T0362417

Peptide and Protein
Drug Delivery

ADVANCES IN PARENTERAL SCIENCES

Editor

JOSEPH R. ROBINSON

Center for Health Sciences
University of Wisconsin
Madison, Wisconsin

Additional Volumes in Preparation

Peptide and Protein Drug Delivery

edited by

Vincent H. L. Lee

University of Southern California School of Pharmacy
Los Angeles, California

MARCEL DEKKER, INC.   NEW YORK · BASEL

Library of Congress Cataloging-in Publication Data

Peptide and protein drug delivery/edited by Vincent H. L. Lee.
      p.   cm. -- (Advances in parenteral sciences; 4)
    Includes bibliographical references.
    Includes index.
    ISBN 0-8247-7896-0 (alk. paper)
    1. Peptides--Therapeutic use--Administration. 2. Proteins-
-Therapeutic use -Administration. 3. Drugs--Dosage forms. I. Lee,
Vincent H. L.   II. Series.
    [DNLM: 1. Peptides--administration & dosage. 2. Proteins-
-administration & dosage. W1 AD714 v. 4/QU 68 P4238]
    RS431.P38P45   1990
    615'.3--dc20
    DNLM/DLC
    for Library of Congress   90-14028
                                                             CIP

This book is printed on acid-free paper.

MARCEL DEKKER, INC.
270 Madison Avenue, New York, New York 10016

Current printing (last digit):
10 9 8 7 6 5 4 3

PRINTED IN THE UNITED STATES OF AMERICA

To Miko and Amanda

Series Introduction

The field of parenterals has seen substantial technical and scientific growth over the past two decades, with the expectation of even greater activity during the last half of the 1980s and through the 1990s. This growth is due, in part, to the expected surge in the number of very potent and sensitive peptides that are arising out of the proliferating genetic engineering programs; an expected increase in the number and type of nutritional products; and the increasing demand for expanded home health care. It is entirely appropriate, therefore, that a scientific/technical series be established not only to report on advances in this field but also to help integrate the various disciplines that impact on parenteral products.

To place the series in perspective it may be useful to provide a personal impression of the level and complexity of the parenteral field. Parenteral activities can arbitrarily be divided into (1) those devoted to making an elegant, safe and effective product, (2) the interface of that product with the route of administration, and (3) the influence of the product on the time course and biological activity of the drug in question. Naturally, all three areas are related and are not easily segregated, especially the latter two.

The technical aspects of parenterals include assurance of stability, sterility, and freedom from particulates as major concerns. These areas have received considerable attention and significant advances have been made in conventional, that is, non-sustained release parenterals. Sustained forms of parenteral products require special consideration from a preparation and quality assurance point of view.

Less well studied is the interaction of the product with the biological interface, for example, biocompatibility, local metabolism and immunological reaction. These areas require a variety of disciplines to fully understand the parenteral product–biological interface. This is a field that has been studied in an uneven manner over the years, providing a less than satisfactory data base.

v

The last area is the influence of the product on the time course and biological activity of the drug, the so-called bioavailability issue. In the absence of a suitable understanding of the product-biological interface, that
is, point number two, it is difficult to provide a thorough mechanistic description.

Thus, there is considerable need for additional work in the area of parenterals. It is hoped that the series "Advances in Parenteral Sciences"
will provide a suitable platform to record advances in this exciting field.

Joseph R. Robinson

Preface

Recent advances in the structural elucidation of numerous natural peptides and proteins, in the understanding of their role in several physiological processes, and in the use of biotechnological techniques in their production have stimulated considerable interest in establishing peptides and proteins as drugs in therapy. Peptides and proteins have been implicated in a wide array of processes including fertility, growth, pain perception, inflammation, blood-pressure regulation, blood clotting, microbial pathogenicity, immunosuppression, and metastasis of neoplasms. The discovery of dual control of other processes by peptides and classical neurotransmitters will undoubtedly alter our approach to treating conditions in which only the autonomic system has heretofore been believed to play a role.

Several challenges confront the delivery of peptide and protein drugs. A major challenge in using peptides and proteins as drugs is preservation of their structural integrity until they reach their sites of action, which are often remote from the site of administration. For some peptides, such inherent metabolic instability has been overcome by developing analogs that are metabolically more stable and that, at the same time, are either as potent or more potent than their parent peptides. But the metabolic instability of proteins has not yet been solved. Another challenge in peptide and protein drug delivery, which is closely related to the first, is to understand the magnitude of the enzymatic barrier in degrading peptides and proteins that are administered parenterally, orally, bucally, rectally, nasally, transdermally, ocularly, or vaginally. An understanding of the nature of this barrier is essential to the development of metabolically stable analogs, to the selection of protease inhibitors to control proteolytic activity, and to the selection of a route for peptide and protein drug delivery.

A third challenge in peptide and protein drug delivery is to overcome the resistance of the mucosal membranes to the penetration by peptide and protein drugs. In the short term, this would require the development of safe and effective absorption promoters. In the long term, this would require a fundamental understanding of the mechanisms of peptide and protein absorption from the mucosal routes. Only then can biochemically based approaches be designed to facilitate peptide and protein absorption. A fourth challenge in peptide and protein drug delivery is the design of delivery systems for each of the unconventional alternative routes as well as for the conventional routes that will mimic the peculiar physiological behavior of certain peptides and proteins as exemplified by their circadian or ultradian secretory patterns. Above all, the ultimate challenge in peptide and protein drug delivery is to understand the requirements to be met in order to successfully deliver peptides and proteins from the oral route, which is expected to be more readily accepted by patients than any of the alternative routes mentioned earlier.

Since this book was first conceived, there have been increasing research efforts dedicated to peptide and protein drug delivery in both academic and industrial laboratories worldwide. Thus far, progress in each of the above-named areas of challenges has been uneven, with more emphasis placed on the modification of existing technologies but less on the development of the truly novel. Moreover, the majority of such studies have emphasized the phenomenological aspects of the processes under investigation. Nonetheless, it is anticipated that, as the field moves forward, more emphasis will be placed on achieving a fundamental understanding of the underlying processes, be they physical or biological. Only then will truly novel drug delivery technologies emerge.

This book was written to fulfill a need for a comprehensive review and assessment of the peptide and protein drug delivery area. To this end, I have organized the book in the following manner. Chapters 1 through 9 provide a description of fundamentals of peptide and protein drug delivery. The topics include synthesis, physical chemistry and biochemistry, analysis, enzymatic and membrane constraints, and pharmacokinetics. Chapters 10 through 16 provide a comprehensive review of approaches taken to optimize peptide and protein drug absorption from the parenteral, oral, buccal, rectal, nasal, vaginal, and transdermal routes. Background information on the anatomy, biochemistry, and physiology of each of these routes has been included. The final chapters of this book, Chapters 17 through 22, focus on formulation challenges unique to peptide delivery, design of controlled release systems, toxicological and immunological aspects of peptide and protein drugs, and quality assurance and regulatory aspects of peptide and protein drugs.

This book will fulfill my expectations if it creates interest in the area of peptide and protein drug delivery and provides a framework for the pharmaceutical or biotechnological scientist to formulate a strategy to deliver peptide and protein drugs to their sites of action. Thus, this book is aimed at those interested in the opportunities and challenges associated with peptide and protein drug delivery.

Vincent H. L. Lee

Contents

Contributors

Abraham Abuchowski, Ph.D. President and CEO, ENZON, Inc., South Plainfield, New Jersey

Reinhold Anders Ph.D. Hoechst AG, Frankfurt, Federal Republic of Germany

Shabbir T. Anik, Ph.D. Director, Technical Services, Syntex Research, Inc., Palo Alto, California

Joffre Baker, Ph.D. Assistant Professor, Department of Biochemistry, University of Kansas, Lawrence, Kansas

Partha S. Banerjee* Research Associate, School of Pharmacy, University of Wisconsin, Madison, Wisconsin

Ronald T. Borchardt, Ph.D. Solon E. Summerfield Professor and Chairman, Department of Pharmaceutical Chemistry, University of Kansas, Lawrence, Kansas

Yie W. Chien, Ph.D. Parke Davis Professor and Director, Controlled Drug Delivery Research Center, Department of Pharmaceutics, College of Pharmacy, Rutgers–The State University of New Jersey, Piscataway, New Jersey

Yuan-Yuan H. Chiu Division of Metabolism and Endocrine Drug Products, Food and Drug Administration, Rockville, Maryland

---

*Current affiliation: Senior Research Scientist, Hercon Laboratories Corporation, Emigsville, Pennsylvania.

Frank F. Davis   Vice President, Research and Development, ENZON, Inc., South Plainfield, New Jersey

Satish Dodda-Kashi   Senior Scientist, Clinical Biology Research Department, Ciba Geigy Pharmaceuticals Division, Summit, New Jersey

Paul D. Gesellchen, Ph.D.   Research Scientist, Biochemistry Research, Lilly Research Laboratories, Eli Lilly and Company, Indianapolis, Indiana

George M. Grass   Staff Researcher II, Department of Dosage Design and Development, Institute of Pharmaceutical Sciences, Syntex Research, Inc. Palo Alto, California

John L. Gueriguian, M.D.   Division of Metabolism and Endocrine Drug Products, Food and Drug Administration, Rockville, Maryland

Ismael J. Hidalgo   Department of Pharmaceutical Chemistry, University of Kansas, Lawrence, Kansas

Ehab A. Hosny   Assistant Professor, University of Cairo, Cairo, Egypt

David M. Johnson, Ph.D.   Institute of Pharmaceutical Sciences, Syntex Research, Inc., Palo Alto, California

Alex Jordan, Ph.D.   Supervisory Pharmacologist, Division of Metabolism and Endocrine Drug Products, Food and Drug Administration, Rockville, Maryland

Glenn M. Kazo, M.S.   Technical Supervisor, ENZON, Inc., South Plainfield, New Jersey

Udaya Bhaskar Kompella, M.Pharm.   Graduate Student. Department of Pharmaceutics, University of Southern California School of Pharmacy. Los Angeles, California

J. R. Kopplin, Ph.D.   Director of Toxicology, Toxicology Department. Cetus Corporation, Emeryville, California

Vincent H. L. Lee, Ph.D.   Gavin S. Herbert Professor and Chairman. Department of Pharmaceutics. University of Southern California School of Pharmacy, Los Angeles, California

Sau-Hung S. Leung, Ph.D.   Senior Research Scientist, Columbia Research Laboratories, Madison, Wisconsin

Yoshiharu Machida, Ph.D.   Associate Professor. Department of Pharmaceutics, Hoshi University, Tokyo, Japan

Thomas R. Malefyt, Ph.D.   Research Section Leader. Pharmaceutical Analysis II, Syntex Research. Inc.. Palo Alto. California

B. J. Marafino, Jr., Ph.D.   Research Scientist, Department of Toxicology, Cetus Corporation, Emeryville, California

Hans P. Merkle, Ph.D.   Professor, Department of Pharmacy, Swiss Federal Institute of Technology, Zurich, Switzerland

Tsuneji Nagai, Ph.D.   Professor and Chairman, Department of Pharmaceutics, Hoshi University, Tokyo, Japan

Tue H. Nguyen, Ph.D.   Scientist, Pharmaceutical Research and Development, Genentech, Inc., South San Francisco, California

Mary L. Nucci   ENZON, Inc., South Plainfield, New Jersey

James Q. Oeswein, Ph.D.   Scientist, Pharmaceutical Research and Development, Genentech, Inc., South San Francisco, California

Hiroaki Okada, Ph.D   Research Head, Pharmaceutics Laboratories, Central Research Division, Takeda Chemical Industries, Ltd., Osaka, Japan

Rodney Pearlman, Ph.D.   Director, Pharmaceutical Research and Development, Genentech, Inc., South San Francisco, California

Susanne Raehs, Ph.D.   Institute of Pharmaceutical Technology, Department of Pharmacy, University of Bonn, Bonn, Federal Republic of Germany

Cynthia S. Randall, Ph.D.   Associate Senior Investigator, Department of Pharmaceutical Analysis, SmithKline Beecham Pharmaceuticals, King of Prussia, Pennsylvania

Joseph R. Robinson, Ph.D.   Professor of Pharmacy, School of Pharmacy, University of Wisconsin, Madison, Wisconsin

Werner Rubas, Ph.D.   Post-Doctoral Fellow, Department of Dosage Design and Development, Institute of Pharmaceutical Sciences, Palo Alto, California

J. Howard Rytting, Ph.D.   Professor, Pharmaceutical Chemistry Department, University of Kansas, Lawrence, Kansas

James M. Samanen, Ph.D.   Assistant Director, Peptide Chemistry Department, SmithKline Beecham Pharmaceuticals, King of Prussia, Pennsylvania

Lynda M. Sanders, Ph.D.   Department Head, Institute of Pharmaceutical Sciences, Syntex Research, Palo Alto, California

Robert F. Santerre, Ph.D.   Research Scientist, Molecular Biology, Lilly Research Laboratories, Eli Lilly and Company, Indianapolis, Indiana

Carlos A. Schaffenburg, M.D.*   Medical Officer, Division of Metabolism and Endocrine Drug Products, Food and Drug Administration, Rockville, Maryland

Steven J. Shire, Ph.D. Senior Scientist, Pharmaceutical Research and Development, Genentech, Inc., South San Francisco, California

Solomon Sobel, M.D.   Director, U.S. Department of Health and Human Services, Center for Drug Evaluation and Research, Food and Drug Administration, Rockville, Maryland

Larry A. Sternson, Ph.D.   Vice President, Pharmaceutical Research and Development, Sterling Drug, Inc., Malvern, Pennsylvania

Kenneth S. E. Su, Ph.D.   Research Scientist, Pharmaceutical Research Department, Lilly Research Laboratories, Eli Lilly and Company, Indianapolis, Indiana

Mitchell E. Taub, B.S.   Graduate Student, Department of Pharmaceutics, University of Southern California School of Pharmacy, Los Angeles, California

Robert D. Traver, B.S.   Graduate Student, Department of Pharmaceutics, University of Southern California School of Pharmacy, Los Angeles, California

Aloys Wermerskirchen   Institute of Pharmaceutical Technology, Department of Pharmacy, University of Bonn, Bonn, Federal Republic of Germany

Gregor Wolany   Institute of Pharmaceutical Technology, Department of Pharmacy, University of Bonn, Bonn, Federal Republic of Germany

*Current affiliation: Consultant, NICHD, National Institutes of Health, Bethesda, Maryland.

part one
FUNDAMENTALS

1

Changing Needs in Drug Delivery in the Era of Peptide and Protein Drugs

Vincent H. L. Lee

University of Southern California School of Pharmacy, Los Angeles, California

I. INTRODUCTION

Hormones, serum proteins, and enzymes have been used as drugs ever since the commercial introduction of insulin, thyroid hormone, and factor VIII from 1920 through 1940. Molecular biology has now given us the tools to expand the range of peptide- and protein-based drugs to combat poorly controlled diseases. Such drugs include synthetic vaccines that promise to offer protection against carcinogens and toxicants [1], and human interferon-αA and monoclonal antibody which, when used in combination, may be effective in overcoming the antigenic heterogeneity of carcinoma cell populations and enhancing the efficacy of monoclonal antibodies in the detection and treatment of carcinoma lesions [2]. The same tools in molecular biology have also given us the opportunity to identify and produce target molecules on which low-molecular-weight drugs act and because of which more specific drugs can be designed [3,4].

While there has been rapid progress in molecular biology, this has not been matched by progress in the formulation and development of peptide and protein drug delivery systems. This is due in part to the lack of appreciation for the unique demands imposed by the physicochemical and biological properties of peptide and protein drugs on routes of delivery as well as on delivery system design and formulation. These properties include molecular size, short plasma half-life, requirement for specialized mechanisms for transport across biological membranes, susceptibility to breakdown in both physical and biological environments, tendency to undergo self-association, complex feedback control mechanisms, and peculiar dose-response relationships [5]. The purpose of this chapter is to review how such properties have led to the development of new routes of drug administration as well as new ways to formulate and test new drug delivery systems.

II. NON-ORAL ROUTES OF PEPTIDE AND
PROTEIN DRUG DELIVERY

The lack of oral efficacy of natural peptides and proteins has prompted
the examination of non-oral routes for peptide and protein drug administration. Such routes include the nasal, buccal, rectal, vaginal, percutaneous, ocular, and, to a limited extent, pulmonary routes. Insulin and
leuprolide, an LHRH analog, are the most studied in this regard. As can
be seen in Table 1, the bioavailability of both peptides is considerably improved when administered by the non-oral routes, but it is still far from
that afforded by the subcutaneous route. Several reasons may be responsible for incomplete bioavailability: resistance of the mucosal membrane to
peptide and protein penetration, degradation of the peptides and proteins
by proteases in the luminal cavity and the cells lining the mucosa, and
rapid clearance of the administered dose from the site of deposition. The
first reason will be discussed below; the second and third will be discussed
in Chapter 7 and in a later section of the present chapter, respectively.

In the G.I. tract, di- and tripeptides are believed to be absorbed by
carriers [6], which are also involved in the transport of antibiotics such
as cefalexin [7], cefatriazine [8], β-lactam antibiotics [9−11], and dipeptide derivatives of L-α-methyldopa [12]. Proteins such as IgG and epidermal growth factor [13,14] are believed to be absorbed by endocytosis,
and all other peptides and proteins are believed to be absorbed by passive
diffusion, if at all. To date, it is not known whether the peptide carriers
and receptors involved in endocytosis are present in non-oral mucosae.
There is, however, indirect evidence for the existence of amino acid carriers in the nasal mucosa of the rat. Tengamnuay and Mitra [15] investigated the mechanism by which amino acids were transported across the
nasal mucosa by monitoring the disappearance of L-tyrosine and L-phenyln-
alanine from 0.25−2.5 mM solutions perfusing the nasal cavity. They proposed that these amino acids are transported across the rat nasal mucosa
on the basis of four lines of evidence: concentration dependence, Na^+-
dependence, stereospecificity, and, in the case of L-phenylalanine, susceptibility to inhibition by ouabain and 2,4-dinitrophenol. The Km and
Vmax values, which had not been corrected for the passive diffusion component, were calculated to be 0.68 mM and 0.44 mM/hr for L-tyrosine and
0.40 mM and 0.39 mM/hr for L-phenylalanine. Since it was the disappearance of amino acids that was monitored, the above findings are best
interpreted as uptake rather than as transport.

There is a perception that the molecular size of peptides and proteins,
which are one to three orders of magnitude larger than that of conventional drug molecules, is a major factor limiting their diffusion across biological membranes. The role of charge on peptide and protein absorption
in the nasal route has been investigated by Maitani et al. [16] using FITC
DEAE-dextran and FITC dextran in the presence of 1% Na glycocholate.
But the results are inconclusive. McMartin et al. [17] analyzed the relationship between the bioavailability and molecular weight of 25 compounds
ranging in molecular weight from 160 (hydralazine) to 34,000 (horseradish
peroxidase) following nasal administration. The peptides that were part of
this analysis included thyrotropin releasing hormone (M.W. 362), metkephamid (M.W. 661), cyclo(-Pro-Phe-D-Trp-Lys-Thr-Phe) (a somastostatin
analog, M.W. 806), oxytocin (M.W. 1,007), lysine vasopressin (M.W. 1,056),
des-amino-D-arginine vasopressin (M.W. 1,069), luteinizing hormone-releasing

Table 1   Percent of Dose Absorbed From Various Routes of Administration

Route	Percent of dose absorbed	
	Insulin (M.W. 6,000)	Leuprolide (M.W. 1,210)
Oral	0.05 [447]	0.05 [452]
Nasal	30 [117]	2−3 [452]
Buccal	0.5 [448]	—
Rectal	2.5 [449]	8 [452]
Vaginal	18 [450]	38 [452]
Subcutaneous	80 [441]	65 [452]

hormone (M.W. 1,182), buserelin (M.W. 1,297), nafarelin (M.W. 1,337), alsactide (M.W. 2,100), secretin (M.W. 3,052), growth hormone releasing factor (M.W. 4,800), and horseradish peroxidase (M.W. 34,000). That analysis revealed that compounds of molecular weight <1,000 were adequately absorbed to the extent of 70 ± 26%. By modeling the percent of drug absorbed as competition between removal of the applied dose by mucociliary clearance from the nasal cavity and drug permeation through the epithelial cell layer, the percent of drug absorbed was found to vary inversely with molecular weight according to the relationship,

$$\% \text{ Absorption} = \frac{100}{[1+a(MW)^{-b}]}$$

where a = 0.003 and b = 1.3 for the rat nasal mucosa; a = 0.001 and b = 1.35 for the human nasal mucosa. Deviations of absorption from values predicted by this model did not appear to correlate with factors such as charge, hydrophobicity, and susceptibility to aminopeptidases. This observation probably reflects the relative homogeneity in charge and hydrophobicity of the compounds examined and their resistance to aminopeptidase action.

Absorption −molecular weight relationships similar to the one obtained for the nasal mucosa are anticipated to exist for the other absorptive mucosae. So far, such information is available for only the oral mucosa, which is about three times more dependent on molecular weight in extent of absorption than is the nasal mucosa. This finding signals the existence of differences in the permeability of the various mucosal routes of administration.

The skin is perhaps the most impermeable of all the mucosae. This is indicated by the experience with thyrotropin releasing hormone (TRH) [18] and vasopressin [19]. Burnette and Marrero [18] showed that TRH did not diffuse across excised nude mouse skin and that iontophoresis was required to facilitate its permeation. Banerjee and Ritschel [19] made similar observations with tritium labeled vasopressin and found that close shaving of the skin, 24 hr before application, or stripping the skin 25 times with cellophane tape, was required to promote its transdermal flux.

Under these conditions, the transdermal flux of vasopressin was improved
5 and 70 times, respectively. Treatment of the skin with Na lauryl sulfate was not effective even at a concentration as high as 20%. In this
study, it was assumed that the tritium label on vasopressin was chemically
and metabolically stable. Further information on the transdermal route of
peptide and protein drug delivery is provided in Chapter 15.

Hersey and Jackson [20] determined that in both the rabbit and the
dog, the nasal mucosa was about 5 times more permeable to water, 9
times more permeable to sucrose, and 16 times more permeable to PEG
5000, when compared with the ileal mucosa. Such differences in permeability are responsible, at least in part, for the different absorptive behavior of peptides and proteins among the mucosal routes. For instance,
the rectal administration of salmon calcitonin (SCT) evoked a sharp peak
of plasma SCT followed by a rapid decrease to less than 10% of the observed peak within 1 hour. The nasal administration of this peptide, on
the other hand, was characterized by its slow passage through the nasal
mucosa to yield sustained peptide levels for several hours. Despite these
kinetic differences, the biological effects were almost similar [21].

Mechanistically, the differences in permeability among the mucosal
routes may be due to differences in lipid composition and porosity. For
instance, the brush border of the intestinal wall possesses unusually large
quantities of glycosylceramide [22,23], the skin contains large quantities
of ceramides and free fatty acids [24], and the buccal mucosa contains
large quantities of phospholipids [25]. While the notion of pores in the
mucosal membranes is a controversial subject [26], estimates of pore size
are available. From the sieving coefficient of antipyrine, Hayashi et al.
[27] calculated a pore size of $7.1-16.0$ Å for jejunum, $5.9-17.0$ Å for
rectum, and $3.9-8.4$ Å for nose. The smaller pore size in the nose was,
however, more than offset by a richer distribution of pores. The net
result was enhanced water influx in the nose when compared with the
jejunum and the rectum.

Undoubtedly, the permeability and metabolic characteristics of mucosal
routes are important factors to consider in the selection of a given route
for the administration of a therapeutic peptide or protein. Several of
these routes have been discussed in some detail in Chapters $10-16$ of this
book. The ocular and pulmonary routes, which have gained some attention
since this book was conceived, will be discussed here.

A. Ocular Route of Peptide and Protein Drug Delivery

The possibility of eliciting a systemic reaction from topically applied peptides to the eye was recognized as early as 1931. At that time Christie
and Hanzal [28] reported dose-dependent lowering in blood glucose after
administration of insulin in a pH 2.5 solution to rabbit eyes. This study,
however, did not stimulate much interest in the ocular route for peptide
and protein drug delivery until a few years ago when it was clear that
some peptides and proteins will become important therapeutic agents.
Among the peptides and proteins that have been investigated for ocular
delivery include enkephalins, thyrotropin releasing hormone, luteinizing
hormone-releasing hormone, glucagon, and insulin [$29-34$]. All of these
peptides and proteins were absorbed into the bloodstream of the albino
rabbit to some extent. Chiou and Chuang [30] noticed sustained peptide
concentrations in plasma following topical solution instillation. These

results are somewhat difficult to interpret, however, since the rabbits were under anesthesia and since the chemical and metabolic stability of the I-125 label was never ascertained. By constrast, Stratford et al. [29] and Yamamoto et al. [34] observed that topically applied [D-Ala^2]metenkephalinamide and insulin reached peak concentrations in plasma within 15−20 min following solution instillation in the fully awake rabbit. The bioavailability was about 36% for the enkephalin and less than 1% for insulin. The bioavailability of insulin was improved to 4−13% in the presence of the penetration enhancers Na glycocholate, Na taurocholate, Na deoxycholate, and polyoxyethylene-9-lauryl ether. Penetration enhancers will be discussed in detail subsequently.

Topically applied ophthalmic drugs reach the bloodstream via the blood vessels in the conjunctival and the nasal mucosae [35]. The relative contribution of these two mucosae to the systemic absorption of topically applied ophthalmic drugs is drug dependent. The nasal mucosa contributes about 4 times more than the conjunctival mucosa to the systemic absorption of ocularly applied timolol [35], inulin [29], and insulin [34]. By contrast, the nasal mucosa contributes equally as the conjunctival mucosa to the systemic absorption of ocularly applied [D-Ala^2]metkephalinamide. By incorporating this peptide in a viscous vehicle, 5% poly(vinyl alcohol), its systemic bioavailability was improved from 36−51% [29]. This is opposite to the situation with timolol, where reduction in systemic absorption was seen [36]. These findings indicate that the permeability of the conjunctival and the nasal mucosae to timolol is different than that of [D-Ala^2]metkephalinamide.

Although ocularly applied peptides can be absorbed into the bloodstream to elicit a pharmacological response, the ocular route is unlikely to be widely accepted for peptide administration, at least at the present. This is primarily because of the innate aversion to instilling drugs to the eye and because of the perceived sensitivity of the eye, notably the corneal epithelium, to external insult, such as when penetration enhancers are included in a formulation to promote the absorption of peptides. Nevertheless, because topically applied drugs will eventually contact the nasal mucosa, a suitable vehicle may some day be developed that will rest in the conjunctival sac while leaking peptides and proteins to the nasal mucosa for absorption.

B. Pulmonary Route of Peptide and Protein Drug Delivery

For decades, medications have been administered to the lungs to control local conditions such as asthma and infections. Unintentionally, such medications are absorbed to varying extents into the bloodstream. Drugs such as sulfanilic acid, p-aminohippuric acid, and phenol red are absorbed at least 8 to 42 times faster from the lungs than from the small intestine, in both rats and humans [37,38]. To improve the safety of lung medications many attempts have been made to minimize the systemic absorption of locally administered drugs. One such attempt is reduction of drug solubility, as has been shown for the leukotriene D$_4$ antagonist, 4-[(3-(4-acetyl-3-hydroxy-2-propylphenoxy)propyl)-sulfonyl]-γ-oxobenzenebutanoate (L-648,051) [39].

The lungs are an attractive site for the systemic delivery of peptides and proteins because of the potentially larger surface area (70 m^2) they have to offer when compared with other mucosal sites such as the nasal, buccal, rectal, and vaginal cavities. Nonetheless, very little information exists on the pulmonary absorption of macromolecules from the lungs.

Leuprolide, insulin, and albumin are perhaps three of the few peptides/ proteins whose absorption from the lungs has been investigated. Vadnere et al. [39a] reported that leuprolide was absorbed to the extent of 18% from the lungs in human volunteers. Yoshida et al. [40] reported that about 40% of the insulin administered in an aerosol to the trachea of the anesthetized rabbit was absorbed. This was better than the 7−16% absorption observed in healthy human subjects and diabetic patients receiving aerosolized insulin in a nebulizer [41]. Loss of the administered dose to areas outside the respiratory tract was thought to be responsible for the smaller extent of insulin absorption in the latter study. Complete absorption of insulin in the rat was seen when penetration enhancers such as 1% Azone, 1% fusidic acid, or 1% glycerol, were incorporated into the aerosol [42]. This was consistent with the favorable action of Na glycocholate (0.01−1 mM), another enhancer, on the pulmonary absorption of intra-bronchally administered FITC-dextran (M.W. 150,000) in rats [43]. There was, however, no information on how well such enhancers were tolerated by the experimental animals.

Albumin was largely absorbed, perhaps via a pinocytosis process [44], within 48 hr upon instillation into the lungs of guinea pigs and dogs [45]. The chemical and metabolic stability of the I-125 label on albumin was not ascertained. In spite of the disparity in molecular size, the permeability coefficient of albumin was only four times smaller than that of sucrose (0.38×10^{-8} cm s^{-1} vs. 1.4×10^{-8} cm s^{-1}) [44].

In order to fully appreciate the potential of the lungs in the systemic administration of peptide and protein drugs, it will be necessary to understand or recognize (1) the permeability characteristics of the lungs, (2) drug absorption mechanisms in the lungs, (3) their drug metabolizing capacity, (4) factors controlling the site of dose deposition in the lungs, and (5) safety issues. These topics will be discussed in turn below.

Permeability Characteristics of the Lungs

About 90% of the absorptive surface area of the lungs is due to the alveoli. The epithelium of the alveoli is comprised of a heterogeneous population of cells [46,47]. About 97% of the alveolar surface area is occupied by squamous of type I cells, while the remainder is occupied by granular cuboidal or type II surfactant-secreting cells. A very few brush cells, called type III cells, are also present as are free alveolar macrophages in the alveolar spaces. Type I cells and macrophages but not type II cells have been shown to be involved in the uptake of horseradish peroxidase by endocytosis [48].

The epithelial cells are generally in intimate contact with the vasculature; the total distance between air and blood is much less than 1 μm [49]. While such an architecture allows the free exchange of gases, it becomes a major barrier to the passage of larger molecules. For instance, horseradish peroxidase (M.W. 40,000) deposited on the air side hardly reached the interstitium [50], although it did reach the interstitium from the blood side [51]. The principal resistance is in the alveolar epithelium, where the cells are tightly interlaced [52]. Wangensteen et al. [53] showed that the alveolar epithelium was 10^3 times less permeable to sucrose than was the microvascular endothelium. Taylor and Gaar [54] and others [44,55] estimated a pore size of 6−10 Å for the alveolar membrane and one of 40–58 Å for the pulmonary capillary membrane. In 1983, Schanker and Hemberger [56] postulated the presence of three populations of pore size in the alveolar epithelium of the rat: (1) numerous small pores that admit urea and guanidine but exclude compounds of M.W. 122 or above, (2) a few medium-size pores that admit compounds with M.W. as high as 1,355 but exclude compounds of M.W. 5,250 or higher, and (3) very few large pores that admit inulin and dextran (M.W. 5,250–20,000) as well as smaller molecules.

In addition to the tight arrangement of the alveolar epithelial cells, the lung surfactant also plays an important role in maintaining the permeability of the alveolar epithelium. The lung surfactant is a complex lipid/protein mixture which coats the air/water interface of the alveoli and alveolar ducts to maintain appropriate surface tension thereby preserving the patency of the airways. Egan et al. [57] showed that goats without detectable surfactant in fetal lung liquid before ventilation allowed the transfer of tracer solutes of all sizes between alveolar liquid and blood, whereas those with detectable surfactant did not allow tracers 14 Å or larger in radius to cross the lung epithelium.

Drug Absorption Mechanisms in the Lung

The lungs are similar to the intestine in that both simple diffusion and carrier-mediated transport mechanisms of drug absorption exist. Simple diffusion is involved in the absorption of a number of drugs ranging in molecular weight from 60 to 75,000 following intratracheal administration in the anesthetized rat. These include neutral molecules (urea, erythritol, mannitol, sucrose, ouabain, dihydroouabain, cyanocobalamin, inulin, and dextran), anionic compounds (fluorescein, carboxyfluorescein, heparin, sulfanilic acid and p-acetylaminohippuric acid), and quaternary amines (tetraethylammonium, diquat and procaine amide ethobromide). Absorption rates are related to molecular size [37,58–62]. The half-time of absorption from the alveolar region ranges from 0.25 min for antipyrine to 26.5 min for mannitol [63]. The absorption half-time was longer for macromolecules, 9.2 hr in the case of heparin and 28 hr in the case of dextran [37,60].

Although the involvement of carriers in the pulmonary absorption of peptides has yet to be determined, carriers are known to exist in several animal species, including rat, mouse, hamster, and guinea pig, and participate in the absorption of 1-aminocyclopentanecarboxylic acid (cycloleucine) [64,65], α-methyl-D-glucose pyranoside [66], and the organic anions phenol red and sodium cromoglycate [67–71]. Carriers do not appear to play a role in the absorption of organic cations except paraquat [38,72]. Conflicting evidence exists in the participation of carriers in the transport of α-aminoisobutyric acid. While this amino acid is taken up by lung slices via a carrier-mediated process [64], its absorption in vivo appears to occur by simple diffusion [65]. No explanation is forthcoming for this inconsistency.

Carrier-mediated transport is usually inferred from several lines of evidence. In the case of α-methyl-D-glucose pyranoside [66], the involvement of a carrier is indicated by sensitivity of the process to inhibition by 0.5 mM phlorizin, 5 mM glucose, 0.5 mM ouabain, and Na^+ depletion. Nonetheless, the extent of drug absorption is independent of drug concentration over the range of 0.01 and 20 mM [71]. In the case of phenol red [69], the involvement of a carrier is indicated by uptake against a concentration gradient by a saturable process (1) that was inhibited by low temperature, anaerobic conditions and certain metabolic inhibitors and (2) that was depressed in the presence of certain anionic dyes, such as chlorphenol red, bromphenol blue, bromthymol blue and bromcresol green, and by various other organic acids, including disodium cromoglycate, probenecid, cephalothin, and benzylpenicillin.

Drug Metabolizing Capacity of the Lungs

The lung contains virtually all of the pathways required for the biotransformation of exogenous chemicals, albeit generally at lower concentrations than in the liver. For instance, bambuterol, a prodrug of terbutaline, underwent oxidative metabolism in the guinea pig isolated lungs [73]. The principal cell type in which metabolism occurs appears to be in the alveolar type II cells, and especially Clara cells, which are enriched in xenobiotic-metabolizing enzymes [74]. On the other hand, the pulmonary endothelial cells are important in the regulation of circulating hormones [75] and certain polypeptides such as bradykinin [76], and angiotensin converting enzyme is known to be present in high activity there [77]. The proteolytic activities elsewhere in the lungs have not been delineated, however. Neither have the extent of metabolism of exogenously applied peptides and proteins and the dominant cell type in which proteolysis occurs.

Factors Controlling Site of Dose Deposition in the Lungs

One of the major challenges in pulmonary drug delivery is ensuring reproducibility in the dose deposition site. Absorption rates are expected to vary at different levels of the respiratory tree owing to the variable thickness of the epithelial lining cells and possibly to other anatomic and physiologic variables. Indeed, drug inhaled as an aerosol generated with an airjet nebulizer (MMAD 2.81 ± 0.08 μm, where MMAD is mass-median aerodynamics diameter) was absorbed roughly two times more rapidly than when administered by intratracheal injection in the rat [63]. To deliver drugs efficiently to that part of the lungs would require overcoming the intense power of the upper airways to filter particles and rapidly remove them by mucociliary action. This, in turn, would require that the particle or droplet size generated from currently used devices be more monodisperse [78]. As an illustration of how the droplet size can affect the percent deposition in the alveoli, a polydisperse aerosol with a MMAD of 3 μm and a geometric standard deviation of 3.5 would yield an alveolar deposition of approx. 30% or less, in contrast to 50% or better for its monodisperse equivalent [79].

Three devices are presently available for administration of medications to the lungs: metered-dose inhaler, nebulizer, and powder inhaler or insufflator. Proper use of the delivery device and appropriate coordination between breathing and dispensing of the aerosol are essential to obtaining a reproducible dose [80]. Sources of variability include accumulation of moisture on the drug particle resulting in significant growth especially in the high humidity of the respiratory tract, evaporation of the propellant resulting in rapid decrease in mass, and the manner of inspiration by the patient and whether he/she holds the breath following inhalation. In the environment of the lung where the relative humidity can be as high as 99%, hydrophilic solids generally begin to dissolve as they absorb water. These solids grow until the aqueous vapor pressure of the droplet equals the pressure in the lung or until deposition occurs, whichever happens first [81]. This undesirable process can be retarded by coating drugs with hydrophobic materials such as lauric acid, oleic acid, lecithin, and Span 85 [83]. The same strategy can be applied to control droplet evaporation kinetics [82]. In this instance, oleic acid was found to be more effective than lecithin and Span 85 due to its ability to form a condensed film at the droplet −air interface.

None of the above devices may be considered as efficient since only 10% of the administered dose reaches the target with the remainder being deposited outside the respiratory tree [80]. Byron et al. [84] felt that this can be improved to about 40% with suitable modifications of the device. Of the dose successfully deposited in the lungs, only 30−40% can reach the alveolar regions from metered-dose inhalers (MDI) and nebulizers, the remainder is left behind in the conducting airways [80,85−89]. Dry powder inhalers are much less efficient than either MDIs or nebulizers [90]. The major factors affecting deposition are listed in Table 2.

Deposition in the alveolar region may be improved by breath-holding if the droplets in the aerosols have aerodynamic diameters in the 1.5−3 µm range [89]. Newman et al. [86] showed that deposition could be maximized in the lower regions of the lung by inhaling slowly and holding the breath for up to 10 seconds. Such maneuvers, however, only minimally affect deposition of the dose in the ciliated airways, which is at a maximum for aerodynamic diameters between 5−9 µm (slow inhalation) and 3−6 µm (fast inhalation) [89]. The time it takes to deplete 99% of the administered dose in the ciliated airways depends on the rate of dissolution in addition to aerosol size and mode of inhalation. In the case of rapidly dissolving solutes, the maximum duration is short (1−2 h). For less soluble particles, it can approach 12 h due to prolonged supply of particles from the alveolar regions [91]. Indeed, drug entrapment in liposomes [92,93] and the use of sparingly soluble compounds [79] or slowly releasing coprecipitates [94,95] have been proposed to achieve sustained drug action following inhalation or intratracheal instillation of therapeutic agents.

The effect of liposomal drug entrapment on drug clearance from the respiratory tract of animals has been investigated [62,92,93]. Woolfrey et al. [96] demonstrated that the absorption of 6-carboxyfluorescein from the lung of the rat was influenced by the charge and dose of liposomes. Absorption was more rapid and extensive from negatively charged than from neutral liposomes. The neutral liposomes were made of dimyristoylphosphatidylcholine (DMPC)/cholesterol (1:1) (105−113 nm) and the negatively charged liposomes were made of DMPC/cholesterol/dicetylphosphate (1:1:0.2) (139−150 nm). Increasing the lipid dose reduced the extent of absorption from both neutral and negative liposomes. In the case of neutral liposomes, the reduction was from 39.5% to 13.6% as the dose was increased from 14−34.3 mg/kg. In the case of negatively charged liposomes, the reduction was from 49.9−32.3% as the dose was increased from 8.6−34.3 mg/kg. This suggests that the higher lipid doses may have induced a biological response which resulted in accelerated removal of the liposome-encapsulated carboxyfluorescein from the absorption site.

Safety Issues in Pulmonary Administration of Peptide and Protein Drugs

A safety issue associated with the use of the lungs in peptide and protein drug delivery is elicitation of an immune response resulting in changes in

Table 2   Major Factors Affecting Dose Deposition in the Respiratory Tree

Particle properties	Diameter
	Density
	Shape
	Charge
	Chemical composition:
	Solubility
	Hygroscopicity
Aerosol properties	Concentration
	Particle size range
	Bolus or continuous cloud
	Velocity of spray
	Evaporation of propellants
Respiratory tract properties	Geometry (variability)
	Preexisting disease
	Humidity
Breathing patterns	Residence time (breath-holding)
	Volumetric flow rate (breathing rate, tidal volume)
	Mouth or nasal breathing

Source: Ref. 453.

pulmonary epithelial permeability.   Indeed, bronchial permeability of the
monkey to macromolecules can be increased upon antigen challenge, due
perhaps to alterations in the tight junctions [97].   Both immunologic and
inflammatory mechanisms control the amounts of inhaled soluble proteins
absorbed.   Braley et al. [98] reported that, in the rabbit, whereas
acute hypersensitivity reaction did not alter the absorption of intact inhaled proteins, chronic hypersensitivity did.   The permeability of the
air−blood barrier to an immunologically unrelated protein was, however,
increased.   Reduction in antigen absorption was attributed partly to enhanced pulmonary metabolism of the antigen [99].

    Alterations in the permeability of the alveolar epithelium can also
result from complications leading to lung edema.   Gardiner and McAnalley
[100] found that lung edema induced by intraperitoneally administered
α-naphthylthiourea increased the absorption of sodium cromoglycate, a
water-soluble drug, while decreasing the absorption of dexamethasone, an
oil-soluble drug.   These results are attributed to increase in the porosity
of the alveolar membrane and to a thickened, fluid-filled interstitium in

the edematous lungs [101]. Another possibility is changes in fluid movement across the alveolar epithelium. The absorption of phenol red from human lungs was found to be more extensive from 0.18% saline than from 0.9% saline [102]. This was attributed to the solvent drag effect as well as to an increase in the concentration gradient for drug absorption due to rapid absorption of water.

Another safety issue associated with pulmonary administration of peptides and proteins is the unlikely event of destabilization of the surfactant film coating the alveolar surface by peptides and proteins. Polypeptides such as ACTH(1-10), porcine β-lipoprotein, α-endorphin, and human fibrinopeptide A have been shown to interact with dimyristoylphosphatidylcholine monolayers resulting in changes in the surface pressure of the film, the efficiency being inversely related to the polypeptide's hydrophobicity [103]. The nature of the polar amino acids and their electrostatic charge also plays a role, as indicated by the interaction of pentagastrin and its basic and neutral analogs with vesicles made of dimyristoylphosphatidylcholine and dimyristoylphosphatidylglycerol [104].

In summary, the lungs are an attractive site for the systemic administration of peptide and protein drugs. But the full potential of the pulmonary route will not be realized unless the various cell types in the conducting and nonconducting airways are characterized with respect to absorptive capacity and mechanisms, type and activities of proteases, and immunological capability. A major challenge to be met in pulmonary drug delivery is ensuring reproducibility in the deposition site of the administered dose.

III. NEED FOR PENETRATION ENHANCERS TO PROMOTE MUCOSAL PEPTIDE AND PROTEIN DRUG ABSORPTION

Penetration enhancers are substances that facilitate the absorption of solutes across biological membranes. Penetration enhancers are often required for the absorption of peptides and proteins in pharmacological amounts. The absorption of vasopressin across the skin of the vasopressin-deficient Brattleboro rat was improved by 3% Azone, resulting in significant reduction in urine volume and increase in osmolality over a 24 hr period. While the percent bioavailability of vasopressin was not determined, the elicitation of antidiuretic response to topically applied vasopressin was consistent with the 15-fold increase in the transdermal flux of vasopressin determined in vitro [105]. Such an increase was not at the upper limit, however, since stripping the skin 25 times with cellophane tape afforded a 70-fold increase [19]. In the presence of palmitoylcarnitine, a cyclic hexapeptide somatostatin analog was nearly completely absorbed rectally in the rat [106]. Moreover, a 180-fold increase in the rectal absorption of [Asu^{1,7}]-eel calcitonin in the Wistar rat was observed in the presence 0.33M phenylalanine enamine of ethylacetoacetate and 0.17M diethylethoxymethylenemalonate. These two adjuvants were more effective than 0.6M Na salicylate and 0.05M sodium p-chloromercurylphenyl sulfate [107].

There are five major types of penetration enhancers: (1) chelators such as EDTA, citric acid, salicylates, N-acyl derivatives of collagen, and enamines (N-amino acyl derivatives of β-diketones), (2) surfactants, such as sodium lauryl sulfate, polyoxyethylene-9-lauryl ether, polyoxyethylene-20-cetyl ether, (3) bile salts, such as sodium deoxycholate, sodium glycocholate, and sodium taurocholate, and their derivatives such as sodium taurodihydrofusidate and sodium glycodihydrofusidate, (4) fatty acids such as oleic acid, caprylic acid, and capric acid, and their derivatives such as acylcarnitines, acylcholines, and mono- and diglycerides, and (5) nonsurfactants such as unsaturated cyclic ureas and 1-alkyl- and 1-alkenylazacycloalkanone derivatives. Few of the above substances were originally designed for penetration enhancement. Today, the design of and search for safe and effective penetration enhancers is an active area of research in peptide and protein drug delivery. The trend appears to be towards enhancers that are based on natural body constituents and that perhaps are readily biodegradable. Examples of the former include glycerides [108,109], acylcarnitines [106], and acylcholines [110]. Examples of the latter include the unsaturated cyclic ureas [111].

The structure–activity relationship of penetration enhancers has been elucidated for both peptide and nonpeptide drugs. In both cases, the enhancer's hydrophobicity has been found to be an important determinant of efficacy. Studies concerning nonpeptide drugs have revealed that for fatty acids and their derivatives, the chain length that maximizes absorption is C12-C18 [112–114]. For N-acyl derivatives of collagen peptide, the most efficient action is observed in the C18 derivative when compared with the C8, C10, and C12 [112]. For glycerides, the rank order of effectiveness is mono- > di- > triglycerides [115].

Results similar to those cited above are also seen in peptide and protein drugs. For instance, the potency of bile salts in enhancing the nasal absorption of insulin in human volunteers was found to correlate with increasing hydrophobicity of the bile salts' steroid nucleus [116]. Specifically, the rank order of effectiveness was deoxycholate > chenodeoxycholate > cholate > ursodeoxycholate. Conjugation of deoxycholate and cholate with glycine and taurine did not affect their enhancement potency. The potency of polyoxyethylene alkyl ethers in enhancing the nasal absorption of insulin in the rat was maximal with the polyoxyethylene 9-lauryl ether, diminishing gradually with smaller or larger numbers of ethylene oxide unit [117]. In the case of acylcarnitines, Fix et al. [106] observed maximal activity with the C12-C18 fatty acid chain length relative to enhancement in the rectal absorption of a cyclic hexapeptide somatostatin analog in the rat. Acylcarnitines with chain lengths shorter than 12 carbons were less effective.

Because enhancers may affect the peptide and protein drugs in addition to the mucosal membrane, one should be cautious in extrapolating structure-activity relationships obtained in conventional drugs to peptide and protein drugs. As a case in point, Duchateau et al. [118] showed that the nasal bioavailability of gentamicin in rabbits increased with the hydrophobicity of trihydroxy bile salts (cholate, taurocholate, and glycocholate) while decreasing with the hydrophobicity of dihydroxy bile salts (deoxycholate, taurodeoxycholate, and glycodeoxycholate). This is not in keeping with the results of Gordon et al. [116] who did not find an inverse relation in absorption promotion between the tri- and dihydroxy bile salts, as mentioned earlier.

A. Mechanisms of Action

The mechanisms by which enhancers act are poorly understood. Enhancers are believed to act by one of three mechanisms: (1) compromising integrity of the mucosal membrane, (2) inhibiting proteolytic activity, or (3) increasing the thermodynamic activity of peptide and protein drugs. Of the three mechanisms, the first has received the most attention and will be discussed in detail here. The second will be discussed in Chapter 7. Very limited attention has been paid to the third until recently. In this regard, Touitou et al. [119] demonstrated that 1.5M Na salicylate increased the solubility of insulin 7875 times, permitting the preparation of an aqueous solution containing 630 mg ml^{-1} of insulin. Such an action of salicylate was also seen with methylene blue, which like insulin underwent self-association [120]. Ion-pair formation between the enhancer and the peptide and protein drug is another interesting possibility by which the thermodynamic activity of the drug may be affected. Hadgraft et al. [121], for instance, proposed that Azone may be capable of forming ion pairs with anionic drugs thereby promoting their permeation across skin. This work, however, was based on membrane filter impregnated with isopropyl myristate. Indeed, the extrapolation of such in vitro results to the biological environment has been questioned [122].

Enhancers may act at either the apical cell membrane (the transcellular pathway) or the tight junctions between cells (the paracellular pathway), or both. Table 3 summarizes the postulated loci of action of some penetration enhancers. The majority of studies designed to investigate the membrane action of enhancers have thus far overlooked the goblet cells and the mucus they secrete as potential targets. This assumption may be incorrect. Fry and Staffeldt [123] showed that the goblet cells in the rat intestinal mucosa were as affected by Na deoxycholate as the columnar cells. Nakanishi et al. [124] noticed the same in the rectum that had been exposed to 5 mM Na deoxycholate, 5 mM Na lauryl sulfate, or 25 mM EDTA. Moreover, goblet cells recovered much more slowly from the effects of the above enhancers than did columnar cells. Whereas columnar cells returned to normal in 2 hr after pretreatment, goblet cells did not completely recover even at 24 hr. There was still a 60% deficiency of goblet cells in the rectal mucosa pretreated with Na lauryl sulfate and Na deoxycholate.

The presence of a mucus layer that coats all epithelial surfaces has also been overlooked in the elucidation of penetration enhancement mechanisms. This is partly because the role of mucus in the absorption of peptide and protein drugs has not yet been established. Limited information does exist on the effect of enhancers on mucus structure and viscosity, however. For instance, Na deoxycholate, Na taurodeoxycholate, Na glycocholate, and lysophosphatidylcholine have been found to reduce the viscosity and elasticity of bronchial mucus and presumably other types of mucus as well [125]. Enhancers that work by the chelation such as N-acylcollagen peptides [112], bile acids [126], and saponins [112] may also affect mucus viscosity since Ca^{2+} and Mg^{2+} play an important role in maintaining mucus layer structure [127]. Whether such mucus thining effect will lead to enhancement in peptide and protein absorption has not yet been shown. Interpretation of data from such studies may be complicated by increase in mucus secretion induced by certain enhancers as is the case with Na taurocholate [128,129].

Transcellular Pathway

Enhancers that increase transcellular permeability to peptide and protein drugs probably do so by affecting membrane lipids and proteins. Evidence for this mechanism is usually indicated by release of lipids and proteins from a given mucosa. The limitation of this approach is that there

Table 3   Postulated Loci of Action of Some Penetration Enhancers

Locus	Enhancer	Reference
Transcellular	Dimethyl maleate	162
	NSAID[a]	124
	Monoglycerides	114
	Na caprylate	140,141
Paracellular	EDTA	163
	Na taurocholate	163
	N-acyl amino acids	454
	N-acyl collagen derivatives	112
	Na laurate	133
	Mixed micelles of Na oleate and Na taurocholate	133
Both	Polyoxyethylated nonionic surfactants	165
	Na caprate	131,133,162
	Salicylates	148,149,455
	Na glycocholate	456
	Na deoxycholate	456

[a]Nonsteroidal anti-inflammatory drugs.

may or may not be a correlation between release of proteins and phospholipids and absorption enhancement potency. Thus, while a good correlation exists between enhancement potency and protein release in the enhancement of nasal insulin absorption by nonionic surfactants [130], the three-fold increase in phospholipid release from the isolated colonic loop is not an important factor in the enhancement of salicylate absorption by caprate [131]. Similarly, Whitmore et al. [132] found that the absorption of salicylate was more sensitive to loss of protein than of phospholipid from the membrane. Other techniques that can be used to determine whether a given enhancer affects the lipid or protein component of mucosal membranes include fluorescence polarization [133], electron spin spectroscopy [134], change in phase transition temperature [135], differential scanning calorimetry, and infrared spectroscopy [136].

Using one of the above techniques, fatty acids and their derivatives have been found to act primarily on the phospholipid component of membranes thereby creating disorder and resulting in increased permeability [137-139]. The extraction of cholesterol out of the epithelial membrane was postulated to be the mechanism by which medium chain monoglycerides glyceryl-1-monooctanoate, glyceryl-1-monodecanoate, and glyceryl-1-monododecanoate promoted rectal cefoxitin absorption [114]. Some fatty acids, however, act on the protein component in membranes. Kajii et al. [140] reported that caprylate caused a decrease in the fluorescence polarization of dansyl chloride, fluorescein isothiocyanate, and eosin maleimide covalently linked to rat small intestinal brush border membrane vesicles. It caused no change in the fluorescence polarization of lipid-soluble fluorescent probes [2-(9-anthroyloxy)stearic acid, 12-(9-anthroyloxy)stearic acid, and 1,6-diphenyl-1,3,5-hexatriene] labeled to the brush border membrane vesicles. Thus, the locus of action of caprylate is at the protein rather than the lipid domain of the brush border membrane vesicles.

Salicylates also act on the protein component of plasma membranes. This is indicated by their effect on the fluorescence polarization of only those small intestinal brush border membrane vesicles labeled with fluorescent dyes with affinity for proteins, such as 8-anilino-1-naphthalene-sulfonic acid and 2',4',5',7'-tetrabromo-5-maleimidofluorescein, but not those labeled with fluorescent dyes with affinity for lipids, such as 1,6-diphenyl-1,3,5-hexatriene and 12-hydroxyoctadecanoic acid 9-anthroate. The additional evidence is that the enhancers released entrapped carboxy-fluorescein from brush border membrane vesicles but not from liposomes made of phosphatidylcholine:cholesterol 1:1 and phosphatidylcholine:cholesterol:phosphatidic acid 1:1:0.1 [141]. Similar results were obtained with red blood cell membranes and rectal brush border membrane vesicles [142,143].

Nonprotein thiols such as glutathione are yet another membrane component where certain enhancers act. These nonprotein thiols are believed to play an important role in maintaining cell integrity [144] and in preventing passive uptake of hydrophilic compounds [145]. Compounds known to affect the nonprotein thiols include oleic acid [146], diethyl maleate [147], and salicylates [148,149]. Nishihata et al. [148] and Suzuka et al. [149] showed that salicylate decreased the levels of nonprotein thiols in intestinal tissues and isolated enterocytes. Murakami et al. [146] showed that the enhancement in carboxyfluorescein penetration across the colonic mucosa by oleic acid solubilized in a nonionic surfactant HCO-60 required intact nonprotein SH groups for its action. This is indicated by the reduction in effectiveness by sulfhydryl agents, such as N-ethylmaleimide, $HgCl_2$, and iodoacetamide. Reduction in mucosal nonprotein thiol by diethylmaleate was associated with increased absorption of cefoxitin, cefmetazole, and phenol red. The administration of cysteamine as an exogenous nonprotein thiol restored nonprotein thiol levels in the mucosa along with the barrier function of the intestinal mucosa.

Paracellular Transport

Promotion of peptide and protein absorption via the paracellular pathway has gained interest recently because of the perception that the paracellular pathway is deficient in proteolytic activity thereby maximizing the amount of peptides and proteins absorbed intact. Increase in the permeability of the paracellular pathway has been depicted as increase in the "pore" size of the mucosal membrane, although the existence of pores in membranes is still a matter of debate. Tomita et al. [133] calculated that the pore size of the colon increased from 8−9 Å to 14−16 Å in the presence of Na caprate (C_{10}), Na laurate (C_{12}) and mixed micelles of Na oleate ($C_{18:1}$) and sodium taurocholate, and to 11−12 Å in the presence of taurocholate and caprylate (C_8).

The barrier to paracellular diffusion of molecules and ions across the epithelial cell layer is the tight junction or zonula occludens (ZO). It is one of the three distinct morphological elements of the epithelial junctional complex, the other two being the zonula adherans (ZA, intermediate junction) and the desmosomes [150]. At the tight junction the plasma membranes are brought into extremely close apposition, but not fused, so as to occlude the extracellular space. Although the degree of permeability of the tight junctions varies in different epithelia, the tight junctions have been reported to be essentially impermeable to molecules with radii ≥ 15 Å [151,152]. In MDCK cells they display a characteristic pattern of cation selectivity which makes them behave as pores with hydrated negative sites [151]. Consequently, anionic substances may not be able to pass through.

Role of Ca^{2+} in regulating paracellular permeability. The integrity of epithelial cell tight junctions has long been known to depend on extracellular Ca^{2+} [153,154]. In fact, the ability of Ca^{2+} to restore the barrier function of a mucosa following exposure to a given enhancer is usually taken as evidence for its paracellular action. The caveat is that one must ascertain that formation of an inactive complex between the enhancer and added Ca^{2+} is not a confounding factor [155]. Gumbiner and Simons [156] observed that within $15-20$ min of incubation, the resistance of MDCK monolayers grown on nitrocellulose filters dropped about 100 times to the value of the bare filter (~ 40 $\Omega.cm^2$). Such Ca^{2+} dependence is now believed to result indirectly from Ca^{2+} effects on other junctional elements rather than from direct effects on the tight junction. The most probable junctional element affected is the Ca^{2+}-dependent cell adhesion molecule uvomorulin (L-CAM) [156-158], which has been localized to the ZA of the small intestinal epithelium by EM immunocytochemistry [159]. This molecule may act in concert with the actin filaments of the cytoskeleton to render the tight junction impermeable [160].

Enhancement in paracellular absorption results not only from expansion in the dimension of the tight junction and the intercellular space but also from increase in water influx through that space [161]. This is the case in the promotion of paracellular transport of phenol red and cefmetazole by EDTA from the rat colon [162], of antipyrine by EDTA [163] and Na caprate [164] from the rat rectal mucosa, and of p-aminobenzoic acid by polyoxyethylated nonionic surfactants from the rat colon [165]. The increase in water flux is Na dependent, as indicated by reduction in its effect by ouabain [164]. This is characteristic of increased water flux in the paracellular pathway when compared with that in the transcellular pathway [162,164]. Increase in water flux in the transcellular pathway can be induced by diethyl maleate which reacts with glutathione in the membrane [166] and by nonsteroidal antiinflammatory drugs such as indomethacin, diclofenac, and phenylbutazone on the rectal absorption of sulfanilic acid [124]. Because the colon is better equipped to conserve water than elsewhere in the GI tract, increase in water flux may not be as pronounced in the upper GI tract. Increase in water influx may affect drug absorption by one of the following mechanisms: (1) increase in concentration gradient for penetration, (2) increase in solvent drag, and (3) increase in blood flow at the absorption site [164].

Other means to enhance paracellular permeability. Lowering extracellular Ca^{2+} concentration is not the only way in which permeability of the tight junction can be enhanced. Other means include (1) enhancing Na^+

transport by increasing osmolality of the dosing solution or by promoting glucose and amino acid transport and (2) altering actin filaments by compounds such as cytochalasins and phorbol esters.

The tight junctions in the jejunum of the rat were found to open in the presence of a hyperosmotic load created by 600 mOsm mannitol [167]. As a result, horseradish peroxidase was found in the intercellular spaces between adjacent absorptive epithelial cells of jejunal villi. The rectal absorption of gentamicin sulfate in rats was also facilitated by the use of high ionic strength aqueous formulations up to 1.2 m [168]. Elevated osmotic pressure was not a major determinant of rectal gentamicin absorption, as indicated by the inability of sorbitol to increase gentamicin bioavailability above control levels. The membrane Na-K pump was suspected to be involved since Na^+ was more effective than K^+. For this reason, the adjuvant effect of Na salicylate, disodium ethylene(dinitrilo)tetraacetate, and Na 5-bromosalicylate was enhanced in the presence of high Na^+ [169].

Pappenheimer and Reiss [170] demonstrated that in the presence of 25 mM glucose fluid flow from the jejunum and upper ileum of the rat was doubled, as was clearance of creatinine (M.W. 113, 3.2 Å), PEG 4000 (M.W. 4,000, 12.4 Å), and inulin (M.W. 5,500, 14 Å). Moreover, Pappenheimer [171] showed that there was a 2- to 3-fold decrease of resistance with a simultaneous increase of membrane surface (capacitance) and width of the intercellular junctions and lateral spaces (conductance). The equivalent pore radius was estimated to be 50 Å. This response was dependent on oxygen tension suggesting the involvement of some energy-dependent contractile process.

According to this mechanism, active transport of gluocse and amino acids, which is coupled to Na^+ transport, across the intestinal mucosa into the intercellular lateral spaces creates an osmotic force for fluid flow. This in turn triggers contraction of the perijunctional actomyosin ring, resulting in increased paracellular permeability [172]. Involvement of actin filaments in this process is indicated by gradual increase in paracellular permeability upon exposure to cytochalasins, drugs that disrupt actin filaments [173] which interact directly with the ZO or indirectly from disruption of the more extensive actin filament interaction with the ZA.

Besides cytochalasins, phorbol esters, through stimulating protein kinase C (a Ca^{2+}/phospholipid-dependent enzyme), induced opening of tight junctions, as judged from increased paracellular permeability of the LLC-PK$_1$ cell monolayers to D-mannitol and PEG 4000 [174]. Positive correlation between transepithelial resistance of cultured intestinal epithelia, their transepithelial mannitol fluxes, and number of strands in their occluding junctions has been established [152]. Indeed, changes in MDCK cytoskeleton induced by the tumor promoter, 12-O-tetradecanoylphorbol-13-acetate, have been shown to give rise to local changes in plasma membrane fluidity [175], even though the active promoters are without specific effect on the membrane fluidity or conductance of artificial lipid membranes [176]. Such changes in membrane fluidity are the consequence of alteration in the other function of the tight junctions, which is to separate the apical from the lateral membrane [177,178] thereby maintaining differences between the two membrane domains with respect to their lipid composition and fluidity [179–181]. Friedlander et al. [182] showed that opening the tight junctions in MDCK cells by incubating the cells in a calcium-free medium had the same effect as benzyl alcohol on membrane fluidity. When the occluding junctions of LLC-PK$_1$ monolayers were disrupted after 24 hr of incubation in Ca^{2+}-free medium there was intermixing of membrane components, as shown by the migration of ectoenzyme alkaline phosphatase [182a] and the Na-cotransport of sugars system [178] from the apical to the basolateral membrane.

B. Factors Influencing Effectiveness

The efficacy of penetration enhancers is influenced by several factors, including physicochemical properties of the peptide and protein drug, site where they are administered, nature of the vehicle in which they are housed, and whether they are used alone or in combination.

Physicochemical Properties of Peptide and Protein Drugs

In selecting a penetration enhancer, the physicochemical properties of the peptide and protein drugs relative to the mode of action of the enhancer must be considered. This point has been emphasized by Guy and Hadgraft [183,184] in their analysis of the mode of action desired of model enhancers in promoting the percutaneous penetration of drugs of varying lipophilicity. These investigators modeled drug diffusion through the skin by consecutive diffusion across the stratum corneum and viable epidermis with a partitioning process at their lipophilic-aqueous phase boundary. Two model enhancers were considered: the first increased only the drug's diffusion coefficient across the stratum corneum, whereas the second decreased the effective stratum corneum-viable tissue partition coefficient of the drug in addition to increasing its diffusion coefficient across the stratum corneum. This analysis revealed that, whereas hydrophilic drugs would benefit from either type of enhancers, lipophilic drugs would benefit only from the second type of enhancer. It seems that since most peptide and protein drugs are hydrophilic, either type of enhancer could be used.

A relating factor is possible interaction between the peptide and protein drugs with the enhancer resulting in loss in activity. This may be the reason for the lesser effectiveness of nonionic surfactants (polyoxyethylene-10-lauryl ether, poloxyethylene-10-cetylether, polyoxyethylene-10-stearylether, and polyoxyethylene-10-nonylphenylether) than the bile salts (Na glycocholate and Na cholate) in promoting the nasal absorption of β-interferon in rabbits, as measured by antiviral activity [16].

Site of Administration

The effectiveness of enhancers varies from site to site. The buccal mucosa and perhaps the skin are among the most resistant to enhancer action. Thus, Na taurocholate was ineffective in lowering the barrier function of the keratinized oral mucosa of the hamster cheek pouch [185]. As seen in Table 4, EDTA and Na salicylate, both at 5%, were less effective in promoting insulin absorption from the buccal as well as the nasal mucosa than from the rectal mucosa.

The above site to site variation in effectiveness of enhancers is also seen within the GI tract. Enhancers like medium chain glycerides [108], dipalmitoylcarnitine [106], and EDTA [186] are more active in the upper than the lower part of the GI tract. Thus, Capmul (medium chain glycerides) were more effective orally than rectally in enhancing the absorption of non-orally available β-lactam antibiotics, such as cefoxitin, moxalactam, and cefamandole [108]. The corresponding rectal and oral

Table 4   The Effects of Enhancers on the Rectal, Nasal, and Buccal
Absorption Efficiency (10 U/kg) Relative to Intramuscular Absorption
Efficiency in the Rat

Enhancer	Rectal	Nasal	Buccal	Sublingual
None	17.0 ± 5.6	0.4 ± 0.2	3.6 ± 2.8	0.3 ± 0.2
Na Salicylate (5%)	41.7 ± 11.3	4.1 ± 2.4	2.9 ± 2.0	—
EDTA.Na (5%)	31.0 ± 6.7	3.5 ± 1.0	0.9 ± 0.4	—
Na Glycocholate (5%)	40.1 ± 8.2	46.8 ± 4.7	25.5 ± 7.5	11.9 ± 7.6
Aprotinin (270 U/ml)	15.1 ± 2.8	9.6 ± 4.3	2.4 ± 1.6	—

Source: Refs. 457 and 458.

bioavailability was 46 and 100% for cefoxitin, 25 and 54% for cefamandole,
and 17 and 40% for moxalactam. Palmitoylcarnitine was more effective in
promoting cefoxitin absorption from the small intestine than from the rectum of the dog [106]. EDTA was more effective in promoting the absorption of fosfomycin in the jejunum than the colon of the rat [186].

Other enhancers, such as polyoxyethylene 9-lauryl ether [186], saponin
[186], and free fatty acids [186], are more active in the lower than the
upper part of the GI tract. Thus, caprylate was more effective in the
colon than in the jejunum relative to fosfomycin absorption [186], even
though brush border membrane vesicles prepared from the rat colon were
less sensitive to the action of caprylate than were the vesicles prepared
from the jejunum [131,140]. This suggests that caprylate's primary site of
action on the mucosa is at the tight junction rather than at the protein
component of the membrane. This finding also emphasizes the importance
of understanding the cellular site of action of the enhancer before selecting
an experimental system to evaluate site to site variation in effectiveness.

The site to site variation in effectiveness of enhancers mentioned above
is probably due to differences in the composition of phospholipids and
other constituents of the epithelial cell membranes of various mucosae.
This is indicated by the role of sphingomyelin content in the erythrocyte
in determining stability of the erythrocytes against emulsion-induced hemolysis [187] and by the role of cholesterol content in determining the susceptibility of multilamellar liposomes made of egg phosphatidylcholine, phos-
phatidic acid, and/or cholesterol to saponins [188]. The same reason is
perhaps responsible for the species differences in sensitivity to a given
enhancer. For instance, sodium taurodihydrofusidate was more effective
in rats (60-fold) than in rabbits (6-fold) relative to enhancing nasal insulin
absorption [188a]. Moreover, 3% Azone is more effective in hairless mouse
skin than in rat skin, even though the hairless mouse skin is less permeable than rat skin to vasopressin in one case [105] and is as permeable as
the rat skin to verapamil in another [189].

Nature of Vehicle

The nature of the vehicle can affect the effectiveness of enhancers because it can affect the enhancer concentration at the absorption site and

its time course relative to that of the peptide and protein drug. Thus, caprate was found to lose its effectiveness in promoting the rectal absorption of hEGF in the rat when the vehicle was changed from Na carboxymethylcellulose to methylcellulose [146]. Moreover, while Tween-type surfactants were effective in promoting the rectal absorption of sulfanilic acid in the rat when administered in an oil or an oil-in-water emulsion vehicle, they were ineffective when administered in aqueous solution [190]. This was attributed to the higher local enhancer concentration achieved in the rectal membranes as a result of partitioning of the surfactant into the small amount of water present in the rectal membranes. Similarly, palmitoylcarnitine was ineffective in promoting rectal insulin absorption when formulated in a PEG 1500/4000 suppository base but was effective (27 ± 7.1%) when formulated in a Witepsol H-15 suppository base [106]. Moreover, Na 5-methoxysalicylate was more effective in propylene glycol than in buffer in enhancing rectal insulin absorption [191]. This was postulated to be due to an increase in the lipophilic characteristics of the enhancer as a result of increase in intramolecular hydrogen bonding and in the apparent pKa of methoxysalicylate in the lower dielectric constant of the vehicle [192]. Such an effect of propylene glycol was first reported by Cooper [193]. This investigator showed that the transport of nonpolar materials across human skin can be increased by an order of magnitude by adding small amounts of fatty acids or alcohols, such as oleic acid and oleyl alcohol, to vehicles containing propylene glycol.

Whether the delivery rate of the enhancer from its vehicle will affect its effectiveness depends in part on its own rate of absorption and therefore accumulation in the mucosal epithelial cells. Salicylate, which is rapidly absorbed, was more effective in promoting the rectal absorption of cefoxitin when given in a bolus than in an infusion [194]. The converse is true of MGK, which is slowly absorbed, in the rectal absorption of cefazolin [113]. MGK is commercially available mixture of glycerol (8%), octanoic acid (3%), glyceryl monooctanoate (57%), glyceryl dioctanoate (29%), and glyceryl trioctanoate (3%). By contrast, the rate of delivery of MGK does not strongly influence its efficacy in enhancing the rectal absorption of des-enkephalin-γ-endorphin, a dodecapeptide, in the rat [109]. This suggests that a lower concentration of MGK is required to promote the absorption of this peptide when compared with the antibiotic.

Combinations

In some instances, enhancers are more effective when used in combination than alone. This was the case for MGK and Na$_2$EDTA in the rectal absorption of des-enkephalin-γ-endorphin. Whereas the rectal bioavailability of this peptide was 0-1% in the presence of EDTA and 8-20% in the presence of MGK, it increased to 10-44% in the presence of the combination [109]. The authors attributed this synergistic effect to the membrane perturbing effect of MGK and to the protease inhibitory effect of EDTA. The protease inhibitory effect of EDTA was inferred from the extension in the clearance of 0.04% peptide from the rectum from 33 ± 7 to 93 ± 45 min in the presence of 0.25% EDTA. Such an effect is not likely to be a major explanation for the synergistic effect, however. This is because even a 90-fold excess of unlabeled peptide was ineffective in further enhancing the bioavailability of the peptide delivered in MGK. Therefore, the synergistic effect of the combination likely results from the concerted effect of MGK on transcellular penetration and of EDTA on paracellular penetration.

Combinations may be a more effective method than increasing the enhancer concentration in enhancing the effectiveness of penetration enhancers. Indeed, increasing the enhancer concentration may not necessarily lead to enhanced drug penetration if the enhancer can form micelles
thereby entrapping drug and reducing its thermodynamic activity. This
occurred when the concentration of Brij 36T was increased to enhance the
transport of hexyl nicotinate across the human skin [195] and when the
concentration of Na lauryl sulfate, cetylpyridinium chloride, and poly-
sorbate 80 was increased to enhance the penetration of salicylic acid across
the keratinized oral mucosa of the hamster cheek puch [185]. On the
positive side, micellar entrapment has been postulated to enhance the intestinal absorption of ergot peptide alkaloids by preventing the drugs
from interacting with mucus [196], which itself may be removed by the
surfactant constituting the micelle.

C. Safety Concerns

Irritation of mucosal tissues by penetration enhancers, the extent of
damage in the mucosal cells they cause, and the rate at which such
damage recovers are three of the most pressing safety concerns associated
with the use of these substances to promote the mucosal absorption of peptide and protein drugs.

Irritation

Irritation is a complex phenomenon involving interaction among the solution
properties of the vehicle, mucosal transport, biological response, and
local drug disposition. In the nose, irritants have been shown to trigger
the release of neuropeptides such as substance P, somatostatin, and calcitonin gene-related peptide from mucosal neurons, epithelial cells, and
leukocytes which may in turn evoke a sneezing reflex and induce nasal
secretion [197,198]. This is akin to the local inflammation and congestion
associated with the common cold or hay fever, whose effect on nasal drug
bioavailability is still uncertain. Larsen et al. [199] determined that
rhinitis induced by nasal application of histamine to the nostrils of 24
healthy subjects had no effect on the nasal absorption of buserelin, a
LHRH agonist. On the other hand, Olanoff et al. [200] reported that
similar treatment improved the nasal absorption of 1-deamino-8-D-arginine
vasopressin (DDAVP).

The irritative response is very subjective. Thus, while mild nasal
discomfort was experienced by patients receiving nasal sprays of salmon
calcitonin containing no enhancers [201], no such local side effects were
reported in a study involving five Paget's disease patients [202]. In
another study on the nasal administration of human pancreatic tumor
growth hormone releasing factor-40 [203], mild burning of the nasal
mucosa was reported in one of the six patients.

Despite its importance as a potentially negative factor limiting the
routine use of penetration enhancers in mucosal peptide and protein drug
delivery, the relationships between irritation and structure of the enhancer
have not been addressed. Such information is, however, available in the
transdermal literature.

In evaluating the relationship between pK_a and acute skin irritation of
a homologous series of benzoic acid derivatives in man, Berner et al. [204]
found that erythema, edema, and the primary irritation index (defined as
the average of the sum of the erythema and edema scores measured on

days 2 and 3) were all linearly correlated with pK_a; erythema at 24 hr appeared to be the most sensitive parameter to variation in pK_a when $pK_a \leq 4$. Similar structure-irritancy relationships were also seen in the 1-alkyl- and 1-alkenylazacycloalkanone derivatives. In this instance, the enhancers with an alkyl chain induced more severe primary irritation (erythema and edema) than those with an alkenyl chain [205].

The relationship between skin irritation and penetration efficacy is not as clear. Aungst [206] reported that the skin irritation potential of fatty acid isomer penetration enhancers correlated poorly with their potency in enhancing the flux of naloxone across human skin. The enhancer was applied as a 10% solution in propylene glycol over a 6 hr period to the mid-back area of albino rabbits. Irritation at the application site was assessed by scoring for edema formation and for erythema and eschar formation. While neodecanoic acid and lauric acid were equally effective as penetration enhancers, lauric acid was more irritating than the former. These results are markedly different from those obtained in 1-alkyl- and 1-alkenylazacycloalkanone derivatives, which were evaluated by Okamoto et al. [205] relative to improvement of 6-mercaptopurine penetration across guinea pig skin. There was good correlation between penetration enhancing effect and skin irritancy of the enhancers. Thus, 1-dodecylazacycloheptan-2-one was the most active, but also the most irritating. 1-Geranylazacyclopentan-2,5-dione was practically nonirritating, but also the least active.

Extent of Membrane Damage

Because all penetration enhancers promote peptide and protein absorption by perturbing membrane integrity, it is inevitable that varying extents of insult would occur to the mucosal tissues which are in intimate contact with the enhancer. The sensitivity of each mucosal tissue to the membrane damaging effect of a given enhancer varies. Thus, the proximal small intestine of the guinea pig appears to be far more sensitive to di-hydroxy bile salts than is the distal mucosa even though far more bile salts are absorbed distally [207]. Moreover, the small intestine appears to be more sensitive to the membranolytic effect of 5-methoxysalicylate than the rectum [155,208]. In addition to structural and functional differences, the greater sensitivity of the proximal region to the membranolytic effect of the enhancers may be due to exacerbation of the effects of a minor lesion by the proteolytic enzymes which are present in larger amounts in the upper than the lower intestine.

Attempts have been made to link penetration enhancement potency with membrane damage, but the results have been conflicting. On the one hand, caprylate and caprate have been found to promote the colonic absorption of cefmetazole in the absence of frank damage to the mucosa [131,133]. On the other hand, dihydroxy bile salts, which are more damaging to the guinea pig and hamster small intestine than trihydroxy bile salts [207], are also more effective than the trihydroxy bile salts in promoting nasal insulin absorption [116]. Similarly, the enhanced absorption of horseradish peroxidase, insulin, and cyclo-(Pro-Phe-D-Trp-Lys-Thr-Phe) (a somatostatin analog) into the portal circulation following the co-administration of 60 mg Na 5-methoxysalicylate in 0.5 ml of saline to the duodenum of the rat was associated with disruption of the epithelial barrier, which occurred as early as 10 min resulting in progressive, severe damage by 1 hr [208]. Such progressive damage may be mediated by luminal proteases that had penetrated the epithelial barrier.   The adjuvant effect appeared to cease after 30 min, probably the result of local shutdown of blood flow in the villus or submucosa associated with extensive mucosal stripping.

Several methods have been used to assess membrane damage; however, the correlation of the results among the various methods has not been attempted.   These methods include morphological examination by scanning or transmission electron microscopy [130,209], determination of extent of hemolysis [130,210], determination of release of cellular constituents [165], measurement of changes in electrical resistance [211], monitoring of changes in permeability to markers [212], and measurement of changes of cilia movement [118].

Perhaps because of its simplicity, hemolysis of erythrocytes has been a popular test to assess the membrane damaging effect of enhancers [130, 210].   Nonetheless, the results obtained in this test may not be extrapolated directly to the mucosal cells exposed to the enhancer.   A case in point is acylcarnitines.   While palmitoylcarnitine at 0.025 mM resulted in 100% lysis of rat erythrocytes in less than 2 min [213,214], no significant mucosal tissue damage was seen in the rat jejunum and colon exposed to palmitoylcarnitine even at 23 mM [106].   This may reflect differences in membrane constituents since membrane proteins, phosphatidylethanolamine, phosphatidylserine, phosphatidylinositol, and sphingomyelin have been shown to protect cells against the lytic activity of amphipathic quaternary ammonium compounds [213,214].

A companion issue is choice of animal model, on which there is very limited information.   Low-Beer et al. [207] noted that the guinea pig ileum was more susceptible to the membrane damaging effects of 1-2 mM Na deoxycholate than the hamster ileum.   Bond and Barry [215] reported that hairless mouse skin showed exaggerated responses to all the penetration enhancer formulations, including Azone, decylmethylsulfoxide, oleic acid, and propylene glycol, when compared with human skin.

Recovery Rate of Mucosal Membrane from Possible
Damaging Effects of Penetration Enhancer

The rate at which mucosal membranes recover from the acute damaging effect of a given penetration enhancer appears to be related to the extent of membrane damages that occur.   But the rate deemed to be safe from the standpoint of exclusion of environmental toxic materials has not yet been defined.   The recovery rate is slow in the case of aggressive enhancers such as Na deoxycholate [20] and polyoxyethylene-9-lauryl ether [117], while it is more rapid in the case of less aggressive ones such as acylcarnitines [106] and Na glycocholate [117].   Fix et al. [106] reported that the increase in rectal membrane permeability afforded by acylcarnitines was rapidly reversible within 60-120 min and occurred at acylcarnitine concentrations which did not appear to alter membrane morphology.   Hirai et al. [117] showed that the barrier function of the rat nasal mucosa recovered within 2 hr after exposure to 1% Na glycocholate but not until 24 hr later in the case of polyoxyethylene-9-lauryl ether [117].   Under scanning electron microscopy, the villi on the nasal mucosa 2 hr after administration of Na glycocholate were observed to be slightly denuded at the tip but gradual recovery was observed with the passage of time and was complete after 24 hr.   By contrast, the alteration of nasal mucosa after administration of polyoxyethylene-9-lauryl ether was significant and complete restoration was not observed at 24 hr.

So far, no deliberate attempts have been proposed to accelerate repair of the mucosal damages caused by penetration enhancers. In this regard, perhaps permeability changes elicited by paracellular enhancers lend themselves more readily to such intervention than those elicited by transcellular enhancers. This is because the permeability of the tight junction is sensitive to extracellular Ca^{2+}. Martinez-Palomo et al. [153] showed that resealing of the tight junctions in MDCK cells required high Ca^{2+} on the extracellular side and low Ca^{2+} concentration of the cytoplasmic compartment. Maximal effect was achieved at very low concentrations, 1 mM. In both this and the study by Gumbiner and Simons [156], the resistance was found to recover rapidly if Ca^{2+} ions were restored within 15−20 min, when the overall disposition of membrane junctional components still remained intact. Recovery was then complete over a 30−60 min period. When Ca^{2+} was not restored until 40−60 min, the resistance recovered much more slowly. Recovery of the electrical resistance across MDCK monolayers was not complete even as late as 24 hr.

There is virtually no information on the chronic effects of penetration enhancers on barrier function. Many years ago, Hirai et al. [117] tested the subchronic effects of the enhancers on nasal morphology by administering the surfactants to rats 3 times a day for one month. Polyoxyethylene 9-lauryl ether was found to cause morphological change in the microvilli of the nasal mucosa whereas Na glycocholate caused no discernible changes. In another study, Nakanishi et al. [190] found that the permeability of the rectal mucosa after repeated treatment was not increased as much as on the first treatment.

Other Safety Concerns

There are three other safety concerns associated with the use of penetration enhancers but which have not been satisfactorily addressed. The first is interference with the normal metabolism and function of the mucosal cells exposed to the enhancer. For instance, perfusing the hamster jejunum with conjugated and unconjugated bile acids resulted in inhibition of water absorption, more so by the di- than the trihydroxy bile acids [216]. Moreover, amino acid metabolism [217] and glucose transport [218] are known to be blocked by unconjugated bile acids. The second safety concern is possible metabolism of the enhancer to toxic or carcinogenic substances in mucosae which are rich in monooxygenases, such as the nasal mucosa [219]. The third safety concern is absorption of the enhancer into the bloodstream. This has been shown for hexamethylene lauramide, which was quantitatively absorbed following topical application to the rat skin [220], and for Azone, which was absorbed to the extent of 0.4% following application to the forearm of the human subjects under 12-hr occlusion [221].

D. Effects Other Than Changes in Barrier Function

In addition to changes in barrier function, changes in membrane fluidity caused by certain penetration enhancers may also affect the activity of numerous membrane-bound proteins, among which are enzymes and transport systems that are known to be affected by membrane fluidity. This possibility, however, has not yet been considered.

Several examples can be cited to illustrate how changes in membrane fluidity may alter the activity of membrane-bound transport proteins and enzymes. Thus, inorganic phosphate uptake and Na-dependent glucose uptake were dramatically impaired upon increasing the fluidity of renal brush-border membrane vesicles and MDCK and LLC-PK$_1$ cells by the local anesthetic drug benzyl alcohol and by deprivation of Ca^{2+} [182,222]. Similar treatments also impaired the activity of adenyl cyclase, a membrane-bound enzyme, in response to hormones such as vasopressin, glucagon, and prostaglandin E$_2$ [223]. On the other hand, the activity of this enzyme in erythrocytes increased following treatment with cis-monoun-saturated acids, but was unaffected by treatment with the corresponding trans-isomers or octadecanoic acid [224]. The order of enzyme activation for the cis acids was cis-11 $\geq$ cis-9 $\geq$ cis-6-octadecenoic, which was also the order of reduction in membrane microviscosity.

E. Alternatives to Penetration Enhancers

Aside from seeking penetration enhancers that are safe and effective, several alternatives to penetration enhancers have been considered. These include use of iontophoresis [18,225,226] and phonophoresis [227,228], altering the formulation pH and tonicity [229–231], and optimizing the vehicle characteristics with respect to deposition and retention at the site of administration [233–238]. The use of iontophoresis and phonophoresis has been discussed in Chapter 15 and elsewhere [228]. Altering formulation pH and tonicity has been found to improve the nasal absorption of insulin [229], secretin [230], and human growth hormone [231]. Nonetheless, unless the companion issue of tissue irritation is resolved, patient acceptance of these two approaches is anticipated to be low.

The most promising alternative in terms of balance between efficacy and patient acceptance is that of optimizing vehicle characteristics with respect to deposition and retention at the administration site. This approach has been emphasized by Harris and coworkers in their work on the nasal absorption of DDAVP. These investigators demonstrated that the nasal absorption of DDAVP was enhanced by administering the dose in the smallest volume possible as sprays than as drops [235,236]. This was attributed to prolonged retention of the dose at the administration site. Illum et al. [237] showed that retention could be further improved by administering the dose in microspheres made of soluble starch, DEAE-Sephadex, or albumin. This led to good absorption of insulin from starch microspheres with a diameter of 45 μm administered as a dry powder to the nasal cavity of the rat [238]. The bioavailability was approximately 30%, as compared with no absorption from an insulin solution. The mechanism for this impressive absorption efficiency was not indicated, however. A possible explanation is loosening of the tight junction as a result of shrinkage of the cells caused by water flux out of the cell into the mucus to replenish what was lost to the deposited microspheres undergoing hydration.

The above positive effect of the vehicle in peptide and protein absorption was also seen in the rectal route. Morimoto et al. [233] demonstrated that the rectal bioavailability of [Asu^{1,7}]-eel calcitonin in the rat was maximized by incorporating it in a 0.1% polyacrylic acid gel than in saline, a triglyceride fatty acid mixture base, or a polyethylene glycol 1000 base. This was attributed to retention of the dose near the site of administration. Following administration, solutions tend to spread away from the rectum into the colon and from there into the portal circulation [239], resulting in reduced peptide and protein bioavailability. By minimizing spreading of

the dose into the colon, short, fat suppositories provide better insulin
bioavailability than long, thin ones [240], as do small suppositories when
compared with larger ones [241].

IV. NEED TO CONSIDER CHRONOPHARMACOLOGY IN
DRUG DELIVERY SYSTEM DESIGN

A variety of endocrine systems are characterized by pulsatile hormone
secretion, with pulse intervals ranging from minutes [242−245] to hours
[246−248]. The biological significance of pulsatile secretion of hormones
such as insulin, glucagon, and growth hormone, as well as the mechanisms
of pulse generation, has been reviewed by Weigle [249]. The mechanisms
governing the circadian shift in pulse frequency seen in hormones such as
luteinizing hormone (LH) [250], prolactin [251], ACTH, growth hormone
(GH) [252,253] remain a matter of speculation. The pineal gland and its
melatonin secretion have, however, been implicated [254]. This is because
melatonin secretion is tightly coupled to both the light-dark and the sleep-
wake cycles, with secretion being increased during darkness and sleep
[255,256]. Melatonin binding sites have been discussed intermittently in
the literature since the early experiments of Wurtman and colleagues [255,
257]. Lang and colleagues [258] observed high melatonin binding site
density in the pituitary gland, hypothalamus, epididymis and adrenals and
lower density in the liver, lung, spleen and heart. Using in vitro autoradiography with ^{125}I-labeled melatonin, Reppert et al. [259] identified
putative melatonin receptors in the suprachiasmatic nuclei of the human
anterior hypothalamus.

Diurnal variations in the concentration of several hormones have been
reported. Thus, plasma cortisol concentration is highest at about the time
of habitual awakening and lowest shortly before sleep [260]. The plasma
concentration of insulin-like growth factor binding protein in children,
aged 7−18 yr, is lowest (0−240 µg/liter) at 1200 hr, rising to a peak of
50−500 µg/liter between 2400 and 0600 hr [261]. Six hypothalmic hormones,
bombesin, cholecystokinin (CCK-8), neurotensin, peptide histidine isoleucine amide, substance P and vasoactive intestinal peptide, are at their
lowest levels around the time of onset of darkness, reaching a peak at
around the beginning of the light phase. Thereafter the hypothalamic
levels of the neuropeptides fall by 45 ± 4% over a 6 hr period [262]. TRH-
like immunoreactivity in the cerebrospinal fluid (CSF) of rhesus monkey
reaches its maximal concentrations (34 ± 2 pg/ml) in the afternoon (1613
hr). Somatostatin-like immunoreactivity in the CSF, on the other hand,
reaches its maximal concentration (82 ± 3 pg/ml) at night (0049 hr) [263].

The circulating levels of gastrointestinal regulatory peptides in healthy
subjects also follow a diurnal pattern. Two such patterns have been observed. One pattern appears to be regulated mainly by meals and is
characteristic of insulin, gastrin, somatostatin, gastric inhibitory peptide,
and pancreatic polypeptide. The second pattern, although also influenced
by meals, is characterized by highest levels of peptides late in the evening or during the night. Peptides that yield such a pattern include vasoactive intestinal peptide, secretin, and cholecystokinin [264].

Basal insulin secretion in healthy subjects shows a circadian rhythm
with a peak time at 1500 hr [265]. Similarly, Hunter et al. [266] reported
that normal mice showed circadian fluctuation in the basal blood glucose

levels with a peak of 112 mg/dl at 1430 hr. Greatest sensitivity to insulin
also occurred at that time, showing a 60% decrease in blood sugar. From
1830 to 1030 hr, insulin produced only 38% decline in glucose. Diabetic
mice showed a circadian variation with phases like that of normal mice,
with basal glucose levels peaking at 438 mg/dl between 1030 and 1430 hr.

The pharmacological effects of several exogenously administered hormones have been observed to be exquisitely sensitive to the pulsatile
secretory pattern of their endogenous counterparts. Thus, pulsatile
glucagon administration was superior to constant rate administration in enhancing glucose production in perifused rat hepatocytes [267]. The optimal pulse interval was between 10 and 20 min, which compared favorably
with the glucagon secretory period of 10 min observed in nonhuman primates [268] and that of 13-20 min observed in humans [269,270]. As
another example, the GH-releasing abilities of growth hormone releasing
factor (GRF) vary markedly according to the time of injection. In the rat,
GH secretion is characterized by an endogenous ultradian rhythm, with
high amplitude GH secretory bursts occurring at 3.3-hr intervals throughout a 24-hr period; in the intervening trough periods, plasma GH levels
are undetectable [271]. The peak GH levels coincide with high levels of
growth hormone releasing factor and low levels of somatostatin [272]. The
weak GRF-induced GH response observed during trough period of the GH
rhythm is due to antagonism by endogenous circulating somatostatin.

The diurnal rhythm of certain endogenous peptides and proteins may
be altered in disease states. Limited information exists in this area, however. Gil-Ad et al. [273] found that in schizophrenics plasma β-endorphin
levels fluctuated randomly throughout the day, ranging within 9-40 pmol/
liter. This pattern was in contrast to the control group, in which β endor-
phin was high in the morning (21 ± 3.5 pmol/liter) and decreased towards
evening. The pattern of plasma human growth hormone level was similar
in both groups. In uncontrolled diabetes mellitus, the circadian rhythms
of hGH and ACTH were disturbed but those of TSH and prolactin were
unaffected [274]. The impairment of certain pituitary hormonal rhythms
was restored after return to normoglycemia. Finally, the mean pulse amplitude, duration, and total fraction of GH secreted in pulses during the 24
hr period have each been observed to be greater in the young but not significantly different between sexes [275]. The mean pulse frequency is not
affected by sex or age. These effects can be accounted for largely by
corresponding variations in endogenous estradiol levels.

A. Influence of Timing of Drug Administration on Drug Efficacy and Toxicity

Because of rhythmic variations in body physiology, reactions to many
drugs may vary at different times during the day [276-278]. This can
result from changes in receptor density [279] or in the absorption, metabolism, and excretion of the drug [280]. Thus, when heparin was infused
at a constant rate, circadian changes in its anticoagulant effect was seen
with a maximum at night 50% higher than the morning value [281]. Moreover, whereas a single dose of theophylline given at 0700 hr resulted in a
plasma drug concentration of 16 µg/ml (within the normally acceptable
range), the same dose given at 1900 hr resulted in a plasma concentration
of only 7 µg/ml [282]. The existence of a circadian rhythm in theophylline
disposition was recently reported by Rackley et al. [283]. These

investigators observed systematic fluctuation in serum theophylline concentrations over a 24-hr period during a 48-hr constant-rate intravenous
infusion of aminophylline in the dog.

Lemmer et al. [284] studied the pharmacokinetics of four beta adrenergic blockers, atenolol, sotalol, metoprolol, and propranolol, in light-dark-
synchronized male rats. After single drug administration, the pharmacokinetic parameters of all drugs depended on light and dark conditions.
Elimination half-lives in plasma and organs were shorter during dark than
during light. This was mainly due to circadian variations in drug elimination with higher hepatic or renal elimination in the activity period of the
rats during dark. Multiple drug dosing abolished the circadian-phase-
dependency in elimination half-lives of the drugs due to the longer lasting
and more pronounced β-receptor blockade over a period of several hours
in dark.

The time of drug administration affects not only the pharmacological
response but also the potential for drug toxicity. This is exemplified in
cis-platinum, which was 25% less nephrotoxic when given at 1800 hr, presumably when urine flow was highest [285]. Moreover, human recombinant
tumor necrosis factor showed a 10-fold variation in toxicity depending on
the time of the day when it was administered relative to the patient's circadian cycle [286]. Similar results are to be expected of other biological
response modifiers such as interferons, leukotrienes, interleukin-2, LAK
cells, growth factors, and monoclonal antibodies. Due to rhythmic changes
in sensitivity of the central nervous system to the drug, intraperitoneally
administered valproic acid in the ICR male mice was more toxic when injected at 1700 hr and least toxic when injected at 0900 hr or 0100 hr [287].
Finally, it is possible to maximize the therapeutic effectiveness of ACTH
(1-17) by optimizing the time of administration. This is because aldosterone is maximally increased when administered at 1400 hr, whereas
both cortisol and testosterone are maximally increased when administered
at 0700 hr [288].

B. Influence of Pattern of Drug Input on Drug Efficacy

There is increasing evidence that pulsed delivery is preferred to constant
delivery in maximizing the effectiveness of therapeutic hormones. Thus,
physiological amounts of triiodothyronine (T_3) given in twice daily injections (0800 and 2000 hr) in female Sprague-Dawley rats were more effective than equal amounts of T_3 administered by constant infusion via the
Alzet osmotic minipump in suppressing TSH secretion in response to
thyroidectomy or to TRH stimulation [289]. Pulsatile delivery of platelet-
derived growth factor (PDGF) in HL-1 serum-free medium at a flow rate
of 2.8 ml/hr for 1.25 hr every 4.75 hr in an atmosphere of 4% CO_2 and
96% air led to lens growth and clarity whereas continuous delivery of this
growth factor resulted in lens opacity and no growth [290]. Pulsed
release is also desirable for vaccines against diseases in which lifelong, or
at least prolonged, immunity can be evoked by exposure to antigens in the
booster. Such diseases include anthrax, cholera, plague, yellow fever,
and malaria. Thus, the initial burst of antigen release over a period of
several days after implantation will elicit the primary immune response.
After a period of weeks, during which little, if any, antigen will be released, the system will spontaneously release a second antigen pulse to
elicit a secondary immune response.

Gonadotropin secretion follows an ultradian rhythm of about 1 hr [246]. Miller et al. [291] demonstrated that pulses of luteinizing hormone-releasing hormone (LHRH) administered every 96 or 120 min to women with hypothalamic amenorrhea activated pituitary-ovarian function with orderly development of a single dominant follicle, a luteinizing hormone surge, and occurrence of ovulation in 20 of the 23 cycles. Similar results were reported earlier by Leyendecker et al. [292] and subsequently by Soules et al. [293]. According to Knobil and his group [294−296], the most effective mode of LHRH administration for restoring and maintaining both LH and FSH secretion was 1 mg/min for a single pulse of 6 min/hr. Three pulses for 20 or 30 min/hr was less effective; five pulses for 12 min/hr was ineffective. The infusion period affected not only the LH and FSH concentrations but also the ratio of these concentrations [297]. A more pronounced effect was seen in serum LH than in serum FSH concentration [297]. In contrast to the results in primates, constant delivery of LHRH to progestin-pretreated anovulatory mares at a rate of 2.5 µg/hr for 7 days via the osmotic minipump surprisingly elevated circulatory mean LH concentrations to those observed in estrous mares. The episodic nature of LH concentrations was preserved, although there was a 22% decrease in the frequency of LH peaks per hour and a 27% longer mean peak lengths [298].

In an attempt to mimic the endogenous bursts of GH, Thorner et al. [298a] administered GHRH by pulsatile injection and demonstrated an increase in GH secretion. By contrast, Rochiccioli et al. [299] found that constant i.v. infusion of GHRH(1-44) at a rate of 0.5 to 1.0 µg/kg·hr from 2000−0800 hr was also effective in correcting impaired nocturnal GH secretion in 16 children with partial GH deficiency, aged 4−14 yr. Moreover, the episodic nocturnal surges of GH secretion were not suppressed, although the number of peaks above 5 ng/ml did increase significantly during the infusions. There were two peaks during the first 4 hr of sleep and a third peak at the end of the night. The reason for restoration of pulsatile GH secretion during constant GHRH infusion is not obvious. It is possible the pulsatile secretion of SRIF is maintained during continuous GHRH infusion thus playing a major role in maintaining pulsatile GH secretion.

In 1980, Cotes et al. [300] reported that hGH administered by continuous infusion via implantable minipumps over 7 days induced greater increase in body weight in hypophysectomized male Wistar rats than hGH administered by daily subcutaneous injections over the same period. Their finding was subsequently refuted in an expanded study by Clark et al. [301]. These investigators found that pulsatile i.v. infusions of human or bovine GH at two doses (12 or 36 mU/day, eight pulses every 3 hr, 5-min duration) produced greater increases in body weight in young hypophysectomized male Sprague-Dawley rats than continuous i.v. infusions of GH at the same daily dose. Continuous infusions of bovine GH produced a lower growth rate in the second of two consecutive 5-day treatment periods, whereas the responses to pulsatile GH did not diminish with time. Both body weight gain and long-bone growth were affected by the frequency of GH pulses; 9 pulses/day were more effective than 3 pulses/day, which in turn produced larger growth responses than 1 pulse/day. Keeping GH pulse frequency constant and varying pulse duration (4, 16, or 64 min) did not affect growth rates. Thus, long-term pulsatile i.v. infusions of GH mimic the endogenous secretory pattern, and are most effective when given at the physiologically appropriate pulse frequency.

Pulsatile administration of human insulin (0.8 mU/kg/min, 7.5 min on/7.5 min off over 4 hr) in normal volunteers has been found to be as effective as continuous administration (0.4 mU/kg/min) in the suppression of endogenous glucose production and in the enhancement of metabolic clearance rate of glucose [302]. Consistent with the findings obtained in isolated perfused rat livers [303], the pulsatile delivery of insulin in a 3-min-on/7-min-off mode at a 40% lower dose is as effective as continuous administration in controlling hepatic glucose production in type I insulin-dependent diabetics [304]. It is proposed that oscillatory insulin replacement is to be preferred since it would help not only to maintain receptor integrity but also to avoid constant hyperinsulinemia with its possible deleterious effects in regard to the development of late diabetic complications [305].

Pulsatile administration is also preferred for glucagon. Weigle et al. [306] demonstrated that over a 90-min period hepatocyte glucose production in response to a fixed total dose of glucagon was enhanced if that glucagon was delivered in a series of six 3-min pulses with a 15-min interpulse interval. The response to continuously administered glucagon exceeded the response to pulsatile glucagon only for very large, but nonphysiological fixed hormone doses. In an effort to explain these observations, Weigle et al. [307] derived a model for the 90-min hepatocyte responses to pulsatile and continuous glucagon delivery based on the waveform of the hepatocyte response to a transient glucagon stimulus. The model demonstrated that the time constant for response decay was an important determinant of the relative efficacy of the two patterns of hormone delivery. The model did not offer insights into the molecular mechanisms for the pulse-enhancement effect by pulsatile glucagon delivery, however.

Whether parathyroid hormone (PTH) stimulates bone formation or resorption is determined by the mode of delivery. Tam et al. [308] demonstrated that continuous infusion of bPTH(1-84) in Sprague-Dawley rats caused an increase in both formation and resorption surfaces and a net decrease in trabecular bone volume. Daily injection of this hormone caused an increase in the formation surface without an increase in the resorption surface, resulting in a net increase in trabecular bone volume. Moreover, Podbesek et al. [309] demonstrated that daily injections of hPTH(1-34) at 1.7 µg/day/kg significantly increased indices of radiocalcium absorption, whereas subcutaneous infusions of hPTH(1-34) at 0.5 µg/day/kg to the same dogs caused no change in the indices of radiocalcium absorption. The lack of effect during continuous infusion was attributed to down-regulation of the PTH receptor, which is known to occur [310]. Indeed, down regulation of receptors has been invoked to explain the loss in efficacy of insulin [311,312], LH [313], GH [314], TRH [315], and LHRH [294] during continuous infusion.

C. Drug Delivery Systems Based on Chronopharmacology

As mentioned above, for peptides and proteins such as luteinizing hormone-releasing hormone, growth hormone, insulin, and glucagon, a pattern of input at different release rates is preferred to that at a constant rate. To this end, several systems are being developed to provide drug on demand or in a pulsed or self-regulating fashion, and they will be discussed next. Implantable infusion pumps which can be preprogrammed to deliver their contents in any pattern has been reviewed by Hrushesky [316] and will not be discussed here.

On-Demand Systems

The first "on-demand" system was reported by Langer et al. [317–318]. This consisted of magnetic beads or cylinders embedded in a ethylene-vinyl acetate matrix containing protein. The rate of protein (bovine serum albumin) release from the matrix increased by a factor of up to 30 in the presence of an oscillating magnetic field and returns to baseline in its absence. A modified version of this system has been extended to insulin using an alginate matrix [319]. The release rate of insulin was about 50 times higher in the presence than in the absence of a 4 Hz magnetic field. When the alginate spheres were cross-linked at the surface by treatment with polyethyleneimine, pulsed delivery of insulin occurred not immediately but during the period just after applying the magnetic field. Other factors influencing peptide release rate include: (1) position, orientation, and strength of the embedded magnets and (2) amplitude and frequency of the applied magnetic field [318]. The mechanism of magnetic modulation is not known. One possibility is that the beads alternately compress and expand the matrix pores in the presence of a magnetic field, thereby forcing out more drug. Other external stimuli under investigation include ultrasound [320] and temperature gradient [321,322].

Pulsed Release Systems

The osmotic pump has been modified by Lynch and co-workers [323] to deliver melatonin, a pineal gland hormone, in an on-off time pattern. The reservoir was filled with a drug-free saline solution and connected to a capillary tubing filled with alternating segments of a melatonin solution and an immiscible melatonin-free mineral oil. Thus, as the contents of the tubing were gradually displaced by saline solution from the pump, the various components of the programmed infusate were discharged sequentially from the opposite end of the tubing. Ewing et al. [324] used this method to produce 0.5 hr pulses of LH and other pituitary hormones every 2 hr, in order to mimic the pulsatile pattern of LH secretion in the rat.
    Alternatively, the approach based on nonuniform drug concentration distribution in either degradable or non-degradable polymeric systems may be considered [325]. Its essential feature is that drug will be loaded in a polymeric matrix nonuniformly to ensure a predetermined time-varying rate of drug release. In 1987, Wise et al. [326] described a pulsed release system of d-norgestrel comprised of blends of coated and uncoated PLGA (90% dl-lactic/10% glycolic copolymer, M.W. 145 kDa) matrices. The profile revealed an initial burst of drug release lasting two weeks, followed by a second burst of greater duration occurring approximately two months after implantation. Although both approaches have not yet been tested in peptides and proteins, they are anticipated to be of practical utility.

Self-Regulating Systems

To date, two major types of polymer-based self-regulating systems have been conceptualized, both of which are geared towards insulin delivery.
    The principle of the first approach is based on competitive binding between blood glucose and glycosylated insulin to lectin (Concanavalin A). This was first tested by Brownlee and Cerami [327] using maltose insulin and subsequently extended by Kim and his co-workers [328,329] who developed a panel of glycosylated insulins with improved affinities for lectin. In addition, this latter group of investigators evaluated the permeability of the glycosylated insulin derivatives across a porous

poly(hydroxyethyl methacrylate) membrane that was freely permeable to both insulin and glucose. The rate and extent of release of these insulin derivatives were found to depend on glucose concentration. When implanted peritoneally in pancreatectomized dogs, the implant produced a blood glucose profile matching that of normal dogs during an intravenous glucose tolerance test [330]. The long-term utility of this system remains to be tested pending investigation of the effect of surface reaction of tissues towards the membrane on insulin and glucose flux.

The principle of the second approach [331] is based on oxidation of glucose to gluconic acid by the glucose oxidase immobilized in a pH-sensitive hydrogel membrane to cause the pH within the membrane to fall and the amine groups in the membrane to become protonated. As a result, the membrane swells, increasing its permeability to the insulin held in a contiguous reservoir. The membrane tested was synthesized by radiation-induced polymerization of frozen solutions containing hydroxyethyl methacrylate, N,N-dimethylaminoethyl methacrylate, tetraethylene glycol dimethacrylate, and ethylene glycol. Thus far, this concept had been confirmed in small drug molecules but not in insulin. Nonetheless, the investigators were optimistic that macroporous glucose-sensitive membranes that are readily permeable to insulin can eventually be developed, thus yielding a self-regulating insulin release system.

A variant of the second approach has recently been described by Fischel-Ghodsian and co-workers [332]. Theirs was based on increase in solubility of a modified insulin derivative with decrease in pH upon oxidation of glucose to gluconic acid by immobilized glucose oxidase in the ethylene-vinyl acetate matrix. The increase in solubility caused a corresponding increase in the release rate of insulin from the matrix upon multiple exposures to buffered glucose solutions over several weeks. Streptozotocin-induced diabetic rats receiving such implants responded as expected to glucose challenge.

Albin et al. [333] cautioned that any glucose sensing device which uses immobilized glucose oxidase, progressive response to glucose concentration in the physiological range is possible only if the enzyme loading is sufficiently low. Equally important is the availability of a mechanism that rapidly removes oxygen, the other product of the reaction, from the membrane. Thus, a membrane that is more permeable to oxygen than to glucose is desirable.

Encapsulated Cells

A fundamentally different approach to achieve nonconstant drug release is transplantation of encapsulated cells. This is exemplified by β-islets, the source of insulin [334]. The iselts were encapsulated in an alginate-polylysine-alginate membrane in order to minimize the incidence of immuno-rejection of the transplant. Of eight streptozotocin-induced diabetic rats who received the transplant intraperitoneally, one had a normoglycemic period of only several days, five had a period ranging from 40–120 days, while two still normoglycemic when sacrificed at 365 and 648 days post-transplantation. By contrast, a single injection of unencapsulated islets was effective for less than 2 weeks. This study demonstrated that the encapsulated cells could continuously secrete insulin and respond to the glucose challenge. The transplant recipients showed normal weight gain and no evidence of cataract development. With subsequent refinements such as prolonging the viability of islets this approach offers the brightest promise of long-term insulin therapy.

An extension of the above approach has been attempted in somatic cell gene therapy. Selden et al. [335] implanted cultured murine fibroblasts containing a human growth hormone fusion gene in mice and found that with appropriate immunosuppression the implants survived for more than 3 months while secreting human growth hormone into the serum. The function of such implants depends on their location; intraperitoneal is superior to subcutaneous. This is because the rate of perfusion and relative surveillance of the implant by cells of the immune system vary with the site of implantation.

V. STABILITY OF PROTEIN DRUGS IN SOLUTION AND OTHER DOSAGE FORMS

Several peptide and protein drug delivery systems have been tested, including nondegradable and degradable polymeric systems, liposomes, and osmotic pumps. Broadly speaking, the selection of a given system depends on intended therapeutic use of a peptide or protein, desired duration of prolongation, and physicochemical properties of the peptide or protein. Among the physicochemical properties is conformational and chemical stability of a peptide or protein during fabrication of its delivery system.

The stability of proteins in solution, the simplest dosage form, varies widely. Some proteins like albumin remain stable at room temperature for days, whereas other proteins like lipoproteins are quite labile and lose their activity quite rapidly when removed from their native environment. Alpha$_1$-antitrypsin rapidly loses activity due to general thermal lability [335] and to oxidation of the exposed methionine residue [336]. A recombinant α_1-antitrypsin injection loses about 40% of its potency after 18 months of storage at 4°C [337]. While crude human gamma interferon induced by antilymphocyte globulin was resistant to inactivation at 56°C for 10 min [338], purified gamma interferons with specific activities of 10^5-10^6 units/mg protein were unstable at 4°C and −20°C even in the presence of human serum albumin [339]. Similarly, while natural human interferon-β derived from fibroblast cells stimulated with viral inducers was stable for 32 months when stored under refrigeration [340], human interferon-β derived from Escherichia coli lost up to 50% of its antiviral activity after only 3 months when similarly stored [341].

Several environmental factors are known to affect protein stability: pH, organic acids, ionic strength, metal ions, detergents, temperature, pressure, interfaces, and agitation. The subject of protein denaturation has been reviewed [342].

The ability of certain solvent additives to stabilize the conformation of proteins such as ribonuclease A, β-lactoglobulin, lysozyme, bovine serum albumin, and α-chymotrypsin is related to preferential hydration of the macromolecules in an aqueous solvent containing these substances as a result of their exclusion from the domain of the protein. Under this condition, the chemical potential of the protein is raised, prompting the system to reduce it by decreasing the area of solvent−protein contact through enhancement of protein self-association [343,344].

The protein conformation stabilizers are sugars (lactose and glucose) [345−347], salts such as NaCl, Na acetate, sodium sulfate, KH_2PO_4,

$(NH_4)_2SO_4$ [348], and amino acids such as glycine, and α- and β-alanine, [349]. In the case of sugars and certain salts, the preferential interactions were found to parallel the increase in surface tension of water by these substances [348]. Nonetheless, increase in surface tension may not necessarily lead to preferential hydration. Even though they lower the surface tension of water, 2-methyl-2,4-pentanediol [350,351], betaine [349], and glycerol [343,344] also stabilize proteins. It is suggested that these compounds stabilize proteins by altering the chemical nature of the protein surface or by assisting the preferential hydration process [349]. The stabilization of protein structure by sugars and polyols is common to both globular and fibrous proteins [352].

A. Manifestations of Protein Instability

Aggregation

Certain proteins tend to undergo self-association resulting in the formation of multimers and, in the extreme, aggregates and precipitates. Indeed, aggregation and gelling of insulin has been a problem in infusion pump therapy [353]. Many contributing factors have been suggested, e.g., metal ions, ionic strength, pH, and temperature [354]. The role of metal ions is well known. Jeffrey [355] demonstrated that heavy metal ions such as Zn^{2+}, Cd^{2+}, Pb^{2+}, and Ca^{2+} catalyzed the self-association of des-(B26-B30)insulin. Zn^{2+} and Cd^{2+} were more effective than Pb^{2+} and Ca^{2+}. Moreover, Zn^{2+} favored hexamer formation, Cd^{2+} favored tetramer formation, whereas Pb^{2+} and Ca^{2+} favored dimer formation [355].

Motion is another contributing factor in the aggregation and gelation of insulin. During motion, shearing and agitation [356] and abrupt changes in flow path [354,357] can occur. Shear has been shown to affect the stability of proteins other than insulin, including catalase, rennet, carboxypeptidase [358], fibrinogen [359], alcohol dehydrogenase [360], urease [361,362], and interferons [363]. The rate of shear is as important as the duration during which the protein is subjected to shear. Charm and Wong [358] showed that if the product of shear rate × exposure time < 10^4, where shear rate is measured in s^{-1} and exposure time in s, there is little or no inactivation. Inactivation probably occurs when a protein molecule's orientation in the shear field causes disruption of its tertiary structure. It is not shear alone that causes the protein to lose its conformation and activity but also shear-associated damage that occurs at the gas-liquid interfaces. Human fibroblast interferon can be stabilized against vortical stability by acid pH, 0.01% Tween 80, 0.01% sodium dodecyl sulfate, 10 mM thioctic acid, bovine serum albumin (0.5 mg/ml) [363–365], and rare earth salts of 0.002M [366], but not by cytochrome c, ovalbumin and human serum albumin.

For insulin, Hutchinson [367] determined that shear rates in a short time caused predominantly aggregation, whereas low shear agitation at 37°C caused predominantly gelation. This investigator also evaluated the role of formulation ingredients such as buffer (0.01M Na acetate and saline) and preservatives (0.02M m-cresol and 8.5 mM phenol) on aggregation and gelation. Although no consistent pattern emerged as to which ingredients promoted or inhibited gelation, the combination of buffer and either or both of the m-cresol and phenol preservatives seemed to favor aggregation, whereas the buffer and preservatives by themselves tended to cause gelation.

Aggregation of insulin can be prevented by the addition of substances like urea [368], dicarboxylic amino acids like aspartic acid and glutamic acid [369], or other agents such as glycerol [370], EDTA, lysine, Tris buffer or bicarbonate buffer [371]. An extensive study of 60 additives and 1125 formulations by Massey and Sheliga [372] revealed that nonionic surfactants such as Pluronic F68 (Poloxamer 188), a polyoxyethylene-polyoxypropylene glycol surfactant, were promising stabilizers. Human insulin appeared to aggregate more readily than porcine or bovine insulin [372].

Multimer formation may not necessarily lead to loss in pharmacological activity. In fact, both recombinant *E. coli* derived and natural interferon-gamma exist as dimers under physiological conditions [373]. At low pH (pH 3.55) and ionic strength dissociation into monomers occurs as a result of increased electrostatic repulsion, leading to denaturation. Recombinant human leukocyte interferon aggregates reversibly, from monomer up to perhaps octamer, as a function of pH and concentration. This self-association appears to be linked to the ionization of three basic residues (Tyr or Lys) with depressed pK values in the aggregated state [374].

Dimer formation is known to occur in two recombinant immune interferons, human and murine, both with Cys-Tyr-Cys-Gln as the amino-terminal residues. Under oxidative conditions, these two cysteine-rich interferons form tetramers and octamers, due to intermolecular disulfide bridge formation [375]. The dimerization tendency of the cysteine-rich human interferon is about three times that of the cysteine-rich murine interferon. This may be related to differences in their surface properties such as hydrophobicity, charged states and presence or absence of the polycationic carboxy-terminal portion [376,377].

Denaturation at Interfaces and Adsorption

Denaturation of proteins at interfaces has been observed under three conditions: shaking protein solutions, adsorption at quiescent interfaces, and compression of spread monolayers. The rates of interfacial denaturation are strongly dependent on the specific protein and on such solution properties, as temperature, pH, and salt concentration [378,379]. Human growth hormone undergoes only limited, and fully reversible, denaturation between pH 1.3 and pH 13 [380,381], while both human choriomammotropin and ovine prolactin undergo substantial and irreversible denaturations above pH 11.0 [382,383].

Under denaturing conditions, disulfide scrambling of interleukin-2 has been observed [384]. Native IL-2 contains three cysteines; two exist in a disulfide bridge (Cys-58 and Cys-105) and the third Cys-125 is a free sulfhydryl. In the presence of 6M guanidine HCl at alkaline pH, IL-2 is converted into three isomers, representing three possible disulfide-linked forms, that can be resolved on a C4 reversed-phase HPLC column. Interferon-α is another disulfide-bridge containing protein (between 1 and 98 and between 29 and 138) that is susceptible to scrambling and intermolecular bridging under redox conditions. This has been shown to lead to the formation of monomers, dimers, trimers, and other oligomers that have little or no activity [385].

Various approaches have been attempted to prevent loss of protein by adsorption to glass and plastic. These include coating polypropylene tubes with bovine serum albumin [386,387], coating glassware with 1% solutions of polyethylene glycol 20,000 in deionized water followed by drying at 110°C

[388], or by adding glycerol and Triton X-100 [389]. Some proteins appear to adsorb to one type of surface but not to others. For instance, human insulin (20 i.u./500 ml dextrose 5% w/v) adsorbed to administration sets made of methacrylate butadine styrene and polybutadiene more extensively than to those made of cellulose propionate and polyvinyl chloride [390]. As much as 96% of the insulin placed in the administration tubings was lost over a 6 hr period. Moreover, dissociated porcine lactic dehydrogenase at pH 2.5 was irreversibly denatured by glass, presumably by adsorption, but was minimally denatured by polypropylene and polycarbonate [391].

Suelter and DeLuca [389] compared the effectiveness of two approaches to prevent adsorption of proteins to glass and plastic surfaces: modification of solvent and modification of the surface. Using bovine serum albumin, luciferase and mitochondrial creatine kinase as model proteins, these investigators found that modifying the solvent by adding Triton X (0.2 mM final concentration) or glycerol (50% final concentration) was more effective than treating surfaces with proteins such as fibrinogen, globulin, ovalbumin, bovine serum albumin, and cytochrome c in reducing protein adsorption to glass and polyethylene. This may be attributed to the ability of the pharmaceutical protein to bind to the proteins coating the surface or to displace the proteins from the coated surface. The type of container used in an experiment is also an important consideration. For plastic containers, glycerol or Triton X-100 at either low or high ionic strength is effective in preventing adsorption of proteins. For glass containers, glycerol at low ionic strength and Triton X-100 at high ionic strength were the preferred solvents. Clearly, the best solvent to use needs to be determined for a specific protein.

B. Influence of Manufacturing Processes on Protein Stability

Temperature

Susceptibility of proteins to thermal inactivation can seriously limit the range of methods that can be used in their sterilization as well as in the fabrication of their delivery systems. Standard heat-melt extrusion techniques for implant preparation could not be used for interferon-β_{ser-17} since such techniques destroyed over 99% of the biological activity of the protein [392]. This led to the development of alternative technology involving the formation of a microemulsion (<1 μm particles) of precipitated protein in a solvent solution of polylactide-glycolide followed by rapid solvent removal by spray drying [393].

At high temperatures native S-S bonds are broken and the new S-S bonds that are formed often lock the protein into a denatured configuration. Jensen et al. [394] found that minute amounts of reducing agents significantly increased the rate of thermal denaturation of proteins containing S-S bonds. Although thermal denaturation can be minimized by avoiding high temperatures during the processing of protein solutions, significant or intense local heating still may occur even when bulk temperatures are controlled. For example, intense, localized adiabatic heating may develop during cavitation [395], a process in which a partial vacuum is formed within a liquid [396]. In an attempt to understand the relationship between protein structure and thermal stability, Komiyama et al. [397]

studied three homologous dimeric inhibitors. They found that an increase
in hydrophilicity of side chains of the α_1-helix was responsible for lowering the temperature stability in this homologous series.

Various approaches have been investigated to stabilize proteins against
thermal inactivation. Low pH, Na dodecyl sulfate, Tween 80, chaotropic
salts, and exogenous proteins have been shown to prevent thermal inactivation of liquid α- and β-interferon preparations [365,366]. Sedmak and
Grossberg [366] reported that salts of the rare earth elements (at. no.
57–71) at concentrations as low as 0.002 M protected partially purified
human β-interferon from inactivation at 37°C. At the same time, the
antiviral activity of β-interferon was enhanced 5 times and that of α-inter-
feron was enhanced to a lesser extent. Ethylene glycol at 30–50% was
very effective in preventing loss of antiviral activity of highly purified
β-interferon preparations [398–400]. Increasing the level of human serum
albumin in recombinant human interferon-β_{ser}-17 results in increased
thermal stability over the range of 25–80°C.

Other approaches to stabilizing proteins include conjugation of proteins
to water-soluble polysaccharides such as dextrans and amylose [401] and
point-specific mutation as is the case for lactate dehydrogenase [402], T4
lysozyme [403], interleukin-2 [404], and human interferon-β [341]. The
introduction of new disulfide bonds into proteins to enhance their thermal
stability is one of the most popular alterations in protein molecules brought
about by site-directed mutagenesis [405–408]. Volkin and Klibanov [409]
questioned the utility of this approach, however. They showed that at
100°C and neutral pH cystine residues in proteins undergo destruction via
two distinct mechanisms: β-elimination and disulfide interchange catalyzed
by thiols formed during β-elimination. The first process is rate-limiting
with a half-life of 1.0 ± 0.4 hr at pH 8 and 12.4 ± 3.4 hr at pH 6. These
half-lives are relatively independent of the primary structure of the proteins examined, including insulin, lysozyme, ovalbumin, papain, peroxidase,
ribonuclease, serum albumin, and transferrin.

Pressure

Proteins are not very sensitive to pressure, and only at extremely large
values of pressure do they exhibit the changes observed in temperature
and pH denaturation [410,411]. Teng and Groves [412] found that urease
subjected to compaction did not lose much activity until the pressure exceeded 474 mPa, above which 50% of the relative activity was lost. Moreover, Truskey et al. [413] demonstrated that the conformational changes
seen in human IgG, bovine insulin, and bovine alkaline phosphatase during membrane filtration were not due to pressure over the range of 0.5–
20 psi. Neither was it due to the pore size of the membrane within the
range of 0.1–0.45 μm. The principal factor appeared to be adsorption or
binding of the proteins to the membranes [414]. Of the three membranes
studied, conformational changes were least extensive and variable in
poly(hydroxypropylacrylate) grafted poly(vinylidene difluoride) membranes,
more in polysulfone membranes, and most in nylon membranes. Nonetheless, pressure has been shown to both increase and decrease protein
activity. Thus, while catheptic proteases are inhibited by pressure treatment [415], about 30% of the proteases examined by Morild [416] exhibited
an increase in enzyme activity with pressure.

Lyophilization

Removal of water from a protein preparation during freeze drying can be
more disruptive to protein function than freeze-thawing or heating [417].
Freezing concentrates protein, buffer salts, and other electrolytes and
may dramatically shift pH. The conditions are therefore favorable for
changes in tertiary and quaternary structure to take place leading to
denaturation, aggregation [418], precipitation, or deaggregation [419].
Indeed, structural changes during freeze drying are known to occur in
carbonic anhydrase [420], lysozyme [421,422], myosin [423], alcohol dehydrogenase [424], and catalase [425-427]. On freeze drying, alcohol dehydrogenase self-associates to high-molecular-weight polymers [424], whereas catalase dissociates into subunits [425-427]. Moreover, L-asparaginase,
which is fully active after freeze thawing, is inactivated by over 80% during freeze drying as a result of extensive dissociation of the enzyme
from active tetramer to inactive monomers [417,419]. Hellman et al. [419]
speculated that dissociation occurred during the drying cycle as a result
of weakening of hydrophobic forces [428]. Hydrogen bonding was apparently not a factor.

Various additives have been included in protein formulations to protect
proteins against changes during lyophilization. Thus, the stability of
partially purified lyophilized human β-interferon greatly enhanced in the
presence of 0.5% bovine albumin, cytochrome C, gelatin, and ovalbumin
[429]. Preparations of human β-interferon freeze dried at pH values 2
and 7 showed essentially the same stability [429]. Similarly, the addition
of trehalose, maltose, or sucrose to a solution of phosphofructokinase purified from rabbit skeletal muscle prior to freeze drying resulted in recovery
of up to 80% of the original activity, when compared with total inactivation
of the enzyme otherwise [430]. Glucose and galactose were ineffective,
however, even at concentrations as high as 500 mM. Addition of Zn^{2+} to
enzyme-sugar mixtures prior to freeze drying greatly enhanced the stabilization imparted by the above sugars, even though this metal ion by itself was ineffective. Other ions found to be effective included Cu^{2+}, Cd^{2+},
Ni^{2+}, Co^{2+}, Ca^{2+}, and Mn^{2+}, but Mg^{2+} was ineffective [430]. Finally, the
addition of sugar or sugar derivatives have been found to protect L-asparaginase against dissociation [419]. Such derivatives fall into three groups
depending on the degree of protection. Members in the most effective
group, i.e., 90-100% tetramer, include glucose, mannose, sorbitol, sucrose,
ribose, tetramethylglucose, and polyvinylpyrrolidone. Members in the next
effective group, i.e., 50-80% tetramer, include α-methylglucoside, penta-
methylglucoside, myo-inositol, glutamine, erythritol, and mannitol. Members
in the least effective group, i.e., less than 50% tetramer, include mannitol,
galactitol, aspartic acid, and NaCl.

A coprecipitation method has recently been developed as an alternative
to lyophilization in preparing dry form of proteins [431]. Coprecipitation
results when an organic solvent such as acetone, ethanol, and isopropanol
is added to a solution of protein in water-soluble starch (M.W. 12,700 and
100,000). The resulting precipitate is filtered, dried in a vacuum for 72
hr at 35°C, and further micronized by milling in a cutting mill.

An important consideration in lyophilization is the amount of residual
moisture that is acceptable for maximal stability [432]. Epidermal growth
factor is a protein whose stability is exquisitely sensitive to moisture. It
loses 62% of its activity in aqueous solution in 48 days except in the

presence of cellulose derivatives such as hydroxypropylmethylcellulose
[433]. A residual moisture of 3% is optimal for stability of the mouse
reference standard.

Irradiation

Gamma irradiation was found to decrease the gel rigidity index of gelatin
depending on the dose and gelatin content. Minimal changes occurred
when the dose was held to 2 Mrad or less [434]. No specific radiolytic
products were detectable when porcine and bovine insulins were subjected
to gamma irradiation (^{60}Co) at doses up to 2.5 mRad at 0°C [435]. The
effects of irradiation on peptides, polypeptides, and proteins has been
reviewed by Garrison [436].

C. Stability of Proteins in the Solid State

Solid-state stability of proteins is a concern for bulk solid storage and
processing, lyophilized formulations, and controlled-release formulations.
The presence of additives such as glycerol, propylene glycol, or other
polyhydric alcohols which are commonly used in protein formulations as
cryoprotectants during lyophilization can affect the water content and
activity in the final formulation and, therefore, amount of available water
for reactant mobilization.

Extensive studies in the hydration of lysozyme [437,438] and other
proteins [439,440] have revealed several critical levels of hydration at
which significant changes in properties of the protein and the bound water
occur. Initial hydration up to 6−8% water content results in very little
change in the properties of tightly bound water; water mobility is about
1/100th that of bulk water. Above 25% water content, the properties of
the bound water are similar to those of bulk water. It is above this
level that significant increases in flexibility of the polypeptide backbone
occur.

The water sorption isotherms for proteins can be roughly divided into
three regions [441]. The first region is binding of water to highly active
sites such as charged and highly polar groups. The water-binding
capacity of the functional group varies with its chemical nature. At a relative humidity of 20%, the moles of water per mole of sorption site are 1.2
for the carboxyl group, 0.17 for the aliphatic hydroxyl group, 0.5 for the
phenolic hydroxyl group, 0.11 for the peptide amide group, and 1.2 for
the amino group [442]. The second region represents binding of water to
weaker sorption sites such as the peptide backbone and the polar surface
groups as well as filling of the voids created by swelling of the protein.
The last region occurs with condensation of water at very weak binding
sites and layering of loosely held water. The sorption isotherm is influenced by the salt form. This results from changes in the ionization of
amino acid side chains and the presence of highly polar counterions
[443,444].

Most decomposition reactions are minimal at or below the monolayer
level of hydration. Sorption of water beyond that generally results in
increased rates of decomposition due to increased conformational flexibility
of the protein and to ability of the less tightly bound water to mobilize
reactants. Indeed, the temperature required for thermal denaturation of
proteins in the solid state decreased with increasing protein hydration

until the water content reaches 50%. At this point the thermal denaturation temperature of the solid state is approximately equal to that in solution [445,446]. The role of moisture in protein stability has been reviewed
by Hageman [441].

VI. CONCLUSION

Although peptides and proteins have been used for decades in the diagnosis and treatment of diseases, it was not until the recent availability of
virtually unlimited quantities of these substances that their realistic potential as drugs was recognized. Already, their emergence on the therapeutic horizon has intensified the investigation of routes of administration
other than oral and parenteral. It has also triggered the development of
new technologies that promise to enhance the permeability of absorptive
epithelia, protect peptide and protein drugs from proteolytic degradation,
prolong retention of the dose at the administration site, and release drug
in a temporal manner consistent with the biochemistry and pathology of a
particular disease state. As a result, new routes of drug administration
have emerged, penetration enhancers are being developed, new drug
delivery systems that release drug in a protracted but nonzero order
manner at its target tissue are being designed and tested, and new ways
to fabricate and assemble drug formulations are being considered. Although
the above technologies are being developed for peptide and protein drugs,
this author is confident that they will benefit the broad field of drug
delivery.

REFERENCES

1. L. K. Sislbart and D. F. Karen, *Science, 243*:1462–1464 (1989).
2. J. W. Greiner, F. Guadagni, P. Noguchi, S. Pestka, D. Colcher,
    P. B. Fisher, and J. Schlom, *Science, 35*:895–898 (1987).
3. B. K. Kobilka, T. S. Kobilka, K. Daniel, J. W. Regan, M. G. Caron,
    and R. J. Lefkowitz, *Science, 240*:1310–1316 (1988).
4. H. A. Lester, *Science, 241*:1057–1063 (1988).
5. V. H. L. Lee, *Pharm. Int., 7*:208–212 (1986).
6. M. L. G. Gardner, *Biol. Rev., 59*:289–331 (1984).
7. J. F. Quay and L. Foster, *Physiologist, 13*:287–290 (1970).
8. P. J. Sinko, M. Hu, and G. L. Amidon, *J. Contr. Rel., 6*:115–121
    (1987).
9. T. Kimura, H. Endo, M. Yoshikawa, S. Muranishi, and H. Sezaki,
    J. Pharm. Dyn., 1:262–267 (1978).
10. T. Kimura, T. Yamamoto, M. Mizuno, Y. Suga, S. Kitade, and
    H. Sezaki, *J. Pharm. Dyn., 6*:246–253 (1983).
11. E. Nakashima, A. Tsuji, S. Kagatani, and T. Yamana, *J. Pharm.
    Dyn., 7*:452–464 (1984).
12. M. Hu, P. Subramanian, H. I. Mosberg, and G. L. Amidon, *Pharm.
    Res., 6*:66–70 (1989).
13. R. Solari, B. Morris, and R. Morris, *Biol. Neonate, 46*:163–170
    (1984).
14. W. Thornburg, L. Matrisian, B. Magun, and O. Koldovsky, *Am.
    Physiol. Soc., 246*:G80–85 (1984).

15. P. Tengamnuay and A. K. Mitra, *Life Sci.*, *43*:585−593 (1988).
16. Y. Maitani, T. Igawa, Y. Machida, and T. Nagai, *Drug Design Deliv.*, *1*:65−70 (1986).
17. C. McMartin, L. E. F. Hutchinson, R. Hyde, and G. E. Peters, *J. Pharm. Sci.*, *76*:535−540 (1987).
18. R. R. Burnette and D. Marrero, *J. Pharm. Sci.*, *75*:738−743 (1986).
19. P. S. Banerjee and W. A. Ritschel, *Int. J. Pharm.*, *49*:189−197 (1989).
20. S. J. Hersey and R. T. Jackson, *J. Pharm. Sci.*, *76*:876−879 (1987).
21. T. Buclin, J. P. Randin, A. F. Jacquet, M. Azria, M. Attinger, F. Gomez, and P. Burckhardt, *Calcif. Tissue Int.*, *41*:252−258 (1987).
22. H. Hauser, K. Howell, R. M. C. Dawson, and D. E. Bowyer, *Biochim. Biophys. Acta*, *602* 567−577 (1980).
23. G. C. Hansson, *Biochim. Biophys. Acta*, *733*:295−299 (1983).
24. G. M. Gray, R. J. White, R. H. Williams, and H. J. Yardley, *Br. J. Dermatol.*, *106*:59−63 (1982).
25. P. W. Wertz, P. S. Cox, C. A. Squier, and D. T. Downing, *Comp. Biochem. Physiol.*, *83B*:529−531 (1986).
26. R. E. Gibson and L. S. Olanoff, *J. Contr. Rel.*, *6*:361−366 (1987).
27. M. Hayashi, T. Hirasawa, T. Muraoka, M. Shiga, and S. Awazu, *Chem. Pharm. Bull.*, *33*:2149−2152 (1985).
28. C. D. Christie and R. F. Hanzal, *J. Clin. Invest.*, *10*:787−793 (1931).
29. R. E. Stratford, Jr., L. W. Carson, S. Dodda-Kashi, and V. H. L. Lee, *J. Pharm. Sci.*, *77*:838−842 (1988).
30. G. C. Y. Chiou and C. Y. Chuang, *J. Ocular Pharmacol.*, *4*:165−177 (1988).
31. G. C. Y. Chiou and C. Y. Chuang, *J. Ocular Pharmacol.*, *4*:179−186 (1988).
32. G. C. Y. Chiou, C. Y. Chuang, and M. S. Chang, *Life Sci.*, *43*: 509−514 (1988).
33. G. C. Y. Chiou, C. Y. Chuang, and M. S. Chang, *J. Ocular Pharmacol.*, *5*:81−91 (1989).
34. A. Yamamoto, A. M. Luo, S. Dodda-Kashi, and V. H. L. Lee, *J. Pharmacol. Exp. Ther.*, *249*:249−255 (1989).
35. S. C. Chang and V. H. L. Lee, *J. Ocular Pharmacol.*, *3*:159−169 (1987).
36. S. C. Chang, D. S. Chien, H. Bundgaard, and V. H. L. Lee, *Exp. Eye. Res.*, *46*:59−69 (1988).
37. S. J. Enna and L. S. Schanker, *Am. J. Physiol.*, *223*:1227−1231 (1972).
38. J. A. Hemberger and L. S. Schanker, *Drug Metab. Disp.*, *11*:73−74 (1983).
39. D. J. Tocco, A. E. W. Duncan, F. A. Deluna, and G. Wells, *Drug Metab. Disp.*, *6*:697−700 (1988).
39a. M. Vadnere, A. Adjei, R. Doyle, and E. Johnson, Evaluation of alternate routes for delivery of leuprolide. Second International Symposium on Disposition and Delivery of Peptide Drugs, Leiden, 1989, Abstract P22.
40. H. Yoshida, K. Okumura, R. Hori, T. Anmo, and H. Yamaguchi, *J. Pharm. Sci.*, *68*:170−171 (1979).
41. F. M. Wigley, J. H. Londono, S. H. Wood, J. C. Shipp, and R. H. Waldman, *Diabetes*, *20*:552−556 (1971).

42. D. A. Creasia, G. A. Saviolakis, and K. A. Bostian, *FABSEB J.*, *2*: A537 (1988).

43. K. Takada, M. Yamamoto, H. Nakae, and S. Asada, *Chem. Pharm. Bull.*, *28*:2806–2808 (1980).

44. D. Wangensteen and R. Yankovich, *J. Appl. Physiol.*, *47*:846–850 (1979).

45. E. A. M. Dominguez, A. A. Liebow, and K. G. Bensch, *Lab. Invest.*, *16*:905–911 (1967).

46. J. Gil, Comparative morphology and ultrastructure of the airways, in *Mechanisms in Respiratory Toxicology*, Vol. 2 (H. Witschi and P. Netlesheim, eds.) CRC, Boca Raton, Florida (1982), pp. 3–25.

47. R. Breezer and M. Turk, *Environ. Health Perspect.*, *55*:3–24 (1984).

48. K. G. Bensch and E. A. M. Dominguez, *Yale J. Biol. Med.*, *43*:236–241 (1971).

49. B. Corrin, Cellular constituents of the lung, in *Scientific Foundations of Respiratory Medicine* (J. G. Scadding and G. Cummings, eds.), W. B. Saunders, Philadelphia (1981), pp. 78–90.

50. E. E. Schneeburger, *Fed. Proc.*, *37*:2471–2478 (1978).

51. E. E. Schneeburger and M. J. Karnovsky, *J. Cell. Biol.*, *37*:781–793 (1968).

52. M. Simeonescu, *Ciba Found. Symp.*, *78*:11–36 (1980).

53. O. D. Wagensteen, L. E. Wittmers, Jr., and J. A. Johnson, *Am. J. Physiol.*, *216*:719–727 (1969).

54. A. E. Taylor and K. A. Garr, Jr., *J. Am. Physiol.*, *218*:1133–1139 (1970).

55. J. T. Gatzy, *Exp. Lung Res.*, *3*:147–161 (1982).

56. L. S. Schanker and J. A. Hemberger, *Biochem. Pharmacol.*, *32*: 2599–2601 (1983).

57. E. A. Egan, R. M. Nelson, and E. F. Beale, *Pediatr. Res.*, *14*:314–318 (1980).

58. R. C. Lanman, R. M. Gillilan, and L. S. Schanker, *J. Pharmacol. Exp. Ther.*, *187*:105–111 (1973).

59. J. Burton, T. H. Gardiner, and L. S. Schanker, *Arch. Environ. Health*, *29*:31–33 (1974).

60. L. S. Schanker, S. J. Ena, and J. A. Burton, *Xenobiotica*, *7*:521–528 (1976).

61. A. R. Clark and P. R. Byron, *J. Pharm. Sci.*, *74*:939–942 (1985).

62. S. G. Woolfrey, G. Taylor, I. W. Kellaway, and A. Smith, *J. Pharm. Pharmacol.*, *38*(Suppl.):34P (1986).

63. R. A. Brown and L. S. Schanker, *Drug Metab. Disp.*, *11*:355–360 (1983).

64. Y. L. Lin and L. S. Schanker, *Biochem. Pharmacol.*, *30*:2937–2943 (1981).

65. Y. J. Lin and L. S. Schanker, *Am. J. Physiol.*, *240*:C215–221 (1981).

66. J. Kerr, A. B. Fisher, and A. Kleinzeller, *Am. J. Physiol.*, *241*: E191–195 (1981).

67. S. J. Enna and L. S. Schanker, *Life Sci.*, *12*:231–239 (1973).

68. T. H. Gardiner and L. S. Schanker, *Xenobiotica*, *4*:725–731 (1974).

69. T. H. Gardiner and L. S. Schanker, *J. Pharmacol. Exp. Ther.*, *196*: 455–462 (1976).

70. J. A. Hemberger and L. S. Schanker, *Proc. Soc. Exp. Biol. Med.*, *161*:285–288 (1979).

71. Y.-J. Lin and L. S. Schanker, *Drug Metab. Disp.*, *11*:75–76 (1983).

72. H. J. Forman, T. K. Aldrich, M. A. Posner, and A. B. Fisher, *J. Pharmacol. Exp. Ther.*, *221* 428 – 433 (1982).
73. A. Ryrfeldt, E. Nilsson, A. Tunek, and L. A. Svensson, *Pharm. Res.*, 5:151 – 155 (1988).
74. T. R. Devereux and J. R. Fouts, *Methods Enzymol.*, 77:147 – 154 (1981).
75. U. S. Ryan, *Ann. Rev. Physiol.*, 44:241 – 255 (1982).
76. U. S. Ryan, *Ann. Rev. Physiol.*, 44:223 – 239 (1982).
77. K. K. F. Ng and J. R. Vane, *Nature,* *216*:762 – 766 (1967).
78. P. R. Byron and A. J. Hickey, *J. Pharm. Sci.*, 76:60 – 64 (1987).
79. I. Gonda, A. F. A. E. Khalik, and A. Z. Britten, *Int. J. Pharm.*, 27:255 – 265 (1985).
80. S. P. Newman, F. Moren, D. Pavia, F. Little, and S. W. Clark, *Am. Rev. Resp. Dis.*, *124*:317 – 320 (1981).
81. P. R. Byron, S. S. Davis, M. D. Bubb, and P. Cooper, *Pestic. Sci.,* 8:521 – 526 (1977).
82. R. Dalby and P. R. Byron, *Pharm. Res.*, 5:36 – 39 (1988).
83. S. P. Newman, D. Pavia, F. Moren, N. F. Sheahan, and S. W. Clark, *Thorax,* *36*:52 – 55 (1981).
84. P. R. Byron, R. N. Dalby, and A. J. Hickey, *Pharm. Res.*, 6:225 – 229 (1989).
85. M. B. Dolovich, R. E. Ruffin, R. Roberts, and M. T. Newhouse, *Chest,* *80*(Suppl.):911 – 915 (1981).
86. S. P. Newman, D. Pavia, N. Garland, and S. W. Clarke, *Eur. J. Respir. Dis.*, *63*(Suppl. 119):57 – 65 (1982).
87. S. P. Newman, *Chest,* *88*:152S – 160S (1985).
88. D. S. Davies, *Eur. J. Resp. Dis.*, *63*:67 – 72 (1982).
89. P. R. Byron, *J. Pharm. Sci.*, 75:433 – 438 (1986).
90. G. Smith, C. Hiller, M. Mazumder, and R. Bone, *Am. Rev. Resp. Dis.*, *121*:513 – 517 (1980).
91. G. M. Green, *Arch. Intern. Med.*, *131*:109 – 114 (1973).
92. H. N. McCullough and R. L. Juliano, *JNCI, 63*:727 – 731 (1979).
93. R. L. Juliano and H. N. McCullough, *J. Pharmacol. Exp. Ther.*, *214*:381 – 387 (1980).
94. A. J. Hickey and P. R. Byron, *J. Pharm. Sci.*, 75:756 – 759 (1986).
95. R. W. Niven and P. R. Byron, *Pharm. Res.*, 5:574 – 579 (1988).
96. S. G. Woolfrey, G. Taylor, I. W. Kellaway, and A. Smith, *J. Contr. Rel.*, 5:203 – 209 (1988).
97. J. C. Hogg, P. D. Pare, and R. C. Boucher, *Fed. Proc.*, *38*:97 – 201 (1979).
98. J. F. Braley, L. B. Peterson, C. A. Dawson, and V. L. Moore, *J. Clin. Invest.*, *63*:1103 – 1109 (1979).
99. J. F. Braley, C. A. Sawson, V. L. Moore, and B. O. Cozzini, *J. Clin. Invest.*, *61*:1240 – 1246 (1978).
100. T. H. Gardiner and B. H. McAnnaley, *Life Sci.*, 23:1827 – 1834 (1978).
101. T. H. Gardiner, *J. Appl. Physiol.: Respirat. Environ. Exercise Physiol.*, 44:576 – 580 (1978).
102. J. L. Maddocks, *Thorax, 30*:333 – 336 (1975).
103. N. A. Williams and N. D. Weiner, *Int.. J. Pharm.*, 50:261 – 266 (1989).
104. W. K. Surewicz and R. M. Epand, *Biochemistry, 23*:6072 – 6077 (1984).
105. P. S. Banerjee and W. A. Ritschel, *Int. J. Pharm.*, 49:199 – 204 (1989).
106. J. A. Fix, K. Engle, P. A. Porter, P. S. Leppert, S. J. Selk, C. R. Gardner, and J. Alexander, *Am. J. Physiol. 251*:G332 – G340 (1986).

107. M. Miyake, T. Nishihata, A. Nagano, Y. Kyobashi, and
     A. Kamada, *Chem. Pharm. Bull.*, *33*:740−745 (1985).
108. J. Unowsky, C. R. Behl, G. Beskid, J. Sattler, J. Halpern, and
     R. Cleeland, *Chemotherapy*, *34*:272−276 (1988).
109. E. J. van Hoogdalem, C. D. Heijligers-Feijen, A. G. de Boer,
     J. C. Verhoef, and D. D. Breimer, *Pharm. Res.*, *6*:91−95 (1989).
110. J. A. Fix, *J. Contr. Rel.*, *6*:151−156 (1987).
111. O. Wong, J. Huntington, R. Konishi, J. H. Rytting, and
     T. Higuchi, *J. Pharm. Sci.*, *77*:967−971 (1988).
112. N. Yata, N. Sugihara, R. Yamajo, T. Murakami, Y. Higashi, and
     H. Kimata, *J. Pharmacobio-Dyn*, *8*:1041−1047 (1985).
113. E. J. van Hoogdalem, A. M. Stijnen, A. G. de Boer, and
     D. D. Breimer, *J. Pharm. Pharmacol.*, *40*:329−332 (1988).
114. Y. Watanabe, E. J. van Hoogdalem, and A. G. de Boer, *J. Pharm.
     Sci.*, *77*:847−849 (1988).
115. M. Sekine, H. Terashima, K. Sasahara, K. Nishimura, R. Okada,
     and S. Awazu, *J. Pharmacobiodyn.*, *8*:286−295 (1985).
116. G. S. Gordon, A. C. Moses, R. D. Silver, J. S. Flier, and
     M. Carey, *Proc. Natl. Acad. Sci.*, *82*:7419−7423 (1985).
117. S. Hirai, T. Yashiki, and H. Mima, *Int. J. Pharm.*, *9*:165−172
     (1981).
118. G. S. M. J. E. Duchateau, J. Zuidema, and F. W. H. M. Merkus,
     Int. J. Pharm., *31*:193−199 (1986).
119. E. Touitou, F. Alhaique, P. Fisher, A. Memoli, F. M. Riccieri,
     and E. Santucci, *J. Pharm. Sci.*, *76*:791−793 (1987).
120. E. Touitou and P. Fisher, *J. Pharm. Sci.*, *75*:384−386 (1986).
121. J. Hadgraft, K. A. Walters, and P. K. Wotton, *J. Pharm. Pharmacol.*,
     37:725−727 (1985).
122. H. H. G. Jonkman and C. A. Hunt, *Pharm. Weekbl. Sci. Ed.*, *5*:
     41−48 (1983).
123. R. J. M. Fry and E. Staffeldt, *Nature*, *203*:1396−1398 (1964).
124. K. Nakanishi, H. Saitoh, M. Masada, A. Tatematsu, and T. Nadai,
     Chem. Pharm. Bull., *32*:3187−3193 (1984).
125. G. P. Martin, C. Marriott, and I. W. Kellaway, *Gut*, *19*:103−107
     (1978).
126. T. Murakami, Y. Sasaki, R. Yamajo, and N. Yata, *Chem. Pharm.
     Bull.*, *32*:1948−1955 (1984).
127. J. F. Forstner and G. G. Forstner, *Biochim. Biophys. Acta*, *386*:
     283−292 (1975).
128. B. L. Slomiany, M. Kosmala, S. R. Carter, S. J. Konturek,
     J. Bilski, and A. Slomiany, *Comp. Biochem. Physiol.*, *87*:657−663
     (1987).
129. F. G. J. Poelma, J. J. Tukker, and D. J. A. Crommelin, *J. Pharm.
     Sci.*, *78*:285−289 (1989).
130. S. Hirai, T. Yashiki, and H. Mima, *Int. J. Pharm.*, *9*:173−184
     (1981).
131. M. Tomita, M. Hayashi, T. Horie, T. Ishizawa, and A. Awazu,
     Pharm. Res., *5*:786−789 (1988).
132. D. A. Whitmore, L. G. Brookes, and K. P. Wheeler, *J. Pharm.
     Pharmacol.*, *31*:277−283 (1979).
133. M. Tomita, M. Shiga, M. Hayashi, and S. Awazu, *Pharm. Res.*, *5*:
     341−346 (1988).

134. C. L. Gay, T. M. Murphy, J. Hadgraft, I. W. Kellaway, J. C. Evans, and C. C. Rowlands, *Int. J. Pharm.*, *49*:39–45 (1989).

135. J. C. Beastall, J. Hadgraft, and C. Washington, *Int. J. Pharm.*, *43*:207–213 (1988).

136. G. M. Golden, D. B. Guzek, R. R. Harris, J. E. McKie, and R. O. Potts, *J. Invest. Dermatol.*, *86*:255–259 (1986).

137. N. Muranushi, N. Takagi, S. Muranishi, and H. Sezaki, *Chem. Phys. Lipids*, *28*:269–279 (1981).

138. K. D. Lillemoe, T. R. Gadacz, and J. W. Harmon, *Surg. Gynecol. Obstet.*, *155*:13–16 (1982).

139. K. Sato, K. Sugibayashi, and Y. Morimoto, *Int. J. Pharm.*, *43*:31–40 (1988).

140. H. Kajii, T. Horie, M. Hayashi, and S. Awazu, *J. Pharm. Sci.*, *77*:390–392 (1988).

141. H. Kajii, T. Horie, M. Hayashi, and S. Awazu, *J. Pharm. Sci.*, *75*:475–478 (1986).

142. T. Nishihata and T. Higuchi, *Biochim. Biophys. Acta*, *775*:269–271 (1984).

143. T. Nishihata, T. Higuchi, and A. Kamada, *Life Sci.*, *34*:437–445 (1984).

144. S. Szabo, J. S. Trier, and P. W. Frank, *Science*, *214*:200–202 (1981).

145. T. Nishihata, H. Takahata, and A. Kamada, *Pharm. Res.*, *1*:307–309 (1985).

146. M. Murakami, K. Takada, T. Fujii, and S. Muranishi, *Biochim. Biophys. Acta*, *939*:238–246 (1988).

147. T. Nishihata, T. Suzuka, A. Furuya, M. Yamazaki, and A. Kamada, *Chem. Pharm. Bull.*, *35*:2914–2922 (1987).

148. T. Nishihata, B. T. Nghiem, H. Yoshitomi, C. S. Lee, M. Dillsaver, T. Higuchi, R. Choh, T. Suzuka, A. Furuay, and A. Kamada, *Pharm. Res.*, *3*:345–351 (1986).

149. T. Suzuka, N. Yata, K. Sakai, and T. Nishihata, *J. Pharm. Pharmacol.*, *40*:469–472 (1988).

150. M. G. Farquhar and G. E. Palade, *J. Cell Biol.*, *17*:375–412 (1963).

151. M. Cereijido, I. Meza, and A. Martinez-Palomo, *Am. J. Physiol.*, *240*:C96–C102 (1981).

152. J. L. Madara and K. Dharmsathaphorn, *J. Cell Biol.*, *101*:2124–2133 (1985).

153. A. Martinez-Palomo, I. Meza, G. Beaty, and M. Cereijido, *J. Cell Biol.*, *87*:736–745 (1980).

154. D. R. Pitelka, B. N. Taggart, and S. T. Hamamoto, *J. Cell Biol.*, *96*:613–624 (1983).

155. G. E. Peters, E. E. F. Hutchinson, R. Hyde, C. McMartin, and S. B. Metcalfe, *J. Pharm. Sci.*, *76*:857–861 (1987).

156. B. Gumbiner and K. Simons, *J. Cell Biol.*, *102*:457–468 (1986).

157. J. R. Sanes, M. Schachner, and O. J. Covault, *J. Cell Biol.*, *102*:420–431 (1986).

158. B. Gumbiner, *Am. J. Physiol.*, *253*:C749–C758 (1987).

159. K. Boller, D. Vestweber, and R. Kemler, *J. Cell Biol.*, *100*:327–332 (1985).

160. B. R. Stevenson and D. A. Goodenough, *J. Cell Biol.*, *98*:1209–1221 (1984).

161. R. W. Freel, M. Hatch, D. L. Earnest, and A. M. Goldner, *Am. J. Physiol.*, *245*:G816−G823 (1983).

162. M. Shiga, M. Hayashi, T. Horie, and S. Awazu, *J. Pharm. Pharmacol.*, *39*:118−123 (1987).

163. M. Shiga, T. Muraoka, T. Hirasawa, M. Hayashi, and S. Awazu, *J. Pharm. Pharmacol.*, *37*:446−447 (1985).

164. M. Shiga, M. Hayashi, T. Horie, and S. Awazu, *Chem. Pharm. Bull.*, *34*:2254−2256 (1986).

165. K. Sakai, T. M. Kutsuna, T. Nishino, Y. Fujihara, and N. Yata, *J. Pharm. Sci.*, *75*:387−390 (1986).

166. E. Boyland and L. F. Chasseaud, *Biochem. J.*, *104*:95−102 (1967).

167. M. Cooper, S. Teichberg, and F. Lifshitz, *Lab. Invest.*, *38*:447−454 (1978).

168. J. A. Fix, P. S. Leppert, P. A. Porter, and L. J. Caldwell, *J. Pharm. Sci.*, *72*:1134−1137 (1983).

169. J. Fix, P. A. Porter, and P. S. Leppert, *J. Pharm. Sci.*, *27*:698−700 (1983).

170. J. R. Pappenheimer and K. Z. Reiss, *J. Membr. Biol.*, *100*:123−136 (1987).

171. J. R. Pappenheimer, *J. Membr. Biol.*, *100*:137−148 (1987).

172. J. L. Madara and J. R. Pappenheimer, *J. Membr. Biol.*, *100*:149−164 (1987).

173. J. L. Madara, D. Barenberg, and S. Carlson, *J. Cell Biol.*, *102*:2125−2136 (1986).

174. J. M. Mullin and T. O'Brien, *Am. J. Physiol.*, *251*:C597−C602 (1986).

175. B. S. Packard, M. G. Saxton, M. J. Bissel, and M. B. Klein, *Proc. Natl. Acad. Sci.*, *81*:449−452 (1984).

176. K. Jacobson, C. E. Wenner, G. Kemp, and D. Papahadjopoulos, *Cancer Res.*, *35*:2991−2995 (1975).

177. P. R. Dragsten, J. S. Handler, and R. Blumenthal, *Fed. Proc.*, *41*:48−53 (1982).

178. C. A. Rabito, *Am. J. Physiol.*, *250*:F734−F743 (1986).

179. G. Van Meer and K. Simons, *EMBO J.*, *1*:847−852 (1982).

180. G. Van Meer and K. Simons, *EMBO J.*, *5*:1455−1464 (1986).

181. G. Carmel, F. Rodrigue, S. Carrière, and C. Le Grimellec, *Biochim. Biophys. Acta*, *818*:149−157 (1985).

182. G. Friedlander, M. Shahedi, C. L. Le Grimellec, and C. Amiel, *J. Biol. Chem.*, *263*:11183−11188 (1988).

182a. C. A. Rabito, J. I. Kreisberg, and D. Wight, *J. Biol. Chem.*, *259*:574−582 (1984).

183. R. Guy and J. Hadgraft, *J. Contr. Rel.*, *5*:43−51 (1987).

184. R. Guy and J. Hadgraft, *Pharm. Res.*, *5*:753−758 (1988).

185. Y. Kurosaki, S. Hisaichi, C. Hamada, T. Nakayama, and T. Kimura, *Int. J. Pharm.*, *47*:13−19 (1988).

186. T. Ishizawa, M. Hayashi, and S. Awazu, *J. Pharm. Pharmacol.*, *39*:892−895 (1987).

187. F. Ishii, Y. Nagasaka, and H. Ogata, *J. Pharm. Sci.*, *78*:303−306 (1989).

188. B. S. Yu, A. Kim, H. H. Chung, W. Yoshikawa, H. Akutsu, and Y. Kyogoku, *Chem.-Biol. Interact.*, *56*:393−319 (1985).

188a. M. J. M. Deurloo, W. A. J. J. Hermens, S. G. Romeijn, J. Verhoef, and F. W. H. M. Merkus, *Pharm. Res.*, 6:853-856 (1989).

189. P. Agrawala and W. A. Ritschel, *J. Pharm. Sci.*, 77:776-777 (1988).

190. K. Nakanishi, M. Masada, and T. Nadai, *Chem. Pharm. Bull.*, 31: 4161-4166 (1983).

191. D. J. Hauss and H. Y. Ando, *J. Pharm. Pharmacol.*, 40:659-661 (1988).

192. K. B. Sloan, K. G. Siver, and S. A. M. Koch, *J. Pharm. Sci.*, 75:744-749 (1986).

193. E. R. Cooper, *J. Pharm. Sci.*, 73:1153-1156 (1984).

194. E. J. van Hoogdalem, H. J. M. van Kan, A. G. de Boer, and D. D. Breimer, *J. Contr. Rel.*, 7:53-60 (1988).

195. P. Ashton, J. Hadgraft, K. R. Brain, T. A. Miller, and K. A. Walters, *Int. J. Pharm.*, 41 189-195 (1988).

196. J. M. Franz and J. P. Vonderscher, *J. Pharm. Pharmacol.*, 33: 565-568 (1981).

197. P. Geppeti, B. M. Fusco, S. Marabini, C. A. Maggi, M. Fanciullacci, and F. Sicuteri, *Br. J. Pharmacol.*, 93:509-514 (1988).

198. K. B. Walker, M. H. Serwonska, F. H. Valone, W. S. Harkonen, O. L. Frick, K. H. Scriven, W. D. Ratnoff, J. G. Browning, D. G. Payan, and E. J. Goetzl, *J. Clin. Immunol.*, 8:108-113 (1988).

199. C. Larsen, M. N. Jorgensen, B. Tommerup, N. Mygind, E. E. Dagrosa, H. G. Griogoleit, and V. Malerczyk, *Eur. J. Clin. Pharmacol.*, 33:155-159 (1987).

200. L. S. Olanoff, C. R. Titus, M. S. Shea, R. E. Gibson, and C. D. Brooks, *J. Clin. Invest.*, 80:546-550 (1987).

201. H. R. D'Agostino, C. A. Barnett, X. J. Zielinski, and G. S. Gordan, *Clin. Orthop.*, 230:223-228 (1988).

202. J. Y. Reginster, A. Albert, and P. Franchimont, *Calcif. Tissue Int.*, 37:577-580 (1985).

203. W. S. Evans, J. L. C. Borges, D. L. Kaiser, M. L. Vance, R. P. Sellers, R. M. MacLeod, W. Vale, J. Rivier, and M. O. Thorner, *J. Clin. Endocrinol. Metab.*, 57:1081-1083 (1983).

204. B. Berner, D. R. Wilson, R. H. Guy, G. C. Mazzenga, F. H. Clarke, and H. I. Maibach, *Pharm. Res.*, 5:660-663 (1988).

205. H. Okamoto, M. Hashida, and H. Sezaki, *J. Pharm. Sci.*, 77:418-424 (1988).

206. B. J. Aungst, *Pharm. Res.*, 6:244-247 (1989).

207. T. S. Low-Beer, R. E. Schneider, and W. O. Dobbins, *Gut, 11*: 486-492 (1970).

208. T. Nishihata, J. H. Rytting, T. Higuchi, and L. Caldwell, *J. Pharm. Pharmacol.*, 33:334-335 (1981).

209. P. Sithigorngul, P. Burton, T. Nishihata, and L. Caldwell, *Life Sci.*, 33:1025-1032 (1983).

210. J. P. Longenecker, A. C. Moses, J. S. Flier, R. D. Silver, M. Carey, and E. J. Dubovi, *J. Pharm. Sci.*, 75:351-355 (1987).

211. M. A. Wheatley, J. Dent, E. B. Wheeldon, and P. L. Smith, *J. Contr. Rel.*, 8:167-177 (1988).

212. T. Nadai, M. Kume, A. Tatematsu, and H. Sezaki, *Chem. Pharm. Bull.*, 23:543-551 (1975).

213. K. S. Cho and P. Proulx, *Biochim. Biophys. Acta, 193*:30−35 (1969).

214. K. S. Cho and P. Proulx, *Biochim. Biophys. Acta, 225*:214−223 (1971).

215. J. R. Bond and B. W. Barry, *J. Invest. Dermatol., 90*:810−873 (1988).

216. M. V. Teem and S. F. Phillips, *Gastroenterology, 62*:261−267 (1972).

217. J. L. Pope, T. M. Parkinson, and J. A. Olson, *Biochim. Biophys. Acta, 130*:218−232 (1966).

218. M. L. Clark, H. C. Lanz, and J. R. Senior, *J. Clin. Invest., 48*: 1587−1599 (1969).

219. P. J. Sabourin, R. E. Tynes, R. M. Philpot, S. Winquist, and A. R. Dahl, *Drug Metab. Disp., 6*:557−562 (1988).

220. D. D. S. Tang-Liu, J. Neff, H. Zolezio, and R. Sandri, *Pharm. Res., 5*:477−481 (1988).

221. J. W. Wiechers, B. F. H. Drenth, J. H. G. Jonkman, and R. A. de Zeeuw, *Int. J. Pharm., 47*:43−49 (1988).

222. B. Carrière and C. Le Grimellec, *Biochim. Biophys. Acta., 857*:131−138 (1986).

223. G. Friedlander, C. Le Grimellec, M.-C. Giocondi, and C. Amiel, *Biochim. Biophys. Acta, 903*:341−348 (1987).

224. M. D. Houslay, I. Dipple, S. Rawal, R. D. Sauerheber, J. A. Esgate, and L. M. Gordon, *Biochem. J., 190*:131−137 (1980).

225. J. C. Liu, Y. Sun, O. Siddiqui, Y. W. Chien, W. M. Shi, and J. Li, *Int. J. Pharm., 44*:197−204 (1988).

226. Y. W. Chien, O. Siddiqui, W. M. Shi, P. Lelawongs, and J. C. Liu, *J. Pharm. Sci., 78*:376−383 (1989).

227. J. Kost, D. Levy, and R. Langer, *Proc. Intern. Symp. Control. Rel. Bioact. Mater., 13*:177−178 (1986).

228. R. Brucks, M. Nanavaty, D. Jung, and F. Siegel, *Pharm. Res., 6*:697−701 (1989).

229. S. Hirai, T. Ikenaga, and T. Matsazawa, *Diabetes, 27*:296−299 (1978).

230. T. Ohwaki, H. Ando, F. Kakimoto, K. Uesugi, S. Watanabe, Y. Miyake, and M. Kayano, *J. Pharm. Sci., 76*:695−698 (1987).

231. A. L. Daugherty, H. D. Liggitt, J. G. McCabe, J. A. Moore, and J. S. Patton, *Int. J. Pharm., 45*:197−206 (1988).

232. K. Morimoto, E. Kamiya, T. Takeeda, Y. Nakamoto, and K. Morisaka, *Int. J. Pharm., 14*:149−157 (1983).

233. K. Morimoto, H. Akatsuchi, R. Aikawa, M. Morishita, and K. Morisaka, *J. Pharm. Sci., 73*:1366−1368 (1984).

234. K. Morimoto, K. Morisaka, and A. Kamada, *J. Pharm. Pharmacol., 37*:134−136 (1985).

235. A. S. Harris, I. M. Nilsson, Z. G. Wagner, and U. Alkner, *J. Pharm. Sci., 75*:1085−1087 (1986).

236. A. S. Harris, M. Ohlin, S. Lethagen, and I. M. Nilsson, *J. Pharm. Sci., 77*:337−339 (1988).

237. L. Illum, H. Jorgensen, H. Bisgaard, O. Krosgaard, and N. Rossing, *Int. J. Pharm., 39*:189−199 (1987).

238. E. Björk and P. Edman, *Int. J. Pharm., 47*:233−238 (1988).

239. E. Wood, C. G. Wilson, and J. G. Hardy, *Int. J. Pharm., 25*:191−197 (1985).

240. G. G. Liversidge, T. Nishihata, K. K. Engle, and T. Higuchi, *Int. J. Pharm., 30*:247−250 (1986).

241. G. G. Liversidge, T. Nishihata, K. K. Engle, and T. Higuchi, *Int. J. Pharm.*, 23:87−95 (1985).

242. C. J. Goodner, B. C. Walike, D. J. Koerker, J. W. Ensinck, and A. C. Brown, *Science*, 195:177−179 (1977).

243. R. W. Reynolds, L. D. Keith, D. R. Harris, and S. Calvano, *Steroids*, 35:305−314 (1980).

244. J. Fox, K. P. Offord, and H. Heath, III, *Am. J. Physiol.*, 30: 247−250 (1986).

245. J. K. Stewart, D. K. Clifton, D. J. Koerker, A. D. Rogol, T. Jaffe, and C. J. Goodner, *Endocrinology*, 116:1−5 (1985).

246. D. J. Dierschke, A. N. Bhattaracharya, L. E. Atkinson, and E. Knobil, *Endocrinology*, 87:850−853 (1970).

247. H. J. Quabble, M. Gregor, C. Bumke-Vogt, A. Eckhof, and I. Witt, *Endocrinology*, 109:513−521 (1981).

248. E. Van Cauter, M. L'Hermite, G. Copinschi, S. Refetoff, D. Desir, and C. Robyn, *Am. J. Physiol.*, 241:E355−E363 (1981).

249. D. Weigle, *Diabetes*, 36:764−775 (1987).

250. B. Zumoff, R. Freeman, S. Coupey, P. Saenger, M. Maskowitz, and J. Kream, *N. Engl. J. Med.*, 309:1206−1209 (1983).

251. P. de Remigis, L. Vianale, A. Damiani, M. D'Angelo, and S. Sensi, *Chronobiologia*, 9:127−132 (1982).

252. T. D. Miller, G. S. Taunenbaum, E. Colle, and H. J. Guyda, *J. Clin. Endocrin. Metab.*, 55:989−994 (1982).

253. R. Durso, C. A. Tamminga, S. Ruggeri, A. Denaro, S. Kuo, and T. N. Chase, *J. Neurol. Neurosurg. Psychiat.*, 46:1134−1137 (1983).

254. A. Miles, *Life Sci.*, 44:375−385 (1989).

255. R. J. Wurtman, J. Axelrod, and L. T. Potter, *J. Pharmacol. Exp. Ther.*, 143:314−318 (1964).

256. A. J. Lewy and D. A. Newsome, *J. Clin. Endocrinol. Metab.*, 56: 1103−1107 (1983).

257. F. Anton-Tay and R. J. Wurtman, *Nature*, 221:474−475 (1969).

258. U. Lang, M. L. Aubert, and P. C. Sizonenko, *Paediatr. Res.*, 15:80−83 (1981).

259. S. M. Reppert, D. R. Weaver, S. A. Rivkees, and E. G. Stopa, *Science*, 242:78−81 (1988).

260. A. R. Glass, A. P. Zavadil, F. Halberg, G. Cornelissen, and M. Schaaf, *J. Clin. Endocrin. Metab.*, 59:161−165 (1984).

261. R. C. Baxter and C. T. Cowell, *J. Clin. Endocrinol. Metab.*, 65:432−440 (1987).

262. G. L. Nicolson and G. Poste, *Curr. Probl. Cancer*, 7:1−42 (1983).

263. M. Berelowitz, M. J. Perlow, H. J. Hoffman, and L. A. Frohman, *Endocrinology*, 109:2102−2109 (1981).

264. R. Jorde and P. G. Burhol, *Scand. J. Gastroenterol.*, 20:1−4 (1985).

265. A. Reinberg, P. Drouin, M. Kolopp, L. Mejean, F. Levi, G. Debry, M. Mechkouri, G. DiCostanzo, and A. Bicakora-Rocher, Pump delivered insulin and home monitored blood glucose in a diabetic patient: retrospective and chronophysiologic evaluation of 3-year time series. Proc. 3rd Int. Conf. Chronopharmacology, Int. Soc. Chronobiology, Nice, France (1988).

266. J. Hunter, J. McGee, J. Saldivar, T. Tsai, R. Feuers, and L. E. Scheving, Circadian variations in insulin and alloxan sensitivity noted in blood glucose alterations in normal and diabetic mice.

Proc. 3rd Int. Conf. Chronopharmacology, Int. Soc. Chronobiology, Nice, France (1988).

267. D. S. Weigle and C. J. Goodner, *Endocrinology, 118*:1606–1613
(1986).

268. C. J. Goodner, D. J. Koerker, B. C. Hansen, and F. Horn,
Cyclic secretion by the islets of Langerhans, in *Carbohydrate
Metabolism* (C. Cobelli, R. N. Bergman, eds.), Wiley, New York
(1981), pp. 37–55.

269. B. C. Hansen, K. L. C. Jen, S. B. Pek, and R. A. Wolfe, *J.
Clin. Endocrinol. Metab., 54*:785–792 (1982).

270. D. A. Lang, D. R. Matthews, M. Burnett, G. M. Ward, and
R. C. Turner, *Diabetes, 31*:22–26 (1982).

271. G. S. Tannenbaum and J. B. Martin, *Endocrinology, 98*:562–569
(1976).

272. G. Tannenbaum and N. Ling, *Endocrinology, 115*:1952–1957 (1984).

273. I. Gil-Ad, Z. Dickerman, S. Amdursky, and Z. Laron, *Psycho-
pharmacology, 88*:496–499 (1986).

274. J. Sieradzki, H. Stanuch, W. Golda, and Z. Szybinski, *Horm.
Metab. Res., 19*:208–211 (1987).

275. K. Y. Ho, W. S. Evans, R. M. Blizzard, J. D. Veldhuis, G. R.
Merriam, E. Samojlik, R. Furlanetto, A. D. Rogol, D. L. Kaiser,
and M. O. Thorner, *J. Clin. Endocrinol. Metab., 64*:51–58 (1987).

276. M. H. Smolensky, A. Reinberg, and J. T. Queng, *Ann. Allergy,
47*:234–252 (1981).

277. A. Reinberg, M. Smolensky, and F. Levi, *Biomedicine, 34*:171–178
(1981).

278. A. Reinberg and M. H. Smolensky, *Clin. Pharmacokin., 7*:401–420
(1982).

279. M. S. Kafka, A. Wirz-Justice, D. Naber, R. Y. Moore, and
M. A. Benedito, *Fed. Proc., 42*:2796–2801 (1983).

280. D. S. Minors, *Clin. Sci., 69*:369–376 (1985).

281. H. A. Decousus, M. Croze, F. A. Levi, J. G. Jaubert,
B. M. Perpoint, J. F. Bonadona, A. Reinberg, and P. M. Queneau,
Br. Med. J., 290:341–344 (1985).

282. G. M. Kyle, M. H. Smolensky, L. G. Thorne, B. P. Hsi,
A. Robinson, and J. P. McGovern, Circadian rhythm in the
pharmacokinetics of orally administered theophylline, in, *Recent
Advances in the Chronobiology of Allergy and Immunology*
(M. H. Smolensky, A. Reinberg, and J. P. McGovern, eds.),
Pergamon Press, Oxford (1980), pp. 95–111.

283. R. J. Rackley, M. C. Meyer, and A. B. Straughn, *J. Pharm.
Sci., 77*:658–661 (1988).

284. B. Lemmer, H. Winkler, T. Ohm, and M. Fink, *Arch. Pharmacol.,
330*:42–49 (1985).

285. W. J. M. Hrushesky, *Ann. Rev. Chronopharmacol., 1*:119–122
(1984).

286. T. Langevin, J. Young, K. Walker, R. Roemeling, S. Nygaard,
and W. J. M. Hrushesky, *Proc. AACCR, 28*:398 (1987).

287. S. Ohdo, S. Nakano, and N. Ogawa, *Chronobiology Int., 6*:229–235
(1989).

288. A. Reinberg, W. Dupont, Y. Touitou, M. Lagoguey, P. Bourgeois,
C. Touitou, G. Muriaux, D. Przyrowsky, S. Guillemant,
J. Guillemant, L. Briere, and B. Zeau, *Chronobiologia, 8*:11–31 (1981).

289. J. M. Connors and G. A. Hedge, *Endocrinology*, *106*:911−917 (1980).
290. B. Brewitt and J. I. Clark, *Science*, 242:777−779 (1988).
291. B. G. Miller, R. Tassell, and G. M. Stone, *Aust. J. Biol. Sci.*, *35*:417−425 (1983).
292. G. Leyendecker, L. Wildt, and M. Hansmann, *J. Clin. Endocrinol. Metab.*, *51*:1214−1216 (1980).
293. M. R. Soules, M. B. Southworth, M. E. Norton, and W. J. Bremner, *Fertil. Steril.*, *46*:578−585 (1986).
294. P. E. Belchetz, T. M. Plant, Y. Nakai, E. J. Keogh, and E. Knobil, *Science*, *202*:631−633 (1978).
295. Y. Nakai, T. M. Plant, D. L. Hess, E. J. Keoch, and E. Knobil, *Endocrinology*, *102*:1008−1014 (1978).
296. E. Knobil, *Recent Prog. Horm. Res.*, *36*:53−88 (1980).
297. B. C. J. M. Fauser, J. M. J. Dony, W. H. Doesburg, and R. Rolland, *Fertil. Steril.*, *39*:695−699 (1983).
298. D. J. Kesler, L. C. Cruz, G. F. Sargent, and J. A. McKenzie, *J. Contr. Rel.*, *8*:55−61 (1988).
299. P. E. Rochiccioli, M.-T. Tauber, F. Uboldi, F.-X. Coude, and M. Morre, *J. Clin. Endocrinol. Metab.*, *63*:1100−1105 (1986).
300. P. M. Cotes, W. A. Bartlett, R. E. Gaines Das, P. Flecknell, and R. Termeer, *J. Endocrinol.*, *87*:303−312 (1980).
301. R. G. Clark, J. O. Jansson, O. Isaksson, and I. C. A. F. Robinson, *J. Endocrinol.*, *104*:53−61 (1985).
302. E. Verdin, M. Castillo, A. S. Luyckx, and P. J. Lefebvre, *Diabetes*, *33*:1169−1174 (1984).
303. M. Komjati, P. Bratusch-Marrain, and W. Waldhausl, *Endocrinology*, *118*:312−319 (1986).
304. P. R. Bratusch-Marrain, M. Komjati, and W. K. Waldhausl, *Diabetes*, *35*:922−926 (1986).
305. R. W. Stout, *Diabetes*, *30*(Suppl. 2):54−57 (1981).
306. D. S. Weigle, D. J. Koerker, and C. J. Goodner, *Am. J. Physiol.*, *247*:E564−E568 (1984).
307. D. S. Weigle, D. J. Koerker, and C. J. Goodner, *Am. J. Physiol.*, *248*:E681−E686 (1985).
308. C. S. Tam, J. N. M. Heersche, T. M. Murray, and J. A. Parsons, *Endocrinology*, *110*:506−511 (1982).
309. R. D. Podbesek, E. B. Mawer, G. D. Zanelli, J. A. Parsons, and J. Reeve, *Clin. Sci.*, *67*:591−599 (1984).
310. C. A. Mahoney, R. A. Nissenson, P. Sarnacki, and K. Pua, *J. Clin. Invest.*, *72*:411−421 (1983).
311. J. R. Gavin, J. Roth, D. M. Neville, P. deMeyts, and D. N. Buell, *Proc. Natl. Acad. Sci.*, *71*:84−88 (1974).
312. A. H. Soll, C. R. Kahn, D. M. Neville, Jr., and J. Roth, *J. Clin. Invest.*, *56*:769−780 (1975).
313. A. J. W. Hsuen, H. L. Dufau, and K. J. Catt, *Proc. Natl. Acad. Sci., U.S.A.*, *74*:592−595 (1977).
314. M. A. Lesniak and J. Roth, *J. Biol. Chem.*, *251*:3720−3729 (1976).
315. P. M. Hinkle and A. H. Tashjian, Jr., *Biochemistry*, *14*:3845−3851 (1975).
316. W. J. M. Hrushesky, *J. Biol. Resp. Mod.*, *6*:587−598 (1987).
317. D. S. T. Hsieh, R. Langer, and J. Folkman, *Proc. Natl. Acad. Sci.*, *78*:1863−1867 (1981).

318. E. R. Edelman, J. Kost, H. Bobeck, and R. Langer, *J. Biomed. Mater. Res.*, *19*:67−83 (1985).

319. O. Saslawski, C. Weingarten, J. P. Benoit, and P. Couvreur, *Life Sci.*, *42*:1521−1528 (1988).

320. S. Miyazaki, C. Yokouchi, and M. Takada, *J. Pharm. Pharmacol.*, *40*:716−717 (1988).

321. A. S. Hoffman, A. Afrassiabi, and L. C. Dong, *J. Contr. Rel.*, *4*:213−222 (1986).

322. L. C. Dong and A. S. Hoffman, *J. Contr. Rel.*, *4*:223−227 (1986).

323. H. J. Lynch, R. W. Rivest, and R. J. Wurtman, *Neuroendocrinology*, *31*:106−111 (1980).

324. L. L. Ewing, T. Y. Wing, R. C. Cochran, N. Kromann, and B. R. Zierkin, *Endocrinology*, *112*:1763−1769 (1983).

325. P. I. Lee, *J. Contr. Rel.*, *4*:1−7 (1986).

326. D. L. Wise, D. J. Trantolo, R. T. Marino, and J. P. Kitchell, *Adv. Drug Del. Rev.*, *1*:19−39 (1987).

327. M. Brownlee and A. Cerami, *Science*, *206*:1190−1191 (1979).

328. S. Y. Jeong, S. W. Kim, M. J. D. Eenink, and J. Feijen, *J. Contr. Rel.*, *1*:57−66 (1984).

329. S. Sato, S. Y. Jeong, J. C. McRea, and S. W. Kim, *J. Contr. Rel.*, *1*:67−78 (1984).

330. S. Y. Jeong, S. W. Kim, D. L. Holmenberg, and J. C. McRea, Self-regulating insulin delivery systems. III. In vivo studies, in *Advances in Drug Delivery Systems* (M. J. Anderson and S. W. Kim, eds.), Elsevier, Amsterdam (1986), pp. 143−152.

331. J. Kost, T. A. Horbett, B. D. Ratner, and M. Singh, *J. Biomed. Mat. Res.*, *19*:1117−1123 (1985).

332. F. Fischer-Ghodsian, L. Brown, E. Mathiowitz, D. Brandenburg, and R. Langer, *Proc. Natl. Acad. Sci. USA*, *85*:2403−2406 (1988).

333. G. W. Albin, T. A. Horbett, S. R. Miller, and N. L. Ricker, *J. Contr. Rel.*, *6*:267−291 (1987).

334. G. M. O'Shea and A. M. Sun, *Diabetes*, *35*:943−946 (1986).

335. J. Lieberman, *Chest*, *64*:579−584 (1973).

336. J. Travis, M. Owen, P. George, R. Carrell, S. Rosenberg, R. A. Hallwell, and P. J. Barr, *J. Biol. Chem.*, *260*:4384−4389 (1985).

337. C. D. Yu, N. Roosdorp, and S. Pushpala, *Pharm. Res.*, *5*:800−802 (1988).

338. R. Falcoff, *J. Gen. Virol.*, *16*:251−253 (1972).

339. M. Y.-S. Chen, M. Wiranowska-Stewart, P. Von Wussaw, and W. E. Stewart, II, *Antiviral Res.*, *1*:24−29 (1981).

340. S. S. Leong, J. S. Horoszewicz, and E. A. Mirand, The stability of human fibroblast interferon (HuIFN-β). 13th International Cancer Congress, Part E, Cancer Management (1983), pp. 327−336.

341. D. F. Mark, S. D. Lu, A. A. Creasey, R. Yamamoto, and L. S. Lin, *Proc. Natl. Acad. Sci. USA*, *81*:5662−5666 (1984).

342. C. Tanford, *Adv. Protein Chem.*, *23*:121−275 (1968).

343. K. Gekko and S. N. Timasheff, *Biochemistry*, *20*:4667−4676 (1981).

344. K. Gekko and S. N. Timasheff, *Biochemistry*, *20*:4677−4686 (1981).

345. J. C. Lee, R. P. Frigon, and S. N. Tamasheff, *Ann. NY Acad. Sci.*, *253*:284−291 (1975).

346. J. C. Lee and S. N. Timasheff, *J. Biol. Chem.*, *256*:7193−7201 (1981).

347. T. Arakawa and S. N. Timasheff, *Biochemistry*, *21*:6536−6544 (1982).
348. T. Arakawa and S. N. Timasheff, *Biochemistry*, *21*:6545−6552 (1982).
349. T. Arakawa and S. N. Timasheff, *Arch. Biochem. Biophys.*, *224*: 169−171 (1983).
350. S. N. Timasheff, J. C. Lee, E. P. Pittz, and N. Tweedy, *J. Colloid Interface Sci.*, *55*:658−663 (1976).
351. E. P. Pittz and S. N. Timasheff, *Biochemistry*, *17*:615−623 (1978).
352. K. Gekko and S. Koga, *J. Biochem.*, *94*:199−205 (1983).
353. W. D. Lougheed, H. Woulfe-Flanagan, J. R. Clement, and A. M. Albisser, *Diabetologica*, *19*:1−9 (1980).
354. W. D. Lougheed, A. M. Albisser, H. M. Martindale, J. C. Chow, and J. R. Clement, *Diabetes*, *32*:424−432 (1983).
355. P. D. Jeffrey, *Biol. Chem. Hoppe-Seyier*, *367*:363−369 (1986).
356. K. Irsigler and H. Kritz, *Diabetes*, *28*:196−203 (1979).
357. W. S. Jackman, W. Lougheed, E. B. Marliss, B. Zinman, and A. M. Albisser, *Diabetes Care*, *3*:322−331 (1980).
358. S. E. Charm and B. L. Wong, *Biotechnol. Bioeng.*, *12*:1103−1109 (1970).
359. S. E. Charm and B. L. Wong, *Enzyme Microbiol. Technol.*, *3*:111−000 (1981).
360. D. J. Fink and V. W. Rodwell, *Biotechnol. Bioeng.*, *17*:1029−1050 (1975).
361. M. Tirrell and S. Middleman, *Biopolymers*, *18*:59−72 (1979).
362. T. J. Narendaranathan and P. Dunnill, *Biotechnol. Bioeng.*, *24*: 2103−2107 (1982).
363. J. J. Sedmak and S. E. Grossberg, *Tex. Reports Biol. Med.*, *41*: 274−279 (1978).
364. J. J. Sedmak and S. E. Grossberg, *Tex. Rep. Biol. Med.*, *35*: 198−203 (1977).
365. S. B. Greenberg, M. W. Harmon, and R. B. Couch, *Modern Pharm. Toxicol.*, *17*:57−87 (1980).
366. J. J. Sedmak and S. E. Grossberg, *J. Gen. Virol.*, *52*:195−197 (1981).
367. K. G. Hutchison, *J. Pharm. Pharmacol.*, *37*:528−531 (1985).
368. S. Sato, C. D. Ebert, and S. W. Kim, *J. Pharm. Sci.*, *72*:228−232 (1983).
369. J. Bringer, A. Heldt, and G. M. Grodsky, *Diabetes*, *30*:83−85 (1981).
370. B. Wigness, F. Dorman, T. Rhode, K. Kernstine, E. Chute, and H. Buchwald, *Diabetes*, *35*:140A (1986).
371. R. Quinn and J. D. Andrale, *J. Pharm. Sci.*, *72*:1472−1473 (1983).
372. E. H. Massey and T. A. Sheliga, *Pharm. Res.*, *3*:26−27 (1986).
373. D. A. Yphantis and T. Arakawa, *Biochemistry*, *26*:5422−5427 (1987).
374. S. J. Shire, *Biochemistry*, *22* 2664−2671 (1983).
375. K. Nagata, T. Izumi, T. S. Kitagawa, and N. Yoshida, *J. Interferon Res.*, *7*:313−320 (1987).
376. P. W. Gray and D. V. Goeddel, *Proc. Natl. Acad. Sci.*, *80*:5842−5846 (1983).
377. T. Arakawa, Y. R. Hsu, C. G. Parker, and P. H. Lai, *J. Biol. Chem.*, *261*:8534−8539 (1986).
378. F. MacRitchie and N. F. Owens, *J. Colloid Interface Sci.*, *29*:66−00 (1969).

379. A. F. Henson, J. R. Michell, and P. R. Mussellwhite, *J. Colloid Interface Sci.*, *32*:162−165 (1970).

380. T. A. Bewley and C. H. Li, *Biochim. Biophys. Acta*, *140*:201−207 (1967).

381. H. Kawauchi, T. A. Bewley, and C. H. Li, *Biochim. Biophys. Acta*, *446*:525−535 (1976).

382. T. A. Bewley and C. H. Li, *Experientia*, *27*:1368−1371 (1971).

383. T. A. Bewley and C. H. Li, *Arch. Biochem. Biophys.*, *144*:589−595 (1971).

384. J. L. Browning, R. J. Mattaliano, E. P. Chow, S.-M. Liang, B. Allet, J. Rosa, and J. E. Smart, *Anal. Biochem.*, *155*:123−128 (1986).

385. S. Pestka, B. Kelder, P. C. Familletti, J. A. Moschera, R. Crowl, and E. S. Kempner, *J. Biol. Chem.*, *258*:9706−9709 (1983).

386. P. L. Felgner and J. E. Wilson, *Anal. Biochem.*, *74*:631−635 (1976).

387. H. C. Beyerman, M. I. Grossman, T. Scratcherd, T. E. Solomon, and D. Voskamp, *Life Sci.*, *29*:885−894 (1981).

388. K. J. Kramer, P. E. Dunn, R. C. Peterson, H. L. Seballos, L. L. Sandburg, and J. H. Law, *J. Biol. Chem.*, *251*:4979−4985 (1976).

389. C. H. Suelter and M. Deluca, *Anal. Biochem.*, *135*:112−119 (1983).

390. J. C. McElnay, D. S. Elliot, and P. F. D'Arcy, *Int. J. Pharm.*, *36*:199−203 (1987).

391. H. Tenenbaum and A. Levitzki, *Biochim. Biophys. Acta*, *445*:261−279 (1976).

392. D. A. Eppstein and J. P. Longenecker, *CRC Crit. Rev. Ther. Drug Carrier Sys.*, *5*:99−139 (1988).

393. D. A. Eppstein, M. A. van der Pas, B. B. Schryver, P. L. Felgner, C. A. Gloff, and K. F. Soike, Controlled release and localized targeting of interferons, in, *Delivery Systems for Peptide Drugs* (S. S. Davis, L. Illum, and E. Tomlinson, eds.), Plenum, New York (1986), pp. 277−283.

394. E. V. Jensen, V. D. Hospelhorn, D. F. Tapley, and C. Huggins, *J. Biol. Chem.*, *185*:411−422 (1950).

395. R. E. Harrington and B. H. Zimm, *J. Phys. Chem.*, *79*:161−167 (1965).

396. R. T. Knapp, J. W. Daily, and F. G. Hammitt, *Cavitation*, McGraw-Hill, New York (1970), pp. 235−256.

397. T. Komiyama, H. Kanno, K. Honbou, and M. Miwa, *Biochim. Biophys. Acta*, *914*:89−95 (1987).

398. J. W. Heine, A. J. Mikulski, E. Sulkowski, and W. A. Carter, *Arch. Virol.*, *57*:185−188 (1978).

399. W. A. Carter and J. S. Horoszewicz, *Pharmac. Ther.*, *8*:359−377 (1980).

400. E. Knight, Jr. and D. Fahey, *J. Biol. Chem.*, *256*:3609−3611 (1981).

401. J. P. Lenders and R. R. Crichton, *Biotech. Bioeng.*, *26*:1343−1351 (1984).

402. D. B. Wigley, A. R. Clarke, C. R. Dunn, D. A. Barstow, T. Atkinson, W. N. Chia, H. Muirhead, and J. J. Holbrook, *Biochim. Biophys. Acta*, *916*:145−148 (1987).

403. R. Wetzel, L. J. Perry, W. A. Baase, and W. J. Becktel, *Proc. Natl. Acad. Sci.*, *85*:401−405 (1988).

404. W. C. Kenney, E. Watson, T. Bartley, T. Boone, and B. W. Altrock, *Lymphokine Res.*, *5*:523−527 (1986).

405. J. E. Villafranca, E. E. Howell, D. H. Voet, M. S. Strobel,
      R. C. Ogden, J. N. Abelson, and J. Kraut, *Science*, *222*:782–788
      (1983).

406. L. J. Perry and R. Wetzel, *Science*, *226*:555–557 (1984).

407. R. T. Sauer, K. Hehir, R. S. Stearman, M. A. Weiss,
      A. Jeitler-Nilsson, E. G. Suchanek, and C. O. Pabo, *Biochemistry*,
      25:5992–5998 (1986).

408. J. A. Wells and D. B. Powers, *J. Biol. Chem.*, *261*:6564–6570
      (1986).

409. D. B. Volkin and A. M. Klibanov, *J. Biol. Chem.*, *262*:2945–2950
      (1987).

410. P. L. Privalov, *Adv. Protein Chem.*, *33*:167–241 (1979).

411. M. J. Kornblatt and G. H. B. Hoa, *Arch. Biochem. Biophys.*, *252*:
      277–283 (1987).

412. C. L. D. Teng and M. J. Groves, *Pharm. Res.*, *5*:776–780 (1988).

413. G. A. Truskey, R. Gabler, A. Dileo, and T. Manter, *J. Parenter.
      Sci. Tech.*, *41*:180–193 (1987).

414. A. M. Pitt, *J. Parenter. Sci. Technol.*, *41*:110–113 (1987).

415. R. H. Locker and D. J. C. Wild, *Meat Sci.*, *10*:207–212 (1984).

416. E. Morild, *Adv. Protein Chem.*, *34*:93–125 (1984).

417. D. I. Marborough, D. S. Miller, and K. A. Cammack, *Biochim.
      Biophys. Acta*, *386*:576–589 (1975).

418. H. Vodenichorova, L. Bozadzhiev, T. Vitanov, V. Velev, and
      B. Zaharia, *Probl. Infect. Paras. Dis. (Sofia)*, *9*:137–141 (1981).

419. K. Hellman, D. S. Miller, and K. A. Cammack, *Biochim. Biophys.
      Acta*, *749*:133–142 (1983).

420. O. De Jesus, E. D. Handel, and H. J. Ache, *Radiochem. Radioanal.
      Lett.*, *41*:133–138 (1979).

421. N.-T. Yu and B. H. Jo, *Arch. Biochem. Biophys.*, *156*:469–474
      (1973).

422. S. Goeteni, J. Raymond, A. Ducastaing, J.-M. Robin, and
      P. Creach, *C. R. Seances Soc. Biol. Ses. Fil.*, *168*:280–285 (1974).

423. T. Yasui and Y. Hashimoto, *J. Food Sci.*, *31*:293–299 (1966).

424. J. B. A. Ross, E. L. Chang, and D. C. Teller, *Biophys. Chem.*,
      10:217–220 (1979).

425. C. Tanford and R. Lovrien, *J. Am. Chem. Soc.*, *84*:1892–1986
      (1962).

426. A. B. Deisseroth and A. L. Dounce, *Arch. Biochem. Biophys.*,
      120:671–692 (1967).

427. A. B. Deisseroth and A. L. Dounce, *Arch. Biochem. Biophys.*,
      131:30–48 (1969).

428. H. A. Sheraga, G. Nemethy, and I. Z. Steinberg, *J. Biol. Chem.*,
      237:2506–2508 (1962).

429. J. J. Sedmak, P. Jameson, and S. E. Grossberg, Thermal and
      vortical stability of purified human fibroblast interferon, in *Human
      Interferon: Production and Clinical Use* (W. Stinebring and
      P. J. Chapple, eds.), Plenum, New York (1978), pp. 133–152.

430. J. F. Carpenter, L. M. Crowe, and J. H. Crowe, *Biochim. Biophys.
      Acta*, *923*:109–115 (1987).

431. L. Randen, J. Nilson, and P. Edman, *J. Pharm. Pharmacol.*, *40*:
      763–766 (1988).

432. P. Jameson, D. Greiff, and S. E. Grossberg, *Cryobiology*, *16*:
      301–312 (1978).

433. A. L. Finkenaur, European Patent 87309748.9, April 1987.

434. A. R. Fasasihi and M. S. Praker, *J. Pharm. Sci.*, 77:876-879 (1988).
435. P. J. M. Salemink, J. C. Roodbeen-Kilman, T. C. J. Gribnau, P. S. L. Janssen, and A. J. Van der Veen, *Pharm. Weekbl. Sci. Ed.*, 9:172-178 (1987).
436. W. M. Garrison, *Chem. Rev.*, 87:381-389 (1987).
437. G. Careri, A. Giansanti, and E. Gratton, *Biopolymers*, 18:1187-1203 (1979).
438. A. Rupley, E. Gratton, and G. Careri, *Trends Biochem. Sci.*, 8:18-22 (1983).
439. N.-T. Yu, B. H. Jo, and C. S. Liu, *J. Amer. Chem. Soc.*, 94:7572-7575 (1972).
440. P. L. Poole and J. L. Finney, *Biopolymers*, 23:1647-1666 (1984).
441. M. J. Hageman, *Drug Devel. Ind. Pharm.*, 14:2047-2070 (1988).
442. J. D. Leeder and I. C. Watt, *J. Colloid Interface Sci.*, 48:339-000 (1974).
443. C. H. Rochester and A. V. Westerman, *J. Chem. Soc. Farad. Trans.*, 72:2753-2768 (1976).
444. M. Ruegg and B. Blanc, *J. Dairy Sci.*, 59:1019-1024 (1976).
445. M. Luescher, M. Ruegg, and P. Schindler, *Biopolymers*, 13:2489-2503 (1974).
446. Y. Fujita and Y. Noda, *Int. J. Peptide Protein Res.*, 18:12-17 (1981).
447. E. Danforth and R. O. Moore, *Endocrinology*, 65:118-123 (1959).
448. M. Ishida, Y. Machida, N. Nambu, and T. Nagai, *Chem. Pharm. Bull.*, 29:810-816 (1981).
449. H. Bar-On, E. M. Berry, A. Eldor, M. Kidron, D. Lichtenberg, and E. Ziv, *Br. J. Pharmacol.*, 73:21-24 (1981).
450. H. Okada, I. Yamazaki, Y. Ogawa, S. Hirai, T. Yashiki, and H. J. Mima, *J. Pharm. Sci.*, 72:75-78 (1983).
451. M. Berger, P. A. Halban, L. Girardier, J. Seydoux, R. E. Offord, and A. E. Renold, *Diabetologia*, 17:97-99 (1979).
452. H. Okada, I. Yamazaki, Y. Ogawa, S. Hirai, T. Yashiki, and H. Mima, *J. Pharm. Sci.*, 71:1367-1371 (1982).
453. J. M. Padfield, Principles of drug administration to the respiratory tract, in *Drug Delivery to the Respiratory Tract* (D. Ganderton and T. M. Jones, eds.), Ellis Horwood, Chichester, England (1987), pp. 77-86.
454. W. M. Wu, T. Murakami, Y. Higashi, and N. Yata, *J. Pharm. Sci.*, 76:508-512 (1987).
455. H. Kajii, T. Horie, M. Hayashi, and S. Awazu, *Life Sci.*, 37:523-530 (1985).
456. E. Hayakawa, A. Yamamoto, and V. H. L. Lee, *Proc. Intern. Symp. Control. Rel. Bioact. Mater.*, 15:458-459 (1988).
457. B. J. Aungst and N. J. Rogers, *Pharm. Res.*, 5:305-308 (1988).
458. B. J. Aungst, N. J. Rogers, and E. Shefter, *J. Pharmacol. Exp. Ther.*, 244:23-27 (1988).

2

Synthesis of Peptides and Proteins by Chemical and Biotechnological Means

Paul D. Gesellchen and Robert F. Santerre

Eli Lilly and Company, Indianapolis, Indiana

I. INTRODUCTION

Interest in peptides and proteins as potential drug candidates increased dramatically in the early 1960s with the development of solid-phase peptide synthesis and was stimulated again in the early 1970s with the advent of recombinant DNA (rDNA) technology. The goal of this chapter is to summarize some of the techniques that are currently utilized by the peptide chemist and molecular biologist to prepare peptides and proteins. We also describe some of the methods that can be utilized to improve the bioavailability of this class of biologically active molecules. This chapter is meant to serve as an overview and we strongly recommend to the reader the various review articles and compendia cited throughout the chapter for an in-depth treatment of any specific topic.

II. SYNTHESIS OF PEPTIDES AND PROTEINS BY CHEMICAL MEANS

A. Historical Perspective: Peptide Synthesis Yesterday and Today

The combination of an acid and an amine with the liberation of water results in the formation of an amide bond (Fig. 1). This deceptively simple reaction is the root of peptide chemistry. When the carboxylic acid and the amino components of the reaction are derived from the class of compounds known as α-amino acids, the product of the reaction is called a peptide, or in the simplest case, a dipeptide (Fig. 2).

The foundation of peptide chemistry is rooted in the first attempts to prepare polymers of α-amino acids late in the nineteenth century. This was rapidly followed by the first crude synthesis of peptides with defined structure [1—8]. The term "peptide" was introduced in 1906 by Emil

Fig. 1 Formation of an amide bond.

Fischer [9]; however, the synthesis of peptides was restricted to a few simple dipeptides.

Clearly, the amino and carboxyl groups that one does not want to participate in the amide-bond-forming reaction must be masked or "protected" or unwanted products will be formed. For example, if the synthesis of L-alanyl—L-leucine were attempted using L-alanine and L-leucine (Fig. 2), the desired dipeptide would be expected to be only one of several products. The other products would include L-leucyl—L-alanine, L-alanyl—L-alanine, L-leucyl—L-leucine, L-alanyl—L-alanyl—L-alanine, and so on. Identification of appropriate protecting groups occupied the peptide chemist for many years. The successful discovery of suitable protecting groups for the amino and carboxyl termini of amino acids and peptides vastly improved the overall synthetic strategy and ease of synthesis of simple peptides.

The controlled, predictable synthesis of peptides was greatly enhanced in 1932 when Bergmann and Zervas reported the development of an amine protecting group (the carbobenzyloxy or Cbz group) which could be selectively removed once the peptide bond was formed [10]. This achievement allowed the removal of the amine protecting group without destroying the all-important peptide bond.

During the next two decades the major advances centered on developing new methodologies for formation of the peptide bond as well as suitable protecting groups for the reactive side chains of amino acids. This progress led to the historic report by du Vigneaud and associates in 1953 on the structure elucidation and total synthesis of the pituitary peptide hormone oxytocin, a nonapeptide [11]. Many mark this date as the beginning of true synthetic peptide research.

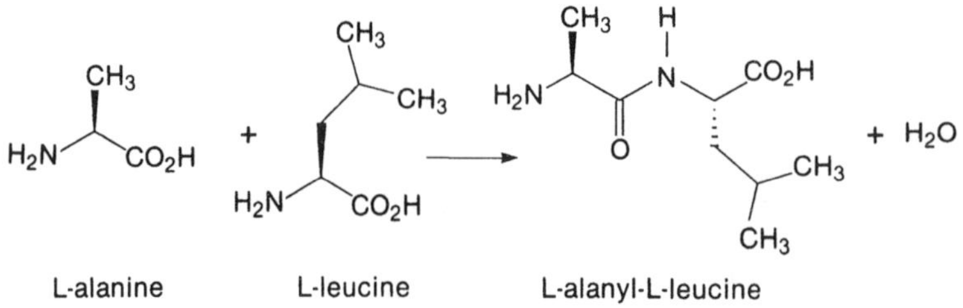

L-alanine   L-leucine   L-alanyl-L-leucine

Fig. 2 Formation of the dipeptide, L-alanyl—L-leucine.

Shortly after the report of the oxytocin synthesis, Sheehan and Hess introduced the use of dicyclohexylcarbodiimide (DCC) as a condensation or "coupling" reagent for the preparation of amide bonds [12]. This reagent quickly became the reagent of choice and still holds that preeminent position.

The next major development in peptide synthesis occurred in 1962, when Robert Merrifield of the Rockefeller Institute described the synthesis of a tetrapeptide (Leu—Ala—Gly—Val) on an insoluble polymeric solid support [13,14]. This accomplishment resulted in an opening of the flood gates in new peptide syntheses which remains unabated today. The technique of synthesis on a solid support eventually led to the development of solid-phase nucleotide synthesis, which has become an integral part of the new technology of rDNA protein synthesis [15]. The far-reaching implications and ramifications of the development of the solid-phase peptide synthesis (SPPS) technique led to the award to Merrifield of the Nobel Prize in Chemistry in 1984.

The term "peptide synthesis" took on a new magnitude in 1969 when two groups independently and simultaneously reported the first total synthesis of a biologically active protein. The enzyme ribonuclease A was synthesized by SPPS [16], while ribonuclease S was synthesized by the classical solution-phase method [17].

B. Nomenclature

The issue of when to refer to a compound as a peptide and when to refer to it as a protein has never been resolved satisfactorily. In this chapter we define all peptides comprised of more than 100 amino acid residues as proteins. In addition, we use the nomenclature proposed by Miklos Bodanszky during the 5th American Peptide Symposium [18]. All peptides larger than 10 residues will be named by an Arabic number corresponding to the number of amino acid residues in the peptide rather than by a Greek prefix. The Greek prefix will be used only in the cases of di- to deca-peptides. This nomenclature has the advantage of reducing uncertainty as to the size of the peptide. Thus methionine enkephalin is a pentapeptide, bombesin is a 14-peptide, and glucagon is a 29-peptide.

C. Peptide Synthesis Methodologies

It has been estimated that there are "about 130 variants of peptide bond formation" [19]. In this chapter we describe briefly only the major methodologies utilized routinely today for the synthesis of peptides. For more details and for a description of the methods not discussed in this chapter, the reader is referred to any of a number of excellent reviews on the subject [20—22].

Classical Solution-Phase Peptide Synthesis

Racemization. All peptide synthesis is based on the reaction of an amine component with an activated carboxyl component to form a peptide bond. In nature this is done with the growing peptide chain being elaborated on the ribosome from the amino terminus toward the carboxyl terminus. The activated carboxyl component in this ribosomal synthesis is

*Racemization

Fig. 3   Potential site of racemization following synthesis of a tripeptide from the amino terminus toward the carboxyl terminus.

always a peptide. When the peptide chemist attempts to mimic nature and synthesize a tripeptide using an activated dipeptide and an amino acid, the resultant tripeptide will be found to have a partially or completely racemized α-carbon atom at the second (penultimate) amino acid residue (Fig. 3).

Racemization occurs due either to proton abstraction followed by base-catalyzed enolization of the parent peptide (Fig. 4) or to formation of the 5(4H)-oxazolone (sometimes referred to as an azlactone) intermediate from the activated peptide (Fig. 5). The latter is at least 100 times more likely to occur than the former [23]. In general, one never activates a peptide acid, because of this potential for racemization [24].

Racemization will also occur if one attempts to form a peptide bond by activation of the carboxylic acid of a simple acylamino acid. Thus acetyl–L-alanine or benzoyl–L-alanine will partially racemize when they are subjected to standard peptide bond formation reactions. This racemization of acyl amino acids can be suppressed or eliminated; however, if one utilizes urethane-protected amino acids (e.g., Cbz or Boc, *tert*-butyloxycarbonyl) (Fig. 6). It was suggested initially that these urethane-protected amino acids (carbamates) cannot form oxazolones [25]; however, later experiments provided evidence that activated Boc-amino acids could indeed form oxazolones, yet they did not undergo the requisite tautomerization and proton abstraction that leads to racemization [26]. The same experiments demonstrated that under appropriate conditions with an excess of a strong base such as 4-dimethylaminopyridine (DMAP), urethane-protected amino acids can and will racemize [27,28].

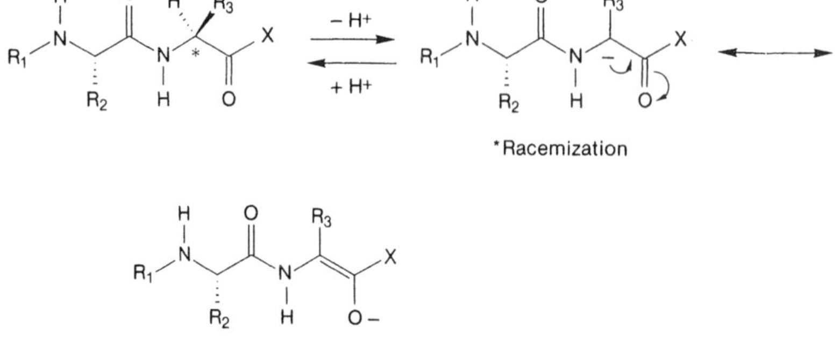

*Racemization

Fig. 4   Racemization of peptides via enolization.

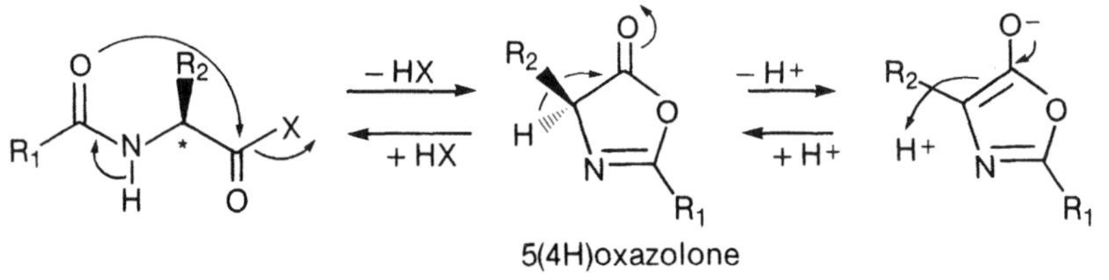

5(4H)oxazolone

*Racemization

Fig. 5   Racemization of peptides via oxazolone formation.

The peptide chemist generally synthesizes peptides from the carboxyl terminus toward the amino terminus one amino acid residue at a time. However, when attempting the synthesis of large peptides, the coupling of peptide fragments, hereafter referred to as segments, can lead to higher yields, greater purity, and a more rapid synthesis.   To avoid racemization, segments generally must terminate in a glycine or a proline residue. The glycine residue has no optical center to racemize and, like proline, also exhibits a reduced tendency to form oxazolones.   However, even acyl-proline derivatives have been shown to racemize under certain conditions [29].   For a detailed description of the phenomenon of racemization, the reader is directed to the excellent reviews by Kemp [23], Kovacs [30], and Benoiton [31].

It is this potential for racemization combined with the presence of re-active functional groups on the side chains of many of the commonly oc-curring amino acids (Arg, Asp, Cys, His, Glu, Lys, Ser, Thr, and Tyr) which makes peptide synthesis significantly more difficult than simple carboxamide formation.   In the following sections we describe the methods and procedures used most often by the peptide chemist to form amide bonds.

Carbodiimide Method.   For a detailed discussion of the carbodiimide method in peptide synthesis, the reader is referred to the review by Rich and Singh [32].   As mentioned previously, Sheehan and Hess introduced the use of DCC as a coupling reagent for the preparation of amide bonds in 1955 [12].   This simple dehydrating reagent has become the de facto standard for peptide bond formation over the ensuing 30 or so years. Amide bond formation proceeds via an O-acylisourea active intermediate. This intermediate is very reactive and cannot be isolated.   It reacts

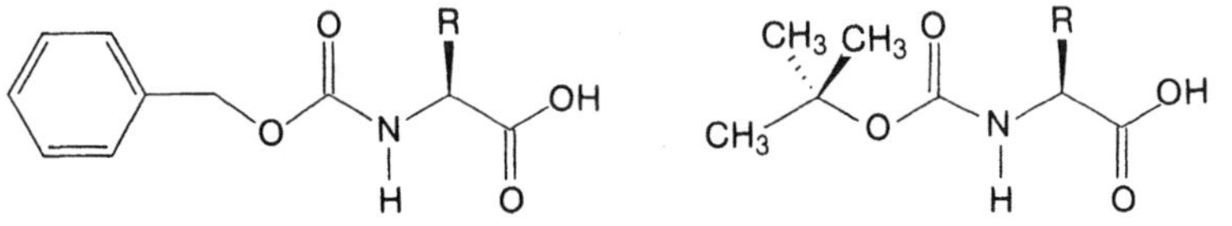

1   2

Fig. 6   Carbobenzyloxy (Cbz) and *tert*-butyloxycarbonyl (Boc) amono acids.

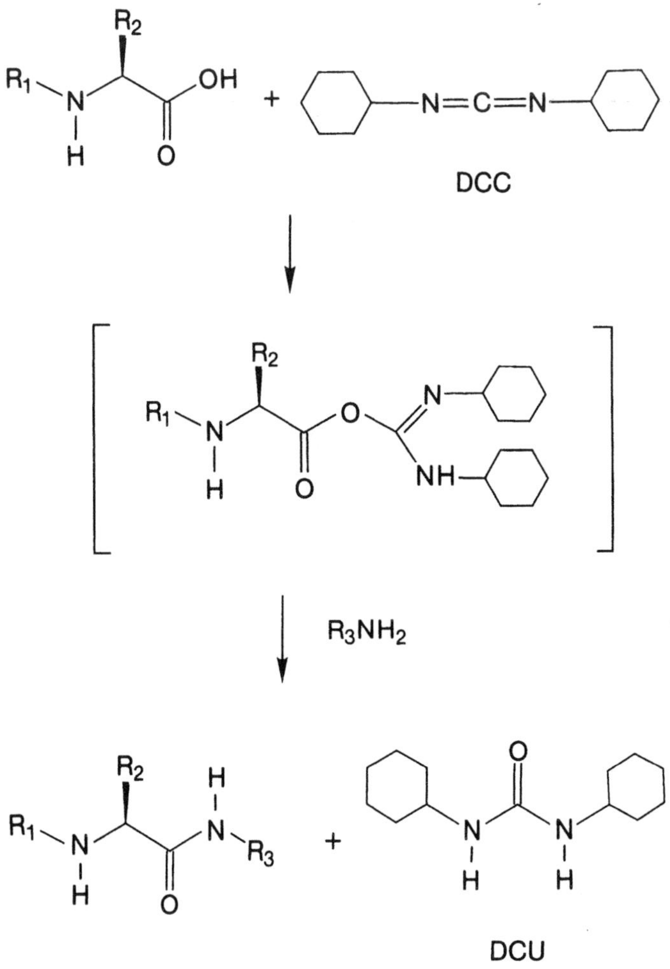

Fig. 7   Formation of an amide bond using dicyclohexylcarbodiimide (DCC) proceeding via an O-acylisourea intermediate.

rapidly with nucleophiles such as amines to form an amide bond, with the concomitant formation of dicyclohexylurea (DCU) as an insoluble by-product (Fig. 7).

The reaction is typically conducted with equimolar quantities of all pertinent reagents in an aprotic solvent such as chloroform, dichloromethane, dimethylformamide, or tetrahydrofuran at an initial temperature of 0°C. The reaction mixture is allowed to warm to room temperature overnight. The reactions are typically done at reduced temperatures to minimize the formation of the N-acylurea, which can be formed when the reaction with nucleophile is slow (Fig. 8). Reaction yields and purity of product are typically in the range of 80 to 100%.

The addition of N-hydroxysuccinimide to DCC-mediated coupling reactions was reported to reduce the extensive racemization that had been observed with DCC alone [33]. The use of N-hydroxysuccinimide, however,

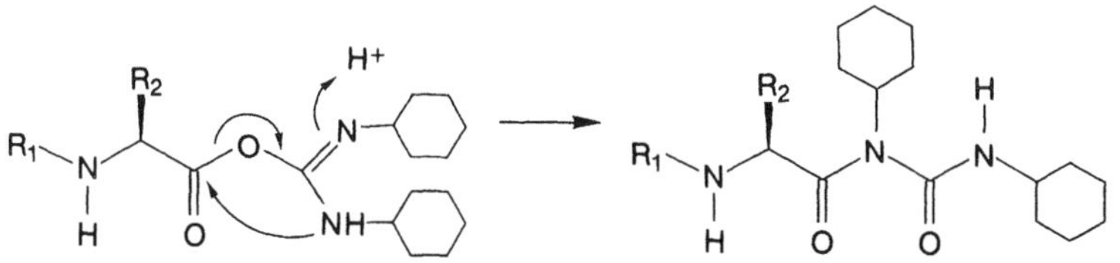

Fig. 8   Rearrangement of an O-acylisourea to an N-acylurea.

can lead to β-alanine-related impurities [34]. In 1970 is was observed that if 1-hydroxybenzotriazole (HOBt) was added to a DCC-mediated peptide synthesis (Fig. 9), the optical and chemical purity of the resultant peptide product was markedly improved [35]. Later it was shown that HOBt could suppress or eliminate the extensive racemization that occurred during DCC-mediated couplings of Boc—His(Bzl) in SPPS [36].

A general trend has developed to include HOBt in the DCC coupling protocol as a racemization suppression agent. The product is usually of higher purity and in those cases where racemization is possible (e.g., segment condensation at a residue other than Gly or Pro) the product will be generally of higher optical integrity. Nonetheless, there have been numerous documented instances of DCC—HOBt-mediated coupling reactions where the resultant peptide is partially racemized [37]. Although additives besides HOBt have been proposed and utilized [32], HOBt is still the most widely used additive for the carbodiimide-mediated synthesis of peptides.

The rationale for the racemization suppression properties of HOBt and related compounds has been debated; however, no definitive explanation has been proposed which satisfactorily accounts for all of the experience with this additive. HOBt apparently forms an active ester intermediate which then reacts with amine nucleophiles to form the amide bond with concomitant regeneration of HOBt (Fig. 9). Although this mechanism implies that HOBt is functioning as a catalyst, it is generally added in equimolar amounts, and in some cases has been added at two or three times the molar level of the acid and amine components of the reaction. The benzotriazole ester that is formed between the acid and HOBt is relatively stable and can be isolated and purified before reaction with the

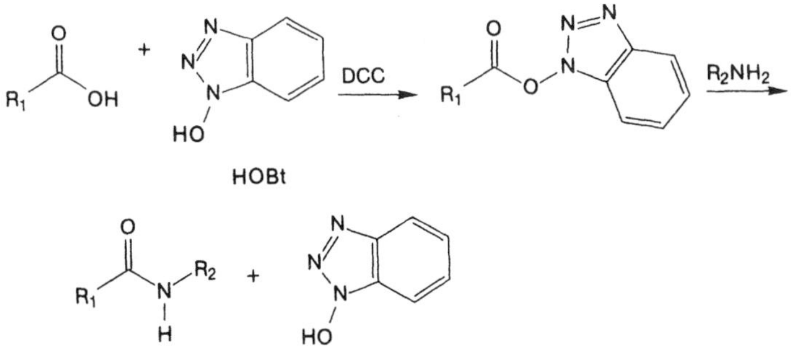

Fig. 9   1-Hydroxybenzotriazole (HOBt)-catalyzed formation of an amide bond.

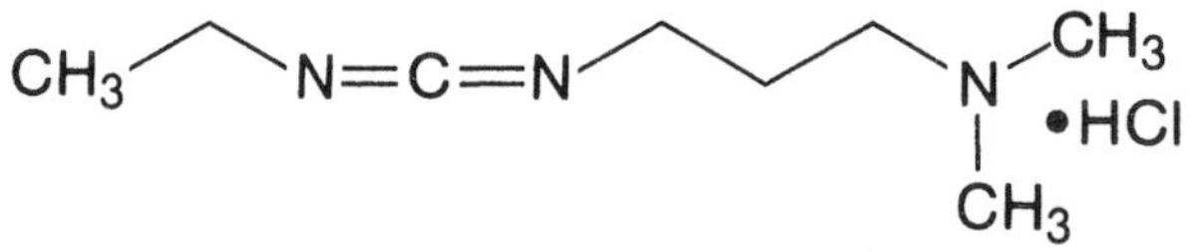

Fig. 10   Water-soluble carbodiimide, 1-ethyl-3-(3-dimethylaminopropyl)
carbodiimide hydrochloride.

amine component (personal observation). Although this is generally not
done by the peptide chemist, it does have the advantage of eliminating the
DCU by-product from the reaction mixture, thus simplifying the purification of the final product.

One of the drawbacks of the DCC method is that the DCU may contaminate the isolated peptide. This problem can be minimized by use of a
water-soluble carbodiimide (Fig. 10). The urea product formed from this
diimide is soluble in water and can readily be separated from the protected
peptide product. Another drawback of this method is that DCC is highly
allergenic. Repeated skin contact can lead to severe and prolonged allergic
reactions. Despite these problems, DCC remains a powerful and popular
tool for the rapid synthesis of peptides.

Active Ester Method. A second major method of peptide synthesis in
solution is via the use of esters. Early attempts by Fischer to prepare
peptides by reaction of amines and simple alkyl esters were largely unsuccessful [38]. Later, electronegatively substituted alkyl esters such as
cyanomethyl esters were introduced and found to possess greatly increased
reactivities toward amine nucleophiles [39]. The major advance in ester
couplings came with the studies by Bodanszky on the reactivities of substituted phenyl esters [40]. From these studies, *para*-nitrophenyl esters
were chosen for general use in the synthesis of peptides. This class of
reagents, more commonly referred to as active esters, generated peptides
in high yields with good purity. These compounds became the first reactive derivatives of protected amino acids to be made commercially
available.

A systematic investigation of other phenyl esters with electronegative
substituents led to the realization that the reactivity of aryl esters was
directly related to the electronegativity of the substituents on the phenyl
ring. It was also shown that steric influences of ortho substituents
could counteract the beneficial effects of electronegative substituents.
Thus the 2,4,5-trichlorophenyl esters were found to be more reactive than
the pentachlorophenyl esters [41]. Pentafluorophenyl esters were reported to exhibit even greater reactivity than other phenyl esters, due
to the high electronegativity of the fluorine atoms coupled with their small
steric bulk [42]. This class of highly reactive agents found ready application in the rapid synthesis of peptide hormones [43].

Kovaks has studied the rates of coupling of various active esters of
cysteine and compared them to their relative rates of racemization [44,45].
The ratio of the coupling rate versus the racemization rate gives a number that can be used to rank these reagents (Table 1). The largest ratio
indicates the most rapid coupling rate with the least racemization. Thus

Table 1   Ranking of Active Esters as a
Function of the Ratio of Their Ability to
Couple (K_c) Versus Their Ability to
Racemize (K_{rac})

R_2	Abbreviation	K_c/K_{rac}
(pentafluorophenyl)	Pfp	122
(2,4-dinitrophenyl)	2,4-Dnp	62.2
(pentachlorophenyl)	Pcp	41.5
(succinimidyl)	Su	11.1
(2,4,6-trichlorophenyl)	2,4,6-Tcp	7.82
(2,4,5-trichlorophenyl)	2,4,5-Tcp	6.11
(2,6-dinitrophenyl)	2,6-Np	5.97
(4-nitrophenyl)	Np	2.67

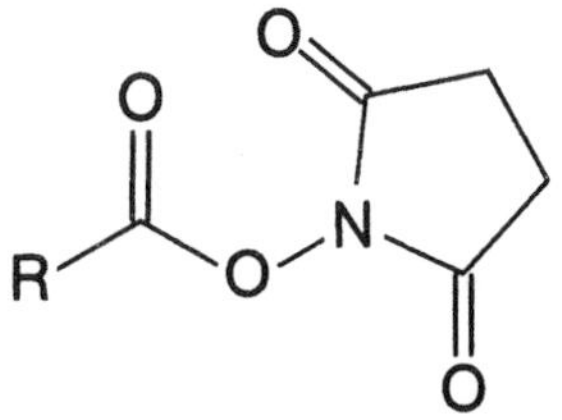

Fig. 11   N-Hydroxysuccinimide active ester.

pentafluorophenyl esters appear to have significant advantage over the
long-used *para*-nitrophenyl esters.

In the early 1960s, the report of the utility of hydroxylamine derivatives as reagents for the synthesis of peptides [46] led to the development
of N-hydroxysuccinimide active esters [47] (Fig. 11). This class of active
esters has retained popularity, and variants of this procedure are used
today as cross-linking reagents for the covalent attachment of peptides to
polymeric supports for use in affinity chromatography and induction of
antigenic responses [48,49] and as protein modification reagents [50,51].

An advantage of this method of peptide synthesis is that the active
esters of most protected amino acids can be synthesized, isolated, crystallized to purity, and stored for later use [52]. The major disadvantage of
the method is the relatively slow aminolysis rate with aryl esters such as
para-nitrophenyl esters and their modest reactivity under conditions of
steric hindrance [53]. However, the additive HOBt can be used to catalyze
the formation of the peptide bond with these slow active esters [54]. For
a detailed review of the active ester method, the reader is directed to
Bodanszky [53].

Mixed Anhydride Method (Classical Method). The mixed anhydride
procedure for the synthesis of peptides was developed in 1951 [55-57].
The method involves the formation of an asymmetric or mixed anhydride
from a protected amino acid and an alkyl chlorocarbonate, followed by reaction with an amine component to generate the peptide bond (Fig. 12).
This reaction was studied in great detail by Anderson and every aspect of
the reaction was optimized [58]. Since mixed anhydrides are among the
most highly activated aminoacyl derivatives, they can undergo numerous
side reactions if the reaction is not conducted under carefully controlled
conditions. For a discussion of these side reactions and a detailed discussion of the mixed anhydride approach, the reader is referred to the
excellent review by Meienhofer [59].

This amide-bond-forming method has the advantage of producing the
volatile and/or readily extractable by-products of CO_2, isobutanol, and
N-methylmorpholine hydrochloride. One disadvantage of this coupling
method is the tendency of hindered amino acids, N-alkylated amino acids,
and proline to react with the "wrong" side of the mixed anhydride, resulting in production of a urethane by-product (Fig. 13). The isobutyloxy-
carbonyl derivative that is generated when using isobutylchloroformate is
neutral and will generally copurify with the desired peptide; however,
this by-product is quite stable to acidiolytic deprotection and is readily

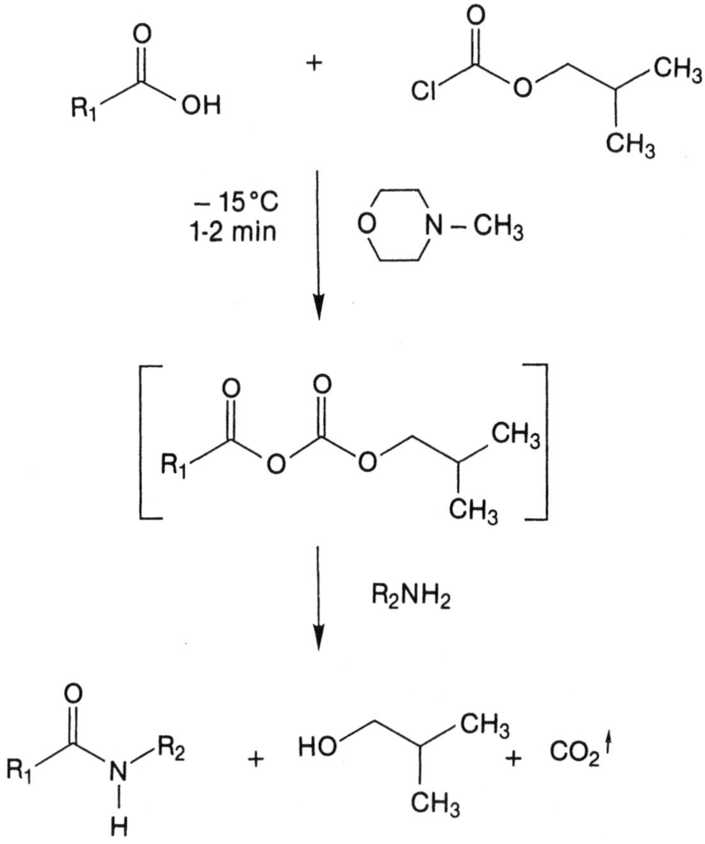

Fig. 12   Formation of an amide bond using the mixed anhydride method.

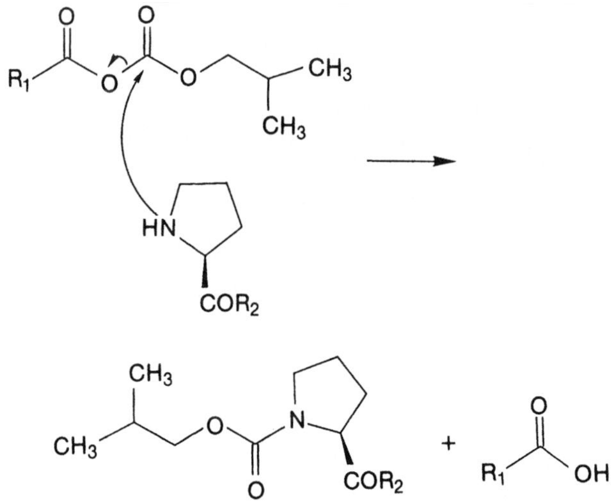

Fig. 13   Urethane (carbamate) formation during the mixed anhydride coupling procedure.

67

separable from the trifluoroacetic acid (TFA) or HCl salt of the peptide following the next deprotection reaction.

Mixed Anhydride Method [Repetitive Excess Mixed Anhydride (REMA) Method]. A basic improvement to the mixed anhydride procedure was developed in 1970 by Tilak [60]. This method involves the repetitive mixed anhydride synthesis of a peptide without intermediate purification. A 10 to 40% excess of the mixed anhydride is utilized at each step to ensure complete acylation of the growing peptide chain. Following completion of the reaction, the excess anhydride is hydrolyzed by a saturated aqueous potassium bicarbonate solution (pH 8) and the peptide is isolated by extraction or precipitation with water. This method was later expanded by Beyerman and collaborators, who embellished it with the acronym REMA (repetitive excess mixed anhydride). Using this method, they rapidly synthesized peptide hormones such as substance P and LH—RH (gonadoliberin); however, the tour de force of this method was their stepwise solution synthesis of the 27-peptide secretin, with full biological activity [61].

Azide Method. The azide coupling procedure, developed by Curtius [62], has been in use since the turn of the century. It involves the reaction of an azide with an amino derivative (Fig. 14). This method gained widespread popularity for the coupling of peptide segments when it was observed that the products retained chiral integrity at the activated residue. Thus whereas the DCC-mediated, active ester, or mixed anhydride methods of peptide synthesis could not be used to couple peptide segments without fear of partial or complete racemization, the azide coupling method was considered to be racemization-free [63]. Subsequent work has shown that this is not always the case [64].

Several procedures have been developed for the prepatation of peptide azides (Fig. 15) [62,65,66]. Following preparation of a peptide hydrazide, the azide is formed by reaction of a nitrite (preferably an organic nitrite such as *tert*-butylnitrite) in the presence of an acid [67]. The azide-coupling procedure is generally performed immediately following azide formation by adjustment of the pH to the range 7 to 8 with a base such as diisopropylethylamine, lowering the temperature to below 0°C, and addition of the amino component. The reaction is typically performed in as high a concentration as possible in a polar solvent such as dimethylformamide. Owing to the fact that azides are only mild acylating reagents, the reaction is often allowed to proceed for several days and on occasion, a week or longer.

A major side reaction that can limit the usefulness of the azide procedure is the classical Curtius rearrangement to form an isocyanate. This

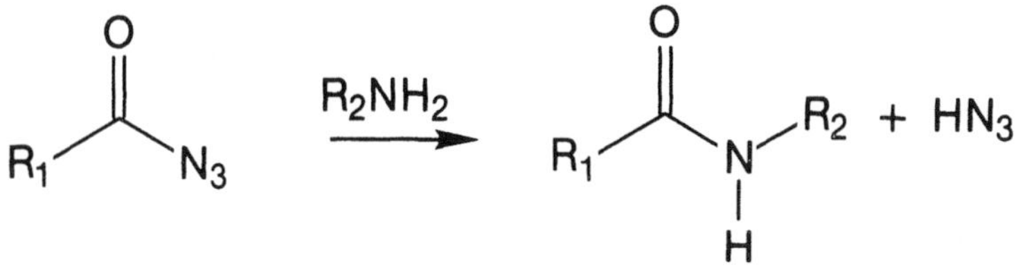

Fig. 14   Formation of an amide bond using the azide procedure.

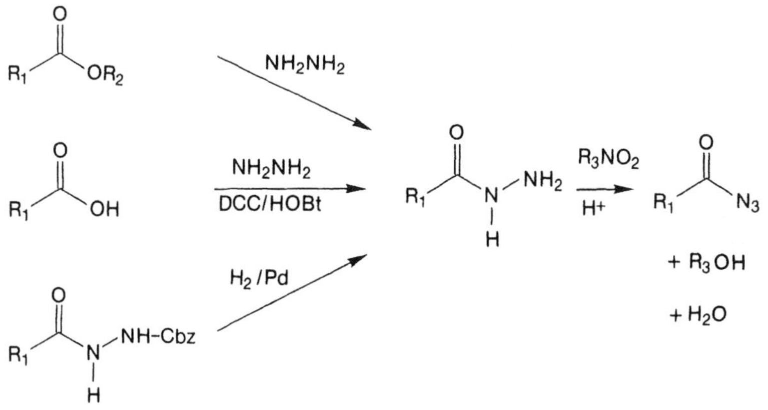

Fig. 15 Preparation of peptide azides from esters, acids, and protected hydrazides.

rearrangement varies depending on the nature of the peptide sequence. Formation of the isocyanate is best monitored by infrared spectroscopy at 2220 cm^{-1} [68]. The isocyanate can undergo further reaction with the components of the reaction mixture to generate symmetric ureas, asymmetric ureas, and urethane derivatives (Fig. 16). These by-products possess properties very similar to those of the desired peptide and are difficult to remove or detect.

An alternative procedure for azide-mediated coupling reactions is now available with the introduction of the reagent diphenylphosphorylazide [69]. Reaction of this reagent with a protected amino acid or peptide followed by

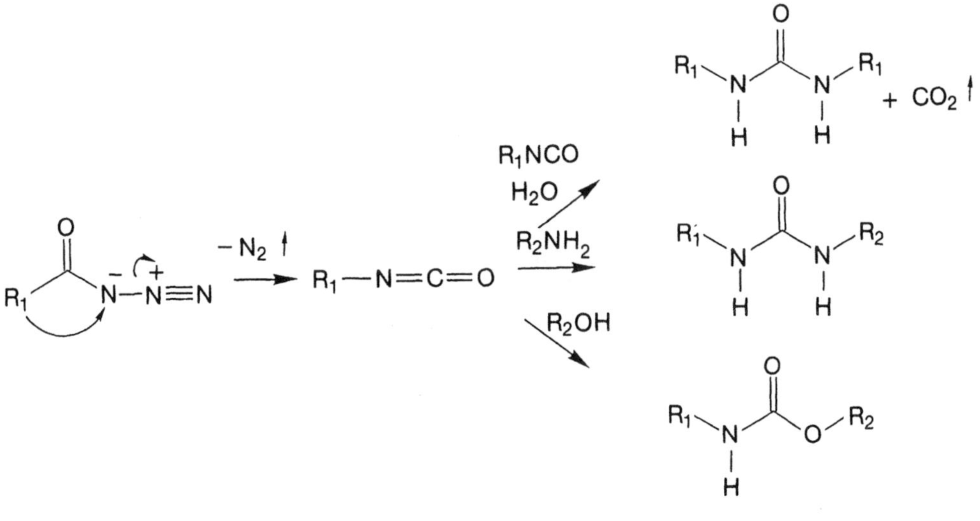

Fig. 16 By-products formed from the Curtius rearrangement of an azide.

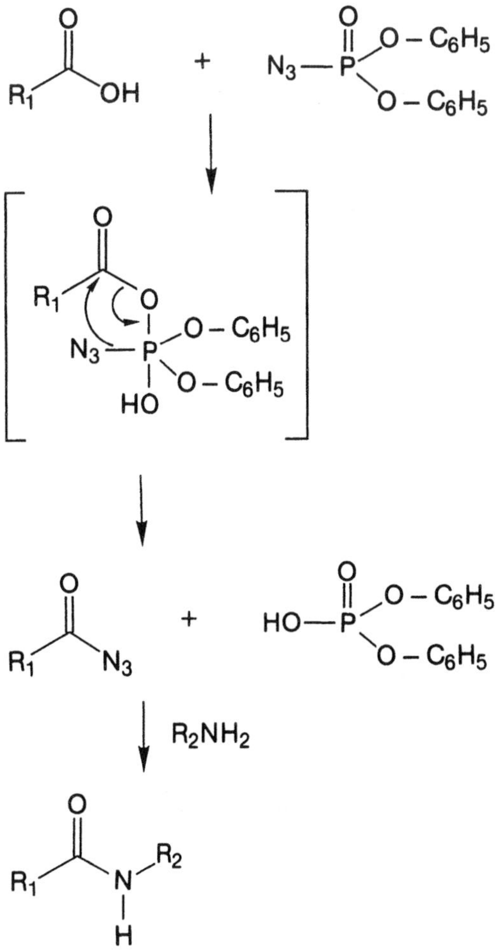

Fig. 17   Formation of an amide bond using diphenylphosphorylazidate.

reaction with an amine leads to peptide bond formation, presumably through the formation of an intermediate azide (Fig. 17).

The advantages of the azide reaction are low racemization rates, the ability to couple peptide segments with minimal side-chain protection (only Lys and Cys residues need to be protected), and the relative ease with which large peptides can be synthesized and subsequently purified. The disadvantages of the azide method are low reactivity rate, thus necessitating long reaction times, the large time investment required to prepare the azide reagent, and the numerous side reactions that have been documented.

Without added proof, use of the azide method must be treated with caution in terms of claims of optical integrity. However, it remains the least racemization-prone method of all known methods of peptide bond forma- tion and is invaluable for the coupling of protected peptide segments as well as for the synthesis of cyclic peptides. For more details of the reaction conditions and a description of the method in general, the reader is referred to the review by Meienhofer [70].

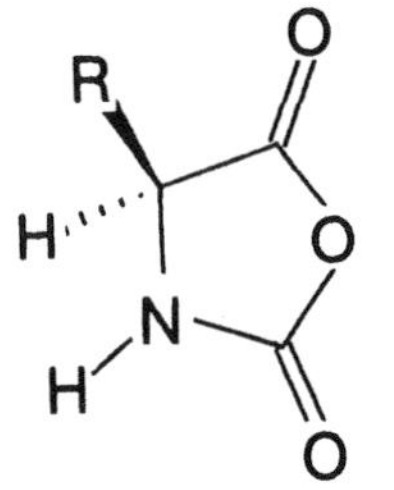

Fig. 18   *N*-Carboxy-α-amino acid anhydride (NCA).

N-Carboxyanhydride Method. A powerful and elegantly simple method for peptide synthesis, the *N*-carboxyanhydride coupling procedure for the synthesis of peptides has had a somewhat checkered history. Although the method has been used for the now classic synthesis of the 104-residue protein bovine ribonuclease A(21−124) (the S protein of RNase [17]), it has not achieved widespread popularity within the research or industrial community.

Although the basic method was utilized for years in the formation of peptide polymers [71], the development of the method into one of general utility for the synthesis of peptides was not achieved until 1967 [72]. *N*-carboxy-α-amino acid anhydrides (NCAs), formerly known as Leuch's anhydrides, are internal anhydrides of amino acids (Fig. 18). This class of compounds has the unique feature of possessing a functionality which serves simultaneously as an internal α-amino protecting group and an α-carboxylic acid activating group. The general procedure for formation of the amide bond involves the addition of an NCA to a well-stirred basic (e.g., pH 10.2) aqueous solution of an amino acid at 0°C. The reaction proceeds rapidly and the sole by-product of the reaction is carbon dioxide (Fig. 19).

This method has the advantages of simplicity, minimal requirement for protection of the side-chain functional groups, and internal protection for the α-amino group. As a result, this method has seen its widest application in the industrial synthesis of peptides, where simplicity of synthesis and cost effectiveness are of paramount importance [73]. The main disadvantages of this method are the somewhat hazardous conditions required to prepare the NCA (generally via phosgene), the relatively short shelf-life of the NCA once it is formed, the necessity to establish the optimal pH conditions for coupling of each different amino acid residue, and the tendency of NCAs to undergo side reactions such as polymerization and

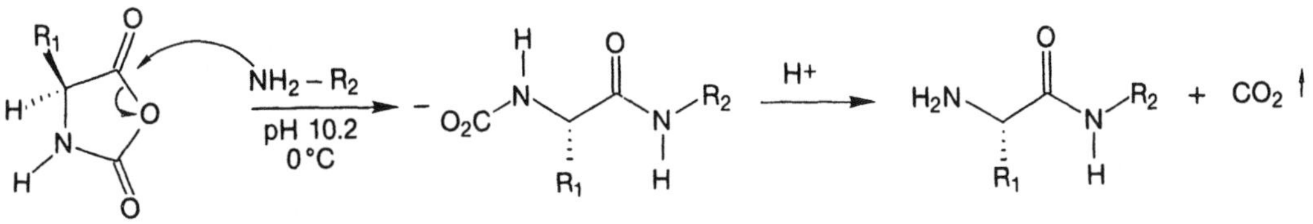

Fig. 19   Formation of an amide bond using the NCA method.

hydantoic acid formation. However, some of the undesirable traits associated with this method of peptide synthesis have been eliminated. For instance, the method of preparing NCA derivatives by reaction of thionyl chloride or phosphorous trihalides with N-siloxycarbonyl amino acid trimethylsilyl esters has eliminated the phosgene hazard [74]. Many of the side reactions associated with NCA reactions have been studied and optimal conditions have been reported [72,75—79].

Removal of Protecting Groups. Following formation of a peptide by one of the preceding methods, one must remove the N^α-protecting group on the peptide so that it may participate in subsequent amide-bond-forming reactions. Although the Cbz protecting group was popular in the 1950s and 1960s, it has been largely replaced by the Boc protecting group. The Boc group is conveniently removed by treatment (30 to 60 min at room temperature or 0°C) with either TFA, 0.1 N HCl in HOAc, or HCl in an organic solvent such as dioxane. The use of TFA is perhaps the most popular method, due to the ease of use and removal of the reagent. Following treatment with TFA, the peptide amine salt typically is precipitated by addition of diethyl ether, filtered, washed, and used without further purification in the next amide-bond-forming reaction.

Following the last amide-bond-forming reaction, the amino terminal, carboxyl terminal, and side-chain protecting groups (if any) are removed by an appropriate method (e.g., catalytic hydrogenolysis, HBr in HOAc, liquid HF, etc.) to produce the desired peptide [80].

Solution Synthesis: Conclusions. From this overview of solution-phase peptide synthesis it should be apparent that there is no one "best" way to synthesize peptides, since success with each method is largely dependent on the specific peptide sequence and length. Nevertheless, the DCC/HOBt method is probably the method of choice for a first attempt and is likely to give satisfactory results. As problems are encountered with this method, the mixed anhydride, active ester, and azide methods should be considered. To assist the peptide chemist in finding the best method for a specific synthesis, a series of manuals has been published which contain the experimental procedures and physical data for thousands of peptides and protected amino acids [80].

Solid-Phase Peptide Synthesis

Basic Method. The basic method of SPPS has remained largely unchanged over the last 25 years. The technique involves the attachment of an α-amino-protected, side-chain-protected amino acid to an insoluble polymeric solid support. The amino protecting group is then chemically removed (deprotection), any salt of the amine component is converted to a free amine by reaction with a weak organic base (neutralization), and an excess (typically 2 to 6 equivalents) of the next amino acid is added and allowed to react with the amine in the presence of an activating reagent such as DCC to form the first amide bond (coupling) (Fig. 20). The solvents and any excess reagents are filtered away from the insoluble peptide-resin and this process of deprotection, neutralization, and coupling is repeated until the final peptide sequence has been completed. The peptide-resin is then treated with a reagent, typically, liquid hydrofluoric acid (HF), to remove the peptide from the solid support and to remove the side-chain protecting groups.

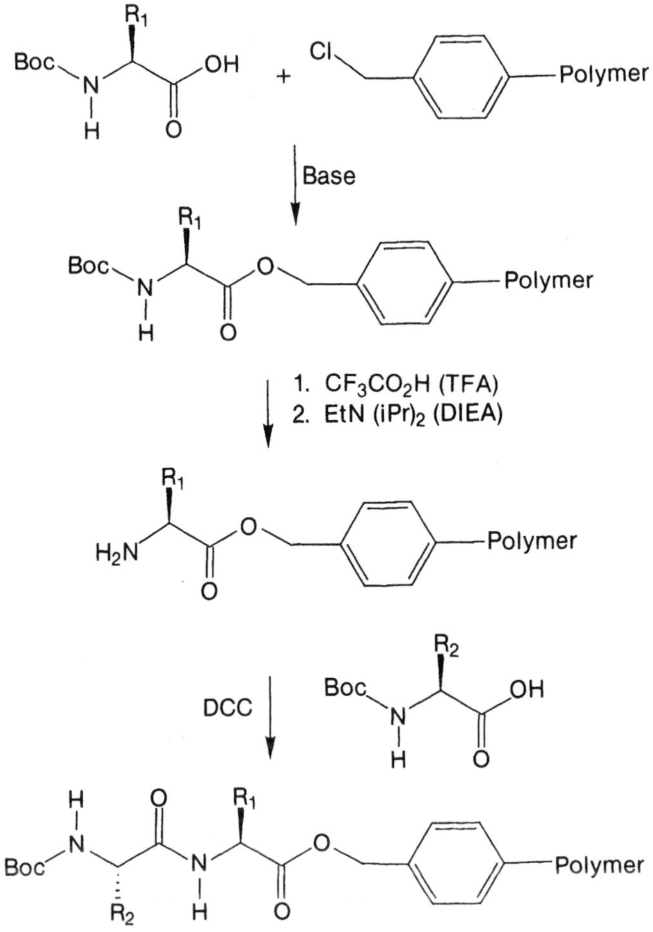

Fig. 20   General scheme for solid-phase peptide synthesis (SPPS).

What has changed since the technique was introduced is the specifics, such as the types of resins that are used, the types of amino terminal and side-chain protecting groups that are used, the chemistries that are employed to form the amide bonds, and the methods of removal of the final peptide from the peptide-resin. In the following section we outline briefly the more recent and popular methods and techniques used with SPPS. For a complete discussion of SPPS, the reader is referred to recent reviews and references cited therein [81–84].

Solid Supports (Polymers/Resins). The original solid support chosen by Merrifield was a copoly(styrene-2%-divinylbenzene) polymer (today, 1% cross-linking is preferred). This type of support (resin) has been used in the vast majority of all solid-phase peptide syntheses reported to date. The resin is characterized as spherical beads that are insoluble in all common solvents but which swell to five to six times their dry volume when placed in organic solvents such as chloroform, dichloromethane, or

dimethylformamide. This swelling allows the reagents and solvents free access to the growing peptide chain, which is held both on the surface and
within the polymer matrix [85]. Although many other types of solid supports have been examined, only the polyacrylamide—gel based solid supports have approached the success rate of the polystyrenes [86].

Since the peptide is usually removed from the solid support by an acid-
based cleavage mechanism, the differential between the acid lability of the
α-amino and side-chain protecting groups and the benzyl ester bond is of
critical importance, especially when faced with a long synthesis. The
classical benzyl ester linkage of the C-terminal amino acid to the functional-
ized polystyrene is sufficiently stable toward the mild acid deprotection
schemes (e.g., 50% TFA) required for removal of the Boc-amino protecting
groups when the peptides are relatively short (5 to 20 residues). However, for longer syntheses, a significant portion of the peptide is cleaved
prematurely from the resin during the course of synthesis.

In 1976, a new peptide-resin linkage strategy was introduced [87]. Instead of a simple benzyl ester, an acetamidomethyl group was inserted between the benzyl ester and the polystyrene resin. This modification [fre-
quently referred to as the PAM (phenylacetamidomethyl) resin] resulted in
a peptide-resin linkage that was 100 times more stable than the standard
benzyl ester to trifluoroacetic acid. This new peptide-resin is best prepared by the DCC-mediated amide bond formation between an aminomethyl-
resin and a 4-methylphenylacetic acid derivative of a Boc-amino acid (Fig.
21). Now that the individual Boc-aminoacyloxymethylphenylacetic acid-
resins are commercially available, the PAM resin is rapidly becoming the
de facto standard for most of the solid-phase syntheses of large peptides.

In another approach to limit the premature cleavage of peptide from the
polymeric support, the concept of "orthogonal" protection schemes was developed. Orthogonal is used here to denote the use of completely selective
techniques for removal of protecting groups [88]. Thus one might envision
the utilization of an α-amino protecting group that is removed by treatment
with base, side-chain protecting groups that are removable by acid treatment, and a peptide-resin linkage that is cleavable by photolysis. The
main advantage of such an orthogonal scheme is that repeated use of one

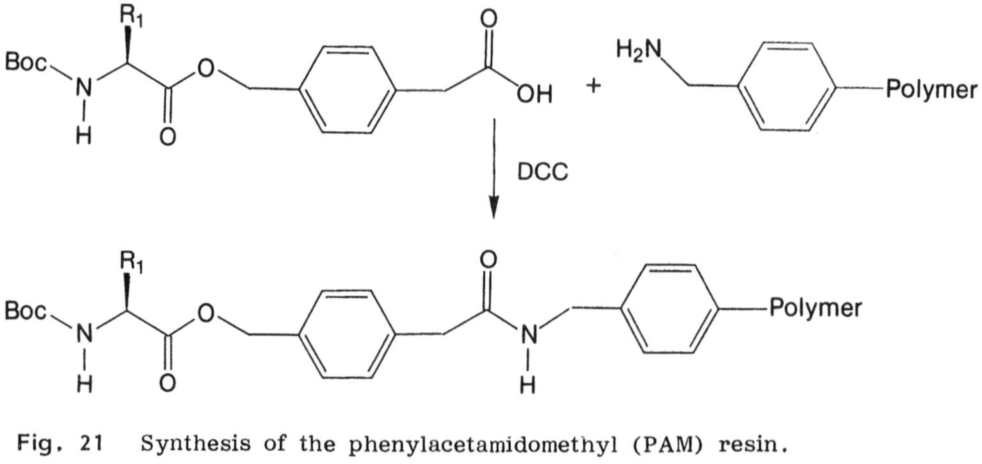

Fig. 21   Synthesis of the phenylacetamidomethyl (PAM) resin.

deprotection scheme (base treatment) throughout the synthesis will not affect the integrity of the photolabile peptide-resin linkage or the acid-labile side-chain protecting groups. This protocol has the added advantage of completely avoiding the use of liquid HF at any stage of the peptide synthesis.

Although there are examples of orthogonal strategies in the literature [88,89], these methodologies have not yet gained widespread acceptance. One strategy that is "semiorthogonal" and appears to be gaining in acceptance and success is the "Fmoc"-based strategy [90,91]. In this protocol, alpha amino protection is based on the fluorenylmethyloxycarbonyl (Fmoc) group, which is base (piperidine) labile (Fig. 22). The side-chain functionalities and the peptide-resin linkages are cleaved by exposure to mild acid [92]. Another attempt to improve the reaction conditions of SPPS is by use of polyamide polymeric support instead of the classical polystyrene support [93]. This modification was made to improve the solvent–peptide interactions in difficult peptide syntheses; however, careful comparisons of the two methods resulted in the conclusion that both are capable of producing excellent results [94].

Recently, a flow-through column procedure for SPPS has been developed [95]. The solid support is a mixed polydimethylacrylamide gel polymerized within a synthetic macroporous kieselguhr matrix. This solid support can withstand applied pressure without compression and allows the free flow of the peptide reagents through the support without a buildup of backpressure. The Fmoc group is used for α-amino protection and the amino acid is coupled to the growing peptide via the very reactive pentafluorophenyl ester [96]. The reagents are passed over the peptide-resin in a column and can be recirculated until the reaction is complete [97]. The effluent from the column can be diverted to an ultraviolet cell and the

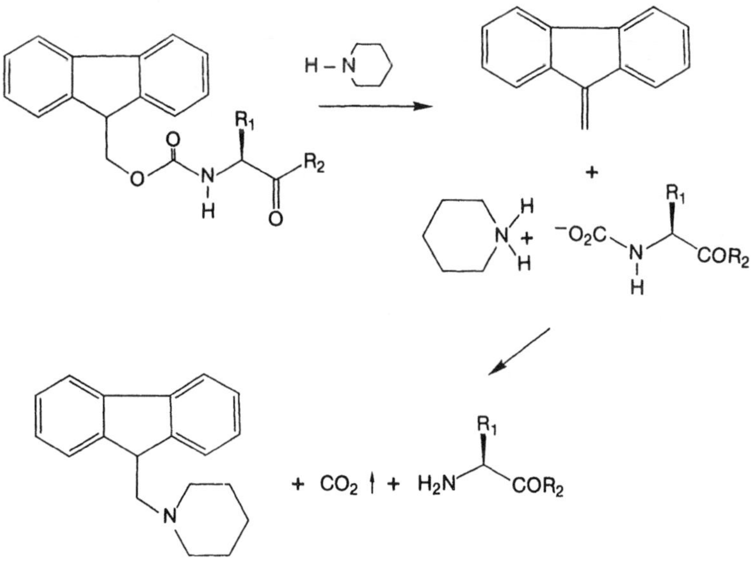

Fig. 22   Piperidine deprotection of a fluorenylmethyloxycarbonyl (Fmoc) peptide.

course of the coupling reaction and the deprotection reaction can be monitored, owing to the high absorbance of the Fmoc moiety and its byproducts.

This process of monitoring the progress of the reaction as it occurs leads us to the first *fully* automated peptide synthesis apparatus. Previously, all automated SPPS machines would automatically add the next amino acid regardless of whether or not the previous coupling reaction was in fact complete. The fully automated apparatus is programmable such that it proceeds with the next coupling step only after a predetermined level of reaction completion has been achieved.

For the synthesis of peptide amides the original procedure of aminolysis of a benzyl ester-linked peptide-resin with saturated methanolic-ammonia [98] has given way to the nearly universal use of the benzhydrylamine class of polystyrene resins. The C-terminal amino acid residue is attached to the benzhydrylamine resin by a standard DCC coupling procedure to create an amide bond. When the peptide-resin is treated with liquid HF, the C-terminal carboxamide is formed due to the formation of the stabilized benzhydryl carbonium ion. Difficulties in removal of C-terminal aromatic and hindered amino acid amides have lead to the development of the *para*-methylbenzhydrylamine (*p*MeBHA) resin (Fig. 23). This resin has become the standard for the production of peptide amides by solid-phase synthesis.

Attachment to Resin. Before the solid-phase synthesis of a peptide can proceed, the C-terminal amino acid of the target peptide must be attached covalently to the polystyrene resin. This was accomplished initially by functionalizing the polystyrene resin with a chloromethyl benzyl residue which was then allowed to react with a triethylamine salt of a suitably protected amino acid to form a benzyl ester linkage (Fig. 20). This method was found to result in incomplete esterification, leaving residual chloromethyl benzyl residues which could later participate in unwanted side reactions. In addition, the reaction of triethylamine with the chloromethyl moiety was documented to form the quaternary ammonium groups, which gave ion-exchange properties to the resin [99].

Alternative methods have been devised which give essentially complete esterification. These involve the use of either a tetramethylammonium salt

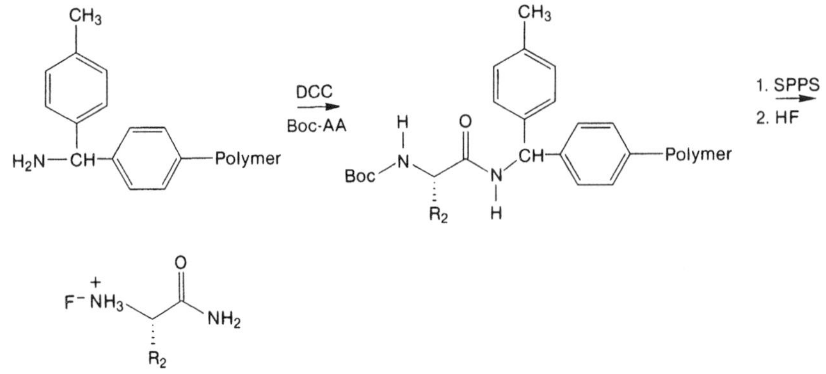

Fig. 23   Synthesis of peptide amides using a *para*-methylbenzhydrylamine resin.

[100], a cesium salt [101], or a potassium salt/18-crown-6 ether complex [102] of the amino acid. These methods function by improving the nucleophilicity of the carboxylate anion and give essentially quantitative esterification. In comparative experiments the crown ether method gave superior results when compared with the cesium salt method [102].

One can avoid the chloromethyl-related problems by use of hydroxymethyl resins. Following esterification of this resin with an amino acid, one can readily acylate any remaining hydroxylmethyl groups to avoid further unwanted acylations [99]. Caution must be exercised, however, with the use of DMAP for the attachment of the first amino acid to the hydroxymethyl resin, due to the reported tendency of this reagent to cause partial racemization [31].

Once the first amino acid is attached to the resin, the degree of substitution or loading level must be determined by amino acid analysis of the resin. The value thus derived is then used to calculate the amounts of reagents to be utilized during the subsequent peptide synthesis. Typically, one uses resins with loading levels between 0.3 and 1.0 mmol of amine content per gram of resin; however, significantly higher and lower loading levels than these have been used with success.

Side-Chain Protection. Owing to (1) the reactivities of the side-chain functional groups in common amino acids, and (2) the requirement of SPPS to drive the coupling reaction to completion by use of two- to sixfold excesses of the acylating amino acid residue, the side chains of all reactive amino acids (e.g., Arg, Asp, Cys, Glu, Lys, Ser, Thr, and Tyr) used in SPPS must, in general, be completely protected. A complete discussion of the "best" side-chain protecting groups is beyond the scope of this review. The reader is referred to the extensive discussion on this topic in a review of SPPS [82].

Coupling Procedures. The use of DCC as a coupling reagent has been the most widely used method of forming peptide bonds in SPPS. It has been proposed that the reactive species present during a typical DCC-mediated solid phase synthesis is to a large extent the symmetric anhydride rather than the O-acylisourea [103]. If symmetric anhydrides of N^α-urethane-protected amino acids (prepared by reaction of 2 equivalents of the amino acid with 1 equivalent of DCC followed by removal of the DCU) are used in SPPS instead of in situ coupling with DCC, the result is rapid peptide synthesis with fewer side reactions. Although symmetric anhydrides of N^α-urethane-protected amino acids can be prepared in crystalline form, they are unstable for extended periods at room temperature [104]. Therefore, the standard procedure has been to preform the anhydride immediately prior to use. Until recently, all of the automated SPPS instruments were designed for use with the DCC protocol. With the popularity and success of the symmetric anhydride procedure, newer instruments are now being produced that will allow automated preparation of preformed symmetric anhydrides [105].

The use of diisopropylcarbodiimide (DIC) (Fig. 24) rather than DCC has been proposed [106] and is beginning to supplant DCC as the coupling reagent of choice. The major advantage of this reagent over DCC is the fact that the urea by-product is soluble in most organic solvents and thus leads to fewer of the problems that have been associated with the relatively insoluble DCU. In addition, DIC is a liquid at room temperature, whereas DCC is a hygroscopic solid.

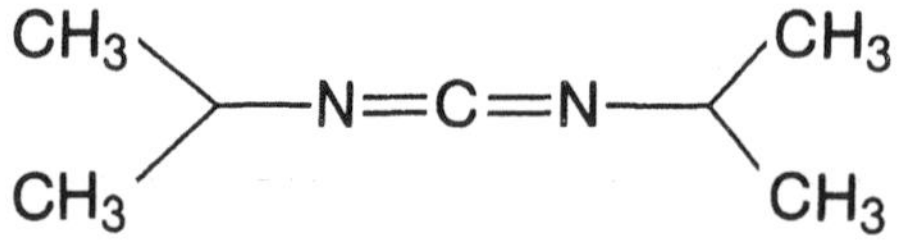

Fig. 24   Diisopropylcarbodiimide (DIC).

More recently, solid-phase instruments have become available which promote the use of pentafluorophenyl esters, in part due to the long-term stability of these active esters relative to the symmetric anhydrides [107].

Deprotection Methods. A major area of concern in the synthesis of
peptides by the SPPS method has been the strong acidiolytic conditions
(liquid HF) which are usually required for the removal of the final peptide
from the solid support. These conditions have been associated with numerous chemical side reactions, which can lead to irreversible modification
of the peptide and losses in the overall yield [82]. Because of the toxic
and corrosive nature of liquid HF, the use of special plastic- or Teflon-
based equipment is required. Despite these drawbacks, the use of liquid
HF remains the most efficient and widely used procedure for cleavage of
the peptide-resin bond and the simultaneous removal of side-chain protecting groups.

Recently, a modification of the classical procedure for the liquid HF-
catalyzed removal of peptides from the solid support was described [108].
This method involves the use of a relatively low initial concentration of
liquid HF (25% by volume) in the presence of a high concentration of the
weak base, dimethyl sulfide (DMS). This low-HF method converts the
benzyl ester cleavage reaction from an SN_1 mechanism to an SN_2 mechanism, with DMS serving as the nucleophile to aid acidolysis.

The SN_1 mechanism has been associated with the production of carbonium ions that can alkylate cysteine, methionine, tryptophan, and tyrosine
residues in the peptide. In addition, the SN_1 mechanism has been associated with the formation of side products at aspartyl and glutamyl residues.
When the reaction is converted to the SN_2 mechanism, a significant lowering
of alkylating side reactions is observed. Any side-chain protecting groups
or linkages that are resistant to cleavage by the low-HF deprotection method can be cleaved subsequently by removing the excess DMS and treating
with the normal high concentration of HF. This procedure has become
known as the "low-high" HF deprotection method for cleavage of peptide-
resins. Recently, this low-acidity protocol has been adapted successfully
to the use of trifluoromethanesulfonic acid/trifluoroacetic acid-based
cleavages of peptide-resins [109]. The interested reader is directed to a
recent review [110] for a detailed discussion of the mechanisms and methods of strong acid deprotection of synthetic peptides.

SPPS: Conclusions. Again, as with solution-phase peptide synthesis,
it is not possible to rank the various SPPS methods. Today, it appears
that many long peptides (20 to 80 residues) are successfully being synthesized on PAM or pMeBHA resins using N^α-Boc protection, symmetric anhydride chemistry, and low—high HF for final cleavage/deprotection. The
procedure composed of N^α-Fmoc protection, pentafluorophenyl active ester
chemistry, and final cleavage/deprotection with TFA is gaining in popularity

with the increasing availability of instruments capable of this type of chemistry.

The advent of SPPS has ushered in a new era in peptide synthesis. There is a trend toward the nonspecialist becoming involved with the synthesis of peptides as the technology for the automated synthesis of peptides on solid supports becomes more and more sophisticated and newer protecting groups and chemistries are developed. The newer automated instruments can now complete the synthesis of a 30-residue peptide in days instead of the months required by solution-phase techniques. In addition, these instruments automatically remove samples of the growing peptide resin for analysis of coupling efficiency [105]. The instruments are controlled by powerful microprocessors, and the manufacturers of these instruments provide the appropriate software for use with the most up-to-date chemistries for forming the amide bond. The necessary protected amino acid derivatives and reagents are commercially available and prepackaged so that all the user has to do is to load the instrument with the correct amino acid sequence, add the necessary reagents and solvents, and start the synthesis. The only thing that the newer instruments do not do is to cleave the peptide from the resin and purify the final peptide!

The capability of newer solid-phase-based methodologies to afford more efficient syntheses with fewer side reactions and products of higher purity has led the peptide chemist to attempt the synthesis of peptides of a size previously unattainable, resulting in new and more challenging purification problems. The purification problems associated with SPPS are discussed below.

Enzymatic Synthesis of Peptides

An exciting new area of peptide synthesis is the use of proteolytic enzymes to catalyze the formation of peptide bonds. The equilibrium between a peptide and its enzymatically generated proteolytic products normally is directed strongly toward the proteolytic products; however, under carefully controlled conditions, the equilibrium can be shifted toward the peptide (Fig. 25). Enzymes that have been used to catalyze the synthesis of peptides include chymotrypsin, trypsin, subtilisin, papain, chymopapain, pepsin, thermolysin, elastase, and carboxypeptidase Y [111].

The potential advantages of this methodology are numerous: (1) enzymes are by nature stereospecific and racemization of the final peptide product is not a concern; (2) the starting amino acids can be racemic and the enzyme-mediated synthesis can be expected to select only the natural L-isomer, resulting in an optically pure peptide, although the enzyme-catalyzed formation of peptides containing D-amino acids has recently been reported [112]; (3) the enzymatic coupling reaction can be carried out

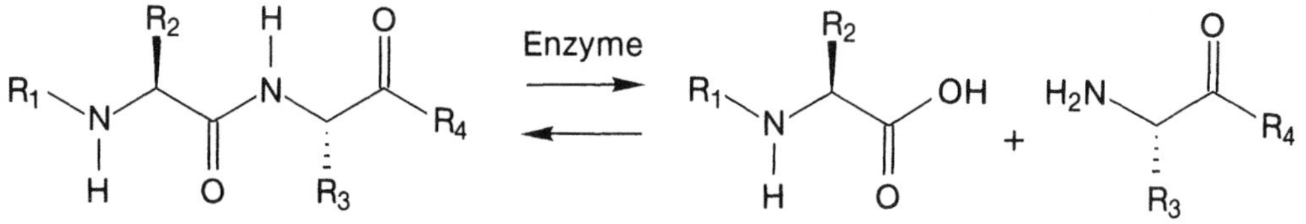

Fig. 25   Enzymatic hydrolysis (or formation) of the amide bond.

under aqueous conditions at or near neutral pH; and (4) one often does
not need to protect side-chain functional groups since enzymes catalyze reactions only at the α-amino and carboxyl groups.

It is therefore not surprising that an increasing number of reports
have been published on the use of proteolytic enzymes to synthesize peptides, such as fragments of substance P [19] [Leu]- and [Met]-enkephalin
[113—115], cholecystokinin (26—33) amide [116], and fragments of mouse
epidermal growth factor [117]. The enzymatic synthesis of the decapeptide
LH—RH on a molar scale has been reported [118].

There are, however, some disadvantages with this method. Each peptide-bond-forming reaction must be optimized with a specific enzyme. The
choice of which enzyme to use for a given peptide bond formation reaction
is critical, as is the purity of that enzyme. The pH conditions, reaction
concentrations, and solubilities of products and reactants all play a critical
role in this methodology and must be optimized for each new reaction.

Efforts are in progress to develop immobilized enzymes and general reaction protocols [19,119]. However, until pure enzymes become rapidly
available and general conditions are established for classes of peptide syntheses, the use of enzymes in the synthesis of peptides will remain limited.
It appears that the most promising situation where enzyme-mediated peptide
synthesis might become widely used is in the semisynthesis of peptides.
Thus one or two problem amide bonds might be generated with the use of
enzymatic catalysis. One notable example of this procedure is the enzyme-
catalyzed transformation of procine insulin into an ester of human insulin
[120—123].

Today, the technique of enzymatic-catalyzed amide bond formation is at
the same approximate stage of development that peptide synthesis was during du Vigneaud's 1953 report of the first chemical synthesis of a peptide
hormone, oxytocin. When one reflects on the recent publication describing
the enzymatic (thermolysin)-catalyzed synthesis of pressinoic acid, the
N-terminal hexapeptide fragment of oxytocin [124], it could be imagined
that history is repeating itself.

Reviews on this method of peptide synthesis have been published [19,
111,125—130]. The reader is strongly urged to consult these references
for an in-depth discussion of this exciting method of peptide synthesis.

Peptide Purification

A major challenge for any peptide chemist is the purification of the synthetic peptide. Peptides are notorious for their failure to crystallize.
Since the early days of peptide synthesis, the purification of the synthetic product has been a major stumbling block to their ready availability.
In addition to impurities arising from side reactions, there was the problem
of racemization. Today, with the advent of rapid peptide synthesis by
SPPS and REMA peptide synthesis, the purification of the final crude synthetic peptide remains as perhaps the single greatest bottleneck for the
rapid analysis of peptide analogs.

In a typical solution-based peptide synthesis, the peptide is purified
at various stages of the synthetic process. It is not unusual for a synthetic pentapeptide to have had as many as five separate chromatographic
procedures performed during the course of the synthesis. However, in the
case of SPPS, purification does not take place until the entire peptide has
been synthesized.

If each step of the coupling process of a typical SPPS does not proceed with 100% efficiency, the final peptide will be contaminated with a mixture of shortened peptides, referred to as deletion peptides. As a simple example, consider the synthesis of a tripeptide, $A_3-A_2-A_1$, beginning with an amino acid-resin. Assume that each step of the amide bond formation reaction is 99% efficient and that the deprotection and neutralization reactions are quantitative. The final crude peptide-resin will contain 98.01% of the correct tripeptide ($A_3-A_2-A_1$), 0.99% of the dipeptide (A_3-A_1) formed by coupling of the third amino acid (with 99% efficiency) to the 1% of the first amino acid that did not couple to the second amino acid, 0.99% of the dipeptide (A_2-A_1) that did not couple to A_3 during the second amide-bond-forming reaction, and 0.01% of the first amino acid that did not react with either A_2 or A_3.

Continuing this process to a 69-peptide, the calculations reveal that with a 99% coupling efficiency the majority of the peptide product would not be the desired 69-peptide (49.88%) but rather, would be a mixture of the various deletion peptides. If the synthesis were to proceed with either a 99.5% or a 99.9% average coupling efficiency, the theoretical yield of the desired peptide would be either 70.8% or 93.3%. Thus, during SPPS of longer peptides, the closer that the synthesis achieves the ideal goal of quantitative coupling, the more likely it will be that the final peptide will not have closely related peptide contaminants and can therefore be purified to homogeneity. This idealistic calculation did not take into account other potential side reactions that can and do occur during a typical synthesis such as chain termination, incomplete deprotection, amino acid insertions, and rearrangements.

As a practical example of the successes and purification problems that are associated with the synthesis of large peptides, consider the recent report of the SPPS of interleukin-3 (IL-3), a 140-residue protein [131]. Although the final crude product was shown to possess physical properties virtually identical to those of the naturally occurring protein, the biological assays indicated that the synthetic material was from 0.5 to 30% as active as the native IL-3. Calculations based on ninhydrin analysis and Edman degradation of the protected peptide chain at various stages throughout the synthesis indicated that 41% of the final crude material possessed the correct sequence. However, the remainder of the material (59%) consisted of a mixture of closely related peptides, primarily those with one internal amino acid missing. No attempts at purification of this crude mixture were reported, and indeed even the use of the most sophisticated purification techniques would not be likely to separate a 140-residue protein from the dozens of contaminating proteins that contain 139, 138, 137, . . . , residues. Nonetheless, the SPPS of the 91-peptide β-lipotropin has been reported to give 0.66% of the correct product after extensive purification [132]. The theoretical yield of this peptide should have been 40.1%, 63.4%, or 91.3% if the average coupling efficiencies were 99%, 99.5%, or 99.9%, respectively. Clearly, with the increase in size of synthetic peptides made possible by SPPS comes an increase in the problems associated with purification and separation of the final peptide from unwanted but very closely related peptide materials.

Fortunately, the development of reversed-phase high-performance liquid chromatography was essentially coincident with these purification challenges. This technique has become the standard method for analysis and purification of peptides [133,134]. Along with the standard procedures of size exclusion, ion exchange, and silica-based chromatography, electrophoresis, and newer developments in countercurrent distribution and affinity chromatography, peptide and protein chemists have kept pace with these purification challenges.

However, great caution must still be exercised when making the claim that a synthetic peptide is "pure." Careful multiple analyses of the final purified peptide must be undertaken to verify that it is indeed a single entity and not contaminated with diastereomeric peptides, deletion sequence peptides, β-aspartyl-linked peptides, and so on. We can learn from the early examples of du Vigneaud and Merrifield, whose landmark reports on the first total synthesis of a peptide hormone in 1953 [11] and the first solid-phase synthesis of an enzyme in 1969 [16] were cautiously entitled: "The Synthesis of an Octapeptide Amide with the Hormonal Activity of Oxytocin" and "The Total Synthesis of an Enzyme with Ribonuclease A Activity," respectively.

Summary of Peptide Synthesis Methodologies

The peptide chemist has the advantage of being able to select from two major classes of peptide synthesis: solution and solid phase. Both methods have their practical advantages and disadvantages, which must be evaluated carefully before proceeding with the synthesis. Neither the solid-phase methodology nor the solution synthesis of peptides can be expected to solve every synthetic problem. Ironically, although solid-phase peptide synthesis appears to be gaining in popularity at the expense of solution synthesis, it has been stated that SPPS might not have been introduced in 1962 without the prior introduction of DCC, the coupling reagent most often used in solution-phase synthesis [135].

SPPS has the advantages of (1) speed, (2) automation, (3) essentially no solubility problems with intermediates during the synthetic process, and (4) the ability to handle readily the synthesis of large peptides. By contrast, the synthesis of peptides in solution has the advantages of (1) ready scale-up to the multigram and kilogram scale, (2) no requirement for expensive machinery, (3) the ability to produce readily peptides with unusual chemical features such as C-terminal aldehydes, (4) few or no requirements to utilize large excesses of expensive or scarce protected amino acid derivatives, (5) simplified purification schemes at the final stage, and (6) the availability of a large repertoire of synthetic protocols.

Generally, solution-phase methods are most successfully employed for peptides of fewer than 10 residues or when the carboxyl terminal residue is of such a nature (e.g., aldehyde, complex amide, alcohol, etc.) that it is not amenable to solid-phase techniques. The use of solid-phase methodologies can be advantageous when one is attempting to produce larger peptides or numerous analogs.

Recent reports have described the solid-phase synthesis of hundreds to thousands of peptides in a period of weeks. This advance in the number of peptides that can be effectively synthesized over a relatively short period now makes it possible to acquire information that previously was available only after years of dedicated work. For example, simply by synthesizing all the possible hexapeptides along the length of a protein sequence and allowing them to interact with the appropriate antibody in an ELISA assay, it was possible to identify the binding epitope of the protein [136,137].

Using these new technologies, it is also now possible to synthesize rapidly all dipeptide combinations from the 40 D- and L-α-amino acids (1600 dipeptides) and to perform an ELISA-type assay on these dipeptides to detect a signal that will denote binding specificity. Following identification of the dipeptide that gives the highest signal, one can make the 40 possible N-terminal or C-terminal tripeptide extensions, measure the binding specificity, and repeat the process. This rapid iterative procedure may allow the determination of a binding epitope of the parent protein or peptide without having to know the primary sequence [138,139]. These and other recent advances are demonstrating the potential for the use of the chemical synthesis of peptides.

The peptide chemist is the beneficiary of two additional methods to prepare peptides. The use of enzymes has been discussed briefly. This method, although potentially of great utility, has been used in a relatively few laboratories. Finally, the advent of protein engineering via the development of rDNA technology adds a powerful method for the preparation of peptides and proteins. This method is discussed in detail in Sec. IV.

III. DESIGN OF PEPTIDE DRUGS

In this section we summarize some of the concepts and approaches that have been utilized to modify a naturally occurring peptide into a compound with an altered pharmacological profile. This modified peptide might have increased receptor binding affinity, improved selectivity for one specific subset or class of receptors, antagonist properties, increased resistance to enzymatic degradation, improved pharmacokinetic properties, or the like. Ideally, this peptide analog would then be a useful therapeutic agent or drug.

The concepts presented in this section are by no means extensive and the reader is encouraged to consult such journals as *Peptide Research*, *Peptide Synthesis*, *International Journal of Peptide and Protein Research*, *Neuropeptides*, *Peptides*, and the like for detailed reports on peptide drug design.

A. Choosing Biological Test Systems

When attempting to develop a peptide into a compound with a therapeutically useful effect in the whole animal, we must be careful not to oversimplify the initial biological test systems. For example, a binding assay might be used as the sole measure of biological activity for a series of analogs only to discover that when the best compounds are tested in vivo or in some other in vitro assay system, they exhibit properties of an antagonist rather than the desired agonist. One might also discover that when administered in vivo the pharmacokinetics of the designed peptide analog has rendered it inaccessible to the site of action, such as the central nervous system (CNS). Finally, the laboratory receptor model system may be found to be a poor model for the in vivo receptor.

The rapid enzymatic degradation of peptides has often been found to be the limitation that prevents them from being useful therapeutic agents. It has become quite popular to assess the potential for degradation of a peptide drug by measurement of the ability of the peptide to survive prolonged exposure to various proteolytic enzymes. Although this may indeed be relevant for selected enzymes, the whole-animal case represents an immense potential source of varied proteolytic degradations which cannot be duplicated accurately by a test tube mixture of commercially available enzymes.

One must therefore be careful to utilize a battery of test systems to evaluate the potential peptide drug candidate [19]. Where possible the assays should be performed simultaneously. Thus chemical modifications that lead to decreases in biological potency will not be eliminated from further studies which may prove that these same structural changes also led to an unpredicted improvement in enzymatic stability, pharmacokinetics, or other physiological or pharmacological changes. When the resources are limited, one must be very careful to decide which biological test systems should be scrutinized at the expense of others in order to maximize the information obtained while minimizing the risks of false or discarded leads.

B. Establishment of Essential Structural Requirements for Biological Activity

One of the first approaches in the study of peptides, as with other non-peptide drug candidates, is the development of a basic structure-activity relationship (SAR). The first step in this process is to establish the essential portions of that peptide necessary for effective biological activity. Again it is important to reiterate that multiple test systems are advisable so that as the peptide is modified and tested for biological activity, one does not lose sight of other factors, such as affinity at the receptor or resistance to enzymatic degradation. In the following sections we describe some of the steps that might be taken when attempting to develop a peptide hormone, neurotransmitter, or other factor into a therapeutically useful drug.

Determination of Minimal Chain Length

A key consideration in the potential of a protein/peptide for drug candidacy is the size of this compound. Specifically, is the complete sequence of the peptide required for full biological potency, or can a smaller version be found to be equally effective? It may even be possible to induce increases in biological activity in this fashion. Such an observation will have important implications for the development of the peptide into a drug candidate.

Enzymatic Digests. When beginning with a relatively large peptide or protein, the first approach to learning about the minimal size requirements for full biological activity involves enzymatic rather than chemical intervention. Typically, the intact biologically active protein is subjected to enzymatic degradation using various proteolytic peptidases. One can then assess the crude enzyme digest and determine if any biological activity remains. If significant biological activity remains, one can proceed to the chromatographic purification of the enzyme digest. The biological activity can hopefully be associated with one peptide which can then be sequenced. Further structural modifications can then be carried out on this peptide fragment.

One typically uses peptidases such as aminopeptidases, carboxypeptidases, trypsin, chymotrypsin, and so on, to trim down the protein and analyze for any changes in biological activity. Alternatively, one can use the cyanogen bromide method of cleavage at methionine residues or other chemical degradative schemes [140] to generate small fragments of proteins for further evaluation.

Chemical Modifications. If the peptide in question is readily amenable to chemical synthesis, the effect of chain length can be evaluated. Generally speaking, we want to make this evaluation quickly, and therefore the technique of SPPS is most frequently employed.

The evaluation of the effect of amino terminal truncation of a peptide is extremely facile by SPPS since synthesis proceeds from the carboxyl terminus toward the amino terminus. The synthesis is begun with a relatively large amount of an α-aminoacyl-resin, and then a sample of the peptide-resin is removed after the coupling of each amino acid. In this fashion a collection of all possible amino terminal truncated forms of the peptide is quickly obtained.

Consider the hypothetical 12-peptide Pro−Asp−Gly−Glu−Ser−Glu−Leu− Leu−Cys−His−Glu−Asn as an example of a potential peptide drug candidate. If we wish to investigate the effect of chain length on biological activity for this peptide, we would begin with a quantity of Boc−Asn−resin for our SPPS and synthesize Glu−Asn, His−Glu−Asn, Cys−His−Glu−Asn, and so on, until we had made all 11 possible amino truncated peptides. These peptides would then be analyzed in our assay system(s) and compared to the "parent" 12-peptide.

Since peptides are synthesized from the carboxyl terminus toward the amino terminus, synthesis of the carboxyl terminally truncated peptides is not as efficient as that of the amino terminally truncated peptides. Although each analog requires a full synthesis, with the progress made in the speed of SPPS, this generally does not pose an insurmountable hardship. Using SPPS and our hypothetical 12-peptide, we would have to use nine separate aminoacyl-resins (Leu-resin is used twice and Glu-resin is used three times). We would thus synthesize Pro−Asp, Pro−Asp−Gly, Pro−Asp−Gly−Glu, and so on, and once again test the peptides produced for biological activity.

The internal deletion of peptides in a fashion similar to that just described for the amino and carboxyl termini has been attempted with many peptides. In general, this procedure has led to inactive analogs. This fact has been rationalized by the assumption there has been a "frame shifting of one or more critical residues within the essential portion of the peptide [141]. Thus synthesis of the glycine deletion 11-peptide Pro−Asp− Glu−Ser−Glu−Leu−Leu−Cys−His−Glu−Asn in our example peptide might result in inactivity because the glycine residue at position 3 may only serve the purpose of placing the Asp side chain at residue 2 and the His side chain at residue 10 of the parent peptide in the correct orientation for maximal receptor interaction.

Once we have determined that a truncated sequence possesses full or significant biological activity, this shortened peptide segment can be used in further structure-activity studies. It is important to realize, however, that further modification on this shortened form of the original peptide will not necessarily parallel the same modification if applied to the parent peptide structure. Conclusions drawn from numerous simultaneous modifications must be interpreted with caution.

Functional Group Requirements

Once we have determined the minimal sequence requirements for biological activity, we are ready to assess how each of the amino acid residues contributes to the overall biological activity. These further elaborations might be designed or random in nature. We would look for improvements in the binding affinity of the peptide to the receptor or in the case of multiple receptors, for improvements in the binding of the peptide to one receptor over the other receptors. In addition, we might concentrate on improvements in the pharmacokinetics of the peptide or reduction of enzymatic degradation of the peptide. In the sections that follow, some of the modifications that have been employed to achieve these goals will be outlined. Note, however, that this process of structural elaboration to assess the functional group requirements has been approached in different ways by many investigators and there is no universally accepted method.

First-Stage Modifications: Side-Chain Substitutions. Again, using our hypothetical 12-peptide as an example, we could proceed through the peptide one residue at a time and replace that residue with either a glycine or an L-alanine. Thus we would synthesize Ala—Asp—Gly—Glu . . . , Pro—Ala—Gly—Glu . . . , Pro—Asp—Ala—Glu . . . , Pro—Asp—Gly—Ala . . . , and so on, and evaluate these analogs for biological activity. Those analogs that retained biological potency would be targets for further chemical modification in the later stages of drug design.

Glycine and alanine are the two most popular residues for substitution experiments. Glycine is popular since it does not possess a side chain. If the glycine analog is active, the implication is that the side chain of the residue that was replaced is not essential for biological activity. One argument against the use of glycine as the substituting amino acid is its lack of chirality. The absence of biological activity of the glycine replacement peptide may indicate a requirement for chirality at that site rather than an absolute requirement for that specific side chain.

To avoid this potential for confusion, alanine is often chosen as the substituting amino acid. Alanine has the advantage of possessing the same chirality as the amino acid that is replaced, and the methyl side chain is small, nonfunctional, and for the most part, devoid of any characteristics that might make it an important factor for binding to a receptor or enzyme.

Stereochemical Requirements. Again, there are two approaches to evaluate the stereochemical requirements for each residue in a given peptide. In one case each residue is replaced with the corresponding D-amino acid residue. Therefore, we could synthesize D—Pro—Asp—Gly—Glu . . . , Pro—D—Asp—Gly—Glu . . . , Pro—Asp—D—Ala—Glu . . . , Pro—Asp—L—Ala—Glu . . . , Pro—Asp—Gly—D—Glu . . . , and so on, and analyze the biological activities of these peptides. Note that in our example the glycine residue is replaced with D- and L-alanine. Although glycine is not chiral, its replacement with a small nonfunctional amino acid such as alanine can provide useful information about conformational tendencies. One argument against this approach, however, is that a specific residue may have no absolute chiral requirements until the side chain becomes large. Thus D—Glu in the fourth position of the foregoing peptide may be completely inactive due to the size of the glutamic acid side chain, while a D—Ala in the same position may be small enough to allow some interaction at the receptor binding site or the enzyme pocket. The second procedure for evaluation of chiral specificity is substitution with D-alanine. One replaces all the

amino acids, one at a time, with a D-alanine (e.g., D—Ala—Asp—Gly—
Glu— . . . , Pro—D—Ala—Gly—Glu— . . . , etc.) and then observes the
effect on biological activity. Those residues that still retain good biological activity after replacement with a D-alanine residue become prime
candidates for further evaluation and modifications.

This practice of substituting D-amino acids for L-amino acids in natural
peptides is widespread and many examples can be cited: LHRH [140,142],
ACTH [143], neurotensin [144], human growth hormone releasing hormone
[145], secretin [146], bradykinin [147], somatostatin [148], and even
insulin [149].

Second-Stage Modifications. Once the amino acid residues that are important for biological activity have been identified, we can concentrate on
changing or modifying individual residues in order to further develop the
SAR. Our initial analysis has identified those residues that are amenable
to change and we could begin our SAR studies with these amino acids.
Thus we might substitute these amenable residues with any or all of the
remaining 18 commonly occurring amino acids and measure the change in
biological potency. Any analogs that exhibited improvements in a biological test system would become targets for further studies.

An alternative approach is the modification of those sites which have
given an indication that they are not amenable to change. These sites
are obviously important for binding, since removal, major modification, or
chirality changes have led to poorly active or inactive analogs. However,
small changes in the structure of these residues might be allowed.

As an example, the phenylalanine at the fourth position of the enkephalins is very important for opiatelike activity in the mouse vas
deferens assay. If this residue is replaced with an L- or D-alanine, a
glycine, or a D-phenylalanine, the resultant analog is two to five orders of
magnitude less active. However, if the phenyl group is substituted at
either the *para* or the *meta* position with a fluorine, chlorine, bromine,
nitro, or trifluoromethyl group, the activity in the mouse vas deferens
assay is improved [150].

The preceding strategies have been utilized with many types of peptides. These changes are made to obtain parameters concerning the structural requirements for this specific peptide. Once these parameters are
known, we can begin to design peptides with improved pharmacokinetic profiles or increased resistance toward proteolytic degradation. These strategies are discussed in more detail in the following sections.

C. Strategies Designed to Improve Absorption of Peptides

Changes in Lipophilicity

Once the basic SAR for the peptide has been established and it has been
determined that the peptide still does not possess adequate biological activity by the desired route of administration, a change in lipophilicity may
be indicated. Peptides that are very polar are not good candidates to
cross lipophilic barriers such as the intestinal membranes. From Hansch-
like analyses of typical drug molecules it is known that increases in lipophilicity, to a point, can improve the ability of the drug to partition
through cell membranes and onto receptors. Conversely, one may wish to
limit the penetration of a given peptide through a biological membrane such

as the blood—brain barrier. In this instance one would want to reduce the overall lipophilicity of the peptide-drug candidate.

Incorporation of Lipophilic Amino Acids. The most logical modification to improve the lipophilicity of a peptide is the substitution of a nonessential amino acid with one of higher lipophilicity. Modern-day peptide chemists have the advantage of being able to utilize lipophilic amino acids other than the naturally occurring amino acid residues which are considered to be the most lipophilic: Ile, Val, Leu, Phe, Cys, Met, Ala, Trp, and Tyr [151]. One can now choose from other synthetic amino acids with significantly increased lipophilicities. A few examples of such lipophilic amino acids are *tert*-butylglycine (Bug) [152], neopentylglycine [153], β-ferrocenylalanine [154,155], α- and β-naphthylalanine (Nal) [156], cyclohexylalanine (Cha) [157,158], *para*-phenylphenylalanine [Phe(Ph)] [157,159], *meta*-phenyl-phenylalanine [Phe(3Ph)] [150], and *para*-iodophenylalanine [Phe(I)] [150]. Other work has focused on increasing the overall lipophilicity of a peptide by alkylating either the amino terminus or the carboxyl terminus with lipophilic long-chain alkyl groups. In one instance with enkephalin peptides, the addition of an alkyl amine to the carboxyl terminus resulted in potent analogs which exhibited prolonged analgesia when injected by the intrathecal route, whereas no analgesia was detected when the compound was injected parenterally. These results were in complete contrast to the unmodified parent enkephalin peptide [160]. In addition to an increase in the ability to partition through lipophilic membranes, highly lipophilic peptides can bind to carrier proteins such as serum albumin or cell membranes and show an increased biological half-life as a consequence [156].

Incorporation of Hydrophilic Amino Acids. To reduce the CNS penetration of peptides one should increase the polarity of the peptide. This can be accomplished by substitution of lipophilic amino acids with hydrophilic ones such as arginine, lysine, glutamic acid, or aspartic acid. Alternatively, one could modify the peptide by addition of amino, guanidine, or carboxyl groups at residues where previous SAR results have indicated that modifications are acceptable without significant adverse effects on receptor binding affinity or efficacy. Recently, enkephalin analogs have been prepared following this scheme and found to have an improved selectivity for peripheral opioid receptors versus those in the CNS. One such analog, Tyr—D—Arg—Gly—Phe(NO_2)—Pro—NH_2, has been carried on to clinical trials [161].

Prodrug Approach. This approach entails the design of a peptide prodrug which is inactive but has an improved absorption profile. Once the peptide prodrug is absorbed, it is converted to an active peptide, usually by an enzyme. Although the prodrug approach to improve oral absorption has had widespread exposure with general organic molecules [162], this procedure has by and large not been exploited with peptides. The most obvious success with a peptide-related drug is in the antihypertensive therapeutic area. Specifically, an angiotensin converting enzyme (ACE) inhibitor (enalapril, Fig. 26) was found to exhibit a delayed onset of activity in lowering blood pressure. This was determined to be due to a postabsorptive enzymatic cleavage of an ethyl ester in enalapril to give the active molecule with a free carboxylic acid moiety [163—165]. The diacid, unlike the starting ethyl ester, has full biological potency as an inhibitor

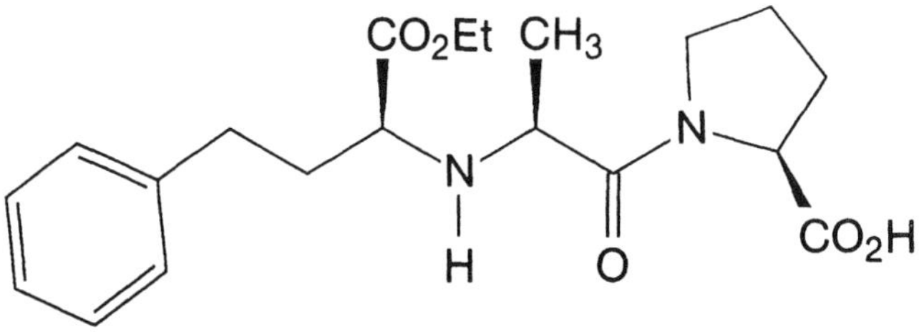

Fig. 26   Prodrug enalapril.

of ACE [165].   Enalapril thus become the prototype for a series of prodrug
ACE inhibitors [166].

Increases in Transport

The transport of peptides across the gut wall is a controversial subject.
The size of intact peptide that can be transported is still debated.   Little
was known about this topic until recently.   Now, however, there are
numerous studies concerning the transport (both active and passive) of
peptides.   The majority of the data seem to favor an upper limit in the
size of the peptide that can penetrate the intestinal wall (thus the desire
to find the smallest fragment of a peptide that still retains biological activity).   There is, however, evidence that peptides of sizes varying from
dipeptides to relatively large peptides the size of insulin can be orally
absorbed to a certain degree.   Another limitation to peptide transport appears to be the presence of proteolytic enzymes which efficiently degrade
the peptide to smaller peptides and amino acids.   Methods to prevent this
enzymatic degradation are outlined below.   In addition, the absorption of
peptides via the lymphatic system, nasal mucosa, and even the skin has
been documented.   Thus the barriers to peptide transport are not absolute.
The reader is directed to the chapters in Section II of this book for detailed information on the complicated topic of peptide transport.

D.   Strategies Designed to Reduce or Eliminate
     Enzymatic Degradation

Much of the problem with the use of peptides as drugs can be traced back
to their relatively short half-lives in vivo.   Most peptide hormones and
neurotransmitters are rapidly degraded in vivo by one or more enzymes
which may or may not be selective for the given peptide.   Some of these
enzymes selectively degrade the peptide as a mechanism to eliminate excess
peptide.   For example, the degradative enzyme referred to as enkephalinase
A has been associated with the metabolic breakdown of methionyl- and
leucyl-enkephalins [167].   Other enzymes degrade an inactive precursor
peptide to generate an active peptide.   As an example, consider the conversion of angiotensin I to angiotensin II by means of angiotensin converting
enzyme [168].

    If one understands the mechanism by which a given peptide is metab-
olized, one can design rational analogs that should limit or eliminate the
enzymatic degradation of that peptide.

N-Terminal Modification

If it is known that the peptide in question is being degraded by amino-peptidases, the first logical step is to introduce an N-terminal modification that will prevent the enzyme from interacting with the peptide. This procedure will be amenable to those peptides that can be modified at the N-terminus without appreciable loss in biological activity.

Since proteolytic enzymes generally prefer the natural L-configuration rather than the unnatural D-configuration, we can substitute a D-amino acid for the L-amino acid at either or both ends of the susceptible amide bond and thereby limit or prevent degradation. For example, the major metabolic degradation of the opioid peptide methionyl-enkephalin, Try—Gly—Gly—Phe—Met, is an aminopeptidase cleavage of the N-terminal amino acid tyrosine. The resultant peptide, Gly—Gly—Phe—Met, is entirely devoid of opioidlike activity. This enzymatic process is primarily responsible for the very short half-life (ca. seconds) of the natural enkephalins in the blood-stream. Early SAR studies established that while the L-tyrosine residue is required for biological activity, the glycine residue in position 2 is not essential. If this glycine residue is replaced with a D-amino acid such as D-alanine, the analog still possesses high binding affinity for the opioid receptor. When tested in vivo, this peptide exhibited a significantly longer half-life, presumably due to the inability of the aminopeptidase to cleave the Tyr—D—Ala amide bond readily.

This "stability" to enzyme degradation is, however, relative rather than absolute. As an example, the opioid peptide metkephamid, Tyr—D—Ala—Gly—Phe—MeMet—NH_2, possesses a half-life on the order of 60 min. A sample was synthesized with a carbon-14 label at the carboxyl group of the tyrosine residue and radiorespirometry studies were performed in rats. Within 10 min after subcutaneous administration of the peptide the animals began to respire $^{14}CO_2$ [169]. These data indicate that the Tyr—D—Ala bond is still quite susceptible to enzymatic cleavage.

In other examples, substance P has been modified at various positions throughout the molecule, resulting in the generation of analogs with potent agonist and antagonist activities. In the case of the antagonists, it has been proposed that coadministration of analogs containing D-amino acid residues causes substrate competition for the degradative peptidases, re-sulting in reduced destruction of the substance P itself and hence larger responses [170].

Some of the other modifications that have been utilized to prevent amino-peptidase degradation are: N^α-acetyl, formyl, pyroglutamic acid, N^α-alkyla-tion, chain extension with other amino acids, and removal of the α-amino group [171]. Nonetheless, some of these "stabilizing" modifications can be circumvented by the appropriate enzyme. The naturally occurring tri-peptide hormone thyrotropin releasing hormone (TRH) has a pyroglutamic acid residue as its amino terminal residue, yet the plasma half-life of TRH is only 4 to 5 min. This short in vivo stability is due in part to the enzymatic attack by a pyroglutamyl aminopeptidase [172].

A cytosolic protease has been isolated that selectively cleaves peptides of the general formula Ac—X—Y—Z . . . to give the products Ac—X + Y—Z . . . . This enzyme is implicated in the processing of nascent pro-teins [173]. Another enzyme, isolated from a *Nocardia* species, has the ability to hydrolyze the peptide bond adjacent to D-amino acids as in D—Leu—D—Leu [174]. These two examples of new enzymes with unexpected

substrate specificities illustrate the fact that our knowledge of the hydrolyzing capabilities of enzymes is still incomplete. Any modification that
we design into a peptide analog in order to limit degradation by a specific
proteolytic enzyme may be circumvented by a different, and as yet unchar-
acterized proteolytic enzyme.

C-Terminal Modification

Carboxypeptidase digestion can lead to the metabolic degradation of peptides from the C-terminal end of the molecule. Again, the use of D-amino
acid replacements at the C-terminal and/or penultimate positions can effectively inhibit carboxypeptidase-mediated degradation. Other strategies
must be employed in cases where the chirality of these two positions cannot
be altered. To eliminate this type of degradation, the carboxylic acid may
be masked or removed. One form of masking the carboxylic acid moiety
is to convert it to an ester. In addition to limiting or preventing carboxypeptidase degradation, formation of an ester results in increased lipophilicity, which in turn may allow improved passage across lipophilic barriers. However, the esterase capacity of many enzymes is well known;
therefore, this method may not always be successful.

A second method that has had some degree of success is conversion of
the C-terminal amino acid residue into an amide. Many of the synthetic
enkephalin peptide analogs possessed a C-terminal primary amide and were
shown to exhibit significantly longer half-lives in vivo [175]. Metkephamid,
$Tyr-D-Ala-Gly-Phe-MeMet-NH_2$, is an example of a peptide that contained both N- and C-terminal modifications designed to limit proteolytic
degradation. It contains a D-alanine designed to limit aminopeptidase
degradation and two modifications (N^α-methylation and a carboxamide) designed to limit carboxypeptidase degradation. While the parent peptide,
methionine enkephalin, $Tyr-Gly-Gly-Phe-Met$, has an in vivo half-life of
seconds, metkephamid has a half-life approaching 1 hr. In addition, in
clinical trials, metkephamid was found to provide pain relief for hours following intramuscular administration [176,177].

Another method that has been utilized to prevent carboxypeptidase-
mediated degradation of the peptide is to convert the C-terminal amino acid
residue to an amino alcohol. This method was successful with enkephalin
peptides wherein the C-terminal methionine was converted to a methioninol
residue. One such peptide was eventually studied in human clinical trials
[178,179].

The modifications described above also impart an increased lipophilicity
to peptides, especially at physiological pH (pH 7.4) where carboxylic acids
would be ionized and esters, amides, and alcohols are neutral. Other possible C-terminal conversions that can reduce or eliminate enzymatic degradation are methyl ketones, nitriles, alkyl amides, tetrazoles, and decarboxy
amino acids [180,181].

Cyclization

One modification that addresses the aminopeptidase and carboxypeptidase
degradation problems simultaneously is the conversion of a linear peptide to
a cyclic analog. In its simplest form this modification will create a peptide
that no longer possesses a free amino terminus or a free carboxyl terminus
which might be susceptible to aminopeptidase or carboxypeptidase degradation, respectively. However, conversion of a linear peptide to a cyclic

analog generally results in major conformational changes that may adversely affect binding to the peptide receptor.

The biological activity of a cyclized peptide can be maintained, however, if one has sufficient conformational information on the parent peptide from physical measurements such as NMR spectroscopy, circular dichroism spectroscopy, or x-ray crystallography. This information can allow rational design of the cyclic analog such that critical conformational parameters are maintained in the cyclic version. An example of the potential success of this method is the development by the Merck group of the cyclic hexapeptide [cyclo(MeAla—Tyr—D—Trp—Lys—Val—Phe)], based on data derived from somatostatin, a linear 14-peptide with one disulfide-based ring. This somatostatin analog was found to possess a high degree of metabolic stability and exhibited a duration of action of more than 4 hr in dogs after oral administration [182].

Internal Modification

D-Amino Acid Substitution. In many instances the enzymatic cleavage sites of peptides are located internally rather than at the termini of the peptide sequence. For example, peptide bonds that involve basic or aromatic amino acid residues can be sensitive to trypsin or chymotrypsinlike enzymes. Many of the modifications described thus far can be incorporated with success at internal residues. For example, the residues on either side of the enzymatically susceptible amide bond can be replaced with a D-amino acid if the earlier replacement studies have indicated that these residues are nonessential for biological activity.

An example of the potential success with this approach can be illustrated by the case of adrenocorticotropic hormone (ACTH). The full ACTH structure contains 39 amino acids and has been used in the treatment of seizures. However, the steroidal effects of ACTH are undesirable when used for treating seizures. Synthetic studies have proven that the heptapeptide (Met—Glu—His—Phe—Arg—Trp—Gly), comprised of residues 4 through 10, is as effective as ACTH for the treatment of seizures, while being devoid of the steroidal effects. The main site of proteolytic attack in this peptide is at the arginyltryptophanyl amide bond.

Synthetic studies established that proteolytic degradation could be reduced and the in vivo half-life increased by oxidation of the N-terminal methionine to the sulfone, removal of the C-terminal glycine residue, replacement of the tryptophan residue with a phenylalanine, and replacement of the L-arginine with a D-lysine. This hexapeptide, Met(O_2)—Glu—His—Phe—D—Lys—Phe (ORG 2766), was also found to exhibit activity after oral administration in human clinical trials [183]. Thus D-amino acid substitution can be a very effective method to limit endopeptidase-mediated degradation.

N^α-Alkyl Amino Acid Substitution. In those cases where the chirality of one or both of the residues that flank the enzymatically labile bond is known to be required for biological activity, alternative strategies may have to be employed to prevent unwanted enzymatic degradation.

Since proteolytic enzymes cleave primary amide bonds, virtually any modification to the amide bond can have an effect on the efficacy of the enzyme. If the C-terminal amino acid residue of the susceptible amide bond is replaced with an N-alkyl amino acid, the resultant peptide becomes resistant to the action of the endopeptidase. This method has been employed successfully in several enkephalin peptides that have progressed to clinical trials. In one case, the degradation of enkephalins by the dipeptidase enkephalinase A was targeted. An analog was synthesized with the Phe residue changed to an N^α-methyl Phe residue. This modification was successful in extending the in vivo half-life of the peptide FK33824, Tyr—D—Ala—Gly—MePhe—Met(O)—ol, versus the parent molecule, methionine enkephalin [175].

There are several general methods available for the synthesis of N^α-alkyl amino acids. A popular method involves the use of sodium hydride and methyl iodide to prepare the N-methyl amino acids. This procedure works well in many situations [184]; however, it fails in the case of N-methylmethionine. N-Alkylation of methionine can be accomplished successfully by treatment of a carbamate analog of methionine with potassium hydroxide and an alkyl halide in the presence of a crown ether (18-crown-6) [185]. Recently, two additional procedures were reported for the N^α-alkylation of amino acids [186,187], thereby highlighting the importance of this type of amino acid to the peptide chemist. The coupling efficiencies of N^α-alkyl amino acids can be significantly lower than that of their unsubstituted parent amino acids. Coupling of one N^α-methyl amino acid to another can be quite difficult [188], while an N^α-ethyl amino acid can be extremely difficult to incorporate in any sequence [personal observations]. Each peptide sequence presents its own degree of difficulty, and in general the relative difficulty cannot be predicted in advance. Recently, the use of bis(2-oxo-3-oxazolidinyl)phosphinic chloride (BOP-Cl) as a reagent for coupling of hindered N^α-alkyl amino acids was reported as an effective solution for this problem [189].

C^α-*Alkyl Amino Acid Substitution*. A modification that has some of the same benefits as N^α-alkylation is the substitution of the α-carbon of an amino acid with any alkyl group. This modification results in a peptide with increased lipophilicity as well as improved enzymatic stability due to steric hindrance in the vicinity of the enzymatically labile amide bond. An angiotensin II analog has been described that possessed an L-α-methyltyrosine in the fourth position. This analog was found to be completely stable when incubated with chymotrypsin for 3 hr [190]. As is the case with N^α-alkylation, this increase in steric hindrance has the undesired side effect of decreasing the coupling yield during the peptide bond formation reactions.

C^α-Alkylation has the potentially undesirable effect of inducing dramatic conformational changes in the peptide into which it is incorporated. For this reason one must be cautious in the utilization and interpretation of the results generated from this approach. In recent years several methods have been reported for the synthesis of C^α-alkylated amino acids [191,192]. These synthetic procedures again highlight the increased interest in obtaining unusual amino acids for incorporation into peptide analogs.

α,β-*Dehydro Amino Acid Substitution*. Another modification to the naturally occurring amino acid residues in peptides which has been exploited in recent years is that of an α,β-dehydroamino moiety. A review describes a number of peptides that contain α,β-dehydroamino acid residues and outlines procedures for synthesis of these amino acids and peptides that contain these amino acids [193].

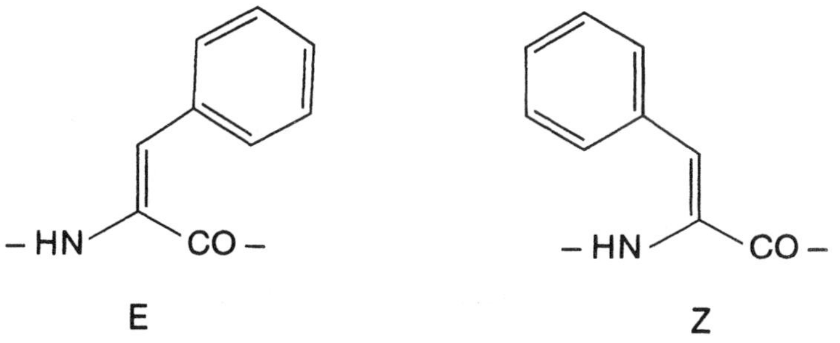

Fig. 27   E and Z conformations of α,β-dehydrophenylalanine.

The incorporation of dehydroamino acids such as α,β-dehydro—Ala, α,β-dehydro—Val, α,β-dehydro—Leu, and α,β-dehydro—Phe into a peptide results in an analog with increased rigidity, lipophilicity, and resistance to enzymatic degradation.   Analogs of bradykinin [194], angiotensin [195], dermorphin [196,197], and enkephalin [198—200] have been made which contain α,β-dehydrophenylalanine.   These analogs have been shown to be resistant to degradation from both the amino terminus and the carboxyl terminus [201].   It has also been established that the conformational isomers (E and Z) (Fig. 27) of α,β-dehydrophenylalanine in a peptide hormone such as the enkephalins can possess markedly different binding affinities for the opioid receptor.   This modification can be extremely useful for mapping the receptor site [200].   Improved synthetic methods for the incorporation of α,β-dehydroamino acids into peptides were recently reported [202,203].

Cyclopropyl Amino Acid Substitution.   Several syntheses of 1-aminocyclopropane-1-carboxylic acid derivatives (more commonly referred to as cyclopropylamino acids) (Fig. 28) have been reported [204,205].   When incorporated into peptides such as the enkephalins [206], these amino acids exhibit many of the same properties as the α,β-dehydroamino acid residues: for example, increased lipophilicity and increased rigidity with conformational selectivity, but without the high chemical reactivity associated with α,β-dehydroamino acids [207].   Consequently, this class of amino acids may prove useful for the moderation of unwanted enzymatic degradations.

Aza Analogs.   This modification to peptide structures, similar to α,β-dehydroamino acid substitutions, leads to a loss of chirality of the amino acid residue that has been replaced.   Since the chiral center of the amino

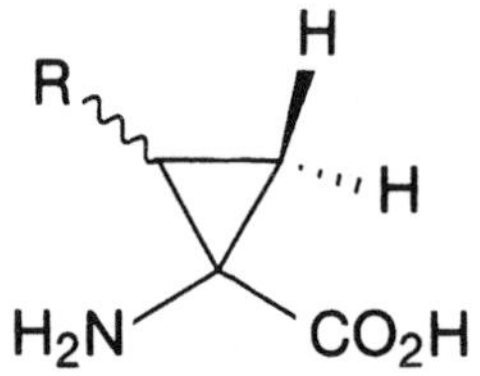

Fig. 28   Cyclopropylamino acid.

Fig. 29   Equilibrium in an aza amino acid leading to a lack of chirality.

acid has been replaced with a trigonal nitrogen which can freely intercon-
vert between conformers, the aza structure can be considered to be intermediate between that of D- and L-amino acids (Fig. 29) [208].   Some aza
analogs have been synthesized and found to be partially resistant to
proteolytic degradation [208].   Aza amino acid substitutions have been reported for LHRH, enkephalins, and tetragastrin peptides [209–211].   An
azaglycine substitution in position 3 of an enkephalin analog resulted in
one of the few, if not the only, potent glycine substitutions yet reported
with the enkephalins [212].   However, in the tetragastrin series, none of
the substitutions is biologically active.   General synthetic strategies have
been reported for aza analogs [213].

Other Non-Protein Amino Acid Substitutions.   The reader is directed to
several compendia on unusual or nonprotein amino acids for excellent resources concerning other amino acids that have been synthesized and incorporated into peptides in order to increase resistance to enzymatic degradation, improve or decrease lipophilicity, and/or study the effects on
structure-activity relationships of peptides [214,215].

Amide Bond Replacement

When a peptide is modified with a D-amino acid, an N-alkylated amino acid,
a C-alkylated amino acid, a dehydro amino acid, or a cyclopropyl amino
acid, the resultant analog is still a peptide.   However, if we replace one or
more of the amide bonds in the peptide, the resulting compound contains a
modified backbone and strictly speaking, can no longer be classified as a
peptide.   Until recently, examples of this class of compounds were rare, due
to the synthetic difficulties inherent in their preparation.

In 1977 an analog of the dipeptide glycylleucine was found to be a competitive inhibitor of aminopeptidase M [216].   More important, this new compound contained a methylene thioether (CH_2S) linkage (Fig. 30) as an

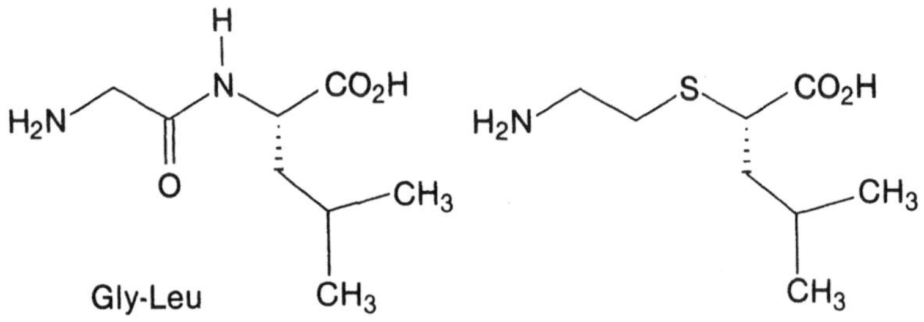

Fig. 30   Methylene thioether pseudodipeptide analog of the dipeptide
glycylleucine.

Table 2 Types of Pseudopeptide Linkages

Linkage	Description	PSI Nomenclature
	Amide	
	Ester	ψ[COO]
	Thioester	ψ[COS]
	Ketomethylene	ψ[COCH$_2$]
	Thioamide	ψ[CSNH]
	Bis Methylene or δ amino acid	ψ[CH$_2$CH$_2$]
	Double Bond Isostere or Olefin Substitution or Alkylidene	ψ[(E) CH = CH] or ψ[(Z) CH = CH]
	Acetylene	ψ[C = C]
	Retro-Inverso	ψ[NHCO]
	Reduced carbonyl	ψ[CH$_2$NH]
	Thioether	ψ[CH$_2$S]
	Thioether Sulfoxide	ψ[CH$_2$–(R)-SO] or ψ[CH$_2$–(S)-SO]
	Thioether Sulfone	ψ[CH$_2$SO$_2$]
	Ether	ψ[CH$_2$O]

isoteric replacement for the amide bond. The authors initially referred to this class of compounds as "peptide-gap" analogs and later introduced the term "pseudopeptides" as an alternative name [217]. Either name, however, was intended to refer only to the $-CH_2S-$ modification. The proliferation of other types of amide bond replacements which appeared shortly thereafter induced Spatola to propose that the term "pseudopeptide" be broadened to include any type of amide bond replacement [218]. While the scope of what constitutes a pseudopeptide is still somewhat nebulous, for purposes of this review we use the term to describe those derivatives that arise from the replacement of an amide bond with an isoteric type of linkage.

Recently, this class of peptide analogs has been thoroughly reviewed [219]. A specific nomenclature (referred to as the psi [ψ], or "psi-bracket" nomenclature) has been proposed and accepted as standard [218, 220]. However, this nomenclature is still not widely recognized and generally should be defined when it is used. The Greek lowercase letter psi [ψ] is used to indicate that a specific amide bond has been replaced with another type of covalent linkage. Thus, in the case of our example peptide, if the amide bond between the two leucine residues were replaced with a methylene thioether bond, the change would be indicated by the written name: Pro—Asp—Gly—Glu—Ser—Glu—Leu—ψ [CH_2S]—Leu—Cys—His—Glu—Asn. In a similar fashion, other isoteric modifications to the amide bond can be designated by the psi nomenclature. Table 2 summarizes the most widely reported types of pseudopeptides in both common name and psi format.

Replacement of the one peptide bond in a dipeptide is referred to as a pseudodipeptide. In actuality, a pseudodipeptide is a γ-amino acid. Thus when reduced to practical aspects, the synthesis of pseudopeptide structures becomes dependent on the ability to synthesize amino acid derivatives that are isosteric with dipeptides. We will use the terms "pseudopeptide" and "peptide isostere" interchangeably throughout the remainder of this chapter.

In the very simple case of a dimethylene amide bond replacement (CH_2CH_2) for the dipeptide glycyl—glycine ($NH_2CH_2CO—NHCH_2CO_2H$), the resultant amino acid is the known compound 5-aminopentanoic acid ($NH_2CH_2CH_2CH_2CH_2CO_2H$). However, replacement of the amide bond in dipeptides that contain one or two chiral amino acids is a synthetic task of not insignificant proportions. One method that has been utilized to simplify this synthetic problem has been to synthesize dipeptide isosteres of Aaa—Gly or Gly—Aaa dipeptides, where Aaa is a chiral amino acid. This reduces the complexity of the isostere synthesis by eliminating one chiral center. The dipeptide isostere may also be incorporated into the peptide as a racemic mixture and the resultant diastereomeric pseudopeptide products may then be separated by suitable chromatographic procedures. A disadvantage of this protocol is that the chirality of the separated diastereomers will be unknown. Since in many peptide analog projects the peptide target does not possess a glycine residue at the metabolically labile site, this approach is of limited utility.

Ideally, the peptide/medicinal chemist would like to have a commercial source for all possible dipeptide isosteres. However, if we consider only the 20 common naturally occurring amino acids, there are 400 possible pseudodipeptides that would be required for each type of amide bond replacement. If we consider that D-amino acids and unusual amino acids are often required in peptide analogs, the number of possible pseudodipeptides

becomes virtually astronomical. Thus for the foreseeable future, each dipeptide isostere will have to be specifically synthesized for an individual
need in a given peptide. Fortunately, synthetic procedures for the stereospecific synthesis of many of the classes of pseudodipeptides are now
being described. This fact reflects the importance of this type of peptide
modification to the medicinal/peptide chemist. In the sections that follow,
the major types of pseudopeptide modifications are briefly described. For
a thorough discussion of these and other pseudopeptide modifications, the
reader is referred to two extensive reviews [219,221].

$\psi[COO]$. The replacement of the amide bond by an ester bond is
perhaps the oldest example of a pseudopeptide-type structure. This type
of pseudopeptide is also commonly referred to as a depsipeptide. Depsipeptides, especially cyclodepsipeptides, are widely distributed in nature
[222]. Depsipeptides are relatively easy to synthesize since they are
formed by reaction between α-hydroxy acids, the latter of which can be
synthesized directly from α-amino acids with retention of the starting
chirality [223,224]. The replacement of an amide bond with an ester bond
increases the overall lipophilicity of a peptide. The effects of this modification on the overall peptide conformation are less predictable and caution must be exercised when attempting to rationalize a change in biological activity based on an isosteric replacement of an amide bond with an
ester bond.

Pseudopeptides that contain ester bonds have been described in the
cases of bradykinin and angiotensin [225,226] and more recently the
enkephalins [227]. No general rule concerning enzymatic stability of depsipeptides can be formulated since reports exist that indicate both increased
and decreased stability toward enzymatic degradation as well as decreased
chemical stability of these esters [226–229].

$\psi[COS]$. Like depsipeptides, the thioester amide bond replacements
are relatively accessible, since they are formed from condensation between
α-amino and α-mercapto acids. However, in general, this class of pseudopeptides has seen limited use, mainly in the enzyme substrate field [230].
Like esters, thioesters are readily hydrolyzed by the same enzymes that
hydrolyze amides and thus may not achieve the desired goal of limiting the
proteolytic degradation of a selected peptide.

$\psi[DOCH_2]$. The ketomethylene amide bond replacement gives an analog
that eliminates the possibility of enzymatic hydrolysis; however, this modification results in a compound that does not possess the rigidity of the
original amide bond. This can result in an analog with greatly modified
conformational properties. However, the carbonyl of the ketone does retain the ability to participate in hydrogen bonds. Until recently the primary example of this class of amide bond replacements was to be found in
the angiotensin-converting enzyme inhibitor field [231]. In addition, a few
peptides have been synthesized with C-terminal methyl ketones, as outlined
above. More recently, the synthesis of ketomethylene pseudopeptide
analogs of substance P has been reported. One analog containing this
modification was shown to possess full agonist properties as well as exhibiting potent inhibition of the substance P degrading activity in rat
diencephalon membranes [232]. Two new procedures have been reported
for the facile synthesis of ketomethylene isosteres of dipeptides [233]. The

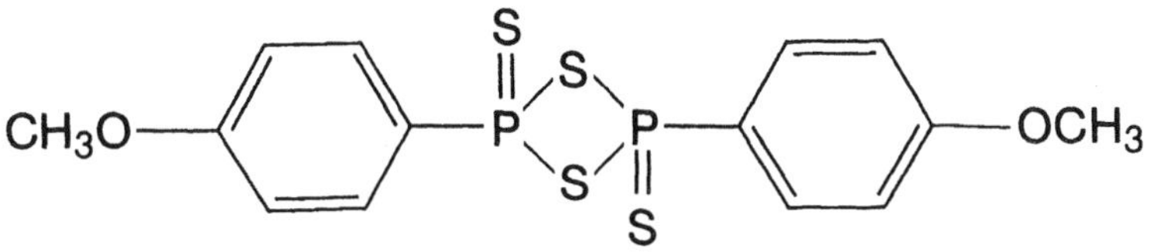

Fig. 31   Lawesson's reagent for the conversion of carbonyl groups into thiocarbonyl groups.

availability of this new chemistry should allow the increased incorporation of this class of amide bond replacements in peptides.

$\psi[CSNH]$. The use of the thioamide replacement of amide bonds has several advantages. Thioamides are isosteric with the amide group they replace. The bond lengths and angles are within 0.01 Å and 0.5° of those values normally found in peptide amide bonds [234]. They are relatively easy to obtain with the success of Lawesson's reagent (Fig. 31) for the conversion of amide carbonyls to their thio equivalents [235]. Interestingly, this reagent has recently been reported to be useful as a racemization-free coupling agent for the synthesis of peptides [236]. Finally, although the thioamide group adopts a trans conformation, the thiocarbonyl group of this amide bond isostere is unable to form strong hydrogen bonds and thus may provide information on the importance of a given amide bond for participation in these conformationally important events [237,238].

$\psi[CH_2NH]$. This class of pseudopeptides is commonly referred to as reduced amide (or peptide) bond analogs, and indeed they can be produced by direct reduction of the amide carbonyl in the target peptide. The reduced carbonyl functionality is often introduced at an early stage in the peptide synthesis and then the peptide is elaborated with the modification in place. Recently, this modification has been successfully incorporated into peptides via SPPS methodology [239,240]. This type of substitution can lead to active agonists as well as antagonists [241−243].

The reduced amide bond modification has several characteristics that make it both suitable and unsuitable for amide bond replacements. First, this modification, like the best amide bond replacements, will effectively prevent enzymatic degradation. However, this replacement is not truly isosteric with the amide bond, owing to the tetrahedral nature of the methylene group. In addition, this amide bond replacement adds an ionic site to the peptide since this residue is a secondary amine. Thus this modification can give rise to a large difference in conformation and ionic charge compared to the original peptide. Consequently, the results of this type of replacement cannot be predicted as readily as with other types of amide bond replacements. Recently, infrared and ^{1}H NMR analysis of protonated and unprotonated forms of reduced amide bond dipeptides has indicated that while the neutral molecule is quite flexible, the protonated form adopts a very stable conformation that resembles a β-turn [244]. These findings may have far reaching implications for this class of pseudopeptides.

$\psi[CH_2S]$. This class of pseudopeptides (a thioether) was first described in 1977 with the preparation of the pseudodipeptide $Gly\psi[CH_2S]Leu$ as a nonhydrolyzable peptide isostere. This compound was found to be a potent inhibitor of aminopeptidase M [216]. Stereospecific synthetic procedures have been developed for the synthesis of pseudodipeptides of the forms $Xxx\psi[CH_2S]Gly$ and $Xxx\psi[CH_2S]Yyy$ using chiral amino acids as the starting materials [245]. This type of pseudopeptide has been measured to be more lipophilic than the parent dipeptide, and the thioether bond is more flexible than the amide bond it replaces [245]. The SAR of leucine enkephalin analogs, wherein each amide bond was successively replaced with the thiomethylene moiety, has been reported; however, only the $[Phe^4\psi[CH_2S]Leu^5]$-leucine enkephalin substitution resulted in an analog with significant biological activity [246]. The $[Gly^3\psi[CH_2S]Phe^4]$-leucine enkephalin analog did exhibit enhanced stability against enzymatic degradation, and this phenomenon was exploited in the development of an effective enkephalin antagonist, $(allyl)_2Tyr-Gly-Gly\psi[CH_2S]Phe-Leu$ [247]. More recently, a procedure has been reported for direct SPPS preparation of peptides with thioether linkages [248].

$\psi[CH_2SO]$. The methylene sulfoxide amide bond replacement is simply an oxidized form of the methylene thioether amide bond replacement and in practice can be prepared directly from the latter class of pseudodipeptides or pseudopeptides. However, this class of pseudopeptides is unique in that the sulfoxide moiety can exist in two configurational states and thus can be classified as chiral. It is possible to use HPLC to separate $\psi[CH_2SO]$ diastereomers in enkephalin and LH—RH analogs [249,250]. The introduction of chirality at the amide bond site holds the promise of introducing interesting new conformational and pharmacokinetic responses. Overoxidation of the thioether starting material with excess oxidizing agent will lead to the thioether sulfone ($\psi[CH_2SO_2]$), which no longer possesses the chirality of the sulfoxide.

$\psi[CH_2O]$. This class of amide bond replacement is similar to the methylene thioether isostere in conformational and lipophilic properties. A substance P analog has been prepared with this substitution which exhibited greater activity than the closest related peptide parent analog [251]. Recently, a novel synthesis of methyleneoxy isosteres of dipeptides with a C-terminal glycine was reported [252].

$\psi[CH_2CH_2]$. The dimethylene amide bond replacement is one of the simplest and, at the same time, most difficult isosteres to utilize. Conceptually, the removal of an amide bond and its replacement with a two-carbon unit in the lowest oxidation state is very straightforward. However, synthesis of the dipeptide isostere with the correct chirality is quite difficult.

The simplest amide bond replacement in this category consists of the δ-amino acid, 5-aminopentanoic acid (5Ape). This amino acid will be isoteric with a glycyl—glycine dipeptide unit; however, the amino acid is considerably more flexible than the dipeptide, which contains an amide bond. In one instance of replacement of a glycyl—glycine dipeptide in the enkephalins with 5Ape, the resultant peptide was more than three orders of magnitude less active than the parent paptide [253]. Use of 5Ape to replace dipeptides that contain chiral amino acids has met with similar lack of success [253]. The most likely instance where this type of replacement

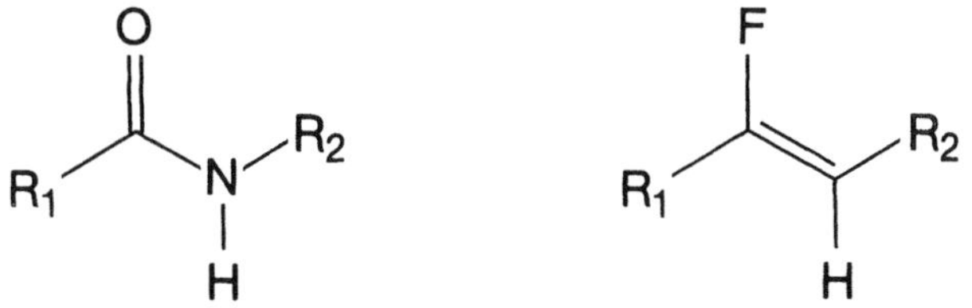

Fig. 32   Fluoroolefins as amide bond isosteres.

will meet with success is in those peptides where the conformation of that particular amide bond is not critical to the overall biological activity.

$\psi[CH{=}CH]$. The alkylidene amide bond replacement is perhaps the most intriguing of the amide bond replacements. Due to the high double-bond character of the amide bond, a trans carbon—carbon double bond should represent a very close isostere with the essentially planar trans amide bond. The bond lengths and angles have been shown to be very similar [254]. It has been calculated that if the degree of vector overlap is established as 0.40 Å the percentage overlap between a trans carbon—carbon double bond and an amide bond is greater than 95% [219].

This amide bond replacement has the advantage of eliminating the enzymatic routes of degradation while maintaining the conformation required by a trans amide bond at that location. Several enkephalin peptides have been synthesized with this isosteric amide bond replacement, and the resultant peptides were found to possess greatly reduced susceptibility to proteolytic enzymes as expected [254].

A major disadvantage of the carbon—carbon double bond isostere is the high degree of difficulty encountered with the synthesis of a dipeptide isostere when both amino acids of the dipeptide model contain chiral centers that are important for biological activity. Recently, several new synthetic methods that address this problem have been published [255,256].

Another potential disadvantage of this isostere is the loss of the polarity inherent with the amide bond. A possible solution to this shortcoming has been suggested recently with the analysis of fluoroolefin geometry. This report indicates that fluorine substitution in olefins will lead to relatively small changes in conformation with an increase in polarity, thus rendering fluoroolefin analogs (Fig. 32) appropriate as isosteric and pseudoisoelectronic amide bond replacements [257].

Two new compounds have been described which are isosteric with the dipeptide Phe—Gly [258]. These dipeptide analogs contain either a ketovinyl or a hydroxyethylidene group in place of the amide bond and the methylene group of the glycine residue (Fig. 33); thus they are not, strictly speaking, pseudodipeptides since they replace more than the amide bond.

$\psi[C{\equiv}C]$. Acetylene pseudopeptides will undoubtedly exhibit major conformational changes relative to the parent dipeptide. To date, only Leu$\psi[C{\equiv}C]$Gly has been synthesized [259] and the inclusion of that pseudodipeptide into a peptide of biological interest has not yet been reported. The addition of this new class of amide bond replacements to the

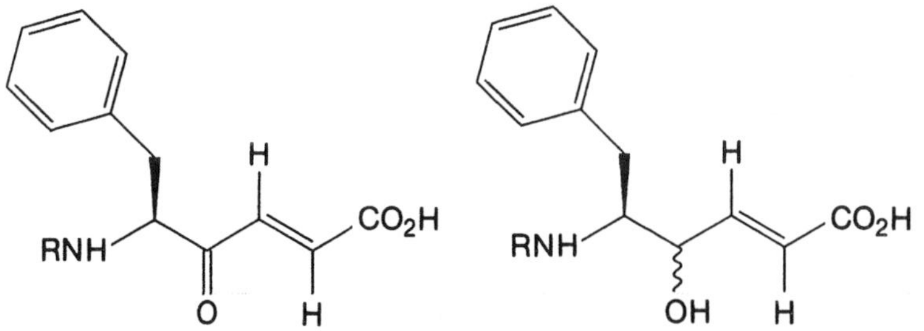

Figure 33 Ketovinyl and hydroxyethylidene dipeptide isosteres.

armamentarium of peptide chemists is indeed welcome; however, their utility remains to be determined.

$\psi[NHCO]$. The simple reversal of the direction of the amide bond (e.g., from —CONH— to —NHCO—) results in a peptide analog that should, in principle, be resistant to proteolysis. When the direction of every amide bond in a peptide is reversed and the chirality of each amino acid residue is reversed, the resultant molecule will be topologically similar to the parent peptide. The peptide that is derived from this type of manipulation is referred to as a retro-enantiomeric or retro-inverso peptide. This has been one of the more popular peptide modifications since the successful synthesis of an active retro-enantiomeric gramacidin S analog [260]. However, the vast majority of retro-enantiomeric peptides that have been synthesized have been biologically inactive. Careful analysis has led to the conclusion that the cumulative effects of small differences in the structural parameters (bond lengths and angles) between a peptide and its retro-enantiomer creates significant differences in side-chain location and resultant differences in overall conformation [261]. This type of modification, like all potential pseudopeptide mimics, has had its greatest success with the synthesis of partial retro-enantiomeric structures. The enkephalins [262] and substance P are examples where single amide bonds have been modified, resulting in biologically active analogs which have shown resistance to enzymatic inhibition [263,264]. This type of pseudopeptide has been reviewed [265,266].

Miscellaneous. All of the amide bond replacements that have been discussed have been isosteric with the amide bond. Other pseudopeptide structures have been reported, such as aza extensions (—NH—CO—NH—) and aminoxy groups (—CO—NH—O—), which place additional atoms into the peptide backbone [267]. While it is more difficult to predict whether an amide bond extension or deletion will provide the desired improvements in biological activity, this approach is gaining wider acceptance as predictive methods of molecular modeling have improved.

Many biologically active peptides have been found to possess a conformation that is dependent on a critical β-turn. If one or more amino acid residues can be replaced with a group that will mimic this β-turn while removing a potentially enzymatically susceptible amide bond from the molecule, the resultant analog may exhibit greater in vivo biological activity. This approach has had its greatest success with the report that a

γ-lactam incorporation into the structure of the peptide LHRH gave an analog with greater potency than the parent peptide when tested both in vitro
and in vivo [268].   An indolizidine nucleus has been used successfully to
mimic the framework of a β-turn in an immunosuppressant peptide [269],
and an eight-membered lactam has been used to mimic a type II β-turn in
the natural product, jaspamide [270].   Another type of turn, variously referred to as a gamma (γ) turn, C-7 turn, or reverse turn, has increasingly been implicated as being important in biologically active peptides.
Recently, a seven-membered lactam has been proposed as a C-7 mimic
structure.   This reverse-turn mimic has been synthesized and incorporated
into a series of enkephalin analogs [271].   Additional work in this area
can be expected based on these early successful applications.

   Conclusions.   Obviously, there are many types of amide bond replacements from which to choose.   When deciding which amide bond or bonds
should be replaced in a peptide drug candidate, the ultimate goal which is
desired for the specific change must first be considered.   For example, if
the oral bioavailability of a given peptide is to be improved, the most important considerations are stability of the peptide to proteolysis and overall pharmacokinetics of the peptide.   We may know from earlier SAR studies
that the susceptible amide bond is important for conformational integrity of
the peptide, and therefore we would choose a replacement that is highly
isosteric with the amide bond, such as the carbon—carbon double bond,
the thioamide, or the ester bond.
   On the other hand, the hydrogen-bond-forming properties of the amide
bond in question may be important, and therefore the replacement should
retain those properties if possible.   We might then choose either the ester
or ketomethylene groups as possible replacements.   If we wanted to increase the lipophilicity of the peptide, we might select from the thioether,
carbon—carbon double bond, dimethylene, ester, or ketomethylene modifications.   Alternatively, we could reduce the overall lipophilicity of the peptide by choosing the reduced carbonyl, the sulfoxide, or the α-aza
replacements.
   As the reader will notice from the preceding examples, the modifications described not only function to limit or eliminate enzymatic degradation, but can also have the secondary benefit of changing the overall
lipophilicity of the molecule.   For example, a molecule that began as an
amino acid peptide could be converted into an N^{α}-acylated peptide amide
with strikingly different ionic properties and lipophilic parameters.   These
new properties must also be considered when accounting for the changes
in biological activity that are observed with a new peptide analog.   Whichever choices we make, it remains important to assess the success or failure
of the synthetic analog in several types of biological assays in order to
measure the potency, the enzymatic stability, and the changes in pharmacokinetics.

E.   Conformational Analysis

Many of the methods and concepts described in previous sections are empirical in nature.   One does not predict a given biological response as
much as look for changes in the response once a new synthetic analog has
been prepared.   To be able to predict the outcome of a given chemical
manipulation, one must first have some concept of the preferred conforma-

tion of the peptide or an idea of how the peptide interacts with its receptor.

The determination of conformational preferences for a given peptide
has developed rapidly over the last decade. In addition to the classical
physical methods of circular dichroism (CD), optical rotatory dispersion
(ORD), x-ray crystallography, and ^{1}H NMR, newer techniques have become increasingly important and useful for conformational analysis. 2D-
NMR spectroscopy has become the method of choice for the analysis of
peptide and even small protein structures [272]. Indeed, the recent side-
by-side publication of the 2D-NMR and x-ray structures of the α-amylase
inhibitor Tendamistat has served as a landmark for the establishment of
the power and validity of this technique [273,274].

With the advent of supercomputers such as the Cray and the development of sophisticated molecular modeling programs that are written for the
express purpose of analyzing peptides, it is now possible to obtain energy-
minimized conformations of peptides as well as "snapshots" of the motion of
these structures that have been generated by molecular dynamics calculations. The physical methodologies coupled with distance geometry, molecular dynamics, and predictive protein/peptide folding programs are beginning to give the chemist an idea of how these compounds may behave in
solution and at the receptor [275-279].

Analysis of the conformation of a given peptide (along with its active
and inactive analogs) by these methods can lead to insights concerning
essential and nonessential features. Application of this knowledge can lead
to analogs which, in the extreme case, may have little resemblance to that
of the original peptide structure. These compounds might then be said to
have been truly "designed." However, to date our ability to generate a
completely nonpeptide drug from the original peptide structure has been
singularly unsuccessful. We remain content for the time to use these tools
as just one additional source of information in our attempts to develop and
design new peptide drugs.

IV. RECOMBINANT DNA TECHNOLOGY: APPLICATIONS IN PROTEIN ENGINEERING AND DRUG DESIGN

A. Introduction

In the past decade certainly one of the brightest stars on the horizon for
human disease therapy has been the promised availability of new drug
entities based on rare human proteins produced by genetic engineering or
rDNA technology. The first realization of this promise came, not surprisingly, with the availability of recombinant human insulin produced by
bacterial fermentation [280] and was rapidly followed by human growth
hormone [281], interferons [282,283], and a hepatitis vaccine [284]. Purified insulins and growth hormone produced from natural sources had previously been available, and thus a large body of knowledge dealing with
drug formulation and delivery had already been developed for these proteins.
Such is not the case for most of the new recombinant proteins now or soon
to be available for clinical use or evaluation (e.g., interferon, subunit
vaccines, tissue plasminogen activator, interleukin-2, etc.). Availability of
this latest crop of recombinant proteins in quantities sufficient to allow
large-scale clinical testing has only recently been realized. These proteins
and a plethora of others under development, such as Factor VIII, protein C,

erythropoietin, tumor necrosis factor, and atrial natriuretic factor, will cause—in fact, already have caused—a great deal of discussion and research activity into development of basic principles and new techniques for production, formulation, and delivery of proteins as therapeutic drug entities [285–288].

Exhaustive studies of serum half-life, clearance, dosing schedules, tissue compartmentalization, absorption rates, delivery modes, and so on, directed toward the use of small molecules in disease therapy have resulted in a large body of knowledge and sets of general guidelines that can be applied to most small-molecule drugs falling into a broad range of chemical classes. Similar parameters and guidelines will now have to be established for the various classes of therapeutic recombinant proteins coming onto the medical scene (Table 3). Delivery modes for proteins is an especially difficult problem since most proteins would be degraded if administered orally or would not be readily absorbed if given by any route other than injection. Dealing with problems of protein delivery will become an even greater issue when therapies move away from life-threatening situations and utilization of recombinant proteins for long-term treatment in otherwise healthy patients becomes more common. The attendant risks and general patient resistance to the injection delivery mode will force the health care industry to become more innovative in its approaches to protein delivery.

Pharmaceutical chemists have developed very sophistated tools for manipulating the chemistry of small-molecule drugs. Frequently, the promise of a newly discovered pharmacological activity is tempered by accompanying undesirable properties or side effects. It is then the pharmaceutical chemist's job to manipulate the physicochemical properties of the "lead" compound to achieve improvements in specific activity, half-life, absorption, clearance, and so on, and to eliminate side effects. Similarly, with proteins used in drug therapy, there will be a need for second- and third-generation compounds to "fine tune" therapies. For the most part chemical approaches toward improving the pharmacologic index of small-molecule drugs which have been so successful in the past will not be applicable to modifying proteins with hundreds of amino acid residues and molecular weights of 10,000 to 200,000 daltons. At least some of the tools for the fine tuning of proteins to make a better drug are inherent in the rDNA technology, which has been the means for large-scale availability. One of the goals of the remainder of this chapter will be to describe how the techniques of genetic engineering can be applied to structure-activity studies of proteins.

B. Gene Cloning: The Basic Principles and Technology

The scientific principles and physicochemical techniques that comprise the basis of recombinant DNA technology or genetic engineering have been derived mainly from the scientific discipline of genetics or molecular genetics. Some of the major discoveries and achievements that have provided the foundation of our present knowledge of molecular genetic mechanisms are listed chronologically in Table 4. An excellent history of discoveries and events leading up to the solving of the structure of DNA has been published [289].

Conceptually, the principles and techniques involved in producing a recombinant protein by isolating a gene from one cell, placing the gene in

Table 3   A Partial List of Recombinant Proteins in Use or
Being Developed for Human Disease Therapy

Hormones and growth factors	Follicle stimulating hormone
	Prolactin
	Angiogenin
	Epidermal growth factor
	Calcitonin
	Erythropoietin
	Thyrotropin-releasing hormone
	Insulin
	Growth hormone
	Insulin-like growth factors 1 and 2
	Skeletal growth factor
	Human chorionic gonadotropin
	Luteinizing hormone
	Nerve growth factor
Cytokines	Interferons
	Interleukin-1
	Colony stimulating factors
	Interleukin-2
	Tumor necrosis factor
Fibrinolytic enzymes	Urokinase
	Kidney plasminogen activator
	Tissue plasminogen activator
Clotting factors	Protein C
	Factor VIII
	Factor IX
	Factor VII
	Antithrombin III
Subunit vaccines	Hepatitis B virus surface antigens
	Influenza virus surface antigens
	Plasmodium surface antigens

Table 3   (Continued)

Subunit vaccines (continued)	Mycobacterium surface antigens
	Schistosoma surface antigens
	Herpes simplex virus surface antigens
	Trypanosoma surface antigens
	Streptococcus surface antigens
	Epstein-Barr virus surface antigens
	HTLV III virus surface antigens
Other proteins	Albumin
	Atrial natriuretic factor
	Renin
	Superoxide dismutase
	α_1-Antitrypsin
	Human lung surfactant protein

Table 4   Major Discoveries, Concepts, and Events in the Evolution of Molecular Genetic Science That Have Provided the Foundations for Recombinant DNA Technology

1865: Mendel	Development of mathematical rules governing gene inheritance in diploid organisms
1869: Miescher	Isolation and chemical characterization of nucleic acids
1911, 1913: Morgan and Sturtevant	Concept of genes as linear arrays on chromosomes and development of techniques for experimental analysis of gene linkage
1927: Muller	Use of x-rays to induce mutations in *Drosophila*; genes are physical entities
1941: Beadle and Tatum	One gene specifies one protein (enzyme)
1943: Avery, MacLeod, and McCarthy	Rigorous proof that genes are DNA
1943: Luria and Delbruck	Fluctuation test proving the heritability of mutations in bacteria

Table 4   (Continued)

1946:	Lederberg and Tatum	Demonstration that genetic recombination in bacteria was a result of conjugation (sexual mating)
1950:	Chargaff	The molar ratios of adenine to thymine and guanine to cytosine in DNA equal 1
1952:	Hershey and Chase	Bacteria viruses inject their DNA into the host cell
1953:	Watson, Crick, Wilkins, and Franklin	Determination of DNA double-helical structure
1958:	Lehman, Bessman, Simms, and Kornberg	Purification of DNA polymerase and in vitro synthesis of DNA
1959:	Weiss and Gladstone	DNA serves as a template to direct RNA synthesis
1960:	Jacob and Monod	Operator theory of gene regulation in bacteria
1961:	Brenner, Jacob, and Meselson	RNA acts as an unstable messenger to carry information from the gene (DNA) to the protein synthetic machinery (ribosome)
1961:	Nirenberg and Matthaei	Discovery of the first triplet genetic codon; uuu = phenylalanine
1970:	Smith, Kelly, and Wilcox	Isolation and characterization of restriction endonuclease HindII
1972:	Temin and Baltimore	Characterization of retrovirus and reverse transcriptase
1972−1975:	Khorana et al.	Chemical synthesis of a gene that codes for tRNA
1974:	Morrow, Boyer, Goodman, Helling, Chang, and Cohen	Transformation of eukaryotic DNA into *E. coli* via a recombinant plasmid vector
1975−1977:	Sanger and Coulson; Gilbert and Maxam	Development of DNA sequencing methods
1977:	Itakura, Hirose, Crea, Riggs, Heyneker, Bolivar, and Boyer	Chemical synthesis and expression of the somatostatin gene in *E. coli*
1980		Recombinant human insulin, produced through a joint venture between Genentech and Eli Lilly, administered to first human subject

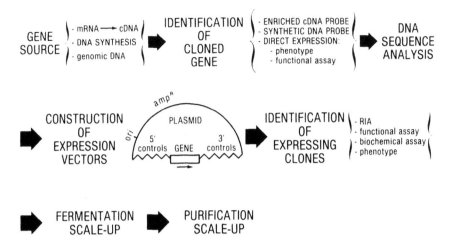

Figure 34   Flow sheet outlining the critical steps in a typical gene cloning and expression project. Examples of various strategies one could employ to proceed from one step to the next are in brackets. Ori denotes a bacterial origin of replication. AmpR denotes ampicillin resistance.

a different cell, and fermenting a culture of the recombinant cells to produce the desired protein are reasonably simple and straightforward. A number of well-written articles and books have been published describing procedures used to clone and express genes [290-292], and the reader who wishes to acquire a more detailed knowledge of rDNA technology is urged to consult these publications.

In its simplest form the underlying genetic principle states that each unique protein is specified by only one gene. Each protein is produced by synthetic machinery in the cell which is able to interpret a set of coded instructions contained in a unique messenger RNA (mRNA) molecule which is in turn a faithful complementary copy of information permanently stored in an inheritance unit, the gene, or DNA of the cell. Thus a genetic engineering project, as outlined in Fig. 34, begins with a consideration of how to purify (or clone) a unique gene from a mixture of genes and other noncoding DNA sequence where the unique stretch of DNA sequence specifying the gene of interest may represent 1 part in 100 to 1 part in 10^6. While it is possible to clone genes directly from total mammalian genomic DNA where the frequency of a unique gene is approximately 1 in 10^6, a more practical route has been to take advantage of the enrichment that one can usually obtain by starting with mRNA. Generally, a particular tissue expresses a limited subset of all the genes that the organism carries. For example, the number of molecules of myosin mRNA may be as high as 1000 per cell in an actively synthesizing muscle cell. This represents a considerable enrichment over the two gene copies per diploid cell.

Starting with a preparation of total mRNA from a particular tissue, one can synthesize faithful double-stranded DNA copies of the mRNAs in this mixture. In the first of a series of enzymatic reactions, avian myeloblastosis reverse transcriptase is used to polymerize deoxynucleoside triphosphates, utilizing an mRNA strand as a template to produce RNA—DNA heteroduplex molecules (Fig. 35). Reverse transcriptase polymerizes deoxynucleoside triphosphates by esterification of the 5'-phosphate (liberating pyrophosphate) to the 3'-hydroxyl group of the growing DNA polymer initiated with a short oligothymidylate primer substrate that has been hybridized to the complementary oligoadenylate tails characteristic of nearly all cytoplasmic mRNA molecules. The enzyme adds deoxynucleotides sequentially, extending the polymer in a 5'-to-3' direction, utilizing the mRNA strand as a template to match dA's with U's, T's with A's, dG's with C's, and dC's with G's. The reaction product is an RNA—DNA heteroduplex molecule in which the newly synthesized DNA strand is a faithful complementary copy of the nucleotide sequence on the mRNA strand.

Next, the mRNA portion of the duplex is degraded with alkali and in a second round of polymerization, a second DNA strand, complementary to the first, is made. This reaction is essentially the same as the first except that the 3'-OH primer is now a short loopback of the first DNA strand hybridizing to itself. Either reverse transcriptase or *Escherichia coli* polymerase can catalyze this second strand reaction. The double-stranded DNA product, termed "copy-" or cDNA, is digested with single strand-specific S1-nuclease to cleave the loop region and trim off overhanging single-stranded ends. This mixture of cDNA genes must now be inserted into vector DNA molecules, which will provide for replication of the cDNA in a host cell. To accomplish this, short, single-stranded deoxynucleotide "sticky" (hybridizable) ends complementary to "sticky" ends of the vector DNA molecules must be added to the cDNA. If, for example, the vector has overhanging 5'-oligo-dG ends, then *E. coli* terminal transferase can be used to add short stretches of dC's to the 3' end of each strand of the cDNA. The overhanging free ends of the vector and cDNA are then allowed to hybridize (anneal) to each other and *E. coli* DNA ligase is used to covalently join the free ends of the vector DNA strands to those of the cDNA. This ligation produces recombinant DNA molecules composed of double-stranded vector DNA sequence and cDNA gene sequence.

Many different types of vector DNA molecules have been developed. The most common have been derived from bacteriophages or plasmids and share several important properties. They are double-stranded DNA molecules that (1) replicate autonomously in bacterial cells, (2) carry marker genes conferring antibiotic resistance, and (3) contain well-characterized cleavage sites for insertion of DNA molecules. One of the best examples of such a vector is the *E. coli* plasmid, pBR322. The entire nucleotide sequence (4363 base pairs) of the plasmid and consequently all the cleavage sites of the known restriction endonucleases are precisely known and can be utilized to insert or remove genes from this plasmid vector.* One common method of inserting a mixture of cDNAs into pBR322 is to add short

---

*Restriction endonucleases represent a very important class of enzymes that can cleave double-stranded DNA molecules at specific sites based on recognition of a unique sequence of nucleotides as shown in Fig. 36. Of equal importance is the fact that enzyme specificity is essentially absolute. Thus a DNA molecule lacking a particular nucleotide sequence will be completely resistant to cleavage by the restriction enzyme(s) that requires

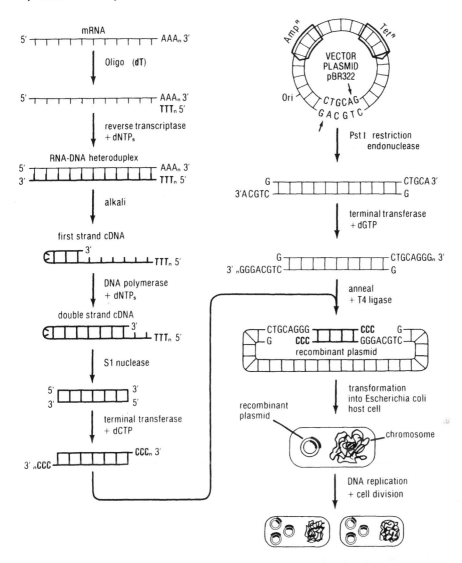

Figure 35 Schematic outline of procedures employed in the synthesis of a cDNA gene copy from a polyadenylated mRNA template, insertion of the cDNA into a bacterial plasmid vector by a homopolymer tailing strategy, and cloning of the recombinant plasmid in an *E. coli* host.

that sequence for activity. The enzymes have been isolated mainly from bacteria, where they can protect the cell against viral attack. Currently, more than 100 unique activities have been isolated and characterized. In rDNA technology restriction enzymes are used singly or in various combinations to fragment and separate DNA molecules or to cleave and linearize circular DNA molecules like plasmids. Many of the enzymes leave short overhanging, single-stranded, "sticky" or hybridizable ends which can be rehybridized to similar or different DNA molecules with complementary ends and then covalently joined with DNA ligase to give "recombined" DNA molecules.

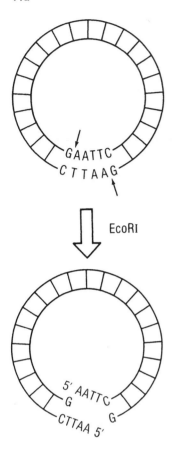

Figure 36   Bacterial plasmid represented by a circular double-stranded
DNA molecule containing the hexanucleotide sequence —GAATTC— which
is recognized and cleaved by a restriction endonuclease, EcoRI, to yield
a linear DNA molecule with "sticky ends."

stretches (tails) of single-stranded dC to cDNA and then hybridize these
dC ends to dG tails generated on linear pBR322.   After covalently joining
free ends of the cDNA to those of the vector, this mixture of recombinant
molecules can be transformed into a culture of *E. coli* cells.   Generally,
each cell will take up only one recombinant plasmid molecule; thus the
mixture of transformed cells represents a living "library" of the mixture
of cDNAs initially synthesized.   When the population of transformed cells
is plated out on agar containing the antibiotic ampicillin, only a cell that
has acquired a plasmid will survive, dividing and replicating its plasmid
DNA.   Thus each resultant colony (or clone) of cells will have been
derived from a single cell containing a single recombinant plasmid.   Of the
thousands of clones on the plate, only one or a few will contain copies of
the vector plasmid with a particular gene of interest.
    At this point the task is to identify the clone of interest from among
the thousands on the plate containing other "uninteresting" genes.   If at

least a portion of the amino acid sequence of the protein of interest is known, a short probe oligonucleotide of 12 to 20 bases can be chemically synthesized. Since many of the amino acids can be specified by more than one DNA triplet codon, in practice it is usually necessary to synthesize a small family of oligonucleotides to cover all the potential gene sequences that could specify a particular amino acid sequence. The radioactively tagged oligonucleotide mix can be used as a hybridization probe to identify which clone contains the gene of interest. Once identified the clone can be expanded and large quantities (milligrams) of the cloned cDNA gene can be isolated from purified plasmid DNA.

Other gene cloning strategies can be equally effective. For example, if the entire amino acid sequence of the protein is known, a gene that would code for that sequence can be chemically synthesized. The practical limit for chemical synthesis of genes is being rapidly extended. Currently, it is reasonable to consider total synthesis for genes of 900 to 1200 base pairs (coding for proteins of 300 to 400 amino acid residues) [293]. Another approach has been insertion and direct expression of the gene in a recombinant host cell. The new protein in the host cell may be identifiable by a functional assay such as interferon antiviral activity [282], a biochemical assay such as enzymatic activity of β-galactosidase [294], or antibody binding [295,296].

Even more subtle cloning strategies are possible. The gene involved in the inherited disorder chronic granulomatous disease (CGD) was recently cloned based primarily on knowledge of its chromosomal location [297]. A large DNA segment from the X chromosome was first identified by comparison with marker DNA segments from CGD patients. Smaller cloned segments of this presumptive CGD gene locus were identified by their absence in patients with gene deletions. Next, mRNA transcripts were compared between normal B lymphocyte cells (the cellular site of expression of the normal CGD gene) and cultured B-cells from CGD patients. One mRNA transcript absent in the cells from CGD patients was shown to be transcribed from a gene lying in the cloned X-chromosome segment. The amino acid sequence of the protein involved in the CGD phenotype was deduced from the DNA sequence of a segment of this genomic clone.

Although one can readily write out a gene cloning strategy employing relatively well characterized techniques and reagents, in practice cloning a gene from a messenger RNA source where the sequence of interest is present at the level of 0.001 to 1% usually represents many hours of benchwork plus a good measure of luck for a multidisciplinary group which would include molecular biologists, protein biochemists, and investigators from other disciplines, such as immunology, cell biology, virology, and animal physiology.

A cDNA clone is often the starting point for a detailed analysis of protein structure/function. The complete nucleotide sequence can be used to confirm and complete the amino acid sequence. Moreover, using a cDNA clone from one species as a hybridization probe it is usually possible to obtain homologous genes from other species or to obtain different but closely related genes within a species. Comparisons of related gene sequences have greatly enhanced our understanding of genetic evolution and can provide insights into the functional role of proteins in cellular physiology [298]. The cDNA clone is also a very specific probe for genomic sequences (frequency 10^{-6}) and can be used to isolate the genomic copy of a gene. The genomic copy can be an important starting point for studies of chromosome structure and mechanisms of gene regulation. On a more practical level cDNA probes are rapidly being developed as medical diagnostic tools. The genetic trace of a disease-causing organism can be an exquisitely sensitive and specific clue for devising therapeutic strategy.

C. Producing Recombinant Proteins in Heterologous Host/Vector Systems

In many cases, however, cloning the cDNA is only the first major hurdle in a project. If the final goal is to produce the protein in sufficient quantity to study its in vitro and/or in vivo function and perhaps ultimately to produce it for commercial use, perhaps in medical therapy, then the gene must be placed in a host/vector system specifically designed to provide for optimal expression of that protein. A wide variety of host cell/expression vector systems have been developed, including prokaryotic systems (e.g., E. coli) and eukaryotic systems such as yeast (e.g., Saccharomyces cerevisiae) and cultured mammalian cells (e.g., Chinese hamster ovary cell lines). A basic concern in expression vector design is to provide for replication of the vector and gene so that dividing cells will inherit the appropriate genetic information. Beyond this the gene must be trimmed to remove noncoding nucleotide sequences. Generally, chemically synthesized adapter DNA molecules must then be added to the front and back of the gene to direct its insertion into the vector in a precise nucleotide register such that vector sequences at the 5' and 3' ends of the gene are placed precisely with respect to the gene coding sequence (Fig. 34). Even single nucleotide insertions, deletions, or exchanges in these control sequences can cause drastic changes in the overall level of protein production [299].

The specific nature of the protein product can influence the choice of cell host. If the protein requires posttranslational modification for activity (e.g., glycosylation, γ-carboxylation, specific disulfide bridging, hydroxylation, acetylation, phosphorylation, etc.), production of an active molecule in a bacterial host may not be possible. For example, tissue plasminogen activator (TPA) normally forms 17 specific intramolecular disulfide bridges. When this protein was expressed in E. coli, only a small fraction of the protein had enzymatic activity probably because of improper disulfide bridging [300]. It was necessary to switch to an animal cell host and expression vector to produce active enzyme in high yield [301]. In contrast, E. coli-produced interleukin-2 has activity equivalent to that of the native molecule [302].

Production of recombinant human insulin by bacterial fermentation was a special case in which the formation of three specific disulfide bridges was required for activity. The initial method of production involved separate fermentations to produce the A-chain and B-chain polypeptides as chimeric C-terminal extensions of a large precursor protein. Unexpectedly, the chimeric proteins formed insoluble granules inside cells. After cell lysis, isolation of the granule fraction proved to be a significant purification step while maintaining the chimeric protein in a physical state relatively resistant to proteolytic degradation. Cyanogen bromide cleavage of methionine was used to liberate A- and B-chain polypeptides. The partially purified S-sulfonylated chains were then mixed and reoxidized, and proper inter- and intrachain disulfide bridges were formed in vitro as part of the overall insulin purification scheme [280]. Proper refolding and reassociation are presumably achieved through the mass action of intrinsic molecular forces driven to seek a minimum free-energy state.

More recently, human insulin production methods have been changed. Human insulin is now produced by refolding and trypsin/carboxypeptidase B cleavage of the recombinant human proinsulin molecule. The advantages of this newer method are that the proinsulin molecule is produced in a single fermentation step, and refolding of the prohormone to generate the native insulin structure is more efficient than refolding and reassociation of the separated A- and B-chains.

Problems associated with large-scale purification of recombinant proteins have also become a significant issue. The observation that chimeric A- and B-chain fusion polypeptides formed insoluble granules inside producing cells proved to be a bit of luck for the purification biochemists. More recently, novel approaches to incorporate purification "handles" into recombinant proteins have been described. Smith et al. described the use of immobilized metal ion affinity chromatography (IMAC) for purification of human proinsulin [303]. A fusion product containing Met—His—Trp-proinsulin was cleaved with CNBr to yield His—Trp-proinsulin. The S-sulfonate derivative of this molecule was bound to a Ni(II)-IDA (iminodiacetic acid) Sephadex column and eluted with 0.1 M acetic acid as a sharp peak at pH 5.12. The His—Trp chelating peptide could be removed by oxidative cleavage of tryptophan. Moks et al. described another approach for affinity purification based on the interaction of staphylococcal protein A and IgG [304]. A synthetic protein A IgG binding domain (Z-domain) was fused by an asparagine-glycine linkage to the N-terminus of human insulin-like growth factor I (IGF-1). The secreted Z-domain-IGF-1 fusion peptide was bound to an IgG-Sepharose column and eluted with 0.5 M acetic acid, pH 2.8. The Asn--Gly bond was then cleaved with hydroxylamine to yield human IGF-1. Clearly, demands for improved yields and simple, rapid purification schemes suitable for large-scale production will stimulate future developments in this area.

D. Protein Engineering and Drug Design

Up to this point the primary concerns have been to optimize production of natural or native proteins. Gene manipulations have been directed toward changing the expression vector or the producing host cell to achieve maximum rates of protein synthesis, optimum stability, and suitable purification properties. An underlying assumption has been that the native protein, produced and delivered in as near native form as possible, will be of therapeutic benefit to the patient. Although this has been demonstrated for some recombinant proteins (insulin, growth hormone, and TPA) and is likely to be so for many others being developed, current therapies are not without problems and side effects. For example, optimum drug payout and patient acceptability are major concerns in the ongoing development of insulin therapy for diabetics. While interferon has clear value in treating certain tumors, some patients experience significant side effects, such as fever, fatigue, and anorexia [305]. TPA is clearly beneficial in restoring patency of blocked coronary arteries in the early phases of acute myocardial infarction; however, initial clinical trials indicate that systemic lytic effects are still a complication [306] even though in vitro studies show that the enzyme has greater fibrin specificity than either streptokinase or urokinase

[307]. While structure-activity studies might be fairly straightforward for a chemical series of small molecules, how are we going to deal with these problems when the drug happens to be a 59,000-dalton protein? Can the power of molecular genetics be applied to yield solutions to these problems as well?

Biotechnology has thus far elegantly solved supply-side problems with these first-generation products. However, issues surrounding delivery modes, stability, clearance, specificity, side effects, and so on, will be a powerful driving force for further improvements fostering, in the not-too-distant future, second-generation recombinant proteins seeking the small increments in pharmacokinetic properties necessary for optimum efficacy. In the past, protein chemists have been limited to modifications such as cleaving away portions of the molecule, adding residues to the carboxy terminus, altering side chains of certain amino acids, and stabilization with various adducts. With access to the gene it is now possible, in theory at least, to change or mutate any nucleotide codon and thus to insert at every amino acid position in a protein any of the 20 naturally occurring amino acids. It may even be possible to develop certain amino acid derivatives that could be incorporated into proteins by the ribosomal machinery. However, this would be a much more difficult problem since first, a transfer RNA synthetase would have to charge its respective transfer RNA with the derivative amino acid before it could be incorporated into protein, and second, the host cell would have to be able to tolerate incorporation of the derivative into its own proteins, perhaps an even greater restriction than the first. Even accepting this limitation to the 20 naturally occurring amino acids, the potential combinations are essentially limitless even for small proteins. The number of combinations is 20^n, where n equals the number of amino acids in the polypeptide, disregarding alterations by amino acid deletion or addition.

Thus, using recombinant DNA techniques, it is possible to exchange any of the 20 amino acids for any other at any position in a polypeptide chain. In addition, by cutting and religating genes, large segments of amino acid sequence can be rearranged, added, or deleted to create proteins that may never have been seen in nature, essentially doing experiments in protein evolution. Techniques such as saturation mutagenesis have been developed and could be coupled to appropriate screening assays to select specific alterations in activity from among thousands of mutant proteins [308]. Selected point mutations have been employed extensively to study the structure and substrate binding relationships of dihydrofolate reductase [309]. Point mutations resulting in amino acid substitutions in the B-chain of human insulin yielded derivatives with more rapid absorption, increased potency, and prolonged action [310,311]. Replacement of two asparagine residues in yeast triosephosphate isomerase with amino acids not susceptible to deamidation (threonine and isoleucine) produced a fourfold increase in the thermostability [312]. Removal of one of three glycosylation sites in TPA (Asn-to-Glu exchange, residue 451) produced a molecule with increased serum half-life [313].

A good deal of protein evolution has been achieved by genetically combining or exchanging functional domains within larger proteins. The serine proteases are a classical example of this natural genetic engineering which has yielded a large series of proteases with closely related amino acid structures at their active sites attached to a variety of functional domains providing unique substrate specificities (e.g., trypsin, chymotrypsin, urokinase, TPA, and plasminogen) [314]. While TPA is clearly a potent thrombolytic agent, its complex secondary and tertiary structure provide fertile grounds for genetic engineering experiments designed to improve its therapeutic index even further. Based on comparisons with related molecules, several functional domains have been recognized. The C-terminal serine protease domain is preceded by 180 residues containing two disulfide-linked kringle structures, one of which is involved in fibrin binding and stimulation of plasminogen activator activity [315]. Preceding the two kringle domains is a short sequence homologous to epidermal growth factor (EGF). The functional contribution of this EGF domain is unclear. The remaining 84 N-terminal residues of the mature protein (527 aa, MW 59,008) comprise a finger domain homologous to regions on fibronectin involved in fibrin binding [316]. Deletion mutagenesis studies have confirmed some of the foregoing assumptions and yielded new molecules with potentially useful biological properties. For example, removal of both kringle structures resulted in a molecule with approximately 50% increased thrombolytic activity that was less susceptible to inhibition by platelet plasminogen activator inhibitor [317]. A TPA derivative in which only the first kringle was deleted gave more rapid lysis of coronary thrombi, improved maintenance of blood flow, and longer plasma half-life in dogs than the native molecule [318]

Fusion of different gene segments has yielded chimeric proteins with novel combinations of properties. For example, fusion of truncated IgG heavy-chain gene to the gene for a staphylococcal nuclease produced a chimeric protein that could combine with light chain to retain antigen binding specificity and had nuclease activity [319]. Chimeric antibodies have been produced by joining gene sequences coding for the hypervariable regions of a mouse antibody binding site to framework and constant regions from a human IgG molecule [320]. The hope is that such "humanized" mouse monoclonal antibodies will be better drugs for passive immunity or antibody therapy in humans. Numerous other gene fusions and amino acid replacements have been reported; for example, production of chimeric human yeast phosphoglycerate kinase [321], fusion of poliovirus antigen sequence to tobacco mosaic virus coat protein [322], and production of thermostable kanamycin nucleotidyltransferase [323].

It is not yet possible to predict accurately the three-dimensional structure of a large polypeptide chain from knowledge of its primary sequence. However, the day may not be far away when one will be able to start with a list of parameters (e.g., a desired chemistry, substrate specificity, solvent interactions, etc.) and based on first principles, design a primary amino acid sequence that will fold to yield a protein exhibiting the desired properties. Protein engineering is a rapidly emerging science, and genetic engineering technology has provided a quantum increase in the quality and quantity of information its practitioners can produce for study.

Algorithms that can predict the folding of a polypeptide chain are not reliable beyond 8 to 10 residues [324]. However, when combined with a cumulative data base generated by studies employing x-ray crystallographic analysis, two-dimensional proton NMR spectroscopy, biochemistry, and comparative analysis of amino acid sequence, sophisticated programs for computer-assisted molecular modeling of proteins can predict with some confidence helical conformation and β-turns in a polypeptide chain [279,325–327]. Beyond this, prediction of β-sheet conformation, for example, is much more difficult because important residues are not contiguous. Prediction

of critical amino acids involved in forming the active site of an enzyme
must still be based mainly on sequence comparisons among related enzymes
to identify highly conserved residues. It may be some time before the
first enzyme designed to carry out some novel chemistry is created. Beyond prediction of what the minimum free-energy state of a globular protein
in aqueous solvent ought to look like, it is yet another matter to create
appropriate conditions enabling a denatured molecule to refold correctly.
It has become clear that polypeptide chain folding must occur along defined
kinetic pathways with short- and long-range interactions playing important
roles at different times to achieve the final higher-order structure. Re-
cent successes in assembly of peptide blocks (α-helices, β-sheets, $\beta\alpha\beta$-
and 4-helix-bundle-folding patterns) on multifunctional carrier or "template"
molecules [328] are an indication that the intractable nature of this problem
may also be yielding to systematic approaches. Difficulties aside, the first
attempts at molecular "tinkering" have already been made [309,323]. Stimulated by both academic and industrial interests, our knowledge of protein
structure-activity relationships is rapidly expanding. Thus, in the foreseeable future it will be possible to make specific changes in protein structure to achieve predictable physicochemical parameters [329,330].

While the prospects of achieving such a goal are enough to get physicians fantasizing about the "ultimate drug," in the real world of protein
drug therapy the patient's immune surveillance system must be considered
a serious limiting factor on how far one can diverge from the native structure in designing a protein to fit the disease. However, even this caveat
does not appear to present an insurmountable barrier. The problem hinges
on which antigenic epitopes of an engineered protein the patient's immune
system will recognize as nonself and how the particular antigen will be presented [331]. Clearly, every amino acid in a protein is not available for
antigenic recognition; determinants are found mainly on the surfaces of
molecules. Certain amino acid changes (even of solvent-exposed residues)
will be antigenically neutral. For engineered proteins that do elicit a significant immune response it may be possible to pretreat the patient to induce immune tolerance. Immunologists and allergists have already made
considerable progress in understanding the cellular and molecular basis of
immunity and the mechanisms of induced immunotolerance [138,332]. Controlling the immune response to engineered protein drugs may be the next
major challenge these scientists will have to tackle.

V. CONCLUSIONS

In this chapter we have attempted to provide the reader with a broad overview of the methods available for the synthesis of peptides and proteins.
These methods vary from chemical synthesis by both the "classical" solution methods and the increasingly popular solid-phase methods, through the
emerging technique of peptide and protein synthesis and semisynthesis
with enzyme-catalyzed amide bond formation, to the expanding arena of recombinant DNA technology.

Although none of the synthetic methods are capable of addressing all
potential synthetic tasks with equal efficiency, all of the methods can be
applied successfully to the synthesis of a given peptide or protein. Thus
human insulin has been prepared by chemical, enzymatic, and recombinant

means. However, in general, each method has its own niche in the overall scheme of the preparation of peptides and proteins. In the realm of production of proteins and enzymes the recombinant methodologies reign supreme. Even the production of peptide hormones as small as insulin may best be accomplished by this powerful technique. The relative ease and cost-effectiveness of scaling-up fermentation technology makes recombinant methodology particularly attractive for larger-scale production.

Enzymatic semisynthesis is a feasible choice if one has a cheap supply of the appropriate peptide/protein and the enzyme with the correct specificity. Chemical synthesis becomes the method of choice when one contemplates the rapid and efficient preparation of peptide analogs that contain unnatural amino acids. The "best" method for the preparation of a given peptide or protein is a function of the sequence and length of the peptide.

We have also provided some general guidelines on how one might begin to design a peptide or protein drug. Among the important factors to consider are (1) the suitability of the biological test systems; (2) pharmacokinetics, including drug absorption and transport; (3) resistance to enzymatic degradation; and (4) conformational analysis by physical or computer modeling techniques. Many of these topics are expanded on in other chapters of this book. Careful application of these guidelines can lead to drug candidates with markedly improved characteristics compared to the original peptide or protein.

These methods are, however, not foolproof and in many cases serve to generate starting points rather than final drug candidates. As a case in point, it is still not possible to predict successfully how to turn a peptide hormone into an antagonist, yet the use of the preceding principles for the design of peptide drugs has generated antagonists in several instances (bradykinin, LH—RH, bombesin, enkephalin, etc.). Similarly, site-specific mutagenesis of proteins and enzymes, although often performed with a specific goal in mind, more often than not has led to products with new and unexpected activities or specificities. Thus the design of peptides and proteins as drugs is still an art as much as an exact science, but with each day and every new journal issue, the "art" in this rapidly expanding field comes one step closer to that of being a predictive science.

ACKNOWLEDGMENTS

The authors wish to express their appreciation to R. A. Gadski, N. D. Jones, R. G. Harrison, J. E. Shields, and S. R. Jaskunas for helpful discussions and criticism. Thanks are also due S. J. Pike for typing Sec. IV of the manuscript.

ABBREVIATIONS

HOBt	1-Hydroxybenzotriazole
3'-OH	3'-Hydroxyl
5Ape	5-Aminopentanoic acid
HOAc	Acetic acid

Ac	Acetyl
ACTH	Adrenocorticotropic hormone
a.a.	Amino acid
ACE	Angiotensin converting enzyme
Bzl	Benzyl
BOP-Cl	Bis(2-oxo-3-oxazolidinyl)phosphinic chloride
Cbz	Carbobenzyloxy
CNS	Central nervous system
CGD	Chronic granulomatous disease
CD	Circular dichroism
cDNA	Copy deoxyribonucleic acid
CNBr	Cyanogen bromide
Cha	Cyclohexylalanine
dA	Deoxyadenylate
dC	Deoxycytidylate
dG	Deoxyguanylate
DNA	Deoxyribonucleic acid
DCC	Dicyclohexylcarbodiimide
DCU	Dicyclohexylurea
DIC	Diisopropylcarbodiimide
DIEA	Diisopropylethylamine
DMAP	Dimethylaminopyridine
DMS	Dimethylsulfide
ELISA	Enzyme-linked immunosorbent assay
EGF	Epidermal growth factor
Fmoc	Fluorenylmethyloxycarbonyl
C-7	Gamma (reverse) turn
LH−RH	Gonadoliberin (leutinizing hormone-releasing hormone)
IGF-I	Human insulin-like growth factor I
HBr	Hydrobromic acid
HCl	Hydrochloric acid
HF	Hydrofluoric acid
IDA	Iminodiacetic acid
IMAC	Immobilized metal ion affinity chromatography

IgG	Immunoglobulin G
IL-3	Interleukin-3
mRNA	Messenger ribonucleic acid
Phe(3Ph)	*Meta*-phenylphenylalanine
Met(O_2)	Methionine sulfone
Met(O)-ol	Methioninol sulfoxide
NCA	N-carboxy-α-amino acid anhydride
MeAla	N^{α}-methylalanine
MeMet	N^{α}-methylmethionine
MePhe	N^{α}-methylphenylalanine
Nal	Naphthylalanine
Ni(II)	Nickel, +2 oxidation state
NMR	Nuclear magnetic resonance
ORD	Optical rotatory dispersion
Phe(I)	*para*-Iodophenylalanine
pMeBHA	*para*-Methylbenzhydrylamine
Phe(NO_2)	*para*-Nitrophenylalanine
Phe(Ph)	*para*-Phenylphenylalanine
PAM	Phenylacetamidomethyl
Pro-NH_2	Proline amide
^{1}H NMR	Proton nuclear magnetic resonance
rDNA	Recombinant DNA
REMA	Repetitive excess mixed anhydride
RNA	Ribonucleic acid
SPPS	Solid-phase peptide synthesis
SAR	Structure-activity relationship
SN_1	Substitution nucleophilic, unimolecular
SN_2	Substitution nucleophilic, bimolecular
Bug	*tert*-Butylglycine
Boc	*tert*-Butyloxycarbonyl
T	Thymidylate
TRH	Thyrotropin releasing hormone
TPA	Tissue plasminogen activator
TFA	Trifluoroacetic acid
2D-NMR	Two-dimensional nuclear magnetic resonance

REFERENCES

1. E. Schaal, *Annalen, 157*:26 (1871).
2. E. Grimaux, *Bull. Soc. Chim., 38*:64 (1882).
3. T. Curtius, *Ber. Dtsch. Chem. Ges., 16*:755 (1883).
4. H. Schiff, *Ber. Dtsch. Chem. Ges., 30*:2449 (1897).
5. P. Schutzenberger, *C. R. Acad. Sci., 106*:1407 (1888).
6. L. Lilienfeld, *Arch. Physiol., 383*:555 (1894).
7. T. Curtius, *J. Prakt. Chem., 24*:239 (1882).
8. E. Fischer and E. Forneau, *Ber. Dtsch. Chem. Ges., 34*:2868 (1901).
9. E. Fischer, *Ber. Dtsch. Chem. Ges., 39*:530 (1906).
10. M. Bergmann and L. Zervas, *Ber. Dtsch. Chem. Ges., 65*:1192 (1932).
11. V. du Vigneaud, J. Ressler, J. M. Swan, C. W. Roberts, P. G. Katsoyannis, and S. J. Gordon, *J. Am. Chem. Soc., 75*:4879 (1953).
12. J. C. Sheehan and G. P. Hess, *J. Am. Chem. Soc., 77*:1067 (1955).
13. R. B. Merrifield, *Fed. Proc., 21*:412 (1962).
14. R. B. Merrifield, *J. Am. Chem. Soc., 85*:2149 (1963).
15. M. H. Caruthers, *Science, 230*:281 (1985).
16. B. Gutte and R. B. Merrifield, *J. Am. Chem. Soc., 91*:501 (1969).
17. R. Hirschmann, R. F. Nutt, D. F. Veber, R. A. Vitali, S. L. Varga, T. A. Jacob, F. W. Holly, and R. G. Denkewalter, *J. Am. Chem. Soc., 91*:507 (1969).
18. M. Bodanszky, in *Peptides: Proceedings of the Fifth American Peptide Symposium* (M. Goodman and J. Meienhofer, eds.), John Wiley & Sons, New York (1977).
19. H.-D. Jakubke, P. Kuhl, A. Konnecke, G. Doring, J. Walpuski, A. Wilsdorf, and N. P. Zapevalova, in *Peptides 1982* (K. Blaha and P. Malon, eds.), Walter de Gruyter, Hawthorne, N.Y. (1983), pp. 43--54.
20. E. Gross and J. Meienhofer, eds., *The Peptides*, Vols, 1, 2, 3, and 5, Academic Press, New York (1979--1983).
21. E. Wunsch, in *Synthese von Peptiden, in Houben-Weyl: Methoden der organischen Chemie*, Vol. 15(1/2) (E. Muller, ed.), Georg Thieme Verlag, Stuttgart, West Germany (1982).
22. M. Bodanszky, Y. S. Klausner, and M. A. Ondetti, eds., *Peptide Synthesis*, John Wiley & Sons, New York (1976).
23. D. S. Kemp, in *The Peptides*, Vol. 1 (E. Gross and J. Meienhofer, eds.), Academic Press, New York (1979), pp. 315–383.
24. E. Gross and J. Meienhofer, in *The Peptides*, Vol. 1 (E. Gross and J. Meienhofer, eds.), Academic Press, New York (1979), p. 39.
25. M. Bodanszky, Y. S. Klausner, and M. A. Ondetti, in *Peptide Synthesis* (G. O. Olah, ed.), John Wiley & Sons, New York (1976), pp. 137–157.
26. N. L. Benoiton and F. M. F. Chen, in *Peptides, Structure and Biological Function* (E. Gross and J. Meienhofer, eds.), Pierce Chemical Company, Rockford, Ill. (1979), pp. 261–264.
27. S. S. Wang, J. P. Tam, B. S. H. Wang, and R. B. Merrifield, *Int. J. Pept. Protein Res., 18*:459 (1981).
28. N. L. Benoiton and F. M. F. Chen, *Can. J. Chem., 59*:384 (1981).
29. J. R. McDermott and N. L. Benoiton, *Can. J. Chem., 51*:2562 (1973).

30. J. Kovacs, in *The Peptides*, Vol. 2 (E. Gross and J. Meienhofer, eds.), Academic Press, New York (1979), pp. 486–539.

31. N. L. Benoiton, in *The Peptides*, Vol. 5 (E. Gross and J. Meienhofer, eds.), Academic Press, New York (1983), pp. 217–284.

32. D. H. Rich and J. Singh, in *The Peptides*, Vol. 1 (E. Gross and J. Meienhofer, eds.), Academic Press, New York (1979), pp. 242–261.

33. F. Weygand, D. Hoffman, and E. Wunsch, *Z. Naturforsch.*, *B21*:426 (1966).

34. H. Gross and L. Bilk, in *Peptides: Proceedings of the Ninth European Symposium* (E. Bricas, ed.), North-Holland Publishing Company, Amsterdam (1968), pp. 156–158.

35. W. Konig and R. Geiger, *Chem. Ber.*, *103*:788 (1970).

36. G. C. Windridge and E. C. Jorgensen, *J. Am. Chem. Soc.*, *93*:6318 (1971).

37. P. Sieber, B. Kamber, A. Hartmann, A. Johl, B. Riniker, and W. Rittel, *Helv. Chim. Acta*, *60*:27 (1977).

38. E. Fischer, *Ber. Dtsch. Chem. Ges.*, *35*:1095 (1902).

39. R. Schwyzer, B. Iselin, and M. Feurer, *Helv. Chim. Acta*, *38*:69 (1955).

40. M. Bodanszky, *Nature*, *175*:685 (1955).

41. J. Pless and R. A. Boissonnas, *Helv. Chim. Acta*, *46*:1609 (1963).

42. L. Kisfaludy, J. E. Roberts, R. H. Johnson, G. L. Mayers, and L. Kovacs, *J. Org. Chem.*, *35*:3563 (1970).

43. L. Kisfaludy, I. Schon, T. Szirtes, O. Nyeki, and M. Low, *Tetrahedron Lett.*, 1785 (1974).

44. J. Kovacs, G. L. Mayers, R. H. Johnson, R. E. Cover, and U. R. Ghatak, *J. Org. Chem.*, *35*:1810 (1970).

45. J. Kovacs, G. L. Mayers, R. H. Johnson, R. E. Cover, and U. R. Ghatak, *J. Chem. Soc. Chem. Commun.*, 53 (1970).

46. G. H. L. Nefkens, *J. Am. Chem. Soc.*, *83*:1263 (1961).

47. G. W. Anderson, J. E. Zimmerman, and F. M. Callahan, *J. Am. Chem. Soc.*, *86*:1839 (1964).

48. P. D. Bragg and C. Hou, *Arch. Biochem. Biophys.*, *167*:311 (1975).

49. J. V. Staros, *Biochemistry*, *21*:3950 (1982).

50. A. E. Bolton and W. M. Hunter, *Biochem. J.*, *133*:529 (1973).

51. R. C. McCarthy and H. Markowitz, *Clin. Chim. Acta*, *132*:277 (1983).

52. M. Bodanszky and V. du Vigneaud, *Biochem. Prep.*, *9*:110 (1962).

53. M. Bodanszky, in *The Peptides*, Vol. 1 (E. Gross and J. Meienhofer, eds.), Academic Press, New York (1979), pp. 105–196.

54. W. Konig and R. Geiger, *Chem. Ber.*, *106*:3626 (1973).

55. T. Wieland and H. Bernhard, *Justus Liebigs Ann. Chem.*, *572*:190 (1951).

56. R. A. Boissonnas, *Helv. Chim. Acta*, *34*:874 (1951).

57. J. R. Vaughan, *J. Am. Chem. Soc.*, *73*:3547 (1951).

58. G. W. Anderson, J. E. Zimmerman, and F. M. Callahan, *J. Am. Chem. Soc.*, *89*:5012 (1967).

59. J. Meienhofer, in *The Peptides*, Vol. 1 (E. Gross and J. Meienhofer, eds.), Academic Press, New York (1979), pp. 263–314.

60. M. A. Tilak, *Tetrahedron Lett.*, 849 (1970).

61. A. van Zon and H. C. Beyerman, *Helv. Chim. Acta*, *59*:1112 (1976).

62. T. Curtius, *Ber. Dtsch. Chem. Ges.*, *35*:3226 (1902).

63. M. Bodanszky and M. A. Ondetti, *Peptide Synthesis*, John Wiley & Sons, New York (1966).

64. P. Sieber, B. Riniker, M. Brugger, B. Kamber, and W. Rittel, *Helv. Chim. Acta*, 53:2135 (1970).

65. K. Hofmann, A. Lindenmann, M. Z. Magee, and N. H. Zahn, *J. Am. Chem. Soc.*, 74:470 (1952).

66. S. S. Wang, I. D. Kulesha, D. P. Winter, R. Makofske, R. Kutny, and J. Meienhofer, *Int. J. Pept. Protein Res.*, 11:297 (1978).

67. J. Honzl and J. Rudinger, *Collect Czech. Chem. Commun.*, 26:2333 (1961)

68. R. Schwyzer and H. Kappeler, *Helv. Chim. Acta*, 44:1991 (1961).

69. T. Shiori and S. Yamada, *Chem. Pharm. Bull.*, 22:849, 855, 859 (1974).

70. J. Meienhofer, in *The Peptides*, Vol. 1 (E. Gross and J. Meienhofer, eds.), Academic Press, New York (1979), pp. 197–239.

71. M. Saware, *Adv. Polym. Sci.*, 4:1 (1965).

72. R. Hirschmann, R. G. Strachan, H. Schwam, E. F. Schoenewaldt, H. Joshua, B. Barkemeyer, D. F. Veber, W. J. Paleveda, T. A. Jacob, T. E. Beesley, and R. G. Denkewalter, *J. Org. Chem.*, 32: 3415 (1967).

73. R. Hirschmann, in *Peptides: Structure and Function* (V. J. Hruby and D. H. Rich, eds.), Pierce Chemical Company, Rockford, Ill. (1983), pp. 1–32.

74. H. R. Kricheldorf, *Chem. Ber.*, 104:87 (1971).

75. R. Hirschmann, H. Schwam, R. G. Strachan, E. F. Schoenewaldt, H. Barkemeyer, S. M. Miller, J. B. Conn, V. Garsky, D. F. Veber, and R. G. Denkewalter, *J. Am. Chem. Soc.*, 93:2746 (1971).

76. Y. Iwakura and K. Uno, *Biopolymers*, 9:1419 (1970).

77. R. Katakai and M. Oya, *Biopolymers*, 10:2199 (1971).

78. R. Katakai and M. Oya, *J. Org. Chem.*, 37:327 (1972).

79. T. J. Blacklock, R. Hirschmann, and D. F. Veber, in *The Peptides*, Vol. 9 (S. Undenfriend and J. Meienhofer, eds.), Academic Press, New York (1987), pp. 39–102.

80. W. Voelter and E. Schmid-Siegmann, *Peptides: Syntheses—Physical Data*, Vols. 1–6, Addison-Wesley Publishing Company, Don Mills, Ontario, Canada (1983).

81. J. M. Stewart and J. D. Young, *Solid Phase Peptide Synthesis*, Pierce Chemical Company, Rockford, Ill. (1984).

82. G. Barany and R. B. Merrifield, in *The Peptides*, Vol. 2 (E. Gross and J. Meienhofer, eds.), Academic Press, New York (1979), pp. 1–284.

83. J. Meienhofer, in *Hormonal Proteins and Peptides* (C. H. Li, ed.), Academic Press, New York (1973), pp. 45–267.

84. M. Bodanszky, Y. S. Klausner, and M. A. Ondetti, eds., *Peptide Synthesis*, John Wiley & Sons, New York (1976), pp. 158–195.

85. R. B. Merrifield, *Science*, 232:341 (1986).

86. E. Atherton, D. L. J. Clive, and R. C. Sheppard, *J. Am. Chem. Soc.*, 97:6584 (1975).

87. A. R. Mitchell, B. W. Erickson, M. N. Ryabtsev, R. S. Hodges, and R. B. Merrifield, *J. Am. Chem. Soc.*, 98:7357 (1976).

88. G. Barany and R. B. Merrifield, *J. Am. Chem. Soc.*, 99:7363 (1977).

89. D. H. Rich and S. K. Gurwara, *J. Am. Chem. Soc.*, 97:1575 (1975).
90. E. Atherton, D. L. J. Clive, and R. C. Sheppard, *J. Am. Chem. Soc.*, 97:6584 (1975).
91. E. Atherton and R. C. Sheppard, in *The Peptides*, Vol. 9 (S. Udenfriend and J. Meienhofer, eds.), Academic Press, New York (1987), pp. 1–38.
92. E. Atherton, R. C. Sheppard, and P. Ward, *J. Chem. Soc. Perkin Trans. 1*, 2065 (1985).
93. E. Atherton, H. Fox, D. Harkiss, and R. C. Sheppard, *J. Chem. Soc. Chem. Commun.*, 539 (1978).
94. S. B. H. Kent and R. B. Merrifield, in *Peptides 1980* (K. Brunfeldt, ed.), Scriptor, Copenhagen (1981), 328–333.
95. A. Dryland and R. C. Sheppard, *Tetrahedron*, 44:859 (1988).
96. E. Atherton, L. R. Cameron, and R. C. Sheppard, *Tetrahedron*, 44:843 (1988).
97. A. Dryland and R. C. Sheppard, *J. Chem. Soc. Perkin Trans. 1*, 125 (1986).
98. M. Bodanszky and J. T. Sheehan, *Chem. Ind. (London)*, 1597 (1966).
99. J. M. Stewart and J. D. Young, *Solid Phase Peptide Synthesis*, W. H. Freeman and Company, San Francisco (1969), p. 7.
100. A. Loffet, *Int. J. Pept. Protein Res.*, 3:297 (1971).
101. B. F. Gisin, *Helv. Chim. Acta*, 56:1476 (1973).
102. R. W. Roeske and P. D. Gesellchen, *Tetrahedron Lett.*, 3369 (1976).
103. J. Rebek and D. Feitler, *J. Am. Chem. Soc.*, 96:1606 (1974).
104. F. M. F. Chen, K. Kuroda, and N. L. Benoiton, *Synthesis*, 929 (1978).
105. S. B. Kent, L. E. Hood, H. Beilan, S. Meister, and T. Geiser, in *Peptides 1984* (U. Ragnarsson, ed.), Almqvist & Wiksell Forlag, Stockholm (1984), pp. 185–188.
106. D. Sarantakis, J. Teichman, E. L. Lein, and R. L. Fenichel, *Biochem. Biophys. Res. Commun.*, 73:336 (1976).
107. E. Atherton, Personal communication.
108. J. P. Tam, W. F. Heath, and R. B. Merrifield, *J. Am. Chem. Soc.*, 105:6442 (1983).
109. J. P. Tam, W. F. Heath, and R. B. Merrifield, *J. Am. Chem. Soc.*, 108:5242 (1986).
110. J. P. Tam and R. B. Merrifield, in *The Peptides*, Vol. 9 (S. Udenfriend and J. Meienhofer, eds.), Academic Press, New York (1987), pp. 185–248.
111. H.-D. Jakubke, in *The Peptides*, Vol. 9 (S. Udenfriend and J. Meienhofer, eds.), Academic Press, New York (1987), pp. 103–165.
112. J. B. West and C. H. Wong, *J. Org. Chem.*, 51:2728 (1986).
113. W. Kullmann, *Biochem. Biophys. Res. Commun.*, 91:693 (1979).
114. W. Kullmann, *J. Biol. Chem.*, 255:8234 (1980).
115. W. Kullmann, *J. Biol. Chem.*, 256:1301 (1981).
116. W. Kullmann, *Proc. Natl. Acad. Sci. USA*, 79:2840 (1982).
117. F. Widmer, S. Bayne, G. Houen, B. A. Moss, R. D. Rigby, R. G. Whittaker, and J. T. Johansen, in *Peptides 1984* (U. Ragnarsson, ed.), Almqvist & Wiksell Forlag, Stockholm (1984), pp. 193–200.

118. A. J. Andersen, F. Widmer, and J. T. Johansen, in *Peptides 1986* (D. Theodoropoulos, ed.), Walter de Gruyter, West Berlin (1986), pp. 183–188.

119. H.-D. Jakubke and A. K. Konnecke, *Methods Enzymol.*, *136*:178 (1987).

120. K. Morihara, T. Oka, and H. Tsuzuki, *Nature*, *280*:412 (1979).

121. A. Jonczyk and H.-G. Gattner, *Hoppe-Seyler's Z. Physiol. Chem.*, *362*:1591 (1981).

122. K. Rose, H. De Piery, and R. E. Offord, *Biochem. J.*, *211*:671 (1983).

123. K. Rose, J. Gladstone, and R. E. Offord, *Biochem. J.*, *220*:189 (1984).

124. V. Cerovsky, *Collect. Czech. Chem. Commun.*, *31*:1352 (1986).

125. J. S. Fruton, *Adv. Enzymol.*, *53*:239 (1982).

126. H.-D. Jakubke and P. Kuhl, *Pharmazie*, *37*:89 (1982).

127. I. M. Chaiken, A. Komoriya, M. Ohno, and F. Widmer, *Appl. Biochem. Biotechnol.*, *7*:385 (1982).

128. R. E. Offord, in *Peptides 1982* (K. Blaha and P. Malon, eds.), Walter de Gruyter, Hawthorne, N.Y. (1983), pp. 31–42.

129. J. D. Glass, in *The Peptides*, Vol. 9 (S. Udenfriend and J. Meienhofer, eds.), Academic Press, New York (1987), pp. 167–184.

130. W. Kullman, *Enzymatic Peptide Synthesis*, CRC Press, Boca Raton, Fla. (1987).

131. I. Clark-Lewis, R. Aebersold, H. Ziltener, J. W. Schrader, L. E. Hood, and S. B. H. Kent, *Science*, *231*:134 (1986).

132. D. Yamashiro and C. H. Li, *J. Am. Chem. Soc.*, *100*:5174 (1978).

133. F. Lottspeich, A. Henschen, and K.-P. Hupe, eds., *High Performance Liquid Chromatography in Protein and Peptide Chemistry*, Walter de Gruyter, Hawthorne, N.Y. (1981).

134. G. L. Hawk, ed., *Biological/Biomedical Applications of Liquid Chromatography IV*, Vol. 20, Marcel Dekker, New York (1982).

135. T. Wieland, in *Perspectives in Peptide Chemistry* (A. Eberle, R. Geiger, and T. Wieland, eds.), S. Karger AG, Basel (1981), pp. 1–13.

136. H. M. Geysen, R. H. Meloen, and S. J. Barteling, *Proc. Natl. Acad. Sci. USA*, *81*:3998 (1984).

137. R. A. Houghton, *Proc. Natl. Acad. Sci. USA*, *82*:5131 (1985).

138. H. M. Geysen, *Immunol. Today*, *6*:364 (1985).

139. H. M. Geysen, S. J. Rodda, and T. J. Mason, *Mol. Immunol.*, *23*:709 (1986).

140. E. Gross and J. Meienhofer, in *The Peptides*, Vol. 1 (E. Gross and J. Meienhofer, eds.), Academic Press, New York (1979), p. 23.

141. J. Rudinger, in *Drug Design*, Vol. II (E. J. Ariens, ed.), Academic Press, New York (1971), pp. 335–336.

142. A. S. Dutta and B. J. A. Furr, in *Annual Reports in Medicinal Chemistry*, Vol. 20 (D. M. Bailey, ed.), Academic Press, New York (1985), pp. 203–214.

143. G. Baumann, A. Walser, P. A. Desaulles, F. J. A. Paesi, and L. Geller, *J. Clin. Endocrinol. Metab.*, *42*:60 (1976).

144. R. Quirion, F. Rious, D. Regoli, and S. St.-Pierre, *Eur. J. Pharmacol.*, *61*:309 (1980).

145. A. Felix, E. P. Heimer, and T. F. Mowles, in *Annual Reports in Medicinal Chemistry*, Vol. 20 (D. M. Bailey, ed.), Academic Press, New York (1985), pp. 185—192.

146. M. Waelbroick, P. Robberecht, P. DeNeef, P. Chatelain, and J. Christophe, *Biochim. Biophys. Acta*, *678*:83 (1981).

147. S. St.-Pierre, P. Gaudreau, J. N. Drouin, D. Regoli, and S. Lemaire, *Can. J. Biochem.*, *57*:1084 (1979).

148. N. R. Voyles, S. J. Bathena, L. Recant, C. A. Meyers, and D. H. Coy, *Proc. Soc. Exp. Biol. Med.*, *160*:76 (1979).

149. F. Marki, M. De Gasparo, K. Eisler, B. Kamber, B. Riniker, W. Rittel, and P. Sieber, *Hoppe-Seyler's Z. Physiol. Chem.*, *360*: 1619 (1979).

150. P. D. Gesellchen, R. T. Shuman, R. C. A. Frederickson, and M. D. Hynes, in *Peptides: Structure and Function* (C. M. Deber, V. J. Hruby, and K. D. Kopple, eds.), Pierce Chemical Company, Rockford, Ill. (1985), pp. 495—498.

151. J. Kyte and R. F. Doolittle, *J. Mol. Biol.*, *157*:105 (1982).

152. J.-L. Fauchere and C. Petermann, *Helv. Chim. Acta*, *63*:824 (1980).

153. J. Pospisek and K. Blaha, *J. Org. Chem.*, *51*:2728 (1986).

154. E. Cuingnet, C. Sergheraert, A. Tartar, and M. Dautrevaux, *J. Organomet. Chem.*, *195*:325 (1980).

155. J. Pospisek, S. Toma, I. Fric, and K. Blaha, *Collect. Czech. Chem. Commun.*, *45*:435 (1980).

156. J. J. Nestor, R. Tahilramani, T. L. Ho, G. I. McRae, and B. H. Vickery, in *Peptides: Structure and Function* (C. M. Deber, V. J. Hruby, and K. D. Kopple, eds.), Pierce Chemical Company, Rockford, Ill (1985), pp. 557—560.

157. J. J. Nestor, T. L. Ho, R. A. Simpson, B. L. Horner, G. H. Jones, G. I. McRae, and B. H. Vickery, in *Peptides: Synthesis, Structure, Function* (D. H. Rich and E. Gross, eds.), Pierce Chemical Company, Rockford, Ill (1981), pp. 109—112.

158. P. D. Gesellchen, R. C. A. Frederickson, S. Tafur, and D. Smiley, in *Peptides: Synthesis, Structure, Function* (D. H. Rich and E. Gross, eds.), Pierce Chemical Company, Rockford, Ill. (1981), pp. 621—624.

159. R. T. Shuman, P. D. Gesellchen, E. L. Smithwick, and R. C. A. Frederickson, in *Peptides, Synthesis, Structure, Function* (D. H. Rich and E. Gross, eds.), Pierce Chemical Company, Rockford, Ill. (1981), pp. 617—620.

160. N. Sole, F. Reig, and J. M. G. Anton, *Int. J. Pept. Protein Res.*, *26*:591 (1985).

161. A. F. Cohen, N. Harkin, and J. Posner, *Br. J. Pharmacol.*, *24*: P267 (1987).

162. S. H. Yalkowsky and W. Morozowich, in *Drug Design*, Vol. IX (E. J. Ariens, ed.), Academic Press, New York (1980), pp. 122—185.

163. A. A. Patchett, E. Harris, E. W. Tristram, M. J. Wyvratt, M. T. Wu, D. Taub, E. R. Peterson, T. J. Ikeler, J. ten Broeke, L. G. Payne, D. L. Ondeyka, E. D. Thorsett, W. J. Greenlee, N. S. Lohr, R. D. Hoffsommer, H. Joshua, W. V. Ruyle, J. W. Rothrock, S. D. Aster, A. L. Maycock, F. M. Robinson, R. Hirschmann,

C. S. Sweet, E. H. Ulm, D. M. Gross, T. C. Vassil, and D. A. Stone, *Nature*, *288*:280 (1980).

164. M. L. Cohen, K. D. Kurz, and K. W. Schenck, *J. Pharmacol. Exp. Ther.*, *266*:192 (1983).

165. D. M. Gross, C. S. Sweet, E. H. Ulm, E. P. Backlund, A. A. Morris, D. Weitz, D. L. Bohn, H. C. Wenger, T. C. Vassill, and D. A. Stone, *J. Pharmacol. Exp. Ther.*, *216*:552 (1981).

166. M. L. Cohen, *Annu. Rev. Pharmacol. Toxicol.*, *25*:307 (1985).

167. J.-C. Schwartz, B. Malfroy, and S. De La Baume, *Life Sci.*, *29*: 1715 (1981).

168. D. W. Cushman and M. A. Ondetti, in *Progress in Medicinal Chemistry*, Vol. 17 (G. P. Ellis and G. B. West, eds.), Elsevier Science Publishing Co., New York (1980), pp. 41–104.

169. P. D. Gesellchen, C. J. Parli, and R. C. A. Frederickson, in *Peptides: Synthesis, Structure, Function* (D. H. Rich and E. Gross, eds.), Pierce Chemical Company, Rockford, Ill. (1981), pp. 637–640.

170. J. S. Dutta, J. J. Gormley, A. S. Graham, I. Briggs, J. W. Growcott, and A. Jamieson, *J. Med. Chem.*, *29*:1163 (1986).

171. J. S. Morley, C. F. Hayward, R. J. Carter, and S. Shuster, *Neuropeptides*, *2*:109 (1981).

172. G. Metcalf, *Brain Res. Rev.*, *4*:389 (1982).

173. T. Orfeo and W. L. Meyer, *Fed. Proc.*, *45*:1712 (1986).

174. M. Sugie, H. Suzuki, and N. Tomizuka, *Agric. Biol. Chem. Tokyo*, *50*:1633 (1986).

175. P. E. Hansen and B. A. Morgan, in *The Peptides*, Vol. 6 (S. Udenfriend and J. Meienhofer, eds.), Academic Press, New York (1984), pp. 269–321.

176. J. F. Calimlin, W. M. Wardell, K. Sriwatanakul, L. Lasagna, and C. Cox, *Lancet 1*, 1374 (1982).

177. S. S. Bloomfield, T. P. Barden, and J. Mitchell, *J. Clin. Pharmacol. Ther.*, *34*:240 (1983).

178. D. Roemer, H. H. Buescher, R. C. Hill, J. Pless, W. Bauer, F. Cardinaux, A. Closse, D. Hauser, and R. Huguenin, *Nature*, *268*:547 (1977).

179. L.-F. Tseng, H. M. Loh, and C. H. Li, *Life Sci.*, *23*:2053 (1978).

180. A. Spatola, in *Chemistry and Biochemistry of Amino Acids, Peptides, and Proteins* (B. Weinstein, ed.), Marcel Dekker, New York (1983), p. 306.

181. J. S. Morley, *J. Chem. Soc. C*, 809 (1969).

182. D. F. Veber, R. Saperstein, R. F. Nutt, R. M. Freidinger, S. F. Brady, P. Curley, D. S. Perlow, W. J. Paleveda, C. D. Colton, A. G. Zacchei, D. J. Tocco, D. R. Hoff, R. L. Vandlen, J. E. Gerich, L. Hall, L. Mandarino, E. H. Cordes, P. S. Anderson, and R. Hirschmann, *Life Sci.*, *34*:1371 (1984).

183. K. Pentella, D. S. Bachman, and C. A. Sandman, *Neuropediatrics*, *13*:59 (1982).

184. S. T. Cheung and N. L. Benoiton, *Can. J. Chem.*, *55*:906 (1977).

185. R. T. Shuman, E. L. Smithwick, D. L. Smiley, G. S. Brooke, and P. D. Gesellchen, in *Peptides: Structure and Function* (V. J. Hruby and D. H. Rich, eds.), Pierce Chemical Company, Rockford, Ill. (1983), pp. 143–146.

186. R. M. Freidinger, J. S. Hinkle, D. S. Perlow, and B. H. Arison, *J. Org. Chem.*, *48*:77 (1983).

187. D. W. Hansen and D. Pilipauskas, *J. Org. Chem.*, *50*:945 (1985).

188. R. D. Tung and D. H. Rich, in *Peptides: Structure and Function* (C. M. Deber, V. J. Hruby, and K. D. Kopple, eds.), Pierce Chemical Company, Rockford, Ill. (1985), pp. 217–220.

189. R. D. Tung, M. K. Dhaon, and D. H. Rich, *J. Org. Chem.*, *51*: 3350 (1986).

190. M. C. Khosla, B. Witczuk, S. Forgac, R. R. Smeby, P. A. Khairallah, and F. M. Bumpus, *Abstr. Pap. Am. Chem. Soc.*, *186*: 23 (1983).

191. U. Schollkopf, J. Nozulak, and U. Groth, *Tetrahedron*, *40*:1409 (1984).

192. M. J. O'Donnell, W. Bruder, K. Wojciechowski, L. Ghosez, M. Navarro, F. Sainte, and J.-P. Antoine, in *Peptides: Structure and Function* (V. J. Hruby and D. H. Rich, eds.), Pierce Chemical Company, Rockford, Ill. (1983), pp. 151–154.

193. K. Noda, Y. Shimohigashi, and N. Izumiya, in *The Peptides*, Vol. 5 (E. Gross and J. Meienhofer, eds.), Academic Press, New York (1983), pp. 284–339.

194. G. H. Fisher, P. Berryer, J. W. Ryan, V. Chauhan, and C. H. Stammer, *Arch. Biochem. Biophys.*, *211*:269 (1981).

195. E. A. Hallinan and R. H. Mazur, in *Peptides: Structure and Biological Function* (E. Gross and J. Meienhofer, eds.), Pierce Chemical Company, Rockford, Ill. (1979), pp. 475–477.

196. S. Salvadori, M. Marastoni, G. Balboni, G. Marzola, and R. Tomatis, *Int. J. Pept. Protein Res.*, *28*:254 (1986).

197. S. Salvadori, M. Marastoni, G. Balboni, G. Marzola, and R. Tomatis, *Int. J. Pept. Protein Res.*, *28*:262 (1986).

198. Y. Shimohigashi and C. H. Stammer, in *Peptides: Synthesis, Structure, Function* (D. H. Rich and E. Gross, eds.), Pierce Chemical Company, Rockford, Ill. (1981), pp. 645–648.

199. Y. Shimohigashi, M. L. English, and C. H. Stammer, *Biochem. Biophys. Res. Commun.*, *104*:583 (1982).

200. T. J. Nitz, Y. Shimohigashi, T. Costa, H.-C. Chen, and C. H. Stammer, *Int. J. Pept. Protein Res.*, *27*:522 (1986).

201. Y. Shimohigashi, H. C. Chen, and C. H. Stammer, *Peptides*, *3*:830 (1982).

202. C. Shin, N. Takamatsu, and Y. Yonezawa, *Agric. Biol. Chem. Tokyo*, *50*:797 (1986).

203. M. Makowski, B. Rzeszotarska, Z. Kubica, G. Pietrzynski, and J. Hetper, *Justus Liebigs Ann. Chem.*, *1986*(6):980 (1986).

204. S. W. King, J. M. Riordan, E. M. Holt, and C. H. Stammer, *J. Org. Chem.*, *47*:3270 (1982).

205. P. K. Subramanian and R. W. Woodward, *J. Org. Chem.*, *52*:15 (1987).

206. C. Mapelli, H. Kimura, and C. H. Stammer, *Int. J. Pept. Protein Res.*, *28*:347 (1986).

207. H. Kimura, C. H. Stammer, Y. Shimohigashi, C. R. Lin, and J. Stewart, *Biochem. Biophys. Res. Commun.*, *115*:112 (1983).

208. A. Spatola, in *Chemistry and Biochemistry of Amino Acids, Peptides, and Proteins* (B. Weinstein, ed.), Marcel Dekker, New York (1983), p. 292.

209. A. S. Dutta, B. J. A. Furr, M. B. Giles, and B. Valcaccia, *J. Med. Chem.*, *21*:1018 (1978).

210. J. S. Morley, *Annu. Rev. Pharmacol. Toxicol.*, *20*:81 (1980).

211. A. S. Dutta and J. S. Morley, in *Peptides 1976* (A. Loffet, ed.), Editions de l'Université de Bruxelles, Brussels (1976), p. 517.

212. A. S. Dutta, J. J. Gormley, C. F. Hayward, J. S. Morley, J. S. Shaw, G. J. Stacey, and M. T. Turnbull, *Life Sci.*, *21*:559 (1977).

213. A. S. Dutta, B. J. A. Furr, and M. B. Giley, *J. Chem. Soc. Perkin Trans. 1*, 379 (1979).

214. D. C. Roberts and F. Vellaccio, in *The Peptides*, Vol. 5 (E. Gross and J. Meienhofer, eds.), Academic Press, New York (1983), pp. 341–449.

215. S. Hunt, in *Chemistry and Biochemistry of the Amino Acids* (G. C. Barrett, ed.), Chapman & Hall, London (1985), pp. 55–138.

216. K.-F. Fok and J. A. Yankeelov, *Biochem. Biophys. Res. Commun.*, *74*:273 (1977).

217. J. A. Yankeelov, K.-F. Fok, and D. J. Carothers, *J. Org. Chem.*, *43*:1623 (1978).

218. A. Spatola, in *Chemistry and Biochemistry of Amino Acids, Peptides, and Proteins* (B. Weinstein, ed.), Marcel Dekker, New York (1983), p. 275.

219. A. Spatola, in *Chemistry and Biochemistry of Amino Acids, Peptides, and Proteins* (B. Weinstein, ed.), Marcel Dekker, New York (1983), pp. 267–357.

220. IUPAC–IUB Joint Commission on Biochemical Nomenclature (JCBN), "Nomenclature and Symbolism for Amino Acids and Peptides, Recommendation 1983," *Eur. J. Biochem.*, *138*:9 (1984).

221. J. M. Samanen, *Polym. Sci. Technol.*, *32*:227 (1985).

222. C. H. Hassall, T. G. Martin, J. A. Schofield, and J. O. Thomas, *J. Chem. Soc. C*, 997 (1967).

223. M. Winitz, *J. Am. Chem. Soc.*, *78*:2423 (1956).

224. G. Losse and G. Bachmann, *Chem. Ber.*, *97*:2671 (1964).

225. E. P. Semkin, A. P. Smirnova, and L. A. Shchukina, *Zh. Obshch. Khim.*, *38*:2358 (1969).

226. G. A. Ravdel, M. P. Filatova, L. A. Shchukina, T. S. Paskhina, M. S. Surovikina, S. S. Trapeznikova, and T. P. Egorova, *J. Med. Chem.*, *10*:242 (1967).

227. P. D. Gesellchen, R. C. A. Frederickson, S. Tafur, and D. Smiley, in *Peptides: Synthesis, Structure, Function* (D. H. Rich and E. Gross, eds.), Pierce Chemical Company, Rockford, Ill. (1981), pp. 621–624.

228. L. Kisfaludy, O. Nyeki, E. Karpati, L. Szporny, K. Sz. Szalay, and G. B. Makara, in *Peptides: Synthesis, Structure, Function* (D. H. Rich and E. Gross, eds.), Pierce Chemical Company, Rockford, Ill. (1981), pp. 225–228.

229. J. S. Morley, *Proc. Roy. Soc. London*, *170*:97 (1968).

230. B. J. McRae, K. Kurachi, R. L. Heimark, K. Fujikawa, E. W. Davie, and J. C. Powers, *Biochemistry*, *20*:7196 (1981).

231. E. W. Petrillo and M. A. Ondetti, *Med. Res. Rev.*, *2*:1 (1982).

232. A. Ewenson, R. Laufer, M. Chorev, Z. Selinger, and C. Gilon, *J. Med. Chem.*, *29*:295 (1986).

233. A. Ervenson, R. Cohen-Suissa, D. Levian-Teitelbaum, Z. Selinger, M. Chorev, and C. Gilon, *Int. J. Pept. Protein Res.*, *31*:269 (1988).

234. T. F. M. LaCour, H. A. Hansen, K. Clausen, and S.-O. Lawesson, *Int. J. Pept. Protein Res.*, *22*:509 (1983).

235. K. Clausen, M. Thorsen, S.-O. Lawesson, and H. Fritz, *Tetrahedron*, *37*:3635 (1981).

236. M. Thorsen, T. P. Andersen, U. Pedersen, B. Yde, S.-O. Lawesson, and H. F. Hansen, *Tetrahedron*, *41*:5633 (1985).

237. M. Kajtar, M. Hollosi, J. Kajtar, and Z. Majer, *Tetrahedron*, *42*:3931 (1986).

238. O. Jensen and A. Senning, *Tetrahedron*, *42*:6555 (1986).

239. Y. Sasaki and D. H. Coy, *Peptides*, *8*:119 (1987).

240. D. H. Coy, S. J. Hocart, and Y. Sasaki, *Tetrahedron*, *44*:835 (1988).

241. P. Vander Elst, M. Elseviers, E. De Cock, M. Van Marsenille, D. Tourwe, and G. Van Binst, *Int. J. Pept. Protein Res.*, *27*:633 (1986).

242. J. Martinez, J.-P. Bali, M. Rodriguez, B. Castro, R. Magous, J. Laur, and M.-F. Lignon, *J. Med. Chem.*, *28*:1874 (1985).

243. D. H. Coy, P. Heinz-Erian, N.-Y. Jiang, Y. Sasaki, J. Taylor, J.-P. Moreau, W. T. Wolfrey, J. D. Gardner, and R. T. Jensen, *J. Biol. Chem.*, *263*:5056 (1988).

244. L. El Masdouri, A. Aubry, C. Sakarellos, E. J. Ganez, M. T. Cung, and M. Marraud, *Int. J. Pept. Protein Res.*, *31*:420 (1988).

245. A. F. Spatola, A. L. Bettag, K.-F. Fok, H. Saneii, and J. A. Yankeelov, in *Peptides: Structure and Biological Function* (E. Gross and J. Meienhofer, eds.), Pierce Chemical Company, Rockford, Ill. (1979), p. 273.

246. A. F. Spatola, H. Saneii, J. V. Edwards, A. L. Bettag, M. K. Anwer, P. Rowell, B. Browne, R. Lahti, and P. Von Voigtlander, *Life Sci.*, *38*:1243 (1986).

247. J. S. Shaw, L. Miller, M. J. Turnbull, J. J. Gormley, and J. S. Morely, *Life Sci.*, *31*:1259 (1982).

248. A. F. Spatola and K. Darlak, *Tetrahedron*, *44*:821 (1988).

249. A. F. Spatola, A. L. Bettag, N. S. Agarwal, H. Saneii, W. W. Vale, and C. Y. Bowers, in *LH—RH Peptides as Female and Male Contraceptives* (G. I. Zatuchni, J. D. Shelton, and J. J. Sciarra, eds.), Harper & Row, New York (1981), p. 24.

250. A. L. Bettag and A. F. Spatola, *1st Int. Workshop HPLC Biochem.*, Washington, D.C. (Nov. 17, 1981).

251. E. Rubini, U. Wormser, D. Levian-Teitelbaum, R. Laufer, C. Gilon, Z. Selinger, and M. Chorev, in *Peptides: Structure and Function* (C. M. Deber, V. J. Hruby, and K. D. Kopple, eds.), Pierce Chemical Company, Rockford, Ill. (1985), pp. 635—638.

252. E. D. Nicolaides, F. J. Tinney, J. S. Kaltenbronn, J. T. Repine, D. A. DeJohn, E. A. Lunney, W. H. Roark, J. G. Marriot, R. E. Davis, and R. E. Voigtman, *J. Med. Chem.*, *29*:959 (1986).

253. D. Hudson, R. Sharpe, and M. Szelke, *Int. J. Pept. Protein Res.*, *15*:122 (1980).

254. M. M. Hann, P. G. Sammes, P. D. Kennewell, and J. B. Taylor, *J. Chem. Soc. Chem. Commun.*, *234* (1980).

255. N. J. Miles, P. G. Sammes, P. D. Kennewell, and R. Westwood, *J. Chem. Soc. Perkin Trans. 1*, 2299 (1985).

256. A. Spaltenstein, P. A. Carpina, F. Miyake, and P. B. Hopkins, *Tetrahedron Lett.*, *27*:2095 (1986).

257. R. J. Abraham, S. L. R. Ellison, P. Schonoholzer, and W. A. Thomas, *Tetrahedron*, *42*:2101 (1986).

258. G. J. Hanson and T. Lindberg, *J. Org. Chem.*, *50*:5399 (1985).

259. M. Van Marsenille, C. Geysen, D. Tourwe, and G. Van Binst, *Bull. Soc. Chim. Belg.*, *95*:127 (1986).

260. M. M. Shemyakin, Y. A. Ovchinnikov, and V. T. Ivanov, *Angew. Chem. Int. Ed. Engl.*, *8*:492 (1969).

261. R. M. Freidinger and D. F. Veber, *J. Am. Chem. Soc.*, *101*:6129 (1979).

262. J. M. Berman, N. Jenkins, M. Hassan, M. Goodman, T. M.-D. Nguyen, and P. W. Schiller, in *Peptides: Structure and Function* (V. J. Hruby and D. H. Rich, eds.), Pierce Chemical Company, Rockford, Ill. (1983), pp. 283–286.

263. M. Goodman and M. Chorev, *Trends Pharm. Sci.*, *463* (1980).

264. E. Rubini, M. Chorev, C. Gilon, Z. Y. Friedman, U. Wormser, and Z. Selinger, in *Peptides: Synthesis, Structure, Function* (D. H. Rich and E. Gross, eds.), Pierce Chemical Company, Rockford, Ill. (1981), p. 593.

265. M. Goodman and M. Chorev, *Acc. Chem. Res.*, *12*:1 (1979).

266. M. Goodman and M. Chorev, in *Perspectives in Peptide Chemistry* (A. Eberle, R. Geiger, and T. Wieland, eds.), S. Karger Publishers, New York (1981), pp. 283–294.

267. A. Spatola, in *Chemistry and Biochemistry of Amino Acids, Peptides, and Proteins* (B. Weinstein, ed.), Marcel Dekker, New York (1983), p. 329.

268. R. M. Freidinger, D. F. Veber, D. S. Perlow, J. R. Brooks, and R. Saperstein, *Science*, *210*:656 (1980).

269. M. Kahn and B. Devens, *Tetrahedron Lett.*, *27*:4841 (1986).

270. M. Kahn and T. Su, in *Peptides: Chemistry and Biology* (G. R. Marshall, ed.), ESCOM Science Publishers B.V., Leiden, The Netherlands (1988), pp. 109–111.

271. W. F. Huffman, J. F. Callahan, D. S. Eggleston, K. A. Newlander, D. T. Takata, E. E. Codd, R. F. Walker, P. W. Schiller, C. Lemieux, W. S. Wire, and T. F. Burks, in *Peptides: Chemistry and Biology* (G. R. Marshall, ed.), ESCOM Science Publishers B.V., Leiden, The Netherlands (1988), pp. 105–108.

272. K. Wuthrich, *NMR of Proteins and Nucleic Acids*, John Wiley & Sons, New York (1986).

273. A. D. Kline, W. Braun, and K. Wuthrich, *J. Mol. Biol.*, *189*:377 (1986).

274. J. W. Pflugrath, E. Wiegand, R. Huber, and L. Vertesy, *J. Mol. Biol.*, *189*:383 (1986).

275. A. M. Groneborn, G. M. Clore, U. Schmeissner, and P. Wingdfield, *Eur. J. Biochem.*, *161*:37 (1986).

276. T. Ohkubo, Y. Kobayaski, Y. Shimonishi, Y. Kyogokui, W. Braun, and N. Go, *Biopolymers*, *25*:S123 (1986).

277. R. E. Bruccole and M. Karplus, *Biopolymers*, *26*:137 (1987).

278. V. J. Hruby, L. F. Kao, B. M. Pettitt, and M. Karplus, *J. Am. Chem. Soc.*, *110*:3351 (1988).

279. F. E. Cohen, P. A. Kosen, I. D. Kuntz, L. B. Epstein, T. L. Ciardelli, and K. A. Smith, *Science*, *234*:349 (1986).

280. B. H. Frank and R. E. Chance, *Muench. Med. Wochenschr.*, *125* (*Suppl. 1*):14 (1983).

281. D. V. Goeddel, H. L. Heyneker, T. Itozumi, R. Arentzen, K. Itakura, D. G. Yansura, M. J. Ross, G. Miozzari, R. Crea, and P. H. Seeburg, *Nature*, *281*:544 (1979).

282. S. Nagata, H. Taira, A. Hall, L. Johnsrud, M. Streuli, J. Ecsodi, W. Boll, K. Cantell, and C. Weissmann, *Nature*, *284*:316 (1980).

283. T. Taniguchi, Y. Fujii-Kuriyama, and M. Muramatsu, *Proc. Natl. Acad. Sci. USA*, 77:4003 (1980).

284. P. Valenzuela, A. Medina, W. J. Rutter, G. Ammerer, and B. D. Hall, *Nature*, *298*:347 (1982).

285. *Stony Brook Symp. Mol. Biol.*, SUNY—Stony Brook, Stony Brook, N.Y. (May 20—25, 1985).

286. *UCLA Symp. Mol. Cell. Biol.*, Keystone, Colo. (Mar. 9—Apr. 4, 1985), *J. Cell. Biochem.*, *Suppl 98*:89—145 (1985).

287. *1st Int. Conf. Protein Eng.*, London (Nov. 21—22, 1985).

288. C. Dempster, *Advances in Enzyme Technology: Artificial, Synthetic, and Designed Enzymes*, Emerging Technology Series, Report No. 12, Technical Insights, Inc., Englewood, N.J. (1984).

289. H. F. Judson, *The Eighth Day of Creation*, Simon and Schuster, New York (1979).

290. W. Gilbert and L. Villa-Komaroff, *Sci. Am.*, 74 (Apr. 1980).

291. T. Maniatis, E. F. Fritsch, and J. Sambrook, *Molecular Cloning: A Laboratory Manual*, Cold Spring Harbor Laboratory, Cold Spring Harbor, N.Y. (1982).

292. R. W. Old and S. B. Primrose, in *Principles of Gene Manipulation*, 3rd ed., *Studies in Microbiology*, Vol. 2 (N. G. Carr, J. L. Ingraham, and S. C. Rittenberg, eds.), Blackwell Scientific Publications, Oxford (1985).

293. L. Ferretti, S. S. Karnik, H. G. Khorana, M. Nassal, and D. D. Oprian, *Proc. Natl. Acad. Sci. USA*, *83*:599 (1986).

294. M. J. Casadaban and S. N. Cohen, *J. Mol. Biol.*, *138*:179 (1980).

295. R. A. Young and R. W. Davis, *Proc. Natl. Acad. Sci. USA*, *80*: 1194 (1983).

296. E. S. Mocarski, L. Pereira, and N. Michael, *Proc. Natl. Acad. Sci. USA*, *82*:1266 (1985).

297. B. Royer-Pokora, L. M. Kunkel, A. P. Monaco, S. C. Goff, P. E. Newburger, R. L. Baehner, F. S. Cole, J. T. Curnutte, and S. H. Orkin, *Nature*, *322*:32 (1986).

298. J. H. Gerlach, J. A. Endicott, P. F. Juranka, G. Henderson, F. Sarangi, K. L. Deuchars, and V. Ling, *Nature*, *324*:485 (1986).

299. B. E. Schoner, R. M. Belagaje, and R. G. Schoner, *Methods Enzymol.*, *153*:401 (1987).
300. D. Pennica, W. E. Holmes, W. J. Kohr, R. N. Harkins, G. A. Vehar, C. A. Ward, W. F. Bennett, E. Yelverton, P. H. Seeburg, H. L. Heyneker, and D. V. Goeddel, *Nature*, *301*:214 (1983).
301. A. D. Levinson, C. C. Simonsen, and E. M. Yelverton, European Patent Application 84,300,299.9 (1984).
302. T. Taniguchi, H. Matsui, T. Fujita, C. Takaoka, N. Kashima, R. Yoshimoto, and J. Hamuro, *Nature*, *302*:305 (1983).
303. M. C. Smith, T. C. Furman, T. D. Ingòlia, and C. Pidgeon, *J. Biol. Chem.*, *263*:7211 (1988).
304. T. Moks, L. Abrahmsen, E. Holmgren, M. Bilich, A. Olsson, M. Uhlen, G. Pohl, C. Sterky, H. Hultberg, S. Josephson, A. Holmgren, H. Jornvall, and B. Nilsson, *Biochemistry*, *26*:5239 (1987).
305. Z. E. Dziewanowska, L. L. Bernhardt, and S. Fein, in *Recombinant DNA Products: Insulin, Interferon and Growth Hormone* (A. P. Bollon, ed.), CRC Press, Boca Raton, Fla. (1984), pp. 115–128.
306. TIMI Study Group, *N. Engl. J. Med.*, *312*:932 (1985).
307. D. Collen, *Thromb. Haemost.*, *43*:77 (1980).
308. R. M. Myers, K. Tilly, and T. Maniatis, *Science*, *232*:613 (1986).
309. E. E. Howell, J. E. Villafranca, M. S. Warren, S. J. Oatley, and J. Kraut, *Science*, *231*:1123 (1986).
310. J. Markussen, I. Diers, A. Engesgaard, M. T. Hansen, P. Hougaard, L. Langkjaer, K. Norris, U. Ribel, A. R. Sørenson, E. Sørenson, and H. O. Voigt, *Protein Eng.*, *1*:215 (1987).
311. J. Brange, U. Ribel, J. F. Hansen, G. Dodson, M. T. Hansen, S. Havelund, S. G. Melberg, F. Norris, K. Norris, L. Snel, A. R. Sørensen, and H. O. Voigt, *Nature*, *333*:679 (1988).
312. T. J. Ahern, J. I. Casal, G. A. Petsko, and A. M. Klibanov, *Proc. Natl. Acad. Sci. USA*, *84*:675 (1987).
313. D. Lau, G. Kuzma, C.-M. Wei, D. J. Livingston, and N. Hsiung, *Biotechnology*, *5*:953 (1987).
314. G. A. Vehar, W. J. Kohr, W. F. Bennett, D. Pennica, C. A. Ward, R. N. Harkins, and D. Collen, *Biotechnology*, *2*:1051 (1984).
315. A. J. van Zonneveld, H. Veerman, and H. Pannekoek, *Proc. Natl. Acad. Sci. USA*, *83*:4670 (1986).
316. T. E. Petersen, H. C. Thøgersen, K. Skorstengaard, K. Vibe-Pedersen, P. Sahl, L. Sottrup-Jensen, and S. Magnusson, *Proc. Natl. Acad. Sci. USA*, *80*:137 (1983).
317. H. J. Ehrlich, N. U. Bang, S. P. Little, S. R. Jaskunas, B. J. Weigel, L. E. Mattler, and C. S. Harms, *Fibrinolysis*, *1*:75 (1987).
318. V. Jackson, T. Craft, J. Sundboom, J. Frank, B. Grinnell, L. Bobbitt, J. Quay, and G. Smith, *FASEB J.*, *2*:A1412 (1988).
319. M. S. Neuberger, G. T. Williams, and R. O. Fox, *Nature*, *312*:604 (1984).
320. L. Riechmann, M. Clark, H. Waldmann, and G. Winter, *Nature*, *322*:323 (1988).
321. M. T. Mas, C. Y. Chen, R. A. Hitzeman, and A. D. Riggs, *Science*, *233*:788 (1986).

322. J. R. Haynes, J. Cunningham, A. Von Seefried, M. Lennick, R. T. Garvin, and S. Shen, *Biotechnology*, *4*:637 (1986).

323. M. Matsumura, S. Yasumura, and S. Aiba, *Nature*, *323*:356 (1986).

324. S. J. Weiner, P. A. Kollman, D. A. Case, U. C. Singh, C. Ghio, G. Alagona, S. Profeta, Jr., and P. Weiner, *J. Am. Chem. Soc.*, *106*:765 (1984).

325. J. Van Brunt, *Biotechnology*, *4*:277 (1986).

326. G. Kolata, *Science*, *233*:1037 (1986).

327. W. F. van Gunsterm, *Protein Eng.*, *2*:5 (1988).

328. M. Mutter, K.-H. Altmann, G. Tuchscherer, and S. Vuilleumier, *Tetrahedron*, *44*:771 (1988).

329. T. L. Blundell, B. L. Sibanda, M. J. E. Sternberg, and J. M. Thorton, *Nature*, *326*:347 (1987).

330. J. S. Fetrow, M. H. Zehfus, and G. D. Rose, *Biotechnology*, *6*:167 (1988).

331. J. D. Sedgwick and P. G. Holt, *Immunology*, *56*:635 (1985).

332. G. Gammon, K. Dunn, N. Shastri, A. Oki, S. Wilbur, and E. E. Sercarz, *Nature*, *319*:413 (1986).

3

Physical Biochemistry of Peptide Drugs: Structure, Properties, and Stabilities of Peptides Compared with Proteins

James M. Samanen

SmithKline Beecham Pharmaceuticals, King of Prussia, Pennsylvania

I. INTRODUCTION

Increasing numbers of pharmaceutical scientists are gaining interest in the therapeutic applications of peptides [1]. They must acquaint themselves with the special structures of peptides, and with the physical and chemical properties of such structures, to better understand the stabilities of such molecules. It is difficult to discuss the properties of peptides without mentioning the properties of proteins, since peptides and proteins both contain chains of amino acids. The main distinction between peptides and proteins is one of size: peptides contain fewer than 100 amino acids (this is a rough demarcation point). Accordingly, in this chapter we contrast the structure, properties, and stability of peptides with proteins. The discussion should show that these properties are often relatable to the amino acid building blocks that make up the structures of the peptides and proteins. One should not enter this chapter with the expectation that quantitative algorithms will be described that relate peptide or protein chemical composition to physical properties, as these are not available. As the demand for such information increases, perhaps the development of such algorithms will be encouraged. To maintain a concise presentation, many of the concepts presented here will not be covered in the depth that some readers may wish; these readers are referred to the references provided.

II. PEPTIDE AND PROTEIN STRUCTURE

A. Elements of Peptide and Protein Structure

All peptides and proteins are composed of *peptide chains*. A peptide chain (Fig. 1) of a mammalian peptide or protein consists of a polyamide *backbone* containing tetrahedral carbon atoms between amide groups, off of

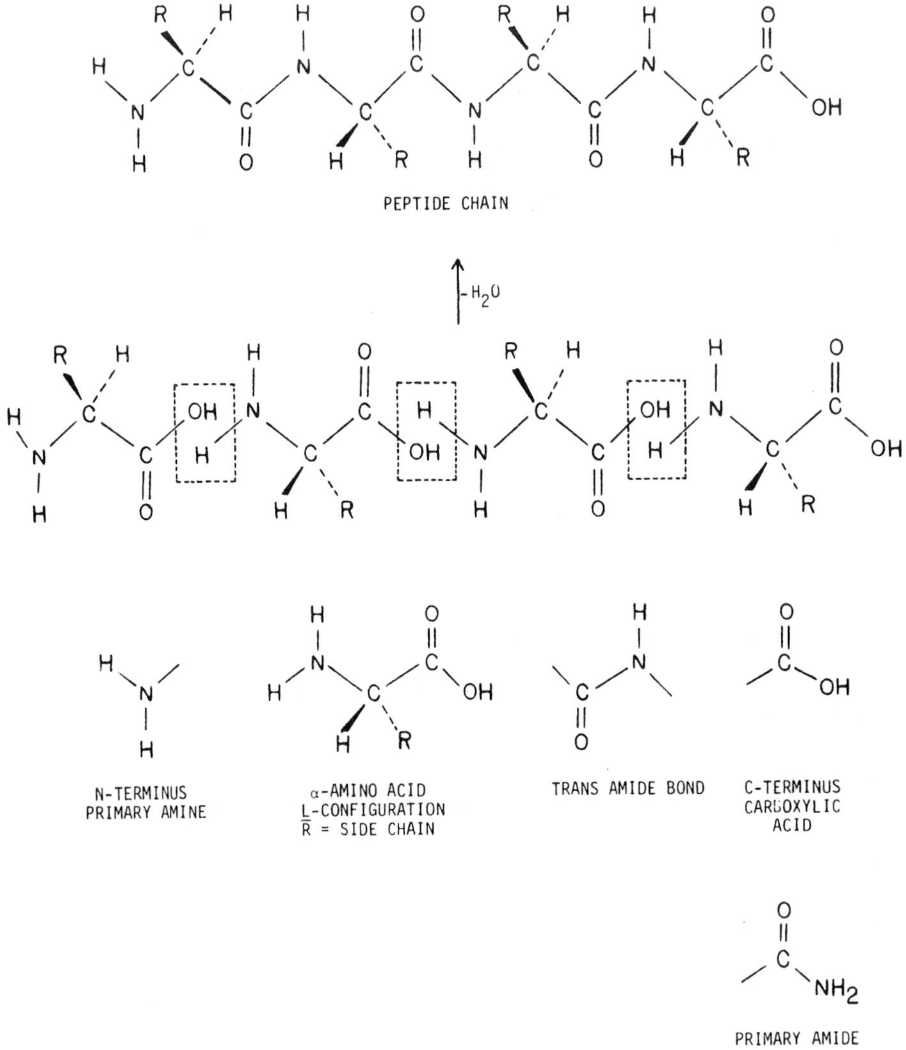

Fig. 1   Structural elements of a peptide chain.

which branch hydrocarbon chains, R, called *side chains*.* The peptide
chain arises from condensations of amino acids. At one end of the chain
is a free amine group, the *N-terminus*. The C-terminus begins with a
free carboxylic acid group, or less frequently with a primary amide. The
internal amide groups are usually *trans* isomers. [†]

The side chains along the backbone come from the amino acids that
make up the chain. The backbone tetrahedral carbon atoms to which the
side chains are connected are "asymmetric" and of the *L absolute configuration* [4]. This chirality is important in determining the *conformation* or
three-dimensional shape of the backbone. The chemical stability of a peptide chain is dependent on the chemistry and conformation of the peptide
backbone and also on the chirality and chemistry of the side chains along
the backbone.

There are 22 amino acids, each with a different side chain in mammalian
peptides and proteins (Fig. 2). [‡] The side chains can either be chemically
reactive (Arg, Asn, Asp, Cys, Glu, Gln, His, Lys, Met, Ser, Thr, Trp,
and Tyr), or inert (Ala, Gly, Ile, Leu, Phe, Pro, and Val). Some are
hydrophobic [Ala, Cys, (Gly), His, Ile, Leu, Met, Phe, Pro, Trp, Tyr,
and Val], while others are hydrophilic [Arg, Asn, Asp, Glu, Gln, Lys,
Ser, and (Thr)]. Side chains can bear ionic charges (Arg, Asp, Glu,
and Lys) and aromatic structures (His, Phe, Trp, and Tyr). Some side
chains have a greater influence on conformation than do others (Cys, Gly,
Ile, Pro, and Val). Due to the abundance of side-chain structures, a
tremendous variety of chemical structures can be constructed in peptide
chains.

Many of these side chains are covalently linked through their side
chains to other groups: sugar residues, phosphate and sulfate groups,
nucleic acids, lipids, metalloporphyrins, flavin, and metals. Each of
these ligands bears its own unique chemical and physical properties which
are important to biological activity and chemical stability.

Peptide chains that bear both positive and negative charges on the
same molecule are called *amphoteric* molecules. A significant property of
amphoteric molecules is the *isoelectric pH* [6], which is defined as the pH
of an aqueous solution at which an amino acid, peptide, or protein bears
a net charge of zero. At this *isoelectric point* the molecule does not
migrate in an electric field (e.g., during electrophoresis) because the
number of cationic groups (positive charges) equals the number of anionic
groups (negative charges). At the isoelectric pH, or pI, amino acids,
peptides, and proteins often become insoluble in aqueous solutions and
precipitate from solution. The isoelectric pH is a useful number to know
if solid amino acids, peptides, or proteins need to be obtained by precipitation. It can readily be calculated simply by adding up the pK values of
the ionizable groups [7] in the molecule and dividing by the number n of
ionizable groups:

$$pI = \frac{pK_1 + pK_2 + \cdots + pK_n}{n} \tag{1}$$

---

*Nonmammalian peptide chains can also include ester linkages and unusual
amino acids as described in Ref. 2.
[†]Small cycle peptides can adapt *cis*-amide bonds; see Ref. 3.
[‡]Peptides in other species contain a bewildering variety of amino acid
structures. These structures are cataloged in Ref. 5.

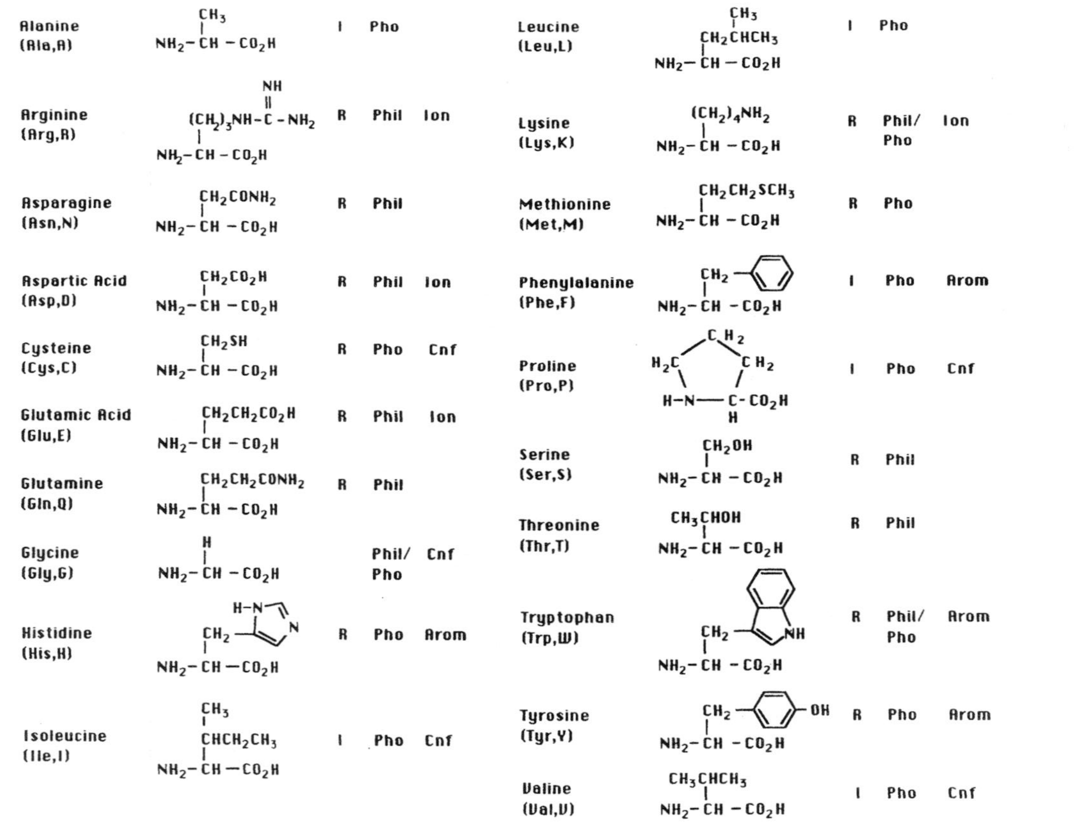

Amino acid	Side chain	Class
Alanine (Ala,A)	CH₃—CH—CO₂H (NH₂)	I Pho
Arginine (Arg,R)	(CH₂)₃NH—C(=NH)—NH₂	R Phil Ion
Asparagine (Asn,N)	CH₂CONH₂	R Phil
Aspartic Acid (Asp,D)	CH₂CO₂H	R Phil Ion
Cysteine (Cys,C)	CH₂SH	R Pho Cnf
Glutamic Acid (Glu,E)	CH₂CH₂CO₂H	R Phil Ion
Glutamine (Gln,Q)	CH₂CH₂CONH₂	R Phil
Glycine (Gly,G)	H	R Phil/Pho Cnf
Histidine (His,H)	CH₂—imidazole	R Pho Arom
Isoleucine (Ile,I)	CHCH₂CH₃ (CH₃)	I Pho Cnf
Leucine (Leu,L)	CH₂CHCH₃ (CH₃)	I Pho
Lysine (Lys,K)	(CH₂)₄NH₂	R Phil/Pho Ion
Methionine (Met,M)	CH₂CH₂SCH₃	R Pho
Phenylalanine (Phe,F)	CH₂—phenyl	I Pho Arom
Proline (Pro,P)	ring	I Pho Cnf
Serine (Ser,S)	CH₂OH	R Phil
Threonine (Thr,T)	CH₃CHOH	R Phil
Tryptophan (Trp,W)	CH₂—indole	R Phil/Pho Arom
Tyrosine (Tyr,Y)	CH₂—phenol(OH)	R Pho Arom
Valine (Val,V)	CH₃CHCH₃	I Pho Cnf

I = inert, R = reactive, Pho = hydrophobic, Phil = hydrophilic, Arom = aromatic, Ion = ionic, Cnf = conformationally significant

Fig. 2   Side-chain structures of the mammalian amino acids.

The manner in which structural variation is built into a peptide or protein is a function of the peptide chain *sequence*, which is the unique ordering of the amino acids in a peptide chain. This unique ordering of amino acid residues in a peptide chain is determined during synthesis. The ordering is determined by genetic code in the biosynthesis of peptide chains, or by the order in which the peptide chemist links amino acids together in peptide-bond-forming reactions in the chemical synthesis of peptide chains.

The uniqueness of sequences is analogous to the uniqueness of words (Fig. 3). One can take the four letters H, S, A, and C and string them together into the sequence C-A-S-H to create the word "CASH," which bears a significant and important meaning to people. The synthetic chemist can take five amino acids: one tyrosine, one phenylalanine, one methionine, and two glycines, and string them together synthetically in the sequence (reading from the amine terminus) Tyr-Gly-Gly-Phe-Met, creating the peptide "enkephalin," a neurotransmitter in the brain that causes analgesia. There are over 5 million possible unique pentapeptide sequences that can be made from 22 amino acids.

Through the construction of sequences of diverse arrangements [1] and lengths, mammals make a variety of peptides and proteins to serve a multitude of functions: hormone, neurotransmitter, neuromodulator, enzyme inhibitor, and structural proteins (Table 1). Structural proteins [8] will not be discussed further; rather, we will focus on peptides (and proteins) that can be employed therapeutically. Notice in Table 1 that peptides and proteins share many functions in the body. As stated earlier, the main distinction between peptides and proteins is one of size: peptides contain fewer than 100 amino acids.

The physiological and physical properties of peptides and proteins arise from their acid-base properties, side-chain chemistry, and conformational properties.

B. Elements of Peptide and Protein Conformation

The peptide chains in peptides or proteins are rarely linear, but adopt a variety of folded patterns or conformations. Conformation affects virtually

COMPONENT POOL:	26 LETTERS	22 AMINO ACIDS
COMPOSITION:	H = 1	TYR = 1
	S = 1	PHE = 1
	A = 1	MET = 1
	C = 1	GLY = 2
SEQUENCE:	C-A-S-H	TYR-GLY-GLY-PHE-MET
WHOLE ITEM:	CASH	ENKEPHALIN (A PEPTIDE)

Fig. 3   Analogy between sequences and words.

Table 1   Physiological Functions of Amino Acids, Peptides, and Proteins

Function	Amino acids	Peptides	Proteins
Hormone	+	+	+
Neurotransmitter	+	+	+?
Neuromodulator	+	+	+?
Enzyme	−	− ?	+
Enzyme inhibitor	+?	+	+
Structural unit	(+)	− ?	+
Molecular weight	75 to 250	132 to 12,000	12,000+

all aspects of the chemistry and biology of peptides and proteins, including (1) physical properties, such as solubility and intrinsic viscosity, and spectral properties, such as circular dichroism (CD) [9]; (2) chemical properties, since folding may sterically shield reactive groups from reagents or stabilize them by hydrogen bonding [10]; (3) biological properties, since the three-dimensional structure places catalytic groups into proper orientation for enzymatic activity or places backbone and side-chain groups into proper orientation for hormone-receptor interaction [1]; and (4) stability to enzymatic cleavage. Amide groups susceptible to proteolysis may be sterically shielded in a folded peptide chain [1].

Peptide chain conformation is determined by the covalently bonded amino acid sequence, by disulfide bridges (Fig. 4) between cysteine residues (creating cyclic structures), and by total conformational energy (Fig. 5). *Total conformational energy* is the sum of all energies of interactions: electrostatic energy, nonbonded energy, hydrogen-bonded energy, and torsional energy. Computer programs have been devised [11] for semiempirical calculation of peptide total conformational energy. Let us consider each component of total conformational energy:

Electrostatic Energy

Electrostatic energy is the repulsive or attractive energy arising from ionic interactions [12]. It is inversely proportional to the dielectric constant of the solvent medium:

$$E = \frac{332}{\varepsilon}\left(\frac{q_1 \cdot q_2}{R_{1-2}}\right) \tag{2}$$

where E is the repulsive or attractive energy (kcal/mol), ε the dielectric constant (e.g., 78.5 for water, 24.3 for ethyl alcohol, and 4.34 for diethyl ether), $q_{1,2}$ the partial charge, and R_{1-2} the distance between partial charges (Å). Ionic interactions may arise between cationic groups (the N-terminal amine, the arginine guanidine group, the lysine amine group, and below pH 7 the histidine imidazole group) and anionic groups

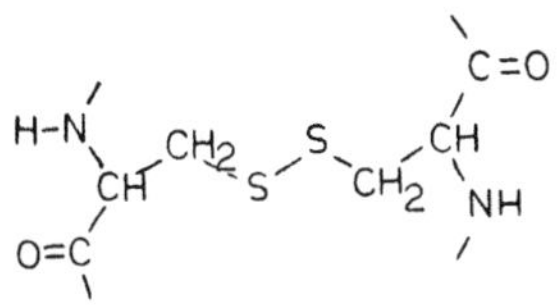

Fig. 4   Disulfide bridge formed between cysteine residues.

(the C-terminal carboxyl groups and the aspartic and glutamic acid carboxyl groups).

In water, which has a high dielectric constant, electrostatic energy is
much smaller than in a nonpolar environment with a low dielectric constant,
such as ethyl ether or the lipid environment of a membrane. In water,
ionic groups remain separated since they are surrounded by solvating water
molecules; consequently, a conformation in water may be quite different in
lipid. Ionic contributions to total energy can be quite large, on the order
of 5 kcal/mol per ion pair.

Hydrophobic Forces

In water, hydrophobic "forces" also contribute to conformational energy.
Hydrophobic forces are the energy gained in removing hydrophobic groups
from water [13]. This energy provides the driving force for the hydrophobic clustering that may be seen between the lipophilic side chains of
many amino acids (Ala, Cyc-Cys, His, Ile, Leu, Met, Phe, Pro, Trp, Tyr,
Val; see Fig. 5). Table 2 lists the additional free energy of transfer
from water to an organic solvent, μ°org-μ°w, that is generated when a
given side chain is substituted for the hydrogen atom of a glycine residue.

Hydrophobic interactions contribute greatly to conformational energy in
water. This orientation of hydrophobic side chains away from water con-
trasts with the tendency for polar side chains to stick out into water. In
a more lipophilic medium, such as ethanol or acetic acid, the hydrophobic
forces are diminished. A peptide chain will no longer prefer to fold in a
way that clusters and buries lipophilic side chains or points ionic groups
out into solvent. As seen in Table 2, the hydrophobic side chains of
amino acids can contribute as much as 3.4 kcal/mol in conformational
energy.

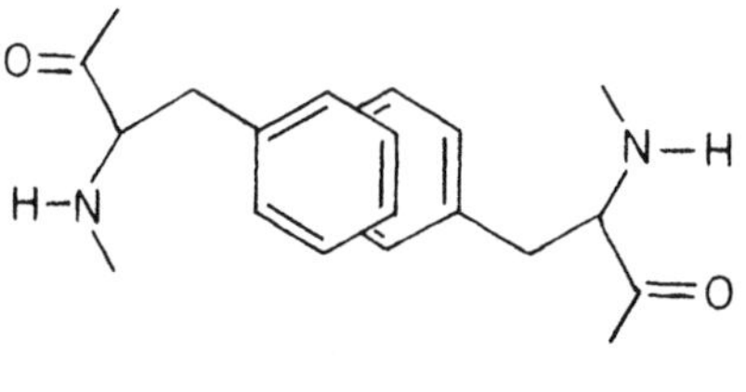

Fig. 5   Hydrophobic interaction between two phenylalanine residues.

Table 2   Hydrophobicity of Protein
Side Chains

Side chain	$\mu°$org-$\mu°$w[a] (kcal/mol)
Tryptophan	−3.4
Norleucine	−2.6
Phenylalanine	−2.5
Tyrosine	−2.3
Leucine	−1.8
Valine	−1.5
Methionine	−1.3
Alanine	−0.5

[a]The additional free energy of transfer
from water to an organic solvent (e.g.,
ethanol) that is generated when the side
chain listed is substituted for the hydrogen atom of a glycyl residue.

Hydrogen Bonds

A final, very important type of interaction, is hydrogen bonding (Fig. 6)
[14]. Hydrogen bonds are intermediate in length between van der Waals
contacts and true covalent bonds. A hydrogen bond requires a donor
atom such as an amine and an acceptor atom such as a carbonyl oxygen.
Each hydrogen bond contributes on the order of 3 kcal to the total conformational energy. As seen in Fig. 6, peptide chains have many opportunities to form hydrogen bonds: between backbone amide groups,
between the many oxygen- and nitrogen-bearing side chains, between side
chains and terminal groups, and between side chains and backbone amide
groups.

The large number of amino acids in proteins lead to large numbers of
hydrogen bonds and nonpolar contacts within the molecules (Table 3) [15].
These interactions contribute significantly to the total conformational
energy. The smaller peptides, such as the cyclic peptide antamanide [16],
typically contain fewer interactions per molecule. Their conformations are
much less stable than those of proteins.

The ways in which the chains fold to bring interacting groups together
fall into four basic types of structure: α-helix, β-structure, reverse
turns, and random coil (Fig. 7). In the α-*helix* the peptide chain coils
in helical fashion with hydrogen bonds between the carbonyl of each
residue and the amide nitrogen of the fourth residue up the chain. Side
chains are all situated outside the helix cylinder. Many native helices are
amphipathic: the polar side chains are clustered on one face of the helix,
with nonpolar side chains clustered on an opposite face. β-*Structure* is a
fully extended, flattened-out portion of chain. β-Chains often link up
with other β-chains running in front and behind, stabilized by many

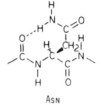

(a) Backbone to backbone

(b) Side chain to side chain (Asp, Asn, Arg, Gln, Glu, His, Lys, Orn, Ser, Thr, Tyr (not Met, or Cys-Cys)

(c) Side chain to terminal   (d) Side chain to backbone
      group (Asn, Thr, Ser)   (Asn, Thr, Ser)

Fig. 6   Types of hydrogen bonds in peptides and proteins. (From Ref. 13.)

hydrogen bonds. The resulting β-*sheet* places side chains above and below the sheet, allowing for hydrophobic interactions between β-sheets. Peptide chains often make sharp turns or *reverse turns*, allowing the chain to fold back on itself and stabilize the resulting structure with hydrogen bonds. Reverse turns are important in bringing β-chains together into sheets. *Random coil* is any segment of peptide chain with no defined structure.

An example of a protein structure is given by the x-ray crystal structure of the globular protein carboxypeptidase (Fig. 8) [17]. This enzyme has been drawn so that the β-chains are represented by arrows and α-helices as coiled ribbon. Carboxypeptidase contains a twisted β-sheet at its core, with helix coils at the edges along with some random coil.

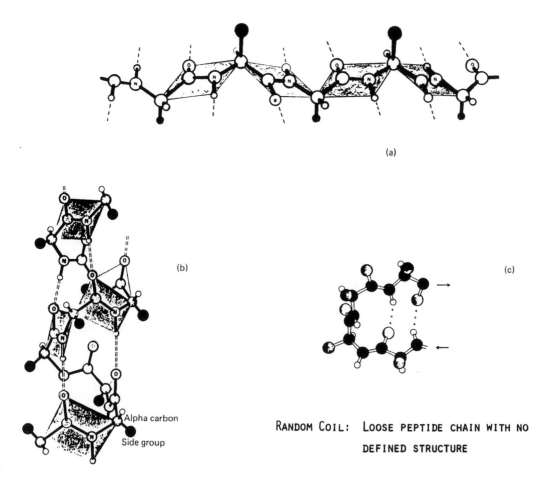

(a)

(b)

(c)

RANDOM COIL: LOOSE PEPTIDE CHAIN WITH NO
DEFINED STRUCTURE

Alpha carbon

Side group

Fig. 7   Types of peptide chain folding. (a) β-structure; (b) α-helix;
(c) reverse turn.

Table 3 Number of Hydrogen Bonds and Nonpolar Contacts in Peptides and Proteins[a]

Protein	MW	N_H	N_a	N_0	N'_0
Ribonuclease	13,600	81	6.0	90	6.6
Parvalbumin	11,500	71	6.2	71	6.2
Lysozyme, egg white	14,300	89	6.2	108	7.5
Lysozyme, T4	18,600	98	5.3	138	7.4
β-Trypsin	23,800	165	6.9	196	8.2
α-Chymotrypsin	25,200	173	6.9	238	9.4
Papain	23,400	139	5.9	204	8.7
Cytochrome c	12,400	70	5.6	136	11.0
Myoglobin	17,900	133	7.4	213	11.9
Antamanide	1,147	2	1.7	4	3.4

Source: Data from Refs. 15 and 16.
[a]MW, molecular weight; N_H, number of hydrogen bonds; N_a, N_H/MW ($\times 10^3$); N_0, number of nonpolar contacts; N'_0, N_0/MW ($\times 10^3$).

Fig. 8   X-ray crystal structure of carboxypeptidase.   [From Jane S.
Richardson, The Anatomy and Taxonomy of Proteins, *Adv. Protein
Chem.*, *34*:295 (1981).]

III. CONFORMATIONAL STABILITY

A. Protein Stability

Chain Folding and Denaturation

Physical studies [18] correlate well with the x-ray structure of carboxypeptidase displayed in Fig. 8.   This globular protein displays one basic
shape or conformation.   Nuclear magnetic resonance (NMR) studies of
proteins show that the various atomic groups in a protein vibrate and
rotate in solution [19], suggesting a number of topologically similar conformational substates.   Nevertheless, globular proteins tend to maintain
one general conformational structure.   Such studies of protein conformation show that although a protein is constantly "flexing" its structure,
overall topology is maintained by:

  1.   Various structural cross-links: the covalent cross-links include
       disulfide bridges and covalently bonded prosthetic groups such as

porphyrins; the noncovalent cross-links include hydrogen bonds, hydrophobic interactions, and interactions with metal ions and ligands.

2. Continuous refolding to the native conformations: there is abundant evidence that proteins fold spontaneously to their native conformations from random conformations in *water* [20].

If one alters the aqueous environment of a protein in aqueous solution by adding an organic solvent, or if one adds reagents that disrupt the noncovalent cross-links in a protein, the native conformation may no longer be the most stable conformation. Under such conditions a protein may undergo conformational change (e.g., it may unfold from a globular shape). As shown in Table 4 [9], the addition of guanidine hydrochloride to protein solutions causes a large increase in intrinsic viscosity. Reduction of disulfide bridges results in still greater intrinsic viscosities. The increase in intrinsic viscosity suggests that the globular proteins are unfolding to loose, expanded random coil peptide chains. Processes such as these are called *denaturation*. Denaturation encompasses any nonproteolytic modification of the unique structure of a native protein that gives rise to definite changes in chemical, physical, and biological properties. Denaturation can be observed by a number of techniques: intrinsic viscosity, optical rotation, optical rotatory dispersion, circular dichroism, ultraviolet difference spectroscopy, and bioassay [9]. Conditions that denature proteins include heat, pH change, added salts, and large surface areas. The environment at the surface of a solution in contact with an ionic glass surface, a hydrophobic polystyrene surface, or an immiscible hydrophobic solvent may stabilize nonnative conformations over native conformations, leading to an equilibrium shift to a denatured state [21]. Chemical agents that denature proteins include polar protic chemicals that are capable of disrupting hydrogen bonds, and surfactants, which involve both hydrophobic disruption and charged group separation, can denature proteins. Figure 9 displays the structures of some common denaturing agents.

Table 4   Increases in Protein Intrinsic Viscosity in Concentrated Guanidinium Hydrochloride (Gu HCl) Solution

Protein	In water	$[\eta]$ (cm^3/g) In Gu HCl solution Intact SS bonds	Broken SS bonds
Ribonuclease	3.3	9.4	16.3
Lysozyme	2.7	6.5	17.1
Serum albumin	3.7	22.9	52.2
Chymotrypsinogen	2.5	11.0	26.8
β-Lactoglobulin	3.4	19.1	22.8

Source: Ref. 21.

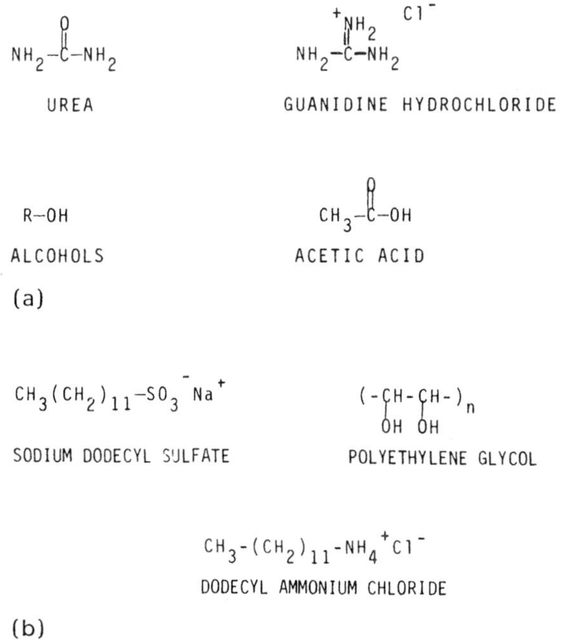

$$NH_2-\overset{\overset{O}{\|}}{C}-NH_2$$

UREA

$$NH_2-\overset{\overset{+NH_2}{\|}}{C}-NH_2 \quad Cl^-$$

GUANIDINE HYDROCHLORIDE

R—OH

ALCOHOLS

$$CH_3-\overset{\overset{O}{\|}}{C}-OH$$

ACETIC ACID

(a)

$$CH_3(CH_2)_{11}-SO_3^-\ Na^+$$

SODIUM DODECYL SULFATE

$$(-\overset{\overset{}{\underset{OH}{|}}}{CH}-\overset{\overset{}{\underset{OH}{|}}}{CH}-)_n$$

POLYETHYLENE GLYCOL

$$CH_3-(CH_2)_{11}-NH_4^+Cl^-$$

DODECYL AMMONIUM CHLORIDE

(b)

Fig. 9   Common denaturing agents.   (a) Polar and protic chemicals (capable of disrupting hydrogen bonds); (b) surfactants (involving both hydrophobic disruption and charge group separation).

Kinetic studies of protein renaturation have been of tremendous importance in illuminating the steps in peptide chain folding.   Proteins that have maintained their disulfide bridges in the denatured state return to their native state fairly rapidly (e.g., less than 1 sec for staphylococcal nuclease and 10 sec for metmyoglobin and ribonuclease) [22–24].   If disulfide bridges are broken, renaturation is slower.   In the case of ribonuclease, it requires 20 min for renaturation.   Stop-flow experiments revealed two phases in the folding of staphylococcal nuclease [22]:   an initial temperature-independent phase with a half-time of only 20 msec, and a subsequent temperature-dependent phase with a half-time of 200 msec. These data suggested to Anfinsen [25] the following protein folding model (Fig. 10):   an initial fast phase of folding, in which the random chain folds to local native formats, for example, α-helix in one section and β-sheet in another, followed by a second slow phase of folding, in which the β-ribbons gather into β-sheets and barrels, and the α-helices gather into bundles around β-structures or other higher-order levels or organization.   Since the latter step involves the translocation of very large molecular substructures, this step is quite temperature sensitive.   In water, the denatured proteins tend to return to their native conformations.   Rose and Roy [26] have shown that initial clustering of hydrophobic residues along the backbone may provide the driving force for initial formation of local secondary structure.

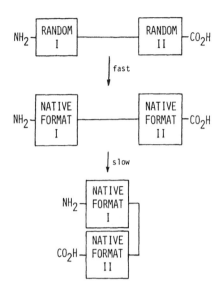

Fig. 10   Anfinsen's protein folding model.   (From Ref. 25.)

Methods of Protein Structure Prediction

Protein chemists have long been interested in developing methods for predicting a protein conformation from its sequence.   With such models the
protein chemist can, for example, search more readily for antigenic determinants in a novel protein.   Two developments were important advances in
protein structure prediction.

Predication of secondary structure.   The advent of refined x-ray crystallographic structures of many proteins has enabled protein chemists to
evaluate statistically the propensity for individual amino acids to appear in
conformational units, such as α-helices, β-sheet, and reverse turns
[27,28].   These data can then be applied to novel proteins, for which
crystal structures are not available, to predict regions of secondary
structure.

Prediction of hydrophobicity and hydrophilicity.   Since hydrophobicity
provides the driving force for placing different sections of sequence
either in the interior or the exterior of a globular protein structure [26],
a method for predicting regions of hydrophobicity (and hydrophilicity)
would be quite useful.   Nozaki and Tanford were the first to establish a
hydrophobicity scale [29] based on the solubility of amino acids in aqueous
ethanol and dioxane solution.   A method for determining hydrophilicity has
appeared since [30].

Applications of predicted hydrophobicity and secondary structure.
With a model of secondary structure, and hydrophobicity or hydrophilicity
data, the protein chemist can then combine the information to predict protein antigenic determinants, which must reside on the presumably hydrophobic surface, from the amino acid sequence of a novel protein [31].

Through the use of conformational energy minimization routines, protein chemists have also sought to use the predictions of hydrophobicity and secondary structure to fold novel protein sequences into reasonable model globular structures [32]. Such models are usually highly speculative. If, however, the novel protein bears a sequence homology to a protein for which the x-ray crystal structure is available, models derived from this structure may be closer to the "real" structure. In this manner a model of the conformational structure of the enzyme renin was recently described [33].

B. Lack of Conformational Stability in Most Peptides

The conservation of native conformation exhibited by globular proteins is not typically observed with peptides, especially noncyclic peptides. For example, many different conformations of the enkephalins have been described. Two crystal structures of the enkephalins have been determined [34,35], both of which have quite different backbone conformations. The more recent structure [35] contains four conformations, which differ slightly in backbone conformation. None of these structures compared to the eight lowest-energy conformations of enkephalins determined by Manavalon and Momany [36]. NMR studies of enkephalins suggest that a number of conformations exist in solution [37]. The general conclusion is that peptides display a multitude of conformations in solution [38]. As discussed earlier, peptides generally lack the many hydrogen bonds and disulfide bridges that stabilize the three-dimensional structure of proteins. We would expect that peptide conformational fluctuations would fall in the "fast phase" of protein folding. Although denaturing agents and denaturing conditions might favor nonnative over native peptide conformations, peptides can reequilibrate to native conformations so rapidly that denaturing reagents and conditions ultimately have no influence on biological activity.

The empirical Chou-Fasman method of protein conformation prediction is of little use in predicting peptide conformation, given the relative lack of stabilizing interactions and the large number of potential conformations. The conformations of small peptides (about 10 residues or less) can be simulated by direct calculations of conformational energy [39]. To a limited extent the influence of solvent on conformation can also be studied. However, since a peptide can adopt a number of different conformations in solution, these calculations are of limited practical value [39].

IV. SOLUBILITY

Proteins were historically classed according to their solubility characteristics [40]. The classes of nonconjugated globular proteins are shown in Table 5. Most globular proteins are soluble in water, dilute acid, and dilute salt. The "globulins" are insoluble in water at their isoelectric pH. Most globular proteins are insoluble in concentrated salt solutions. The "albumins" typically require more salt (e.g., 2 N ammonium sulfate) for precipitation from solution. Whereas albumins and globulins are soluble in ammonium hydroxide solution, the "histones" and "protamines" are insoluble because they contain large percentages of the basic amino acids lysine and arginine.

Table 5   Solubility of Nonconjugated Globular Proteins

Class	Concentrated salt	Dilute salt	Isoelectric pH H_2O	Dilute acid	Dilute base	Aqueous alcohol
Albumins	*Fairly sol.*	Sol.	Sol.	Sol.	Sol.	Insol.
Globulins	Insol.	Sol.	*Insol.*	Sol.	Sol.	Insol.
Histones (Lys, Arg)	Insol.	Sol.	Sol.	Sol.	*Insol.*	Insol.
Protamines (Lys, Arg, His)	Insol.	Sol.	Sol.	Sol.	*Insol.*	Insol.

Source: Ref. 40.

Peptides display a wide range of solubilities. Table 6 gives some examples of peptides soluble in dimethyl sulfoxide (DMSO), ammonium acetate, alcohols, and glacial acetic acid [41]. The shorter peptide sequences lend themselves to rough estimates of solubility based on side-chain lipophilicity or hydrophilicity. Many peptides bear fewer ionic groups per molecule than do proteins, leaving them less water soluble than proteins. Peptides can also be prone to intermolecular interactions leading to aggregate formation, decreasing solubility. Thus peptides often dissolve better in denaturing solvent systems [e.g., aqueous acetic acid, trifluoro-acetic acid (TFA), N,N-dimethylformamide (DMF), DMSO, and ethanol]. Aggregate dispersal is best achieved by initially forming a concentrated peptide solution in a solvent mixture with a high ratio of organic solvent to water. The solution can then be diluted out to physiological concentrations with water to avoid denaturation of proteins that may be exposed to the peptide solution.

Table 6   Examples of Peptide Solubility

Peptide	Soluble in:
Pepstatin	DMSO
Human gastrin	Ammonium acetate buffer
Growth hormone releasing peptide: Tyr-D-Trp-Ala-Trp-D-Phe-NH_2	DMSO, methanol, ethanol
Chemotactic peptide: For-Met-Leu-Phe	DMSO
Substance P, angiotensin I, II, or III and analogs	Acetic acid
β-Endorphin, somatostatin, TRH, LHRH	Water or acetic acid

Source: Ref. 41.

It may be desirable to obtain a peptide or protein as a solid from aqueous solution, but the peptide or protein may resist precipitation. Since many peptide chain degradation reactions are temperature sensitive, one should not evaporate peptide or protein solutions above room temperature. It is better to degas, freeze, and lyophilize (freeze-dry) the solution to minimize oxidation. Since the weight of peptide or protein may be only 60 to 80% of the total powder weight due to bound water and salts, one should determine the peptide content by amino acid analysis or elemental analysis if accurate amounts of peptide or protein are to be measured out from the lyophilized solid. Lyophilized powders may be hygroscopic; they may absorb moisture from the air, resulting in decreases in percent peptide weight. Even worse, the powder may wet to a transparent oil, leading one to believe that no peptide is present in the vial. Lyophilized powders are best kept dry over desiccant. These are discussed further in Chap. 16.

V. CHIRAL STABILITY

The stereoisomerism exhibited by the native amino acids in peptides and proteins leads to the subtle problem of chiral stability. Except for glycine, each of the mammalian amino acids occurs as one of two *stereoisomers* [4]. These stereoisomers are called *enantiomers* or *optical antipodes*, which are nonsuperimposable chemical structures (i.e., they are mirror images of each other). The enantiomers of alanine are displayed in Fig. 11, the L and D forms, as related to L- and D-glyceraldehyde. These structures are asymmetric because the α-carbon bears four different groups in a tetrahedral arrangement. Enantiomers are chemically equivalent, that is, most physical properties are identical except for one, optical rotation: L-alanine displays a specific rotation of −14° and D-alanine +14° [4].

In a dipeptide such as Val-Phe, there are four optical isomers (four combinations of the L- and D-amino acids). Stereochemically, they fall into two groups of enantiomers, displayed in Fig. 12: L-Val-L-Phe is the mirror image of D-Val-D-Phe; and L-Val-D-Phe is the mirror image of D-Val-L-Phe. The stereoisomers that are not enantiomers of each other are called *diastereomers* [4]. Most important, diastereomeric peptides have different physical properties. Conversion of an L-amino acid to a D-amino acid in a peptide chain can dramatically change the properties of the peptide chain.

If L-valine is heated in 6 N HCl at 110°C, a mixture of L-valine and D-valine is obtained (Fig. 13). After 8 days, the mixture is 10% D-valine;

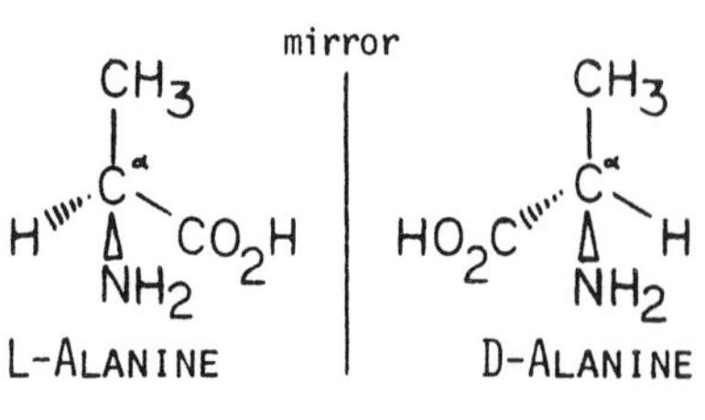

Fig. 11   Amino acid stereoisomers.

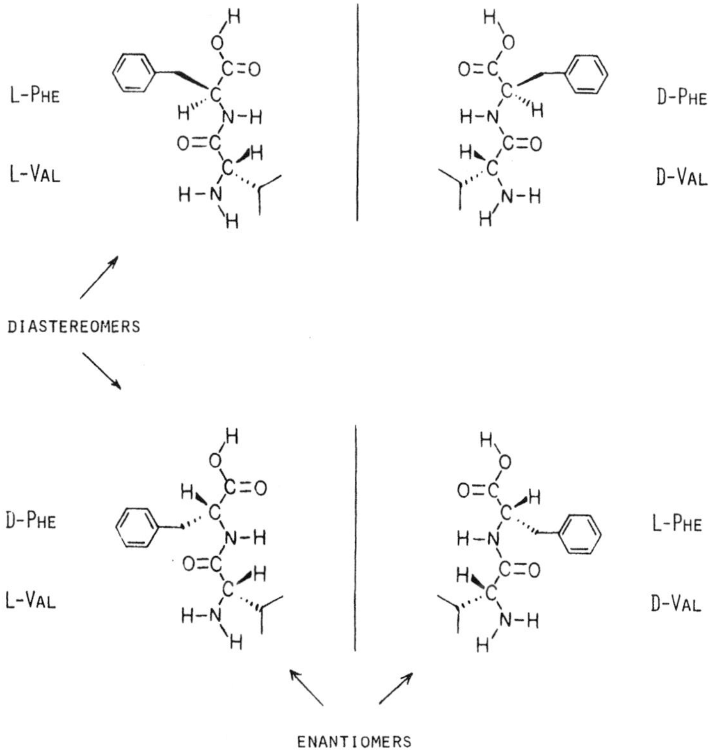

Fig. 12   Optical isomers of Val-Phe.

and after 42 days the mixture is 50% L-valine and 50% D-valine [42]. This chemical alteration of L- and D-amino acid to a 50:50 mixture of L- and D-amino acids is called *racemization*. It is important to note that a racemization reaction only goes to a 50:50 mixture of L- and D-amino acids. At this point a *racemic mixture* is obtained, that is, the optical rotation (α_D) is zero.

At room temperature and neutral pH, amino acids are quite stable to racemization. It would take about 100 years to obtain 0.1% conversion of L-valine to D-valine at room temperature [42].

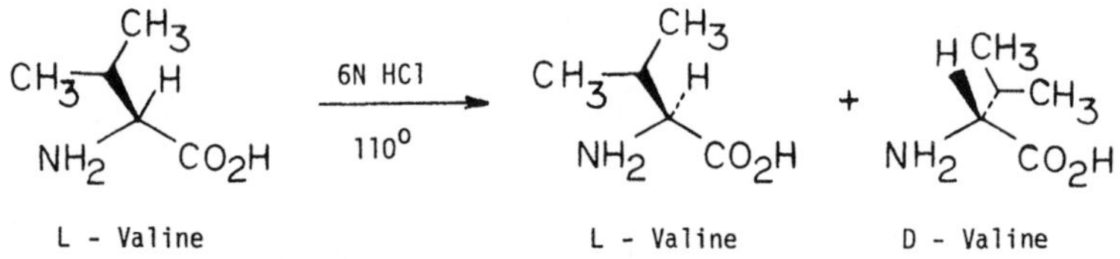

Fig. 13   Racemization.   (From Ref. 42.)

Racemization can produce diastereomers of a native peptide chain with different physical properties such as different retention times in reversed-phase high-performance liquid chromatography (HPLC). For example, five of the diastereomers of oxytocin [43] had been individually prepared for biological evaluation, each displaying peaks in reversed-phase HPLC that are baseline resolvable from other peaks. These particular diastereomers can be removed from the desired native peptide by chromatographic purification. Any peptide exposed to racemization conditions should be checked afterward for diastereomer formation by HPLC.

Introduction of a D-amino acid into a native peptide can alter the way the resulting peptide folds and its ability to interact with a biological receptor. For example, angiotensin II is an octapeptide hormone that elevates blood pressure in mammals. An equimolar dose of the D-His-6 analog of angiotensin II displays only 4% of the pressor activity of angiotensin II [44]. Thus racemization may be detrimental to bioactive conformation and biological activity.

The mechanism for racemization of intact peptide chains for unactivated amino acids is *enolization* [43]. By this mechanism (Fig. 14) the racemizing amino acid rearranges to the enol form with a planar trigonal C_α carbon atom, losing its asymmetry. On reprotonation of the C_α carbon, asymmetry is restored. However, both the L and D forms result in a 50:50 mixture in the absence of an asymmetric catalyst. Thus the amount of racemic amino acid formed equals the amount of enol generated by the racemization reaction. Factors that promote enolization are acidic or basic pH; increased temperature; electron-withdrawing groups in the side chain (e.g., Asn, Asp, Cys, Gln, Glu, His, Phe, Ser, Thr, Trp, and Tyr); and position of the amino acid in the interior of the sequence (the N- and C-terminal amino acids do not racemize as readily) [43].

To repeat, peptide chains that have been exposed to these conditions should be monitored for diastereisomer formation by HPLC. If peptide solutions are kept cool and near neutral pH, racemization should not present a problem.

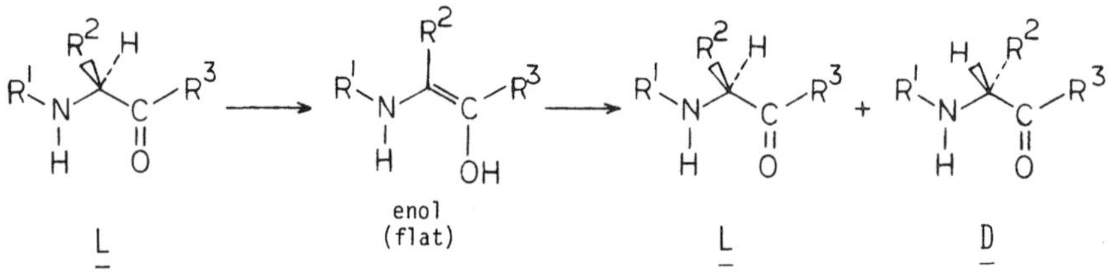

Fig. 14   Racemization mechanism for unactivated amino acids and peptides. (From Ref. 43.)

VI. CHEMICAL STABILITY

The extent to which a chemical degrades a peptide or protein depends on
several factors: temperature, length of exposure, and the amino acid
composition, sequence, and conformation of the peptide or protein. Conformation is important since groups that typically react with a chemical
may be buried and shielded from contact with the reagent. The stability
of conformation may also be important: for example, chemical reaction
could result in a lowering of the conformational energy of a peptide in a
highly strained conformation, or an elevation of the conformational energy
of a peptide in an unstrained conformation.

A vast array of chemicals can modify peptide chains. These have
received ample attention [10,45−47]. Thus it is more relevant to consider
here those degradation reactions that arise from the inherent chemical
reactivities of certain amino acids in peptide chains. The reagents that
promote these reactions include air, water, acid, and base, reagents that
are inevitably employed in typical handling of peptides and proteins. If
conditions harsher than these are contemplated, the synthetic literature
listed above should be consulted.

Quite often the products of these degradation reactions are difficult to
characterize or are unstable, undergoing further degradation during isolation. Most often these degradation products represent a minor percentage
of total peptide weight. Consequently, peptide and protein chemists find
it easier to delineate methods for avoiding degradation of a particular protein or peptide than to analyze a degradation reaction thoroughly. Consequently, this discussion will not contain many specific examples from the
literature, since only a few have been reported. Instead, reaction
mechanisms and factors affecting degradation rates that have been well
characterized are presented. The stability of any particular protein or
peptide is best studied by monitoring degradation by analytical gel electrophoresis or HPLC.

A. Diketopiperazine Formation

The N-terminus of a peptide chain can suffer rearrangement with loss of
the N-terminal dipeptide as a cyclic *diketopiperazine* (Fig. 15). This reaction is inhibited in acid and promoted in base [46]. Certain amino acids,
especially proline and glycine in the N-terminal dipeptide, promote the
reaction.

B. Imide Formation

Asparagine, glutamine, aspartic, and glutamic acids tend to cyclize back
onto the peptide backbone, forming *aspartimide* or *glutarimide* (Fig. 16)
[46]. Asparagine and glutamine lose a molecule of ammonia in the
process. The imides are sensitive to hydrolysis and can reopen in the
presence of water along two reaction pathways: transpeptidation, which
gives a β-aspartyl peptide or a C-glutamyl peptide, or side-chain hydrolysis, which gives the normal α-aspartyl or α-glutamyl peptide.

Imides form slowly on storage in water. Heat, low pH, and high pH
promote imide formation. Adjacent lysine or arginine residues autocata-
lyze imide formation. Rates of imide formation vary widely with sequence,

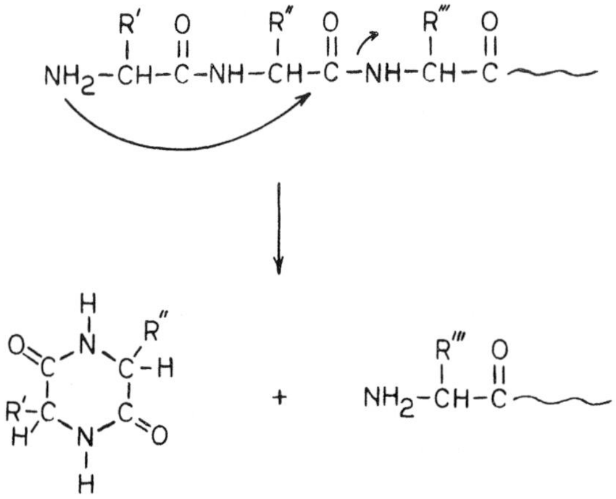

Fig. 15   Diketopiperazine formation.   (From Ref. 47.)

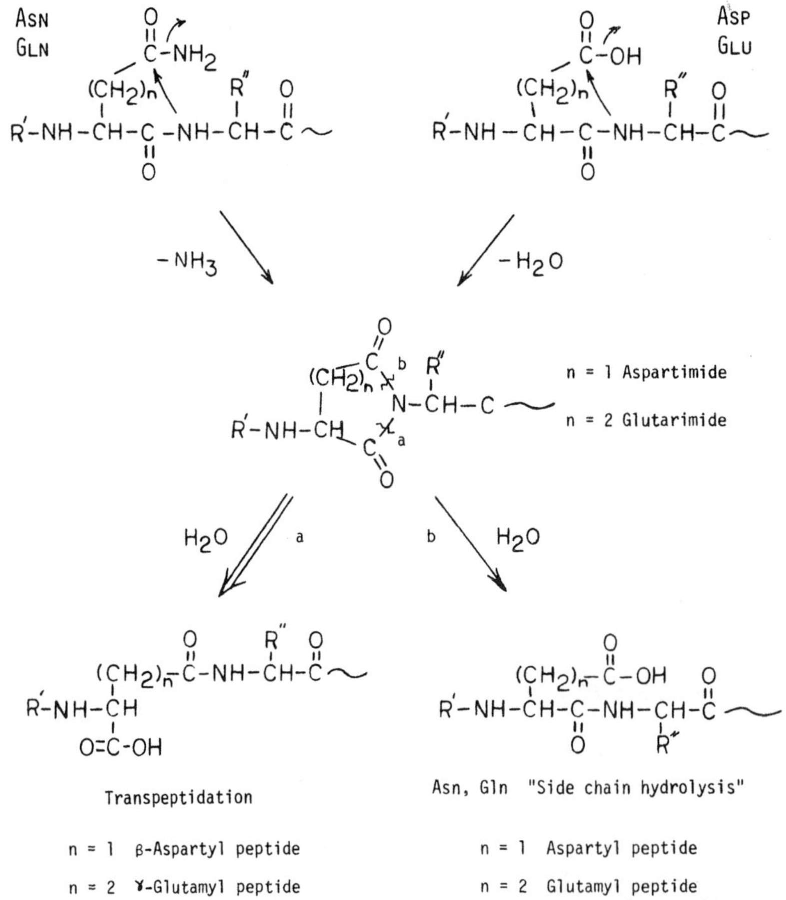

Fig. 16   Imide formation.   (From Ref. 47.)

with a half-time varying from as little as 18 days to several years at neutral pH [46].

C. N-Terminal Asparagine Cleavage

A second side reaction is possible with asparagine at the N-terminus [46]. The side-chain amide nitrogen can attack its own α-carboxyl group, forming an unsubstituted aspartimide. In the process, asparagine cleaves from the peptide chain. This reaction is acid catalyzed.

   The propensity for degradation of asparaginyl peptides is shown in the study by Riniker and co-workers (Fig. 17) [48]. [Asparagine-1, Valine-5]angiotension II was kept in solution at 50°C. After 6 months, none of the original asparaginyl peptide remained. The product mixture included 81% of the β-aspartyl peptide, 9% of the side-chain-hydrolyzed peptide and 10% of the peptide lacking an N-terminal asparagine. At room temperature these degradation products should constitute only minor impurities.

D. Oxidation of Amino Acids

Major degradation of peptides and proteins occurs by oxidation of certain susceptible amino acids. Cystine is stable to air oxidation. Air alone can oxidize tryptophan, methionine, and cysteine (Table 7) [46]. These oxidations are acid or base catalyzed and may be blocked if an oxidation scavenger is present in solution. Exclusion of air by degassing solvents and by avoiding violent stirring reduces air oxidation. Since cysteine rarely exists in the reduced form in native peptides but is usually linked to

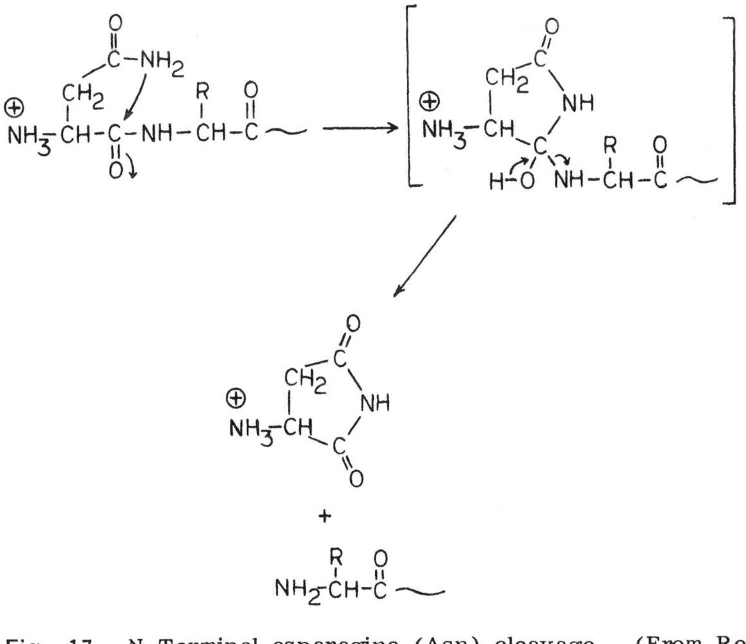

Fig. 17   N-Terminal asparagine (Asn) cleavage.   (From Refs. 47 and 48.)

Table 7   Oxidation of Amino Acids in Peptides and Proteins

Amino acid	Oxidation products	Air (O_2)	H_2O_2	BR_2, H_2O	Catalysis	Oxidation scavenger
Tryptophan (Trp)	Oxindoylalanine, etc.	Yes	Yes	Yes	Acid	Indole, $(CH_3)_2S$
Methionine (Met)	Methionine sulfoxide	Yes	Yes	Yes	Base	$(CH_3)_2S$
	Methionine sulfone	No	Yes	Yes	Base	$(CH_3)_2S$
Cysteine (Cys)	Cystine (disulfide)	Yes	Yes	Yes	Base	$(CH_3)CH_2SCH_3$
Cystine (Cys-Cys)	Cysteic acid ($-SO_3H$)	Yes	Yes	Yes	Base	$(CH_3)CH_2SCH_3$
Tyrosine (Tyr)	DOPA	No	Yes	No	Base	Phenol, $(CH_3)_2S$

Source: Ref. 47.

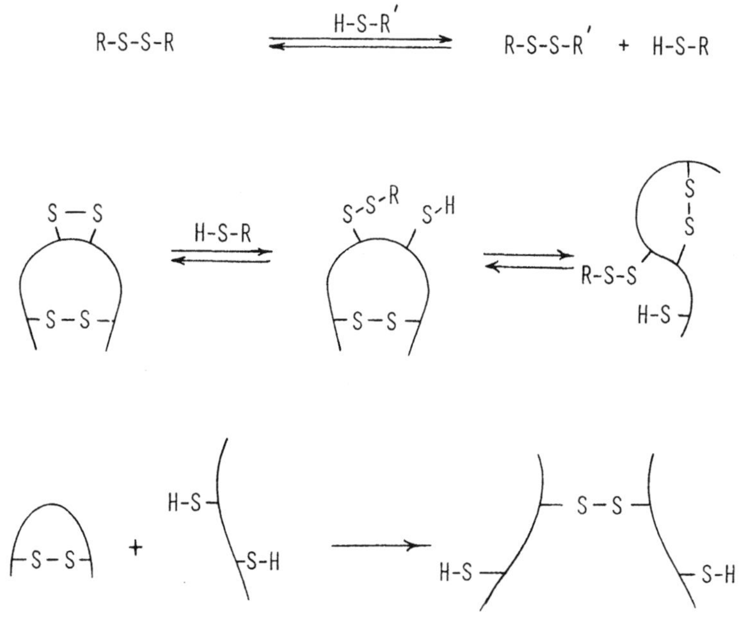

$$R-S-S-R \xrightleftharpoons{\text{H-S-R}'} R-S-S-R' + H-S-R$$

Fig. 18   Cystine disulfide exchange.   (From Ref. 46.)

another cysteine residue as a cystine disulfide, this reaction of cysteine is
of interest more to the synthetic chemist. Stronger oxidizing agents, such
as hydrogen peroxide or bromine water, degrade not only tryptophan and
methionine but also cystine and tyrosine.

E.   Cystine Disulfide Exchange

A more common problem with cystine is disulfide exchange. In the presence of another mercaptan, HSR', a disulfide, R—S—S—R, interchanges
mercaptan groups to give a mixed disulfide, R—S—S—R' (Fig. 18) [45].
The mercaptan can be a free cysteine in a partially reduced peptide or
protein, or an agent such as mercaptoethanol mistakenly added as an
antioxidant.
     As seen in the second reaction of Fig. 18, a peptide chain containing
more than one disulfide can enter into disulfide exchange reactions, resulting in a scrambling of disulfide bridges and a change in conformation. As
seen in the third reaction, one peptide chain with free cysteines can add
to the disulfide of another peptide chain to give a dimeric disulfide.
Trimers and tetramers can form by analogous reactions.
     The disulfide exchange reaction is base catalyzed. Thus cystine-
containing peptides should not be stored in alkaline solution. Mercaptans
such as mercaptoethanol or mercaptoacetic acid promote the reaction.
Other oxidation scavengers, such as ethylmethyl sulfide, should be employed instead. The reaction is concentration dependent, especially for
oligomer formation. One should avoid storage of concentrated solutions of
cystinyl peptides. The oligomers formed appear at low R_f values on thin-
layer chromatography (TLC) and are readily removed by gel filtration.

VII. SUMMARY OF PEPTIDE/PROTEIN DEGRADATION CONDITIONS AND EXAMPLES OF PEPTIDE STABILITY

Table 8 summarizes the conditions that degrade peptides and proteins containing certain amino acids. Introduction of air into a solution, especially by shaking, can cause the oxidation of methionine and tryptophan. Low pH promotes transpeptidation of asparagine and glutamine, as well as side-chain hydrolysis. Low pH also promotes racemization. In contrast, high pH promotes transpeptidation of aspartic acid and glutamic acid, and transpeptidation and side-chain hydrolysis of asparagine and glutamine. High pH also promotes cystine disulfide exchange and oxidation of methionine and tryptophan, as well as racemization of many amino acids. Disulfide exchange of cystine residues is promoted in concentrated solution. Finally, heat or long-term storage at room temperature will increase all of the degradation reactions mentioned.

The extent to which these reactions affect sample purity and biological activity is heavily sequence dependent. For example, the β-Asp-1 derivative of angiotensin I is more active than is angiotensin I itself [48]. The glutamine and asparagine side chains in substance P slowly hydrolyze at room temperature with no apparent effect on activity, while methionine oxidation is detrimental to substance P [49]. Methionine oxidation lowers activity in [Met-5]-enkephalin [50]. Oligomers of somatostatin form slowly under most conditions, resulting in decreased activity [41]. The 1–34

Table 8 Conditions That Degrade Peptides and Proteins Containing Certain Amino Acids

	Air (shaking)	Low pH	High pH	Conc'd. sol'n.	Heat/RT long term	Potential problem[a]
Asp			×		×	TP
Glu			×		×	TP
Asn		×	×		×	TP/Hyd
Gln		×	×		×	Hyd
Pro[b]					×	DKP
His[b]					×	DKP
Gly[b]					×	DKP
(Cys)$_2$			×	×	×	DE
Met	×		×		×	OX
Trp	×		×		×	OX
All amino acids		×	×		×	Rac

[a]TP, transpeptidation; Hyd, side-chain hydrolysis; DKP, diketopiperazine formation; DE, disulfide exchange; OX, oxidation; Rac, racemization.
[b]In N-terminal dipeptide.

fragment of bovine parathyroid hormone, on the other hand, contains one aspartic acid, two asparagines, two glutamines, one methionine, three serines, three glutamic acids, and one tryptophan. Yet this peptide is stable during normal handling [41].

VIII. CONCLUSIONS

Although many peptides and proteins have been studied for decades, their solubility and stability problems are rarely a subject of publication. Solubility problems generally disappear with increased experience in handling peptides. Degradation typically produces minor impurities which the physiologist has come to live with but which are important to the pharmaceutical scientist and to the FDA for product specifications. An examination of amino acid composition will indicate which degradation problems may arise for a particular peptide or protein. One must then determine impurity generation under typical handling conditions by analytical gel electrophoresis or HPLC. Few stability problems should be expected if peptide or protein samples are kept cool, near neutral pH, and with minimal exposure to air. A number of peptides have already been approved by the FDA for use in humans [51]. A larger group of peptides have displayed therapeutically useful actions in humans (Table 9) [1], and several of these actions can be achieved by intranasal or oral administration. Thus the problem of chemical stability is clearly surmountable. As pharmaceutical scientists familiarize themselves with the ways to circumvent these problems, the industry should be able to bring an even larger group of peptide pharmaceutical agents to market.

Table 9 Unmodified Peptides Displaying Therapeutically Useful Actions in Humans

Peptide	Condition[a]
Adrenocorticotropic hormone (ACTH)	(IV) Hypercalcemia[b], inflammation[b]
Bacitracin	(topical) Bacterial infection[b]
Bestatin	(oral) Cancer therapy
Calcitonin (CT)	(SC or IM) Paget's disease,[b] osteoporosis[b]
	(IV) Hypercalcemia[b]
	(oral) Gastric secretion
	(SA) Intractable pain
Cholecystokinin	(IN) Chronic pancreatitis
	(IV) Appetite
	(IV) Postoperative paralytic ileus
Chorionic gonadotropin (hCG)	(IV) Cruptorchidism[b], induction of ovulation[b]

Table 9 (Continued)

Peptide	Condition[a]
Cydosporine	(PO, IV) Immunosuppression in allogenic transplants[b]
Delta sleep-inducing peptide (DSIP)	(IV) Insomnia
β-Endorphin	(IV, ITH) Cancer pain
	(ITH) Childbirth pain
	(IV) Narcotic abstinence syndrome
Glucagon	(IV, IM) Hypocalcemia[b]
Growth hormone	(IM) Short stature[b]
Gramicidin	(topical) Bacterial infections[b]
Insulin	(IV) Diabetes mellitus[b]
Luteinizing hormone	(IV) Amenorrhea
Releasing hormone (LHRH)	(IN) Cryptorchidism
Melanocyte inhibiting factor-I (MIF-I)	(oral) Depression
	(oral) Tardive dyskinesia
Neurotensin (NT)	(IV) Gastric juice secretion
Oxytocin (OT)	(IV) Postpartum bleeding,[b] labor
	(IN) Lactation[b]
Parathyroid hormone (PTH 1−34)	(SC) Osteoporosis
Somostatin (SS)	(IV) Gastric ulcers
Terprotide	(PAR) Hypertension
Serum thymide factor (FTS)	(IV) Immune deficiencies
Crude thymosin	(PAR) Autoimmune disorders
	(IM) Collagen vascular disease
	(IV) Chemotherapy
	(IV) Rheumatoid arthritis
Thyrotropin releasing hormone (TRH)	(oral) Lactation
Arginine^8-vasopressin and lysine^8-vasopressin	Diabetes insipidus[b]

Source: Ref. 1, 51.

[a]IV, intravenous; SC, subcutaneous; IM, intramuscular; SA, IN, intranasal; ITH, intrathecal; PAR, parenteral.

[b]Application approved for humans by the U.S. Food and Drug Administration.

REFERENCES

1. J. Samanen, in *Bioactive Polymeric Systems* (C. G. Gebelein and C. E. Carraher, Jr., eds.), Plenum Press, New York (1985), pp. 279–344.
2. E. Schroeder and K. Lubke, *The Peptides*, Academic Press, New York (1965).
3. G. N. Ramachandran and V. Sasisekharan, *Adv. Protein Chem.*, 23:283 (1968).
4. J. P. Greenstein and N. Winitz, *Chemistry of the Amino Acids*, Vol. 1, John Wiley & Sons, New York (1961), Chap. 1.
5. Royal Society of Chemistry, *Amino Acids, Peptides, and Proteins*, Royal Society of Chemistry, London, Vol. 1 (1969) through present.
6. A. L. Lehninger, *Biochemistry*, Worth Publishers, New York (1970), p. 75.
7. J. P. Greenstein and N. Winitz, *Chemistry of the Amino Acids*, Vol. 1, John Wiley & Sons, New York (1961), Chap. 4.
8. P. Bornstein and W. Traub, *The Proteins*, Vol. 4, Academic Press, New York (1979), pp. 412–632.
9. R. W. Woody, in *The Peptides: Analysis, Synthesis, Biology*, Vol. 7 (S. Undefriend, J. Meienhofer, and V. J. Hruby, eds.), Academic Press, New York (1985), pp. 16–115.
10. G. E. Means and R. E. Feeney, *Chemical Modification of Proteins*, Holden-Day, San Francisco (1971).
11. S. S. Zimmerman, in *The Peptides: Analysis, Synthesis, Biology*, Vol. 7 (S. Undefriend, J. Meienhofer, and V. J. Hruby, eds.), Academic Press, New York (1985), pp. 166–213.
12. G. E. Schulz and R. H. Schirmer, *Principles of Protein Structure*, Springer-Verlag, New York (1979), pp. 30–31.
13. C. Tanford, *The Hydrophobic Effect, Formation of Micelles, and Biological Membranes*, 2nd ed., John Wiley & Sons, New York (1980), pp. 139–145.
14. G. E. Schulz and R. H. Schirmer, *Principles of Protein Structure*, Springer-Verlag, New York (1979), pp. 33–36.
15. P. L. Privalov, *Adv. Protein Chem.*, 33:167 (1979).
16. I. L. Karle, *Proc. Natl. Acad. Sci. USA*, 76:1532 (1979).
17. J. S. Richardson, *Adv. Protein Chem.*, 34:168 (1981).
18. G. N. Reese, J. A. Harstuck, M. L. Ludwig, F. A. Quiocho, T. A. Steitz, and W. N. Lipscomb, *Proc. Natl. Acad. Sci. USA*, 58:2220 (1967).
19. R. N. Gurd and T. M. Rothgels, *Adv. Protein Chem.*, 33:73 (1979).
20. G. E. Schulz and R. H. Schirmer, *Principles of Protein Structure*, Springer-Verlag, New York (1979), Chap. 8, pp. 149–165.
21. F. Marcritchie, *Adv. Prot. Chem.*, 32:283 (1978).
22. H. F. Epstein, A. N. Schecter, R. F. Chen, and C. B. Anfinsen, *J. Mol. Biol.*, 60:499 (1971).
23. L. L. Shen and J. Hermans, Jr., *Biochemistry*, 11:1836 (1972).
24. P. J. Hagerman and R. L. Baldwin, *Biochemistry*, 15:1462 (1976).
25. C. B. Anfinsen, *Science*, 181:223 (1973).
26. G. D. Rose and S. Roy, *Proc. Natl. Acad. Sci. USA*, 77:4643 (1980).
27. P. Y. Chou and G. D. Fasman, *Biochemistry*, 13:211–222, 222–245 (1974).
28. F. R. Maxfield and H. A. Scheraga, *Biochemistry*, 45:5138 (1976).

29. Y. Nozaki and C. Tanford, *J. Biol. Chem.*, *246*: 2211 (1971).
30. M. Levitt, *J. Mol. Biol.*, *104*: 59 (1976).
31. T. P. Hopp and K. R. Woods, *Proc. Natl. Acad. Sci. USA*, *78*: 3824 (1981).
32. C. Auffray and J. Novotny, *Hum. Immunol.*, *15*: 381 (1986).
33. T. Blundell, B. L. Silbanda, and L. Pearl, *Nature*, *304*: 273 (1983).
34. G. D. Smith and J. F. Griffin, *Science*, *199*: 1214 (1978).
35. I. Karle, J. Karle, D. Mastropaolo, A. Camerman, and N. Camerman, in *Peptides: Structure and Function, Eighth American Peptide Symposium* (V. J. Hruby and D. H. Rich, eds.), Pierce Chemical Company, Rockford, Ill. (1983), pp. 291–294.
36. P. Manavalan and F. A. Momany, *Int. J. Pept. Protein Res.*, *18*: 256 (1981).
37. S. Premilat and B. Margret, *J. Phys. Chem.*, *84*: 293 (1980).
38. V. J. Hruby and H. I. Mosberg, *Peptides*, *3*: 329 (1982).
39. A. T. Hagler, in *Chemistry of the Amino Acids*, Vol. 1 (J. P. Greenstein and N. Winitz, eds.), John Wiley & Sons, New York (1961), pp. 219–300.
40. H. L. Fevold, in *Amino Acids and Proteins* (D. M. Greenburg, ed.), Charles C Thomas Publisher, Springfield, Ill. (1951), pp. 257–311.
41. D. Chan, Beckman Instruments, Inc., SPINCO Division, personal communication, and author's personal observations.
42. D. S. Kemp, in *The Peptides: Analysis, Synthesis, and Biology*, Vol. 1 (E. Gross and J. Meienhofer, eds.), Academic Press, New York (1979), Chap. 7.
43. B. Larsen, V. Viswantha, S. Y. Chang, and V. J. Hruby, *J. Chromatogr. Sci.*, *16*: 207 (1978).
44. M. C. Khosla, M. M. Hall, R. R. Smeby, and F. M. Bumpus, *J. Med. Chem.*, *16*: 829 (1973).
45. E. Schroeder and K. Lubke, *The Peptides, Methods of Synthesis*, Vol. 1, Academic Press, New York (1965).
46. M. Bodanzky, *Synthesis*, 5 333 (1981).
47. B. Witkop, *Adv. Protein Chem.*, *16*: 221 (1961).
48. B. Riniker, M. Brunner, and R. Schwyzer, *Angew. Chem. Int. Ed. Engl.*, *1*: 405 (1962).
49. J. Stewart, *Peptide Pointers*, Beckman Instruments, SPINCO Division, Palo Alto, Calif. (1981), p. 1.
50. J. S. Morley, *Annu. Rev. Pharmacol. Toxicol.*, *20*: 81 (1980).
51. E. R. Barnhart, *Physician's Desk Reference*, 42nd ed., Medical Economics Company, Oradell, N.J. (1988).

4

Physical Biochemistry of Protein Drugs

James Q. Oeswein and Steven J. Shire

Genentech, Inc., South San Francisco, California

I. INTRODUCTION

Although most traditional pharmaceutical agents have been small organic molecules, the advent of biotechnology, especially recombinant DNA techniques, has made available bioactive proteins which are usually found in extremely low concentrations in vivo. These proteins provide a greater formulation and drug delivery challenge than do smaller organic molecules because of their greater structural complexity. In particular, the large number of functional groups in these macromolecules makes them very susceptible to a variety of degradation and inactivation pathways. The complex three-dimensional structure of a protein is also very dependent on solution conditions, and the wrong choice of manufacturing process steps, conditions, or formulation may lead to improperly folded polypeptide chains which are biologically inactive. For these reasons, great care must be taken in defining and choosing the appropriate conditions under which a protein pharmaceutical may be assured maximal stability. Further complicating this situation, however, is the fact that there are presently a limited number of acceptable administration routes for proteins (mainly parenteral). Consequently, in designing formulations and processes for alternative delivery systems and administration routes, one must weigh those stability concerns against the protein's bioavailability following administration.

In this chapter we discuss the chemical and physical properties of proteins that are responsible for their biological functions and those which make them unique as therapeutic agents. Since these properties are varied and complex, we also detail some of the potential routes of inactivation, as well as the effects of processing and formulation conditions on the final product.

II. LEVELS OF PROTEIN STRUCTURE

A. Primary to Quaternary Structure

Proteins, once thought to be loosely associated aggregates of small multimers, are, instead, polymers of amino acids connected via amide linkages
commonly referred to as peptide bonds. The sequence of amino acids in a
peptide or protein is known as the primary structure and is determined
genetically by the sequence of nucleotides in DNA. Within living organisms
this protein sequence may involve any or all of the 20 naturally occurring
common L-amino acids, as well as some uncommon metabolic variants [1].
The peptide bonds linking them are fairly rigid and planar and almost exclusively trans in configuration, although cis configurations have been
observed [2].

   The secondary structure of a peptide or protein is the respective arrangement of the individual amino acids along the polypeptide backbone,
sometimes resulting in very specific and well-ordered structures such as
helices, β-strands, loops, and β (or reverse)-turns. These structures
are defined by two torsional angles: those made between the α-carbon
and the amino group (Φ) and those made between the α-carbon and the
carboxyl group (Ψ) of each amino acid within that particular portion of
the molecule. Theoretically, many combinations of Φ and Ψ angles are
possible for all the amino acids within a peptide or protein, except for
the cyclic imino acid proline. However, steric constraints made by the
various side (R) groups greatly limit the angles that are favorable. For
example, a fully extended β-chain ($\Phi = -180°$; $\Psi = +180°$) with a dipeptide
repeat distance of 7.2 Å can occur only for polyglycine [3]. For β-struc-
tures within proteins containing amino acids other than glycine, the chains
must become more compact through appropriate rotation of the Φ and Ψ
angles.

   Other secondary structures seen within peptides and proteins when Φ
and Ψ are approximately $-60°$ are helices [4,5]. The right-handed α-helix
is most common. It has a repeat distance or helical pitch of 5 to 5.5 Å
per turn, depending on the R groups involved, with each turn containing
approximately 3.6 amino acids. This structure is more compact than
β-structures due to intrachain hydrogen (H) bonding involving amino and
carboxyl groups of the chain backbone 3 residues apart within the helix.
Left-handed α helices are not as stable and have never been observed in
native proteins. Other helices, rarely seen in proteins except near the
ends of right-handed α-helices, are the 3_{10}-helix (3 residues per turn)
and the π-helix (4.4 residues per turn).

   Illustrations and examples of helices and β-structures may be found in
any of several biochemistry texts [3,4,6,7]. The remaining portions of a
protein that do not fall into any of the categories mentioned above are
referred to as random structure. More recently, however, a new category
of secondary structure—the loop—has been described by Leszczynski and
Rose [8]. Loops, whose idealized shapes each "resemble a Greek omega
(Ω)," may represent greater than 20% of a protein's structure on the
average. They are usually found at the protein's surface and therefore
may play "important roles in molecular function and biological recognition" [8].

Tertiary structure* is the three-dimensional arrangement of a single protein molecule and refers to the way that specific secondary structural elements interact with each other as well as with random portions of the molecule to form stable domains. Four broad categories of tertiary structures have been very elegantly classified by Richardson [9] as "antiparallel α, parallel α/β, antiparallel β, and small S—S rich or metal rich." Each category is defined according to the arrangement of multilayered domains (approximately 100 distinctly different domains are known) which form the central core or dominant stabilizing features of a particular protein. For example, a parallel or antiparallel β-sheet may be formed when a single polypeptide chain of sufficient length folds back upon itself using a β-turn. β-Sheets are stabilized through hydrogen bonding of amino and carboxyl groups on the two adjacent portions of the chain forming the sheet, with the R groups projecting above or below the planes defined by each of the peptide bonds therein [3]. Such arrangements are the most common among proteins. Examples of each of the four broad categories of tertiary structure are illustrated in Fig. 1.

Disulfide (S—S) bonds are also a common feature of many proteins and even some peptides, occurring within the same chain or between two chains. The S—S bonds have a preferred geometry, usually with a dihedral rotational angle of approximately ±90°, which once formed, obviously impart a great deal of stability to a protein's overall structure by limiting the number of conformational transitions that it may undergo.

In addition to S—S bonds and metals, tertiary interactions are affected by other posttranslational modifications, both covalent and noncovalent, which may occur naturally or synthetically. These modifications include alkylation, acylation, hydroxylation, glycosylation, phosphorylation, cross-linking, and binding of prosthetic groups (e.g., heme groups). Some of these modifications are discussed further in subsequent sections.

Quaternary structure is the final form or conformation of a protein as it exists in the solid state or in solution and refers to the noncovalent interaction or spatial arrangement of individual protein subunits (monomers) with each other to form oligomers, as well as the interaction of monomers and oligomers with the solvent and other solute components that affect the protein's molecular weight distribution.

In self-associating protein systems, monomers and all oligomeric species are in dynamic equilibrium with each other in solution, although one

---

*Although in the past the terms "peptide" and "protein" have been used interchangeably or at best only vaguely differentiated, it is appropriate to make a distinction when one is concerned with purifying or formulating a polypeptide based on its physicochemical properties or behavior. Therefore, sequences of amino acids for which the chain length is not sufficient to produce the higher levels of tertiary and/or quaternary structure (as discussed in this chapter) should be designated as "peptides," while those exhibiting such characteristics should be designated as "proteins."

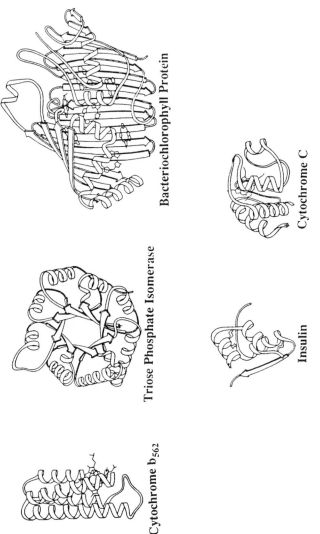

Cytochrome b$_{562}$

Triose Phosphate Isomerase

Bacteriochlorophyll Protein

Insulin

Cytochrome C

Fig. 1   Four categories of tertiary structure.   The examples are representative of subcategories of the major types of tertiary structures as defined in the book.   The spiral ribbons represent helices, the arrows β-strands, and the ropes random structure and turns.   Cytochrome b$_{562}$ is a subcategory of the antiparallel α category with up-and-down helix bundles.   Triose phosphate isomerase is from the parallel α/β category and contains a singly wound β-barrel.   Bacteriochlorophyll Protein is an open-face sandwich β-sheet from the antiparallel β category.   The small disulfide-rich and metal-rich category are represented by insulin and cytochrome c, respectively.   (Adapted from Ref. 9.)

species is usually preferred. An example is the equilibrium distribution
of monomers, dimers, hexamers, and higher oligomers in a solution of
pork insulin at neutral pH [10]. In the absence of divalent metals the
dimer form is preferred. The addition of Zn^{2+}, Cu^{2+}, Cd^{2+}, Co^{2+}, or
Ni^{2+} under appropriate conditions results in the coordination of two or
more metal atoms with the B5 and/or B10 histidine residues of three insulin
dimers to form a stabilized hexamer [11–13].

It has also been observed that the secondary and tertiary structure
of a protein may be altered significantly via quaternary interactions,
thereby affecting its biological function. Examples of such quaternary
structures are those produced by the supramolecular assembly of proteins
with other biological compounds, such as other proteins, lipids, carbohydrates, nucleic acids, or other prosthetic groups. A specific example of
interaction with prosthetic groups is the binding of the heme group to
apohemoglobin, affecting the tertiary structure of the apoglobin chains as
well as the subunit assembly of hemoglobin into its final quaternary structure [14,15]. Figure 2 illustrates the levels of structure involved (primary to quaternary) in the folding and assembly of hemoglobin. Another
example of quaternary interaction is the self-assembly of proteins with
nucleic acids to form viruses.

B. Thermodynamic and Kinetic Considerations

Protein Folding as a Self-Assembly Process

In Sec. II.A we discussed the complex structures that polypeptide chains
can attain. Even more remarkable than the ability to fold into a particular
conformation is the fact that the folding of a randomly coiled polypeptide
chain (generally referred to as the denatured state of the protein) into a
compact tertiary structure is a self-assembly process. Essentially all the
information required to fold the chain into a globular structure is provided by the primary structure, as demonstrated by the classical unfolding-refolding experiments with ribonuclease-A by Anfinsen et al. [16].
These experiments show that an unfolded polypeptide chain can be refolded
spontaneously to a functional three-dimensional structure, without the addition of energy sources or catalysts, simply by reverting to solvent conditions in which the protein was folded into its "native" structure (Fig. 3).
It has been estimated that a random search of structures would require a
protein the size of ribonuclease-A (approximately 100 amino acid residues)
10^{77} years to refold to a particular conformation [17]. Since proteins
typically refold on a millisecond-to-hours time scale, it appears that the
refolding process must be a well-directed one that avoids a random
sampling of all possible conformations. The implication of this observation
is that the native globular structure for a protein need not be the most
stable conformation but rather, a "metastable" conformation that is attained
easily via a particular kinetically controlled pathway [18].

Protein Folding as a Concerted Process

The majority of proteins tend to unfold and refold within narrow ranges
of denaturing conditions. Thus the folding of a polypeptide chain appears
to be a highly concerted process governed by long-range, highly cooperative interactions. The importance of these interactions is exemplified by
recent discoveries that similar amino acid sequences in different proteins

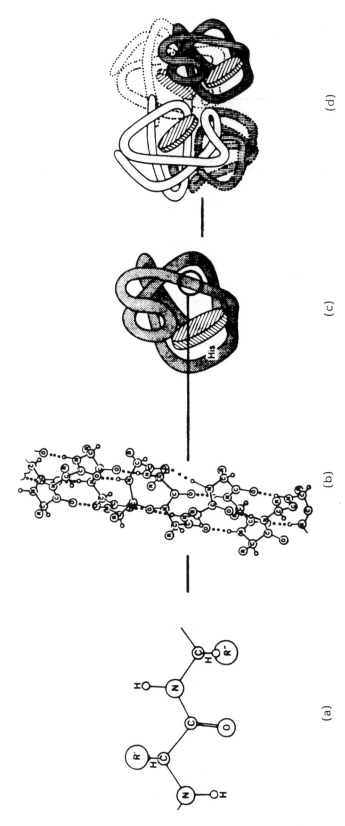

(a)   (b)   (c)   (d)

Fig. 2   Levels of structural organization in hemoglobin.   (a) Primary, peptide chain backbone;   (b) secondary, a representative helical portion;   (c) tertiary, folding of the β-chain helix;   (d) quaternary, association of α- and β-chain subunits. (From Ref. 153.)

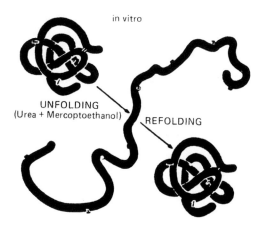

in vitro

UNFOLDING
(Urea + Mercoptoethanol)

REFOLDING

Fig. 3   Schematic drawing showing the in vitro unfolding of a small globular native protein such as ribonuclease to an extended polypeptide chain
that has no biological activity. In this example the protein is unfolded in
the presence of urea and β-mercaptoethanol to reduce all the disulfide
cross-links. Subsequent removal of urea and reducing agent results in a
refolded molecule that regains its biological activity. (From Ref. 154.)

have been found to have very different secondary structures [19,20].
This observation demonstrates how difficult it is to predict structure (and
ultimately function) based on short amino acid sequences without considering the remaining interactions from the other folded sequences in a polypeptide chain.

The highly cooperative nature of protein folding transitions has allowed
researchers to model protein folding as a two-state transition from a denatured to a native structure without the presence of a significant concentration of intermediates. To ascertain if a protein can be analyzed by
such a model, researchers have typically investigated the reversible unfolding and refolding of a protein using spectroscopic or hydrodynamic
methods. The various methods probe different regions of a protein's
structure. As an example, spectroscopic techniques such as fluorescence
and ultraviolet absorption spectroscopy at about 260 to 290 nm can be
used to probe the environment of aromatic amino acid residues such as
tryptophan, tyrosine, and phenylalanine; while circular dichroism measurements in the far-ultraviolet (UV) range (about 200 to 230 nm) probe
changes in secondary structure. Viscosity and ultracentrifugation can be
used to examine overall changes in shape and size. The experimental
data obtained from the various methods may be used to construct a transition curve with plateau regions on each side. These plateau regions are
used to normalize the curve to give a plot of fraction denatured as a function of pH, temperature, or concentration of an added chemical denaturant
(Fig. 4). If the data from the various methods results in essentially one
transition, this is considered to be a good indication that the protein
folds in a two-state manner. However, it has been pointed out that the
only way to demonstrate that the thermodynamics of folding for a particular
protein can be analyzed with a two-state model is to measure the change
in enthalpy for the refolding process calorimetrically (ΔH_{cal}) and to compare this value to the more indirect van't Hoff enthalpy (ΔH_{vH}), obtained

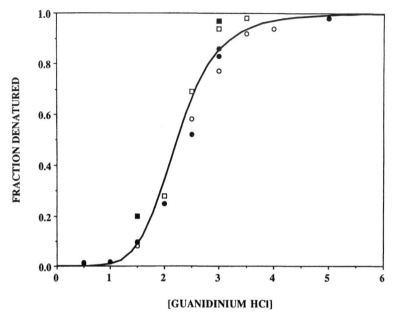

[GUANIDINIUM HCl]

Fig. 4   Reversible unfolding and refolding in GuHCl of recombinant-DNA-derived bovine α-interferon I. The fraction denatured was monitored by circular dichroism at 220 nm (open circles, unfolding; solid circles, refolding) and fluorescence emission at 336 nm after excitation at 292 nm (open squares, unfolding; solid squares, refolding). The solid line is a fit to the two-state model for protein folding. (From S. Shire and J. Maher, unpublished data.)

by the temperature dependence of the equilibrium constant for the unfolding-refolding transition [21]. The two-state model is considered applicable if the enthalpies are equivalent. This comparison has been made for a number of small proteins and the average value for $\Delta H_{cal}/\Delta H_{vH}$ is 1.05 ± 0.03, thus demonstrating for the small proteins studied that the deviation from a two-state process is less than 5% [22].

Recent work on larger multidomain proteins indicates that the separate domains may fold independently [23,24], and in some cases these separate foldings still appear to follow a two-state process. Despite the great success in interpreting folding of small proteins (range 12,000 to 20,000 daltons) and domains of larger multidomain proteins by the two-state model, kinetic analysis has demonstrated the presence of intermediates in the folding pathway. In particular, Baldwin and co-workers [25] have shown that for a sequential folding model, the presence of as much as 20% folding intermediates would result in ΔH_{cal} essentially in agreement with ΔH_{vH} because of baseline errors in the calorimetric measurements. Wetlaufer has discussed the protein folding problem in terms of a modular process which also involves the presence of intermediates. The oxidative refolding of bovine serum albumin is discussed in terms of this model. The refolding kinetics were monitored using circular dichroism as well as disulfide regeneration and antibody, bilirubin, and palmitate binding. The experiments demonstrated that the secondary structure as monitored by circular dichroism was attained before the disulfide regeneration and formation of antigenic determinants, which in turn occurred well before the binding of bilirubin and palmitate [18]. These results strongly suggest that after a rapid protein refolding step the important binding sites in the protein are formed as a final finishing process. For more detailed discussions and references concerning the kinetic aspects of protein folding, the reader is referred to a recent treatise on this subject [26].

The possible presence of significant amounts of intermediates in the protein folding pathway has obvious ramifications for designs of formulations for protein pharmaceuticals and would complicate the choice of formulation components, especially if intermediates were stabilized which prevented the protein from attaining its final native conformation.

Interactions Involved in Stabilizing Protein Structure

A great many of the conformations theoretically attainable by a polypeptide chain cannot be attained simply because of steric restrictions imposed by the backbone folding and the respective geometric placements of the functional side chains of the amino acid residues. The respective energies of the permissible conformations are determined by a delicate balance of intramolecular interactions which result in the highly cooperative process of protein folding. These interactions include electrostatic, hydrogen bonding, and hydrophobic interactions.

Electrostatic interactions. The properties and conformations of proteins are very dependent on the ionic strength as well as the pH of the solutions in which the proteins are stored [27,28]. The reasons for this are very likely due to the important role that electrostatic interactions play in determining the conformation of a protein. The charge interactions can involve fully ionized functional groups of side chains of some of the amino acids in a protein or may involve an interaction of partial charges. The partial charges in a protein are often due to the variety of polar covalent bonds, an important example of which is the hydrogen bond. Since hydrogen bonding is a common occurrence in proteins as well as having great importance in stabilizing protein conformations, it is treated in a separate subsection.

The variety of electrostatic interactions involves ionizing functional groups and can be described by Coulomb's law, where the energy between two charges q_1 and q_2 separated by distance r is given by

$$E = \frac{-q_1 q_2}{\varepsilon r}$$

The quantity ε is the dielectric constant, which is 1.0 in a vacuum and is 78.5 in water at 25°C. In hydrophobic organic solvents such as pentane the value is approximately 2, and therefore the interactions between the charges will be greater than in water. The dielectric constant is actually a macroscopic property of the solvent treated as a continuous medium in which the charges are immersed. Use of the dielectric concept should be valid when the number of solvent molecules between charges on a protein is large enough to be considered bulk solvent.* The choice of an appropriate dielectric constant for the interior of proteins has been a matter of great debate. The traditional view is that proteins are like oil droplets, with the polar amino acid side chains on the surface of the protein and highly hydrophobic residues in the interior. Many researchers have used macroscopic models with dielectric constants less than 4 to estimate electrostatic interactions. It is very likely that the interactions have been overestimated since recent work suggests that the apparent dielectric constant representing the interior of proteins may be considerably greater than for nonpolar liquids [32,33]. For example, measurements of redox potentials in cytochrome c suggest an apparent dielectric constant of about 55 when charges are separated by a distance of 12 Å [33].

The interactions between charges in proteins are long-range interactions since the interaction energy falls off as $1/r$, as opposed to short-range van der Waals interactions, which decrease more rapidly as $1/r^6$. In addition to interactions between charges, interactions can occur between charges and dipoles as well as between dipoles. Dipoles are simply asymmetric charge distributions whereby the dipole moment $\mu = zd$, where d is the distance separating charges within the dipole of equal and opposite magnitude, z. The interaction energy, E, between a dipole and an ion varies as $1/r^2$, whereas a dipole-dipole interaction varies as $1/r^3$.

Because this interaction energy involving dipoles falls off more rapidly than do charge-charge interactions, it probably does not contribute as much. However, the large number of permanent dipoles in a protein, which includes the peptide bond, can make significant contributions to the overall conformational energy of the protein. It has been suggested that α-helices in proteins possess a dipole moment that is equivalent to a separation along the helical axis of 0.5 formal charge units [34,35]. However, estimates of the interaction energy were performed with the protein in a vacuum ($\varepsilon = 1$), and therefore the formal charges may be overestimated.

Hydrogen bonds. Many of the chemical bonds within amino acids are polar covalent bonds where the charge distribution is asymmetric along the bond and the atoms involved in the bond can be represented by partial charges. A sharing of a hydrogen atom between two polar atoms is referred to as a hydrogen bond. Although the exact nature of this bond is not clear, it is extremely important in maintaining protein conformation, and the typical types of hydrogen bonds in proteins are shown in Fig. 5. The formation of hydrogen bonds between the carboxyl and amino functional groups of the peptide bond stabilizes secondary structures such as α-helices and β-sheets. Some side-chain functional groups can also participate in hydrogen bonding and therefore stabilize tertiary structure in proteins. In particular, residues with —OH and —NH functional groups can serve as hydrogen donors, whereas almost any oxygen atom and nonprotonated nitrogen atom can serve as a hydrogen acceptor (Fig. 5). Typical values for the interaction energies of these bonds have been estimated to be between -3 and -6 kcal/mol. Although these types

---

*Many electrostatic models for proteins use so-called macroscopic models, but microscopic models have also been proposed. The breadth of this topic is too detailed for this review but has been covered in the reviews by Warshel and Russell [29], Matthew [30], and Hoenig et al. [31].

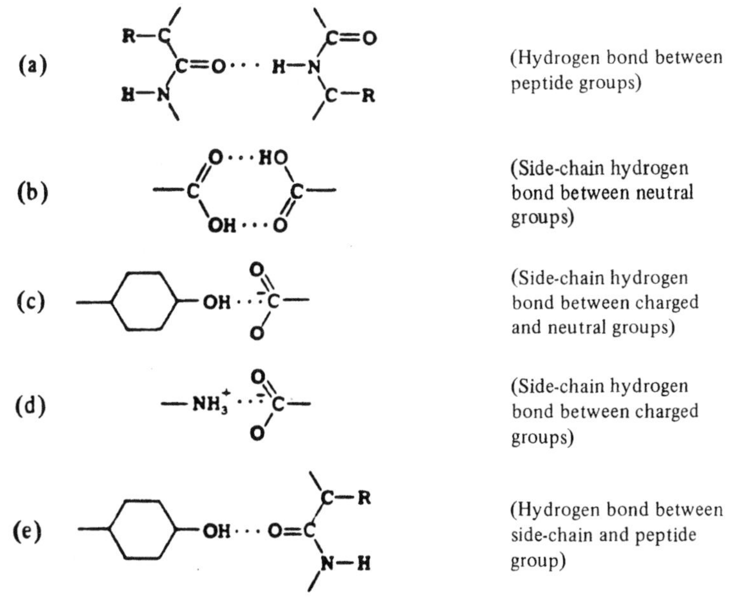

(a) (Hydrogen bond between peptide groups)

(b) (Side-chain hydrogen bond between neutral groups)

(c) (Side-chain hydrogen bond between charged and neutral groups)

(d) (Side-chain hydrogen bond between charged groups)

(e) (Hydrogen bond between side-chain and peptide group)

Fig. 5   Types of hydrogen bonds found in proteins.   (From Ref. 155.)

of bonds are considerably weaker than the usual covalent bond, the large number of them in a protein can make a significant contribution to a protein's stability. An extreme version of the hydrogen bond involves the side chains between charged functional groups, where the hydrogen donor has virtually a full formal positive charge. This type of interaction can be considered an ion pair, generally referred to as a salt bridge, and it undoubtedly stabilizes structures in proteins and may even be essential in the regulation of function by ligands in proteins [36].

Hydrophobic interactions. The forces that are responsible for the immiscibility of water with hydrocarbons are major contributors to the maintenance of protein structure. These forces are actually a property of water and its unusual ability to hydrogen-bond extensively [37,38]. The properties of liquid water, including its high boiling point and especially its decreased density upon freezing, are due to the extensive network of hydrogen bonds, where each water molecule can participate as both an acceptor and a donor. The highly ordered tetrahedral geometry that results from this bonding leads to a greater volume being occupied by the water molecules and hence a decreased density upon freezing. The degree of hydrogen bonding in water is highly dependent on the thermal fluctuations of the water molecules. The insertion of a hydrocarbon chain into the aqueous phase would require the breaking of hydrogen bonds, leading to an unfavorable (positive) enthalpy contribution to the free energy. To compensate for this, the water molecules would attempt to increase the degree of hydrogen bonding by becoming more ordered around the hydrocarbon chain. This ordering of water molecules may be substantially greater than that in the bulk water and can result in a large unfavorable decrease in entropy. The association of the hydrocarbon into a separate phase [39,40] would prevent this large decrease in entropy and thus is an entropically driven process.

Proteins are polymers with hydrophilic and hydrophobic side groups, and in general, the hydrophobic amino acids will tend to be removed from the aqueous phase and become folded into the interior of the protein. This is effectively the same as decreasing the surface area of the hydrophobic regions of a protein since a folded polypeptide chain will have a decreased surface area accessible to the solvent. In fact, correlations have been found to exist between the accessible surface area of amino acid side chains and the free energy of transfer of the amino acids into water [41]. This dependency of hydrophobicity on surface area has the interesting consequence that the hydrophobicity of a large protein is approximately equivalent to the sum of its constituent parts. Hydrophobicity scales as summarized by Rose et al. [42], based on various properties of amino acids, such as solubility in water or vapor pressure measurements, have allowed researchers to estimate the hydrophobicity of proteins. Self-association of some proteins has also been found to be an entropically driven hydrophobic process, where the accessible surface area of the protein is decreased upon association [43]. A variety of theories have also been proposed to account for the thermodynamic properties of proteins on the basis of this hydrophobic effect [44,45].

Strong denaturants such as urea and guanidinium hydrochloride are believed to unfold a protein by decreasing the hydrophobic interactions. These agents are often referred to as chaotropic agents because of their

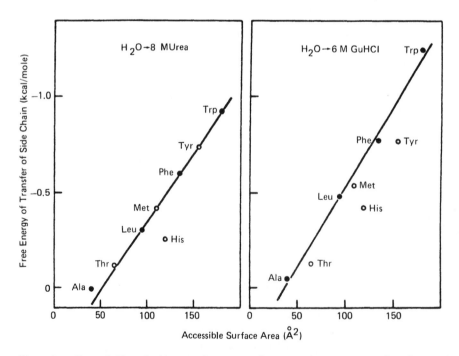

Fig. 6   Correlation between the accessible surface areas of amino acids and their free energies of transfer from water to either 8 M urea or 6 M GuHCl. (From Ref. 156.)

ability to decrease ordered water structure, which, as explained, is the driving force for the interaction. Figure 6 shows the correlation between the free energy of transfer of amino acid side chains into either 8 M urea of 6 M guanidinium hydrochloride and the accessible surface area of the amino acid side chain. Other salts also have an effect on hydrophobic interaction, depending on the ions comprising the salts. The rank ordering of the ions with respect to their ability to increase hydrophobic interactions is given by the Hofmeister series:

$$\text{Cations:} \quad Mg^{2+} > Li^{+} > Na^{+} > K^{+} > NH_4^{+}$$

$$\text{Anions:} \quad SO_4^{2-} > HPO_4^{2-} > CH_3COO^{-} > Cl^{-} > NO_3^{-} > ClO_3^{-} >$$

$$I^{-} > ClO_4^{-} > SCN^{-}$$

This ordering of ions, as discussed by Von Hippel and Schleich [46], was discovered over 100 years ago by Hofmeister [47] in studies of the effectiveness of salts to precipitate serum globulins at a given ionic strength. This series correlates well with the effect of the salts on the surface tension of water [48]. These changes in surface tension may arise via increases in ordered water structure, an effect opposite to that described above for chaotropic agents. Alternatively, the salt ions may undergo a preferential hydration, resulting in the reduction of the hydration layer surrounding the protein molecules [49]. In either case, changes in the solvation layer around the protein will undoubtedly affect the hydrophobic interaction of neighboring protein molecules.

III. CHARACTERIZATION OF PROTEINS AS PHARMACEUTICALS

A. Criteria for Characterization

Recombinant DNA technology will allow pharmaceutical chemists to develop an almost unlimited new class of drugs. Despite the optimistic outlook for the future, a problem that will be encountered concerns the final form of the protein that is formulated compared to its form in vivo. Initial research into the development of a protein pharmaceutical involves isolating the protein and demonstrating a particular activity. However, large polypeptide chains could be made as precursors in vivo, which ultimately are processed into the smaller active forms, such as the conversion of proinsulin to insulin [50] and the processing of polypeptide hormones [51]. In vivo proteolytic cleavages could also result in multiple active forms, as for example the two-chain and single-chain forms of thrombolytic agents such as plasminogen activators and urokinases [52,53]. A great many of the protein pharmaceuticals that are being developed are also glycoproteins, and heterogeneity exists in the number of carbohydrate chains attached, as well as the composition and sequence of the carbohydrate chains.

The proteins that are chosen to be pharmaceuticals will have to be characterized, and safety and efficacy demonstrated. Since the isolated proteins may have properties that differ from the proteins produced in vivo, it will be very difficult, if not impossible, to introduce these proteins into the patient in exactly the way that the body does. The

appropriate choice of formulation and/or administration design may minimize these differences and result in a safe and efficacious product.

    Compounding this problem is the fact that most protein pharmaceuticals
of the future will be produced by recombinant DNA technology. Differences between the natural and recombinant protein can occur as a result
of posttranslational modifications to the protein that are dependent on the
host organism that is used to express the recombinant-DNA-derived protein. Some of these modifications are summarized below. However, unless
it is demonstrated that the safety and/or efficacy of the final product are
compromised, it may be unnecessary and even "impractical to remove all
detectable components that are not identical" to the main (active) component
[54].

Glycosylation

Carbohydrates can be attached to proteins via the hydroxyl groups of
serine (ser) or threonine (thr) residues (O-linked) or via the amide
nitrogen atom of asparagine (asn) (N-linked). The linkage to the asn
residues occurs in the sequence -asn-X-ser- or asn-X-thr. The details
of the biosynthesis as well as basic structures of the carbohydrate chains
that are added to proteins are summarized nicely by Sharon and Lis [55].
The carbohydrate chains attached to proteins can be rather complex and
the particular kinds of carbohydrate added appear to be dependent on
the host organism used for expression of the recombinant protein [56].
In particular, proteins that are normally glycosylated will not be glycosylated when expressed in bacteria such as *Escherichia coli*. Recently it has
been demonstrated that glycosylation is also dependent on the tissue source
of isolated proteins [159], and therefore it is likely that the choice of cell
lines from the same organism for expression of proteins will determine the
nature and disposition of oligosaccharides on the protein. The purpose of the
glycosylation is not well understood but probably affects the folding of the
protein, its immunogenicity, and its solubility. It has also been shown that
glycosylation can be related to the biological half life and activity of a protein
[57,160,161], as well as the routing of the protein to its final destination [58].
The differences in details of the carbohydrate structure between recombinant
proteins (e.g., those expressed in a mammalian tissue culture system or in
yeast) and the native glycoprotein could have a great impact on these properties, and thus careful comparisons between native proteins and recombinant
proteins should be made whenever possible.

Specific Modification of Amino Acid Residues

Posttranslational modifications of amino acids can take place and include
hydroxylation of proline and lysine (e.g., in collagen), phosphorylation of
hydroxyl groups of serine or threonine, and methylation of the α-amino
groups or side chains of lysine, arginine, histidine, and glutamic acid [59].
These types of modifications affect the secretion, targeting, and possibly the
longevity of proteins.

    Modifications to the N and C termini of a protein that occur in vivo may
not occur in bacterial hosts. Particular examples of these modifications are
acetylation of the N-terminal amine and amidation of the C-terminal carboxyl.
These modifications may be essential for pharmacological activity, as, for
example, in the hormone falcitonin [60]. Another N-terminal modification that
can occur in recombinant proteins produced in prokarvotes is the extension
of the N-terminal sequence with a methionine (met) residue. The triplet

codon used for this residue (AUG) is required for initiation of protein synthesis, and the met may not be cleaved off by the bacterial host in which it is expressed. While recent evidence suggests that the intracellular half-life of a protein in vivo may be greatly influenced by the amino acid at its N-terminus [61], the effects of any modifications or changes in sequence on the circulating half-life and/or pharmacological activity of a protein may be negligible. Therefore, these effects should be assessed on a case-by-case basis.

Arrangement of Disulfide Bonds

It has been suggested by Garvin that the choice of expression host may influence the arrangement of disulfide bonds in recombinant proteins [62]. In particular, it has been noted that an enzyme which promotes disulfide bond formation is bound to the rough endoplasmic reticulum (RER), which E. coli do not possess. Furthermore, it appears that E. coli as well as other prokaryotes do not have disulfide interchange activity but do possess the thioredoxin system, which is known to be active in the reduction of protein disulfides. This suggests that proteins produced in E. coli are in a reducing environment. However, it has been possible to produce successfully fully active, relatively small recombinant proteins in E. coli, such as human alpha interferon (MW 19,000), which are dependent on proper disulfide bond formation for their biological activity [63]. Undoubtedly, larger proteins with a great many disulfide bonds may be more difficult to refold with the proper disulfide cross-links. Expression in other systems such as yeast or mammalian cell lines have helped circumvent this difficulty.

B.   Physical Properties

Size and Shape

The size and shape of a protein pharmaceutical can be determined by a number of analytical methods. One of the most common methods for size determination has been the use of sodium dodecyl sulfate polyacrylamide gel electrophoresis (SDS-PAGE), discussed in greater depth in Chap. 6. This method assumes that the binding of SDS to proteins results in essentially the same charge density for all proteins and therefore that the mobility through the gel matrix is related to the hydrodynamic volume and shape of the protein with bound detergent. The problem with this type of characterization is that this method denatures the protein and is capable only of giving the apparent size of the protein monomer or covalently linked monomers (such as through disulfides), which may not be the size of the active protein. Nondenaturing techniques, such as gel permeation chromatography or analytical ultracentrifugation, can be used to determine the size of the protein in solution, usually in the presence of formulation components. The combined use of both methodologies ascertains whether a protein produced by recombinant DNA techniques has the size expected from the DNA sequence and at the same time determines if the protein exists as an oligomer in solution. In this manner it was shown that recombinant-DNA-derived human γ-interferon exists as a dimer under most solution conditions [64].

  The shape of the protein as well as the size can also be characterized by using a variety of hydrodynamic techniques, including viscometry, analytical ultracentrifugation, and laser light scattering. Mathematical expressions are available for ellipsoids of revolution that can be used to compute hydrodynamic properties such as the frictional coefficient [65].

These computed frictional coefficients can be used to evaluate the diffusion or sedimentation coefficient, which can then be compared to experimentally determined values. The changes in size and shape may be small, but careful measurements will detect the differences. As an example, small changes in sedimentation coefficient have been used successfully to distinguish between tetrahedral and square planar geometry for the hemoglobin tetramer [66].

Conformation

A protein's conformation is the most elusive of its physical properties in that it may be the most difficult not only to ascertain but to maintain. Although the specific interactions that give rise to a protein's unique three-dimensional structure impart a certain degree of stability, there may still be more than one allowable conformation even under relatively "mild" conditions. Previously, protein chemists may not have been concerned with minor conformational "impurities," as they would normally have been considered to be statistically irrelevant. However, when one is developing a protein formulation, particularly one intended for human use, the nature of other conformers, even those which are present at levels of less than 1%, becomes very important indeed. For example, one must establish that other conformational forms present have no undesirable chemical, physical, or biological effects.

The structural features that determine a protein's conformation may be elucidated in several ways. The simplest approaches are those based on spectroscopic properties, which for a protein arise not only from the strong absorption of peptide bonds in the far-UV region but also from the absorption of the aromatic chromophores tryptophan, tyrosine, and phenylalanine in both the far-UV and near-UV regions. As the electronic transitions that these chromophores undergo are very sensitive to their environment, the employment of any of several spectroscopic techniques [UV, infrared (IR), Raman, fluorescence, and especially circular dichroism (CD)] on a protein in two or more solvent systems may yield information on its conformation [67-74].

Magnetic resonance techniques have also been employed to yield detailed information on physical changes at particular sites within proteins as a result of local or global conformational changes, particularly near the active site of enzymes, yet data analysis for complex proteins is very difficult [75-78].

By far the most valuable and telling physical technique for determining protein conformation is x-ray crystallography. However, although there have been vast improvements in optics and computer technology, crystallization of a purified protein under appropriate conditions remains the limiting factor in obtaining its native three-dimensional structure. Even once obtained, it is not an easy task to establish a correlation between the crystal structure and that in solution [78-83]. Yet this technique still provides valuable clues as to the protein's physical and chemical stability and may yield important information on structure-function relationships as well.

In any case the maintenance of a protein's conformation is of utmost importance, as incorrectly folded structures may not only be inactive themselves but may be more susceptible to covalent modification and/or may aggregate and precipitate [84].

Solubility and Charge

The solubility of a protein is a useful indicator of changes that may occur
in the protein conformation. If a protein denatures, more hydrophobic
groups may be exposed to the aqueous phase from the interior of the
protein. To prevent a decrease in entropy from the ordering of water
molecules, the contact of hydrophobic groups with the aqueous phase must
be minimized. If the protein cannot refold into a conformation that minimized this contact, an alternative mechanism must be found. Generally, a
minimization of surface area can be realized by the association of protein
molecules, in much the same way that hydrocarbons tend to associate into
a separate phase when placed into water. As the level of association increases, protein solubility decreases and precipitation may occur. This
decrease in solubility is a result of hydrophobic interactions. However,
decrease in solubility could also occur due to the balance of charge interactions and hydrophobic interactions. Proteins tend to be least soluble at
their isoelectric point, when the net charge on the protein is zero. Increasing the net charge (i.e., via an increase in pH) will result in favorable interactions with the polar water molecules and may also increase the
repulsive interaction between the protein macromolecules, preventing them
from associating and ultimately increasing their solubility.

C. Biological Properties

Antigenicity and Immunogenicity

The manner in which the immune system responds to an antigen such as a
protein, triggering the production of antibodies, is referred to as immunogenicity and is not well understood. The recognition of particular sites
(epitopes) on the antigen and binding by antibodies is referred to as
antigenicity.

Proteins made by a particular organism are recognized by the immune
system as "self" proteins and normally do not elicit an immune response.
However, misfolded or denatured forms of proteins may be immunogenic.
The regions of a protein that are recognized by an antibody are referred
to as antigenic determinants [85]. A determinant comprising a contiguous
sequence of amino acids is referred to as a continuous determinant or epitope. An antigenic region formed from amino acids located at different
points of the polypeptide chain brought together in close proximity to each
other is known as a discontinuous determinant (Fig. 7). A discontinuous
determinant will be very dependent on the folding of a peptide chain, and
a continuous determinant not present on the properly folded molecule may
become accessible to solvent in a misfolded form of the protein. The immunogenicity of a protein may also be enhanced due to self-association of
the protein. Recent theories and experiments suggest how the molecular
weight and antigen density of a polymer might affect the production of
antibodies [86,87].

Protein pharmaceuticals produced by recombinant DNA technology may
also possess a different set of antigenic determinants because of the
differences between the recombinant and natural protein. Differences in
glycosylation and amino acid modifications were discussed previously.
Alterations of carbohydrates could result in altered antigenic determinants.
Usually, determinants are found on the surface of the proteins, with

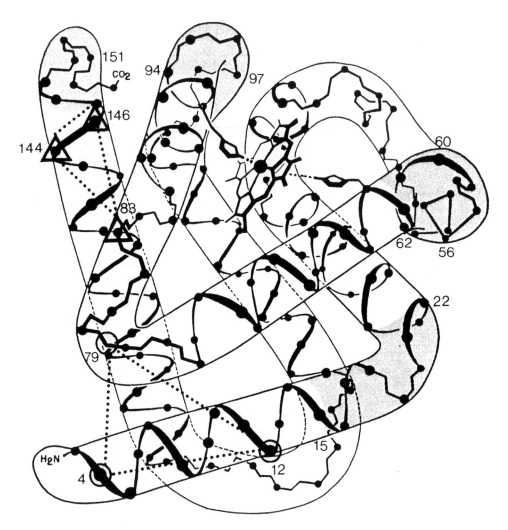

Fig. 7   Some of the reported antigenic determinants for sperm whale myoglobin. For clarity, only the α-carbon backbone is shown. Sequential
antigenic determinants at residues 15–22, 56–62, 94–99, and 146–151 are
shown shaded. Amino acid residues that contribute to a discontinuous
determinant are labeled either with a shaded triangle (residues 83, 144,
and 145) or a shaded circle (residues 4, 12, and 79). Dotted lines are
drawn from the centers of the α carbon positions solely to demonstrate
the proximity of the residues to each other in the tertiary structure,
although far apart in the primary sequence. (Adapted from Ref. 157.)

increases in hydrophilicity generally resulting in increased antigenicity
[88]. Lack of a carbohydrate chain may expose regions of the protein to
solvent which were previously inaccessible to solvent and thus produce a
new set of antigenic determinants. The alteration of antigenic determinants
on the protein may result in the protein being treated as a foreign antigen
by the immune system, with the concomitant production of antibodies.
Some of these (neutralizing) antibodies may cross-react with the native
protein and ultimately decrease the activity and/or alter the clearance
time from serum of the protein drug. Some formulations may prevent
immunogenicity problems by maintaining the molecule in the properly
folded conformation as well as decreasing protein self-association, whereas
immunogenicity problems caused by chemical differences between a recombinant DNA produced pharmaceutical protein and the natural protein may
be more difficult to resolve.

Biological Activity

Biological activity is obviously the most important property of a protein
pharmaceutical. Methods of measurement of biological activity are as
variable as proteins themselves. Activity is often assessed in vivo using
one or more response variables in an appropriate animal model, with
activity given as USP or International Units per milliliter based on an
official reference standard (e.g., USP, WHO). Specific examples are the
USP rabbit hypoglycemic assay for insulin [89] and the rat weight gain
bioassay for human growth hormone (hGH) [90]. However, with respect
to using biological assays to assess a protein's stability under various con-
conditions, or in defining structure-function relationships, many in vitro
assays serve just as well or better. For example, radioreceptor assays
(RRA) may be used to relate binding of a protein hormone (e.g., hGH) to
a cellular receptor and may even be used to assess activity of protein
fragments, produced via enzymatic degradation [91,92]. Biological activity
as a measure of protein stability may also be assessed using a radioimmunoassay (RIA), with losses in antibody binding traced to cleavage of peptide
bonds or other physical alterations [91–93].
     The important point to remember, however, is that even though a pro-
tein's function or full biological activity is strongly dependent on its
native three-dimensional structure, sometimes major changes in physical or
chemical properties are well tolerated, producing only minor losses in assay
activity, while in other cases very subtle physical alterations result in
significant loss of biological activity. Therefore, in characterizing and
formulating proteins for pharmaceutical use it is imperative to use a wide
variety of physical, chemical, and biological techniques to assure safety
and efficacy.

IV. PROTEIN STABILITY AND FORMULATION

A. Potential Routes of Inactivation

A protein may undergo a number of chemical or physical alterations with
time, depending on storage conditions. The changes that occur may or
may not affect the safety and efficacy of the protein as a drug. In
particular, the understanding of the relationship of protein primary structure to function is not sufficiently well understood that one can predict a

priori the effect of an alteration on stability or function. Therefore, the alterations in a particular protein pharmaceutical must be investigated on a case-by-case basis to determine if the alterations truly affect the performance and safety of the drug.

Chemical Alterations

Deamidation. The amide side chains of glutamine (gln) and asparagine (asn) can deamidate to glutamate (glu) and aspartate (asp) residues. At one point this chemical reaction was considered essentially a nuisance that occurred during the isolation and purification of a protein. More recent data suggest that the rates of deamidation may be correlated with the biological half-life of the protein [94]. The activity of a protein as well as its physical parameters may be greatly altered by the deamidation of gln and asn [95]. Thermal inactivation of proteins has also been related to the deamidation of these residues [96]. However, in some cases deamidation may not affect the biological potency or immunogenicity of a protein. For example, certain formulations of bovine or porcine insulin, when exposed to elevated temperatures, undergo significant deamidation and aggregation. Yet in clinical studies, these samples have shown no significant loss of biological potency or increase in immunogenicity [97,98].

The effects of buffers, salts, temperature, ionic strength, and pH on the rate of deamidation reactions have been studied extensively using small peptides containing asn or gln [99]. The deamidation rate of one model peptide is in good agreement with the measured deamidation rate of the same peptide sequence in an intact protein [100], and thus peptides may be good model systems for the study of deamidation reactions in some proteins. Increasing temperature increases the rate of deamidation in peptides and the reaction appears to be acid and base catalyzed, with the slowest rates of deamidation occurring near pH 6. The rates of these reactions, which are first order in peptide concentration, are also highly dependent on the anion used in the buffer preparation, as shown in Table 1. In particular, phosphate buffers are most prone to increase deamidation rates in proteins. These types of effects can be controlled by the pharmaceutical chemist in order to increase the stability of the product. The rates of deamidation of polypeptides have also been proposed to be very dependent on the sequence of the amino acids neighboring the asn or gln side chains (as summarized in Table 2). Asn residues tend to be more labile than gln residues and the rates of the reaction can vary almost 3000-fold in small peptides (Table 2), depending on the neighboring residues.

Recently, it has been demonstrated that the mechanism of deamidation involves the formation of a succinimide intermediate which, depending on the bonds cleaved during hydrolysis, can also result in transpeptidation or racemization [162,163]. The formation of such a cyclic structure will undoubtedly be influenced by the folding of the polypeptide chain around the deamidation site. Kossiakoff has shown that in trypsin, the propensity toward deamidation of specific asn side chains is related to "distinct local conformation and hydrogen-bonding structure" and not to neighboring residues [101]. Indeed, it is possible that for most proteins the rate of deamidation is more influenced by tertiary and quaternary interactions than by primary sequence as seen in small peptides. In any case, the influence of sequence on tertiary structure and deamidation obviously cannot be controlled by the formulation chemist but would require careful engineering of recombinant-DNA-derived protein pharmaceuticals.

Table 1 Variation of Deamidation Rate of Asparaginyl Peptides with Different Anions at 37°C

Peptide	Buffer type	pH	Ionic strength	Half-time of deamidation reaction
GlyThrAsnGluGly	Borate	10	1.0	34 hr
	Carbonate	10	1.0	11 hr
	Phosphate	10	1.0	5 hr
	Borate	7.4	0.15	27 days
	Phosphate	7.4	0.15	12 days
GlyArgAsnArgGly	Tris-HCl	7.4	0.15	115 days
	Phosphate	7.4	0.15	20 days

Source: From Ref. 148.

Oxidation and reduction. Several side groups in proteins are susceptible to either oxidation or reduction. Proteins containing free cysteine (cys) are especially prone to oxidation, as the thiol (SH) group ($pK_a \sim 8.3$ for the free amino acid) is the most reactive R group near neutral pH. Not only is the thiolate anion a potent nucleophile susceptible to alkylation, addition across double bonds, and complexation with heavy metals, but it also undergoes spontaneous oxidation in air to disulfide (—S—S—), especially in the presence of trace metals [102]. The resultant oxidation product of two cys residues in proteins is called cystine. When both free cys and cystine are present in proteins, they may undergo exchange at alkaline pH, leading to a variety of covalent isomers. Disulfides, whether natural or process induced, may be reduced by β-mercaptoethanol (β-ME) or dithiothreitol (DTT), although complete reduction usually requires denaturing conditions. Disulfide bonds may also be reduced by excess hydroxide to hydroxysulfides. The sulfur atom of methionine (met) is also prone to alkylation or reversible oxidation to a sulfoxide. More vigorous treatment of the sulfoxide results in the further irreversible reduction to a sulfone.

Other R groups in proteins are not nearly as susceptible to oxidation or reduction as cys, although the carboxyl groups of aspartic acid (asp), glutamic acid (glu), and the C-terminus may be reduced to alcohols under certain conditions, and the indole ring of tryptophan (trp) is somewhat susceptible to irreversible oxidation [102]. The oxidation of these R groups may also be enhanced upon exposure to light of certain wavelengths [103].

Covalent aggregation/cross-linking. The variety of reactions that the N-terminus, C-terminus, R groups, and other groups (e.g., carbohydrate, lipid, etc.) undergo makes possible the formation of covalent cross-links within proteins, especially at elevated temperatures. Such cross-linking can lead to significant losses in activity. The most common cross-link is via the disulfide bond, formed within or between protein subunits by free SH groups or rearrangement of existing S—S bonds. At alkaline pH, intermolecular S—S interchange can result in extensive aggregation. In addition, hydrolysis of some disulfides produces dehydroalanine, which

Table 2 Deamidation Half-Times for Peptides[a]

Peptide	$t_{\frac{1}{2}}$ (days)	Peptide	$t_{\frac{1}{2}}$ (days)
GlySer*Asn*HisGly	6[b]	GlyThr*Asn*AlaGly	68[b]
GlyThr*Asn*ArgGly	16[c]	GlyAsp*Asn*IleGly	75[b]
GlyThr*Asn*Glu	16[b]	GlyMet*Asn*AlaGly	78[c], 77[b]
GlyArg*Asn*AlaGly	18[b]	GlyGlu*Asn*ProGly	80[b]
GlyIle*Asn*SerGly	18[c]	GlyTyr*Asn*AlaGly	85[b]
GlyArg*Asn*ThrGly	28[c]	GlyGly*Asn*AlaGly	87[b]
GlyCys*Asn*AspGly	28[c]	GlyTrp*Asn*AlaGly	87[b]
GlyAsp*Asn*AlaGly	44[b]	GlyLys*Asn*LysGly	94[b]
GlyHis*Asn*AlaGly	45[b]	GlyAla*Asn*AlaGly	95[b]
GlyPhe*Asn*AlaGly	47[b]	GlyCys*Asn*IleGly	100[c]
GlyGlu*Asn*AlaGly	49[b]	GlyPro*Asn*AlaGly	100[b]
GlySer*Asn*AlaGly	52[b]	GlyVal*Asn*AlaGly	111[b]
GlyAla*Asn*LysGly	54[b]	GlyArg*Asn*LeuGly	113[c]
GlyArg*Asn*ArgGly	71[c], 38[b]	GlyPhe*Asn*ThrGly	123[c]
GlyLys*Asn*AlaGly	61[b]	GlyGlu*Asn*ValGly	145[b]
GlyCys*Asn*AlaGly	68[b]	GlyLeu*Asn*AlaGly	217[b]
GlyPro*Asn*LeuGly	277[b]	GlyGly*Gln*AlaGly	418[b]
GlyIle*Asn*AlaGly	507[b]	GlyVal*Gln*LysGly	421[b]
GlyHis*Gln*AlaGly	96[b]	GlyAla*Gln*AlaGly	538[b]
GlyMet*Gln*AlaGly	102[b]	GlyPhe*Gln*GlyGly	553[b]
GlyAla*Gln*CysGly	113[b]	GlyLeu*Gln*AlaGly	663[b]
GlyPro*Gln*GlyGly	144[b]	GlyTyr*Gln*AlaGly	689[b]
GlyAla*Gln*LysGly	157[b]	GlyTrp*Gln*AlaGly	713[b]
GlyAla*Gln*ArgGly	188[b]	GlyIle*Gln*GlyGly	735[b]
GlyAsp*Gln*AlaGly	209[b]	GlyTyr*Gln*LeuGly	884[b]
GlyLys*Gln*ArgGly	223[b]	GlySer*Gln*AlaGly	889[b]
GlyGlu*Gln*AlaGly	226[b]	GlyPhe*Gln*AlaGly	1060[b]
GlyLys*Gln*LysGly	251[b]	GlyIle*Gln*AlaGly	1087[b]
GlyLys*Gln*AlaGly	280[b]	GlyAla*Gln*IleGly	1094[b]
GlyArg*Gln*ArgGly	285[b]	GlyPro*Gln*AlaGly	1114[b]
GlyArg*Gln*GlyGly	305[b]	GlyVal*Gln*AlaGly	3278[b]
GlyArg*Gln*AlaGly	389[b]	GlyThr*Gln*AlaGly	3409[b]

[a]Data are for 1 mM peptides at 37°C.
[b]pH 7.4 phosphate buffer, ionic strength 0.2.
[c]pH 7.5 phosphate buffer, ionic strength 0.15.
Source: Adapted from Ref. 99.

may subsequently react with the amino groups of lysine (lys) or the
N-terminus, producing another type of cross-link [104]. In other cases,
cross-links may be formed by esterification of the carboxyl groups of asp,
glu, or the C-terminus with the hydroxyl groups of serine (ser) or
threonine (thr), or by amidation of these same carboxyl groups with the
free amino groups of lys or the N-terminus. Additionally, some carbohydrate groups of glycoproteins, such as those with a free anomeric carbon,
may also form cross-links with the amino groups of lys or the N-terminus.
There are other types of cross-links that occur naturally in some proteins, such as those between lys and aldehydes or amides [59,105], although these are not important degradation pathways.

Some of the different types of cross-links may be distinguishable by a
combination of electrophoretic techniques, such as SDS-PAGE (under both
reducing and nonreducing conditions) and isoelectric focusing (IEF).

The prevention of covalent aggregation in protein formulations is obviously of utmost importance since such processes are not only irreversible
but invariably result in the production of inactive species which may also
be immunogenic.

Carbamylation. Production of proteins by recombinant DNA technology
using bacterial hosts often results in bacteria with one or two optically
dense bodies when viewed by phase contrast microscopy. The extraction
of the protein from these dense bodies requires a strong denaturant, such
as guanidinium hydrochloride or urea [106]. Urea is a preferred solvent
since the extracted protein can then be chromatographed using ion-exchange
chromatography. However, urea contains small amounts of cyanate which
can react with N-terminal and ϵ-amino groups of lys to form the stable
carbamyl form of the amino acid [107]. This reaction is highly pH dependent and involves reaction of the unprotonated amine with electrically
neutral cyanic acid. Whether the N-terminal amino group and/or ϵ-amino
groups are carbamylated will depend on the pH of the urea buffer since
the pK_a values for the N-terminal amino group and ϵ-amino groups can
differ by 2 to 3 pH units. Carbamylation should be avoided since the
modifications of the amino groups may have profound effects on the immunogenicity as well as the biological activity of the protein [108–110].
Modification of the N-terminal amino group by cyanate also prevents the
amino acid sequence of the polypeptide chain from being determined with
the standard Edman chemistry [107].

Proteolysis. The hydrolysis of peptide bonds within a protein backbone can obviously reduce or destroy its activity. This may occur when
the protein is subjected to harsh conditions, such as prolonged exposure
to extremes of pH or high temperatures [104] or when the protein is exposed to proteolytic enzymes, which may be derived from several sources.
The most obvious source is through bacterial contamination, which may be
avoided by storing the protein in the cold under sterile conditions. Proteases may also be introduced during the isolation and recovery of recombinant proteins through copurification from cell extracts or culture
fluid. The control of proteolysis will depend on the particular proteases
that copurify with the recombinant protein. Manipulation of the solution
conditions during purification and/or addition of protease inhibitors can
minimize this problem [111]. Furthermore, some proteins possess auto-
proteolytic activity (i.e., they have the ability to cleave themselves).
Although this situation may be necessary in controlling a protein's level

or function in vivo, it is highly undesirable for a pharmaceutical, especially if cleavage results in loss of biological activity. Therefore, one
must choose formulation conditions that minimize the protein's autocatalytic
activity while maintaining its desired integrity and functionality.

Physical Alterations

Conformational changes. Induced changes in conformation can occur
frequently within proteins. Indeed, it has been stated that "native proteins are only marginally stable" [112]. Alterations in environment may
lead to many conformational forms, often referred to as the "denatured
state" of a protein, although they may be reversible to the native form.
Conformational changes may be induced through a variety of mechanisms,
including changes in temperature, pH, ionic strength, or solvent composition. In fact, the "alloplastic" effects of solvent constituents on conformation have been described by Klotz as being of two types: conformational
changes brought about by direct interaction of the solvent constituent
with the protein (e.g., O_2 binding to hemoglobin) and those produced via
changes in chemical potential or structure of the solvent [113]. In any
case, some of the resultant conformers may themselves be inactive, and
conditions leading to their formation should be avoided on that basis
alone. However, even subtle changes such as partial unfolding are cause
for concern when they increase the probability of subsequent chemical
and other physical alterations, which in turn lead to irreversible loss of
activity [84,104]. Very often the rates of such subsequent processes are
too fast to allow isolation or detection of the various conformational states
that have been postulated to exist as intermediates or nuclei for the noncovalent or covalent aggregation that follows.

Noncovalent aggregation (self-association). The self-association of
protein subunits (either native or misfolded) may occur readily under
certain conditions and can lead to precipitation and loss of activity. In
fact, noncovalent aggregation is one of the primary mechanisms of protein
degradation. This process is dependent on the formulation conditions
chosen and is a result of hydrophobic interactions.

In some cases, protein aggregation may be limited and the oligomers
produced may remain in solution (e.g., as for insulin hexamers). Such
soluble aggregates may affect (increase or decrease) the stability of a
protein pharmaceutical. Therefore, protein self-association processes
should be well characterized. One of the most accurate techniques for
assessing the concentration distribution of monomer and oligomer species
in solution is analytical ultracentrifugation using the sedimentation
equilibrium method [114].

In other cases, protein aggregation may become quite extensive,
resulting in the production of insoluble oligomers. Such a process may
be rationalized in terms of the dynamic interplay between the thermodynamic and kinetic parameters involved. Thus the rate of favorable
conformational change concomitant with the addition of each monomer unit
slows with each successive step, due to the increasing global constraints
made by the growing aggregate. Eventually, the surface area occupied
by hydrophobic residues becomes overwhelming, and the rate of precipitation exceeds the rate of association.

In the case of insulin, many types of noncovalent aggregates have
been observed, depending on the stress used to induce their formation.

These include both the soluble forms and insoluble forms, ranging from microcrystals to globular clumps of elongated fibrils produced by isoelectric precipitation, freeze-heat-acid, or mechanical agitation [115]. Such variations in morphology are common among other proteins as well, and it is likely that the types of monomers present (i.e., native versus clipped or unfolded or refolded forms) dictate the final morphology of the precipitate. Finally, the relationship between noncovalent and covalent aggregation processes is unknown, although it is very likely to be protein and formulation dependent.

B. Processing and Formulation Conditions

Temperature

Most proteins can be reversibly denatured with increases in temperature and very often go through a sharp transition over a small temperature range. Calorimetric measurements have shown that for some proteins a pre-denaturational transition can occur, presumably related to small conformational changes prior to large-scale unfolding of the protein structure [22]. Continued increase in temperature can result in further denaturation, which is not reversed by subsequent cooling. This type of irreversible denaturation has been studied in enzymes [104]. In some cases the studies indicate that the protein unfolds and refolds into a new structure that is different from the original native structure. Even after lowering the temperature, the protein remains in this misfolded form because of a high kinetic energy barrier which prevents refolding to the native structure. In other cases it appears that the protein undergoes chemical modifications such as those described in the preceding section. Freeze-drying of the protein in the presence of appropriate buffer excipients can minimize these problems but does not necessarily eliminate them. Evidence has been accumulating which demonstrates that exposure to low temperature can also result in a cold denaturation, which occurs as a two-state process but with inverse enthalpic and entropic effects compared to the denaturation of protein with increasing temperatures [116]. Thus careful analysis of physical properties and biological activity as a function of storage temperature is extremely important.

pH

The pH of the formulation buffer can have a profound impact on protein solubility, folding, and rates of reactions which chemically alter the amino acid residues. The large number of reactions that can occur may have different pH profiles, and thus it may not be possible to formulate the protein at a particular pH that eliminates all the modification reactions while maintaining high solubility as well as the proper conformation of the protein. Usually, a trade-off will have to be accepted where the rates of the deleterious chemical reactions are minimized as much as possible while still maintaining protein conformation. In some cases it may be possible to formulate the protein under conditions where chemical modifications and/or proteolysis (as in the case of proteolytic enzyme drugs) are minimized while maintaining the protein in a misfolded conformation which is readily and rapidly refolded to the "native" form upon administration of the drug.

Ionic Strength

The concentration of counterions is known to be an important property in mediating electrostatic interactions in molecules [28,117] and can have large effects on the stability of a protein as well as its solubility. Basic Debye-Hückel theory of electrolytes suggests that the important parameter for assessing the effects of counterion concentration on electrostatic interactions is the ionic strength, $I = \frac{1}{2} \Sigma \, c_i z_i^2$, where c_i is the molar concentration of the ith counterion with charge z_i. Despite the great success of this basic theory and its applications in analyzing electrostatic interactions in proteins [28,118–120] and stabilization of proteins [121,122], the effects of particular counterions are not addressed by the use of the ionic strength parameter.

Previously, we discussed the Hofmeister series, where various ions are ranked according to their ability to increase hydrophobic interactions at high ionic strength and thus affect protein solubility. However, sometimes overlooked is the "salting-in" phenomenon that these ions produce at low concentrations. That is, there exists for each ion, holding other factors constant, an optimal concentration at which a given protein will be maximally soluble. It is believed that this effect is produced through shifts in the pK_a values of ionizable R groups [123]. Therefore, in the development of a formulation buffer for proteins, it is especially crucial that attention be paid to the types of counterions used as well as the ionic strength. For example, guanidinium hydrochloride is a strong denaturant of proteins, whereas guanidinium sulfate at equivalent concentrations can stabilize proteins.

Effects of Other Formulation Components

The development of a suitable pharmaceutical formulation usually involves the screening of any of a number of physiologically acceptable buffers, salts, chelators, antioxidants, surfactants, solvents, cosolvents, preservatives, dispersants, bulking agents, and so on. A fairly comprehensive list of excipients used in parenteral products may be found in a review by Wang and Kowal [124]. One must be assured that the combination of ingredients used is not deleterious to the active compound. This task is all the more difficult when formulating proteins, since the effects of multicomponent systems on the physicochemical properties of proteins are highly diverse and not well understood. Timasheff and co-workers have attempted to relate the preferential interaction parameters between proteins and salts, sugars, and other compounds to their effectiveness as stabilizers against thermal denaturation [125–129]. For example, the presence of cosolvents such as glycerol or polyethylene glycol may stabilize protein structure by a mechanism of preferential hydration, where an increase in the system entropy is affected by decreasing the protein surface area in contact with the solvent. This decrease in surface area can be obtained by folding the polypeptide chain into a compact structure. However, in some situations this decrease can also be accomplished by protein self-association. Alteration of the solvent polarity and dielectric constant will also affect electrostatic interactions and may influence the association of the protein as well.

Yet when systems involving four or more components are used (as is often the case for protein formulations), predictions on structure cannot be made, since the chemical potentials of all solvent components are interdependent [113]. In fact, the final choice of a clinical or market formulation for a protein pharmaceutical will very often involve a trade-off in which those factors considered most detrimental to product stability will be minimized, while other degradation pathways may actually be favored and will have to be tolerated. In such cases, the resultant degradation products should be well characterized and their clinical significance understood. For example, the addition of phenolic preservatives to neutral regular insulin solutions is necessary to prevent microbial growth, yet it has been suggested that they may alter the protein's tertiary structure as well as increase its propensity toward noncovalent aggregation [130,131]. However, these aggregates remain soluble and the phenolic preservatives may act to stabilize insulin protomers, preventing more extensive aggregation. Furthermore, even though the insulin-preservative complex may have an altered tertiary structure, the presence of the preservatives has not been shown to have any adverse effects on insulin administration [130,131]. Also, these physicochemical effects are expected to be fully reversible upon dilution of the protein at the site of administration, since insulin will dissociate from hexamers to monomers whose tertiary structure is unaltered in the presence of phenolic preservatives.

In other more extreme cases, however, one may have to choose a formulation which, for a hypothetical example, prevents noncovalent aggregation and loss of activity due to precipitation, yet results in the production of a small amount of deamidated protein which may be immunogenic. Therefore, as with the administration of the drug itself, one must consider the risk versus benefit when choosing formulation components that alter the physicochemical properties of the protein, especially in the absence of safety and efficacy data on degradation products.

Surface Interactions

The various steps involved in processing and formulating a protein bring it into contact with many types of surfaces, such as air, glass, rubber, plastics, and other synthetic materials. The potential for interaction with and even adsorption to such surfaces is of great concern with regard to maintaining native protein structure. The type and degree of interaction depend not only on the physical and chemical properties of both the protein and the surface, but also on the stresses placed on the system.

In general, proteins interact with surfaces via the mechanisms already discussed that stabilize protein structure (i.e., electrostatic, hydrogen bonding, hydrophobic, and van der Waals interactions) [132]. The physical characteristics of the surface are also important. Some proteins may adsorb more to a rough surface (i.e., one with many microscopic irregularities such as crevices, holes, pockets, etc.) than to a smooth one. In any case, surface interactions can result in protein conformational changes, the degree and reversibility of which are dependent on time, temperature, agitation, and type of surface. For example, some plasma proteins have been shown to undergo conformational changes at the air-liquid interface [133]. Additionally, Bohnert and Horbett showed that different proteins adsorb to synthetic materials at different rates proportional to temperature and further, that adsorption was typically irreversible, as the surface-bound protein could not be completely eluted with detergents, organic solvents, or denaturants under varying conditions [134]. Agitation of a protein solution by vigorous shaking or stirring can also exacerbate surface-affected protein denaturation and even cause precipitation. The mechanisms involved

in such processes are not well understood, nor are the processes themselves well studied. It has been speculated that the interaction of proteins with surfaces involves the formation of bonds between the protein
and surface via one of two mechanisms: through multiple bonds which
increase with time as the protein unfolds, exposing more surface area, or
through very strong, slowly forming bonds which do not require protein
denaturation [134]. Another group suggests further that the "interfacial
forces" responsible for the interaction of proteins with low-energy surfaces
are mainly Lifshitz/van der Wall interactions (long range) and hydrogen
bonding (short range) [135].

Of the various types of surface materials studied, polyethylene and
especially Teflon have proven most acceptable in terms of low adsorption-
interaction potential. Other materials, such as glass and silastic, may
have high adsorption-interaction potential [134]. When working with large
amounts of protein over reusable surfaces (as in many processing and
manufacturing stages), losses due to adsorption may not be as critical,
especially since the surface will become saturated very quickly [134,136],
although surface-induced denaturation of the bulk protein solution may
still be a concern. However, when working with small amounts of protein,
losses may be quite dramatic. In either case, assuming that the interaction is primarily hydrophobic in nature, it may be possible to minimize
or even prevent adsorption (as well as surface-induced aggregation)
through the use of a nondenaturing surfactant in the protein solution [137].

V. INTERACTION WITH OTHER BIOMOLECULES

A. Protein-Protein Interactions

The biological functioning of a protein often depends on interactions with
other molecules. These interactions may involve small molecules, such as
oxygen binding to hemoglobin, or other large proteins, such as the case
for binding of bovine pancreatic trypsin inhibitor with the enzyme trypsin.
The forces involved are similar to those responsible for maintaining protein conformation, yet the interactions may be highly specific in nature.
For example, binding constants for specific associations of IgG immunoglobulins (antibodies) with protein antigens, or proteins with cell receptors can
be as high as 10^{11} M^{-1}. Despite these high association constants the protein interactions may be extremely selective, as in the case of an antibody
raised against one protein antigen that will not bind tightly with a different
antigen. Protein conformation is also crucial in these interactions, since
antibodies raised against denatured protein will not necessarily cross-react
with the native protein. The specificity of binding is often dictated by an
active-site region on the protein, and variations of a few amino acids in
this region can determine this specificity. This is demonstrated nicely in
the serine proteases, where the binding sites are highly homologous [138].
Variations of single amino acids within the binding sites determine the
specificity of the binding with the polypeptide chain to be cleaved:
trypsin cuts with high specificity at arg and lys residues, whereas
chymotrypsin is more specific for aromatic residues such as phe, tyr,
and trp.

Although the binding between proteins may be highly specific, nonspecific binding may also occur. This binding is due to weaker protein-
protein interactions, the degree of which is dependent on the relative

concentrations of the proteins that are interacting. An example of non-specific interactions are those that can occur when using immuno-blot techniques, commonly referred to as Western blotting [139]. Nonspecific interactions of protein antigens with antisera can lead to false positive results. These interactions generally are a result of long exposure times of high concentrations of antibody to antigen.

Both specific and nonspecific interactions need to be considered in dealing with the development of a protein as a pharmaceutical. Nonspecific interactions of the protein with other protein components during purification may have an impact on the route of purification. In particular, recombinant-DNA-derived proteins expressed in an organism such as *E. coli* may interact with proteins from the organism, resulting in binding to ion-exchange columns under conditions where binding was not expected [106]. To purify and ultimately characterize the protein pharmaceutical, it may be necessary to disrupt these interactions using either detergents or strong denaturing agents such as urea or guanidinium hydrochloride. However, the methods used to disrupt these interactions may also affect the conformation of the protein being purified. Protein pharmaceuticals such as growth hormone and lymphokines are designed to interact specifically with cell receptors, and therefore the proteins may have to be refolded after purification, since these specific interactions are required for bioactivity.

Generally, solvent interactions and the presence of other solutes can affect protein function as well as association. For example, hemoglobin exists as a tetrameric protein consisting of two different types of poly-peptide chains. The interactions between these chains are highly dependent on the state of oxygen binding as well as the presence of small effector molecules, which include organic phosphates, bicarbonate, and chloride [140,141]. Therefore, changes in solvent composition that may appear to be innocuous for nonprotein drugs could have profound effects on the function and state of association of a protein drug.

B. Protein-Lipid Interactions

Proteins are known to associate with a variety of lipids, both specifically and nonspecifically. Examples of specific interactions are the binding of free fatty acids to albumin [142] and the binding of prothrombin to acidic phospholipids in membranes [143]. Such binding is associated with a specific biological function and as such is reversible. Examples of non-specific interactions are the binding of other proteins, such as albumin and fibrinogen, to membrane phospholipids, which are akin to the inter-action of proteins with other surfaces. Such nonspecific binding is usually irreversible.

Sackmann et al. [144] have classified protein-lipid interactions into five types based on the degree of penetration of proteins into biological membranes: electrostatic, hydrophilic (i.e., hydrogen bonding, dipole-dipole, etc.), hydrophobic, combined hydrophobic/hydrophilic, and com-bined hydrophobic/electrostatic. The effects of such interactions on pro-tein structure are critical. For example, integral membrane proteins such as cytochrome c oxidase or cytochrome c reductase have altered conforma-tions and lose activity in the absence of lipid [145,146]. Other proteins, such as the antenna protein of *Rhodopseudomonas sphaeroides*, may aggre-gate in the presence of certain phospholipids [147]. Still others, such as the apolipoproteins A-I and A-II, may interact hydrophobically with lipids with little or no conformational change [148,149]. The presence of lipid may also mediate protein-protein interactions, as suggested by the formation of the enzyme complexes of the electron transport system in mitochondrial membranes [144].

In any case, an understanding of protein-lipid interactions is important in the design of delivery systems involving the use of lipids. For example, liposomes have been investigated by several groups for the incorporation and delivery of proteins (especially lymphokines) to certain target tissues. Unfortunately, studies of the physicochemical effects on proteins of the lipids and/or processes used for incorporation are extremely limited. However, in one study of the use of multilamellar vesicles (MLVs) of varying composition to deliver recombinant murine interferon-γ to murine L929 cells, Eppstein and co-workers [150] demonstrated that as much as 50% (depending on preparation conditions) of the protein could be associated with positively or negatively charged phospholipids on the liposome surface rather than being incorporated into the internal aqueous volume, as many other investigators had assumed. Characterization of the distribution of the protein in or on the liposome is critical, since the types of interactions involved not only affect the protein's physicochemical properties, but will also affect its release rate from the liposome following administration.

VI. CONCLUSIONS

The complex composition and physical properties of proteins make the task of formulation much more difficult than for small-molecule drugs. The physical properties of proteins as discussed in this chapter may be closely related, in that changes in one property have profound affects on other properties. As an example, alterations in primary structure such as deamidation may result in conformational changes in the protein, which in turn may result in self-association of the protein. Although the genetic code for the creation of a linear amino acid polymer is known, the complex set of rules (in a sense the second half of the genetic code) for determining protein conformation, properties, and bioactivity from the primary structure is poorly understood [151]. Hence a great deal of empirical work is required to fold and stabilize a protein properly as a drug. Often, many of the degradative pathways in proteins, such as proteolysis, deamidation, oxidation or self-association, will be subject to a diverse set of solution conditions. In fact, it may not be possible to produce a formulation that will eliminate all the potential routes of inactivation. Therefore, it may be necessary to optimize all conditions to essentially minimize overall the impact on the protein of the various degradation routes deemed to be adverse.

In dealing with the designs of future formulations for proteins, new methodologies in drug delivery systems will be developed, including alternative routes of administration, drug targeting, and sustained-release technologies which utilize nonerodible and erodible polymers [152]. The use of certain materials and solvent systems, as well as the stresses placed on a protein in the preparation of these delivery systems, will require thorough knowledge of the interaction of the system components with the protein drug. In particular, the design of an appropriate delivery system should be based on the physical properties of the protein drug (e.g., as in the effects on the protein of the solvent systems used in the preparation of a controlled-release polymer matrix). Indeed, from the formulation chemist's point of view, it may be a more straightforward and successful task to design a delivery system around the properties of the protein rather than to alter those properties to fit an existing delivery system. However, with advances in biotechnology it may soon be possible to design active proteins with the necessary properties to give the desired biodisposition when administered alone or the appropriate stability when incorporated into a particular delivery system. To be successful, such an effort must involve the close collaboration of the molecular biologists who will engineer the protein and the pharmaceutical chemists who will formulate it.

REFERENCES

1. A. L. Lehninger, *Biochemistry*, 2nd ed., Worth Publishers, New York (1975), pp. 71–77.
2. G. N. Ramachandran and A. K. Mitra, *J. Mol. Biol.*, 107:85 (1976).
3. A. L. Lehninger, *Biochemistry*, 2nd ed., Worth Publishers, New York (1975), pp. 127–135.
4. D. E. Metzler, *Biochemistry: The Chemical Reactions of Living Cells*, Academic Press, New York (1977), pp. 63–69.
5. A. T. Hagler, D. J. Osguthorpe, P. Dauber-Osguthorpe, and J. C. Hempel, *Science*, 227 1309 (1985).
6. L. Stryer, *Biochemistry*, 2nd ed., W. H. Freeman and Company, San Francisco (1981), pp. 27–32.
7. R. E. Dickerson and I. Geis, *The Structure and Action of Proteins*, Harper & Row, New York (1969), pp. 24–43.
8. J. F. Leszczynski and G. D. Rose, *Science*, 234:849 (1986).
9. J. S. Richardson, *Adv. Protein Chem.*, 34:167 (1981).
10. A. H. Pekar and B. H. Frank, *Biochemistry*, 11:4013 (1972).
11. J. Schlichtkrull, *Acta Chem. Scand.*, 10:1455 (1956).
12. B. K. Milthorpp, L. W. Nichol, and P. D. Jeffrey, *Biochim. Biophys. Acta*, 495:195 (1977).
13. E. J. Dodson, G. G. Dodson, D. C. Hodgkin, and C. D. Reynolds, *Can. J. Biochem.*, 57:469 (1979).
14. Y. K. Yip, M. Waks, and S. Beychok, *J. Biol. Chem.*, 247:7237 (1972).
15. M. Waks, Y. K. Yip, and S. Beychok, *J. Biol. Chem.*, 248:6462 (1973).
16. C. B. Anfinsen and E. Haber, *J. Biol. Chem.*, 236:1361 (1961).
17. T. E. Creighton, *Proteins, Structures and Molecular Properties*, W. H. Freeman and Company, San Francisco (1983), p. 287.
18. D. B. Wetlaufer, in *The Protein Folding Problem* (D. B. Wetlaufer, ed.), Westview Press, Boulder, Colo. (1984), pp. 29–46.
19. A. M. Lesk, *Bioessays*, 2 213 (1985).
20. W. Kabsch and C. Sander, *Proc. Natl. Acad. Sci. USA*, 81:1075 (1984).
21. C. Tanford, *Adv. Protein Chem.*, 23:121 (1968).
22. P. L. Privalov, *Adv. Protein Chem.*, 33:167 (1979).
23. J. M. Betton, M. Desmadril, A. Mitraki, and J. M. Yon, *Biochemistry*, 23:6654 (1984).
24. H. Nojima, A. Ikai, T. Oshima, and H. Noda, *J. Mol. Biol.*, 116:429 (1977).

25. T. Y. Tsong, R. L. Baldwin, and P. McPhie, *J. Mol. Biol.*, *63*:453 (1972).

26. C. Ghelis and J. Yon, *Protein Folding*, Academic Press, New York (1982).

27. J. B. Matthew, F. R. N. Gurd, M. A. Flanagan, K. L. March, and S. J. Shire, *CRC Crit. Rev. Biochem.*, *18*:91 (1985).

28. C. Tanford, *Adv. Protein Chem.*, *17*:69 (1962).

29. A. Warshel and S. T. Russell, *Q. Rev. Biophys.*, *17*:283 (1984).

30. J. B. Matthew, *Annu. Rev. Biophys. Biophys. Chem.*, *14*:387 (1985).

31. B. H. Hoenig, W. L. Hubbell, and R. F. Flewelling, *Annu. Rev. Biophys. Biophys. Chem.*, *15*:163 (1986).

32. E. L. Mehler and G. Eichele, *Biochemistry*, *23*:3887 (1984).

33. D. C. Rees, *J. Mol. Biol.*, *141*:323 (1980).

34. W. G. J. Hol, P. T. van Duijnen, and H. J. C. Berendsen, *Nature*, *273*:443 (1978).

35. W. G. J. Hol, L. M. Halie, and C. Sander, *Nature*, *294*:532 (1981).

36. E. Antonini and M. Brunori, *Hemoglobin and Myoglobin in Their Reactions with Ligands*, North-Holland Publishing Company, Amsterdam (1971), pp. 373–377.

37. D. Eisenberg and W. Kauzman, *The Structure and Properties of Water*, Oxford University Press, New York (1969).

38. F. H. Stillinger, *Science*, *209*:451 (1980).

39. T. E. Creighton, *Proteins: Structures and Molecular Properties*, W. H. Freeman and Company, San Francisco (1983), p. 139.

40. C. Tanford, *The Hydrophobic Effect*, John Wiley & Sons, New York (1980).

41. F. M. Richards, *Annu. Rev. Biophys. Bioeng.*, *6*:151 (1977).

42. G. D. Rose, A. R. Geselowitz, G. J. Lesser, R. H. Lee, and M. H. Zehfus, *Science*, *229*:834 (1985).

43. M. A. Lauffer, *Entropy Driven Processes in Biology: Polymerization of Tobacco Mosaic Virus Protein and Similar Reactions*, Springer-Verlag, New York (1975).

44. G. Nemethy and H. A. Scheraga, *J. Phys. Chem.*, *66*:1773 (1962).

45. R. B. Hermann, *J. Phys. Chem.*, *76*:2754 (1972).

46. P. Von Hippel and T. Schleich, in *Structure and Stability of Biological Macromolecules* (S. N. Timasheff and G. D. Fasman, eds.), Marcel Dekker, New York (1969), p. 417.

47. F. Hofmeister, *Arch. Exp. Pathol.*, *24*:247 (1888).

48. T. E. Creighton, *Proteins: Structures and Molecular Properties*, W. H. Freeman and Company, San Francisco (1983), p. 151.

49. E. E. Saffen, Jr. and P. W. Chun, *Biophys. Chem.*, *9*:329 (1979).

50. B. H. Frank, J. M. Pettep, R. E. Zimmerman, and P. J. Burck, in *Peptides: Synthesis, Structure, Function* (D. H. Rich and E. Gross, eds.), Pierce Chemical Company, Rockford, Ill. (1981).

51. S. W. J. Lamberts and R. Oosterom, *Front. Horm. Res.*, *14*:137 (1985).

52. D. C. Rijken, M. Hoylaerts, and D. Collen, *J. Biol. Chem.*, *257*:2920 (1982).

53. S. Kasai, H. Arimura, M. Nishida, and T. Suyama, *J. Biol. Chem.*, *260*:12377 (1985).

54. W. F. Bennett, "Criteria for Identification and Assessment of Purity of Proteins Produced by Recombinant DNA," *Therapeutic Agents Produced by Genetic Engineering "Quo Vadis?" Symposium*, Sanofi Group, Toulouse-Labege, France (1985), p. 111.

55. N. Sharon and H. Lis, *Chem. Eng. News*, *59(13)*:21 (1981).

56. R. Kornfeld and S. Kornfeld, *Annu. Rev. Biochem.*, *45*:217 (1976).

57. A. G. Morrell, G. Gregoriadis, H. Scheinberg, J. Hickman, and
    G. Ashwell, *J. Biol. Chem.*, *246*:1461 (1971).

58. J. Green, G. Griffiths, D. Louvard, P. Quinn, and G. Warren,
    J. Mol. Biol., *152*:663 (1981).

59. T. E. Creighton, *Proteins: Structures and Molecular Properties*,
    W. H. Freeman and Company, San Francisco (1983), pp. 79–86.

60. P. Seiber, M. Brugger, B. Kamber, B. Rineker, M. Rittel, R. Maier,
    and M. Staehelin, in *Calcitonin 1969, Proceedings of the 2nd International
    Symposium* (S. Taylor, ed.), Heinemann, London (1970),
    pp. 28–33.

61. A. Bachmair, D. Finley, and A. Varshavsky, *Science*, *234*:179 (1986).

62. R. T. Garvin, private communication (1986).

63. H. Morehead, P. D. Johnston, and R. Wetzel, *Biochemistry*, *23*:2500
    (1984).

64. D. A. Yphantis and T. Arakawa, *Biophys. J.*, *49*:499a (1986).

65. F. Perrin, *J. Phys. Radium*, 7:1 (1936).

66. C. R. Cantor and P. R. Schimmel, *Biophysical Chemistry—Part II:
    Techniques for the Study of Biological Structure and Function*,
    W. H. Freeman and Company, San Francisco (1980), p. 611.

67. S. Beychok, *Science*, *154*:1288 (1966).

68. S. Beychok, *Proc. Natl. Acad. Sci. USA*, *53*:999 (1965).

69. J. W. S. Morris, D. A. Mercola, and E. R. Arquilla, *Biochim.
    Biophys. Acta*, *160*:145 (1968).

70. B. Myers II and A. N. Glazer, *J. Biol. Chem.*, *246*:412 (1971).

71. N. Greenfield and G. D. Fasman, *Biochemistry*, *8*:4108 (1969).

72. E. H. Strickland, *CRC Crit. Rev. Biochem.*, *2*:113 (1974).

73. M. J. Ettinger and S. N. Timasheff, *Biochemistry*, *10*:824 (1971).

74. Y. A. Lazerav, B. A. Grishkovsky, and T. B. Khromova, *Biopolymers*,
    24:1449 (1985).

75. M. N. G. James, A. Sielecki, F. Salituro, D. H. Rich, and T. Hofmann,
    Proc. Natl. Acad. Sci. USA, *79*:6137 (1982).

76. P. G. Schmidt, M. S. Bernatowicz, and D. H. Rich, *Biochemistry*,
    21:1830 (1982).

77. P. G. Schmidt, M. S. Bernatowicz, and D. H. Rich, *Biochemistry*,
    21:6710 (1982).

78. K. L. Williamson and R. J. P. Williams, *Biochemistry*, *18*:5966 (1979).

79. J. Goldman and F. H. Carpenter, *Biochemistry*, *13*:456 (1974).

80. S. P. Wood, T. L. Blundell, A. Wollmer, N. R. Lazarus, and R. W. J.
    Neville, *Eur. J. Biochem.*, *55*:531 (1975).

81. E. H. Strickland and D. A. Mercola, *Biochemistry*, *15*:3875 (1976).

82. A. Wollmer, W. Strassburger, E. Hoenjet, U. Glatter, J. Fleischhauer,
    D. A. Mercola, R. A. G. de Graaf, E. J. Dodson, G. G. Dodson,
    D. G. Smith, D. Brandenburg, and W. Danho, in *Insulin: Chemistry,
    Structure, and Function of Insulin and Related Hormones* (D. Brandenburg
    and A. Wollmer, eds.), Walter de Gruyter, N.Y. (1980),
    pp. 27–35.

83. H. Renscheidt, W. Strassburger, U. Glatter, A. Wollmer, G. G. Dodson,
    and D. A. Mercola, *Eur. J. Biochem.*, *142*:7 (1984).

84. A. Light, *BioTechniques*, *3*:298 (1985).

85. D. C. Benjamin, J. A. Berzofsky, I. J. Past, F. R. N. Gurd,
    C. Hannum, S. J. Leach, E. Margoliash, J. G. Michael, A. Miller,
    E. M. Prager, M. Reichlin, E. E. Sercarz, S. J. Smith-Gill, P. E. Todd,
    and A. C. Wilson, *Annu. Rev. Immunol.*, *2*:67 (1984).

86. H. M. Dintzis, R. Z. Dintzis, and B. Vogelstein, *Proc. Natl. Acad. Sci. USA*, *73*:3671 (1976).

87. R. Z. Dintzis, B. Vogelstein, and H. M. Dintzis, *Proc. Natl. Acad. Sci. USA*, *79*:884 (1982).

88. T. P. Hopp and K. R. Woods, *Proc. Natl. Acad. Sci. USA*, *78*:3824 (1981).

89. United States Pharmacopeial Convention, *United States Pharmacopeia*, 21st ed., Mack Publishing Company, Easton, Pa. (1985), p. 1180.

90. C. H. Li, in *Hormonal Proteins and Peptides*, Vol. IV (C. H. Li, ed.), Academic Press, New York (1977), p. 1.

91. A. E. Wilhelmi, *Hormone Drugs: Proceedings of the FDA-USP Workshop on Drug and Reference Standards for Insulins, Somatotropins, and Thyroid-axis Hormones*, United States Pharmacopeial Convention, Rockville, Md. (1982), pp. 369–381.

92. R. E. Chance, E. P. Kroeff, J. A. Hoffmann, and B. H. Frank, *Diabetes Care*, *4*:147 (1981).

93. A. J. S. Jones and J. V. O'Connor, *Hormone Drugs: Proceedings of the FDA-USP Workshop on Drug and Reference Standards for Insulins, Somatotropins, and Thyroid-axis Hormones*, United States Pharmacopeial Convention, Rockville, Md. (1982), pp. 335–351.

94. A. B. Robinson, *Proc. Natl. Acad. Sci. USA*, *71*:885 (1974).

95. T. Flatmark, *J. Biol. Chem.*, *242*:2454 (1967).

96. T. J. Ahern and A. M. Klibanov, *Science*, *228*:1280 (1985).

97. B. V. Fisher and P. B. Porter, *J. Pharm. Pharmacol.*, *33*:203 (1981).

98. J. Schlichtkrull, M. Pingel, L. G. Heding, J. Brange, and K. H. Jorgensen, in *Handbook of Experimental Pharmacology*, Vol. 32 (A. Hasselblatt and F. von Bruchhausen, eds.), Springer-Verlag, New York (1975), p. 729.

99. A. B. Robinson and C. J. Rudd, *Curr. Top. Cell. Regul.*, *8*:247 (1974).

100. J. H. McKerrow and A. B. Robinson, *Science*, *183*:185 (1974).

101. A. A. Kossiakoff, *Science*, *240*:191 (1988).

102. T. E. Creighton, *Proteins: Structures and Molecular Properties*, W. H. Freeman and Company, New York (1983), pp. 20–24.

103. G. E. Means and R. E. Feeney, *Chemical Modification of Proteins*, Holden-Day, San Francisco (1971), pp. 165–169.

104. A. M. Klibanov, *Adv. Appl. Microbiol.*, *29*:1 (1983).

105. D. E. Metzler, *Biochemistry: The Chemical Reactions of Living Cells*, Academic Press, New York (1977), p. 665.

106. S. J. Shire, L. Bock, J. Ogez, S. Builder, D. Kleid, and D. M. Moore, *Biochemistry*, *23*:6474 (1984).

107. G. E. Means and R. E. Feeney, *Chemical Modification of Proteins*, Holden-Day, San Francisco (1971), pp. 84–89.

108. G. R. Stark, W. H. Stein, and S. Moore, *J. Biol. Chem.*, *235*:3177 (1960).

109. D. G. Smyth, *J. Biol. Chem.*, *242*:1579 (1967).

110. S. Rimon and G. E. Perlmann, *J. Biol. Chem.*, *243*:3566 (1968).

111. M. Laskowski, Jr. and I. Kato, *Annu. Rev. Biochem.*, *49*:593 (1980).

112. R. D. Schmid, *Adv. Biochem. Eng.*, *12*:41 (1979).

113. I. M. Klotz, *Arch. Biochem. Biophys.*, *116*:92 (1966).

114. D. C. Teller, *Methods Enzymol.*, *27*:346 (1972).

115. D. S. Schade, R. P. Eaton, J. DeLongo, L. C. Saland, A. J. Ladman, and G. A. Carlson, *Diabetes Care*, *5*:25 (1982).

116. P. L. Privalov, Y. V. Griko, S. Y. Venyaminov, and V. P. Kutyshenko, *J. Mol. Biol.*, *190*:487 (1986).

117. S. J. Shire, G. I. H. Hanania, and F. R. N. Gurd, *Biochemistry*, *13*:2974 (1974).

118. K. Linderstrom-Lang, *C. R. Trav. Lab. Carlsberg*, *15*:1 (1924).

119. S. J. Shire, G. I. H. Hanania, and F. R. N. Gurd, *Biochemistry*, *14*:1352 (1975).

120. C. R. Hartzell, R. A. Bradshaw, K. D. Hapner, and F. R. N. Gurd, *J. Biol. Chem.*, *243*:690 (1968).

121. S. H. Friend and F. R. N. Gurd, *Biochemistry*, *18*:4612 (1979).

122. M. A. Flanagan, E. B. Garcia-Moreno, S. H. Friend, R. J. Feldmann, H. Scouloudi, and F. R. N. Gurd, *Biochemistry*, *22*:6027 (1983).

123. A. L. Lehninger, *Biochemistry*, 2nd ed., Worth Publishers, New York (1975), p. 162.

124. Y. J. Wang and R. R. Kowal, *J. Parenter. Drug Assoc.*, *34*:452 (1980).

125. K. Gekko and S. N. Timasheff, *Biochemistry*, *20*:4667 (1981).

126. T. Arakawa and S. N. Timasheff, *Biochemistry*, *21*:6536 (1982).

127. T. Arakawa and S. N. Timasheff, *Biochemistry*, *21*:6545 (1982).

128. T. Arakawa and S. N. Timasheff, *Biochemistry*, *24*:6756 (1985).

129. T. Arakawa and S. N. Timasheff, *Biophys. J.*, *47*:411 (1985).

130. J. Q. Oeswein and Z. A. Bergstedt, "The Physicochemical Effects of Phenolic Preservatives on Human Insulin," *Abstracts of 190th ACS National Meeting*, Chicago (1985), p. BIOL 0052.

131. J. Q. Oeswein, "Aggregation of Proteins in Solution—Behavior of Insulin," *Abstracts of 133rd APhA Annual Meeting*, San Francisco (1986), p. 11.

132. J. D. Andrade, in *Surface and Interfacial Aspects of Biomedical Polymers*, Vol. 2, *Protein Adsorption* (J. D. Andrade, ed.), Plenum Press, New York (1985), pp. 1–80.

133. L. Vroman, A. L. Adams, and M. Klings, *Fed. Proc.*, *30*:1494 (1971).

134. J. L. Bohnert and T. A. Horbett, *J. Colloid Interface Sci.*, *111*:363 (1986).

135. C. J. Van Oss, R. J. Good, and M. K. Chaudhury, *J. Colloid Interface Sci.*, *111*:378 (1986).

136. W. G. Pitt, K. Park, and S. L. Cooper, *J. Colloid Interface Sci.*, *111*:343 (1986).

137. A. S. Chawla, I. Hinberg, P. Blais, and D. Johnson, *Diabetes*, *34*:420 (1985).

138. B. S. Hartley, *Philos. Trans. R. Soc. London*, B257:77 (1980).

139. W. N. Burnette, *Anal. Biochem.*, *112*:195 (1981).

140. S. J. Edelstein, *Annu. Rev. Biochem.*, *44*:209 (1975).

141. D. H. Atha and A. Riggs, *J. Biol. Chem.*, *251*:5537 (1976).

142. A. L. Lehninger, *Biochemistry*, 2nd ed., Worth Publishers, New York (1975), p. 837.

143. J. W. Corsel, G. M. Willems, J. M. M. Kop, P. A. Cuypers, and W. T. Hermens, *J. Colloid Interface Sci.*, *111*:544 (1986).

144. E. Sackmann, R. Kotulla, and F. Heiszler, *Can. J. Biochem. Cell Biol.*, *62*:778 (1984).

145. C. A. Yu, S. H. Gwak, and L. Yu, *Biochim. Biophys. Acta, 812*: 656 (1985).

146. J. C. Salerno, S. Yoshida, and T. E. King, *J. Biol. Chem.*, *261*: 5480 (1986).

147. W. M. Heckl, M. Losche, H. Scheer, and H. Mohwald, *Biochim. Biophys. Acta*, *810*:73 (1985).

148. R. J. King, M. C. Carmichael, and P. M. Horowitz, *J. Biol. Chem.*, *258*:10672 (1983).

149. B. W. Shen, *J. Biol. Chem.*, *260*:1032 (1985).

150. D. A. Eppstein, Y. V. Marsh, M. Van der Pas, P. L. Felgner, and A. B. Schreiber, *Proc. Natl. Acad. Sci. USA*, *82*:3688 (1985).

151. G. Kolata, *Science*, *233* 1037 (1986).

152. R. Langer, *Technol. Rev.*, *83*:26 (1981).

153. T. P. Bennett and E. Frieden, *Modern Topics in Biochemistry*, Macmillan Publishing Company, New York (1968), p. 24.

154. C. J. Epstein, R. F. Goldberger, and C. B. Anfinsen, *Cold Spring Harbor Symp. Quant. Biol.*, *28*:439 (1963).

155. H. A. Scheraga, *The Proteins*, Vol. 1, 2nd ed., Academic Press, New York (1963).

156. T. E. Creighton, *Proteins, Structures and Molecular Properties*, W. H. Freeman and Company, San Francisco (1983), p. 151.

157. R. E. Dickerson, *The Proteins*, Vol. 2, 2nd ed., Academic Press, New York (1966), p. 634.

158. J. H. McKerrow and A. B. Robinson, *Anal. Biochem.*, *42*:565 (1971).

159. R. B. Parekh, R. A. Dwek, J. R. Thomas, G. Opdenakker, T. W. Rademacher, A. J. Wittwer, S. C. Howard, R. Nelson, N. R. Siegel, M. G. Jennings, N. K. Harakas, and J. Feder, *Biochemistry*, *28*:7644 (1989).

160. A. J. Wittwer, S. C. Howard, L. S. Carr, N. K. Harakas, J. Feder, R. B. Parekh, P. M. Rudd, R. A. Dwek, and T. W. Rademacher, *Biochemistry*, *28*:7662 (1989).

161. R. P. Parekh, R. A. Dwek, P. M. Rudd, J. R. Thomas, T. W. Rademacher, T. Warren, T.-C. Wun, B. Hebert, B. Reitz, M. Palmier, T. Ramabhadran, and D. C. Tiemeier, *Biochemistry*, *28*:7670 (1989).

162. S. Clarke, *Int. J. Peptide Protein Res.*, *30*:808 (1987).

163. R. C. Stephenson and S. Clarke, *J. Biol. Chem.*, *264*:6164 (1989).

5

Approaches to the Analysis of Peptides

Cynthia S. Randall

SmithKline Beecham Pharmaceuticals, King of Prussia, Pennsylvania

Thomas R. Malefyt

Syntex Research, Inc., Palo Alto, California

Larry A. Sternson

Sterling Drug, Inc., Malvern, Pennsylvania

I. INTRODUCTION

Peptides represent a pharmacologically and therapeutically diverse chemical class of compounds. Their development as drugs has been facilitated by advances in recombinant DNA technology and in chemical synthesis approaches that have made peptides and proteins more readily available. The full therapeutic value of peptides has yet to be clearly realized, but possible applications include the modulation of cardiovascular, central nervous system, and endocrine (e.g., growth) functions. These potential drugs are structurally similar to one another, differing only in the specific amino acids or in the sequence of the amino acids that define the oligomer. Although the first peptides evaluated for therapeutic purposes consisted entirely of natural amino acids, recent attempts have been directed toward improving the physical and chemical stability, pharmacological activity, and biological transport properties of peptides through structural alterations of the component amino acids. The approaches have included terminal N- and C-blockage, N-alkylation, substitution of D-amino acids for the naturally occurring L-isomers, and various side-chain modifications

The development and production of peptides as drugs present daunting scientific challenges. The rigor of the analytical methods that are adopted to characterize and evaluate this group of substances will not only help define our understanding of their chemical and biological properties, but will probably affect their regulatory approval as well as the success and safety of the resulting products.

Several properties of peptides and proteins present particular challenges to their analysis. The biological potency characteristic of these molecules requires methods that are exquisitely sensitive. Additionally, the striking similarity in structure among analyte, degradation products, impurities, and matrix components places tremendous demands for selectivity on analytical methodology. Analysis of peptides is further complicated by their inherent reactivity. As discussed in Chap. 3, the most common reaction is hydroly- sis of amide linkages, although oxidation reactions (of tryptophan, methio- nine, cysteine, or tyrosine), racemization, and modifications of the amino acid side chain are also frequently observed. Physical changes in peptide structure due to aggregation or adsorption of sample material onto container surfaces, as discussed in Chap. 17, can also result in an apparent loss of analyte. Although adherence onto container surfaces does not present a significant problem in analyzing microgram or greater amounts of peptide, the loss due to adsorption can become significant in samples containing trace levels (i.e., submicrogram) of such analytes. Attempts to circumvent surface adsorption by chemically coating glassware (e.g., siliconization) or by addition of competing adsorptive amines to the test solution can intere- fere with subsequent analysis of the peptide of interest.

In addition to these "traditional" problems, analytical methods developed for peptides and proteins must also consider the relationship between the purity and activity of the analyte. Physical-chemical identity and purity assessment are sufficient to guarantee therapeutic potency for small molecules; however, many peptides and all proteins, due to their complexity, require an additional, independent evaluation of activity. Purity must reflect not only chemical composition, but also chirality, conformation, and physical states of aggregation of the analyte. Consideration must be given to the fact that in some cases, loss of purity measured in terms of these criteria may not affect biological activity, whereas in other cases, limitations on the scope of available analytical methodology cause impurities that do affect bio- logical activity not to be recognized.

Peptide analysis is an active and varied area of study. It is our in- tention to provide an overview of current analytical methodology, especial- ly as relates to peptides of therapeutic interest. Focus is directed primarily on peptides containing 50 or fewer amino acids, although many of the methods reviewed here are also applicable to larger peptides and proteins. The reader is referred to Chap. 6 for a detailed discussion of the specific tech- niques employed in protein analysis. For organizational purposes, the methods of peptide analysis discussed herein have been placed into four categories: measurement of biological activity, evaluation of purity and stability, quantitative detection, and structural characterization.

II. MEASUREMENT OF BIOLOGICAL ACTIVITY

A. Assays

Bioassays measure biological response to a particular compound; as such, they are considered to give the best indication of a drug's therapeutic po- tency. Potency determinations for current U.S. Food and Drug Administra- tion (FDA)—approved peptide drugs uniformly involve animal bioassays, which are summarized in Table 1. Details and guidelines for these assays can be found in the *United States Pharmacopeia* (USP) [1]. Bioassays can be conducted in vitro as well as in vivo. An example of the former is an assay for corticotropin releasing factor (CRF), which involves measuring the corticotropin (ACTH) released by rat pituitary cells upon exposure to

Table 1   USP Bioassays for Peptide Drugs

Peptide	Basis of assay
Corticotropin	Decreased ascorbic acid concentration in adrenal glands of hypophysized rats
Glucagon	Increased blood sugar concentration in cats
Insulin	Decreased blood sugar concen- in rabbits
Chorionic gonadotropin	Increased weight of rat uterus
Follicle stimulating hormone	Decreased ascorbic acid concentration in rat ovaries
Luteinizing hormone	Decreased ascorbic acid concentration in rat ovaries
Oxytocin	Decreased blood pressure in chickens
Vasopressin	Increased blood pressure in rats

CRF; representative data are shown in Fig. 1. This assay can detect as little as 5 to 10 femtomoles of added CRF and can also be used for activity assays of CRF agonists and antagonists [2].

Such in vitro or cytochemical bioassays offer certain advantages of their in vivo counterparts. The former assays involve a less complex biological system than a living animal and are more easily controlled in the laboratory. Cytochemical bioassays can be carried out several times using tissue from one animal, which diminishes "interanimal" variability. In vitro bioassays are generally considerably less time consuming than in vivo assays. However, it is important to remember that the in vitro bioassay measures a secondary or biochemical phenomenon, whereas the in vivo bioassay measures a characteristic biological function, making it the most relevant indicator of drug efficacy.

Despite the importance of bioassays, there are significant drawbacks associated with their use on a routine basis. In general, the necessary procedures are cumbersome, costly, time consuming, and labor intensive. Variability within test tissue cultures or laboratory animals, coupled with interference by other substances, such as impurities and peptide degradation products, often gives rise to assay variabilities in excess of ±20%. This causes difficulty in the assignment of drug shelf life and may lead to severe underestimation of drug stability. Clearly, methods other than a bioassay are needed for reliable estimation of drug purity and stability. At the same time, there remains the need to establish and verify drug potency. The implementation of analytical methodology that satisfies both needs and provides high degrees of selectivity and sensitivity will be extremely important in the development of commercially feasible peptide drugs.

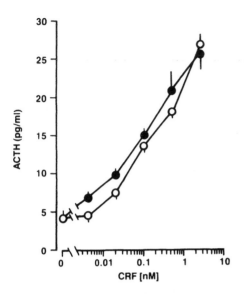

Fig. 1   Effect of synthetic rat corticotropin releasing factor (dark circles) and ovine CRF (open circles) on secretion of ACTH by cultured rat anterior pituitary cells. (From Ref. 2.)

There is now evidence that this goal is within reach for some compounds. In a study of corticotropin stability, samples of the peptide stored in ampules at 20 and 37°C for 15 years were compared to standards kept in the dark at 20°C. The activity of corticotropin as measured with two common bioassays showed good correlation with the amount of native corticotropin present on isoelectric focusing gels [3]. In the case of oxytocin and other nonapeptides in solid and liquid dosage forms, a good correlation was also found between results of bioassays and a high-performance liquid chromatographic (HPLC) method [4]. Perhaps most significant, an HPLC assay for insulin purity is now carried out in conjunction with the rabbit hypoglycemic assay as described in the USP; the requirement for concordance is that the total chromatographic potency differs by not more than ±6% of the biological potency [1]. From these examples, the use of chromatographic and/or electrophoretic methods to supplement bioassays would appear to be a feasible approach for potency determination of other peptide drugs.

B. Radioreceptor Assays (RRAs)

Biological responses such as those described in Table 1 are initiated by the binding of peptide to a receptor in the cell membrane or within the cell interior. Thus characterization of the peptide-receptor interaction can also be considered as a measure of biological activity. In practice, this characterization is carried out through a competitive assay. The nonradioactive sample of interest is mixed with a "tracer" (most often a highly purified radiolabeled peptide) and the mixture is incubated with a known concentration of cultured cells that possess the peptide receptor complementing the peptide analyte. Substances that bind to the cell receptors will precipitate upon centrifugation, while unbound material will remain in the supernatant. The ability of the sample to compete with the peptide tracer for binding sites is then assessed from the measurement of receptor-bound radioactivity. A variation of this method is to use a nonpeptide tracer in conjunction with a peptide standard. For example, Chan et al. [5] have shown that the displacement of radiolabeled naloxone from opiate receptors can be used to evaluate potency of various analgesic peptides and their degradation products.

RRAs have been developed for other peptides, including insulin [6], human growth hormone [6], lutenizing hormone [7], and chorionic gonadotropin [8]. Currently, the greatest limitation to more widespread application of RRA is probably the preparation of suitable peptide tracers. Because the number of receptors per cell is small, on the order of 10^4 to 10^5, a peptide of high specific radioactivity is often required. For RRA, there is the additional requirement that the radiolabel itself not interfere with the peptide's receptor binding ability. This is a particular problem for ^{125}iodine labels, due to their large size, which has resulted in some peptides losing their bioactivity upon iodination. Although tritium (^{3}H) can be used as an alternative, its detection is more difficult and time consuming and can be adversely affected by exchange of label with the solvent medium and quenching effects induced by the matrix.

Other limitations of RRA are similar to those encountered in bioassays: variability within and among cell receptor preparations and interference by other substances in the incubation mixture. Despite such problems, results of RRAs have generally correlated very well with those of in vitro bioassays and fairly well with in vivo ones. This is to be expected, since in vitro bioassay results are more heavily weighted by the affinity of the peptide for its receptor and are less influenced by distribution and disposition phenomena. However, there are situations where RRA and in vitro bioassay results necessarily diverge. This happens with peptides that are partial agonists or antagonists. These compounds may exhibit strong affinities for the natural peptide receptors, but show diminished or nonexistent biological responses when subjected to the full complement of materials present in the living system.

The underlying principles of RRA are similar to those of other competitive assays involving radioactive ligands, such as the widely applied radioimmunoassay (RIA). The key difference between RRA and RIA is one of biological specificity. Whereas RRA measures an interaction related to biological activity, RIA measures one related to antibody-binding affinity, which need not, and probably will not, coincide with biological response. RRA may thus have an important application in clinical analyses of peptide drugs in cases where quantitative determinations by RIA are at variance with clinical observations [6]. Some aspects of RIA are discussed further in Sec. IV; for more information on the merits of RRA versus RIA, the reader is directed to a pertinent review [9].

III. EVALUATION OF PURITY AND STABILITY

A. Chromatography

Chromatography encompasses a wide variety of separation methods based on the relative affinities of a solute for mobile and stationary phases. In the past, several chromatographic techniques have proven useful in the characterization of pharmaceutical peptides, including thin-layer chromatography

(TLC), gas-liquid chromatography (GLC), and column liquid chromatography. TLC has been particularly valuable for monitoring batch impurities in synthetic peptide preparations, such as those found in salmon calcitonin [10]. Applications of TLC to characterize other peptides of therapeutic interest, notably angiotensins, glucagon, and insulin, have been described by Lepri et al. [11]. GLC has been used mainly for identification of small-molecule impurities and cannot be used to identify peptides containing more than a few residues, due to their nonvolatility and thermal lability. However, GLC does have an important application in peptide sequence determination when carried out in conjunction with mass spectrometry, which is discussed in Sec. V.

Over the past decade, TLC, GLC, and conventional column liquid chromatography have become overshadowed by high-performance liquid chromatography (HPLC). The successful application of HPLC in peptide analysis, and in analysis of other pharmaceuticals and biomacromolecules, has been due in large part to significant technical improvements in the preparation of inert, pressure-resistant stationary-phase materials. Today, HPLC is probably the most versatile and widely applied method for separation and identification of peptides, replacing or at least supplementing more time-consuming and laborious methods. The key advantages of HPLC are speed and simplicity; moreover, unlike most other peptide separation methods, HPLC procedures can be automated. The varied uses of HPLC for peptide analysis have been the subject of numerous articles; the interested reader is referred to recent reviews [12,13] for an introduction to the theoretical and practical considerations involved. The present discussion focuses on the applications of three HPLC techniques relevant to separation of pharmaceutical peptides: reversed-phase, ion-exchange, and gel permeation (size exclusion) chromatography.

Reversed-Phase HPLC

Reversed-phase HPLC (RP-HPLC) discriminates among peptides primarily on the basis of hydrophobic character. In the presence of a polar mobile phase, column retention is related to the strength of interactions occurring between the hydrophobic portion of a peptide, the surface of a nonpolar stationary phase (reverse packing), and components of the eluting phase. HPLC reversed-phase packings usually consist of hydrocarbon chains chemically bonded to microparticulate silica gels; chain lengths frequently employed are the octyl (C_8) and octadecyl (C_{18}) forms. Chemically bonded silicas are superior to the physically adsorbed stationary phases used in conventional liquid chromatography, as the former are compatible with gradient elution systems and permit much greater flexibility in the selection of mobile-phase compositions. A significant limitation is that silica supports are chemically unstable above pH 7.5. Below this pH, however, they can be used reliably with a wide range of eluents and flow velocities. It is this versatility that has made HPLC a popular as well as a powerful tool for peptide analysis.

Meek and Rosetti [14] have carried out a detailed study of the factors affecting RP-HPLC retention and resolution of small peptides (i.e., those containing 20 or fewer residues). Results from their experiments confirm the hypothesis that hydrophobicity of the component amino acids is the major determinant of retention behavior. Retention can be predicted by summing the hydrophobic contributions of a peptide's amino acid constituents, yielding a "retention coefficient." Amino acids with aromatic or aliphatic side chains contribute positively to retention, while those with polar side chains either have no effect or contribute a small negative effect on peptide retention. Prediction of retention becomes more complicated for peptides larger than 20 residues, due to the existence of long-range intramolecular interactions between various amino acids brought about by folding. Indeed, there are frequent instances where observed retention behavior deviates from theoretical hydrophobicity predictions. A primary cause of anomalous retention behavior is the presence of reactive free silanol groups at the surface of the stationary phase, promoting "silanophilic" interactions with cationic solutes. Such interactions may result in chromatographic peak tailing and an inability to achieve desired selectivity of separation. Mobile-phase buffer components can also alter predictions of retention based on assumed reversed-phase behavior, due to dynamic ion-exchange and ion-exclusion effects during elution [15].

While very polar peptides can be eluted from alkyl-silica-bonded stationary phases with a purely aqueous solvent system, in most cases an organic modifier is added to the mobile phase. Peptide retention behavior is quite sensitive to mobile-phase composition, and efficient isocratic separation often can occur only in a narrow range of organic solvent percentages. For smaller peptides, retention on the reversed-phase silica generally decreases with increasing volume fraction of organic solvent (ϕ_S) in the mobile phase over the range $0 < \phi_S < 0.4$, and peptides will be eluted in order of increasing hydrophobicity [15]. Separation of very hydrophobic and/or larger peptides may require mobile phases of higher organic solvent content; such systems sometimes exhibit reversals of retention order, referred to as normal-phase behavior. The properties of the organic modifier chosen can also affect chromatographic behavior. For instance, broader peaks were observed for smaller peptides when 2-propanol was used in the mobile phase compared to the same peptides eluted with systems containing methanol or acetonitrile; this effect was ascribed to the greater viscosity of 2-propanol [15].

Column selectivity can also be enhanced by adding an ion-pairing agent to the mobile phase. Such agents are charged species that associate with peptides carrying opposite charges. Common anions added as ion-pairing agents include phosphate, sulfonate, sulfate, and acetate; frequently used cations include quaternary n-alkyl ammonium ions. By careful consideration of the ion-pairing agent's charge, its hydrophobicity, and its associated counterion polarity, one can promote selective retention of a hydrophilic peptide on the reversed phase. Consider the elution of an arginine-containing peptide on a C_{18} column in 1:1 water/methanol at pH 3.0. The presence of a cationic species in the mobile phase would decrease retention due to electrostatic repulsion with the positively charged arginine side chain. The retention time can be slightly increased by employing a hydrophilic anionic species ($H_2PO_4^-$); retention can be significantly increased by employing a hydrophobic anionic species such as dodecyl sulfate [16].

In addition to hydrophobicity, peptide molecular weight also affects RP-HPLC retention. Specifically, column efficiency decreases with increasing solute molecular weight, presumably due to reduced diffusivity, which results in band spreading. A similar reduction in chromatographic efficiency occurs with increasing mobile-phase flow rate. Thus high-resolution isocratic separations are generally achieved employing a low flow rate (≤ 1 ml/min). Resolution in HPLC separations using organic modifier gradients depends on the gradient profile, becoming optimal at very slow rates of

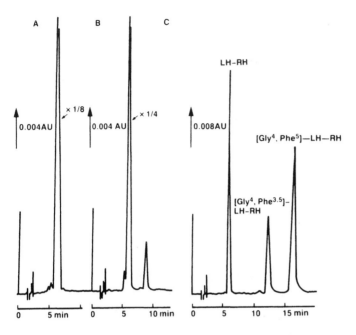

Fig. 2   RP-HPLC chromatograms of LH-RH and analogs.   Content of acetonitrile in the mobile phase: 21%. A, LH-RH synthesized by the solid
phase method (1.7 µg); B, LH-RH synthesized by the solution method
(0.95 µg); C, a mixture of LH-RH, $[Gly^4, Phe^{3,5}]$-LH-RH and $[Gly^4,$
$Phe^5]$-LH-RH. (From Ref. 17.)

change in solvent composition (0.5%/min) [14]. Thus one often is faced
with a trade-off for practical applications: short analysis time at the expense of maximum resolution.

     A common use of RP-HPLC in pharmaceutical analysis has been in the
determination of peptide homogeneity. This is particularly important for
peptides produced by synthetic methods, since various chemical reactions
(e.g., racemization) can alter the desired biological activity of the final
product. Successful applications of RP-HPLC for separation of peptide diastereomers and other closely related structural analogs are well documented
[12]. A representative example involving LH-RH is given in Fig. 2. The
RP-HPLC chromatogram of LH-RH synthesized by a solid-phase method indicates a highly pure product, whereas the LH-RH prepared from solution
synthesis contains noticeable impurities. In addition, the chromatogram of
a mixture containing LH-RH and two analogs indicates that all three compounds are well resolved, with no significant evidence of contamination by
other peptides.

     RP-HPLC is also useful for examining peptide stability in vitro as well
as in vivo. In the former situation, Corran and Calam [10] have described
such an assay for a set of tetracosatrin samples in which degradation was
induced by exposrue to elevated temperatures for different time intervals.
The amount of native tetracosatrin remaining according to the HPLC chromatograms agreed well with bioassay data. However, the authors noted
some evidence of nonparallelism with respect to the native peptide standard,

suggesting that the degradation products themselves possessed a certain amount of bioactivity.

The development of strategies for assessing in vivo stability of peptide drugs is an active area of research, complicated by the challenges of achieving adequate selectivity and sensitivity. Here RP-HPLC has been useful in assessing tissue peptidase activity toward different peptides. In one such study, tissue extracts of rat hypothalamus were incubated with LH-RH; HPLC was then applied to the samples to examine the extent of LH-RH degradation as a function of incubation time [18]. The various degradation products were subsequently isolated to identify the sites of peptide bond cleavage in LH-RH. Identification of chromatographic peaks was confirmed by amino acid analysis and by comigration with peptide standards of known sequence.

Ion-Exchange HPLC

Ion-exchange chromatography separates molecules principally on the basis of charge, although hydrophobic and polar interactions between analyte and stationary phase also contribute to the observed retention behavior. The molecules to be separated are reversibly associated with the solid phase (ion exchanger) through electrostatic interactions. The column is eluted with buffers of varying pH and/or ionic strength; the buffers compete with the ion exchanger for molecular binding sites. It has been pointed out that in some instances, ion-exchange HPLC and RP-HPLC are complementary (i.e., peptides that cannot be well resolved by one method may be resolved by the other) [19]. Thus the combined use of ion-exchange HPLC and RP-HPLC enhances the probability of achieving complete separation of a given peptide mixture. An impressive example of such a "tandem" system for separation of extremely complex peptide mixtures has been described by Takahashi et al. [20].

The presence of an organic modifier is often also required for optimal separation and elution in ion-exchange chromatography. Kato et al. [21] have reported on the application of a resin known as TSK gel-5PW, a hydrophilic polymer-based cation exchanger containing sulfonylpropyl groups, to separate a mixture of brain peptides. In the absence of organic solvent in the mobile phase, some peptides were eluted very slowly as broad peaks, probably due to hydrophobic interactions between these peptides and the ion exchanger. This effect could be eliminated by the addition of 20 to 40% acetonitrile to the mobile phase. Furthermore, the relative elution positions of several peptides were sensitive to the percentage of acetonitrile employed, providing a means for improving selectivity.

The widespread use of ion-exchange HPLC to peptide separation has been hindered by a lack of suitable stationary phases. Probably the most commonly used ion exchangers are prepared from polystyrene-divinylbenzene (PS-DVB) copolymers. These resins tend to display low rates of mass exchange, and elevated temperatures (>60°C) are often needed to achieve adequate column efficiency. Many PS-DVB resins are also susceptible to deformation at high flow velocities, a drawback that has largely been overcome by chemically bonding the ion exchanger to microparticulate silica gels. The major disadvantage of silica-based systems is again their limited chemical stability above pH 7.5. The practical consequence of this for ion-exchange HPLC is that buffer gradients are needed to elute solutes over a wide pH or pI range. Alternative packing materials (e.g., ion exchangers based on hydrophilic polyether resins) may become important in the future, as they appear to possess good chemical stability and high loading capability [12].

Recently, Dizdaroglu [19] reviewed the applications of a silica-bonded weak anion exchanger to the HPLC characterization of several endogenous peptides, including bradykinins, angiotensins, and neurotensins, as well as their diastereomers and other synthetic analogs; the peptides studied contained from 6 to 11 amino acid residues. This chromatographic system, referred to as weak anion-exchange HPLC (WAE-HPLC), uses a stationary phase known as MicroPak Ax-10, which carries the functional group $-CH_2-$ $-CH_2-CH_2-NH_2^+-CH_2-NH_3^+$. For elution, a gradient of pH 6.0 triethylammonium acetate buffer (TEAA) into acetonitrile was utilized. Good separation could be achieved for closely related peptides, some differing by only one amino acid residue, together with high sensitivity and efficient recovery of the separated peptides (80% or greater).

The selectivity of the WAE-HPLC stationary phase was found to be influenced by the column temperature, which could be varied to optimize the separation of different peptide mixtures. For example, the retention times of five sarcosine-containing angiotensins decreased as temperature was increased from 26°C to 50°C, while all other angiotensins studied exhibited longer retention times at higher temperatures. In the case of the neurotensins, retention times for all derivatives increased with increasing temperature. But the magnitude of the change in retention differed among the individual peptides, allowing enhanced selectivity of separation at a given temperature. The pH of the eluting buffer is also an important separation parameter in WAE-HPLC. In the aforementioned elution procedure, good separation of dipeptide mixtures could be carried out at pH 4.3. However, for the neuropeptides cited previously, pH 6.0 was preferred in terms of retention, peak symmetry, and resolution.

Gel (Size Exclusion) HPLC

Separation by gel chromatography is based on molecular size. Molecules in a mixture which are larger than the pores in the gel cannot diffuse into them, and continue to move down the column. Molecules smaller than the pores will travel a more tortuous path, moving in and out of the pores with a probability inversely related to molecular size, thereby retarding movement along the column. Molecules will thus be eluted from the column in order of decreasing molecular size (or molecular weight) if shape is constant. Although size exclusion is the primary factor governing retention, secondary effects such as ionic interactions and adsorption to the gel also affect elution behavior.

Classical gel chromatography has been used extensively for peptide and protein purification. However, the use of gel permeation HPLC for peptides with molecular weights under 5000 has been limited to date. This is due primarily to a lack of suitable stationary-phase materials. The traditional cross-linked gels (dextran, agarose, etc.) are often subject to deformation under the experimental conditions of HPLC. Newer materials based on hydrophilic silica supports and cross-linked organic polymers (e.g., Spheron, which consists of polymeric methylmethacrylate ester matrices) are still unsatisfactory in terms of column efficiency.

Despite current technical limitations, gel HPLC of peptides promises to be more valuable in the future as an analytical tool. It should be especially useful for characterizing peptides derived from natural sources, wherein high-molecular-weight polypeptides may be present as contaminants. Gel

HPLC can also be used to monitor the extent of peptide aggregation in solution; conversely, it can be used to examine dissociation of peptides under
denaturing conditions, as has been done for insulin [22]. Separation based
on both a reversed-phase and a gel permeation mechanistic component may
also be achieved through use of high-porosity reversed-phase supports.
In fact, such a combination of HPLC techniques has been applied to separate several highly hydrophobic peptides [23].

    In conclusion, it should be reemphasized that successful application of
any HPLC method to peptide analysis will depend on several factors. These
include the nature of the column packing, particle size, porosity, composition of the mobile phase, length and diameter of the column, operating temperature, and properties of a given peptide, such as molecular weight, isoelectric point, and hydrophobicity.

B. Electrophoresis

Electrophoresis can be defined as the controlled movement of charged species
through a solvent in the presence of an electric field. It has been an effective method for separation and characterization of macromolecules on the
basis of their net charge or molecular size. In general, electrophoresis is
more suitable for routine analysis of proteins and large peptides (molecular
weights $\geqslant 10,000$) rather than the smaller peptides which are the primary
focus of this paper. Nevertheless, with proper choice of experimental conditions, electrophoresis can often be a valuable tool for evaluation of purity
and stability of such peptides. Three electrophoretic methods useful in
peptide analysis are zone electrophoresis, isoelectric focusing, and isotachophoresis.

Zone Electrophoresis

This method involves mounting the sample on a supporting medium to which
a voltage in the range of 3500 to 12,000 V/m is applied. Common stabilizing
supports include paper, cellulose acetate, and gels such as polyacrylamide.
After development is complete, the support is removed and stained, typically
with a peptide-specific dye, to detect the sample components (zones). The
results can be compared with those of reference standards electrophoresed
under identical conditions in order to evaluate sample molecular weight and
purity. Polyacrylamide gel electrophoresis (PAGE) typically gives good resolution with high reproducibility, and the presence of a single stained band
is considered a reliable indicator of a sample's homogeneity.

    Probably the most widespread use of zone electrophoresis has been in
"fingerprinting," a method of protein sequence analysis involving initial
enzymatic digestion of the analyte, followed by characterization using electrophoretic and chromatographic procedures. Zonal electrophoresis has
also been used to assess purity and stability of peptide drug substances.
The *British Pharmacopeia* (BP) now includes such a peptide sequencing
test for identification of tetracosatrin [10]. Other applications described
in the current BP monographs are cellulose acetate electrophoresis (for
tetracosatrin) and polyacrylamide gel electrophoresis (PAGE) (for glucagon).
In the latter test, PAGE is used to limit levels of desamidoglucagon (which
is significantly less active than native glucagon), as well as other contaminating peptides. A similar PAGE procedure can be carried out on insulin
to assess contamination by desamidoinsulin and proinsulin.

Peptide molecular weight is an important consideration when choosing
the supporting medium and associated reagents for zonal electrophoresis.
Paper and cellulose supports can be used successfully for peptides with
molecular weights up to 2000, but higher-molecular-weight peptides tend to
interact strongly with these supports, as manifested by tailing. On the
other hand, PAGE is best suited for peptides with molecular weights above
2000; the effective pore diameter within the polyacrylamide matrix is generally too large for resolution and detection of smaller peptides. This limitation of PAGE can sometimes be overcome by carrying out electrophoresis
in the presence of ionic detergents such as sodium dodecyl sulfate (SDS),
and/or urea, or by increasing the polyacrylamide content of the gel. West
et al. [24] have recently described the use of such strategies in PAGE
methods capable of separating nonderivatized peptide fragments ranging in
length from 2 to 19 amino acids. The PAGE resolution of peptides containing disulfide linkages may be further improved by carrying out electrophoresis under reducing conditions, which promote the solubilizing effect
of SDS. If one is interested in separating a peptide preparation containing
both high- and low-molecular-weight species, it is also possible to combine
PAGE with either isoelectric focusing or isotachophoresis in a two-dimensional electrophoresis system.

Detection of separated peptides following electrophoresis can be accomplished through various methods, many of which involve treatment of the
supporting medium with agents that react with peptide functional groups
to produce a characteristic color. Ninhydrin, which reacts with free amino
groups to form colored products (see Sec. IV), is commonly applied to detect peptides on cellulose and paper supports. However, ninhydrin is
relatively insensitive, and cannot reliably detect peptide amounts smaller
than 1 mg on these supports. Sensitivity can often be improved significantly by use of amino acid-specific reagents, such as the Ehrlich's reagent,
which can detect microgram sample amounts of tryptophan. For visualization of proteins and some peptides on polyacrylamide gels, staining with
the dye Coomassie Blue is popular. Although this method can detect sample
amounts below 1 μg, it is not reliable for identification of peptides containing less than 20 residues. In this instance, silver staining, which can detect nanogram amounts of peptides, is preferred. These and other chemical methods used in the electrophoretic detection of peptides have been
summarized by Prusik [25].

One disadvantage of conventional zonal electrophoresis methods relates
to the existence of temperature gradients generated by passage of electric
currents. These gradients cause convective flow, which in turn leads to
irregularities and broadening of the separated zones. Although the use of
stabilizers helps to diminish convection, it does not eliminate the source of
convection (i.e., the temperature gradients). Furthermore, stabilizers themselves promote broadening effects, as in the case when solutes within a sample mixture adsorb to the supporting gel.

A more promising development is capillary zone electrophoresis, where
electrophoresis is performed in a tube of small internal diameter (<100 μg)
without the addition of stabilizers at very high field strengths (about
30,000 V). This is based on the observation that heat can be dissipated
by reducing the distance perpendicular to the axis. Compared with conventional zone electrophoresis, the capillary method offers greater speed
and simplicity as well as high resolution. Capillary zone electrophoresis
can be carried out in less than 30 min, whereas the conventional zone

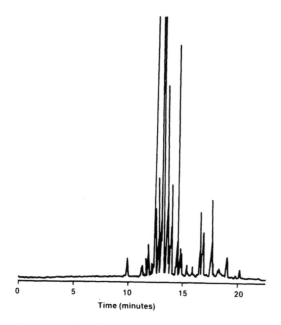

Fig. 3 Capillary zone electrophoresis separation of fluorescamine-labeled peptides obtained from a tryptic digest of reduced and carboxymethylated egg white lysozyme. (From Ref. 26.)

method requires a minimum of several hours to complete. An indication of the high separation efficiency that can be achieved with capillary zone electrophoresis is given in Fig. 3.

Isoelectric Focusing

Isoelectric focusing separates components based on their isoelectric points (pIs) in the presence of a preformed pH gradient. The pH gradient, typically in the pH range 3 to 10, is established with a mixture of low-molecular-weight synthetic polyaminopolycarboxylic acids. Such substances, which carry both positively and negatively charged functional groups, are commonly known as carrier ampholytes. The optimal ampholyte mixture should have component pIs closely spaced over the desired pH range for peptide analysis. During electrolysis, current moves in the direction of increasing pH; each ampholyte will migrate to a point matching its own pI, thus establishing a pH gradient. A peptide that carries a net positive charge in regions where pH < pI travels with the current toward its isoelectric zone (i.e., the region where the peptide net charge is zero). In regions where pH > pI of the peptide, it moves against the current, back toward the isoelectric zone. Eventually, all the peptide's cationic and anionic forms become condensed or "focused" in one sharp isoelectric zone. Isoelectric focusing is most often carried out on thin slabs of polyacrylamide; a detailed description of this methodology is given by Righetti et al. [27].

In contrast with zonal electrophoresis, resolution of isoelectric focusing is not significantly affected by peptide molecular weight under the influence

of an imposed electric field. However, "de-focusing" (spreading) of peptide zones due to diffusion can occur after the current is turned off and
is inversely related to peptide size. The size of the carrier ampholytes
must also be considered; ampholytes with molecular weights comparable to
those of the peptides being separated may interfere with the latter's detection by some chemical staining procedures. Despite such potential problems, isoelectric focusing does offer an important advantage in that it can
be carried out using a relatively simple experimental apparatus, yet is extremely sensitive to slight charge differences within a sample mixture. It
is now possible to resolve components whose pIs differ by only 0.02 pH
unit with isoelectric focusing. Thus isoelectric focusing may often be used
to separate, identify, and detect peptides that are not resolved by other
methods.

As a practical illustration of the resolving capabilities of this technique,
consider the application of isoelectric focusing to synthetic human calcitonin
which was radioiodinated by two different procedures [28]. In both cases,
isoelectric focusing gels showed that the reaction products were heterogeneous, exhibiting more acidic pI values compared to unlabled calcitonin. Other
methods either failed to detect heterogeneity (chromatography) or gave ambiguous indications (irreproducible radioimmunoassay data). Isoelectric
focusing also has potential as the basis for stability assays; as noted in
Sec. II, amounts of native corticotropin present after long-term storage as
determined by isoelectric focusing agreed well with bioassay data [3].

Isotachophoresis

Isotachophoresis separates ionic species according to their effective mobilities, which are defined by the magnitude of the velocity of a given species
(i.e., distance traveled per unit time) divided by the magnitude of the
strength of an applied electric field. Isotachophoresis is simple, accurate,
and requires only a small amount of sample, on the order of 0.1 to 1.0 nmol
of peptide. Like capillary zone electrophoresis, isotachophoresis has an
advantage of rapid sample analysis time, on the order to 10 to 15 min,
making it much less time consuming than PAGE or isoelectric focusing analyses.

In the techniques of analytical capillary isotachophoresis, the sample is
introduced in a capillary tube between an electrolyte of higher mobility
(leading electrolyte) and one of lower mobility (terminating electrolyte).
The leading, terminating, and sample ions all have net charges of the same
sign (i.e., either all positive or all negative). After applying an electric
field, sample components of different mobilities will move with correspond-
ingly different velocities, with those of highest mobility traveling the greatest distance in a given period. Eventually, the system reaches a steady
state in which the various sample ions become clearly separated between the
leading and terminating ions. The sample zones can be detected with various methods, including thermometry (each zone generates a characteristic
heat), electrical conductivity, and ultraviolet (UV) absorbance. A more
detailed description of the principles and practice of isotachophoresis can
be found in a review by Everaerts and Mikkers [29].

As an illustration, a basic peptide such as secretin is typically characterized in protonated form by cationic isotachophoresis at pH 5.1, with potassium cations as the leading electrolyte and β-alanine as the terminating
electrolyte [30]. Acidic and neutral peptides are best separated using
anionic isotachophoresis. For example, a 30-amino acid fragment of synthetic

human growth hormone was analyzed at pH 7.6 with chloride as the leading ion and glycine as the terminating species [31].

Although analytical isotachophoresis is not yet a routine tool in peptide analysis, it appears to have great potential value, particularly with respect to analyzing peptides generated by synthetic routes. In one such application, isotachophoresis was used as an analytical monitor during synthesis of a bombinin fragment, Gln-His-Phe-Ala-Asn [32]. Isotachophoretic analyses revealed that stages of synthesis up to the Phe-Ala-Asn tripeptide gave pure products. However, addition of the histidine residue to the peptide chain resulted in the introduction of considerable impurities, which were not detected with TLC.

Considering all three electrophoretic methods, isotachophoresis may be most useful for characterization of synthetic peptides, while isoelectric focusing and zonal electrophoresis promise to be equally important analytical tools for quality control of recombinant-DNA derived peptides. Indeed, the latter methods have already been used for this purpose with respect to analysis of human insulin [33]. However, it should be noted that like HPLC, electrophoretic methods may not always be applicable to analysis of peptide formulations, due to interference in their separation by excipients. Other limitations to routine use of zonal electrophoresis and isoelectric focusing are their unsuitability for automation, and the difficulty of removing separated components from supports for further characterization. Also, at the present time, the chemical staining methods commonly used for electrophoretic detection generally do not offer adequate sensitivity. Alternative detection systems with enhanced sensitivity may require radiochemical, fluorometric, or immunochemical markers. The principles underlying the use of such methods for quantitative detection of peptides are discussed in Sec. IV.

IV. QUANTITATIVE DETECTION

The availability of methodology for quantitative monitoring of peptides is essential to their pharmaceutical development. Such methods are needed to (1) optimize the isolation and purification processes for the putative drug, (2) study the kinetics of drug degradation in vitro, (3) assess content uniformity of dosage forms, and (4) determine the fate of peptide drugs in biological systems (i.e., laboratory animals and humans). In the case of in vivo studies, quantitative assessment of biological disposition is needed to describe a drug's absorption, distribution, metabolism (clearance), and excretion. This information can then be used to optimize delivery of a given peptide, or to facilitate the design and selection of improved peptide analogs.

One is faced with two significant challenges in developing methodology for such purposes. First, peptides are often present at very low concentrations in vivo (below the nanogram level), necessitating methods of high sensitivity. An additional complication is that the peptide drug may exist in a matrix containing other components with similar structural features (e.g., a biological matrix will have endogenous peptides present as well as their metabolites). Therefore, useful quantitative methods must be sufficiently specific as well as sensitive for the peptide of interest.

On the basis of convenience and rapid sample analysis time, HPLC is a desirable method for peptide separation and isolation. However, the standard HPLC detector, which utilizes ultraviolet absorbance of the peptide bond at about 210 nm, often lacks the requisite specificity and sensitivity for quantitative analysis at or below the nanogram level; in this situation, alternative detection systems must be considered. The methodology and analytical strategies described below can be applied either as part of a batch procedure, or as one step in an analytical sequence that includes a chromatographic component. In many cases, the detection systems presented herein have been interfaced successfully with HPLC, providing on-line capability for rapid identification and quantitation of separated species.

A. Radiochemical Methods

A convenient and potentially sensitive means of detection involves using radiolabeled analytes for investigative studies and then monitoring the label's emitted radiation. Means of detection depend on the type of radiation emitted from the isotope label; a liquid scintillation counter is used for β-emitters (e.g., ^{3}H or ^{14}C), while a gamma counter analyzes γ-emitters such as ^{125}I. Sensitivity is determined by the specific activity of the radiolabeled compound, as well as by the emission energy profile and intensity of the radionuclide. The excellent sensitivity that can often be achieved with radiochemical methods has contributed to their widespread use in quantitation of picomole amounts of labeled material. However, radiochemical detection is based on monitoring the label, and is not responsive to other molecular parameters of the labeled compound. Thus radiochemical detection has an inherent lack of specificity which must be considered in potential applications. For example, consider a solution containing radiolabeled drug that has been degraded, producing fragments some of which are also radiolabeled. Measurement of radioactive emission by itself will not provide information regarding the extent of degradation nor the identity of the degradation products. However, it is possible to obtain this information by introducing a separation step (e.g., chromatography) prior to either on- or off-line radiochemical detection. Implementation of such procedures has been facilitated by the commercial availability of flow-through liquid scintillation detectors for HPLC with sophisticated data processing systems.

Radiochemical detection methods have often been applied to the in vivo monitoring of peptides. In one such study, Bennett and McMartin [34] examined the distribution and degradation of two ^{3}H-labeled corticotropin analogs after intravenous administration in rats. Characterization of radioactivity revealed that both peptides were rapidly cleared from plasma, mainly entering muscle and skin tissue, where they were rapidly degraded. However, for one analog that was blocked at its amino terminus, a measurable amount of intact peptide persisted (20 to 30%, as ascertained by HPLC) and returned to the circulation. The authors proposed that protection of the amino terminal residue would confer increased resistance to degradation, resulting in enhanced drug potency and duration of action. This hypothesis was supported by bioassay data.

Although peptide radiolabeling is most often used for quantitative determinations, it can sometimes be extended to characterize biological activity as well. An example of such an application is a radioassay for nonoxidized methionine residues, which involves labeling a methionine residue of interest with iodo [2-^{14}C]acetic acid (oxidized methionines fail to react with this reagent). The derivatized material is separated from other matrix components by chromatographic or electrophoretic procedures, and analyzed for

radioactivity. The amount of radioactivity in the derivatized product can
be directly related to the amount of reduced methionine in the peptide,
which in turn can be related to the extent of biological activity as determined previously from bioassays. The radioassay can be carried out on
up to 30 samples at one time, with a lower detection limit of less than 1
nmol of methionine per sample. This assay is especially useful for peptides
whose biological activity depends on their sole methionine residue, and has
been applied successfully to several therapeutically important peptides, including corticotropin, glucagon, and calcitonin [25].

Choice of the peptide radiolabel is an important consideration in any
radiochemical detection method. For best sensitivity (detection at or below
the picomolar level, ^{125}I or other high-energy γ-emitters are the preferred
radiolabels. In the most commonly applied procedures for iodination of peptides, the iodine radiolabel is attached to the activated aromatic ring of a
tyrosine residue. However, there are some important endogenous peptides
that do not contain tyrosin (e.g., somatostatin); iodination of such peptides
requires more sophisticated synthetic strategies [36]. Furthermore, the lipophilicity of bound iodine moieties can alter the biological, immunological, and
physical-chemical properties characteristic of the unlabeled peptide.

An additional disadvantage of using ^{125}I is that many radioiodination
procedures alter the integrity of the peptide being labeled, resulting in a
heterogeneous population of [^{125}I]peptides, as was described for calcitonin
in Sec. III [28]. The extent of product heterogeneity is a function of
several factors, including pH of the reaction medium and molar ratio of
radiolabel to peptide [37]. Use of a heterogeneous peptide preparation in
in vivo systems can lead to errors in the determination of a peptide's pharmacokinetic profile and in the evaluation of bioassay data. Heterogeneity
will also complicate analysis of peptide-ligand associations and dissociations,
which are the basis of the RRA and RIA methods. For these reasons,
heterogeneity in the radioiodinated peptide must be ascertained and avoided
at the synthetic level; chromatographic and electrophoretic methods are
often helpful for this purpose.

If the peptide of interest cannot be successfully iodinated, labels such
as ^{3}H, ^{14}C, or ^{15}S may be utilized; this is done by incorporating labeled
amino acids during peptide synthesis. However, the aforementioned isotopes
are β-emitters and have relatively low energies compared with γ-emitters
much as ^{125}I. A consequence of this is that often, not all emitted particles
reach the detector, thus compromising sensitivity. Furthermore, the synthesis of such peptides can be time consuming, as each synthetic step is
likely to produce labeled compounds other than the desired product, necessitating cumbersome separation procedures. Even if a highly purified radiolabeled compound is obtained, its degradation can be accelerated over time
by the presence of radiolabel. This can happen when one emitted particle
ionizes another, leading to chemical rearrangement, or when the emitted
particle ionizes the solvent, producing reactive species that attack the
parent peptide. The latter process is referred to as radiolysis, and frequently occurs with radiochemicals, particularly those that are stored in
aqueous solution.

B. Immunochemical Methods

The high affinity and specific association of an antibody with an antigenic
determinant can be exploited to detect very low concentrations (about

10^{-13} M) of peptide antigens in a variety of biological matrices. Immunochemical methods are thus particularly well suited for detection of peptide
drugs within biological tissues and fluids. Three types of immunoassays
have been commonly employed for this purpose: the radioimmunoassay
(RIA), the immunoradiometric assay (IRMA), and the enzyme immunoassay
(EIA).

Radioimmunoassays

RIA is based on competition for specific antibody binding sites between an
unknown amount of nonradiolabeled peptide antigen and a known quantity
of the same antigen which has been radiolabeled (tracer). Quantitation is
achieved by measuring the amount of radiolabeled antigen bound to antibody
relative to the amount remaining in solution. Sensitivity of RIA depends on
the reaction of an antigen with primary antibody binding sites:

$$Ag + Ab \rightleftarrows Ag - Ab$$

$$K_{eff} = \frac{[Ag - Ab]}{[Ag][Ab]}$$

A high-sensitivity assay requires a large K_{eff} value. For RIAs, K_{eff}
values are on the order of 10^{12} M^{-1} and allow detection of peptide antigens
at or even below the picomolar level. Specificity of RIA depends on the
extent of recognition between antigen and antibody, which is essentially
defined by structural determinants in both macromolecules. An RIA of optimal specificity requires highly selective recognition with minimal interference from other components within the sample matrix. In practice, this
selectivity has proven difficult to achieve. The reasons for this are discussed herein, together with proposed strategies for improving RIA selectivity.

Meaningful application of an RIA requires satisfying three important conditions, which have been sumarized by Eckert [38]. First, the unlabeled
antigen to be determined must be immunochemically identical to the radiolabeled antigen used as tracer. That is, K_{eff} must be the same for complexes of the antibody with either substance. Second, the amount of radiolabeled antigen added to the test solution should exceed the number of antibody binding sites available. In other words, unbound labeled antigen will
exist along with the tracer-antibody complex. Third, the concentration of
tracer and amount of antibody present in a given experiment must be kept
constant for all test runs. This ensures that the concentration of unlabeled
antigen is the sole experimental variable. To measure an unknown concentration of antigen, one first carries out RIAs using known concentrations
of the antigen, typically over a 100-fold dilution range. The resulting data
are used to prepare a calibration curve. The concentration of the analyte in
in samples can then be determined by relating the ratio of bound radiolabeled antigen to total radioactivity (bound plus unbound radiolabeled antigen) to points on the standard curve. A representative standard curve is
shown in Fig. 4. This curve was used in RIAs of motilin, a 22-residue
gastrointestinal peptide; the assay could detect native motilin in dog plasma
at concentrations below 2 pg/ml.

To carry out an RIA, the appropriate antiserum must first be obtained.
Since small peptides (<20 residues) normally are not immunogenic, the

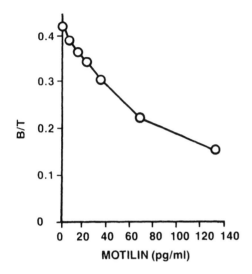

Fig. 4   Standard curve of motilin radioimmunoassay; B/T = number of
bound counts per total counts.   (From Ref. 39.)

peptide of interest (hapten) is usually coupled with a larger immunogenic
macromolecule (carrier) such as bovine serum albumin (BSA), in order to
stimulate antibody production.   The site at which the hapten is conjugated
to its carrier is important, as conjugation may alter the antigen-antibody
recognition and binding processes.   For example, antisera used in the RIA
of the tripeptide thyrotropin-releasing hormone (TRH) are commonly raised
against an immunogen produced by conjugation of BSA via the peptide's
histidine residue.   However, the resulting antisera are only weakly immunogenic, due apparently to the "masking" of histidine from the antibody by
the bulky BSA molecule [36].

In animals, antibody response is generally achieved by injecting the
antigen in the form of an emulsion made with Freund's complete adjuvant.
This adjuvant is a mixture of mineral oil and emulsifier containing heat-
treated mycobacteria to stimulate the general immune response.   Such emulsions promote a slow and steady release of the immunogen into the test animal.   Slow antigen release is desirable in order to produce polyclonal antibodies with high affinity; excessive administration of antigen is a common
cause of failure to induce high-affinity binding antibodies [40].   It should
also be noted that antisera quality can vary with the animal species injection technique and immunization schedule chosen.

In addition to antisera, a highly purified radiolabeled antigen is an essential component of the RIA.   Although RIAs can sometimes be carried out
with ^{3}H-labeled peptides, achieving the full sensitivity potential of RIA generally requires ^{125}I as the peptide radiolabel.   This is due to the latter
radionuclide's high efficiency of detection, as noted in Sec. IV. A.   But
as also noted therein, iodination procedures can result in a heterogeneous
population of radiolabeled peptides, leading to decreased affinity of antibodies

for the labeled antigen preparation and, in turn, reduced assay sensitivity. The site(s) of iodination on the peptide of interest must also be considered. In certain instances, the incorporation of ^{125}I may sterically hinder antibody binding or may alter the antigen-antibody association via electronic effects.

Another key requirement of RIA is a method of separating antibody-bound tracer from free tracer at the end of the incubation period. Normally, this separation cannot be accomplished directly by centrifugation, as the low concentrations of antigen and antibody used in RIA do not result in quantitative precipitation of the antigen-antibody complex. Rather, separation is achieved using differences in adsorption behavior between the free and bound antigen forms (i.e., free antigen is adsorbed onto solid phases, whereas antibody-bound antigen remains in solution). Common solid-phase materials used for this purpose include charcoal, talc, and ion-exchange resins. Charcoal is probably the most widely used adsorbent but has certain disadvantages with respect to selectivity. For instance, it can bind labeled peptide fragments resulting from proteolytic degradation in addition to binding the intact free peptide [41].

An alternative means of separation is the second antibody method (also known as the double antibody method). A "second antibody" is one raised against the primary antibody in a species other than the one in which the primary antibody has been raised. The second antibody is added to the assay medium after the primary antibody incubation period. After the second incubation period, the primary antibody-antigen complex can be precipitated by centrifugation; decantation is used to separate the supernatant (which contains the unbound antigen) and the precipitate (which contains the bound antigen). The relative amounts of tracer in each phase can then be determined using a gamma counter or a scintillation counter, as appropriate.

One limitation of the aforementioned RIA separation procedures is that they require a significant amount of time and effort to complete. Furthermore, fresh standard curves must be prepared for each analytical run due to variability of reagents, experimental conditions, and quality of separation. From the standpoints of convenience and marketability, it would be preferable to employ methods that eliminate the need for physical separation of free and bound fractions, instead relying on observed changes induced in the antigen label upon binding to its antibody. Such homogeneous assays have been developed using fluorescent probes, spin labels, or enzyme reporter groups. Fluorescence polarization immunoassays (FPIA), for example, are based on the differential rates of fluorescence depolarization for small (i.e., free molecule) and large (i.e., associated molecular complex) species, which allow the ratio of the free and unbound labeled antigens to be determined [42]. However, this approach may not be widely applicable to macromolecular analytes, as their depolarization rates in free solution often do not differ significantly from their depolarization in the bound form.

A potentially serious concern with RIA is the existence of peptide heterogeneity in vivo (i.e., the presence of peptides immunologically similar to the antigen being assayed). Such peptides can include higher-molecular-weight precursors (e.g., prohormones) and metabolic degradation products, as well as otherwise unrelated peptides containing sequence homologies with the antigen. If the antiserum used reacts to a significant extent with any of these components, the ability of RIA to specifically quantitate the peptide of interest may be greatly diminished.

Yalow and Straus [43] have discussed the problem of in vivo heterogeneity with respect to RIA measurements of gastrointestinal hormones,
many of which can be divided into two families (gastrin-related and secretin-
related) containing significant homologies. For example, the C-terminal
pentapeptide epitope of gastrin is shared by cholecystokinin (CCK). Thus
an antiserum directed against this portion of CCK is likely to cross-react
strongly with gastrin. Both peptides are normally present in plasma, which
introduces complications for the RIA of CCK using the aforementioned antiserum. The presence of gastrin must be determined independently with an
antiserum not sensitive to CCK and the subtracted from the apparent total
immunoreactivity as determined with the CCK-sensitive antiserum. This
procedure obviously adds to the overall analysis time, but more important,
it represents an additional source of error which may become large if the
CCK and gastrin levels are comparable in the plasma being evaluated.

Immunoradiometric Assays (IRMAs)

RIA involves radiolabeling of an antigen, which is then added in excess to
the test solution, competing with unlabeled antigen for antibody binding
sites. In contrast, an immunoradiometric assay (IRMA) utilizes a radiolabeled antibody reagent, which is added to a sample in an amount exceeding
the number of antigen molecules. After removal of the unbound antibody,
the remaining labeled antibody can be quantitatively related to the presence
of the antigen. Although not as widely applied as RIAs, IRMAs have some
advantages over the former. Unlike small peptide antigens, where radioiodination can be synthetically difficult and lead to altered antibody recognition, the much larger antibody molecule (which is likely to contain many
more potential iodination sites) can easily be iodinated to produce a highly
stable radiolabeled product that retains full immunoreactivity.

A comparison of RIA versus IRMA has been made for human growth
hormone using a common specific antibody to the antigen [44]. The IRMA
showed a 13-fold increase in sensitivity and a 6-fold increase in assay
range over the RIA procedure. However, the IRMA does have disadvantages.
Most notably, the relatively slight difference in size between the free antibody and the antigen-antibody complex makes their separation difficult.
This problem has necessitated the use of immunoadsorbents in which the
antigen is bound to a polymeric carrier. Washing steps carried out to separate free antibodies form those bound to the immobilized antigen can result
in losses of material, requiring the use of relatively large amounts of expensive antiserum.

Enzyme Immunoassays (EIAs)

Enzyme immunoassay (EIA) methods utilize enzyme-catalyzed reactions for
identification of immunoreactants. Typically, an enzyme is conjugated to
one of the reactants (either the antigen or the antibody) and serves to
amplify the occurrence of antigen-antibody binding through conversion of
enzymatic substrates to products. A "homogeneous" EIA can be performed
in a single step and involves covalent coupling of a peptide hapten with an
enzyme. The conjugation is done in such a way that the enzyme-substrate
interaction required for catalytic activity is inhibited when the hapten forms
a complex with its antibody. Thus the magnitude of enzyme activity (determined by kinetic analysis of known products) can be related to the amount
of free (i.e., nonantibody-bound) hapten present. However, most current

EIA methods are heterogeneous, requiring one or more washing steps to separate unbound reactants and often utilizing more than one type of antibody. These EIAs can be subdivided into several categories (competitive versus noncompetitive; direct versus indirect) depending on the relationship of the enzyme indicator to the compound being assayed. The principles and procedures involved in them have been reviewed by Trivers et al. [45] and are summarized in Table 2.

Common enzymes used in immunoassays include alkaline phosphatase, β-galactosidase, and horseradish peroxidase, all of which can easily be conjugated to antibodies or antigens. Conjugation can be achieved with covalent linkages (e.g., using glutaraldehyde) [36,40], or through noncovalent means such as the avidin-biotin system [45]. In addition to ease of conjugation, enzymes as well as their respctive substrates should be inexpensive, stable for long-term storage, and readily available in highly purified form. Moreover, quantitation of the enzymatic reaction products in biological samples necessitates detection methods of high sensitivity. This generally entails the use of compounds which are either chromogenic, fluorescent, or radioactive. EIAs for peptides using bioluminescence and chemiluminescence as detection modes have also been described [46,47].

EIAs have comparable sensitivity to RIAs and offer particularly good sensitivity with radioactive reagents. EIAs utilizing radiolabeled (as opposed to chromogenic) substrates have now been developed to detect femtomole amounts of selected substrates [48]. However, some of the problems cited for RIAs have analogous counterparts in enzyme immunoassays. One area of concern relates to enzyme-antibody conjugation procedures and the possibility of enzyme activity alterations due to the introduction of contaminants. Before using a conjugate in subsequent EIA analyses, one should first ascertain its immunological specificity, antigen binding affinity, and physical-chemical stability relative to the catalytic activity of the enzyme moiety. The absence of unreacted enzyme and of free antibody in the conjugate preparation must also be verified. With respect to the immobilization of antibodies on a solid phase, the possibility of antibody denaturation must be considered. Furthermore, the number of antibody molecules adsorbed is often subject to variability and can change over the course of repeated analyses.

A significant barrier to the optimization of immunochemical methods has been the use of antisera that are polyclonal. A related problem has been the limited availability of highly purified antisera along with the batch-to-batch variations in quality. However, these limitations may be largely overcome in the future with increased production and use of monoclonal antibodies, which are produced by a colony of cells derived from fusion of spleen and myeloma cells to produce hybridomae. After immunizing a laboratory animal with the immunogen of interest, spleen cells are removed and fused with myeloma cells. Selection of the "best" hybridoma is based on observed ability to grow rapidly in culture and on the production of homogeneous antibody populations. Such monoclonal antibodies recognize a single antigenic determinant and have only one affinity constant for that determinant, providing optimal specificity as well as sensitivity for immunochemical assays. The hybridoma technology allows one to screen and select the most desirable antibody for a particular experiment and furnishes a continuous, chemically homogeneous supply of that antibody. Immunochemical methods utilizing monoclonal antibodies thus offer a considerable advantage in providing sensitive detection of a variety of analytes from complex

Table 2 Types of Solid-Phase Enzyme Immunoassays[a]

Assay type	Sequence of reagent addition and measurement after washing the immunoadsorbent				
	1	2	3	4	5
Immobilized antigen					
1. NC/D	$*Ab_1$	Substrate	Measure products	—	—
2. N/ID	Ab_1	$*Ab_2$	Substrate	Measure products	—
3. CI/D	$*Ab_1 + Ag$	Substrate	Measure products	—	—
4. CI/ID	$Ab_1 + Ag$	$*Ab_2$	Substrate	Measure products	—
5. C/D	$*Ab_1 + Ab_1$	Substrate	Measure products	—	—
Immobilized antibody					
6. NC/D	Ag	$*Ab_1$	Substrate	Measure products	—
7. NC/ID	Ag	Ab_1	$*Ab_2$	Substrate	Measure products
8. CI/D	$*Ag + Ab_1$	Substrate	Measure products	—	—
9. C/D	$*Ag + Ag$	Substrate	Measure products	—	—

Source: Adapted from Ref. 45.

[a]NC, noncompetition; C, competition; CI, competitive inhibition; DI, direct measurement; ID, indirect measurement; *Ab, enzyme-conjugated antibody; *Ag, enzyme-conjugated antigen.

matrices. By combining such techniques with a chromatographic separation step, additional assurances of selectivity can be made.

C. Optical Methods

Ultraviolet Absorbance Measurements

As noted previously, ultraviolet absorbance at 210 nm provides a universal HPLC detection system for peptides; and amide bond moiety has an absorptivity on the order of 10,000 liter M^{-1} cm^{-1} in this spectral region. However, this absorbance often is not sensitive enough for many applications. Furthermore, the exact position and intensity of the amide spectral feature can vary with the length and composition of the peptide, as well as with the surrounding environment (e.g., solution pH). Absorbance measurements in the wavelength region 200 to 220 nm can also be greatly affected by the presence of UV-absorbing impurities in the mobile phase, leading to baseline drift and peak artifacts.

Components within a sample matrix (e.g., formulation excipients) can also limit the ability to detect and quantitate peptides. The use of rapid-scanning UV detectors based on photodiode arrays appears promising for the elimination of such background interference [49]. Another strategy is to choose a longer wavelength for peptide detection, either 254 nm (characteristic absorbance of aromatic amino acids) or 280 nm (characteristic absorbance of tryptophan); solvent impurities pose less of a problem in this region. However, these absorbances are inherently less intense than peptide bond absorbance, reducing sensitivity at least 10-fold. Moreover, detection is limited to peptides containing aromatic chromophores.

Fluorescence Measurements

Use of fluorescence for peptide detection can improve sensitivity by two orders of magnitude. However, only peptides containing tyrosine and/or tryptophan residues are naturally fluorescent. Peptides that are not intrinsically fluorescent or which fluoresce only weakly must be chemically derivatized if they are to be detected fluorometrically. Derivatization of a peptide preparation can be carried out prior to chromatographic separation (precolumn labeling) or after separation (postcolumn labeling). The latter type of derivatization is often done as a prelude to amino acid composition analysis. A well-known reagent employed for this purpose is dansyl cloride (see Sec. V).

Two other reagents commonly employed for fluorescent labeling of peptides are fluorescamine and o-phthalaldehyde (OPA). Fluorescamine reacts with the α-amino group at the N-terminus of peptides, as well as with the ε-amino group of lysine residues:

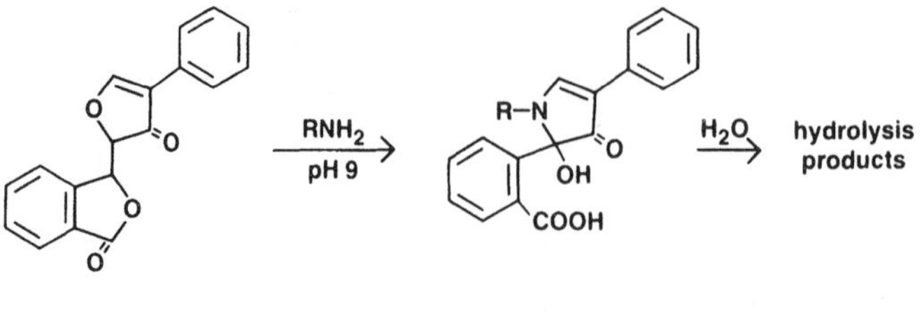

Reaction occurs rapidly at pH 9 (reaction time $\ll$ 1 min), yielding a highly fluorescent product that is stable over several hours: fluorescamine itself

and the reaction hydrolysis by-products are nonfluorescent. The resulting
fluorescence signal is thus proportional to the quantity of labeled amine
and is capable of detecting picomoles of peptides [50]. (It should be noted,
however, that fluorescamine cannot label a peptide amino terminus that is
blocked, nor will it react with secondary amines such as proline.) It is now
possible to combine fluorescamine derivatization procedures with HPLC separation of peptides in an automated system known as "fluorescamine-stream sampling" [51]. In this system, small aliquots of column effluent are injected with
a stream of water and transported to mixing tees, where fluorescamine and
supporting media are added. The reaction mixture then flows through a filter fluorometer; fluorescence is plotted versus time using a chart recorder.

Another reagent commonly applied for fluorescent derivatizatin of amino
acids and peptides is o-phthalaldehyde (OPA). In the presence of reagents
such as 2-mercaptoethanol, OPA reacts with free ε-amino groups of lysine
residues to form relatively stable, strongly fluorescent products [52]:

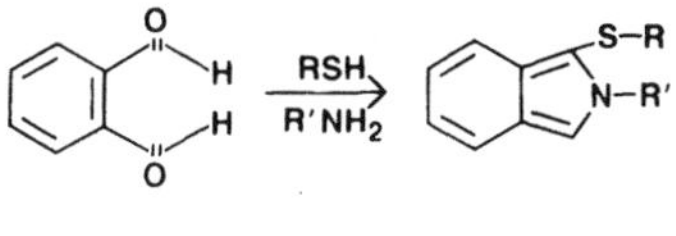

The sensitivity of OPA measurements is comparable to that obtained with
fluorescamine; moreover, OPA is less costly to use. But like fluorescamine,
OPA does not label proline residues and thus cannot be used to detect this
amino acid. The fluorescent intensity of N-terminal OPA-peptide products
diminishes dramatically with increasing reactant peptide chain length. For
this reason, OPA produces fluorescent products only from single amino
acids or from peptides containing free lysine residues.

Recent studies carried out with the naphthalene analog of OPA suggest
that OPA reacts with the N-terminus of peptides, but that the Schiff base
intermediate formed initially in the reaction goes on to form a nonfluorescent
product, rather than the luminescent isoindole formed with amino acids [53].
As illustrated by the following reaction pathway, the nucleophilicity of the
peptide amide nitrogen in the Schiff base intermediate (B and C in the reaction scheme shown below) is sufficient to allow for preferential intramolecular

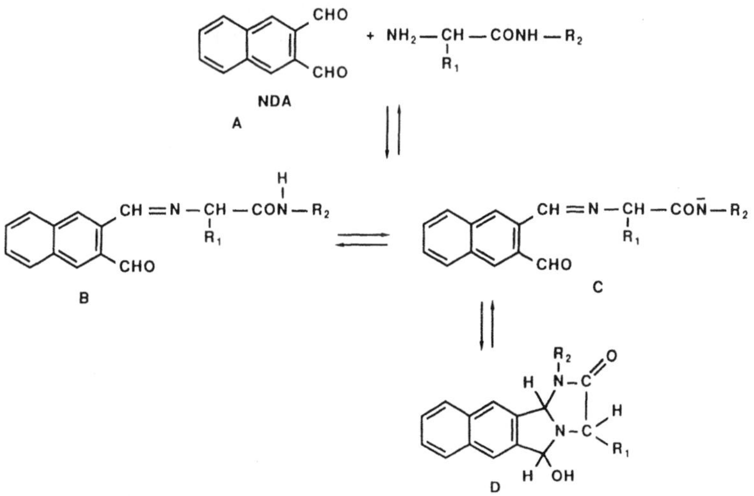

cyclization to occur, forming a nonfluorescent tetrahydrominidazol-4-one (D in the reaction scheme shown below). This nonproductive pathway is not available to amino acids (due to the absence of stable internal nucleophile), accounting for the efficient production of fluorescent isoindoles with amino acids but not with peptides. It is likely that a similar reaction course occurs with OPA.

Alternative fluorometric reagents have also been developed which may be applicable to peptide derivatization. One reagent that appears promising is 5-[(4,6-dichlorotriazin-2-yl)amino]fluorescein (DTAF):

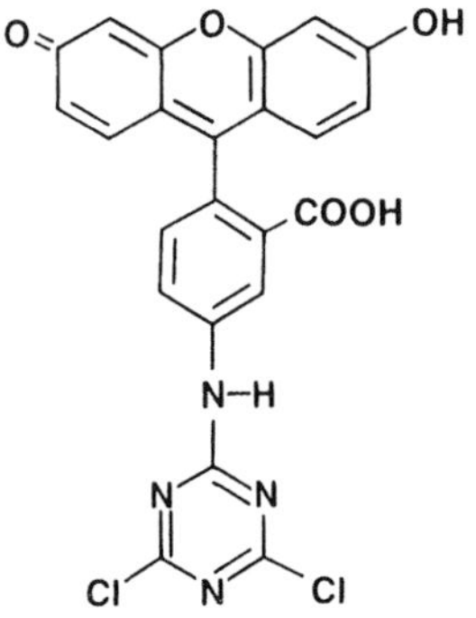

In aqueous alkaline media, DTAF reacts with amine functions to produce stable fluorescent products. The derivatization reaction proceeds via replacement of one of the chloro substituents of the triazine ring with the nucleophilic amine species, although details of the reaction mechanism remain unclear. In contrast with OPA and fluorescamine, DTAF reacts quantitatively with proline groups [54]. However, unlike the other two reagents, DTAF itself is fluorescent, necessitating separation of reagent and reagent degradation products from the derivatized product before measurement of peptide fluorescence is carried out.

Visible Absorbance Measurement

Peptides can also be derivatized to give colored products whose absorbance can be measured spectrophotometrically. Many early methods of peptide detection involved color formation, and some remain in routine use today (e.g., the Lowry method) [55]. However, such methods often are not suitable for analysis of peptides within complex matrices (formulations and biological samples), due to interference by other substances. Sensitivity is also a limitation. Consider the postcolumn derivatization of peptides with ninhydrin:

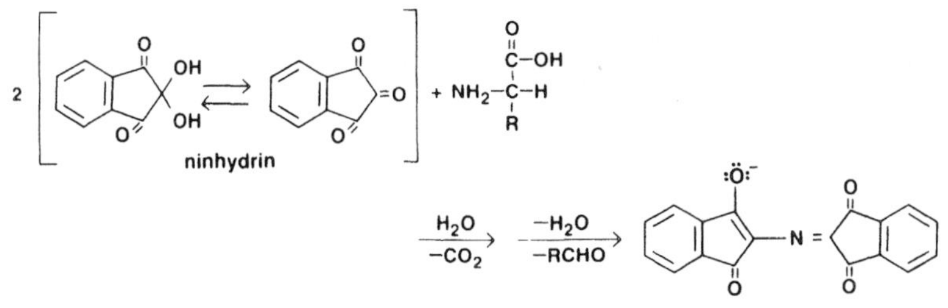

The ninhydrin reaction yields colored products that absorb strongly in the visible region, but detection sensitivity is still 10 to 100 times lower than that achievable with fluorescamine or OPA [56]. However, in recent years, chromophoric reagents capable of visualizing picomole amounts of peptides have been developed for precolumn derivatization purposes. These reagents, whose characteristic color arises from the presence of a dimethylaminoazo function, have been reviewed by Chang et al. [57].

Colorimetry can also be a valuable tool in the quality control of peptide pharmaceuticals. An example is the colorimetric assay developed for a synthetic sleep-inducing peptide with the structure

This assay is specific for the peptide glycinamide group, which reacts with 3,5-dibromosalicylaldehyde (DBSA) in the presence of piperidine to form a red-colored product. The characteristic absorbance of the reaction product was used as the basis for content uniformity determinations of the peptide in raw material samples [58]. Peptide amounts on the order of 10 mg could be quantitated with a standard deviation of less than ±2%. A potential impurity formed by ring closure of the original peptide was also tested and found not to interfere with the assay.

D. Electrochemical Methods

Electrochemical (amperometric) detection of compounds involves their oxidation or reduction to produce current, which can then be amplified and measured. The use of HPLC with amperometric monitoring (LCEC) has often provided detection limits in the low-picogram range, and in some cases has been extended into the femtogram range (for compounds with extremely small oxidation potentials). LCEC thus has sensitivity comparable to that achieved with immunochemical methods, but the former has additional advantages of simple, inexpensive instrumentation which can be easily operated and maintained. The use of LCEC to analyze ascorbate and neurotransmitter amines and their metabolites has been well established [59]. In recent years, this methodology has been successfully adapted to study peptides [60].

The primary requirement for successful electrochemical detection is that the compound of interest be electroactive within the potential range of the electrode material in the chosen solvent medium. Sensitivity is generally limited by the redox potential of the electroactive species: the lower the E value, the less likely will be interference due to concomitant electrolysis of

sample contaminants, and accordingly, the higher the sensitivity that can
be achieved.   Common electrode materials used for LCEC include carbon,
mercury, and platinum.   The factors involved in electrode selection have
been discussed.   If electrochemical transducers are to be interfaced with
HPLC, one must also consider the effects of the mobile phase (pH, identity
and amount of organic modifier, and ionic strength) on the expected electrochemical reaction.

The constituent amino acids in peptides normally possessing electroactivity are tyrosine (from the phenol group), tryptophan (from the indole
ring), and cysteine (from the thiol group).   Examples of tyrosine-containing
peptides amenable to electrochemical detection include oxytocin, vasopressin,
leu-enkephalin, and angiotensin II.   White [61] has discussed the use of an
LCEC system to analyze these peptides.   This system can detect sample peptide amounts as small as 40 pmd, making it comparable in sensitivity to UV
detection at 220 nm.   However, the LCEC method is more specific for the
peptides of interest and less affected by the presence of other matrix components.

In the case of peptides that do not possess inherent electroactivity,
electrochemical detection may still be possible through chemical derivatization procedures.   For example, derivatization of amino acids with OPA yields
products that are anodically electroactive [62].   Another useful reagent is
3,6-dinitrophthalic anhydride (DPNT).   Derivatization of small oligopeptides
with DPNT resulted in electroactive products that could be detected after
reduction at -0.24 V [63]:

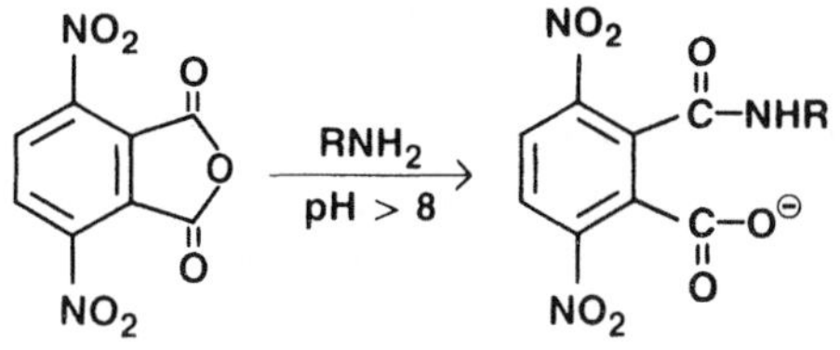

The detection limit for the electroactive derivative was on the order of
1 pmol, compared to the limit of 500 pmol for underivatized peptides as
measured by UV absorbance at 210 nm.   However, it should be noted that
cathodic processes are generally of limited value, particularly in HPLC systems, due to the large signal produced by O_2 reduction in samples that
have not been vigorously deoxygenated.

V.   STRUCTURAL CHARACTERIZATION

In this section the focus is on methods for determination of peptide amino
acid composition and the linear sequence of amino acids within a peptide
chain (i.e., the primary structure).   (Methods of conformational analysis
concerned with aspects of secondary and tertiary structure, such as circular dichroism, are discussed in Chap. 6.)   Methods for identifying constituent amino acids in peptides often involve partial or complete digestion of
the peptide sample using chemical, thermal, or enzymatic means.   Amino
acid composition analysis, N- and C-terminal analysis, and mass spectrometry
fall into this category.   There are also nondestructive techniques that allow

recovery of the original sample, such as nuclear magnetic resonance (NMR) spectroscopy.

A. Amino Acid Composition

Determination of the identity and number of constituent amino acids requires that the peptide bonds be completely hydrolyzed to liberate free amino acids. Hydrolysis can be accomplished in aqueous acid, aqueous base, or enzyme catalysis. Acid hydrolysis is the most commonly applied hydrolytic procedure. This typically involves dissolving the peptide in 6 N HCl and heating the acid solution in a sealed tube at 110°C for 24 h. To detect cysteine groups accurately in such peptides, they must first be oxidized to cysteic acid by treatment with performic acid. In addition to cysteine sensitivity, another limitation of acid hydrolysis is that it generally leads to almost total destruction of tryptophan residues. However, it is often possible to estimate tryptophan content through use of base hydrolysis. This typically entails heating a solution of the peptide in 2 N NaOH at 100°C for several hours. This procedure does not harm tryptophan but does result in complete or partial degradation of cysteine, serine, threonine, and arginine, as well as promoting amino acid racemization. Finally, proteolytic enzymes can also be used to promote peptide hydrolysis. Most of these enzymes cleave peptides at specific sites and thus cause only partial hydrolysis of the original peptide. Enzymatic hydrolysis methods offer the advantage of mild, nondestructive reaction conditions. However, they have a disadvantage in that proteases tend to undergo measureable self-digestion, which can introduce contaminants into the peptide-derived amino acid mixture.

Once a peptide has been completely hydrolyzed, an acidic solution of the amino acid products is applied to an ion-exchange column and eluted using a buffer gradient of increasing pH and ionic strength. The most acidic amino acids will be eluted first, followed by neutral and then basic amino acids. An eluted amino acid is then mixed on-line with ninhydrin to form a colored product whose absorbance can be measured spectrophotometrically. The resulting absorbance is proportional to the amount of a particular amino acid in the hydrolytic mixture, allowing calculation of the relative numbers of component amino acids in the original peptide. These procedures have been successfully automated and form the basis of the commercially available amino acid analyzer [64].

An amino acid analysis of the type outlined above can be carried out on hydrolyzate from as little as 0.1 μmol of peptide. The detection limit can be even further reduced by use of fluorescence detection instead of the ninhydrin reaction. For example, Bohlen and Schroeder [65] have described the use of an automated fluorescence amino acid analyzer with o-phthalaldehyde (OPA) as the derivatizing agent; this system can detect picomole quantities of peptides. The authors note that optimization of sensitivity depends on careful sample handling and use of highly purified reagents to minimize background contamination. Sensitivity in the low-femtomolar range can be achieved with another fluorescent derivatizing reagent, naphthalene-2,3-dicarboxaldehyde (NDA). The improvement in sensitivity over OPA is attributable to the larger cross-sectional area of the NDA reaction product, resulting in a higher quantum capture and emission efficiency [66].

The variability of amino acid analysis is normally between 5 and 10% for each amino acid. For this reason, small observed deviations with

respect to predicted composition may not be meaningful. Also, it is often
not possible to discern the extent of side-chain modifications in the original
peptide, as these features may disappear upon hydrolysis (e.g., glutamine
and asparagine are liberated as glutamic acid and aspartic acid, respectively).
Amino acid analysis is perhaps most useful for detecting and identifying
impurities present in an isolated peptide of known amino acid sequence. If
a peptide's sequence is unknown, or if side-chain modification is of concern,
other methods of structural characterization must be employed.

B. Primary Sequence Determination: Chemical and
    Enzymatic Methods

N-Terminal Reactions

The reactivity of a free α-amino group can be exploited to identify a peptide's N-terminal amino acid. One such strategy involves the use of 2,4-
dinitrofluorobenzene (Sanger's reagent). This reagent reacts with the
α-amino moiety to form a yellow-colored derivative that is stable to acid hydrolysis and can thus be separated from the unmodified amino acids liberated upon hydrolysis [67]:

N^{α}–(2,4–dinitrophenyl) derivative

Although Sanger's reagent can also react with other free amines (e.g.,
the ε-amino group of lysine), these derivatives are generally less stable
and do not interfere with isolation of the α-amino derivative. The derivatized α-amino acids remain in the aqueous phase. Chromatographic comparison of the derivatized amino acid with N-(2,4-dinitrophenyl)amino acid standards allows the identity of the N-terminal residue to be established.

The sensitivity of amino acid detection can be enhanced by derivatizing
the α-amino group to produce a fluorescent compound stable to acid hydrolysis. A reagent commonly employed for this purpose is 1-dimethylamino-
naphthalene-5-sulfonyl chloride (dansyl chloride), which under mildly alkaline conditions reacts at the peptide α-amino terminus to form a sulfonamide
[68]:

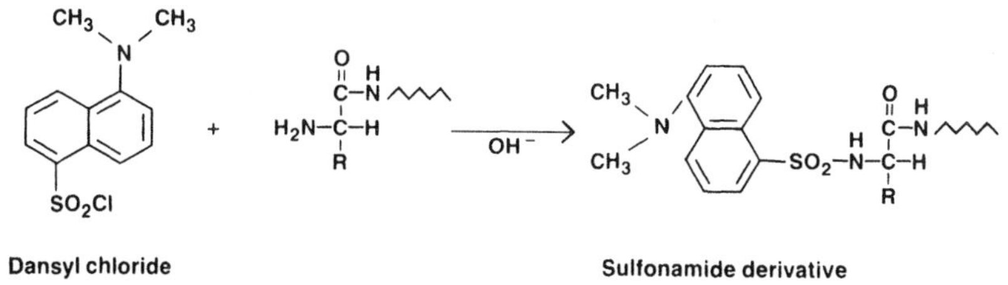

Dansyl chloride   Sulfonamide derivative

The separation of the amino-terminal residue is accomplished in a manner similar to that described previously for dinitrophenyl-labeled peptides. The use of such a fluorescence detection system typically provides a 100-fold increase in sensitivity over Sanger's reagent, and can be used to analyze nanomole amounts of amino acids by TLC.

The main limitation of the reactions described above is that they identify only the N-terminal amino acid; information relating to the sequence of other amino acids becomes lost upon hydrolysis. Obviously, it would be desirable to have a method that modifies and separates the amino terminal residue while leaving the remainder of the peptide intact for further derivatization reactions. The Edman degradation is the best known example of such a method [69]. It utilizes phenylisothiocyanate (or related reagent), which reacts with the α-amino group to form a thiazolinone derivative. Under mildly acidic conditions, the thiazolinone-modified amino acid will be cleaved from the rest of the peptide chain, which remains unhydrolyzed and now has a "new" amino terminus that can be modified in the same fashion. Thus each thiazolinone-modified group can be successively isolated and identified. Usually, identification is accomplished through conversion of the thiazolinone to the more stable phenylthiohydantoin (PTH) derivative, which is compared chromatographically with PTH derivatives of amino acid standards. The reaction scheme is as follows:

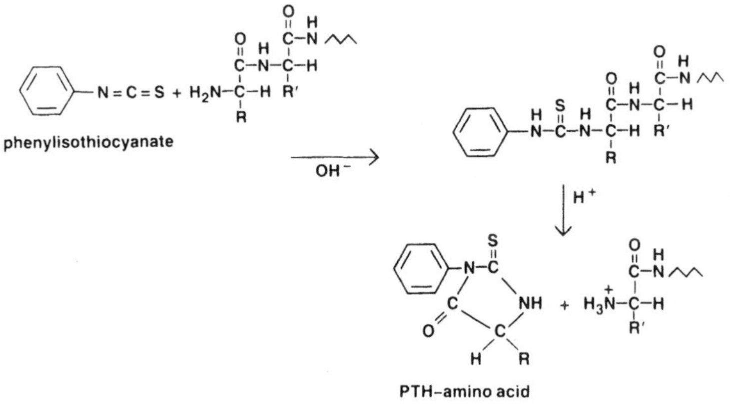

phenylisothiocyanate

PTH-amino acid

The Edman degradation procedures have been successfully automated to sequence micromole amounts of peptides containing up to 60 residues. In recent years, "microsequencing" procedures have been developed that can detect picomole amounts of PTH-amino acids [70]. The sensitivity of Edman degradation can also be enhanced by utilizing dansyl instead of PTH as the amino acid derivative [71]. However, Edman degradation sequencing of larger peptides may become prohibitively time consuming, as only about 15

residues can be determined in a 24-hr period. Furthermore, there are
small losses of peptide (typically 5%) on each turn of the cycle, which further limit the length and amount of peptide that can be sequenced unambiguously.

Enzymatic cleavage can also be used to identify N-terminal residues.
For example, the enzyme leucine aminopeptidase catalyzes the hydrolytic
cleavage of leucine as well as several other amino acids from the N-terminus
of a peptide, which can be identified through chemical derivatization and/or
chromatographic methods. However, application of such enzymes is complicated by the fact that upon cleavage of the original N-terminal group, a
new N-terminus forms from the remaining peptide; this residue may also be
susceptible to peptidase action. If the enzyme has greater reactivity toward
the new N-terminal group compared with the previous group, both amino
acids will be released in comparable amounts, and it will be difficult to determine which residue corresponds to the original N-terminus. Moreover,
products from autoproteolysis of the enzyme can be a source of artifacts.
Finally, it should be noted that the aminopeptidase reactions, as well as the
other N-terminal reactions discussed here, are readily applicable only to
peptides with an unblocked amino terminus. (Free amino termini are normally present in naturally occurring peptides, but may often be absent in synthetic peptide analogs. In many such compounds, the terminal amino group
is deliberately modified to confer enhanced resistance to proteolytic degradation in vivo.)

C-Terminal Reactions

Perhaps the most useful method for determining peptide C-terminal residues
involves hydrazinolysis, that is,

This reaction can be carried out under conditions that convert all peptide
residues to hydrazine derivaties with the exception of the C-terminal amino
acid. The resultant free amino acid can be separated from the generated
hydrazides and identified chromatographically.

Another useful method involves the reduction of the peptide carboxyl
group to an aldehyde, commonly known as the Loudon procedure [72]:

3.

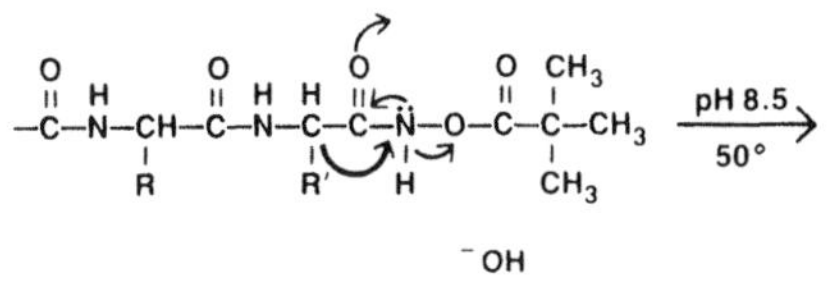

$$-C-N-CH-C-N-C-C-N-O-C-C-CH_3 \xrightarrow[50°]{pH\ 8.5}$$

^-OH

4.

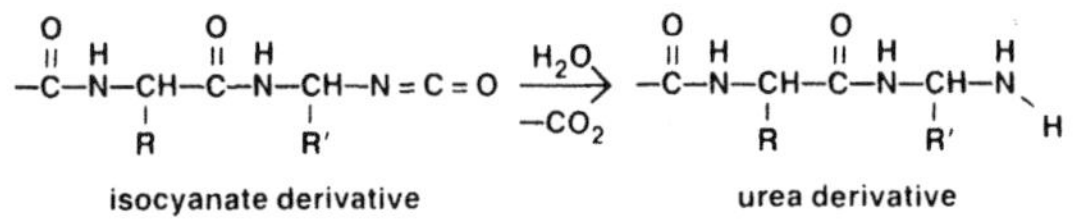

isocyanate derivative   urea derivative

$$\xrightarrow[-CO_2]{H_2O}$$

5.

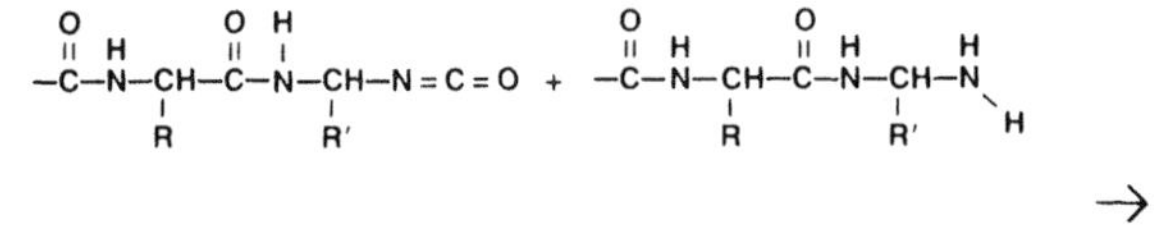

$\rightarrow$

6.

6 M HCl
110°C
24 hrs

7.

$$2NH_3 + 2CO_2 + \text{free amino acids} + R'-C-H$$

aldehyde

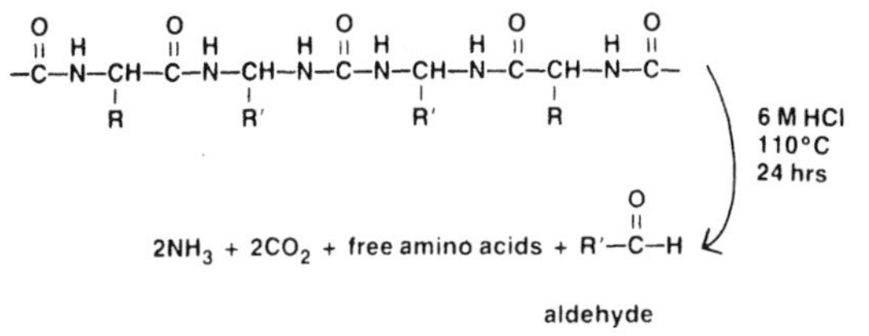

The resulting aldehyde is stable to extended acid treatment, whereas the rest of the peptide will be hydrolyzed to give free amino acids. The alde-hyde-modified residue can be separated from the hydrolysis products and derivatized with 2,4-nitrophenylhydrazine to produce a yellow-colored com-pound for chromatographic characterization.

There are relatively few chemical reactions specific to the peptide car-boxyl terminus. For this reason, it has proven difficult to develop C-ter-minal sequencing methods comparable to the Edman degradation method for N-terminal sequencing. A procedure that has been applied with some suc-cess is the Stark method [73]. This method involves conversion of the terminal carboxyl group into an aryl isothiocyanate, which undergoes intra-molecular rearrangement upon warming to form a thiohydantoin derivative.

$$\xrightarrow{(CH_3CO)_2O} \xrightarrow{NCS}$$

$$\xrightarrow{warm} \xrightarrow{OH^-}$$

isothiocyanate derivative

use to repeat cycle

thiohydantoin derivative

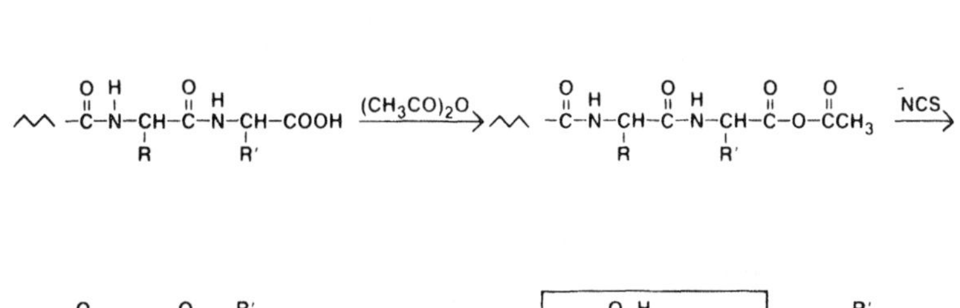

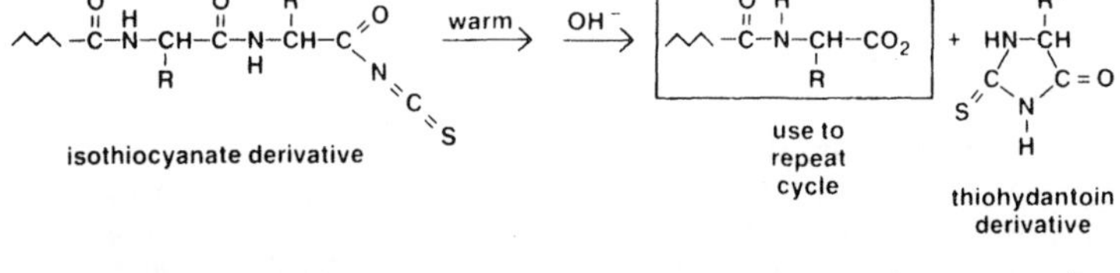

Mild alkaline hydrolysis releases the thiohydantoin-modified residue, which can then be identified chromatographically. The rest of the peptide chain remains intact and thus is available for subsequent derivatization at its new carboxyl terminus. In practice, however, the Stark method is often much less satisfactory than the Edman degradation, and can typically be run only for two or three cycles. Moreover, some C-terminal residues, notably aspartic acid and proline, cannot easily be removed with this procedure.

A C-terminal sequencing method that appears more promising involves modifications of the Loudon procedure, in which the initial step involves coupling of the peptide to a polymeric solid support (P) via its α-amino terminus [74]:

$$\text{(P)} \sim\!\!\sim \underset{R'}{\overset{O\ H\ \ \ \ O}{C-N-CH-C-NCH-COOH}} \longrightarrow \text{isocyanate derivative} \xrightarrow[\substack{DMF\\70°}]{\substack{CH_2OH\\ \\OCH_3}}$$

$$\text{(P)} \sim\!\!\sim \underset{R'}{\overset{O\ H\ \ \ \ O\ \ H}{C-N-CH-C-N-CH-N-C-OCH_2-\!\!\bigcirc\!\!-OCH_3}} \xrightarrow[\substack{(heat, pH\ 7\\80\ min.)}]{\text{mild hydrolysis}}$$

$$\text{(P)} \sim\!\!\sim \underset{R'}{\overset{O\ H\ \ \ \ O}{C-N-CH-C-NH_2}} + CO_2 + RCHO + NH_3 + \begin{matrix}CH_2OH\\ \\OCH_3\end{matrix}$$

$$\longrightarrow \text{(P)} \sim\!\!\sim \underset{R'}{\overset{O\ \ \ H}{C-N-CH-NH_2}} \xrightarrow{\text{mild hydrolysis}} \boxed{\text{(P)} \sim\!\!\sim \overset{O}{C-NH_2}} + R'CHO + NH_3$$
$$\text{use to repeat cycle}$$

This sequencing procedure has been applied successfully through several cycles with fairly good yields of peptide (typically 80%) after each cycle. Note, however, that it cannot be applied to peptides with synthetically blocked N-termini, since these peptides cannot readily be coupled to the solid support.

Enzymatic cleavage can also be applied to C-terminal sequencing (e.g., by use of carboxypeptidases). Isobe et al. [75] have utilized carboxypeptidase digestion in conjuction with RP-HPLC and Edman degradation methods to identify the C-terminal region of proteins. However, carboxypeptidase methods have the same problems associated with use of aminopeptidases for N-terminal sequencing: the possibility that more than one amino acid will be released in significant amounts after initial incubation with the enzyme, as well as possible contamination of the hydrolysate by products of enzyme autoproteolysis. Furthermore, larger peptides may have a high degree of ordered structure, thus conferring resistance to enzymatic attack. For example, human chorionic gonadotropin is not degraded by carboxypeptidases even in the presence of urea [76]. Also, like all the C-terminal reactions discussed here, the carboxypeptidase method is not directly applicable to peptides with synthetically blocked carboxyl termini.

Site-Specific Cleavage of the Peptide Chain

Clearly, the applicability of the Edman degradation and related procedures to carry out peptide sequencing will diminish with increasing chain length.

If a peptide is too large to be sequenced conveniently using this methods, it is often possible to selectively degrade the peptide into smaller fragments for direct sequencing. Selective cleavage can be induced with enzymes, for example, trypsin (which cleaves peptide bonds on the carboxyl side of lysine or arginine) or chymotrypsin (most reactive toward peptide bonds on the carboxyl side of tyrosine, tryptophan, or phenylalanine). Selective cleavage can also be obtained by treating the peptide with reagents such as cyanogen bromide, which cleaves peptides on the carboxyl side of their methionine residues [77]. The use of other amino-acid-specific reagents has been reviewed by Dence [78]. As noted in Sec. III, such site-specific cleavage can be combined with electrophoresis and chromatography to provide a unique "fingerprint" of a given peptide or protein, thus serving as an identity test.

Selective cleavage processes also have utility in the sequencing of smaller peptides with blocked amino and/or blocked carboxyl terminal residues. As mentioned previously, peptides with modified termini cannot easily be chemically derivatized or cleaved by exopeptidases at the blocked positions. But by inducing cleavage at other (i.e., unblocked) residues within the peptide chain, direct sequencing can be carried out on the resulting fragments. It should be pointed out that use of site-specific cleavage requires some a priori knowledge of the original peptide sequence to identify appropriate reagents and their likely sites of cleavage in the peptide chain. Furthermore, the cleavage procedures must be optimized to produce satisfactory yields of fragments that are readily separable for individual sequence determinations.

C. Primary Sequence Determinations: Mass Spectrometry

Despite the commercial availability of Edman degradation instrumentation, and the continuing progress in development of related sequencing schemes, these methods are generally inadequate for the structural characterization of peptides extracted from complex matrices (e.g., biological fluids and tissues). The small amounts of peptides normally present in biological extracts may preclude their analysis by degradative sequencing methods, even with the use of microsequencing techniques. Furthermore, a biological sample often contains a mixture of closely related peptides. Some of these may be endogenous contaminants, others may be peptide degradation products, while still others may reflect side-chain modifications of the native peptide via deamidation, oxidation-reduction, or racemization. Thus methodology is needed to separate the various peptide components, as well as to carry out their individual sequence determinations. In practice, this can be accomplished by combining chromatographic separation techniques (either GLC or HPLC) with mass spectrometry (MS) for sequence analysis.

GLC-MS

In gas-liquid chromatography, separation occurs through the partitioning of sample components between an inert gaseous mobile phase and a thin layer of nonvolatile liquid bonded to a solid support. Initially, the sample is injected into a heated block, where it becomes rapidly vaporized before being carried to the column inlet by a stream of inert gas. Each solute within the sample travels down the column, at a rate determined by its relative propensity to associate with the stationary phase. Upon elution from the column, separated solutes enter a detector, in this case a mass spectrometer. The mass spectrometer produces charged particles consisting of the parent ion and various fragment ions, which are subsequently separated according to their mass/charge ratio.

As noted previously, peptides are not well suited to GLC analysis, due to their nonvolatility and thermal lability. However, the lack of volatility may be overcome through derivatization to create peptides that are more easily vaporized. A common synthetic strategy involves trifluoroacetylation of the N-terminal residue, followed by reduction of amide bonds, and tri-silylation to protect the C-terminal moiety:

MBTFA = N-methylbis-trifluoroacetamide
BSTFA = N,O-bis(trimethylsilyl)-trifluoroacetamide

N-, S-, and O-permethylation can also be carried out to enhance thermal stability and improve the volatility of peptides. The use of such derivatization procedures has permitted GLC-MS analyses of nanogram amounts of peptides containing up to five amino acids; biomolecules successfully sequenced by GLC-MS include enkephalin fragments [79] and various oligopeptides present in human urine [80] and various peptide antibiotics [81]. For small peptides (or larger peptides which can be efficiently degraded into sufficiently small fragments), GLC-MS offers a rapid, relatively inexpensive means of sequence determination. But for larger peptides which are less efficiently degraded, nonvolatility precludes direct application of this technique.

HPLC-MS

The on-line combination of HPLC with mass spectrometry provides an additional dimension to analysis of peptide mixtures, allowing reliable identification of component peptides, as well as insight into sequence differences between them. For example, Yu et al. [82] applied HPLC-MS to analyze C-methylation artifacts associated with the permethylation derivatization of tri- and dipeptides. HPLC separated the N-, O-, C-methylated peptides from the N, O-permethylated derivatives prior to MS detection, thus facilitating mass spectral interpretations. Representative HPLC chromatograms and mass spectra from each type of peptide are shown in Fig. 5.

Until recently, MS methods for producing ions were not readily adaptable to peptides. Conventional ionization procedures (e.g., electron or chemical ionization) resulted in extensive fragmentation, difficult to interpret for all but the smallest peptides. This problem has been substantially overcome with the development of less destructive ionization methods, such as fast-atom bombardment (FAB) and field desorption (FD), which minimize the production of fragment ions. Desiderio [83] has discussed the applications of FAB and FD in the off-line HPLC-MS detection of picogram amounts of neuropeptides. To sequence these peptides, another method, known as collision-activated dissociation-linked field scanning (CAD-B/E), is used. This involves the use of a double-focusing mass spectrometer with three fields: accelerating voltage (V), electric field (E), and magnetic field (M). The V field is used to produce parent $(M + H)^+$ ions, which then undergo fragmentation. All fragment ions arising from a selected precursor ion are collected by scanning the B and E fields at a constant B/E ratio. In effect, a mass spectrum is obtained from one ion selected from the original mass spectrum. CAD-B/E thus analyzes the sequence of an ion specific to a given peptide, virtually guaranteeing that the substance being quantified is in fact the peptide of interest. Development of microcomputer-based instrument operation and data acquisition modes for this system is in progress, which will further improve sensitivity and diminish potential matrix effects.

D. Nuclear Magnetic Resonance (NMR)

In recent years, development of high-field NMR spectrometers coupled with the use of Fourier transform processes in data acquisition have extended the capability of NMR to analyze constituents of macromolecules, including amino acids within peptides. A significant advantage of NMR is that it represents a nondestructive method of amino acid identification, allowing recovery of the original peptide after completion of analysis. Furthermore, it improves the quantitation of amino acids that might be partially destroyed by hydrolytic procedures or which cannot easily be derivatized for sensitive detection. Such analyses were carried out on angiotensin I by Margolis and Coxon [84]. In this study, chemical shifts of protons in the NMR spectrum at 400 MHz (Fig. 6) were used to determine the amino acid composition of the peptide sample. The NMR data were then compared to data obtained from chromatographic analysis of the angiotensin I hydrolysate. Both methods gave similar results with comparable variability (±10% or better).

A significant disadvantage of proton NMR with respect to characterization of macromolecules is that the relatively small range of ^1H chemical shifts

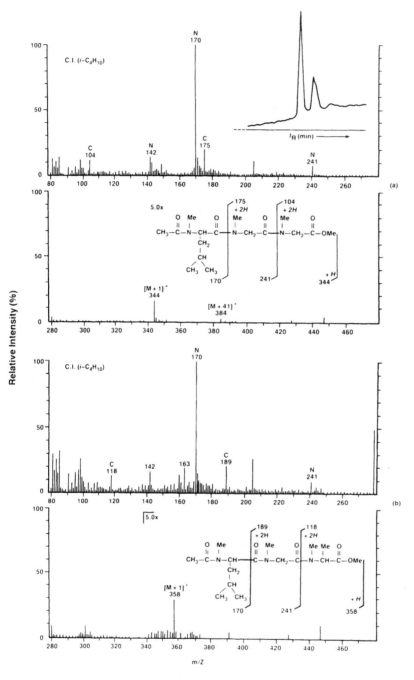

Fig. 5 (a) Mass spectrum of *N, O*-permethylated-leu-gly-gly. HPLC/UV
(λ = 214 nm) chromatogram of mixture from short reaction time in inset.
(b) Mass spectrum of *N, O, C*-methylated product from Leu-Gly-Gly. (From
Ref. 82.)

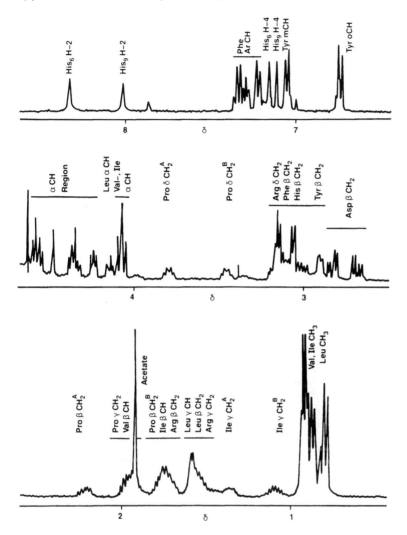

Fig. 6   Partial proton NMR spectrum at 400 MHz of a solution of angiotensin I (3.6×10^{-4} mol/liter) and acetate (5.22×10^{-4} mol/liter in D_2O). The water region is not shown. (From Ref. 84.)

(0 to 10 ppm) coupled with the large peak widths (5 to 30 Hz) result in broad bands of overlapping resonances. The ability to identify and assign each amino acid diminishes significantly with the size and complexity of the peptide sequence. In the future, however, the capability of proton NMR to identify specific amino acids within a peptide molecule is likely to be greatly improved by applications of two-dimensional J-resolved spectroscopy. This technique enhances NMR spectral dispersion by spreading spin-coupled multiplets into a second dimension, as well as by suppressing the line broadening due to inhomogeneity. The combination of two-dimensional J-resolved spectroscopy with other techniques [e.g., spin echo correlated spectroscopy (SECSY)] should further facilitate assignment of peptide proton resonances [85].

^{13}C NMR is potentially more valuable than ^{1}H NMR for studying individual amino acids in a peptide, since the ^{13}C chemical shift range is much wider (0 to 200 ppm). An elegant application of ^{13}C NMR for this purpose has been described by Hruby et al. [86]. In this study, vasopressin, oxytocin, and a series of related analogs were synthesized, with substitution of deuterium for hydrogen in the C—H bonds of selected amino acids. This isotopic substitution helped facilitate the unambiguous assignments of chemical shift values to specific carbon atoms (i.e., the carbons directly bonded to hydrogen in the native peptides). However, the low natural abundance of the ^{13}C isotope (1.108%) necessitates sample concentrations even higher than those required for ^{1}H NMR (typically 10^{-2}M for ^{13}C versus 10^{-3}M for ^{1}H), as well as long data acquisition times to obtain useful information. Although this limitation can be overcome with ^{13}C isotopic enrichment of samples, such procedures can be prohibitively costly and time consuming.

There is no doubt that NMR utilizing ^{1}H, ^{13}C, and other nuclei (e.g., ^{15}N and ^{17}O) can be valuable in peptide analysis. However, the relatively large amounts of highly purified samples needed for meaningful spectra, coupled with requisite long data acquisition times and the cost of maintaining high-field spectrometers, limit the use of NMR to specialized (i.e., nonroutine) applications.

VI. CONCLUSIONS

The inherent complexity of peptide drug substances demands suitable analytical methodology for determining their purity, stability, and potency and for correlating this information with biological activity. Considerable progress has been achieved in the development of physical-chemical methods for the separation, identification, and quantitation of peptides. However, additional efforts will be needed to further enhance the selectivity and sensitivity of such methods and to allow scientists to better understand the chemistry of these molecules, both in vivo and in vitro.

The high sensitivity demands anticipated for biological fluid analysis are predicated on the general observation that peptides are of high potency and have short biological half-lives, resulting in extremely low plasma concentrations of drug. A high degree of selectivity is also required of analytical methods for peptides, due to the proteinaceous nature of the matrices from which these molecules are to be analyzed and because of their proclivity toward biodegradation or modification. Similar concerns must be raised in in vitro systems and even in formulations, where high levels of sensitivity and selectivity must be achieved for definitive characterization of degradation profiles. Such information is needed to formulate these molecules and to define the shelf-life and storage conditions of the resulting drug products. It is also needed to gain insight into molecular mechanisms of degradation, as this will affect the drug discovery process (i.e., the rational design of improved second-generation agents).

Analytical methodology suitable for various aspects of peptide drug development is complicated by the size and complexity of the analyte and the yet-to-be resolved definition of purity as related to biological activity. These complexities will probably require that future analytical strategies include components of each of the types of methods described in this chapter, as well as others not discussed or yet to be identified. Furthermore, panels of such analytical tests will probably be needed to assess all aspects pertinent to the description of identity and purity of peptide drugs and their degradation products in a variety of matrices (fermentation broths, formulations, biological fluids, etc.). If results obtained with these methods can be correlated successfully to those of traditional bioassays, the understanding of peptide drug behavior will be greatly enhanced. This increased understanding will in turn permit more rational approaches to the synthesis and formulation of therapeutically useful peptides.

REFERENCES

1. United States Pharmacopeial Convention, *United States Pharmacopeia*, 21st ed., Mack Publishing Company, Easton, Pa. (1985).
2. W. Vale, J. Vaughan, G. Yamamoto, T. Bruhn, C. Douglas, D. Dalton, C. Rivier, and A. Rivier, *Methods Enzymol.*, *103*:565 (1983).
3. P. Storring, R. Gaines-Das, R. Tiplady, B. E. Stenning, and Y. Mistry, *J. Endocrinol.*, *85*:533 (1980).
4. K. Krummen and R. W. Frei, *J. Chromatogr.*, *132*:429 (1977).
5. J. S. D. Chan, N. G. Seidah, M. Ikeda, J. Gutkowska, R. Boucher, J. Genest, and M. Chretien, *Can. J. Physiol. Pharmacol.*, *59*:811 (1981).
6. J. Roth, *Methods Enzymol.*, *37*:66 (1975).
7. M. Schmidt-Gollwitzer, J. Eiletz, U. Sackmann, and J. Nevinney-Stickel, *J. Clin. Endocrinol. Metab.*, *46*:92 (1978).
8. S. M. Shahani, P. P. Kulkarni, and K. L. Patel, *Contraception*, *18*: 543 (1978).
9. H. Schluesner and P. Kotulla, *Nuklearmedizinn (Suppl.)*, *17*:776 (1981).
10. P. H. Corram and D. H. Calam, in *Recent Developments in Chromatography and Electrophoresis* (A. Frigerio and L. Renoz, eds.), Elsevier Science Publishing Co., Amsterdam (1979), p. 341.
11. L. Lepri, P. G. Desideri, and D. Heimler, *J. Chromatogr.*, *211*:29 (1981).
12. M. T. W. Hearn, in *High-Performance Liquid Chromatography: Advances and Perspectives*, Vol. 3 (C. Horvath, ed.), Academic Press, New York (1983), p. 87.
13. H. Parvez, Y. Kato, and S. Parvez, ed., *Progress in HPLC*, Vol. 1, *Gel Permeation and Ion-Exchange Chromatography of Proteins and Peptides*, VNU Science Press, De Meern, The Netherlands (1985).
14. J. L. Meek and Z. L. Rosetti, *J. Chromatogr.*, *211*:15 (1981).
15. M. T. W. Hearn and B. Grego, *J. Chromatogr.*, *218*:497 (1981).

16. M. T. W. Hearn, G. Grego, and W. S. Hancock, *J. Chromatogr.*, *185*: 429 (1979).
17. S. Terabe, R. Konaka, and K. Ionouye, *J. Chromatogr.*, *172*:163 (1979).
18. J. E. Krause and J. F. McKelvy, *Methods Enzymol.*, *103*:539 (1983).
19. M. Dizdaroglu, *J. Chromatogr.*, *334*:49 (1985).
20. N. Takahashi, N. Ishioka, Y. Takahashi, and F. W. Putnam, *J. Chromatogr.*, *326*:407 (1985).
21. Y. Kato, K. Nakamura, and T. Hashimoto, *J. Chromatogr.*, *294*:207 (1984).
22. B. Welinder, *J. Liq. Chromatogr.*, *3*:1399 (1980).
23. A. K. Taneja, S. Y. M. Lau, and R. S. Hodges, *J. Chromatogr.*, *317*:1 (1984).
24. M. P. H. West, R. S. Wu, and W. M. Bonner, *Electrophoresis*, *5*:133 (1984).
25. Z. Prusik, *J. Chromatogr. Libr.*, *18*(pt. B):81 (1983).
26. J. W. Jorgenson and K. D. Lukacs, *J. High Res. Chromatogr. Chromatogr. Comm.*, *4*:230 (1981).
27. P. Righetti, E. Gianazza, and A. B. Bosisio, in *Recent Developments in Chromatography and Electrophoresis* (A. Frigerio and L. Renoz, eds.), Elsevier Science Publishing Co., Amsterdam (1979), p. 1.
28. W. C. Dermody, A. G. Levy, P. E. Davis, and J. K. Plowman, *Clin. Chem.*, *25*:939 (1979).
29. F. M. Everaerts and F. E. P. Mikkers, in *Biochemical and Biological Applications of Isotachophoresis* (C. Schots, ed.), Elsevier Science Publishing Co., Amsterdam (1980) p. 1.
30. L. Arlinger, *J. Chromatogr. Libr.*, *18*(pt. A):363 (1979).
31. A. Kopwillem, F. Chillemi, A. Bosisio-Righetti, and P. G. Righetti, *Protides Biol. Fluids Proc. Colloq.*, *22*:697 (1975).
32. K. Friedel and C. J. Holloway, *Electrophoresis*, *2*:116 (1981).
33. R. E. Chance, E. P. Kroepf, J. A. Hoffman, and B. H. Frank, *Diabetes Care*, *4*:147 (1981).
34. H. P. J. Bennett and C. McMartin, *J. Endocrinol.*, *82*:33 (1979).
35. P. L. Storing and R. J. Tiplady, *Anal. Biochem.*, *141*:43 (1984).
36. W. W. Youngblood and J. S. Kizer, *Methods Enzymol.*, *103*:435 (1983).
37. V. Pingoud and I. Trautschold, *Anal. Biochem.*, *140*:305 (1984).
38. H. G. Eckert, *Angew. Chem. Int. Ed. Engl.*, *15*:525 (1976).
39. T.-M. Chang and W. Y. Chey, in *Gastrointestinal Hormones* (G. B. J. Glass, ed.), Raven Press, New York (1980), p. 797.
40. L. Hennes and W. E. Stumpf, *Methods Enzymol.*, *103*:448 (1983).
41. E. Straus and R. Yalow, *J. Lab. Clin. Med.*, *87*:292 (1976).
42. M. E. Jolley, *J. Anal. Toxicol.*, *5*:236 (1981).
43. R. Yalow and E. Straus, in *Gastrointestinal Hormones* (G. B. J. Glass, ed.), Raven Press, New York (1980), p. 751.
44. M. A. Wilson, Y. Kao, and L. Miles, *Nucl. Med. Commun.*, *2*:68 (1981).
45. G. E. Trivers, C. C. Harris, C. Rougeot, and F. Dray, *Methods Enzymol.*, *103*:409 (1983).
46. T. O. Baldwin, T. F. Holzman, P. S. Satoh, and F. S. C. Yein, European Patent Application (1984).
47. J. L. Brockelbank, J. B. Kim, G. J. Bernard, W. P. Collins, B. Gaier, and F. Kohen, *Ann. Clin. Biochem.*, *21*:284 (1984).
48. C. C. Harris, R. H. Yolken, and I. C. Hsu, *Proc. Natl. Acad. Sci. USA*, *76*:5336 (1979).

49. A. F. Fell, G. J. Clark, H. P. Scott, *J. Chromatogr.*, *297*:203 (1984).
50. S. Udenfriend, S. Stain, P. Bohlen, W. Dairman, W. Leimgruber, and M. Weigele, *Science*, *178*:871 (1972).
51. S. Stein and J. Moschera, *Methods Enzymol.*, *79*:7 (1981).
52. T. M. Joys and H. Kim, *Anal. Biochem.*, *94*:371 (1979).
53. L. A. Sternson, A. J. Repta, and J. F. Stobaugh, *Anal. Biochem.*, *144*:233 (1985).
54. P. de Montigny, J. F. Stobaugh, R. S. Givens, R. G. Carson, K. Srinivasachar, L. A. Sternson, and T. Higuchi, *Anal. Chem.*, *59*: 1096 (1987).
55. O. H. Lowry, N. J. Rosebrough, A. L. Farr, and R. J. Randall, *J. Biol. Chem.*, *193*:265 (1951).
56. R. L. Cunico and T. Schlabach, *J. Chromatogr.*, *266*:461 (1983).
57. J. Y. Chang, R. Aebersold, T. Gautter, G. Rosenfelder, and D. G. Braun, *Protides Biol. Fluids Proc. Colloq.*, *32*:955 (1985).
58. R. Ikenishi, T. Kitagawa, and E. Hirai, *Chem. Pharm. Bull.*, *32*:748 (1984).
59. I. N. Mefford, *Methods Biochem. Anal.*, *31*:221 (1985).
60. G. W. Bennett, M. P. Brazell, and C. A. Marsden, *Life Sci.*, *29*:1001 (1981).
61. M. W. White, *J. Chromatogr.*, *262*:420 (1983).
62. L. A. Allison, G. S. Mayer, and R. E. Shoup, *Anal. Chem.*, *56*:1089 (1984).
63. J. L. Meek, *J. Chromatogr.*, *266*:401 (1983).
64. S. Moore, D. H. Spackman, and W. H. Stein, *Anal. Chem.*, *126*:144 (1982).
65. P. Bohlen and R. Schroeder, *Anal. Biochem.*, *126*:144 (1982).
66. L. A. Sternson, T. Higuchi, P. Demontigny, A. Repta, and J. Stobaugh, U. S. Patent Application (1985).
67. F. Sanger, *Biochem. J.*, *39*:507 (1945).
68. G. Schmer and G. Krell, *J. Chromatogr.*, *28*:458 (1967).
69. P. Edman, *Acta Chem. Scand.*, *4*:283 (1950).
70. J. J. L'Italien and J. E. Strickler, in *High-Performance Liquid Chromatography of Proteins and Peptides* (M. T. W. Hearn, F. E. Regnier, and C. T. Wehr, eds.), Academic Press, Orlando, Fla. (1983), p. 195.
71. J. M. Walker, in *Methods in Molecular Biology*, Vol. 1 (J. M. Walter, ed.), Humana Press, Clifton, N.J. (1984), p. 213.
72. M. J. Miller and G. M. Loudon, *J. Am. Chem. Soc.*, *97*:5295 (1975).
73. G. R. Stark, *Biochemistry*, *7*:1796 (1968).
74. G. M. Loudon and M. E. Parham, *Tetrahedron Lett.*, 437 (1978).
75. T. Isobe and T. Okuyama, *Anal. Biochem.*, *155*:135 (1986).
76. V. G. Hum, H. G. Botting, and K. F. Mori, *Endocr. Res. Commun.*, *4*:205 (1977).
77. W. B. Lawson, E. Gross, C. M. Foltz, and B. Witkop, *J. Am. Chem. Soc.*, *83*:1509 (1961).
78. J. B. Dence, *Steroids and Peptides*, John Wiley & Sons, New York (1980), p. 307.
79. E. Peralta, H.-Y. Tang, J. Hong, and E. Costa, *J. Chromatogr.*, *190*:43 (1980).
80. W. Steiner and A. Niederwieser, *Clin. Chem. Acta*, *92*:431 (1979).
81. W. A. Koenig, in *Glass Capillary Chromatography in Clinical Medicine and Pharmacology* (N. Jaeger, ed.), Marcel Dekker, New York (1975), p. 551.

82. T. J. Yu, B. L. Karger, and P. Vouros, *Biomed. Mass Spectrum*, *10*:633 (1983).

83. D. M. Desiderio, *Adv. Chromatogr.*, *22*:1 (1983).

84. S. A. Margolis and B. Coxon, *Anal. Biochem.*, *141*:355 (1984).

85. K. Nagayama and K. Wuthrich, *Eur. J. Biochem.*, *114*:365 (1981).

86. V. J. Hruby, K. K. Deb, A. F. Spatola, D. A. Upson, and D. Yanomoto, *J. Am. Chem. Soc.*, *101*:202 (1979).

6

Analysis of Protein Drugs

Rodney Pearlman and Tue H. Nguyen

Genentech, Inc., South San Francisco, California

I. INTRODUCTION

In this chapter we describe methods and approaches for the analysis of protein-based pharmaceuticals. While excellent texts [1,2] and individual methods reviews [3–5] on the analysis of proteins are available, they are not presented from a pharmaceutical perspective. This perspective differs from typical bioanalytical approaches because issues such as purity, stability, and characterization have unique and well-defined meanings for a pharmaceutical product. Also, we are often analyzing proteins in dosage forms or biological fluids, and thus an understanding of both the assay and any influence the matrix may have on the analysis is essential.

The increased size and complexity of proteins compared to those of conventional drugs pose additional challenges for the analyst. To exert their biological activity, proteins often must exist in a precise conformation which also should be amenable to analytical scrutiny.

We will first describe some major differences between proteins and conventional drugs and the impact of such differences on their analyses. Then a compilation of some important analytical techniques for proteins is presented. Although many of the examples cited involve nontherapeutic proteins, an effort has been made to refer to protein drugs.

For the purposes of the chapter, we refer to small proteins as polypeptides whose molecular weights are in the range of a few thousand to 50,000 daltons, and to large proteins as those with molecular weights in excess of 50,000 daltons. This nomenclature is arbitrary but will help in distinguishing these size differences. Also, this review deals with proteins—as distinct from peptides, which were covered in Chapter 5.

A. Problems in Protein Analysis

Size

Proteins are orders of magnitude larger than conventional pharmaceuticals,
and this feature alone accounts for many of the difficulties in their analysis.
For example, many common drugs possess molecular weights on the order of
hundreds of mass units, whereas proteins such as insulin (ca. 6000 daltons),
human growth hormone (ca. 22,000 daltons), and the blood clotting agent
Factor VIII (ca. 360,000 daltons) are considerably larger. This increased
size complicates many of the traditional analytical techniques, such as extraction, separation, derivatization, and functional group analysis.

Thus some of the most fundamental assays are ones that determine the
size of the protein. An electrophoretic technique, sodium dodecyl sulfate
polyacrylamide gel electrophoresis (SDS-PAGE) [6], is probably the most
common method for molecular weight determination. Gel permeation chromatography is another widely used method for the determination of molecular
size and size distribution [7]. Although these techniques suffer from
rather imprecise results and insensitivity to small changes in the protein,
they are certainly starting points in the analytical spectrum. Often, surprisingly large amounts of information can be deciphered from a single gel.

Complexity

Proteins are linear polymers built up from amino acids linked via amide
bonds and folded into a three-dimensional structure [8]. This entire structure is held together by a variety of interactions, including hydrogen bonds,
disulfide bonds, charge interactions, and "hydrophobic bonding". Table
1 lists the 20 common amino acids and some of their properties. Immediately
apparent is the diversity and heterogeneity of these building blocks of proteins. Proteins contain acidic and basic groups, hydrophilic and hydrophobic residues, charged and neutral species, chromophores, fluorophores,
and neutral moieties. Such diversity in functional groups in any one analyte
requires multiple methodologies; on the other hand, it helps to "fingerprint"
each protein and is therefore a useful source for protein identification.

An additional level of complexity arises when studying glycoproteins, in
which complex carbohydrates are linked to the peptide backbone via amino
or alcohol groups. Not only can the presence of such groups affect the
physical properties and pharmacokinetics of proteins, but various glycosylation patterns can lead to a form of microheterogeneity in the analyte. Microheterogeneity due to carbohydrate composition could be described as a
population of molecules possessing the identical primary sequence, with
varying degrees and structure of carbohydrate constituents.

Polymeric Nature

Proteins are hetero polymers in that they are made up of different individual building blocks rather than a repeating unit. Privalov [9] argues that
this repeating sequence cannot be of a random monomeric nature and still
yield globular, soluble proteins. Polymer theory has made important contributions in our understanding of a variety of protein properties [10,11].
Consideration of the polymeric properties of proteins has also aided in
their analysis, especially in the areas of ultracentrifugation [12] and gel
filtration [7]. Information about the primary sequence of biopolymer is
obtained by controlled enzymatic digestion at specific amide bonds [13,14]

Table 1   The 20 Common Amino Acids, Their One- and Three-Letter
Symbols and Molecular Weights, and the pK_a Values of Any Ionizable
Side Chains[a]

Amino acid	Three-letter symbol	One-Letter symbol	Mass (daltons)	pK_a
Glycine	Gly	G	57.06	—
Alanine	Ala	A	71.08	—
Serine	Ser	S	87.08	—
Proline	Pro	P	97.12	—
Valine	Val	V	99.14	—
Threonine	Thr	T	101.11	—
Cysteine	Cys	C	103.14	9.1–9.5
Leucine	Leu	L	113.17	—
Isoleucine	Ile	I	113.17	—
Asparagine	Asn	N	114.11	—
Aspartate	Asp	D	115.09	4.5
Glutamine	Gln	Q	128.14	—
Lysine	Lys	K	128.18	10.4
Glutamate	Glu	E	129.12	4.6
Methionine	Met	M	131.21	—
Histidine	His	H	137.15	6.2
Phenylalanine	Phe	F	147.18	—
Arginine	Arg	R	156.12	~12
Tyrosine	Tyr	Y	163.18	9.7
Tryptophan	Trp	W	186.12	—

[a]Data compiled from Ref. 8.

as well as complete hydrolysis to the constituent amino acids [15]. Polymer
theory has also played a vital role in sizing proteins and studying their
aggregation behavior by means of various light-scattering techniques [16].

Surface Properties

Because of their large size and three-dimensional structure, proteins possess
a flexible "surface," with measurable properties. Early models of proteins
comprised of a hydrophobic core (i.e., hydrophobic amino acid side chains)
surrounded by a coat of hydrophilic residues have given way to ideas of a
mosaic surface [17]. This surface contains both polar and nonpolar groups

and is therefore accessible to analysis by a variety of techniques. Analyses based upon charge separations (ion-exchange chromatography) and hydrophobicity (both reversed-phase and hydrophobic interaction high-performance liquid chromatography) have been used to characterize human growth hormones and their variants [18]. However, the surface heterogeneity of proteins gives rise to problems during analysis, including aggregation [19] and adsorption phenomena [20].

B. Characterization Versus Quantitation

The characterization of small organic molecules is a relatively straightforward task. Various spectroscopic methods, such as NMR and mass spectrometry, can generally supply unambiguous information about the structure of the molecule. Because of their complexity, the characterization of proteins requires the use of many different assays, each probing a particular structural or functional feature. Thus some form of characterization of the protein is necessary before any quantitation may be performed.

The type of assays used for protein characterization include the determination of amino acid sequence, isoelectric point (pI), molecular size, and glycosylation pattern, to name a few. They are not typically amenable to quantitative analysis. By contrast, quantitative assays, such as bioassays, colorimetric protein concentration assays, and enzyme-linked immunosorbent assays, are rarely sensitive or specific enough to distinguish small perturbations in protein structure. In the case of bioassays, large assay variability and the inability to distinguish between two similarly acting proteins further complicate the analysis of proteins. Such observations have, at the very least, important ramifications in the pharmaceutical development of proteins, even if they only lead to the use of more assays at each stage of the process. These difficulties in analysis can become particularly evident in the course of stability studies, and a clear set of assays for describing stability must be established.

C. Inadequacy of Any Single Analytical Method

From the previous discussion it should be apparent that no single assay will be definitive for the analysis of any particular protein. A repertoire of assays is therefore required to yield sufficient analytical information. Such a multiplicity of assays make stability studies a very complex procedure.

The choice of assays is important, because different assays yield vastly different information on the state of the analyte. We use the term "orthogonal assays" to describe assays that probe different aspects of protein structure or chemistry, and strongly advise the use of such assays in the analysis of proteins. Such an orthogonal approach would employ a panel of assays, based on different separation and/or quantitation techniques. For example, one might use gel filtration to obtain size information about a protein as well as ion-exchange chromatography to separate charged species; or, if they are available, multiple monoclonal antibody probes, each "seeing" a different portion of the protein's structure.

Such assays may also be performed sequentially on the same sample; by collecting peaks from one type of chromatography, or eluting protein samples out of a gel matrix, and then analyzing them via a second analytical method. This approach has several drawbacks, due to destruction of the sample or dilution of the sample to below analytical sensitivity during the first analysis.

II. LEVELS OF STRUCTURAL ANALYSIS

A. Sequence

The determination of amino acid composition involves total cleavage of all peptide linkages in the molecule and the separation and quantitation of the amino acids released. A complete amino acid sequence is often required to unequivocally establish the chemical identity of the protein. Identification of the amino acids at the N- or C-terminus, or partial sequencing of the first few residues, is routinely performed to evaluate product purity. The presence of more than one amino acid at the terminal position tends to indicate that the sample is contaminated by an extraneous protein, that only partial cleavage of the molecule has occurred, or that the protein has undergone some form of proteolysis.

Amino Acid Composition

Total hydrolysis. Complete hydrolysis of proteins can be achieved by heating the sample in acid or base. Heating a protein sample in 13 M NaOH at 110°C for 30 min completely hydrolyzes all peptide bonds, but the procedure invariably results in the racemization of all amino acids. Such harsh conditions result in the destruction of cystine, cysteine, serine, and threonine residues, which must be analyzed by a different method. Base-catalyzed hydrolysis has been used primarily for the determination of tryptophan, which tends to be destroyed in acidic medium.

The most common approach for total protein hydrolysis employs 6 N HCl at 100°C. The reaction is carried out in an evacuated, sealed glass container for 12−72 hr. Serine and threonine degrade slowly under such conditions and their concentrations are estimated based on the extrapolation of the data to initial time. Isoleucine and leucine linkages are quite resistant, and often 72 hr is required for the reaction to go to completion.

Amino acid analysis. Typically, the amino acids in the protein hydrolysate are separated by a chromatographic method, then derivatized for spectrophotometric quantitation. Thin-layer chromatographic techniques that detect phenyl thiohydantoin derivatives at nanomolar levels [21] and tritiated dansyl derivatives of amino acids at picomole range have been described [22]. Amino acid analysis by gas chromatography has also attained sensitivity at the picomole level [23].

The method of choice, however, is the liquid chromatographic technique first developed by Moore and Stein [15,24]. Subsequent improvements and automation, and its ability to offer precise quantitation, have made it the standard method for amino acid analysis [25,26]. The most widely used reagent for amino acid analysis is still ninhydrin. With specially designed systems, detection of amino acids at the nanomole level can be achieved [27]. Fluorogenic reagents such as fluorescamine and O-phthalaldehyde produce more stable baselines and can increase the sensitivity another one or two orders of magnitude [28−30]. The reagent, however, does not react with secondary amines. Consequently, proline and hydroxyproline are not detected. Conversion of the secondary amines to primary amines by N-chlorosuccinimide or oxidation with hypochlorite is required prior to derivatization [31].

With most commercially available automatic analyzers today, baseline separation of almost all common amino acids can be achieved in a single run.

N-Terminal Analysis

Sanger and his collaborators, in their work on insulin, have developed several sequencing methods which are the basis for current protein sequence determination [32].   The most important technique involves the derivatization of the N-terminus peptide with 1-fluoro-2,4-dinitrobenzene.   Acid-catalyzed hydrolysis of the protein is then performed and the dinitrophenyl amino acid from the N-terminus can be identified by partition chromatography.   Fluorodinitrobenzene was later replaced with dansyl chloride as the reagent for the N-terminal amino acid.   The fluorescent derivative increases the sensitivity of the method by approximately two orders of magnitude [33].   The procedure, however, requires large amounts of sample. More efficient methods are now available.   These are based on the derivatization of the N-terminus followed by the sequential degradation of the polypeptide chains pioneered by Edman [34-36].   The reaction involves coupling of the N-terminal amino acid with phenyl isothiocyanate, cleavage of the derivatized terminal residue in acidic medium, followed by the quantitation of the released amino acid derivative.   The cycle is then repeated for the next residue.

The coupling reaction shown in Scheme 1 is performed at pH 8 to 9. Excess reagent and by-products are removed by solvent extraction.   The sample is then dried at 50°C under vacuum.   Anhydrous trifluoroacetic acid, hydrochloric acid in nitromethane, hydrochloric acid in acetic acid, or concentrated hydrochloric acid [35-37] is used in cleavage of the terminal amino acid.   The anilinothioazolinone derivative is unstable under the experimental conditions, and reaction with methylamine yields a more stable

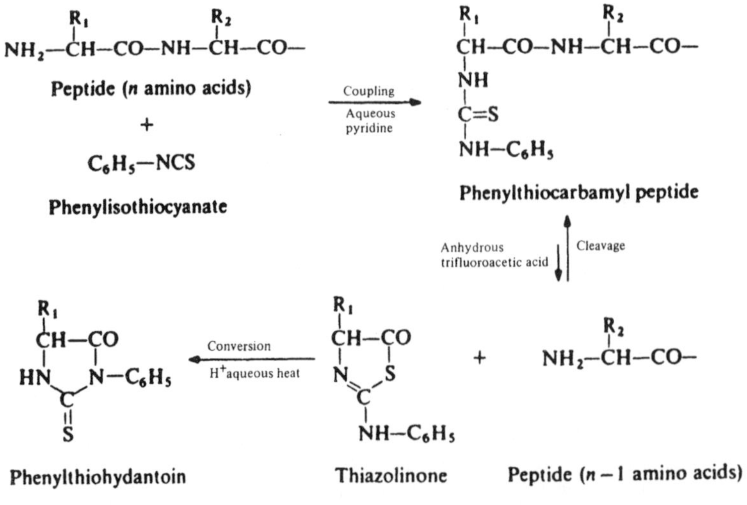

$$NH_2-\overset{R_1}{\underset{}{CH}}-CO-NH-\overset{R_2}{\underset{}{CH}}-CO-$$

Peptide (*n* amino acids)

+

C₆H₅—NCS

Phenylisothiocyanate

Coupling
Aqueous pyridine

$$\overset{R_1}{\underset{}{CH}}-CO-NH-\overset{R_2}{\underset{}{CH}}-CO-$$
$$NH$$
$$C=S$$
$$NH-C_6H_5$$

Phenylthiocarbamyl peptide

Anhydrous trifluoroacetic acid | Cleavage

$$\overset{R_1}{\underset{}{CH}}-CO$$
$$HN \quad N-C_6H_5$$
$$C$$
$$\parallel$$
$$S$$

Phenylthiohydantoin

Conversion
H⁺ aqueous heat

$$\overset{R_1}{\underset{}{CH}}-CO$$
$$N \quad S$$
$$C$$
$$NH-C_6H_5$$

Thiazolinone

+

$$NH_2-\overset{R_2}{\underset{}{CH}}-CO-$$

Peptide (*n* − 1 amino acids)

Scheme 1

product which can be extracted into ethyl acetate. The solvent is then evaporated under nitrogen.

Quantitation of the amino acid is performed chromatographically. Gas chromatography was perhaps the most widely used method for the quantitation of phenylthiohydantoin amino acids. Recent technological advances in high-performance liquid chromatography have made this technique the current method of choice. Several methods have been reported [38−44], all of which are based on reversed-phase high-performance liquid chromatography (RP-HPLC). The separation is performed most commonly on an octadecylsilane C_{18} column using a gradient of increasing concentration of a polar organic solvent, such as methanol or acetonitrile.

The introduction of automatic sequencers has allowed rapid and efficient determination of the primary structure of proteins. The liquid-phase sequencer was first described by Edman and Begg [45]. The protein is dissolved in heptafluorobutyric acid and introduced in a spinning cup. A vacuum is then applied to dry the protein to a thin film around the sides of the cup. Phenylisothiocyanate is added, followed by the coupling buffer. The coupling reaction is allowed to proceed at 50°C for 30 min, the excess reagent is extracted with benzene and ethyl acetate, and the solvents are removed under vacuum. Fluorobutyric acid is added to the dried film of derivatized protein to cleave the terminal anilinothiazolinone amino acid, which is subsequently extracted with chlorobutane and collected. The sample is ready for the next cycle.

Improvements in instrumentation [46−49], coupled with the use of highly purified reagents and sensitive HPLC detection [50], have allowed the effective sequencing of proteins with up to 150 residues. Loss of hydrophobic amino acids to the extracting solvent is still a problem to be overcome.

An alternative approach is the covalent attachment of the protein to an insoluble support, first proposed by Laursen [51,52]. In this case the extraction can be performed effectively without risk of sample loss. The protein molecules may be attached to the support through the C-terminal amino acid and sequential degradation can thus be performed from the N-terminus by Edman degradation. The polypeptide chain can also be anchored onto the support via the ε-amino groups or sulfhydryl moieties. After the coupling step, excess reagents are washed through with methanol and the terminal amino acid anilinothioazolinum derivatives are cleaved with trifluoroacetic acid and collected. The column is washed with methanol and the next cycle started.

C-Terminal Analysis

The methods available for C-terminus sequencing are based primarily on cleavage of the terminal amino acid residue by exopeptidases. Carboxypeptidase A cleaves nonpolar amino acids rapidly from the C-terminus, while asparagine, serine, glycine, and other acidic residues are cleaved more slowly. The enzyme is not active against C-terminal proline and arginine residues [53]. Carboxypeptidase B is particularly active against C-terminal lysine and arginine [54]. Carboxypeptidase C appears to be the most universal, as it cleaves all of the common amino acids from the carboxy chain ends [55]. Carboxypeptidase Y [56] is also successfully used in C-terminal sequencing; that is, a mixture of carboxypeptidases A and B is employed when lysine or arginine are at or near the C-terminus. The amino acid arrangement is deduced from a plot of the amino acids released versus time. Contamination of the enzymes by endopeptidases is

a troublesome problem, and misinterpretation of the data is not uncommon
[57].

Partial Cleavage of the Protein

For many proteins of biological interest having between 200 to 300 residues,
it is not always possible to determine the amino acid sequence directly.
The polypeptide chains have to be cleaved into smaller fragments either
enzymatically or chemically, which are then chromatographically separated,
purified, and sequenced. The primary structure of the whole protein molecule is then derived from overlapping the sequences of different sets of
peptide digests.

Enzymatic digests. Enzymatic digestion is often performed on the denatured and reduced molecule. Trypsin is the most widely used and is
perhaps the most specific of the proteases employed. It catalyses the hydrolysis of lysyl and arginyl peptide bonds except lys-pro or arg-pro.
Succinylation or maleylation of the lysine residues results in the cleavage
of the protein only at the arginyl sites. Other enzymes frequently used
are, in order of decreasing specificity, chymotrypsin, pepsin, thermolysin,
papain, and subtilisin. Table 2 describes the properties of some of these
enzymes.

The tryptic digest, once chromatographically separated, yields a pattern of peptide fragments unique to the protein. These patterns can be
used as "fingerprints" for the identification of the molecule, as first suggested by Ingram [13]. Several variants of hemoglobin were identified
and characterized by Ingram using gel electrophoresis and two-dimensional
chromatographic analysis of the tryptic digests of these proteins.

More recent work has made extensive use of reversed-phase HPLC.
Schroeder et al. [58,59] have identified all the peptide fragments of hemoglobin A, separated on an Ultrasphere octadecyl silane column employing a
phosphate buffer/acetonitrile mobile phase. Application of tandem HPLC,
combining ion-exchange and reversed-phase modes, proved to be a powerful separation method [60]. Four variants of human serum albumin were
identified using this technique [61].

Specific cleavage by chemical reaction. Selective cleavage of the peptide
bonds can also be achieved chemically. Care must be taken, however, to

Table 2 Partial Cleavage of Proteins by Proteolytic Enzymes

Enzyme	Amino acid (1) − amino acid (2)
Trypsin	Amino acid (1): Arg or Lys
Chymotrypsin	Amino acid (1): Phe, Trp, or Tyr
Pepsin	Amino acid (1): Phe, Tryp, Tyr, or between two adjacent hydrophobic residues
Thermolysin	Amino acid (2): Leu, Ile, Val, Met, Phe
Subtilisin	Relatively nonspecific
Papain	Relatively nonspecific

minimize the degradation of amino acid side chains, especially via deamidation of asparagine and glutamine.

Fragmentation of the protein at tryptophan residues is performed on the reduced and S-carboxymethylated samples with o-iodosobenzoic acid in concentrated acetic acid and guanidine hydrochloride [62]. Cleavage of the polypeptide chain between asparagine and glycine has been reported [63]. The protein is reacted with 0.2 M hydroxylamine in 6 M guanidine and potassium carbonate buffer at pH 9.0. It is believed that a cyclic imide ring is formed between asparagine and glycine residues. Hydroxylamine cleaves the imide ring, and asparagine is converted to a mixture of α- and β-aspartyl hydroxamate [64].

By far the most widely used reaction is cleavage of the protein with cyanogen bromide (CNBr) under acidic conditions [65]. Cleavage occurs at methionine residues, forming a mixture of C-terminal homoserine and homoserine lactone [66].

Cyanogen bromide is also used to cleave the protein backbone at cysteine residues. Conversion of cysteine to S-cyanocysteine by 2-nitro-5-thiocyanobenzoic acid is first performed [67,68]. Incubation in basic medium cleaves the peptide bond, accompanied by the formation of the iminothiazolidine-4-carboxyl residue from S-cyanocysteine (Scheme 2).

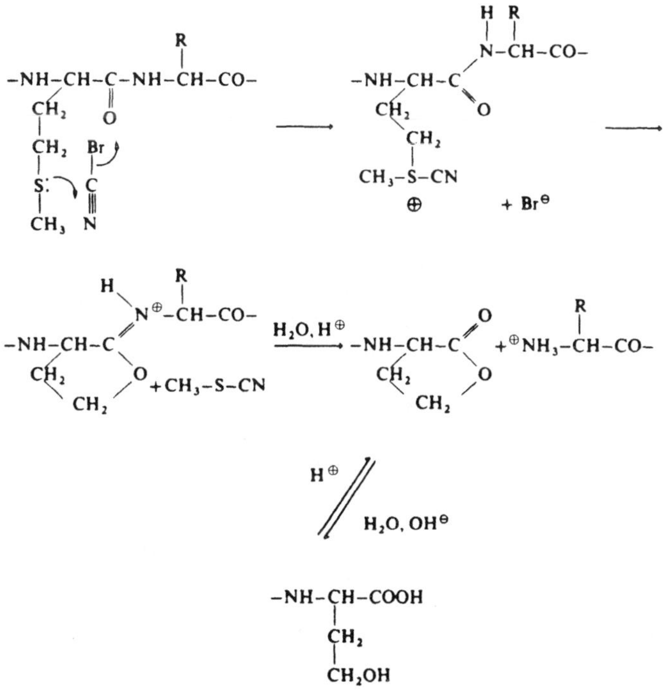

Scheme 2

B. Conformation, Size, and Shape

Spectrophotometric Techniques

Spectroscopic experiments often are the most rapid and most convenient
means of obtaining certain structural information about a protein, and often they require only small amounts of sample. In general, the methods
are nondestructive, and proteins can be observed in their surrounding
medium with minimum perturbation. Spectroscopy can also be applied to
dynamic rather than static situations. Thus reaction kinetics and molecular
interactions may be accurately and conveniently measured.

Ultraviolet absorption spectroscopy. The typical ultraviolet (UV) absorption spectrum of a protein can be divided into three distinct regions:
the far-UV region, below 210 nm; the region between 210 nm and 240 nm;
and the near-UV region, above 250 nm.

In proteins, absorption of the peptide bond occurs at around 190 nm.
Because changes in absorption in this region are ascribed to conformational
changes of the backbone of the molecule, they are potentially of great importance. In a detailed study on well-characterized model compounds such
as poly(L-lysine), Rosenbeck and Doty [69] have observed significant
changes in absorption spectra following conformational rearrangement of the
polypeptide chains. The transition from random coil to β-sheet in poly-
(L-lysine) results in a 50-nm shift of the absorption spectrum toward longer
wavelengths. When the random coil to an α-helix transition occurs, a sharp
decrease in molar absorptivity and the appearance of a shoulder at 205 nm
are observed. Extending the technique, the authors have determined the
α-helical content of several proteins and obtained results that are in good
agreement with literature data [69].

Despite these encouraging results, absorption spectroscopy in this region has not found widespread usage because of many technical difficulties.
In addition to oxygen absorption in the gas phase of the optical path length,
dissolved oxygen contributes significantly to the sample absorbance. Common inorganic buffer ions, including chloride and hydroxide, also have
measurable absorption around 200 nm [70].

Sulfur-containing amino acids, such as cysteine, cystine, and methionine,
and the histidine imidazole chain, all absorb appreciably between 210 nm
and 240 nm. However, little information can be extracted from UV absorption spectra in this region, due to the strong absorbance of aromatic residues.

The absorption bands in the region of long-wavelength radiation are
assigned to the transfer of electrons between the π-orbitals of aromatic
amino acids. The maximum absorption peaks center around 295, 275, and
260 nm for tryptophan, tyrosine, and phenylalanine, respectively. Since
the molar absorptivity of phenylalanine is approximately an order of magnitude lower than those of tryptophan and tyrosine, its contribution to
the spectrum is often insignificant.

Among the aromatic amino acid monomers, tyrosine is the only residue
possessing an acidic functional group. Ionization of the phenolic group
results in more intense absorption bands and a 20-nm red shift of the maximum absorbance peak [71]. In general, protein absorption of long-wavelength radiation is maximal at around 280 nm.

Although the molar absorbance of a protein is determined by the number of aromatic residues in the 280-nm region, a precise spectrum cannot

be predicted from the spectra of individual amino acids. This is because the pattern observed reflects the environment around such residues, which in turn depends on the molecule's structure and conformation. The aromatic amino acids in proteins often absorb at slightly longer wavelengths and higher intensity than do the spectra of free amino acids [71]. This is due to the fact that some of the chromophores are buried in the core of the protein and thus are exposed to an environment of higher refractive index. Intramolecular bonding and charge-charge interactions between nearby residues also contribute to observed differences in spectra.

The dependence of the absorption spectrum on the environment offers opportunities to study the interaction between proteins and their surroundings. The changes produced by environmental perturbations are relatively small, and difference spectroscopy is often used for a direct (and hence more accurate) comparison of protein spectra in two different environments.

Difference spectroscopy and the solvent perturbation technique. The environment of a chromophore situated on the surface of a protein can be varied by altering solvent composition. As long as the addition of co-solvent does not result in the denaturation of the protein, the number of aromatic residues exposed to the solvent can be estimated by difference spectroscopy. The experiment is based on the premise that only external residues are exposed to solvent and contribute to the difference spectra. Herkovits and Sorensen [72] have developed equations for assessing the degree of exposure of tyrosine and tryptophan residues. Comparison is made between the difference spectra of proteins, with those generated from the model compounds, N-acetyltyrosine ethyl ester and N-acetyltryptophan ester, using the same perturbant.

$$\Delta\varepsilon_{\lambda 1}(p) = a\Delta\varepsilon_{\lambda 1}(trp) + b\Delta\varepsilon_{\lambda 1}(tyr)$$

$$\Delta\varepsilon_{\lambda 2}(p) = a\Delta\varepsilon_{\lambda 2}(trp) + b\Delta\varepsilon_{\lambda 2}(tyr)$$

where a and b are the number of exposed tryptophan and tyrosine residues in the protein, and $\Delta\varepsilon$ is the molar absorptivity difference between the protein and the free tryptophan and tyrosine model compounds at the wavelength λ of interest. Since the difference spectrum of tryptophan shows positive peaks at 293 and 284 nm and those corresponding to tyrosine are at 288 and 277 nm, the data at long wavelengths are used to estimate a, assuming that the contribution of tyrosine in this range is negligible. Thus

$$a \approx \frac{\Delta\varepsilon_{\lambda 1}(p)}{\Delta\varepsilon_{\lambda 1}(trp)}$$

The value of b is approximated using the value of a at a new wavelength. The two equations are then reiterated to refine a and b.

Measurement can be carried out similarly at 288 nm. At this wavelength the difference spectrum of tyrosine shows a peak, but the difference absorption coefficient of tryptophan is zero. Thus only changes in the tyrosine residues are recorded. Figure 1 is a plot of experimental data obtained by Herkovits and Laskowski [73] from their studies on ribonuclease. The ratio of $\Delta\varepsilon$ at various pH values to the $\Delta\varepsilon$ of the native protein at

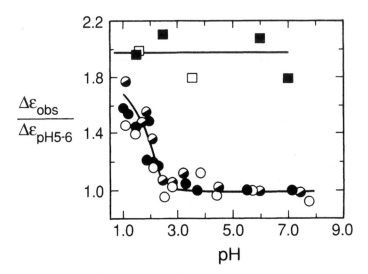

Fig. 1   Difference absorbance spectrum of ribonuclease at 287 nm in aqueous solutions containing 20% polyhydroxy solvents relative to native ribonuclease in solution at pH 5 to 6. Filled circle, 20% sucrose; open circle, 20% ethylene glycol; half-filled circle, 20% glycerol; filled square, 20% glycerol (reduced ribonuclease); open square, 20% ethylene glycol (reduced ribonuclease). (From *J. Biol. Chem.*, *235*:56.)

pH 6 is plotted against the pH of the solution. As the pH decreased, $\Delta \varepsilon_{obs} / \Delta \varepsilon_{pH6}$ increases and approaches 2, due to the loss of tertiary structure. Since ribonuclease contains six tyrosine residues, the results indicate that three of these tyrosines are buried in the core of the native protein, and as the protein denatures, the remaining three residues become accessible to the solvent.

Similar changes were discovered when ethylene glycol or sucrose was used as a perturbant. In another experiment with serum albumin, the perturbation caused by ethylene glycol affected a larger number of tyrosine residues than did sucrose, suggesting that some tyrosine residues are located in small pockets or crevices and are accessible only to small perturbants [74].

UV absorption spectroscopy has also been used to monitor the reversible denaturation of proteins. The unfolding of a globular protein molecule is often accompanied by changes in absorbance properties of buried aromatic residues which become exposed to the external medium. The resulting decrease in absorbance and shift of the peak of maximum absorption toward shorter wavelengths can readily be monitored by difference spectroscopy. Figure 2 is shown as an example of a two-state process. Here the folding-unfolding of ovalbumin was monitored by following the absorption of tryptophan at 293 nm and tyrosine at 288 nm [75]. The experimental data obtained from denaturation and renaturation experiments lie on the same curve, indicating that the process is completely reversible.

Interactions between protein chains are also amenable to analysis by UV difference spectroscopy. Frequently, multiunit proteins have aromatic

acids at the interface between the subunits. Upon dissociation, these residues become more accessible to solvent molecules, resulting in decreased
absorption intensity. Using difference spectroscopy, Pauly and Pfleiderer
[76] demonstrated that tryptophan is involved in the reversible association
of glucose dehydrogenase subunits, and that dissociation of the polymer
induces a blue shift in the spectrum with a concomitant decrease in absorption intensity.

Circular dichroism. Normal light sources consist of a collection of randomly oriented waves. When light is passed through an object that filters
out all but one wave oscillating in a single plane, the transmitted radiation
is termed plane polarized. Circularly polarized light results from the super-
imposition of two plane-polarized waves in mutually perpendicular directions
(i.e., vertical and horizontal). Polarization is achieved in such a way as
to give a difference in phase of one-quarter wavelength. The tip of the
vector sum of the two waves describes a circular spiral as the radiation
propagates. The polarized light is right-hand circularly polarized if the
vector sum rotates in a clockwise direction; conversely, a counterclockwise
rotation results in left-hand circularly polarized sources.

The combination of a right (RC) and a left (LC) circularly polarized
waves of equal amplitude produces plane-polarized light. However, if the
intensities of the two circularly polarized waves are different, the vector
sum will follow an elliptical path and such light is called elliptically polarized.

When a beam composed of two circularly polarized light sources, one
RC and one LC, passes through a solution of asymmetric molecules, RC
and LC are absorbed to different extents. The emerging radiation is

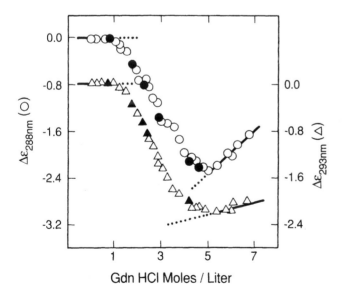

Fig. 2   Denaturation (open circle and triangle) and renaturation (filled
circle and triangle) of ovalbumin by guanidine HCL at 25°C and pH 7.0 as
followed by ultraviolet absorption spectroscopy at 288 nm (open circle) and
293 nm (open triangle). The protein concentration was 1 to 2 mg/ml. (From
Biochem., *15*:5168.)

consequently elliptically polarized and is measured by the difference between the molar extinction coefficients of LC (ε_1) and RC (ε_r).

Thus the circular dichroism (CD) is

$$\Delta\varepsilon = \varepsilon_1 - \varepsilon_r$$

The observed circular dichroic absorbance is defined by the equation

$$\Delta A = \Delta\varepsilon\, CL = A_1 - A_r$$

where C is the concentration of the chromophore and L the light path
length, and A_1 and A_r are the absorbance of RC and LC.

Another common way to express circular dichroism is in terms of molar
ellipticity θ. This is the angle with a tangent equal to the ratio of the
minor and major axes of the ellipse.

The relationship between θ and $\Delta\varepsilon$ is given by

$$\theta = 3300\,\Delta\varepsilon$$

The mean residue ellipticity θ_m is defined as

$$\theta_m = \frac{\theta}{\text{mean residue molecular weight}}$$

Most amino acids, with the exception of glycine, are asymmetric but
display only weak CD signals [77]. The CD spectrum of a protein results
primarily from the relative spatial asymmetric arrangement of its constituent amino acids.

Thus circular dichroism measured in the far-UV region reflects the
organization of the polypeptide backbone (i.e., secondary structure), while
the spectrum in the near-UV region is more closely related to the environment around the aromatic side chains and their relative orientation (tertiary
structure). The main contribution of cystinyl disulfide bonds to the CD
spectra of proteins arises from the dihedral angle of the disulfide bond,
the C—S—S bond angle, and its interaction with vicinal residues [78,79].
In general, however, disulfide CD bands are diffuse and difficult to interpret, so that their relationship to the protein structure has not yet
been well characterized.

1. *CD in the far-UV region and protein secondary structure.* Measurement performed on synthetic homopolypeptides in the far-UV region between
180 nm and 250 nm, particularly on poly(L-lysine) and poly(L-glutamic
acid), have helped established a set of characteristic spectra for α-helix
and β-sheet conformations [80,81]. In the fully charged state, these
polymers have the hydrodynamic properties of an expanded random coil,
and their spectra were attributed to the random arrangement of the chromophores [80,81].

Representative reference spectra are shown in Fig. 3. The α-helix
spectrum is characterized by two pronounced minima at 209 and 222 nm
and a maximum at around 190 nm. The CD absorption profile corresponding to the β-pleated sheet form has a peak at 194 nm and a minimum at
217 nm, whereas the random coil conformation exhibits a broad positive
band that peaks at 218 nm and a negative band with a minimum at 197 nm.

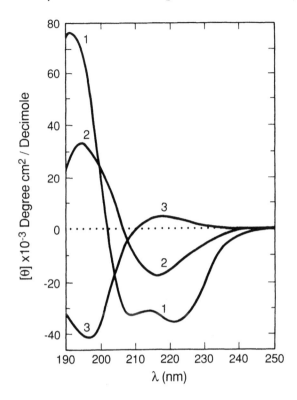

Fig. 3   Circular dichroism spectra of poly(L-lysine).   Curve 1, α-helix; curve 2, β-sheet; curve 3, random coil conformation.   (From Biochem., 8:4108.)

These models have been extended to describe large globular proteins by assuming the additivity of component structures as expressed in the equation

$$\theta_\lambda = a(\alpha)_\lambda + b(\beta)_\lambda + c(R)_\lambda$$

where $(\alpha)_\lambda$, $(\beta)_\lambda$, and $(R)_\lambda$ are the spectra of α-helix, β-sheet, and random coil, respectively, and a, b, and c are the fractional contents of each of these forms.   Computerized curve-fitting routines are used to estimate best values of a, b, and c.   Figure 4 shows an example of a best-fit line for myoglobin, generated from poly(L-lysine) basic spectra, which agrees very well with crystallographic data.   In many instances, however, the difference between structures estimated by CD spectroscopy and by x-ray diffraction analyses are substantial [82].

Comparing the secondary structure of large globular proteins to that of synthetic homopolypeptides may prove to be an oversimplification.   Unlike these model compounds, which have extended α-helix regions, globular proteins often contain many short helical segments.   Because the mean contribution of each residue within an α-helix increases with increasing helix length [83—85], the equation tends to overestimate the helical content of

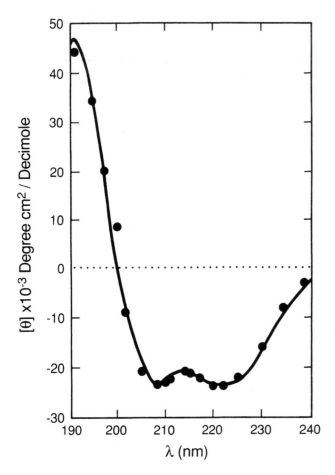

Fig. 4   Circular dichroism spectrum of myoglobin at pH 7.0.   Solid line, experimental curve; filled circle, data calculated from poly(L-lysine) reference spectra corresponding to 68.3% α-helix, 4.7 β-structure, and 27.0% random coil.   (From Biochem., 8: 4108.)

these proteins.   The β-form in a protein is more complex than the well-known pleated-sheet model.   CD spectra of the β-sheet conformation are sensitive to the surrounding medium, the number and arrangement of the strands, and the number of amino acids residues per strand [86].   The unordered segments in globular proteins are usually held in place in compact, rigid regions, the CD spectra of which should be relatively different from those of the extended loosely coiled polypeptide chains in solution. Finally, β-turns, which were found to be present in a significant portion of native proteins, are not accounted for.

The availability of x-ray crystallographic data on many proteins has led to another approach, credited to Chen et al [82], in which the reference spectra are based on a set of proteins of known structure.   Similar to the previous method, the ellipticity at any wavelength λ is a combination of reference values for the α-helix (α), β-sheet (β), and the remainder (r) forms:

$$\theta_\lambda = f\alpha[\alpha_\lambda] + f\beta[\beta_\lambda] + fr[r_\lambda] \quad \text{with} \quad 0 \leqslant f \leqslant 1 \quad \text{and} \quad \Sigma f = 1$$

where $f\alpha$, $f\beta$, and fr are the corresponding fractional content of each of these forms. The basic spectra for the α-helix and the β-sheet conformations generated according to this procedure closely resemble those obtained for the homopolypeptides. The intensity of the signals is, however, significantly smaller at all wavelengths measured. More pronounced differences are found between spectra assigned to the unordered region. The magnitude of the minimum at 194 nm is approximately one-third that of synthetic poly(L-lysine) at 197 nm, and absent are the negative bands near 240 nm and 225 nm and the positive band at 218 nm. Efforts to account for the contribution of β-turns and helix length have also been published by Chang et al. [87].

A more recent approach was proposed by Provencher and Gloekner [88], in which the CD spectra of a number of proteins were analyzed as linear combinations of the spectra of 16 other proteins. The detailed structure of the 16 proteins has been characterized by x-ray crystallography. The method obviates the need to define a unique reference spectrum for each conformation, thus allowing more flexibility in the analysis. A direct comparison of the application of the Chen-Yang-Martinez and the Provencher-Gluekner approaches on CD data of interleukin-2 can be found in a paper by Cohen et al. [89]. It appears that the latter method better describes the experimental data.

Considering the complexity of the problem, then, it appears that different sets of references should be applied in different situations. Thus homopolypeptide model compounds are probably more suitable for the characterization of small to medium-sized peptides in solution. More accurate estimations of the structure of large globular proteins are likely to come from analogy to standard proteins of similar structure.

2. *CD in the near-UV region.* The phenylalanyl, tryptophanyl, and tyrosinyl side chains all absorb in the region near 280 nm. Studies conducted on simple amino acid derivatives demonstrate that the CD signals of these residues are relatively weak. The more intense spectra of native globular proteins in this region are attributable to the immobilization of the aromatic side chains buried in the core of the molecule or the restriction of the rotational freedom of the surface side chains due to specific interactions with other residues or the solvent molecules [90]. Thus CD spectra in the near-UV region tend to be more sensitive to subtle environmental changes than do CD spectra of the peptide backbone recorded in the far-UV region.

The CD signals of phenylalanyl side chains are relatively weak, being obscured by much more intense tyrosine and tryptophan CD bands. The CD of horseradish peroxidase (HRP) is one of the few examples where distinctive fine structures between 255 and 270 nm can be observed [91,92]. In this region, the other two aromatic side chains tend to give only broad, featureless bands. The characteristic CD absorption spectrum of HRP is attributed to the presence of 23 phenylalanines and the interaction among these residues in the native state of the protein.

Tryptophan exhibits strong transitions between 290 and 305 nm. In proteins containing several tryptophans, such as carboxypeptidase A and chrymotrypsin, intense CD fine structures are visible between 295 and 305 nm, a region often identified with this residue [93,94]. The contribution of tyrosine to the CD spectrum is most conveniently studied in proteins in which the aromatic side chain is comprised solely of this residue. The near-UV spectra of proteins such as insulin, ribonuclease A, and ribonuclear S [95—97] are all characterized by intense bands between 275 and 289 nm.

Most proteins, however, contain all three aromatic side chains, and in many cases, several disulfide bridges. Their CD spectra in the near-UV region are thus too complex to be resolved into overlapping and canceling individual absorption profiles. On the other hand, CD spectra in this region reflect the fine organizational structure of the molecule in its native state. These signals are extremely sensitive to the environment around the aromatic side chains, and thus constitute a valuable tool in the study of protein-protein interaction [98], enzyme-substrate binding [99,100], or protein solvation phenomena [101].

3. *Protein folding/unfolding and CD spectra*. Circular dichroism is a particularly useful tool in monitoring the properties of both aromatic and peptide chromophores during protein folding/unfolding experiments. In many cases, direct evidence of intermediates can be derived from comparison of the near- and far-UV CD spectra. The influence of guanidine hydrochloride on the denaturation of penicillinase has been studied by Carrey and Pain [102]. CD spectra of the aromatic residues disappeared rapidly at denaturant concentrations as low as 0.84 M, at which point the peptide signals still remain largely unchanged. This was interpreted as evidence of an intermediate with a hydrodynamically expanded structure, in which the side-chain mobility is greatly increased.

The influence of temperature on the conformational change of insulin in solution has been followed by both near- and far-UV circular dichroism [103]. It is clear from the data presented that the structural organization in the side-chain domain is more susceptible to increases in temperature than is the polypeptide backbone which, even at 70°C, retains a large degree of ordered structure.

Fluorescence spectroscopy. The aromatic side chains of tyrosine and particularly tryptophan are relatively strong fluorophores when subjected to UV light of wavelengths in the vicinity of their absorption maxima (between 280 and 290 nm). The emission spectrum of tryptophan has a maximum at around 250 nm, but changes in the environment surrounding the amino acid lead to large changes in the intensity of the emitted radiation and shift in the spectrum. As a general rule, a lowering in the polarity of the medium results in an increase in fluorescence intensity accompanied by a shift toward radiation of shorter wavelengths [104].

Thus the addition of 3 M guanidine hydrochloride to horse muscle phosphoglycerate kinase is followed by a marked decrease in fluorescence intensity, with a 10-nm red shift of the emission spectrum peak [105]. Apparently, as the molecule unfolds, the buried tryptophan residues become fully exposed to the more polar environment of the aqueous solution. Specific interactions leading to radiationless transitions also result in the quenching of fluorescence light. In some cases an actual increase in fluorescence intensity was observed upon unfolding of the molecules [106].

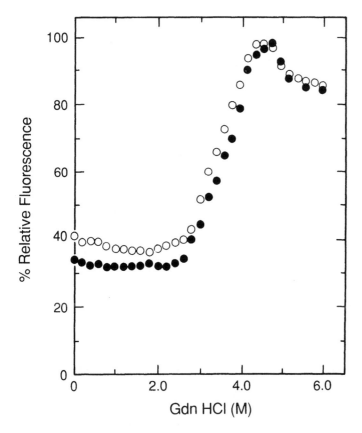

Fig. 5   Denaturation of bovine growth hormone (bGH) as followed by
fluorescence spectroscopy.   Open circle, pituitary bGH; filled circle, recombinant bGH.   The percent fluorescence was expressed relative to that
of an equimolar concentration of N-acetyltryptophanamide.   The protein
concentration was 0.4 to 0.5 mg/ml.   The emission spectrum has a maximum
of 325 nm at an excitation wavelength of 283 nm.   (From Biochem., *24*:
7662.)

An example of the use of fluorescence spectroscopy in following changes
in the environment of tryptophan upon denaturation of bovine growth hormone can be found in Fig. 5.   In this case, an increase in fluorescence
intensity accompanies the denaturation of the protein.   This increase was
attributed to a specific interaction between tryptophan and a proximal
lysine group in the protein's native conformation, which is relieved upon
unfolding of the peptide backbone of the molecule.

A major drawback of fluorescence spectroscopy is that much of the
theory explaining fluorescence and quenching of the radiation is not well understood.   A direct correlation between the fluorescence spectra and the environment of the fluorophore is difficult to establish in view of the many overlapping effects.   The method is, however, nondestructive and is a sensitive
and convenient means of recording and comparing conformational changes of
the molecules.

X-Ray Diffraction

This is one of the most powerful methods available for the study of protein structure. X-ray diffraction is basically a scattering phenomenon.
The utility of the method is due to the fact that the wavelengths of scattered radiation are comparable to the spacings of molecular structure.
This results in reinforcement of some of the refracted radiation, when observed from a particular angle. Thus when an x-ray beam is passed
through a protein crystal, it is scattered by the atoms and molecules, producing a pattern of dots of varying intensities that can be stored as a
photographic image in computer memory. Such patterns yield information
on the crystal lattice periodicity and geometry, the organization within a
unit cell of the crystal, and ultimately a detailed structure of the molecule.

If large enough crystals of the pure protein are available, the three-
dimensional conformation of the molecule can be described unequivocally.
With the latest improvements in optics and computer technology, the resolution of the method can reach as low as 1.5 to 0.5 Å in some cases. At
these levels, the precise spatial location of all the residues may be defined.

Once the coordinate data have been established, they may be down-
loaded into various computer graphics packages, permitting visualization
of the molecule from different spatial angles. Enzyme-substrate binding,
interactions between macromolecules, and conformational changes due to
mutation of residues are just some of the types of information available
through such analysis.

Protein crystals typically contain large amounts of solvent (i.e., water),
as much as 50% [107]. In the solid state, it was found that protein molecules are only touching each other at a few places, with solvent-filled
channels in between [107–109]. This picture is strongly supportive of
the assumption that the three-dimensional structure of proteins as determined by crystallography is basically similar to that of their native structure in solution. Such an assumption justifies the use of x-ray information
as a frame of reference for other structural methods, such as circular
dichroism. In some cases, solvent molecules in the immediate proximity of
the macromolecules become quite ordered, and their contributions to the
x-ray diffraction pattern of the crystal can then be used to explore the
solvent-solute interactions of proteins in solution.

Light Scattering

Most proteins of biological origin tend to fold on themselves and assume a
compact globular conformation [9]. When a solution of a macromolecule is
illuminated with a light source of wavelength away from any absorbing region, the amount of light scattered by the macromolecule can be measured.
In classical static light scattering, the intensity of the scattered light is
measured as a function of the angle of observation and solute concentration.
The results yield direct information on the molecular weight and dimensions
of the macromolecule. The fluctuation of the scattered radiation can also
be recorded, which may then be used to determine the diffusion coefficient
of the molecule. This information may be used to calculate the size of the
molecule by invoking the Stokes-Einstein equation.

Static light scattering. Macromolecules in solution scatter light strongly,
and random Brownian motion of these solute molecules induces instantaneous
inhomogeneity in the solution. These fluctuations of the solute concentration on a microscopic scale permit part of the scattered light to escape

destructive interference and be detected. Mathematical expressions exist which relate the intensity of the incident light to the angle of observation and the polarizability of the molecule [110]. When light of intensity i_0 is directed at a solution of macromolecules, the intensity of the scattered radiation i', observed at an angle θ and a distance r, is described by the Rayleigh ratio:

$$R_\theta = K \frac{i'}{i_0} \cdot \frac{r^2}{1 + \cos^2\theta}$$

with the constant quantity K being

$$K = \frac{2\pi^2 n_0^2 (dn/dC)^2}{N\lambda^4}$$

where n_0 is the refractive index of the solvent, n the refractive index of the solution of concentration C, N is Avogadro's number, and λ the wavelength of the incident beam.

The scattering intensity can be expressed as

$$\frac{KC}{R_\theta} = \frac{1}{M}\left(1 + C\frac{\delta \ln y}{\delta C}\right)$$

By expressing the equation in terms of its virial expansion;

$$\frac{KC}{R_\theta} = \frac{1}{M} + 2B_2 C + 3B_3 C^2 + \cdots$$

one can measure the intensity of light scattered from solutions of varying solute concentrations, then extrapolate the data to zero concentration, to obtain the absolute molecular weight of the molecule.

For particles with dimensions larger than $\lambda/10$, the intensity of the scattered light is usually smaller than that predicted by the equations above. The path difference between light scattering from two foci on the same particle may be large enough to produce a phase difference that results in destructive interference. A structure factor is added to the expression to account for this effect:

$$R'_\theta = \frac{1}{P(\theta)} R_\theta$$

Since $P(\theta)$ can be related to the radius of gyration of the molecule, r_g [110], the Rayleigh ratio measured from different angles contains information on the shape of the molecule in solution.

$$\frac{KC}{R_\theta} = \left[1 + \frac{16\pi^2 r_g^2}{3\lambda^2} \sin^2\theta\right]\left(\frac{1}{M} + 2BC + \cdots\right)$$

To obtain M, it is necessary to extrapolate KC/R_θ to zero concentration and zero angle. This can be elegantly done by a method devised by Zimm [111].

Dynamic light scattering. In the preceding section, the average intentensity of the scattered light is of primary interest. Its dependence on the solute concentration and the scattering angle is exploited to arrive at the molecular weight and dimensions of the solute. In dynamic light-scattering experiments, it is the time dependency of scattering that is recorded. The fluctuations of scattering intensity are directly related to the random motion of the molecules in solution. Large molecules diffusing slowly produce slow fluctuations, while small molecules that move faster cause more rapid fluctuations in the scattered light. These fluctuations can be followed and analyzed by means of the autocorrelation function [112,113].

The difference between the scattering intensity at a time t and the average intensity i is

$$\Delta i(t) = i(t) - i$$

At some later time $t + \tau$,

$$\Delta i(t + \tau) = i(t + \tau) - i$$

The average of the product of these values over many observations is described by the equation

$$A(\tau) = \Delta i(t)\Delta i(t + \tau)$$

If τ is very large, $A(\tau)$ will approach zero as the negative and positive fluctuations cancel each other out. If τ is small enough, the product is positive since within this period of time the fluctuations are likely to be of the same direction. The rate at which $A(\tau)$ decays as τ increases depends on how fast the fluctuations are. For a monodisperse system, $A(\tau)$ is described by the function

$$A(\tau) = A(0)e^{-(\tau/\tau_0)}$$

where $\tau_0 = 1/2K^2D$; K is a geometric factor dependent on the wavelength λ, the solvent refractive index, and the scattering angle θ; and D is the diffusion coefficient. The radius of gyration of the molecule is calculated from the Stokes-Einstein equation,

$$Rg = \frac{kT}{6\pi\eta D}$$

where k is the Boltzmann constant, T the absolute temperature, and η the solvent viscosity.

By using the powerful argon ion laser light sources currently available, changes in molecular size of a few nanometers can be recorded accurately. Using a 100-mW laser, Nicoli and Benedek [114] determined that the hydrodynamic radii of lysozyme and chymotrypsinogen increase by 18% and 25%, respectively, upon thermal denaturation. Similarly, the dynamics of protein aggregation in solution have been studied by application of laser light-scattering techniques [115].

Electrophoresis

In aqueous solution, a protein carries a net charge at any pH value other than its isoelectric point. When subjected to an electric field, the molecule will orient itself and migrate. The velocity with which the molecule travels is determined by a balance between the acceleration caused by the application of the field and the opposing frictional force arising from the viscosity of the medium. The mobility, U, of a protein undergoing electrophoresis is defined as its velocity per unit of electrical field and is given by the expression

$$U = \frac{v}{E} = \frac{Ze}{6\pi\eta r}$$

where v is the velocity of the molecule, E the electric field, Z the net charge of the molecule, e the electron unit charge, and η and r are the viscosity of the medium and the Stokes radius of the particle, respectively. Thus the velocity of the protein is proportional to its charge and inversely related to its size and the viscosity of the medium. At pH values below its isoelectric pH the molecule moves toward the cathode, as its average charge is positive. Conversely, it will migrate to the anode at pH values more basic than its isoelectric point. In reality, technical difficulties in minimizing band broadening due to diffusion and convection have limited the application of electrophoresis in low-viscosity media, such as in solution.

Gel electrophoresis is very widely used in practice. In this method, a supporting matrix is introduced to stabilize the medium against convection, and in most cases the gel also acts as a molecular sieve separating molecules on the basis of their size. Almost all gel electrophoresis of proteins is presently performed on agarose or polyacrylamide gels. These matrices can be prepared over a wide range of concentrations, or in the case of polyacrylamide, the density of the cross-linking agent can be varied to alter the separation characteristics of the gel.

A resolving polyacrylamide gel is prepared by mixing a solution typically containing 10 to 15% of the acrylamide monomer in Tris-glycine buffer pH 8.3, with solutions of a cross-linking agent (bis-N,N'-methylene acrylamide) and initiators (ammonium persulfate and tetramethylethylenediamine). The mixture is then poured between two glass plates shimmed to a desired thickness. Polymerization is usually completed within 3 to 5 hr. A number of wells for sample application are cast on top of the gel by inserting a comb during the polymerization process. Sample resolution is greatly improved by the introduction of a more porous stacking gel and running the separation in a discontinuous buffer system [116]. In this case a layer of polyacrylamide at lower monomer concentration, usually 4 to 5%, in Tris-HCl pH 6.7 is poured on top of the resolving gel and wells are cast into this gel layer.

Nondenaturing gel electrophoresis. Electrophoresis under nondenaturing conditions is particularly useful when intact proteins, especially oligoproteins, are to be recovered for later assessment of biological activity. The migration rate of the molecule is dependent on its charge, size, and shape [164]. Separation by size is accomplished by varying the acrylamide monomer concentration and the amount of cross-linker. Performing the experiment at a known pH that allows maximal charge differences between samples in a mixture offers an opportunity to separate isozymes on the basis of their net charge. An empirical expression was developed by Ferguson and others [117—119] to relate the mobility of a protein to the gel concentration in the supporting matrix:

$$\log U = \log U^\circ - kC$$

where

$$U = \frac{U_i}{U_d} = \frac{d_i}{d_d}$$

In the expressions above, U, the relative mobility, is defined as the ratio of the distance traveled by the protein to the distance traveled by a dye marker of high mobility, such as bromophenol blue; C is the concentration of the gel; and U° is the relative mobility extrapolated to zero gel concentration. This value is assumed to be equal to the mobility of the molecule in free solution. k, the slope of the plot of log U versus C, was found to be directly related to the molecular weight of the protein. Accordingly, a small but highly charged molecule will have a large value of U° but a small k value.

Large neutral molecules will result in a small U° and large k values, whereas a macromolecule that carries several charges will have a large intercept as well as a steeper slope when log U is plotted against gel concentration.

Electrophoresis of a set of isozymes should typically result in a set of parallel straight lines. For molecules existing in solution as long rods with net charge proportional to their length, it can be shown that U°, the free solution mobility, is essentially independent of molecular size and charge. Thus, in native gel electrophoresis, these molecules are separated primarily by the molecular sieving effect. In this case the slope of the Ferguson plot would be propotional to the molecular size of the species.

Sodium dodecyl sulfate polyacrylamide gel electrophoresis (SDS-PAGE). A rapid method for the estimation of molecular weight is electrophoresis on polyacrylamide gel in the presence of sodium dodecyl sulfate [6]. The protein is first denatured with SDS by heating at about 90°C for several minutes, and the separation is carried out on polyacrylamide gel in contact with a buffer containing SDS. A reducing agent such as β-mercaptoethanol (BME) or dithiothreitol (DTT) can be added to cleave all disulfide bonds, yielding additional structural information.

In the denaturation process, SDS binds to the protein molecules, the amount bound being proportional to the size of the unfolded protein. The negatively charged polar end of the surfactant swamps the net charge on the protein, and elongated micelles with a uniform negative surface charge are formed with the protein in the core.

The "coated" proteins are applied to the cathode and migrate toward the anode, through the gel matrix. The separation mechanism should therefore be due purely to the molecular sieving effect, similar to the case described above [109]. Indeed, the mobility of the proteins was found to be

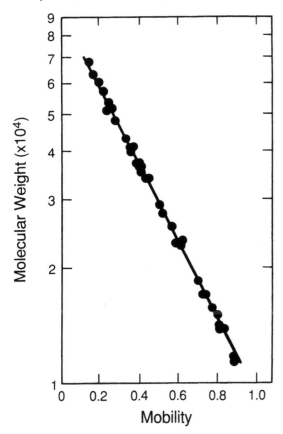

Fig. 6   Relationship between protein molecular weight and electrophoretic mobility of SDS-PAGE in the molecular weight range from 11,000 to 70,000 daltons.   (From J. Biol. Chem., 244:4406.)

totally dependent on their molecular weights, and the influence of net charge is rather insignificant.   A relationship between relative mobility and molecular weight is shown in Fig. 6.   In practice, the molecular weight of the protein is estimated by calibrating the gel with molecules of known size.   There are, however, a number of artifacts, which may lead to mis-interpretation of the data.   Some proteins can bind more or less SDS per unit length than reference compounds, branched sugars in glycosylated proteins may cause atypical migration, and the high charge on some proteins may not be negligible when compared to the bound SDS.

The protein, once separated, can either be eluted from the gel and recovered or by visualized by staining.   The two most widely used staining methods are silver staining [120–122] and Coomassie blue staining [123–126].   Silver staining is more sensitive and is used for the detection of trace amounts of protein in the nanogram range.   A recent paper describes the coloration effects due to silver grains bound to protein in the gels [126a].   Coomassie blue staining, although more rapid, is less sensitive, requiring microgram quantities of sample for good visualization.

Since Coomassie blue binds to proteins in a stoichiometric manner within a certain concentration range, with careful calibration [144], the amount of protein in each band can be estimated conveniently by laser densitometry. Other types of electrophoresis exist which separate proteins based on surface charge, and these techniques are described in the section on surface properties of proteins.

Ultracentrifugation

The exact molecular weight of proteins can be determined based on their amino acid composition, or in the case of glycoproteins, by a combination of amino acid and carbohydrate composition data. These data, however, may not accurately reflect the apparent molecular weight of the protein in solution. Many proteins exist in multichain polymeric forms rather than the monomeric form attributed to their sequence. Also, storage under adverse conditions may lead to noncovalent and covalent aggregation of the molecule, or conversely, to the cleavage of the peptide chain into smaller fragments. Since the complete primary structure of many large proteins is not always readily available, an alternative means of determining molecular size, based on experimental methods, is necessary. Sedimentation behavior of proteins under a centrifugal field offers an excellent alternative for studying molecular weight, the state of molecular aggregation, and the shape of molecules in their native environments, with minimal perturbation.

Sedimentation velocity determination. In this technique, the rate of sedimentation of the molecule in solution is followed under the influence of a centrigugal force created by subjecting the sample to high-speed rotation (normally 40,000 to 60,000 rpm). Under such conditions, the macromolecules are subjected to three types of oppositely directed forces. First, a centrifugal force F_c, which is proportional to the angular velocity (ω) of the rotor, the distance from the center of rotation to the molecule (r), and the mass of the molecule (m). As the molecule migrates through the solution, a frictional force F_f arises. F_f is proportional to the frictional coefficient (f) and the velocity (v) of the protein. The movement of the molecules also displaces some solution, creating an opposing force (f_d) similar to F_c wherein m is replaced by m_0, the mass of the displaced solution [127].

$$F_c = \omega^2 rm$$

$$F_f = -fv$$

$$F_d = -\omega^2 rm_0$$

At steady state,

$$F_c + F_f + F_d = 0$$

$$\omega^2 rm - fv - \omega^2 rm_0 = 0$$

Upon rearrangement,

$$rm\omega^2(1 - \nu\rho) = fv = 0$$

since $m_0 = m\nu\rho$, where ν is the partial specific volume of the protein and ρ is the density of the solution. Multiplying both sides by Avogadro's number N yields

$$\frac{M(1 - \nu\rho)}{Nf} = \frac{v}{\omega^2} = S$$

Since $f = kT/ND$, where k is Boltzmann's constant, T the temperature, and D the diffusion coefficient of the molecule, the equation can be expressed on a molar basis as

$$S = \frac{v}{\omega^2 r} = \frac{DM(1 - \nu\rho)}{RT}$$

where R is the gas constant. The sedimentation coefficient S, which is directly proportional to the molecular weight and diffusion coefficient of the protein, represents the mobility of the molecule per unit centrifugal field [128]. The unit for S is Svedberg, and 1 Svedberg is set equal to 1×10^{-13} sec. Table 3 lists typical values for some common proteins.

When a centrifuge field is applied to a solution, solute molecules near the meniscus migrate toward the cell bottom and thus form a moving boundary, the movement of which can be followed by a Schlieren optical system which monitors the concentration gradient dc/dr. The area under the peak is therefore proportional to the concentration of the protein, and S can be derived from the measured rate of migration dr/dt of this boundary as follows:

$$S = \frac{v}{\omega^2 r} = \frac{dr/dt}{\omega^2 r}$$

$$dr = S\omega^2 dt$$

Integration between initial time 0 and t yields

$$\ln r(t) - \ln r(t_0) = S\omega^2(t - t_0)$$

S can be determined from the slope of a plot of ln r versus time.

The frictional coefficient of the protein can be calculated from knowledge of its molecular weight and sedimentation coefficient. This should provide an approximate estimation of the molecular shape and dimensions based on models developed by Kirwood [129] and Bloomfield et al [130].

The sedimentation behavior of proteins in many cases is strongly influenced by intermolecular interaction. Highly extended molecules in concentrated solutions tend to entangle with each other as they migrate,

Table 3   Sedimentation and Diffusion Data of Some Proteins in Water at 20°C[a,b]

Protein	$S_{20,W}$ $(10^{13}$ sec$)$	$D_{20,W}$ $(10^{7}$ cm^{2}/sec$)$	$V_{20,W}$ $(cm^{3}$/g$)$	MW
Lysozyme	2.19	11.92	0.721	15,000
Serum albumin	4.6	4.31	0.733	69,000
Plasminogen	4.2	4.31	0.715	81,000
IgG	6.6−7.2	4.0	0.739	153,000
Factor XIII	9.9	2.5	0.730	320,000
IgM	18−20	1.71−1.75	0.723	1,000,000

[a]Data from G. D. Fasman, ed., *The Handbook of Biochemistry and Molecular Biology*, 3rd ed., Chemical Rubber Co., Cleveland, Ohio (1976).

[b]$S_{20,W}$, Sedimentation coefficient; $D_{20,W}$, diffusion coefficient; V_{20}, partial specific volume; M, molecular weight.

resulting in increased frictional coefficient, hence a reduced sedimentation coefficient. On the other hand, an increase in the value of S with solute concentration is often attributed to the reversible association of the sedimenting species.

In a polydisperse system, if the molecular weight of the species is sufficiently different and the molecules are noninteractive, multiple boundaries may appear whose individual S values can be determined. However, some interaction often does occur and the sedimentation coefficients of the different species become mutually dependent. In this situation, difficulties in interpreting the data arise which sometimes lead to substantial errors.

Sedimentation equilibrium. Instead of monitoring the rate of migration of the molecule, the rotation speed can be reduced and the experiment all allowed to proceed toward equilibrium. As long as the rotor speed and temperature are held constant, there is no net flux of matter from one point of the cell to another. The movement of the molecules toward the cell bottom creates a concentration gradient, which, in turn, augments the diffusional migration of the molecules toward the opposite direction.

At equilibrium:

$$S = \frac{M(1 - \nu\rho)}{Nf} \quad \text{and} \quad D = \frac{RT}{Nf}$$

Since J, the net flux of matter can be expressed as

$$J = \frac{dM}{dt} = S\omega^{2}rC - D\frac{dC}{dr} = 0$$

Replacing S and D by their values gives

$$\frac{dM}{dt} = \frac{M(1 - \nu\rho)\omega^2 rC}{Nf} - \frac{RT}{Nf}\frac{dC}{dr} = 0$$

or

$$\frac{dC}{C} = \frac{M(1 - \nu\rho)\omega^2 r\,dr}{RT}$$

The equation can be integrated between the meniscus a and r, giving

$$\ln C(r) - \ln C(a) = \frac{\omega^2 M(1 - \nu\rho)}{2RT}\,(r^2 - a^2)$$

Accordingly, a plot of ln C against r^2 yields a straight line with a slope equal to $\omega^2 M(1 - \nu\rho)/2RT$, and since ν and ρ can be obtained from density measurements, sedimentation experiments yield direct information on the protein molecular weight. For a polydisperse system, the value obtained represents the weight average molecular weight of the mixture.

The concentration gradient established in the cell offers a convenient and powerful means for studying the association/dissociation equilibrium of proteins. The process can be investigated as a function of distance from the axis of rotation, and thus data can be collected at a series of concentrations in a single experiment [131].

The aggregation behavior of hepatitis B immunoglobulin in a vaccine formulation was characterized by analytical ultracentrifugation [140]. A stability study of the vaccine was performed, and the results were in good agreement with gel permeation data performed on the same samples.

The use of analytical centrifugation is an important tool in the analysis of proteins of pharmaceutical interest, because molecular size and aggregation can readily be determined in solution formulations. However, because the equipment is expensive and cumbersome and the procedure is time consuming, it has serious limitations as a tool for routine analysis.

Gel Filtration

Gel filtration (or size exclusion, as it is sometimes called) involves the separation of macromolecules based on their size. It has been used as a preparative chromatographic technique for proteins for almost three decades [132]. Its analytical application has been more recent [133], and the advent of new packing materials has permitted the development of high-performance chromatography, thereby extending its analytical utility [132]. The analytical uses of gel filtration include protein molecular weight determination, characterization of higher-order aggregates in protein samples, and the determination of equilibrium constants for self-association.

Separation of macromolecules in gel filtration occurs because different-sized molecules diffuse into the column matrix to differing extents during their passage along the column [133,135]. The stationary phases employed as column packings are porous matrices that permit the ready diffusion of macromolecules into the matrix up to a specified cutoff. Because smaller

molecules can enter the pores of the stationary phase more readily, they
elute more slowly than do larger species. Molecules whose size is greater
than the cutoff of the gel do not even enter the matrix, and thus elute
with the void volume of the column. Early work in gel filtration was performed with soft porous gels prepared from dextrans cross-linked with
epichlorohydrin, the most notable being the Sephadex brand from Pharmacia
[136]. Other types of stationary phases include beaded agarose, cross-
linked agarose, modified silica gel [137], and polyether matrices. The
latter three column packing materials, being relatively rigid, can be used
in the high-pressure (and generally, fast flow) applications necessary in
high-performance gel filtration.

Since separation in gel permeation chromatography is based on exclusion, any specific interactions between the analyte and the column material
need to be greatly minimized [134]. This is achieved by performing analyses away from extremes in pH and by the inclusion of salts in the mobile
phase to reduce charge-charge interactions, or by the addition of semi-
polar solvents to decrease hydrophobic interactions.

It is common to refer to the separation of proteins by gel filtration to
be based on the "size" of the analytes, when in fact separation also depends on the shape of the protein. This is because shape also determines
entry of the protein into the gel matrix. Thus for similarly shaped globular
proteins, the simplification of separation based on relative size is employed—
but should be done so with caution. Generally, gel filtration is performed
under native conditions. In some instances, denaturing agents such as
urea and sodium dodecyl sulfate are added to improve separations or to
gain more accurate estimates of molecular weights [138,139].

Before a separation is performed on a particular column, the column
should be calibrated with a standard mixture of proteins whose molecular
weights cover the range of interest in the separation [134]. Such mixtures should contain proteins of similar shape and they should be used
routinely to verify the performance of the column. The eluting proteins
are most commonly monitored by UV spectroscopy. It is useful to check
the absorption reading at two or more wavelengths, particularly if aggregates are present in a sample. By so doing, a rapid check is made that
the absorbance of the peak in question is due to absorption rather than
scattering of light.

One application of the analytical use of gel filtration was performed
by Sackett and West [140]. These authors used HPLC gel filtration with
a TSK G 3000 SW column to separate aggregates of hepatitis B immunoglobulin from its monomeric form. They measured the amounts of monomer,
dimer, and trimer occurring in 17 stability samples and found them to be
in excellent aggreement with measurements from an analytical ultracentrifuge.

The separation of various growth factors was performed by Sullivan
et al. [141] using gel filtration in conjunction with anion-exchange chromatography. They were able to separate chrondrosarcoma growth factor
(CHSA GF), human milk growth factor (HMGF), retinal-derived growth
factor (RDGF), and mouse epidermal growth factor (EGF) covering molecular
weights from 6000 to 18,000 daltons.

Becker et al. [142] described the presence of a noncovalent dimer of
human growth hormone in samples of pituitary-derived and recombinant
hormone. Under nondenaturing conditions of gel filtration, the material
was collected preparatively and characterized by tryptic mapping.

The homogeneity of several insulin preparations was determined by gel filtration and compared with reversed-phase HPLC (RP-HPLC) by Welinder and Andresen [143]. Both techniques were necessary, since each revealed different information about the sample. The gel filatration method readily quantitates higher-molecular-weight species, whereas the RP-HPLC can separate human beef and pork insulins and detect small fragments of the protein.

C. Surface Properties

Surface properties of proteins are the means by which proteins interact with their surroundings. Because of the diverse groups that reside on the surface of proteins, a variety of techniques are employed to study these properties. Both electrophoretic and chromatographic methods rely on surface interactions to effect the separation and analysis of proteins.

Electrophoretic methods include native polyacrylamide gel and isoelectric focusing techniques. The former method involves PAGE under native conditions at a pH away from the pI of the protein, such that the mobility of the protein is a result of a combination of both its size and surface charge. In contrast, isoelectric focusing is perfomed across a pH gradient, and proteins migrate until they attain charge neutrality at their pI. Such techniques are useful in the pharmaceutical analysis of proteins because they are important characterization tools and can yield information on the degradation pathways of proteins.

High-performance liquid chromatography (HPLC) has become a vital tool in the analysis of small organic molecules. Chromatographic methods have also been popular techniques for proteins, but until recently, their use was mainly preparative rather than analytical. In general, preparative chromatography involves separations at low and medium pressures, employing larger-particle-size packing materials, whereas HPLC is most commonly used as an analytical technique.

Electrophoretic Techniques

Isoelectric focusing (IEF) is an extremely useful electrophoretic technique that separates proteins by their differences in surface charge under a pH gradient. Righetti has written an excellent laboratory guide on the subject of IEF [144]. Although IEF can be used as both a preparative and an analytical took, only the analytical applications are described here.

The principle behind IEF is that a pH gradient is established across a gel matrix (in the presence of an applied voltage) by means of a mixture of amphoteric substances dissolved in the gel [145]. The type of gel most commonly used is formed by polymerization of acrylamide and N,N-methylenebisacrylamide with ammonium persulfate. The gels are either purchased as precast plates available in fixed pH ranges, or prepared in the laboratory as plates or tube-type gels [146]. Agarose-based gel systems have also been used, especially in preparative applications [147].

The amphoteric substances, referred to as carrier ampholytes, are typically mixtures of low-molecular-weight polymeric materials, containing both acidic and basic functional groups with relatively close pKa values. To maintain an effective pH gradient across the gel, the individual ampholytes should possess a pK less than 1.5 pH units from their pI values [144].

Such a requirement led to the synthesis of dedicated substances containing polyamines and carboxyl groups. When a variety of these carrier ampholytes are mixed in set proportions, a predictable pH range is achieved. The two most common products are the Ampholines from LKB Produkter AB [148] and the Pharmalytes from Pharmacia [149], both being available for specified pH intervals. pH ranges across a gel can be as large as pH 3 to 10 to as small as fractions of a pH unit [144]. Also, LKB supplies precast plates of polyacrylamide, 1 mm thick, for a variety of pH ranges. Typical pH ranges from precast plates are pH 3.5 to 9.5, 4 to 6.5, 5.5 to 8.5, and similar ranges.

A recent innovation in the area of IEF is the use of ampholytes which are derivatives of acrylamide that are chemically bonded to the gel matrix. These ampholytes are known by the trade name Immobilines. Unlike conventional IEF, these are able to maintain very narrow (<0.1 pH units across a gel) and stable pH gradients [150].

When a voltage is applied to the conventional IEF gel, the randomly dispersed ampholytes migrate in the matrix until they attain charge neutrality. In so doing, the various ampholytes create a pH gradient across the gel, with low pH values at the anode and high pH values at the cathode. The pH range over which the gel is run may be selected by choosing the desired range of ampholytes of the system.

When a protein is applied to such a system, it will migrate until it reaches a state of net charge neutrality, that is, its pI. The actual separation of proteins that occurs during isoelectric focusing is a function of the mobility of the proteins in a local buffer. In the case of gel plates, protein solutions are applied to the gel by adding known volumes of the solution to paper wicks positioned along the length of the gel. The position of the applicators along the charge gradient is independent of the pI of the protein loaded, since IEF is an equilibrium method. To ensure that the system has attained equilibrium, samples can be loaded onto opposite ends of the gel. The measured pIs should be identical, independent of the site of loading. One concern in the analysis of proteins in complex matrices (e.g., a pharmaceutical formulation) is the distortion of the separated bands due to high salt concentrations. The simplest solution is to desalt the analyte prior to application by using a disposable gel filtration column, such as the PD 10 from Pharmacia. The influence of salts on IEF patterns is discussed by Righetti [144].

The electrodes used in IEF are placed in contact with the gel via wicks soaked in an appropriate buffer: acidic buffers for the anode and basic buffers for the cathode. The nature of the buffer varies depending on the pH range of the gel. Isoelectric focusing gels are run under constant power, at least in the initial stages of a separation. This is because the current and voltage are changing in response to migration of the carrier ampholytes. The current will decrease and the voltage will increase throughout a run [151]. Once the gel has focused, constant voltage is often imposed to improve separation and band sharpening [152]. Typical power ratings across a 10-cm-wide gel are on the order of 10 to 30 W with voltages as high as 3000 V. Such high voltages require the electrophoresis to be performed with a means of cooling the gel to temperatures usually less than 10°C to avoid overheating. Runs of almost 2 hr are required. Longer run times may cause shifts in the pH gradient and drifting of the pH, especially at the cathode [153,154].

Determination of the pH values across the gel is most commonly made by running pI markers as standards in one lane of the gel. These markers are mixtures of pure proteins of known pI values and are available for a range of pI values, depending on the pH range of the gel. Another technique is the use of a surface electrode to physically measure the pH across the gel just after completion of the run [156]. Examples of the use of isoelectric focusing include the separation of various forms of hemoglobin [157,158], identification of deamidated forms of human growth hormone [159,160], and numerous other examples contained in previously cited references [144,156].

By combining IEF with SDS-PAGE, the technique of two-dimensional gel electrophoresis is obtained [161]. This technique yields information on the size of proteins separated initially by charge. An excellent text edited by Celis and Bravo [162] describes this technique and gives numerous examples of separations performed by two-dimensional gel electrophoresis. Biosynthetic methionyl human growth hormone was assayed by two-dimensional gel electrophoresis and the identity of the bands determined by Western blot [163]. Western blotting [163a] determines the identity of protein bonds by measuring the immunoreactivity of the bands to antibodies raised to the protein of interest (e.g., growth hormone). After running the gel, the protein bands are electroeluted onto a membrane that is "blotted" with antibodies. Visualization of the reactive bands is often made using an enzyme marker linked to a second antibody.

Chromatography

The modes of HPLC fall into four general categories, based on the type of separation involved [165]. *Size exclusion chromatography* (or gel permeation/filtration chromatography) has already been described. *Ion-exchange chromatography* is classified as anion or cation exchange, depending on the charge of the species being separated. An extension of ion-exchange chromatography is the useful technique of chromatofocusing, which separates proteins based on their pI values. *Reversed-phase HPLC* (RP-HPLC) separates proteins primarily by their degree of surface hydrophobicity and often involves elution of the anlyte often under denaturing conditions. A more recently developed technique is *hydrophobic interaction chromatography* (HIC), which has among its advantages that of being performed under "native" conditions.

The general principles of these techniques and some recent examples will be described; however, the interested reader is referred to the text of Hancock and Sparrow [166] and the two-volume book edited by Hancock [167] for more thorough treatments of the topic. Also, the *Journal of Chromatography* publishes annual proceedings of each International Symposium on High-Performance Liquid Chromatography of Proteins, Peptides and Polynucleotides [168]. It is, therefore, an excellent resource for the latest developments in a rapidly changing field.

Ion-exchange chromatography. The analytical use of ion-exchange chromatography evolved from the preparative use of this technique in the purification of proteins. Because proteins possess charged amino acid residues, they are readily amenable to ion-exchange separations. Since the loading capacity of an ion-exchange resin is large compared with other chromatographic methods, this technique has gained widespread use on a preparative scale. Both cation- and anion-exchange separations may be used on proteins; the choice of technique depends on the pI of the protein and the pH of the separation.

The ion-exchange matrix is comprised of ionic groups covalently bound to a support. The ion-exchange groups are bound to solid supports of porous material such as silica gel, polymethacrylates, microcrystalline cellulose, or cross-linked dextran [170]. Requirements of the support material used in protein separations include a hydrophilic nature, a large pore size (300 Å), and in high-performance systems, structural rigidity. The term "weak" or "strong" ion exchanger refers to the strength of the ion-exchange group with respect to its acidity or basicity. It does not describe the strength of the interaction between the exchanger and the analyte. Examples of weak exchangers are tertiary amines for anion exchange and carboxyl groups for cation exchange. Examples of strong ion exchangers are quaternary amines for anion exchange and sulfonic acid derivatives for cation exchange [169].

The principle of the separation involves binding a charged protein mixture to the ion-exchange matrix in a low ionic strength medium, and eluting the various species with a gradient of increasing ionic strength [169]. Solvent selection and optimization are crucial to the quality of the separation and are described in detail in a publication by Kopaciewica and Regnier [171]. The gradient can be generated by a two-pump system or by means of proportioning valves connected to a single pump. Other variations include isocratic elution [172] and elution by means of a pH gradient. Detection of protein eluent is made by a variety of techniques, the most common being by ultraviolet spectroscopy at 214- or 280-nm wavelength.

Welinder and Linde [173] employed ion-exchange chromatography in the separation and quantitation of insulin and some insulin derivatives. Their method used an ion-exchange column and a buffered mobile phase (pH 8.6) containing 7 M urea to prevent insulin aggregation. The salt gradient was formed using sodium chloride, up to 0.3 M. In fact, salt gradients of 1 M or steeper are not uncommon in the ion-exchange chomatography of proteins. However, the analyst must be aware of the potential problems caused by high salt buffers. Halide salts at such high concentrations can exert their deleterious effects on stainless steel components of the HPLC system. A recent report compares reversed-phase, gel filtration, and anion-exchange HPLC in the analysis of human growth hormones [174]. The authors concluded that with the sequential use of two of the three techniques, they were able to separate all forms of the hormones, including aggregates, deamidated hormone, and forms that had undergone conformational changes.

Chromatofocusing. The technique of chromatofocusing is a form of ion-exchange chromatography that separates proteins by charge, due to a pH gradient rather than a salt gradient [175,176]. In its simplest form, chromatofocusing involves the use of weak anion-exchange resins and multicomponent buffers. The sample is loaded onto a column already equilibrated in a start buffer at a pH higher than the pI of the protein. An elution buffer is next run through the column to establish a pH gradient. This buffer is at a lower pH than the start buffer and contains multiple amphoteric buffer species similar to those employed in IEF. Proteins will remain bound on the column until the pH nears their pI, whereupon they elute.

Although chromatofocusing can be performed using ordinary anion-exchange columns, its utility is significantly improved when specially designed stationary phases are employed [177]. These contain a mixture of ionizable groups rather than a single species. This range of ionization is achieved by amplifying primary, secondary, and tertiary amines bonded to the support material [177].

Because the analytes are required to bind to the column to effect separation, chromatofocusing must be performed in low-ionic-strength buffers. Also, the column must be regenerated with start buffer prior to the next run, a process that requires almost as long to achieve as the elution phase. This problem can be circumvented by running two columns in tandem, such that one column is performing the elution whilst the other is being regenerated [177]. This method requires two pumps and the appropriate valving.

While chromatofocusing is being used principally as a purification tool, it has also been used to separate components of serum [178], interferons [179], and a variety of other species where a difference in pI occurs [177]. We have employed this technique in our laboratories to study the deamidation of growth hormone and to determine the pI of various proteins, including recombinant human tumor necrosis factor (TNF). Interestingly, the pI of TNF as determined by isoelectric focusing is about 6.7, whereas the pI measured by chromatofocusing is around 5.5. This difference, in excess of 1 pH unit, is rather surprising and may be due to some specific interactions occurring in the chromatofocusing column.

Reversed-phase HPLC (RO-HPLC). The prominent attention that RP-HPLC has recently received is due to several factors, including a better understanding of the theory of protein interactions in such systems [165, 180], improved stationary phases [181], and improvements in pump design and data reduction methods. Some concerns over the use of reversed-phase techniques for the separation of proteins stem from the potential denaturing effects caused by the use of hydrophobic stationary phases, the need for organic solvents, and the typically low pH values used to effect separation [182].

Because of the complex nature of proteins and of their ability to assume different conformational forms with different surface properties, their behavior in reversed-phase chromatography is very different from that encountered by small organic molecules [180,183]. Separation by RP-HPLC involves interaction of the surface hydrophobic areas (or "patches") of proteins with the alkyl-bonded silica stationary phase. Elution of adsorbed proteins is produced by an increasing gradient of organic modifier, such as propanol or acetonitrile. Because the chromatography of proteins by reversed phase is extremely sensitive to the amount and type or organic modifier used in the mobile phase [166,192], stringent requirements are imposed on the pumping systems. Thus even small perturbations in the gradient may have a significant influence on the analysis.

An ion pairing agent is often added in the mobile phase to minimize ionic interactions between the protein and unreacted silanol groups on the stationary phase [166], thereby sharpening peaks and improving separations. Typical ion-pairing agents include trialkyl ammonium phosphate (TAAP), phosphoric acid, and trifluoroacetic acid (TFA) in buffers of various pH values. Because the nature of the ion-pairing agent can greatly affect the separation, experimentation with various agents is advised [184].

In the course of binding to the column, some unfolding of the protein may occur, which alters the physical properties of the molecule. The mechanisms of retention and elution of proteins in RP-HPLC are quite complex [185–187] and the kinetics of the adsorption, desorption, and conformational changes are subjects of considerable research [180,188]. The interested reader is encouraged to study these references and the review chapters of Hancock [165] and Hancock and Harding [182] before embarking on reversed-phase chromatography of proteins.

The types of column materials employed in RP-HPLC include C_4, C_8, C_{18}, as well as aryl-conded silica and other common packings. The smaller alkyl chain packings are preferred for protein separations because of better recovery of the protein at lower organic modifier concentrations [182]. The particle size range of packing material of less than 10 μm is suitable for analytical work, and stationary phases should possess wide pore sizes (diameter about 300 A angstroms and greater) for the separation of large proteins [189]. Such large pore sizes are necessary to permit the adequate diffusion of proteins with molecular weight of 50,000 daltons and greater into the matrix.

While denaturation of proteins prepared and analyzed by RP-HPLC can occur, regeneration of native conformations upon removal of the organic solvent has been reported [190,191]. The ability of a protein to withstand RP-HPLC depends on many factors, but it cannot be predicted a priori. The unfolding of proteins during chromatography raises problems for their pharmaceutical analysis, especially if the amount of unfolded protein in a sample must be known. It is often necessary to determine the stabilizing (or destabilizing) influence of formulations on protein structure, and consequently, an accurate measure of the effect of the analysis is valuable. Karger and co-workers [180.188] have measured the degree of reversible unfolding that occurs during RP-HPLC of ribonuclease A under a variety of chromatographic conditions, using a photodiode array spectrophotometer. By monitoring the different spectra of the eluting peaks, the amount of unfolded protein was quantitated, and the rate at which the protein refolds was determined.

Reversed-phase HPLC has been used for the analysis of insulins [193, 194] and can be used to separate insulin from desamido insulin, higher aggregates, and other derivatives. The USP RP-HPLC method is an acceptable alternative to the rabbit bioassay for insulin, except in the case of highly purified insulins. Examples of other protein drugs assayed by RP-HPLC include interleukin-2 derivatives [195], interferons [196], human growth hormone [174], and additional examples in references already cited [165–167].

Hydrophobic interaction chromatography (HIC). The basic principle behind HIC is the adsorption of proteins to a hydrophobic stationary phase in the presence of high concentrations (1 M or greater) of antichaotropic salts, which tend to order the structure of water [197,198]. Examples include ammonium sulfate, sodium sulfate, potassium citrate, and other salts that are high on the Hofmeister series [199]. By reducing the interactions of the water and the protein, these compounds promote increased interactions of the hydrophobic surface patches of the protein with the stationary phase. Elution of the protein is achieved by a gradient of *decreasing* concentration of the salt.

Like other forms of chromatography, the initial utility of HIC was preparative rather than analytical. Because of the interest in HIC, a number of manufacturers have produced new column support materials and pumping systems better suited to the high-salt mobile phases involved in this technique. Stationary phases used in HIC were originally prepared from various-length alkyl "spacer arms" attached to agarose beads, for preparative dffinity chromatography [197]. More recently, dedicated column materials have been developed, based on a range of supports, including a variety of hydrophobic groups [200–202]. Also, hydrophobic groups have been attached to either silica (via ether linkages) or rigid, cross-linked agarose gels, permitting true high-performance liquid chromatography HIC. An example of the former is Bio-Rad's Bio-Gel TSK Phenyl-5SW, and an example of the latter is Pharmacia's Phenyl-Superose. Casein proteins have been separated on a Phenyl-Superose column using a sodium phosphate gradeint from 0.8 to 0.05 M, in the presence of urea [203]. The separation characteristics of a number of proteins on Bio-Gel TSK Phenyl-5SW is described by Goheen and Englehorn [205]. These authors found that a gradient of ammonium sulfate starting at 1.7 M and concluding with 0.1 M sodium phosphate in a pH 7.0 buffer was suitable for chromatographing a range of proteins, with retention times less than 60 min. Many papers report methods for the purification of proteins by HIC, and such techniques may be adapted to analytical separations.

Although both RP and HIC employ hydrophobic groups to effect separation, the density of the exposed hydrophobic groups is one to two orders of magnitude greater in the case of reversed phase. One result of this difference in ligand density is that RP-HPLC often requires high concentrations of organic solvents for the elution of proteins, whereas HIC is usually carried out under fully aqueous conditions. The harsh conditions of RP-HPLC are thus avoided, and unfolding of proteins is less likely to occur during HIC. Nonetheless, strongly hydrophobic proteins will often require the addition of small amounts of organic solvent, particularly from the more hydrophobic HIC columns (e.g., TSK-phenyl).

A possible drawback of HIC is that column selection and solvent optimization can be difficult until some experience is gained with the system. Also, prediction of retention order and separation based on RP-HPLC elution profiles or other measures of relative hydrophobicity is generally not successful [203]. In an interesting article that compares HIC with RP-HPLC of proteins, the authors describe differences in retention patterns between the two systems and the overall higher recovery of enzyme activity in HIC [204].

III. QUANTITATIVE ANALYSIS

A. Concentration

Routine laboratory practice often requires a sensitive and rapid assay for the determination of protein concentration. Although many techniques are available, colorimetric and spectroscopic assay are most commonly used. These assays are convenient, requiring no specialized instrumentation. On the negative side, these methods will not give unambiguously correct results unless they are calibrated against a solution of identical protein composition. Furthermore, they do not supply identity information as immunoassays do.

Colorimetric Assays

Biuret assay. The biuret is one of the earliest colorimetric methods available for determining protein concentration [206]. It involves a strong basic copper sulfate reagent, which produces a purple color upon reaction with proteins. The reaction is rapid and equilibrium is usually reached within 10 min.

The absorption spectrum of the copper-protein complex has a maximum at 750 nm and an absorptivity of approximately 23 per mole of nitrogen [207]. The sensitivity of the assay is therefore relatively low, and several milligrams of sample must be sacrificed for a reliable measurement. Since copper binds to the polypeptide backbone rather than to the side-chain residues, there is little variation in color yield from protein to protein

[208]. Consequently, dosage-form excipients should be reviewed prior to
analysis. Possible interferences include ammonia, tris buffer, and glycerol
[206,209,210].

Lowry assay. Wu [211], and later Lowry et al. [207], discovered that a
combination of the biuret and Folin-Ciocalteau phenol reagent increased the
color intensity of the assay almost 15-fold. Typically, the sample is reacted with copper sulfate in an alkaline medium, then incubated with Folin
reagent for 30 min. The absorptivity measured at 750 nm may vary from
1145 per mole of nitrogen for gelatin to 3600 for trypsin [207]. Thus accurate measurements can only be made for pure proteins, after proper
standardization of the method.

The reaction is sensitive to many common buffers and reagents used
in protein purification. These include ammonium and magnesium salts,
TRIS buffer, ethylenediaminetetracetic acid (EDTA), reducing agents, carbohydrates, and glycerol [212—214]. Many modifications of the method
have been reported, chiefly to circumvent these interferences [215].
Nevertheless, because of its sensitivity, the Lowry assay has been the
method of choice for protein determination in many laboratories.

Bradford assay. More recently, Coomassie brilliant blue G-250 has
been used successfully in the quantitation of microgram amounts of protein
in solution [216]. The dye, when dissolved in perchloric acid, turns to
a red-brown color. Bound to protein in this same medium, its original
blue color is restored. This is a one-step procedure where the protein
solution is simply mixed with the reagent and incubated for 2 to 5 min before measuring the absorbance at 595 nm. The reaction is relatively insensitive to inorganic ions, ethanol, or ammonium salts. Interferences are
observed with reducing agents, sucrose and glycerol, but they are small
and can be eliminated by running proper controls along with the samples.
The presence of large amounts of sufactant, such as sodium dodecyl sulfate or Triton X-100, is the source of error in many instances. The binding affinity of the dye to different proteins may vary slightly [216,217].
Overall, this method is sensitive, faster, and more accurate than the Lowry
assay.

Spectrophotometric Assays

The near-UV absorption spectrum of proteins has a maximum at around
280 nm. In this region, the major contribution to the spectrum is from
tryptophan and tyrosine residues, which have absorptivities of about 5700
and 1300 M^{-1} cm^{-1}, respectively. Thus if the number of these residues
is known, the absorptivity of the protein can be estimated. The average
absorptivity of some 14 proteins chosen at random was found to be 1.1 ±
0.5 [218], and this value has been used for rough estimates of protein
concentration. Although this method is nondestructive and convenient, it
suffers from having a sensitivity of about one order of magnitude lower
than that of standard colorimetric assays.

Greater sensitivity can be obtained in the far-UV region, where peptide absorption occurs. The absorptivity of proteins at 230 nm is about
300. This translates to an absorbance of 3 for a 1-mg/ml solution. Unfortunately, many common solvents also absorb in this region and far-UV
spectrometry is thus limited in many respects. Nonetheless, provided that

these absorbing species are accounted for, knowledge of the absorptivity will permit the rapid determination of protein concentration in solution.

Protein solutions containing significant amounts of aggregates that are capable of scattering light can complicate simple absorbance at 280 nm. Here, some of the light that would normally reach the photomultiplier tube is scattered rather than absorbed by the protein [109]. A measure of the true absorbance at 280 nm can be obtained by plotting the log of the absorbance versus the log of the wavelength, beyond the absorbing region around 280 nm. Once the degree of scatter is determined, it can be extrapolated back to 280 nm and the amount of 280-nm "absorbance" due to scatter may be corrected for.

In addtion to its chromophore content, the higher structural organization of the protein also contributes to its absorptivity in the near-UV region. For globular proteins, conformational induced red-shifting and hyperchromicity may contribute significantly to their total absorbance [208,219]. Enzymatic digestions of proteins can remove all conformational contributions to the UV spectrum. However, cleavage next to aromatic residues will produce ionizable functional groups which may also alter the spectrum [220]. In a clever method proposed by Bewley [221], the conformational contribution is removed by enzymatic digestion. The absorptivity is then measured at two pH values, 1.5 and 13, and the molar concentration of the native protein is calculated based on the difference in absorptivities of ionized and unionized tyrosine and on the density of this residue on the polypeptide chain.

Immunological Assays

Immunological assays are based on the reaction between an antigen (the protein analyte) and an antibody raised to that antigen. Because of their great specificity and sensitivity, immunological assays are often used for analysis of proteins in complex mixtures, such as in fermentation media and blood. Although immunoassays may provide valuable identity information, they are not necessarily indicative of the protein's biological activity. The antigen-antibody reaction is dependent on recognition of an epitope on the surface of the protein, which may or may not be sensitive to other changes occurring in the molecule. To this end, bioassays are still required. The first use of an immunoassay was reported by Yalow and Berson [222,223] for the analysis of insulin in plasma. The term "immunological assay" refers to a large number of different techniques, all of which share the common basis of an antigen-antibody reaction at some stage of their execution. However, numerous means of reporting analyte levels and methods of separating the antigen-antibody complex from the sample exist. For example, Yalow and Berson used a radioimmunoassay (RIA) which employed a radiolabled antigen to measure insulin levels. Other types of immunoassay effect quantitation by means of fluorescence [224], colorimetric enzymatic reactions [225], chemiluminescence [226], and a variety of other methods [227].

The antibodies in immunoassay can be produced in the sera of several animal species through immunization [228]. Different species produce antibodies of differing specificity, and these antibodies are polyclonal in nature. The advent of hybridoma techniques has led to the availability of monoclonal antibodies, which can be employed in immunoassays to probe specific regions of the protein analyte [229].

We will briefly describe only two types of immunoassay: the RIA and the enzyme-linked immunosorbent assay (ELISA), which are the ones most commonly in use. RIA, which is often sensitive below the nanogram level, employs a specially prepared antigen (or ligand) which is identical to the analyte except that it has a radiolabel attached to it. The first radio-labeled ligand used was insulin in which tyrosines were iodinated with ^{131}I [222].

RIA. In this method, the protein ("cold" ligand) to be analyzed is exposed to its corresponding antibodies in the presence of a known amount of radiolabeled ("hot") ligand. The concentration of protein is determined by measurement of the unbound, hot ligand after separation from the bound ligand. Separation of the antigen-antibody complex is often performed by adsorption of the unbound ligand onto charcoal or other adsorbents [230], or by precipitation of the complex with an additional antibody against the first antibody [231]. The antibody precipitation technique employs a gamma globulin against the first antibody, which is raised in a different species to the first antibody. An essential part of development and use of RIA is the production of a standard curve over the concentration range of interest.

In many cases, labeling of the antigen may alter the structure of the protein sufficiently to change its binding characteristics with the antibody. Since the basis for the assay is equivalent binding of the hot and cold antigens, labeling must be such that binding is unaltered by the process, and less severe labeling techniques, such as iodination via chloramine T, have been developed [232].

Examples of the use of RIA for proteins include its use in the analysis of insulin [222], ACTH [235], insulin like growth factors [236], and growth hormone [237].

ELISA. The ELISA assay may be used for the analysis of antigens or antibodies themselves [233]. The assay requires several types of antibody, one of which contains the linked enzyme from which the name of the assay is derived [225]. As illustrated in Fig. 7, an antibody to the analyte antigen is bound to a solid support, such as that in a sample vial. An unknown amount of antigen is added, followed by rinsing the vial to remove potentially interfering substances. A second antibody to the antigen is then added, which binds to the already complexed antigen. Finally, an antibody that recognizes the last antibody, and which contains the covalently linked enzyme, is added. The amount of original antigen present is quantitated by measuring the activity of the enzyme reporter present in the final complexed antibody upon exposure to a substrate for the enzyme. Substrates that produce colorimetic or fluorescent products are used. Examples of the latter include β-D-galactoside [232] and coumarin derivatives [234]. The analysis of angiotensin has been performed down to levels of 10^{-16} mol using the ELISA method, with fluorometric detection [234]. A review of advances in ELISA is contained in the chapter by Schall and Tenoso [232].

B. Activity

Bioactivity and Potency

For many protein-based drugs, potency is the frame of reference from which the activity of the compound is gauged. It is measured by a

A   **B**

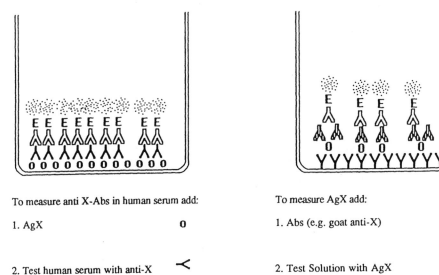

To measure anti X-Abs in human serum add:

1. AgX   **0**

2. Test human serum with anti-X   ⤙

3. E-Ab (e.g. alkaline phosphatase **E**⤙
   covalently linked to goat anti-
   human Igs)

4. Substrate for E

To measure AgX add:

1. Abs (e.g. goat anti-X)   ⤙

2. Test Solution with AgX   **0**

3. A standard amount of a standard   ⤙
   anti-X serum (e.g. human anti-X)

4. E-Ab (same as in assay at A)   **E**⤙

5. Substrate for E

Fig. 7   Schematic description of two applications of the enzyme-linked
immunosorbent assay (ELISA). The method can be used to measure antibodies (Abs) as shown in panel (a) or antigens (Ag) in panel (b). The
test relies on the ability of certain plastics (polystyrene and polyvinyl
derivatives) to strongly adsorb protein antibodies and antigens. Only the
component added in the first step is bound to the plastic, which is then
coated with unrelated proteins to ensure that the substances added subsequently will stick only specifically to reactants, rather than nonspecifically
to the plastic. (From *Microbiology*, 3rd edition, Harper & Row, 1980.)

bioassay unique for each class of proteins. Bioassays include in vitro enzymatic assays, assays performed in cell cultures, and assays measuring chemical or biological changes occurring in organs and whole animals. The aim of such assays is to compare a sample of unknown biological activity against a reference sample and thus determine the potency of the unknown sample.

One example of a bioassay is the USP Insulin Assay, which measures the ability of insulin to lower rabbit blood glucose levels. The USP sets a limit on the potency of insulin: not less than 26.0 USP insulin units per milligram. This measure of potency per unit weight is referred to as the specific activity of the protein, and is a useful estimate of purity of the compound.

An example of a whole-animal bioassay for growth hormone measures the weight gain in hypophysectemized ("hypoxed") rats as a function of dose of the hormone [241,242]. Since hypoxed rats have essentially no endogenous pituitary function, they do not grow unless supplemented by an external source of growth hormone. In essence, this assay measures the potency of a growth hormone sample relative to a reference of designated potency. Obviously, the ability to quantify such data is extremely difficult, and such assays are of little value in obtaining information about absolute activity of the hormone [243]. Other examples of bioassays include an antiviral assay for α-interferon [239] and a measure of killing efficacy of tumor necrosis factor against cultured lung tumor cells [240].

Bioassays are usually complex and are prone to large variability. As such, they are not preferred in measuring the stability of a compound over a period of time. Thus, although bioassays may be "stability indicating" for a proteinaceous drug, their lack of precision renders them unsuitable for any quantitative extrapolation, especially that required for stability studies. A monograph on the design and analysis of biological assays appears in the USP XXI [238], as do a number of monographs of official bioassays.

Enzymatic Assays

Enzymes have been used as therapeutic agents for many years. Early preparations of these proteins often consist of crude extracts of plants or animal organs. Recently, several enzymes have been made available in purified form, because of improved purification processes and the advent of biotechnology. These pure proteins have been subjected to more rigorous physicochemical characterization. However, for such proteins, regardless of physical measurements, the catalytic activity is still the ultimate measure of quality.

Two unique features of enzymatic reactions are their catalytic efficiency and specific reactivity. Most enzymatic reactions can be interpreted in the framework of the classical Michaelis-Menten kinetic model.

$$E + S \underset{k_{-1}}{\overset{k_1}{\rightleftharpoons}} ES \overset{k_2}{\longrightarrow} P + E$$

The equation describing the reaction velocity has the form

$$\frac{dP}{dt} = \frac{dS}{dt} = \frac{k_2ES}{[(k_{-1} + k_2)/k_1] + S}$$

or, more familiarly,

$$\frac{dP}{dt} = \frac{k_2ES}{Km + S}$$

where $K_m = (k_{-1} + k_2)/k_1$ is the Michaelis constant, S and P are the concentrations of the substrate and the product, respectively, and E is the enzyme activity. When S is much larger than 10 times K_m, the initial rate of reaction can be approximated by a first-order process according to the equation

$$\frac{dP}{dt} = k_2E$$

since Km + S $\simeq$ S. Consequently, under these conditions, the velocity of the reaction is directly proportional to the activity of the enzyme. The specific activity of the enzyme is defined as its catalytic activity per unit mass of protein.

Enzymatic reactions are particularly sensitive to experimental conditions. Whereas the temperature, pH, and ionic strength of the buffer can easily be controlled, the nature of the substrate is often the main source of variation; for example, lipase activity is determined based on the digestion of vegetable oil [244]. The type and quality of the oil as well as the droplet size of the emulsified lipid will undoubtedly affect the experimental results.

For many proteolytic and peptidolytic enzymes, synthetic substrates are ideal, as these compounds are specific and commercially available in pure form. p-Toluenesulfonyl L-arginine methyl ester is used for the quantitation of chymotrypsin [245], H-D-valyl-L-leucyl-L-lysine-p-nitroanilide dihydrochloride is specific for plasmin [246,247], and H-D-isoleucyl-L-prolyl-L-arginine-p-nitroanilide dihydrochloride is commonly employed in the measurement of tissue plasminogen activator (t-PA) activity [248,249]. Hydrolytic cleavage of peptides catalyzed by these enzyme releases the chromophore, which can conveniently be monitored spectrophotometrically.

Although these assays are accurate and reproducible with coefficient of variation rarely exceeding 5%, the physiological relevance of the assay is an important consideration, especially for pharmaceutical enzymes. For example, t-PA is involved in the activation of plasminogen to plasmin, a fibrinolytic enzyme. An assay directly measuring the activation of plasminogen by t-PA is obviously more indicative of its biological activity than that based on synthetic substrates. Assays such as an enzymatic assay for t-PA based on plasminogen activation have been designed to follow the formation of plasmin from plasminogen by coupling two reactions [250], as shown in Scheme 3. As expected, the reliability of this assay is strongly dependent on the quality of plasminogen (e.g., plasmin free). Moreover, since t-PA has a different specificity for Glu-plasminogen than for Lys-plasminogen [250], it is imperative that this substrate be thoroughly characterized.

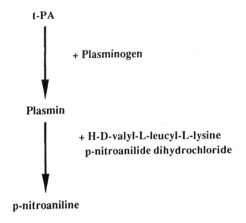

t-PA

+ Plasminogen

Plasmin

+ H-D-valyl-L-leucyl-L-lysine
   p-nitroanilide dihydrochloride

p-nitroaniline

Scheme 3

Another level of complexity arises when the ultimate therapeutic indication is considered. t-PA is now being used as a blood-clot-specific thrombolytic agent. In vivo, t-PA binds to the fibrin clot and cleaves entrapped plasminogen to plasmin, which in turn, dissolves the blood clot. Since its fibrin binding domain resides in a different location on the molecule than the catalytic site, preservation of proteolytic activity may not reflect the integrity of the whole molecule.

Assays in which plasminogen containing fibrin clots are formed in vitro can be designed so that upon addition of t-PA, the rate of dissolution of the clot can be monitored photometrically or by measuring the radioactive fibrin fragment release into the supernatant medium [250,251,251a]. This assay is obviously more meaningful, but as the degree of complexity increases and more variables are introduced, the accuracy and reproducibility of the assay may suffer.

IV. ANALYSIS OF PROTEINS IN COMPLEX ENVIRONMENTS

The analysis of proteins of pharmaceutical interest is further complicated by the presentation of the protein in a dosage form, or the need for analyses in biological samples. In addition, information about the stability and degradation mechanisms of the protein upon storage is also required for the development of a protein-based pharmaceutical.

A. Dosage Forms and Stability Indicating Assays

By far the most common route for the administration of therapeutic proteins is by injection. This will be the standard route of administration in the foreseeable future because of the difficulties in transport of proteins across membranes and their great lability toward metabolic degradation. Thus proteins will not be found in as many diverse types of dosage forms as will conventional drugs.

Nevertheless, the analytical challenges in separating the protein from its formulation will still require ingenuity and good scientific rationale.

Many protein dosage forms are presented as lyophilized solids containing the protein along with bulking agents, buffers, and various stabilizers. Such excipients could adversely affect electrophoretic systems, which typically require low-ionic-strength samples, or interfere with spectroscopic analyses, amino acid determinations, and protein digests. Thus the analyst needs to be aware of the interactions of such excipients with the assays utilized.

Of particular concern are the so-called stability indicating assays. In the design of a dosage form, numerous studies are undertaken to provide information on the intrinsic stability profile of the active agent and the influence of various excipients and processing conditions on that profile. Excipients may have a profound effect on the properties of the protein because of the potentially large number of opportunities for interactions to occur.

Since biological activity is the ultimate measure of the proteins' activity, one might argue that this is the benchmark assay for determining stability. Because of the notorious imprecision of most bioassays, this cannot be the case. Furthermore, it has been our experience that one biochemical assay or another will detect extremely subtle changes occurring in the protein far in advance of any signal from the bioassay. The skill is in determining just which assay that might be for a particular protein.

B. Biological Matrices

One of the greatest analytical challenges in employing protein drugs is the analysis of the drug in vivo. Since many proteins are active in doses far lower than those of conventional drugs, new levels of sensitivity are required, namely, in the picogram to femtogram range and beyond. To complicate the issue further, the biological matrix possesses numerous interfering substances that are themselves proteins. In the analysis of conventional drugs from blood or tissue, often one of the preliminary steps is the removal (usually by precipitation) of much of the protein load in the sample to make the analysis possible. Obviously, such an approach is not feasible when analyzing proteins. Thus a great reliance has been placed on protein specific assays that work in the presence of the biological matrix. Bioassays (with all of their attendant problems) have been used for the analysis of insulin, interferons, and tumor necrosis factor, as referenced above.

Another approach is the use of immunological assays in the case of insulin [222] and human growth hormone [237]. An article by Aston et al. compared growth hormone preparations in an in vivo bioassay with an in vitro cell proliferation assay and a solid-phase RIA [252]. This study demonstrated that although immunological potency can be retained, bioactivity can be markedly reduced.

Another problem that is observed in the analysis of proteins, particularly when low levels are being assayed, is the potential for adsorption of the protein onto surfaces. Several articles dealing with the adsorption of insulin to intravenous injection sets exist [253,254]. Inhibition of adsorption by the inclusion of competing proteins such as human serum albumin and partially hydrolyzed gelatin have been used [254]. Where ELISA and bioassays are used, the obvious limitations due to the addition of a potential interfering substance to an assay have to be considered.

V. CONCLUSIONS

In this chapter we have described many of the principal methods for the analysis of proteins, with particular reference to protein drugs. In attempting to present a comprehensive survey of current analytical methodology, a number of techniques had to be omitted from the discussion. Most notable among those are mass spectrometry (MS), in particular fast atom bombardment MS (FAB-MS) [255] and two-dimensional nuclear magnetic resonance (2D-NMR) spectroscopy [256], which are both proving to be powerful techniques for the determination of structural information of proteins. Although the work of Privalov [9] was referenced, no description of the technique of scanning microcalorimetry was made.

The necessity of employing multiple analytical techniques in the development of protein-based pharmaceuticals must be emphasized. The need for increased analytical scrutiny occurs during all phases of the development of a protein pharmaceutical and is felt especially when monitoring stability of the agent. Along with the vast improvements that have occurred in analytical technology over the past decade or so, a higher degree of analytical precision is now available. Such analytical sophistication has already raised and, will undoubtedly continue to raise, many more questions than it answers.

ACKNOWLEDGMENTS

The authors wish to thank Dr. Richard Jones for his guidance and advice, and Dr. William Hancock for his comments and suggestions on the manuscript.

REFERENCES

1. I. Kerese, ed., *Methods of Protein Analysis*, Halsted Press, New York (1984).
2. C. R. Cantor and P. R. Schimmel, *Biophysical Chemistry-Part II: Techniques for the Study of Biological Structure and Function*, W. H. Freeman and Company, San Francisco (1980).
3. M. T. W. Hearn F. E. Regnier, and C. T. Wehr, eds., *High Performance Liquid Chromatography of Proteins and Peptides*, Academic Press, Orlando, Fla. (1983).
4. D. B. Wetlaufer, *Adv. Protein Chem.*, *17*:703 (1962).
5. M. W. Hunkapiller, J. E. Strickler, and K. J. Wilson, *Science*, *226*: 304 (1984).
6. U. K. Laemmli, *Nature*, *227*:680 (1970).
7. G. K. Ackers, *Adv. Protein Chem.*, *24*:343 (1970).
8. C. R. Cantor and P. R. Schimmel, *Biophysical Chemistry-Part I: The Conformation of Biological Macromolecules*, W. H. Freeman and Compnay, San Francisco (1980).
9. P. L. Privalov, *Adv. Protein Chem.*, *33*:167 (1979).
10. P. J. Flory, *Proc. Natl. Acad. Sci. USA*, *79*:4510 (1982).
11. K. A. Dill, *Biochemistry*, *24*:1501 (1985).
12. H. K. Schachman, *Ultracentrifugation in Biochemistry*, Academic Press, New York (1959).
13. N. M. Ingram, *Nature*, *178*:792 (1956).

14. C. A. Bishop, W. S. Hancock, S. D. Brennan, R. W. Carrell, and M. T. W. Hearn, *J. Liq. Chromatogr.*, *4*:599 (1981).
15. S. Moore and W. H. Stein, *J. Biol. Chem.*, *192*:663 (1951).
16. K. E. Van Holde, *Physical Biochemistry*, Prentice-Hall, Englewood Cliffs, N.J. (1985).
17. C. H. Chothia, *Nature*, *248*:338 (1974).
18. R. L. Patience and L. H. Rees, *J. Chromatogr.*, *352*:241 (1986).
19. W. D. Lougheed, A. M. Albisser, H. M. Martindale, J. C. Chow, and J. R. Clement, *Diabetes*, *32*:434 (1981).
20. E. W. Kraegen, L. Lazarus, H. Miller, L. Campbell, and Y. O. Chia, *Br. Med. J.*, *3*:464 (1975).
21. M. R. Summers, G. W. Smythers, and S. Orosland, *Anal. Biochem.*, *53*:624 (1973).
22. S. R. Burzynski, *Anal. Biochem*, *65*:93 (1975).
23. R. W. Zumwalt, K. Kuo, and C. W. Gerhke, *J. Chromatogr.*, *57*:193 (1971).
24. D. H. Spackman, W. H. Stein, and S. Moore, *Anal. Chem.*, *30*:1190 (1958).
25. P. B. Hamilton, *Anal. Chem.*, *30*:914 (1958).
26. P. B. Hamilton, *Anal. Chem.*, *35*:2055 (1963).
27. T. H. Liao, G. W. Robinson, and J. Salmikov, *Anal. Chem.*, *45*:2286 (1973).
28. P. E. Hare, *Methods Enzymol.*, *47*:3 (1977).
29. A. M. Felix and G. Terkelsten, *Anal. Biochem.*, *56*:610 (1973).
30. J. R. Benson and P. E. Hare, *Proc. Natl. Acad. Sci. USA*, *72*:619 (1975).
31. M. Roth and A. Hampai, *J. Chromatogr.*, *83*:353 (1973).
32. F. Sanger, *J. Biochem.*, *39*:507 (1945).
33. W. R. Gray and B. S. Hartley, *Biochem. J.*, *89*:379 (1963).
34. P. Edman, *Acta Chem. Scand.*, *4*:283 (1950).
35. P. Edman, *Acta Chem. Scand.*, *7*:700 (1953).
36. P. Edman and A. Henschen, in *Protein Sequence Determination* (S. B. Needleman, ed.), Springer-Verlag, Berlin (1975), p. 232.
37. G. E. Tarr, *Methods Enzymol.*, *47*:335 (1977).
38. A. S. Brown, J. E. Mole, A. Weissinger, and J. C. Bennett, *J. Chromatogr.*, *156*:35 (1978).
39. A. S. Brown, J. E. Mole, A. Weissinger, and J. C. Bennett, *J. Chromatogr.*, *148*:532 (1978).
40. C. L. Zimmerman, E. Apella, and J. J. Pisano, *Anal. Biochem.*, *77*:569 (1977).
41. J. E. River, in *Peptide Structure and Biological Function, Proceedings 6th American Peptide Sympodium* (E. Gross and J. Meinhofer, eds.), Pierce Chemical Company, Rockford, Ill. (1980).
42. I. N. Macheva, R. N. Nikolov, and X. Pfetschinger, *J. Chromatogr.*, *213*:99 (1981).
43. C. L. Zimmerman, *Anal. Biochem.*, *75*:77 (1976).
44. K. L. Stone and K. R. Williams, *J. Chromatogr.*, *359*:203 (1986).
45. P. Edman and G. Begg, *Eur. J. Biochem.*, *1*:80 (1967).
46. B. Wittmann-Liebold, H. Gradffunder, and H. Kohls, *Anal. Biochem.*, *3*:426 (1975).
47. B. Wittmann-Liebold, H. Gradffunder, and H. Kohls, *Anal. Biochem.*, *75*:621 (1976).
48. M. W. Hunkapiller and L. E. Hood, *Biochemistry*, *17*:2124 (1978).

49. M. W. Hunkapiller and L. E. Hood, *Science,* 207:523 (1980).
50. R. Somack, *Anal. Biochem.,* 104:464 (1980).
51. R. A. Laursen, *J. Am. Chem. Soc.,* 88:5344 (1966).
52. R. A. Laursen, ed., *Solid Phase Methods in Protein Sequence Analysis,* Pierce Chemical Company, Rockford, Ill. (1975).
53. R. P. Ambler, *Methods Enzymol.,* 25:143 (1972).
54. J. E. Folk, *Methods Enzymol.,* 19:504 (1970).
55. H. Tscheske and S. Kupfer, *Eur. J. Biochem.,* 26:33 (1972).
56. R. Hayashi, *Methods Enzymol.,* 47:84 (1977)
57. L. Rask, H. Amundi, and P. A. Peterson, *FEBS Lett.,* 104:55 (1979).
58. W. A. Schroeder and T. H. J. Huisman, *The Chromatography of Hemoglobin,* Marcel Dekker, New York (1980).
59. W. A. Schroeder, J. B. Shelton, and J. R. Shelton, in *Advances in Hemoglobin Analysis* (S. M. Hanash and G. J. Brewer, eds.), Alan R. Liss, New York (1981).
60. N. Takahashi, N. Ishioka, Y. Takahashi, and F. W. Putnam, *J. Chromatogr.,* 326:181 (1986).
61. N. Takahashi, Y. Takahashi, N. Ishioka, and B. S. Blumberg, *J. Chromatogr.,* 359:181 (1986).
62. W. C. Mahoney and M. A. Hedmodson, *Biochemistry,* 18:3810 (1979).
63. W. T. Butler, *J. Biol. Chem.,* 244:3415 (1969).
64. P. Bornstein, *Biochemistry,* 9:2408 (1970).
65. E. Gros and B. Witkop, *J. Am. Chem. Soc.,* 83:1510 (1961).
66. R. P. Ambler, *Biochem. J.,* 96:32p (1965).
67. N. Catsimpoolas and J. C. Wood, *J. Biol. Chem.,* 241:1790 (1966).
68. Y. Degani and A. Patchornik, *Biochemistry,* 13:1 (1974).
69. K. Rosenbeck and P. Doty, *Proc. Natl. Acad. Sci. USA,* 29:633 (1961).
70. R. P. Buck, S. Singhadeja, and L. B. Rogers, *Anal. Chem.,* 26:1240 (1954).
71. O. B. Wetlaufer, *Adv. Protein Chem.,* 17:303 (1962).
72. T. T. Herkovits and S. R. Sorensen, *Biochemistry,* 7:2523 (1968).
73. T. T. Herkovits and M. Laskowski, Jr., *J. Biol. Chem.,* 235:56 (1960).
74. T. T. Herkovits and M. Laskowski, Jr., *J. Biochem.,* 237:2481 (1962).
75. F. Ahman and A. Salahudin, *Biochemistry,* 15:5168 (1976).
76. H. E. Pauly and G. Pfleiderer, *Biochemistry,* 16:4599 (1977).
77. C. Torriolo, *J. Phys. Chem.,* 74:1390 (1970).
78. T. A. Bewley and C. H. Li, *Arch. Biochem. Biophys.,* 138:338 (1970).
79. E. R. Simons, in *Spectroscopy in Biochemistry* (J. E. Bell, ed.), CRC Press, Boca Raton, Fla. (1981).
80. N. Greenfield, B. Davidson, and G. D. Fasman, *Biochemistry,* 6:1630 (1967).
81. N. Greenfield and G. D. Fasman, *Biochemistry,* 8:4108 (69).
82. Y. H. Chen, J. T. Yang, and H. M. Martinez, *Biochemistry,* 11:4122 (1972).
83. R. W. Woody, *Biopolymers,* 8:669 (1969).
84. J. N. Vournakis, J. T. Yan, and H. R. Sheraga, *Biopolymers,* 6: 1531 (1968).
85. J. R. Parrish, J. R., and E. R. Blout, *Biopolymers,* 10:1491 (1971).
86. R. W. Woody, *Biopolymers,* 8:669 (1969).

87. C. T. Chang, C. C. Wu, and J. T. Yang, *Anal. Biochem.*, *91*:13 (1978).

88. S. W. Provencher and J. Gloekner, *Biochemistry*, *20*:33 (1981).

89. F. E. Cohen, P. A. Kosen, I. D. Kuntz, L. B. Ebstein, T. L. Ciardelli, and K. A. Smith, *Science*, *234*:349 (1986).

90. E. H. Strickland, CRC *Crit. Rev. Biochem.*, *2*:113 (1974).

91. E. H. Strickland, E. Kay, L. M. Shannon, and J. Howitt, *J. Biol. Chem.*, *243*:3560 (1968).

92. E. H. Strickland, E. Kay, and L. M. Shannon, *J. Biol. Chem.*, *245*: 1233 (1970).

93. L. Fretto and E. H. Strickland, *Biochim. Biophys. Acta*, *235*:473 (1971).

94. M. J. Gorbunoff, *Biochemistry*, *8*:2591 (1969).

95. M. J. Ettinger and S. N. Timasheff, *Biochemistry*, *10*:824 (1971).

96. J. Horwitz and E. H. Strickland, *J. Biol. Chem.*, *246*:3749 (1971).

97. J. Horwitz, E. H. Strickland, and Billups, *J. Am. Chem. Soc.*, *92*:2119 (1970).

98. W. B. Gratzer and G. H. Beaven, *J. Biol. Chem.*, *244*:6675 (1969).

99. D. A. Holowka, A. D. Storsberg, J. W. Kimball, E. Haber, and R. E. Cathon, *Proc. Natl. Acad. Sci. USA*, *69*:3399 (1972).

100. L. Fretto and E. H. Strickland, *Biochim. Biophys. Acta*, *235*:489 (1971).

101. M. J. Ettinger and S. N. Timasheff, *Biochemistry*, *10*:831 (1971).

102. E. Carrey and R. Pain, *Biochim. Biophys. Acta*, *533*:12 (1978).

103. M. J. Ettinger and S. N. Timasheff, *Biochemistry*, *10*:824 (1971).

104. J. E. Bell, in *Spectroscopy in Biochemistry* (J. E. Bell, ed.), CRC Press, Boca Raton, Fla. (1981).

105. J. M. Betton, M. Desmadril, A. Mitraki, and J. M. Yon, *Biochemistry*, *23*:6654 (1984).

106. D. N. Brems, S. M. Plaisted, H. A. Havel, E. W. Kaufman, J. D. Stodola, L. C. Eaton, and R. D. White, *Biochemistry*, *24*:7662 (1985).

107. C. R. Cantor and P. R. Schimmel, in *Biophysical Chemistry-Part II: Techniques for the Study of Biological Structure and Function* (C. R. Cantor and P. R. Schimmel, eds.), W. H. Freeman and Company, San Francisco (1980).

108. J. L. Smith, W. A. Hendrickson, R. B. Honzatko, and S. Sheriff, *Biochemistry*, *25*:5018 (1986).

109. K. E. Van Holde, in *Physical Biochemistry* (K. E. Van Holde, ed.), Prentice-Hall, Englewood Cliffs, N.J. (1985).

110. C. R. Cantor and P. R. Schimmel, in *Biophysical Chemistry-Part II: Techniques for the Study of Biological Structure and Function* (C. R. Cantor and P. R. Schimmel, eds.), W. H. Freeman and Company San Francisco (1980).

111. B. H. Zimm, *J. Chem. Phys.*, *16*:1093 (1948).

112. D. E. Koppel, *J. Chem. Phys.*, *57*:4814 (1972).

113. N. Mazer, B.S. thesis, Massachusetts Institute of Technology, Cambridge, Mass. (1973).

114. D. F. Nicoli and G. B. Benedek, *Biopolymers*, *15*:2421 (1976).

115. J. R. Cohen, J. A. Jedziniak, and G. B. Benedek, *Proc. R. Soc. London*, *A345*:73 (1975).

116. A. Chrambach and D. Rodbard, *Science*, *172*:440 (1971).

117. K. A. Ferguson, *Metabolism*, *13*:985 (1964).

118. K. Weber and M. Osborn, *J. Biol. Chem.*, *244*:4406 (1969).
119. K. E. Van Holde, in *Physical Biochemistry* (K. E. Van Holde, ed.), Prentice-Hall, Englewood Cliffs, N.J. (1985).
120. R. C. Switzer, C. R. Merril, and S. Shifrin, *Anal. Biochem.*, *98*: 231 (1979).
121. B. L. Oakley, D. R. Kirsch, and N. R. Morris, *Anal. Biochem.*, *105*: 361 (1980).
122. J. H. Morrissey, *Anal. Biochem.*, *117*:307 (1981).
123. S. Fazekas de St. Groth, R. B. Webster, and A. Datyner, *Biochim. Biophys. Acta*, *71*:377 (1963).
124. T. S. Meyer and B. L. Lambert, *Biochim. Biophys. Acta*, *100*:144 (1965).
125. A. H. Reiner, *Methods Enzymol.*, *104*:439 (1984).
126. V. Neuhoff, R. Stamm, and H. Eibl, *Electrophoresis*, *6*:427 (1985).
126a. C. R. Merril, M. E. Bisher, M. Harrington, and A. C. Stereng, *Proc. Natl. Acad. Sci. USA*, *85*:453 (1988).
127. C. R. Cantor and P. R. Schimmel, in *Biophysical Chemistry-Part II: Techniques for the Study of Biological Structure and Function* (C. R. Cantor and P. R. Schimmel, eds.), W. H. Freeman and company, San Francisco (1980).
128. K. E. Van Holde, in *Physical Biochemistry* (K. E. Van Holde, ed.), Prentice-Hall, Englewood Cliffs, N.J. (1985).
129. J. G. Kirwood, *J. Polym. Sci.*, *12*:1 (1954).
130. V. Bloomfield, W. O. Dalton, and K. E. Van Holde, *Biopolymers*, *5*: 135 (1967).
131. D. C. Teller, *Methods Enzymol.*, *27*:346 (1972).
132. J. Porath and P. Flodin, *Nature*, *183*:1657 (1959).
133. G. K. Ackers, *Adv. Protein Chem.*, *24*:343 (1970).
134. K. Unger, *Methods Enzymol.*, *104*:154 (1984).
135. F. E. Regnier, *Science*, *222*:245 (1983).
136. Pharmacia Fine Chemicals, *Gel Filtration, Theory and Practice*, Pharmacia Fine Chemicals, Uppsala (1982).
137. Y. Kato in *Handbook of HPLC for the Separation of Amino Acids, Peptides and Proteins*, Vol. 2 (W. S. Hancock, ed.), CRC Press, Boca Raton, Fla. (1984), p. 363.
138. T. I. Pristoupil, *J. Chromatogr.*, *19*:64 (1965).
139. T. Imamura, K. Konishi, M. Yokohama, and K. Konishi, *J. Biochem.*, *86*:639 (1979).
140. P. H. Sackett and J. West, *J. Chromatogr.*, *360*:433 (1986).
141. R. C. Sullivan, Y. W. Shing, P. A. D'Amore, and M. Klagsbrun, *J. Chromatogr.*, *266*:301 (1983).
142. G. W. Becker, R. R. Bowsher, W. C. MacKellar, M. L. Poor, P. M. Tackitt, and R. Riggin, *Biotech. Appl. Biochem.*, *9*:478 (1987).
143. B. S. Welinder and F. H. Andresen, *Hormone Drugs, Proceedings of the FDA-USP Workshop on Drug and Reference Standards for Insulins, Somatotropins, and Thyroid-Axis Hormones* (J. L. Gueriguian, E. D. Branscome, and A. S. Outschoorn, workshop organizers), United States Pharmacopeial Convention, Rockville, MD (1982), p. 163.
144. P. G. Righetti, in *Isoelectricfocusing: Theory Methodology and Applications. Laboratory Techniques in Biochemistry and Molecular Biology* (T. S. Work and R. H. Burdon, eds.), Elsevier/North-Holland Biomedical Press, Amsterdam (1983).

145. P. G. Righetti, E. Gianzza, and A. B. Bosisio, in *Recent Developments in Chromatography and Electrophoresis* (A. Frigerio and L. Renoz eds.), Elsevier/North-Holland Biomedical Press, Amsterdam (1979), p. 1.

146. T. Tanaka, *Sci. Am.*, *224*:124 (1981).

147. P. G. Righetti, B. C. W. Brost, and R. S. Snyder, *J. Biochem. Biophys. Methods*, *4*:347 (1981).

148. O. Vesterberg, in *Isoelectric Focusing* (J. P. Arbuthnott and J. A. Beeley, eds.), Butterworth & Company (Publishers), London (1976).

149. J. W. Gelseman, C. L. De Lignyand, and N. G. Van Der Veen, *J. Chromatogr.*, *173*:33 (1979).

150. B. Bjellqvist, B. Ek, P. G. Righetti, E. Gianazza, A. Gorg, W. Postel, and R. Westermeier, *J. Biochem. Biophys. Methods*, *6*:317 (1982).

151. H. E. Schaffer and F. M. Johnson, *Anal. Biochem.*, *51*:577 (1973).

152. O. Vesterberg and H. Svensson, *Acta Chem. Scand.*, *20*:820 (1966).

153. G. R. Finlayson and A. Chrambach, *Anal. Biochem.*, 40292 (1971).

154. P. G. Righetti and J. W. Drysdale, *Ann. N.Y. Acad. Sci.*, *209*:163 (1973).

155. O. Vesterberg, *Biochim. Biophys. Acta*, *257*:11 (1972).

156. Pharmacia Fine Chemicals, *Isoelectric Focusing. Principles and Methods.* Pharmacia Fine Chemicals, Uppsala (1982).

157. H. F. Bunn and J. W. Drysdale, *Biochim. Biophys. Acta*, *229*:51 (1971).

158. P. Bassett, F. Braconnier, and J. Rosa, *J. Chromatogr.*, *227*:267 (1982).

159. U. J. Lewis, R. N. P. Singh, L. F. Bonewald, and B. K. Seavey, *J. Biol. Chem.*, *256*:11645 (1981).

160. F. Talamantes, J. Lopez, U. J. Lewis, and C. B. Wilson, *Acta Endocrinol.*, *98*:8 (1981).

161. P. H. O'Farrell, *J. Biol. Chem.*, *250*:4007 (1975).

162. J. E. Celis and R. Bravo, eds., *Two-Dimensional Gel Electrophoresis of Proteins*, Academic Press, Orlando, Fla. (1984).

163. A. J. S. Jones and J. V. O'Connor, *Hormone Drugs, Proceedings of the FDA-USP Workshop on Drug and Reference Standards for Insulins, Somatotropins, and Thyroid-Axis Hormones* (J. L. Gueriguian, E. D. Branscome, and A. S. Outschoorn, workshop organizers), United States Pharmacopeial Convention, Rockville, Md. (1982), p. 335.

163a. W. N. Burnette, *Anal. Biochem.*, *112*:195 (1981).

164. J. Margolis and K. G. Kendrick, *Biochem. Biophys. Res. Commun.*, *27*:68 (1967).

165. W. S. Hancock and D. R. K. Harding, *Hormone Drugs, Proceedings of the FDA-USP Workshop on Drug and Reference Standards for Insulins, Somatotropins, and Thyroid-Axis Hormones* (J. L. Gueriguian, E. D. Branscome, and A. S. Outschoorn, workshop organizers), United States Pharmacopeial Convention, Rockville, Md. (1982), p. 452.

166. W. S. Hancock and J. T. Sparrow, eds., *HPLC Analysis of Biological Compounds: A Laboratory Guide*, Marcel Dekker, New York (1984).

167. W. S. Hancock, ed., *Handbook of HPLC for the Separation of Amino Acids, Peptides and Proteins*, CRC Press, Boca Raton, Fla. (1984).

168. 5th International Symposium on High-Performance Liquid Chromatography of Proteins, Peptides and Polynucleotides, Toronto, Nov. 4–6, 1985, *J. Chromatogr.*, *359*:1 (1986).

169. Pharmacia Fine Chemicals, *Ion Exchange Chromatography Principles and Methods*. Pharmacia Fine Chemicals, Uppsala, (1982).

170. F. E. Regnier, *Anal. Biochem.*, *126*:1 (1982).

171. W. Kopaciewica and F. E. Regnier, *Anal. Biochem.*, *133*:251 (1983).

172. R. M. David and Z. K. Shihabi, *J. Liq. Chromatogr.*, *7*:2874 (1984).

173. B. S. Welinder and S. Linde, in *Handbook of HPLC for the Separation of Amino Acids, Peptides and Proteins*, Vol. 2 (W. S. Hancock, ed.), CRC Press, Boca Raton, Fla. (1984), p. 347.

174. R. L. Patience and L. H. Rees, *J. Chromatogr.*, *352*:241 (1986).

175. L. A. Sluyterman and O. Elgersma, *J. Chromatogr.*, *150*:17 (1978).

176. L. A. Sluyterman and J. Wijdenes, *J. Chromatogr.*, *150*:31 (1978).

177. Pharmacia Fine Chemicals, *FPLCR Ion Exchange and Chromatofocusing. Principles and Methods*, Pharmacia Laboratory Separation Division, Uppsala (1985).

178. L. G. Fägerstam, J. Lizana, U. B. Axiö-Fredriksson, and L. Wahlström, *J. Chromatogr.*, *266*:523 (1983).

179. J. Friedlander, D. G. Fischer, and M. Rubinstein, *Anal. Biochem.*, *137*:115 (1984).

180. S. A. Cohen, K. Benedek, Y. Taphui, J. C. Ford, and B. L. Karger, *Anal. Biochem.*, *144*:275 (1985).

181. I. Halász, in *Handbook of HPLC for the Separation of Amino Acids, Peptides and Proteins*, Vol. 1 (W. S. Hancock, ed.), CRC Press, Boca Raton, Fla. (1984), p. 23.

182. W. S. Hancock and D. R. K. Harding, in *Handbook of HPLC for the Separation of Amino Acids, Peptides and Proteins*, Vol. 2 (W. S. Hancock, ed.), CRC Press, Boca Raton, Fla. (1984), p. 303.

183. J. E. Rivier, *J. Liq. Chromatogr.*, *1*:343 (1978).

184. W. S. Hancock, C. A. Bishop, R. L. Prestige, D. R. K. Harding, and M. T. W. Hearn, *Science, 200*:1168 (1978).

185. C. S. Horváth, W. Melander, and I. Molnar, *J. Chromatogr.*, *125*:129 (1976).

186. R. A. Barford, B. J. Sliwinski, A. C. Breyer, and H. L. Rothbart, *J. Chromatogr.*, *235*:281 (1982).

187. M. T. W. Hearn, *Methods Enzymol.*, *104*:190 (1984).

188. X. M. Lu, K. Benedek, and B. L. Karger, *J. Chromatogr.*, *359*:19 (1986).

189. R. V. Lewis and A. S. Stearn, in *Handbook of HPLC for the Separation of Amino Acids, Peptides and Proteins*, Vol. 2 (W. S. Hancock, ed.), CRC Press, Boca Raton, Fla. (1984), p. 313.

190. A. J. Sadler, R. Micanovic, G. E. Katzenstein, R. V. Lewis, and C. R. Middaugh, *J. Chromatogr.*, *317*:93 (1984).

191. M. P. Strickler, M. J. Gemski, and B. P. Doctor, *J. Liq. Chromatogr.*, *4*:1765 (1981).

192. M. J. O'Hare and E. C. Nice, *J. Chromatogr.*, *171*:209 (1979).

193. E. P. Kroeff and R. E. Chance, *Hormone Drugs, Proceedings of the FDA-UPS Workshop on Drug and Reference Standards for Insulins, Somatotropins, and Thyroid-Axis Hormones* (J. L. Gueriguian, E. D. Branscome, and A. S. Outschoorn, workshop organizers), United States Pharmacopeial Convention, Rockville, Md. (1982), p. 148.

194. D. J. Smith, R. M. Venable, and J. Collins, *J. Chromatogr. Sci.*, *23*:81 (1985).

195. M. Kunitani, P. Hirtzer, D. Johnson, R. Halenback, A. Boosman, and K. Koths, *J. Chromatogr.*, *359*:391 (1986).

196. M. Rubinstein, S. Rubinstein, P. C. Familetti, R. S. Miller, A. A. Waldman, and S. Pestka, *Proc. Natl. Acad. Sci. USA*, *74*:3052 (1977).
197. S. Shaltiel, *Methods Enzymol.*, *104*:69 (1984).
198. W. R. Melander and C. Horváth, *Arch. Biochem. Biophys.*, *183*:200 (1983).
199. W. R. Melander, D. Corradini, and C. Horváth, *J. Chromatogr.*, *317*:67 (1984).
200. D. L. Gooding, M. N. Schmuck, and K. M. Gooding, *J. Chromatogr.*, *296*:107 (1984).
201. Y. Kato, T. Kitamura, and T. Hashimoto, *J. Chromatogr.*, *266*:49 (1983).
202. N. T. Miller, B. Feibush, and B. L. Karker, *J. Chromatogr.*, *316*:519 (1985).
203. L. C. Chaplin, *J. Chromatogr.*, *363*:329 (1986).
204. J. L. Fausnaugh, L. A. Kennedy, and F. E. Regnier, *J. Chromatogr.*, *317*:141 (1984).
205. S. C. Goheen and S. C. Engelhorn, *J. Chromatogr.*, *317*:55 (1984).
206. A. G. Gornall, C. J. Bardawill, and M. M. David, *J. Biol. Chem.*, *177*:751 (1949).
207. O. H. Lowry, N. J. Rosebrough, A. L. Farr, and R. J. Randall, *J. Biol. Chem.*, *193*:265 (1951).
208. M. M. Rising and P. S. Yang, *J. Biol. Chem.*, *99*:755 (1932).
209. M. Robson, D. E. Goll, and M. J. Temple, *Anal. Biochem.*, *24*:339 (1964).
210. M. D. Zishka and J. S. Nishimura, *Anal. Biochem.*, *34*:291 (1970).
211. H. Wu, *J. Biol. Chem.*, *51*:33 (1922).
212. C. G. Vallejo and R. Lagunas, *Anal. Biochem.*, *36*:207 (1967).
213. H. Kuno and H. K. Kiharo, *Nature*, *215*:974 (1967).
214. C. Lo and H. Stelson, *Anal. Biochem.*, *45*:331 (1972).
215. G. L. Peterson, *Anal. Biochem.*, *100*:201 (1979).
216. M. M. Bradford, *Anal. Biochem.*, *72*:248 (1976).
217. M. M. De Moreno, J. F. Smith, and R. V. Smith, *J. Pharm. Sci.*, *75*:9 (1986).
218. C. R. Cantor and P. R. Schimmel, in *Biochemical Chemistry-Part II: Techniques for the Study of Biological Structure and Function* (C. R. Cantor and P. R. Schimmel, eds.), W. H. Freeman and Company, San Francisco (1980).
219. M. Lakowski, Jr., S. J. Leach, and H. R. Sheraga, *J. Am. Chem. Soc.*, *82*:571 (1960).
220. J. W. Donovan, in *Physical Principles and Techniques of Protein Chemistry Part A* (S. J. Leach, ed.), Academic Press, New York (1968).
221. T. A. Bewley, *Anal. Biochem.*, *123*:55 (1982).
222. S. A. Berson and R. S. Yalow, *J. Clin. Invest.*, *38*:1966 (1959).
223. R. S. Yalow and S. A. Berson, *J. Clin. Invest.*, *39*:1157 (1960).
224. E. F. Ullman, M. Schwartzberg, and K. E. Rubenstein, *J. Biol. Chem.*, *251*:4172 (1976).
225. E. Engvall and P. Perlmann, *Immunochemistry*, *8*:871 (1971).
226. B. Velan and M. Halmann, *Immunochemistry*, *15*:331 (1978).
227. R. M. Nakamura, W. R. Dito, and E. S. Tucker, eds., *Immunoassays. Clinical Laboratory Techniques for the 1980s*, Alan R. Liss, New York (1980), p. 127.
228. B. A. L. Hurn, *Br. Med. Bull.*, *30*:26 (1974).

229. B. S. Wilson, A. K. Ng, V. Quaranta, and S. Ferrone, in *Current Trends in Histocompatibility* (R. A. Reisfeld and S. Ferrone, eds.), Plenum Press, New York (1980).

230. V. Herbert, K. S. Lau, C. W. Gottlieb, and S. J. Bleicher, *J. Clin. Endocrinol. Metab.*, *25*:1375 (1965).

231. C. N. Hales and P. J. Randle, *Biochem. J.*, *88*:137 (1967).

232. R. F. Schall and J. J. Tenoso, in *Immunoassays. Clinical Laboratory Techniques for the 1980s* (R. M. Nakamura, W. R. Dito, and E. S. Tucker, eds.), Alan R. Liss, New York (1980), p. 127.

233. B. D. Davis, R. Dulbecco, H. N. Eisen, and H. S. Ginsburg, eds., *Microbiology*, *3rd ed.*, Harper & Row, New York (1980).

234. T. Aikawa, S. Suzuki, M. Murayama, K. Hashiba, T. Kitagawa, and I. Ishikawa, *Endocrinology*, *105*:1 (1979).

235. D. N. Orth, in *Methods of Hormone Radioimmunoassay* (B. M. Jaffe and H. R. Behrman, eds.), Press, New York (1974).

236. I. Zangger, J. Zapf, and E. R. Froesch, *Acta Endocrinol.*, *114*:107 (1987).

237. R. Oslapas, in *Radiobioassays*, Vol II (F. S. Ashkar, ed.), CRC Press, Boca Raton, Fla. (1983).

238. United States Pharmacupical Convention, *United States Pharmacopeia*, *21st ed.*, Mack Publishing Company, Easton, Pa. (1985).

239. R. J. Wills, H. E. Speigel, and K. F. Soike, *J. Interferon Res.*, *4*: 399 (1984).

240. B. Spofford, R. A. Daynes, and G. A. Granger, *J. Immunol.*, *112*: 2111 (1974).

241. W. Marx, M. E. Simpson, and H. M. Evans, *Endocrinology*, *30*:1 (1942).

242. A. E. Wilhelmi, in *Methods in Investigative and Diagnostic Endocrinology* (S. A. Berson and R. S. Yalow, eds.), North-Holland Publishing Company, Amsterdam (1973).

243. C. H. Li, in *Hormonal Proteins and Peptides*, Vol. IV (C. H. Li, ed.), Academic Press, New York (1977).

244. United States Pharmacopeial Convention, "Pancreatin," *United States Pharmacopeia 21st ed.*, Mack Publishing Company, Easton, Pa. (1985).

245. United States Pharmacopeial Convention, "Chymotrypsin," *United States Pharmacopeia*, *21st ed.*, Mack Publishing Company, Easton, Pa. (1985), p. 217.

246. P. Friberger, *Haemostasis*, *7*:138 (1978).

247. P. Friberger, *Scand. J. Clin. Lab. Invest.*, *42*:162 (1982).

248. P. Friberger, *Scand. J. Clin. Lab. Invest.*, *42*:58 (1982).

249. KabiVitrum, Determination of Tissue Plasminogen Activator in Purified Preparations, laboratory instructions accompanying S2288 reagent, KabiVitrum, Stockholm.

250. M. Hoylaerts, D. C. Rijken, H. R. Lijnen, and D. Collen, *J. Biol. Chem.*, *257*:2912 (1982).

251. J. C. Unkeless, A. Tobia, L. Ossowski, J. P. Quigley, D. B. Rifkin, and E. Reich, *J. Exp. Med.*, *137*:85 (1973).

251a. R. H. Carlson, R. L. Garnick, A. J. S. Jones, and A. M. Meunier, *Anal. Biochem.*, *168*:428 (1988).

252. R. Aston, A. M. Whitaker, A. L. Baldwin, and A. T. Holder, *Dev. Biol. Stand.*, *64*:227 (1986).

253. C. Petty and N. L. Cunningham, *Anesthesiology*, *40*:400 (1974).

254. E. W. Kraegen, L. Lazarus, H. Meier, L. Campbell, and Y. O. Chia, *Br. Med. J.*, 3:464 (1975).
255. M. E. Hemling, *Pharm. Res.*, 4:5 (1987).
256. K. Wüthrich, *NMR of Proteins and Nucleic Acids*, John Wiley & Sons, Chichester, West Sussex, England (1986).

7
Enzymatic Barriers to Peptide and Protein Drug Delivery

Vincent H. L. Lee, Robert D. Traver, and Mitchell E. Taub

University of Southern California School of Pharmacy, Los Angeles, California

Potential peptide and protein drugs are subject to degradation by numerous enzymes or enzyme systems throughout the body. This degradation can come in two forms: (1) hydrolytic cleavage of peptide bonds by proteases, such as insulin-degrading enzyme, enkephalinases, angiotensin-converting enzyme, and renin and (2) chemical modification of the protein, such as oxidation by xanthine oxidase or glucose oxidase and phosphorylation by kinases, causing the protein to denature and aggregate or fragment. But proteolysis is by far the more common. Usually there exists a rate-limiting step in the degradation so that there is no accumulation of intermediate degradation products. Proteins are rapidly degraded to amino acids and small peptides following the initial event.

Therefore, a major challenge in peptide and protein drug delivery is to overcome the enzymatic barrier that limits the amount of peptide and protein drugs from reaching their targets. Degradation usually begins at the site of administration and can be extensive. Even when the subcutaneous or intramuscular route is used, less than complete bioavailability is often observed. For instance, the subcutaneous and intramuscular bioavailability of thyrotropin-releasing hormone, a tripeptide, in mice is only 67.5% and 31.1%, respectively [1]. Parathyroid hormone and its analogs, bPTH(1-34) and hPTH(1-34), as well as calcitonin are also incompletely absorbed following subcutaneous administration, since their absorption improved when coadministered with aprotinin, a serine protease inhibitor [2,3]. Although its bioavailability was not quantitated, proinsulin was found to degrade extensively to four metabolites in subcutaneous tissues, none of which was insulin [4]. This is to be contrasted with the fate of proinsulin in secretory vesicles. Here, proinsulin is converted to insulin during the maturation of secretory vesicles by the action of two proteases which are tryptic and carboxypeptidase-like enzymes [5,6]. This process is inhibited by ionophores that disrupt intracellular H^+ gradients maintained by proton pumps found in coated vesicles [7,8], endosomes [9], and lysosomes

[10,11] in the endocytic pathway and in the Golgi apparatus [12,13] and secretory vesicles [14,15] in the exocytic pathway. The different physico-chemical properties of proinsulin and insulin have been suggested to play a role in proinsulin processing within the β cell. Halban et al. [16] observed that whereas proinsulin is rapidly degraded by lysosomal proteases, insulin in its crystal (not soluble) form is degraded at a remarkably slower rate.

I. ROLE OF PROTEOLYSIS IN CELLULAR FUNCTION

It is widely accepted that enzymes act to catalyze chemical reactions by lowering the energy of the transition-state intermediates of biological reactions [17,18]. Proteases are certainly no different, although the exact means by which this is accomplished is still the subject of much research and speculation. Like other enzymes, the activity of proteases are defined by their affinity for the substrate (described in terms of the Michaelis-Menten constant, Km) and the speed at which catalysis can be elicited (described in terms of kcat, or the turnover number of product formed per molecule of enzyme per minute under defined conditions). These values are typically determined by using purified enzyme preparations in vitro, and they can give us a clue as to the preference of enzymes for substrates and the efficiency of catalysis under specific conditions of pH, temperature, and buffers. Both values are important to know because they will be used to determine the ultimate rate of enzymatic reaction under various substrate concentrations. For example, if an enzyme shows a low Km and a small kcat, the rate of hydrolysis must then be very low. But if the enzyme has a high Km and a large kcat, then it is considered a poor substrate at low concentrations and a good substrate at high concentrations. A good way to express specificity of the substrate is in terms of kcat/Km, which takes into account both rate and binding terms.

The kinetics and specificity of proteases in vivo are likely to be quite different from those in vitro [19]. For example, cathepsin activity against substrates is usually measured using catalytic amounts of enzyme (micrograms, <1 µM), while in the lysosome where these enzymes act, the concentrations of the cathepsins are estimated to be in the range of 1 mM (25 mg/ml). Also, when enzymes are isolated from membranes, harsh treatments are often used which can alter the activity of enzymes against specific substrates.

Proteolysis is an irreversible reaction which, when considered with lack of absolute specificity, introduces the potential for severe damage. Consequently, it is important that the activity of this class of enzymes be stringently controlled. Many biological events require the spatial restriction of proteolysis. Examples include the localized activation of plasma coagulation and the release of proteases at a site of inflammation, where cells migrate to the site of action and release proteases at the site. Intracellularly, proteases are denied free access to many potential substrates by compartmentalization, by the action of inhibitors, or by the loss of freedom due to membrane association [19].

There are certain naturally occurring protease inhibitors which are usually large peptides or proteins capable of suppressing the activity of more than one protease [20 – 23]. These inhibitors are exclusively proteins with M.W. of at least 5000, and those described to date are all directed toward endopeptidases. Such inhibitors include the recently identified cystatins [24], which are a family of proteins having a molecular weight of about 11,000 – 13,000 and many function to control inappropriate hydrolysis by cathepsins.

The aspartic proteases, renin, cathepsin D, and pepsin, are all restricted in their action, either by pH limitations/intolerance to their surrounding medium or by extremely restricted substrate specificity. This is assumed to be a protective function. Protease activity can also be restricted by the simple physicochemical inactivation of the enzyme in a foreign environment. The best known example of this is denaturation of pepsin in the neutral-alkaline environment of the duodenum of the GI tract [19].

Many proteases are initially synthesized as inactive precursors called zymogens which are subsequently activated at the functional site. This is the case for the cathepsins, such as cathepsins B, D, E, and F. Unrestricted action of a protease has the potential to deplete the pool of substrate to a great extent in some cases. Thus, the rate of synthesis, secretion, or uptake may control the effect of a protease that has this potential to rapidly deplete a given pool of substrate [19].

Continuous turnover of intracellular proteins and other various macromolecules is a basic cellular process that assists in the regulation of cytoplasmic content and provides amino acids for the cell during stages of nutrient deprivation [25]. The rates of breakdown and regulation are quite variable among the different cell types. For example, high absolute rates of proteolysis and regulatory effects that diminish during times of starvation are seen in the rat hepatocytes, while such responses move in the opposite direction in the skeletal and cardiac myocytes [26 – 29]. Overall, the process of intracellular breakdown of proteins is highly active. More protein is broken down in individual cells than in the lumen of the GI tract [19]. One other site of extensive proteolysis is the proximal tubule of the kidney. Much of the protein that enters the glomerular filtrate is hydrolyzed and recovered as amino acids or small peptides [19].

Proteases are important in the pathogenesis of a great number of diseases, such as insulin-dependent diabetes, cancer, nephrotic syndrome, septicemia, and postsurgical trauma. Abnormal or increased degradation of myelin basic protein has been reported in multiple sclerosis [30], while the failure of normal protein degradation has been implicated as a major contributor to the debilitating effects of Alzheimer's disease [30].

There is considerable evidence to suggest that collagen- and proteoglycan-degrading proteases are involved in cartilage degradation in osteoarthritis [31 – 33]. Pelletier et al. have proposed that interleukin-I (IL-I) plays a role in chronic inflammatory diseases such as rheumatoid arthritis [34 – 36]. But its exact role in these diseases remains to be clarified. IL-I has been shown to enhance protease production in vitro and to stimulate the production of rabbit chondrocyte-derived collagenase and proteoglycanase [37]. This IL-I-mediated effect itself can be enhanced by fibroblast growth factor, which in turn may allow the cells to express additional receptors for IL-I [38]. All the above diseases elicit a general wasting of tissues (cachexia), which involves extensive entracellular protein catabolism. This is not surprising, since this class of enzymes participates so ubiquitously in normal cellular function. In some cases the disease itself can be attributed to the deficiency of a protease alone, for example, Christmas disease [19].

Interleukin-1B is a key hormone of the immune system. It plays an important role in hematopoiesis, inflammation, and wound healing [19]. Mature IL-1B contains C-terminal 153 residues of an inactive 33-kD precursor. Unlike most secretory proteins, this precursor lacks a signal sequence and is not associated with membrane-bound compartments [40]. IL-1B is produced primarily by macrophages [39]. The maturation and subsequent release of this hormone do not proceed by conventional mechanisms. In a pulse-chase experiment, Hazuda et al. [41] have shown that mature IL-1B is generated by a single cleavage of the precursor. This reaction has been shown by Black et al. [42] to be mediated by a single protease of high specificity. Evidence that this protease is required for activation of the prohormone by monocytic cells includes the correlation of its appearance with that of mature IL-1B following stimulation of the cells [41,43]. Since no mature IL-1B is found in the cytoplasm [41,42], processing must have occurred during or following release of the precursor. The maturation pathway is highly unusual due to the absence of any typical processing sites near the N-terminus and the inability of fibroblasts to process the prohormone [43].

II. CHARACTERISTICS OF PROTEASES

Proteases are essentially hydrolases; hence they have the ability to cleave peptide bonds with the addition of water. Proteases rarely show absolute specificity in their action, hence any protease has the potential to hydrolyze more than one substrate [44]. Thrombin, for example, can hydrolyze Arg−Gly bonds in fibrinogen, Arg−Thr in prothrombin, Arg−Ile in Factor VII, and Lys−Ala in actin.

Proteases are globular proteins that are between 20,000 and 60,000 daltons. The minimal requirements for a substrate binding site can be met with a protease of approximately 25,000 daltons. Even when one considers the very large proteases, the binding site or catalytic domain is roughly the same size [19]. There is increasing evidence that the large size of certain proteases is due to gene fusion [45], an event in the evolution of proteases from simple digestive enzymes to regulatory enzymes. For example, human blood coagulation Factor IX and protein C are quite similar, as both contain similar pre−pro leader sequences, activation peptide domains, catalytic domains, growth factor domains, and Ca^{2+} binding domains [46].

There are many examples of mammalian proteases that do not fit into the generalization of small, monomeric, stable enzymes. The endopeptidase calpain [47], which is present in the cytosol of many cells, has a molecular weight of 110,000 and is a heterodimer with subunits of 80,000 and 30,000. These large proteases are not restricted to cellular enzymes, and there are many examples of large, complex, extracellular proteases as well [48]. For example, kallikrein, the protease that converts kininogen to bradykinin, is active as a 90-kD protein.

The stability of proteases also encompasses quite a wide range. There exist a large number of proteases that are quite unstable and are very difficult to purify. Many of the large, cytosolic proteases fall into this category. One feature common to many cytosolic endopeptidases that may relate to their instability is the presence of cysteine residues at the active site as well as in several other areas of the protein. The problem with the cysteine residues is the potential for oxidation or their reactivity with certain metal ions [19].

A. Kinetic Properties of Proteases

Proteases are classified as exo- or endopeptidases, and each is listed under the formal enzyme classification (EC) system. Exopeptidases, which are listed under the subcategories 3.4.11−3.4.19, include amino- and carboxypeptidases and dipeptidases. Exopeptidases act on the N- or C-end termini of a protein, so long as it is not blocked by any substitution. Examples of exopeptidases are aminopeptidase (EC 3.4.11), dipeptidase (EC 3.4.13), diaminopeptidase (EC 3.4.14), dipeptidyl carboxypeptidase (EC 3.4.15), serine carboxypeptidase (EC 3.4.16), metallocarboxypeptidase (EC 3.4.17), and cysteine carboxypeptidase (EC 3.4.18). Endopeptidases cleave the peptide bonds internally in accordance with their mechanism rather than substrate specificity [49]. Examples of endopeptidases are serine proteinase (EC 3.4.21), cysteine proteinase (EC 3.4.22), aspartic proteinase (EC 3.4.23), and metalloproteinase (EC 3.4.24). The properties of endopeptidases will be described below [50].

Serine Proteases

These are the most abundant proteases of those that have been discovered to date. Members in this group include trypsin, chymotrypsin, elastase, coagulation factors that are proteases, plasmin, thrombin, and plasminogen activator. Chymotrypsin is probably the most extensively investigated member of this group. Using the active site directed inhibitor diisopropyl flavophosphate (DFP), Spencer and Sturtevant [52] demonstrated that a serine residue at position 195 at the active site of chymotrypsin was necessary for activity. This serine residue was the only one of the 28 serine residues in chymotrypsin which reacted with DFP. Schoellman and Shaw [53] showed that tosyl-L-phenylalanyl chloromethane (TPCK) reacted with an active histidine residue on chymotrypsin but that it would not react with inactive forms of chymotrypsin (including denatured chymotrypsin and chymotrypsinogen) or histidine alone in solution.

Cysteine Proteases

These proteases are also abundant in nature and are found in plant, bacterial, and animal cells [54]. This group includes papain and chymopapain, ficin, bromelain, actinidin, lysosomal cathepsins, calpains, and high-molecular-weight cytosolic proteases of mammalian cells. Papain is the best characterized of the cysteine proteases, having a single polypeptide chain and a molecular weight of 23,400. Some cysteine proteases possess both endo- and exopeptidase activity. Papain is a single polypeptide chain with a M.W. of 23,400 [54]. It is fully inactivated by thiol-blocking agents such as iodoacetic acid. A histidine residue, His-159, increases the reactivity of the active-site thiol. Acylation of the enzyme

involves attack of the negatively charged sulfhydryl of Cys-25 on the
substrate.

The comparison of the amino acid sequences of five cysteine proteases —
papain, actinidin, rat cathepsins B and H, and chicken cathepsin L — demonstrates a striking homology among their sequences [55]. As shown by
X-ray diffraction studies, papain and actinidin possess a clearly defined
double domain structure. Each domain has a core of nonpolar side chains,
which are retained in cathepsins B, H, and L, with the exception of the
nonpolar residue 203 of the core, which is replaced by glutamic acid in
cathepsin B. The main ordered structures in papain and actinidin are
most likely retained in cathepsins B, H, and L [56]. The differences
occur essentially in the middle region, a place where sequences display the
lowest homologies and which is far removed from the active site.

Aspartic Proteases

Formerly called acid proteases because of their activity at low pH values,
aspartic proteases are less abundant than serine and cysteine proteases
[50]. Aspartic proteases are synthesized only by eukaryotic organisms.
Members in this group include pepsin, chymosin, renin, and cathepsins D
and E. Pepsin, one of the most well-known proteases in this class, is a
single polypeptide chain with a molecular weight of 34,644. The substrate
binding site interacts with about 4 or 5 amino acid residues on different
substrates. Characteristic of all aspartic proteases, this enzyme has a
preference for hydrophobic amino acid residues on both sides of the
scissile bond. For pepsin, Asp-32 and Asp-215 are the two catalytically
active residues and there is evidence that they are hydrogen-bonded.
The details of this mechanism are as yet unknown, but it does not appear
to involve the formation of covalent enzyme-substrate intermediates. General acid-base catalysis is most likely involved in this case [19].

Metalloproteases

This class of proteases occurs widely in nature and is found in prokaryotes
and eukaryotes. Yet relatively few of them have been purified and well
characterized [57]. Members of this class include thermolysin, collagenases, gelatinases, and plasma membrane peptidases. Most of these metalloproteases contain zinc as the metal ion that is essential for proper activity
of the protease. Unlike metal-activated proteases, which bind metal loosely and are easily dissociable, metalloproteases contain metals as an important and integral part of their structure. Most of the metalloproteases
are secreted from cells or are bound to the plasma membrane of cells,
contain disulfide bridges, and are not exposed to the reducing environment of cells [58,59].

Thermolysin is a bacterial metalloprotease with a M.W. of 34,600. Its
three-dimensional structure [60] reveals one zinc atom per enzyme molecule
at the active site that is important for catalysis. Each molecule of thermolysin also contains four calcium ions, which help enhance stability of the
enzyme. This enzyme contains an extended cleft that binds the substrate
polypeptide. Binding at the active site involves a complex between zinc,
histidine, and glutamate residues. The role of zinc is to enhance the
reactivity of water or hydroxyl ions and polarize the peptide bond prior
to nucleophilic attack [19].

Collagenases, another type of metalloprotease, exhibit a very high degree of specificity for native collagen and have very little proteolytic activity against most other substrates [58,59]. Nevertheless, these proteases have been implicated in the degradation of proinsulin to yield two intermediates [61]. One intermediate resulted from cleavage at residues 23 and 24 in the main portion of proinsulin. The other intermediate resulted from cleavage at residues 55 and 56 in the connecting peptide retion. Unlike collagenases, thermolysin and meprin are general peptidases that hydrolyze a broad range of polypeptide substrates. Intrasubunit disulfide bridging is also found in other classes of proteases (i.e., in the serine peptidase chymotrypsin and the cysteine peptidase cathepsin B). At least two metallopeptidases, meprin and enteropeptidase, contain intersubunit disulfide bridges resulting in the formation of covalently linked subunits [58,59].

With respect to the different proteases, many similarities exist between the mechanisms of hydrolytic cleavage of proteases in the same class. The serine proteases, for example, all have the three amino acids aspartic acid, histidine, and serine in close proximity in the catalytic sites. These amino acids react with the peptide or protein to be hydrolyzed when the hydroxyl oxygen in the active-site serine residue attacks the carbonyl carbon of the peptide bond. The amino terminus of the peptide bond is a good leaving group and is cleaved to produce the acyl-enzyme intermediate. The deacylation process then is mechanistically just the opposite of the acylation process, except that it is a water molecule which adds across the carbonyl bond when the hydroxyl group of the water molecule attacks the carbonyl carbon, releasing the hydrolyzed product and returning the enzyme to its original form [17,18]. Based on sophisticated, yet simple, free energy calculations for the transition-state intermediates for subtilisin and trypsin, Warshel et al. [18] concluded that serine proteases act by electrostatically complementing the charge distribution changes occurring in the transition state of the catalyzed reaction. This is accomplished when the catalytic site aspartic acid and histidine residues stabilize the charges of the serine-peptide transition state, which accounts for the overall rate acceleration associated with the enzymes. This recently proposed mechanism is to be preferred to the alternative mechanisms of double-proton transfer and semimacroscopic models widely presented in textbooks.

Cysteine proteases possess a thiol group on cysteine in the active site which acts in the same way as the hydroxyl group of serine proteases, forming a covalent carbonyl-substrate bond. Such a bond is then deacylated as water adds across it, restoring the enzyme and liberating the new hydrolyzed product. Here a histidine residue is also important in stabilizing the active-site thiol. The acylation of the enzyme involves the attack of a negatively charged sulfhydryl group on a cysteine residue in the enzyme on the substrate. The histidine and cysteine residues are both in the enzymatic cleft, although they appear on opposite sides of the thiol group and function to stabilize the transition-state intermediates in a manner apparently similar to the better-studied serine proteinases. Like serine proteases, most cysteine proteases are not demanding of the residues occupying the positions on either side of the scissile bond and they, in fact, allow for the linkage of nonpeptidic moieties in these positions [62−64].

Much less is known about the mechanism of catalysis of aspartic proteases. Formation of a covalent bond between the enzyme and the hydrolyzed substrates does not appear to occur. There is evidence and speculation for hydrogen bonding and/or acid-base catalysis at the active site, which contains two aspartic acid residues and which can accommodate at least seven amino acid residues of the peptide substrate. The enzymes of this class show a preference for hydrophobic amino acids (alanine, valine, methionine, isoleucine, leucine, phenylalanine, tryptophan, or proline) on both sides of the hydrolyzed bond [65-67].

The catalytic reaction of metalloproteases involves the formation of a complex between the active site metal ion (typically zinc) and surrounding histidine and glutamate residues on the enzyme which serves to polarize, and therefore activate, the substrate-peptide bond to nucleophilic attack by water and subsequent hydrolysis [68-70]. The entire family of metalloproteases appears to be related through specificity for Gly-Ile or Ala-Leu bonds. Neither the metallo- nor the aspartic proteases, however, will tolerate nonpeptidic groups on either side of the scissile bond. Consequently, this property can be used to produce protein or peptide analogs with nonpeptidic substitutions around the scissile bond that are resistant to degradation by proteolytic enzymes in these two classes [71].

Among the factors that influence the rate of proteolytic reactions are pH and substrate concentration.

pH is one of the several factors influencing enzymatic activity, and many of the endogenous proteases operate in cellular and extracellular compartments where the pH is optimal for activity. An example is pepsin, which works best in the acidic pH of the stomach. Other examples include trypsin and chymotrypsin, which possess peak activity in the alkaline (pH = 8) environment of the small intestine, and the cathepsins, which are concentrated in the lysosome and hydrolyze polypeptides best in an acidic range (pH = 3-6) [19].

Another important consideration for the action of proteases is concentration of the substrate. For example, several proteases may have the ability to hydrolyze neuropeptides that are present in the plasma and in the external environment of cells, but at the low concentrations that these peptides are present, only specific proteases are able to degrade the neuropeptides, and interactions with receptor proteins might possibly hold precedence over those with proteases. Conversely, it is possible that receptor proteins can increase the concentration of peptides in the membrane, and this may increase the availability of the substrate to the proteases. The presence of inhibitors and other modulators of protease activity (such as calcium ions) may also have an effect on protease activity in vivo. There are a considerable number of endogenous protease inhibitors, such as cystatins, protease nexins, calpastatin, and antithrombin, which may fully inhibit proteolysis at inappropriate times or regions and also may interfere with normal hydrolysis.

B. Tissue Distribution of Proteases

There is considerable heterogeneity in both the types and amount of proteases associated with different types of tissues, cells, and subcellular organelles. For example, many of the proteases present in the plasma, such as thrombin, plasmin, renin, and angiotensin-converting enzyme, are synthesized in and secreted by the liver. By contrast, other cells such as mast cells and leukocytes express specific proteases, such as cathepsin G, elastase, and chymase, which are associated with their special roles in defense against foreign matter.

Most proteases are widely distributed among various organs. A case in point is dipeptidyl aminopeptidase IV (DAP IV) [71–75]. This plasma-membrane-bound protease is most prominent in the pancreas, liver, lung, heart, and thymus. In the pancreas, DAP IV is found principally in the plasma membrane of the intercalated duct cells. In the liver, it is found in the plasma membrane at the biliary pole of hepatocytes. In the lung, it is associated with the plasma membrane of the endothelium of all segments of the capillary bed. In the heart, it is most prominent in the surface membrane of the endothelium of special segments of the capillary bed and in that of endocardial endothelial cells. In the thymus, DAP IV is present in the cortical and medullary thrombocytes. The brush border of the duodenum, jejunum, and ileum enterocytes also contains DAP IV.

Compared with DAP IV, the protease angiotensin-converting enzyme (ACE) has a more restricted distribution. ACE has, in fact, been used as a convenient marker for endothelial cells. While the presence of ACE does not guarantee that a cell is endothelial, its absence in a putative endothelial cell is a definite cause for concern. To date, ACE has been located in the aorta [76], umbilical vein [77], pulmonary artery and vein [78], the brain capillaries [79], adrenal capillaries [80], alveolar macrophages, enterocytes, and proximal tubules of the kidney [81].

Angiotensin-converting enzyme in the lung is believed to play an important physiological role in inactivating bradykinin [82], a nonapeptide with hypotensive properties. Under normal circumstances, almost all of the bradykinin that enters the lung via the pulmonary artery is cleaved and inactivated during passage through the lung [83]. Cleavage occurs at the Pro^7-Phe^8 bond of bradykinin (Arg–Pro–Pro–Gly–Phe–Ser–Pro–Phe–Arg). In addition to ACE, post-proline-cleaving enzyme (PPCE) can also inactivate bradykinin [84]. The hydrolytic products are $Arg^1-Pro^2-Pro^3$, $Gly^4-Phe^5-Ser^6-Pro^7$, and Phe^8-Arg^9, as a result of cleavage at the Pro^3-Gly^4 and Pro^7-Phe^8 bonds of bradykinin [85,86]. PPCE is an endopeptidase with a molecular weight approaching 70,000 daltons [87] that cleaves peptide bonds where proline is the carboxy residue constituent of the bond, and which is other than an N-terminal or penultimate N-terminal residue. PPCE does not cleave the bond between two proline residues, however [83]. This protease has been found in the rat brain [88], lamb brain [89], rabbit brain [90], bovine kidney [91], human placenta [92], and a Flavobacterium [93].

The role of proteases in inactivating neuropeptides in neural tissues is well known. For instance, aminopeptidase and dipeptidyl carboxypeptidase have been shown to play a very important role in the cerebral metabolism of endogeneously and exogeneously administered enkephalin [94–96]. Cerebral microvessel isolation, immunoelectrophoresis, and enzyme assays have clearly shown the binding of aminopeptidase M to cerebral microvessels [97]. In addition to modulating local cerebral blood flow, cerebral microvascular aminopeptidase M may also play an important role in preventing circulating enkephalins from crossing the blood-brain barrier [98,99]. Despite the possibility of contamination by other aminopeptidases, the characteristics of the protease involved in microvascular enkephalin hydrolysis are comparable in catabolic activity to those reported for vascular aminopeptidase M, renal aminopeptidase M, and cerebral membrane aminopeptidase [100–102].

There is increasing evidence that proteases may also play a role in the processing, deposition, and maintenance of memory [103]. Siman et al. [104,105] reported that synaptic membrane preparations of hippocampal tissue responded to the administration of exogenous calcium by stimulating calpain in addition to increasing the number of glutamate receptors, which are considered to be essential in the maintenance of neuronal interactions in memory deposition [104]. The activated calpain subsequently degraded fodrin, a cytoskeletal protein that is sensitive to proteolytic attack [103].

The kidneys play a major role in the degradation of insulin and glucagon, contributing nearly one-third of the metabolic clearance of these hormones [106,107]. Metabolism occurs primarily subsequent to glomerular filtration, followed by intraluminal or intracellular degradation. In the rabbit, maximum degradation occurs in the proximal nephron and is similar in proximal convoluted and straight tubules. The degradation of both hormones by the distal segments of the nephron averages one-quarter and one-third of that measured in the proximal convoluted tubule and pars recta, respectively [108]. Unlike insulin, however, most of the glucagon-degrading activity in the kidney homogenates is due to a brush-border neutral peptidase [109].

In addition to glucagon and insulin, the kidney also degrades insulin-like growth factors [110,111]. These are growth hormone-dependent polypeptides having insulin-like activity [112,113]. Specific binding of IGF to cell surface receptors has been reported in several rat tissues in addition to the kidney, including placenta, liver, adipocytes, and lymphoid tissues [114,115]. The receptors have been characterized by Cheranausek et al. [116] and Massague and Czech [117] as type I and II IGF receptors, with IGF-I weakly binding insulin and IGF-II not binding insulin at all.

Several distinct differences exist between IGF-I and IGF-II. First, the binding Km values for IGF-I and -II are 625 nM and 500 nM, respectively. These values are approximately 900 and 400 times higher than the concentration of IGF-I (0.7 nM) and IGF-II (1.3 nM) needed to reduce the binding of labeled IGF-I and IGF-II by 50%, respectively. Second, the relative specificities of IGF binding and degrading systems are different. Third, insulin that recognizes IGF-I receptor does not affect the inhibition of labeled IGF-I degradation, but insulin that does not recognize IGF-II receptor does inhibit the degradation of labeled IGF-II at very high concentrations. These observations have led Bhaumick and Bala to conclude that the IGF binding and degrading activities of the kidney membrane are two independent systems with specificity for either IGF-I or -II. These investigators have also established that the characterized IGF-degrading system enzymes are different from the previously described insulin protease [118].

Gossrau et al. [119] have shown that in rats, acute renal failure (ARF) caused metabolic alterations in various organs, such as the liver, skeletal muscle, and lung, and increased protease activities in plasma and bronchial lavage fluid, which could be detected histochemically. The organs that showed a clear-cut response were the extraorbital gland, thymus, and lung. The other organs did not show as definite a response. In the affected organs, activities of plasma membrane-associated and lysosomal proteases were either increased or decreased.

The absorptive mucosa for which the enzymatic barrier has been extensively studied is the intestinal mucosa. Much of the information in this regard was obtained in the 1960s and 1970s using peptides such as Gly – Gly, Gly – Leu, Gly – Pro, Gly – Sar, Ala – Phe, and β – Ala – L – His [120 – 122]. The focus then was on its role in nutrient absorption. Even to this date, there is scantly information on the gastrointestinal hydrolysis of biologically active peptides.

Compared with the oral route, much less is known about the nature of the enzymatic barrier to peptide and protein absorption in the nasal, buccal, rectal, and vaginal routes. As a first step in characterizing the proteolytic barrier, the proteolytic activity in various mucosal tissues can be determined by incubating a peptide or protein in mucosal tissue homogenates. Lee and his co-workers [123,124] demonstrated that the proteolytic activities in homogenates of the mucosal tissues of the albino rabbit against methionine enkephalin, [D – Ala^2] methionine enkephalinamide, substance P, insulin, and proinsulin were, surprisingly, comparable to those in the ileum. The proteolytic activities against the small peptides YGGFM, YAGFM, and substance P do not differ as much among the mucosal routes of delivery as do the activities against the proteins insulin and proinsulin. Moreover, the rank order of proteolytic activities against the peptides is different from that against the proteins. For small peptides, the rectal route is the most active, followed by the buccal, nasal, and vaginal. By contrast, for proteins, the nasal route is far more active than the rectal, vaginal, and buccal routes, in that order. It remains to be seen whether these trends hold for other peptides and proteins.

The proteases constituting the enzymatic barrier at each mucosal route have not been isolated and characterized. Based on the pattern of cleavage of enkephalins, substance P, insulin, and proinsulin, both exo- and endopeptidases are present in each mucosal tissue. These include aminopeptidases, diaminopeptidases, post-prolyl-cleaving enzyme, angiotensin-converting enzyme, endopeptidase 24.11 (a metalloproteinase), and thiol proteinases [123 – 125]. What distinguishes one mucosa from another is probably the relative proportion of these proteases as well as their subcellular distribution. The former has yet to be determined, while the latter has been characterized only for the aminopeptidases. Subcellular distribution of proteases will be amplified upon in the next section.

The subcellular distribution of aminopeptidase activity based on methionine and leucine enkephalins as substrates reveals that aminopeptidases are distributed throughout the cell to degrade peptides and proteins both during and after absorption into the cell. The subcellular aminopeptidase distribution was determined by subjecting a low-speed (1000× g) supernatant of a mucosal tissue homogenate to high-speed contrifugation (250,000× g) to obtain membrane and cytosolic fractions, followed by incubating each fraction with aminopeptidase substrates over a period of time. Of all the mucosae, the nasal is the most similar to the ileal in subcellular distribution of aminopeptidases. Specifically, almost half the aminopeptidase activity in the nasal mucosa of the albino rabbit is membrane-bound, in comparison to the 80% seen in the ileal mucosa. By contrast, the aminopeptidase activity is principally cytosolic in the buccal, rectal, and vaginal mucosae. This amounts to 85 – 88%, 88 – 90%, and 79 – 80%, respectively.

C. Subcellular Distribution of Proteases

Different proteases are associated with various subcellular structures. The diversity, amount, and compartmentalization of proteases provide yet another way to facilitate the role of this class of enzymes in the mediation of important biological processes. The lysosome contains a very high concentration of cathepsins to complete the degradation of endocytosed proteins (heterophagy) and intracellular proteins (autophagy) [19]. The endoplasmic reticulum contains special proteases that are used in the processing of nascent proteins. These proteases are not as well characterized as the lysosomal proteases owing to their low concentrations in the cell [126,127].

The mitochondria contain proteases that participate in the processing and degradation of some mitochondrial proteins. The precursors of most mitochondrial proteins imported into the mitochondria, such as ornithine aminotransferase (OAT) [128], have been shown to be converted to their mature forms by cleavage of their extrapeptides in the mitochondrial matrix [129]. In the case of OAT, this degradation process does not require any energy [130].

Proteases are present on cell surfaces and can hydrolyze peptides and proteins in the extracellular environment, as may be the case in the hydrolysis of exogenous leucine enkephalin in the nasal cavity of the rat [131]. Examples of such proteases are meprin and aminopeptidase M, both of which are metalloproteases. These are plasma membrane proteins with their active domains expressed at the extracellular surface [132]. Meprin appears to be a kidney-specific protease. It is a glycoprotein with large subunits (M.W. 90,000) which tend to aggregate to a higher-molecular-weight complex upon purification. Its dimer, tetramer, and possibly octamer are active under certain conditions [133,134].

Cell surface proteases may play a key role in the growth and differentiation of many cellular systems by modulating the activity of peptide factors and regulating their access to adjacent cells [135–138]. In the central nervous system, endopeptidase-24.11 is located in some axonal and synaptic membranes to inactivate substance P and enkephalins [139]. Renal and brush-border membranes are quite abundant sources of membrane peptidases, but these particular enzymes are also found in small amounts on the surfaces of several other cell types [135]. Their functions are likely to vary with their cellular locations. In the intestine, such proteases are associated with the final steps of digestion, but they may be quite powerful elsewhere in the inactivation of peptide signals [140].

There is much evidence that lysosomes are the major degradation site of extracellular proteins which are brought into the cell via endocytosis [141–143]. Aside from cathepsins B, H, and L [144–147], little is known about proteases and cellular locations that are involved in the degradation of intracellular proteins [58,148]. While there is little doubt that lysosomes play a role in the breakdown of intracellular proteins in certain situations (i.e., nutritional deprivation and pathological states), further study is necessary in order to further clarify these mechanisms [149–152].

The lysosomes are by far the best-characterized intracellular proteolytic system that plays a very important role in basal proteolysis [153,154]. Nutritional deprivation accelerates protein degradation occurring in lysosomes [155], and this is mediated by bulk uptake of portions of cytoplasm into autophagic vacuoles [156]. There are two routes by which cytoplasm may enter the autophagic vacuolar apparatus. The first is macroautophagy, which involves the segregation of distinct portions of cytoplasm, including entire organelles, within membrane-bound structures known as autophagic vacuoles [157]. Such vacuoles are believed to acquire hydrolytic enzymes mostly by fusion with dense bodies to subsequently initiate intralysosomal digestion. The second pathway is microautophagy [158–161], which involves invagination of the lysosomal membrane to form intralysosomal vesicles. The membranes of these vesicles are then digested, thus exposing the internalized material to the hydrolytic enzymes [162]. Microautophagy differs from macroautophagy in that the cytoplasmic "bite" is smaller and sequestration is not acutely regulated. It is, however, adaptively decreased during starvation in parallel with absolute rates of basal turnover.

Factors affecting macroautophagy include the stringent omission of amino acids [163], ratio between sequestered protein and quantity of enzyme acquired by fusion [164], and sensitivity to osmotic shock [165]. In the hepatocyte, the amino acid control of macroautophagy is mediated by a small group of direct inhibitors (leucine, tyrosine, phenylalanine, glutamine, proline, methionine, tryptophan, and histidine) and by the permissive effect of alanine. By contrast, in myocytes and adipocytes only leucine is involved. In hepatocytes, macroautophagy is stimulated by glucagon, cyclic AMP, and beta agonists. In skeletal and cardiac myocytes [25], however, the same factors exert an inhibitory effect.

The lysosome is by no means the only site capable of extensive degradation of cellular proteins. It is now clear that significant proteolysis occurs in the cytosolic compartment of cells and that this is important in the activation, processing, and initiation of degradation of proteins [166]. Microinjection studies have shown that some proteins can be fully degraded in the cell cytosol. In fact, in cells such as mammalian red blood cells and bacteria that have no lysosomes, all proteolysis must be accomplished extralysosomally [167]. There are several exo- and endopeptidases in the cytosol [168,169], such as ATP-dependent proteases, other high-molecular-weight proteases, and calpains [170,171]. Most are cysteine proteases and are rather large, ranging from M.W. 80,000 to 750,000 [172]. A ubiquitin-dependent, ATP-dependent proteolytic system has been described in reticulocytes [173], whereas an alkaline ATP-dependent protease that does not require ubiquitin has been elucidated in soluble liver extracts [174]. In the latter case, ATP is not hydrolyzed during activation of the protease. Several other nucleotides (ADP, GTP, GDP, and CTP) can activate proteolysis, and these compounds could be acting as allosteric activators.

Beinfeld et al. [175] have determined that the rat brain cytosol contains a neutral metalloprotease of about M.W. 80,000 that is capable of cleaving substrates containing the site at which mammalian prosomatostatin is cleaved to generate somatostatin. This represents a cleavage on the carboxyl side of a single arginine residue at an Arg–Ser bond. This enzyme is, however, unable to cleave several other substrates containing single arginine residues or two substrates containing an Arg–Lys or Lys–Arg pair.

Thus, the initial events in the breakdown of proteins may occur outside the lysosome, but the final stages are inherently intralysosomal. The route taken by a given protein may vary according to cell type,

nutritional status, hormonal status, and environmental conditions. Within
one cell type, alternate routes and rates of degradation may exist, but
there is probably one single underlying pathway for final degradation [19].

III. STRUCTURE–PROTEOLYSIS RELATIONSHIPS

Different properties of peptides and proteins have been studied in an attempt to relate their physical characteristics to mechanism and rate of
hydrolytic cleavage. In both bacterial and eukaryotic cells, relatively
long-lived proteins (half-lives circa cell generation time) coexist with proteins whose half-lives can be less than 1% of the cell generation time
[176]. Relative to the life of the cell, the half-lives of proteins in mammalian cells are typically quite short. For example, mitochondrial proteins
have half-lives of 1 hr to 8 days, and the average protein half-life in a
liver cell is approximately 3 days. Although the specific functions of
selective protein degradation are still unknown in most instances, it is
clear that many regulatory proteins, such as RNA polymerase, insulin,
growth hormone, and corticotropin, are extremely short-lived in vivo
[176,177]. This would provide a mechanism for enhanced control by these
proteins over their action. Wide fluctuations in their concentration could
then be achieved by increases or decreases in protein synthesis. This
has been found to be the case in general for regulatory enzymes [178–
181]. How exactly these enzymes are recognized as being "regulatory" in
nature, however, is not fully understood.

    Some correlation seems to exist between higher turnover rate and
peptides and proteins of larger size (glutamate dehydrogenase, M.W. =
1,000,000; gamma globulin, M.W. = 149,900), acidic pI values (pepsin,
pI $<$ 1.0; serum albumin, pI = 4.9; urease, pI = 5.0) or greater (hydrophobicity as measured by affinity for (i.e., longer elution time on) an
octyl-Sepharose column (lactate dehydrogenase, $t_{1/2}$ = 84 hr, $<$ glyceralde-
hyde-3-phosphate dehydrogenase, $t_{1/2}$ = 75 hr, $<$ glucose-6-phosphate
dehydrogenase, $t_{1/2}$ = 15 hr) [182–184]. The interdependence of all
these characteristics, however, does not allow for the singling out of any
one characteristic as being of primary importance, nor does it provide an
adequate explanation for the recognition of regulatory proteins, which
come in many sizes, hydrophobicities, and charges.

    What all these characteristics might better point out is that thermo-
dynamic stability (or in the case of hydrophobicity, acidic pI, and size —
instability) may allude to the in vivo stability of a protein. This thermodynamic instability, then, suggests that the rate of protein degradation
correlates with the amount of protein in the unfolded, denatured state,
making it more vulnerable to attack by proteolytic enzymes [185].

    Bond [186] has shown that proteins having short half-lives in vivo
are significantly more sensitive to proteolytic attack in vitro. This is a
property generally associated with large rather than small proteins. It is
also often stated that "abnormal" or "aberrant" proteins are degraded
rapidly by cells [187,188]. There is some evidence for this hypothesis,
which again is indicative that protein structure is an integral factor in
the determination of degradation rate. Factors such as ligand binding,

covalent modification, and other reversible events could play an important
role in the rate of entry of a protein into the catabolic pathway [189].
For example, it could be possible that substrates or allosteric activators
could stabilize a protein in vivo. A high level of either ligand could be
indicative of, and signal the need for, an increase in the intracellular
concentration of active sites. On the other hand, product accumulation or
allosteric inhibition could be viewed as a signal for accelerated proteolysis
of the protein. Enzymes that catalyze the covalent modification of a protein may be essential in the initial stage of degradation, by "destabilizing,"
"marking," or "branding" a protein for degradation [173]. It is possible
that this is the step in which protein half-life is determined, and thus
eventually leads to extensive degradation [189].

Rogers et al. observed that proteins that are rapidly degraded in
eukaryotic cells contain regions rich in proline, glutamic acid, serine, and
threonine (the PEST hypothesis) [190]. This observation was confirmed
by computer analysis of the sequences of 10 rapidly degraded eukaryotic
proteins of known sequence, such as c−myc ($t_{1/2}$ = 0.5 hr), ornithine
decarboxylase ($t_{1/2}$ = 0.5 hr), hydroxymethyl glutaryl−CoA reductase
($t_{1/2}$ = 1.5−3 hr), and alpha- and beta-casein ($t_{1/2}$ = 2−5 hr). These
PEST regions are often flanked by clusters of several positively charged
amino acids such as lysine, arginine, and histidine. Moreover, many are
characterized by clustering of negatively charged residues such as glutamic
acid and aspartic acid and of residues such as serine, threonine, and
tyrosine, which can, through phosphorylation, become negatively charged.
This is exemplified by the prominent phosphorylation of alpha- and beta-
casein at their serine residues. Of the proteins studied, only c−myc and
EIA are thought to share common ancestry. No correlations seem to exist
that would suggest a common compartmentation or functional activity. To
further test this hypothesis, radiolabeled caseins were introduced into
cultured HeLa cells by means of red blood cells-mediated microinjection and
their half-lives monitored. These were found to be less than 5 hr.

The occurrence of the sequences has also been analyzed to estimate
the probability that these regions arise randomly. The occurrence of
PEST regions (13−14 regions per 50−100 amino acids) far in excess of the
amount expected to arise randomly (one in every 6570−18,000 residues)
suggests that the probability that these regions are the result of random
distribution is strikingly remote. How these regions might confer rapid
degradation of the proteins that contain them is not known. Several
mechanisms have been suggested, however. As polyproline and poly-
glutamic acid sequences form extended left-hand helices, enrichment with
these two amino acids could present conformations or specific regions that
are readily recognized by proteolytic enzymes. A more complicated pathway suggests that phosphorylation of PEST regions generates calcium ion
binding sites which could then elicit degradation by calcium-activated
proteases such as calpains. As some stable proteins such as aspartic
aminotransferase, dihydrofolate reductase, and pyruvate kinase were found
to contain PEST regions, other factors such as intracellular location and
conformational masking of these regions must also influence the control of
proteolytic degradation by these regions. Thus far, experiments to lengthen
the intracellular half-life of proteins by removing PEST sequences have not
been conducted.

IV. EFFECT OF CHEMICAL CHANGES ON PROTEOLYTIC STABILITY OF PROTEINS

Not surprisingly, several chemical changes in the substrate protein have also been identified as having an effect on the rate and site of hydrolysis catalyzed by proteases. Lewis et al. found that human growth hormone (hGH) can undergo chemical deamidation prior to hydrolytic cleavage by the serine protease subtilisin and that this deamidation can affect the site of subsequent hydrolysis on the molecule [191]. The desamido form can be produced by incubation in alkaline medium. Deamidation of hGH may occur at two positions on the peptide (Asn-152 and Gln-137). hGH undergoes hydrolysis by subtilisin at positions 139 and 149 or positions 139 and 146. Deamidation of hGH at Asp-152 results in hGH that is hydrolyzed only at positions 139 and 146. Deamidation of the peptide at Gln-137 results in a form of hGH that undergoes hydrolysis by subtilisin at positions 95–127 in addition to cleavage at positions 139 and 149. Deamidation at this site seems to produce greater resistance to hydrolysis by subtilisin in the resulting peptide, as indicated by decreased peptide production seen on a tryptic peptide map, which was not assessed quantitatively, however, for the degree of decrease in peptide production. This finding demonstrates that deamidation can alter the location of hydrolytic cleavage of human growth hormone.

Glycation is a common postribosomal modification of body proteins. Glycation of proteins involves the covalent attachment of aldose or ketose to free amino acid groups of proteins. Glucose reacts with protein to initially form a Schiff's base, which can then undergo Amadori rearrangement to form the corresponding ketamine. Since the Amadori rearrangement is essentially irreversible, any protein modified in this manner stays in the body for its complete lifespan and is eventually degraded [192].

Chemical changes such as carbamylation, denaturation, phosphorylation, oxidation, and ubiquitization are thought to act as markers on the peptides or proteins which make the enzyme more susceptible to proteolytic attack [193].

Carbamylation of glutamate dehydrogenase renders this enzyme at least twice as susceptible to hydrolytic degradation in 1-ml incubation mixtures containing a phosphate buffer and 100 μg chymotrypsin, pronase, or trypsin. After incubation for 0.5 hr at 37°C, chymotrypsin had hydrolyzed 2.5 times as much carbamylated glutamate dehydrogenase as native protein, trypsin had hydrolyzed two times as much, and pronase had hydrolyzed 2.3 times as much of the carbamylated enzyme. It is possible that carbamylation may lead to decreased stability of other proteins to proteolysis by a number of proteolytic enzymes [194].

Glutathione disulfide, cysteine, and cysteamine inactivate, destabilize, and enhance the proteolytic susceptibility of rabbit-muscle fructose-1,6-biphosphate aldolase to chymotrypsin in vitro. Increasing glutathione disulfide concentration from 2 to 12 mM in an incubation mixture containing Tris buffer and aldolase at 37°C reduced the half-life of aldolase from 91 to 19 min. Addition of ammonium sulfate to this mixture, however, abolished the effect of the added glutathione disulfide. Similarly, addition of cysteamine to the incubation mixture at a concentration of 1–5 mM reduced the half-life from 30 to 11 min. These results suggest that inactivation is caused by the formation of intramolecular disulfide bonds and enzyme-glutathione mixed disulfide linkages, thereby thermodynamically destabilizing the enzyme, as indicated by a lowered transition temperature and enthalpy of denaturation. Thus, covalent modification of aldolase by biological disulfides plays an important role in modulating enzyme stability and vulnerability to proteinases, hence enzyme activity [195].

Phosphorylation by AMP-dependent protein kinase increases proteolytic degradation of bovine cardiac troponin by calcium-activated neutral protease in vitro. The rate of troponin degradation is dependent on the degree of phosphorylation, unphosphorylated enzyme being degraded at only one-third the rate of the completely phosphorylated (1.1 moles $P/10^5$ g protein) enzyme and only one-half the rate of the partially phosphorylated (0.6 mole $P/10^5$ protein) enzyme [196].

Numerous studies have shown increased hydrolysis following oxidative denaturation of a wide variety of peptides and proteins, including transferrin, human serum albumin, superoxide dismutase, peroxidase, trypsin, and chymotrypsin in vitro [178,197–199]. Glutamine synthetase, which has been studied most extensively as a model of oxidative denaturation, can be oxidized in vitro by the enzymatic systems cytochrome P-450 and P-450 reductase, NADH oxidase, xanthine oxidase, peroxidase, or glucose oxidase and by the nonenzymatic systems ascorbate, molecular oxygen, and Fe(III), or molecular oxygen and Fe(II) [200]. All these systems are capable of generating superoxide anions or hydroxyl radicals which then oxidize nearby amino acids of the enzyme [201]. Oxidative modification of glutamine synthetase takes place on a single histidine residue per subunit and possibly other amino acids, thus rendering the protein catalytically inactive and susceptible to degradation by endogenous proteolytic enzymes calpain I, cathepsin D, and alkaline protease, which exhibit little activity on the nonoxidized, native form of the protein [200–202].

There is recent biochemical and genetic evidence that, in eukaryotes, covalent conjugation of ubiquitin to short-lived intracellular proteins is essential for their selective degradation [176]. Ubiquitin is a 76-residue protein abundant in all eukaryotic cells with an amino acid sequence identical in both humans and insects, making it one of the most highly conserved sequences known. Ubiquitin occurs in cells either unbound or covalently linked to proteins on the epsilon-amino groups of lysine residues [203]. Wilkinson et al. showed that a specific factor required for ATP-dependent nonlysosomal proteolysis in reticulocyte extracts isolated earlier by Ciechanover was indeed ubiquitin [62]. Hershko et al. elucidated the steps in the pathway of ubiquitin conjugation to proteins. These steps include the adenylation of the carboxy terminus of ubiquitin proteins followed by linkage via a thiolester bond to ubiquitin-activating enzyme, which is then transesterified, forming a complex that links ubiquitin (or a multiplicity of ubiquitin moieties) to the acceptor protein via an isopeptide bond [63]. Ciechaonver et al. subsequently showed that the cell cycle mutant ts85 cells, unlike the parental FM3A cells, failed to degrade otherwise short-lived intracellular proteins such as L-threo-alpha-amino-beta-chlorobutyric acid (a valine analog) and thialysine (a lysine analog). The primary lesion in these cells was loss of their ubiquitin-protein ligase system [64,203]. These results demonstrate that the degradation of most short-lived intracellular proteins in the eukaryotic cell proceeds through a ubiquitin-dependent pathway. Indeed, Hershko and Ciechanover have proposed that ubiquitin binding acts as a signal for attack by proteinases specific for ubiquitin-protein conjugates [204]. On the other hand, as proteolytic pathways have been found that are not energy dependent, such as PEST sequences, deamidation, or carbamylation, the ubiquitin-mediated pathway must not be responsible for the degradation of all intracellular proteins.

An intriguing correlation between the nature of amino-terminus amino acid residues and protein half-life has been shown for the beta-galacto-sidase protein [205]. When a gene encoding a ubiquitin-beta-galactosidase fusion protein is expressed in yeast, ubiquitin is first cleaved off, ex-posing the N-terminus amino acid. By altering the genetic sequence to produce beta-galactosidase proteins, which are otherwise identical except for their amino terminus residues, the in vivo half-lives of these proteins can be dramatically altered. The amino acid residues can then be ordered to correlate with the half-life of the beta-galactosidase enzyme to which they are attached and used to predict the half-life (from several minutes to 20 hr) of the enzyme following cleavage of the ubiquitin protein (the N-end rule). By measuring the enzymatic activity of beta-galactosidase in crude galactose-free extracts containing 10% glucose, the half-lives of beta-galactosidase analogs with N-terminal amino acids such as methionine, serine, and glycine were shown to be greater than 20 hr. By this meth-od and in vivo pulse-chase techniques, analogs containing tyrosine and glutamic acid at the amino terminus displayed half-lives of only 10 min and analogs with N-terminal arginine residues had half-lives of only 2 min. In this way, the turnover rate of a protein can be encoded for genetically by different amino acid residues at the N-terminus of the protein. Additional-ly, peptides or proteins of interest might be manipulated to exhibit a slower or more rapid turnover rate in vivo by choosing and attaching dif-ferent amino acids to the amino terminus.

All the above chemical modifications tend to increase the rate of proteolysis, and this may at least partly account for the regulation of the protein turnover rate. Each chemical change represents a modification that alters the allosteric conformation and even completely denatures the target peptide or protein. In this way, all the above changes can be viewed as simple chemical modifications which act to expose and thereby increase hydrolysis of susceptible peptide bonds by peptidases and pro-teinases. It is therefore likely that conditions used in the fabrication of peptide and protein drug delivery systems which may cause conformational changes or denaturation similar to the modifications mentioned above may potentially "mark" a peptide or protein drug for rapid hydrolytic degrada-tion.

V. NATURE OF ENZYMATIC BARRIER: DEGRADATION OF SELECTED PEPTIDE AND PROTEIN DRUGS

The enzymatic barrier has three essential features, as follows.

First, because proteases and other proteolytic enzymes are ubiquitous, peptides and proteins are usually susceptible to degradation in multiple sites, including the site of administration, blood, liver, kidney, and vascular endothelia. Consequently, peptides and proteins must be pro-tected against degradation in more than one anatomical site for them to reach their target sites intact.

Second, almost all the proteases capable of degrading a given peptide or protein are likely to be present in a given anatomical site where the peptide or protein is located [206-209]. For instance, diaminopeptidase

IV, angiotensin-converting enzyme, and several other proteases that participate in angiotensin II (DRVYIHPFHL) degradation are all present in the vascular endothelium [207,208]. The implication is that protecting a peptide or protein from degradation by one protease may not necessarily lead to marked increases in its stability or in the amount of peptide or protein reaching its site of action [125,210].

Third, a given peptide or protein is usually susceptible to degradation at more than one linkage within the backbone; each locus of hydrolysis is mediated by a certain protease. Often, even when one linkage is modified to circumvent one protease, the rest of the peptide molecule is still vulnerable to the other proteases. This usually manifests itself as a shift in the relative proportion of the various degradation products of a given peptide. For example, when the penultimate N-terminus glycine residue of methionine enkephalin (YGGFM) is substituted by D-alanine to render this peptide more resistant to aminopeptidases, the role played by enkephalinase in degrading the resultant analog, Tyr−D−Ala−Gly−Phe−Met−NH$_2$ (YAGFM), assumes greater quantitative importance [211]. This enzyme was found to be responsible for about 88% of the hydrolysis of YAGFM, as compared with a 4% contribution to the hydrolysis of YGGFM, in the nasal homogenates of the albino rabbit [125].

To illustrate the above three features of the enzymatic barrier, the degradation of cholecystokinin (CCK), insulin, luteinizing hormone-releasing hormone (LHRH), substance P, and thyrotropin-releasing hormone will be described.

A. Cholecystokinin

Cholecystokinin (CCK) was discovered as a gastrointestinal peptide in the duodenal mucosa. It was originally isolated in two molecular forms, one with 33 amino acid residues and the other with 39. In both forms the carboxyl-terminals are amidated [212]. The principal physiological functions of CCK are stimulation of the secretion of pancreatic enzymes into the duodenum and contraction of the gallbladder [213]. CCK has also been found to inhibit gastric emptying and to augment stimulation of bicarbonate secretion from the pancreas by secretin [214]. Like a number of other gut peptides, CCK also occurs in the brain, where it has been implicated as a neurotransmitter or neuromodulator [215]. CCK has now been shown to exist in multiple molecular forms in both the gut and the brain. The major form in neurons is the C-terminal octapeptide, Asp−Tyr(SO$_3$H)−Met−Gly−Trp−Met−Asp−Phe−NH$_2$. The intestinal forms, on the other hand, are more heterogeneous in size. In fact, the C-terminal tetrapeptides, CCK-4 as well as sulfated and unsulfated CCK-12 and CCK-8, have been found in the rat gut mucosa and gut muscle. It appears that the small C-terminal peptides are the circulating forms and that CCK-33 and larger forms are the biosynthetic precursors [216].

The actions of CCK in the gut on pancreatic secretion and gut motility and in the brain on satiety are mediated by at least two CCK receptor subtypes, termed CCKa and CCKb. CCKa receptors are found primarily in tissues such as pancreas, gallbladder and colon, although isolated regions have been found in the central nervous system (CNS). The minimum fully potent endogenous ligand for CCKa receptor for the carboxy-terminal octapeptide CCK-8 [Asp−Tyr(SO$_3$H)−Met−Gly−Trp−Met−Asp−Phe−NH$_2$]. The primary CCK receptor subtype in the CNS is

CCKb, and the minimum ligand for this receptor is tetragastrin (CCK-4) [212]. In vivo, endogenous CCK-8 has been found by radioimmunoassay to be stable in humans for 5-7 min [214].

In 1982, Koulischer et al. investigated the in vitro half-lives of different fragments of cholecystokinin in human plasma using high-performance liquid chromatography (HPLC) [215]. CCK-10 (Asp-Arg-Asp-Tyr-Met-Gly-Trp-Met-Asp-Phe-NH$_2$), CCK-8 (Asp-Tyr-Met-Gly-Trp-Met-Asp-Phe-NH$_2$), and CCK-4 (Trp-Met-Asp-Phe-NH$_2$) were added to either rat or human plasma and incubated for 15-45 min at 37°C. HPLC was used to separate and identify the degradation products. CCK tetrapeptide showed a half-life of 13 min in human plasma while its cleavage in rat plasma occurred at a very high rate (<1 min). The kinetics of sulfated CCK-8 [Asp-Tyr(SO$_3$H)-Met-Gly-Trp-Met-Asp-Phe-NH$_2$] in human plasma showed a half-life of 18 min, while unsulfated CCK-8 (Asp-Tyr-Met-Gly-Trp-Met-Asp-Phe-NH$_2$) had a prolonged half-life of 50 min. This degradation was inhibited by bestatin (0.1 mM) and puromycin (1 mM), suggesting that aminopeptidases play a major role in the breakdown of this peptide. CCK-10 exhibited a serum half-life of 30-45 min, giving rise to CCK-8 in vitro. This degradation was inhibited by aprotinin (5000 U/ml), suggesting that trypsin-like or kallikrein-like enzymes are involved in the plasma metabolism of this peptide [215].

Subsequently, McDermott et al. studied the degradation of CCK-8 by intact synaptosomes isolated from rat cortex and hypothalamus [217]. They showed that the majority of the degrading activity was present in the cytoplasmic fraction and a small amount (7%) was membrane-bound. The degradation products were isolated by HPLC and characterized by amino acid analysis. The Met-Gly bond was the primary site of cleavage, giving rise to Asp-Tyr(SO$_3$H)-Met and Gly-Trp-Met-Asp-Phe-NH$_2$, which were further degraded by sequential removal of amino-terminal residues. The Asp-Tyr and Asp-Phe bonds were also sites of cleavage although the enzymes involved were not isolated. This degradation of CCK-8 was most strongly inhibited (90% inhibition) by p-chloromercuribenzoate (PCMB 1 mM), indicating that thiol proteases played a major role in degrading the peptide in synaptosomes while inhibitors of metallopeptidases (Hg^{2+}, Cu^{2+}, Zn^{2+} all 2 mM) reduced the formation of the secondary degradation products by 75-100% but had little effect on the initial degradation of CCK-8 [217].

Deschodt-Lanckman and Strosberg investigated the hydrolysis of CCK-8 by membrane-bound enzymes from synaptic rat membranes [218]. These enzymes were solubilized by homogenization of the animals' brain and centrifugation at 50,000× g for 15 min, followed by resuspension of the pellet in Tris buffer (pH 7.7) and incubation at 37°C for 45 min. The solubilized enzymes were then obtained after centrifugation at 100,000× g for 60 min. Two aminopeptidases of broad specificity gave rise to free tryptophan as the major fluorescent degradation product. These enzymes cleaved the N-terminal tyrosine residue off enkephalins, the N-terminal tryptophan residue off CCK-4, and sequentially all the peptide bonds of CCK-8. Additionally, a metallopeptidase that cleaved CCK-8 at the Try-Met bond appeared indistinguishable from enkephalinase A on the basis of (1) chromatographic behavior on DEAE-cellulose columns, (2) sensitivity to the inhibitors 1.5 mM 1,10-phenanthroline (complete inhibition), 0.2 mM phenobarbital (IC$_{50}$ = 60 μM), and 0.03 mM phosphoramidon (IC$_{50}$ = 23 nM), and (3) similar affinity of this enzyme for Met- and

Leu-enkephalin. These findings suggest that CCK-8 is inactivated by the same peptidases that degrade the enkephalins. In characterizing the metallopeptidase responsible for cleaving the Trp—Met bond, it also became obvious that the presence of the negatively charged sulfate group enhanced the rate of cleavage of the peptide (two- to threefold increase for CCK-8 and three- to fourfold increase for CCK-7), compared to its unsulfated form [218].

The involvement of a different major inactivating enzyme has been suggested for the degradation of endogenous CCK-8 in rat cerebral cortex [219]. By studying the degradation of CCK-8 released by depolarization of brain slices, Rose et al. [219] observed that a variety of inhibitors of metallopeptidases (bestatin, amastatin, puromycin, thiorphan, captopril, or o-phenanthroline), thiol peptidases (leupeptin or antipain), and carboxypeptidases (pepstatin) had no effect on the recovery of CCK-8 degradation products. By contrast, serine-alkylating reagents such as diisopropyl fluorophosphate or phenylmethylsulfonyl fluoride markedly increased (two- to threefold each reagent) the amount of CCK-8 recovered intact by immunoassay from the brain slices following depolarization, suggesting the involvement of a serine peptidase in the endogenous inactivation of cholecystokinin in rat brain. This endopeptidase is more elastase-like than trypsin-like or chymotrypsin-like, which is far more common. It cleaves the two peptide bonds of CCK-8 where the carboxyl group is donated by a methionine residue and is a major inactivation ectoenzyme for this neuropeptide [219]. HPLC analysis of endogenous cholecystokinin recovered in the media of depolarized brain slices indicates that endogenous CCK-5 was the most abundant fragment and that its formation was strongly decreased in the presence of an elastase inhibitor. CCK-8 released by depolarization of rat cerebral cortex slices showed only low immunoreactivity [220], indicating that endogenous CCK-8 was highly sensitive to hydrolysis by one or more tissue endopeptidases. This study suggests that metabolic pathways of neuropeptides studied exogenously may not reliably reflect those responsible for endogenous neuropeptide inactivation, and as such, the study of the fate of endogenous neuropeptides released by depolarization of brain slices allows the physiologically relevant peptidases to be identified.

Recently, Charpentier et al. [221], using conformational studies of the N-terminal folding of the CCK tyrosine-sulfated octapeptide, produced two cyclic cholecystokinin analogs by conventional peptide synthesis. The binding characteristics of the peptides Boc—D—Asp—Try(SO_3H)-2-amino-hexanoic acid-D—Lys—Trp-aminohexanoic acid-Asp—Phe—NH_2 (compound I) and Boc-gamma-D-Glu—Tyr(SO_3H)-aminohexanoic acid-D-Lys—Trp-aminohexanoic acid-Asp—Phe—NH_2 (compound II) were then investigated on brain cortex membranes and pancreatic acini of guinea pigs. Both compound I (Ki = 5.1 nM) and compound II (Ki = 0.49 nM) showed a high affinity for central CCK receptors in competitive inhibition assays using the synthetic ligand t-butoxycarbonyl-2-aminohexanoic acid-CCK hexapeptide and were only weakly active in stimulation of amylase release from pancreatic acini, while they were unable to induce contractions in the guinea pig ileum. These two analogs, therefore, appear to be synthetic ligands exhibiting both high affinity and high selectivity for central CCK binding sites.

A literature review reveals an unfortunate lack of examples of the use of protease inhibitors either in vivo or in vitro to protect CCK from

proteolytic attack. These studies will undoubtedly follow a clearer understanding of the proteases involved and the degradative pathway at different sites of CCK release.

B. Luteinizing Hormone-Releasing Hormone

Luteinizing hormone-releasing hormone (LHRH) is a decapeptide (pGlu –
His –Trp –Ser –Tyr –Gly –Leu –Arg –Pro –Gly –NH$_2$) that regulates the secretion of both luteinizing hormone and follicle-stimulating hormone from
the anterior pituitary. The neurohormone is synthesized in the hypothalamus and carried via the hypophysial portal system to the anterior
pituitary where it exerts its stimulatory effects on hormone release [222].

LHRH degrading-enzyme activity is the focus of considerable interest
because it may represent one of the control mechanisms that determines
the concentration of active LHRH in the hypothalamus and at the level of
the pituitary gonadotroph. Gonadal steroids capable of modulating LHRH
degrading activity both in vivo and in vitro have been reported [223].
Thus, the time period preceding gonadotropin surge, with its rapidly
changing steroidal milieu, appears to be a likely interval during which to
examine LHRH-degrading activity.

The possibility of cyclic fluctuations in LHRH-degrading activity in
female rat hypothalami and pituitaries was investigated by Lapp and
O'Conner [223]. These tissues were collected at selected time points during the 4-day estrous cycle, homogenized, and centrifuged at $100,000 \times g$.
Supernatants were incubated with synthetic LHRH, the reactions terminated,
and the decapeptide and by-products were separated by reversed-phase
HPLC. Degradation of LHRH incubated with active cytosol was estimated
by comparison of integrated LHRH peak area with that from incubations
with heat-inactivated cytosol. Hypothalamic LHRH degradation was depressed from 4.6 to 2.6 ng/min/mg protein during the latter hours of
diestrus two (1800 – 2400 hr), a time during which the LHRH content in the
hypothalamus has been reported to be increasing [224]. From diestrus two
2400 hr to proestrus 1500 hr, there was significant increase in degrading
activity from 2.6 to 4.3 ng/min/mg protein. This was then followed by a
statistically insignificant decline (4.3 to 4.0 ng/min/mg protein) from
1500 to 1800 hr proestrus. At the time of the LHRH surge, the activity
had not undergone significant increase in comparison to 1800 hr. Pituitary
LHRH degradation was significantly increased from 3.2 to 4.3 ng/min/mg
protein during the 6-hr period preceding the surge (1200 – 1800 hr), but
was significantly depressed from 5.0 to 3.1 ng/min/mg protein at the surge.
The hypothalamic reduction in activity associated with diestrus two as well
as the hypothalamic and pituitary reductions associated with proestrus may
represent a permissive effect allowing increased LHRH accumulation in the
hypothalamus and its prolonged action in the pituitary. These cyclic
variations in net LHRH degradation may arise from modulation of one or
several peptidases that degrade the hormone and support the hypothesis
that the ovulatory cycle may in part be strongly influenced by LHRH degrading activity [223].

In the rat, LHRH is degraded primarily in the pituitary gland, ovaries,
and peripheral tissues [224,225]. Berger et al. [226] have demonstrated
that in the ovary about 1% of the LHRH inactivation occurs in the plasma
membrane. The remainder is mediated by intracellular proteases leaked
out from the cell.

In 1981, degradation of LHRH by purified plasma membranes from rat pituitaries was investigated by Elkabes et al. [222]. Synthetic LHRH (0.5 mg/ml) was incubated at 37°C for 20 min with pituitary plasma membranes (750 µg protein/ml). The reaction was stopped by centrifugation at 4°C and the degradation products were isolated by HPLC using a reversed-phase column. Amino acid analysis of the degradation products indicated that the N-terminal tripeptide (pGlu – His – Trp) and the N-terminal hexapeptide (pGlu – His – Trp – Ser – Tyr – Gly) sequences of LHRH are the main degradation products. Thus, the main cleavage sites of LHRH by the pituitary plasma membrane-bound enzymes are the $Trp^3 - Ser^4$ and $Gly^6 - Leu^7$ bonds of the neurohormone [222].

In 1988, Molineaux et al. used specific, active-site-directed endopeptidase inhibitors to identify the enzyme involved in LHRH degradation by hypothalamic and pituitary membrane preparations and by an intact anterior pituitary tumor cell line (AtT 20) [227]. Incubation of LHRH with pituitary and hypothalamic membrane preparations led to the formation of LHRH tripeptide (pGlu – His – Trp) as the main reaction product. Under the same conditions, captopril, an inhibitor of angiotensin-converting enzyme, led to the accumulation of LHRH pentapeptide (pGlu – His – Trp – Ser – Tyr) and, to a lesser extent, LHRH hexapeptide (pGlu – His – Trp – Ser – Tyr – Gly). The degradation of LHRH and the formation of the N-terminal tri- and pentapeptides were blocked by N-[1-(R,S)-carboxy-3-phenylpropyl]-Ala – Ala – Phe-p-aminobenzoate (cFP – AAF – pAB), a specific active-site-directed inhibitor of endopeptidase-24.15. Some inhibition of LHRH degradation (only 55% of the LHRH was degraded) and formation of the N-terminal hexapeptide were also obtained in the presence of N-1-carboxy-2-phenylethyl]-Phe – p-aminobenzoate (cFE – F – pAB), an inhibitor of endopeptidase-24.11. Similar results were obtained with AtT 20 cell membranes and with intact AtT 20 cells in monolayer culture. When LHRH was injected directly into the third ventricle of rats, the presence of cFP – AAF – pAB inhibited LHRH degradation, indicating that LHRH degradation was primarily initiated by the membrane-bound form of endopeptidase-24.15 to yield pGlu – His – Trp – Ser – Tyr and to a lesser extent by endopeptidase-24.11 to yield pGlu – His – Trp – Ser – Tyr – Gly. These findings indicate that the principal enzyme involved in the degradation of the Try – Gly bond of LHRH in both hypothalamic pituitary membranes is the zinc-containing metalloendopeptidase EC 3.4.24.15 [227].

LHRH degrading activity in the median eminence has been investigated by Advis et al. [228]. Incubation of LHRH with median eminence tissue supernatant produced pGlu – His – Trp – Ser – Tyr (LHRH 1-5) and Gly – Leu – Arg – Pro – Gly – NH_2 (LHRH 6-10) degradation fragments, as detected by HPLC, suggesting $Try^5 - Gly^6$ cleavage of the decapeptide. Because these fragments were also present after incubation of LHRH with α-chymotrypsin, the possibility that the irreversible inhibitor of alpha-chymotrypsin, N-tosyl-L-phenylalanine chloromethyl ketone (TPCK), might inhibit LHRH degrading activity and effect LHRH release was examined [229]. The irreversible inhibitors of trypsin proteases N-alpha-p-tosyl-L-lysine chloromethyl ketone and phenylmethylsulfonylfluoride were used as controls. LHRH degrading activity in aliquots of median eminence supernatant was determined by monitoring the loss of synthetic LHRH in the presence or absence of the inhibitors. Moreover, LHRH release from median eminence fragments was assessed by radioimmunoassay after incubating the tissue with the inhibitors in Krebs-Ringer bicarbonate buffer.

LHRH degrading activity in both the incubation medium and the median
eminence tissue was determined at the end of the incubation.

N-Tosyl-L-phenylalanine chloromethyl ketone inhibited LHRH degrading
activity in median eminence tissue supernatant in a dose-dependent manner
over the range of 0.5–100 μM. In contrast, this inhibitor was able to
suppress LHRH release from intact median eminence only at doses of 25,
50, and 100 μM, suggesting relative inability of the inhibitor to reach endopeptidase pools in the intact tissue. There was a corresponding increase
in LHRH release from the incubated median eminence. No evidence of tissue damage after 100 μM TPCK was detected as assessed by the absence of
lactate dehydrogenase levels in the incubation medium. Neither LHRH degrading activity nor LHRH release was affected by the irreversible trypsin
inhibitors N-alpha-p-tosyl-L-lysine chloromethyl ketone and phenylmethylsulfonyl fluoride. These results indicate that TPCK is a potent inhibitor
of an endopeptidase which degrades LHRH at its $Try^5–Gly^6$ bond, and
that TPCK releases LHRH from the median eminence not by damaging the
nerve terminals or entirely by inhibiting endopeptidase activity, but rather
through activation of a cell-membrane-dependent signal transduction mechanism [228].

The observation that proteolysis of the Try–Gly bond is the most important step in terminating the activity of LHRH has led to the search for
LHRH analogs with greater biological activity.

The first success was achieved by Fujino et al. [230] in 1973, who
found that substituting ethylamine for the glycineamide residue in position
10 resulted in an analog, [Des–Gly–NH$_2$]-proethylamide LHRH, which
was approximately five times as active as LHRH in an ovulation-inducing
assay in adult Sprague-Dawley diestrous rats following subcutaneous administration as well as in constant-estrus rabbits following intravenous administration. In both cases, examination for the presence of ova in the
oviduct was made on the next day and the results were compared with
those of synthetic LHRH. In addition, the analog was incubated with hemi-
sected male rat pituitaries and the media was assayed for LH and FSH release by both bioassay and radioimmunoassay. The analog was shown to
possess over 2.5 times the potency of LHRH in release of both LH and
FSH in vitro.

Subsequently, Monahan et al. [231] showed that substituting a
D-alanine for L-Gly in position 6 also increased the potency of LHRH.
This analog, [D-ala^6]LHRH, was found to be 3.5–4.5 times more potent
than LHRH in elevating plasma LH 15 min after an intravenous injection of
the test compound in chronically ovariectomized rats [231].

In 1974, Coy et al. produced a nonapeptide ethylamide analog with
D-alanine at position 6 [232]. This analog had a very high potency and
long-acting properties, and as such, it was suggested that it could have
important clinical and veterinary applications. The nonapeptide analog,
[D-Ala^6,des-Gly–NH$_2$^{10}]-LHRH ethylamide, was prepared by solid-phase
methodology and assayed for elevation in LH and FSH levels following infusion into immature male rats. The analog showed LH-releasing activity
of 1600% and FSH-releasing activity of 1200% that of LHRH [232].

Buserelin, pGlu –His – Trp –Ser –Tyr –D –Ser(tertbutyl) –Leu –Arg –
Pro-ethylamide, is a highly bioactive analog synthesized with a D amino acid
at the sixth position of the LHRH decapeptide to resist proteolytic cleavage
of this bond. Compared with LHRH, buserelin is 27, 3, and 5.5 times more
stable toward hydrolysis in pituitary plasma membranes and homogenates

of the pituitary and liver, respectively. This enhanced stability probably leads to a 20–40 times gain in potency at releasing gonadotropins
with greatly prolonged duration of gonadotropin release (up to 8 hr) in
ewes in vivo. This suggests that degradation of LHRH and its analogs by
the gonadotroph surface membrane may well be of physiological importance
and may be a major factor in determining the potency and prolonged action
of LHRH analogs [233]. The pituitary data further suggest that pituitary
plasma membrane LHRH degrading enzyme is different from pituitary homogenate enzyme.

Swift and Crighton then compared three analogs, des–Gly–NH_2^{10}LHRH,
[D–Ser^6]des–Gly–NH_2^{10} LHRH ethylamide, and [D–Ser(Bu^t)^6]des–Gly–
NH_2^{10} LHRH ethylamide, relative to their ability to release LH and FSH
and their rates of elimination and degradation in various tissues [234].
When injected intravenously into mature ewes as a single dose (30 µg),
des–Gly–NH_2^{10} LHRH was found to be the least potent with no significant
difference in FSH or LH release than seen with LHRH itself, whereas
[D–Ser(Bu^t)^6] des–Gly–NH_2^{10} LHRH ethylamide, buserelin, was the most
potent, causing a 3.5-fold greater increase in LH and FSH release. All
three analogs were eliminated from plasma at rates similar to LHRH.

In the same way, the incubation of 500 ng of LHRH with tissue extracts
followed by radioimmunoassay of the residual LHRH revealed that extracts
of the hypothalamus (97% degradation) and anterior pituitary gland (91%)
degraded LHRH to a similar extent. Both the hypothalamic and anterior
pituitary gland extracts degraded more LHRH than did lung extract (62%),
which in turn destroyed more LHRH than did extracts of kidney (44%) or
liver tissue (35%). The degradative abilities of kidney and liver extracts
did not differ significantly from each other. Plasma failed to degrade
LHRH or the analogs. By contrast, extracts of liver and lung were incapable of catabolizing any of the anlogs. Of the three analogs, des–Gly–
NH_2^{10} LHRH was the only analog that was degraded (20% by hypothalamic
extract in vitro). The anterior pituitary gland and kidney extracts degraded [D–Ser(Bu^t)^6] des–Gly–NH_2^{10} LHRH ethylamide (81 and 25%, respectively) and [D–Ser^6] des–Gly–NH_2^{10} LHRH ethylamide (25 and 12%,
respectively) less extensively than LHRH. The less extensive catabolism
of these analogs of LHRH by the anterior pituitary gland may explain their
increased LH- and FSH-releasing potency.

As more information becomes available on the mechanism and extent to
which LHRH levels are controlled by proteolytic enzymes in these tissues,
a better understanding of the ideal LHRH analog characteristics and the
most appropriate method for their administration will undoubtedly arise.

C. Substance P

Substance P is an 11-peptide found in the CNS as well as the GI tract.
Although the function of substance P as a neurotransmitter is most likely
terminated by enzymatic degradation in the synapse [90,93], the enzymes
that are involved have yet to be fully identified. It has been proposed
that membrane-bound proteases are involved in the degradation of substance P in the same fashion as acetylcholinesterase degrades acetylcholine
in the synapse. Several membrane-bound proteases in the brain have been
reported to be integral in the degradation of substance P, including substance P-degrading enzyme [91], endopeptidase-24.11 [87,92], angiotensin-
converting enzyme [235], and post-proline-cleaving enzyme [236].

On the basis of experimental results using neuroblastoma N-18 cells,
Endo et al. [237] proposed that metalloendopeptidases distinct from either
endopeptidase-24.11 or ACE may play a key role in the degradation of substance P by neurons. This unidentified metalloendopeptidase was successfully isolated and purified from the rat brain, and was thought to be
a substance P-degrading endopeptidase distinct from all other proteases
known to cleave substance P. While this enzyme is able to cleave substance P in the glia, it does not appear to be very active in the brain
synaptic membranes [238].

D. Thyrotropin-Releasing Hormone

Thyrotropin-releasing hormone is a hypothalamic regulatory tripeptide
(pyro-Glu-His-Pro-NH$_2$) that stimulates the release of thyrotropin,
prolactin, and growth hormone from the pituitary. It is used to test
thyroid and pituitary-hypothalamic function. TRH is protected against
classical exopeptidase attack by having a pyroglutamyl residue at its amino
terminus and an amidated carboxyl-terminus [239].

The catabolism of TRH is believed to proceed through two different
pathways: removal of the pyroglutamyl moiety by pyroglutamyl aminopeptidase followed by deamidation of the terminal amide or, to a lesser
extent, initial deamidation followed by removal of the pyroglutamyl moiety
[240-247].

Pyroglutamyl aminopeptidase (E.C. 3.4.11.8), which is responsible for
degrading TRH to pGlu and His-ProNH$_2$, has a molecular weight of
28,000 daltons, a Km for TRH of 45 µM, and total enzyme activity of 600
nmol/hr/rat brain [248]. The resulting dipeptide cyclizes spontaneously
at neutral pH to form the histidyl-proline diketopiperazine, cyclo(His-
Pro), which is stable to further degradation [249]. This enzyme appears
to be involved in the removal of the pGlu residue from other neuropeptides, such as LHRH and neurotensin, as well [250]. It is activated by
EDTA and dithiothreitol and inhibited by Hg^{2+}, N-ethylmaleiimide, and
2-iodoacetamide, indicating the presence of sulfhydryl groups in the active
site of the enzyme [251]. Pyroglutamyl aminopeptidase has been found in
the rat colon and other intestinal tissues [252,253].

The inactivation of TRH in serum is catalyzed by an enzyme that also
hydrolyzes TRH into pyroGlu and His-Pro-NH$_2$ and which exhibits chemical characteristics distinctly different from pyroglutamyl aminopeptidases
[254,255]. It is inhibited by EDTA and dithiothreitol but not by
2-iodoacetamide and N-ethylmaleinimide. Its specificity is also much different than that of pyroglutamyl aminopeptidases in that it causes rapid,
stereospecific cleavage of the pyroGlu-His bond of TRH only and will not
cleave LHRH, neurotensin, or gastrin. This property has given rise to
the hypothesis that it might be involved in the regulatory mechanisms
governing TRH [256-259]. In 1988, however, Iversen reexamined this
hypothesis by measuring the TRH-degrading activity in the serum of adult
males and females, aged 23-40, at different times and concentrations of
TRH following intravenous administration (10 nmol/min over 60 min) [260].
TRH was assayed by radioimmunoassay (RIA). Serum TRH-degrading activity was found to be independent of sex, phase of female menstral cycle, time of day, or concentration of TRH used, suggesting that the serum
peptidase responsible for specific TRH degradation has little regulatory
control of serum concentration of TRH. This also implies that the plasma

levels of this peptide are regulated by means other than fluctuations in proteolytic degradation.

Prolyl endopeptidase (E.C. 3.4.21.26), which is responsible for de-aminating TRH and cyclo(His−Pro), has a molecular weight of 76,000 daltons, a Km for TRH of 34 μM, and total enzyme activity of 3500 nmol/hr/rat brain [248]. This protease, also known as post-prolyl-cleaving enzyme, is highly specific for cleavage of peptides with the structure Y−Pro−X, where Y is a peptide, N-protected amino acid, amide, or ester, and X is an amino acid or ester [261]. It produces TRH−OH from TRH, but it is not specific for its peptide as it has been shown to cleave other proline-containing neuropeptides such as neurotensin, substance P, oxy-tocin, vasopressin, and angiotensin [262]. The enzyme is inhibited by di-isopropyl fluorophosphate, N-ethylmaleimide, and 2-iodoacetamide but is activated by EDTA, indicating that —SH and —OH residues represent essential functional groups of the enzyme [261,263]. Once formed, TRH−OH is rapidly degraded by soluble rat brain peptidases in the lateral cerebral ventricle, to its constituent amino acids ($t_{1/2}$ = 3 min) [264].

The presence of TRH−OH and cyclo(His−Pro) in the pituitary sug-gests the presence of the same degrading enzymes found in the brain [252]. Indeed, proline endopeptidase has been purified and character-ized from bovine pituitaries by Bauer and Kleinkauf and found to be similar to the brain enzyme in properties [253]. Prolyl endopeptidase is present in the liver, kidney, heart, lung, pancreas, skeletal muscle, and ileum of the rat [252,265].

An aminopeptidase that cleaves the His−Pro bond of TRH and its me-tabolite, cyclo(His−Pro), has been isolated from porcine brain by Matsui et al. [245]. Fractionation of an extract of porcine brain acetone powder by DEAE cellulose chromatography produced a fraction with an enzyme that cleaved His−Pro−NH$_2$ to histidine and proline when incubated with His−Pro−NH$_2$ in Tris buffer (7.7) at 37°C for 30 min. Competition studies using a variety of compounds containing histidine or proline [pGlu−His, His−Pro, pGlu−His−Pro−OH, prolineamide, TRH, and cyclo-(His−Pro)] suggest that the best substrates for this enzyme contain a free alpha amino acid on histidine and a blocked carboxyl group on proline. The enzyme has a sensitive sulfhydryl group and a pH optimum of 7−7.5 and appears to be identical to the prolyl endopeptidase of kidney that hydrolyzes the prolyl-glycinamide bond of LHRH and the prolyl-glutamine bond of substance P [266]. There is little support, however, for this degradative step being important in the initial loss of biological activity of TRH [248].

Several TRH analogs, which have been used to study the sites and mechanisms of TRH degradation and to test thyroid and pituitary-hypo-thalamic function, have been synthesized [249]. [^{3}Me−His]TRH (pGlu−^{3}methyl−His−ProNH$_2$), a hyperactive analog, has a slower plasma clear-ance rate than TRH, presumably because of its increased resistance to serum degradation by the enzyme that cleaves the protected pyroGlu−His bond [267]. In cerebral cortex, hypothalamus, and thalamus, however, this particular analog is readily deamidated and forms a diketopiperazine after pyroglutamyl peptidase action. Its enhanced activity may, there-fore, be due to higher affinity for TRH receptors than TRH itself [268]. Similarly, pGlu−His-amphetamine and pGlu−His−Pro-amphetamine are re-sistant to degradation by gastrointestinal and liver enzymes as well as brain peptidases and serum enzymes [269]. HPLC and radioimmunoassay techniques have revealed the cleavage of pGlu and amphetamine from these analogs in vitro. Unfortunately, these pseudopeptides and their derivatives are devoid of TRH-releasing properties. Indeed, they were initially studied as possible antidepressants as radioimmunoassay revealed the release of amphetamine from these analogs [269].

Modification of the pGlu residue of TRH has also produced several long-acting, centrally active TRH analogs such as CG3509, orotyl–His–ProNH$_2$, and CG3703, 6-methyl-5-oxothiomorpholinyl-3-carbonyl–His–ProNH$_2$, which are completely stable to the action of the pyroglutamyl aminopeptidase in the brain and serum, but which may still be sensitive to deamidation [267]. Similarly, MK771, L-pyro-2-aminoadipyl-His-thiazolidine-4-carboxide, with modifications in both the pGlu- and ProNH$_2$ residues of TRH, can be readily deamidated by brain prolyl endopeptidase but is stable to pyroglutamyl aminopeptidase [267]. Of all the TRH analogs studied, the most stable TRH analog and the one with the most therapeutic potential is RX77368 (pGlu–His-3-monomethyl-ProNH$_2$), which has enhanced stability to inactivation by brain, serum, gut, liver, and kidney [267].

Several competitive inhibitors, which are highly effective in preventing the breakdown of synthetic TRH and which appear to be useful as specific inhibitors of TRH degradation in vitro, have also been synthesized [261]. PyroGlu–NH$_2$ (Ki 185 µM) and pyroGlu–His–OCH$_3$ (Ki 16.5 µM) are effective competitive inhibitors of pyroglutamyl aminopeptidase, while N-acetyl-Gly–Pro–NH$_2$ (Ki 1143 µM) and N-benzoxylcarbonyl-Gly–Pro–NH$_2$ (Ki 442 µM) are effective inhibitors of prolyl endopeptidase. Because of the extremely high rate of TRH catabolism by tissue homogenates, utilization of these specific inhibitors of TRH breakdown facilitates the in vitro study of TRH synthesis by avoiding the addition of nonspecific inhibitors such as bacitracin [261].

E. Insulin

The most widely used protein, insulin, is only 80% absorbed in the pig following subcutaneous administration [270] and is variably absorbed in diabetics with a wide range of 19–104% [271]. Aside from drug loss during diffusion through the connective tissue and vascular membranes to reach the bloodstream, degradation of this protein at the site of injection has been implicated as a factor contributing to the less than complete bioavailability [272–275]. In homogenates of the rat skeletal muscle, insulin was degraded rapidly, losing 40% of its activity within 5 min [276]. It is therefore not surprising that coadministration of insulin with certain protease inhibitors results in improved absorption following subcutaneous administration. Hori et al. [277] tested four such inhibitors and found that benzyloxycarbonyl-Gly–Pro–Leu–Gly, benzyloxycarbonyl-Gly–Pro–Leu, and dinitrophenyl-Pro–Leu–Gly were effective, whereas benzyloxycarbonyl-Gly–Pro was not.

Degradation of insulin at the site of injection probably occurs before transport across the vascular endothelium. This is because more than 80% of the transported insulin, which is taken into the cell by receptor-mediated process, has been found at the transluminal side intact [278–280]. A possible explanation for the relatively small amount of degradation of the insulin could be that the endothelial cells, unlike other cells, do not degrade insulin to any appreciable degree since the

internalized insulin may not be channeled through a lysosomal pathway
[281]. Consequently, a large portion of the molecular transport across
the capillary wall could be due to micropinocytosis with the formation of
vesicles that shuttle between the luminal and external sides of the vessel
wall [282,283].

In cells other than endothelial cells, internalized insulin or insulin
receptor complexes can be found in a variety of subcellular compartments,
including the nucleus, endoplasmic reticulum, Golgi, and cytoplasm as well
as lysosomes [284]. A small, undefined amount of internalized insulin is
transported back to the membrane and released from the cell, intact, in a
process called retroendocytosis. Normally, insulin, which is bound in
membrane-coated pits and internalized via an endosome, is degraded in the
early endosome [286]. The degradation products consist of intact A chains
with cleavage of B chain at B16-17, 24-25, and 25-26 while the insulin is
still bound to the receptor [285]. Subsequent processing of the internalized insulin then occurs, which ultimately results in multiple cleavages in
both the A and B chains. Unfortunately, the enzyme involved in this
intracellular cleavage has yet to be established, but it has been shown by
Yonezawa et al. [285] that the lysosome has a minor or secondary effect
on the degradation of insulin.

Two principal enzymes, glutathione insulin transhydrogenase (EC
1.8.4.2) and insulin-degrading enzyme (EC 3.4.22.11), have been implicated in the degradation of insulin. The former cleaves insulin at the
disulfide bridges [287], whereas the latter cleaves insulin principally at
the $Tyr^{16}-Leu^{17}$ bond in the B chain [288]. The relative role of these
two enzymes in the degradation of insulin is still controversial. At one
time initial disulfide bond reduction followed by proteolysis was felt to be
the only mechanism of insulin degradation [289]. This resulted in the
formation of A and B chains, which were then proteolytically degraded to
amino acids. Then another possibility was proposed, in which insulin-
degrading enzyme was envisioned to play a key role in initiating degradation. This resulted in a molecule with three peptide chains held together
by disulfide bonds, which was further degraded by nonspecific proteases
[289]. Recently, Hamel et al. [276] observed the formation of 11 products during the degradation of insulin in rat skeletal muscle homogenates.
The products, however, were not identified. At about the same time,
Yonezawa et al. [290] reported the formation of seven products from insulin as a result of action by insulin-degrading enzyme from pig skeletal
muscle. The primary product appeared within 30 sec of incubation, reaching a peak within 10 min. Interestingly, three of the seven products
were able to bind to the insulin receptor in IM-9 lymphocytes and to react
with antiinsulin antibodies. This suggests that only minor changes had
occurred to the insulin molecule. Subsequent work by Goldfine et al.
[291] on insulin degradation in isolated mouse pancreatic acini suggests
that glutathione insulin transhydrogenase and insulin-degrading enzyme
cooperate in an as yet undetermined manner in degrading insulin. Such a
scenario may be possible at the site of injection and elsewhere in the body.

A few comments on the properties of glutathione insulin transhydrogenase and insulin-degrading enzyme are in order. Glutathione insulin
transhydrogenase is a glycoprotein with a molecular weight of 48−50 K
daltons, has a pH optimum between 7 and 8, and requires glutathione as a
cofactor [287,292,293]. It is found on the cell membrane but in much reduced quantities in the cell cytoplasm [294]. The Km for insulin in the

liver is 43 µM. Because of its affinity toward thiol disulfide bonds in peptides and proteins other than insulin, glutathione insulin transhydrogenase has also been called thiol protein disulfide oxidoreductase.

Insulin-degrading enzyme, on the other hand, is a cysteine proteinase with a molecular weight of 135 K daltons [295]. It is therefore susceptible to inhibition by N-ethylmaleimide and p-hydroxymercuribenzoate and to activation by glutathione and dithiothreitol. Other activators include Ca^{2+} and Mg^{2+}. In contrast, it is inactivated by Zn^{2+} and Cu^{2+} [296]. Its pH optimum is between 7.5 and 7.7 [297]. The Km for insulin in a variety of tissues is 70–400 nM [297–301]. Because this is lower than the Km determined for glutathione insulin transhydrogenase, it is possible that insulin-degrading enzyme plays a more prominent role than glutathione insulin transhydrogenase in degrading insulin extracellularly. The relative roles of these two enzymes in processing insulin in the intact cell are, however, not known, and the subcellular location of these two enzymes is still controversial [301–305].

Insulin-degrading enzyme is by no means a specific enzyme since it also degrades other peptides and proteins, albeit at a much reduced rate. These include insulin-like growth factors I and II [274,306], proinsulin and its intermediates, insulin derivatives with blocked N- and C-termini, and insulin without the C-terminal octapeptide on the B chain [197]. However, insulin-degrading enzyme has no activity against insulin A and B chains, glucagon, human growth hormone, globin, and bovine serum albumin. When this enzyme was first isolated and purified, it was found primarily in the cytosol fraction and only in trace quantities in the plasma membrane [296,297,307]. Subsequent work, however, revealed the existence of this enzyme in the plasma membrane of a variety of cell types as well [291,308–311]. In human fibroblasts [312], rat adipocytes [313–315], and rat liver hepatocytes [316], insulin-degrading enzyme is virtually 100%, greater than 75%, and about 50% plasma membrane-bound, respectively. Goldfine et al. [291] suggested that the insulin-degrading enzyme located on the plasma membranes may contribute to insulin degradation in three ways: while both insulin and enzyme are present on the plasma membrane, by shedding of enzymes into the bathing medium where insulin is digested extracellularly, and cointernalization of the enzyme with insulin and its receptor to form an endosome within which insulin is degraded.

In addition to the glutathione insulin transhydrogenase, insulin-degrading enzyme, and other contributing proteases indigenous at the site of injection [317], these enzymes can also be recruited to the area via monocytes and polymorphonuclear leukocytes in which these enzymes are housed [318–321]. These blood cells are usually present whenever there is an acute inflammatory response, as would be the case following a subcutaneous or intramuscular injection. In time, these cells disintegrate, liberating the proteolytic enzymes, which then can degrade insulin and other peptides and proteins.

VI. APPROACHES TO CIRCUMVENT ENZYMATIC BARRIER

Clearly, in order to promote the absorption of peptides and proteins from any route of administration, the many components of the enzymatic barrier must be controlled. This can be achieved to some extent by modifying the peptide or protein structure, through coadministration of protease inhibitors, or by using the formulation approach. Each of these strategies will be discussed in turn. The focus will be on the oral route since it is the best studied in this regard.

A. Modification of Peptide and Protein Structure: Development of Orally Active Renin Inhibitors

For practical reasons, modifying the amino acid composition to improve enzymatic stability appears to be more amenable to peptides than to proteins, simply because of the complexity of the latter when compared with compounds of the former. Several methods have been used to improve the stability of peptides to proteases [322]. These methods include: (1) substituting unnatural amino acids, including D-amino acids, for L-amino acids in the primary structure, (2) introducing conformational constraints, (3) reversing the direction of the peptide backbone and inverting the chirality of each amino acid, the so-called retroenantiomer analogs, and (4) acylating or alkylating the N-terminus or altering the carboxy terminus by reduction or amide formation. The advantages and disadvantages of each method have been reviewed by Samanen [323]. The experience with renin inhibitors will be described to illustrate the chemical modification approach.

The clinical success of angiotensin-converting enzyme inhibition in the treatment of hypertension has stimulated research on other modes of blockade of the renin-angiotensin system. Most recent efforts have been focused on inhibitors of the renin-angiotensin reaction, the most specific and rate-limiting step in the reaction sequence [324,325]. Renin is an aspartic acid protease (E.C. 3.4.9.19) produced in the juxtaglomerular apparatus of the kidney. Other members of the aspartic protease family include pepsin, cathepsin D, and gastricsin. Renin catalyzes the first and rate-limiting step of the enzyme cascade that exists for the biosynthesis of angiotensin II. This protease selectively cleaves angiotensinogen, a glycoprotein synthesized in the liver, to the decapeptide intermediate, angiotensin I. Angiotensin I is then cleaved by angiotensin-converting enzyme to yield the biologically active octapeptide product, angiotensin II. This octapeptide causes vasoconstriction and stimulates the secretion of aldosterone and catecholamines. In this way, the renin-ACE cascade has been implicated in the regulation of blood pressure and electrolyte balance as well as in pathophysiological states related to various forms of hypertension. Inhibition of renin may, therefore, be therapeutically important in the development of novel antihypertensive agents [326,327].

About 30-40 novel renin inhibitors have been synthesized and tested but so far none is available for use as an antihypertensive drug. The specificity of these inhibitors for renin, as opposed to other aspartic proteinases, was determined by comparing their inhibitory potency against a range of human aspartic proteinases [328]. Homogeneous preparations of human pepsin, human gastricsin, and human cathepsin D were obtained and their ability to hydrolyze chromogenic peptide substrates at pH 3.1 with an ionic strength of 0.1 M was followed spectrophotometrically at 300 nm. Kinetic parameters for substrate hydrolysis by each of the enzymes were obtained using five different concentrations of the respective substrates. For all enzymatic reactions, Michaelis-Menten kinetics were observed, and the best values for the kinetic parameters were obtained by

computer fitting of the measured velocities to those predicted by the
Michaelis-Menten equation [328]. Thus far, the major obstacles to the
design of renin inhibitors have been lack of oral activity and short duration of action, due to proteolysis in the GI tract and beyond.

Early attempts to synthesize renin inhibitors were based on analogs of
the N-terminal sequence in angiotensinogen. The octapeptide comprised of
the amino acid sequence from 6 through 13 of angiotensinogen (His—Pro—
Phe—His—Leu—Val—Ile—His) was the shortest substrate fragment that was
still cleaved by renin [328]. Substitution of the amino acids in position 10
and 11 (Leu—Val), the scissile bond hydrolyzed by renin, with His—Phe—
Phe prevented cleavage and resulted in a weakly competitive inhibitor
(Ki = 2.0 mM) [329,330]. Subsequently, more potent inhibitors were synthesized by substituting the amino acids at the scissile bond with the non-
cleavable dipeptide analogs statine, (4-amino-3-hydroxy-6-methyl-heptanoic
acid)-Val, or a reduced Leu—Val bond ($-CH_2NH-$ in place of $-CONH-$).
These analogs are believed to act as transition-state analogs since they resemble the tetrahedral conformation assumed by the scissile peptide bond
upon binding of the substrate to the enzyme and being distorted for
cleavage [331]. This approach brought significant advantages in potency
of inhibitors in vitro (Ki = 0.19 μM and 20 nM, respectively). Unfortunately, these inhibitors were rapidly cleared from the circulation, rendering them ineffective in lowering blood pressure in sodium-deficient dogs
unless administered by continuous intravenous infusion [332—334].

In 1986, Thaisrivongs et al. developed tert-butoxycarbonyl-Pro—Phe-
N-alpha-methyl-His—Leu—psi[$CHOHCH_2$]Val—Ile-2-pyridylmethylamine as a
potential renin inhibitor with good specificity and extended half-life [335].
The N-alpha-methylhistidine imparted resistance to proteolytic degradation due to increase in steric hindrance, as indicated by stability of the
inhibitor in rat liver homogenate during a 60-min incubation. In addition,
the hydroxyethylene isostere at the scissile site produced a potent renin
inhibitor at physiological pH (IC_{50} = 0.39 nM at pH 7.4), as demonstrated
by decreased angiotensin I generation in human plasma reconstituted from
lyophilized form in vitro. Following intravenous administration of this peptide over 17 min to sodium-depleted cynomolgus monkeys, the plasma renin
activity was essentially zero (ng angiotensin I/ml/hr) for nearly 200 min.
This led to profound reductions in blood pressure (−15 ± 5 mm Hg) for
more than 3 hr postinfusion [335].

In 1988, Hiwada et al. developed an orally active renin inhibitor,
N-[(2R)-3-morpholinocarbonyl-2-(1-naphthylmethyl)propionyl]-(4-thiazolyl)-
L-alanyl-cyclostatine-(2-morpho-linoethyl)amide, which contained no natural
amino acids and was therefore resistant to proteolytic degradation, at
least for 1 hr when incubated with homogenates of the rat liver, kidney,
pancreas, and small intestine [329]. Moreover, the C- and N-termini were
derivatized to confer properties on the renin molecule to make it active.
Specifically, the morpholinocarbonyl group at the N-terminus rendered the
resulting molecule more lipophilic while the C-terminus morpholinoethyl
group rendered the molecule more water soluble. The compound was also a
potent competitive inhibitor of human renin (Ki = 7.3 nM), but much more
so than against renins from pig (Ki = 130 nM), dog (Ki = 35 nM), goat
(Ki = 200 nM), rabbit (Ki = 91 nM), and rat (Ki = 1400 nM). In addition,
this inhibitor showed good specificity for renin in vitro. This was indicated by its lack of inhibitory effect against cathepsin D, pepsin, trypsin,
chymotrypsin, angiotensin-converting enzyme, or urinary kallikrein at a

high concentration of 50 μM. Oral administration of the drug (30 mg/kg) to sodium-depleted marmosets produced significant lowering of blood pressure and almost complete inhibition of plasma renin activity which persisted for 5 hr. Although one-third of the drug was lost to hepatic first-pass metabolism, the above findings indicate that it is feasible to design an orally effective renin inhibitor for control of blood pressure [329].

By attempting to improve existing or create new inhibitor-enzyme interactions, Kleinert et al. [336] designed a novel class of potent (Ki = 0.3–7 nM) renin inhibitors with a dihydroxyethylene replacement for what is usually the Leu–Val amide bond of angiotensinogen. Starting with a dipeptide attached to the nonpeptide replacement for the scissile Leu–Val amide bond (t-butyloxycarbonyl-Phe–His-isobutyl-leucine), shown earlier to be a weak renin inhibitor, the inhibitory potency was increased 100 times by changing the side chain of the scissile bond from isobutyl to cyclohexylmethyl and introducing a hydroxyl group side chain on the other side of the scissile bond, resulting in t-butyloxycarbonyl-Phe–His-cyclo-hexylmethyl-leucinol. This was indicated by the change in IC_{50} from 1500 nM to 1.5 nM. Inhibitor potency was measured in vitro by monitoring for angiotensin I produced in the presence of human renal renin and human angiotensinogen in a maleate buffer containing 0.135 M maleate, 1% DMSO, 0.44% BSA, 1.4 nM PMSF, and 3 mM EDTA using RIA. By replacing the acid-labile t-butyloxycarbonyl group with beta-valine, the potency was maintained while the water solubility was increased from <0.026 mg/ml to >10 mg/ml at 37°C. Finally, to overcome degradation of the chemical by chymotrypsin, the Phe residue was replaced with 4-OCH_3Phe, which increased the resistance to hydrolysis by over 100-fold. The resulting compound, beta-valine-4-OCH_3-Phe–His-cyclohexylmethyl-leucinol, was shown to be potent (IC_{50} = 0.6 nM), water soluble (>10 mg/ml at physiological temperature), stable toward degradation by chymotrypsin (in vitro half-life of 820 min), and active for 1–2 hr following intravenous injection in a primate model (15% drop in mean arterial pressure in association with complete inhibition of plasma renin activity) [336].

To further improve the oral bioavailability of renin inhibitors more recent efforts have been focused on the synthesis of smaller molecules with fewer peptidic bonds. In 1989, Wood et al. introduced a low-molecular-weight renin inhibitor, CGP38560, with the potency and specificity of the larger molecules, thereby improving its oral activity to being the best produced to date [328]. CGP38560 is a synthetic inhibitor of human renin that contains only one natural amino acid. It is a potent inhibitor of human renin (Ki = 0.4 nM) with high specificity as indicated by its over 1000-fold greater affinity for renin than for human pepsin, gastricsin, and cathepsin D. In addition, due to its low molecular weight (M.W. = 730), it shows one of the lowest doses of a renin inhibitor (1–10 mg/kg) reported to be effective after oral administration to marmosets. This dose is, nevertheless, 10 times higher than the i.v. dose (0.1–1 mg/kg) required to produce the same decrease in blood pressure lowering of 23 mm Hg and complete inhibition of plasma renin activity. This low oral bioavailability is attributed to a sizable first-pass effect. For this reason as well as its 2-hr duration of action, the usefulness of CGP38560 as an antihypertensive agent in humans is far from clear [328].

The radioimmunoassay for angiotensin I used in the above experiments was developed by Haber et al. in 1969 [337]. The initial antigen was synthesized by coupling succinylated poly-L-lysine to angiotensin I to render it immunogenic. Amino acid analysis indicated a substitution of one of four lysine-succinate residues or 160 moles of peptide per mole of poly-L-lysine. This complex was then injected in an emulsion in Freund's adjuvant in distilled water containing 2 mg/ml into the toepads of New Zealand white rabbits and blood samples were obtained at weekly intervals. The antibodies with maximum binding activity were collected at 9-16 weeks following primary immunization, and the plasma from the animals with the highest binding activity was selected for use.

To produce the labeled peptide, angiotensin I was iodinated and the immunoreactive fraction isolated by high-voltage paper electrophoresis. The ^{125}I-angiotensin I has an average specific activity of 100 mCi/mg as determined by tracer displacement from antibody.

In preparing incubation mixtures for radioimmunoassay, 5 µl antiserum and 50 µg of labeled peptide (about 5000 cpm) were added to 1 mL 0.1 M Tris acetate buffer (pH 7.4) containing 1 mg/ml lysozyme and either unknowns or standard solution of angiotensin I. Equilibrium is reached in this system in 24 hr at 4°C. Antibody-bound and free peptide are then separated by dextran-coated charcoal.

To determine plasma renin activity, blood samples from the test subjects (dogs, monkeys, and marmosets) were drawn into EDTA tubes, chilled on ice, and the cells separated by centrifugation at 0°C. One milliliter of the resulting plasma containing 2.6 mM EDTA, 1.6 mM dimercaptol, and 3.4 mM 8-hydroxyquinoline is incubated for 3 hr at 37°C. A similar aliquot is held at 0°C. Fifty microliters of each plasma sample is then added to the radioimmunoassay incubation mixture. Displacement of labeled peptide from antibody in the generated sample is corrected by displacement observed in the chilled control sample. The quantity of angiotensin I present is estimated by comparison with standards. Since each sample is corrected by its own nonincubated control, no correction is needed for possible nonspecific effects of plasma on the radioimmunoassay or for cross-reaction with renin substrate [337].

With widespread use of the above radioimmunoassay for plasma renin activity, no chemical assays have been developed to quantitate renin inhibitor levels in plasma. While such an analytical procedure is cost saving, it has inherent inaccuracies. Cross-reaction of angiotensin I antibodies with angiotensinogen, renin inhibitor, and possibly renin itself must be evaluated. Species differences in angiotensin I could also lead to over- or underestimation of plasma renin activity in assays for renin activity in dog, marmoset, or human using rabbit antibodies.

Additionally, a species specificity for the renin-angiotensin reaction is known in which human angiotensinogen can be hydrolyzed only by human renin, which can, however, hydrolyze the angiotensinogens of other species [338]. Morris' studies on the tertiary structure of human renin suggest that this effect is due to the presence of a valine residue on the C-terminal scissile bond of primate angiotensinogen, whereas in nonprimate mammalian species, this residue at this position is leucine [338]. Morris suggested that this could lead to weaker binding between nonprimate renin and primate angiotensinogen. A recent study by Burton and Quinn showed that replacement of this C-terminal residue in a "human" synthetic substrate with leucine produced a threefold increase in Kcat and fourfold increase in Km in a reaction with dog renin in vitro [339].

The above results raise additional concerns about the applicability of inhibitor studies that rely on radioimmunoassay in different nonprimate animal models to human renin inhibition. Such results also accentuate the difficulty in comparing studies of renin inhibitors using different experimental models or "mixed" in vitro experimental models in which the angiotensinogen, renin, and angiotensin I antibodies may come from different species. The use of entirely human in vitro model and stringent testing of antibody cross-reactivity should greatly increase the ability to compare different renin inhibitor studies and predict the effects of these inhibitors on the human renin-angiotensin system.

B. Coadministration with Protease Inhibitors: Protease Inhibitors as Drugs

The second strategy, coadministration with protease inhibitors, has been successful in promoting the oral absorption of the peptides pentagastrin and PHPFHLFVF (a nonapeptide renin inhibitor) and of the proteins insulin and RNAase. The absorption of pentagastrin from the duodenum of the rat was enhanced 11.5 times when the pancreatic duct was ligated to exclude the pancreatic digestive enzymes from the small intestine [340]. This procedure was three times more effective than bypassing the liver, indicating that pancreatic proteases played a larger role than hepatic proteases in the degradation of this peptide.

In the case of the nonapeptide renin inhibitor, PHPFHLFVF, it was found to cross the adult rabbit jejunum 90% intact in the presence of phosphoramidon, a metalloproteinase inhibitor [341]. The permeability coefficient was 4.44×10^{-6} cm/sec. Pepstatylglutamic acid, an aspartyl proteinase inhibitor, was also effective, while aprotinin, a serine proteinase inhibitor, was not. In the absence of protease inhibitors, however, over 55% of the nonapeptide was degraded in 5 min and none was left by 30 min. The main metabolite was Phe, although small quantities of the dipeptide, VF, and the pentapeptide, HLFVF, were also detected.

Insulin is the protein that has gained the most attention in terms of enhancing oral absorption via control of proteolytic activity. As early as 1959, Danforth and Moore [342] demonstrated that diisopropyl fluorophosphate, a serine proteinase inhibitor, was effective in protecting insulin from proteolysis, thereby slightly enhancing its absorption across the jejunum of the rat. Soybean trypsin inhibitor, at a dose of 100 mg, was ineffective. Subsequently, Kidron et al. [343] contradicted this finding by showing that the same inhibitor was effective in enhancing the absorption of insulin from the ileum of the rat at a much lower dose (3 mg). Another effective inhibitor was aprotinin at a dose of 3000 KIU/ml, which also protected RNAase from proteolysis [344]. The effectiveness of both aprotinin and soybean trypsin inhibitor in enhancing insulin absorption is augmented by coadministration with the bile salts deoxycholate and cholate, which when used alone were less effective than the protease inhibitors themselves in promoting oral insulin absorption [344]. This suggests that the enzymatic barrier is more of an impediment to insulin absorption than the membrane transport barrier. The implicit assumption, of course, is that the protease inhibitors did not affect membrane integrity.

Partly because the enzymatic barrier has not been recognized as being important in mucosal peptide and protein absorption until recently, there is virtually no information on the influence of protease inhibitors on mucosal peptide and protein bioavailability. In vitro results indicate that aminopeptidases can be controlled by bestatin, puromycin [124], and

α-aminoboronic acid derivatives [131] and that endopeptidase 24.11 and cysteine proteinase can be controlled by 1,10-phenanthroline and p-hydroxymercuribenzoate, respectively (Inagaki and Lee, unpublished results). Hussain et al. [131] reported that α-aminoboronic acid derivatives, which are reversible inhibitors of aminopeptidases, markedly inhibited the hydrolysis of leucine enkephalin in the nasal cavity of the rat. Boroleucine was more potent than borovaline and boroalanine, in that order. Compared with bestatin and puromycin, boroleucine was 100 and 1000 times more potent, respectively, and was effective at 0.1 µM. Whether protection of leucine enkephaliln from aminopeptidase action would lead to therapeutic concentrations of the peptide in the blood was not determined in that study.

In addition to the known protease inhibitors, sodium glycocholate, a penetration enhancer, also inhibits leucine aminopeptidase activity and protects insulin from proteolysis in rat nasal homogenates [345]. This is advantageous since the peptides would be protected from the otherwise inaccessible cytosolic proteases. Sodium glycocholate is, however, not as potent as the most potent aminopeptidase inhibitors bestatin and amastatin [346]. Protection of insulin from proteolysis by sodium glycocholate is not limited to the nasal mucosa, but is extended to the buccal, rectal, vaginal, and ileal mucosae as well [123]. Moreover, the extent of protease inhibition is dependent on both the nature [346] and concentration of the substrate [51]. This and other bile salts appear to inhibit proteolytic activity by denaturing the enzyme and preventing the enzyme-substrate complex formed to undergo the necessary conformational change which aligns the catalytic site on the protease with the susceptible bond of the substrate [346]. Entrapment of the peptide or protein in micelles of the surfactant-type penetration enhancers, thereby protecting it from the proteolytic enzymes, is another possibility that needs to be explored.

A concern with using protease inhibitors to promote peptide and protein absorption is their possible effect on physiological processes following absorption into the bloodstream. Nevertheless, there are several therapies based on the administration of protease inhibitors. Such therapies include anticoagulant therapy [347], treatment of hypertension [348], and septic shock [19]. Angiotensin-converting enzyme inhibitors such as catopril and enalapril are being used to treat hypertension. Anticoagulant therapy is another target for antiprotease intervention. Excessive activation of the coagulation cascade can lead to disseminated intravascular coagulation; insufficient activity of coagulation or increased fibrinolysis can lead to imbalances in either direction. Aprotinin is an effective inhibitor of plasmin, but it is not orally effective because it is a protein and is therefore vulnerable to digestion [349]. Eglin C is another protease inhibitor investigated for use as a potential antiproteolytic drug [349a]. Undoubtedly, protease inhibitors have been recognized as having great potential as drugs. As information in this field accumulates, the number of protease inhibitors that can potentially be used as drugs is certain to grow.

Selective inhibition of a number of serine proteases would potentially be useful as a basis for therapeutic agents. For instance, Kettner and Shenvi [350] have shown that peptide boronic acids are effective inhibitors of the serine proteases, chymotrypsin, pancreatic elastase, and leukocyte elastase. The kinetic parameters for these peptide boronic acids in their interaction with alpha-lytic proteases were elucidated and were found to be similar to those of other serine proteases. Alpha-lytic

proteases are able to hydrolyze substrates having either alanine or valine
in the P1 site and have a preference for substrates having a P1 alanine.
The kinetic parameters of the tri- and tetrapeptide analogs correlate with
the mechanism for slow-binding inhibition, while the less effective inhibitors are simply competitive inhibitors [351].

In the following section, α-1-antitrypsin (A1AT) and an inhibitor of
the HIV protease will be discussed to illustrate the therapeutic application
of protease inhibitors.

Alpha-1-Antitrypsin

Individuals with a deficiency in alpha-1-antitrypsin (A1AT), the major
endogenous inhibitor of neutrophil elastase, are susceptible to premature
development of emphysema. Thus, a greater understanding of this serine
proteinase inhibitor has been a major objective of research on the pathogenesis of emphysema. Clinically, A1AT deficiency is associated with
emphysema in adults and, less commonly, liver disease in neonates [352].

A1AT is a 52-kD, 394-amino-acid, single-chain glycoprotein normally
present in serum at 150−350 mg/dl. The A1AT gene, composed of seven
exons dispersed over 12 kb of chromosomal segment 14q31-32.3, is expressed in hepatocytes and mononuclear phagocytes. The A1AT protein,
a member of the class of protease inhibitor proteins known as serpins
(serine protease inhibitors), is a globular molecule composed of nine alpha-
helicies and three beta-pleated sheets. The major function of A1AT is to
inhibit neutrophil elastase. It does so through an active site centered
around Met^{358} contained within an external stressed loop on the surface
of the molecule. A1AT is a highly pleomorphic protein with greater than
75 variants determined at the protein and/or gene level. These variants
can be categorized into four groups according to their serum A1AT level
and function: normal, deficient, dysfunctional, and absent. There are
two important salt bridges within the A1AT molecule ($Glu^{342}-Lys^{263}-$
Lys^{387}) and a mutation in the A1AT gene causing disruption of either salt
bridge causes distinct molecular pathology resulting in reduced serum
A1AT levels. Clinically relevant variations can be distinguished by a
combination of isoelectric focusing of serum, restriction-fragment-length
analysis of genomic DNA, oligonucleotide probes, and direct sequencing of
the variant A1AT genes [353].

A1AT acts competitively by allowing its target enzymes to bind directly to a substrate-like region near its carboxy-terminal. A peptide bond
in the inhibitor is hydrolyzed during formation of the enzyme-inhibitor
complex. Hydrolysis of this reactive-site peptide bond, however, does
not proceed to completion. An equilibrium close to unity is established
between complexes in which the reactive-site peptide bond of A1AT is
intact and those in which this peptide bond is cleaved. The complex of
A1AT and serine protease is a covalently stabilized structure that is resistant to dissociation by denaturing compounds such as sodium dodecyl
sulfate and urea. The interaction between A1AT and serine protease is
suicidal in that the modified inhibitor is no longer able to bind or inactivate enzyme. During complex formation and hydrolysis of the reactive-
site peptide bond in vitro, a 4-kD carboxy-terminal fragment of the inhibitor is released. But it is not known whether this peptide fragment is
actually released under physiological conditions, since it remains attached
to modified A1AT by tenacious hydrophobic association during isolation
[352].

The functional activity of A1AT in complex biological fluids may be modified by several factors. First, the reactive site methionine of A1AT may be oxidized, rendering it inactive. This is precisely the fate met by A1AT when exposed to the oxidants released by activated neutrophils and alveolar macrophages in cigarette smokers. Whether the above reaction occurs in vivo is unknown. Second, the functional activity of A1AT may be modified by proteolytic inactivation. Examples of such proteases include a metalloprotease secreted by mouse macrophages, a human neutrophil-derived metalloenzyme, thiol protease C, cathepsin L, and *Pseudomonas* elastase [352].

Perhaps the most important reason underlying the interest in human A1AT is the possible role of its absence in plasma in the early development of emphysema. Indeed, in persons deficient in A1AT, the circulating levels of the protein are approximately 10% of that in normal persons. This is primarily because of abnormal glycosylation of an aberrant form of the inhibitor during posttranslational processing so that most of the synthesized protein accumulates as inclusion bodies in the liver [354].

A deficiency of A1AT results in insufficient antielastase protection in the lower respiratory tract, thus allowing neutrophil elastase to destroy alveolar structures. The goal of A1AT augmentation therapy in A1AT deficiency is to raise lung A1AT levels and anti-neutrophil elastase capacity to levels that will provide adequate protection against neutrophil elastase, thereby preventing the lung from further elastase-mediated degradation [355].

A1AT was isolated via large-scale fractionation methods from pooled human plasma for use as a therapeutic agent by Fournel et al. in 1988 [356]. A concentrated preparation of this protein, both in a highly purified form and in a composition identical to that intended for clinical use, was tested in rabbits and cynomolgous monkeys relative to pharmacokinetics and pharmacodynamics. Both unlabeled and ^{125}I-labeled A1AT isolated from pooled human plasma were used. A significant increment in A1AT in bronchial-alveolar lavage fluid was detected. A catabolic half-life of 48.5 hr was obtained for the labeled material in rabbit plasma, which agreed well with the antigenic decay (35.5 hr) measured with a specific enzyme-linked immunosorbent assay and with the functional activity decay (38.1 hr) measured antigenically by the ability of resident human A1AT to complex with human neutrophil elastase. The catabolic half-life in monkeys was 55.45 hr. No unusual tissue distribution was observed at the first, 24th, or 168th hr of sacrifice. Safety studies assessing acute physiological response and both acute and subacute toxicity presented no significant adverse effects [356].

Studies with intravenous administration of human A1AT (60 mg/kg at weekly intervals) were then preformed by Hubbard and Crystal at NIH [355]. Serum A1AT levels increased from an average 33 mg/dl preinfusion to a steady-state trough level of 117 mg/dl, well above the threshold protective serum level of A1AT. The infused A1AT diffused into the lung, significantly augmenting the epithelial lining fluid A1AT levels from an average 0.44 μM to 2.62 μM. This led to significant augmentation of epithelial lining fluid anti-neutrophil elastase capacity and restoration of the lung anti-elastase protection. In the 800 weekly infusions administered no significant adverse reactions occurred, demonstrating that long-term augmentation therapy with weekly infusions of A1AT is rational, safe, and biochemically effective [355].

The efficacy of replacement therapy of A1AT in patients suffering from moderate emphysema has been reported by Moser et al. [357]. All were administered a single intravenous infusion of A1AT in a dose of 60 mg/kg followed by 300 μCi of ^{131}I-labeled A1AT over a 30-min period [357]. No acute toxicity was observed. Compared with baseline data, significant elevations of serum A1AT, measured both antigenically and as anti-elastase activity, occurred with serum half-life of 110 hr. Bronchoalveolar lavage fluid, obtained 48 hr after infusion, reflected a significant increase in A1AT concentration versus baseline bronchoalveolar lavage fluid values. Serial gamma camera images of the lungs confirmed persistence of enhanced lung radioactivity for several days.

In a 6-month, multicenter feasibility and safety study conducted by Schmidt et al. [358], 20 patients, all of whom had a congenital deficiency of A1AT accompanied by chronic obstructive lung disease, were treated with human plasma-derived A1AT. A weekly dose of 60 mg/kg administered intravenously was shown to be sufficient to maintain patient serum levels above the threshold limit of 35%, the serum level of healthy persons with the normal A1AT phenotype. The global concentration in serum or bronchiolar lavage fluid A1AT including active and inactivated A1AT was measured immunologically by rate nephelometry and radial immunodiffusion. The functional activity of A1AT, expressed as free inhibitor activity against trypsin and leukocyte elastase, confirmed that the infused A1AT remained mostly in its active form in the circulation. Reported adverse reactions were moderate and did not require alteration of the schedule of the infusions and/or the dose and rate of administration. Antibodies to A1AT measured by the Ouchterlony method did not develop, and physical signs of possible hepatitis virus contamination were not observed. Consequently, long-term replacement therapy with this protease inhibitor appears to be safe and effective.

Inhibitor of HIV Protease

The etiological agent of acquired immune deficiency syndrome (AIDS) is the human immunodeficiency retrovirus, HIV. HIV proteins, like other retroviral proteins, are initially translated as the large precursor polyproteins *gag*, *pol*, and *env*.* These polyproteins, in turn, are proteolytically processed to generate structural proteins such as p6, p7, p17, and p24, enzymes such as protease, reverse transcriptase, and integrase, and the envelope proteins gp120 and gp41. Processing of the *gag* and *pol* polyproteins involves a virtually encoded protease, as is found in avian myleoblastosis, bovine leukemia, and murine leukemia retroviruses [359].

In 1988, Kohl et al. examined the role of the human immunodeficiency virus protease in the viral replication cycle using site-directed mutagenesis in the protease gene [360]. The HIV protease gene product was expressed in *Escherichia coli* and observed to cleave HIV gag p55 to gag p24 and

---

*The *gag*, *pol*, and *env* proteins are encoded for on the retroviral RNA genome, which is reverse-transcribed to DNA, transcribed to RNA, and translated to produce the three proteins. The *gag*, *pol*, and *env* proteins are then cleaved proteolytically by a peptidase also encoded for on the viral genome to core proteins, polymerases, and envelope glycoproteins, respectively.

gag p17 in vitro. Substitution of aspartic acid residue 25 (Asp-25) of
this protein with an asparagine residue did not affect the expression of
the protein, but it eliminated detectable in vitro proteolytic activity
against HIV gag p55. When SW480 human colon carcinoma cells were transfected with an Asn-25 mutant, proviral DNA produced virions that contained gag p55 but not gag. By contrast, virions from cells transfected
with the wild-type DNA contained both gag p55 and gag p24. The mutant
virions were not able to infect Mt-4 lymphoid cells, whereas the wild-type
virions were able to. The above findings demonstrate that the HIV protease is an essential viral enzyme. Consequently, proteolytic enzyme encoded by the virus that facilitates this processing becomes a potential
target for therapeutic intervention to block HIV infection.

In 1989, Darke et al. expressed the human immunodeficiency virus
protease in *E. coli* and purified it to apparent purity [359]. Recombinant
DNA procedures were performed in which the 5' portion of the *pol* gene of
the NY5 strain of HIV-1 was subcloned into the HindIII/KpnI site of M13
mp8 and mp9. The resulting plasmid containing a promoter and gene fragment coding for the protease was completely sequenced on both strands
and transformed into *E. coli* strain HB101 for expression. The *E. coli*
containing these plasmids were grown in Luria broth at 37°C overnight
and centrifuged. The cells were then resuspended in medium containing
20 µg/ml beta-indolacrylic acid and incubated for 2.5 hr, harvested by
centrifugation, and frozen at −70°C. The cell paste was thawed and suspended in 50 mM Tris HCl (pH 7.5), homogenized, and centrifuged at
12,000 rpm for 20 min. The clear supernatant was applied to and eluted
from a DEAE-Sephadex A-25 column in 50 mM Na HEPES buffer (pH 7.8).
Fractions containing peptide hydrolysis activity were pooled and concentrated in dialysis bags and the volume was reduced two- to threefold.
The dialyzed sample was applied to a phosphocellulose column and eluted
in PC buffer with a gradient of 0−0.1 M NaCl. Active fractions were
pooled and dialyzed and protein concentrations were measured with a dye
binding assay kit.

The activity of HIV-1 protease expressed in bacteria was determined
by incubating HIV-1 gag p55 with extracts from protease-expressing cells
followed by immunoblot analysis of the products. Protease activity as a
function of pH was measured using 10 mM substrate (Val−Ser−Gln−Asn−
Tyr−Pro−Ile−Val) in buffers ranging in pH from 3 to 7. The enzyme
expressed as a nonfusion protein exhibited proteolytic activity with a pH
optimum of 5.5 and was inhibited by aspartic protease inhibitor pepstatin
(Ki = 1.1 µM). Replacement of the conserved residue Asp-25 with an Asn
residue through genetic engineering of a point mutation on the transformed plasmid resulted in complete loss of proteolytic activity of the resulting purified protein supporting the classification of the human immunodeficiency virus protease as an aspartic protease [358].

Later that year, Richards et al. examined the blocking of the HIV-1
proteinase activity with different inhibitors of aspartic inhibitors [361].
Pepstatin is most commonly described as the isovaleryl-derivative (isovaleryl-
Val−Val−Sta−Ala−Sta); however, other forms of the molecule, such as
acetyl-pepstatin (acetyl-Val−Val−Sta−Ala−Sta) and lactoyl-pepstatin
(lactoyl-Val−Sta−Ala−Sta), have been reported and characterized as inhibitors of classical aspartic proteinases. The effect of these compounds
as well as the synthetic aspartic proteinase inhibitor H-261 (tert-butyl-
oxycarbonyl-His−Pro−Phe−His−Leu−CHOH−CH$_2$−Val−Ile−His) on the

viral proteinase was examined in detail. HIV-1 proteinase was similarly prepared and purified from *E. coli*. The peptide substrate, Tyr –Val – Ser –Gln –Asn –Phe –Pro –Ile –Val –Gln –Asn –Arg, was incubated with the recombinant HIV-1 proteinase produced, resulting in specific cleavage at the Phe –Pro bond. Initial rates of hydrolysis were determined by removal of samples at three different time points followed by analysis by reversed-phase HPLC/FPLC with quantitation of the cleavage product, Tyr –Val –Ser –Gln –Asn –Phe. Lactoyl-pepstatin was a rather poor inhibitor of HIV-1 proteinase (Ki = 6 μM) whereas acetyl-pepstatin was much more effective (Ki = 35 nM) and H-261 was still more potent (Ki = 15 nM). These findings imply that this enzyme, essential for replication of the AIDS virus, may be classified unequivocally as belonging to this proteinase family.

An identical study was carried out by the same group on the recently characterized HIV-2 proteinase [362]. Since it would be desirable to develop proteinase inhibitors that would be effective against enzymes from different strains and species of virus, the susceptibility of the proteinase from HIV-2 to a variety of available inhibitors of aspartic proteinases was examined in detail. Isovaleryl-pepstatin (Ki = 1.5 μM), lactoyl-pepstatin (Ki = 16 μM), and H-261 (Ki = 35 nM) were weaker inhibitors of the HIV-2 proteinase than the HIV-1 proteinase, while acetyl-pepstatin (Ki = 5 nM) was a better inhibitor of HIV-2. The finding that the enzymes from HIV-1 and HIV-2 are both relatively susceptible to inhibition by acetyl-pepstatin suggests that a single inhibitor would be effective against both viral enzymes [362].

Recently, McQuade et al. prepared a synthetic peptidomimetic substrate of the HIV-1 protease with nonhydrolyzable pseudodipeptidyl insert at the protease cleavage site [363]. This inhibitor, U-81749, contains a hydroxyethylene isostere [CH(OH)CH$_2$] as a nonhydrolyzable, synthetic replacement of the scissile amide bond at the cleavage site, a cyclohexylalanine residue on the amino- and valine-residue on the carboxyl terminal side of the scissile bond, an amino-terminal tert-butylacetyl, and a carboxy-terminal aminomethylpyridine, tert-butylacetyl-cyclohexylalanine-CH(OH)CH$_2$-Val –Ile –Amp. This inhibitor competitively inhibited recombinant HIV-1 in vitro (Ki = 70 nM) and exhibited good selectivity toward the HIV-1 enzyme, as indicated by a >100 times lower potency toward human renin, another aspartic protease. In the presence of 1 μM of this protease inhibitor, the level of HIV p24, a structural protein of the virion core, in culture supernatants of human peripheral blood lymphocytes already infected 3 or 4 days with HIV virus was reduced 70% compared with control infected cells (8 versus 25 ng/ml and 18 versus 58 ng/ml of p24 at 3 and 4 days after infection, respectively).

To investigate whether the HIV protease was the actual target of U-81749 in cells, processing of the HIV p55 and the maturation of HIV-like particles in recombinant virus-infected CV-1 cells was studied. Infection of CV-1 cells with a recombinant vaccinia virus (vVK-1) engineered to express the HIV *gag –pol* genes resulted in the synthesis and processing of *gag –pol* precursors. Initially, the most prominent HIV polypeptides synthesized were p55, p46, and p41; with time, however, these proteins were further processed to the mature viral proteins p17 and p24. vVK-1-infected CV-1 cell lysates were analyzed for *gag –pol* precursors and p24 at various times after infection by protein immunoblotting. The p24 product was first detected 6 –9 hr after infection, and by 24 hr it accounted for 31% of the immunoreactive proteins as determined by densitometric analysis of the immunoblots. Addition of 2.5, 5, or 10 μM U-81749 to infected cultures reduced the levels of p24 in cell lysates at this time 32, 42, and 84%, respectively. Similarly, the levels of P17 were also reduced 16, 34, and 49% at these concentrations of U-81749. These data indicate that U-81749 inhibits the processing of p55 by the HIV-1 protease in these mammalian cells.

Cells transfected with HIV proviral DNA encoding a protease mutagenized in the presumed active site produced noninfectious virions composed of p55 but not p24. This protease appears to be essential for the maturation of viral particles to infectious virus. In a study to determine whether this inhibitor prevented the maturation of the HIV-like particles found in the medium of vVK-1-infected cells, the medium was collected 16 hr after infection of CV-1 cells, and the HIV-like particles recovered by centrifugation. The protein components of the particles were then analyzed by protein immunoblotting. HIV-like particles recovered from control medium contained almost exclusively p24 (>97% of immunoreactive proteins) and no detectable p55 or other *gag* precursors. In contrast, particles recovered from drug-treated cultures contained almost exclusively p55 and p46 (>90% of the immunoreactive protein) and only trace levels of p24 (<9% of the immunoreactive protein). These data indicate that U-81749 is present within the HIV-like particle, where it directly inhibits the protease and thus the subsequent maturation of the particle.

There is, therefore, encouraging evidence that a chemical compound, specifically targeted against the HIV-1 protease, can inhibit viral replication in cell culture. This optimism is further supported by the recent findings of Roberts et al. [364], who demonstrated that the more active peptide derivatives based on the transition-sate mimetic concept were able to inhibit both HIV-1 and HIV-2 proteinases in the nanomolar range without causing cytotoxicity in C8166 and JM cells at 10 and 5 μM, respectively. As it is still early in their development, there is as yet no information on the pharmacokinetics and pharmacodynamics of the above inhibitors.

C. Formulation Approach

A third strategy to circumvent the enzymatic barrier is the formulation approach. Here, the peptide or protein drug is housed within a delivery system that is designed not only to protect the drug from contact with the luminal proteases, but also to release the drug only upon reaching an area favorable to its absorption, be it due to low protease content or enhanced permeability characteristics.

Over the years several formulations have been explored to protect insulin from proteolysis in the GI tract, including water-in-oil-in-water emulsions [365 – 367], liposomes [368 – 370], nanoparticles [371], and soft gelatin capsules coated with polyacrylic polymers with pH-dependent properties [372]. The results have been mixed. Thus far, the most elegant delivery system designed for this purpose is that of Saffran et al. [373]. The heart of their system is an azo-cross-linked copolymer of styrene and hydroxyethymethacrylate that coats the protein and remains intact throughout the GI tract until arrival at the ileocecal junction. The azoreductase elaborated by the indigenous bacteria proceeds to reduce the $R - C_6H_4 - N = N - C_6H_4 - R$ bond, causing the polymer to disintegrate and the encapsulated protein to be released and absorbed. While this system releases insulin at the site where it is least likely to be degraded, very little insulin may be absorbed into the liver where it acts. Cho and Flynn [374] attempted to solve this problem by administering insulin in a water-in-oil microemulsion composed of cholesterol, lecithin, and an esterified fatty acid in a proportion mimicking that in the chylomicrons, that is, 1:2:14. An unspecified amount of aprotinin, a protease inhibitor, was also added to guard against insulin degradation. At a dose of 1 U/kg, three patients, all with a long history of diabetes, responded positively, showing significant reduction in blood glucose. While these results are encouraging, additional information is needed on the droplet size in the microemulsion, the mechanism of drug release from the microemulsion, the fate of the microemulsion in the GI tract and beyond, the bioavailability of insulin from this formulation, and total duration of blood glucose reduction.

REFERENCES

1. T. W. Redding and A. V. Schally, *Neuroendocrinology*, 9:250–256 (1972).
2. J. A. Parsons, B. Rafferty, R. W. Steveson, and J. M. Znaell, *Br. J. Pharmacol.*, 59:489P–490P (1977).
3. J. A. Parsons, B. Rafferty, R. W. Stevenson, and J. M. Zanelli, *Br. J. Pharmacol.*, 66:25–27 (1979).
4. B. Given, R. Cohen, B. Rrank, A. Rubenstein, and H. Tager, *Clin. Res.*, 33:569A (1985).
5. K. Docherty, R. Carroll, and D. F. Steiner, *Proc. Natl. Acad. Sci. USA*, 79:4613–4617 (1982).
6. S. O. Emdin, G. G. Dodson, J. M. Cutfield, and S. M. Cutfield, *Diabetologia*, 19:174–182 (1974).
7. M. Forgac and L. Cantley, *J. Biol. Chem.*, 259:8101–8105 (1984).
8. M. Forgac, L. Cantley, B. Wiedenmann, L. Altstiel, and D. Brnaton, *Proc. Natl. Acad. Sci. USA*, 80:1300–1303 (1983).
9. C. J. Galloway, G. E. Dean, M. Marsh, G. Rudnick, and I. Mellman, *Proc. Natl. Acad. Sci. USA*, 80:3334–3338 (1983).
10. S. Ohkuma, Y. Moriyama, and T. Takano, *Proc. Natl. Acad. Sci. USA*, 79:2758–2762 (1982).
11. L. Orci, M. Ravazzola, M. Amherdt, O. Madsen, A. Perrelet, J.-D. Vassalli, and R. G. W. Anderson, *J. Cell Biol.*, 103:2273–2281 (1986).
12. J. Glickman, K. Croen, S. Kelly, and Q. Al-Awqati, *J. Cell. Biol.*, 97:1303–1308 (1983).
13. F. Zhang and D. L. Schneider, *Biochem. Biophys. Res. Commun.*, 114:620–625 (1983).
14. J. C. Hutton, *Biochem. J.*, 204:171–178 (1982).
15. D. Scherman, J. Nordmann, and J.-P. Henry, *Biochemistry*, 21:687–694 (1982).
16. P. A. Halban, R. Mutkoski, G. Dodson, and L. Orci, *Diabetologia*, 30:348–353 (1987).
17. J. Kraut, *Annu. Rev. Biochem.*, 15:331–358 (1977).
18. A. Warshel, G. Nary-Szabo, F. Sussman, and J. K. Hwang, *Biochemistry*, 28:3629–3637 (1989).
19. J. S. Bond and R. J. Beynon, in *Proteolysis and Physiological Regulation* (H. Baum, ed.), Pergamon Press, Oxford (1987), pp. 173–285.

20. A. J. Barrett and G. Salvesen, *Proteinase Inhibitors*, Elsevier-North Holland Biomedical Press, Amsterdam (1986).

21. T. Murachi, *Biochem. Soc. Trans.*, *13*:1015–1018 (1985).

22. I. T. Carney, C. G. Curtis, J. Kay, and N. Birket, *Biochem. J.*, *185*:423–434 (1980).

23. E. G. Afting, in *Proteinase Inhibitors: Medical and Biological Aspects* (N. Katunuma, H. Umezawa, and H. Holzer, eds.), Springer-Verlag, Berlin (1983), pp. 191–200.

24. D. D. J. Green, A. A. Kembhari, M. E. Davies, and A. J. Barrett, *Biochem. J.*, *218*:939–946 (1984).

25. G. E. Mortimore and A. R. Poso, *Annu. Rev. Nutr.*, *7*:539–564 (1987).

26. N. J. Hutson and G. E. Mortimore, *J. Biol. Chem.*, *257*:95448–95454 (1982).

27. G. E. Mortimore, N. J. Huston, and C. A. Surmacz, *Proc. Natl. Acad. Sci. USA*, *80*:2179–2183 (1983).

28. G. Soberon and Q. E. Sanchez, *J. Biol. Chem.*, *239*:1602–1606 (1961).

29. D. J. Millward, D. O. Nnanyelugo, W. P. T. James, and P. J. Garlick, *Br. J. Nutr.*, *32*:127–142 (1974).

30. A. Pope and R. A. Nixon, *Neurochem. Res.*, *9*:291–323 (1984).

31. J. F. Woessner and W. H. Azzo, *J. Rheumatol.*, *14*:S36–S37 (1987).

32. K. Deshmukh-Phadke, M. Lawrence, and S. Nanda, *Biochem. Biophys. Res. Commun.*, *85*:490–496 (1978).

33. M. K. B. McGuire-Goldring, J. E. Meats, D. D. Wood, E. J. Ihrie, N. M. Ebsworth, and R. G. G. Russell, *Arthritis Rheum.*, *27*:654–662 (1984).

34. J. P. Pelletier, J. Martel-Pelletir, D. S. Howell, L. G. Mnaymne, J. E. Enis, and J. F. Woessner, *Arthritis Rheum.*, *26*:63–68 (1983).

35. J. Marel-Pelletier, J. P. Pelletier, J. M. Clouther, D. Howell, L. Ghandur-Mnaymneh, and J. F. Woessner, *Arthritis Rheum.*, *26*:63–68 (1983).

36. D. D. Wood, E. J. Ehrie, C. A. Dinarello, and P. L. Cohen, *Arthritis Rheumat.*, *26*:975–983 (1983).

37. R. D. Pasternak, S. J. Hubbs, R. C. Caccese, R. L. Marks, J. M. Conaty, and G. DiPasquale, *Agents–Actions*, *21*:328–330 (1987).

38. K. Phadke, *Biochem. Biophys. Res. Commun.*, *142*:448–453 (1987).

39. J. J. Oppenheim, E. J. Kovac, K. Matsushima, and S. K. Durum, *Immunol. Today*, *7*:45–56 (1986).

40. G. Braedt, V. Price, S. Gillis, C. S. Henney, S. R. Kronheim, K. Grabstein, P. J. Conlon, T. P. Hopp, and D. Cosman, *Nature*, *315*:641–647 (1985).

41. D. J. Hazuda, J. C. Lee, and P. R. Young, *J. Biol. Chem.*, *263*:8473–8479 (1988).

42. R. A. Black, S. R. Kronheim, and P. R. Sleath, *FEBS Lett.*, *247*:386–390 (1989).

43. P. R. Young, D. J. Hazuda, and P. L. Simon, *J. Cell Biol.*, *107*:447–456 (1988).

44. P. Elodi, I. Kiss, G. Cs-Szabo, and M. Pozsgay, in *Proteinase Action* (P. Elodi, ed.), Akademiai Kiado, Budapest (1984), pp. 81–98.

45. H. Neurath, *Science, 224*:350–357 (1984).
46. K. Katayama, L. H. Ericsson, D. L. Enfield, K. A. Walsh, H. Neurath, E. W. Davie, and K. Titani, *Proc. Natl. Acad. Sci. USA, 76*:4990–4994 (1979).
47. T. Murachi, *Biochem. Soc. Trans., 13*:1015–1018 (1985).
48. C. M. Jackson and Y. Memerson, *Annu. Rev. Biochem., 49*:765–811 (1980).
49. J. K. McDonald, *Histochem. J., 17*:173–185 (1985).
50. A. J. Barrett and J. K. McDonald, *Mammalian Proteases: A Glossary and Bibliography. Vol. 1: Endopeptidases*, Academic Press, London (1980).
51. E. Hayakawa, A. Yamamoto, Y. Shoji, and V. H. L. Lee, *Life Sci., 45*:167–174 (1989).
52. T. Spencer and J. M. Sturtevant, *J. Am. Chem. Soc., 81*:1874–1881 (1959).
53. G. Schoellman and E. Shaw, *Biochemistry, 2*:252–255 (1963).
54. L. Lorand, *Methods in Enzymology*, Academic Press, New York (1981), p. 80
55. E. Dufour, *Biochimie, 70*:1335–1342 (1988).
56. E. D. Harris and E. C. Cartwright, in *Proteinases in Mammalian Cells and Tissues* (A. J. Barrett, ed.), Elsevier-North Holland Biomedical Press, Amsterdam (1977), pp. 249–283.
57. J. S. Bond and R. J. Beynon, *Int. J. Biochem., 14*:565–574 (1985).
58. J. S. Bond and R. J. Beynon, *Curr. Top. Cell. Regul., 28*: 263–290 (1986).
59. J. J. Kiepnieks and A. Light, *J. Biol. Chem., 254*:1677–1683 (1979).
60. B. W. Mathews, P. M. Colman, J. N. Jansonius, K. Titani, K. A. Walsh, and H. Neurath, *Nature, 238*:42–43 (1972).
61. W. C. Duckworth, D. E. Peavy, F. G. Hamel, J. Liepnieks, M. R. Brunner, R. E. Heiney, and B. H. Frank, *Biochem. Jpn., 255*: 277–284 (1988).
62. K. D. Wilkinson, M. K. Urban, and A. L. Haas, *J. Biol. Chem., 255*:7529–7532 (1980).
63. A. Hershko, H. Heller, S. Elias, and A. Chiechanover, *J. Biol. Chem., 258*:8206–8214 (1983).
64. A. Ciechanover, D. Finley, and A. Varshavsky, *Cell, 37*:57–66 (1984).
65. J. Kay, *Biochem. Soc. Trans., 13*:1027–1029 (1985).
66. W. Carlson, M. Karplus, and E. Haber, *Hypertension, 7*:13–26 (1985).
67. A. M. Hemmings, S. I. Foundling, B. L. Sibanda, et al., *Biochem. Soc. Trans., 13*:1036–1041 (1985).
68. J. S. Bond and R. J. Beynon, *Curr. Top. Cell. Regul., 28*:263–290 (1986).
69. J. S. Bond and R. J. Beynon, *Int. J. Biochem., 14*:565–574 (1985).
70. B. W. Matthews, J. N. Colman, J. N. Jansonius, et al., *Nature, 238*:41–43 (1972).
71. Y. Tahara, K. Shima, M. Hirota, H. Ikegami, A. Tanaka, and Y. Kumahara, *Biochem. Biophys. Res. Commun., 113*:340–347 (1983).
72. H. Kaneda, T. Manaka, T. Kaminura, T. Katagiri, and H. Sasaki, *Diabetologia, 26*:392–396 (1984).

73. H. Tsubouchi, E. Gohda, H. Miyazaki, A. Kamibeppu, H. Oka,
    J. Nagahama, S. Hirono, K. Fujisaki, O. Nakamura, Y. Daikuhara,
    and S. Hashimoto, *J. Clin. Invest.*, *49*:837–848 (1976).
74. R. Gossrau, *J. Histochem. Cytochem.*, *29*:464–480 (1981).
75. J. K. McDonald and A. J. Barrett, *Mammalian Proteases: A Glossary
    and Bibliography, Vol. 2: Expeptidases*, Academic Press, London
    (1986).
76. L. W. Hays, C. A. Goguen, S. F. Ching, and L. L. Stakey,
    Biochem. Biophys. Res. Commun., *82*:1147–1153 (1978).
77. A. R. Johnson and E. G. Erdos, *J. Clin. Invest.*, *59*:684–695
    (1977).
78. R. R. Johnson, *J. Clin. Invest.*, *65*:841–850 (1980).
79. P. D. Bowman, A. L. Betz, D. Ar, J. S. Wolinsky, J. B. Penney,
    R. R. Shivers, and G. W. Goldstein, *In Vitro*, *17*:353–362 (1981).
80. R. S. Bar, *Annu. NY Acad. Sci.*, *401*:150–162 (1982).
81. P. R. B. Caldwell, B. C. Seegal, K. C. Shu, M. Das, and R. L.
    Soffer, *Science*, *191*:1050–1051 (1976).
82. E. Erdos, *Handbook Exp. Pharmacol.*, *25*:428–468 (1979).
83. J. Ryan, U. Ryan, A. Chung, and G. Fisher, *Adv. Exp. Med.
    Biol.*, *156*:775–781 (1983).
84. S. L. Stephenson and A. J. Kenny, *Biochem. J.*, *241*:237–247
    (1987).
85. J. Szechinski, W. Hsia, and F. Behal, *Enzyme*, *29*:21–31 (1983).
86. W. Sidorowicz, J. Szechniski, P. Canizar, and F. Behal, *Proc.
    Soc. Exp. Biol. Med.*, *175*:503–509 (1984).
87. T. Kato, M. Okada, and T. Nagatsu, *Mol. Cell. Biochem.*, *32*:
    117–121 (1980).
88. T. Kato, T. Nakano, K. Kojima, T. Nagatsu, and S. Sakakibara,
    J. Neurochem., *35*:527–535 (1980).
89. T. Yoshimoto, W. Simmons, T. Kita, and D. Tsuru, *J. Biochem.*,
    90:325–334 (1980).
90. M. Orlowski, S. Wilk, S. Pearce, and S. Wilk, *J. Neurochem.*, *33*:
    461–469 (1979).
91. L. Hersch, *J. Neurochem.*, *37*:172–189 (1981).
92. S. Mizutani, S. Sumsi, O. Suzuki, O. Narita, and Y. Tomada,
    Biochim. Biophys. Acta, *786*:113–117 (1984).
93. T. Yoshsimoto and D. Tsuru, *Agric. Biol. Chem.*, *42*:2417–2419
    (1978).
94. J. M. Mambrook, B. A. Morgan, M. J. Rance, and C. F. C. Smith,
    Narure, *262*:782–783 (1976).
95. T. Hazato, M. Shimamura, T. Katayama, A. Kasama, S. Nishioka,
    and K. Kaya, *Life Sci.*, *33*:443–448 (1983).
96. B. Giros, C. Gros, B. Solhonne, and J. C. Schwartz, *Mol.
    Pharmacol.*, *29*:281–287 (1986).
97. L. Churchill, H. H. Bausback, M. E. Gerritsen, and P. E. Ward,
    Biochim. Biophys. Acta, *923*:35–41 (1987).
98. C. Gross, B. Giros, and J. C. Schwartz, *Biochemistry*, *24*:2179–2185
    (1983).
99. M. Kobari, F. Gotoh, Y. Fukuuchi, T. Amano, N. Suzuki,
    D. Uematsu, K. Obara, I. Gogolak, and P. Sandor, *J. Cereb. Blood
    Flow Metab.*, *5*:34–39 (1985).

100. W. M. Pardridge and L. J. Mietus, *Endocrinology, 109*:1138–1143 (1981).

101. A. L. Betz, *Fed. Proc., 44*:2614–2615 (1985).

102. F. E. Palmieri, P. M. Luckett, S. A. Stalcup, and P. E. Ward, *Fed. Proc., 43*:651 (1984).

103. R. Siman, M. Baudry, and G. Lynch, *Proc. Natl. Acad. Sci. USA, 81*:3572–3576 (1984).

104. R. Siman, M. Baudry, and G. Lynch, *Nature, 313*:225–228 (1985).

105. R. Siman, M. Baudry, and G. Lynch, *Nature, 313* :225–228 (1985).

106. D. S. J. B. Emmanouel, A. H. Jaspan, A. H.-J. Rubenstein, E. Huen, and A. I. Katz, *J. Clin. Invest., 62*:6–13 (1978).

107. A. I. Katz and A. H. Rubenstein, *J. Clin. Invest., 52*:1113–1121 (1973).

108. R. Nakamura, M. Hayashi, D. S. Emmanouel, and A. I. Katz, *Am. J. Physiol., 250*:F144–F150 (1986).

109. W. C. Duckworth, *Biochim. Biophys. Acta, 437*:531–542 (1976).

110. A. J. D'Ercole, L. E. Underwood, J. J. Van Wyk, D. J. Decedue, and D. B. Fonshee, in *Proceeding of the Third International Symposium on Growth Hormone and Related Peptides* (A. Pecile and E. E. Miller, eds.), Excerpta Medica, Amsterdam (1976), p. 220.

111. J. A. D'Ercole, D. J. Decedue, R. W. Furlanetto, L. E. Underwood, and J. J. Van Wyk, *Endocrinology, 101*:577–586 (1977).

112. E. Runderknecht and E. Humbel, *Proc. Natl. Acad. Sci. USA, 73*:2365–2369 (1976).

113. M. E. Svoboda, J. J. Van Wyk, D. G. Knapper, R. E. Fellows, F. E. Grissom, and R. J. Schlneter, *Biochemistry, 19*:790–797 (1980).

114. B. Bhaumick, H. J. Goren, and R. M. Bala, *Horm. Metab. Res., 13*:515–518 (1981).

115. J. Zapf, U. Schoenle, and E. R. Froesch, *Eur. J. Biochem., 87*:285–296 (1978).

116. S. D. Cheranausek, S. Jacobs, and J. J. Van Wyk, *Biochemistry, 20*:7345–7350 (1981).

117. J. Massague and M. P. Czech, *J. Biol. Chem., 257*:5038–5045 (1982).

118. B. Bhaumick and R. M. Bala, *Endocrinology, 120*:1439–1448 (1987).

119. R. Gossrau, A. Heidland, and J. Haunschild, *Adv. Exp. Med. Biol., 240*:351–360 (1988).

120. D. M. Matthews and J. W. Payne, *Curr. Top. Memb. Transp., 14*:331–425 (1980).

121. D. M. Matthews, *Physiol. Rev., 55*:537–608 (1975).

122. M. G. Gardner, *Biol. Rev., 59*:289–331 (1984).

123. V. H. L. Lee, S. Dodda Kashi, R. H. Patel, E. Hayakawa, and K. Inagaki, *Proc. Int. Symp. Control. Rel. Bioact. Mater., 14*:23–24 (1987).

124. R. E. Stratford and V. H. L. Lee, *Int. J. Pharm., 30*:73–82 (1986).

125. S. Dodda Kashi and V. H. L. Lee, *Life Sci., 38*:2019–2028 (1986).

126. S. Pontremoli and E. Melloni, *Annu. Rev. Biochem., 55*:455–481 (1986).

127. J. S. Bond and P. E. Butler, *Annu. Rev. Biochem., 56*:333–364 (1987).

128. H. Ono and S. Tuboi, *J. Biol. Chem., 263*:3188–3193 (1985).

129. M. Yang, R. E. Jensen, M. P. Yaffe, W. Oppliger, and G. Schatz, *EMBO J.*, *7*:3857–3862 (1988).

130. W.-P. Ren, H. Ono, and S. Tuboi, *Biochem. Biophys. Res. Commun.*, *163*:215–219 (1989).

131. M. A. Hussain, A. B. Shenvi, S. M. Rowe, and E. Shefter, *Pharm. Res.*, *6*:186–189 (1989).

132. A. J. Kenny and A. J. Turner, eds., in *Mammalian Ecoenzymes*, Elsevier, Amsterdam (1987), pp. 67–84

133. A. J. Barrett, ed., in *Proteinases in Mammalian Cells and Tissues*, Elsevier, North Holland, Amsterdam (1977).

134. R. J. Beynon, J. D. Shannon, and J. S. Bond, *Biochem. J.*, *199*: 591–598 (1981).

135. A. J. Kenny, S. L. Stephenson, and A. J. Turner, in *Mammalian Ectoenzymes* (A. J. Kenny and A. J. Turner, eds.), Elsevier, Amsterdam (1987), pp. 169–210.

136. R. Zolfaghari, C. R. F. Baker, P. C. Canizaro, M. Feola, A. Amirgholami, and F. J. Behal, *Enzyme*, *36*:165–178 (1986).

137. K. A. Thomas and R. A. Bradshaw, *Enzymology*, *80*:609–620 (1981).

138. K. A. Thomas, N. C. Baglan, and R. A. Bradshaw, *J. Biol. Chem.*, *266*:9156–9166 (1981).

139. G. M. Littlewood, L. L. Iversen, and A. J. Turner, *Neurochem. Int.*, *12*:383–389 (1988).

140. A. J. Kenny and S. L. Stephenson, *FEBS Lett.*, *232*:1–8 (1988).

141. R. T. Dean and A. J. Barrett, *Essays Biochem.*, *12*:1–40 (1976).

142. C. de Duve, *Eur. J. Biochem.*, *137*:391–397 (1983).

143. J. S. Bond and N. N. Aronson, Jr., *Arch. Biochem. Biophys.*, *227*:367–372 (1983).

144. A. J. Barrett and H. Kirschke, *Methods Enzymol.*, *80*:535–561 (1981).

145. J. S. Huang, S. S. Huang, and J. Tang, in *Enzyme Regulation and Mechanism of Action* (P. Mildner and B. Ries, eds.), Pergamon, Oxford (1980), pp. 289–306.

146. N. Katunuma and E. Kominami, *Curr. Top. Cell Regul.*, *22*:71–101 (1983).

147. J. S. Bond and A. J. Barrett, *Biochem. J.*, *189*:17–25 (1980).

148. R. J. Mayer and F. Doherty, *FEBS Lett.*, *198*:181–193 (1986).

149. G. E. Mortimore and A. R. Poso, *Fed. Proc.*, *43*:1289–1294 (1984).

150. E. Kominami, S. Hashida, E. A. Khairallah, and N. Katunuma, *J. Biol. Chem.*, *258*:6093–6100 (1983).

151. J. Ahlberg and H. Glaumann, *Exp. Mol. Pathol.*, *42*:78–88 (1985).

152. H. Glaumann and F. J. Ballard, *Lysosomes: Their Role in Protein Breakdown*, Academic Press, New York (1986).

153. R. Dean, *Nature*, *257*:414–416 (1975).

154. F. Henell and H. Gluamann, *Exp. Cell Res.*, *158*:257–262 (1985).

155. L. Marzell and H. Glaumann, *Int. Rev. Exp. Pathol.*, *25*:239–279 (1983).

156. G. E. Mortimore, *Nutr. Rev.*, *40*:1–12 (1983).

157. C. De Duve and R. Wattiaux, *Annu. Rev. Physiol.*, *28*:435–492 (1966).

158. A. B. Novikoffand and W.-Y. Shin, *Proc. Natl. Acad. Sci. USA*, *75*:5039–5044 (1978).

159. L. Marzella, J. Ahlberg, and H. Glaumann, *Exp. Cell Res.*, *129*: 460–467 (1980).
160. J. M. Dayer, S. M. Krane, R. G. G. Russell, and D. R. Robinson, *Proc. Natl. Acad. Sci. USA*, *73*:945–949 (1976).
161. J. Ahlberg and H. Glaumann, *Exp. Mol. Pathol.*, *42*:78–82 (1985).
162. R. J. Beyon, E. J. Cookson, and P. E. Butler, *Biochem. Soc. Trans.*, *13*:1005–1026 (1985).
163. C. M. Schworer, K. A. Shiffer, and G. E. Mortimore, *J. Biol. Chem.*, *256*:7652–7658 (1981).
164. C. A. Surmacz, A. R. Poso, and G. E. Mortimore, *Biochem. J.*, *242*:253–258 (1987).
165. A. N. Neely, P. B. Nelson, and G. E. Mortimore, *Biochim. Biophys. Acta*, *338*:458–472 (1974).
166. I. A. Rose, J. V. Warms, and A. Hershko, *J. Biol. Chem.*, *254*: 8135–8138 (1979).
167. S. Bigelow, R. Hough, and M. Rechsteiner, *Cell*, *25*:83–93 (1981).
168. A. C. M. Camargo, R. Shapanka, and L. J. Greene, *Biochemistry*, *12*:1838–1844 (1973).
169. M. Orlowski, C. Michaud, and T. G. Chu, *Eur. J. Biochem.*, *135*: 81–88 (1983).
170. J. H. Hanko and J. E. Hardebo, *Eur. J. Pharmacol.*, *51*:295–297 (1978).
171. P. Sandor, I. T. Demchenko, Y. N. Morgalyov, Y. E. Moskalenko, and A. G. B. Kovach, *Acta Physiol. Acad. Sci. Hung.*, *61*:155–161 (1983).
172. E. A. Khairallah, J. S. Bond, and J. W. C. Bird, *Intracellular Protein Catabolism*, Liss, New York (1985).
173. A. Hershko and A. Ciechanover, *Annu. Rev. Biochem.*, *51*:335–364 (1982).
174. G. N. DeMartino and A. L. Goldberg, *J. Biol. Chem.*, *254*: 3712–3715 (1979).
175. M. C. Beinfeld, J. Bourdais, A. Morel, P. F. Kuks, and P. Cohen, *Biochem. Biophys. Res. Commun.*, *160*:968–976 (1989).
176. S. Pontremoki and E. Melloni, *Annu. Rev. Biochem.*, *55*:455–481 (1986).
177. T. Curan, A. D. Miller, L. Zokas, and I. Verma, *Cell*, *36*:259–261 (1984).
178. F. J. Ballard, in *Principles of Metabolic Control in Mammalian Systems* (R. H. Herman, R. M. Cohn, and P. D. McNamara, eds.), Plenum Press, New York (1980), pp. 221–263.
179. C. M. Berlin and R. T. Schimke, *Mol. Pharmacol.*, *1*:149–156 (1965).
180. A. L. Goldberg and A. C. St. John, *Annu. Rev. Biochem.*, *45*: 747–803 (1976).
181. F. T. Kenney and K. L. Lee, *Bioscience*, *32*:181–184 (1982).
182. J. F. Dice and A. L. Goldberg, *Arch. Biochem. Biophys.*, *170*: 213–219 (1975).
183. W. E. Duncan, M. K. Offermann, and J. S. Bond, *Arch. Biochem. Biophys.*, *199*:331–341 (1980).
184. D. F. Mann, K. Shah, D. Stein, and G. A. Snead, *Biochim. Biophys. Acta*, *788*:17–22 (1984).
185. R. J. Beyon and J. S. Bond, *Am. J. Physiol.*, *251*:C141–C152 (1986).

186. J. S. Bond, *Biochem. Biophys. Res. Commun.*, *43*:333–339 (1971).
187. A. L. Goldberg and J. F. Dice, *Annu. Rev. Biochem.*, *43*:835–869 (1974).
188. A. L. Goldberg and A. C. St. John, *Annu. Rev. Biochem.*, *45*: 747–803 (1976).
189. R. J. Beyon and J. S. Bond, *Am. J. Physiol.*, *251*:C141–C152 (1986).
190. S. Rogers, R. Wells, and M. Rechsteiner, *Science*, *234*:364–368 (1986).
191. U. J. Lewis, R. N. P. Singh, L. F. Bonewald, and B. K. Seavey, *J. Biol. Chem.*, *256*:11645–11650 (1981).
192. H. M. Johnson, M. P. Langford, B.-S. Lakhchura, T.-S. Chan, and G. J. Stanton, *J. Immunol.*, *129*:2357–2359 (1982).
193. E. R. Stadtman, *Trends Biochem. Sci.*, *11*:11–13 (1986).
194. W. Hood, E. de la Morena, and S. Grisolia, *Acta Biol. Med. Germ.*, *36*:1667–1672 (1977).
195. M. K. Offermann, M. J. Mckay, M. W. Marsh, and J. S. Bond, *J. Biol. Chem.*, *259*:8886–8891 (1984).
196. T. Toyo-Oka, *Biochem. Biophys. Res. Commun.*, *7*:44–50 (1982).
197. K. J. A. Davies, *J. Biol. Chem.*, *262*:9895–9901 (1987).
198. K. J. A. Davies and S. W. Lin, *Free Rad. Biol. Med.*, *4*:225–236 (1988).
199. O. Marcillat, Y. Zhang, S. W. Lin, and K. J. A. Davies, *Biochem. J.*, *254*:677–683 (1988).
200. E. R. Stadtman and M. E. Wittenberger, *Arch. Biochem. Biophys.*, *239*:379–387 (1985).
201. R. L. Levine, C. N. Oliver, R. M. Fulks, and E. R. Stadtman, *Proc. Natl. Acad. Sci. USA*, *78*:2120–2124 (1981).
202. J. A. Rivett, *J. Biol. Chem.*, *260*:300–305 (1985).
203. D. Finley, A. Ciechanover, and A. Varshavsky, *Cell*, *37*:43–55 (1984).
204. A. Hershko and A. Ciechanover, *Annu. Rev. Biochem.*, *51*:335–364 (1982).
205. A. Bachmair, D. Finley, and A. Varshavsky, *Science*, *234*:179–186 (1986).
206. F. E. Palmieri and P. E. Ward, *Biochim. Biophys. Acta*, *755*: 522–525 (1983).
207. P. E. Ward, *Biochem. Pharmacol.*, *33*:3183–3193 (1984).
208. F. E. Palmieri, J. J. Petrelli, and P. E. Ward, *Vas. Biochem. Pharmacol.*, *34*:2309–2317 (1985).
209. T. Najdovski, N. Collette, and M. Deschodt-Lanckman, *Life Sci.*, *37*:827–834 (1984).
210. V. H. L. Lee, L. W. Carson, S. Dodda Kashi, and R. E. Stratford, *J. Ocular Pharmacol.*, *2*:345–352 (1986).
211. S. Dodda Kashi and V. H. L. Lee, *Invest. Ophthalmol. Vis. Sci.*, *27*:1300–1303 (1986).
212. R. M. Freidinger, *Trends Pharmacol. Sci.*, *10*:270–274 (1989).
213. H. Tuskamoto, M. A. Kiefer, G. Ananda Rao, E. C. Larkin, C. Largman, and H. Sankaran, *Life Sci.*, *37*:1359–1365 (1985).
214. R. F. Harvey, L. Dowsett, M. Hartog, and A. E. Read, *Lancet*, *2*:826–828 (1973).

215. D. Koulischer, L. Moroder, and M. Deschodt-Lanckman, *Reg. Peptides*, *4*:127–139 (1982).

216. S. Ryder, J. Ing, E. Straus, and R. S. Yalow, *Gastroneterology*, *81*:267–275 (1981).

217. J. R. McDermott, P. R. Dodd, J. A. Edwardson, J. A. Gardy, and A. I. Smith, *Neurochem. Int.*, *5*:641–647 (1983).

218. M. Deschodt-Lanckman and A. D. Strosberg, *FEBS Lett.*, *152*: 109–113 (1983).

219. C. Rose, A. Camus, and J. C. Schwartz, *Proc. Natl. Acad. Sci. USA*, *85*:8326–8330 (1988).

220. K. A. Zuzel, C. Rose, and J. C. Schwartz, *Neuroscience*, *15*: 149–158 (1985).

221. B. Charpentier, D. Pelaprat, C. Durieux, et al., *Proc. Natl. Acad. Sci. USA*, *85*:1968–1972 (1988).

222. S. Elkabes, M. Fridkin, and Y. Koch, *Biochem. Biophys. Res. Commun.*, *103*:240–248 (1981).

223. C. A. Lapp and J. L. O'Conner, *Neuroendocrinology*, *43*:230–238 (1986).

224. J. W. Simkins, P. S. Kalra, and S. P. Kalra, *Endocrinology*, *107*: 573–583 (1980).

225. S. Elkabes, M. M. Fridkin, and Y. Koch, *Biochem. Biophys. Res. Commun.*, *103*:240–248 (1981).

226. H. Berger, R. Pliet, L. Mann, and B. Mehlis, *Peptides*, *9*:7–12 (1988).

227. C. J. Molineaux, A. Lusdun, C. Michaud, and M. Orlowski, *J. Neurochem.*, *51*:624–633 (1988).

228. J. P. Advis, A. M. Contijoch, H. F. Urbanski, and S. R. Ojeda, *Neuroendocrinology*, *47*:102–108 (1988).

229. J. E. Krause, J. P. Advis, and J. F. McKelvy, *Biochem. Biophys. Res. Commun.*, *108*:1475–1481 (1982).

230. M. Fujino, S. Shinagawa, I. Yamazaki, S. Kobayashi, M. Obayasi, T. Fukuda, and R. Nakayama, *Arch. Biochem. Biophys.*, *154*: 488–489 (1973).

231. M. W. Monahan, M. S. Amoss, H. A. Anderson, and W. Vale, *Biochemistry*, *12*:4616–4620 (1973).

232. D. H. Coy, E. J. Coy, A. V. Schally, J. Valchez-Martinez, Y. Hirotsu, and A. Arimura, *Biochem. Biophys. Res. Commun.*, *57*: 335–339 (1974).

233. R. N. Clayton and R. A. Shakespear, *J. Endocrinol.*, *77*:34P (1978).

234. A. D. Swift and D. B. Crighton, *J. Endocrinol.*, *80*:141–152 (1979).

235. N. M. Hooper and A. J. Turner, *Biochem. J.*, *241*:625–633 (1987).

236. B. O'Connor and G. O'Cuinn, *Eur. J. Biochem.*, *104*:999–1006 (1988).

237. S. Endo, H. Yokosawa, and S. Ishii, *Biochem. Biophys. Res. Commun.*, *129*:684–700 (1985).

238. S. Endo, H. Yokosawa, and S. Ishii, *Neuropeptides*, *14*:31–37 (1989).

239. H. P. J. Bennett and C. McMartin, *Pharmacol. Rev.*, *30*:247–292 (1979).

240. K. Bauer and F. Lipmann, *Endocrinology*, *99*:230–242 (1976).

241. E. C. Griffiths and J. A. Kelly, *Mol. Cell Endocrinol.*, *14*:3–17 (1979).
242. E. C. Griffiths, J. A. Kelly, N. White, and S. L. Jeffcoate, *Acta Endocrinol.*, *93*:385–391 (1980).
243. L. B. Hersh and J. F. McKelvy, *Brain Res.*, *168*:553–564 (1979).
244. M. S. Kreider, A. Winodur, and N. R. Krieger, *Neur. Endocr. Lett.*, *3*:115 (1981).
245. T. Matsu, C. Prasad, and A. Peterdofsky, *J. Biol. Chem.*, *254*: 2439–2445 (1979).
246. C. Prasad and A. Peterkosfky, *J. Biol. Chem.*, *251*:3229–3234 (1976).
247. W. L. Taylor and J. E. Dixon, *Biochim. Biophys. Acta*, *444*:428–434 (1976).
248. W. H. Busby, W. W. Youngblood, and J. S. Kizer, *Brain Res.*, *242*: 261–270 (1982).
249. E. C. Griffiths and J. R. McDermott, *Mol. Cell Endocrinol.*, *33*:1–25 (1983).
250. P. Browne and G. O'Cuinn, *Eur. J. Biochem.*, *137*:75–87 (1983).
251. A. Szewzuk and J. Swiatkowska, *Eur. J. Biochem.*, *15*:92–96 (1970).
252. M. Mori, J. Pergues, C. Prasad, R. M. Edwards, and J. F. Wilber, *Biochem. Biophys. Res. Commun.*, *109*:982 (1982).
253. K. Bauer and H. Kleinkauf, *Eur. J. Biochem.*, *106*:107–117 (1980).
254. K. Bauer and P. Nowak, *Eur. J. Biochem.*, *99*:239–246 (1979).
255. W. L. Taylor and J. D. Dixon, *J. Biol. Chem.*, *253*:6934–6940 (1979).
256. K. Bauer, *Nature*, *259*:591–593 (1976).
257. N. White, S. L. Jeffcoate, E. C. Griffiths, and K. C. Hooper, *J. Endocrinol.*, *71*:13–19 (1976).
258. J. T. Neary, J. D. Kieffer, P. Federico, et al., *Science*, *193*: 403–405 (1976).
259. C. Oliver, C. R. Parker, and J. C. Porter, *J. Endocrinol.*, *74*: 339–340 (1977).
260. E. Iversen, *J. Endocr.*, *118*:511–516 (1988).
261. J. H. Rupnow, W. H. Taylor, and J. E. Dixon, *Biochemistry*, *18*: 1206–1212 (1979).
262. S. Wilk and M. Orlowski, *J. Chromatogr.*, *249*:121–129 (1982).
263. P. C. Andrews, C. M. Hines, and J. E. Dixon, *Biochemistry*, *19*: 5494–5500 (1980).
264. E. R. Spindle, M. Lakher, and R. J. Waurtman, *Brain Res.*, *216*: 343–350 (1981).
265. M. Saffran, C. F. Wu, and C. H. Emerson, *Endocrinology*, *110*: 2101–2106 (1982).
266. L. J. Hersh, *J. Neurochem.*, *37*:172–178 (1981).
267. C. Brewster, P. W. Dettmar, and B. Metcalf, *Neuropharmacology*, *20*:497–503 (1981).
268. E. C. Griffiths, J. R. McDermott, and A. I. Smith, *Neurosci. Lett.*, *28*:61–65 (1982).
269. E. H. Morier, K. K. Han, L. Patsouris, O. Mareau, and R. Rips, *Int. J. Peptide Protein Res.*, *18*:513–515 (1981).
270. M. Berger, P. A. Halban, L. Girardier, J. Seydoux, R. E. Offord, and A. E. Renold, *Diabetologia*, *17*:97–99 (1979).

271. T. Lauritzen, O. K. Faber, and C. Binder, *Diabetologia*, *17*:291−295 (1979).

272. E. P. Paulsen, J. W. Courtney, and W. C. Duckworth, *Diabetes*, *28*: 640−645 (1979).

273. G. Freidenberg, N. White, S. Cataland, T. O'Dorisio, J. Sotos, and J. Santiago, *N. Engl. J. Med.*, *305*:363−368 (1980).

274. R. I. Misbin, E. C. Almira, and M. W. Cleman, *J. Clin. Endocrinol. Metab.*, *52*:177−180 (1981).

275. G. F. Maberly, G. A. Wait, J. A. Kilpatrick, E. G. Loten, K. R. Gain, R. D. H. Stewart, and C. J. Eastman, *Diabetologia*, *23*: 333−336 (1982).

276. F. G. Hamel, D. E. Peavy, M. P. Ryan, and W. C. Duckworth, *Endocrinology*, *118*:328−333 (1986).

277. R. Hori, F. Komada, and K. Okumura, *J. Pharm. Sci.*, *72*:435−439 (1983).

278. K. D. Dernovsek and R. S. Bar, B. H. Ginsberg, and M. N. Lioubin, *J. Clin. Endocrinol. Metab.*, *58*:761−763 (1984).

279. M. P. Carson, S. W. Peterson, M. E. Moynhan, and P. Shepro, *In Vitro*, *19*:833−840 (1983).

280. N. Kaiser, I. Vlodarsky, A. Tur-Sinai, Z. Fuks, and E. Cerasi, *Diabetes*, *31*:1077−1083 (1982).

281. G. L. King and S. M. Johnson, *Science*, *227*:1583−1586 (1985).

282. N. Simionescu, M. Simionescu, and G. E. Palade, *J. Cell Biol.*, *64*: 586−607 (1985).

283. G. E. Palade, M. Simionescu, and N. Simionescu, *Acta Physiol. Scand.*, *463*:11−32 (1979).

284. I. D. Goldine, *Endocrinol. Rev.*, *8*:235−255 (1987).

285. K. Yonezawa, K. Yokono, and K. Shii, *Biochem. Biophys. Res. Commun.*, *150*:604−614 (1988).

286. W. C. Duckworth, F. G. Hamel, and D. E. Peavy, *Am. J. Med.*, *85*:71−76 (1988).

287. P. T. Varandani, *Biochim. Biophys. Acta*, *286*:126−135 (1972).

288. W. C. Duckworth, F. B. Stentz, M. Heinemann, and A. E. Kitabachi, *Proc. Natl. Acad. Sci. USA*, *76*:635−639 (1979).

289. P. T. Varandani, L. A. Shroyer, and M. A. Nafz, *Proc. Natl. Acad. Sci. USA*, *69*:1681−1684 (1972).

290. K. Yonezawa, K. Yokono, S. Yaso, J. Hari, K. Amano, Y. Kawase, T. Sakamoto, K. Shii, Y. Imamura, and S. Baba, *Endocrinology*, *118*:1989−1996 (1986).

291. I. D. Goldfine, J. A. Williams, A. C. Bailey, K. Y. Wong, Y. Iwamoto, K. Yokono, S. Baba, and R. A. Roth, *Diabetes*, *33*: 64−72 (1984).

292. H. H. Tomizawa, *J. Biol. Chem.*, *237*:428−432 (1962).

293. P. T. Varandani, *Biochim. Biophys. Acta*, *305*:642−659 (1973).

294. P. T. Varandani, D. Raveed, and M. A. Nafz, *Biochim. Biophys. Acta*, *538*:343−352 (1978).

295. S. Baba, H. Sakai, Y. Imamura, K. Yokono, and S. Smi, in *Proinsulin, Insulin, C Peptide* (Y. Kanazawa, M. Ikeuchi, M. Hayashi, M. Kasai, and K. Kosaka, eds.), Excerpta Medica, Amsterdam (1979), p. 270.

296. G. A. Burghen, A. E. Kitabchi, and J. S. Brush, *Endocrinology*, *91*:633−642 (1972).

297. J. S. Brush, *Diabetes*, *20*:140–145 (1971).
298. W. C. Duckworth and A. E. Kitabchi, *Diabetes*, *23*:536–543 (1974).
299. W. C. Duckworth and A. E. Kitabchi, *Endocrinol. Rev.*, *2*:210–215 (1981).
300. B. J. Goldstein and J. N. Livingston, *Metabolism*, *30*:825–835 (1981).
301. N. Kaiser, I. Vlodasvsky, A. Tur-sinai, Z. Fuks, and E. Cerasi, *Diabetes*, *31*:1077–1083 (1982).
302. K. Dernovsek, R. Bar, B. Ginsberg, and M. Lioubin, *J. Clin. Endocrinol. Metab.*, *58*:761–763 (1984).
303. I. Jialal, G. L. King, S. Buckwald, C. R. Kahn, and M. Crettaz, *Diabetes*, *33*:794–800 (1984).
304. K. D. Dernovsek and B. S. Bar, *Am. J. Physiol.*, *248*:E244–E251 (1985).
305. G. L. King and S. Johnson, *Science*, *227*:1583–1586 (1985).
306. J. A. D'Ercole, C. J. Decedue, R. W. Furlanetto, L. E. Underwood, and J. J. Van Wyk, *Endocrinology*, *101*:577–586 (1977).
307. J. S. Brush and A. E. Kitabchi, *Biochim. Biophys. Acta*, *215*:134–144 (1970).
308. W. C. Duckworth, *Endocrinology*, *104*:1758–1764 (1979).
309. K. Yokono, Y. Imamura, H. Sakari, and S. Baba, *Diabetes*, *28*:810–817 (1979).
310. K. Yokono, R. Imamura, K. Shii, N. Mizuno, H. Sakai, and S. Baba, *Diabetes*, *29*:856–864 (1980).
311. B. R. Zetter, *Diabetes*, *30*:24–28 (1981).
312. T. Kooistra and J. B. Lloyd, *Int. J. Biochem.*, *17*:805–811 (1985).
313. S. Marshall and J. M. Olefsky, *J. Biol. Chem.*, *254*:10153–10160 (1979).
314. G. T. Hammons, R. M. Smith, and L. Jarrett, *J. Biol. Chem.*, *257*:11563–11570 (1982).
315. S. Bishayee and M. Das, *Arch. Biochem. Biophys.*, *214*:425–430 (1982).
316. J. F. Caro, G. Muller, and J. A. Glennon, *J. Biol. Chem.*, *257*:8459–8466 (1982).
317. L. J. Ignarro and W. J. George, *J. Exp. Med.*, *140*:225–238 (1974).
318. M. L. Chandler and P. T. Varandani, *Diabetes*, *23*:232–239 (1974).
319. A. C. Powers, S. S. Solomon, and W. C. Duckworth, *Diabetes*, *29*:27–32 (1980).
320. J. H. Im, C. J. Frangakis, W. J. Rogers, S. W. Puckett, H. R. Bowden, C. E. Rackley, E. Meezan, and H. D. Kim, *J. Mol. Cell Cardiol.*, *18*:157–168 (1986).
321. V. Trischitta, L. Benzi, A. Burnetti, P. Cecchetti, P. Marchetti, R. Vigneri, and R. Navalesi, *J. Clin. Endocr. Metab.*, *64*:914–920 (1987).
322. M. J. Wyvratt and A. A. Patchett, *Med. Res. Rev.*, *5*:483–531 (1985).
323. J. M. Samanen, in *Polymeric Materials in Medication* (C. G. Gebelein and C. E. Carraher, eds.), Plenum Press, New York (1985), p. 227.
324. J. M. Wood, J. L. Stanton, and K. G. Hofbauer, *J. Enzyme Inhib.*, *1*:169–185 (1987).
325. W. J. Greenlee, *Pharm. Res.*, *4*:364–374 (1987).
326. E. D. Freis, B. J. Materson, and V. Flamenbaum, *Am. J. Med.*, *74*:1029–1041 (1983).

327. E. J. Haber, *J. Hypertension*, 2:223–230 (1984).
328. J. M. Wood, L. Criscione, M. de Gasparo, et al., *J. Cardiovasc. Pharmacol.*, *14*:221–226 (1989).
329. K. Hiwada, T. Kokubu, E. Murakami, S. Muneta, Y. Morisawa, Y. Yabe, H. Hoike, and I. Yasuteru, *Hypertension*, *11*:708–712 (1988).
330. J. Burton, R. J. Cody, J. A. Herd, and E. Haber, *Proc. Natl. Acad. Sci. USA*, 77:5476–5479 (1980).
331. T. T. Ngo and G. Tunnicliff, *Gen. Pharmacol.*, *12*:129–138 (1981).
332. M. Szelke, B. J. Leckie, M. Tree, et al., *Hypertension*, 4(suppl. 2): 59–69 (1982).
333. J. M. Wood, P. Forgiarini, and K. G. Hofbauer, *J. Hypertension*, *1*(suppl. 2):189–191 (1983).
334. E. H. Blaine, T. W. Schorn, and J. Boger, *Hypertension*, 6(suppl. 1):111–118 (1984).
335. S. Thaisrivongs, D. T. Pals, D. W. Harris, W. M. Kati, and S. R. Turner, *J. Med. Chem.*, *29*:2088–2093 (1986).
336. H. D. Kleinert, J. R. Luly, P. A. Marcotte, et al., *FEBS Lett.*, *230*:38–42 (1988).
337. E. Haber, T. Koerner, L. B. Page, B. Kliman, and A. Purnode, *J. Clin. Endocrinol.*, *29*:1349–1355 (1969).
338. B. J. Morris, *J. Hypertension*, 7(suppl. 2):S9–S14 (1989).
339. B. J. Morris, D. F. Catanzaro, and J. Hardman, *Clin. Exp. Pharmacol. Physiol.*, *11*:369–374 (1984).
340. H. M. Jennewein, F. Waldeck, and W. Konz, *Arzneim. Forsch.*, *24*: 1225–1228 (1974).
341. F. Takaori, J. Burton, and M. Donowitz, *Biochem. Biophys. Res. Commun.*, *137*:682–687 (1986).
342. E. Danforth and R. O. Moore, *Endocrinology*, *65*:118–123 (1959).
343. M. Kidron, J. Bar-on, E. M. Berry, and E. Ziv, *Life Sci.*, *31*: 2837–2841 (1982).
344. E. Ziv, O. Lior, and M. Kidron, *Biochem. Pharmacol.*, *36*:1035–1039 (1987).
345. S. Hirai, T. Yashiki, and H. Mima, *Int. J. Pharm.*, *9*:173–184 (1981).
346. V. H. L. Lee, D. Gallardo, and J. P. Longenecker, *Proc. Intern. Symp. Control. Rel. Bioact. Mater.*, *14*:55–56 (1987).
347. M. Verstraete, *Drugs*, *29*:236–261 (1985).
348. M. A. Ondetti and D. W. Cushman, *CRC Crit. Rev. Biochem.*, *16*: 381–411 (1985).
349. M. Jochum, J. Witte, K.-H. Duswald, D. Inthorn, H. Welter, and H. Fritz, *Behring Inst. Mitt.*, *79*:121–130 (1986).
349a. G. L. Snider, E. C. Lucey, T. G. Christensen, P. J. Stone, J. D. Calore, A. Catanese, and C. Branzblau, *Am. Rev. Respir. Dis.*, *129*:155–160 (1984).
350. C. Kettner and A. B. Shenvi, *J. Biol. Chem.*, *259*:15106–15114 (1984).
351. C. A. Kettner, R. Bond, D. A. Agard, and W. W. Bachovchin, *Biochemistry*, *27*:7682–7688 (1988).
352. D. H. Perlmutter and J. A. Pierce, *Am. J. Physiol.*, *257*:L147–L162 (1989).

353. M. Brantly, T. Nukiwa, and R. G. Crystal, *Am. J. Med.*, *84* (suppl. 6a):13–31 (1988).

354. J. Travis, *Am. J. Med.*, *84*(suppl. 6a):37–42 (1988).

355. R. C. Hubbard and R. G. Crystal, *Am. J. Med.*, *84*(suppl. 6a): 52–62 (1988).

356. M. A. Fournel, J. O. Newgren, C. M. Betancourt, and R. G. Irwin, *Am. J. Med.*, *84*(suppl. 6a):43–47 (1988).

357. K. M. Moser, R. M. Smith, R. G. Spragg, and G. M. Tisi, *Am. J. Med.*, *84*(suppl. 6a):70–74 (1988).

358. E. W. Schmidt, B. Rasche, W. T. Ulmer, N. Konietzko, M. Becker, J. P. Fallise, J. Lorenz, and R. Ferlinz, *Am. J. Med.*, *84* (suppl. 6a):63–69 (1988).

359. P. L. Darke, C. T. Leu, L. F. Davis, J. C. Heimbach, R. E. Diehi, W. S. Hill, R. A. F. Dixon, and I. S. Sigal, *J. Biol. Chem.*, *264*:2307–2312 (1989).

360. N. E. Kohl, E. A. Emini, W. A. Schleif, L. J. Davis, J. C. Heimbach, R. A. F. Dixon, E. M. Scolnick, and I. S. Sigal, *Proc. Natl. Acad. Sci. USA*, *85*:4686–4690 (1988).

361. A. D. Richards, R. Roberts, B. M. Dunn, M. C. Graves, and J. Kay, *FEBS Lett.*, *247*:113–117 (1989).

362. A. D. Richards, A. V. Broadhurst, A. J. Ritchie, B. M. Dunn, and J. Kay, *FEBS Lett.*, *253*:214–216 (1989).

363. T. J. McQuade, A. G. Tomasselli, L. Liu, V. Daracostas, B. Moss, T. K. Sawyer, R. L. Heinrikson, and W. G. Tarpley, *Science*, *247*: 454–456 (1990).

364. N. A. Roberts, J. A. Martin, D. Kinchington, A. V. Broadhurst, J. C. Craig, I. B. Duncan, S. A. Galpin, B. K. Handa, J. Kay, A. Kröhn, R. W. Lambert, J. H. Merrett, J. S. Mills, K. E. B. Parkes, S. Redshaw, A. J. Ritchie, D. L. Taylor, G. J. Thomas, and P. J. Machin, *Science*, *248*:358–361 (1990).

365. R. H. Engel, S. J. Riggi, and M. J. Fahrenbach, *Nature*, *219*: 856–857 (1968).

366. M. Shichiri, Y. Shimizu, Y. Yoshida, R. Kawamori, M. Fukuchi, Y. Shigeta, and H. Abe, *Diabetologia*, *10*:317–321 (1974).

367. M. Shichiri, R. Kawamori, M. Yoshida, N. Etani, M. Hoshi, K. Izumi, Y. Shigeta, and A. Hiroshi, *Diabetes*, *24*:971–976 (1975).

368. H. M. Patel, N. G. L. Harding, F. Logue, C. Kesson, A. C. MacCuish, J. C. MacKenzie, B. E. Ryman, and I. Scobie, *Biochem. Soc. Trans.*, *6*:784–785 (1978).

369. H. M. Patel, R. W. Stevenson, J. A. Parsons, and B. E. Ryman, *Biochim. Biophys. Acta*, *716*:188–193 (1982).

370. J. F. Arrieta-Molero, K. Aleck, M. K. Sinha, C. M. Brownscheidle, L. J. Shapiro, and M. A. Sperling, *Horm. Res.*, *16*:249–256 (1982).

371. R. C. Oppenheim, N. F. Stewart, L. Gordon, and H. M. Patel, *Drug Dev. Ind. Pharm.*, *8*:531–546 (1982).

372. E. Touitou and A. Rubinstein, *Int. J. Pharm.*, *30*:95–99 (1986).

373. M. Saffran, G. S. Kumar, C. Savariar, J. C. Burnham, F. Williams, and D. C. Neckers, *Science*, *233*:1081–1084 (1986).

374. Y. W. Cho and M. Flynn, *Lancet*, *30*:1518–1519 (1989).

8

Intestinal Epithelial and Vascular Endothelial Barriers to Peptide and Protein Delivery

Joffre Baker, Ismael J. Hidalgo, and Ronald T. Borchardt
University of Kansas, Lawrence, Kansas

I. INTRODUCTION

To produce a meaningful therapeutic effect, drugs must be able to traverse several biological membranes interposed between the administration site and the target site(s). The epithelial and vascular endothelial membranes play important roles in the delivery of drugs from the site of administration to the site of action, which for most drugs is located extravascularly. Despite vigorous research, selective and efficient drug delivery across epithelial and vascular endothelial barriers remains a formidable challenge for pharmaceutical scientists. This situation will be further complicated as the number of peptides and proteins exhibiting therapeutic potential increases. Through rational drug design, drugs (including peptide and protein drugs) may be developed with molecular characteristics suitable for interaction with a target site or a specific receptor. However, the rational drug design approach does not necessarily lead to a compound with the required molecular properties for delivery across the biological barriers between the site of administration and the target site.

Although several epithelia are being considered for peptide and protein delivery, we focus on the intestinal epithelia since they are the best characterized in terms of peptide and protein absorption. Thus relevant morphological and biochemical aspects of the intestinal epithelial and vascular endothelial membranes are described. In particular, recent studies illustrating the difficulties and promises of peptide and protein drug delivery across these membranes are discussed in greater detail.

II.   INTESTINAL EPITHELIAL BARRIER

A.   Organization and Architecture of the Small Intestinal Mucosa

In an average adult human, the small intestine is approximately 280 cm long
and 4 cm in diameter.   Three anatomical modifications increase the effective
area of the small intestine in a dramatic way.

First, *plicae circulares* or folds of Kerckring are spiral or circular concentric folds up to 1 cm in height and 5 cm in length.   These folds vary
in number along the small intestine and are most prominent in the distal
duodenum and proximal jejunum [1].   Second, superimposed on these folds
and in spaces between, numerous microscopic mucosal villi (Fig. 1) increase the absorptive surface area between 7- and 14-fold [2,3].   These
villi, which may number as many as 40 per square millimeter, are present
throughout the small intestine.   They exhibit variations in size and shape
from the duodenum to the ileum [1].   For example, in humans, villi of the
distal duodenum and proximal jejunum may be leaf- or finger-shaped and
about 0.5 to 0.8 mm in height.   In contrast, the villi of the ileum are usually finger-shaped under 0.5 mm in height [1].   Third, each villus is covered by a layer of columnar epithelial (absorptive) cells and a few goblet
cells (Fig. 1).   The absorptive cells have a series of projections called

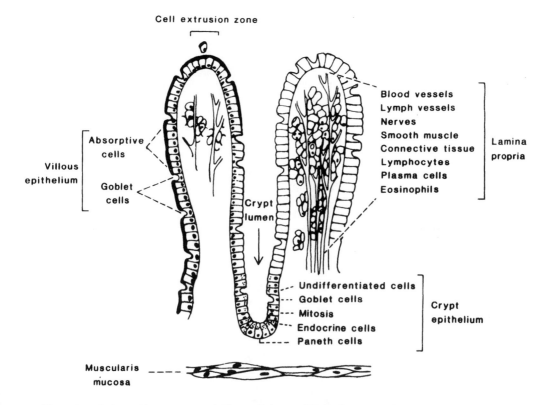

Fig. 1   Schematic representation of two villi and a crypt.

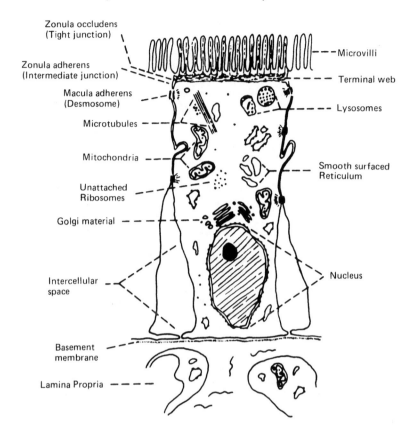

Zonula occludens
(Tight junction)

Microvilli

Zonula adherens
(Intermediate junction)

Terminal web

Macula adherens
(Desmosome)

Lysosomes

Microtubules

Mitochondria

Smooth surfaced
Reticulum

Unattached
Ribosomes

Golgi material

Nucleus

Intercellular
space

Basement
membrane

Lamina Propria

Fig. 2   Schematic representation of the organization of the intestinal epithelial absorptive cell.

microvilli at their apical (luminal) end (Fig. 2). Collectively, the microvilli, also known as the brush border, increase the luminal surface area by 14- to 40-fold [4,5].

The mucosa of the small intestine consists of three main layers: the muscularis mucosa, the lamina propria, and the epithelial cells (Fig. 1). The muscularis mucosa is a continuous sheet of smooth muscle, 3 to 10 cells thick, composed of inner circular and outer longitudinal layers, at the boundary between the mucosa and the submucosa. It has been proposed that this muscle layer may help the functioning of the intestine by virtue of its contractile potential, and that it may facilitate the emptying of the crypt luminal contents by exerting luminal compression [1].

The middle layer, or lamina propria, is the connective tissue inside the villus and surrounding the crypt epithelium (Fig. 1). Numerous defensive cells, such as lymphocytes, plasma cells, and macrophages, are found in the lamina propria of the small intestine, where they react with foreign substances that enter this layer through the gastrointestinal tract.

Additional structures present in the lamina propria are eosinophils, mast cells, blood and lymph vessels, and fibroblasts. Particles entering the lamina propria of the villus are drained by a centrally located lymphatic duct, the central lacteal, which constitutes the beginning of the lymphatics of the mesentery.

The lamina propria performs a variety of functions, including important immunological functions [6]. In addition, it provides structural support to the epithelial cells (whose basal surface resides directly on the lamina propria) and, through its blood vessels, nourishes the epithelial cells and allows the transport of absorbed material to the systemic circulation.

B. Epithelial Cells of the Small Instestine

Absorptive (Villus) Cells

The absorptive cells of the small intestine are highly polarized, columnar cells, whose major function is absorption. Under the light microscope, the apical pole exhibits a characteristic striated border. This border consists of numerous closely packed microvilli, approximately 0.5 to 1.5 μm in length and 0.1 μm in width, depending on the mammalian species studied [1]. Immediately below there is a narrower area named the terminal web, which is clear in appearance and lacking in cytoplasmic organelles (Fig. 2). The nuclei of the absorptive cells are located toward the basal membrane (Fig. 2). The cytoplasm contains Golgi material, mitochondria, smooth and rough endoplasmic reticuli, ribosomes, microtubules, and lysosomes [7].

Microvillus membrane. The width of the microvillus membrane, 10 to 11 nm, is substantially larger than most eukaryotic plasma membranes [8–10]. Although the reason for this peculiar structural feature is not known. it may be related to the unique biochemical composition of such membranes.

An important specialization of the apical portion of the absorptive cell is the glycocalyx or "fuzzy coat," a fine filamentous surface coat directly connected to the outer leaflet of the apical plasma membrane. The material of the fuzzy coat consists of weakly acidic sulfated mucopolysaccharides [11], synthetized and extruded by the cells to which they are connected and histochemically different from the goblet cells' mucus [12].

Electron microscopic analysis of freeze-fracture preparations of small intestinal epithelial cells reveals that the microvillus membrane contains amounts of intramembrane proteins, fractions of proteins, or protein assemblies which exceed those found in many other plasma membrane surfaces [13,14]. These studies also indicate that as in most biological membranes, the protein distribution between the two fractured faces of the microvillus membrane is asymmetrical [15].

There is evidence that the numerous proteins of the microvillus membrane differ in their degree of anchorage in the hydrophobic center of the plasma membrane. For example, sucrase-isomaltase is only partially embedded in the hydrophobic core, with most of the molecule located outside the membrane in the hydrophilic glycocalyx [16]. By contrast, other membrane proteins, such as alkaline phosphatase, ATPase, and glucose transport protein, are more deeply associated with the hydrophobic core of the membrane [17].

Another feature of the microvillus membrane is the high (1.7:1) protein-to-lipid molar ratio and an uncommon abundance of cholesterol and glycolipids [18,19]. In the rat microvillus, the molar ratio of cholesterol to phospholipids has been found to be 1.26:1 [18]. In the mouse microvillus, the glycolipid-to-phospholipid ratio has been formed to be as high as 1:1 [20]. These ratios are in contrast with cholesterol/phospholipid

and glycolipid/phospholipid ratios of less than 0.5:1 in basolateral membrane—enriched preparations of absorptive cells [20,21]. It thus appears
that the low membrane fluidity of the microvillus membrane reflects the
high protein-to-lipid ratio and the high cholesterol and glycolipid content
[22].

Although the biochemical composition of the microvillus membrane may
vary between different regions of the small intestine [23], significant morphological dissimilarities have not yet been identified.

The absorptive cells constantly renew their brush border membrane
proteins and glycoproteins. Autoradiography studies have shown that the
renewal rate for the vast majority of the protein and glycoprotein components of the microvillus membrane is faster than the renewal rate of the
villus cells [24].

Basolateral membrane. The basolateral membrane of the absorptive epithelial cells differs from the microvillus membrane in morphology, biochemical composition, and functions [28], although the two membranes renew
their glycoprotein contents at similar rates [29]. The absorptive cell's basolateral membrane is approximately 7 nm wide [1]. As shown in Fig. 2,
there are intercellular spaces between adjacent cells. These spaces vary
in width, depending on the villus level at which the cells are located and
in the degree of hydration of the epithelial mucosa. The lateral plasma
membranes are about 15 to 30 nm apart along their entire lengths under
conditions of low hydration (net fluid secretion) but are widened, often to
2 or 3 μm, under conditions of high hydration (net fluid absorption) [25].
This expansion in intercellular space is more pronounced toward the basal
half of the cell and decreases in the apical direction up to the tight junctions. The spaces existing between neighboring cells are substantially decreased by desmosomes, or points at which adjacent cells attach to each
other, and by thin sheetlike cytoplasmic projections which approach those
of adjacent cells (Fig. 2).

The basolateral membrane of the absorptive cells rests directly on a
continuous basal lamina or basement membrane, approximately 30 nm wide,
which serves to bridge the gaps between adjacent cells. Sometimes, the
basal lamina membrane shows gas. These are sufficiently large to allow the
passage of chylomicrons and other lipoproteins from the epithelial intercellular space to the lamina propria [26] and migration of lymphocytes from
the lamina propria to the lateral intercellular space between epithelial cells.

Many aspects of the dynamics of the basal lamina remain unclear, including the stability or rate of renewal of the components of this membrane.
Moreover, it is not known whether the basal lamina migrates from the crypt
to the villus tip along with the epithelial cells which it underlies, or remains in place as the epithelium slides over its surface.

As is the case with the microvillus membrane of the absorptive epithelial
cells, the biochemical composition of the basolateral membrane reflects the
specialized functions of this structure in intestinal transport processes.
For instance, Na^+-K^+-ATPase has been identified in this membrane [21,27,
28].

Junctional complexes. Epithelial cells of the intestinal mucosa are joined
at intercellular attachment zones or junctional complexes, which are 0.5 to
2 μm wide. The elements of this complex are known as zonula occludens or
tight junctions, zonula adherens or intermediate junctions, and macula adherens or spot desmosomes [30]. The zonula occludens or tight junctions

are located at the apical end of the lateral membrane of adjacent cells (Fig.
2), thereby obliterating the intercellular space over a variable distance
[30]. The structure of the tight junctions varies with the region, cell
type, and position along the crypt-villus axis. For example, the tight junctions of ileal absorptive cells are deeper and have more strands (chain of
closely spaced intramembrane particles assumed to represent integral membrane proteins) than their jejunal counterparts. In addition, while tight
junctions between villus cells are similar in depth and strand number, those
of the goblet cells and undifferentiated crypt cells exhibit variability in
these parameters.

The zonula adherens or intermediate junctions are positioned directly
below the tight junctions (Fig. 2) [30]. It is characterized by intercellular
spaces 15 to 20 nm wide which are partially occupied by a homogeneous,
finely fibrillar material [30]. This junction appears to constitute the chief
site of membrane insertion for the filaments of the terminal web. The last
element of the junctional complex, macula adherens or spot desmosomes, is
located approximately 0.2 μm or more from the basal end of the intermediate
junction (Fig. 2). Unlike the tight and the intermediate junctions, the
spot desmosomes do not form a continuous belt around the cell, but rather
a discontinuous band of circular or oval-shaped plaques 100 to 500 nm in
diameter. Desmosomes of epithelial cells consist of two straight plaques of
dense material running parallel to the inner leaflet of the corresponding
plasma membrane and being separated by a clear space of about 8 nm (Fig.
2) [30]. They are located not only in the junctional complex region but
also along the lateral membrane, thereby stabilizing the cytoskeletal network
of the epithelial cells [31].

Undifferentiated Crypt Cells

The major functions of these cells are proliferation and secretion, not absorption. After their formation in the crypt, many undifferentiated daughter
cells undergo differentiation into the villus cells as they migrate from the
crypt to the villus. The undifferentiated crypt cells can also differentiate
into goblet cells, Paneth cells, and endocrine cells [32], which are primarily
secretory in function.

Undifferentiated cells are columnar in shape with the nuclei located
towards the basal half of the cell. During mitosis, the nuclei migrate toward
the luminal half of the cell where the formation of chromosome masses gives
the cells a more oval shape. There are substantial structural differences
between the undifferentiated crypt cells and the differentiated absorptive
cells. Electron micrographs reveal that in the undifferentiated cells, the
microvilli are shorter and more sparse and the glycocalyx less abundant.
Furthermore, the protein content in the microvilli of undifferentiated cells
is less than one-half that present in the differentiated absorptive cells [33].
Similarly, tight junctions of undifferentiated cells show less depth and
complexity than those of the differentiated cells. In addition to the structural differences mentioned above, substantial biochemical differences have
been reported. Unlike the differentiated cells, the undifferentiated crypt
cells contain less alkaline phosphatase, disaccharidase, and dipeptidase [34].
Thus it appears that these enzymes are developed during the differentiation
process and therefore are suitable enzyme markers for differentiated absorptive epithelial cells.

M (mircrofold) Cells

Peyer's patches are groups of subepithelial, lymphoid follicles distributed throughout the small intestine. These patches, generally oval or rectangular, are usually encountered on the antimesenteric wall of the intestine. However, sometimes they may be randomly distributed around the intestinal wall. Peyer's patches may be as long as 28 cm in human adults and may contain as few as 5 or as many as 980 individual follicles [35]. They are most prominent in the ileum and are characterized by the presence of specialized epithelial cells called M cells. These cells are intermittently distributed among the absorptive cells covering the follicular dome. Unlike the absorptive cells, whose apical membranes are modified to form microvilli, the apical membranes of M cells bear numerous microfolds. These folds are not only substantially shorter than the microvilli but also are more sparse [35] (Fig. 3). In addition, the glycocalyx covering the M cells is less abundant than that of the neighboring columnar absorptive cells.

The cell apex appears under the electron microscope as a thin cytoplasmic band that connects the absorptive cells surrounding it. Thus the M cells share tight junctions with adjacent absorptive cells. From the apex, the lateral membrane projects down along the lateral membrane of adjacent absorptive cells, creating a deep invagination that normally enfolds lymphocytes (or macrophages), which migrate in and out of the intercellular space of the epithelium (Fig. 3).

The M cell nucleus is located in the basal cytoplasm. Additional subcellular structures contained in the cytoplasm include mitochondria, rough endoplasmic reticulum, and golgi complexes. It has been hypothesized that M cells perform sampling and transport of undegraded luminal particles and macromolecules into lymphoid follicles for immunologic surveillance and initiation of appropriate immunologic responses [35].

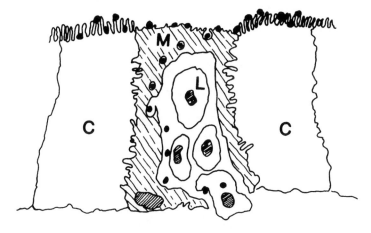

Fig. 3   Graphic depiction of the structure of the M cell.

Goblet Cells

The small intestinal epithelial microvilli are covered by a mucus layer located outside the glycocalyx or fuzzy coat. This layer is secreted by
specialized cells known as goblet cells, which are found with increasing
frequency along the small intestine, being most abundant in the ileum [10].
Goblet cells are characterized by their dense basophilic cytoplasm and their
basally located nuclei. The presence of mucus-secreting granules gives them
their characteristic goblet shape as a result of filling and distending the
apices of the cells. Unlike the absorptive cells, the goblet cells' microvilli
are sparse and irregular in size and shape [36]. In addition, the protein
content of the microvillous membrane is low and both the supporting terminal
web and the microvilli-associated glycocalyx are not well developed [36].

There is evidence that goblet cells are derived from undifferentiated
crypt cells [37,38]. They differentiate deep in the crypt, and then migrate
upwardly, becoming gradually filled with mucus granules. The mucus secreted by goblet cells is chiefly glycoproteins, whose molecular weight may
be up to 2×10^6 daltons and consists of approximately 80% carbohydrates
and 10% proteins [39,40]. Mucus secretion is carried out through a series
of exocytosis of intracellular mucus granules.

Although the exact function of mucus in the small intestine is not known,
the fact that mucus secretion is accelerated by *E. coli* and *Vibrio cholerae*
enterotoxins [41], and by chemical or physical irradiation of the intestinal
mucosa [42], appears to suggest that goblet cells' mucus may protect the
small intestinal epithelial mucosa by acting as a barrier that binds macromolecules, coats bacteria, and cushions particulate matter [7]. The importance of the goblet cells in the transpithelial transport of macromolecules is
not yet clear. However, the mucus layer, which is a functional part of
the unstirred water layer overlying the intestinal epithelium, appears to
represent a potential barrier to the mucosal transport of macromolecules.

C. Methodology to Study Intestinal Transport and Metabolism

Numerous experimental approaches have been utilized to investigate intestinal transport and metabolism. The methods used may be classified into
three categories: (1) in vitro, (2) in situ, and (3) in vivo.

In Vitro Systems

The in vitro systems include the everted intestinal sac [43], everted intestinal ring [44], mucosal sheets [45], brush border membrane visicles
[46], and basolateral membrane vesicles [47]. Due to their simplicity and
controllability, these systems have been widely used and have provided important information on the transport and metabolism of drugs and nutrients.
In recent years, isolated intestinal epithelial cells are gaining acceptance as
useful tools in the study of mucosal binding, uptake, and transport [48].
The main limitation of all these systems is their short viability, which clearly
restricts their potential value as investigative tools.

While still in their early stages of development, the CACO-2 and HT
cell lines appear to offer hope in circumventing the limited viability of isolated cell systems. The CACO-2 cell line undergoes spontaneous differentiation in culture [49], whereas the HT-29 cell line differentiates when galactose
is substituted for glucose [50]. Like the normal enterocytes, the differentiated CACO-2 and HT-29 cell lines are joined by tight junctions and

express a brush border. Thus it would appear that growing these cells to monolayer densities on a microporous substrate would provide a potentially valuable system for evaluating protein and peptide drugs transport across intestinal epithelial-like membranes.

Osiecka et al. [46] have recently compared brush border membrane vesicles, everted intestinal rings, and isolated intestinal epithelial cells in terms of their suitability for drug uptake studies. They concluded that the intestinal ring was a good potential system for drug uptake studies and that the brush border membrane vesicles were not. In addition, they found that despite some advantages, the usefulness of the isolated cells system in this type of studies was not clearly validated.

In Situ Systems

In situ systems have also been utilized to study mucosal transport and metabolism. These systems include the modified Doluisio method [51] and the through-and-through method [52]. In these two methods, a segment of the small intestine is perfused through the luminal cavity with a physiological solution containing a compound of interest. Subsequently, the rate of drug disappearance from the lumen or its appearance in the portal venous circulation is carefully monitored. These methods have provided valuable information on the relationship among the physical, chemical, and biological determinants of mucosal drug transport. These include perfusion flow rate, thickness of the aqueous boundary layer, biomembrane permeability, solute lipophilicity, pH, and reserve length of the intestinal segment.

In Vivo Systems

Intestinal mucosal transport is also investigated in vivo. Here, the drug is orally administered to animals or humans and some indicator of absorption (e.g., biological activity) may be monitored. For an extensive review on peptides and proteins biologically active through the oral route, see Ref. 62 and Chap. 16. An alternative vascular autoperfusion technique has been described by Barr and Riegelman [43] and Windmeuller and Spaeth [53]. This permits the investigation of drug transport in nearly normal intestinal segments, where the arterial, neural, and lymphatic connections of the intestinal segment are maintained intact and the venous outflow is totally collected.

D. Mechanisms for Peptide and Protein Transport Across the Intestinal Epithelium

Passive and Carrier Mediated

It is well known that di- and tripeptides are absorbed extensively from the small intestine of humans and animals. Although it appears that these molecules are absorbed by both active transport and simple diffusion, the latter appears to be of minimal importance [54]. In addition, there is little evidence that peptides with more than three or four amino acid residues are transported across the intestinal mucosa by the peptide transport system.

Dipeptide absorption is affected by factors such as stereoisomerism, side-chain length, and N- or C-terminal substitution. For example, the rate of absorption of L-Ala-L-Phe and L-Leu-L-Leu from the jejunum was much greater than that of the DD isomers [55]. In some instances D- and L-enantiomorphs of dipeptides are transported by the same system but with different affinities [56].

The presence of a larger side chain in either the N-terminal or C-terminal appears to favor dipeptide absorption. Adibi and Soleimanpour demonstrated that the substitution of a Gly residue in diglycine with Leu in either the N- or C-terminal position resulted in bulkier dipeptides which were better absorbed [57]. Furthermore, there was no difference in the absorption rate of Leu-Gly and Gly-Leu.

In addition to being affected by stereoisomerism and side-chain length, the absorption of dipeptides is influenced by other structural factors. Basic (Lys-Lys) and acidic (Glu-Glu) dipeptides appear to possess a lower affinity for the peptide transport than neutral dipeptides [56]. Similarly, acetylation and methylation of N-terminal amino groups, esterification of C-terminal carboxyl groups, presence of Γ-linkage, and incorporation of β-amino acids also decrease the affinity of dipeptides for the peptide transport system [56]. Furthermore, an increase in the number of amino acid residues from two (i.e., Gly-Sar) to three (i.e., Gly-Sar-Sar) produces an increase in the K_m value [58].

A substantial amount of work has been carried out with the purpose of increasing our present understanding of the peptide transport system. Since there is strong evidence that amino acids are absorbed by different amino acid transport systems, it has been suggested that dipeptides containing acidic, basic, or neutral amino acids may be transported by different carrier systems [59]. However, detailed studies have failed to demonstrate the existence of multiple carrier systems [54,60]. Therefore, it would appear that unlike amino acids, which are absorbed by heterogeneous transport systems, dipeptides are transported by the same carrier system regardless of their net charge.

Endocytosis and Transcytosis

There is firm evidence that endocytotic uptake and transport of large protein molecules takes place in the neonatal intestine. As the epithelial membrane of the small intestine matures, there is a pronounced decrease in the permeability of the intestinal mucosa to such molecules. Although the amounts of intact protein taken up by the small intestine may not be sufficiently large to be nutritionally significant, some are biologically significant [61]. An extensive list of small peptides with biological activity after oral administration has been compiled by Samanen [62]. It follows that the intestinal mucosa constitutes a potentially important site for the delivery of peptides and proteins to the systemic circulation.

Cellular internalization of peptides and proteins by endocytosis is an important biological process whereby peptides and proteins too large to be absorbed by the di- and tripeptide transport systems may be taken up. Two different pathways of endocytosis are (1) fluid-phase type (nonspecific endocytosis, pinocytosis), and (2) adsorptive or receptor-mediated type (specific endocytosis).

Fluid-phase endocytosis (FPE) (Fig. 4a) is a process by which macromolecules dissolved in the extracellular fluid are incorporated by bulk transport into the fluid phase of endocytic vesicles. It begins with a "pinching off" of the plasma membrane to form a vesicle or pinosome, which migrates inwardly. The vesicle transports the ligand to an endosome (prelysosomal vesicle), which subsequently migrates to the perinuclear region, where it fuses with a lysosome. Here the ligand is degraded and the membrane recycled back to the plasma membrane [63].

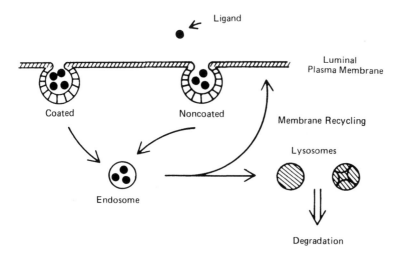

(a)

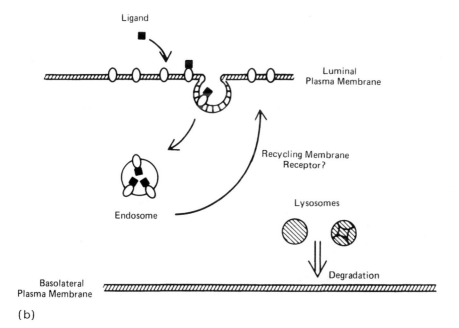

(b)

Fig. 4   Cellular uptake mechanisms. (a) Fluid-phase endocytosis; (b)
receptor-mediated endocytosis; and (c) transcytosis.

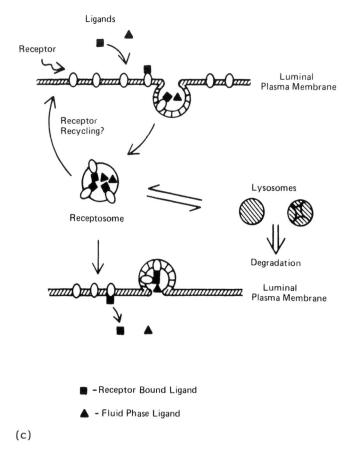

Luminal
Plasma Membrane

Luminal
Plasma Membrane

■ – Receptor Bound Ligand

▲ – Fluid Phase Ligand

(c)

Figure 4   (Continued)

Table 1 lists some peptides and proteins that are known to traverse the mucosa of the small intestine, albeit in small amounts. Although there is some evidence that the mucosal uptake of some of these compounds is mediated by endocytic processes, [74,79], in most instances, the mechanism(s) responsible for such uptake is unclear. The potential of the endocytic pathway has been demonstrated in the past. For example, using horseradish peroxidase (HRP) as an enzyme marker, Cornell et al. [66] found that following intraluminal administration of HRP into ligated segments of jejunum and ileum, the enzyme accumulated on the apical surface membrane and within membrane-bound cytoplasmic canalicular vesicular and vacuolar structures. In addition, extracellular absorbed HRP was found in the intercellular spaces between absorptive cells, crossing the basement membrane, and in the lamina propria [66].

Adsorptive endocytosis involves the binding of macromolecules to the plasma membrane before being incorporated into endocytic vesicles. The process commonly known as receptor-mediated endocytosis (RME) is a subclass of adsorptive endocytosis, where the membrane binding site is a specific receptor for the macromolecule involved (Fig. 4b). In this instance, the specific binding of molecules to membrane receptors is quickly followed by clustering of the receptor-ligand complex into clathrin-coated pit regions of the plasma membrane. Subsequently, the complex is endocytosed. After internalization, a process known as "sorting" controls the fate of the elements of the complex (i.e., receptor and ligand). Normally, the ligand-receptor complex is transported to endosomes (where the low pH may cause the complex to dissociate), and then the ligand is sent to lysosomes for degradation, while the receptors may be degraded (e.g., epidermal growth factor receptor) or recycled back to the plasma membrane (e.g., transferrin receptor and asialoglycoprotein receptor).

In polarized cells, such as the intestinal epithelial cells, some endosomes carrying the ligand or the receptor-ligand complex bypass the lysosomes and migrate toward the basolateral membrane, resulting in the release of the ligand in the extracellular space bounded by the basolateral membrane. This process, known as transcytosis, combines elements of membrane protein sorting and endocytosis and represents a potentially useful and important pathway for the mucosal transport of peptides and proteins (Fig. 4c). Although both types of endocytosis have been demonstrated in epithelial cells, FPE appears to be of little importance compared with RME.

The small intestine epithelial mucosa is located in the strategic position in the sense that it constitutes a barrier to the permeation of macromolecules. However, endocytotic uptake provides ways to circumvent this barrier. RME allows the absorptive cells of the intestinal epithelium to select and transport specific molecules while excluding undesirable or potentially harmful ones (e.g., bacterial enterotoxins and endotoxins). Thus, depending on their molecular structure, endogenous and foreign macromolecules may interact with the intestinal epithelial mucosa receptors and subsequently undergo internalization. Indeed, some peptides and proteins are known to enter intestinal mucosal cells through endocytosis [74,79,84]. Furthermore, a few peptides and proteins have been reported to reach blood vessels

Table 1   Representative List of Protein and Large Peptides Transported Across the Intestinal Mucosa

Protein or peptide	Species used in study	References
Insulin	Human (adult)	64
	Rat	65
HRP	Rat	66
Egg albumin	Human (infant)	67,68
Trypsin and chymotrypsin	Human	69
Chymotrypsin	Rat	70
Γ-globulin	Neonatal pig	68,71
	Neonatal mouse	72
Native ferritin	Rat	72
IgG	Rat	73
EGF	Rat	84
NGF	Rat	79

in the lamina propria and the portal venous circulation. Three important examples are immunoglobulin G (IgG), nerve growth factor (NGF), and epidermal growth factor (EGF) [79,80–84].

It is widely known that in mammals, mothers can transfer passive immunity to newborns through the intestinal absorption of immunoglobulins (Ig) from colostrum [73]. One of the Igs best characterized in terms of its intestinal mucosal transport is IgG. An early study by Kraehenbuhl and Campiche [74] showed that antiferritin IgG and antiperoxidase IgG were appreciably endocytosed by jejunal absorptive cells of newborn rats, rabbits, and pigs. Regional differences in gamma-globulin uptake along the small intestine have been reported in two separate studies. Rodewald [75] showed that more IgG is transported across the jejunum than across the ileum, and Jones and Waldman [76] found that IgG binds to brush border membrane preparations from newborn rat jejunum but not to membrane preparations from newborn rat ileum. The finding that Fc fragments (but not the Fab fragments) inhibited the mucosal transport of IgG in young rats supports the involvement of the receptor-mediated endocytosis in the uptake of IgG. It also indicates that the Fc fragment contains the active binding site responsible for the interaction with the receptor [77].

After undergoing receptor-mediated internalization, receptor-bound IgG molecules are transported across the cell and exocytosed across the basal membrane [74]. One study showed that when endosomes containing receptor-bound IgG and free proteins fused with lysosomes, the free protein molecules were degraded but the IgG molecules were not [78]. Therefore, it appears that when bound to specific receptors, IgG molecules are protected from lysosomal degradation [73]. In contrast, nonspecifically endocytosed peroxidase is almost completely delivered to the lysosomes for enzymatic degradation. Indeed, Kraehenbuhl and Campiche [74] detected the appearance of moderate amounts of intact gamma globulin in the portal venous blood of newborn rats and pigs following intrajejunal administration of the immunoglobulin.

Certain growth factors constitute valuable model peptides for elucidation of mechanisms involved in the transepithelial transport of macromolecules. For example, Siminoski et al. [79] found that after 45 to 60 min of incubation of [^{125}I]NGF with fixed ileal epithelial sheets, radioactivity accumulated in microvilli and apical invaginations of absorptive cells, and that 78% of the radioactivity could be diplaced following the administration of an excess amount of NGF [79]. These results appear to suggest the existence of NGF receptors in the ileal epithelial absorptive cells. The intraluminal administration of [^{125}I]NGF into ileal loops of suckling rats resulted in detectable amounts of the intact radiolabeled peptide in brain, kidneys, liver, and lungs, and to a lesser extent in other tissues [70]. These results indicate that transepithelial transport may play a role in the mucosal transport of NGF.

EGF has also been shown to be absorbed following oral administration to suckling rats [80]. Moreover, specific receptor-mediated binding and endocytosis of EGF was demonstrated in intestinal epithelial cells of suckling rats [81], mouse [82], and subcellular fractions of rat small intestine enriched in microvillus membrane [83]. More recently, Gonella et al. [84] found that [^{125}I]EGF bound to microvillus and basolateral membranes in fixed isolated epithelial sheets. In addition, [^{125}I]EGF was taken up by absorptive (villus) cells, but the binding or uptake of the radioactive peptide by crypt cells was minimal [84]. These results are consistent with the possibility that on crypt cells EGF receptors are mainly located on the basolateral membrane [84]. As coadministration of 1000-fold unlabeled EGF resulted in a 44% reduction in radioactivity in basal regions on absorptive cells and 38% reduction in TCA precipitable serum radiolabel, it would appear that the transepithelial transport of EGF is at least in part mediated by specific receptors [84]. A smaller fraction of the dose is believed to be absorbed by a nonspecific adsorptive mechanism, similar to that reported for cationic ferritin [84]. Following endocytic uptake most EGF molecules are directed to the lysosomal vacuole, where they are degraded, but a small portion of unchanged EGF enters the transcytotic (transepithelial) pathway [84]. Some EGF fragments appear to be generated by limited metabolism of EGF in prelysosomal vesicles. While some of these fragments reach the basolateral membrane by transcytosis [80], the exact site at which they enter the transepithelial pathway is not known.

Hepatocyte receptors bind EGF rapidly and extensively, resulting in removal of 90% of the peptide entering the liver (i.e., extraction ratio is 0.9). Thus the lack of appreciable amounts of intact [^{125}I]EGF in serum, reported by Gonella et al. was probably due to a virtually complete removal of the radiolabeled peptide during the first pass through the liver [84], so that the intact peptide never reached the systemic circulation. However, EGF fragments, which in some instances retain biological activity, have been detected in blood and liver after oral administration of [^{125}I]EGF to suckling rats [80].

Paracellular Movement

It is widely recognized that junctional complexes are important determinants of intestinal epithelial permeability because they control the paracellular flow of water and ions and prevent the passage of macromolecules. Although the paracellular movement of water and ions plays an important role in the absorption of water from the intestinal lumen, the exact mechanism by which these materials pass between the cells joined by tight junctions is unknown. Permeability characteristics differ from one epithelium to another. Thus it appears that the passive flux of cations is almost exclusively through tight junctions [85], with "leaky" epithelia (i.e., intestinal epithelium) having higher transport rates than "tight" epithelia. Conceivably, the passage of water across the tight junction could carry dissolved drugs or facilitate the transport of macromolecules that would not otherwise traverse the apical membrane. Munck and Rasmussen found that a glucose-induced increase in mucosal water transport carried [^{3}H]methoxyinulin and [^{14}C]polyethylene glycol 4000 across isolated mucosa of rat small intestine [86]. In the absence of such increase in water transport, the transmucosal movement of these compounds (which normally do not enter cells) was zero [86]. Therefore, it has been suggested that they cross the intestinal mucosal epithelium through the tight junctions [86]. These results emphasize the potential role of paracellular movement in the transmucosal transport of macromolecules.

There is a strong relationship between the structure of the tight junction and its permeability. For example, disruption of tight junctions of toad urinary bladder mucosa is associated with decreased electrical resistance and permeation of tight junction by macromolecules that would be excluded under normal conditions [88,89]. Rhodes and Karnovsky [87] found that following surgical trauma to small intestine of adult guinea pigs, HRP traversed intercellular tight junctions of epithelial cells without being taken up to any appreciable extent.

Within local regions of the small intestinal epithelium morphological differences have been found that might indicate differences in epithelial permeability. It is important to note that the tight junctions between undifferentiated crypt cells are more lossely organized and composed of fewer strands than those between the absorptive (villus) cells. This suggests that the crypt epithelium may be of a "leakier" nature than the villus epithelium.

III. CAPILLARY ENDOTHELIAL BARRIER

A. Organization and Structure of Capillary Endothelium

Extravascular tissues are bathed in a filtrate of plasma rather than whole plasma because capillaries, which account for most of the interfacial surface that separates plasma and interstitial tissues [90], are a selective barrier to the exchange of solutes. The luminal surface of capillaries consists of the plasma membrane surface of a monolayer of endothelial cells which are joined together by more or less continuous tight or occluding junctions [91]. From tissue to tissue, the endothelial cells can vary in thickness from 0.1 to 1.0 μm. Their nuclei often bulge into the capillary lumen. Scanning electron micrographs show a large number of fingerlike 300- to 3000-nm projections extending from the cells [92], which greatly increase the amount of cell surface exposed to plasma. The luminal surfaces of the endothelial cells are coated with a glycocalyx, a fuzzy matrix of glycoproteins, and glycoaminoglycans [93,94].

A prominent feature of many capillary endothelial cells is the presence of numerous vesicles about 60 nm in diameter, many of which appear to be continuous with the plasma membrane and open to the extravascular space [95,96]. Most of these open bowl-shaped plasmalemmal vesicles (caviolae) are spanned at their 20 to 40-nm openings by a fuzzy sheet of material called a diaphragm. Electron micrographs often reveal a central knob extending out of the diaphragm. These features of plasmalemmal vesicles, plus the absence of an underlying coat, distinguish them from the coated vesicles, which are present in fewer numbers [94].

The abluminal face of the endothelium is ensheathed by a dense 30 to 50-nm-thick basement membrane or basal lamina [97], in which are often embedded pseudopodia of extravascular cells called pericytes. In certain cases thin fibers extend from the endothelial cells into the basement membrane, presumably serving as an anchor [98]. The basement membrane contains the fibrous cell adhesion proteins laminin and fibronectin, and a larger amount of collagen (types IV and V [99,100]). Both faces of this membrane contain clusters of anionic sites, which may be patches of the sulfated glycosaminoglycan heparin sulfate [101].

The structure and permeability properties of capillaries vary enormously between tissues. Capillaries are conventionally categorized into three major classes: sinusoidal (or discontinuous), fenestrated, and continuous. The first type, present in liver, spleen, and bone marrow, contains frequent large gaps (>100 nm) that are devoid of both endothelial cells and basement membrane [90,102]. Fenestrated capillaries tend to be located in tissues through which there is rapid movement of fluid (e.g., kidney glomerulus, intestinal mucosa, and fat pads of synovial membranes). Fenestrations are

60- to 80-nm cell-free openings that are often but not always spanned by
a highly anionic heparin sulfate-containing diaphragm membrane that lacks
a lipid bilayer [103]. The continuous capillaries serve the lung, brain,
muscle, and connective tissue and account for most of the total capillary
surface area. These capillaries contain no gaps and have few fenestrations.
It should be emphasized that there is considerable variation within these
categories. For example, the continuous capillaries of tongue have a density of fenestrations that is intermediate between muscle and renal glomerulus
[104]. The renal glomerulus capillaries have an extraordinarily thick basement membrane [105].

Brain capillaries differ from other continuous capillaries in both their
structure [95] and permeability properties. Whereas other continuous epithelia are at least moderately permeable to proteins the size of albumin,
brain capillaries are extraordinarily resistant to the passage of even very
small polar solutes [106]. This "blood-brain barrier" maintains a low extracellular level of K^+ in brain which is necessary for generation of nerve im-
pulses [107]. The price of this exquisitely well-sealed barrier is that exchange of solutes between brain and blood has required the evolution of
specialized, often energy-requiring, transport systems in the capillary endothelium (see below).

The morphological, physical, and biochemical properties of brain capillaries are consistent with their barrier and transport functions. No gaps
or fenestrations are observed in the capillaries that form the blood-brain
barrier. The tight junctions between the cells are so well sealed that the
electrical resistance of brain capillaries is equivalent to that observed in
tight epithelia [108]. Brain capillary endothelial cells often fuse to generate a seamless endothelium [109], a structure that is not observed in the
peripheral capillaries. In addition, in contrast to other capillaries, brain
capillaries contain very few intracellular vesicles, which, as discussed below, are potential vehicles for the transport of proteins. Brain capillary
endothelial cells contain a number of enzymes found in actively transporting
epithelium (e.g., glutamyl transpeptidase, monoamine oxidase, and Na^+/K^+
ATPase) (97,110,111). These cells also have an unusually large concentration of mitochondria [108], consistent with intense active transport and
metabolic activity.

Microvascular structure varies not only from tissue to tissue but also
between arteriole, capillary, and venule segments [94]. Transmission electron microscopy and freeze-fracture analysis indicate that the arteriole endothelium contains large gap junctions (which allow communication between
the cytoplasms of the participating cells) and extensive tight junctions.
Capillaries generally contain few gap junctions and tight junctions that are
morphologically distinct from those of arterioles. The intercellular junctions
of pericytic (postcapillary) vesicles are also distinct. In freeze-fracture
replicas their junctions appear to be an assembly of "scattered ridges and
groves" [112]. The morphology of the microvascular segments differs somewhat between tissues. Figure 5 depicts certain key features of the capillary
endothelium.

B. Methods for Measuring Capillary Permeability

A number of independent procedures have been used to measure peptide
or protein exchange across capillaries. These include measurement of the
rate of disappearance of radiolabeled solutes from plasma or their

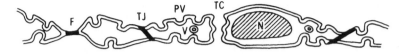

Fig. 5   Structureal features of capillary endothelium. The endothelial cell monolayer, shown in cross section, is a principal barrier to the exchange of solutes between plasma and extravascular tissues. F, fenestral diaphragm; TJ, tight junctions; PV, plasmalemmal vesicles; V, vesicles; TC, transmembrane channel; N, cell nucleus.

accumulation in tissues, comparison of the plasma and lymph concentrations of solutes, use of fluorescence microscopy to trace the penetration of solutes tagged with fluorescent dyes [113], and osmotic-transient methods. In the latter case a hyperosmotic solution is injected into the arterial plasma of a tissue. Determination of solvent flux and the solute osmotic pressure gradient allows measurement of the capacity of a solute to be transported conductively across a membrane [115]. The osmotic reflection coefficient varies from zero, when the membrane is not a barrier, to 1, when the membrane is totally nonpenetrable. Each of the methods noted above has pitfalls. Measurements of solute disappearance from plasma or leakage of fluorescent solutes from capillaries provide only qualitative indications of capillary permeability. Accurate estimation of several parameters required for the osmotic-transient approach may be difficult [116].

Several in vitro models of capillaries have recently been employed to study peptide and protein transport. These include cultured brain endothelial cell monolayers [117] and isolated brain microvessels [118]. Junctional integrity in these systems, which can be assessed by assaying penetration of an inert polymer (e.g., inulin), is often great. The standard reservations regarding extrapolation of in vitro data to in vivo cell behavior apply.

C.   Membrane, Metabolic, and Junctional Components of the Endothelial Barrier

To cross the capillary endothelium solutes must either traverse the endothelial cells themselves or pass between the cells. The nature of the cellular barrier has been most clearly defined by studies of the blood-brain barrier because the extracellular transport of solutes is exquisitely restricted in brain capillaries. The blood-brain barrier has, essentially, the permeability properties of phospholipid bilayer membranes. Thus the rate of nonspecific (i.e., nonsaturable) solute exchange correlates relatively well with solute lipid solubility [119], as measured by octanol/H_2O, or olive oil/ H_2O partition coefficients, and also varies inversely with solute size. The latter effect is manifested through changes in the diffusion coefficient, which varies as the square root of molecular weight [120].

Solutes that are able to traverse the endothelial cell plasma membrane may be modified or metabolized by cytoplasmic enzymes. The endothelial cells thus pose a possible metabolic or enzymatic barrier to solute passage. Cellular metabolism accounts for the failure of circulating dopamine, fatty acids, and ammonia to enter the brain [121–123].

In most capillaries, endothelial tight junctions serve as the major extracellular barrier to solute exchange. These junctions stop the transcapillary
movement of macromolecular tracers injected into plasma or interstital fluid
[124]. Conditions that open tight junctions relax both the high transendothelial electrical resistance and the barrier to solute exchange [125,126].
In mammals, glial or astrocyte extensions ensheath the brain capillaries.
At one time this suggested that junctions between the astrocytes, rather
than between the endothelial cells, prevent the movement of solutes from
brain to plasma. That this is not the case is demonstrated by (1) the
finding that horseradish peroxidase injected into cerebrospinal fluid is
stopped by the endothelial cell junctions [127], and (2) the presence of a
blood-brain barrier in animals (amphibians) that lack a glial sheath around
their brain capillaries [128].

The first electron microscopic analysis of tight junction ultrastructure
showed that the plasma membranes of adjoining cells fused, creating a 2-
to 3-nm-thick electron dense line [8]. Freeze-fracture studies revealed an
anastomosing meshwork of interbilayer strands. However, the nature of
the freeze-fracture images varies considerably depending on preparation or
fixation procedures. In unfixed tissue, irregular particles occupy the exoplasmic face tight junction contact sites, whereas in specimens fixed with
glutaraldehyde offset cylindrical structures are found [129]. Until a few
years ago the prevailing view was that the particulate morphology was
physiologically correct, and that the particles were junctional proteins.
Subsequently, Pinto da Silva and Kachar [130] and Kachar and Reese [131]
provided evidence that it is instead the cylindrical structures that are non-
artifactual. They showed that in rapidly forzen unfixed tissue, cylinders
rather than particles predominated [131]. The cylinders resembled the
inverted micelle hex II phase structures that are observed under certain
conditions in synthetic phospholipid membranes [132] and which are thought
to exist transiently during cell or liposome fusion [133]. The tight junction
cylinders were destroyed by agents, such as acetone, that extract lipids
but not proteins. An inverted micelle model of tight junctions has been
proposed [131] and is shown in Fig. 6. The junction is considered to form
by fusion of exoplasmic membrane leaflets. This arrangement would restrict
not only the movement of solutes between adjoining cells, but also isolate the
apical exoplasmic leaflets from the basal side of the plasma membrane. The
restricted diffusion of fluorescent membrane probes indicates that this membrane partitioning occurs [134].

Another feature of the model discribed above is that the plasma membrane exoplasmic leaflets are continuous between the joined cells [131].
However, lipids or other molecules intercalated into the plasma membrane
exoplasmic leaflet have been reported not to diffuse between the cells joined
by tight junctions [134]. The inverted micelle model could still be fundamentally correct if proteins, perhaps recruited by the inverted micelle
strands, impose the barrier between the exoplasmic leaflets of neighboring
cells.

Several groups have detected proteins in tight junctions. Stevenson
and Goodenough isolated tight junctional complexes from mouse liver using
subcellular fractionation methods [135]. Detergent extraction of the junctional structures yielded an anastomosing network of fibrous proteins, which
on SDS gels migrated as major 37- and 47-kD bands and several minor
bands. Using a line of cultured epithelial cells, Gumbiner and Simons have
identified a cell surface antigen that is required for the formation of tight

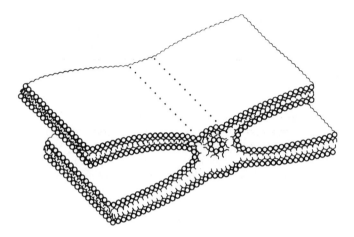

Fig. 6   Model of the tight junction based on a hypothetical intramembrane cylinder (see Refs. 130 and 131). A single cylinder is shown. Electron micrographs suggest that the junctions may contain two such strands, opposed and offset. In cross section the cylinders resemble inverted phospholipid micelles.

junctions [136]. Intriguingly, the antigen is similar or identical to uvomorulin, a calcium-dependent cell adhesion molecule.

The freeze-fracture morphology of tight junctions may vary between tissues [131,137]. Thus it should be emphasized that the term "tight junction" may actually describe a set of junctions that differ somewhat between tissues.

D. Mechanisms of Solute Transit Across Vascular Endothelium

Passive, Nonfacilitated

All vascular endothelium is permeable to water, gases, and low-molecular-weight lipid-soluble solutes. Diffusion of charged or polar solutes across capillaries varies profoundly between capillary types. Sinusoidal capillaries permit free exchange of even very large ($>1 \times 10^6$ M_r) molecules, whereas the blood-brain barrier restricts the passage of metal ions [138]. Continuous capillaries of the peripheral tissues are remarkably permeable to proteins that are about 70 kD or even larger [138].

In the early 1950s, Pappenheimer and colleagues found that the permeability of (what are now known to be) continuous capillaries to solutes of various sizes could be accounted for by the existence of two sets of rigid water-filled pores: 120 Å in diameter and 600 Å in diameter, at densities of 12 and 0.05 per square micrometer of capillary surface, respectively [139]. This model leads to the prediction that in the peripheral continuous capillaries, pores account for about 0.1% of the capillary surface area [140]. Subsequently, variations in pore radii have been proposed to accommodate variations in measured solute permeability or reflection coefficients. These differences are in part attributable to variations between capillaries of different tissues [138]. In addition, the simplest form of the pore theory has been found to be inadequate to explain solute exchange properties within certain capillary types [141]. For example, working with frog mesenteric capillaries, Michel found that the measured filtration coefficients of small hydrophilic molecules yielded an estimated pore radius that was almost twice as large as that calculated from the measured osmotic reflection coefficients of proteins [142]. Such discrepancies have led to the reintroduction of a "fiber matrix" pore model, in which pore permeability is determined chiefly by the mesh size of a network of glycoproteins that inhabit the pores [141, 143].

For decades electron microscopists have sought to visualize the cellular structures that allow diffusion of solutes across continuous capillaries. In the case of small (<2 nm diameter) charged or polar solutes (e.g., metal ions or sugars) the primary passageway seems likely to be small clefts located in capillary tight junctions [144,145; see Ref. 140 for a comprehensive review). In contrast, in the case of macromolecules this issue is highly controversial. Vesicles, intercellular junctions, and transendothelial channels have all been advocated as possible protein transport pathways [94, 137]. It is now generally conceded that intercellular junctions which open to macromolecular solutes are rare or nonexistent in arteriolar or capillary endothelia, although approximately on-fourth of the junctions are probably open to about ≥20 Å diameter in postcapillary venules [146,147]. The latter openings may be a major route for macromolecular exchange [148]. Leakage of macromolecular tracers has been observed to be somewhat restricted to the venular segment when the tracer molecular radius is larger than 30 Å [149]. (The radius of serum albumin, for example, is 36 Å.) A number of investigators have suggested that the passageways corresponding to the "large pore" system are concentrated at the venous ends of capillaries [149–151], but this has been disputed [94].

Whether transmembrane channels play a significant role in macromolecular exchange is unclear because their frequency of occurrence is still uncertain. It is difficult to enumerate these structures because their positive identification requires either a rare section that is in the plane of the channel across the entire width of the cell, or stereoscopic reconstruction.

A number of investigators believe that the plasmalemmal vesicles are the primary mediators of macromolecular exchange between plasma and interstitial fluid [94,152]. The strongest evidence comes from experiments in which the transendothelial passage of various macromolecular tracers has been followed at the EM level as a function of time (e.g., see Refs. 153–155). The sequence of events usually inferred following injection of a macromolecule probe into the bloodstream is uptake of the probe into vesicles at the luminal plasma membrane of the capillary endothelial cells followed by migration across the endothelial cell and release of the tracer on the abluminal side. This process, which has been termed capillary-trancytosis, appears to function bidirectionally [154,155]. These types of studies generally suggest that passage of macromolecule tracers also occurs, albeit to a lesser extent, through endothelial channels (which may be formed by fusion of the vesicles) but not through capillary junctional sites [94].

Unlike other bulk-phase pinocytosis or receptor-mediated endocytosis, trancytosis does not seem to utilize coated pits or culminate in vesicle fusions with lysosomes. Double-labeling experiments employing distinct plasma membrane and fluid-phase markers suggest that in trancytosis, unlike in other forms of endocytosis, the transporting vesicles do not arise by pinching-off from

the plasma membrane. That is, the vesicles seem to be composed of a pool
of membrane that does not mix with the plasma membrane pool [156].

The theory that trancytosis is responsible for much of the exchange of
macromolecules across capillaries has been criticized on grounds that it is
difficult to reconcile vesicular transport with physiological data that suggests pore-mediated solute exchange. For example, most physiological
measurements suggest that convection (a function of solvent volume flow)
makes a much larger contribution to macromolecular exchange than does
diffusion (a function of transcapillary solute gradient) [157,158]. That
solute transport would increase with volume flow is intuitively obvious in
the case of the pore theory but not in the case of a vesicular transport
pathway. Asymmetric vesicle exchange could accommodate coupling between
macromolecule flux and volume flow [159]. It should also be noted that
the relative contributions of convection and diffusion to macromolecule exchange is not without controversy (see Ref. 138 for a review). A recent
theoretical review [141] indicates that both pore and vesicle transport systems are plausible and that it is at present difficult to estimate the relative
contributions of these pathways to the flux of macromolecules across capillaries.

Electron microscopy, which is the source of virtually all the evidence
for the trancytosis pathway, is clearly subject to artifacts [160]. Based
on three-dimensional reconstruction of serial sections [161] and other
findings [162], Bundgaard and colleagues concluded that the plasmalemmal
vesicles observed in capillary endothelium are actually open at one or the
other cell surface. These investigators proposed that these structures are
stationary and therefore do not conduct fluid across the cells. However,
high-voltage electron microscopy, which provides a depth of view that is
many vesicle diameters deep, has indicated the existence of vesicular structures that are free in the cytoplasm [163]. The juxtaposition of free
vesicles with the distal ends of the caviolae in some cases suggested a
fusion or fission relationship, consistent with the concept of trancytosis.

The potential modes of solute transport above pertain to fenestrated
and discontinuous capillaries as well as to the continuous capillaries, with
gaps and fenestrated openings making additional contributions to solute
flux. The basement membrane−free 1000 to 10,000-Å gaps of discontinuous
epithelium account for the high permeability of discontinuous capillaries to
macromolecules. In fact, the large size of the gaps suggests that macromolecule filtration rate should be even faster than it is. The rate-limiting
barrier in the case of discontinuous capillaries may be the interstitial extracellular matrix [164].

Fenestral openings, despite their relatively large diameter, are not all
open to large macromolecules. Simionescu et al. determined that although
horseradish peroxidase (about 30 Å in radius) passed through all intestinal
capillary fenestrae, glycogen (about 300 Å in radius) passed through a
small fraction of the fenestrae [165]. This observation is compatible with
the findings that fenestrated intestinal capillaries restrict the passage of
certain macromolecules even more effectively than do the continuous capillaries of other tissues [166,167]. A primary function of fenestral openings
may be to enhance the transendothelial flow of water and low-molecular-
weight solutes [167,168]. The impermeability of some but not all fenestrae
to large macromolecules may be related to observed heterogeneity in fenestral
ultrastructure. Most, but not all fenestrae possess diaphragms that limit

the passage of molecules that are >100 **A** in diameter. The subset of open
fenestrae may be responsible for the "large pore" properties of fenestrated
endothelium [138]. Fenestral diaphragms contain heparan sulfate and hence
are negatively charged [169], which may explain findings that diaphragmed
fenestrae are a greater barrier to anionic proteins than to cationic proteins
[94].

Carrier-Mediated Transport Pathways

Specific solute transport pathways permit the movement across capillaries of
solutes that otherwise would not penetrate. Several of the transport systems for the passage of nutrients and ions are now relatively well characterized. Both uni- and bidirectional transport systems have been identified. The brain capillary endothelial Na^+/K^+ ATPase (Na/K^+ pump) is an
example of the former type. Its concentration at the abluminal half of the
plasma membrane [121] accounts for the fact that K^+ moves more rapidly
from brain to blood [172] than in the reverse direction [173]. This polar
active transport not only serves to maintain a low K^+ concentration in brain
interstitial fluid, but also provides a path for the movement of water from
blood to brain [171].

   At present, eight independent nutrient transport systems have been
identified in brain capillary endothelium: (1) a hexose carrier; (2) a mono-
carboxylic acid carrier; (3) a neutral amino acid carrier, which functions
in the transport of 14 neutral amino acids; (4) a carrier for lysine, arginine,
and ornithine; (5) a choline carrier; (6) an adenine/guanine carrier; (7)
a porter of purine and uracil nucleosides; and (8) a carrier for aspartic
and glutamic acids (see Ref. 170 for recent references and a brief recent
review). These are saturable and generally stereospecific [174].

   The neutral amino acid carrier is perhaps of greatest interest with regard to drug delivery (see below). This carrier in certain respects is
similar to the Na-independent leucine-preferring L system of the peripheral
tissues [175]. Its very low K_m, which is roughly 10% as large as the K_m
of the carrier in the peripheral tissues, renders it extraordinarily sensitive
to competition effects [176]. The high affinity of the brain neutral amino
acid carrier may resolve the findings that, although the concentration of
neutral amino acids is 10-fold higher in plasma than in brain fluid, the
rates of neutral amino acid influx and efflux are similar [176]. Unfortunately, the biochemical nature of the neutral amino acid carrier is not yet defined [177].

   In the last few years it has become increasingly apparent that specific
transport systems carry polypeptides across the capillary endothelium. The
endothelial trancytosis described above is a nonspecific process. In other
epithelial systems specific receptor—mediated trancytosis of proteins has
been demonstrated and in certain cases extensively characterized (see
above).

   The most persuasive evidence for receptor-mediated trancytosis by
vascular endothelium comes from analysis of insulin interaction with cultured
endothelial cell monolayers and with isolated brain microvessels [178,179].
Bovine aortic endothelial cell monolayers transfer [^{125}I]-insulin from the
side of the monolayer that corresponds to the luminal face to the side that
corresponds to the abluminal face [178]. Transfer in the opposite direction
does not occur. (Interestingly, unidirectional transport is a feature of
receptor-mediated trancytosis of immunoglobulins across intestinal epithelium

[180].) High-affinity insulin receptors present on the endothelium seem to mediate this in vitro transport of insulin because the process is inhibited by nonlabeled insulin or by anti-insulin receptor antibody. Microvessels isolated from human brain (22 to 36 hr postmortem) bind insulin with high affinity, internalize it, and transport it, intact, to culture medium [179]. Experiments that monitor insulin transport across capillary endothelial cells in vivo are crucial to determine whether receptor-mediated trancytosis is an important physiological mechanism for insulin delivery. Significantly, intracarotid injection of [^{125}I]insulin into suckling rabbits results in this radiotracer rapidly entering the brain by a saturable and therefore, presumably, receptor-mediated pathway [181]. Tentative evidence suggests that endothelial receptor-mediated trancytosis pathways also exist for other polypeptides (e.g., transferrin, insulin-like growth factor I (B. Keller and R. T. Borchardt, unpublished data), and β-lipotropin [182]).

E. Basement Membrane Barrier

In addition to the epithelial barrier, capillaries also impose a basement membrane barrier to solute transport. This structure is absent only in the sinusoidal gaps of discontinuous endothelium. Elsewhere it is elaborated in varying densities and thicknesses. Glomerular capillaries, which posses a particularly thick basement membrane, provide clear evidence for basement membrane—mediated retardation of solute movement. In this case the basement membrane is apparently the limiting barrier for capillary permeation by macromolecules [183]. It limits dextrans according to size. Moreover, its filtration properties are exquisitely sensitive to charge. Thus fractional solute clearance of 30-Å-diameter dextran is twice as great for cationic dextran as for neutral dextran, and about 10-fold as great for neutral dextran as for anionic dextran [184,185]. The charge selectivity can be accounted for by the fact that the basement membrane surface is anionic. Retardation of macromolecules by basement membranes has also been detected in intestinal capillaries [186].

F. Perturbation or Circumvention of the Vascular Barrier

The capillaries pose a major obstacle to the delivery of certain drugs. Delivery to the brain is the extreme case. The blood-brain barrier is regarded as "a major obstacle to the development of pharmaceutically useful peptides" [187]. Although the continuous capillaries that feed most other tissues allow the translocation of even relatively large proteins, the solute filtration rate is nevertheless proportional to solute size. Relative to low-molecular-weight solutes, proteins are thus subject to prolonged residence time in the circulation and consequently increased vulnerability to clearance by the reticuloendothelial cells. The latter are mononuclear phagocytic cells of the liver and spleen which filter out foreign or altered molecules that enter the sinusoids of these tissues [188]. In certain cases even native physiological proteins injected into the bloodstream are cleared at extraordinary rates [189,190]. Pharmacological or physiological conditions that temporarily relax the capillary barrier may indicate new strategies for drug delivery. Several examples are noteworthy.

In the case of low-molecular-weight drugs the substitution of nonpolar for polar groups may allow penetration of lipid bilayers and hence passage through capillary endothelial cells. Nature provides an example of this type of modification. Heroin and morphine differ at two positions, at which the former has non-hydrogen-bonding acetyl groups and the latter has hydroxyl groups. Heroin enters the brain about 100-fold more rapidly than does morphine [191]. In the case of di- and tripeptides, closure of the polar amino and carboxyl termini (which results in the formation of diketopiperazines) increases the rate of peptide entry into the brain by 10- to 100-fold. This modification has yielded CNS-active analogs of several peptide hormones [192,193].

The specific endothelial transport systems for both low-molecular-weight and polypeptide solutes are promising pthways for drug delivery. Because of its relatively low specificity, the neutral amino acid carrier mediates transport of certain drugs (e.g., dopa and melphalan) [194,195]. However, it seems doubtful that molecules of the size of peptides or larger can be delivered by this carrier. The ability of insulin to penetrate the blood-brain barrier suggests that it may be possible to deliver peptides or even proteins across brain capillaries by cross-linking these molecules to insulin or to the domain of insulin that is recognized by insulin receptors [196]. This type of approach, utilizing insulin or other ligands that undergo trancytosis across capillary endothelium, will undoubtedly be the subject of intense investigation over the next several years.

An alternative, less solute-specific means of increasing transcapillary movement of peptides or proteins is to perturb the endothelial cell membranes or their junctions. Injection of hypertonic sugar, NaCl, or urea solutions into the carotid aretery causes temporary opening of the blood-brain barrier [125,197]. This procedure, which increases cerebrovascular permeability more than 20-fold, facilitates delivery of antitumor drugs to the brain [198,199]. Measurement of solute clearance rates [200] and electron microscopic examination of tracer penetration [201] have usually indicated that hyperosmotic solutions act by opening tight junctions, but certain studies suggest that the major effect is to increase the rate of vesicular transport [202,203].

The endothelial barrier is modulated by various physiological parameters. Glucocorticoids tighten the blood-brain barrier [204], perhaps by decreasing vesicular transport [205]. Inflammatory agents, certain hormones, vasoactive agents [206], or hypertension [207] relax the endothelial barrier. Abundant evidence indicates that histamine, bradykinin, serotonin, estrogen, and angiotensin increase vascular permeability by opening large gaps in the endothelial junctions of postcapillary venules [208,209; see Ref. 138 for a review]. Numerous investigators have suggested that the gaps are created by stimulation of cactomyosin (contractile elements in the endothelial cells, possibly utilizing Ca^{2+} as a second messenger [210]). This hypothesis is as yet unproved [211].

All of the mechanisms of opening interendothelial junctions noted above either cause gross changes in the capillary tissue (e.g., perfusion of hyperosmotic solutions) or undesired side effects (e.g., inflammatory agents). Progress in defining the nature and regulation of tight junctions, at a biochemical level, brings with it the hope of developing a drug that will specifically open the junctions, for example, a monoclonal antibody against a junction protein.

IV. SUMMARY

Although some amino acids may cross the intestinal epithelial and some vascular endothelial membranes (at least partly) by simple diffusion, peptides and proteins are commonly transported by carrier-mediated processes. As attempts to deliver peptide and protein drugs across these biological barriers have met with limited success, it is clear that a more fundamental understanding of the composition and properties of these membranes is necessary before meaningful advances can be made in the area of peptide and protein drug delivery. Recently, cultured microvessel endothelial and intestinal epithelial cell systems have been employed in uptake and/or transport studies. Because such systems allow more detailed characterization of specific transport processes at a cellular or subcellular level, their utilization is expected to increase in the near future.

REFERENCES

1. J. S. Trier and J. L. Madara, in *Physiology of the Gastrointestinal Tract* (L. R. Johnson, ed.), Raven Press, New York (1981), p. 926.
2. F. Verzar and E. J. McDougall, *Absorption from the Intestine*, Longmans, Green & Co., London (1936), p. 9.
3. F. Moog, *Sci. Am.*, *245*:154 (1981).
4. A. L. Brown, Jr., *J. Cell Biol.*, *12*:623 (1962).
5. H. Zetterqvist, *The Ultrastructural Organization of the Columnar Absorbing Cells of the Mouse Jejunum*, Monograph, Aktiebolaget Godvil, Stockholm (1956).
6. M. F. Kagnoff, in *The Role of the Gastrointestinal Tract in Nutrient Delivery* (M. Green and H. L. Greene, eds.), Academic Press, Orlando, Fl. (1984), p. 239.
7. M. R. Neutra and H. A. Padykula, in *Histology: Cell and Tissue Biology* (L. Weiss, ed.), Elsevier/North-Holland Biomedical Press, New York (1983), p. 659.
8. M. G. Farquhar and G. E. Palade, *J. Cell Biol.*, *17*:375 (1973).
9. F. S. Sjostrand, *J. Ultrastruct. Res.*, *8*:517 (1963).
10. J. S. Trier, in *Handbook of Physiology*, Section 6, Vol. 3 (C. F. Code, ed.), American Physiological Society, Bethesda, Md. (1968), p. 1125.
11. J. P. Revel and S. Ito, in *The Specificity of Cell Surfaces* (B. D. Davis and L. Warren, eds.), (1967), p. 211.
12. S. Ito, *J. Cell Biol.*, *27*:475 (1965).
13. H. H. Edwards, T. J. Mueller, and M. Morrison, *Science*, *203*:1343 (1979).
14. P. Pinto da Silva, P. S. Moss, and H. H. Fudenberg, *Exp. Cell Res.*, *81*:127 (1973).
15. D. Branton, *Proc. Natl. Acad. Sci. USA*, *55*:1048 (1966).
16. J. Brunner, H. Hauser, H. Braun, K. J. Wilson, H. Wacker, B. O'Neill, and G. Semenza, *J. Biol. Chem.*, *254*:1821 (1979).
17. I. A. Brasitus, D. Schachter, and T. G. Mamounes, *Biochemistry*, *18*: 4136 (1979).
18. A. Eichholz, *Biochim. Biophys. Acta*, *163*:101 (1967).
19. J.-F. Bouhours and R. M. Glickman, *Biochim. Biophys. Acta*, *487*:51 (1977).
20. K. Kawai, M. Fujita, and M. Nakao, *Biochim. Biophys. Acta*, *369*:222 (1974).

21. A. P. Douglas, R. Kerley, and K. J. Isselbacher, *Biochem. J.*, *128*: 1329 (1972).

22. D. Schachter and M. Schinitzky, *J. Clin. Invest.*, *59*:536 (1977).

23. M. E. Etzler and M. L. Branstrator, *J. Cell Biol.*, *62*:329 (1974).

24. B. Messier and C. P. Leblond, *Am. J. Anat.*, *106*:247 (1960).

25. D. R. Dibona, L. C. Chen, and G. W. G. Sharp., *J. Clin. Invest.*, *53*:1300 (1974).

26. G. N. Tytgat, C. E. Rubin, and D. R. Saunders, *J. Clin. Invest.*, *50*:2065 (1971).

27. C. E. Stirling, *J. Cell Biol.*, *53*:704 (1972).

28. M. M. Weiser, M. M. Neumeier, A. Quaroni, and K. Kirsch, *J. Cell Biol.*, *77*:722 (1978).

29. A. Quaroni, K. Kirsch, and M. M. Weiser, *Biochem. J.*, *182*:203 (1979).

30. M. G. Farquhar and G. E. Palade, *J. Cell Biol.*, *17*:375 (1963).

31. B. E. Hull and L. A. Staehelin, *J. Cell Biol.*, *81*:67 (1979).

32. W. D. Troughton and J. S. Trier, *J. Cell Biol.*, *41*:251 (1969).

33. J. L. Madara, J. S. Trier, and M. R. Neutra, *Gastroenterology*, *78*: 963 (1980).

34. C. Nordstrom and A. Dahlqvist, *Scand. J. Gastroenterol.*, *8*:407 (1973).

35. J. S. Cornes, *Gut*, *6*:225 (1965).

36. H. Moe, *Int. Rev. Cytol.*, *4*:299 (1955).

37. J. Merzel and C. P. Leblond, *Am. J. Anat.*, *124*:281 (1969).

38. A. Bella, Jr., and Y. S. Kim, *Arch. Biochem. Biophys.*, *150*:679 (1972).

39. J. F. Forstner, I. Jabbae, and G. G. Forstner, *Can. J. Biochem.*, *51*:1154 (1973).

40. H. W. Moon, S. C. Whipp, and A. L. Baetz, *Lab. Invest.*, *25*:133 (1971).

41. M. F. Sullivan, E. V. Hulse, and R. M. Mole, *Br. J. Exp. Pathol.*, *46*:235 (1965).

42. J. J. Skillman, S. A. Gould, R. S. K. Chung, and W. Silen, *Ann. Surg.*, *172*:564 (1970).

43. W. H. Barr and S. Riegelman, *J. Pharm. Sci.*, *59*:154 (1970).

44. R. D. Shaw, K. Li, J. W. Hamilton, A. L. Shug, and W. A. Olsen, *Am. J. Physiol.*, *245*:G376 (1983).

45. C. Fernandez-Tejero and M. Gilles-Baillien, *J. Physiol.*, *266*:35P (1977).

46. I. Osiecka, P. A. Porter, R. T. Borchardt, J. A. Fix, and C. R. Gardner, *Pharm. Res.*, *2*:293 (1985).

47. A. K. Mircheff, C. H. Van Os, and E. M. Wright, *J. Membr. Biol.*, *52*:83 (1980).

48. L. M. Pinkus, in *Methods in Enzymology*, Vol. 77 (W. B. Jakoby, ed.), Academic Press, New York (1981), p. 154.

49. M. Pinto, S. Robine-Leon, M.-D. Appay, M. Kedinger, N. Triadou, E. Dussaulx, B. Lacroix, P. Simon-Assmann, K. Haffen, J. Fogh, and A. Zweibaum, *Biol. Cell*, *47*:323 (1983).

50. A. Zweibaum, M. Pinto, G. Chevalier, E. Dussaulx, N. Triadou, B. Lacroix, K. Haffer, J.-L. Brun, and M. Rousset, *J. Cell Physiol.*, *122*:21 (1985).

51. N. F. H. Ho, J. Y. Park, W. Morozowich, and W. Higuchi, in *Design of Biopharmaceutical Properties Through Prodrugs and Analogs*

(E. B. Roche, ed.), American Pharmaceutical Association/American
Physiological Society, Washington, D.C. (1977), p. 136.

52. I. Komiya, J. Y. Park, A. Kamani, N. F. H. Ho, and W. Higuchi,
    Int. J. Pharm., *4*:249 (1980).

53. H. G. Windmueller and A. E. Spaeth, *Arch. Biochem. Biophys.*, *171*:
    662 (1975).

54. D. M. Matthews, R. H. Gandy, E. Taylor, and D. Burston, *Clin. Sci.*,
    56:15 (1979).

55. A. M. Asatoor, A. Chedra, M. D. Milne, and D. I. Prosser, *Clin.
    Sci. Mol. Med.*, *45*:199 (1973).

56. J. M. Addison, D. Burston, J. A. Dalrymple, D. M. Matthews, J. W.
    Payne, M. H. Sleisenger, and S. Wilkinson, *Clin. Sci. Mol. Med.*, *49*:
    313 (1975).

57. S. A. Adibi and M. R. Soleimanpour, *J. Clin. Invest.*, *53*:1368 (1974).

58. M. H. Sleisenger, D. Burston, J. A. Dalrymple, S. Wilkinson, and
    D. M. Matthews, *Gastroenterology*, *71*:76 (1976).

59. V. J. Gupta and K. D. G. Edwards, *Clin. Exp. Pharmacol. Physiol.*,
    3:511 (1976).

60. E. Taylor, D. Burston, and D. M. Matthews, *Clin. Sci.*, *58*:221
    (1980).

61. E. Danforthand and R. D. Moore, *Endocrinology*, *65*:118 (1959).

62. J. Samanen, in *Bioactive Polymeric Systems* (C. G. Gebelain and C. E.
    Carraher, Jr., eds.) Plenum Press, New York (1985), p. 279.

63. P. J. Jacques, in *Lysosomes in Biology and Pathology*, Vol. 2 (J. T.
    Dingle and H. B. Fell, eds.), North-Holland-Amsterdam Publishing
    Comapny, (1969), p. 395.

64. C. W. Crane and G. R. W. N. Luntz, *Diabetes*, *17*:625 (1968).

65. M. Kidron, H. Bar-On, E. M. Berry, and E. Ziv, *Life Sci.*, *31*:2837
    (1982).

66. R. Cornell, W. A. Walker, and K. J. Isselbacher, *Lab. Clin. Invest.*,
    25:42 (1971).

67. F. L. Gruskay and R. E. Cooke, *Pediatrics*, *16*:763 (1955).

68. S. L. Clark, Jr., *J. Biophys. Biochem. Cytol.*, *5*:41 (1959).

69. J. L. Ambrus, H. B. Lassman, and J. J. DeMarchi, *Clin. Pharmacol.
    Ther.*, *8*:362 (1966).

70. H. Moriya, C. Moriwaki, S. Akimote, K. Yamaguchi, and M. Iwadare,
    Chem. Pharm. Bull., *15*:1662 (1967).

71. J. G. Lecce, *J. Physiol.*, *184*:594 (1966).

72. J. R. Casley-Smith, *Experentia*, *15*:370 (1967).

73. F. W. R. Brambell, R. Halliday, R. Brierley, and W. A. Hemmings,
    Lancet, *1*:694 (1954).

74. J. P. Kraehenbuhl and M. A. Campiche, *J. Cell Biol.*, *42*:345 (1969).

75. R. Rodewald, *J. Cell Biol.*, *45*:635 (1970).

76. E. A. Jones and T. A. Waldman, *J. Clin. Invest.*, *51*:2916 (1972).

77. I. G. Morris, *Proc. R. Soc. London, B157*:160 (1963).

78. F. W. R. Brambell, *Lancet*, *2*:1087 (1966).

79. K. Simonski, P. Gonella, J. Bernanke, L. Owen, M. R. Neutra, and
    R. A. Murphy, *J. Cell Biol.*, *103*:1979 (1987).

80. W. Thornburg, L. Matrisian, B. Magnun, and O. Koldousky, *Am. J.
    Physiol.*, *246*:680 (1984).

81. R. K. Rao, W. Thornburg, M. Korc, L. M. Matrisian, B. E. Magun,
    and O. Koldovsky, *Am. J. Physiol.*, *250*:G850 (1986).

82. N. Gallo-Payet and J. S. Hugon, *Endocrinology*, *116*:194 (1985).

83. J. F. Thompson, *Gastroenterology*, *90*:1664 (1986).

84. P. Gonella, R. Murphy, L. Owen, J. Bernanke, and M. Neutra, *J. Cell Biol.*, *99*:283a (1984).

85. J. M. Diamond, *Fed. Proc.*, *33*:2220 (1974).

86. B. G. Munck and S. N. Rasmussen, *J. Physiol.*, *271*:473 (1977).

87. R. S. Rhodes and M. J. Karnovsky, *Lab. Invest.*, *25*:220 (1971).

88. M. M. Civan and D. R. DiBona, *J. Membr. Biol.*, *38*:359 (1978).

89. J. B. Wade, J.-P. Revel, and V. A. DiScala, *Am. J. Physiol.*, *224*: 407 (1973).

90. N. Simionescu and M. Simionescu, in *Histology* (L. Weiss, ed.), Elsevier Science Publishing Co., New York (1983), p. 371.

91. A. R. Muir and A. Peters, *J. Cell Biol.*, *12*:443 (1962).

92. U. Smith, J. W. Ryan, D. D. Michie, and D. S. Smith, *Science*, *173*: 925 (1971).

93. U. S. Ryan, J. W. Ryan, and D. J. Crutchley, *Fed. Proc.*, *44*:2603 (1985).

94. M. Simionescu and N. Simionescu, in *Handbook of Physiology*, Section 2, *The Cardiovascular System*, Vol. 4, *Microcirculation*, Part I, American Physiological Society, Bethesda, Md. (1984), p. 41.

95. H. S. Bennett, J. H. Luft, and J. C. Hampton, *Am. J. Physiol.*, *196*:381 (1959).

96. G. E. Palade, *Anat. Rec.*, *136*:254 (1960).

97. R. R. Brune and G. E. Palade, *J. Cell Biol.*, *37*:244 (1968).

98. W. E. Stehbens, *J. Ultrastruct. Res.*, *15*:389 (1966).

99. N. A. Kefalides, J. D. Cameron, E. A. Tomichek, and M. Yanoff, *J. Biol. Chem.*, *251*:730 (1976).

100. L. Biempica, R. Morecki, C. H. Wu, M. A. Giambrone, and M. Rojkind, *Am. J. Pathol.*, *98*:591 (1980).

101. M. P. Cohen and C. J. Ciborowski, *Biochim. Biophys. Acta*, *674*:400 (1981).

102. H. J. Geuze, J. W. Slot, G. J. A. M. Strous, H. F. Lodish, and A. L. Schwartz, *Cell*, *32*:277 (1983).

103. F. Clementi and G. E. Palade, *J. Cell Biol.*, *41*:33 (1969).

104. J. R. Maynard, *Am. J. Anat.*, *100*:409 (1957).

105. M. A. Farquhar, *Kidney Int.*, *8*:197 (1975).

106. M. W. B. Bradbury, *Circ. Res.*, *57*:213 (1985).

107. R. D. Tschirgi, in *Handbook of Physiology*, Section 3, *Neurophysiology*, Vol. 3, American Physiological Society, Bethesda, Md. (1960), p. 1865.

108. C. Crone and S. P. Olesen, *Brain Res.*, *241*:49 (1982).

109. J. R. Wolff and T. Bar, *Brain Res.*, *41*:17 (1972).

110. A. Meister, *Science*, *180*:33 (1973).

111. F. M. Lai, S. Udenfriend, and S. Spectro, *Proc. Natl. Acad. Sci. USA*, *72*:4622 (1975).

112. M. Simionescu, N. Simionescu, and G. E. Palade, *J. Cell Biol.*, *67*: 863 (1975).

113. E. M. Landis, *Ann. N.Y. Acad. Sci.*, *116*:765 (1964).

114. F. Vargas and J. A. Johnson, *J. Gen. Physiol.*, *47*:667 (1964).

115. O. Kedem and A. Katchalsky, *J. Gen. Physiol.*, *45*:143 (1961).

116. D. N. Granger, J. P. Granger, R. A. Bruce, R. E. Parker, and A. E. Taylor, *Circ. Res.*, *44*:335 (1979).

117. K. L. Audus and R. T. Borchardt, *Ann. N.Y. Acad. Sci.*, in press (1987).

118. D. Pardridge, J. Eisenberg, and J. Yung, *J. Neurochem.*, *44*:1771 (1985).

119. W. H. Oldendorf, *Proc. Soc. Exp. Biol. Med.*, *147*:813 (1974).

120. J. D. Fenstermacher and S. I. Rapoport, in *Handbook of Physiology*, Section 2, *The Cardiovascular System*, Vol. 4, *Microcirculation*, Part II, American Physiological Society, Bethesda, Md. (1984), p. 969.

121. J. E. Hardabo, L. Edvinsson, B. Fulck, M. Lindvall, C. Owman, E. Rosengren, and N. A. Svengaard, in *The Cerebral Vessel Wall* (J. Cervos-Navarro and F. Matukis, eds.), Raven Press, New York (1976), p. 233.

122. A. J. L. Cooper, J. M. McDonald, A. S. Gelbard, R. F. Gledhill, and T. E. Duffy, *J. Biol. Chem.*, *254*:4982 (1979).

123. J. E. Cremer, H. M. Teal, D. F. Heath, and J. B. Cavanaugh, *J. Neurochem.*, *28*:215 (1977).

124. N. Simionescu, M. Simionescu, and G. E. Palade, *Microvasc. Res. J.*, *15*:17 (1978).

125. S. I. Rapaport, D. S. Bachman, and H. K. Thompson, *Science*, *176*:1243 (1972).

126. K. Dorovini-Zis, P. D. Bowman, A. L. Betz, and G. W. Goldstein, *Brain Res.*, *302*:383 (1984).

127. M. W. Brightman and T. S. Reese, *J. Cell Biol.*, *40*:648 (1969).

128. H. F. Cserr and M. Bundgaard, *Am. J. Physiol.*, *246*:R277 (1984).

129. L. A. Staehelin, *J. Cell Sci.*, *13*:763 (1973).

130. P. Pinto da Silva and B. Kachar, *Cell*, *28*:441 (1982).

131. B. Kachar and T. S. Reese, *Nature*, *296*:464 (1982).

132. A. J. Verkleij, *Proc. Electron Microsc. Sci. Am.*, *38*:688 (1980).

133. A. J. Verkleij, C. Mombers, W. J. Gerritsen, L. Leunissen-Bijvelt, and P. R. Cullis, *Biochim. Biophys. Acta*, *555*:358 (1979).

134. G. van Meer, B. Grumbiner, and K. Simons, *Nature*, *322*:639 (1986).

135. B. R. Stevenson and D. A. Goodenough, *J. Cell Biol.*, *98*:1209 (1984).

136. B. Gumbiner and K. Simons, *J. Cell Biol.*, *102*:457 (1986).

137. E. Raviola, D. A. Goodenough, and G. Raviola, *J. Cell Biol.*, *87*:273 (1980).

138. A. E. Taylor and D. N. Granger, in *Handbook of Physiology*, Section 2, *The Cardiovascular System*, Vol. 4, *Microcirculation*, Part I, American Physiological Society, Bethesda, Md. (1984), p. 467.

139. J. R. Pappenheimer, E. M. Renkin, and L. Borrero, *Am. J. Physiol.*, *167*:13 (1951).

140. C. Crone and D. L. Levitt, in *Handbook of Physiology*, Section 2, *The Cardiovascular System*, Vol. 4, *Microcirculation*, Part I, American Physiological Society, Bethesda, Md. (1984), p. 411.

141. F. E. Curry, in *Handbook of Physiology*, Section 2, *The Cardiovascular System*, Vol. 4, *Microcirculation*, Part I, American Physiological Society, Bethesda, Md. (1984), p. 309.

142. C. C. Michel, *J. Physiol.*, *309*:341 (1980).

143. F. E. Curry, *Fed. Proc.*, *44*:2610 (1985).

144. J. R. Casley-Smith, N. S. Green, J. L. Harris, and P. J. Wadey, *Microvasc. Res. J.*, *10*:43 (1975).

145. M. A. Perry, *Microvasc. Res.*, *19*:142 (1980).

146. G. E. Palade, M. Simionescu, and N. Simionescu, *Acta Physiol. Scand. Suppl.*, *463*:11 (1979).

147. T. S. Reese and M. J. Karnovsky, *J. Cell Biol.*, *34*:207 (1969).

148. A. Bill, *Acta Physiol. Scand.*, *73*:511 (1968).

149. G. Hauck, *Bibl. Anat.*, *15*:202 (1977).

150. P. Rous, H. P. Gilding, and F. Smith, *J. Exp. Med.*, *51*:807 (1930).

151. Y. Nakamura and H. Wayland, *Microvasc. Res.*, *9*:1 (1975).

152. M. J. Poznansky and R. L. Juliano, *Pharm. Rev.*, *36*:277 (1984).

153. N. Simionescu, *J. Histochem. Cytochem.*, *27*:1120 (1979).

154. B. R. Johansson, *Microvasc. Res.*, *16*:354 (1978).

155. B. R. Johansson, *Microvasc. Res.*, *16*:362 (1978).

156. N. Simionescu, in *Internal Cell Biology*, 1980–1981 (H. G. Schweiger, ed.), Springer-Verlag, Berlin (1981), p. 657.

157. B. Rippe, A. Kamiya, and B. Folkow, *Acta Physiol. Scand.*, *105*:171 (1979).

158. E. M. Renkin, W. L. Joyner, C. H. Sloop, and P. D. Watson, *Microvasc. Res.*, *14*:191 (1977).

159. E. M. Renkin, *Physiologist*, *23*:57 (1980).

160. C. Crone, *Fed. Proc.*, *45*:77 (1986).

161. M. Bundgaard, P. hagmar, and C. Crone, *Microvasc. Res.*, *25*:358 (1983).

162. M. Bundgaard, J. Frokjaer-Jensen, and C. Crone, *Proc. Natl. Acad. Sci. USA*, *76*:6439 (1979).

163. R. C. Wagner and C. S. Robinson, *Microvasc. Res.*, *28*:197 (1984).

164. D. N. Granger, T. Miller, R. Allen, R. E. Parker, J. C. Parker, and A. E. Taylor, *Gastroenterology*, *77*:103 (1979).

165. N. Simionescu, M. Simionescu, and G. E. Palade, *J. Cell Biol.*, *57*: 424 (1973).

166. H. J. Granger, *Adv. Biomed. Eng.*, *7*:1 (1979).

167. E. M. Renkin, *Circ. Res.*, *41*:735 (1977).

168. E. M. Renkin, *Microvasc. Res. J.*, *15*:123 (1978).

169. M. Simionescu, N. Simionescu, J. E. Silbert, and G. E. Palade, *J. Cell Biol.*, *90*:614 (1981).

170. W. M. Pardridge, *Fed. Proc.*, *45*:2047 (1986).

171. A. L. Betz, *Fed. Proc.*, *45*:2050 (1986).

172. M. W. B. Bradbury and B. Stulcova, *J. Physiol.*, *208*:415 (1970).

173. A. J. Hansen, H. Lund-Anderson, and C. Crone, *Acta Physiol. Scand.*, *101*:438 (1977).

174. W. M. Pardridge and W. H. Oldendorf, *J. Neurochem.*, *28*:5 (1977).

175. H. M. Christensen, *Fed. Proc.*, *32*:19 (1973).

176. W. M. Pardridge and T. B. Choi, *Fed. Proc.*, *45*:2073 (1986).

177. K. L. Audus and R. T. Borchardt, in *Theory and Applications of Bioreversible Carriers in Drug Design* (E. B. Roche, ed.), Plenum Press, New York, in press (1988).

178. G. L. King and S. M. Johnson, *Science*, 227:1583 (1985).

179. W. M. Pardridge, J. Eisenberg, and J. Yang, *J. Neurochem.*, 44: 1771 (1985).

180. K. E. Mostov and N. E. Simister, *Cell*, *43*:389 (1985).

181. K. R. Duffy, T. B. Chi, and W. M. Pardridge, *Clin. Res.*, *34*:57A (1986).

182. W. M. Pardridge, *Endocr. Rev.*, *7*:314 (1986).

183. J. P. Caulfield and M. G. Farquhar, *J. Cell Biol.*, *63*:883 (1974).

184. R. L. S. Chang, W. M. Deen, C. R. Robertson, and B. M. Brenner, *Kidney Int.*, *8*:212 (1975).

185. M. P. Bohrer, P. C. Baylis, D. Humes, R. J. Glassock, C. R. Robertson, and B. M. Brenner, *J. Clin. Invest.*, *61*:72 (1978).
186. N. Simionescu, M. Simionescu, and G. E. Palade, *J. Cell Biol.*, *53*: 365 (1972).
187. G. Meisenberg and W. H. Simmons, *Life Sci.*, *32*:2611 (1983).
188. D. L. Knook and E. Wisse, *Sinusoidal Liver Cells*, Elsevier/North-Holland, Amsterdam (1982).
189. H. E. Fuchs, H. G. Trapp, M. J. Griffith, H. R. Roberts, and S. V. Pizzo, *J. Clin. Invest.*, *73*:1696 (1984).
190. H. E. Fuchs, H. Berger, Jr., and S. V. Pizzo, *Blood*, *65*:539 (1985).
191. W. H. Oldendorf, S. Hyman, L. Braun, and S. Z. Oldendorf, *Science*, *178*:984 (1972).
192. P. L. Hoffman, R. Walter, and M. Bulat, *Brain Res.*, *122*:87 (1977).
193. J. S. Peterson, P. W. Kaliras, and C. Prasad, *Soc. Neurosci. Abstr.*, *10*:1123 (1984).
194. W. M. Pardridge, in *Directed Drug Delivery* (R. T. Borchardt, A. J. Repta, and V. J. Stella, eds.), Humana Press, Clifton, N.J. (1985), p. 83.
195. W. M. Pardridge, *J. Neurochem.*, *28*:103 (1977).
196. W. M. Pardridge, *Annu. Rep. Med. Chem.*, *20*: Chap. 33 (1985).
197. S. I. Rapopart, M. Hori, and I. Klatzo, *Am. J. Physiol.*, *223*:323 (1972).
198. E. A. Neuwelt, E. Balalan, J. Diehl, S. Hill, and E. Frenkel, *Neurosurgery*, *12*:662 (1983).
199. E. A. Neuwelt, J. Minna, E. Frenkel, P. A. Barnett, and C. I. McCormick, *Am. J. Physiol.*, *250*: 875 (1986).
200. W. G. Mayhan and D. D. Heistad, *Am. J. Physiol.*, *248*:H712 (1985).
201. K. Korovini-Zis, P. D. Bowman, A. L. Betz, and G. W. Goldstein, *Brain Res.*, *302*:383 (1986).
202. E. Westergaard, E. B. Van Deurs, and H. E. Bronsted, *Acta Neuropathol.*, *37*:141 (1977).
203. C. L. Farrell and R. R. Shrivers, *Acta Neuropathol.*, *63*:179 (1984).
204. J. B. Long and J. W. Holaday, *Science*, *227*:1580 (1985).
205. E. T. Hedley-Whyte and D. W. Hsu, *Ann. Neurol.*, *19*:373 (1986).
206. G. J. Grega, *Fed. Proc.*, *45*:75 (1986).
207. H. Hansson, B. B. Johansson, and C. Blomstrand, *Acta Neuropathol.*, *32*:187 (1975).
208. A. Kahn and E. Brachet, *Bibl. Anat.*, *15*:452 (1977).
209. E. Svensjo and G. J. Grega, *Fed. Proc.*, *45*:89 (1986).
210. G. Manjo, S. M. Shea, and M. Leventhal, *J. Cell Biol.*, *42*:647 (1969).
211. F. Hammersen, E. Hammersen, and U. Osterkamp-Baust, *Prog. Appl. Microcirc.*, *1*:1 (1983).

9
Pharmacokinetics of Peptide and Protein Drugs

Udaya Bhaskar Kompella and Vincent H. L. Lee

University of Southern California School of Pharmacy, Los Angeles, California

I. INTRODUCTION

The effects of therapeutic peptides and proteins are diverse, ranging from metabolism of glucose (insulin), calcium (calcitonin and parathyroid hormone), and sodium (atrial natriuretic factor) to immunoregulation (interferons, interleukins, thymic hormones, and tuftsin), cytostasis, and cytotoxicity (interferons, neocarzinostatin, and immunotoxins). Other effects include inhibition of proteinases (aprotinin, α_1-antitrypsin, and α_2-macroglobulin), on the one hand, and activation of proteinases involved in the fibrinolytic system (tissue-type plasminogen activator), on the other.
The molecular size of peptides and proteins is very heterogeneous, ranging from about 1 to 160 kD. Moreover, while proteins such as insulin, interferon-α, tumor necrosis factor-α(TNF-α), and interleukin-1β are not glycosylated, many others, such as interferon-β_1, -β_2, and -Γ, TNF-β, colony stimulating factors, erythropoietin, and most of the interleukins, are glycosylated.
　　Commercial production of peptide and protein drugs is now possible due to the availability of sophisticated synthetic and recombinant DNA techniques. To date, more than 150 recombinant proteins have reached phase I clinical trials or beyond, and some of them have received FDA approval. The purpose of this chapter is to highlight the pharmacokinetic aspects relevant to peptide and protein drugs, including clearance mechanisms, design of pharmacokinetic studies, and structure-pharmacokinetic relationships.

II. RELEVANT PHARMACOKINETIC PARAMETERS

Pharmacokinetic studies describe what the body does to the drug and may be described as the mathematical relationship that exists between the dose

of drug and its concentrations in readily accessible sites in the body, such
as blood and urine. The data are usually analyzed using either compartmental or noncompartmental methods. The important parameters are
clearance, apparent volume of distribution, half-life, and bioavailability.

A. Clearance

Drug clearance may be defined as the volume of fluid that is cleared of
drug per unit time. Total clearance of a drug from the body almost always involves more than one organ. The clearance sites may include liver,
kidney, lung, saliva, sweat, and gut, of which the kidney and the liver
are the most important. Thus, systemic or total clearance is the sum of
all individual organ clearances:

$$Cl_s = Cl_{renal} + Cl_{liver} + Cl_{gut\ wall} + \cdots$$

Measurement of clearances of various organs is well described [1].

Once in the blood, small molecules such as thyroid hormones, antibiotics, and other drugs are more or less extensively bound to albumin or
globulins. Cyclosporine binds to erythrocytes in a reversible manner [2].
Plasma protein binding has also been reported for calcitonin [3] and
β-endorphin (β-EP) [4], among others. Sato et al. [4] demonstrated that
about 36% and 35−44% of ^{125}I-β-EP was protein bound in human and rat serum,
respectively. ^{125}I-β-EP was bound partly to albumin and partly to α_2-
macroglobulin, β_2-macroglobulin, and lipoproteins. Binding of the drug to
blood constituents limits the availability of drug to the metabolic or
excretory sites.

Depending on the biological fluid in which the drug concentration is
measured, clearance is expressed as either blood clearance (Cl_b) or plasma
clearance (Cl_p). Blood clearance can be determined from plasma clearance
if the drug equilibrates between plasma and blood cells very rapidly according to the following relationship:

$$Cl_p/Cl_b = 1 + H(C_{RBC}/C_p - 1)$$

where H is the hematocrit and C_{RBC} and C_p are drug concentrations in
red blood cells and plasma, respectively.

All theories of pharmacokinetics generally assume linearity. Linearity
implies that for a given route and method of administration, the time
course of drug present in or eliminated from the body, normalized to the
administered dose, superimposes for all doses regardless of size and frequency. If the dose-normalized curves do not superimpose after appropriate corrections for variability, it can be construed that some kind of
nonlinearity exists. This is normally referred to as dose-dependent
kinetics. In this situation, the rate of a given process increases with the
concentration driving it, but approaches a limiting value.

Time dependency of pharmacokinetic parameters is one cause of nonlinearity. As described in Chap. 1, time-dependent pharmacokinetics may

result from circadian rhythms in absorption, distribution, and elimination or from auto- and heteroinduction of metabolic enzymes [5]. For instance, the 19% decrease in apparent clearance during the resting (p.m.) vs. activity (a.m.) period in five recipients of pancreatic allografts was thought to be responsible for the 23% increase in the amount of orally administered cyclosporine absorbed into the bloodstream [6]. Most dose-dependent situations can, however, be attributed to the saturability of one or more of the processes involved in drug absorption, distribution, metabolism, or excretion. Saturable kinetics are of therapeutic concern, because slight alterations in dosing rate may result in disproportionate changes in steady-state drug concentration. Additional causes for the dose dependency include: (1) physical properties such as low solubility, as is the case in cyclosporine [7], (2) altered renal handling due to diuresis or nephrotoxicity caused by the drug, and (3) altered hepatic handling due to hepatotoxicity, enzyme induction, or increased blood flow caused by the drug.

B. Apparent Volume of Distribution

Apparent volume of distribution is a proportionality constant relating drug concentration in blood or plasma to amount of drug in the body. It should be called as "apparent" volume of distribution since it cannot be related to any physiological space within the body. The relationship between amount of drug in the body (A), concentration of drug in the plasma (C), and apparent volume of distribution (V) may be given by the equation:

$$A = VC$$

Unlike drugs that follow one-compartment kinetics, those with multi-compartment kinetics usually do not equilibrate with various tissues instantaneously. As a result, the apparent volume of distribution can be used for some a priori comments concerning drug dosing. These comments should be made in conjunction with consideration of other parameters, such as amount of drug in the various compartments and elimination constants for drug removal from these compartments.

C. Half-Life

Half-life, $t_{1/2}$, is an expression of the relationship between volume of distribution and clearance. For a drug that exhibits monoexponential plasma disappearance, the half-life is given by the relationship:

$$t_{1/2} = 0.693 \ V/Cl$$

As disease state can affect either of the parameters of volume of distribution (V) or clearance (Cl), $t_{1/2}$ may not necessarily reflect the expected change in drug elimination. Nevertheless, it can be used as an indicator of the time required to reach and decay from steady-state

conditions, after administration of a drug at a definite rate. $t_{1/2}$ deline-
ates the maximum and minimum blood concentrations obtained for a particular dosage regimen, which is important in defining the pharmacodynamic response to a drug.

Most drugs exhibit multiexponential plasma disappearance curves, leading to many half-lives. A number of factors must be considered in defining the relevance of a particular half-life in drug therapy [8], including: (1) concentration (or amount)—effect (or toxicity) relationship,
(2) half-time for equilibration between measured plasma concentration and
pharmacodynamic effect, (3) fractional clearance related to a particular
half-life as defined by area under the curve (AUC), (4) increase in
amount of drug in the body at steady state which may be related to the
terminal half-life, (5) dosing interval with respect to the terminal half-life,
and (6) total period of drug dosing with respect to the terminal half-life.

D. Bioavailability

Bioavailability refers to the rate and extent of drug absorption. The rate
of absorption is more important for drugs given as a single dose. It may
be obtained either by measuring the rate constant for absorption or by
comparing the peak concentration and peak time. The rate at which a
peptide or a protein reaches the plasma compartment depends on the route
of administration. It will be instantaneous after intravenous (i.v.) or
intraarterial (i.a.) administration. It will take minutes to hours after
intramuscular (i.m.), subcutaneous (s.c.), intraperitoneal (i.p.), intra-
pleural (i.pl.), intrathecal (i.t.), rectal, or nasal administration. In all
cases except the i.t. route, proteins can reach the systemic circulation
either directly by venous blood capillaries or indirectly via the lymphatic
circulation. The relative amounts absorbed depend on the molecular size
of the protein, the anatomical area, and the nature of the pharmaceutical
vehicle. Protein molecules up to the size of insulin (6 kD) are probably
absorbed equally well by capillaries and lymphatics, while proteins over
20 kD are absorbed preferentially or exclusively via lymphatics [9]. Indeed, the very low, short-lived plasma levels and the very poor bioavailability of intramuscularly administered interferon β (IFN-β) previously
reported [10–12] may now be attributed to its lymphatic uptake. Thus,
contrary to popular belief, bioavailability of IFN-β cannot be simply estimated by the plasma AUC as its absorption through lymphatic system can
result in important immunomodulatory effects.

The extent of absorption (F) may be defined as the fraction of unchanged drug reaching the systemic circulation from a given route of administration. It is frequently calculated by dividing the area under the
concentration-time curve (AUC) obtained after administering the drug by
a particular route by the AUC of a separate, equally sized intravenous
dose. This method assumes that the overall clearance, defined as the
amount of drug eliminated (not including first-pass elimination, if any),
divided by AUC, is the same for both doses. Bioavailability (F) values
less than unity can be attributed to one of the following reasons:
(1) incomplete absorption, (2) metabolism at the site of administration,
(3) metabolism in the liver prior to its entry into the systemic circulation,
and (4) incomplete reabsorption after enterohepatic cycling on oral administration. The F value for a drug, which is completely absorbed and

eliminated only by the liver, will be decreased by hepatic extraction ratio (ER) when given orally. This can be expressed as:

$$F = 1 - ER = 1 - (Cl_{liver}/Q_{liver})$$

where Cl_{liver} and Q_{liver} represent the hepatic clearance and hepatic blood flow rate, respectively.

Drugs with saturable metabolism and a medium-to-high hepatic extraction ratio are expected to show an increase in bioavailability on increasing the dose or dosing rate. In this case, bioavailability becomes more variable due to the influence of absorption rate on the fraction escaping first-pass metabolism at any point of time. Thus, food, drugs, and other factors that affect the rate of absorption can influence the concentration entering the liver and the corresponding degree of saturation. Bioavailability estimation methods for drugs showing Michaelis-Menton kinetics have been reviewed by Martis and Levy [13].

III. CLEARANCE MECHANISMS OF PEPTIDE AND PROTEIN DRUGS

Clearance of peptide and protein drugs from the systemic circulation generally begins with passage across the capillary endothelia. The transendothelial passage of proteins is determined by such physicochemical properties of the molecule as size, shape, and charge [14,15], as well as by the ultrastructural and physicochemical characteristics of the capillaries themselves [16,17]. For instance, cytokines such as interferons (IFNs), interleukins (ILs), tumor necrosis factors (TNFs), and colony-stimulating factors (CSFs), which range from 17 to 26 kD in molecular weight depending on the extent of glycosylation, may either be partially taken up by the liver or be filtered by the kidneys with no return into the circulation.

Capillary endothelia are of three types: continuous, fenestrated, and discontinuous [18,19]. Their roles as barriers to the transport of macromolecules have been reviewed by Taylor and Granger [19].

Macromolecules may pass through continuous endothelia by one of three mechanisms: (1) capture by pinocytotic vesicles with an internal diameter of 500−1000 Å, (2) diffusion across intercellular junctions of 20−60 Å in width, and (3) passage through transendothelial channels formed by one or more vesicles that have opened simultaneously on either side of the endothelium. The basal lamina surrounding continuous capillaries is formed by a layer of fine, fibrous networks of collagen, glycoproteins, and probably fibronectin produced by the endothelial cells themselves. Particles with diameters between 50 Å and 110 Å can readily pass through the basal lamina [19]. This type of continuous epithelium is typical of capillaries in muscle, central nervous system, lung, subcutis, and bone tissue.

Fenestrated capillaries are permeable to macromolecules in the size range range 50−300 Å. These macromolecules can pass through fenestrated capillaries by the pathways of pinocytotic vesicles, diaphragm fenestrae, open fenestrae, intercellular junctions, and basal lamina [19]. This type of endothelium is typical of capillaries surrounding renal glomeruli, intestinal villi, skin, synovial tissue, and endocrine glands.

The discontinuous or sinusoidal endothelium has large pores ($10^3 - 10^4$ Å) in the intercellular junctions and is the most permeable of all micro-vascular beds. These pores and pinocytotic vesicles are the two potential pathways for the transport of macromolecules in this type of capillary. Discontinuous capillaries are found in liver, spleen, bone marrow, and at the level of postcapillary venules of lymph nodes.

A. Liver

Being well perfused and equipped with discontinuous capillaries, the liver plays an important role in the removal of macromolecules from the systemic circulation. The liver is composed of several cell types, including hepato-cytes, endothelial cells, Kupffer cells, and fat-storing cells. Their con-tributions to the total liver volume are 78%, 2.8%, 2.1%, and 1.4%, respec-tively. The remainder is constituted of sinusoids (10.6%), space of Disse (4.9%), and bilary tree (0.4%) [20,21]. Cell surface receptors for differ-ent proteins exist in the hepatocytes, Kupffer cells, endothelial cells, and fat-storing cells [22–27]. Hepatocytes recognize and internalize IgA and many galactose- and N-acetylgalactosamine-terminated glycoproteins. These cells also contain recognition sites for transferrin [28] and low-density lipoprotein (LDL) [23,29], as well as receptors for insulin, glucagon, growth hormone, and epidermal growth factor [30]. Kupffer cells endocy-tose particulate material to which galactose groups are connected and, to-gether with endothelial cells, also recognize immunoglobulin G (IgG) and fucose-, mannose-, or N-acetylglucosamine-terminated glycoproteins. In addition, endothelial cells possess a scavenger receptor, which binds and internalizes negatively charged proteins [31]. The fat-storing cells also recognize mannose-6-phosphate-terminated proteins.

Uptake of peptides and proteins from plasma by hepatocytes occurs by two distinct, yet not entirely separable processes: receptor-mediated endocytosis and nonselective pinocytosis. In receptor-mediated endocytosis, plasma-derived proteins become internalized following specific recognition and binding by hepatocyte receptor proteins located within the plasma membrane. These receptors are typically integral membrane glycoproteins possessing nonembedded, exposed domain located on the extracellular side of the hepatocyte membrane [32]. Receptor-mediated endocytosis is op-erative in hepatocytes for several proteins, including the hormones insulin, glucagon, growth hormone, and epidermal growth factor, as well as for glycoproteins, lipoproteins, immunoglobulins, intestinal and pancreatic pep-tides, and metallo- and hemoproteins [30]. It also occurs in nonparen-chymal hepatic cells which can recognize mannose-containing glycoproteins such as horseradish peroxidase (HRP) [30].

A requisite for the demonstration of receptor-mediated endocytosis in vivo is delay in the plasma disappearance rate of labeled peptides in the presence of excess unlabeled ligand. Using this so-called isotope dilu-tion method, Kim et al. [33] demonstrated that increasing the dose of un-labeled epidermal growth factor (EGF) from 0.035 nmol/kg to 22.7 nmol/kg reduced the half-time of disappearance of ^{125}I-EGF from plasma during the first 3 min following i.v. administration in the rat.

Endocytosis begins with invagination of a portion of the plasma mem-brane forming a vesicle on the order of 1000 Å in diameter, thus capturing the ligand, which may or may not be associated with the receptor. The fate of the receptor and ligand portions after receptor-mediated endocytosis has been reviewed [23,30,34].

Two major intracellular pathways of polypeptide internalization by hepatocytes have been identified. In the direct shuttle pathway, the endocytotic vesicle formed at the sinusoidal surface traverses the cell to the peribiliary space, where it fuses with the bile canalicular membrane and releases its contents by exocytosis into bile. The shuttle vesicle does not appear to interact with other compartments of the cell and bypasses the GERL area (Golgi, endoplasmic reticulum, lysosome) entirely. Polymeric immunoglobulin A is secreted by this pathway [35,36].

In the second pathway, the ligand-receptor complexes proceed through coated vesicles to another intracellular compartment, known as endocytotic vesicle or compartment of uncoupling receptor and ligand (CURL) [22]. Both the endocytotic compartment and the coated vesicles, which deliver the ligand-receptor complex, are acidified by proton pumps that exist within the membranes of these compartments. The acidic nature of CURL controls the final destinations of the ligand and/or receptors by regulating receptor-ligand binding affinity and hence ability of receptor proteins to recycle. Receptor molecules known to undergo recycling to the cell surface are the LDL, asialoglycoproteins, and transferrin receptors. Some receptors, such as the interferon receptor, undergo degradation and subsequent down-regulation. Finally, some receptors may show mixed behavior, such as the insulin and epidermal growth factor receptors.

Of note is the asialoglycoprotein receptor, which is present exclusively in the liver. This receptor binds serum glycoproteins whose carbohydrate side chains terminate in galactose, thus rapidly clearing them from the circulation [24,25]. The protein-receptor complex thus formed is transported into the cell in an endosomal compartment from which the asialoglycoprotein is transported to lysosomes for degradation [23,24] and the receptor is recycled to the cell membrane. Two hundred and fifty such circuits may be made in a single receptor's lifetime. The distribution, recycling kinetics, and biosynthesis of this receptor protein have been described in the Hep G2 cell line derived from human hepatoma [37]. In general, the receptor-mediated endocytosis process is both of high capacity because of receptor recycling and of a high affinity due to the highly specific recognition.

Even though specific binding proteins are absent on the hepatocyte plasma membrane, proteins may still gain access to the cytoplasm of hepatocytes by another process known as nonselective pinocytosis [38]. These proteins are internalized according to their concentration within plasma. The amount of proteins internalized by this pathway, however, probably accounts for only a small portion of the total protein internalized by hepatocytes. Albumin [39], certain antigen-antibody complexes [40], and some pancreatic proteins [41,42], as well as some glycoproteins such as the plant glycoprotein horseradish peroxidase [43], are examples of proteins removed from plasma by hepatocytes by a nonreceptor-mediated process.

B. Kidney

The kidney plays an important role in the clearance of proteins, polypeptides, small peptides, and amino acids [K22]. Indeed, following the injection of radiolabeled growth hormone or gonadotropin releasing hormone (GnRH) in rats, the highest tissue concentration of radioactivity was localized in the kidney [44,45]. It is therefore not surprising that in patients with advanced renal disease the metabolic clearance rates of growth hormone and GnRH were reduced [46,47] and that in nephrectomized rats their circulating halftimes were increased [48,49].

Many peptides are filtered by the glomerulus and excreted. Such irreversible uptake followed by degradation appears to be an important process in the clearance of some peptides. Up to 20% of peptides in plasma with molecular weights less than 30 kD can be cleared in a single pass in this way unless they are protein bound. In the case of iodinated growth hormone this value is as high as 67% [50].

Both the charge and size of proteins can affect the efficacy of glomerular filtration. Because the glomerular filter is negatively charged, anionic molecules are repelled and consequently do not filter as well as cationic ones. This is indeed the case for anionic fractions of IFN-β, IFN-γ, and TNF-α [51]. Peptides and small proteins (less than 5 kD) pass into the ultrafiltrate almost like inulin or creatinine, which are typical markers for assessing the glomerular filtration rate (GFR). Thus, the protein clearance/inulin clearance ratio ranges from almost 0 for albumin (69 kD) and IgG (160 kD) to 1 for small peptides. A related index, called the glomerular sieving coefficient (GSC), behaves similarly. This is defined as the ratio of solute concentration in the ultrafiltrate to that in plasma. Its value ranges from 0 for albumin and larger proteins to about 0.5 and 0.8 for anionic and cationic proteins of 20 kD, respectively, and to about 1 (for peptides) [52]. Compared with their monomeric forms, dimeric or trimeric forms of IFN-β, IFN-γ, and TNF-α are filtered at a slower rate [51].

After glomerular filtration, a peptide can be excreted unchanged (melanostatin [53]), degraded to products that are excreted (oxytocin and vasopressin [54]), or reabsorbed by the proximal tubular cells followed by degradation by digestive enzymes in the lysosomes [55,56] (corticotropin analogs [55], calcitonin [57], glucagon [58], insulin [59], growth hormone [60], and glycopeptide hormones [60]).

The proximal tubule reabsorbs polypeptides and proteins filtered at the glomerulus by luminal endocytosis [50,61–65] and reabsorbs amino acids by a carrier-mediated, energy-dependent transport mechanism operative at the luminal membrane. Several reabsorptive systems have been described for tubular transport of varying classes of amino acids [66]. Generally, L-isomers are more effectively transported than D-isomers [66]. The transport of amino acids across the plasmalemma is dependent on a Na^+ gradient [66].

Renal tubular cells also possess an active mechanism to transport di- and tripeptides [67]. Peptide transport is unique among the transport systems known to be present in the kidney in that it is energized by a H^+ gradient rather than by a Na^+ gradient [67]. An inward-directed H^+ gradient exists across the brush-border membrane of proximal tubular cells under physiological conditions as a result of the concerted action of $Na^+-K^+-ATPase$ in the basolateral membrane and the Na^+-H^+ exchanger in the brush-border membrane. Since the exchanger is electroneutral, the magnitude of the H^+ gradient should be at least equal to that of the Na^+ gradient. Under physiological conditions, this active transport of peptides prevents the urinary loss of amino nitrogen in the form of peptides and also regulates the concentration of small peptides in the circulation. Certain aspects of renal peptide transport system remain to be investigated, including: (1) stoichiometry between H^+ and peptide, (2) relative contribution of membrane potential and chemical H^+ gradient to the uphill transport of peptides, (3) molecular characterization of the transporter, and (4) functional characterization of the transport system along the proximal tubule [67].

Several studies indicate that circulating peptides are hydrolyzed by the kidney. Oxytocin [45,68 – 71], vasopressin [69], parathyroid hormone [72], and calcitonin [73] were degraded during incubation with renal tissue. Oxytocin and vasopressin were hydrolyzed in the isolated perfused rat kidney, liberating glycine amide [59]. Angiotensin II was largely degraded in a single passage through the dog kidney, with recovery of most of the labeled material in the renal vein [74]. Electron microscopic autoradiography indicated that GH [75], lysozyme [62], and insulin [59] were reabsorbed in the proximal nephron and degraded within lysosomes.

More direct evidence for the renal transport and hydrolysis of peptides came from the studies of Gottschalk et al. [76], Burg et al. [77], and Carone et al. [78]. In these studies, the renal transport and hydrolysis of radiolabeled angiotensin I, angiotensin II, bradykinin, oxytocin, glucagon, insulin, and LHRH were investigated. Several techniques were used, including in vivo microinfusion of surface tubules in rats, arterial infusion in filtering and nonfiltering rat kidneys in vivo and microperfusion of isolated rabbit nephron segments. Reabsorption of radiolabeled material was measured, and the intact peptide or its metabolites were identified and quantified in urine, renal venous blood, bathing medium, and/or collection fluid. In addition, peptides were incubated with isolated renal membrane preparations to identify a probable cellular site of hydrolysis. These studies revealed that in the proximal, but not the distal, tubules, radiolabeled angiotensin I, angiotensin II, bradykinin, glucagon, and LHRH were hydrolyzed by brush-border enzymes at the luminal membrane to various small peptide fragments, which were then reabsorbed and further degraded intracellularly. Peptides such as insulin, oxytocin, and vasopressin that contain disulfide bridges, by contrast, were not hydrolyzed at the luminal brush border of the proximal tubule. In vivo sequestration and slow degradation of insulin by rat tubules suggested that this peptide was reabsorbed by endocytosis and degraded in lysosomes. Thus, as the molecular complexity or size of a peptide increases, the mechanism for renal tubular degradation shifts from luminal membrane hydrolysis to lysosomal digestion following endocytosis.

Katz and Emmanouel [79] and Talor et al. [80] suggested that significant peritubular hydrolysis also occurs in several peptides, such as glucagon, insulin, parathormone, and C-peptide [79,52]. This is indicated by renal arteriovenous differences that are greater than the filtration fraction and by significant extraction which occurs in the nonfiltering kidney. Talor et al. [80] have shown that semipurified contraluminal membranes derived from rat renal cortex are capable of degrading glucagon. Thus, it appears that for several peptides, hydrolysis may occur by mechanisms in addition to those involving filtration and tubular reabsorption.

Membrane hydrolysis was first described in the small intestine [82,82]. The intestinal brush border is rich in hydrolytic enzymes. Brush-border enzymes are either sequestered in the external surface glycocalyx or bound in the cell membrane, with part of their molecular domain exposed to the cell exterior [81,83,84]. The external location of these hydrolases is supported by several observations. Electron microscopy has revealed

knob-like projections on the outside of brush-border membranes that were
removed by exposure to proteases resulting in solubilization of enzymatic
activity [85]. Moreover, peroxidase-labeled antibody to aminopeptidase
exhibited binding to essentially the same number of antigenic sites when
the enzyme was either bound to the membrane or solubilized [86].

Membrane hydrolysis in the intestine yields a mixture of amino acids,
dipeptides, and tripeptides, of which the latter predominate [81]. The
enterocyte possesses active transport system for amino acids and small
peptide fragments [87]. The peptide carrier system has a higher maximal
rate of uptake than the amino acid carrier system. It favors peptides with
lipophilic amino acids in both N- and C-terminal positions, and it is stereochemically specific, preferring peptides composed of L-amino acids [81].
Hydrolysis follows the absorption of dipeptides and tripeptides [81].

The epithelial cells of the renal proximal tubule are similar to those of
the small intestine morphologically and, in some respects, functionally.
Both cell types possess a brush border rich in proteolytic enzymes, and
both contain carriers for the uptake of amino acids, dipeptides, and tripeptides [88]. In the intestine, proteins are degraded by enzymes in the
digestive fluid, and liberated peptides are further degraded after contact
with the luminal membrane. In the renal proximal tubule there is no evidence for secretion of hydrolytic enzymes, but proteins and polypeptides
are reabsorbed by endocytosis and subsequently hydrolyzed within lysosomes.
The liberated amino acids, di-, and tripeptides are absorbed by active
transport in proximal renal tubular cells, where further hydrolysis can
occur.

C. Conjugation of Protein Drugs to Macromolecules to Circumvent Hepatic and Renal Clearance

As discussed in detail in Chapter 21, a number of proteins have been conjugated to macromolecules in an attempt to circumvent hepatic and renal
clearance, thereby extending plasma half-life. These macromolecules include soluble and compatible polymers, such as dextran [89–91], polyethylene glycol [92,93], and polystyrene-maleic acid copolymer [94], and
proteins, such as albumin [95,96] and monoclonal antibodies [97]. Two
examples will be cited to illustrate the usefulness of such an approach.

L-Asparaginase

L-Asparaginase is an enzyme with antitumor effects against Gardener
lymphoma [98,99]. Hypersensitivity reactions ranging from mild allergic
reactions to anaphylactic shock were reported in approximately 5–20% of
cancer patients treated with L-asparaginase [100,101]. Park et al. [102]
suggested that the administration of PEG-L-asparaginase prepared by covalently linking PEG to L-asparaginase might lower the potential for
anaphylactic reactions and extend its plasma half-life. Such a concept has
been investigated by Ho et al. [103], who administered PEG-L-asparaginase
($500-800$ U/m^2) over 60 min in cancer patients. Plasma levels of enzyme
were determined by a spectrophotometric assay of the rate of NADH oxidation at 340 nm in a series of coupled enzyme reactions [104,105]. The
plasma disappearance curve of the conjugate was monoexponential for all
the doses studied. The volume of distribution of either L-asparaginase or
PEG-L-asparaginase was comparable to that of plasma volume, indicating
little or no tissue binding for either. The corresponding mean values of

$t_{1/2}$, apparent volume of distribution, and total clearance were 357 ± 243 hr, 2093 ± 643 ml/m^2, and 128 ± 74 ml/m^2·day. The mean AUC normalized to a dose of 1000 U/m^2 was 10.2 ± 2.6 U/ml·day. By comparison, the corresponding values for unconjugated asparaginase over a dose range of $16,500 - 100,000$ U/m^2 were 20 ± 6 hr, 2336 ± 663 ml/m^2, 2196 ± 1098 ml/ m^2·day, and 0.4 ± 0.1 U/ml·day [106]. These favorable pharmacokinetic results were supported by the pharmacodynamic results based on the time course of L-asparagine levels in plasma following the i.v. administration of the two forms of the enzyme [107]. Thus, the coupling of L-asparaginase to PEG increased both the plasma and pharmacological $t_{1/2}$ of L-asparaginase significantly. The extended half-life is partly due to reduced renal clearance of the conjugate and partly due to reduced susceptibility of the conjugate to systemic proteases. Despite the favorable alterations in the pharmacokinetic parameters of L-asparaginase following conjugation, anaphylactic reaction was still observed in 3 of 31 patients studied.

Soybean Trpysin Inhibitor

Soybean trypsin inhibitor is useful in the treatment of trypsin-induced shock and acute pancreatitis. Takakura et al. [108] studied the pharmacokinetics and pharmacodynamic aspects of soybean trypsin inhibitor (STI) and its conjugates with dextran (STI-D) and polyethylene glycol (STI-PEG). Dextran having a molecular weight of 9.9 kD was covalently attached to the STI molecule by periodate oxidation. Activated PEG having a molecular weight of 10 kD was reacted with STI to synthesize STI-PEG conjugates.

Mice were injected intravenously with saline solution of STI or conjugates at a dose of 250 U/kg. Urine and plasma samples were collected over a period of 60 min and assayed for inhibitory activity on tryptic hydrolysis of synthetic substrate [109]. STI showed a rapid decrease in trypsin inhibitory activity in plasma with a half-life of only 2 min, and 60% of the dose was excreted in urine within 1 hr after injection. By contrast, considerable activity was detected in plasma even at 60 min after administration with these conjugates; 20% of the activity of the amount injected was recovered in urine within 1 hr.

The pharmacological effects of STI and its conjugates were evaluated in two animal experimental models: trypsin-induced shock in mice and acute pancreatitis in rats. In mice, shock induced by i.v. injection of trypsin (25 mg/kg) was inhibited by i.v. pretreatment with STI (100 U/kg), but the effect lasted for only 60 min. The STI-D conjugate showed superior inhibitory effect at the same dose for up to 5 hr. A similar effect was observed in mice given an i.v. injection of STI-PEG. In rats with acute pancreatitis, no significant therapeutic effect was observed with i.v. STI (25 U/kg) or saline treatment. By contrast, i.v. treatment with STI-D at the same dose lowered the mortality of the rats (23.1 vs. 53.8%). The improved activity, persistence of plasma concentration, and reduced urinary excretion of conjugates can be attributed to decreased glomerular filtration of STI-D (127 kD) and STI-PEG (182 kD).

IV. DESIGN OF PHARMACOKINETIC STUDIES

Pharmacokinetic studies of peptide and protein drugs are important in understanding their biological action and in the rational design of their

dosing regimens. Interpretation of these observations in the context of
clearance mechanisms is, however, complex. Plateau blood concentration
reached after a period of constant infusion at a particular rate yields information on total body clearance under conditions where distribution has
achieved an equilibrium or where distribution and metabolism have reached
steady state. Study of total clearance at different infusion rates can
establish whether clearance is dose-dependent.

Pharmacokinetic studies may yield invaluable information on interactions
between peptide/protein drugs and other coadministered drugs. For instance, TNF significantly inhibited drug metabolism in the rat, as indicated by 30% and 25% decreases in the average total clearance of antipyrine
and diazepam, respectively [110]. The mechanism of this interaction is
probably due to suppression of hepatic cytochrome P-450-dependent mono-
oxygenase activities responsible for antipyrine and diazepam metabolism.
Similar interactions have been reported between interferons and antipyrine
[111,112]. This may be clinically relevant if a similar effect occurs in
humans following administration of therapeutically effective doses of these
biological response modifiers.

Investigation of the pharmacokinetics of peptides and proteins is
complicated by a number of factors. The available assays suffer from limitations in acquiring absolutely specific information about their concentrations and identity. There are also difficulties in the selection of appropriate experimental animal models and sampling intervals. This section
focuses on parameters important in the design of pharmacokinetic studies
for peptide and protein drugs.

Choice of animals may have a great influence on the outcome of peptide and protein pharmacokinetics. The animal species often used in
pharmacokinetic studies on peptide and protein drugs include *Cynomolgus*
monkeys, *Rhesus* monkeys, New Zealand rabbits, Sprague-Dawley rats,
Listar rats, mongrel dogs, and others. Obviously, extrapolation from one
animal species to another needs to be made with caution as the different
animal species may differ in their metabolic clearance rates and proteolytic
activities. Also, the therapeutic peptide or protein should have the desired pharmacological effect in the animal species chosen. For example,
human interferon-γ (HuIFN-γ) is not active in rats. Hence to acquire knowledge about the disposition of IFN-γ in rats, murine IFN-γ should be used.
Despite the fact that IFN-γ possesses species restricted activity, pharmacokinetic studies have been conducted with HuIFN-γ in rabbits [113], monkeys,
and guinea pigs [114]. As pointed out by Cantell et al. [114], intramuscularly
administered natural human and recombinant rat IFN-Γ resulted in lower
blood levels in those species in which they were biologically inactive.

Because peptides and proteins have relatively short plasma half-lives,
it is imperative that samples be collected at short intervals during the
first $10-15$ min, after allowing $2-3$ min for drug mixing in the circulation.
During this first phase rapid equilibration of drug into a pool with an apparent volume greater than plasma occurs. Absence of early phase for
many peptides and proteins can often be attributed to insufficient number
of samples collected. A case in point is the prolonged monoexponential
disposition ($t_{1/2}$, 8.2 hr) observed by Macdougall et al. [115] for recombinant human erythropoietin instead of the more established biexponential disposition for this glycoprotein [115]. This long plasma half-life
is probably related to the inadequate number of samples, totalling 13,
collected over 48 hr.

Following the first phase, drug levels usually decline exponentially for 20-40 min. The rate of decline is nearly two to five times slower when compared to the first phase. Often the half-life obtained from this phase is reported as the drug's half-life. This may range from a few minutes to a few hours, depending on the chemical nature of the peptide and protein drug. It has been suggested that measurements should be made five times per half-life for four or five half-lives of decay to get an accurate estimate of $t_{1/2}$ [116].

The distribution of peptide and protein drugs in tissues can be elucidated by autoradiography or by measuring drug concentrations. The presence of labeled drug or drug fragments in a tissue does not, however, indicate where or when it was formed. It cannot be concluded that only those tissues that accumulate the label are involved in metabolism, because the drug can undergo degradation during transit to various tissues. Only for those drugs whose metabolites do not circulate over the period under consideration is it possible to interpret tissue uptake and degradation patterns in terms of the roles played by tissues involved.

Provided that care is taken to avoid damaging the perfused tissue and the subsequent release of peptidases into the perfusion medium, isolated perfused organs and tissues are useful in (1) elucidating the mechanisms of clearance and degradation by a particular organ or tissue and (2) determining whether the drug is accumulating or being degraded in transit. Even then, the results of these studies cannot be related quantitatively to metabolism in vivo.

Comparison of therapeutic ranges and pharmacokinetic parameters of a peptide or protein drug should be made with careful attention given to the method of sample collection and treatment prior to assay. A case in point is cyclosporine. As this peptide is highly lipophilic, it may adhere to plastic cannulae [117] or to indwelling polyurethane catheters [118]. Blifeld and Ettenger [117] administered cyclosporine to renal allograft patients as an i.v. infusion over 4 hr. Cyclosporine levels were assayed in specimens obtained simultaneously from the indwelling catheter and a peripheral site. Catheter levels were more than twice the peripheral levels. This was observed despite the fact that other solutions (total peripheral nutrition, Intralipid, and dextrose) were infused for over 20 hr before specimen withdrawal. It is therefore recommended that samples not be withdrawn from an intravenous cannula through which the peptide has been administered [117].

As is the case of some nonpeptide and protein drugs, whether plasma or blood is used may affect the numerical values of the derived pharmacokinetic parameters. Consider again the example of cyclosporine. This peptide and many of its metabolites concentrate preferentially in erythrocytes in a temperature-dependent manner [2,119]. This makes plasma concentration measurements highly variable unless due precautions are taken in the separation of plasma. With a temperature decrease from 37 to 21°C, about 50% of cyclosporine diffuses from plasma to red blood cells, where it binds to hemoglobin. This process is reversible upon reequilibration at 37°C for 2 hr [2,119]. Obviously, since cyclosporine has erythrocyte-binding properties [2,120,121] and since it binds to all clones of circulating human lipoproteins [2,120], the patient's hematocrit and lipid profile must be taken into account, along with the type of assay, sample matrix, and concomitant drug therapy, in deciding the therapeutic concentration of cyclosporine. Even the type of anticoagulant used may influence the assay results. Specifically, EDTA is preferable to heparin as anti-coagulant since the use of heparin may result in clots large enough to af-fect the assay result [121,123].

Precautionary measures must also be taken to reduce degradation of peptide and protein drugs in biological specimens prior to assay. Degra-dation of peptide and protein drugs can readily be determined in vitro using either plasma or blood, provided care is taken not to activate pep-tidases of the clotting mechanism during and after sample collection. In vitro studies using tissue homogenates or slices are easy to conduct, but the results must be interpreted with caution. In tissue homogenates, the drug may be exposed to enzymes that are never accessible to it in vivo. Similarly, in studies using tissue slices, the drug may be ex-posed to some irrelevant enzymes released from the surface.

The situation is quite opposite in the case of rt-PA. Binding of this protein to proteinase inhibitors in plasma was found to be responsible for detection of only 65−75% of rt-PA activity in thawed plasma samples [124] by the chromogenic [125] and fibrin plate assays [126]. By contrast, there was no interference from the proteinase inhibitors in immunoradio-metric assay (IRMA) [127] and enzyme-linked immunosorbent assays (ELISA) for rt-PA [128]. Such differences in assay results may, in fact, be responsible for the different numerical values of the area under the concentration time curve of rt-PA shown in Table 1.

As mentioned in Chapters 5 and 6, various assays can be used in pharmacokinetic studies, including bioassays, radioimmunoassays, and chromatographic assays. Although bioassays are useful for acquiring information about the concentration of biologically active peptide or protein drugs in samples, they do not give information about inactive products and may fail to indicate conversion of a peptide or a protein to active metabo-lites. Bioassays can be very sensitive, accurate, and practical.

Table 1   Influence of Assay Methods on Derived Pharmaco-kinetic Parameters of Recombinant Tissue Plasminogen Activator Following a 30-Min Intravenous Infusion of 0.25 mg/kg of the Protein to Healthy Volunteers

Parameter	Chromogenic	ELISA[a]	ELISA[b]
$t_{1/2\alpha}$ (min)	3.4 ± 0.6	3.3 ± 0.4	3.5 ± 0.4
$t_{1/2\beta}$ (min)	34 ± 12	26 ± 12	35 ± 4
AUC (ng.h/ml)	313 ± 65	530 ± 73	455 ± 64
Vd_{ss} (L/kg)	7.6 ± 1.9	7.2 ± 1.0	10.7 ± 1.2
Clearance (ml/min)	1235 ± 170	687 ± 63	810 ± 66

[a]Without PPACK treatment.

[b]With PPACK treatment.
Source: Ref. 124.

Like bioassays, radioimmunoassays lack specificity, but they are usually sensitive and reproducible. An unpredictability with radioimmunoassay is the cross-reactivity of inactive products, congeners, or even some totally unrelated structures, which may lead to overestimation of the immunoactive peptide or protein. Also, the presence of impurities in either the drug sample or tracer may result in overestimation of the drug concentration. If the tracer is not radiochemically pure, then the binding characteristics of the impurities may affect the assay. The consequence of such artifacts for pharmacokinetic studies is most marked at times of low circulating peptide analog levels when positive assay bias induces systematic errors in pharmacokinetic analysis. Indeed, this explains the artifactually high apparent basal endogenous GnRH immunoreactivity seen in studies that measured GnRH in unextracted serum or plasma samples [129–133]. Extraction is therefore essential before assays for peptide and protein druts, particularly those that are analogs of endogenous substances, to remove some of the blank effects mentioned above [130].

Isotopic labeling of peptides and proteins makes their quantitative assay relatively simple, particularly since it minimizes the interference from endogenous peptides and proteins. Ideally, the labeled molecule should be essentially identical to the unlabeled drug in its physicochemical and biological properties. This can be achieved by using ^{14}C or ^{3}H labels, but such products are difficult to prepare. Radioactive iodine labeling results in material of high specific activity. Iodine-labeled drugs are, however, chemically distinct analogs and the labeling process may change the drug chemically. Chromatographic separation is usually required to distinguish between the original compound and metabolites [134,135].

Because of its high resolution and applicability to a wide variety of samples, reversed-phase HPLC (RP-HPLC) has become the most widely used separation method. As many as 60–70% of all recent HPLC applications involve the use of RP-columns [136]. The various columns available for RP-HPLC include octadecyl, octyl, and cyano columns. Acetonitrile is by far the most widely used organic solvent for RP-HPLC. Other popular solvents include 1-propanol and 2-propanol. Acetone gives very good resolution for tryptic peptides, but because of its high UV absorption it has only been used with radiolabeled samples.

Retention of peptides in RP-HPLC can be controlled by varying the mobile phase, by altering the concentration of the organic modifier, by changing the pH, or by adding salts or ion-pairing agents [137]. The fundamental mechanism governing the retention of peptides on reversed-phase supports is hydrophobic interaction. Since basic amino acids are also hydrophilic, very poor retention in RP-columns is to be expected for peptides which, in addition to basic residues, contain mainly hydrophilic amino acids. In such cases, the presence of basic groups can be exploited successfully with the use of ion-pair chromatography [138,139]. A wide variety of ionic modifiers have been investigated in an attempt to combine the desirable characteristics of suppression of silanol interactions, UV transparency, and volatility. Volatile acids include formic, trifluoroacetic, heptafluorobutyric, acetic, and hydrochloric acids. Volatile amine salts include ammonium acetate and chloride, pyridinium acetate and formate, and ammonium bicarbonate. Also, amine phosphates have been widely used because of their excellent chromatographic properties [136].

V. PHARMACOKINETICS OF SELECTED PEPTIDE AND PROTEIN
DRUGS: STRUCTURE-PHARMACOKINETIC RELATIONSHIPS

A. Cardiovascular Peptide and Protein Drugs

Atrial Natriuretic Factor

Atrial natriuretic factor (ANF) is stored in specific granules of mammalian
cardiocytes as a 126-amino-acid precursor of molecular weight 13.6 kD,
ANF(Asn[1]-Tyr[126]). Three atrial natriuretic factors, α, β, and γ, have
been isolated from human atria. Of the three, α-hANF is the circulating
and biologically most active form of ANF in human subjects [140,141].
β-hANF is 56 amino acids long and comprises an antiparallel dimer of
α-hANF, whereas Γ-hANF is composed of 126 amino acids containing the
28-amino-acid sequence of α-hANF at its carboxy terminus. α-hANF is
four and five times more potent than β-hANF and γ-hANF, respectively
[141]. The major circulating form of ANF in the rat has been identified
as ANF (Ser[99]-Tyr[126]) [142].

ANF can be considered a cardiac hormone with potent natriuretic, diuretic, and vasorelaxant properties [142,143]. Furthermore, ANF has
been shown to inhibit the secretion of aldosterone, plasma renin, and
vasopressin [142]. Release of ANF into plasma occurs in response to
acute volume expansion [144] and water immersion [145]. ANF release by
these stimuli appears to be associated most closely with elevations in right
atrial pressure. Stimuli that raise plasma ANF levels also cause a decrease in atrial stores of pro-ANF.

Receptors for ANF have been identified in the kidney, adrenal gland,
small intestine, colon, certain brain areas, and blood vessels using autoradiographic and receptor binding techniques [142]. There is a high concentration of receptors in the glomeruli of the kidney and a lower concentration in the basa recta bundles in the inner part of the outer medulla.
This observation supports the concept that ANF exerts its major renal effects by modulating glomerular function and medullary blood flow. Additionally, ANF receptors in moderate density have been demonstrated in the
inner renal medulla, of which about half the grains are located on collecting ducts. This, in conjunction with the observation of ANF receptor
sites in cultured renal tubular cells, implies that the natriuretic effect of
the hormone may, at least in part, be mediated by renal tubular mechanisms and by effects on the intrarenal vasculature [142].

There is evidence that two classes of ANF receptors exist, namely,
B-receptor and C-receptor. Binding of ANF to the former leads to expression of pharmacological activity, whereas binding of ANF to the latter
does not. The C-receptors were proposed to exist by Maack and co-
workers based on the binding behavior of des[Gln[18],Ser[19],Gly[20],Leu[21],
Gly[22]] rat ANF(102-121)-NH$_2$ [C-ANF(102-121)], a COOH-terminal ring
deleted analog of ANF, toward the majority of the specific binding sites of
ANF(99-126) in kidney cortex, isolated glomeruli, and cultured endothelial and vascular smooth muscle cells [146-148]. In the isolated rat
kidney, C-ANF(102-121) at perfusate concentrations that led to maximal
occupancy of its renal binding sites did not have functional effects on its
own and did not antagonize any of the known renal vascular, hemodynamic,
or excretory effects of ANF(99-126) [147,148]. Thus, C-ANF(102-121)
does not interact with the biological receptors proper of the hormone
(B-ANF receptors), but binds to a distinct class of ANF receptors that
are biologically silent (C-ANF receptors). On the other hand, biologically

active ANF(99−126) binds with similar high affinity to B-ANF and C-ANF receptors [146,148]. Thus, B- and C-receptors of ANF are binding sites that are functionally and biologically distinct.

Murthy et al. [149] evaluated blood as a possible site for the processing of prohormone in rat, as the processing site of pro-ANF(Asn[1],Tyr[126]) is as yet unknown. ^{125}I-ANF(Asn[1]-Tyr[126]) was incubated with whole rat blood, plasma, or serum for different time intervals. No significant activation was observed in either whole blood or plasma. Incubation with serum, however, resulted in the formation of an 11-kD and another 3-kD peptide, which corresponded, respectively, to the N-terminal and C-terminal parts of the propeptide. These results suggest that hydrolysis of the propeptide in serum is brought about by enzymes that might have been stimulated only during coagulation. Presumably this process occurred rapidly since no pro-ANF had been found in the circulation.

Murthy et al. [149] also analyzed plasma for pro-ANF at different intervals following intravenous administration using RIA. Two antibodies were utilized. One antibody was directed against the C-terminus of the ANF while the other was directed toward the N-terminus of the pro-ANF molecule. The results indicate that pro-ANF was hydrolyzed to multiple immunoreactive forms of ANF. The disappearance curves obtained following the injection of ^{125}I-ANF (Asn[1]-Tyr[126]) in rats [149] revealed two components with half-lives of 2.1 ± 0.4 min and 52.5 ± 8.4 min, respectively. Metabolic clearance and initial volume of distribution were 1.49 ± 0.22 ml/min and 47.4 ± 8 ml, respectively.

Pharmacokinetic studies of ANF have been conducted in rabbits [150], dogs [151], rats [157,159−161], and humans [152,155]. The kinetic parameters observed in different species are shown in Table 2. Nakao et al. [152] studied the pharmacokinetics of synthetic α-hANF following i.v. bolus (100 µg) injection in healthy male volunteers. Disappearance of ANF from plasma was rapid and obeyed biexponential kinetics. The plasma

Table 2   Pharmacokinetic Parameters of Atrial Natriuretic Factors Following Intravenous Bolus Administration

Parameter	ANF(103−126)		ANF(99−126)
Subject	Wistar rats (normal)	Wistar rats (anephric)	Humans
Dose (µg)	0.25	0.25	100
$t_{1/2\alpha}$ (min)	0.44	0.95	1.7
$t_{1/2\beta}$ (min)	−	−	13.3
AUC (ng/ml.sec)	32.4	78.6	−
Vd (ml)	352	345	3731
Cl (ml/min)	500	206	1520

Source: Refs. 152 and 160.

half-times of the fast and slow components were 1.7 ± 0.07 min and 13.3 ±
1.69 min, respectively. The mean plasma clearance of ANF was 25.4 ml/
min·kg. Such rapid disappearance of α-hANF from human plasma and the
finding that the volume of the central compartment was approximately twice
that of plasma volume may be explained in part by the binding of α-hANF
to specific receptors in blood vessels and other target organs such as the
kidney. The above kinetic data are consistent with the rapid and short
action of α-hANF in humans. Renal and hypotensive responses commenced
at 5 min, reached a peak at 15 min, and were almost complete by 30 min
[153]. Since the dose used (100 µg) was high enough to exert natriuretic,
diuretic, and hypotensive actions in humans [153,154], it is possible that
the hemodynamic changes thus induced may have contributed to the kinetic
data observed.

The effect of aging on the pharmacokinetics of α-hANF has been
studied in healthy men [155]. The disappearance of α-hANF from plasma
following intravenous administration (2 µg/kg) obeyed biexponential decay
curve in young (25−28 years) and aged (71−77 years) subjects. The
initial phase was about the same in both age groups, the half-life being
1.2 ± 0.2 min in the young group and 2.9 ± 2.0 min in the aged group.
The second phase was, however, significantly prolonged in the aged group
(34.3 ± 3 min) when compared with the young group (17.3 ± 3.9 min).
Consequently, the metabolic clearance rate in the aged was markedly
diminished, 13.6 ± 1.3 ml/min·kg vs. 21.1 ± 2.2 ml/min·kg. This diminished metabolic clearance in the aged may explain the high plasma hANF
values observed in this age group [156].

Brier et al. [157] studied the kinetics and pharmacodynamics of hANF
in freshly isolated perfused Wistar rat kidney. hANF (1 µg) was added to
the perfusate in a bolus. Urine was collected during nine 10-min intervals,
and the perfusate was withdrawn at the midpoint of each urine collection
interval and analyzed for hANF using RIA. Perfusate hANF concentrations decreased rapidly with an elimination half-life of 12.4 min, yielding a
clearance rate of 5.71 ml/min. Urine flow was significantly increased.
hANF increased the fractional excretions of sodium (FE_{Na}), lithium (FE_{Li}),
and potassium (FE_K), which were calculated by dividing the individual
clearances by the creatinine clearance. Using lithium as a marker of
proximal tubular sodium reabsorption, proximal tubular Na reabsorption
was found to be reduced. At the peak natriuretic effect, FE_{Na} was increased by 3.9% whereas FE_{Li} was increased by 18.8%. ANF did not affect creatinine clearance, renal vascular resistance, renal perfusion flow
rate, or renal perfusion pressure, however. The fractional excretion of
Na as a function of ANF concentration traced out a counterclockwise
hysteresis loop, which is indicative of lack of instantaneous equilibration
between ANF concentrations in the perfusate and at its receptor site
[158]. A more linear response should ensue by infusing ANF to steady
state. This model, however, does not indicate whether the action of ANF
on sodium reabsorption was direct or indirect.

Katsube et al. [159] studied the plasma disappearance of ANF(103-126)
and a N-terminal fragment of proANF [ANF(13-104)]. ANF(103-126) is a
24-amino-acid peptide which was also found in the mammalian circulatory
system (1). ANF(13-104) lacks most of the biologically active C-terminal
portion of the prohormone but retains the majority of the remaining
N-terminal region. This peptide, while sharing 87% homology with the

corresponding rat sequence, is biologically inactive in natriuretic, diuretic, and vasorelaxant assays.

Following a continuous infusion (300 ng/min) of ANF(103-126) for 15 min in rats, an elimination $t_{1/2}$ of 31 ± 4 sec was observed. This value was similar to that (26.5 sec) reported by Luft et al. [160] following an i.v. bolus injection of 250 ng of ANF(103-126). Nephrectomizing the rat 1 and 24 hr before the infusion increased the $t_{1/2}$ values for the disappearance of exogenous ANF(103-126) to 48 ± 4 and 64 ± 12 sec, respectively.

The half-life of hANF(13-104) infused at a rate of 1 μg/min in normal rats was much longer (287 ± 38 sec) than that of ANF(103-136). In rats nephrectomized 1 hr before infusion, the $t_{1/2}$ was 60 min, suggesting that the majority of clearance was renal dependent. Plasma immunoreactivity of the N-terminal fragment of the prohormone increased 35-fold by 26 hr after nephrectomy in comparison with a fivefold increase in the immunoreactivity of hANF(103-126) in a similar experiment.

Dose-dependent decrease in blood pressure was observed in the rat following i.v. bolus administration of hANF(103-126) over the range $0.067-1.0$ ng/kg [160]. Heart rate also showed an increase, but not in a dose-dependent manner [160].

Krieter and Trapani [161] studied the pharmacokinetics and extraction of rat ANF (103-126) by various organs in conscious rats. Conscious rats were used in order to minimize the confounding effects of anesthetics on reduction of regional blood flow to the kidney [162], gut [163], and muscle [164]. Indeed, the half-life in anesthetized rabbits was somewhat longer (1.2 min) [150] following the i.v. infusion of ANF(103-126) (0.97 ml/min for 15 min). In conscious rats, the half-life of ANF-like immunoreactivity was 35 ± 5 sec following i.v. infusion of ANF(103-126) for 25 min at a rate of 1250 ng/kg·min. The volume of distribution was slighyly greater than the blood volume, and the total clearance was 1/3 to 1/2 of the cardiac output. The kidney, intestine, muscle, and sex organs, but not the liver, heart, and lungs, contributed to this clearance. Specifically, the extraction ratios were 0.55 and 0.61, respectively, at the two infusion rates for kidneys, 0.44 and 0.27 for the intestines, and 0.43 and 0.54 for both muscle and sex organs. Thus, kidneys accounted for 1/6 to 1/3 of the total clearance while the intestines accounted for 1/6 of the total clearance.

Cernacek et al. [151] investigated the dose-plasma level-response relationships with increasing doses of ANF(99-126) administered by constant infusion in conscious dogs. Plasma immunoreactive ANF increased by 12, 19, 23, and 35 times during 45 min consecutive infusions of 50, 75, 125, and 175 ng/kg.min of this peptide, respectively. Over this pharmacological range, natriuresis increased linearly with the infused dose to a maximum response of 1550%. This occurred despite the significant, gradual fall in blood pressure, which attained a minimum of 83 mm Hg (-26%) at 125 ng/kg·min. There was no change of glomerular filtration rate or renal plasma flow at any dose. Metabolic clearance rate during the infusion was 1.09 ± 0.19 L/min. The postinfusion decay curve of plasma immunoreactive ANF was best described by a biexponential function. The $t_{1/2}$ values of the fast and slow components were 1.44 min and 10.3 min, respectively. These results showed that: (1) natriuretic response to ANF in the pharmacological range studied is dose dependent and occurs despite pronounced hypotension, (2) increase in glomerular filtration rate is not a prerequisite of ANF-induced natriuresis, and (3) ANF is rapidly eliminated from the circulation, suggesting intensive uptake and/or degradation in the target tissues.

Rapid removal of ANF from plasma could be due to one or more of a number of clearance processes, including: (1) receptor-mediated binding, uptake, and metabolism by target tissues; (2) degradation by proteases and other processes at plasma membranes; and (3) excretion into non-plasma fluids such as urine. All three processes exist in the kidney. Being a major target organ for the natriuretic and vascular actions of ANF, the kidney has been proposed to play an important role in the clearance and metabolism of ANF [160]. As mentioned earlier, specific receptors for ANF have been demonstrated in the tubules, glomeruli, and vasa recta within the kidney [142]. Moreover, proteases are present in the kidney and can potentially degrade ANF.

To quantify the role of the kidney in the clearance of ANF from plasma, Woods [165] infused synthetic α-hANF(99-136) at a rate of 200 ng/min to steady-state conditions in conscious dogs. The right kidney of the dogs was removed surgically before the start of the experiment and the left renal artery was fitted with indwelling catheters, a renal artery flow meter, and an inflatable occluding cuff. Clearances were measured in dogs with one normally filtering kidney as well as in those whose glomeruler filtration rate had been reduced to close to zero by inflating a cuff around the renal artery. Such a procedure resulted in minimal urine production and reduction of renal blood flow to 59% of the resting level. In normal dogs MCR was 1090 ± 134 ml/min, with renal clearance rate contributing only 13.9%. After the cuff was fixed around the renal artery, the metabolic clearance rate (MCR) fell to 64 ± 151 ml/min, due in part to a fall in renal clearance rate (−41.5 ± 12.9 ml/min). The reduced glomerular filtration rate accounted for virtually all the fall in renal clearance rate. This finding indicates that filtration across glomerular membranes plays an important role in the renal clearance. Plasma ANF half-life was 59.6 ± 7.9 sec. Such a rapid rate of clearance was close to the cardiac output. While the extraction of exogenous and endogenous ANF across the kidney was high, 50−70%, renal clearance contributed only 14% to the total body clearance. Removal of ANF by non-renal tissues, which accounted for most of the whole body clearance, was influenced by changes in regional vascular tone such as those occurring in response to reduced renal perfusion pressure.

Almeida et al. [166] investigated the function of C-ANF receptors as clearance receptors for circulating ANF(99-126) by evaluating the effects of carboxy-terminal ring deleted analog of ANF [C-ANF(102-121)], a specific ligand of C-ANF receptors, on the pharmacokinetics and hydrolysis of ^{125}I-ANF(99-126) in anesthetized rats. Radioactivity in plasma was measured by trichloroacetic acid solubility and by HPLC. Intravenous infusion of C-ANF(102-121) at two different rates (1 and 10 μg/min.kg) led to marked dose-dependent increases in initial plasma concentration of administered ^{125}I-ANF(99-126) (0.5 μci/rat) and decrease in its volume of distribution at steady state (V_{ss}), MCR, and appearance of hydrolytic products ([^{125}I] monoiodotyrosine and free ^{125}I) in plasma. At the higher dose, C-ANF(102-121) decreased the V_{ss} from 97 ± 12 to 36 ± 2 ml/100 g body weight, MCR from 50 ± 4 to 12 ± 1 ml/min 100 g, and hydrolytic products in plasma from 54 ± 8 to 11 ± 2% of initial plasma ^{125}I-ANF(99-126). These results demonstrated that C-ANF receptors are mainly responsible for the very large volume of distribution and fast MCR of ANF in the rat.

In this way, C-ANF receptors are likely to play an important role in the homeostasis of circulating ANF.

Endothelins

Endothelins are potent vasoconstrictor peptides produced by the endothelial cells of blood vessels and may be secreted in a way similar to some paracrine hormones [167,168]. Three structurally and pharmacologically distinct endothelin isopeptides encoded by separate human genes have been characterized: ET-1, formerly porcine and human ET; ET-2, a new human ET; and ET-3, formerly rat ET. Their biological coexistence may be associated with distinct and/or complementary biological functions [169,170].

ET-1 is comprised of 21-amino-acid residues with a molecular weight of 2692 daltons and two interchain disulfide bonds between cysteines at positions 1 and 15 and at positions 3 and 11. The four cysteine residues at positions 1, 3, 11, and 15, the carboxy-terminal tail portion including the last tryptophan residue, and the cluster of alternating charges, Asp^8–Lys^9–Glu^{10}, are highly conserved among all the three peptides. ET-2 is (Trp^6, Leu^7)ET-1 and ET-3 is $(Thr^2, Phe^4, Thr^5, Tyr^6, Lys^7, Tyr^{14})$ET-1.

ET-1, ET-2, and ET-3 have been synthesized by solid-phase chemistry [171]. All three synthetic peptides exhibited strong vasoconstrictor activity at nanomolar concentrations in vitro, characteristic of endothelin. The time course for the development of contractile tension in porcine coronary artery strips was about the same for all three peptides. All three peptides produced strong pressor responses in anesthetized, chemically denervated rats.

Following i.v. bolus injection, a transient depressor response lasting 1–2 min always preceded the increase in blood pressure. Two distinct phases of the pressor effect were typically seen in response to ET-1 and ET-2. The early phase immediately followed the depressor response and lasted for the next 3–10 min, whereas the next phase developed 10–20 min after the injection and lasted for more than 1 hr. The peak increases in blood pressure are not significantly different between the two phases. These two separate phases were not clearly seen in the responses to ET-3. Similar triphasic response was reported for ET-1 and ET-3 in a study by Watanabe et al. [172]. The peak pressor effect was smaller with ET-3 (51 ± 4 mm Hg), but it was not significantly different between ET-1 (66 ± 5 mm Hg) and ET-2 (72 ± 3 mm Hg). The time required for the blood pressure to return to baseline was the longest with ET-2 (nearly 4 hr) and shortest with ET-3 (nearly 1.5 hr). Based on these findings, the biological potency of endothelins seems to be in the order: ET-2 > ET-1 > ET-3. Such a rank order appears to correlate with the hydrophobicity of the peptides. ET-2, due to its Trp^6 residue, is the most hydrophobic, whereas ET-3, due to its Lys^7 residue, is the most polar.

Following i.v. bolus injection of 30 pmol/kg of ^{125}I-ET-1 into the femoral vein of urethane-anesthetized rats, the total radioactivity in the right atrial blood decayed rapidly in a biexponential manner, with a half-life of 7 min for the fast component [173]. The half-life of the slow component was not calculated, however. Five minutes after peptide administration, the radioactivity was found mainly in the parenchyma of the lungs, kidneys, and liver (406-, 296-, and 128-thousand cpm/g, respectively). No degradation products were detected in plasma by reversed phase HPLC as late as 60 min postinjection.

In a similar study by Anggard et al. [174], the disappearance of ^{125}I-labeled ET-1 and ET-3 from blood followed the bolus injection of 0.2 μCi into the left ventricles of rats. Biphasic disappearance of injected radioactivity with an initial fast phase of less than 1 min was observed for both ET-1 and ET-3. The half-life of the slower phase was 46 min for ET-1 and 35 min for ET-3. Whether this radioactivity represented intact or degraded ET was not reported. Forty minutes after the injection of labeled peptides, the tissue uptake of radioactivity was highest in the lung (4.6 cpm/g) followed by the kidney (2.8 cpm/g) and the liver (1.5 cpm/g) for both ET-1 and ET-3. Isolated guinea pig-lung perfusate studies using labeled ET indicated retention of 64 ± 1.9% in one passage of 0.2 pmol of ET-1 over a period of 5 min. Homogenization and differential centrifugation followed by radioactivity measurement indicated the accumulation of over 93% of the radioactivity in the sediment. This portion contained the intracellular organelles, indicating internalization of the peptide [174]. Thus, the rapid elimination of labeled ET from blood can be attributed to its rapid distribution to different tissues, particularly the lung, kidney, and liver [173–175].

Erythropoietin

Erythropoietin (EPO) is a glycoprotein hormone produced primarily in the kidneys [176] and to a lesser extent in the liver [177]. The physiological function of EPO is to regulate the proliferation and differentiation of erythroid progenitor cells [178]. Interestingly, EPO produced in *Escherichia coli* or yeast lacks the carbohydrate portion and is inactive or weakly active in vivo. On the other hand, EPO produced in COS cells or Chinese hamster ovary cells resembles the natural EPO and is fully active in vivo. This form of recombinant EPO [179,180] has been used successfully in end-stage renal disease patients undergoing chronic hemodialysis [181–185].

rEPO is a single-chain polypeptide with a molecular weight of 30.4 kD, 40% of which is accounted for by carbohydrates. These carbohydrates have been shown to consist of one O-linked and three N-linked oligosaccharides [186,187]. The majority of rEPO oligosaccharides are tetra-antennary core saccharides comprised of N-acetyllactosamine repeating units [186]. Oligosaccharides with one lactosamine repeat account for 21.1%, those with two lactosamine repeats account for 16.5%, and those with three lactosamine repeats account for 4.7% of the total saccharides [186]. EPO is resistant to denaturation by heat, alkali, or reducing agents [178]. It is very hydrophobic and hence adsorbs readily to glass surfaces. Such adsorptive losses can be prevented by adding a detergent or a carrier protein such as serum albumin to the formulation.

The half-life of endogenous EPO in rats has been reported to be in the range of 2–7 hr [188–190]. The large variation in the values of the half-life is partly due to the lack of a sensitive assay for EPO. Plasma EPO titers in these studies were expressed as percent of ^{59}Fe incorporation in polycythemic mice and in arbitrary units with reference to a dilution curve of active rat plasma.

The bioavailability of subcutaneous rEPO (21.5%) in humans was seven times greater than that of intraperiotoneal EPO (2.9%) [115], but was substantially lower than that of subcutaneously administered insulin (50–80%),

heparin (40%), and growth hormone (71%) [191−193]. This can be attributed to impeded absorption of EPO due to its size (30.4 kD) or degradation of EPO by peptidases in the skin, as is the case for insulin [194].

^{125}I-rEPO (2 × 10^5 cpm/kg) has been reported to disappear biexponentially in the plasma of dogs following i.v. infusion over 15 min
[195]. The distribution of radioactivity in different organs in the rat at
30 min after infusion was: liver, 64%, kidney, 24%; lung, 9%, and spleen,
3%. In the dog, the mean distribution half-life is 23.7 ± 5.0 min, which is
about three times shorter than that (75.3 ± 21.2 min) of native EPO (nEPO)
extracted by concentration on a hollow fiber filter from the urine or patients with aplastic anemia. nEPO was administered by i.v. infusion at a
dose of 25 U/kg and assayed by a sensitive and specific radioimmunoassay
with a sensitivity of 5 mU/ml [196,197]. Fu et al. [195] attributed the
shorter distribution half-life of rEPO to a larger intercompartmental clearance (Cl_{ic}, an index of drug distribution between the central and peripheral compartment). This value was 0.068 ± 0.018 L/kg.hr for rEPO and
0.018 ± 0.006 L/kg.hr for nEPO. The smaller Cl_{ic} for nEPO may in turn
be due to binding of nEPO to plasma constituents, aggregation of nEPO
into large molecular complexes during or after extraction, or both. Gel
filtration chromatography indicated stability of the I-125 label on rEPO during the 24 hr after infusion of ^{125}I-rEPO in the dog. Consequently, the
shorter distribution half-life of rEPO when compared with nEPO cannot be
an artifact.

Except for the difference in distribution half-life just noted, the native
and recombinant forms of EPO were rather similar in other pharmacokinetic
parameters, including volume of distribution of the central and peripheral
compartments, elimination half-life, and elimination clearance (Cl_e) [195].
The apparent volume of distribution of nEPO (0.102 ± 0.016 L/kg) and
rEPO (0.149 ± 0.014 L/kg) exceeded plasma volume and approximated the
extracellular fluid space (approximately 0.2 L/kg). The estimated apparent
volume of distribution was similar to that of warfarin (0.11 ± 0.01 L/kg)
[198] and thyroxine (0.15 ± 0.014 L/kg) [199], both of which are highly
protein bound, and smaller than that of gallamine (0.23 ± 0.05 L/kg) and
inulin (0.22 ± 0.03 L/kg), which are both confined to the extracellular
space [200]. The volume of the central compartment was larger than that
of the peripheral compartment.

Even though EPO acts primarily on the bone marrow eryhthroid progenitor
cells, it is distributed mainly to the liver and kidney [201]. The role
played by the kidney in the clearance of EPO is indicated by the prolongation in the mean elimination half-life of ^{124}I-rEPO in dogs that had been
rendered anephric by bilateral nephrectomy 16−18 hr prior to the start of
the experiment [195]. The elimination half-life of rEPO in anephric dogs
was found to be 13.8 ± 1.4 hr, compared with 9.0 ± 0.6 hr in dogs with
functional kidneys. The observed 30% reduction in total clearance after
nephrectomy, however, exceeded the reported contribution of less than
10% of renal clearance to total clearance. To explain this disproportionally
larger reduction in total clearance following bilateral nephrectomy, Fu
et al. [195] invoked reduction in nonrenal (metabolic) clearance of EPO
due to azotemia caused by the above surgical procedure.

The hepatic clearance of EPO is facilitated by removal of the large
number of sialic acid residues on the oligosaccharide side chains [186,202],

thereby exposing galactose residues which can then be recognized by the galactose-binding proteins in the hepatocytes [203]. Indeed, ^{125}I-asialoerythropoeitin, produced by treating rEPO with sialidase, disappeared almost completely within 6 min, whereas untreated rEPO was retained in the circulation longer with a $t_{1/2\alpha}$ of 10 min and a $t_{1/2\beta}$ of 108 min. Spivak and Hogans [204] also reported that 96% of the asialoerythropoeitin disappeared from rat plasma with a $t_{1/2}$ of 2.0 min following i.v. bolus administration. These observations suggested that the sialic acid residues prevent the more rapid removal of EPO from circulation by the liver.

To investigate the contribution of N-acetyllactosamine repeats to the clearance of EPO, Fukuda et al. [205] compared the plasma kinetics of ^{125}I-EPO that had been fractionated on tomato lectin – Sepharose, which binds carbohydrates with more than three N-acetyllactosamine repeats, against unfractionated EPO. Following i.v. administration, the tomato lectin-bound EPO was cleared more rapidly and to a greater extent than was the unbound fraction. The organ distribution of radioactivity at 30 min was in the order of liver, 67%; kidney, 18%; lung, 10%; and spleen, 5%.

To determine whether lactosamine repeats are a general determinant in the removal of glycoproteins by liver, polylactosaminoglycans isolated from erythrocytes were further tested. These glycosylated peptides were found to be cleared very rapidly from the circulation, with more than 90% of the dose being captured by the liver. This clearance pattern was the same whether the erythrocyte polylactosamines were intact or desialyated and whether they were branched or linear. Since clearance was inhibited by asialo-α_1-acid glycoprotein, it is possible that polylactosaminoglycans and asialoglycoproteins are cleared by the same galactose binding receptors in the liver. The implication is that glycoproteins including EPO with more than three lactosamyl repeat units are preferentially cleared by the liver.

Tissue Plasminogen Activator

Tissue plasminogen activator (t-PA) is a naturally occurring serine protease which catalyzes the conversion of plasminogen to plasmin [206,207]. Plasmin, in turn, proteolytically degrades the fibrin network associated with blood clots. The activation of plasminogen by t-PA is markedly increased in the presence of fibrin [208,209]. t-PA is therefore considered a promising thrombolytic agent in patients suffering from thromboembolic diseases.

t-PA is a single-chain glycoprotein with 530 amino acids. It can be spliced proteolytically into a two-chain protein, consisting of a heavy (M_r = 38,000) and a light (M_r = 31,000) chain [209 – 211]. Three N-glycosylation sites have been identified: on amino acid Asn-117 forming the oligomannose type of glycan and on the amino acids Asn-184 and Asn-448 forming the N-acetyllactosamine type of glycan [212]. Cloning and genomic mapping studies of the t-PA gene have led to the assignment of five putative structural domains in t-PA based on polypeptide homologies. These are the firbonectin finger-like domain (Cys-6 through Ser-50), the epidermal growth factor (EGF) domain (Cys-51 through Asp-87), the kringle 1 domain, the kringle 2 domain, and the serine protease domain. These domains represent the first reported example of exon transfer between genes [213]. The light chain of t-PA contains the active serine protease domain. The heavy chain contains the finger domain, the EGF

domain, and the two kringle domains. t-PA derived from various sources
may vary in the extent of their glycosylation, carbohydrate structure,
chain content (one-chain versus two-chain), and in the amino acid
sequence.

Studies conducted with t-PA derived from various sources have demonstrated that exogenous t-PA is eliminated very rapidly from the systemic
circulation with plasma half-lives of $1-4$ min in various animal species
[214-216] and $1.4-9$ min in humans [217-220]. These initial reports on
t-PA pharmacokinetics had two major drawbacks, however. First, important pharmacokinetic parameters like clearance and volume of distribution were not reported. Second, most of these studies employed immunoassays that were not specific, so that the measured t-PA antigen levels did
not necessarily represent pharmacologically active t-PA concentrations.

Due to the rapid disappearance of t-PA from plasma, intravenous infusion of large doses (100 mg) of t-PA is often required to treat patients
with myocardial infarction. Fong et al. [221] reported dose-related increases in steady-state two-chain t-PA plasma concentration (Css) and
total AUC following intravenous infusion in a number of anesthetized,
open-chested mongrel dogs undergoing reperfusion. But the increase was
not linear throughout the entire dose range. Due probably to a 30% reduction in systemic clearance of t-PA, a twofold increase in dose from 4
to 8 µg/kg·min led to a 2.5-fold increase in Css and AUC. There is,
therefore, dose-dependent pharmacokinetics at optimum thrombolytic doses
of two-chain t-PA.

In male Wister rats, human t-PA is cleared primarily via the liver
[215,222-224]. Kuiper et al. [225] reported that over 80% of the injected
dose of ^{125}I-tPA was found in the liver of rats at 6 min following i.v. ad-
miniatration of 50 ng of the labeled peptide. Uptake is believed to be
mediated through interaction of t-PA with specific cell surface receptors in
the parenchymal cells as well as by mannose receptors in the endothelial
cells [225], leading to internalization and lysosomal degradation [222,226].
Since the heavy chain of t-PA ($t_{1/2}$ = 1.0 min) is cleared almost six times
faster than the light chain ($t_{1/2}$ = 5.7 min), it is possible that intact t-PA
is recognized by the liver primarily through the heavy chain and much
less through the light chain [227].

Recent work by Larsen et al. [228] based on variant forms of t-PA
suggests that the polypeptide sequence encoding the finger and EGF domains in the heavy chain is responsible for rapid clearance by liver.
Seven forms of t-PA were tested: (1) natural or glycosylated wild-type
t-PA; (2) nonglycosylated wild type t-PA; (3) ΔF t-PA, which lacked fibronectin fingerlike domain (Cys-6 through Ser-50); (4) ΔE t-PA, which
lacked the epidermal growth factor domain (Cys-51 through Asp-87); (5)
ΔFE t-PA, which lacked both the finger and EGF domains; (6) $\Delta FE3X$
t-PA, a form of ΔFE t-PA in which Asn-linked glycosylation was prevented at all known glycosylation sites (Asn-117, 184, and 448 through
replacement by Gln); and (7) $\Delta FE1X$ t-PA, a form of ΔFE t-PA in which
high-mannose-type glycosylation was prevented at Asn-117. Approximately
1 µg of each of these proteins was injected as a bolus into the tail vein
of previously anesthetized rats. Blood samples were collected at 10 time
points within 15 min for all the proteins studied.

Both the glycosylated and nonglycosylated wild-type t-PA disappeared
from rat plasma with an initial α-phase $t_{1/2}$ of 0.8 and 1.9 min, respectively. The corresponding $t_{1/2\beta}$ values for these two forms of t-PA were 11.9

and 11.7 min, respectively. This finding indicates that the carbohydrate is not the primary mediator of rapid t-PA clearance. All other t-PA forms studied, except ΔE t-PA, demonstrated monophasic clearance, with $t_{1/2}$ ranging between 12 and 27 min. ΔE t-PA, however, was cleared in a biphasic manner, with $t_{1/2\alpha}$ of 2.1 min and $t_{1/2\beta}$ of 9.2 min.

These results indicate that hepatic clearance of t-PA is mediated in two distinct ways. The first involves a primary determinant responsible for rapid clearance of t-PA which appears to reside within the polypeptide sequence encoding the finger and EGF domains, with major contribution from finger domain as indicated by an approximately 20-fold increase in $t_{1/2}$ following the deletion of the finger region of t-PA. The second contribution to clearance is attributed to the presence and type of glycosylation which was more apparent in variants lacking the finger and EGF domains. For instance, the $\Delta FE1X$ t-PA and $\Delta FE3X$ t-PA variants disappeared from rat plasma less rapidly than the wild-type t-PA. The corresponding $t_{1/2}$ was 19.3 and 11.8 min, respectively. The more rapid clearance of $\Delta FE3X$ t-PA than $\Delta FE1X$ t-PA was attributed to its reduced size and charge. $\Delta FE3X$ t-PA was also found to have a markedly longer plasma half-life than natural t-PA in dogs [229]. The pharmacokinetic parameters of this variant in dogs were: $t_{1/2\alpha}$, $14-18$ min; $t_{1/2\beta}$, $72-125$ min; and plasma clearance, $1.26-2.16$ L/hr.

To elucidate the structure-pharmacokinetic and structure-pharmacodynamic relationships for t-PA, Colleen et al. [230] have studied the pharmacokinetics and thrombolytic properties of three deletion mutants in rabbits: (1) ΔFE t-PA, a form of t-PA lacking both the finger domain and the EGF domain; (2) $\Delta FE1X$ t-PA, a form of ΔFE t-PA in which high-mannose-type glycosylation is prevented at Asn-117; and (3) $\Delta FE3X$ t-PA, another form of ΔFE t-PA in which all Asn-glycosylation is prevented at all known glycosylation sites (Asn-117, -184, and -448) by substituting Gln for Asn. These proteins were constructed by deletion mutagenesis, expressed and amplified in Chinese hamster ovary cells, and purified from conditioned medium by affinity chromatography using erythrina trypsin inhibitor-agarose.

Following intravenous infusion for 4 hr into rabbits with jugular vein thrombosis at doses ranging between 0.13 and 0.75 mg/kg, 50% fibrinolytic activity was observed with 0.40 mg/kg of wild-type t-PA, 0.37 mg/kg of ΔFE t-PA, 0.20 mg/kg of $\Delta FE1X$ t-PA, and 0.40 mg/kg of $\Delta FE3X$ t-PA. The plateau t-PA antigen levels in plasma were 0.055, 2.1, 0.6, and 0.5 µg/ml, respectively. At 50% lysis, the residual fibrinogen 30 min after the end of infusion was 100%, 81%, 100%, and 85% of baseline and the residual α_2-antiplasmin was 82%, 55%, 85%, and 90%, respectively. The significantly greater decrease of the α_2-antiplasmin levels with ΔFE t-PA suggests that this variant, when infused at a dose required to produce 50% clot lysis, is somewhat less fibrin specific than intact t-PA, $\Delta FE1X$ t-PA or $\Delta FE3X$ t-PA. After the end of the infusion, t-PA-related antigen disappeared from plasma with an initial $t_{1/2}$ of 4 min for intact t-PA, 25 min for ΔFE t-PA, 42 min for $\Delta FE1X$ t-PA, and 14 min for $\Delta FE3X$ t-PA. Collectively, these findings indicate that the structures responsible for the clearance of t-PA reside in a different domain of t-PA than those responsible for its fibrin affinity and specificity. It is therefore possible to independently modify the pharmacokinetic and thrombolytic properties of t-PA through structural modification.

Vasopressin and Desmopressin

Arginine vasopressin (AVP) is a neurohypophysial nonapeptide hormone with direct antidiuretic effect on the kidney [231]. Many kinetic studies have been conducted with AVP and the results are summarized below.

Following i.v. administration, vasopressin disappears in a biphasic manner [150,232,233]. In rabbits anesthetized with a combination of urethane and barbital, the plasma half-lives of the fast and slow phases were 0.9 min and 5.4 min, respectively [150]. In rabbits anesthetized with urethane, the corresponding values were 0.6 and 5.9 min. In conscious dogs, such values were 1.4 and 4.1 min [233]. $t_{1/2}$ values of 0.9–8 min in the rat, 5–8 min in the dog, and 2.21 min [232] as well as 26.1 min [234] in humans have also been reported. The above differences in the $t_{1/2}$ may reflect differences in species and the anesthetics used.

Following intravenous infusion of AVP at different rates in humans [235] and conscious dogs [236], a more or less linear relationship was observed between plasma concentration of AVP and its urinary metabolic clearance rates as well as urinary excretion rates. These findings suggest that measurement of urinary AVP is a useful way to assess pituitary AVP secretion because the urinary excretion rate of AVP represents an integral value of plasma AVP levels.

Vasopressin is stable in normal plasma in vitro [237]. In pregnant women, oxytocinase-like activity, which inactivates vasopressin by cleavage of the pentapeptide ring between the cysteine and tyrosine residues, appears in the plasma [238]. AVP is extensively degraded in the proximal renal tubule in the rat releasing the COOH-terminal glycinamide residue (position 9) in the urine [54]. In renal cortical homogenates, free phenylalanine (position 3), hexapeptide(1-6), heptapeptide(1-7), and the octapeptide were formed from AVP. The same products were detected in proximal tubules microperfused in vivo with labeled AVP [239], indicating endocytotic uptake and rapid lysosomal degradation of AVP in the proximal tubule. Considerable variability was seen in the rate at which these degradation products were formed from tubule to tubule and between rats.

The metabolic clearance rate (MCR) of AVP appears to vary with the stage of pregnancy. Davison et al. [236] measured the MCR in five women before conception, during gestational 7–8 weeks (early), 22–24 weeks (middle), and 36–38 weeks (late pregnancy), and again 10–12 weeks postpartum. The MCR of AVP was similar before conception, in early pregnancy, and postpartum. Values at midpregnancy and late pregnancy increased three- to fourfold. Plasma oxytocinase-like activity, undetectable at 7–8 weeks gestational period, increased markedly by mid- and slightly more by late gestation. Also, the relationships between AVP levels in plasma and urine osmolality were similar before, during, and after pregnancy. These results indicate that the marked increments in the MCR of AVP between gestational weeks 7 and 8 and midpregnancy coincide with rise in both trophoblastic mass and plasma vasopressinase.

1-Desamino-8-D-arginine vasopressin (DDAVP, desmopressin), a synthetic analog of vasopressin [240], has potent antidiuretic and hemostatic activities. It stimulates the release of Factor VIII and von Willebrand factor and also causes fibrinolytic activity [241–243]. Desmopressin is resistant to aminopeptidase and trypsin-like activities [244,245]. It produces prolonged antidiuretic effect on the kidneys of

patients with diabetes insipidus. This is a reflection of its metabolic
stability and probably of increased receptor affinity [245-247].

In a randomized cross-over trial in 10 healthy volunteers, the influence
of three routes of administration, i.v. infusion, s.c. injection, and i.n.
(intranasal) dosing, on the efficacy of desmopressin in the treatment of
bleeding disorders was evaluated [248]. In the i.v. series 0.4 µg/kg
body-weight dose of desmopressin was dissolved in 100 ml saline and was
administered over a period of 30 min. In the s.c. series 0.4 µg/kg of the
same preparation was injected. In the i.n. series a total dose of 300 µg
was slowly applied into both nasal cavities. Factor XII and high-molecular-
weight kininogen levels increased only slightly after desmopressin administration. The mean increase of Factor VIII was 3.1- (i.v.), 2.3- (s.c.),
and 1.3- (i.n.) fold over baseline. Ristocetin cofactor (von Willebrand
factor antigen) increased 3.1-, 2.0-, and 1.2-fold over baseline mean values after i.v., s.c., and i.n. desmopressin, respectively. Thus, i.v.
and s.c. administration of desmopressin resulted in appropriate stimulation
of hemostasis. The half-disappearance-time of Factor VIII and von
Willebrand factor after desmopressin was about 5 and 8 hr, respectively.
The mean increase of fibrinolytic activity was more pronounced after i.v.
administration.

In a subsequent study, the pharmacokinetics of desmopressin were
evaluated in addition to hematological effects following i.v., s.c., and i.n.
administration of desmopressin [249]. Here, desmopressin was administered
at a dose of 0.4 µg/kg body weight via i.v. and s.c. routes. A total dose
of 300 µg of desmopressin was administered intranasally. Two different
concentrations (4 µg/ml and 40 µg/ml) were studied via s.c. route. The
effect of desmopressin on Factor VIII was similar to that reported in the
early study [250]. Also tissue-type plasminogen activator values were increased by 1.9- (i.v.), 1.3- (s.c. 4 µg/ml), 3.1- (s.c. 40 µg/ml), and
1.2- (i.n.) fold over baseline values. The elimination of desmopressin from
plasma in all three routes of administration obeyed first-order kinetics, the
half-life ranging from 2.7 to 4.6 hr. There was no significant difference
in AUC between i.v. and either method of s.c. dosing. Also, there was
no significant difference between the methods of s.c. administration in the
maximum plasma concentration or the time to reach the maximum concentration, although there was less variability after injection of the more concentrated solution.

Absorption after i.n. administration was poor, with high interindividual
variation. The bioavailability was only 2%, when compared with a bioavailability of 112% and 94% for the low and high subcutaneous doses.
There was, however, two- to threefold improvement in the bioavailability
as well as the biological response when the nasal dose was administered as
sprays rather than as drops [251]. Consistent with earlier observations
[249], the maximal increase in Factor VIII and von Willebrand factor was
independent of desmopressin concentration above 300 pg/ml.

B. Centrally Acting Peptide and Protein Drugs

Delta Sleep-Inducing Peptide

Delta sleep-inducing peptide (DSIP, Trp-Ala-Gly-Gly-Asp-Ala-Ser-
Gly-Glu) was first discovered as a humoral hypnotic factor in the rabbit
[252]. DSIP-like immunoreactivity (LI) has been demonstrated in the

central nervous system of the rat and in the pituitary and adrenal glands of humans, pig, mouse, and rat. DSIP-LI has also been found in several peripheral organs of the rat, pig, and humans, particularly in the stomach and small intestine [253,254].

Besides sleep-inducing properties [255], DSIP has been claimed to exert several other actions. These actions include (1) modulation of the release of somatostatin from the hypothalamus [256] and several hormones from the pituitary, including adrenocorticotropin (ACTH) [257,258], growth hormone [259], and luteinizing hormone [260]; (2) reduction of blood pressure [251]; and (3) interaction with adrenergic transmission [262]. It also has antinociceptive properties [263], probably owing to its indirect action on opioid receptor through stimulating the release of immunoreactive methionine enkephalin [264].

DSIP in plasma disappears rapidly with a $t_{1/2}$ of about 6–8 min in humans and even faster in other species [254,255,265,266]. DSIP is rapidly degraded in both plasma and tissue extracts. The major metabolite is des-Trp-des-Ala (DSIP 3-9) [266,267]. Various metabolically stable analogs have now been synthesized in an attempt to obtain longer circulation half-life than DSIP. They include des-Trp-DSIP, D-Ala^3-DSIP, D-Ala^4-DSIP, and D-Ala^4-DSIP-NH$_2$. Their respective n-octanol/saline partition coefficients are 0.00638, 0.0170, 0.0103, and 0.0335, compared with 0.00927 for DSIP. Only the D-Ala^4-DSIP and D-Ala^4-DSIP-NH$_2$ analogs afford the expected increase in circulation half-life. Even then, the increase is very modest. In addition, there are subtle changes in the volume of distribution and percent of protein binding (Table 3).

Banks et al. [268,269] reported that DSIP and its analogs mentioned above entered the CSF to a measurable degree following i.v. administration in anesthetized mongrel dogs. RIA was used to measure DSIP in plasma and cerebrospinal fluid. This assay used an antibody which was highly specific for DSIP and which required at least eight of the nine amino acids for cross-reactivity. Column chromatographic studies revealed that the intact peptide accounted for almost all of the immunoactivity present in the CSF following DSIP administration [269].

Table 3   Pharmacokinetic Parameters of DSIP and Its Analog Following Intravenous Bolus Administration of 0.1 mg/kg of Peptide to Mongrel Dogs

Peptide	$t_{1/2}$ (min)	Vd (L/kg)	% Bound in plasma
DSIP	7.24 ± 0.77	0.671 ± 0.117	9.6
des-Trp-DSIP	7.55 ± 0.52	0.488 ± 0.043	50.6
D-Ala^3-DISP	8.19 ± 0.30	0.617 ± 0.050	0
D-Ala^4-DSIP	12.9 ± 0.83	0.234 ± 0.019	12.6
D-Ala^4-DSIP-NH$_2$	10.2 ± 0.69	0.280 ± 0.022	32.1

Source: Ref. 268.

Banks et al. [269] found that the increase in CSF levels correlated positively with basal plasma levels, plasma half-life, and octanol/saline partition coefficient. This is consistent with the proposed nonsaturable mode of entry of DSIP into rat brain [270]. By contrast, blood-brain-barrier permeability did not correlate with protein binding and molecular weights of peptides studied. Here, plasma protein binding was measured by incubating the different peptides with pooled dog plasma for 10 min. It was speculated that the in vitro values thus obtained could be different from those occurring in vivo, thus accounting for the lack of correlation between protein binding and CSF levels. Moreover, the lack of correlation between molecular weight and CSF levels is not surprising given that the molecular weights of the peptides studied were distributed over a narrow range (664–879 daltons).

β-Endorphin

β-Endorphin (β-EP) [271], a straight-chain polypeptide containing 31 amino acid residues, is an endogenous peptide that exhibits potent analgesis activity when injected either cerebroventricularly [272] or intravenously [273]. β-EP has been shown to depress gastrointestinal mobility [274] as well as cardiovascular [275] and respiratory functions [276].

Many pharmacokinetic studies of β-endorphin have been performed [277–282] and the data analyzed exclusively by noncompartmental or two-(three-)compartmental methods. Aronin et al. [281] studied the pharmacokinetics of synthetic β-EP and its precursor protein, native purified β-lipoprotein, in humans following i.v. bolus administration at doses of 100 µg and 250–370 µg, respectively. Plasma concentrations of β-EP and β-lipoprotein were measured using radioimmunoassay. The $t_{1/2}$ of β-EP and β-lipoprotein was 44 and 49 min, respectively. The corresponding mean volumes of distribution were 29.6 L and 21.7 L. The average MCR of β-EP was 0.468 L/min and that of β-lipoprotein was 0.329 L/min. The observed mean MCR/kg for β-EP (7.8 ml/min.kg) was higher than that reported (2.1 to 4.8 ml/min.kg) by Foley et al. [278] following a much higher initial dose (5 or 10 mg). In another study, Bertanga et al. [283] observed that endogenous immunoreactive-β-EP disappeared from the plasma of a subject with Addison's disease three times faster than immunoreactive-β-lipoprotein. In yet a third study, Sato et al. [282] evaluated the pharmacokinetics of exogenously administered β-EP using a rapid radioreceptor assay in rats. The method was based on the inhibition by β-EP of ^{3}H-naloxone binding to the specific receptors on rat brain membranes. These studies indicate biexponential decline for human β-EP following i.v. bolus (168 µg/kg) administration in male Wistar rats. The $t_{1/2\alpha}$ and $t_{1/2\beta}$ were 2.6 ± 0.5 min and 6.2 ± 1.6 hr, respectively. The volume of the central compartment and the steady-state volume of distribution were 67 ± 16 and 480 ± 75 ml/kg, respectively. The total body serum clearance was 2.1 ± 0.9 ml/min.kg. These observations were similar to those reported by Houghten et al. [279] in rats following the administration of tritiated β-EP.

A physiologically based pharmacokinetic model incorporating diffusional transport of the peptide across the capillary membrane has been proposed by Sato et al. [284] to simulate the distribution and elimination of ^{125}I-β-EP after i.v. injection in rats. Schematically it is shown in Fig. 1. The model consists of nine tissue and blood compartments. It is assumed that ^{125}I-β-EP is metabolized and excreted by liver and kidney, respectively.

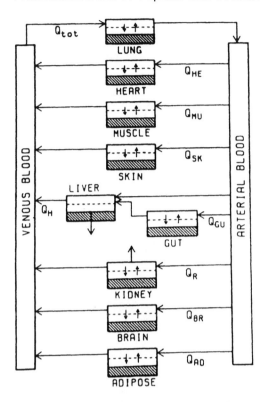

Fig. 1   A physiologically based pharmacokinetic model incorporating the transcapillary diffusion barrier for the distribution, metabolism, and excretion of $125I$-β-endorphin in rats.   (From Ref. 284.)

Each compartment is further subdivided anatomically into three fluid compartments, namely, capillary bed, interstitial fluid (ISF), and intracellular space.   There are five assumptions: (1) each compartment constituting a whole organ is well stirred; (2) binding equilibrium of $125I$-β-EP in the capillary bed and ISF is instantaneously established; (3) the distribution of the labeled peptide is restricted only within extracellular fluids, i.e., blood and ISF; (4) passage of the labeled peptide is limited by the diffusion process across the capillary membrane in each tissue except the liver; and (5) only unbound $125I$-β-EP is capable of undergoing transcapillary diffusion, metabolism, and excretion.

The incorporation of a diffusional barrier across the capillary membrane in the above model is justified, because rapid equilibrium between the capillary bed and interstitial fluid cannot be assumed for relatively high-molecular-weight substances, e.g., inulin [285,286], dextrans, and albumin [287,288].   The liver is, of course, an exception.   Here, it is expected that free exchange of materials between the capillary bed and extracellular fluid (in the space of Disse) would occur through its open fenestrae [289], and filtration effects can be expected when particles of about the size of the fenestrae (approximately 0.1 μm) arrive in the liver [290].

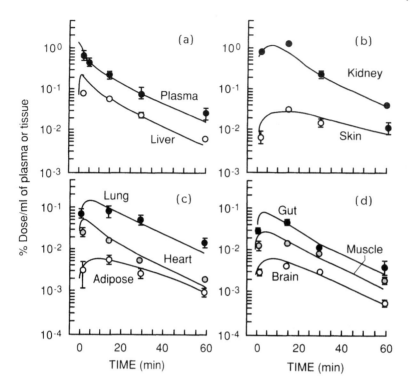

Fig. 2   Observed and simulated concentration profiles of ^{125}I-β-endorphin in plasma, liver, kidney, lung, gut, brain, heart, skin, muscle, and adipose tissue (a–d) after i.v. bolus injection of tracer doses of ^{125}I-β-endorphin (8 μg/kg) in rats. Each point and vertical bar represents the mean and SE of three individual rats. Solid lines show the simulated curves based on the physiological pharmacokinetic model incorporating the transcapillary diffusion-barrier for ^{125}I-β-endorphin. (From Ref. 284.)

The above pharmacokinetic model appears to describe the observed data reasonably well, as shown in Fig. 2a–d. Thus, the rapid decline of plasma concentration of ^{125}I-β-EP after i.v. injection could be attributed to transcapillary diffusion into the tissue interstitial fluids, followed by the postdistributive elimination phase.

The expression of CNS activities by peripherally administered β-EP [273] has prompted Rapoport et al. [291] to investigate the cerebrovascular permeability of four synthetic radioactive opioid peptides derived from lipoprotein. Because of the tight junctions between the cerebrovascular endothelial cells, the blood-brain barrier is generally viewed as essentially impermeable to peptides, proteins, as well as water-soluble electrolytes and nonelectrolytes [292,293]. Consequently, it has been suggested that circulating peptides enter the brain via the choroid plexus and cerebrospinal fluid or, if produced in the pituitary gland, through routes such as the pituitary portal system [294].

The four peptides evaluated by Rapoport et al. [191] were: (1) [D-Ala^2]methionine enkephalinamide; (2) β-[D-Ala^{62},^{14}C-Homoarg^{69}]-lipoprotein(61-69); (3) α-[D-Ala^2,^{14}C-Homarg]-endorphin; and (4) β-[D-Ala^2, ^{14}C-Homoarg]-endorphin. They were analogs of methionine enkephalin, β-lipotropin(61-69), α-endorphin, and β-endorphin, respectively. All shared a common amino acid terminal sequence, Tyr-D-Ala-Gly-Phe-Met, in which the D-alanine residue was placed in position 2 to increase resistance to degradation without significantly reducing biological activity. Each peptide was administered at a dose of 2-10 μCi intravenously in conscious and partially restrained adult male rats with indwelling femoral artery and vein catheters. Rats were decapitated 3 min after the injection of [D-Ala^2]-methionine enkephalinamide; 3 or 10 min after β-[Ala^{62},^{14}C-Homoarg^{69}]-lipoprotein (61-69); 4, 8, 20, or 30 min after α-[D-Ala^2,^{14}C-Homoarg]-endorphin; and 5 or 10 min after β-[D-Ala^2,^{14}C-Homoarg]endorphin. The brain was removed, dissected into 13 regions, and assayed for radioactivity. Brain parenchymal concentration was measured by subtracting intravascular radioactivity from net measured regional radioactivity. Intravascular radioactivity was obtained by multiplying the whole blood concentration with regional blood volume, which ranged from 0.015 to 0.035 ml/g of brain. The resulting cerebrovascular permeability coefficient was 2.5 × 10^{-6} cm/sec for met-[D-Ala^2]enkephalinamide, 1.4 × 10^{-6} cm/sec for β-[D-Ala^{62},^{14}C-Homoarg^{69}]lipoprotein(61-69), 2.3 × 10^{-6} cm/sec for α-[D-Ala^2,^{14}C-Homoarg^9]endorphin, and 3.9 × 10^{-6} cm/sec for β-[D-Ala^2, ^{14}C-Homoarg]endorphin. These permeability coefficients were about the same as those of nonelectrolytes glycerol, formamide, and thiourea, which have similar octanol-water partition coefficients (between 0.003 and 0.1) [295]. The n-octanol/water partition coefficients of the above peptides were 0.066, 0.012, 0.0017, and 0.013, respectively. For comparison, very permeable, lipid-soluble substances like caffeine and antipyridine exhibit permeability coefficients two orders of magnitude higher (10^{-4} cm/sec) than those of the peptides, whereas poorly permeable, water-soluble compounds like sucrose and erythritol have permeability coefficients equal to 10^{-8} to 10^{-7} cm/sec [293,295].

Enkephalins

Met-enkephalin (Tyr-Gly-Gly-Phe-Met) and leu-enkephalin (Tyr-Gly-Gly-Phe-Leu) are two naturally occurring opiate-like peptides, which are widely distributed in the brain and gastrointestinal tract of various animal species. These peptides are concentrated in the synaptosomal fraction of brain homogenates and are also present in the peripheral nervous system. Enkephalins bind to opiate receptors with high affinity. Parenteral administration of enkephalins elicits many morphine-like effects, including analgesia, secretion of growth hormone and prolactin, and inhibition of neurotransmitter release. Enkephalins possibly have a role as neurotransmitters or neuromodulators in the CNS and in the periphery [296,297].

The reported plasma half-lives of enkephalins ranged from 2-4 sec in rats [298] to 9 min in human plasma [299]. Met- and Leu-enkephalins are unstable in blood in vitro. The half-life is about 2 min in rat blood. The major metabolite arising from incubation of [^{3}H-Tyr^1] enkephalin in blood or brain homogenates is [^{3}H]tyrosine [230], as a result of aminopeptidase action on the Tyr^1-Gly^2 bond [301,302]. Cleavage of enkephalin at the

3−4 amide bond by synaptic membranes yields the metabolites Tyr−Gly−
Gly and Phe−Leu or Phe−Met [303,304]. The generation of Tyr−Gly−Gly
represents less than 25% of enkephalin metabolism in washed synaptic membranes. The responsible enzyme, enkephalinase, is membrane-bound. It
is unevenly distributed in various regions of the brain and appears to be
closely associated with opioid receptors [305]. Almenoff et al. [306] suggested that this enkephalinase may be identical to the previously identified
class of zinc metalloendopeptidases, which are strongly inhibited by thiols
and metal chelators [305,307], rather than being a dipeptidylcarboxy-
peptidase, as suggested by Schwartz et al. [305]. Dupont et al. [298]
studied the plasma disappearance of [^{3}H] Met-enkephalin in male Sprague-
Dawley rats. In these studies, 0.5 ml of 23 μM Met-enkephalin in 0.9%
NaCl was injected intrajugularly. Blood samples were collected in a capillary tube at 30-sec intervals during the first 7 min and then at 8, 10, 15,
and 20 min after injection of the labeled peptide. Only 5% of total plasma
radioactivity comigrated with intact Met-enkephalin on Sephadex G-25 15 sec
after injection and no detectable intact pentapeptide remained 2 min after
injection. A half-life on the order of 2−4 sec was suggested for the intact
peptide. Thirty seconds after the injection, the apparent volume of distribution and metabolic clearance rate of tritium label were 53 ml and 10
ml/min, respectively.

Buscher et al. [308] studied the analgesic effects of Met- and Leu-
enkephalins using the tail flick test in mice. Morphine (0.56−3.2 μg/
mouse) had a slow onset of action following intracerebroventricular (i.c.v.)
administration, peak value being reached after 15 min. By contrast, Met-
and Leu-enkephalin showed maximum activity after 2 min and had a duration of action of about 5 min. This observation is consistent with the
rapid in vitro degradation reported for the enkephalins [300]. The ED$_{50}$
values of analgesic activity following i.c.v. administration were 75 and 240
μg/mouse for Met-enkephalin and Leu-enkephalin, respectively. Naloxone
(0.1 mg/kg) administered subcutaneously 20 min before the measurement of
analgesia inhibited the analgesic effects of Met-enkephalin and morphine
administered intracerebroventricularly. The implication is that these two
compounds acted on the same system in a similar manner. By contrast,
the analgesic effect of Leu-enkephalin was not inhibited by pretreatment
with naloxone except at a very high subcutaneous dose (10 mg/kg). Following i.v. administration, Met-enkephalin exhibited a transient effect at
very high doses (ED$_{50}$ = 170 mg/kg intravenously) [308]. Analgesic activity was detected only within the first 15 sec of administration. Leu-
enkephalin had weak analgesic activity (4 of 10 mice showed analgesia) following i.v. administration of 320 mg/kg. These findings indicate that
Met-enkephalin is more active than Leu-enkephalin.

C. Immunoactive Peptide and Protein Drugs

Cyclosporine

Cyclosporine (CsA), a neutral, hydrophobic cyclic peptide composed of 11
amino acid residues, is a third-generation immunosuppressive agent that
has been used successfully in organ transplantation in humans since 1981
[309−311]. The first-generation immunosuppressive agents include azathioprine, steroids, and antilymphocyte globulin derivative. An example
of second-generation immunosuppressive agent is cyclophosphamide.
Cyclosporine depresses cellular and humoral immunity with a preferential

and rapidly reversible action against T lymphocytes. The drug also has generated much interest as a useful tool in various autoimmune conditions in humans. The popularity of cyclosporine is due to its lack of bone marrow depression, which often limits the use of azathioprine and cyclophosphamide. Nonetheless, cyclosporine may cause nephrotoxicity [309], hepatotoxicity [312], oncogenicity [313], and other occasional side effects [314].

A clear understanding of the pharmacokinetics of CsA and the subsequent therapeutic drug monitoring is essential for various reasons. First, wide variations in CsA pharmacokinetic parameters have been observed following oral and intravenous administration. Variations were observed in bioavailability, peak serum concentrations, time to reach peak concentration, absorption half-life, blood-to-plasma ratio, and volume of distribution. Such variations in the kinetics of CsA are related to the patient's disease state, type of organ transplant, age of the patient, and therapy with other drugs that interact with CsA. Second, maintaining blood concentration of CsA is required to prevent rejection of the transplanted organ. Third, it is essential to minimize drug toxicity by maintaining trough concentration below that at which toxicity is most likely to occur. Fourth, it is necessary to ensure patient compliance, since patient noncompliance with drug regimens is a significant reason for graft loss after 60 days [315].

Pharmacokinetic studies of cyclosporine can be performed using different biological fluids (serum, plasma, and whole blood) and different analytical techniques [RIA (6,7) and HPLC (8-10, 17-19)]. RIA provides a rapid and sensitive means for estimating cyclosporine in different biological fluids. The sensitivity of the assay is 25 µg/L requiring only a limited amount (1 ml) of blood or plasma. Concentrations above the linear range of the standard curve are frequently observed in the course of pharmacokinetic studies, and in such a case, the samples must be diluted with blank blood or plasma prior to analysis. The assay is linear in the range of 100−2500 µg/L on a log-logit plot. The assay, however, has two drawbacks. First, kits that use tritium-labeled drug suffer from low counting efficiency and limited kit stability. Second, the antibody in RIA analysis cross-reacts with the metabolites of cyclosporine and therefore is unable to differentiate between the parent compound and its metabolites [316]. This problem is solved to a great extent with the availability of a monoclonal antibody for CsA [317]. Quesniaux et al. [317] showed that monitoring of cyclosporine could be improved by using monoclonal antibodies of restricted specificity instead of polyclonal antisera that recognize both cyclosporine and its unmodified metabolites. This monoclonal antibody required 15- to 1000-fold higher concentrations of different metabolites studied to inhibit the cyclosporine−monoclonal antibody reaction by 50%. As the immunosuppressive [318,319] and toxic [320] properties of CsA are controversial issues, it has been recommended [321] that routine drug monitoring should include at least one immunoassay with a specific antibody detecting the unchanged CsA, and a supplementary immunoassay with a nonspecific antibody detecting a composition of cross-reacting metabolites plus the unchanged substance.

Many high-performance liquid chromatography (HPLC) procedures are available for estimation of cyclosporine in blood, involving either extensive sample extraction or a column-selective isolation technique that requires special instrumentation [322−324]. The extraction of cyclosporine from biological fluids involves solid−liquid or liquid−liquid extraction. In the

solid–liquid method, cyclosporine is separated from endogenous substances
using columns such as Baker cyano columns. In the liquid–liquid extraction method, cyclosporine is extracted with ether. In some procedures
further sample purification is carried out using hexane [324]. Isolated
cyclosporine can be chromatographed using a C-18 or cyano column maintained at a temperature of 55–75°C. The mobile phase generally consists
of various combinations of acetonitrile, methanol, and water. Cyclosporine
usually elutes within 12 min. Late peaks may occasionally be seen in samples prepared by liquid–liquid extraction, thereby lengthening the chromatographic time to as long as 25–30 min. HPLC is specific for unchanged
CsA, and it is sensitive to concentrations as low as 10 µg/L in 1 ml of
blood. The assay is linear over the range of 25–4000 µg/L and is very
easily adapted to specifically measure the metabolites of the cyclosporine
family [325]. Obviously, HPLC is more demanding of laboratory time,
equipment, and expertise.

Concentrations of cyclosporine obtained by HPLC are consistently lower
than those obtained by RIA with polyclonal antisera because of the cross-
reactivity of cyclosporine metabolites with the antiserum used in RIA kits
[316,326]. Cyclosporine is extensively metabolized in humans and animals
[327,328]. More than 90% of the administered intravenous dose of CsA is
excreted as metabolites in bile in humans, dogs, and rats [329]. Urinary
excretion of the metabolites accounts for less than 6% of the dose administered [330]. The metabolites retain the cyclic structure of the parent
compound [312]. The metabolites of cyclosporine include monohydroxylated
(metabolites 1 and 17), dihydroxylated (metabolites 8, 9, 10, and 16),
N-dimethylated (metabolite 21), and tetrahydrofuran ether products of
cyclosporine.

The RIA:HPLC ratio of cyclosporine concentration in blood, plasma,
and serum is highly variable [316,326,331]. It depends on the patient's
liver function, time of blood sampling in reference to the time of drug administration, absolute cyclosporine concentration, and whether the patient
is receiving other drugs that may alter the enzymes involved in hepatic
metabolism. Average value of RIA:HPLC ratio of cyclosporine in healthy
subjects is 2:1 [118]. Typical RIA:HPLC ratios less than 2.7:1 in renal
patients and a mean value of 3.9:1 for heart and liver transplant recipients have been reported [332]. The highest ratios were found shortly
after liver transplantation and a fall was observed as liver function improved [118,333,334].

After intravenous administration, cyclosporine exhibits multicompart-
mental behavior [331,335]. There are two phases of distribution for cyclosporine in humans. The half-life was 0.1 hr for the initial phase and 1.1
hr for the terminal phase. Rapid initial distribution can be attributed to
the high lipid solubility of cyclosporine and its ability to cross some biological membranes. Consistent with its lipophilic characteristics, cyclosporine accumulates in body fat [336]. Studies using intravenous tritiated
cyclosporine in rats showed the highest accumulation of radioactivity in the
liver, followed by fat, kidney, reticuloendothelial system, endocrine sys-
stem, and blood, in that order [328].

Estimates of the total volume of distribution of cyclosporine appear to
vary with patient condition. In renal transplant patients mean steady-state
volumes of distribution of 8.7 ± 6.2 L/kg and 4.5 ± 3.6 L/kg were reported
by RIA [327] and HPLC [337] methods, respectively. Patients with liver
disease showed a steady-state volume of distribution of 3.9 ± 1.8 L/kg

(measured by HPLC in whole blood), whereas normal volunteers, children with heart failure, and patients following heart transplantation reportedly have a smaller volume of distribution of 1.2, 0.9, and 1.3 L/kg, respectively (HPLC method) [315]. The smaller volume of distribution may be attributed to a higher hematocrit in normals and heart transplant recipients compared with other transplant recipients. Rosano [338] studied the effect of hematocrit on CsA concentration in whole blood and plasma. Both RIA and HPLC estimates indicated an inverse correlation between hematocrit and plasma fraction of cyclosporine in the circulating blood. Specifically, a 10% increase in hematocrit would result in a decrease in the portion of cyclosporine in the plasma by 12–14%.

The volume of distribution at steady state (Vd_{ss}) appears to vary not only with patient condition, but also with the dosage form. Venkataram et al. [339] reported that the Vd_{ss} of intravenously administered cyclosporine in the commercial formulation (2.7 ± 0.2 L/kg) was significantly lower than that of either Intralipid (10.6 ± 2.7 L/kg) or liposomes constituted of egg phosphatidylcholine and cholesterol (7.4 ± 2.3 L/kg) in the albino rabbit. The terminal disposition half-life, which ranged from 400 to 475 min, was not statistically different among the three groups. Similar observations were made by Vadiei et al. [340], who evaluated the pharmacokinetics of liposomal cyclosporine administered intravenously over a 10-day period in mice. The liposomes were constituted of dimyristoylphosphatidylcholine and stearylamine in mice. More importantly, these investigators found that liposomal cyclosporine did not affect the glomerular filtration rate whereas the same dose of commercial cyclosporine formulation impaired renal function by about 50%.

In blood, cyclosporine is highly bound to erythrocytes and plasma proteins, especially lipoproteins. The relative distribution of cyclosporine in blood is a function of drug concentration, hematocrit, temperature, and lipoprotein concentration [120]. At a blood concentration of 500 µg/L, 58% of the drug is associated with erythrocytes, 4% with granulocytes, 5% with lymphocytes, and the remaining 33% is distributed within plasma. In normal subjects, the blood-to-plasma ratio of cyclosporine is approximately 2:1, indicating a greater affinity for red blood cells than for plasma proteins. In liver and kidney transplant patients the blood-to-plasma ratios observed were 1.32 and 1.36, respectively. This lower ratio is probably a consequence of low hematocrit observed in these patients. Low hematocrit values may be observed in transplant patients due either to renal failure, liver failure, or blood loss associated with surgery. The plasma concentration of cyclosporine increases linearly with whole blood concentration up to 1000 µg/L. Above this concentration, the distribution of cyclosporine between blood and plasma becomes nonlinear. Blood cells appear to be saturated with cyclosporine at concentrations above 5 mg/L [2]. Temperature can affect the distribution of cyclosporine between red blood cells and plasma. With a temperature decrease from 37 to 21°C, about 50% of cyclosporine diffuses from plasma to red blood cells. This process is reversible upon reequilibration at 37°C for 2 hr [2,119].

In plasma, about 80% of cyclosporine is bound to lipoproteins while very little is bound to other plasma proteins such as albumin and globulins [2,120]. The binding of cyclosporine to plasma proteins is independent of concentration in the range of 0.2–20 mg/L. However, binding is markedly influenced by temperature. About 70% of the drug is bound at 4°C, 93% at 20°C, and 98% at 37°C [2,341]. Of the different lipoproteins in plasma, the high density lipoprotein (HDL) binds 57% of the drug compared with 25% by low-density lipoproteins (LDL), and approximately 2% by the very-low-density lipoproteins (VLDL). The amount of cyclosporine bound to chylomicrons is negligible [2].

Cyclosporine does not cross the blood-brain barrier in significant amounts [342]. Cefalu and Pardridge [343] reported that extraction of [^{3}H]cyclosporine by rat brain following carotid artery injection in ketamine-anesthetized rats was 2.9 ± 0.5%. By contrast, cyclosporine A appears to pass through the blood-aqueous barrier more readily. Palestine et al. [344] studied the penetration of cyclosporine into the anterior chamber of the eye following oral administration in patients suffering from uveitis. Aqueous humor was obtained by paracentesis 24 hr after the last dose but before cataract surgery, and plasma samples were also collected for assay of cyclosporine by RIA. The mean ratio of anterior chamber concentration to plasma concentration was 0.40 ± 0.24. The concentrations in the anterior chamber ranged from 19 to 125 ng/ml. (Surprisingly, cyclosporine concentration in the amniotic fluid was similar to that in mother's peripheral blood and CsA was also detected in breast milk [345].)

Cyclosporine is poorly absorbed after intramuscular administration [330,346]. Studies by Beveridge et al. [330] and Keown et al. [346] revealed that even after the administration of 20 mg/kg in Miglyol (100 g/L), the maximum concentration of cyclosporine in plasma as measured by HPLC was less than 100 μg/L. For this reason, cyclosporine is administered either orally or intravenously. Although administration of cyclosporine via subcutaneous route resulted in a consistent absorption pattern in the rat [337], this route has not been tried in humans.

Following oral administration, cyclosporine is absorbed from the upper part of the small intestine. Factors that affect the oral absorption of cyclosporine include elapsed time after surgery [118], dose administered [7,347], gastrointestinal dysfunction [348–351], external bile drainage [352], liver disease [353], and food [354–356]. The absorption process is slow, incomplete, and extremely variable [337,347,357]. In transplant patients, the bioavailability of orally given cyclosporine ranges from 1% [327] to 89% [315]. In renal transplant patients, peak concentration is observed at various times ranging from 1 to 8 hr, with an absorption half-life ranging from 0.6 to 2.3 hr [330,357] and peak serum concentration ranging from 240 to 1250 μg/L following a 600-mg dose [330] and from 1800 to 3300 μg/L following a 17.5 mg/kg daily dose [346]. Renal transplant patients receiving 17.5 mg/kg with breakfast had peak whole blood concentrations of 862 to 3431 μg/L, as measured by HPLC [354].

Interferon-α

Human interferon-α (IFN-α) is a family of more than 26 species with molecular weights ranging from 16,000 to 27,000 daltons [358–360]. The majority of human IFN-α species are nonglycosylated [361]. While these IFN-α species share more than 50% homology in their amino acid sequences, subtle sequence differences do exist. For instance, IFN-αA (IFN-α$_{2a}$) has a lysine at position 23 while IFN-α$_2$ (IFN-α$_{2b}$) has an arginine at this position. The molecular weights of the recombinant human IFN-α species range from 19,241 to 19,745 daltons. Most of these species have a signal peptide sequence of 23 amino acid residues and a mature amino acid

Dose

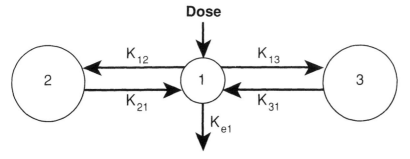

Fig. 3   Mammillary model describing the distribution of recombinant
human interferon-α_2 after IV administration in rabbits.   (From Ref.366.)

sequence of 166 amino acid residues.   The exceptions are human IFN-α_2
and human IFN-α_{11}, which have 165 and 172 residues, respectively.   Such
differences in the composition of natural IFN-α result in differences in
physicochemical, biological, and metabolic properties [362].   For instance,
rHUIFN-α_2 and leukocyte IFN circulate for a longer time than recombinant
IFN-α_1.   Also, analogs or hybrid IFNs have a shorter $t_{1/2}$ than natural
IFN-α [363].

    After i.v. administration in the rabbit, both human lymphoblastoid IFN
(natural human IFN-α) and recombinant IFN-α_2 disappear from the systemic circulation in a triexponential fashion.   The final slope in the concentration-time curve is rather steep and plasma levels become undetectable
fairly rapidly.   Their distribution has been described by the mammillary
model shown in Fig. 3.   In patients from whom a good number of blood
samples in the first few minutes after the administration of the protein
were not obtained, only biphasic plasma disappearance was observed.   In
Fig. 1, compartment 1 corresponds to the plasma pool, compartment 2 probably corresponds to intestinal, muscular, CNS, subcutaneous, and pulmonary interstitial spaces, and compartment 3 corresponds to the hepatic,
splenic and bone-marrow interstitial spaces.   The elimination rate constant
(k_{el}) is far greater than the other rate constants, indicating the IFN is
mostly lost through the kidneys.   The half-life of the metabolic phase was
4.0 and 4.6 hr for lymphoblastoid IFN and rIFN-α_2, respectively [364–366].

    rIFN-α_2 is the most studied of the recombinant IFN-α's.   The elimina-
tion $t_{1/2}$ was under 2 hr in cancer patients [367,371] after i.v. bolus injection
and about 6 hr after i.v. infusion in African green monkeys and in patients
with leukemia [368,369].   Probably during a slow and prolonged infusion, more
IFN was distributed into body fluids, acting as reservoirs.   Intramuscularly
administered rIFN-α_2 was virtually completely absorbed [368,370,371].   The
time of maximum serum concentrations (t_{max}) occurred at 4.6 ± 2.5 hr at a
3-MU dose and 6.1 ± 3.6 hr at a 72-MU dose [372].   The t_{max} occurred a
little later (7.3 ± 0.8 hr) following subcutaneous injection [368].   Depending on the dose, intramuscularly or subcutaneously administered rIFN-α_2

could still be detected after 24−36 hr [368,373]. A fairly good correlation between the maximum serum concentration and dose was observed [371−373].

Very little IFN-α was detectable in the CSF of cancer patients even after continuous i.v. administration of high doses [374−376]. The serum/CSF concentration ratio was found to be 30:1 after i.m. administration of natural leukocyte IFN-α in monkeys [377]. This problem can be overcome in one of three ways: (1) by modifying the blood-brain barrier, (2) by infusing IFN via carotid artery [378], or (3) by injecting IFN directly into the ventricular CSF and thus obtaining a high IFN level throughout the CSF pathway without many side effects [378].

Conventional routes of administration (i.v., i.m., and s.c.) tend to confine IFNs to the plasma pool and consequently may cause toxicity without a corresponding clinical effect [378]. Many cytokines can be considered paracrine hormones and thus act physiologically in microenvironment rather than as circulating proteins [380]. This is certainly true when these proteins are used as immunoenhancers. In this case, the administration of IFN via the lymphatic route has been advocated, thereby producing a lymph-to-plasma gradient similar to the one observed in physiological conditions [381]. This can be achieved by increasing the colloid osmotic pressure of interstitial fluid through the administration of IFN with high concentrations of human albumin in small-volume injections in multiple sites [382].

In unanesthetized, unrestrained rabbits, multisite subcutaneous administration or rHuIFN-α_2 in 10% human albumin (ALB) solution resulted in plasma C_{max} values significantly lower than those in the absence of albumin [383]. Both the mean residence time and the release time of IFN increased linearly with albumin concentration over the range of 4−10%. By contrast, in the presence of 75 U hyaluronidase, both residence and release times were signifcantly lower.

Unlike i.m. administration, intradermal administration of rHuIFN-α_2, in 13% albumin in multiple sites resulted in no changes in the concentration-time profile when compared with the control [384]. This is probably due to the fact that i.d. absorption of IFN already occurs via the lymphatics in the dermis. Moreover, intradermally administered IFN-α_2 appeared to be more extensively distributed in the body and more slowly eliminated. The V_d was sixfold larger and the half-time of the elimination phase was over six times longer. Intradermally administered rHuIFN-α_2 caused no visible inflammation even at 8−24 hr, as confirmed by histological examination of skin biopsies [385].

Interferon-β

Natural human interferon-β (IFN-β) is a glycoprotein with a molecular weight of about 20 kD which has been purified to homogeneity from human diploid fibroblasts [386]. It is comprised of a signal peptide of 21 amino acid residues and a mature protein of 166 amino acid residues [387]. Human IFN-β has three cysteines at positions 14, 17, and 31. Recombinant IFN-β, [Ser[17]] IFN-β, is a cloned, nonglycosylated variant that has been stabilized by the substitution of serine for cysteine in the 17 position [388]. [Ser[17]]IFN-β is similar to naturally produced IFN-β in specific activity as well as spectrum of antiviral, antiproliferative, and immunodulatory activities [388,389].

Following i.v. bolus (6 mU) administration in rabbits, natural human IFN-β disappeared rapidly from the systemic circulation [390]. Only 25% of the dose remained after 10 min. Passage of IFN-β from the plasma into the CSF and vice versa is extremely limited, while IFN-β levels persist in the CSF for some time after intralesional and intrathecal administration [391]. Bocci et al. [392] reported that intravenously administered IFN-β equilibrated fairly rapidly with the lymph, yielding a lymph AUC/plasma AUC ratio of 0.35 ± 0.15. Nonetheless, this value is lower than expected [380] on the basis of the leakiness of the discontinuous and fenestrated capillaries in the liver and gut, respectively [17], conditions that favor the rapid and almost free passage of IFN-β from plasma to lymph. Restricted transendothelial passage is probably due to the hydrophobic nature of β-interferon and its tendency to dimerize.

The lymph AUC/plasma AUC ratio rose to 3.0 ± 2.0 following intramuscular administration of IFN-β in rabbits, indicating that IFN-β was preferentially absorbed from the muscle via lymphatics rather than via blood capillaries [392]. Indeed, marked and sustained lymph levels were observed as late as 12–24 hr following injection. Unlike subcutaneous administration (see below), coadministration of intramuscularly administered IFN-β with 13% albumin afforded no further enhancement through the lymphatics, while actually delaying absorption and causing a fall in the lymph AUC/plasma AUC ratio to 1.3 ± 0.8. This observation remains unexplained.

A lymph AUC/plasma AUC ratio of 1.2 ± 1.3, which was intermediate to those obtained from i.v. and i.m. administration, was obtained following s.c. administration of IFN-β in saline in rabbits [390]. Addition of 13% albumin to the original dose of IFN-β enhanced absorption after s.c. administration. There was also a fourfold increase in systemic bioavailability (40 ± 18%), reduced clearance (5 ± 2 ml/min.kg), and enlarged volume of distribution (21 ± 2 L/kg). The concentration in lymph remained higher than controls during the 12–26-hr interval. This was attributed to local edema associated with the interstitial fluid expanding properties of albumin, which favors lymphatic uptake not only directly, but also indirectly by delaying the entry of IFN-β into the plasma pool. The net result was an increase in the lymph/plasma ratio to 2.5 ± 1.8.

Interferon-γ

Interferon-γ has a variety of antiproliferative, immunomodulatory, and antiviral effects, which suggest a role for it in cancer treatment. In vitro, IFN-γ is cytocidal to human tumor cell lines [394,395] and increases the natural killer (NK) cytotoxicity [395]. It is also a powerful monocyte/macrophage activating factor, as indicated by the induction of increased tumor cell cytostasis [397], Fc receptor-mediated phagocytosis [398], and increased cell membrane expression of class II major histocompatibility antigens [397,399,400]. Pharmacokinetic studies of IFN-γ have been conducted using recombinant human interferon-γ (rHuIFN-γ) [401–411] or recombinant murine interferon-γ (rMuIFN-γ) [412–414] in mice [412–414], rabbits [415,416], monkeys [409], and humans [401–408,410,411]. Unlike the naturally occurring IFN-γ, rIFN-γ is not glycosylated. Nonetheless, rIFN-γ exhibits the same high antiproliferative activity against HeLa cells as IFN-γ in supernatants from PHA-stimulated lymphocyte cultures [417].

Because IFN-γ activity is species specific, rMuIFN-γ has been an invaluable tool in efforts to characterize the central role of IFN-γ in the regulation of the immune system [418]. Specifically, the enhancement of peritoneal exudate-derived natural killer (NK) cell function in mice treated with rMuIFN-γ has been studied extensively [414,418]. Cellular internalization studies using cultured L929 mouse fibroblasts indicated that ^{125}I-rMuIFN-γ entered cells at a somewhat slower rate than natural ^{125}I-MuIFN-γ [419]. Nonetheless, the total amount of IFN taken into the nucleus of the cells within 2–3 min was almost the same. Unlike natural IFN-γ, rMuIFN-γ was found in substantial amounts in pinocytotic vesicles. These findings suggested that the carbohydrate found on natural MuIFN-γ may influence molecular conformation and hence cellular binding and processing characteristics.

Ferraiolo et al. [409] designed a three-period crossover study to determine the pharmacokinetic parameters of rHuIFN-γ after i.v., i.m., and s.c. administration using 12 healthy adult rhesus monkeys. A washout period of 2 weeks was allowed between doses. Four additional monkeys received 0.25 mg/kg rHuIFN-γ by i.v. route alone. Following the intravenous administration of 0.10 mg/kg of rHuIFN-γ, 5 of 11 monkeys displayed a biexponential disposition. The distribution and elimination half-lives were 12.3 min and 130 min, respectively [409]. The remaining six monkeys exhibited a monoexponential disposition with a mean $t_{1/2}$ of 13.4 min. The clearance and V_{ss} values following 0.10 mg/kg dose were 27.4 ml/min.kg and 510 ml/kg, respectively. Following an i.v. dose of 0.25 mg/kg in four monkeys, three monkeys exhibited a monoexponential disposition with a mean $t_{1/2\alpha}$ of 15.2 min. One monkey exhibited biphasic plasma disappearance $t_{1/2}$ and $t_{1/2\beta}$ of 10.6 and 30 min, respectively. The clearance and V_{ss} were 18.5 ml/min.kg and 312 ml/kg, respectively. There was no significant difference in clearance, half-life, or steady-state volume of distribution at the two different doses (0.10 and 0.25 mg/kg) administered intravenously, indicating linear kinetics.

Following the intramuscular and subcutaneous administration of 0.25 mg/kg of rHuIFN-γ in rhesus monkeys, peak serum concentrations of 50.7 and 52.3 ng/ml, respectively, were achieved at 80 min. The corresponding bioavailabilities were 109 and 90%. Bioavailability values above 100% may be attributed to an underestimated intravenous AUC. In these studies blood samples were assayed for the presence of antibodies at zero time and before the administration of each dose and 7 days after the final dose by a radioimmunoprecipitation method. The anti-rHuIFN-γ antibody titer was expressed as the $\log_{10}$ of the reciprocal of the dilution that was two times the nonspecific binding value. The antibodies were also characterized by a neutralization assay employing the A549 cell line as a target for the encephalomyocarditis virus. The neutralizing antibody titer was expressed as the $\log_{10}$ of the reciprocal of the dilution that reduced 10 U/ml of rHuIFN-γ activity to 1 U/ml. A titer of less than 1.5 was considered to be negative in this assay. Nearly all the monkeys (10/11) had detectable antibodies to rHuIFN-γ, 1 week after receiving the final dose. At that time, only three animals had developed neutralizing antibodies, and the titers ranged from 1.7 to 2.6. Neither type of antibodies affected the pharmacokinetic parameters in a consistent way or discernible pattern.

Pharmacokinetic studies of rHuIFN-γ in humans have shown serum $t_{1/2}$ values of 25–35 min after i.v. bolus injection [401,402,404]. Kurzrock et al. [402] studied the pharmacokinetics and biological activity of rHuIFN-γ

in patients with metastatic cancer. In these studies, rHuIFN-γ was given at doses of $0.01-2.5$ mg/m^2 by alternating i.m. and i.v. bolus injections with a minimum intervening period of 72 hr. After i.v. administration, rHuIFN-γ was cleared monoexponentially with a short half-life of $25-35$ min as determined by bioassay and enzyme immunoassay. Following i.m. administration, rHuIFN-γ was apparently well absorbed, with the percentage of dose absorbed ranging from 33 to 77%. Serum titers of rHuIFN-γ were detected by ELISA at dose levels greater than or equal to 0.1 mg/m^2. Administration of rHuIFN-γ intramuscularly resulted in prolonged serum $t_{1/2}$ of $227-462$ min.

Subcutaneous administration of rHuIFN-γ in cancer patients resulted in C_{max} values ranging from 1.9 ng/ml at the 0.5 MU/m^2 dose to 17.1 ng/ml at the 4.0 MU/m^2 dose within $6.7-13$ hr [405]. The C_{max} increased linearly with dose up to 4.0 MU/m^2, but this relationship disappeared at higher doses (6.0 and 8.0 MU/m^2). This was attributed to the slow release of rHuIFN-γ from the site of injection at the higher dose. The serum elimination half-life was $2-3$ hr and the absorption half-life was $9-20$ hr. No antibodies were detected against rHuIFN-γ.

Although the intravenous bolus injection of rMuIFN-γ in female and male C57BL/6 and CBA/2 mice also showed biphasic disappearance of antiviral activity in plasma, as was the case for rHuIFN-γ in the monkey, its $t_{1/2\beta}$ was much shorter, $19-32$ min [413]. After intramuscular and subcutaneous injection of rMuIFN-γ, antiviral activity in serum could be detected from 30 to 270 min. Plateau activity was observed from 65 to 135 min after s.c. and from 84 to 143 min after i.m. injection. In both routes, biphasic elimination kinetics were observed. The same was the case in the BALB/c mice [414].

Interleukin-2

Interleukin-2 (IL-2) is a secretory product of activated T lymphocytes that induces the proliferation of antigen-primed T cells [420,421]. IL-2 has been shown to induce the secretion of other lymphokines like gamma interferon [422] and B-cell growth factor [423] and augments the cytolytic activity of natural killer (NK) cells [424] and cytotoxic T cells [425]. It also induces the differentiation of inactive precursors of a population of cytotoxic lymphocytes [lymphokine-activated killer (LAK) cells] which are capable of lysing a wide range of malignant cells, including NK-resistant cell lines and autologous tumors cells [426-428].

Native human IL-2 is a glycosylated protein with a molecular weight of 15 kD [429]. Welte et al. [430] showed that purified natural human IL-2 could exist in three molecular forms with molecular weights up to 26 kD. The pI of these molecular forms ranges from 6.7 to 8.1. IL-2 is rather hydrophobic and contains an intramolecular disulfide bridge linking cysteines in positions 58 and 105. Recombinant IL-2 (rIL-2) has been purified from genetically engineered *E. coli* as an unglycosylated protein with biological activities equivalent to those of the native protein [431].

Pharmacokinetic studies of IL-2 in cancer patients [432] and mice [433] have been reported. Single doses of recombinant human IL-2 ranging from 0.1 to 30×10^6 units were administered to cancer patients as intravenous infusions [432]. The serum samples were analyzed using a sandwich capture antibody bioassay, which utilized a monoclonal antibody specific for rIL-2. These studies indicated rapid elimination of rIL-2 with half-lives

ranging from 0.24 to 3.3 hr. Following subcutaneous administration, higher elimination half-lives, ranging from 2.7 to 12.2 hr, were obtained.
These values may not represent the true elimination half-life, as slow absorption from injection site could result in flip-flop kinetics. When administered intravenously, the serum concentrations increased in an apparently
dose-proportional manner. Following the subcutaneous administration of
0.1, 0.3, 1.0, and 3.0 MU of IL-2, the peak serum concentrations were
5.6, 10.5, 25.1, and 40.8 U/ml, respectively. The corresponding t_{max}
values were 3.6, 3.5, 3.9, and 2.7 hr, respectively.

In C57BL/6 mice the pharmacokinetics and biodistribution of radioiodinated recombinant interleukin-2 (^{125}I-IL-2) were studied in comparison
to radioiodinated (^{131}I) bovine β-lactoglobulin after intravenous injection
[433]. β-Lactoglobulin is inactive, having no known enzymatic or hormonal functions, and has a molecular weight of 30 kD. After simultaneous
i.v. administration of 0.04−0.1 µg of these two proteins, mice were killed
by cervical dislocation at specified times, and blood and various organs
were weighed and the radioiodine content was determined by a gamma
scintillation counter. Both macromolecules showed highest blood levels
(23−34% injected dose/g) 1 min after injection. Both were readily cleared
from the blood since relatively low blood levels remained at 1 min. By
5 min, blood values of both macromolecules were less than half of that at
1 min and slowly declined over the next hour. Significant differences in
biodistribution of the two macromolecules occurred in liver and spleen.
Liver values of radiolabeled IL-2 at 1 min were very high (41% injected
dose/g) and remained high for 5−15 min. These values started to decrease after 15 min, reaching control protein values only at 60 min post
i.v. injection. By contrast, accumulation of radiolabeled β-lactoglobulin in
liver decreased from 8% injected dose/g at 1 min to 3% injected dose/g at
60 min. The splenic uptake of ^{125}I-IL-2 followed a similar pattern. Uptake was not as great as that seen for the liver and the time course was
delayed, with the peak accumulation (32% injected dose/g) attained at 15
min post i.v. injection. At 1 min after i.v. injection, the accumulation of
^{125}I-IL-2 in kidney was 24% injected dose/g and increased to 48% at 5 min.
By contrast, labeled β-globulin accumulated in kidney to the extent of 71%
at the end of 1 min. Following i.p. injection of either ^{125}I-IL-2 or
^{131}I-β-lactoglobulin, no significant differences were seen in their organ
distribution. Blood levels of both proteins were similar and relatively
constant during the 3 hr of study (7 and 4−6% injected dose/g at 5 min
and 3 hr, respectively). Peak blood levels were observed at 15 min and
were the same (10% injected dose/g) for both proteins. No organ preferentially accumulated IL-2 over β-lactoglobulin. Peak kidney values (about
10% injected dose/g) were observed for both proteins after 15 min.

The biodistribution of ^{125}-I-IL-2 in rat has also been evaluated by
Gennuso et al. [434] using gamma camera imaging, autoradiography, and
organ well counting [434]. Both camera imaging and organ well counting
demonstrated that liver and kidney were the principal accumulators of
radioactivity with peak liver uptake at 10 min and peak kidney uptake at
20 min from the onset of infusion. Autoradiographic assessment of selected
organs (kidney, adrenal glands, liver, lung, and brain) revealed marked
heterogeneity of uptake in the kidney and adrenal gland with preponder-
ance of radioactivity in the cortex of these organs. More homogeneous
uptake of the label was noted in liver, lung, and brain parenchyma.

Ohnishi et al. [435] administered ^{125}I-IL-2 intravenously in mice and measured the tissue distribution of radioactivity at 2, 5, 10, 30, and 60 min postinjection. Kidney was found to be the organ accumulating the most radioactivity throughout the first three measurements. Accumulation of radioactivity in urine reached a peak at 60 min. In comparison, in animals receiving the same dose of Na^{125}I, the radioactivity accumulated by kidney was less than 12% of that accumulated when ^{125}I-IL-2 was injected. This indicates that the distribution of ^{125}I-IL-2 to kidney was an apparent and significant accumulation of ^{125}I-IL-2 by the kidney. Following the simultaneous i.v. injection of ^{125}I-IL-2 (3×10^5 cpm, 4 ng protein) and excessive cold IL-2 (0.1 mg protein), the kidney no longer accumulated the majority of radioactivity. Rather, the spleen accumulated the major portion of the radioactivity, followed by the lung. Secretion of the radio-activity into the urine was also significantly increased in specimens collected at 30 and 60 min.

Both serum and urine samples collected from animals injected with ^{125}I-IL-2 alone were subjected to gel filtration on Sephadex G-50. The samples analyzed included the serum specimens collected at 2, 5, and 10 min and urine specimens collected at 30 and 60 min. Similar to that of initial ^{125}I-IL-2, only two radioactive peaks were detected in serum throughout the first 10 min when most of the circulating ^{125}I-IL-2 was cleared. The first peak represented ^{125}I-IL-2 while the second represented free ^{125}I-iodine and inorganic ^{125}I-iodide. However, analysis of urine samples revealed single chromatographic peak located at a position near or at the size of ^{125}I-iodide. Thus, ^{125}I-IL-2 was metabolized and rapidly cleared by the kidney.

The incubation of radiolabeled drug in vitro with either kidney homogenate or the cytosol fraction of the kidney was shown to hydrolyze ^{125}I-IL-2 maximally at pH 4.0 [435]. The reactivity of the kidney cytosol fraction with ^{125}I-IL-2 was inhibited to the extent of 93% by pepstatin, an acid protease inhibitor, but not by the serine protease inhibitors TLCK or TPCK, indicating the possible involvement of an acid protease in the breakdown of IL-2. A heat-treated kidney cytosol fraction plus ^{125}I-IL-2 was further incubated with either or both of the two major acid proteases present in the kidney, namely, cathepsin D and renin. Incubation with renin did not result in the degradation of IL-2. On the other hand, incubation with cathepsin D resulted in the degradation of ^{125}I-IL-2, the degree of degradation being proportional to the concentration of cathepsin D used. Maximal degradation by cathepsin D was found to occur at pH 4.8. These findings indicate that cathepsin D played a major role in the renal catabolism of IL-2.

Tumor Necrosis Factor-α

Tumor necrosis factor-α (TNF-α) is a cytokine produced by activated macrophages [436–438], natural killer cells [439,440], natural cytotoxic cells [441], and mast cells [442]. It performs a variety of biological functions [443,444], including induction of interleukin-1 production [445,446], simulation of class I major histocompatibility complex antigen expression [447], augmentation of polymorphonuclear neutrophil functions [448–450], inhibition of lipoprotein lipase activity [451], stimulation of fibroblast growth [452], and regulation of immune function [453]. TNF has cytostatic and/or cytotoxic activity against a variety of tumors in vitro [452,

454−458] and in vivo [455,458−461]. This cytokine has been cloned and expressed. The resultant recombinant human TNF (rHuTNF) has been shown to have in vitro and in vivo activities similar to those of natural numan TNF [436]. rHuTNF has shown some antitumor effects in cancer patients in phase I clinical trials [462,463].

TNF-α is 157 amino acid residues long, contains one disulfide bond, has a pI of 5.3, and has a mol. wt. of about 17 kD in its monomeric form. The molecule has a strong tendency to aggregate into dimers and trimers, as determined by gel permeation chromatography [464,465].

TNF does not disappear from the circulation as rapidly as other lymphokines such as interferons and interleukins. This is partly attributed to slower renal filtration [466] and to limited dilution in the interstitial fluids [467]. In mice, I-125 labeled natural mouse TNF was found to have a half-life of 6 min [466], whereas rHuTNF-α was found to have a $t_{1/2}$ of 18.5−19.2 min [468]. This difference in half-lives was probably due to the glycosylated nature of natural mouse TNF. The rank order of accumulation of the radiolabel (H-3 or I-125) in the major organs (% dose per organ over 1440 min) was: liver > kidney > lung > heart > spleen [468]. Pharmacokinetic studies in rabbits and monkeys indicated biphasic clearance of TNF-α [467]. In rabbits, $t_{1/2\alpha}$ and $t_{1/2\beta}$ values were 0.89 hr and 1.99 hr, respectively. In monkeys, these values were 1.29 hr and 4.52 hr, respectively. This species difference in half-life can be attributed to species difference in metabolic rates. The smaller the animal, the higher the metabolic rate [469]. Specifically, metabolic rates decrease in the order: mouse > rabbit > monkey.

The pharmacokinetics of TNF-α appear to be dose dependent. In rhesus monkeys, whereas first-order kinetics was observed during intravenous infusion of high doses (120 µg/kg for 30 min and 54−325 µg/kg for 6.5 hr) of rHuTNF-α yielding a plasma $t_{1/2}$ of 1.2−2.1 hr [470], non-first-order kinetics was observed with infusions of low doses (10−30 µg/kg for 30 min and 22 µg/kg for 6.5 hr). At such low doses, the elimination rate increased steadily, showing dependence on concentration and time. Such time-dependent increases in elimination rate at low dose of TNF-α (22 µg/kg) were abolished upon simultaneous long-term infusion of TNF-β (200 µg/kg), which is similar to TNF-α in several aspects. Based on these results, two different elimination mechanisms of TNF-α in the rhesus monkey have been postulated: a specific, saturable mechanism and a nonspecific, nonsaturable mechanism. The saturable mechanism, which predominates at low doses, involves binding of TNF-α to its receptor sites, followed by internalization and degradation. The nonsaturable mechanism, which predominates at high doses, involves primarily glomerular filtration.

Tumor Necrosis Factor-β

Tumor necrosis factor-β (TNF-β) or lymphotoxin has been purified from human OKT 4⁺ lymphocytes upon photohemagglutinin stimulation [471]. It is biochemically and immunologically similar [472] to factors isolated from a lymphoblastoid 1788 cell line [473−475]. TNF-β is a glycoprotein with an isoelectric point of 5.8 and a molecular weight of about 25 kD [474,475]. It has a strong tendency to polymerize and a molecular mass of about 70 kD has been measured by molecular sieve chromatography. Human TNF-β has now been produced in E. coli using recombinant DNA technology [476]. Although the recombinant product is nonglycosylated, it shows cytotoxic

activity in murine and human tumor cell lines in vitro and causes necrosis
of certain murine sarcomas in vivo. Like TNF-α, TNF-β can induce
necrosis in Meth-A-tumors (methylcholanthrene-induced sarcomas) [476–
478].

TNF-α and TNF-β are similar in three aspects. First, their amino
acid sequences share about 50% homology. Second, both TNF-α and
TNF-β bind to various cell types via a single class of high-affinity receptors, which can be up-regulated by interferons and lectins, leading to
synergistic and antagonistic responses, respectively [479]. Third,
TNF-α and TNF-β exhibit antiviral effects in infections caused by either
DNA or RNA viruses [479]. In spite of these similarities, there are important differences between TNF-α and TNF-β, in that TNF-α contains a
single disulfide bridge between cysteines 69 and 101, does not contain
methionine, has no carbohydrates, and has its own antigenic site [475].

Following intraperitoneal administration of radioiodinated TNF-β to
tumor-bearing or normal BALB/c mice, lytic activity reached its peak
between 1 and 2 hr and became undetectable by 5 hr [480]. The $t_{1/2}$ of
disappearance of TNF-β bioactivity was approximately 42 min. Tissue distribution measurements indicated localization of radioactivity in the kidneys
within minutes, reaching the peak value after 30 min. There was, however, no selective accumulation of TNF-β in any other organ, including
the Meth-A-tumor. No obvious differences in tissue distribution were observed in normal and tumor-bearing mice.

D. Metabolic Peptide and Protein Hormones

Calcitonin

Calcitonin is a hypocalcemic and hypophosphatemic polypeptide hormone
secreted by parafollicular cells (C-cells) of the thyroid gland [481,482].
Calcitonin causes hypocalcemia by inhibiting the release of calcium from
bone and by stimulating urinary calcium excretion [483]. Synthetic human
calcitonin and salmon calcitonin preparations are being used in the treatment of osteoporosis and Paget's disease. Of the two forms of calcitonin,
salmon calcitonin [sCT(1-32)] is 30 times more potent in hypocalcemic activity in rats and 10–100 times more potent in Paget's disease patients
[484,485].

The higher potency of sCT when compared with hCT is due in part to
its greater resistance to metabolic degradation and in part to its slower
total clearance. Differences in total clearance between sCT and hCT have
been evaluated in humans by Huwyler et al. [486]. Following i.v. infusion of 0.04 mg of either sCT or hCT over 240 min, the plasma concentrations of hCT, as indicated by radioimmunoassay (RIA), were lower than
those of sCT. This was due to the 2.4 times higher MCR for hCT than
for sCT. The higher MCR of hCT was consistent with its shorter half-life
(16.6–34.4 hr) determined in vitro when compared with sCT (43–72 hr).

The inactivation of sCT and pCT in specific organs in anesthetized
dogs has been studied by Singer et al. [487]. Natural or synthetic sCT
and natural porcine calcitonin (pCT) were infused for 60 min into the right
atria of mongrel dogs anesthetized with phenobarbital and which had been
catheterized in the left ventricle, abdominal aorta, pulmonary artery,
hepatic vein, renal vein, and femoral vein. Blood samples were collected
simultaneously from arterial and venous catheters and centrifuged to yield

Table 4   Arteriovenous Differences in Calcitonin Concentrations Across
Organs in the Dog

Calcitonin	Lung	Kidney	Liver	Muscle and/or bone (hindquarter)	Ref.
Salmon	−3%	−17%	−3%	−2%	487
Porcine	−1%	−26%	−22%	−32%	487
Human	N.A.	−30%	−8%	N.A.	488

plasma and stored at −20°C until analysis by RIA.   The arteriovenous differences found across the lung, kidney, liver, muscle, and/or bone are
summarized in Table 4.   pCT was inactivated by the liver, kidney, muscle,
and/or bone, whereas sCT was inactivated only in the kidney.   The data
in Table 4 indicate that all three calcitonins share a common fate in the
kidney.   The pulmonary circulation plays little part in reducing salmon or
porcine calcitonin levels while the hindquarters and liver show considerable
ability to degrade or sequester porcine calcitonin.   However, these clearance studies do not differentiate between degradation and sequestration.
Clark et al. [488] reported results for hCT similar to pCT results in dogs.
In this instance, the arteriovenous (a.v.) differences across the kidney
and the liver were 30% and 8%, respectively (Table 4).

     Forslund et al. [489] have studied the distribution of exogenous hCT
in rats using whole body autoradiography.   One minute following the i.v.
injection of ^{3}H-hCT (0.7 µCi/g) and ^{125}I-hCT (0.3 µCi/g), distinct
localization of radioactivity, somewhat higher than that of blood, was observed in the epiphysis of the long tubular bones, such as femur.   Five
minutes later, radioactivity was observed in the fasciae of skeletal muscles,
the walls of large blood vessels, and the renal cortex.

     The kidneys play a prominent role in the clearance of calcitonin.   This
is indicated by the fivefold increase in mean metabolic clearance of calcitonin in patients with renal failure than in normal subjects [490].   High
plasma concentration of endogenous immunoreactive calcitonin was observed
in patients with renal failure.   Clark et al. [488] showed that the disappearance of human calcitonin from the circulation of nephrectomized dogs
was slower ($t_{1/2\alpha}$, 5 min) than that of intact animals ($t_{1/2\alpha}$, 3 min).

     In order to delineate the pathways of renal clearance of calcitonin,
Simmons et al. [491] perfused the isolated rat kidney with either immunoreactive hCT (25−50 ng) or ^{125}I-sCT (2 µCi) 10 min after inulin was
added to the perfusion medium.   The disappearance of calcitonin in the
perfusion medium was monoexponential, far exceeding that of inulin, which
was also added to the perfusion medium.   The total renal immunoreactive
hCT clearance averaged 1.96 ± 0.18 ml/min, which was greater than
glomerular filtration rate (GFR) (0.68 ± 0.05 ml/min).   This finding indicates that glomerular filtration can account for at most 41% of the total
calcitonin cleared by the isolated perfused kidney and that filtration-
independent clearance accounts for the balance.   Since urinary clearance
of immunoreactive calcitonin (0.018 ± 0.004 ml/min) was only 2.6% of the
GFR, avid tubular removal of filtered calcitonin must have occurred.   The

renal clearance, filtration-independent clearance, and urinary clearance of 125_I-sCT averaged 1.7 ± 0.32, 0.70 ± 0.05, and 0.019 ± 0.003 ml/min, respectively.   Gel filtration chromatography of 125_I-sCT revealed that the principal degradation products were low-molecular-weight forms coeluting with monoiodotyrosine.   Intermediate-size products were not detected. Such a low contribution of urinary clearance to total MCR has also been reported by Huwyler et al. following i.v. infusion of hCT and sCT in humans [486].

Subcellular fractionation studies by Simmons et al. [491] indicate that at pH 7.5, 42%, 9%, and 29% of the total calcitonin degrading activity resided in the brush-border membranes, basolateral membranes, and cytosol, respectively.   At pH 5.0, calcitonin-degrading activity was limited to the lysosomal fractions.   The total activity of tubular homogenate at this pH was considerably less than that at pH 7.5.

The management of disorders by calcitonin is dependent on full conservation of its biological effects.   Loss of CT activity has been observed both in vitro [492] and in vivo during continuous administration to rats [493].   Furthermore, the injection of large doses of sCT in the long-term treatment of postmenopausal osteoporosis has led to cessation of response to the hormone [494].   These findings suggest that there is loss or down-regulation of CT receptor sites during treatment with CT.   A heterogeneous distribution of 125_I-sCT binding sites has been demonstrated by autoradiography in the kidney, with high densities in both the superficial layer of the cortex and the outer medulla [495].

Tashjian et al. [496] observed a reduction in the number of CT binding sites in cultured bone cells in the presence of excess CT, and they attributed the loss of available sites to occupancy of the receptor sites by a poorly dissociable hormone [496].   Bouizar et al. [495] investigated whether a similar loss of receptor sites would occur in the rat kidney following exposure to excessive amount of hormone.   The effects of sCT infusion via minipumps on plasma calcium concentrations were tested over a period of 6 days.   During the first day, plasma calcium levels significantly decreased at both infusion rates studied (32 mU/hr and 2000 mU/hr). This decrease was no longer observed after 2 days of treatment with 32 mU/hr infusion.   With 2000 mU/hr infusion, the hypocalcemic effect gradually decreased after day 1 and reached normal levels by day 6.   Infusions of 32 mU/hr of sCT also decreased inorganic phosphate (Pi) and enhanced Mg plasma levels during the first day.   This effect was no longer observed after 4−6 days.

Autoradiograms of rat kidney sections infused with 32 mU/hr of sCT for 7 days revealed a reduction in 125_I-sCT binding in both the outer medulla and the cortex.   Following infusion of sCT at a rate of 2000 mU/hr for 7 days, a greater decrease in the density of 125_I-sCT binding sites was observed.   The greatest decrease was mainly observed in the outer medulla.   This observation is consistent with the binding of the hormone at the level of thick ascending limb of the loop of Henle, where CT stimulates adenylate cyclase activity [497,498] and alters reabsorption of water and electrolytes such as calcium, Pi, and Mg [499−501].

To determine whether the loss of hypocalcemic effect after several days of infusion was due to loss of biological activity of the hormone, the same minipumps with 32 mU/hr releasing capacity used in the initial studies were transplanted to age-matched control rats.   The significant decrease in plasma calcium concentrations (−2.2 ± 0.2 mg/dl) observed 6 hr later and the significant hypophosphatemic (-2.8 ± 0.2 mg/dl) and Mg effects ($+0.29 \pm 0.05$ mg/dl) demonstrated that sCT was still biologically active in the minipumps at the end of the experiment.

Incubation of rat kidney sections with ^{125}I-sCT revealed only a single class of binding sites for ^{125}I-sCT [495]. After 7 days of infusion of sCT (32 mU/hr), the number of binding sites decreased by 15%. Infusion of 2000 mU/hr for 7 days caused an 80% decrease in the number of binding sites. There was no modification in the dissociation constant with either dose of sCT. The resistance to high doses of sCT may involve down-regulation of CT-sensitive kidney and bone receptor sites.

Gonadotropin-Releasing Hormone and Its Analogs

Gonadrotropin-releasing hormone (GnRH) is a decapeptide (pGlu – His – Trp – Ser – Tyr – Gly – Leu – Arg – Pro – Gly – NH_2) secreted by the hypo-thalamus in a pulsatile fashion and stimulates the release of both luteiniz-ing hormone and follicle-stimulating hormone from the anterior pituitary. Pituitary gonadotropes exposed to GnRH pulses outside the physiological range of $0.5-1$ pulses/hr fail to sustain gonadotropin output [502]. Ana-logs of GnRH (either agonists or antagonists) are intended to suppress gonadal function via pituitary desensitization as a result of sustained pituitary overexposure to GnRH effects by continuous or quasicontinuous administration [503,505]

The kidneys play an important role in clearing GnRH from the systemic circulation [135,505]. This is indicated by rapid, extensive renal uptake of GnRH after tracer administration [44,506,507] and by the recovery of substantial portions of tracer [44,508] as well as immunoreactive [509,510] and bioactive [511] GnRH in urine. Moreover, it is chronic renal, not hepatic, failure which greatly prolongs the apparent plasma half-time and reduces the metabolic clearance rate of GnRH [129]. Perfusion studies utilizing isolated renal tubules by Flouret et al. [134] and Stetler-Stevenson et al. [135] revealed that intact GnRH underwent glomerular reabsorption.

Hepatic metabolism of exogenous GnRH and analogs seems minor since hepatic uptake [44,506] and biliary excretion [134] are low. Metabolites observed after administration of exogenous GnRH [130,505,509] and analogs include the products of pyroglutamate aminopeptidase (specificity >Glu^1 – His^2) and neutral endopeptidase (specificity Tyr^5-Gly^6) [512]. The other sites of cleavage of GnRH by endogenous peptidases are the Gly^6-Leu^7 and Pro^9-Gly^{10} bonds [134,513]. Many GnRH agonists contain a D-amino acid at position 6 to retard degradation by proteases, most of which are specific for L-amino acids. The accumulation of GnRH analogs is partly responsible for desensitization of pituitary receptors. Also, the increased hydrophobicity of GnRH agonists and antagonists has been suggested to cause prolonged duration of action due to retention of these analogs in the body [514,515] through protein binding or accumulation in adipose tissues [516].

Plasma disappearance of GnRH following i.v. bolus administration is best described by a two-compartment open model with clearance from the central compartment [133,510,511,517–521]. The first phase is rapid ($t_{1/2}$ $2-8$ min), representing rapid mixing throughout the vascular and extracellular fluid spaces. During this phase, negligible uptake of GnRH by the liver, lung, muscle, and brain occurs [134,135]. The second phase is slower ($t_{1/2}$ $15-60$ min), probably due to saturable clearance mechanisms or return of GnRH from slowly equilibrating remote tissue pools. Plasma protein binding of GnRH is, however, negligible [516,522]. During this phase, accumulation of GnRH in the kidney, liver, skin, intestine, and pituitary occurs [44,506]. Noncompartmental analysis indicated steady-state volume of distribution of 12−20 L, metabolic clearance rate of 750−1500 ml/min, and mean transit time of 20−30 min, depending on the physiological states of the subjects [523,524]. Following i.v. continuous infusion to steady state, the metabolic clearance rate was slower [520] and the postinfusion plasma half-times (5−8 min) were longer [129, 525,528]. The metabolic clearance rates of 6- and/or 10-position substituted analogs are similar to native GnRH in animals as determined by tracer [529] or RIA [512,518,530,532]. In humans, the plasma clearance rates of [D-Trp^6]GnRH and [D-Trp^6,Pro^9]GnRH NEt obtained after steady-state infusion are slower than those observed for native GnRH [525,526].

Plasma levels tend to remain elevated for longer periods with delayed return to baseline following subcutaneous administration when compared to i.v. bolus [133,519,520,533]. Similar delay in achieving steady state by continuous s.c. when compared to i.v. infusion indicates a retarding, depot-like effect of the s.c. tissues on GnRH absorption. Consequently, plasma GnRH pharmacokinetics via the s.c. route is best described by an absorption-rate-limiting, flip-flop model [534]. Prolonged absorption of subcutaneously administered hormones has also been demonstrated for other peptides, including insulin, glucagon, growth hormone, and corticotropin-releasing factor. The role of s.c. degradation in limiting GnRH bioavailability appears minimal since bioavailability is high (75−90%) after both s.c. bolus and infusion. The bioavailability of subcutaneously (1-mg-dose) administered leuprolide (a potent analog of GnRH) was nearly complete, about 94% [535]. The bioavailability of subcutaneously administered detirelix, a synthetic analog (GnRH), was only 50 ± 20%, however [536].

In addition to differences in concentration-time profiles, differences in pharmacokinetic parameters, such as apparent volume of distribution of the central compartment and mean residence time between subcutaneous and intravenous administration, have also been reported for leuprolide, a potent GnRH analog [535]. The mean residence time averaged 3.7 hr following i.v. dosing in male human volunteers and averaged 4.3 hr following subcutaneous administration, indicating an average mean residence time at the s.c. injection site of 1.2 hr. The mean volume of distribution at steady state from the intravenous subcutaneous doses was 26.5 and 37.1 L, respectively. The AUC(0-infinity) values following i.v. and s.c. administration were similar (125.8 ± 32.9 ng.hr/ml vs. 118.6 ± 34.7 ng.hr/ml). In addition, the plasma drug clearance was similar for either route with a clearance of 139 ± 30 ml/min for i.v. route and 151 ± 43 ml/min for s.c. route.

The development of antibodies to GnRH or its analogs has been reported following chronic s.c. [537] but not i.v. [538] administration. Antibodies occurred in 3% of s.c.-treated patients with either congenital or acquired GnRH deficiency [537]. Depending on the circulating levels and affinity of the antisera, these antibodies can neutralize the hormone or can bind the hormone, thus creating a large pool of loosely bound ligands [539,540]. In either case, the pharmacokinetic consequence would be reduced clearance and altered volumes of distribution.

Buserelin ([D-Ser(t Bu)^6,Pro^9-NHET]GnRH), a highly potent agonist of GnRH, has high affinity for GnRH receptors and is resistant to

enzymatic degradation [541]. Pharmacokinetics of i.v. and intranasally
administered buserelin in patients with endometriosis have been reported
by Kiesel et al. [542]. Following i.v. bolus administration, multiexponential disappearance was observed. The major metabolite was buserelin
(5-9) pentapeptide, which accounted for 32% of the total dose recovered in
the urine between 6 and 24 hr.

Detirelix, a synthetic decapeptide containing five D-amino acids,
$[N-Ac-D-Nal(2)^1, D-PCl-Phe^2, D-Trp^3, D-hArg(Et_2)^6, D-Ala^{10}]$-
GnRH, is a very potent GnRH antagonist [536]. The plasma elimination
half-lives of detirelix were 1.6 hr and 7 hr following i.v. administration of
300 µg/kg in rat and 80 µg/kg in monkey, respectively. The plasma $t_{1/2}$
of s.c. detirelix was dose dependent, being longer with larger doses.
Long $t_{1/2}$ values of 18.7 hr and 31.6 hr were observed after single 0.2
and 1.0 mg/kg s.c. doses in the monkey. Several factors may contribute
to the long plasma $t_{1/2}$ of detirelix [543]. Detirelix has three hydrophobic
amino acid substituents like $Nal(2)^1$, $pCl-Phe^2$, and Ala^{10}. It was suggested that the increased hydrophobicity may enhance the binding of
detirelix to membrane receptors on the pituitary gonadotrophs. Binding to
the receptor is further reinforced by interaction between the positively
charged diethyl homoarginine residue at position 6 of detirelix with the
sialic acids on the receptors [544]. The increased affinity of detirelix
with membrane receptors may prolong the plasma $t_{1/2}$ and its biological
activity. Another important factor contributing to the long $t_{1/2}$ of
detirelix is reduced susceptibility to enzymatic degradation due to the
presence of hydrophobic residues. The major metabolite of detirelix,
isolated from rat bile, was formed by the cleavage of Ser^6-Tyr^5 bond.
This is reasonable since the Ser^4-Tyr^5 bond is one of the only two peptide bonds in detirelix not containing a D-amino acid. The 1-8-octa-
peptide fragment was not detected in these studies, however. This is
probably due to further cleavage of 1-8-octapeptide fragment at the
Ser^4-Tyr^5 bond.

Human Chorionic Gonadotropin and Derivatives

Human chorionic gonadotropin (hCG) is secreted by the placenta into the
circulation and excreted in urine throughout pregnancy. Serum and
urinary titers reach a peak in the first trimester and diminish thereafter
[545,546]. hCG is also found in the urine of patients with certain malig-
nancies [547]. Clinically, hCG is used in selected programs for induction
of ovulation and in in vitro fertilization protocols as a surrogate for a
luteinizing hormone surge, which often does not occur in gonadotropin-
induced ovulatory cycles.

In addition to hCG, the urine of pregnant women contains free β-subunit known as β-core fragment [548−550], accounting for up to 70% of
the total β-immunoreactivity in the urine [551]. Following the administration of hCG-β, β-core fragment was found in the urine of nonpregnant
individuals with various forms of trophoblastic disease and malignancy
[552,553]. While some nontrophoblastic malignant disease patients had
positive RIA hCG-β levels in both urine and serum, many of them were
positive for RIA hCG-β only in the urine. Thus the β-core is of interest
as a normal metabolite of pregnancy urine and as a potential marker for
malignancy.

The β-core fragment has been purified from pregnancy urine by Blithe et al. [554] using Sephadex G-100, concanavalin A (ConA)-Sepharose, DEAE-Sephacel, and Sephadex G-75 chromatography. The purified material, called pregnancy-related β-core or P-core, had an apparent molecular weight of 17 kD, in order to compare the carbohydrate moieties of β-core with those of native hCG-β, a trypsin fragment of hCGβ (T-core) was prepared. This T-core retained the β-core conformational immunological determinant recognized by the SB6 antiserum. It, however, had lost the carboxy-terminal immunological determinant recognized by the R529 antiserum. This T-core has been suggested to contain the N-linked oligosaccharides that are present in the amino-terminal portion of the molecule, but not the O-linked oligosaccharides of the β-subunit that are associated with carboxy-terminal peptide portion of hCG. About 84% and 86% of the P-core and T-core molecules, respectively, were bound to ConA, but failed to bind DEAE under conditions in which intact hCG-β would bind. The lectin-binding data indicated that the antennae on P-core did not contain appreciable sialic acid or galactose, in contrast to the antennae on T-core, which contain both.

Lefort et al. [555] demonstrated that infusions of hCG-β, desialylated hCG, or intact hCG in rats resulted in the accumulation of a large quantity of small-molecular-weight peptides in the kidney. These low-molecular-weight peptides also lacked the immunological determinants of the carboxy-terminal peptide of the β-subunit. In this light, it was speculated that the P-core in pregnancy urine may be a product of metabolism of hCG or hCG-β in the kidney. β-Core fragments have been documented in the urine of normal subjects infused with hCG [556] or hCG-β [557], clearly indicating that degradative pathways exist for the production of urinary β-core fragments in humans. Since there are no measurable β-core fragments in the serum of pregnant women [548,549], the source of these molecules in pregnancy urine must be derived from metabolism of hCG-β in the renal paraenchyma.

Wehmann and Nisula [556] have determined the MCR and renal clearance rate (RCR) of pure hCG in healthy young men and women following continuous intravenous infusion. The MCR was the same at both the low infusion rate (1.83 ml/min·m^2 at 0.2 μg/min) and the higher infusion rate (1.95 ml/min·m^2 at 0.8 μg/min). The mean MCR of all subjects was 1.88 ml/min·m^2. The MCR was not significantly different between men (2.04 ml/min·m^2) and women (1.76 ml/min·m^2). The RCR also did not vary between low and high infusion rates (0.40 ml/min·m^2). There was no significant difference in RCR between men (0.42 ml/min·m^2) and women (0.39 ml/min·m^2).

A mean $t_{1/2}$ of 2.32 days has been reported by Damewood et al. [558] for hCG following i.m. administration in 34 patients undergoing ovarian stimulation in an in vitro fertilization program [558]. Serum hCG was measured by an immunoenzymetric assay in which serum was reacted with plastic beads coated with a monoclonal antibody directed toward a specific antigenic site on the hCG molecule. A second enzyme-labeled monoclonal antibody directed to a different antigenic site was then added to the beads. After the production of beads/hCG/enzyme-labeled antibody sandwich, the beads were washed with buffer and incubated with p-nitrophenyl phosphate as the enzyme substrate. The amount of substrate turnover was determined colorimetrically by measuring the rate of change in absorbance at 405 nm.

To evaluate changes in carbohydrate and/or polypeptide structure on the plasma pharmacokinetics of glycoprotein hormones, Liu et al. [559] compared the initial volume of distribution, rate constants, and metabolic clearance rates of several highly purified human chorionic gonadotropin analogs in monkeys. The analogs studied were hCG, deglycosylated hCG (dg-hCG), desialylated hCG (ds-hCG), and β-core fragment of hCG.

β-Core fragment was purified from pregnancy urine as described above. It had no terminal sialic acid or galactose residues but retained its ConA-binding site (trimannosyl core) and fucose core. Its molecular weight was 10 kD, and its biological activity was nil [550]. dg-hCG was prepared from purified hCG by treatment with anhydrous hydrogen fluoride, which removed 70% of the carbohydrate residues. It was then subjected to Sephadex G-100 chromatography and further purified by chromatography on ConA-Sepharose 4B, which removed hCG contaminants as well as incompletely deglycosylated hCG. dg-hCG had a molecular weight of 27 kD and retained 89% of the receptor binding affinity of hCG [560]. Its maximal agonist activity was indistinguishable from that of hCG in terms of testosterone production by rat Leydig cells in vitro. The potency of dg-hCG was, however, reduced to 1/39th of that of hCG [561].

Desialylated hCG was prepared by incubating hCG in sodium acetate solution [560] with 5 units of neuraminidase for 1 hr at 37°C. Following overnight dialysis against sodium acetate (pH 5.0), the incubation was repeated with an additional 5 units of neuraminidase. The final solution was dialyzed against 1% acetic acid and lyophilized. After removal of neuraminidase with an agarose-insolubilized N-(p-aminophenyl)oxamic acid reagent, ds-hCG was further purified by chromatography on Sephadex G-100 and DEAE-Sephadex A-50. Warren's thiobarbituric acid assay [562] showed greater than 98% desialylation of hCG. The ds-hCG thus obtained had a molecular weight of about 32 kD and retained its biological activity in vitro [563] and in vivo [561,564].

Following intravenous bolus administration of pure hCG (93 µg), dg-hCG (63 µg), ds-hCG (118 µg), and β-core (59−103 µg) to cynomolgus monkeys [559], hCG, dg-hCG, and β-core were found to disappear from plasma biexponentially, whereas ds-hCG displayed monoexponential disappearance. The metabolic clearance rates of dg-hCG, β-core, and ds-hCG were 15, 47, and 152 times, respectively, that of hCG. Such increases can be explained entirely by increases in the β-rate constant, since the initial volumes of distribution remained unchanged.

Thus, despite profound variations in molecular weight (10−37 kD) and carbohydrate composition, all the hCG analogs tested had initial volumes of distributions indistinguishable from that of hCG [559]. These findings are consistent with the observed similarity in the initial volume of distribution of hCG, ds-hCG, hCG-α-subunit, and hCG-β subunit (1.665, 1.786, 1.729, and 1.958 L/m^2, respectively) in humans [556,560,565]. These findings support the notion that the volume of initial distribution of glycoprotein hormones, which is principally confined to the plasma compartment, is quite independent of the carbohydrate composition. This is so despite the wide differences in the disposal sites of hCG derivatives after carbohydrate or polypeptide modifications. Although renal clearance accounts for 20% of the total disposal of intact hCG [556], it contributes to 1% of the disposal of hCG-α, hCG-β [556], and ds-hCG in humans [560]. In rats, deglycosylation increases the renal accumulation of hCG five- to six-fold [567], whereas desialylation increases the hepatic accumulation of hCG 14-fold [568].

Interestingly, the MCRs of the hCG derivatives showed no distinct relationship to the molecular weight or sialic acid content of the hormone. For example, β-core, with the lowest molecular weight (10 kD), had an MCR value intermediate between those of the two other larger hCG analogs, dg-hCG and ds-hCG. The three molecular species ds-hCG, dg-hCG, and β-core, which were equally devoid of terminal sialic acid, had MCRs differing by as much as 10-fold [559].

Insulin

Insulin is secreted by the β-cells of the islets of Langerhans in the pancreas in response to a number of stimuli, including elevated blood glucose concentration [569]. Insulin has a well-defined three-dimensional structure and is composed of two polypeptide chains, A-chain with 21 amino acid residues and B-chain with 30 amino acid residues, which are joined by disulfide linkages. Human insulin has been prepared by several means, including extraction from human cadaver pancreas, complete chemical synthesis, enzymatic conversion of porcine insulin (semisynthetic insulin), and use of recombinant DNA technology (biosynthetic insulin) [570]. Porcine and bovine insulins are closely related structurally to human insulin, and these preparations have been used in many experimental studies. Porcine insulin differs from human insulin in having alanine instead of threonine at position 30 of the insulin B-chain. Bovine insulin has alanine at positions A-8 and B-30 instead of threonine at the same positions in human insulin.

In most tissues, insulin is involved in the metabolism of sugars, lipids, amino acids, and ions. Liver, muscle, and fat tissues are the major physiological targets of insulin. The action of insulin is initiated by interaction with its receptor on the plasma membrane [571]. Following one or more unknown transmembrane signaling mechanisms, regulation of three major types of metabolic actions occurs: (1) enhancement of membrane events such as glucose transport, (2) enzyme activation in the cytoplasm and in organelles, and (3) regulation of DNA and RNA synthesis in the nucleus [572,573]. Gavin et al. [574] showed that the interaction of insulin with its receptor led to loss of insulin binding to target cells (down-regulation of receptors). Insulin may down-regulate its receptors by two possible mechanisms: (1) by accelerating receptor degradation following internalization, and (2) by inhibiting the biosynthesis of its own receptor [575].

The in vitro potency of semisynthetic or biosynthetic insulin is not significantly different from pancreatic human or purified porcine insulin, as indicated by a variety of assays [576–581]. Such assays include receptor binding, extent of glucose oxidation, and stimulation of lipogenesis in adipocytes. Similarly, the hypoglycemic effect of i.v. human insulin in healthy subjects or in diabetic patients [582–588] is not significantly different from that produced by equal doses of purified porcine insulin [584–588]. A more prompt and pronounced hypoglycemic effect observed with soluble and Isophane formulation of biosynthetic human insulin [582], compared with equivalent formulations of purified porcine insulin, can be attributed to the different absorption rates of the two forms of insulin.

While small differences in secretion of counterregulatory hormones such as epinephrine, cortisol, glucagon, and growth hormone following the administration of biosynthetic or semisynthetic human insulin and purified porcine insulin have been reported, the clinical significance of such differences is still unknown [570].

Human insulins as well as porcine and bovine insulins are immunogenic. Nonetheless, the production of IgG antibodies occurs less frequently with human insulin than with purified porcine insulin and, in turn, with purified bovine insulin [570]. Both Fineberg et al. [589] and Heding et al. [590] reported a lower incidence of antibody production with biosynthetic human insulin (44%) than with purified porcine insulin (59−60%) [589].

The absorption kinetics of subcutaneously injected insulin has been extensively evaluated [591−596], either by monitoring the disappearance rate of H-3 [597] or I-125 [592] at the site of injection or by monitoring the time course of insulin concentration in the circulation after suitably correcting for (1) endogenous insulin in healthy subjects and (2) the presence of circulating antibodies in diabetic subjects [598].

Suppression of endogenous insulin production can be verified by radioimmunoassay of C-peptide, a biologically inactive fragment of proinsulin. Two methods have been used to suppress endogenous insulin in healthy subjects. One such method involves the infusion of somatostatin [599−601]. Somatostatin i.v. infusion rates ranging from 100 μg/hr to 500 μg/hr were used throughout these studies. The other method involves the maintenance of blood glucose at a constantly suppressed level of 3−3.5 mmol/L using a hypoglycemic clamp [602,603], an adaptation of the euglycemic clamp technique [604]. In the latter technique, i.v. infusion of insulin at 1−1.4 U/hr and glucose (variable rate) is used to clamp blood glucose level between the desired limits. This infusion commences 60 min before the s.c. administration and continues throughout the study.

The disappearance of ^{125}I-NPH insulin from the subcutaneous tissue in diabetic patients was monoexponential [594]. In spite of the transient (60 min) stimulation of blood flow caused by insulin, there was a delay period of 1.5 ± 0.8 hr from injection to start of radioactivity disappearance [605]. Deviation from monoexponential kinetics has been observed, however. This effect is probably due to the release of histamine, serotonin, and other vasoactive compounds in the subcutaneous tissues, which presumably would reduce the capillary membrane area in the terminal vascular bed [606], thereby reducing the rate of absorption [593,596,606]. Possible stimuli triggering the release of vasoactive substances include fluids that are not physiological with respect to ion composition and pH, trauma of the needle, and injection pressure [606].

In diabetic patients with palpable lipohypertrophy at the injection site, insulin absorption was significantly prolonged ($t_{1/2}$, 11.2 ± 3.1 hr) when compared to that observed in patients without lipohypertrophy ($t_{1/2}$, 6.8 ± 1.8 hr). This prolonged absorption might be due to considerable tissue changes due to lipohypertrophy [594]. Patients with low serum bicarbonate levels (10−18 mmol/L) showed faster absorption ($t_{1/2}$, 3.9 ± 2.3 hr) compared to that ($t_{1/2}$, 7.0 ± 1.8 hr) observed in patients with high serum bicarbonate levels (24−31 mmol/L). Lean diabetics had a faster absorption ($t_{1/2}$, 6.2 ± 1.9 hr) than normal-weight diabetics ($t_{1/2}$, 7.5 ± 2.0 hr). The overall interindividual coefficient of variation for insulin absorption in nonketotic patients was 27.4%. Considerable intrapatient day-to-day variation was also found (24.5%).

Considerable intra- and interpatient variations on the order of 25% exist in the absorption of subcutaneously administered insulin [594,607]. Variable absorption has been observed in subjects with lipodystrophy [596], ketoacidosis [608], or low body weight [596]. Pramming et al. [609] studied the subcutaneous absorption of insulin in seven insulin-dependent diabetic patients 16−50 years of age with a known history of diabetes of 4−24 years. Each patient received subcutaneous injections of 0.1 IU/kg body weight of 40 IU/ml radiolabeled purified porcine insulin. Disappearance of radioactivity from the injection site was monitored over an 8-hr period by external counting. One patient was found to have a considerably faster absorption than the rest. Another patient was found to have a delayed absorption.

Factors that may contribute to variations in absorption but which are difficult to control include variable blood flow at the injection site [610], ambient temperature effects [611], local heating or massage [591], and exercise of the injected limb [612]. Factors that can readily be controlled include concentration, regional area, and depth of injection. Absorption from the abdominal area is favored by shallow injections of dilute solutions [592] as well as by jet injections [613], which tend to disperse insulin over a wider area.

A better understanding of the absorption process of insulin can significantly improve metabolic control in insulin-requiring diabetic patients. In spite of considerable research, the detailed mechanisms of the absorption process are still unknown. As an illustration, the following variations in the absorption of soluble insulin are as yet unexplained:

1. For a typical bolus injection of 0.2 ml of 40 IU/ml of soluble insulin, the absorption rate does not follow a simple monoexponential time course [593,609,612]. Instead, absorption is relatively slow in the beginning, reaching the maximum rate only after 3−4 hr.
2. The absorption rate is inversely related to insulin concentration in the injected solution [593,595,614,615]. Thus, the same volume of 0.1 ml is absorbed considerably faster if injected as 4 IU/ml than as 40 IU/ml.
3. At very low concentrations and injection volumes, a tail is observed in the disappearance curve. This amounts to about 40% of the injected insulin being absorbed slowly [593].

A mathematical model has been developed by Mosekilde et al. [616] to describe the kinetics of subcutaneously administered insulin. The model assumes the existence of three forms of insulin at the subcutaneous depot: (1) a low-molecular-weight form (dimeric insulin), (2) a higher-molecular-weight form (hexameric insulin), and (3) an immobile form in which the molecules are bound in the tissue or precipitated as polymeric insulin. Only monomeric and dimeric insulin can be absorbed according to a calculation by Vinder [583] based on restricted capillary pore area and diffusion radii for insulin [617]. In addition, the model assumes that:

1. The depot is cleared by the absorption of dimeric insulin molecules into the vascular bed, and the depot is widened through diffusion of insulin molecules in the tissues.

2. Local degradation of insulin in the subcutaneous tissue is relatively insignificant [593,618].

3. Small amounts of insulin can either be reversibly bound in the tissue or become precipitated as insoluble polymerized insulin. In both cases, absorption will be inhibited.

The simulated absorption curves based on this model agree very well with the experimental data on the influence of changes in injection volume and insulin concentration on subcutaneous insulin absorption. Moreover, using this model, the values of the various absorption parameters can be calculated, including the effective diffusion constant D for insulin in sub-cutis, the absorption rate constant B for the low-molecular-weight form of insulin, the equilibrium constant Q between high- and low-molecular-weight insulin, the binding capacity C for insulin in the tissue, and the average lifetime T for insulin in its bound state. The typical values for a bolus injection in the thigh of fasting type I diabetics are $D = 0.9 \times 10^{-4}$ cm^2/min, $B = 1.3 \times 10^{-2}$ IU/min, $Q = 0.13$ (ml/IU)^2, and $C = 0.05$ IU/cm^2 with atypical average lifetime in the bound state of 800 min.

Growth Hormone

Growth hormone (GH) is present in large amounts in the anterior pituitary and has a wide range of functions, including stimulation of protein synthesis and promotion of skeletal and general body growth [619]. Two forms of human GH (hGH) have been produced by recombinant DNA technology. One form has an amino acid sequence identical to that of the naturally occurring hormone (rhGH), while the other form has an additional N-terminal methionine (Met–hGH). Both forms have been found to be equipotent based on weight gain and changes in femur length and width of the proliferative zone in the tibial epiphysis following multiple injections over 8 days in hypophysectomized rats [620].

The half-time of disappearance of endogenous growth hormone from serum has been studied using physiological effectors to stimulate and then suppress GH release [621]. Stimulation of GH secretion can be achieved by a single i.v. injection of GHRH (1 µg/kg) followed by an i.v. bolus dose (250 µg) after 45 min. Suppression of GH secretion can be achieved by a 2.5-hr infusion (500 µg/hr) of somatostatin. Under these conditions, the $t_{1/2\alpha}$ and $t_{1/2\beta}$ values of endogenous GH were found to be 3.5 and 20.7 min, respectively. The $t_{1/2}$ of endogenous GH in acromegalic patients after resection of a GH-secreting adenoma or during SRIF infusion ranged from 21 to 50 min [622–624]. Values of 15–45 min have been reported for the $t_{1/2}$ of exogenously administered radiolabeled GH in normal subjects [116,622,625,626].

The pharmacokinetics of rhGH and Met–hGH was compared following i.v. administration in the cynomolgus monkey [620]. Serum GH concentrations were determined in duplicate using an immunoradiometric assay and checked for potential interference by anti-hGH antibodies using a radioimmunoprecipitation assay. No antibodies to hGH were detected. Both forms of recombinant hGH were kinetically indistinguishable. This was indicated by statistically similar volumes of distribution (180 ± 40 ml for Met–hGH and 200 ± 60 ml for rhGH), AUC values normalized to standard dose (125 µg/kg) (38,000 ng.min/ml for Met–hGH and 34,000 ng.min/ml for rhGH), and clearance (13 ml/min for Met–hGH and 15 ml/

min for rhGH). The same conclusion was reached following subcutaneous administration. In this case, no significant differences were observed in C_{max}, t_{max}, and AUC. The bioavailability was 59% for Met–hGH and 72% for rhGH. The similar pharmacokinetic behavior of the two forms of recombinant hGH is consistent with their similar biological potencies mentioned earlier.

The MCR of Met–hGH is dependent on sex and age. Rosenbaum and Gertner [627] reported that the mean MCR was significantly greater in adult men aged 24.8 ± 0.6 years (125.2 ml/min·m^2) than in adult women aged 26.2 ± 0.9 years (89.9 ml/min·m^2) or in prepubertal children aged 11.0 ± 0.9 years (66.8 ml/min·m^2). The smaller difference in MCR between adult women and prepubertal children in this study suggests that circulating androgens in adult men may in some way enhance the MCR of GH. Indeed, androgens are known to inhibit the synthesis of binding proteins, including sex-hormone-binding globulin [628]. Bauman et al. [629] reported that a minimum of 15–18% of circulating GH is bound under normal physiological conditions. Consequently, if androgens exert their inhibitory effect on the GH-binding protein, an increase in GH clearance would result.

The liver plays an important role in the clearance of hGH. Picard et al. [630] demonstrated that in male rats 24% of an i.v. dose of I-125 hGH was captured by the liver in 15 min. The hormone was internalized by hepatocytes following interaction with specific somatogenic binding sites [631], which are also present in human IM-9 lymphocytes [632], and rat adipocytes [633,634].

Subcellular fractionation studies on sucrose density gradients by Husman et al. [635] revealed that after initial plasma membrane association of ^{125}I-bGH, the ligand was transported in two successive endocytotic compartments before reaching the lysosomes. The molecular weights of the somatogenic binders in the rat liver involved in the internalization of ^{125}I-bGH were determined to be 95, 64, 55, 43, and 35 kD, assuming a 1:1 binding of the hormone to the binder. The presence of several binders may be a reflection of receptor heterogeneity in ligand internalization or the proteolytic cleavage of receptor and/or ligand during internalization. These binders were also seen in both endosomes and lysosomes, suggesting that growth hormone was transported to lysosomes in a complex with its receptor.

In order to determine the integrity of ^{125}I-bGH in the membranes participating in growth hormone endocytosis, ^{125}I-bGH was incubated with plasma membranes, different endosome-enriched fractions (E-I, early endosome fraction; and E-II, late endosome fraction), and lysosomes at pH 7.4 and 24°C for 16 hr [635]. After the incubation mixture was analyzed by SDS-PAGE under reducing conditions, a fragment with molecular weight 15 kD as well as intact ^{125}I-bGH with molecular weight 23 kD were seen in all fractions analyzed. More 15-kD fragments were generated in the plasma membranes, early endosomes (E-I), and lysosomes than in the late endosome fraction (E-II). Addition of excess unlabeled bGH led to decreased formation of the 15-kD fraction, suggesting that degradation of ^{125}I-bGH was receptor-dependent. When incubation was repeated at pH 5.8, further degradation of the 15-kD fragment into smaller fragments occurred in the lysosomes. More of the 23-kD intact hormone was metabolized to the 15-kD fragments in the endocytotic vesicles E-I and E-II.

To identify the proteases involved in the formation of the 15-kD fraction, ^{125}I-bGH was incubated in the presence of various protease inhibitors, including leupeptin, PMSF, aprotinin, TLCK, and α_1-antitrypsin inhibitor. Since only leupeptin, PMSF, and aprotinin inhibited the formation of the 15-kD fragment, while TLCK and α_1-antitrypsin inhibitor were without effect, a non-trypsin-like serine protease was proposed to be the principal protease acting on bGH.

As is the case for other small proteins, the therapeutic effectiveness of GH has been limited by its rapid clearance by the kidneys. Extensive renal filtration (glomerular sieving coefficient = 0.6) and subsequent tubular reabsorption (the ratio of renal clearance to glomerular filtration rate was less than 1%) of I-125 rat GH were seen in an isolated, stable, functional, perfused rat kidney preparation [50]. The reabsorption process was inhibited by iodoacetate. The reabsorbed ^{125}I-rat GH was catabolized and a detectable amount of ^{125}I-monoiodotyrosine was returned to the circulation. By contrast, in the nonfiltering kidney preparation, renal accumulation, extraction, and catabolism of ^{125}I-rat GH from the peritubular side were found to be minimal when compared to those occurring from the luminal side.

The filtered growth hormone may lead to varying degrees of renal toxicity. This is indicated by elevation in blood urea nitrogen (BUN) levels [636] and creatinine levels [637] following i.v. administration of high doses of the hormone (200 µg–1 mg) in the rat. In an effort to circumvent this drawback and to prolong its half-life, porcine GH was conjugated with serum albumin using glutaraldehyde as the cross-linking agent [636]. On average, the resulting conjugate, 180 kD in molecular weight, consists of two molecules of albumin and six molecules of GH. Upon i.v. injection into anesthetized male Sprague-Dawley rats, a half-life of 2–3 hr was obtained, which was significantly longer than the 5 min observed in the native hormone. Moreover, the conjugate was not cleared by the kidneys when compared with 25% of the free hormone being cleared in 5 min. There was no evidence of renal complications even at doses as high as that which would cause renal complications by the free hormone [636]. The major pathway of clearance was shifted to the liver, which accumulated 22% of the injected dose at the end of 2 hr. In addition to the above favorable pharmacokinetic properties, the GH-albumin conjugate was also 2–10 times more potent than the native hormone on the basis of insulin-like growth hormone activity (as indicated by the amount of $^{14}CO_2$ produced from [^{14}C]glucose in the adipose tissue) and of lipolysis (as indicated by the amount of glycerol produced from tissue fragments of epididymal fat pads).

Growth-Hormone-Releasing Hormone and Analogs

Growth-hormone-releasing hormone (GHRH) is a peptide hormone containing 40 or 44 amino acids [638,639]. GHRH stimulates linear growth in growth-hormone-deficient patients [640,641]. The biological activity of the native peptide resides in the N-terminal region. The amidated sequence of GHRH 1-29 [GHRH(1-29)-NH$_2$] exhibits full growth-hormone-releasing activity in several species, including humans [642,643]. GHRH(1-29)NH$_2$ administered by continuous s.c. infusion has been shown to restore pulsatile GH secretion in children with short stature and low 24-hr GH secretion [644].

GHRH(1-29)-NH$_2$ was rapidly cleared from the systemic circulation following i.v. administration in urethane-anesthetized rats. The plasma decay curve between 10 and 30 min gave a $t_{1/2}$ of 6.2 ± 0.7 min [645]. Such a rapid rate of clearance is consistent with the susceptibility of this peptide to cleavage at its NH_2 terminus in plasma [646]. Surprisingly, D-amino acid analogs of GHRH(1-29)NH_2, namely (D –Ala-2) -, (D –Asp-3)-, (D –Asn-8)-, (D –Tyr-10)-, and Ac-(D –Tyr-1,D –Ala-2)GHRH(1-29)NH_2, were equally rapidly cleared from the systemic circulation [645]. The half-life ranged from 4.7 to 7.4 min.

Subcutaneously administered GHRH(1-29)-NH_2 or analogs were rapidly absorbed in urethane-anesthesized rats [645]. Rise in plasma GHRH occurred within 1 min, reaching a peak between 5 and 10 min. After approximately 20 min, the rate of decline was similar to that seen after i.v. administration. At 60 min postinjection, GHRH was undetectable. The bioavailability of the analogs ranged from 4.6 to 7.2%, compared to 5.1% observed with the parent peptide. These findings indicate that while modification of GHRH(1-29)-NH_2 resulted in increased potency in rats and pigs [643], it did not confer any stability on the molecule in vivo. Thus, the analogs were as susceptible as the parent peptide to proteolytic degradation both at the site of injection and in the blood [647]. Consequently, the increased biological activity of the analogs is probably due to enhanced receptor affinity rather than to increased resistance to proteolytic degradation.

Somatostatin and Octreotide

Somatostatin (SRIF), a cyclic tetradecapeptide first isolated from bovine and porcine hypothalamus, is a potent growth hormone inhibitory factor [648]. It is present also in other areas of the central nervous system and in the specialized D-cells of the stomach, small intestine, and pancreas [649]. It delays glucose utilization and acutely inhibits the secretion of growth hormone (GH), insulin, glucagon, and stimulated thyrotropin (TSH) [650]. In the brain and in the gastroenteropancreatic area, SRIF may play a physiological role as both a hypophysiotrophic and a gastrointestinal regulatory hormone [651].

Sheppard et al. [652] studied the metabolic clearance rate and plasma half-disappearance time of exogenously administered somatostatin in normal volunteers, patients with chronic liver failure, and patients with chronic renal failure. In these studies, synthetic somatostatin was administered intravenously as a constant infusion for 30 min. Normal subjects received 250 or 500 µg of the peptide. Patients with liver disease and renal failure received 500 and 250 µg of the peptide, respectively. Immunoreactive somatostatin in plasma was measured by a sensitive and specific RIA. The antiserum used in the RIA was directed against the core of the somatostatin molecule and did not measure the C- or N-terminal fragments. Mean MCR value of somatostatin in normal subjects was 28.4 ± 4.2 ml/min·kg body weight (BW). Patients with chronic liver disease exhibited a mean MCR of 25.4 ± 3.4 ml/min·kg BW, which was similar to that observed in normal subjects. However, patients with chronic renal failure showed a highly significant lowering ($p < 0.001$) of the MCR (7.8 ± 0.6 ml/min·kg BW). The plasma disappearance was reported to be biphasic in the above three groups. $t_{1/2\alpha}$ values ranged from 1.1 to 3.0 min, 1.2 to 4.8 min, and 2.6 to 4.9 min in normal subjects, patients with liver disease, and patients with renal failure, respectively. However, the significant decrease in MCR observed in renal failure patients did not correlate with the observed small increment in $t_{1/2\alpha}$. The very short half-life of somatostatin makes this substance unsuitable for long-term clinical use. Search for long-acting analogs of somatostatin has led to the synthesis of a useful cyclic octapeptide analog, octreotide [SMS 201-295, D −Phe −Cys −Phe −D −Trp − Lys −Thr −Cys −Thr(ol)] [653]. Octreotide possesses the essential pharmacophore of natural somatostatin (Phe −Trp −Lys −Thr), with a dextroisomer replacing Trp in position 8 and a preserved Cys −Cys disulfide bridge. The aromatic side chain occupies the conformational space of the essential residue Phe at position 6 in the native peptide, and the molecule is protected at the C-terminus by an aminoalcohol. This synthetic analog is longer acting, more potent, and more specific in pharmacological action than somatostatin [653,654].

In healthy subjects octreotide produces a mild transient postprandial hyperglycemic action, probably related to the postprandial suppression of insulin release. This suppressin appears to override the other actions of octreotide (namely, glucagon suppression and delayed nutrient absorption) which would tend to improve glucose tolerance [655]. Results from studies in patients with acromegaly generally confirm the effects of octreotide on growth hormone secretion in normal volunteers, with octreotide producing a potent and prolonged inhibitory effect [655]. Octreotide at a dose of 50 µg appears to provide maximal inhibition, although there is evidence that a higher dose (100 µg) causes a prolongation of the period of inhibition. Additionally, no rebound hypersecretion of growth hormone is observed when the suppressive effect of octreotide has worn off [655].

Octreotide produces a clear symptomatic improvement in flushing episodes and diarrhea, with a concomitant significant reduction in serotonin levels in patients with carcinoid tumors. In addition, the reduction in urinary excretion of 5-hydroxyindole acetic acid (a metabolite of serotonin) in patients with carcinoid tumors following octreotide administration appears to be dose-related [655]. Administration of octreotide to patients with vasoactive intestinal peptide-producing tumors rapidly controls secretory diarrhea and causes a marked reduction in mean plasma vasoactive peptide levels, although these values remain well above the upper limit of normal. This discrepancy between symptomatic relief and hormone response may be explained by the presence of larger, possibly less biologically active forms of circulating vasoactive intestinal peptide during octreotide therapy [655].

Results from intestinal perfusion experiments suggest that octreotide-induced increases in fluid and electrolyte absorption from the jejunum and ileum may be associated with a direct action of octreotide on intestinal fluid, pancreatic fluid, and electrolyte transport or a secondary response to a reduction in the release of vasoactive intestinal peptide and other peptides from the tumor mass [655]. Short-term studies performed in patients with glucagonomas have demonstrated that octreotide dramatically reduces elevated plasma glucagon levels and produces rapid symptomatic improvement in the necrolytic migratory erythmea associated with this syndrome. Similarly, plasma gastrin levels and basal acid output are rapidly suppressed, following octreotide administration to patirnts with gastrinomas. Sustained improvement in hypoglycemia in association with a marked increase in plasma glucose levels have been reported in patients with insulinomas who were administered single doses of octreotide [655].

Kutz et al. [656] studied the pharmacokinetics of octreotide after i.v. and s.c. administration in eight healthy subjects. Each subject was given octreotide acetate in separate doses of 25, 50, 100, and 200 µg as 3-min i.v. infusions. The resulting serum concentration vs. time curves showed

biexponential characteristics. The calculated serum $t_{1/2\alpha}$ ranged from
9 min (25 µg) to 14 min (200 µg). The serum $t_{1/2\beta}$ values ranged from
72 min (25 µg) to 98 min (200 µg). These values were not significantly
different for the four doses studied. There was no evidence for saturation kinetics, as indicated by a dose-dependent increase in $AUC_{0-\infty}$ from
153 ng/ml.min (25 µg) to 1244 ng/ml.min. The volume of the central compartment ranged between 5.2 L (25 µg) and 10.2 L (200 µg).

Following s.c. administration in doses of 50, 100, 200, and 400 µg,
octreotide acetate was rapidly absorbed and the peak plasma concentrations
were reached within 30 min. The absorption half-life ranged from 5.3 min
(50 µg) to 11.7 min (400 µg) [656]. The serum disposition showed monoexponential behavior with a half-life of 88 min (50 µg) to 102 min (400
µg). As was the case with i.v. administration, the peak serum concentrations increased in a dose-dependent manner from 2.4 ng/ml (50 µg) to
23.7 ng/ml (400 µg). The $AUC_{0-\infty}$ also increased in a dose-dependent
manner from 320 ng/ml.min (50 µg) to 3163 ng/ml.min (400 µg). The
absolute bioavailability of octreotide was calculated to be about 100% (50
µg) or more (higher doses). There were, however, large interindividual
variations in the absolute subcutaneous bioavailability of octreotide.

del Pozo et al. [657] studied the pharmacodynamics of octreotide in a
group of 35 normal subjects following s.c. administration of separate doses
of 50 µg and 100 µg. The GH increment was markedly suppressed by
octreotide within 15 min (1488 ± 607 vs. 203 ± 82 ng/ml.min, $p < 0.001$).
Insulin was also significantly inhibited within 15 min (1146 ± 466 vs. $454 \pm
32$ ng/ml.min, $p < 0.02$). Arginine stimulation (50% arginine HCl, 0.5
g/kg.30 min) tests ($n = 6$) were conducted 15 min and 3 hr after s.c. injection of 50 µg of octreotide. The stimulatory effect of arginine on GH
and insulin was counteracted by the peptide at $p < 0.001$ and $p < 0.02$
significance levels, respectively. Arginine stimulation 3 hr after injection
of ocreotide led to persistent blockade of GH release ($p < 0.02$), whereas
recovery of the insulin response was observed. Thus, octreotide affords
a different duration of action on GH and insulin suppression. Plasma
glucagon increments following a standard protein meal ($n = 10$) were significantly ($p < 0.001$) inhibited by previous s.c. injection of 50 µg and
100 µg of octreotide. Previous treatment with 50 µg and 100 µg of
octreotide ($n = 9$) inhibited ($p < 0.001$) the stimulatory effect of TRH
(200 µg i.v.) on TSH without modifying basal levels. The injection of 100
µg of the drug during sleep completely abolished the noctornal GH peak.
Unlike somatostatin, there was no rebound rise after decline of the suppressive action of GH [658,659].

The efficacy of octreotide on gastric functions following 25 µg and 100
µg of the peptide given three times daily for 7 days subcutaneously was
investigated in healthy male subjects [660]. Serum concentrations of the
drug were well reproducible within 1 week. Octreotide significantly raised
24-hr median intragastric pH on day 1, but no longer on day 6. Peptone-
stimulated gastric acid and volume secretion were less suppressed by
octreotide on day 7 compared with those observed on day 2. Peptone-
stimulated gastrin release was abolished on days 2 and 7, as was peptone-
stimulated insulin release. Blood glucose was altered in a biphasic manner
on days 2 and 7. All effects of octreotide were without clear-cut dose-
response relationship.

The reason for the diminished action of octreotide on exocrine glands
noted above is unknown. Nonetheless, it may be attributed to receptor

down-regulation. A model of desensitization has been proposed for the
adenylate-cyclase-coupled β-adrenergic receptor by Struclovici et al.
[661]. Thus, chronic administration of an agonist causes the number of
plasma membrane receptors to diminish (down-regulation). As suggested
by Hollenberg [662] for antagonism, an opposite mechanism might provoke
receptor recruitment on the cell surface with hypersensitization for
agonist stimulation (up-regulation). This hypersensitization could explain
the rebound effects observed in patients with VIPoma and the restoration
of the effector cell response to octreotide upon cessation of drug administration [663]. Receptor binding studies before and after several days
of treatment with octreotide could provide further information on this
question.

Thyrotropin-Releasing Hormone

Thyrotropin-releasing hormone (TRH, L-pyroglutamyl-L-histidyl-L-prolin-
amide) is a hypothalamic hormone that stimulates the secretion of thyrotropin [664,665] and the release of prolactin and growth hormone from the
pituitary [666,667]. In addition to its endocrine functions, TRH may
exert direct actions on the central nervous system (CNS). For example,
TRH administered centrally or peripherally antagonizes the sedation and
hypothermia produced by CNS depressants such as reserpine, chlor-
promazine, and various barbiturates [668,669]. Indeed, TRH has been
suggested for use as an antidepressant [670,671].
    TRH is rapidly degraded in serum and in many tissues. This has
been studied by Brewster and Waltham [672] using tissues from a variety
of animal species including humans. Degradation of TRH in pooled plasma
followed first-order kinetics in all species studied. The degradation
rates were, however, highly species dependent. The degradation half-
life in rabbit plasma was over 2 hr compared with 39 min in human plasma.
TRH degradation in pig plasma was 40 times faster than in guinea pig
plasma. Surprisingly, in either male or female dog plasma, little, if any,
degradation was observed (<5% within 6 hr). Such stability to degradation
was responsible, at least in part, for the longer plasma half-life and reduced MCR in dogs when compared with the rat and human [673]. The
rate of degradation of TRH in mouse plasma was approximately 10-fold
slower than that observed in rat plasma [672]. This is interesting since
in reversing chlorpromazine-induced hypothermia, TRH was 20 times more
potent in mouse than in rat [669]. In three different strains of male rats,
significant differences in TRH-degrading activity were observed (Table 5).
There was, however, no influence of sex or phase of the female menstrual cycle on TRH degradation in plasma in human subjects [674].
    Species differences in TRH-degrading activity were observed in other
tissues as well. The liver was an active tissue in the rat and in the
mouse, but it was less active in humans and was inactive in the dog (<5%
change within 6 hr). TRH-degrading activity was low in all gut tissues
(rat, human, mouse) and, again, absent in dog. In all four species examined (mouse, rat, dog, and human), the brain appeared consistently
rich in TRH-degrading activity. Of the dog tissues examined (plasma,
liver, gut, brain), the brain was the only tissue containing TRH-degrading
activity.
    Two pathways of degradation have been described for TRH. In the
first pathway, deamino-TRH (TRH-acid, TRH−OH) is generated from TRH

Table 5   Degradation of [^{14}C-His]-TRH by Plasma of Different Strains of Rat

Strain	% TRH remaining after 20 min[a]	Metabolites detected
Sprague-Dawley	62.8 ± 1.5	His only
New Wistar	51.9 ± 0.8	His, trace TRH acid, trace cyclo (*His−Pro*)
Listar hooded	39.8 ± 2.5	His only

[a]Plasma diluted with an equal volume of phosphate buffer (pH 7.4). Results represent mean values ± SEM for plasma samples of 5 individual male animals.
Source: Ref. 672.

by a deamidase enzyme. The second pathway of TRH degradation involves a pyroglutamyl peptidase which yields the products pGlu and His−Pro−NH$_2$. The latter is sensitive to peptidases in serum and several tissues and can also transform nonenzymatically to the stable, biologically active cyclo(His−Pro) [675]. The pathways of TRH metabolism vary more with the tissue type than with species [672]. For instance, following incubation of [^{14}C-His]-TRH in plasma, histidine was the only significant product, whereas in liver and gut both TRH acid and histidine were formed. In brain tissue histidine was the predominant metabolite but His−Pro was also formed. This is especially so in humans.

Using radioimmunoassay, TRH was found to disappear very rapidly from the systemic circulation with a t$_{1/2}$ of 5.3 min following i.v. administration of 400 µg in humans. Over 90% of the TRH was removed within the first 20 min. The mean distribution volume was 15.7 ± 3.8 L. TRH was readily detected in urine for 90 min, most being found during the first 30 min. The amount of peptide in the urine during this period represented 84.9% of the recovered TRH. Similar pharmacokinetic behavior was reported by Leppaluoto et al. [676] using a T-3 TRH bioassay. The half-life of TRH in vivo in the rat after i.m. injection of H-3 or C-14-labeled peptide was estimated to be 2.2 min [677,678].

Iverson [674] studied the kinetics of TRH following the infusion of TRH into antecubital vein for 60 min at two different rates in normal male subjects. These rates were (1) 1 nmol/min initiated once at 1100 hr and again at 2300 hr and (2) 10 nmol/min initiated at 1100 hr. Steady-state levels of TRH were reached at about 40 min. The time of dosing, morning vs. evening, did not significantly affect the plasma profiles of TRH. This finding indicates that the plasma clearance rates and t$_{1/2}$ of TRH do not undergo diurnal variations. The plasma clearance rate and plasma t$_{1/2}$ values were observed to be 1532 ± 423 ml/min and 6.6 ± 1.5 min, respectively. The infusion of 10 nmol/min of TRH resulted in 10 times higher plasma TRH values, indicating lack of dose-dependent pharmacokinetics.

Jackson et al. [679] reported that the thyroid status affected the metabolic clearance of TRH in rat. In that study, TRH (2 µg) was administered i.v. to hypothyroid, hyperthyroid, and euthyroid controlled

animals and the blood and urine samples were assayed for TRH using RIA.
The plasma disappearance curves showed biexponential kinetics in the
hypothyroid and control groups. The half-lives of the fast component
were similar in both groups (2.1 − 2.2 min), while the second component
was prolonged in hypothyroid rats (4.8 min) compared to the euthyroid
rats (3.6 min). The disappearance rate was more rapid in hyperthyroid
animals with a $t_{1/2}$ of 1.6 min. MCR and renal clearance were increased
in hyperthyroid rats (57.5 ml/min and 5.8 ml/min, respectively) but were
reduced in hypothyroid animals (11.4 ml/min and 1.14 ml/min, respectively) compared to the control group (20.2 ml/min and 2.53 ml/min, respectively). Despite alterations in renal clearance, the amounts of TRH excreted were similar (14.5 − 16% of dose administered) in all groups. This
finding was consistent with plasma TRH levels that were higher in hypothyroid but lower in hyperthyroid animals after TRH injection. Incubation
of TRH with serum in vitro [679] showed good agreement with the in vivo
findings. Specifically, compared with control group ($t_{1/2}$, 3.6 min),
serum derived from hyperthyroid rats caused enhanced breakdown of TRH
($t_{1/2}$, 2.4 min), while that from the hypothyroid group had reduced degrading activity ($t_{1/2}$, 6 min) compared with control group ($t_{1/2}$, 3.6
min). Collectively, these findings suggest that thyroid status affects the
MCR of TRH primarily through its effect on the rate of TRH degradation
in plasma.

Safran et al. [680] evaluated and compared the pharmacokinetics of
TRH and its metabolite deamido-TRH (TRH −OH) in rats administered as a
60 − 90-min infusion. TRH or TRH −OH in plasma extracts was quantitated
by specific RIA. The cross-reaction of TRH in the TRH −OH RIA was
0.4%, and the cross-reaction of TRH −OH in the TRH RIA was 0.34%. The
serum concentrations of both TRH and TRH −OH decreased biexponentially
after the cessation of infusion. The $t_{1/2}$ values were 2.4 and 3.9 min for
the initial phase of disappearance and 14.1 and 20.6 min for the latter
phase of disappearance for TRH and TRH −OH, respectively. The volume
of distribution of TRH was 28.5% of total body weight and that of TRH −OH
was 59% of total body weight. The MCR of TRH averaged 4 ml/min,
whereas that of TRH −OH was 6.5 ml/min. The fractional conversion of
TRH to TRH −OH, based on serum TRH −OH concentrations measured during TRH infusions, was 0.7 − 1.4%. The rapid clearance of TRH −OH and
low fractional conversion of TRH to TRH −OH in serum suggested that
little TRH secretion will be reflected in serum TRH −OH concentrations.

Morley et al. [681] measured the plasma clearance rate and $t_{1/2}$ of
TRH and methyl-TRH in normal subjects following an i.v. infusion at a
rate of 3.82 µg/min for 1 hr. Methyl-TRH, [pyroglutamyl-N^{3im}-methyl-
histidyl-prolineamide], is a potent synthetic analog of TRH relative to
stimulation of TSH and prolactin release [682 − 685]. The plasma clearance
rate of TRH (1500 ± 329 ml/min) was significantly greater than that of
methyl-TRH (783 ± 96 ml/min). The $t_{1/2}$ was 6.2 min for TRH and 11.5
for methyl-TRH. Dvorak and Utiger [686] demonstrated that the immunoreactivity of methyl-TRH incubated in serum was destroyed about 50%
faster than TRH. Based on this, the observed slower clearance of methyl-
TRH [681] can be attributed to its increased resistance to degradation by
serum enzymes compared to TRH.

REFERENCES

1. M. Gibaldi and D. Perrier, *Pharmacokinetics*, 2nd ed., Marcel Dekker, New York (1982), pp. 319−354.
2. W. Niederberger, M. LeMaire, G. Maurer, K. Nussbaumer, and O. Wagner, *Transplant. Proc.*, 15:2419−2421 (1983).
3. Y. Sato, M. Yasuhara, K. Okumara, and R. Hori, *Biochem. Pharmacol.*, 34:3543−3546 (1985).
4. H. Sato, Y. Sugiyama, Y. Sawada, T. Iga, and M. Hanano, *Life Sci.*, 37:1309−1318 (1985).
5. R. H. Levy, in: *Topics in Pharmaceutical Sciences* (D. D. Breimer and P. Speiser, eds.), Elsevier Science Publishers B.V., Amsterdam (1985), pp. 161−178.
6. R. J. Cipolle, D. M. Canafax, J. Rabatin, L. D. Bowers, D. E. R. Sutherland, and W. J. M. Hrushesky, *Pharmacotherapy*, 8:47−51 (1988).
7. J. P. Reymond, J. L. Steimer, and W. Neiderberger, *J. Pharmacokin. Biopharm.*, 16:331−353 (1988).
8. L. Z. Benet, *Eur. J. Resp. Dis.*, 65(suppl. 134):45−61 (1984).
9. W. R. Hein and A. Supersaxo, *Immunology*, 64:69−474 (1988).
10. J. Treuner, G. Dannecker, K. E. Joester, A. Hettinger, and D. Niethammer, *J. Interferon Res.*, 1:373−380 (1981).
11. J. R. Quesada, J. U. Gutterman, and E. M. Hersh, *J. Interferon Res.*, 2:593−599 (1982).
12. A. Billiau, *Med. Oncol. Tumor Pathopharmacother.*, 1:87−96 (1986).
13. L. Martis and R. H. Levy, *J. Pharmacokin. Biopharm.*, 1:283−294 (1973).
14. B. M. Brenner, T. H. Hostelter, and H. D. Humes, *N. Engl. J. Med.*, 298:826−833 (1978).
15. D. E. Griffin and J. Giffels, *J. Clin. Invest.*, 70:289−295 (1982).
16. H. S. Bennett, J. H. Luft, and J. C. Hampton, *Am. J. Physiol.*, 196:381−390 (1959).
17. G. Majno, in: *Handbook of Physiology* (W. F. Hamilton and P. Dow, eds.), American Physiology Society, Washington, DC (1965), pp. 2293−2375.
18. M. Simionescu and N. Simionescu, in: *Handbook of Physiology*, Vol. 4 (E. M. Renkin and C. C. Michel, eds.), American Physiological Society, Bethesda, MD (1984), pp. 41−101.
19. A. E. Taylor and G. N. Granger, in: *Handbook of Physiology*, Vol. 4 (E. M. Renkin and C. C. Michel, eds.), American Physiological Society, Bethesda, MD (1984), pp. 467−520.
20. E. R. Weibel, W. Staubli, H. R. Gragi, and F. A. Hess, *J. Cell. Biol.*, 42:68−91 (1969).
21. A. Blouin, R. P. Bolender, and E. R. Weibel, *J. Cell. Biol.*, 72:441−455 (1977).
22. P. Stahl and A. L. Schwartz, *J. Clin. Invest.*, 77:657−662 (1986).
23. J. L. Goldstein, M. S. Brown, R. G. W. Anderson, D. W. Russell, and W. J. Schneider, *Annu. Rev. Cell. Biol.*, 1:1−39 (1985).
24. C. J. Steer and G. Ashwell, in: *Progress in Liver Diseases*, Vol. 8 (H. Popper and F. Schaffner, eds.), Grune & Stratton, NY (1986), pp. 99−123.

25. A. L. Schwartz, *CRC Crit. Rev. Biochem.*, *16*:207−223 (1984).

26. T. C. Wileman, C. Harding, and P. Stahl, *Biochem. J.*, *232*:1−14 (1985).

27. G. Ashwell and J. Harford, *Annu. Rev. Biochem.*, *51*:531−544 (1982).

28. J. van Renswoude, K. R. Bridges, J. B. Harford, and R. D. Klausner, *Proc. Natl. Acad. Sci. USA*, *79*:6186−6190 (1982).

29. M. S. Brown and J. L. Goldstein, *Science*, *232*:34−47 (1986).

30. A. L. Jones, R. H. Renston, and S. T. Burwen, in: *Progress in Liver Diseases*, Vol. 7 (H. Popper and F. Schaffner, eds.), Grune & Stratton, NY (1982), pp. 51−69.

31. G. Ashwell and C. J. Steer, *JAMA*, *246*:2358−2364 (1981).

32. E. A. Jones, J. M. Vierling, C. J. Steer, and J. Reichen, in: *Progress in Liver Diseases*, Vol. 6 (H. Popper and F. Schaffner, eds.), Grune & Stratton, New York (1979), pp. 43−80.

33. D. C. Kim, Y. Sugiyama, H. Sato, T. Fuwa, T. Iga, and M. Hanano, *J. Pharm. Sci.*, *77*:200−207 (1988).

34. Y. Sugiyama and M. Hanano, *Pharm. Res.*, *6*:192−202 (1989).

35. J. M. Schiff, M. M. Fisher, A. L. Jones, and B. J. Underworn, *J. Cell Biol.*, *102*:920−931 (1986).

36. R. H. Renston, A. L. Jones, W. D. Christiansen, and G. T. Hradek, *Science*, *208*:1276−1278 (1980).

37. A. L. Schwartz, A. Bolognesi, and S. E. Fridovich, *J. Cell. Biol.*, *98*:732−738 (1984).

38. J. Kaplan, *Science*, *212*:14−20 (1981).

39. R. Weisiger, J. Gollan, and R. Ockner, *Science*, *211*:1048−1051 (1981).

40. D. J. Socken, E. S. Simms, B. R. Nagy, M. M. Fisher, and B. J. Underdown, *Mol. Immunol.*, *18*:345−348 (1981).

41. J. H. Grendell and S. S. Rothman, *Gastroenterology*, *80*:1164 (1981).

42. G. Scheele and G. Rohr, *Gastroenterology*, *80*:1274 (1981) (abstract).

43. R. H. Renston, D. G. Maloney, A. L. Jones, G. T. Hradek, K. Y. Wong, and I. D. Goldfine, *78*:1373−1388 (1980).

44. T. W. Redding and A. V. Schally, *Life Sci.*, *12*:23−32 (1973).

45. M. Koida, J. D. Glass, I. L. Schwartz, and R. Walter, *Endocrinology*, *88*:633−643 (1971).

46. D. P. Cameron, H. G. Burger, E. Gordon, and J. M. Watts, *Metabolism*, *21*:895−904 (1972).

47. N. Samaan and R. M. Freeman, *Metabolism*, *19*:102−113 (1970).

48. A. L. C. Wallace and B. D. Stacy, *Horm. Metab. Res.*, *7*:135−138 (1975).

49. P. J. Collipp, J. R. Patrick, C. Goodheart, and S. A. Kaplan, *Proc. Soc. Exp. Biol. Med.*, *121*:173−177 (1966).

50. V. Johnson and T. Maack, *Am. J. Physiol.*, *233*:F185−F196 (1977).

51. V. Bocci, *Adv. Drug Del. Rev.*, *4*:149−169 (1990).

52. T. Maack, V. Johnson, S. T. Kau, J. Figueiredo, and D. Sigulem, *Kidney Int.*, *16*:251−270 (1979).

53. T. W. Reding, A. J. Kastin, D. G. Barzena, D. H. Coy, Y. Hirotsu, J. Ruelas, and A. V. Schally, *Neuroendocrinology*, *16*:119−126 (1974).

54. R. Walter and R. H. Bowman, *Endocrinology, 92*:189−193 (1973).
55. J. R. J. Baker, H. P. J. Bennett, R. A. Christian, and
    C. McMartin, *J. Endocrinol., 74*:23−35 (1977).
56. H. P. J. Bennett and C. McMartin, *J. Endocrinol., 82*:33−42 (1979).
57. E. P. Hicks, C. W. Cooper, and W. J. Waddell, *J. Dent. Res., 50*:
    1307−1331 (1971).
58. H. T. Narahara, N. B. Everett, B. S. Simmons, and R. H.
    Williams, *Am. J. Physiol., 192*:227−231 (1985).
59. J. E. K. Bourdeau, E. R. Y. Chen, and F. A. Carone, *Am. J.
    Physiol., 225*:1399−1404 (1973).
60. O. Vilar, B. Alvane, O. Davidson, and R. E. Mancini, *J. Histochem.
    Cytochem., 12*:621−627 (1964).
61. J. E. Bourdeau and F. A. Carone, *Nephron., 13*:22−34 (1974).
62. E. I. Christensen and A. B. Maunsbach, *Kidney Int., 6*:396−407
    (1974).
63. T. Maack, *Am. J. Med., 58*:57−64 (1975).
64. A. B. Maunsbach, in: *MTP International Review of Science, Kidney
    and Urinary Tract Physiology*, Vol. 11 (K. Thurau, ed.),
    University Park Press, Baltimore (1976), pp. 145−167.
65. F. A. Carone and D. P. Peterson, *Am. J. Physiol., 238*:F151−F158
    (1980).
66. K. J. Ullrich, *Annu. Rev. Physiol., 41*:181−195 (1979).
67. V. Ganapathy and F. H. Leibach, *Am. J. Physiol., 25*:F945−F953
    (1986).
68. W. H. Sawyer, *Proc. Soc. Exp. Biol. Med., 87*:463−465 (1954).
69. M. W. Smith and H. Sachs, *Biochem. J., 79*:663−669 (1961).
70. N. A. Thorn and N. B. S. Willumsen, *Acta Endocrinol., 44*:563−569
    (1963).
71. R. Walter and H. Shlank, *Endocrinology, 89*:990−995 (1971).
72. S. Fang and A. H. Tashjian, Jr., *Endocrinology, 90*:1177−1184
    (1972).
73. A. L. Bailey, S. B. Baylin, G. V. Foster, and P. Bard, *Fed. Proc.,
    32*:285 (1973) (abstract).
74. S. Oparil and M. D. Bailie, *Circ. Res., 33*:500−507 (1973).
75. B. D. Stacy, A. L. C. Wallace, R. T. Gemmell, and B. W. Wilson,
    J. Endocrinol., 68:21−30 (1976).
76. C. W. Gottschalk, F. Morell, and M. Myelle, *Am. J. Physiol., 209*:
    173−178 (1965).
77. J. J. Burg, M. Grantham, M. Abramow, and J. Orloff, *Am. J.
    Physiol., 210*:1293−1298 (1966).
78. F. A. Carone, D. R. Peterson, and G. Flouret, *J. Lab. Clin. Med.,
    100*:1−14 (1982).
79. A. Katz and D. Emmanouel, *Nephron, 22*:69−80 (1978).
80. Z. Talor, D. S. Emmanouel, and A. I. Katz, *Clin. Res., 29*:779A
    (1981).
81. D. M. Matthews, *Physiol. Rev., 55*:537−608 (1975).
82. K. Elliott and M. O'Connor (eds.), *Peptide Transport and Hydrolysis*
    (CIBA Foundation Symp. 50), Elsevier-North Holland, New York
    (1977).
83. D. H. Alpers and B. Seetharam, *N. Engl. J. Med., 296*:1047−1050
    (1977).

84. A. M. Ugolev, N. M. Timofeeva, L. F. Smirnova, P. DeLaey, A. A. Grudzdkov, N. N. Iezuitova, N. M. Mityushova, G. M. Roshchina, E. G. Gurman, V. M. Gusev, V. A. Tsvetkova, and G. G. Shcherbakov, in: *Peptide Transport and Hydrolysis* (CIBA Foundation Symp. 50) (K. Elliott and M. O'Connor, eds.), Elsevier-North Holland, New York (1977), pp. 221–243.

85. D. Maestracci, *Biochim. Biophys. Acta, 433*:469–481 (1976).

86. D. Louvard, M. Semeriva, and S. Marous, *J. Mol. Biol., 106*: 1023–1035 (1976).

87. L. Josefsson, H. Sjostrom, and O. Noren, in: *Peptide Transport and Hydrolysis* (CIBA Foundation Symp. 50) (K. Elliott and M. O'Connor, eds.), Elsevier-North Holland, New York (1977), pp. 199–207.

88. S. Silbernagl, J. A. Foulkes, and P. Deetjen, *Rev. Physiol. Biochem. Pharmacol., 74*:105–167 (1975).

89. A. D. Inglot, B. Kisielow, and O. Inglot, *Arch. Virol., 60*:43–50 (1979).

90. T. Wileman, R. L. Foster, and P. N. C. Elliot, *J. Pharm. Pharmacol., 38*:264–271 (1986).

91. R. G. Melton, C. N. Wiblin, R. L. Foster, and R. F. Sherwood, *Biochem. Pharmacol., 36*:105–112 (1987).

92. Y. Kamisaki, H. Wada, T. Yagura, A. Matsushima, and Y. Inada, *J. Pharmacol. Exp. Ther., 216*:410–414 (1981).

93. N. V. Katre, M. J. Knauf, and W. J. Laird, *Proc. Natl. Acad. Sci. USA, 84*:1487–1491 (1987).

94. H. Maeda, T. Matsumoto, T. Konno, K. Iwai, and M. Ueda, *J. Protein Chem., 3*:181–193 (1984).

95. M. J. Poznansky, in: *Methods of Drug Delivery* (G. M. Ihler, ed.), Pergamon, Oxford (1986), pp. 59–82.

96. A. Traub, B. Payess, S. Reuveny, and A. Mizhrai, *J. Gen. Virol., 53*:389–392 (1981).

97. M. G. Rosenblum, B. W. Unger, J. U. Gutterman, E. M. Hersh, G. S. David, and J. M. Frncke, *Cancer Res., 45*:2421–2424 (1985).

98. J. G. Kidd, *J. Exp. Med., 98*:583–606 (1953).

99. L. T. Mashburn and J. C. Wriston, Jr., *Biochem. Biophys. Res. Commun., 12*:50–55 (1963).

100. R. L. Capizzi, J. R. Bertino, and R. E. Handschumacher, *Annu. Rev. Med., 21*:433–444 (1970).

101. A. Khan and J. M. Hill, *J. Lab. Clin. Med., 73*:846–852 (1969).

102. Y. R. Park, A. Abuchowski, S. Davis, and F. Davis, *Anticancer Res., 1*:373–376 (1981).

103. D. H. Ho, N. S. Brown, A. Yen, R. Holmes, M. Keating, A. Abuchowski, R. A. Newman, and I. H. Krakoff, *Drug Metab. Dispos., 14*:349–352 (1986).

104. D. A. Cooney, R. L. Capizzi, and R. E. Handschumacher, *Cancer Res., 30*:929–935 (1970).

105. A. Abuchowski, T. van Es, N. C. Palczuk, J. R. McCoy, and F. F. Davis, *Cancer Treat. Rep., 63*:1127–1132 (1979).

106. D. H. Ho, H. Y. Yap, N. Brown, G. R. Blumenschein, and G. R. Bodey, *Clin. Pharmacol., 21*:72–78 (1981).

107. D. H. W. Ho, C. Y. Wang, J.-R. Lin, N. Brown, R. A. Newman, and I. H. Krakoff, *Drug Metab. Dispos., 16*:27–29 (1988).

108. Y. Takakura, T. Fujita, M. Hashida, H. Maeda, and H. Sezaki, *J. Pharm. Sci.*, *78*:219−222 (1989).

109. Y. Takakura, Y. Kaneko, T. Fujita, M. Hashida, H. Maeda, and H. Sezaki, *J. Pharm. Sci.*, *78*:117−121 (1989).

110. Li Duan, P. Ghezzi, I. Conti, R. Tridico, M. Bianchi, and S. Caccia, *J. Biol. Resp. Mod.*, *7*:365−370 (1988).

111. S. J. Williams and G. C. Farrell, *Br. J. Clin. Pharmacol.*, *22*: 610−612 (1986).

112. G. Taylor, B. J. Maratino, Jr., J. A. Moore, V. Gurley, and T. F. Blaschke, *Drug Metab. Dispos.*, *13*:459−463 (1985).

113. V. Bocci, A. Pacini, G. P. Pessina, L. Paulesu, M. Muscettola, and G. Lunghetti, *J. Gen Virol.*, *66*:887−891 (1985).

114. K. Cantell, H. Schellekens, L. Phyala, S. Hirvonen, P. H. Van Der Meide, and A. J. De Reus, *Interferon Res.*, *5*:571−582 (1985).

115. I. C. Macdougall, D. E. Roberts, P. Neubert, A. D. Dharmasene, G. A. Coles, and J. D. Williams, *Lancet*, *1*:425−427 (1989).

116. B. J. Boucher, *Nature*, *210*:1288−1289 (1966).

117. C. Blifeld and R. B. Ettenger, *N. Engl. J. Med.*, *317*:509−511 (1987).

118. Task Force on Cyclosporine Monitoring, *Clin. Chem.*, *33*: 1269−1288 (1987).

119. J. Smith, J. Hows, and E. C. Gordon-Smith, *Transplant. Proc.*, *15*: 2422−2425 (1983).

120. M. Le Marie and J. P. Tillement, *J. Pharm. Pharmacol.*, *34*:715−718 (1982).

121. R. P. Agarwal, R. A. McPherson, and G. A. Threatte, *Transplantation*, *42*:627−632 (1986).

122. R. Prasad, M. S. Muddux, M. F. Mozes, N. S. Biskup, and A. Maturen, *Transplantation*, *39*:667−669 (1985).

123. J. M. Potter and H. Self, *Ther. Drug Monit.*, *8*:122−125 (1986).

124. E. Seifried, P. Tanswell, D. C. Rijken, M. M. Barrett-Bergshoeff, C. A. P. F. Su, and C. Kluft, *Arzneim. Forsch./Drug Res.*, *38*: 418−422 (1988).

125. J. H. Verheijen, E. Mulleart, G. T. G. Chang, C. Kluft, and G. Wijngaards, *Thromb. Hemostas.*, *48*:266−269 (1982).

126. T. Astrup and S. Muellertz, *Arch. Biochem. Biophys.*, *40*:346−351 (1952).

127. D. C. Rijken, I. Juhan-Vague, F. de Cock, and D. Collen, *J. Lab. Clin. Med.*, *101*:274−284 (1983).

128. P. Holvoet, H. Cleemput, and D. Collen, *Thromb. Hemostas.*, *54*: 684−687 (1985).

129. B. Primstone, S. Epstein, S. M. Hamilton, D. Le Roith, and S. Hendricks, *J. Clin. Endocrinol. Metab.*, *44*:356−360 (1977).

130. T. M. Nett and G. D. Niswender, in: *Methods of Hormone Radioimmunoassay*, 2nd ed. (H. R. Behrman and R. B. Jaffe, eds.), Academic Press, New York (1979), pp. 57−85.

131. C. A. Huseman and R. P. Kelch, *J. Clin. Endocrinol. Metab.*, *47*: 1325−1331 (1978).

132. J. P. Fauconnier, B. Teuwissen, and K. Thomas, *Gynecol. Obstet. Invest.*, *9*:229−237 (1978).

133. V. Menon, W. R. Butt, R. N. Clayton, R. L. Edwards, and S. S. Lynch, *Clin. Endocrinol. (Oxf)*, *21*:223–232 (1984).

134. G. Fluoret, M. A. Stetler-Stevenson, F. A. Carone, and D. R. Peterson, in: *LHRH and Its Analogs — Contraceptive and Therapeutic Applications* (B. H. Vickery, J. J. Nestor, Jr., and E. S. E. Hafez, eds.), MTP Press, Boston (1984), pp. 397–410.

135. M. A. Stetler-Stevenson, G. Fluoret, S. Nakamura, B. Gulczynski, and F. A. Carone, *Am. J. Physiol.*, *244*:F628–F632 (1983).

136. R. L. Prestidge, in: *Handbook of HPLC for the Separation of Amino Acids, Peptides, and Proteins*, Vol. II (W. S. Hancock, ed.), CRC Press, Orlando, FL (1985), pp. 279–285.

137. E. Lundanes and T. Greibrokk, in: *Handbook of HPLC for the Separation of Amino Acids, Peptides, and Proteins*, Vol. II (W. S. Hancock, ed.), CRC Press, Orlando, FL (1985), pp. 49–52.

138. U. Ragnarsson, B. Fransson, and O. Zetterqvist, in: *Handbook of HPLC for the Separation of Amino Acids, Peptides, and Proteins*, Vol. II (W. S. Hancock, ed.), CRC Press, Orlando, FL (1985), pp. 75–88.

139. J. Crommen, B. Fransson, and G. Schill, *J. Chromatogr.*, *142*:283–297 (1977).

140. K. Kangawa and H. Matsuo, *Biochem. Biophys. Res. Commun.*, *118*:131–139 (1984).

141. K. Kangawa, A. Fukuda, and H. Matsuo, *Nature (London)*, *313*:397–400 (1985).

142. R. E. Lang, T. Unger, and D. Ganten, *J. Hypertension*, *5*:255–271 (1987).

143. P. Needelman, S. P. Adams, B. R. Cole, M. G. Currie, D. M. Geller, M. L. Michener, C. B. Saper, D. Schwartz, and D. G. Standaert, *Hypertension*, *7*:469–482 (1985).

144. R. E. Lang, H. Tholken, D. Ganten, F. C. Luft, H. Ruskoaco, and T. Unger, *Nature (London)*, *314*:264–266 (1985).

145. N. Katsube, D. Schwartz, and P. Needleman, *Biochem. Biophys. Res. Commun.*, *133*:937–944 (1985).

146. T. Maack, M. Suzuki, F. A. Almeida, D. Nussenzveig, R. M. Scarborough, G. A. McEnroe, and J. A. Lewicki, *Science*, *238*:675–678 (1987).

147. D. Nussenzveig, R. Scarborough, J. Lewicki, and T. Maack, *Proc. Annu. Meet. Am. Soc. Nephrol. 20th* (1987), pp. 132A.

148. R. M. Scarborough, G. McEnroe, L. Kang, A. Arfsten, K. Schwartz, M. Suzuki, F. Almeida, T. Maack, and J. Lewicki, *Proc. World Congr. on Biol. Active Atrial Peptides. II*, New York (1987), pp. 190.

149. K. K. Murthy, G. Thibault, and M. Cantin, *250*, 665–670 (1988).

150. K. A. King, C. A. Courneya, C. Tang, N. Wilson, and J. R. Ledsome, *Endocrinology*, *124*:77–83 (1989).

151. P. Cernacek, E. Maher, J. C. Crawhall, and M. Levy, *Am. J. Physiol.*, *255*:R929–R935 (1988).

152. K. Nakao, A. Sugawara, N. Morii, M. Sakamoto, T. Yamada, H. Itoh, S. Shiomo, Y. Saito, K. Nishimura, T. Ban, K. Kangawa, and H. Matsuo, *Eur. J. Clin. Pharmacol.*, *31*:101–103 (1986).

153. A. Sugawara, K. Nakao, N. Morii, M. Sakamoto, K. Horii,
     M. Shimokura, Y. Kiso, K. Nithimura, T. Ban, M. Kihara,
     Y. Yamori, K. Kangawa, H. Matsuo, and H. Imura, *Hypertension*,
     8(Suppl. 1):151–155 (1986).
154. A. M. Richards, M. G. Nicholls, H. Ikram, M. W. Webster, T. G.
     Yandle, and E. A. Espiner, *Lancet*, *1*:545–548 (1985).
155. M. Ohashi, N. Fujio, H. Nawata, K. Kato, H. Ibayashi, K. Kangawa,
     and H. Matsuo, *Regulatory Peptides*, *19*:265–272 (1987).
156. M. Ohashi, N. Fujio, H. Nawata, K. Kato, H. Ibayashi, K. Kangawa,
     and H. Matsuo, *J. Clin. Endocrinol. Metab.*, *64*:81–85 (1987).
157. M. E. Brier, R. A. Brier, F. C. Luft, and G. R. Aronoff, *J.
     Pharmacol. Exp. Ther.*, *243*:868–983 (1987).
158. H. G. Holford and L. B. Sheiner, *Pharmacol. Ther.*, *16*:143–166
     (1982).
159. N. Katsube, D. Schwartz, and P. Needleman, *J. Pharmacol. Exp.
     Ther.*, *239*:474–479 (1986).
160. F. C. Luft, R. E. Lang, G. R. Aronoff, H. Ruskoaho, M. Toth,
     D. Ganten, R. B. Sterzel, and T. Unger, *J. Pharmacol. Exp.
     Ther.*, *236*:416–418 (1986).
161. P. A. Krieter and A. J. Trapani, *Drug Metab. Dispos.*, *17*:14–19
     (1989).
162. L. A. Walker, M. Gellai, and H. Valtin, *J. Pharmacol. Exp. Ther.*,
     236:721–728 (1986).
163. W. C. Segde, L. McGowan, N. Lund, B. Duling, and D. E.
     Longnecker, *Am. J. Physiol.*, *249*:H164–H173 (1985).
164. H. Sugimoto, W. W. Monato, and S. G. Eliasson, *Am. J. Physiol.*,
     251:H1211–H1216 (1986).
165. R. L. Woods, *Am. J. Physiol.*, *255*:E934–E941 (1988).
166. F. A. Almeida, M. Suzuki, R. M. Scarborough, J. A. Lewicki, and
     T. Maack, *Am. J. Physiol.*, *256*:R469–R475 (1989).
167. T. Masaki, *J. Cardiovasc. Pharmacol.*, *13*(Suppl. 5):S1–S4 (1989).
168. M. Yanagisawa, H. Kurihara, S. Kimura, Y. Tomobe, M. Kobayashi,
     Y. Mitsui, K. Goto, and A. T. Masaki, *Nature (London)*, *332*:
     611–615 (1988).
169. Y. Ito, M. Yanagisawa, S. Ohkubo, C. Kimura, T. Kosaka,
     A. Inoue, N. Ishida, Y. Mitsi, H. Onda, M. Fujino, and T. Masaki,
     FEBS Lett., *231*:440–444 (1988).
170. M. Tanagisawa, A. Inoue, T. Ishikawa, Y. Kasuya, S. Kimura,
     S. Kumagaye, K. Nabajime, T. X. Wakanabe, S. Sababisara,
     K. Goto, and T. Masae, *Proc. Natl. Acad. Sci. USA*, *85*:6964–6967
     (1988).
171. A. Inoue, M. Yanagisawa, S. Kimura, Y. Kasuya, T. Miyauchi,
     K. Goto, and T. Masaki, *Proc. Natl. Acad. Sci. USA*, *86*:2863–2867
     (1989).
172. T. X. Watanabe, S. I. Kumagaye, H. Nishio, K. Nakajima,
     T. Kimura, and S. Sakakibara, *J. Cardiovasc. Pharmacol.*, *13*
     (Suppl. 5):S207–S208 (1989).
173. R. Shiba, M. Yanagisawa, T. Miyauchi, Y. Ishii, S. Kimura,
     Y. Uchiyama, T. Masaki, and K. Goto, *J. Cardiovasc. Pharmacol.*,
     13(Suppl. 5):S98–S101 (1989).
174. E. Anggard, S. Galton, G. Ral, R. Thomas, L. McLoughlin,
     G. de Nucci, and J. R. Vane, *J. Cardiovasc. Pharmacol.*, *13*
     (Suppl. 5):S46–S69 (1989).

175. J. Pernow, A. Hemsen, and J. M. Lundberg, *Biochem. Biophys. Res. Commun.*, *161*: 667-653 (1989).

176. L. O. Jacobson, E. Goldwasser, M. Fried, and L. Plzak, *Trans. Assoc. Am. Physicians*, *70*:305–317 (1957).

177. W. Fried, *Blood*, *40*:671–677 (1972).

178. J. L. Spivak, *Int. J. Cell Cloning*, *4*:139–166 (1986).

179. K. Jacobs, C. Schoemaker, R. Rudersdorf, S. D. Neil, R. M. Kaufman, A. Mufson, J. Sechra, S. S. Jones, R. Hewick, E. F. Rich, M. Vawakila, T. Shimizu, and T. Miyake, *Nature*, *313*:806–810 (1985).

180. F. K. Lin, S. Suggs, C. H. Lin, J. K. Browne, R. Smalling, J. C. Egrie, K. K. Chen, G. M. Fox, F. Martin, Z. Stabinsky, S. M. Badrawi, P.-H. Lai, and E. Goldwasser, *Proc. Natl. Acad. Sci. USA*, *82*:7580–7584 (1985).

181. C. G. Winearls, M. J. Pippard, M. R. Downing, D. O. Oliver, C. Reid, and P. M. Cotes, *Lancet*, *2*:1175–1177 (1986).

182. J. W. Eschbach, J. C. Egrie, M. R. Downing, J. K. Browne, and J. W. Adamson, *N. Engl. J. Med.*, *316*:73–78 (1987).

183. S. Casati, P. Passerini, M. R. Campise, G. Graziani, B. Cesana, M. Perisic, and C. Ponticelli, *Br. Med. J.*, *295*:1017–1010 (1987).

184. J. Bommer, C. Alexiou, U. Muller-Buhl, J. Eifert, and E. Ritz, *Nephr. Dialysis Trans.*, *2*:238–242 (1987).

185. J. W. Eschbach and J. W. Adamson, *Kidney Int.*, *33*:189 (1988) (abstract).

186. H. Sasaki, B. Bothner, A. Dell, and M. Fukuda, *J. Biol. Chem.*, *262*:12059–12079 (1987).

187. P. H. Lai, R. Everett, F. F. Wang, T. Arakawa, and E. Goldwasser, *J. Biol. Chem.*, *261*:3116–3121 (1986).

188. F. Stohlman, Jr. and D. Howard, in: *Erythropoiesis* (L. O. Jacobson and M. Doyle, eds.), Grune & Stratton, New York (1962), pp. 120–124.

189. K. R. Reissman, D. A. Diederich, K. Ito, and J. W. Schmaus, *J. Lab. Clin. Med.*, *65*:967–975 (1965).

190. J. P. Naets and M. Wittek, *Am. J. Physiol.*, *217*:297–301 (1969).

191. J. A. Galloway, C. T. Spradlin, R. L. Nelson, S. M. Wentworth, J. A. Davidson, and J. L. Swarner, *Diabetes Care*, *4*:66–376 (1981).

192. R. M. Emanuele and J. Fareed, *Thromb. Res.*, *48*:591–596 (1987).

193. P. Wilton, L. Widlund, and O. Guilbaud, *Acta Paediatr. Scand.*, *337*:118–212 (1987).

194. E. P. Paulsen, J. W. Courtney, and W. C. Duckworth, *Diabetes*, *18*:640–645 (1979).

195. J. S. Fu, J. J. L. Lertora, J. Brookins, J. C. Rice, and J. W. Fisher, *J. Lab. Clin. Med.*, *111*:669–676 (1988).

196. J. B. Sherwood and E. Goldwasser, *Blood*, *54*:885–893 (1979).

197. J. C. Egrie, P. M. Cotes, J. Lane, R. E. Gaines Das, and R. C. Tam, *J. Immunol. Meth.*, *99*:235–241 (1987).

198. R. A. O'Reilly, P. G. Wetting, and J. G. Wagner, *Thromb. Diath. Haemorrh.*, *25*:178–186 (1971).

199. P. R. Larsen, A. J. Atkinson, Jr., H. N. Wilman, and R. E. Goldsmith, *J. Clin. Invest.*, *49*:1266–1279 (1970).

200. T. K. Henthorn, M. J. Avram, M. C. Fredriksen, and A. J. Atkinson, Jr., *J. Pharmacol. Exp. Ther.*, *22*:389–394 (1982).
201. S. E. Steinberg, J. Maldenovie, G. R. Matzke, and J. F. Garcia, *Blood*, *67*:646–649 (1986).
202. M. S. Dordal, F. F. Wang, and E. Goldwasser, *Endocrinology*, *116*:2293–2299 (1985).
203. A. G. Morrell, K. A. Irvine, I. Sternlieb, I. H. Scheinberg, and G. Ashwell, *J. Biol. Chem.*, *243*:155–159 (1968).
204. J. L. Spivak and B. B. Hogans, *Blood*, *73*:90–99 (1989).
205. M. N. Fukuda, H. Sasaki, L. Lopez, and M. Fukuda, *Blood*, *73*:84–89 (1989).
206. D. Collen, *Thromb. Haemostas.*, *43*:77–89 (1980).
207. D. C. Rijken and D. Collen, *J. Biol. Chem.*, *256*:7035–7041 (1981).
208. O. Matsuo, D. C. Rijken, and D. Collen, *Nature*, *291*:590–591 (1981).
209. P. Wallen, N. Bergsdorf, and M. Ranby, *Biochim. Biophys. Acta*, *709*:318–328 (1982).
210. D. C. Rijken, G. Wijngaards, M. Zaal-de Jong, and J. Welbergen, *Biochim. Biophys. Acta*, *580*:140–153 (1979).
211. P. Wallen, G. Pohl, N. Bergsdorf, M. Ranby, T. Ny, and H. Jornwall, *Eur. J. Biochem.*, *132*:681–686 (1983).
212. G. Pohl, L. Kerne, B. Nilsson, and M. Emarason, *Eur. J. Biochem. (Tokyo)*, *170*:69–75 (1987).
213. J. Rogers, *Nature*, *315*:458–459 (1985).
214. C. Konninger, J. M. Stassen, and D. Collen, *Thromb. Haemostas.*, *46*:658–661 (1981).
215. H. E. Fuchs, H. Berger, and S. V. Pizzo, *Blood*, *65*:539–546 (1985).
216. J. J. Emeis, C. M. Van de Hoogen, and D. Jense, *Thromb. Haemostas.*, *54*:661–664 (1985).
217. T. Nilsson, P. Wallen, and O. Mellbring, *Scand. J. Haematol.*, *33*:49–53 (1984).
218. M. Verstraete, H. Bounameaux, F. De Cock, F. De Werf Van, and D. Collen, *J. Pharmacol. Exp. Ther.*, *235*:506–512 (1985).
219. A. J. Tiefenbrunn, A. K. Robinson, P. B. Kiernik, P. A. Lundbrook, and B. R. Sobel, *Circulation*, *70*:110–116 (1985).
220. H. D. Garabedian, H. K. Gold, R. C. Leinbach, T. Yasuda, J. A. Johns, and D. Collen, *J. Am. Coll. Cardiol.*, *9*:599–607 (1987).
221. K. L. L. Fong, C. S. Crysler, B. A. Mico, K. E. Boyle, G. A. Kopia, L. Kopaciewicz, and R. K. Lynn, *Drug Metab. Disp.*, *16*:201–206 (1988).
222. C. Bakhit, D. Lewis, R. Billings, and B. Malfroy, *J. Biol. Chem.*, *262*:8716–8720 (1987).
223. S. Nilsson, M. Emarsson, S. Ekvarn, L. Hoggroth, and C. Mattson, *Thromb. Res.*, *39*:511–521 (1985).
224. H. Bounameaux, J. M. Stassen, C. Seghers, and D. Collen, *Blood*, *67*:1493–1497 (1986).
225. J. Kuiper, M. Otter, D. C. Rijken, and T. J. C. Van Berkel, *J. Biol. Chem.*, *263*:18220–18224 (1988).
226. D. A. Owensby, B. E. Sobel, and A. L. Schwartz, *J. Biol. Chem.*, *263*:10587–10594 (1988).

227. D. C. Rijken and J. J. Emeis, *Biochem. J.*, *238*:643−646 (1986).

228. R. G. Larsen, M. Metzger, K. Henson, Y. Blue, and P. Horgan, *Blood*, *73*:1842−1850 (1989).

229. P. Cambier, F. van de Werf, G. R. Larsen, and D. Collen, *J. Cardiovasc. Pharmacol.*, *11*:468−472 (1988).

230. D. Collen, J.-M. Stassen, and G. Larsen, *Blood*, *71*:216−219 (1988).

231. D. M. Gibbs, *Psychoneuroendocrinology*, *11*:131−140 (1986).

232. H. D. Lauson, in: *Handbook of Physiology and Endocrinology*, Sect. 7, Part 1, Vol. 4.1 (R. O. Greep and E. B. Astwood, eds.), American Physiological Society, Washington, DC (1974), pp. 287−393.

233. R. E. Weitzman and D. A. Fisher, *Am. J. Physiol.*, *235*:E591−E597 (1978).

234. G. Baumann and J. F. Dingman, *J. Clin. Invest.*, *57*:1109−1116 (1976).

235. J. L. Sondeen and J. R. Claybaugh, *Am. J. Physiol.*, *256*: R291−R298 (1989).

236. J. M. Davison, E. A. Sheills, W. M. Barron, A. G. Robinson, and M. D. Lindheimer, *J. Clin. Invest.*, *83*:1313−1318 (1989).

237. K. Adamsons, S. L. Engel, and H. B. Van Dyke, *Endocrinology*, *63*:679−687 (1958).

238. H. Tuppy and H. Nesvadba, *Monatschr. Chem.*, *88*:977−988 (1957).

239. F. A. Carone, E. J. Christensen, and G. Flouret, *Am. J. Physiol.*, *253*:F1120−F1127 (1987).

240. P. M. Mannucci, Z. M. Ruggeri, F. I. Pareti, and A. Capitano, *Lancet*, *1*:869−872 (1977).

241. A. M. Gader, J. DaCosta, and J. D. Cash, *Lancet*, *2*:1417−1418 (1973).

242. J. D. Cash, A. M. Gader, and J. DaCosta, *Br. J. Haematol.*, *15*: 363−364 (1974).

243. P. M. Mannucci, F. I. Pareti, L. Holmberg, I. M. Nilsson, and Z. M. Ruggeri, *J. Lab. Clin. Med.*, *88*:662−671 (1976).

244. M. Zaoral, J. Kolc, and F. Sorm, *Collect. Czech. Chem. Commun.*, *32*:1250−1257 (1967).

245. C. R. W. Edwards, M. J. Kitau, T. Chard, and G. M. Besser, *Br. Med. J.*, *3*:275−278 (1973).

246. A. S. Aronson, K. E. Andersson, C. G. Bergstrand, and I. L. Mulder, *Acta Paediatr. Scand.*, *62*:133−140 (1973).

247. J. P. Rado, J. Marosi, L. Szende, L. Borbely, J. Tako, and J. Fischer, *J. Clin. Pharmacol. Biopharmacol.*, *13*:199−209 (1976).

248. M. Kohler, P. Hellstern, C. Miyashita, G. von Blohn, and E. Wenzel, *Thromb. Haemostas.*, *55*:108−111 (1986).

249. M. Kohler and A. Harris, *Eur. J. Clin. Pharmacol.*, *35*:281−285 (1988).

250. P. M. Mannucci, M. T. Canciani, L. Rota, and B. S. Donovon, *Br. J. Haematol.*, *47*:283−293 (1981).

251. A. S. Harris, I. M. Nilsson, Z. G. Wagner, and U. Alkner, *J. Pharm. Sci.*, *75*:1085−1088 (1986).

252. M. Monnier, T. Koller, and S. Graber, *Exp. Neurol.*, *8*:264−277 (1963).

253. A. Bjartell, R. Ekman, J. Hedenbro, K. Bjolund, and F. Sundler,
      Peptides, *10*:163–170 (1989).
254. M. V. Graf and A. J. Kastin, *Proc. Soc. Exp. Biol. Med.*, *177*:
      197–204 (1984).
255. M. V. Graf and A. J. Kastin, *Peptides*, *7*:1165–1187 (1986).
256. K. S. Iyer and S. M. McCann, *Neuroendocrinology*, *46*:93–95
      (1987).
257. M. V. Graf, A. J. Kastin, D. H. Coy, and A. J. Fischman,
      Neuroendocrinology, *41*:353–356 (1985).
258. T. Okajima and G. Hertting, *Horm. Metab. Res.*, *18*:497–498
      (1986).
259. K. S. Iyer and S. M. McCann, *Peptides*, *8*:45–48 (1987).
260. A. Sahu and S. P. Kalra, *Life Sci.*, *40*:1201–1206 (1987).
261. M. V. Graf, A. J. Kastin, and G. A. Schoenenberger, *Pharmacol.
      Biochem. Behav.*, *26*:1797–1799 (1986).
262. M. V. Graf and G. A. Schoenenberger, *Peptides*, *7*:1001–1006
      (1986).
263. A. Nakamura, N. Nakashima, T. Sugao, H. Kanemoto, Y. Fukumura,
      and H. Shiomi, *Eur. J. Pharmacol.*, *155*:247–253 (1988).
264. A. Nakamura, M. Nakashima, K. Sakai, M. Niwa, M. Nozaki, and
      H. Shiomi, *Brain Res.*, *481*:165–168 (1989).
265. M. V. Graf and A. J. Kastin, *Neurosci. Biobehav. Res.*, *8*:83–93
      (1984).
266. N. Marks, F. Stern, A. J. Kastin, and D. H. Coy, *Brain Res.
      Bull.*, *2*:491–493 (1977).
267. J. T. Huang and A. Lajtha, *Res. Commun. Chem. Pathol.
      Pharmacol.*, *19*:191–199 (1978).
268. W. A. Banks, A. J. Kastin, D. H. Coy, and E. Angulo, *Brain Res.
      Bull.*, *17*:155–158 (1986).
269. W. A. Banks, A. J. Kastin, and D. H. Coy, *Pharmacol. Biochem.
      Behav.*, *17*:1009–1014 (1982).
270. W. A. Banks, A. J. Kastin, and D. H. Coy, *Brain Res.*, *301*:
      201–207 (1984).
271. C. H. Li and D. Chung, *Proc. Natl. Acad. Sci. USA*, *73*:1145–1148
      (1976).
272. H. H. Loh, L. F. Tseng, E. Wei, and C. H. Li, *Proc. Natl. Acad.
      Sci. USA*, *73*:2895–2898 (1976).
273. L. F. Tseng, H. H. Loh, and C. H. Li, *Nature*, *263*:239–240
      (1976).
274. T. P. Davis, A. J. Culling, H. Schoemaker, and J. J. Galligan,
      J. Pharmacol. Exp. Ther., *277*:699–507 (1983).
275. J. W. Holaday, *Biochem. Pharmacol.*, *32*:573–585 (1983).
276. J. Florez, A. Mediavilla, and A. Pazos, *Brain Res.*, *199*:197–206
      (1980).
277. W. C. Chang, A. J. Rao, and C. H. Li, *Int. J. Protein Res.*, *11*:
      93–94 (1978).
278. K. M. Foley, I. A. Kourides, C. E. Inturrisi, R. F. Kaiko, C. G.
      Zaroulis, J. B. Posner, R. W. Houde, and C. H. Li, *Proc. Natl.
      Acad. Sci. USA*, *76*:5377–5381 (1979).
279. R. A. Houghten, R. W. Swann, and C. H. Li, *Proc. Natl. Acad.
      Sci. USA*, *77*:4588–4591 (1980).

280. R. L. Reid, J. D. Hoff, S. S. C. Yen, and C. H. Li, *J. Clin. Endocrinol. Metab.*, *52*:1179–1184 (1981).

281. N. Aronin, M. Wiesen, A. S. Liotta, G. C. Schussler, and D. T. Krieger, *Life Sci.*, *29*:1265–1269 (1981).

282. H. Sato, Y. Sugiyama, Y. Sawade, T. Iga, and M. Hanano, *Life Sci.*, *35*:1051–1059 (1989).

283. X. Y. Bertanga, W. J. Stone, W. E. Nicholson, C. D. Mount, and D. N. Orth, *J. Clin. Invest.*, *67*:124–133 (1981).

284. H. Sato, Y. Sugiyama, Y. Sawada, T. Iga, and M. Hanano, *Drug Metab. Disp.*, *15*:540–550 (1987).

285. L. E. Wittmers, M. Bartlett, and J. A. Johnson, *Microvasc. Res.*, *11*:67–78 (1976).

286. N. Ito, Y. Sawada, Y. Sugiyama, T. Iga, and M. Hanano, *Jpn. J. Physiol.*, *35*:291–299 (1985).

287. L. J. Nugen and R. K. Jain, *Am. J. Physiol.*, *246*:H129–H137 (1989).

288. W. P. Paaske, *Microvasc. Res.*, *25*:101–107 (1983).

289. C. A. Goresky, P. M. Huet, and J. P. Villeneuve, in: *Hepatology: A Textbook of Liver Disease* (D. Zakim and T. Boyer, eds.), W. B. Saunders, Philadelphia (1982), pp. 32–63.

290. E. Wisse, R. De Zanger, K. Charles, P. V. D. Smissen, and R. S. McCuskey, *Hepatology (Baltimore)*, *5*:683–692 (1985).

291. S. I. Rapoport, W. A. Klee, K. D. Pettigrew, and K. Ohno, *Science*, *207*:84–86 (1980).

292. T. S. Reese and M. J. Karnovsky, *J. Cell Biol.*, *34*:207–217 (1967).

293. S. I. Rapoport, *Blood-Brain-Barrier in Physiology and Medicine*, Raven Press, New York (1976).

294. D. DeWied, *Annu. NY Acad. Sci.*, *297*:263–274 (1977).

295. S. I. Rapoport, K. Ohno, and K. D. Pettigrew, *Brain Res.*, *172*:354–359 (1979).

296. O. H. Viveros, E. J. Diliberto, Jr., E. Hazum, and K. J. Chang, in: *Natural Peptides and Neuronal Communication* (E. Costa and M. Trabucchi, eds.), Raven Press, NY (1980).

297. J. Hughes, *Nature*, *39*:17–24 (1983).

298. A. Dupont, L. Cusan, M. Garon, G. A. Urbina, and F. Labrie, *Life Sci.*, *21*:907–914 (1977).

299. L. G. Roda, G. Roscetti, R. Possenti, F. Venturelli, and F. Vita, in: *Enkephalins and Endorphins. Stress and the Immune System*, Plenum Press, New York and London (1986), pp. 17–33.

300. J. M. Hambrook, B. A. Morgan, M. J. Rance, and C. F. C. Smith, *Nature*, *262*:782–783 (1976).

301. M. A. Coletti-Previero, H. Mattras, B. Decomps, and A. Previero, *Biochim. Biophys. Acta*, *657*:122–127 (1981).

302. L. B. Hersh and J. F. McKelvy, *J. Neurochem.*, *36*:171–178 (1981).

303. B. Malfroy, J. P. Swerts, A. Guyon, B. P. Roques, and J. C. Schwartz, *Nature (London)*, *276*:523–526 (1978).

304. S. Sullivan, H. Akil, and J. D. Barchas, *Commun. Psychopharmacol.*, *2*:525–531 (1978).

305. J. P. Schwartz, B. Malfroy, and S. De la Baume, *Life Sci.*, *29*:1715–1740 (1981).

306. J. Almenoff, S. Wilks, and M. Orlowski, *Biochem. Biophys. Res. Commun.*, *102*:202–214 (1981).

307. M. Benuck and N. Marks, *Biochem. Biophys. Res. Commun.*, *95*: 822–828 (1980).

308. H. H. Buscher, R. C. Hill, D. Romer, F. Cardinaux, A. Closse, D. Hauser, and J. Pless, *Nature*, *261*:423–425 (1976).

309. S. M. Flechner, *Urol. Clin. North Am.*, *10*:263–275 (1983).

310. T. E. Starzl, S. Iwatsuki, D. H. Van Thiel, J. C. Gartner, B. J. Zitelli, J. J. Malatack, R. R. Schade, B. W. Shaw, Jr., T. R. Hakaka, T. Rosenthal, and K. A. Porter, *Hepatology*, *2*:614–636 (1982).

311. Preliminary results of a European multicentre trial, *Lancet*, *2*:57–60 (1982) (editorial).

312. T. E. Starzl, R. Weill, S. Iwatsuki, G. Klintmalm, G. P. J. Schroter, L. J. Koep, Y. Iwaki, P. I. Terasaki, and K. A. Porter, *Surg. Gynecol. Obstet.*, *151*:17–26 (1980).

313. S. Thiru, R. Y. Calne, and J. Nagington, *Transplant. Proc.*, *13*: 359–364 (1981).

314. A. Laupacis, P. A. Keown, R. A. Ulan, N. McKenzie, and C. R. Stiller, *Can. Med. Assoc. J.*, *126*:1041–1046 (1982).

315. R. J. Ptachcinski, R. Venkataramanan, and G. J. Burckart, *Clin. Pharmacokinetics*, *11*:107–132 (1986).

316. P. Donatsch, E. Abisch, M. Homberger, R. Trabar, M. Trapp, and R. Voger, *J. Immunoassay*, *2*:19–32 (1981).

317. V. Quesniaux, R. Tees, M. H. Schreier, G. Maurer, and M. H. V. Van Regenmortel, *Clin. Chem.*, *33*:32–37 (1987).

318. B. M. Freed, T. G. Rosano, and N. Lempert, *Transplantation*, *43*: 123–127 (1987).

319. B. Ryffel, P. Hiestand, P. Foxwell, P. Donatsch, H. J. Boelsterli, G. Maurer, and M. J. Mihatsch, *Transplant Proc.*, *6*(Suppl. 5): 41–45 (1986).

320. K. Leunissen, F. Bosman, G. Beuman, and van Hoff, *Second Intl. Cong. on Cyclosporine, Book of Abstracts*, Washington, DC (Nov. 4–7, 1987), p. 239.

321. U. Kunzendorf, J. Brockmoller, F. Jochimsen, F. Keller, I. Roots, G. Waltz, and G. Offermann, *Klin. Wochenschr.*, *67*:438–441 (1989).

322. G. C. Yee, D. J. Gmur, and M. S. Kennedy, *Clin. Chem.*, *28*: 2269–2271 (1982).

323. R. L. Kates and R. Latini, *J. Chromatography*, *309*:441–447 (1984).

324. R. J. Sawchuck and L. L. Cartier, *Clin. Chem.*, *27*:1368–1371 (1981).

325. D. J. Freeman, A. Laupacis, P. A. Keown, C. R. Stiller, and S. G. Carruthers, *Br. J. Clin. Pharmacol.*, *18*:887–893 (1984).

326. E. Abisch, T. Beveridge, A. Gratwohl, W. Niederberger, K. Nussabaumer, P. Schaub, B. Speck, and M. Trapp, *Pharm. Weekbl. Sci. Ed.*, *4*:84–86 (1982).

327. B. D. Kahan, *Transplantation*, *40*:457–476 (1985).

328. G. Maurer, H. R. Loosli, E. Schreier, and B. Keller, *Drug Metab. Dispos.*, *12*:120–126 (1984).

329. T. Beveridge, in: *Cyclosporin A* (D. J. G. White, ed.), Elsevier Biomedical, New York (1982), pp. 5–17.

330. T. Beveridge, A. Gratwohl, F. Michot, W. Niederberger, E. Nuescha, K. Nussabaumer, P. Schaul, and B. Speck, *Current Ther. Res.*, *30*: 5–18 (1981).

331. G. C. Yee, M. S. Kennedy, R. Storb, and E. D. Thomas, *Transplantation, 38*:511–513 (1984).

332. G. Hamilton, F. Muhlbacher, E. Roth, I. Wolf, F. Piza, M. Havel, A. Lockovics, J. Schindler, and W. Wolosczuk, *Transplant Proc., 19*:1706–1708 (1987).

333. S. Aziz, P. Y. Dyer, and R. E. Kates, *J. Clin. Pharmacol., 26*: 652–657 (1986).

334. P. E. Wallemacq, M. Lesne, and J. B. Otte, *Clin. Transplantation, 1*:132–137 (1987).

335. F. Follath, M. Wenk, S. Vozeh, G. Thiel, F. Brunner, R. Loertscher, M. LeMaire, K. Nussabaumer, W. Niederbager, and A. Wood, *Clin. Pharmacol. Ther., 34*:638–643 (1983).

336. M. R. Waters, J. D. M. Albano, V. L. Sharman, and G. Venkat Raman, *Clin. Sci., 71*(Suppl. 15):2P (1986).

337. R. J. Ptachcinski, R. Venkataramanan, J. T. Rosenthal, G. J. Burckart, R. J. Taylor, and T. R. Hakala, *Clin. Pharm. Ther., 38*: 296–300 (1985).

338. T. G. Rosano, *Clin. Chem., 31*:410–412 (1985).

339. S. Venkataram, W. M. Awni, K. Jordan, and Y. E. Rahman, *J. Pharm. Sci., 79*:216–219 (1990).

340. K. Vadiei, R. Perez-Soler, G. Lopez-Berestein, and D. R. Luke, *Int. J. Pharm., 57*:125–131 (1989).

341. W. Mraz, R. A. Zink, and A. Graf, *Transplant. Proc., 15*: 2426–2429 (1983).

342. J. K. Fazakerley and H. E. Webb, *Lancet, 2*:889–890 (1985).

343. W. T. Cefalu and W. M. Pardridge, *J. Neurochemistry, 45*: 1954–1956 (1985).

344. A. G. Palestine, R. B. Nussenblatt, and C. C. Chan, *Am. J. Ophthalmol., 99*:210–211 (1985).

345. S. M. Flechner, A. R. Katz, A. J. Rogers, C. Van Buren, and B. D. Kahan, *Am. J. Kidney Dis., 5*:60–63 (1985).

346. P. A. Keown, C. R. Stiller, R. A. Ulan, N. R. Sinclair, W. J. Wall, G. Carruthers, and W. Howson, *Lancet, 1*:686–689 (1981).

347. C. T. Ueda, M. LeMaire, G. Gsell, P. Misslin, and K. Nussbaumer, *Biopharm. Drug Dispos., 5*:141–144 (1984).

348. C. T. Ueda, M. LeMaire, G. Gsell, and K. Nussbaumer, *Biopharm. Drug Dispos., 4*:113–124 (1983).

349. K. Atkinson, J. C. Biggs, K. Britton, R. Short, R. Mrongovius, A. Concannon, and A. Doddls, *Br. J. Haematol., 56*:223–231 (1984).

350. N. K. Wadhwa, T. J. Schroeder, E. O'Flaherly, A. J. Pesce, S. A. Myre, and M. R. First, *Transplant. Proc., 19*:1730–1733 (1987).

351. N. K. Wadhwa, T. J. Schroeder, A. J. Pesce, S. A. Myre, C. W. Clardy, and M. R. First, *Ther. Drug Monit., 9*:399–406 (1987).

352. B. Z. Ericzon, S. Todo, S. Lynch, I. Kam, R. J. Ptachcinski, G. J. Burckart, D. H. Van Thiel, T. E. Starzl, and R. Venkataramanan, *Transplant. Proc., 19*:1248–1249 (1987).

353. R. Venkataramanan, F. E. Starzl, S. Yang, G. J. Burckart, R. J. Ptachcinski, B. W. Shaw, S. Iwatsuki, D. H. Van Thiel, A. Sanghui, and H. Seltman, *Transplant. Proc., 17*:286–289 (1985).

354. R. J. Ptachcinski, R. Venkataramanan, J. T. Rosenthall, G. J. Burckart, R. J. Taylor, and T. R. Hakala, *Transplantation*, 40: 174–176 (1985).

355. A. Johnston, J. T. Marsden, K. K. Hla, J. A. Henry, and D. W. Holt, *Br. J. Clin. Pharmacol.*, 21:331–333 (1986).

356. B. Ota, *Transplant. Proc.*, 17:1252–1255 (1983).

357. J. Newberger and B. D. Kahan, *Transplant. Proc.*, 15:2413–2415 (1983).

358. S. Pestka, *Meth. Enzymol.*, 119:3–14 (1986).

359. C. Weissmann and H. Weber, *Prog. Nucl. Acids Res. Mol. Biol.*, 33: 251–300 (1986).

360. K. C. Zoon, R. Q. Hu, D. Zur Nedden, and N. Nguyen, in: *The Biology of the Interferon System 1984* (H. Kirchner and H. Schellekens, eds.), Elsevier, Amsterdam (1985), pp. 61–67.

361. K. C. Zoon, *Interferon*, 9:1–12 (1987).

362. G. Bodo and G. R. Adolf, in: *The Biology of the Interferon System* (E. D. Meyer and H. Schellekens, eds.), Elsevier, Amsterdam (1983), pp. 113–118.

363. G. Bodo and I. Fogy, in: *The Interferon System* (F. Dianzani and G. B. Rossi, eds.), Serono Symposia 24, Raven Press, New York (1985), pp. 23–27.

364. V. Bocci, G. P. Pessina, A. Pacini, L. Paulesu, M. Muscettola, and G. Lunghetti, *Gen. Pharmacol.*, 16:277–279 (1985).

365. V. Bocci, A. Pacini, E. Maioli, M. Muscettola, and L. Paulesu, *IRCS Med. Sci.*, 14:360–361 (1986).

366. V. Bocci, *Pharmacol. Ther.*, 34:1–49 (1987).

367. G. Emodi, M. Just, R. Hernandez, and H. R. Hort, *J. Natl. Cancer Inst.*, 54:1045–1049 (1975).

368. R. J. Wills, H. E. Speigel, and K. F. Soike, *J. Interferon Res.*, 4: 399–409 (1984).

369. R. J. Wills and H. E. Speigel, *J. Clin. Pharmacol.*, 25:616–619 (1985).

370. R. J. Wills, S. Dennis, H. E. Spiegel, D. M. Gibson, and P. I. Nalder, *Clin. Pharmacol. Ther.*, 35:722–727 (1984).

371. L. D. Bornemann, H. E. Spiegel, Z. E. Dziewanaowska, J. E. Krown, and W. A. Colburn, *Eur. J. Clin. Pharmacol.*, 28:469–471 (1985).

372. J. U. Gutterman, S. Fine, J. Quesada, S. J. Horning, J. F. Levine, R. Alexanian, L. Bernhardt, M. Kramer, H. Spiegel, W. Colburn, P. Trown, T. Merigan, and Z. Dziewanowski, *Annu. Intern. Med.*, 96:549–556 (1982).

373. S. A. Sherwin, J. A. Knost, S. Fein, P. G. Abramas, K. A. Foon, J. J. Ochs, C. Schoenberger, A. E. Maluish, and R. K. Oldhamal, *JAMA*, 248:2461–2466 (1982).

374. A. Z. S. Rohatiner, F. R. Balkwill, D. B. Griffin, J. S. Malpas, and T. A. Lister, *Cancer Chemother. Pharmacol.*, 9:97–102 (1982).

375. M. S. Mahley, M. B. Urso, R. A. Whaley, M. Blue, T. E. Williams, A. Guaspari, and R. G. Selker, *J. Neurosurg.*, 63:719–725 (1985).

376. R. A. Smith, F. Norris, D. Palmer, L. Bernhardt, and R. J. Wills, *Clin. Pharmacol. Ther.*, 37:85–88 (1985).

377. D. V. Habif, R. Lipton, and K. Cantell, *Proc. Soc. Exp. Biol. Med.*, *149*:287–289 (1985).

378. V. Bocci, *Cancer Drug Del.*, *1*:337–351 (1984).

379. A. M. Salazar, G. J. Gibbs, D. C. Gajdust, and R. A. Smith, in: *Interferons and Their Applications* (P. E. Came and W. A. Carter, eds.), Springer Verlag, Berlin (1984), pp. 471–497.

380. V. Bocci, *J. Biol. Resp. Mod.*, *4*:340–352 (1985).

381. V. Bocci, *Immunol. Today*, *6*:7–9 (1985).

382. V. Bocci, M. Muscettola, G. Grasso, Z. S. Magyar, A. Naldini, and G. Szabo, *Experientia*, *42*:432–433 (1986).

383. V. Bocci, M. Muscettola, A. Naldini, E. Bianchi, and G. Segre, *Gen. Pharmacol.*, *17*:93–96 (1986).

384. V. Bocci, M. Muscettola, and A. Naldini, *Int. J. Pharmaceutics*, *32*: 103–110 (1986).

385. G. M. Scott, in: *Interferon*, Vol. 5 (L. Gresser, ed.), Academic Press, New York (1983), pp. 85–114.

386. E. Knight, Jr., *Proc. Natl. Acad. Sci. USA*, *73*:520–523 (1976).

387. T. Taniguchi, M. Sakai, Y. Fuji-Kuriyama, M. Muramatsu, S. Kobayashi, and T. Sudo, *Proc. Jpn. Acad.*, *B55*:464–469 (1979).

388. D. V. Mark, S. D. Lu, A. Creasey, R. Yamamoto, and L. S. Lin, *Proc. Natl. Acad. Sci. USA*, *81*:5662–5666 (1986).

389. D. Mark, R. Drummond, A. Creasey, L. Lin, B. Khosrovi, B. Edwards, D. Groveman, J. Joseph, M. Hawkins, and E. Borden, *Interferons, Proc. Intl. Symp. Interferons*, Kyoto, Japan (1983), pp. 167–172.

390. V. Bocci, M. Muscettola, and A. Naldini, *Gen. Pharmacol.*, *17*: 445–498 (1986).

391. Y. Koyama, in: *Interferons* (T. Kishida, ed.), Japan Convention Services, Osaka (1983), pp. 189–195.

392. V. Bocci, G. P. Pessina, L. Paulesu, M. Muscettola, and A. Valeri, *J. Interferon Res.*, *8*:633–640 (1988).

393. G. Sarna, M. Pertchek, R. Figlin, and B. Ardalan, *Cancer Treat. Rep.*, *70*:1365–1372 (1986).

394. J. Blalock, M. Georgiades, M. Langford, and H. Johnson, *Cell. Immunol.*, *49*:390–394 (1980).

395. B. Y. Rubin and S. L. Gupta, *Proc. Natl. Acad. Sci. USA*, *77*: 5928–5932 (1980).

396. D. Weigent, G. Stanton, and H. Johnson, *Biochem. Biophys. Res. Commun.*, *111*:525–529 (1983).

397. K. Pfitzenmaier, H. Bartsch, P. Schemich, B. Saliger, V. Ucer, K. Vehmeyer, and G. Nagel, *Cancer Res.*, *45*:3503–3509 (1985).

398. D. Fertsch and S. Vogel, *J. Immunol.*, *132*:2436–2439 (1984).

399. G. Gastl, C. Marth, E. Leiter, C. Gattringer, I. Mayer, G. Daxenbichler, R. Flener, and C. Huber, *Cancer Res.*, *45*: 2957–2961 (1985).

400. J. L. Virelizer, N. Perez, F. Arenzana-Seisdedos, and R. Devos, *Eur. J. Immunol.*, *14*:106–108 (1984).

401. K. A. Foon, S. A. Sherwin, P. Abrahms, H. Stevenson, P. Holmes, A. Maulish, R. Oldham, and R. Herberman, *Cancer Immunol. Immunother.*, *20*:193–197 (1985).

402. R. Kurzrock, M. G. Rosenblum, S. A. Sherwin, A. Rios, M. Talpaz, J. Quesada, and J. Gutterman, *Cancer Res.*, 45:2866-2872 (1985).

403. M. Van der Burg, M. Edelstein, L. Gerlis, C.-M. Liang, M. Hirschi, and A. Dawson, *J. Biol. Response Mod.*, 4:264-272 (1985).

404. S. Vadhan-Raj, C. F. Nathan, S. A. Sherwin, H. F. Oettgen, and S. E. Krown, *Cancer Treat. Rep.*, 70:609-614 (1986).

405. J. A. Thompson, W. W. Cox, C. G. Lindgren, C. Collins, K. A. Neraas, E. M. Bonnem, and A. Fefer, *Cancer Immunol. Immunother.*, 25:47-53 (1987).

406. J. J. Rinehart, L. Malspeis, D. Young, and J. A. Neidhart, *J. Biol. Response Mod.*, 5:300-308 (1986).

407. R. Kurzrock, J. R. Quesada, M. G. Rosenblum, S. A. Sherwin, and J. U. Gutterman, *Cancer Treat. Rep.*, 70:1357-1364 (1986).

408. J. Wagstaff, D. Smith, P. Nelmes, P. Loynds, and D. Crowther, *Cancer Immunol. Immunother.*, 25:54-58 (1987).

409. B. L. Ferraiolo, G. B. Fuller, B. Burnett, and E. Chan, *J. Biol. Response Mod.*, 7:115-122 (1988).

410. B. M. Bowman, M. M. Gagen, E. Bonnem, J. A. Ajani, S. Schmidt, I. W. Dimery, J. Golando, and J. Neidhart, *J. Biol. Response Mod.*, 7:438-446 (1988).

411. H. C. Lane, R. T. Davey, Jr., S. A. Sherwin, H. Masur, A. H. Rook, J. F. Manischewitz, G. V. Quinnan, P. D. Smith, M. E. Easter, and A. S. Fauci, *J. Clin. Immunol.*, 9:351-361 (1989).

412. S. L. Gonias, S. V. Pizzo, and M. Hoffman, *Cancer Res.*, 48:2021-2024 (1988).

413. I. Rutenfranz and H. Kirchner, *J. Interferon Res.*, 8:573-580 (1988).

414. S. A. Chen, M. R. Shalaby, D. R. Crase, M. A. Palladino, Jr., and R. A. Baughman, *J. Interferon Res.*, 8:597-608 (1988).

415. K. Cantell, W. Fiers, S. Hirvonen, and L. Pyhala, *J. Interferon Res.*, 4:291-293 (1983).

416. K. Cantell, S. Hirvonen, L. Pyhala, A. De Reus, and S. Schellekins, *J. Gen. Virol.*, 64:1823-1826 (1983).

417. R. Eife, T. Hahn, and M. De Taverna, *J. Immunol. Methods*, 47:339-347 (1981).

418. M. R. Shalaby, L. P. Svedersky, P. A. McKay, B. S. Finkle, and M. A. Palladino, Jr., *J. Interferon Res.*, 5:571-582 (1985).

419. V. M. Kushnaryou, H. S. MacDonald, J. J. Sedmak, and S. E. Gossberg, *Biochem. Biophys. Res. Commun.*, 157:109-114 (1988).

420. D. A. Morgan, F. W. Ruscetti, and R. Gallo, *Science*, 193:1007-1008 (1976).

421. J. Gootenberg, F. Ruscetti, J. Mier, A. Gazarz, and R. C. Gallo, *J. Exp. Med.*, 154:1403-1418 (1981).

422. J. K. Yamamoto, W. L. Farrar, and H. M. Johnson, *Cell Immunol.*, 66:333-341 (1982).

423. M. Howard, L. Matis, T. R. Malek, et al., *J. Exp. Med.*, 158:2024-2034 (1983).

424. C. S. Henney, K. Kuribayashi, D. E. Kern, and S. Gillis, *Nature*, 291:333-337 (1981).

425. S. Gillis and K. Smith, *Nature*, 268:154-155 (1977).

426. E. A. Grimm, A. Mazumdar, H. Z. Zhang, and S. A. Rosenberg, *J. Exp. Med.*, 155:1823-1841 (1982).

427. E. A. Grimm, K. Ramsey, A. Mazumdar, D. J. Wilson, J. Y. Djeu,
     and S. A. Rosenberg, *J. Exp. Med.*, *157*:884 – 897 (1983).
428. E. Grimm, R. J. Robb, J. A. Roth, L. M. Neckers, L. B.
     Lachman, D. J. Wilson, and S. A. Rosenberg, *J. Exp. Med.*, *158*:
     1356 – 1361 (1983).
429. R. J. Robb, *Methods Enzymol.*, *116*:493 – 525 (1985).
430. K. Welte, C. Y. Wang, R. Mertelsmann, S. Venuta, S. P. Feldman,
     and M. A. S. Moore, *J. Exp. Med.*, *156*:454 – 464 (1982).
431. S. A. Rosenberg, E. A. Grimm, M. McGrogan, M. Doyle,
     E. Kawasaki, K. Koths, and D. F. Mark, *Science*, *223*:1412 – 1415
     (1984).
432. L. E. Gustarson, R. W. Nadeau, and N. F. Oldfield, *J. Biol.
     Resp. Mod.*, *8*:440 – 449 (1989).
433. H. Sands and S. E. Loveless, *Int. J. Immunopharmacol.*, *11*:
     411 – 416 (1989).
434. R. Gennuso, M. K. Spigelman, S. Vallabhajosula, F. Moore, R. A.
     Zappulla, J. Nieves, J. A. Strauchen, P. A. Paciucci, L. I. Malis,
     S. J. Goldsmith, and J. F. Holland, *J. Biol. Resp. Modifiers*, *8*:
     375 – 384 (1989).
435. H. Ohnishi, J. T. Y. Chao, K. K. M. Lin, H. Lee, and T. M.
     Chu, *Tumor Biol.*, *10*:202 – 214 (1989).
436. D. Pennica, G. E. Nedwin, J. S. Hayflick, P. H. Seeburg,
     R. Derynck, M. A. Palladino, W. J. Kohr, B. B. Aggarwal, and
     D. V. Goeddel, *Nature*, *312*:724 – 729 (1984).
437. B. Beutler, D. Greenwald, J. D. Hulmes, M. Chang, Y.-C. E. Pan,
     J. Mathison, R. Ulevitch, and A. Cerami, *Nature*, *316*:552 – 554
     (1985).
438. L. J. Old, *Science*, *230*:630 – 632 (1986).
439. G. Degliantoni, M. Murphy, M. Kobayashi, M. K. Francis,
     B. Perussia, and G. Trinchieri, *J. Exp. Med.*, *162*:1512 – 1530
     (1985).
440. J. P. Scedersky, G. E. Nedwin, T. X. Bringman, M. R. Shalaby,
     J. A. Lamott, D. V. Goeddel, and M. A. Palladino, Jr., *Fed.
     Proc.*, *44*:589 (1985) (abstract).
441. J. R. Ortaldo, L. H. Mason, B. J. Mathieson, S.-M. Liang, D. A.
     Flick, and R. B. Herberman, *Nature*, *321*:700 – 702 (1986).
442. P. M. Peters, J. R. Ortaldo, M. R. Shalaby, L. P. Svedersky, G. E.
     Nedwin, T. S. Bringman, P. E. Hals, B. B. Aggarwal, R. B. Heberman,
     D. V. Goeddel, and M. A. Palladino, Jr., *J. Immunol.*, *137*:2592 – 2598
     (1986).
443. M. R. Shalaby, D. Pennica, and M. A. Palladino, Jr., *Semin.
     Immunopathol.*, *9*:33 – 37 (1986).
444. D. V. Goeddel, B. B. Aggarwal, P. W. Gray, D. W. Leung, G. E.
     Nedwin, M. A. Palladino, J. S. Patton, D. Pennica, H. M. Shepard,
     B. J. Sugarman, and G. H. W. Wong, *Cold Spring Harbor Symp.
     Quant. Biol.*, *51*:597 – 609 (1986).
445. C. A. Dinarello, J. G. Cannon, S. M. Wolff, H. A. Bernheim,
     B. Beutler, A. Cerami, I. S. Figari, M. A. Palladino, Jr., and
     J. V. O'Connor, *J. Exp. Med.*, *163*:1433 – 1450 (1986).
446. P. P. Nawroth, I. Bank, D. Handley, J. Cassimeris, L. Chess, and
     D. Stern, *J. Exp. Med.*, *163*:1363 – 1375 (1986).
447. T. Collins, L. A. Lapierre, W. Fiers, J. L. Strominger, and J. S.
     Pober, *Proc. Natl. Acad. Sci. USA*, *83*:446 – 450 (1986).

448. M. R. Shalaby, B. B. Aggarwal, E. Rinderknecht, L. P. Svedersky, B. S. Finkle, and M. A. Palladino, Jr., *J. Immunol.*, *135*:2069–2073 (1985).

449. S. J. Klebanoff, M. A. Vadas, J. M. Harlan, L. H. Sparks, J. R. Gamble, J. M. Agosti, and A. M. Waltersdorph, *J. Immunol.*, *136*: 4220–4225 (1986).

450. M. R. Shalaby, M. A. Palladino, Jr., S. E. Hirabayashi, T. E. Eessalu, G. D. Lewis, H. M. Shepard, and B. B. Aggarwal, *J. Leukocyte Biol.*, *41*:196–204 (1987).

451. F. M. Torti, B. Dieckmann, B. Beutler, A. Cerami, and G. M. Ringold, *Science*, *229*:867–869 (1985).

452. B. J. Sugarman, B. B. Aggarwal, P. E. Hass, I. S. Figari, M. A. Palladino, Jr., and H. M. Shepard, *Science*, *230*:943–945 (1985).

453. M. R. Shalaby, S. E. Hirabayashi, L. P. Svedersky, and M. A. Palladino, Jr., *Fed. Proc.*, *44*:569–571 (1985) (abstract).

454. L. Helson, S. Green, E. Carswell, and L. J. Old, *Nature*, *258*: 731–732 (1975).

455. L. Helson, C. Helson, and I. Gree, *Exp. Cell Biol.*, *47*:53–60 (1975).

456. K. Haranaka and N. Satomi, *Jpn. J. Exp. Med.*, *51*:191–194 (1981).

457. K. Nakano, S. Abe, and Y. Sohmura, *Int. J. Immunopharmacol.*, *8*: 347–355 (1986).

458. M. A. Palladino, Jr., M. R. Shalaby, S. M. Kramer, A. B. DeLe, D. A. Baughman, and J. S. Patton, *J. Immunol.*, *138*:4023–4032 (1987).

459. E. A. Carswell, L. J. Old, R. L. Kassel, S. Green, N. Fiore, and B. Williamson, *Proc. Natl. Acad. Sci. USA*, *72*:3666–3670 (1975).

460. K. Haranaka, N. Satomi, and A. Sakurai, *Int. J. Cancer*, *34*: 263–267 (1984).

461. N. Watanabe, Y. Niitsu, H. Sone, H. Neda, I. Urushizaki, A. Yamamoto, M. Nagamuta, and Y. Sugawara, *J. Immunopharmacol.*, *8*:271–283 (1986).

462. M. Blick, S. A. Sherwin, M. Rosenblum, and J. Gutterman, *Cancer Res.*, *47*:2986–2989 (1987).

463. B. Widenmann, P. Reichardt, U. Rath, L. Theilmann, B. Schule, A.-D. Ho, E. Schlick, J. Kempeni, W. Hunstein, and B. Kommerell, *J. Cancer Res. Clin. Oncol.*, *115*:189–192 (1989).

464. B. B. Aggarwal, *Drugs Future*, *12*:891–898 (1987).

465. P. Wingfield, R. H. Pain, and S. Craig, *FEBS Lett.*, *211*:179–184 (1987).

466. G. P. Pessina, A. Pacini, V. Bocci, E. Maioli, and A. Naldini, *Lymphokine Res.*, *6*:35–44 (1986).

467. V. Bocci, A. Pacini, G. P. Pessina, E. Maioli, and A. Naldini, *Gen. Pharmacol.*, *18*:343–346 (1987).

468. B. L. Ferraiolo, J. A. Moore, D. Crase, P. Gribling, H. Wilking, and R. A. Baughman, *Drug Metab. Dispos.*, *16*:270–275 (1988).

469. A. C. Allison, *Nature*, *188*:37–40 (1960).

470. A. Greischel and G. Zahn, *J. Pharmacol. Exp. Ther.*, *251*:358–361 (1989).

471. O. Pichyangkul, J. E. Miller, S. Waldrop, and A. Khan, *Clin. Immunol. Immunopathol.*, *35*:22–34 (1985).

472. D. S. Stone-Wolff, Y. K. Yip, H. C. Kelker, J. Le, D. Henriksen-Destefano, B. Y. Rubin, E. Rinderknechct, B. B. Aggarwal, and J. Vilecek, *J. Exp. Med.*, *159*:828–843 (1984).

473. S. Pichyangkul, N. O. Hill, and A. Khan, in: *Human Lymphokines* (A. Khan and N. O. Hill, eds.), Academic Press, New York (1982), pp. 173–183.

474. B. B. Aggarwal, B. Moffat, and R. N. Harkins, *J. Biol. Chem.*, *259*:686–691 (1986).

475. B. B. Aggarwal, W. J. Kohr, P. E. Hoss, B. Moffat, S. A. Spencer, W. J. Henzel, T. S. Bringman, G. E. Nedwin, D. V. Goeddel, and R. N. Harkins, *J. Biol. Chem.*, *260*:2345–2354 (1985).

476. P. W. Gray, B. B. Aggarwal, C. V. Benton, T. S. Bringman, Λ. J. Henzel, J. A. Jarrett, D. W. Leung, B. Moffat, P. Ng, L. P. Svedersky, M. A. Palladino, and G. E. Nedwin, *Nature (London)*, *312*:721–724 (1984).

477. E. Jeffes, B. Averbook, T. Ulich, and G. Ganger, *Lymphokine Res.*, *6*:141–149 (1987).

478. P. F. Torrence, in: *Lymphotoxins: A Multicomponent Family of Cell-Lytic and Growth Inhibitory Proteins* (G. Granger, R. Yamamoto, and S. Orr, eds.), Academic Press, New York (1985), pp. 293–306.

479. B. B. Aggarwal, R. A. Aiyer, D. Pennica, P. W. Gray, and D. V. Goeddel, *CIBA Found. Symp.*, *131*:39–51 (1987).

480. B. J. Averbook, E. B. Jeffes, R. S. Yamamoto, I. Masunata, M. Kobayashi, and G. A. Granger, *J. Biol. Resp. Mod.*, *8*:344–350 (1989).

481. P. F. Hirsch and P. L. Munson, *Physiol. Rev.*, *49*:548–622 (1969).

482. K. E. W. Melvin, A. H. Tashjian, Jr., and H. H. Miller, *Recent Prog. Horm. Res.*, *28*:399 (1972).

483. J. Moran, W. Hunziker, and J. A. Fischer, *Proc. Natl. Acad. Sci. USA*, *75*:3948–3988 (1978).

484. F. E. Newsome, R. K. O'Dor, C. O. Parkes, and D. H. Copp, *Endocrinology*, *92*:1102–1106 (1973).

485. L. Galante, G. F. Joplin, I. MacIntire, and N. J. Y. Woodhouse, *Clin. Sci.*, *44*:605–610 (1973).

486. R. Huwyler, W. Born, E. E. Ohnhaus, and J A. Fischer, *Am. J. Physiol.*, *5*:E15–E19 (1979).

487. F. R. Singer, J. F. Habener, E. Green, P. Godin, and J. T. Potts, Jr., *Nature New Biol.*, *237*:269–270 (1972).

488. M. B. Clark, C. C. Williams, B. M. Nathanson, R. E. Horton, H. I. Glass, and G. V. Foster, *J. Endocrinol.*, *61*:199–210 (1974).

489. K. Forslund, P. Slanina, M. Stridsberg, and L. E. Appelgren, *Acta Pharmacol. Toxicol.*, *46*:398–400 (1980).

490. R. Ardaillou, F. Paillard, C. Savier, and A. Bernier, *Rev. Eur. Etud. Clin. Biol.*, *16*:1031–1036 (1971).

491. R. E. Simmons, J. T. Hjelle, C. Mahoney, L. J. Deftos, W. Lisker, P. Kato, and R. Rabkin, *Am. J. Physiol.*, *254*:F595–F600 (1988).

492. L. G. Raisz, J. S. Brand, W. Y. U. Au, and I. Niemann, in: *Parathyroid Hormone and Thyrocalcitonin* (R. V. Talmage and L. F. Belanger, eds.), Excerpta Medica, Amsterdam (1968), pp. 370–380.

493. J. F. Obie and C. W. Cooper, J. Pharmacol. Exp. Ther., 209: 422−428 (1979).

494. C. H. Chesnut, H. E. Gruber, J. L. Ivey, M. Matthews, K. Sisom, W. B. Nelp, H. M. Juggart, B. A. Roos, and D. J. Baylink, in: Human Calcitonin (A. Canniggia, ed.), CIBA-GEIGY, Stresa, Italy (1983), pp. 136−140.

495. Z. Bouizar, W. H. Rostene, and G. Milhaud, Proc. Natl. Acad. Sci. USA, 84:5125−5218 (1987).

496. A. H. Tashjian, D. A. Wright, Jr., J. L. Ivey, and A. Pont, Rec. Prog. Horm. Res., 34:285−332 (1978).

497. D. M. Chabardes, M. Imbert-Teboul, A. C. B. Montegut, and F. Morell, Proc. Natl. Acad. Sci. USA, 73:3608−3612 (1976).

498. F. Morell, Am. J. Physiol., 240:F159−F164 (1983).

499. G. A. Quamme, Am. J. Physiol., 238:E573−E578 (1980).

500. J. M. Elalouf, N. Roinel, and C. de Rouffiniac, Am. J. Physiol., 246:F213−F220 (1984).

501. C. de Roffiniac and J. M. Elalouf, Am. J. Physiol., 245:F506−F511 (1983).

502. P. E. Belchetz, T. M. Plant, Y. Nakai, E. J. Keogh, and E. Knobil, Science, 202:631−633 (1978).

503. A. V. Schally, A. Arimura, and D. H. Coy, Vitam. Horm., 38: 257−323 (1980).

504. R. S. Swerdloff and D. Heber, Annu. Rev. Med., 34:491−500 (1983).

505. E. C. Griffiths and J. R. McDermott, Mol. Cel. Endocrinol., 33: 1−15 (1984).

506. A. Dupont, F. Labrie, G. Pelletier, R. Puviani, D. H. Coy, E. J. Coy, and A. Schally, Neuroendocrinology, 16:65−73 (1974).

507. J. Sandow and W. Konig, J. Endocrinol., 81:175−182 (1979).

508. T. W. Redding, A. J. Kastin, D. Gonzalez-Barcena, D. H. Coy, E. J. Coy, D. S. Schalch, and A. V. Schally, J. Clin. Endocrinol. Metab., 37:626−631 (1973).

509. J. P. Bourguignon, C. Hoyoux, A. Reuter, P. Franchimont, C. Leinartz-Dourcy, and Y. Brindts-Gevaert, J. Clin. Endocrinol. Metab., 48:78−84 (1979).

510. S. L. Jeffcoate, R. H. Greenwood, and D. T. Holland, J. Endocrinol., 60:305−314 (1974).

511. P. Virkunnen, H. Lybek, J. Partanen, T. Ranta, J. Leppaluoto, and M. Seppala, J. Clin. Endocrinol. Metab., 39:702−705 (1984).

512. J. Sandow, R. N. Clayton, and H. Kuhl, in: Endocrinology of Human Infertility: New Aspects. Proceedings of the Serono Clinical Colloquia on Reproduction No. 2 (P. G. Crosignani and B. L. Rubin, eds.), Academic Press, London (1981), p. 221.

513. R. L. Chan and C. A. Nerenberg, in: LHRH and Its Analogs — Contraceptive and Therapeutic Applitations, Part 2 (B. H. Vickery and J. J. Nestor, Jr., eds.), MTP Press, Lancaster (1987), pp. 577−593.

514. J. J. Nestor, Jr., T. L. Ho, R. A. Simpson, B. L. Horner, G. H. Jones, G. I. McRae, and B. H. Vickery, J. Med. Chem., 25: 795−801 (1982).

515. J. J. Nestor, Jr., R. Tahilramani, T. L. Ho, G. I. McRae, and B. H. Vickery, J. Med. Chem., 27:1170−1174 (1984).

516. R. L. Chan and M. D. Chaplin, Biochem. Biophys. Res. Commun., 127:673−679 (1985).

517. L. E. Seyler, *J. Clin. Endocrinol. Metab.*, *41*:1155–1160 (1975).
518. A. D. Swift and D. B. Chrichton, *J. Endocrinol.*, *80*:141–152 (1979).
519. N. Lahlou, M. C. Feinstein, B. Kerdelhue, M. Roger, G. Schiason, and R. Scholler, *Pathol. Biol. (Paris)*, *31*:649–651 (1983).
520. D. J. Handelsman, R. P. S. Hansen, L. M. Boylan, J. A. Spaliviero, and J. R. Turtle, *J. Clin. Endocrinol. Metab.*, *59*:739–746 (1984).
521. G. Fink, G. Gennser, P. Liedholm, J. Thorell, and J. Mulder, *J. Endocrinol.*, *63*:351–360 (1974).
522. L. Tharandt, H. Schulte, G. Benker, K. Hackenberg, and D. Reinwein, *Horm. Metab. Res.*, *11*:391–394 (1979).
523. J. Roth, C. R. Kahn, M. A. Lesniak, P. Gorden, P. De Meyts, K. Megyesi, D. M. Neville, J. R. Gavin, A. M. Soll, P. Freychet, I. D. Goldfine, R. S. Bar, and J. A. Archer, *Recent Prog. Horm. Res.*, *31*:95–134 (1975).
524. C. R. Kahn, K. L. Baird, J. S. Flier, C. Grunfeld, J. T. Harmon, L. C. Harrison, F. A. Karlsson, M. Kasuga, G. L. King, V. C. Lang, J. M. Podskalnt, and E. Vanb Oberggen, *Recent Prog. Horm. Res.*, *37*:477–538 (1981).
525. J. L. Barron, R. P. Millar, and D. Searle, *J. Clin. Endocrinol. Metab.*, *54*:1169–1173 (1982).
526. J. L. Barron, E. Griffiths, G. Tsalacopoulos, and R. P. Millar, in: *LHRH and Its Analogs — Contraceptive and Therapeutic Applications* (B. H. Vickery, J. J. Nestor, and E. S. E. Hafez, eds.), MTP Press, Lancaster, England (1984), p. 411.
527. L. Tharandt, C. Rosanowski, R. Windeck, G. Benker, K. Hackenberg, and D. Reinwein, *Horm. Metab. Res.*, *13*:277–281 (1981).
528. K. Chikamori, F. Suehiro, T. Ogawa, K. Sato, H. Mori, I. Oshima, and S. Saito, *Acta Endocrinol. Copenh.*, *96*:1–6 (1981).
529. J. J. Reeves, G. R. Tarnavsky, S. R. Becker, D. Coy, and A. V. Schally, *Endocrinology*, *101*:540–547 (1977).
530. I. Yamazaki and H. Okada, *Endocrinol. Jpn.*, *27*:593–605 (1980).
531. I. Yamazaki, *Endocrinol. Jpn.*, *29*:415–421 (1982).
532. I. Yamazaki, *J. Reprod. Fertil.*, *72*:129–136 (1984).
533. A. Miyake, Y. Kawamura, and T. Aono, *J. Endocrinol. Invest.*, *5*:255–257 (1982).
534. D. J. Handelsman and R. S. Swerdloff, *Endocr. Rev.*, *7*:95–105 (1986).
535. L. T. Sennello, R. A. Finley, S. Y. Chu, C. Jagst, D. Max, D. E. Rollers, and K. G. Tolman, *J. Pharm. Sci.*, *75*:158–160 (1986).
536. J. J. Nestor, Jr., T. L. Ho, R. Tahilramani, B. L. Horner, R. A. Simpson, G. H. Jones, G. I. McRae, and B. H. Vickery, in: *LHRH and Its Analogs—Contraceptive and Therapeutic Applications* (B. H. Vickery, J. J. Nestor, Jr., and E. S. E. Hafex, eds.), MTP Press, Boston (1984), pp. 22–23.
537. J. L. Meakin, E. J. Keogh, and C. E. Martin, *Fertil. Steril.*, *43*:811–815 (1985).
538. G. Leyendecker and L. Wildt, *J. Reprod. Fertil.*, *69*:397–409 (1983).
539. J. G. Wagner, *Fundamentals of Clinical Pharmacokinetics*, Drug Intelligence Publication, Hamilton, IL (1975).

540. A. B. Kurtz and J. D. N. Nabarro, *Diabetologia*, *19*:329–334 (1980).

541. J. Sandow and R. H. Clayton, in: *Hormone Biochemistry and Pharmacology*, Vol. 2 ((M. Briggs and A. Corgin, eds.), Eden Press, Montreal (1988), pp. 63–106.

542. L. Kiesel, J. Sandow, K. Bertges, et al., *J. Clin. Endocr. Metab.*, *28*:1167–1173 (1989).

543. R. L. Chan, W. Ho, A. S. Webb, J. La Fargue, and C. A. Nerenberg, *Pharm. Res.*, *5*:335–340 (1988).

544. E. Hazum, *Mol. Cell. Endocrinol.*, *26*:217–222 (1982).

545. R. B. Jaffe, in: *Reproductive Endocrinology* (S. S. C. Yen and R. B. Jaffe, eds.), Saunders, Philadelphia (1986), p. 758.

546. J. R. Marshall, C. B. Hammond, G. T. Ross, A. Jacobsen, P. Rayford, and W. D. Odell, *Obstet. Gynecol.*, *32*:760–769 (1968).

547. G. D. Braunstein, J. L. Vaitukaitis, P. O. Carbone, and G. T. Ross, *Ann. Intern. Med.*, *78*:39–45 (1973).

548. P. Franchimont, U. Gaspard, A. Reuter, and G. Heynen, *Clin. Endocrinol. (Oxf.)*, *1*:315–319 (1972).

549. A. Good, M. Ramos-Uribe, R. J. Ryan, and R. D. Kempers, *Fertil. Steril.*, *28*:846–850 (1977).

550. G. S. Taliadouros, S. Amr, J. P. Louvet, S. Birken, R. E. Canfield, and B. C. Nisula, *J. Clin. Endocrinol. Metab.*, *54*:1002–1009 (1982).

551. H. R. Schroeder and C. M. Halter, *Clin. Chem.*, *29*:667–671 (1983).

552. P. D. Papapetrou and S. C. Nicopoulou, *Acta Endocrinol. (Copenh.)*, *112*:415–422 (1986).

553. P. D. Papapetrou, N. P. Sakarelou, H. Braouzi, and P. H. Fessas, *Cancer*, *45*:2583–2592 (1980).

554. D. L. Blithe, A. H. Akar, R. E. Wehmann, and B. C. Nisula, *Endocrinology*, *122*:173–180 (1988).

555. G. P. Lefort, J. M. Stolk, and B. C. Nisula, *Endocrinology*, *119*:924–931 (1986).

556. R. E. Wehmann and B. C. Nisula, *J. Clin. Invest.*, *68*:184–194 (1981).

557. R. E. Wehmann and B. C. Nisula, *J. Clin. Endocrinol. Metab.*, *51*:101–105 (1980).

558. M. D. Damewood, W. D. Schlaft, et al., *Fertil. Steril.*, *52*:398–400 (1988).

559. L. Liu, J. L. Southers, J. W. Cassels, Jr., S. M. Banks, R. E. Wehmans, D. L. Blithe, H. C. Chen, and B. C. Nisula, *Am. J. Physiol.*, *256*:E721–E724 (1989).

560. C. Rosa, S. Amr, S. Birken, R. E. Wehmann, and B. C. Nisula, *J. Clin. Endocrinol. Metab.*, *59*:1215–1219 (1984).

561. L. Liu, J. L. Southers, S. M. Banks, D. L. Blithe, R. E. Wehmann, J. H. Brown, H. C. Chen, and B. C. Nisula, *Endocrinology*, *124*:175–180 (1989).

562. L. Warren, *J. Biol. Chem.*, *234*:1971–1975 (1959).

563. M. L. Dufau, K. J. Catt, and T. Tsuruhara, *Biochem. Biophys. Res. Commun.*, *44*:1022–1029 (1971).

564. L. Liu, R. E. Wehmann, C. Rosa, S. Birken, and B. C. Nisula, *J. Androl.*, *9*:62–66 (1988).

565. R. E. Wehmann and B. C. Nisula, *J. Clin. Endocrinol. Metab.*, *48*: 753–759 (1979).

566. R. E. Wehmann and B. C. Nisula, *J. Clin. Endocrinol. Metab.*, *50*: 674–679 (1980).

567. N. K. Kalyan and O. P. Bhal, *J. Biol. Chem.*, *258*:67–74 (1983).

568. G. P. Lefort, J. M. Stolk, and B. C. Nisula, *Endocrinology, 115*: 1551–1557 (1984).

569. T. Blundell, G. Dodson, D. Hodgkin, and D. Mercola, *Adv. Protein Chem.*, *26*:279–402 (1972).

570. R. N. Brogden and R. C. Heed, *Drugs, 34*:350–371 (1987).

571. S. Jacobs and P. Cuatrecasas, *Endocrinol. Rev.*, *2*:251 (1981).

572. I. D. Goldfine, *Endocrinol. Rev.*, *3*:235–255 (1987).

573. I. D. Goldfine, in: *Effects of Insulin on Intracellular Functions in Biochemical Actions of Hormones* (G. Litwack, ed.), Academic Press, New York (1981), pp. 273–305.

574. J. R. Gavin, J. Roth, D. M. Neville, Jr., P. De Meyts, and D. N. Buell, *Proc. Natl. Acad. Sci. USA*, *71*:84–88 (1974).

575. Y. Okabayashi, B. A. Maddux, A. R. McDonald, C. D. Logsdon, J. A. Williams, and I. D. Goldfine, *Diabetes, 38*:182–187 (1989).

576. M. Fehlmann, Y. L. Marehand-Brustel, J. Dolais-Kitabgi, O. Morin, and P. Freychet, *Diabetes Care, 4*:223–227 (1981).

577. R. D. Fussgaenger, H. H. Ditschuneit, H. Martini, et al., *Diabetes Care, 4*:228–234 (1981).

578. L. M. Keefer, M. A. Piron, P. De Meyts, *Diabetes Care, 4*:209–214 (1981).

579. S. Gammeltoft, *Diabetes Care, 4*:235–237 (1981).

580. W. Bachmann, C. Sieger, F. Lacher, N. Lotz, and H. Mchnert, *Diabetes Care, 4*(Suppl. 2):215–219 (1981).

581. B. Vialettes, E. Zotian, M. C. Simon, and P. H. Vague, *Diabetologia, 23*:208 (1982) (abstract).

582. A. Ebihara, K. Kondo, K. Ohashi, K. Kosaka, T. Kuzuyz, and A. Matsuda, *Diabetes Care, 5*(Suppl. 2);35 (1982).

583. K. Federlin, H. Lause, and H. G. Velcovsky, *Diabetes Care, 4*: 170–174 (1981).

584. J. A. Galloway, M. A. Root, C. T. Spardlin, M. A. Root, and S. E. Fineberg, *Diabetes Care, 4*:183 (1981).

585. M. Klier, W. Kerner, A. A. Torres, and E. F. Pfeiffer, *Diabetes Care, 4*:193–195 (1981).

586. J. C. Pickup, R. W. Bilous, G. C. Viberti, H. Keen, R. J. Jarrett, A. Glynne, J. Claudwell, M. Root, and A. H. Rubenstein, *Diabetes Care, 5*(Suppl. 2):29–34 (1982).

587. C. Rosak, S. H. Althoff, W. Fassbinder, and K. Schoffling, 11th IDF Congress, Excerpta Medica International Congress Series No. 577 (1982), p. 56 (abstract).

588. K. J. Schluter, K. G. Petersen, J. Sotheimer, F. Enzmann, and L. Kerp, *Diabetes Care, 5*(Suppl. 2):78 (1982).

589. S. E. Fineberg, J. A. Galloway, N. S. Fineberg, M. J. Rathbun, and S. Hufferd, *Diabetes Care, 5*:107 (1982).

590. L. G. Heding, M. O. Marshall, B. Persson, G. Dahlquist, B. Thalme, F. Lindgren, H. K. Akerbloom, A.-R. Iva, M. Knip, J. Ludvigsson, L. Stenhemmer, L. Stromberg, O. Sovik, H. Baevre, K. Wefring, J. Vidnes, J. J. Kjaergard, D. Bro, and P. H. Kaad, *Diabetologia, 27*: 96 (1984).

591. M. Berger, H. J. Cuppers, H. Hegner, V. Jorgens, and P. Berchtold, *Diabetes Care, 5*:77–91 (1982).

592. C. Binder, T. Lauritzen, O. Faber, and S. Pramming, *Diabetes Care*, 7:188–199 (1984).
593. C. Binder, *Acta Pharmacol. Toxicol.*, 27(Suppl. 2):1–84 (1969).
594. E. W. Morre, M. L. Mitchell, and T. C. Chalmers, *J. Clin. Invest.*, 38:1222–1227 (1959).
595. J. A. Galloway, C. T. Spardlin, R. L. Nelson, S. M. Wentworth, J. A. Davidson, and J. L. Swarner, *Diabetes Care*, 4:366–376 (1981).
596. K. Kolendorf, J. Bosgen, and J. Deckert, *Horm. Metab. Res.*, 15:274–278 (1983).
597. M. Berger, P. A. Halban, J. P. Assal, R. E. Offord, M. Vranic, and A. E. Renold, *Diabetes*, 28(Suppl. 1):53–57 (1979).
598. A. B. Kurtz and J. D. N. Nabarro, *Diabetologia*, 19:329–334 (1980).
599. P. D. Home, J. C. Pickup, H. Keen, K. G. M. M. Albert, J. A. Parsons, and C. Binder, *Metabolism*, 30:439–442 (1981).
600. D. R. Owens, M. K. Jones, T. M. Hayes, L. G. Heding, K. G. M. M. Albert, P. D. Home, and J. M. Burrin, *Lancet*, 2:118–122 (1981).
601. W. K. Waldhaus, P. R. Bratusch-Marrian, H. Vierhapper, and P. Nowotng, *Metabolism*, 32:478–486 (1983).
602. E. W. Kraegen and D. J. Chrisholm, *Diabetologia*, 26:208–213 (1984).
603. D. J. Chisholm, E. W. Kraegen, and G. S. Zelenka, *Diabetes Care*, 4:265–268 (1981).
604. R. A. De Fronzo, D. J. Tobin, and R. Andres, *Am. J. Physiol.*, 237:E214–E223 (1979).
605. G. Williams, J. Pickup, A. Clark, S. Bowcock, E. Cook, and H. Keen, *Diabetes*, 32:466–473 (1983).
606. J. Schou, *Pharmacol. Rev.*, 13:441–464 (1964).
607. D. A. Henry, J. M. Lowe, D. Citrin, and W. G. Manderson, *Lancet*, 2:741 (1978).
608. K. Kolendorf, J. Bosjen, and T. Deckert, *Horm. Metab. Res.*, 15:274–278 (1983).
609. S. Pramming, T. Lauritzen, B. Thorsteinsson, K. Johansen, and C. Binder, *Acta Endocrinol.*, 105:215–220 (1984).
610. T. Lauritzen, C. Binder, and O. K. Faber, *Acta Paediatr. Scand.*, 283(Suppl.):81–85 (1980).
611. V. A. Koivisto, S. Fortney, R. Hendler, and P. Felig, *Metabolism*, 30:402–405 (1981).
612. V. A. Koivisto and P. Felig, *N. Engl. J. Med.*, 298:79–83 (1978).
613. R. Taylor, P. D. Home, and K. G. M. M. Alberti, *Diabetes Care*, 4:377–379 (1981).
614. K. Kolendorf, J. Bosjen, and T. Deckert, *Diabetes Care*, 6:6–9 (1983).
615. A. Murat and G. Slama, *Metabolism*, 34:120–123 (1985).
616. E. Mosekilde, K. S. Jensen, C. Binder, S. Pramming, and B. Thorsteinsson, *J. Pharmacokin. Biopharm.*, 17:67–87 (1989).
617. C. Binder, in: *Artificial Systems for Insulin Delivery* (P. Brunetti K. G. M. M. Alberti, A. M. Albesser, K. D. Hepp, and M. M. Bendetri, eds.), Raven Press, New York (1988), pp. 53–57.
618. T. Deckert, B. Hansen, K. Kolendorf, J.-S. D. Poulsen, and M. Smith, *Acta Pharmacol. Toxicol.*, 51:30–37 (1982).

619. L. A. Frohman, L. Burek, and M. E. Stachura, *Endocrinology, 91*: 262–269 (1972).

620. J. A. Moore, C. G. Rudman, N. J. MacLachlan, G. B. Fuller, B. Burnett, and J. W. Frane, *Endocrinology, 122*:2920–2926 (1988).

621. A. C. S. Faria, J. D. Veldhuis, M. O. Thorner, and M. L. Vance, *J. Clin. Endocrinol. Metab., 68*:535–541 (1989).

622. S. Refetoff and P. H. Sonksen, *J. Clin. Endocrinol. Metab., 30*: 386–392 (1970).

623. S. M. Glick, J. Roth, and E. T. Lonergan, *J. Clin. Endocrinol. Metab., 24*:501–505 (1964).

624. S. S. C. Yen, T. M. Siler, and G. W. DeVane, *N. Engl. J. Med., 290*:935–938 (1974).

625. D. Ownes, M. C. Srivastava, C. V. Tompkins, J. D. N. Nabarro, and P. H. Sonksen, *Eur. J. Clin. Invest., 3*:284–294 (1973).

626. M. L. Parker, R. D. Utiger, and W. H. Daughaday, *J. Clin. Invest., 41*:262–268 (1962).

627. M. Rosenbaum and J. M. Gertner, *J. Clin. Endocrinol. Metab., 69*: 821–824 (1989).

628. S. R. Plymate, J. M. Leonard, C. A. Paulsen, B. L. Fariss, and A. E. Karpas, *J. Clin. Endocrinol. Metab., 57*:645–247 (1983).

629. G. Bauman, M. W. Stolar, K. Amburn, C. P. Barsano, and B. C. De Vries, *J. Clin. Endocrinol. Metab., 62*:134–141 (1986).

630. F. B. Picard, M. C. P. Vinay, and C. Keyser, *J. Endocrinol., 121*: 19–25 (1989).

631. B. Husman, G. Andersson, G. Norstedt, and J. A. Gustafsson, *Endocrinology, 116*:2605–2611 (1985).

632. J. P. Hughes, J. S. A. Simpson, and H. G. Friesen, *Endocrinology, 112*:1980–1985 (1983).

633. C. Carter-Su, J. Schwartz, and G. Kikuchi, *J. Biol. Chem., 259*: 1099–1104 (1984).

634. E. Gorin and H. M. Goodman, *Endocrinology, 114*:1279–1286 (1984).

635. B. Husman, J. A. Gustafsson, and G. Andersson, *Mol. Cell. Endocrinol., 59*:13–25 (1988).

636. M. J. Poznansky, J. Halford, and D. Taylor, *FEBS Lett., 239*: 18–22 (1988).

637. M. D. Groesbeck, A. F. Parlow, and W. H. Daughaday, *Endocrinology, 120*:1963–1975 (1987).

638. J. River, J. Spiess, M. Thorner, and W. Vale, *Nature, 300*: 276–278 (1982).

639. W. B. Wehrenberg and N. Ling, *Biochem. Biophys. Res. Commun., 115*:525–530 (1983).

640. R. J. M. Ross, S. Tsagarakis, A. Grossman, M. A. Preece, C. Rodda, P. S. W. Davies, L. H. Rees, M. O. Savage, and G. M. Besser, *Lancet, 1*:5–8 (1987).

641. P. Rochiccioli, M. T. Tauber, F. X. Coude, M. Arnone, M. Niorre, F. Uboldi, and C. Barbeau, *J. Clin. Endocrinol. Metab., 65*:268–274 (1987).

642. A. Grossman, M. O. Savage, N. Lytras, M. A. Preece, J. Suieras-Diaz, D. H. Coy, L. H. Rees, and G. M. Besser, *Clin. Endocrinol., 21*:321–330 (1984).

643. V. A. Lance, W. A. Murphy, J. Sueiras-Diaz, and D. H. Coy, *Biochem. Biophys. Res. Commun., 119*:265–272 (1984).

644. M. T. Tauber, C. Pienkowski, F. Landier, R. Gunnarsson, and P. Rochiccioli, *Acta Paediatr. Scand.*, *349*(Suppl.):117–122 (1989).

645. B. Rafferty, D. H. Coy, and S. Poole, *Peptides*, *9*:207–209 (1988).

646. L. A. Frohman, T. R. Downs, T. C. Williams, E. P. Heimer, Y.-C. E. Pan, and A. M. Felix, *J. Clin. Invest.*, *78*:902–913 (1988).

647. B. Rafferty, S. Poole, R. Clarke, and D. Schulster, *J. Endocrinol.*, *107*:R5–R8 (1985).

648. N. Fleischer and R. Guillemin, in: *Peptide Hormones* (J. A. Parsons, ed.), Macmillan, London (1976), pp. 317–332.

649. J. M. Polak, L. Grimelius, A. G. E. Pearse, S. R. Bloom, and A. Arimura, *Lancet*, *2*:1220–1222 (1975).

650. I. S. Gottesmann, L. J. Mandarino, and J. E. Gerich, *Spec. Top. Endocrinol. Metab.*, *4*:177–243 (1982).

651. E. D. Pozo, *Hormone Res.*, *29*:89–91 (1988).

652. M. Sheppard, B. Shapiro, and B. Pimstone, *J. Clin. Endocrinol. Metab.*, *48*:50–53 (1979).

653. W. Bauer, U. Briner, W. Doepfner, R. Haller, R. Huguenin, and P. Marbach, *Life Sci.*, *31*:1131–1141 (1982).

654. M. D. Katz and B. L. Erstad, *Clin. Pharm.*, *8*:255–273 (1989).

655. P. E. Battershill and S. P. Clissold, *Drugs*, *38*:658–702 (1989).

656. K. Kutz, E. Nuesch, and J. Rosenthaler, *Scand. J. Gastroenterol.*, *21*(Suppl. 119):65–72 (1986).

657. E. del Pozo, M. Neufeld, K. Schlüter, F. Tortosa, P. Clarenbach, E. Bieder, L. Wendel, E. Nuesch, P. Marbach, and H. Cramer, *Acta Endocrinol.*, *111*:433–439 (1986).

658. M. Peracchi, E. Reschini, and L. Cantalamessa, *Metabolism*, *23*:1009–1015 (1974).

659. R. H. Unger, A. M. Eisentraut, M. S. McCall, and L. L. Madison, *J. Clin. Invest.*, *41*:682–689 (1962).

660. W. Londong, M. Angerer, K. Kutz, R. Landgraf, and V. Londong, *Gastroenterology*, *96*:713–722 (1989).

661. B. Struclovici, J. M. Stadel, and R. J. Lefkowitz, in: *Mechanisms of Receptor Regulation* (G. Poste and S. T. Crooke, eds.), Plenum Press, New York, London (1985), pp. 279–294.

662. M. D. Hollenberg, in: *Mechanisms of Receptor Regulation* (G. Poste and S. T. Crooke, eds.), Plenum Press, New York, London (1985), pp. 295–322.

663. A. Kolez, M. Kraenzlin, K. Gyr, V. Meier, S. R. Bloom, P. Heitz, and H. Stalder, *Gastroenterology*, *92*:527–531 (1987).

664. R. G. Guillemin and R. Burgus, *Sci. Am.*, *11*:24–73 (1972).

665. C. Y. Bower, H. G. Friesten, P. Hwang, H. J. Guyda, and K. Folkers, *Biochem. Biophys. Res. Commun.*, *45*:1033–1041 (1971).

666. L. S. Jacobs, P. J. Snyder, J. F. Wilber, R. D. Utiger, and W. H. Daughaday, *J. Clin. Endocrinol. Metab.*, *33*:996–998 (1971).

667. M. Irie and T. Tsushima, *J. Clin. Endocrinol. Metab.*, *35*:97–100 (1972).

668. G. R. Breeze, J. M. Colt, B. R. Cooper, A. J. Prange, Jr., M. A. Lipton, and N. P. Plotinkoff, *J. Pharmacol. Exp. Ther.*, *193*:11–22 (1975).

669. H. Kruse, *J. Pharmacol. (Paris)*, *6*:249–268 (1975).

670. A. J. Prage, Jr., I. C. Wilson, P. P. Lara, L. B. Alltop, and G. R. Breeze, *Lancet*, *2*:999−1002 (1972).

671. A. J. Kastin, R. H. Ehrensing, D. S. Schalch, and M. S. Anderson, *Lancet*, *2*:740−742 (1972).

672. D. Brewster and K. Waltham, *Biochem. Pharmacol.*, *30*:619−622 (1981).

673. C. Liberman, T. Nogimori, C. F. Wu, T. V. Buren, J. Wilson, T. Miller, E. Smith, and C. H. Emerson, *Acta Endocrinologica (Copenh.)*, *120*:134−142 (1989).

674. E. Iverson, *J. Endocrinol.*, *118*:511−516 (1988).

675. M. Safran, C. F. Wu, R. Matys, S. Alex, and C. H. Emerson, *Endocrinology*, *115*:1031−1037 (1984).

676. J. Leppaluoto, P. Virkkunen, and H. Lybek, *J. Clin. Endocrinol. Metab.*, *35*:477−478 (1972).

677. A. Dupont, F. Labrie, L. Levasseur, and A. V. Schally, *Can. J. Physiol. Pharmacol.*, *52*:1012−1019 (1974).

678. T. W. Redding and A. V. Schally, *Neuroendocrinology*, *9*:250−256 (1972).

679. I. M. D. Jackson, P. D. Papapetrou, and S. Reichlin, *Endocrinology*, *104*:1292−1298 (1979).

680. M. Safran, C. F. Wu, R. Matys, S. Alex, and C. H. Emerson, *Endocrinology*, *115*:1031−1037 (1984).

681. J. E. Morley, T. J. Garvin, A. E. Pekary, R. D. Utiger, M. G. Nair, and C. M. Baugh, *J. Clin. Endocrinol. Metab.*, *48*:377−380 (1979).

682. W. Vale, J. River, and R. Burgus, *Endocrinology*, *89*:1685−1688 (1971).

683. J. River, W. Vale, M. Monahan, N. Ling, and R. Burgus, *J. Med. Chem.*, *15*:679−681 (1972).

684. J. R. Sowers, J. M. Hershaman, A. E. Pekary, M. G. Nair, and C. M. Baugh, *J. Clin. Endocrinol. Metab.*, *43*:741−748 (1976).

685. J. R. Sowers, J. M. Hershman, H. E. Carlson, A. E. Pekary, M. G. Nair, and C. M. Baugh, *Endocrinol. Metab.*, *43*:749−755 (1976).

686. J. C. Dvorak and R. D. Utiger, *J. Clin. Endocrinol. Metab.*, *44*:582−587 (1977).

part two
ROUTES OF DELIVERY

10

Parenteral Delivery of Peptide and Protein Drugs

Partha S. Banerjee*

University of Wisconsin, Madison, Wisconsin

Ehab A. Hosny

University of Cairo, Cairo, Egypt

Joseph R. Robinson

University of Wisconsin, Madison, Wisconsin

I. INTRODUCTION

Peptides and proteins have received much attention in recent years as drug candidates [1,2]. The rapid progress in peptide and protein pharmacology along with large-scale manufacture of these compounds by recombinant DNA technology and sophisticated synthetic techniques have understandably prompted this optimism. Peptides and proteins play an important role in all biological processes. The significance and remarkable scope of their functions includes enzymatic catalysis, immune protection, and control of growth and differentiation. This large functional diversity [3] makes them a very important and ubiquitous class of therapeutic agents unlike any other class of known compounds.

Unfortunately, the large-scale manufacture of peptide and protein drugs has advanced far more rapidly than our ability to deliver these molecules systemically in a convenient and effective delivery system. This is due largely to the unique requirements and restrictions that peptides pose to the formulation chemist compared to conventional drug compounds. These properties include molecular size, susceptibility to proteolytic breakdown, rapid plasma clearance, sometimes unusual dose-response curve, immunogenicity, biocompatibility, as well as the tendency of a peptide to undergo aggregation, adsorption, and denaturation [2].

The demand for effective delivery systems for peptides has brought about a tremendous thrust in recent years in both the scope and complexity of drug delivery technology. Two important aspects of drug delivery are

---

Current affiliation: Hercon Laboratories Corporation, Emigsville, Pennsylvania.

route and mode of delivery. Oral route is by far the most convenient and popular. However, with the few exceptions of small and cyclic peptides such as cyclosporin, ACTH (4–9), TRH, captopril, DDAVP, and muramyl dipeptide [2,4] most peptide drugs have low oral activity. This is due largely to degradation by proteolytic enzymes in the gastrointestinal tract and poor permeability of the intestinal mucosa to high-molecular-weight substances. Consequently, most of the new proteinaceous therapeutics are administered by parenteral routes such as intravenous, intramuscular, and subcutaneous. This form of delivery has traditionally been poorly accepted by patients, except those suffering from life-threatening diseases.

One approach to alleviate this problem has been to deliver them by alternative routes, such as nasal, pulmonary, rectal, buccal, vaginal, transdermal, and ocular routes [5]. However, in the absence of an acceptable penetration enhancer, the absorption of peptides and proteins from these routes is much less than after parenteral administration. Depending on the route, this could be due to poor membrane permeability and/or metabolism at the absorption site. One recent report indicates that protease activities in the homogenates of the nasal, buccal, rectal, and vaginal mucosa of rabbit are substantial and quite comparable to those in the intestinal mucosa [6]. Even the nasal route, which is most promising in terms of permeability, shows large differences in the bioavailabilities of peptides with similar molecular weights and almost insignificant bioavailabilities for proteins composed of more than 27 amino acids [7]. Low bioavailability by these routes coupled with the high cost of peptides [4] make these routes less than satisfactory for delivery of peptides. Although future advances in this direction will make some of these routes potentially useful, in the near future parenteral administration will remain the route of choice, particularly for large peptides and proteins.

Another approach has been to improve existing parenteral dosage forms to increase patient compliance. These include new devices for parenteral administration, polymeric controlled-release implants, implantable pumps, biodegradable microspheres, liposomes, and conjugation with carriers. Many of these advanced systems are primarily in the research stage with future potential.

The parenteral mode of delivery is also important for peptide drugs. The body provides these agents in pulse rather than continuous form to a particular receptor site [8]. Pulsed delivery rather than continuous administration is the preferred mode for many of these agents, due to down-regulation of receptors after continuous administration. Moreover, since many peptide drugs are highly potent, it is desirable to target the drug to specific receptors to reduce toxicity. Considering the complexity of the aforementioned processes, the parenteral route will be preferred for this purpose.

The main emphasis of this chapter is to consider the possibilities and problems for parenteral delivery of peptide and protein drugs for a biopharmaceutic, pharmacokinetic, and formulation viewpoint. The prospects and limitations of advanced delivery systems and their impact on parenteral delivery of peptide drugs are also discussed.

II. BIOPHARMACEUTICS AND PHARMACOKINETICS OF PARENTERAL PEPTIDE DELIVERY SYSTEMS

There are two key factors that determine the pharmacokinetics and bioavailability of peptide and protein drugs after parenteral administration. These are the route of delivery and the mode of delivery. Subsequent discussion will elaborate on those aspects from a comparative viewpoint with reference to different peptide drugs.

A. Route of Delivery

The word "parenteral" means other than enteral (i.e., administration of a drug other than through the intestine). In the medical literature, parenteral administration is commonly understood as the injection of a dosage form into the body by a sterile syringe or some other mechanical device such as an infusion pump. The major routes of parenteral administration for most peptide and protein drugs are intravenous, intramuscular, and subcutaneous, the last two being more frequently used in clinical situations. Other parenteral routes for peptides include intraperitoneal, intraspinal, intrathecal, intracerebroventricular, intraarterial, subarachnoid, and epidural.

Intravenous

In intravenous administration the drug is directly introduced into one of the superficial veins, usually in the form of an aqueous solution, either as a single injection or as continuous infusion at a specified rate. Veins in the lower forearm are preferred because they are usually large and easily accessible. It is better to avoid the antecubital area, where veins and arteries lay in such close proximity that they are sometimes confused and it is possible to inject a solution into an artery instead of a vein.

   This route has the advantage of most rapid onset of action. In emergency situations, when rapid onset of action is desired, this is the route of choice for most peptide and protein drugs. This was apparent in a recent unblinded, large-scale trial (11,806 patients) of intravenous streptokinase in early acute myocardial infarction [9]. The beneficial effect of steptokinase was a function of the time from onset of pain to streptokinase infusion (Table 1), being most striking in patients treated within 3 hr of an attack.

   Intravenous drug delivery offers selective distribution of particles to different organs depending on size. Particles larger than 7 μm are trapped in the lung, smaller than 0.1 μm accumulate in bone marrow, and particles with diameter between 0.1 and 7 μm are taken up by liver or spleen [10]. This approach may be useful for targeting of peptide drugs.

   Intravenous administration is the only available avenue of delivering peptide and protein drugs in situations where the drug is excessively metabolized and/or bound at the site of intramuscular or subcutaneous injection or where drug absorption is very low. Since in theory none of the the drug is lost by intravenous administration, smaller doses may be given than by other extravascular routes of administration. This is important for peptide drugs because most of these substances are highly potent and at the same time have a short half-life. For example, the threshold dose necessary for a significant electromyographic response in mice for thymopentin, a pentopeptide, is 0.03 mg/kg for intravenous injection and 0.3 mg/kg for both intraperitoneal and subcutaneous injection [11,12]. When large doses need to be administered to maintain optimum blood levels of antibody (e.g., with gamma globulin) [13], intravenous therapy is superior

Table 1 Effectiveness of Intravenous Thrombolytic Therapy with Streptokinase (SK) in Acute Myocardial Infarction[a,b]

Hours	SK [%(deaths/n)]	C [%(deaths/n)]	p	RR (95% CI)	Total [%(deaths/n)]
<1	8.2(52/635)	15.4(99/642)	0.0001	0.49(0.34—0.69)	11.8(151/1277)
<3	9.2(278/3016)	12.0(369/3078)	0.0005	0.74(0.63—0.87)	10.6(657/6094)
3—6	11.7(217/1849)	14.1(254/1800)	0.03	0.80(0.66—0.98)	12.9(471/3649)
6—9	12.6(87/693)	14.1(93/659)	NS	0.87(0.64—1.19)	13.3(180/1352)
9—12	15.8(46/292)	13.6(41/302)	NS	1.19(0.75—1.87)	14.6(87/594)

Source: Ref. 9.

[a]Mortality by hours from onset of symptoms.

[b]C, control; n, number of patients; RR, relative risk estimate; NS, not significant; CI, confidence interval.

to intramuscular because of muscle mass volume limitations with the latter route.

Some disadvantages of intravenous delivery include local reactions due to trauma to the vessel wall, such as thrombophlebitis, extravasation, tissue necrosis, and so on, and systemic effects such as infection, pulmonary embolism resulting from air or particulate emboli, reactions due to leachables from the delivery system, metabolic complications, volume overload, hemolysis, hypersensitivity reactions, and speed shock [14]. Previously intravenous administration of gamma globulin often led to severe systemic reactions [15,16], which were attributed to the anticomplementary activity of aggregated IgG, the latter being formed as a result of the removal of stabilizing serum proteins during the Cohn fractionation process [17]. Modification of the immune globulin preparation, where the Cohn fraction II was treated with dithiothreitol to reduce selected interchain disulfide bonds, improved the therapeutic advantage of the intravenous route significantly [18].

Immunogenicity probably represents the most serious concern with parenteral administration of peptides and proteins [14]. As a general rule, molecules smaller than molecular weight 10,000 are only weakly immunogenic or not immunogenic at all [19]. The most potent immunogens are macromolecular proteins with molecular weights greater than 100,000. Chemical complexity also contributes to immunogenicity. Homopolymers are poor immunogens regardless of size, whereas copolymers of two amino acids contribute more to immunogenicity than nonaromatic residues. Impurities during manufacturing and formulation (e.g., foreign bacterial proteins) probably represents one prime source of immunogenic substances [20]. Immunogenic response to a protein may be more severe following intravenous delivery than corresponding intramuscular or subcutaneous delivery. For example, the time to death in guinea pigs sensitized to modified gelatin was significantly shorter when the animals were challenged by the intravenous route than by the intraperitoneal route (Table 2). This may be because of the sudden massive antigen-antibody reaction that can occur following intravenous delivery. When the drug is given by other parenteral routes, the access to antibody molecules is slower. The immunogenicity of the peptide or protein drug can be reduced by conjugation with a polymer or macromolecule [21]. The conjugation of polymers with peptides is discussed in a subsequent section and in Chap. 21.

In intravenous administration, since the drug solution is placed directly into the blood stream, no absorption is involved and thus the onset of action is rapid even though the drug may have to cross the capillaries to reach the site of action. The rapidity of intravenous circulation times between various sites of the human body as measured by a dye dilution method is shown in Table 3. When the half-life of the peptide is in the order of seconds (e.g., with bradykinin and angiotensin) [23,24], the circulation time is of importance in selecting the site of administration of the drug because significant drug loss can occur before the drug reaches the site of action.

Another factor of importance in the onset of action is the capillary permeability. Capillary permeability of endogenous serum albumin and a graded series of dextrans of molecular weight 8000 to 500,000 was studied in dogs [25]. Analysis of the transport data suggested that the principal barrier to blood-lymph transport is at the blood capillary wall, not at the lymph capillary or within the interstitial space. The free diffusion coefficient of dextrans decline linearly with increasing molecular weight. In

Table 2   Immunogenicity of Modified Gelatin (MG) in Guinea Pigs: Effect
of Route of Administration on Time to Death

Group	Number of animals	Immunization	Challenge dose	Number of deaths			
				0-15 min	1 hr	24 hr	Total
1	10	IP—MG	1 g/kg MG—IP	2	0	6	8
2	10	IP—saline	1 g/kg MG—IP	0	0	0	0
3	10	IP—MG	0.1 g/kg MG—IV	9	0	1	10
4	10	IP—saline	3 ml/kg saline—IV	0	0	0	0

Source: Ref. 14.

the range between 10,000 and 80,000, capillary diffusion of dextran molecules is approximately six times slower than predicted by reduction in free
diffusivity.   Above 80,000, however, there is little further restriction.
The mechanism of capillary permeation is not explained satisfactorily.   It is
hypothesized that few large pores (radius = 800 Å) or pinocytotic vesicles
(radius = 250 Å) carry larger molecules [25].

Visual evidence of vascular permeability is difficult to quantitate.   Intravenous injection of fluorescent-tagged plasma proteins or dextrans of graded
molecular weight colored the perivenular tissue diffusely and then spread
into the interstitium until after some 20 to 35 min it was no longer evident,
depending on the amount of macromolecule that was injected [26].   If the
blood flow to the vascular bed was tempararily interrupted, the dye-colored

Table 3   Circulation Times Following Intravenous
Administration

Distance in body	Circulation time (sec)
Arm to ear	8-14
Lung to ear	3-5
Arm to finger	17.5 ± 4.4
Arm to toe	24.8 ± 4.4
Right ventricle to left ventricle	2-4

Source: Ref. 22.

albumin or fluorescent-tagged protein did not appear in the perivenular tissue in sufficient concentrations to be detectable [27]. In 1984, Nuget and Jain [28] developed a photometric technique to measure fluorescent intensity-time response in both intravascular and interstitial regions of the rabbit ear chamber capillary bed. Interstitial transport was adequately described by a one-dimensional diffusional model. Interstitial diffusion coefficients decreased progressively with Stokes-Einstein radius, with values for albumin being significantly reduced from that for a dextran of equivalent hydrodynamic radius. This was explained by the electrostatic repulsion of negatively charged albumin by negative charges on the walls of membrane pores.

Conventionally, bioavailability by any route is defined as the relative amount available compared to the intravenous route, assuming that intravenous administration makes the drug completely bioavailable. However, this may not be true for peptide drugs, such as bradykinin, angiotensin, vasopressin, and oxytocin, because they are rapidly degraded by aminopeptidases or carboxypeptidases in the plasma [23,29,30]. For some of these hormones clearance by tissues such as kidney, liver, and lung is even faster than degradation in the blood [23,24]. Massive insulin resistance was observed in a patient with insulin-dependent diabetes, even by the intravenous route, apparently due to rapid clearance of circulating insulin [31]. Enkephalins and endorphins are active analgesics when administered intraventricularly [32] and β-endorphin is also active when administered intravenously [33]. Since β-endorphin is 31 residues long it is apparent that peptides of moderate size can cross the blood-brain barrier. The natural enkephalins are almost inactive when administered intravenously [34], but this does not necessarily indicate inability to cross the blood-brain barrier since enkephalins are more unstable than β-endorphin in blood and in brain preparations. In fact, a stabilized analog of enkephalin has been reported to have considerable analgesic activity after parenteral and even after oral administration [32].

Pharmacokinetics. The pharmacokinetics of peptide hormones and related drugs have been reviewed in detail elsewhere [35,36], including Chapter 9 of this book. Much of the available information about endogenous peptides and proteins is in the form of half-life following intravenous injection rather than a formal pharmacokinetic analysis. The pharmacokinetics of peptide and protein drugs are ultimately dependent on reliable assay methodology. These methodological problems are all the more serious because peptides are complex molecules capable of giving rise to many products, and most of the existing assay methods lack the sensitivity and resolving power needed to characterize their disposition. Half-lives of most of the peptide and protein drugs are very short, on the order of few minutes or shorter. A bolus injection may yield shorter half-lives than intravenous infusion because of rapid distribution outside the vascular compartment [37]. According to Bennett and McMartin [36], peptide plasma concentration decay follows three phases, an early rapid distribution phase lasting less than 2 min, a second more prolonged decline for 20 to 40 minutes, and a third phase characterized by a much slower decline of plasma levels. Bradykinin and angiotensin have a potent local effect and exceptionally short half-lives (few seconds). For these hormones it would therefore not be easy to distinguish the first two phases. Moreover, the third phase sometimes may appear as an assay artifact.

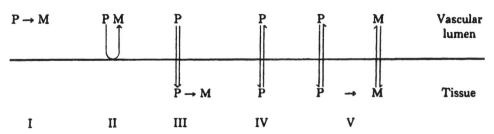

Fig. 1   Processes observed in peptide metabolism. P, peptide; M, metabolites; I, degradation in blood; II, degradation in transit; III, irreversible
uptake and degradation; IV, equilibration; V, equilibration and degradation.
(From Ref. 36.)

The clearance and distribution of peptide hormones usually occur by
five major pathways (Fig. 1): degradation in blood (I), degradation in
transit (II), irreversible uptake and degradation (III), equilibration (IV),
and equilibration and degradation (V). Degradation in transit through a
tissue bed occurs in the lung, kidney, and probably liver. Peptide analogs
that resist degradation in blood and in transit should be longer acting and
more potent. Irreversible uptake followed by degradation appears to be a
common process in kidney and may also be important for some peptides in
liver. The rate-limiting step for clearance by this type of mechanism is uptake and the degradation process is of secondary importance. Equilibration
and degradation refer to a process in which a peptide hormone can enter
and leave sites within a tissue at which degradation occurs. If equilibration
is fast, the process will become indistinguishable from degradation in transit.
If equilibration is slow and degradation is not too extensive, the tissue will
act as a depot from which intact peptide hormone as well as peptide fragments will enter the circulation [36]. Clearance estimates from steady-state
infusion is usually independent of peptide concentration except for insulin
[38,39] and vasopressin [40]. The reason for the concentration dependence
seen in insulin and vasopressin is not clear.

Extravascular

The two major extravascular routes of administration are intramuscular and
subcutaneous. In intramuscular administration the drug is injected deep
into the skeletal muscles, principally in the deltoid or gluteal region. Depending on the amount of fluid administered, the injection site can be selected according to its capability to hold a certain volume of fluid. The
lateral thigh region holds up to 15 ml of fluid, in contrast to 6 to 8 ml in
the case of the gluteal muscle [41]. Intramuscular administration is often
preferred because of the sustained action it provides compared to intravenous administration. The onset of action after intramuscular administration
is faster than subcutaneous delivery, probably due to faster blood flow in
the muscle [42]. Pain following injection, slow absorption, degradation or
binding of peptides at the site of injection, and muscle mass volume limitations are some of the problems associated with intramuscular injection of
peptides and proteins [14].

In subcutaneous injection, the drug solution is administered beneath the
skin into the subcutaneous tissue. Although the subcutaneous tissues are

somewhat loose, permitting the administration of moderate amounts of fluid (0.5 to 2 ml) [43], the normal connective tissue matrix prevents indefinite lateral spread of the injected solution [44]. In general, the absorption rate from a subcutaneous depot is slower than from the intramuscular site, and hence it is better suited for long-term therapy. However, absorption sites must be changed frequently to avoid local tissue damage and accumulation of unabsorbed drug. Besides having the drawbacks of intramuscular administration, subcutaneous delivery often lowers the potency of a peptide or protein drug due to degradation or incomplete absorption [11,12,14,45].

The major barrier to absorption from the intramuscular or subcutaneous site is believed to be the capillary endothelial membrane or cell wall [44], whcih is formed by a single layer of flat cells approximately 0.0025 mm thick. The capillaries themselves have a diameter of approximately 0.02 mm [46]. The farthest capillary is within approximately 0.125 mm of any cell. The biopharmaceutics of subcutaneous and intramuscular administration has been reviewed by Ballard [47]. The subcutaneous region is well supplied by capillary and lymphatic vessels [48]. Muscle tissue also has a rich supply of capillary vessels and the adjoining connective tissue has an abundant supply of lymphatic vessels. The intercellular junction in lymphatics opens more readily in active or injured regions than in motionless regions [49]. It is shown that particles and fluid contacting one endothelial surface enter cells in small vesicles and then leave the opposite surface from the same organelle by a process called cytopemphis [49], discharging their contents randomly on the other side of the cell. Thus cytopemphis permits the net transport of material proportional to its concentration difference on either side of the endothelial tissue. There are theoretically only four possible paths through the endothelial tissue: intercellular, intracellular in organelles, cytoplasmic matrix, and fenestral paths [50]. It appears that all four of these paths are used to some extent in various endothelial tissues.

Although it has not been determined whether the microscopic absorption route is via capillaries, lymphatics, or both, many drugs in solution injected into subcutaneous or intramuscular sites behave as if their absorption were taking place by passive diffusion. The permeation rate of the drug from the injection site can be described by

$$\frac{dN}{dt} = \frac{DAK}{h}(C_s - C_b)$$

where dN/dt is the permeation rate, N the amount permeating the tissue, D the mean diffusion coefficient of the drug in the membrane, A the area of the absorbing membrane exposed to the solution, K the partition coefficient of the drug between the membrane phase and delivery system at the injection site, h the thickness of the thin subcutaneous or intramuscular membrane, and C_s and C_b are the drug concentration at the injection site and in body fluids (e.g., blood and lymph), respectively. Some primary factors influencing the rate and extent of absorption after intramuscular or subcutaneous administration are discussed below.

Molecular size. In a diffusion-controlled process, large molecules would be expected to have slower penetration rates than smaller ones. A prime factor in the slow absorption of insulin is presumed to be the molecular weight of insulin (MW 6000) [51], which is close to the upper limit for transcapillary absorption [44]. Growth hormone with a molecular weight of 22,000

Table 4   Carbohydrate Absorption from Rat Muscle

Substance	Molecular weight	Aqueous diffusion coefficient $\times 10^6$ (cm^2/sec)	Fraction of drug cleared 5 min after intramuscular injection
D-Mannitol-1-[14]C	182	8.7	0.7
Sucrose-[14]C	342	7.5	0.6
Inulin-methoxy-[3]H	3000–4000	2.1	0.2
Inulin-carboxyl-[14]C	3000–4000	2.1	0.2
Dextran-carboxyl-[14]C	60,000–90,000	0.5	0.07

Source: Ref. 53.

is absorbed more slowly than insulin when administered subcutaneously [52]. There was a 10-fold difference (Table 4) in clearance rates of labeled carbohydrates, with molecular weight ranging from 182 to 90,000 from rat muscle [53]. However, for small molecules, the absorption rate is virtually the same and is probably limited by blood flow [54]. It appears that molecules of low molecular weight (< 20,000) are absorbed primarily via the capilaries, while molecules having high molecular weights are absorbed primarily via lymph vessels [55].

   Body movement. Body movement has a profound effect on the rate at which molecules are abosrobed via the lymphatic route. Muscular movement increases lymph flow, whereas at rest there is little or no lymph flow [48]. Following subcutaneous injection of black tiger venom (MW > 20,000) into rabbits, control animals lived no longer than 150 min after the injection, while animals whose injected limb was immobilized by a plaster cast lived more than 8 hr [55]. Immobilization of the limb had no effect on the absorption of a low-molecular-weight (MW < 5000) cobra venom. Since lymph flow was reduced during immobilization, it was assumed that high-molecular-weight compounds are absorbed primarily via lymphatic channels. To explain sudden anaphylactic deaths that occurred occasionally in humans after subcutaneous injections of macromolecules such as diptheria antitoxin, Lewis [56] injected horse serum subcutaneously in dogs. Table 5 shows that the serum was absorbed slowly from the site, but massage of the wheal produced by the injection and the use of high injection pressures resulted in a more rapid appearance of the serum in the thoracic duct lymph. Drug absorption is increased by rubbing the skin around the injection site, and also by exercise. There was a substantial increase in the absorption rate of [3H]insulin due to leg exercise following subcutaneous injection in rats and humans, leading to markedly elevated plasma levels of exogenous insulin [57]. The effect is more pronounced when insulin is injected into an exercised limb than when it is injected at another site [58].

   Viscosity. Viscosity may also delay or impede diffusion of the drug into body fluids. Coles et al. [59] studied the effect of aluminum monostearate-paraffin gels on the antigen response to *Clostridium welchii* type D toxoid. When these formulations were injected subcutaneously into rabbits and guinea pigs, the level and duration of the antitoxin titer in the blood was directly related to viscosity of the vaccine formulations studied. When gonadotropin-releasing hormone (GnRH) was intramuscularly injected into rats, the addition of dextran to the injected peptide solution led to a prolongation of the GnRH plasma level at the expense of the peak value (Fig. 2). This decreased absorption rate can be explained by the decreased diffusion coefficient of GnRH due to obstruction effect and microviscosity in the polymer solution [60].

Area. The local distribution of solution injected subcutaneouly or intramuscularily is of interest because the permeation rate of the drug depends in part on the geometry and the resulting area of the depot exposed to the tissue [61]. It is difficult, if not impossible, to control the area of an injected solution, suspension, or emulsion in contact with the tissues [62]. If other factors such as metabolic and excretion rates are equal, one would expect that higher serum levels of drug would result when the solution has a larger area exposed to the tissue, since drug penetration rate is directly proportional to this area. Hyaluronidase, an enzyme, enhances drug absorption by increasing the surface area of the drug solution in contact with the tissue [63,64]. It hydrolyzes hyaluronic acid, a component of the tissue ground substance which limits the spreading of fluids at the injection site. After hydrolysis the area of drug distribution in the tissue is increased and as a result, the absorption rate is also increased.

Anatomical site. The anatomical characteristics of the extravascular injection site significantly influences the rate of absorption of a drug. In a study in children with diabetes mellitus, Nora and associates [65] found that the mean absorption half-life for ^{131}I-labeled lente insulin was 224 min and 314 min after arm and thigh injections, respectively. In a similar study [66], gamma globulin uptake and blood levels differed according to the site of intramuscular injection in humans. This difference could be due to regional muscle blood flow. The resting blood flow is greater in the deltoid, intermediate in the thigh, and least in the buttock [67]. However, Cockshott et al. [68] suggested that most of the intramuscular injections have actually been intralipomatous (i.e., in the fat), due to the size of the needle and fat tissue present at particular site. Thus the amount of fat

Table 5   Appearance of Subcutaneously Administered Horse Serum in Lymph and Blood of Dogs Under Various Conditions

Condition	Time to detect presence of horse serum in:	
	Thoracic duct lymph	Blood
No massage at site	40 min	3.5 hr
Massage at site	15–20 min	2 hr
High-pressure injection	<5 min	40 min

Source: Ref. 56.

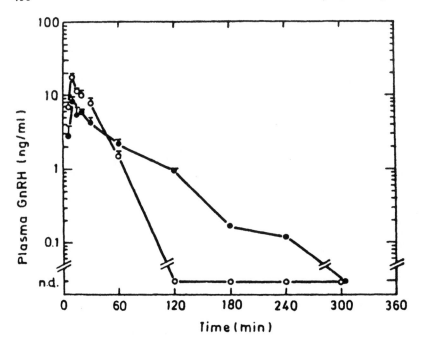

Fig. 2   Plasma levels of GnRH after intramuscular administration of 10 µg of GnRH (open circle) and 10 µg of GnRH in 10% dextran T500 (filled circle) into male rats.   Each point represents the mean ± SEM (n.d., not detectable: <0.1 ng/ml). (From Ref. 60.)

tissue at the site may cause differences in absorption, although the benefit or disadvantage for injections to be given in fat rather than in muscle is not clearly established.

Tissue condition.   The condition of the tissue at the site of injection can also affect peptide absorption rate; for example, subcutaneous absorption of insulin was retarded in diabetics when injected into regions where induration or an insulin pad had formed.   After a recovery period the tissue once again regained its ability to abosrb insulin rapidly [69].

Degradation at the site.   Degradation or metabolism at the site of injection is another major determinant of peptide and protein drug absorption from intramuscular or subcutaneous injection.   The enzymatic barrier to peptide absorption has been reviewed recently [6].   Insulin [70,71], interferon [72], and gamma globulin [17] are shown to degrade at the site of intramuscular injection, resulting in low bioavailability.   Clinical laboratory studies in a diabetic patient showed that insulin resistance by subcutaneous route was due to excessive insulin degradation in adipose and muscle tissue and could be reversed by aprotinin, a serine protease inhibitor [70].   Berger et al. [73] also reported that aprotinin increased the absorption rate of subcutaneously injected insulin and amplified its biological effects by inhibiting local degradation of exogenous insulin at the injection site.   Subcutaneous insulin sensitivity improves temporarily following prolonged continuous intravenous or intramuscular insulin infusion, suggesting that depot subcutaneous insulin may induce its own degrading enzyme [70]. On the contrary, the possibility of misinterpreting the slow subcutaneous absorption of insulin as evidence of degradation has been pointed out [74]. The dynamic nature of the degradation and absorption of insulin at the subcutaneous injection site was described by Hori et al. [75]. They used four protease inhibitors to improve subcutaneous bioavailability in rats. Benzoxyl carbonyl-gly-pro-leu-gly was found to inhibit insulin degradation markedly. In addition, contrary to previous findings [73], aprotinin was ineffective. The rationale underlying the selection of the amino acid sequence in the inhibitors was not provided and the proteolytic enzymes affected were not identified.

Subcutaneous and intramuscular bioavailability of thyrotropin releasing hormone (TRH), a tripeptide, in mice is only 67.5% and 31.4%, respectively [76], presumably due to degradation at the injection site. Parathyroid hormone and its analogs, bPTH(1-34) and hPTH(1-34), as well as calcitonin are also incompletely absorbed due to degradation following subcutaneous administration in chicks, since their absorption improved when coadministered with aprotinin [77,78].

Often, a correct balance between the rate of absorption and that of clearance at the site of injection is required to achieve therapeutic blood levels. Cyclosporin A, a water-insoluble cyclic endecapeptide, achieved this balance presumably by improved absorption and maintained adequate blood levels when injected intramuscularly in chacma baboons after diluting the ethanolic solution with Intralipid 20% [79].

Other factors. Absorption from the extravascular site also depends on blood flow, pH of the vehicle, volume and concentration of injection, osmolarity of the solution, disease state, age of the patient, polymorphic form, particle size, solubilities, and solvents [43,47,53,54,80]. Agents that affect blood flow (e.g., vasoconstrictors) can delay absorption [81]. Also because of arteriovenous anastomoses, regional capillary blood flow is different than total blood flow through the muscle. Absorption after intramuscular injection of tissue plasminogen activator in rabbits was enhanced by hydroxylamine and electrical stimulation (improves blood flow) without deleterious local effects [82].

Pharmacokinetics. The pharmacokinetix of insulin after extravascular administration has been studied extensively [51,83] and various models have been proposed. The processes by which insulin is absorbed are extremely complex, involving both spatial and temporal factors [83]. There is considerable difficulty at the present time in precisely determining the factors likely to be involved, such as local diffusion rates in the tissue, insulin binding in the tissue, and local blood flow. Insulin seems to be absorbed directly into the blood stream rather than via the lymphatics, irrespective of whether the insulin is injected in a dissolved, amorphous, or crystalline state [84]. A precise determination of the amount of insulin cleared from the injection site per unit time requires measurement of the blood and lymph flow through the tissue and of the arteriovenous concentration difference. This is impracticable in humans and difficult in animal models [83]. By labeling the insulin with a γ-emitting isotope, the remaining amount of insulin at the injection site can be determined indirectly by external counting of the nonabsorbed radioactivity, provided that in every respect the labeled substance behaves like the unlabeled, and that the radioactivity counted externally is proportional to the nonabsorbed amount of radioactivity. After the injection, insulin remains for an appreciable time in a micromilieu at 35 to 37°C containing all sorts of enzymes, buffers, and ions—the influence of which on insulin is more or less unknown. It has been postulated that these conditions are conducive to polymerization of the longer-acting insulins especially [85].

If circulating insulin-binding antibodies are present, the pharmacokinetics of insulin are considerably more complicated [86,87]. The circulating insulin antibodies only seem to act as a carrier protein. The antibodies are heterogenous and of IgG type, comprising a mixed population of high-affinity and low-affinity antibodies. This division is somewhat arbitrary, the binding of insulin to antibodies being complex, with a large variation between patients [88]. Even in the same patient, considerable time-to-time variations are seen [89], presumably related to the quality of insulin preparation.

A simplified model has been developed to describe the general peptide pharmacokinetics after subcutaneous or intramuscular administration [90] (Fig. 3). The single-pool model (Fig. 3) assumes that all the drug at the injection site is immediately available for transcapillary absorption. However, following injection of insulin, there is some evidence of an initial delay phase prior to the monoexponential disappearance phase [84]. The second, split-pool model allows for such an initial delay phase, where drug absorption takes place from a pool secondary to the injection pool. Both models assume a fractional rate of systemic delivery K_{sp} and a degradation rate constant K_d from the respective tissue pool. The process labeled as degradation may include drug that is irreversibly bound to the tissue as well as drug that is physically degraded. No distinction can be made mathematically between these possibilities. If the rate constant K_{12} is much higher than K_{sp}, the first pool has little influence beyond the early period. To identify parameters in these models, it is necessary to estimate independently the metabolic clearance rate and total distribution volume of the drug. Some used a multipool model to represent insulin kinetics [91,92]. Others found a single pool adequate [93]. Sometimes, however, the single-pool model may be an oversimplification. For example, there may be some local reversible binding of insulin, delaying absorption at low infusion rates, with saturation of the binding at high infusion rates. Also in the case of circulating antibodies, the single-pool model may not be adequate [86,87].

One kinetic study of insulin showed ingenuity by using a three-compartment model with infusion not to steady-state concentration but to steady-state glucose levels [91]. Compartment 1 is the plasma space. The other two compartments are extravascular; compartment 2 is small and equilibrates rapidly with plasma, and compartment 3 is large and equilibrates slowly with plasma. It was concluded that compartment 3, which could not be sampled but for which concentrations of insulin could be calculated, was the compartment pertinent to glucose utilization.

Other Routes

Intraperitoneal administration, in which the drug is injected into the peritoneal cavity, is infrequently used for peptide and protein drugs. For patients with resistance to insulin subcutaneously injected, intraperitoneally it can gain direct access to the lymphatic system and then return slowly to the vascular compartment [48]. Thus this route is especially useful for targeting drugs to the lymphatic system [95]. Intraperitoneal route is often less potent than intravascular route for peptide drugs [11,12].

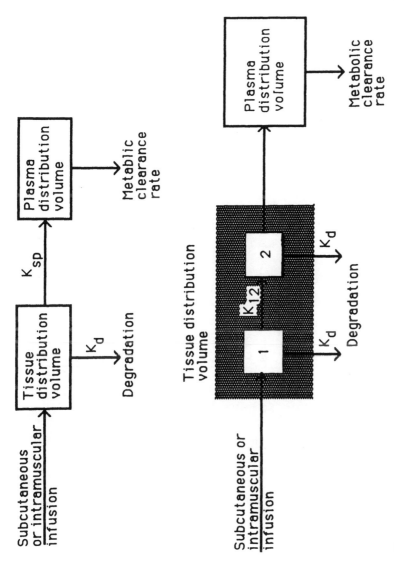

Fig. 3 Single-pool (top) and split-pool (bottom) models of subcutaneous or intramuscular peptide pharmacokinetics. (Modified from Ref. 90.)

Table 6　Comparative Efficacy of Peptide/Protein Drugs by Different Parenteral Routes of Administration

Peptide/protein	Dose[a]	Route of administration	Species	Relative efficacy[b]	Response measured	References
ACTH (1–24)	24 µg/kg	Intracerebro-ventricular(ICV)	Rat	++	Antishock effect	104
	24 µg/kg	Intravenous		+		
Calcitonin	3 µg	Epidural	Humans	+	Pain relief	105
	1 µg	Subarachnoid		+		
Corticotropin-releasing hormone	3 µg/kg	Intravenous	Humans	+	Cortisol and ACTH levels	106
	30 µg/kg	Subcutaneous		+		
DDAVP	0.4 µg/kg	Intravenous	Humans	+	Clotting factors	107
	0.4 µg/kg	Subcutaneous		+		
Dermorphin	NS	Intracerebro-ventricular	Mouse	++	Analgesic activity	108
		Subcutaneous		+		
GHRH	0.003–0.1 µg/kg	Intravenous	Humans	+	GH secretion	45
	1–10 µg/kg	Subcutaneous		±|		
	1 µg/kg	Subcutaneous	Children	+		109
Growth hormone	0.1 U/kg	Subcutaneous	Children	+	Height velocity, somatomedin level	52
	0.1 U/kg	Intramuscular		+		
Influenza vaccine	0.1 ml	Intradermal	Humans	+	Immunization titer	99,100
	1 ml	Subcutaneous		+		
Interleukin-1β	100 pg/kg	Intravenous	Mouse	+++	Immune response	110
	20 ng/kg	Subcutaneous		+		
	20 ng/kg	Intraperitoneal		+		
	80 ng/kg	Oral		±|		

Drug	Dose	Route	Species	Efficacy[b]	Response	Ref.
163–171 fragment of interleukin-1β	10 µg/kg	Intravenous	Mouse	+++	Immune response	110
	33 mg/kg	Oral		+±+		
	100 mg/kg	Subcutaneous		+		
	100 mg/kg	Intraperitoneal		+		
LHRH	NS	Intraperitoneal	Rat	+++	Ovulation induction	111
		Subcutaneous		++		
		Intravenous		+		
		Intramuscular		+		
Met-enkephalin	0.01–0.1 mg/kg	Intravenous	Guinea pig	+	Cell-mediated immunity	14
		Subcutaneous		++		
Peptidoglycan	1.0 mg	Intravenous	Mouse	++	Immune response	112
	1.0 mg	Subcutaneous		+		
Satietin	10–20 mg/kg	Subcutaneous	Rat	+	Food intake, running-wheel activity	113
	25 µg	ICV Consecutive		+		
		Alternate day		–		
Thymopentin	0.03 mg/kg	Intravenous	Mouse, guinea pig	+	Electromyographic response	11, 12
	0.3 mg/kg	Subcutaneous		+		
	0.3–0.6 mg/kg	Intraperitoneal		+		
TRH	1 mg	Intramuscular	Mouse	++	Brain concentration	114
	1 mg	Intravenous		+		
	1 mg	Intraperitoneal		+		
Fibrin degradation product	20 µg	Intrasplenic	Mouse	++	Monoclonal antibody produced	115
	50 µg	Subcutaneous		+		
	50 µg	Intraperitoneal		+		

[a]NS, not specified.

[b] –, little or no effect; ±, low and/or controversial efficacy; +, effective; ++, more effective than +; +++, much more effective than +.

Intraarterial infusion increases drug delivery to the area supplied by the infused artery while reducing the access of the drug to the systemic circulation [96]. One example is selective intraarterial infusion of vasopressin in treating massive upper gastrointestinal tract hemorrhage [97]. Intradermal injections are used predominantly for local effects, but the potential of this route for systemic activity, particularly for vaccines, have been illustrated in several studies [98—101]. Intradermal injection is made into the upper layers of skin, just beneath the epidermis, the usual site being the anterior surface of the forearm. The volume of solution that may be administered in this manner is only about 0.1 ml [99]. Intradermal route often produces comparable or superior effect than the subcutaneous route [99—101].

Sometimes in acute conditions when the peptide drugs is not effective intravenously, it is injected directly at the site of action or into the body fluid near it (e.g., intraspinal, intrathecal, intracerebroventricular, etc.). Since no absorption step is involved, the onset of action is instantaneous and the drug effect is predictable following injection or infusion. For many central nervous system (CNS) drugs, intravenous route is not effective due to the presence of the blood-brain barrier, and thus local administration is necessary. For example, the analeptic properties of the tripeptide TRH and analogs are 100,000-fold greater following intracerebroventricular administration [102] than from the intravenous route. Continuous intraspinal or intraventricular infusion of somatostatin has also been used to alleviate cancer pain [103]. Because of the special techniques involved and necessity of trained personnel, such routes are of limited use in long-term clinical therapy and are commonly used as the last resort when all other routes become ineffective in a life-threatening disease.

Comparative Features

As mentioned before, detailed pharmacokinetic analysis of peptide and protein drugs are intimately related to sensitive and specific assay methodology. Comparison between different parenteral routes of administration thus relied primarily on some kind of biological response. Table 6 lists the comparative efficacy of some representative peptide drugs administered by different parenteral routes. With growth hormone releasing hormone (GHRH) in adults the subcutaneous route was 30-fold less effective than the intravenous route [45]. This was in contrast to a significant effect in growth hormone-deficient children when given subcutaneously [109]. While the adult subjects received GHRH dissolved in a 1-ml vehicle, the children received their doses in 50-μl vehicle. Whether the absorption of the hormone is more complete and/or degradation of the peptide less pronounced as a function of a smaller volume of vehicle is unknown.

Sometimes the species difference in the activity of a drug depends on the route of administration. Threshold dose of response for thymopentin by intravenous injection is similar in both mouse and guinea pig. However, for intraperitoneal and subcutaneous routes there appear to be species differences that may relate to permeability of vessels in peritoneum and skin, rates of diffusion in these tissues, and the presence of local proteolytic enzymes [11,12].

Despite the analytical problems encountered, some studies have reported pharmacokinetic (Table 7) and bioavailability (Table 8) data after administration by different routes. Interferon kinetics after intravenous, intramuscular, and subcutaneous injection (Fig. 4) best illustrates the relative merits

Table 7 Pharmacokinetic Parameters of Peptide/Protein Drugs by Different Parenteral Routes of Administration[a]

Peptide/protein	Route of administration	Species	$t_{1/2}$ (abs)	C_{max}	T_{max}	$t_{1/2}$	CL (liters min)	References
Des-enkephalin-γ-endorphin	Intravenous	Rat	—	—	—	6.3 min	20.3	116
GnRH	Intravenous	Humans	—	400 pg/ml	—	6.6 min	1.0	37,117
	Subcutaneous			93.5 pg/ml	—	—	—	
IgG	Intramuscular	Humans	1.6 days	44%	3.5 day	22–46 days	—	66
	Subcutaneous		3.1 days	33%	6 day	—	—	
Insulin	Intravenous	Humans	—	6.9 n mol/ liter	2 min	30.9 min	1.0	118
	Intramuscular			0.2 n mol/ liter	60 min	—	—	
Interferon	Intravenous	Humans	—	13.9 ng/ml	5.1 h	—	0.2	72
	Intramuscular		—	2.0 ng/ml	3.8 h	2.3 hr	—	
	Subcutaneous		—	1.7 ng/ml	7.3 h	3.5 hr	—	
Leuprolide	Intravenous	Humans	—	—	—	2.9 hr	0.14	119
	Subcutaneous		4.2 min	32.3 ng/ml	0.6 h	3.6 hr	0.15	
LHRH	Intravenous	Humans	—	—	—	7.8 min	1.8	120
D-Trp[6]LHRH	Intravenous	Humans	—	—	—	19 min	0.5	121
	Subcutaneous		—	1.85 ng/ml	38 min	7.6 min	0.16	
α-MSH	Intraperitoneal	Rat	7.3 min	14.1 n mol/ liter	16.2 min	26 min	—	122
	Subcutaneous Acid-saline		5.6 min	8.3 n mol/ liter	12.5 min	23.2 min		

Table 7 (Continued)

Peptide/protein	Route of administration	Species	$t_{1/2}$ (abs)	C_{max}	T_{max}	$t_{1/2}$	CL (liters min)	References
α-MSH	Subcutaneous ZnPO$_4$ complex	Rat	17.7 min	4.8 n mol/liter	26.6 min	19.3 min	—	
Tumor necrosis factor	Intravenous	Rabbit	—	—	—	—	—	123
	Intramuscular		—	3700 U/ml	2.5 h	8.8 hr	—	
	Subcutaneous		—	1930 U/ml	2.5 h	4.5 hr	—	
	Intraperitoneal		—	2605 U/ml	6.0 h	5.8 hr	—	
[125]I-tumor necrosis factor	Intravenous	Rabbit	—	—	—	2.97 hr	—	123
	Intramuscular		—	8.8 cpm/μl	3.0 h	10.4 hr	—	
	Subcutaneous		—	8.8 cpm/μl	2.5 h	6.8 hr	—	
	Intraperitoneal		—	15.1 cpm/μl	6 h	7.2 hr	—	

[a] —, data not available.

Table 8   Bioavailability of Peptide/Protein Drugs by Different Parenteral Routes of Administration

Peptide/protein	Bioavailability[a] (%)			Assay[b]	Species	Reference
	Im	Sc	Ip			
Cyclosporine[c]	100	58.5	57.6	RIA	Rat	124
Des-enkephalin-γ-endorphin	8.5	30.9	–	HPLC	Rat	116
Insulin	48.2[d]	54.1[d]	–	RIA	Humans	118
		84.4	41.6	RIA	Humans	125
Interferon	82.9	90.3	–	EIA	Humans	72
Leuprolide	–	94.3	–	RIA	Humans	119
α-MSH	–	2.8	7.0	RIA	Rat	122
Metkephamid (enkephalin analog)	–	90.6	–	HPLC	Rat	126

[a]–, Data not available.

[b]RIA, radioimmunoassay; HPLC, high-pressure liquid chromatography; EIA, enzyme immunoassay.

[c]IV data not available, hence IM was assumed to be 100%.

[d]AUC calculated over 0 to 360 min compared to 0 to 240 min for IV.

of these three routes. Adverse effects, which appear to be a function of exceeding and maintaining serum concentrations above a threshold level, were dependent on the route of administration. The clinical adverse experiences were least severe and of shortest duration after intravenous infusion, whereas they were more severe and of longer duration after the intramuscular injection and most severe and of longest duration after the subcutaneous injection [72]. Interferons (IFN) show different patterns of distribution depending on the route of distribution. Thus primarily, the highly diffusible IFNs-α are administered intramuscularly and rarely by intravenous infusion or bolus injection. The poorly diffusible IFN-β are administered primarily by brief intravenous infusion and rarely by intramuscular injection [127]. It was suggested that to improve the efficacy of interferon after parenteral routes of administration, it must be shifted to the lymph pool minimizing direct absorption into the blood. It appears that the increased colloid osmotic pressure of subcutaneous interstitial fluid obtainable after injection of an IFN solution containing serum albumin is an important driving force for a major opening and absorption of IFN via lymphatics. This new route is considered lymphatic, even though it does not require an injection directly into a lymphatic duct [127].

Cyclosporine, an extremely useful immunosuppressant in organ transplantation, exhibits multicompartment behavior following intravenous

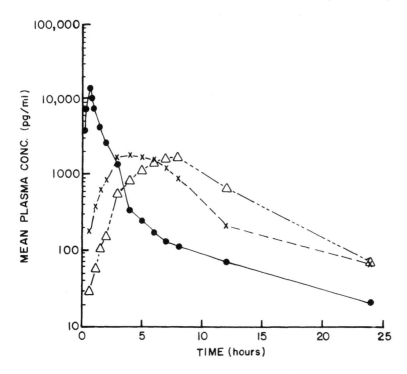

Fig. 4   Mean serum interferon concentrations after a single 36×10^6 U dose as an intravenous infusion (open circle) or an intramuscular (cross) or subcutaneous injection (open triangle). (From Ref. 72.)

administration in humans [128].   The influence of the route of administration and dosage form on cyclosporine pharmacokinetics in rat was recently reported [124].   The intramuscular route gave variable but highest plasma levels, whereas the subcutaneous route provided reproducible and steady levels over a 24-hr period.   Intraperitoneal administration of cyclosporine resulted in an intravenous-like pattern [129] with early, very high peaks followed by a steady decrease.

Bioavailability data for oligopeptides after subcutaneous, particularly after intramuscular injection, are scarce and vary considerably.   Only a few percent of drug has been found for α-MSH after subcutaneous administration in rats [122], whereas high (but not numerically expressed) bioavailability has been claimed for a somatostatin analog [130] and LHRH-related peptides [111].   Bioavailability after intramuscular injection (8.5%) of des-enkephalin-γ-endorphin (DEγE) was significantly smaller than that after subcutaneous injection (30.9%).   It is noteworthy, however, that in various behavioral paradigms, DEγE has been shown to be active for hours or even days after subcutaneous injection in rats [131,132], even though the blood levels of DEγE are below the limit of detection.   Therefore, bioavailability might be considered only to a certain extent as a measure of the biological effect of neuropeptides.   It is known that subcutaneous administration of ACTH-related peptides is more effective in behavioral tests than in intravenous injection [133].   This illustrates particularly the difficulty in correlating bioavailability and biological effectiveness for neuro-peptides.

The importance of the route of delivery is clearly apparent in one inter-esting study, where substance P fragment elicited precisely opposite effects in rabbits when administered into the cerebral ventricles and when adminis-tered intravenously (Fig. 5) [134]. Following intravenous injection the arterial blood pressure decreased in conscious rabbits, whereas when ad-ministered into the cerebral ventricles, the hexapeptide caused an increase in blood pressure. The underlying mechanism of this effect is unknown.

The preceding discussion showed that judicious choice of the route of parenteral delivery is important for most peptides. It depends on the par-ticular effect desired and the rate of absorption and/or degradation at the injection site. It appears that the subcutaneous route achieves prolonged therapeutic effect for most peptides, although the plasma levels are lower than for other routes. Patients also indicated an overwhelming preference for subcutaneous injection over intramuscular injection [52]. Since most of the studies compared direct or indirect pharmacological effect rather than bioavailability based on pharmacokinetic analysis, a quantitative com-parison between different routes is not possible.

B. Mode of Delivery

The main goal of any therapy is to maintain the therapeutic effect for a desired period. This will be better achieved by continuous infusion rather than conventional multiple daily injections, since most peptide drugs have a very short half-life. However, the relationship between pharmacokinetics and pharmacodynamics for many peptide and protein drugs is often not linear. Pulsatile secretion of a number of peptide hormones, including luteinizing hormone [135], growth hormone [136], vasopressin [137], insu-lin, and glucagon [138,139], is well documented in humans. Oscillating plasma hormone concentrations are thought to be of major biological impor-tance in that down-regulation of receptors is avoided and thus hormone action is enhanced. One interesting example is gonadotropin-releasing hor-mone and its analogs, in which the pulsatile dosing and dosing to steady state achieve opposite pharmacological effects—mimicking the normal physio-logical condition and achieving receptor desensitization, respectively [140]. Basal insulin secretion in humans also shows a circadian rhythm with a peak time at 1500 hr. This physiologic circadian baseline concentration can best be achieved by a programmable pump [141,142]. Circadian variation in absorption of peptide hormones has been reported [143]. There are three major modes of delivery:

1. Conventional multiple daily injections
2. Continuous infusion
3. Pulsatile delivery
   a. Open-loop
   b. Closed-loop

Insulin has been widely studied with respect to all these modes of de-livery [144]. Comparison of the modes of delivery for insulin (Fig. 6) shows that optimal use of multiple daily injections results in glycemic con-trol equivalent to that achieved with closed-loop intravenous insulin

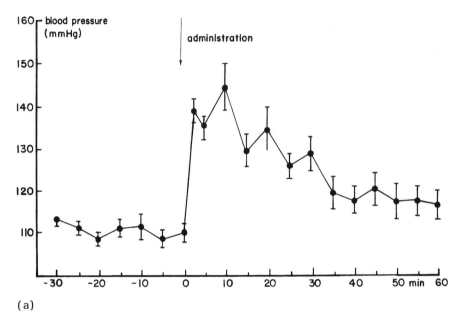

(a)

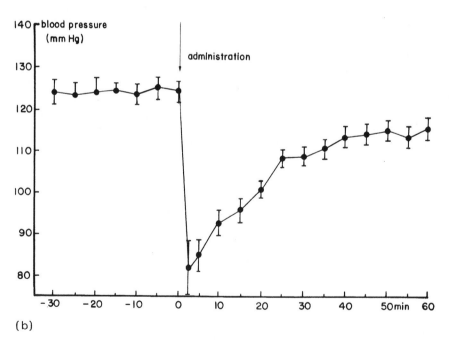

(b)

Fig. 5 Arterial blood pressure before and after SP_{6-11} administration (a) into the cerebral ventricles and (b) into the marginal ear vein of rabbit. Filled circles and vertical bars denote mean ± SEM. (From Ref. 134.)

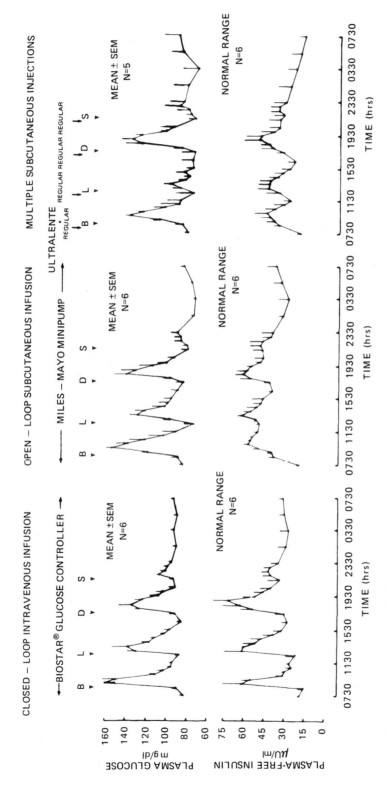

Fig. 6 Concentrations of plasma glucose and free insulin in insulin-dependent diabetic subjects treated with three modes of therapy. B denotes breakfast, L lunch, D dinner, and S snack. (From Ref. 144.)

infusion. Besides inconvenience, the primary drawback of multiple daily injections is the delay inherent in changing the basal insulin concentrations.

Closed-loop delivery involves feedback of the response to control the infusing system; for example, in the case of insulin, blood glucose level is monitored and the infusion rate is changed accordingly to achieve the desired glycemic profile. Typically, these systems have three characteristics: a method to measure blood glucose continuously, an algorithm to determine insulin infusion rates, and a pump to infuse insulin [145–147]. All these are designed to respond not only to the prevailing concentration of glucose but also to the rate and direction of change [148]. Theoretically, closed-loop systems are ideal in normalizing the plasma glucose concentration. But they are not yet practical for long-term use, given their size and the need for an extracorporeal glucose sensor. This has led to the development of so-called "open-loop" systems consisting of an insulin pump and a program for insulin delivery [149–153]. The program for delivery ranges from a single basal infusion rate, which is briefly increased at the time of meal ingestion, to a multicomponent program, including multiple basal rates and complex postprandial waveforms. The system is called open-loop because the prevailing glucose concentration does not influence the insulin infusion. In continuous infusion, the drug is infused only at one constant rate. Most of the pumps currently in use are programmable and thus have the flexibility for either use.

A number of studies reported the effectiveness of pulsatile administration for peptide drugs [154–167]. Some showed that pulsatile administration is more effective than continuous delivery for peptide hormones such as insulin [154,166], gonadotropin-releasing hormone [155–157], parathyroid hormone [164], and glucagon [165,166]. Continuous GnRH infusion was ineffective regardless of the infusion rate in monkeys (Fig. 7). In sharp contrast, however, long-term restoration of gonadotropin secretion was achieved in the same animals by the intermittent administration of GnRH (Fig. 7). Change from pulsatile to continuous mode, without a change in the infusion rate, results in declining gonadotropin levels, which reverses when pulsatile administration is reinsituted (Fig. 8). For insulin, receptor binding study showed a relatively enhanced insulin binding with pulsatile insulin or a relative down-regulation with continuous insulin [154]. Thus while the underlying cellular mechanism is not clear, it appears that down-regulation of receptors during continuous administration partly explains the higher effectiveness of pulsatile mode of administration. In a recent study of growth hormone in rats, single injection gave a higher area under the curve than that of continuous infusion (Table 8) [168]. The authors did not provide any explanation.

Successful pulsatile therapy depends on the dose per pulse, the frequency of pulses, and the route of administration [162]. Lack of difference in metabolic effects after pulsatile and continuous delivery of insulin [169] was explained by the higher insulin levels maintained in those subjects. Pulsatile insulin appears to be more effective under close-to-basal conditions. It has been pointed out that assuming that insulin receptors are directly influenced by the ambient insulin concentration, pulsatile variations in plasma levels must be present at the receptor site as well [169]. Changes in pulse frequency represent a mechanism of LH regulation alternative to changes in pulse amplitude, and it appears that there may be a frequency threshold that allows GnRH pulses to the pituitary to either turn on or turn off gonadotropin release in vivo [170].

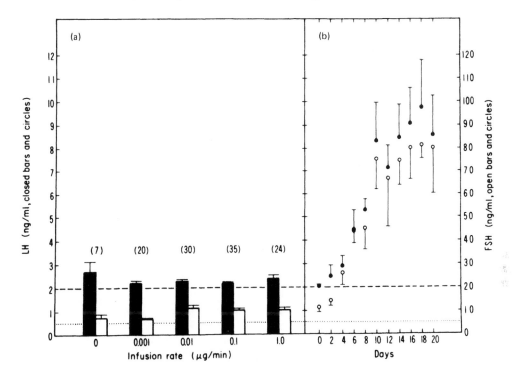

Fig. 7   Gonadotropin secretion in ovariectomized rhesus monkeys bearing hypothalamic lesions after (a) continuous and (b) pulsatile (1 μg/min for 6 min once per hour) intravenous infusion of GnRH. In plot (a), each bar represents the mean ± standard error (SE) of the number of observations in parentheses obtained during the last 5 days of the infusion period. In plot (b), each point is the mean ± SE of three to five observations. The horizontal dotted and dashed lines show the sensitivity limits of the FSH and LH assays, respectively. (From Ref. 155.)

Three major routes of long-term infusion are intravenous, subcutaneous, and intraperitoneal. The intravenous route offers excellent dynamics, a change in infusion rate being accompanied immediately by a change in circulating drug concentration. The difficulty in maintaining venous access remains the major disadvantage of intravenous infusion. Both infection and catheter obstruction have also proved formidable obstacles.

Subcutaneous routes, although useful for long-term use, has been quite controversial in terms of efficacy [117,159,162,164,171–173]. Lack of difference in pulsatile and continuous mode of administration of GnRH [172] and growth hormone releasing factor [174] could be due to the subcutaneous route chosen. The pharmacokinetic differences due to a combination of delayed absorption and irreversible losses of the drug when injected into subcutaneous tissues produces a damping of the plasma profile. This converts a pulsatile waveform into a more sinusoidal pattern and may underlie the lower efficiency of this route in pulsatile therapy of some drugs [117]. Higher doses may be needed by the subcutaneous route to achieve equivalent effect of the intravenous route [158,159]. Development of hematoma

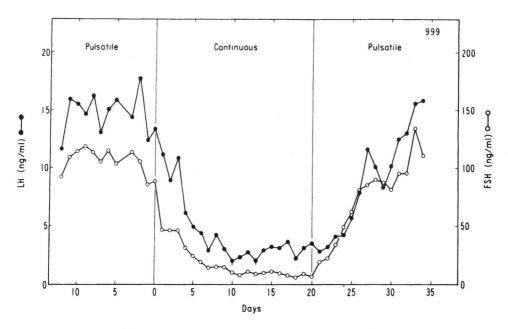

Fig. 8   Gonadotropin secretion in ovariectomized rhesus monkeys after pulsatile (1 µg/min for 6 min once per hour) and continuous (1 µg/min) infusion of GnRH.   (From Ref. 155.)

at the cannula site may lead to local degradation of GnRH [175].   Frequent changes of site every 1 to 2 days and avoidance of heparin are recommended for successful use of the subcutaneous route.   Absorption can be quite variable by the subcutaneous route [176].   Knowledge of the kinetics of drug absorption can help in deciding the dose per pulse and frequency of pulse that will achieve the physiological pulsatile profile in plasma.

The kinetics of absorption of insulin from the peritoneal space are excellent, more closely mimicking those observed after intravenous infusion. The primary limitation of the intraperitoneal route is related to access. Chronic infusion can be complicated by peritonitis [144].

III.   DELIVERY SYSTEM DESIGN

A.   Conventional Formulations

Conventional parenteral formulations have been reviewed before [177,178]. The greater chemical and structural complexity of protein molecules may present some unusual problems in their formulation design.   Efficiency and safety of protein drugs are of utmost concern in the design of parenteral formulations.   To assume that these goals are met, the chemical, physical, and biological integrity of the peptide or protein must be maintained up to and including delivery to its site of action in the body [179].   Since most parenteral formulations are liquid, stability of the protein drug in solution for the required period of time is important.   Stability problems can involve either integrity of the primary structure or alterations in secondary or tertiary structure (i.e., denaturation). Denaturation is usually the more serious problem, since it can be induced by a variety of conditions encountered in normal processing or use: heat, pH extremes, hydrophobic surfaces, and shear or agitation during filtration or reconstitution.

Human alpha interferon in solution undergoes concentration and pH-dependent self-association, whereas bovine interferon does not appear to self-associate [179]. However, at higher temperatures (up to 37°C), bovine interferon forms a substantial precipitate. The rate of precipitate formation is extremely temperature dependent and is probably due to conformational changes in the protein at higher temperatures. Aggregation has also been observed with small peptides such as delta sleep-inducing peptide (DSIP) [180] and LHRH analogs [181]. Insulin aggregation remains a fundamental obstacle to the long-term application of many insulin infusion systems [182–184]. Some of the factors that affect insulin aggregation are metal ion type and amount, ionic strength, pH, temperature, and agitation [183].

Most peptide or protein drugs have a significant hydrophobic component. Thus they have a tendency to adsorb to manufacturing or delivery equipment [185–188]. Losses are particularly significant at low concentrations, and this can result in a lower dose for the patient. This property may also result in physical instability in solution, gelling [181,189], or a gross increase in viscosity to the point of noninjectability. The phenomena of adsorptive loss and gelling are both influenced by the presence of excipients, and materials for pH control and tonicity must therefore be carefully selected. In general, electrolytes exacerbate the situation and a suitable solvent may alleviate it [181].

Aggregation and surface adsorption of insulin can be prevented by urea [190], glycerol [191], dicarboxylic amino acids [192], or anionic and nonionic surfactants [182]. Contact with different materials in dosing devices, particularly silicone rubber [182], other design-related factors, and motion appear to be chemically more detrimental than storage in glass vials at the same temperature [193]. Surface-active polyethylene-polypropylene glycol-type stabilizer increases the physical stability remarkably [194]. A recent study of a new insulin formulation [195], however, showed that insulin precipitated in glass vials but remained clear and active in polyethylene reservoirs, catheters, and pumps, although only 83 to 90% of insulin was delivered chemically intact. Heparin can be added to insulin formulations to prevent catheter obstruction due to fibrin deposition. An extensive study of 60 additives and 1125 formulations [196] reported that Pluronic F68, a polyoxyethylene–polypropylene glycol surfactant, appears to be a promising stabilizer.

Filtration has been used to sterilize drugs and proteins that cannot be heat sterilized. The effect of membrane filtration on protein conformation has been studied recently with alkaline phosphatase, insulin, and immunoglobulin [197]. The applied pressure and filter pore size did not appear to affect protein conformation of filtered samples. Hydrodynamic variables in the form of shear stress are probably not the cause of any conformational changes either, while the membrane's ability to bind protein seems to be more closely related to conformational changes. Thus membrane composition is important.

The stability problem of many protein drugs, except for some small peptides, are serious enough that they are commonly formulated as a solid usually by lyophilization [198] and reconstituted with a sterile diluent just prior to administration. Even so, stabilizers such as amino acids, proteins, or other polymers are often required to achieve an adequate shelf life. Their precise mechanism of action is unknown but can include the following: (1) competition with drug for potential reactants, (e.g., water), (2) lowering the activity of the drug by providing multiple interaction sites, (3) competition with drug for irreversible adsorption sites (e.g., container), and (4) providing a nonreactive carrier to which drug adsorbs until dissolution. In designing a suitable manufacturing process for a lyophilized protein product, several issues are of particular concern: adsorption onto contact surfaces, shear and heat sensitivity, loss of activity during freeze-drying, and moisture sorption by the product after drying.

Injections

The syringe and the needle constitute part of the injection delivery system. Pain after intramuscular or subcutaneous injection is one of the major reasons for patient noncompliance with injections. Injection pain can arise from three factors [199,200]: (1) local irritation due to either the antiseptic used on the skin or the properties of the parenteral formulations, (e.g., pH, tonicity, or chemical action on the tissues), (2) mechanical trauma due to introduction of the needle to the injection site or to the sudden distension of the tissues resulting from a too rapid discharge of the syringe contents, and (3) abnormal sensitivity of tissue at the injection site. Factors (1) and (2) should be considered in designing the delivery system. Delayed pain can be caused by infection, aseptic irritation and necrosis, antigenic reaction, and reactions to pyrogens.

To eliminate skin pain, a sharp needle, free of barbs, having the smallest diameter compatible with viscosity of the solution should be used. With the widespread availability of "superfine" needles for injection and some novel methods, the discomfort of injection can be largely reduced. A new jet-type injector (Preci-jet 50) was recently developed for insulin [201]. A jet injector may improve patient compliance to multiple daily injection protocols. With these high-pressure devices, the drug is ejected through a very fine hole (diameter: 0.008 in.). At high velocity, the shock wave generated in front of the insulin bolus pierces the skin and the insulin spreads into the subcutaneous area without the need to use a needle. It is small, simple in design, and is capable of mixing two types of insulin, which are not available in the earlier jet injection technology. The method results in relatively painless, reproducible injections. Moreover, because of the wider subcutaneous dispersion of insulin, absorption is faster than when injected with a conventional needle [202].

Novopen is another new device for convenient and accurate injection of insulin [203]. It is like a fountain pen in size and design. The insulin is delivered through a detachable, disposable 27-gauge superfine needle by a button on the pencap which, on full depression, delivers 2 U of insulin. In a multicenter trial, this device vastly improved patient compliance compared to conventional injections [204]. A sprinkler needle for insulin injection has also been described recently [205].

B. Advanced Delivery Systems

Besides improvement in injection techniques, another approach to alleviate the pain and discomfort of frequent injections is either to use a programmable infusion pump or to prolong release or minimize the proteolytic degradation of the peptide. Successful development of a peptide drug also involves reduction of dose cost and enhancement of selectivity, which can be achieved by targeting the drug to the desired site of action. These two concerns, minimizing the dosing frequency and targeting, have led to a number of investigational advanced delivery systems. Some of these systems are discussed in the following section.

Pumps

The use of a portable pump for continuous subcutaneous insulin infusion was first reported by Pickup and colleagues in 1978 [153]. Since then a large number of pumps have been used to deliver peptide hormones, insulin being the most common. The prospects and limitations of pumps with respect to insulin, LHRH, and other drugs have been the subject of many recent reviews [206-212].

A pump can be distinguished from other diffusion-based system in that the primary driving force for delivery by a pump is pressure difference rather than concentration difference of the drug between the formulation and the surroundings. The pressure difference can be generated by pressurizing a drug reservoir, by osmotic action, or by direct mechanical actuation. The pump can be also externally portable or implantable. Implantable. Implantable pumps are technically more complex. The ideal pump should have the following characteristics [207]:

1. The pump must deliver drug at the prescribed rate(s) for extended periods.
    a. Wide range of delivery rate
    b. Accurate, precise, and stable delivery
    c. Reliable pump and electrical components
    d. Drug compatible with pump internals
    e. Simple means to monitor the status and performance of the pump
2. The pump must be safe.
    a. Biocompatible exterior if implanted
    b. Overdose protection, no leakage, fail-safe mechanism
    c. For implantable pumps, sterilizable interiors, and exteriors
3. The pump must be convenient.
    a. Reasonably small size and inconspicuous
    b. Long reservoir life
    c. Easy programmability

Mechanical pumps. Most portable pumps in use for insulin delivery are syringe driven. There are two drive principles: lead screw and direct drive [211]. In leadscrew drive the motor drives the leadscrew through a reduction gearbox. Rotary motion of the leadscrew is transferred via a carriage assembly to linear movement, thus driving the syringe plunger. The size of this drive system makes the currently available leadscrew devices extremely large.

In direct-drive systems, the drive is directly transmitted to the cylindrical plunger of the syringe by means of a serrated or spliced roller which forms teeth in the plastic plunger as it turns. The advantages of this drive system are its compact size, freedom from backlash, and ease of loading the syringe assembly. Some examples of syringe driven infusion devices are AutoSyringe, Milhill infuser, Penpump, Microjet, Pharmajet, and CPI-9100 [206-212].

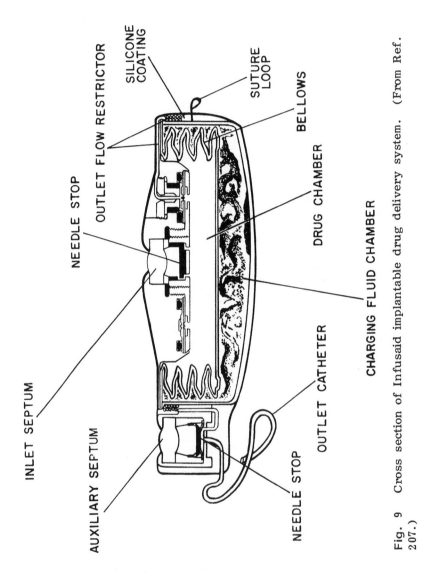

Fig. 9   Cross section of Infusaid implantable drug delivery system.   (From Ref. 207.)

Another widely used principle is roller peristaltic. This uses a disposable bag and silicone outlet tube. The tube is stretched around the roller mechanism, and as the roller turns, the contents of the tube are expelled from the cannula. This type of drive is compact but requires careful filling to avoid air inclusion [211]. Examples include Zyklomat, Siemens Promedos, Sandia, and the Medtronic system.

The infusaid implantable pump (Fig. 9) developed by Blackshear et al. [213] is a purely mechanical pump. It uses no external energy at constant rate of delivery. It consists of a hollow titanium disk divided into two chambers by a freely movable titanium bellows. The outer chamber contains a flurocarbon liquid which exerts a vapor pressure well above atmospheric pressure at 37°C. This forces the drug solution in the inner chamber through a capillary flow resistor into a suitably placed catheter. According to the Poiseuille equation, flow through the pumps is inversely related to the capillary length or drug viscosity, and the flow rate can be modified by changing one of those two parameters. Without the infusion regulator, these pumps are sensitive to ambient temperature and pressure, since they are implanted near the skin.

Typically, these devices consist of a pump, a reservoir with capacity ranging from 1 to 30 ml, rechargable nickel-cadmium battery or long-life Li-thionyl chloride battery, a program for both basal rate and bolus delivery, and an alarm system for low battery, pump runaway, reservior empty, motor fault, and computer fault [207-209,212].

Portable infusion devices are currently in fairly wide use for open-loop insulin delivery [171,206,210] and for pulsatile therapy of LHRH [157-163, 170,173,211,212,214]. The primary limitations of these devices are patient acceptance because they are somewhat bulky and uncomfortable to wear, although smaller devices are becoming available [214]. The implanted system, however, is both more convenient and more socially acceptable. Portable pumps are usually less costly. Refilling or replacing the reservoir and recharging the battery are convenient in a portable device, while reservoir and battery life assume more importance in an implanted device since access to the device is much more restricted. Implantation of a drug reservoir necessitates that a large quantity of drug be implanted. The potential leakage of this drug becomes an overwhelming safety concern that is not present in a portable device. Smaller size, biocompatible materials, and reliability are more important in an implanted pump. For practical reasons, infusion pump use is restricted to subcutaneous delivery of insulin, because routine use of the intravenous and intraperitoneal routes is too hazardous [215]. Infusion pumps have also been used for somatostatin [103, 216] and peptide hormones [208] such as growth hormone, vasopressin, calcitonin, and glucagon. A variable-rate implanted pump was used to deliver cyclosporine to nephrectomized dogs receiving an allografted kidney [217].

The two main advantages of infusion pumps are the potential for achieving physiological levels of the drug, and more freedom and flexibility for the patient when compared to conventional multiple injections. Pumps can deliver the peptide in several different wave forms. There is a large body of information [206,209,210,218] on continuous insulin delivery systems which demonstrated that at least on a short-term basis, various metabolic parameters can be returned to normal and glycemic control was significantly better by pump than by conventional therapy. However, it is argued that the pump itself may not be the only reason; other factors, such as intensive diabetes education, motivation, blood sugar monitoring, and diet intake, are also important. Patients with brittle diabetes (wide swings in glucose levels) have benefited most from insulin pumps [210]. Long-term (36 months) studies [219,220] indicated that not only better metabolic control, but also positive patient motivation, was achieved by long-term pump therapy [219]. Despite some concerns, a recent study showed that use of continuous subcutaneous insulin-infusion pumps is not associated with excess mortality [221].

The disadvantages of pump therapy may or may not be pump-specific. The potential for electrical or mechanical failure is high: for example, failure of the battery charger, slippage of the syringe, accidental withdrawal of the needle, and unnoticed leakage of the plastic catheter [206]. The pump delivery systems are not yet sufficiently convenient or easy to use and their cost is quite high ($900 to $3300 per pump in 1987) [210]. Dermatological complications such as skin infection at the site of needle insertion or local reaction and formation of fibrotic capsule around the implanted pump may occur [222]. Aggregation of insulin in the catheter [182,183] or capillary flow restrictor in the Infusaid device [223] can be a significant problem because many of the pumps rely on a linear relationship between volumetric flow rate and actual drug delivery rate. Thus reliability of delivery must refer to mass and not volumetric delivery. Some of these problems are detectable and correctable [224]. In the case of needle-site infections, needle site should be changed more frequently.

Osmotic pumps. The Alza osmotic minipump (Alzet) [225] have been used extensively in research to deliver short half-life agents such as peptides in various animals [8]. The minipump (Fig. 10) consists of a flexible, impermeable diaphragm surrounded by a sealed layer containing an osmotic agent (e.g., salt) at a particular concentration which, in turn, is contained within a cellulose ester semipermeable membrane. The pump is filled with sterile solution by the investigator with a separate filling tube and a flow moderator is inserted into the orifice. When the filled pump is placed in an aqueous environment (e.g., tissue), water diffuses into the osmotic agent chamber at a rate determined by the permeability of the surrounding membrane and the osmotic agent concentration (i.e., osmotic pressure). The absorbed water generates a hydrostatic pressure that acts on the flexible lining to force drug through the pump orifice. The reservoir volume ranges from 200 to 2 ml. The pump can be implanted subcutaneously. Pumps are available with a variety of delivery rates between 0.5 and 10 µl per hour and delivery durations between 3 days and 4 weeks. The in vitro and in vivo volume delivery rates of this pump are within 5% of the labeled rate [226] and rate control administration is possible. Osmotic pumps have been used with a large number of peptides and protein drugs in animals. Examples include ACTH, angiotensin, bombesin, calcitonin, endorphins, enkephalins, erythropoietin, growth hormone, insulin, LHRH, neurotensin, parathyroid hormone, and vasopressin [226]. Stability of the drug at 37°C within the pump for the entire infusion period and compatibility with the internal pump components are critical features of this pump. To the authors' knowledge, an osmotic pump has not been used clinically in humans.

Controlled-release micropump. This is a novel implantable pump (Fig. 11) that has been used in dogs for intraperitoneal delivery of insulin [227, 228]. Without an external power source, the concentration difference between the drug reservoir and the delivery site causes diffusion of the drug

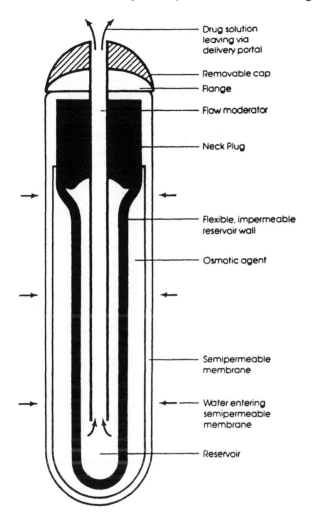

Fig. 10 Cross section of functioning Alza Alzet osmotic minipump. (From Ref. 207.)

to the delivery site to provide basal delivery. Current applied through the solenoid coil causes the piston to compress the foam disk repeatedly and augmented delivery is achieved. Interruption of the current causes the membrane to relax, drawing more drug into the foam disk for the next compression cycle. The basal rate is determined by the magnitude of the concentration difference and by the permeability of the rate-controlling membrane and other diffusion resistances between the reservoir and the outlet. The augmented rate is a function of the elastic properties of the foam, the force applied by the solenoid piston, and the frequency of compression.

Recently, a piezoelectric controlled micropump was developed [229]. This was composed of a piezoelectric disk bender and a cellulose acetate microporous membrane located at the opposite sides of the insulin reservoir. Augmented delivery was obtained by applying a square-wave voltage (100 to 1000 Hz, 20 to 90 V dc) to the piezoelectric bender. Its advantage over the solenoid-driven pump was that greater augmentation factors were obtained with significantly lower power inputs.

Absence of an outlet catheter, valves, and motors makes controlled-release micropumps less susceptible to limiting problems such as catheter blockage and mechanical failure. Lack of a valve system and presence of the membrane prevents accidental overdosage in case of failure. The long-term stability of the membrane to repeated compression is important, although in vitro results indicate that a life of 1 year can be expected under normal use [230]. The need to minimize diffusion resistances limits the location of the implant and thus introduces greater emphasis on the biocompatibility of the pump exterior.

Other electrochemical principles have been proposed for implantable pumps. One group used electroosmosis through a cation-exchange membrane [231]. The membrane was placed between two Ag/AgCl electrodes and aqueous NaCl solution. In response to electric potential hydrated Na^+ ions passed through the membrane accompanied by the movement of volume of fluid. A constant-current power supply which reversed the direction of the current every 10 min caused a to-and-fro transport of fluid through the membrane. The potential advantages of this pump are low power requirement, high reliability, and very low flow rates, in the microliters per day range [231]. It would be particularly suited for highly concentrated drug solution that remains stable for several years. However, the problem of the consumption of the electrodes has not been solved satisfactorily, and the necessary valve and reservoir arrangements have not yet been devised. An electrolytic pump, utilizing the pressure of gases evolved at the electrodes of an electrolytic cell, has been proposed for insulin delivery [232]. However, it is not clear how the evolved gases will be collected, vented, or recombined to control the driving pressure and to slow the pump.

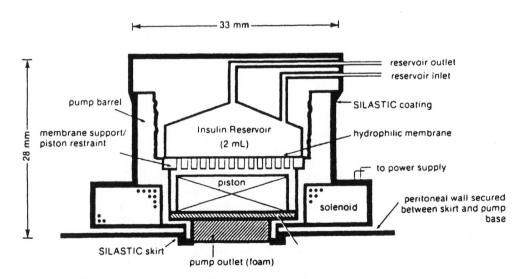

Fig. 11   Schematic illustration of prototype VIII of controlled-release micropump. (From Ref. 207.)

Mechanical pumps are technically simple and rugged and can deliver the peptide in several different waveforms. Nonetheless, the potential for mechanical failure, high power requirement, and relatively large size are some major limitations of these pumps. In contrast, the controlled-release micropumps and the electrochemical pumps are technically more sophisticated and they can deliver the drug reliably at a low rate for a long time with very little power consumption. Some of the technical details and biocompatibility issue need to be worked out before their widespread use. Osmotic pumps are simple in principle, small, and reliable, but they have less flexibility in the delivery program and the reservoir volume is low compared to mechanical pumps, thus necessitating frequent replacement.

In comparison to many other drug delivery systems, infusion pumps for variable rate delivery of peptides are at a crude stage of development. Some researchers consider them as a research tool. Future research should be aimed at developing simple and reliable actuations, without the need for refillable reservoires and little if any external energy. In insulin therapy, the development of a portable glucose sensor will represent a major breakthrough.

Liposomes

Liposomes are spherical vesicles formed when phospholipids are allowed to swell in an aqueous media. They consist of one or a number of concentric bilayers surrounding aqueous phases. Hydrophilic or lipophilic drugs can be encapsulated in their aqueous or lipid phase, respectively [233]. Recent reviews have dealt extensively with the structural and biological aspects of liposomes [234,235] and their usefulness as drug carrier systems [236-238].

Liposomes have three distinct applications in parenteral peptide and protein delivery. First, subcutaneously, intraperitoneally, or intramuscularly injected liposomes can act as a "depot" for the peptide drug at the injection site, the drug being released slowly as the liposomes are degraded by local hydrolytic enzymes and by neutrophills. Encapusulation of a hydrophilic peptide (*P-18, MW < 5000) in liposomes resulted in significantly sustained release after intramuscular injection in rats (Fig. 12). Intramuscular injection of liposomal interferon in mice resulted in significantly increased localized retention of interferon [239,240]. Intramuscular or subcutaneous administration of insulin and growth hormone in a liposome collagen matrix gave similar prolongation [241]. A maximum of 3 to 5 day release for insulin and 14-day release for growth hormone was observed. Enhanced sequestration of liposomes with the callagen can be achieved by modifying the liposome surface with fibronectin. The release of the drug from the liposome can be modulated by magnetic field [242] or microwave [243]. Liposomes can also be designed to release an entrapped drug preferentially at higher temperatures attainable by mild local hyperthermia [244-246].

Second, liposome entrapment of peptides can protect them from enzymatic degradation after intravenous administration. Weingarten et al. [247] reported that insulin entrapped in multilamellar liposomes was more stable against proteolytic enzymes in vitro. Liposome-encapsulated cyclosporin [248] reduced its nephrotoxicity, presumably due to metabolic breakdown products, without altering the immunosuppressive effect in rats. Liposomes have been used as intravenous carriers for enzymes such as amylglycosidase, β-fructofuranosidase, and neuraminidase [249-252]. The elimination

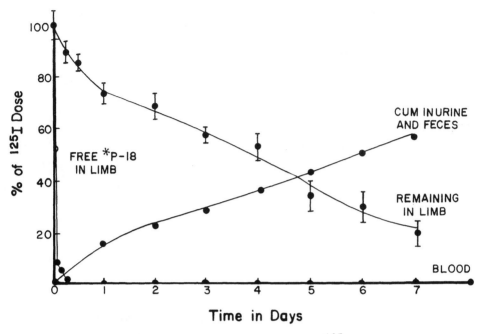

Fig. 12   Pharmacokinetics of liposome containing [^{125}I] *P-18 after subcutaneous injection into hind limbs of rats.   (From Ref. 238.)

and distribution of liposomes are controlled by the size, charge, and composition of the liposome [253,254].

Third, liposomes can be used for targeting the peptide to a particular
site of action [255].   Two limiting factors, localization of liposomes in macrophages and the second, more important problem of gaining access to target
cells from blood vessel walls, must be overcome to achieve this objective.
The localization of liposomes in macrophages can offer a convenient method
for targeting drugs to these cells.   Immunopotentiating agents such as
muramyl dipeptide (MDP) can be encapsulated into liposomes for specific delivery to macrophages.   The macrophages then become activated and in animal models they have been shown to control metastases of experimental
lung tumors in mice and rats [256,257] and to protect mice against fatal
herpes virus infection [258].   The inclusion of muramyl dipeptide and its
derivative into liposomes resulted in a 700- to 7000-fold increase in efficacy
compared to free MDP.   Freeze-dried liposomes were 50,000 to 100,000 times
more effective [256].   Another example is enhancement of calcitonin effects
in rats after liposome entrapment [259].   The size, composition, and route
of administration each seem to influence this action.   Large multilamellar
vesicles were more potent than small unilamellar vesicles after intravenous
administration.   Cholesterol inclusion in the large vesicles prolonged the hypocalcemia.   However, with intramuscular administration, the cholesterol-free
liposomes were more potent than their cholesterol-containing counterparts
regardless of size.   The mechanism of this effect has not been delineated.

Antibodies can be coupled covalently to liposomes [260] to enhance
their cell specificity [261−263].   The present approach is to target the

antibody-directed liposomes to specific cell types or subset of cells within the vascular system [255]. The potency of the liposomes with a given antibody is proportional to their relative binding and endocytosis by the cells, and to the reactivity of the particular antibody with the cell [262]. This is a very active area of research at present. Recently also, magnetic liposomes [264] have been studied as a targeting device for drugs. Liposomes containing ultrafine magnetite were injected in rats having implanted tumor. In the presence of an external magnetic field of 4000 G, a very small amount of the liposomes (0.3%), but significantly more than control, was trapped at the tumor tissue. It was speculated that the holding ability of the magnet may be related to the flow rate and viscosity of the blood and a stronger magnet would be necessary.

The advantages of liposomes are their flexibility in structure, size, and shape, ability to encapsulate both hydrophilic and lipophilic compounds, and relative nontoxicity. The presence of water inside liposomes makes them an excellent delivery system for the proteins, whose tertiary and quaternary structures are sensitive to irreversible damage caused by dehydration as often occurs with other alternative carrier systems [238]. However, their widespread use in peptide and protein delivery can be limited by their capacity as immunoadjuvants [265] and by the tendency of their constituent phospholipids to interact with the peptide being encapsulated [266]. The latter can affect the release kinetics of the peptide from the liposomes as well as the shelf life of the liposomal formulation [267]. Proteins prone to aggregation may affect liposomal stability by initial fusion [268] due to hydrophobic interaction, followed by aggregation [269]. A novel method for encapsulation of macromolecules such as hemoglobin and alkaline phosphatase has recently been reported [270]. The main advantage of this method is that it does not require organic solvents, detergent dialysis, or freezing, which are central features of other liposome preparation methods. Absence of detergents or organic solvents makes it the method of choice for encapsulation of macromolecules that are sensitive to these compounds. It is also easy to scale up. A limitation of the method is that the protein may bind to the exterior surface of the liposome. Liposomes are also useful in coencapsulating hydrophilic and hydrophobic peptides to exploit their interdependence and synergism [257,271]. Despite its potentials, the commercial production of liposomes for clinical use is beset with numerous technological problems [272].

Emulsions

The use of emulsions for parenteral delivery of peptides is rather limited to date. As a parenteral drug carrier system [273], emulsions can be used to protect hydrophilic or hydrophobic drugs from direct contact with body fluids, while being delivered slowly over a prolonged period of time. They can be also used to deliver drugs to the reticuloendothelial system. Release of the drug can be further prolonged by using multiple emulsion systems [274]. Emulsion delivery leads to prolonged and higher antibody levels after administration of diptheria toxoid [275] and influenza vaccine [276]. Continuous, slow release of the antigen from the emulsion was suggested as the probable mechanism [277,278]. Intradermal injection of nonliving mycobacteria cell wall in emulsion to guinea pigs produces less intense inflammatory reaction to similarly administered free cell walls [279]. Simple mixtures of cell wall and mineral oil droplets were ineffective in tumor suppression. Other oils, (e.g., vitamin A) were also ineffective

[279]. Optimal oil concentration is also important [280]. Similar behavior
has been shown when emulsions have been used as carriers for muramyl
dipeptide (MDP) and their synthetic analogs. Subcutaneous administration
of MDP in water-in-oil emulsion to the hind paw of mice prolonged the effect of the peptide significantly [281]. More than 96% of the administered
dose remained in the injected paw at 30 min with 45% present after 24 hr.
The figures for free MDP in saline were 8% and <1%, respectively. Mineral oil emulsion as a carrier for the coadministration of muramyl dipeptide
and trehalose dimycolate in the emulsion was essential for tummor regression [282]. Emulsion adjuvant may also orient the molecule in a biologically
effective form [283].

Cyclosporine in Intralipid emulsion when administered intramuscularly
in baboons achieved therapeutic levels [79]. Marketed cyclosporine emulsion for infusion uses Cremophor EL, a polyethoxylated castor oil, as the
surfactant [284]. Although it is relatively nontoxic and the surfactant of
first choice worldwide [285], anaphylactic reactions associated with parenteral cyclosporine is most likely due to Cremophor [286]. Since intravenous
fat emulsions such as Intralipid cause a rise in serum triglyceride level
and cyclosporine is known to bind with lipoproteins in plasma, the effects
of the concomitant use of cyclosporine and Intralipid was studied in dogs
recently [287]. It was found that cyclosporine pharmacokinetics was unchanged and use of fat emulsion does not require a change in dosage regimen.

Magnetic emulsions for targeting anticancer agents to tumor tissue
have also been reported [288]. After intravenous injection in rats about
twice the amount was localized in lung compared to control in the presence
of a magnetic field. However, there was no good correlation between emulsion levels and magnetic field intensity. The advantages of emulsion systems are: they are clinically acceptable and technologies exist for their
large-scale production. Future investigations may succeed in realizing the
potential of emulsions as a parenteral delivery system for peptides.

Microspheres

Microspheres are small soild, spherical particles that range in size from
few tenths of a micrometer up to several hundred micrometers in diameter.
They can be prepared by various polymerization processes and encapsulation
[289]. The drug molecules are dispersed in either solution or crystalline
form. Microspheres have potential application in controlled release and
targeting of drugs [290]. Subcutaneous delivery of an ACTH(4-9) analog
from saline solution, osmotic minipump, and biodegradable microsphere consisting of 1:1 copolymer of lactic acid and glycolic acid demonstrated that
microspheres and minipumps were equally effective [291]. Microspheres
have the advantage over minipumps that they are biodegradable and thus
do not have to be removed. These microspheres released the peptide for
14 days. In patients with prostate cancer, intramuscular injection of
[D-trp6)LHRH in aqueous suspension of poly d,l-lactide-coglycolide microspheres achieved therapeutic plasma concentration for 4 weeks [121].
Crystallized carbohydrate spheres have been used for slow release of insulin, interferon, and growth hormone and for immunization with entrapped
antigens and immunomodulators [292]. The advantage of this concept was
that the matrix degradation was not enzymatic but dependent on the pH
and ionic strength, which are stable in a human body. Microspheres can
be used as enzyme carriers in vivo (293). Depending on the route of

administration, they may or may not act as immunological adjuvants to enhance the immune response to the bound or trapped enzyme. In case of L-asparaginase–polyacrylamide microbeads, intramuscular/subcutanous route is the most effective in mice in terms of both reducing plasma L-asparagine levels and minimizing the immune response to the enzyme [294]. Intraperitoneal and intravenous injections of particles produced significantly higher antibody levels and this was not satisfactorily explained.

Targeting of microspheres can be to a particular organ, a specific part of the organ, or selective intracellular delivery. Targeting can be passive, by occlusion (size dependent), cellular uptake, or local injection. It can be active by conjugating with antibodies, the so-called immunomicrospheres [295], by incorporating magnetic particles [296], or by a combination of both, the so-called magnetic immunomicrospheres [297]. Magnetic microspheres were developed to minimize reticuloendothelial clearance and to increase target site specificity. Significant target site specificity has been achieved by this carrier [296], localization being controlled by the intensity of the external magnetic field.

The advantages of microspheres are that they can be prepared inexpensively if the correct encapsulation method is chosen. They can be injected subcutaneously, intramuscularly, or intraperitoeally and thus implantation is not necessary. However, release of high-molecular-weight compounds may be difficult and drug loading can be restricted (20 to 30%). A prerequisite for success of this microparticulate delivery systems is their ability to pass successfully through biological barriers such as blood, endothelium, the reticuloendothelial system, and cellular barriers. Interaction and complexation of blood components with drug carrier systems may obviate their effectiveness [289].

Cellular Carriers

When erythrocytes are lysed and resealed, exchange of intracellular and extracellular solutes will occur and the drug present in the medium will be encapsulated. Enzymes and/or drugs can be encapsulated in erythrocytes for targeting or slow release of the drug [298]. The methods of encapsulation include hemolysis, dialysis, and electric field breakdown [299]. After administration, the carrier erythrocytes usually exhibit a bimodal survival curve with a rapid loss of the cells in the first 24 hr postinjection and a much slower loss of cells afterward. The early loss represents cells damaged by in vitro handling procedure. Damaged erythrocytes are removed from circulation by phagocytosis, and thus these carriers can be exploited for targeting to liver or spleen. The release of the drug following phagocytosis occurs by simple diffusion or by a specific transport system [299]. Charged molecules are retained longer than uncharged molecules. Examples of enzymes encapsulated in erythrocytes include glycosidase enzymes for lysosomal storage disease, L-asparaginase for leukemia, aminolevulinic acid dehydratase in prophyrias, and arginase in hyperarginemia [298].

Erythrocytes have the advantages that they are biodegradable, nonimmunogenic, can circulate up to 4 months, are easily obtainable, and large (20 to 90% of extracellular concentration) quantities of material can be entrapped in a small volume of cells and afford immunological and enzymatic protection. Although they are interesting tools for experimental manipulation in enzyme replacement therapy, clinical utility of carrier erythrocytes

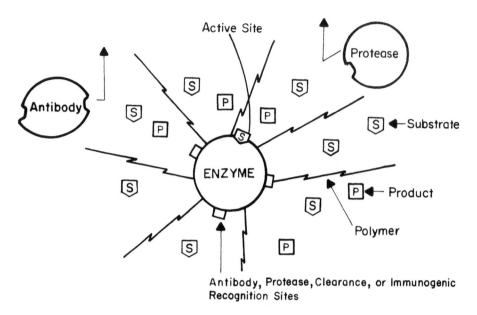

Fig. 13 Proposed model to explain the enzymatic and biological properties of poly-DL-alanylasparaginases. (From Ref. 306.)

has not been proven. Their long-term storage is a problem and they are permeable to various drugs [298,299].

The use of mammalian cells as a delivery vehicle for ricin toxin has recently been reported [300]. Immobilized enzymes covalently bound to some polymers [301,302] are shown to have increased stability in circulation, decreased ability to provoke undesirable complications, and the possibility of simplified administration. Another approach is to encapsulate the cell within semipermeable aqueous microcapsule [303] to impart more control over the release and prevent immune rejection of the cell. Microencapsulated pancreatic islet of Langerhans cells indicate that they can provide a long-term insulin delivery system comparable to pump treatment [304].

Polymeric Conjugates

The peptide or protein drug, conjugated with a polymer or macromolecule, can decrease immunogenicity, improve protease stability, and achieve selected or targeted drug delivery [305]. Immunological properties of bovine serum albumin can be reduced by conjugation with polyethylene glycol [21]. The chemical modification of asparaginase by a DL-alanine-N-carboxyanhydride polymerization technique produced modified enzyme which had greater protease stability, retained most of its catalytic activity, and demonstrated a 7- to 10-fold prolongation in plasma clearance properties in rats and mice [306]. The polymers probably act as a steric barrier (Fig. 13), which retards or prevents the interaction of macromolecules (proteases, antibodies, immunogenic and/or clearance recognition receptors) with the native enzyme, while micromolecular substances (substrates, products,

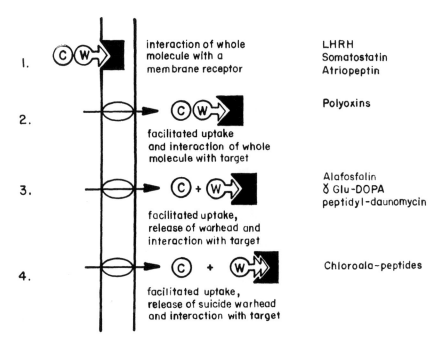

Fig. 14   Scheme showing mechanisms for targeting peptide drugs.   (From Ref. 307.)

hydrogen ions, coenzymes, etc.) can still interact with the enzyme with little difficulty. Polymer modification may (1) mask antigenic determinants, (2) mask immunogenic recognition sites, (3) mask protease-susceptible sites, (4) mask clearance recognition signals, (5) allow free access to low-molecular-weight substrates, (6) maintain systemic injectability, and (7) alter pH optimum by changing microenvironment. The nature as well as the extent of polymerization are important factors [306].

Protein molecules often act as carriers for targeting of other peptide or protein drugs [307,308]. Targeting of peptide drugs can involve four mechanisms (Fig. 14): (1) interaction with surface receptor, (2) facilitated uptake and interaction of intact molecules with intracellular target site, (3) facilitated uptake and intracellular release of a warhead moiety, and (4) facilitated uptake and release of a suicide warhead, which is activated only after interacting with its target [307]. There are naturally occurring multisubunit proteins of toxins which also act as conjugates. The subunits have different roles; one subunit mediates binding and the other is the pharmacologically active entity. The peptide or protein drug, by its conjugation to the binding subunit of the carrier protein or toxin, can be targeted to the receptor for the carrier subunit. One widely used carrier is the B chain of ricin toxin [309]. Its use with interferon [309] and insulin [310] has been reported. The interferon-ricin B conjugate appears to possess interferon activity mediated through the receptor for the ricin B chain. The specific binding and the antiviral activity of the conjugate were inhibited in the presence of galactose, an inhibitor of ricin B chain binding [309]. Conversely, the ricin A chain can be conjugated to binding subunit of

other proteins or monoclonal antibody to tumor antigen for cytotoxic effect
in tumor cells [311–316]. Synthetic polypeptides have also been used as
carriers for ligands such as muramyl dipeptide [317]. The conjugation in
these cases is often multiple and results in increased potency, increased
receptor affinities, enhanced resistance to enzymatic degradation, and prolonged durations of action. The subject of macromolecular-drug conjugate
has been reviewed in detail recently [318] and in Chap. 21. Targeting of
peptide drugs is an actively growing research field with potential for further future development. The major obstacle to the carrier approach is
their susceptibility to removal by the reticuloendothelial system and limited
ability to cross capillary endothelium [318].

Polymeric Controlled-Release Systems

Controlled delivery systems such as microcapsules, implants, gels, or vis-
cosity-controlled systems for peptides have been dealt with extensively in
Chap. 19. Some more sophisticated systems are described here.

On-demand systems. There are a number of situations where externally
augmented delivery on demand could be beneficial. An example would be
delivery of insulin for patients with diabetes mellitus. Such a system with
magnetic modulation has been designed [319–321]. It consists of magnetic
beads or cylinders embedded in a ethylene-vinyl acetate matrix containing
a model protein, bovine serum albumin. This matrix, containing the protein
and magnets, can release up to 30 times more drug when exposed to oscillating external magnetic field, and release rates return to normal when the
magnetic field is discontinued. This was confirmed in vivo in rats with
insulin [321]. Factors influencing the release rate include (1) the position,
orientation, and strength of the embedded magnets; (2) the amplitude and
frequency of the applied magnetic field, and (3) the mechanical properties
of polymer matrix. The mechanism of magnetic modulation is not yet known.
One possibility is that the beads alternately compress and expand the matrix pores in the presence of a magnetic field, thereby facilitating drug release. Recently, external regulation of release by ultrasound (20 kHz for
20 min) has been studied [322] in rats after subcutaneous implantation of
a model drug, *para*-aminohippuric acid (PAH), in a polyanhydride matrix.
The effect of ultrasound on the PAH concentration in the urine is pronounced during the exposure and mainly just after the exposure. One
possible explanation of this phenomenon is that the ultrasound enhances the
penetration of water into the bioerodible polyanhydride matrix, exposing
more linkages for hydrolysis and therefore higher degradation and release
rate.

Self-regulated systems. The best recognized need for self-regulated
systems is the need for a device that would deliver insulin in response to
blood glucose concentration for diabetic patients, and two polymer-based
systems have been conceptualized for this purpose. One is based on a
pH-sensitive bioerodible polymer [323,324]. A glucose-sensitive membrane
is fabricated from a cationic hydrogel polymer with immobilized glucose
oxidase. As glucose diffuses into the polymer, glucose oxidase catalyzes
its conversion to gluconic acid, thereby lowering the microenvironment pH
within the membrane and protonating the amine groups in the membrane.
As a result, the membrane swells, thereby increasing its permeability to
the insulin held in a contiguous reservoir. Thus far, this concept appears
to hold promise only for small peptide molecules. A modification of this

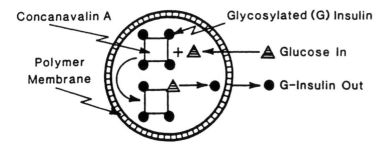

Fig. 15   Schematic representation of a self-regulating insulin delivery system.   (From Ref. 329.)

approach is based on the increase in solubility of a modified insulin derivative with a decrease in pH in the membrane and corresponding increase in release [325].   Another method used a biochemical approach based on competitive binding between glucose and glycosylated insulin (G-insulin) to concanavalin A (lectin) [326–329].   As the glucose level increases, the influx of the glucose to the pouch increases, displacing G-insulin from the concanavalin A substrate (Fig. 15).   Increasingly, displaced G-insulin in the pouch results in efflux of G-insulin to the body.

   Temperature sensitive systems.   Some polymers (e.g., polyacrylamide derivatives exhibt thermosensitive swelling behaviors [330,331], which results in temperature-dependent release.   This can be exploited in pulsatile or on-off delivery of peptides.   The temperature sensitivity of insulin permeation through a poly-N-isopropylacrylamide-butylmethacrylate copolymer membrane varied with the hydrophobic component of the copolymer [332].   Temperature was changed in a stepwise manner from 37 to 27°C.   The thermosensitive permeation demonstrated reversibility without noticeable lag times.   Similar temperature-dependent release due to swelling was observed for bovine serum albumin in ethylene–vinyl acetate matrix [333] and myoglobin in N-isopropylacrylamide gel [331].   This type of gel can also have application in immobilizing enzymes for feedback control [334].

IV.   CONCLUSIONS

Although peptide and protein drugs have been used for decades in the diagnosis and therapy of a number of diseases, it was not until recently that these compounds have become available in large quantities in pure form.   Also in the last 10 years clinical interest in peptides has escalated. Lack of adequate delivery systems for these drugs has challenged the pharmaceutical scientist to reevaluate the existing drug delivery technologies and to explore new approaches.   Considering the impermeability and metabolic barrier of the major mucosal routes, parenteral routes will continue to be the route of choice for most large peptides and proteins.   The major drawbacks of parenteral peptide drug delivery are the inconvenience and immunogenicity of some drugs.   Development of painless, reproducible injection devices will undoubtedly improve patient compliance in the future. Notwithstanding its benefits, the success of pump therapy will depend on miniaturization of the device, reliable and simple design, and versatility. Magnetic microspheres and conjugates with antibody or protein appear very promising for targeted delivery of peptides.

The three major formulation challenges posed by peptide and protein drugs are pulsatile or multirate delivery, biofeedback or self-regulated delivery, and targeted or site-specific delivery. All these approaches are anticipated to expand in the near future. The acceptance and success of parenteral delivery systems for peptide and protein drugs are contingent on the extent to which they can meet and adapt to these sophisticated delivery approaches.

REFERENCES

1. W. Sadee, *Pharm. Res.*, *3*:3 (1986).
2. V. H. L. Lee, *Pharm. Int.*, *7*:208 (1986).
3. E. L. Smith, R. L. Hill, I. R. Lehman, R. J. Lefkowitz, P. Handler, and A. White, *Principles of Biochemistry: General Aspects*, McGraw-Hill Book Company, New York (1983).
4. J. M. Samanen, in *Polymeric Materials in Medication* (C. G. Gebelein and C. E. Carraher, eds.), Plenum Press, New York (1985), p. 227.
5. O. Siddiqui and Y. W. Chien, *CRC Crit. Rev. Ther. Drug Carrier Syst.*, *3*:195 (1987).
6. V. H. L. Lee, *CRC Crit. Rev. Ther. Drug Carrier Syst.*, *5*:69 (1988).
7. W. A. Lee and J. P. Longnecker, *Biopharm. Manuf.*, *1*:30 (1988).
8. J. Urquhart, J. W. Fara, and K. L. Willis, *Annu. Rev. Pharmacol. Toxicol.*, *24*:199 (1984).
9. F. Rovelli, C. de Vita, G. A. Feruglio, The study was done by a group of 70 people so it was not possible to give all names. *Lancet*, *1*:397 (1986).
10. L. Illum and S. S. Davis, *J. Parenter. Sci. Technol.*, *36*:242 (1982).
11. T. Audhya and G. Goldstein, *Int. J. Pept. Protein Res.*, *22*:187 (1983).
12. T. Audhya and G. Goldstein, *Surv. Immunol. Res.*, *4* (*Suppl. 1*):17 (1985).
13. R. H. Buckley, *J. Clin. Immunol.*, *2*:155 (1982).
14. L. Martis, *28th Ann. Nat. Ind. Pharm. Res. Conf.*, Wisconsin (1986).
15. S. Barandun, P. Kistler, F. Jeanetand, and H. Isliker, *Vox Sang.*, *7*:157 (1962).
16. E. Merier, F. S. Rosen, S. Salmon, and J. D. Crain, *Vox Sang.*, *13*: 102 (1967).
17. R. A. Good, *J. Clin. Immunol.*, *2* (*Suppl.*):5S (1982).
18. B. Pirofsky, S. M. Campbell, and A. Montanaro, *J. Clin Immunol.* *2* (*Suppl.*):7S (1982).
19. J. W. Goodman, in *Basic and Clinical Pharmacology* (D. P. Stites, J. D. Stobo, and J. V. Wells, eds.), Lange Medical Publications, Los Altos, Calif. (1987), p. 20.
20. T. E. van Metre, *In Vitro*, Monograph 6, (1985), p. 172.
21. A. Abuchowski, T. van Es, N. C. Palczuk, and F. F. Davis, *J. Biol. Chem.*, *252*:3578 (1977).
22. W. A. Ritschel, *Pharm. Ind.*, *35*:273 (1973).
23. S. H. Ferreira and J. R. Vane, *Br. J. Pharmacol. Chemother.*, *30*: 417 (1967).

24. R. L. Hodge, K. K. F. Ng, and J. R. Vane, *Nature*, 215:138 (1967).
25. D. G. Garlick and E. M. Renkin, *Am. J. Physiol.*, 219:1595 (1970).
26. G. Hanck, *Angiologica*, 8:236 (1971).
27. S. Witte, D. M. Goldenberg, and K. T. Schricker, *Z. Gesamte Exp. Med.*, 148:72 (1968).
28. L. J. Nuget and R. K. Jain, *Am. J. Physiol.*, 246:H129 (1984).
29. D. Regoli, B. Rinniker, and H. Brunner, *Biochem. Pharmacol.*, 12:637 (1963).
30. H. Tuppy and H. Nesvadha, *Monatsh. Chem.*, 88:977 (1957).
31. G. Williams, J. C. Pickup, and H. Keen, *Am. J. Med.*, 82:1247 (1987).
32. D. Roemer, H. H. Buescher, R. C. Hill, J. Pless, W. Bauer, F. Cardinaux, A. Closse, D. Hauser, and R. Huguenin, *Nature*, 268:547 (1977).
33. L. F. Tseng, H. H. Loh, and C. H. Li, *Nature*, 263:239 (1976).
34. H. H. Buescher, R. C. Hill, D. Roemer, F. Cardinaux, A. Closse, D. Hauser, and J. Pless, *Nature*, 261:423 (1976).
35. M. J. Humphrey and P. S. Ringrose, *Drug Metab. Rev.*, 17:283 (1986).
36. H. P. J. Bennett and C. McMartin, *Pharmacol. Rev.*, 30:247 (1979).
37. B. Pimstone, S. Epstein, S. M. Hamilton, D. LeRoith, and S. Hendricks, *J. Clin. Endocrinol. Metab.*, 44:356 (1977).
38. J. R. M. Franckson and H. A. Ooms, *Postgrad. Med. J.*, 49 (Suppl):931 (1973).
39. P. H. Sonksen, C. V. Tompkins, M. C. Srivastava, and J. D. N. Nabarro, *Clin. Sci. Mol. Med.*, 45:633 (1973).
40. J. B. Little, L. M. Kievay, E. P. Radford, and R. B. McGandy, *Am. J. Physiol.*, 211:786 (1966).
41. Pfizer, *Science for the World's Well Being*, 4th ed., Pfizer, Inc., New York (1967) p. 50.
42. A. C. Guyton, *Textbook of Medical Physiology*, 6th ed., W. B. Saunders Company, Philadelphia (1981), p. 232.
43. F. L. S. Tse and P. G. Welling, *J. Parenter. Drug Assoc.*, 34:484 (1980).
44. J. Schou, *Pharmacol. Rev.*, 13:441 (1961).
45. W. S. Evans, M. L. Vance, D. L. Kaiser, R. P. Sellers, J. L. C. Borges, T. R. Downs, L. A. Frohman, J. River, W. Vale, and M. O. Thorner, *J. Clin. Endocrinol. Metab.*, 61:846 (1985).
46. B. W. Zweifach, *Sci. Am.*, 200:54 (1959).
47. B. E. Ballard, *J. Pharm. Sci.*, 57:357 (1968).
48. J. M. Yoffey and F. C. Courtice, *Lymphatics, Lymph and the Lymphomyeloid Complex*, Academic Press, New York (1970).
49. J. R. Casley-Smith, *Ann. N.Y. Acad. Sci.*, 116:803 (1964).
50. E. M. Landis, *Ann. N.Y. Acad. Sci.*, 116:765 (1964).
51. E. W. Kraegen and D. J. Chisholm, *Clin. Pharmacokinet.*, 10:303 (1985).
52. L. Russo and W. V. Moore, *J. Clin. Endocrinol. Metab.*, 55:1003 (1982).
53. R. B. Sund and J. Schou, *Acta Pharmacol. Toxicol.*, 21:313 (1964).
54. J. Bederka, A. E. Takemori, and J. W. Miller, *Eur. J. Pharmacol.*, 15:132 (1971).
55. J. M. Barnes and J. Trueta, *Lancet*, 1:623 (1941).
56. J. H. Lewis, *J. Am. Med. Assoc.*, 76:1342 (1921).
57. M. Berger, P. A. Halban, J. P. Assal, R. E. Offord, M. Vranie, and A. E. Renold, *Diabetes*, 28:53 (1979).

58. V. A. Koivisto and P. Felig, *N. Engl. J. Med.*, *298*:77 (1978).

59. C. L. J. Coles, K. R. Heath, M. L. Hilton, K. A. Lees, P. W. Muggleton, and C. A. Walton, *J. Pharm. Pharmacol.*, *17* (*Suppl.*):87S (1965).

60. N. Heinrich, D. Lorenz, H. Berger, K. Fechner, H. E. Schmidt, H. Schaefer, and M. Burkhard, *Pharm. Res.*, *2*:198 (1985).

61. S. Zelman, *Am. J. Med. Sci.*, *241*:563 (1961).

62. B. E. Ballard and E. Menczel, *J. Pharm. Sci.*, *56*:1476 (1967).

63. H. Sekijima and L. D. Fink, *Proc. Soc. Exp. Biol. Med.*, *80*:158 (1952).

64. K. K. Keawn, S. M. Fisher, D. F. Downing, and P. Hitchcock, *Anaesthesiology*, *18*:270 (1957).

65. J. J. Nora, D. W. Smith, and J. R. Cameron, *J. Pediatr.*, *64*:547 (1964).

66. G. N. Smith, D. Mollison, B. Griffiths, and P. L. Mollison, *Lancet*, *1*:1208 (1972).

67. E. F. Evans, J. D. Proctor, M. J. Fratkin, J. Velandia, and A. J. Wasserman, *Clin. Pharmacol. Ther.*, *17*:44 (1975).

68. W. P. Cockshott, G. T. Thompson, L. J. Howlett, and E. T. Seeley, *N. Engl. J. Med.*, *307*:356 (1982).

69. H. F. Root, J. W. Irvine, Jr., R. D. Evans, L. Reiner, and T. M. Carpenter, *J. Am. Med. Assoc.*, *124*:84 (1944).

70. G. F. Maberly, G. A. Wait, J. A. Kilpatrick, E. G. Loten, K. R. Gain, R. D. H. Stewart, and C. J. Eastman, *Diabetologia*, *23*:333 (1982).

71. E. P. Paulsen, J. W. Courtney, and W. C. Duckworth, *Diabetes*, *28*: 640 (1979).

72. R. J. Wills, S. Dennis, H. E. Spiegel, D. M. Gibson, and P. I. Nadler, *Clin. Pharmacol. Ther.*, *35*:722 (1984).

72. M. Berger, H. J. Cuppers, P. A. Halban, and R. E. Oxford, *Diabetes*, *29*:81 (1980).

74. E. W. Kraegen, D. J. Chisholm, and M. J. Hewett, *Diabetes Care*, *6*:118 (1983).

75. R. Hori, F. Kamada, and K. Okumura, *J. Pharm. Sci.*, *72*:435 (1983).

76. T. W. Redding and A. V. Schally, *Life Sci.*, *12*:23 (1970).

77. J. A. Parsons, B. Rafferty, R. W. Stevenson, and J. M. Zanelli, *Br. J. Pharmacol.*, *59*:489P (1977).

78. J. A. Parsons, B. Rafferty, R. W. Stevenson, and J. M. Zanelli, *Br. J. Pharmacol.*, *66*:25 (1979).

79. D. Novitzky, D. K. C. Cooper, and W. N. Wicomb, *S. Afr. Med. J.*, *68*:737 (1985).

80. F. L. S. Tse and P. G. Welling, *J. Parenter. Drug Assoc.*, *34*:409 (1980).

81. S. Kety, *Am. Heart J.*, *39*:321 (1949).

82. B. E. Sobel, L. E. Fields, A. K. Robinson, K. A. A. Fox, and S. J. Sarnoff, *Proc. Natl. Acad. Sci. USA*, *82*:4258 (1985).

83. C. Binder, T. Lauritzen, O. Faber, and S. Pramming, *Diabetes Care*, *7*:188 (1984).

84. C. Binder, *Acta Pharmacol. Toxicol.*, *27* (*Suppl. 2*):1 (1969).

85. A. B. Kurtz and J. D. N. Nabarro, *Diabetologia*, *19*:329 (1980).

86. S. A. Berson, R. S. Yalow, M. A. Bauman, A. Rothschild, and K. Newerly, *J. Clin. Invest.*, *35*:170 (1956).

87. R. E. Bollinger, J. H. Morris, F. G. McKnight, and D. A. Diedrich, *N. Engl. J. Med.*, *270*:767 (1964).

88. S. A. Berson and R. S. Yalow, *J. Clin. Invest.*, *38*:1996 (1959).

89. K. Dixon, *Clin. Chem.*, *20*:1275 (1974).

90. E. W. Kraegen and D. J. Chisholm, *Diabetologia*, *26*:208 (1984).

91. R. S. Sherwin, K. J. Kramer, D. J. Tobin, P. A. Insel, J. E. Liljenquist, M. Berman, and R. Andres, *J. Clin. Invest.*, *53*:1481 (1974).

92. K. G. Tranberg and H. Dencker, *Am. J. Physiol.*, *235*:E577 (1978).

93. R. N. Bergman, Y. Z. Ider, C. R. Bowden, and C. Cobelli, *Am. J. Physiol.*, *236*:E667 (1979).

94. P. Dandona, V. Fonseca, O. Fernando, R. K. Menon, J. Weerakoon, A. Kurtz, and R. Stephen, *Diabetes Res.*, *5*:47 (1987).

95. K. Takeda, H. Yoshima, N. Shibata, Y. Masuda, H. Yoshikawa, S. Muranishi, T. Yasumura, and T. Oka, *J. Pharmacobio-Dyn.*, *9*: 156 (1986).

96. W. W. Eckman, C. S. Patlak, and J. D. Fenstermacher, *J. Pharmacokinet. Biopharm.*, *2*:257 (1974).

97. A. Mallory, J. W. Schaefer, J. R. Cohen, S. A. Holt, and L. W. Norton, *Arch. Surg.*, *115*:30 (1980).

98. M. I. Marks and J. J. Eller, *Am. Rev. Respir. Dis.*, *103*:579 (1971).

99. M. D. Sanger, *Ann. Allergy*, *17*:173 (1959).

100. M. L. Clark, H. Reinhardt, M. C. Miller III, and R. Wilson, *J. Lab. Clin. Med.*, *66*:34 (1965).

101. W. Halperin, W. I. Weiss, R. Altman, M. A. Diamond, K. J. Black, A. W. Iaci, H. C. Black, and M. Goldfield, *Am. J. Public Health*, *69*:1247 (1979).

102. G. Metcalf, P. W. Dettmar, A. Lynn, D. Brewster, and M. E. Havler, *Regul. Pept.*, *2*:277 (1981).

103. K. Weigl, F. Mundinger, and J. Chrubasik, *Acta Neurochir.* *39*:163 (1987).

104. S. Guarini, A. V. Vergoni, and A. Bertolini, *Pharmacol. Res. Commun.*, *19*:255 (1987).

105. C. E. Fiore, F. Castorina, L. S. Malatino, and C. Tamburino, *Int. J. Clin. Pharm. Res.*, *3*:257 (1983).

106. C. R. DeBold, W. R. Sheldon, Jr., G. S. DeCherney, R. V. Jackson, W. E. Nicholson, D. P. Island, and D. N. Orth, *J. Clin. Endocrinol. Metab.*, *60*:836 (1985).

107. M. Koehler, P. Hellstern, B. Reiter, G. von Biohn, and E. Wenzel, *Klin. Wochenschr.*, *62*:543 (1984).

108. S. Salvadori, M. Marastoni, G. Balbani, E. D. Uberti, and R. Tomatis, *Peptides*, *6*:127 (1985).

109. M. O. Thorner, J. Reschke, J. Chitwood, A. D. Rogol, R. Furlanetto, J. Rivier, W. Vale, and R. M. Blizzard, *N. Engl. J. Med.*, *312*:4 (1985).

110. L. Neneioni, L. Villa, A. Tagliabue, and D. Boraschi, *Lymphokine Res.*, *6*:335 (1987).

111. I. Yamazaki, in *LHRH and Its Analogues: Basic and Clinical Aspects* (F. Labrie, A. Belanger, and A. Dupont, eds.), Excerpta Medica, Amsterdam (1984), p. 77.

112. Z. Valinger, B. Ladesie, I. Hrsak, and J. Tomasie, *Int. J. Immunopharmacol.*, *9*:325 (1987).

113. V. E. Mendel, L. L. Bellinger, F. E. Williams, and R. A. Iredale, *Pharmacol. Biochem. Behav.*, *24*:247 (1986).

114. T. Mitsuma and T. Nogimori, *Experientia*, *39*:620 (1983).

115. R. Thorpe, M. J. Perry, M. Callus, P. J. Gaffney, and M. Spitz, *Hybridoma*, *3*:381 (1984).

116. J. C. Verhoef and H. M. van den Wildenberg, *Regul. Pep.* *14*:113 (1986).

117. D. J. Handelsman, R. P. S. Jansen, L. M. Boylan, J. A. Spaliviero, and J. R. Turtle, *J. Clin. Endocrinol. Metab.*, *59*:739 (1984).

118. D. R. Owens, *Human Insulin. Clinical Pharmacological Studies in Normal Man,* MTP Press, Lancaster, England (1986), p. 98.

119. L. T. Sennello, R. A. Finley, S.-Y. Chu, C. Jagst, D. Max, D. E. Rollins, and K. G. Tollman, *J. Pharm. Sci.*, *75*:158 (1986).

120. J. L. Barron, R. P. Miller, and D. Searle, *J. Clin. Endocrinol. Metab.*, *54*:1169 (1982).

121. K. Drien, J. P. Devissagnet, R. Duboistesselin, F. Dray, and E. Ezan, *Prog. Clin. Biol. Res.*, *243A*:435 (1987).

122. A. Wright and J. F. Wilson, *Peptides*, *4*:5 (1983).

123. A. Pacini, E. Maioli, V. Bocci, and G. P. Pessina, *Cancer Drug Delivery*, *4*:17 (1987).

124. R. Wassef, Z. Cohen, and B. Langer, *Transplantation*, *40*:489 (1985).

125. D. S. Schade, R. P. Eaton, N. Friedman, and W. Spencer, *Diabetes*, *28*:1069 (1979).

126. K. S. E. Su, K. M. Campanale, G. A. Kerchner, and C. L. Gries, *J. Pharm. Sci.*, *74*:394 (1985).

127. V. Bocci, *Cancer Drug Delivery*, *1*:337 (1984).

128. A. J. Wood, G. Mauer, W. Niederberger, and T. Beveridge, *Transplant. Proc.*, *15*:2409 (1983).

129. K. Nooter, F. Schultz, and P. Sooneveld, *Res. Commun. Chem. Pathol. Pharmacol.*, *43*:407 (1984).

130. J. R. J. Baker, B. H. Kemmenoe, C. McMartin, and G. E. Peters, *Regul. Pep.*, *9*:213 (1984).

131. D. DeWied, G. M. van Ree, and H. M. Greven, *Regul. Pep.*, *14*:113 (1986).

132. O. Gaffori and D. DeWied, *Eur. J. Pharmacol.*, *85*:115 (1982).

133. A. Witter and D. DeWied, in *Handbook of the Hypothalamus*, Vol. 2 (P. J. Morgane and J. Panksepp, eds.), Marcel Dekker, New York, (1980), p. 307.

134. B. Meltzer, W. Z. Traczyk, and J. Kubicki, *Arzneim-Forsch.*, *35*: 1374 (1985).

135. H. R. Nakin and P. Troen, *J. Clin. Endocrinol. Metab.*, *33*:558 (1971).

136. B. E. Spiliotis, G. P. Angust, W. Hung, W. Sonis, W. Mendelson, and B. B. Bercu, *J. Am. Med. Assoc.*, *152*:2223 (1984).

137. B. Koch and B. Lutz-Bucher, *Endocrinology*, *116*:671 (1985).

138. D. A. Land, D. R. Mathews, M. Burnett, G. M. Ward, and R. C. Turner, *Diabetes*, *31*:22 (1982).

139. B. C. Hansen, K. L. C. Jen, S. B. Pek, and R. A. Wolfe, *J. Clin. Endocrinol. Metab.*, *54*:785 (1982).

140. D. J. Handelsman and R. S. Swerdloff, *Endocr. Rev.*, *7*:95 (1986).

141. J. Hunter, J. McGee, J. Saldivar, T. Tsai, R. Feuers, and L. E. Scheving, *Proc. 3rd. Int. Conf. Chronobiol.* (1988).

142. A. Reinberg, P. Drovin, M. Kolopp, L. Mejean, F. Levi, G. Debry, M. Mechkouri, G. DiCostanzo, and A. Bicakora-Rocher, *Proc. 3rd. Int. Conf. Chronobiol.* (1988).

143. B. Tarquini, V. Cavallini, A Cariddi, M. Checchi, V. Sorice, and M. Cecchettin, *Chronobiol. Int.*, *4*:199 (1987).

144. R. A. Rizza, *Clin. Chem.*, *32*:B97 (1986).

145. A. M. Albisser, B. S. Leibel, T. G. Ewart, Z. Davidovac, C. K. Botz, W. Zingg, H. Schipper, and R. Gander, *Diabetes*, *23*:397 (1974).

146. W. Kerner, C. Thum, G. Tamasjun, W. Beischer, A. H. Clemens, and E. F. Pfeiffer, *Horm. Metab. Res.*, *8*:256 (1976).

147. A. Lambert, M. Buysselaert, E. Marchard, M. Pierard, S. Wojeck, and L. Lambotte, *Diabetes*, *27*:825 (1978).

148. E. Kraegan, D. Chisholm, and M. McNamara, *Horm. Metab. Res.*, *12*:365 (1981).

149. S. Genuth and P. Martin, *Diabeies*, *26*:571 (1977).

150. T. Deckert and B. Lorup, *Diabetologia*, *12*:573 (1977).

151. J. Mirouze, J. Selam, T. Pham, and D. Chenon, *Acta Diabetol. Lat.*, *17*:103 (1980).

152. K. Perlman, R. Ehrlich, R. Filler, and A. Albisser, *Diabetes*, *30*: 710 (1981).

153. J. C. Pickup, H. Keen, J. A. Parsons, and K. G. M. M. Alberti, *Br. Med. J.*, *1*:204 (1978).

154. D. R. Mathews, B. A. Naylor, R. G. Jones, G. M. Ward, and R. C. Turner, *Diabetes*, *32*:617 (1983).

155. P. E. Belchetz, T. M. Plant, Y. Nakai, E. J. Keogh, and E. Knobil, *Science*, *202*:631 (1978).

156. T. F. Davies, A. Pan-Gomez, M. J. Watson, C. Q. Mountijoy, J. P. Hauker, G. M. Besser, and R. Hall, *Clin. Endocrinol.*, *6*:213 (1977).

157. B. C. J. M. Fauser, J. M. J. Dony, W. H. Doesburg, and R. Rolland, *Fertil. Steril.*, *39*:695 (1983).

158. V. Menon, W. R. Butt, R. N. Clayton, R. L. Edwards, and S. S. Lynch, *Clin. Endocrinol.*, *21*:223 (1984).

159. B. Couzinet, N. Lahlou, N. Lestrat, P. Bouchard, M. Roger, and G. Schaison, *J. Endocrinol. Invest.*, *9*:103 (1986).

160. G. Leyendecker, L. Wildt, and M. Hansman, *J. Clin. Endocrinol. Metab.*, *51*:1214 (1980).

161. M. M. Siebel, M. Kamrava, C. McArdle, and M. L. Taymor, *Obstet. Gynecol.*, *61*:292 (1983).

162. A. Loucopoulos, M. Ferin, R. L. V. Wiele, I. Dyrenfurth, D. Linkie, M. Yeh, and R. Jewelewicz, *Obstet. Gynecol.*, *148*:895 (1984).

163. G. Amodeo, R. Palermo, M. Gabrielli, and A. Girasolo, *Acta Eur. Fertil.*, *18*:113 (1987).

164. R. D. Podbesek, E. B. Mawer, G. D. Zanelli, J. A. Parsons, and J. Reeve, *Clin. Sci.*, *67*:591 (1984).

165. D. S. Weigle, D. J. Koerker, and C. J. Goodner, *Am. J. Physiol.*, *247*:E564 (1984).

166. M. Komjati, P. Bratusch-Marraim, and W. Waldhause, *Endocrinology*, *118*:312 (1986).

167. N. Santoro, M. E. Wierman, M. Filicori, J. Waldstreicher, and W. F. Crowley, Jr., *J. Clin. Endocrinol. Metab.*, *62*:109 (1986).

168. J. A. Moore, H. Wilking, and A. L. Daugherty, in *Delivery Systems for Peptide Drugs* (S. S. Davis, L. Illum, and E. Tominson, eds.), Plenum Press, New York (1986), p. 317.

169. E. Verdin, M. Castillo, A. S. Luyckx, and P. J. Lefebvre, *Diabetes*, 33:1169 (1984).

170. A. Souvatzoglou, Z. Voulgaris, R. Charitopoulou, and E. Rapt, *J. Endocrinol. Invest.*, 9:325 (1986).

171. M. Gulan, K. Perlman, A. M. Albisser, J. Pyper, and B. Zinman, *Diabetes Care*, 10:453 (1987).

172. B. C. J. M. Fauser, R. Rolland, J. M. J. Dong, and R. S. Corbey, *Andrologia*, 17:143 (1985).

173. M. R. Soules, M. B. Southworth, M. E. Norton, and W. J. Bremmer, *Fertil. Steril.*, 46:578 (1986).

174. K. Takano, N. Hizuka, K. Shizuma, N. Honda, and N. C. Ling, *Acta Endocrinol.*, 108:11 (1985).

175. G. Skarin, S. J. Nillius, L. Wibell, and L. Wide, *J. Clin. Endocrinol. Metab.*, 55:723 (1982).

176. T. Lauritzen, S. Pramming, T. Decker, and C. Binder, *Diabetologia*, 24:326 (1983).

177. K. E. Avis, in *Remingtons' Pharmaceutical Sciences*, 17th ed. (A. R. Gennaro, ed.), Mack Publishing Company, Easton, Pa. (1985), p. 1815.

178. K. E. Avis, L. E. Lachman, and H. A. Lieberman, *Pharmaceutical Dosage Forms, Parenteral Medications*, Marcel Dekker, New York (1986).

179. S. J. Shire, *Proc. Pharm. Tech. Conf.*, (1987), p. 47.

180. A. J. Kastin, P. F. Castellanos, A. J. Fischman, J. K. Proffitt, and M. V. Graf, *Pharmacol. Biochem. Behav.*, 21:969 (1984).

181. A. Rogerson and L. M. Sanders, *Proc. Int. Symp. Controlled Release Bioact. Mater.*, 14:97 (1987).

182. W. D. Lougheed, A. M. Albisser, H. M. Martindale, J. C. Chow, and J. R. Clement, *Diabetes*, 32:424 (1983).

183. W. D. Lougheed, H. Woulfe-Flanagan, J. R. Clement, and A. M. Albisser, *Diabetologia*, 19:1 (1980).

184. L. Pachucki, P. Gotch, J. Shaker, and J. Findling, *Ann. Intern. Med.*, 107:781 (1987).

185. S. T. Anik and J. Y. Hwang, *Int. J. Pharm.*, 16:181 (1983).

186. J. C. McElnay, D. S. Elliott, and P. F. D'Arcy, *Int. J. Pharm.*, 36:199 (1987).

187. P. S. Adams, R. F. Haines-Nutt, and R. Town, *J. Pharm. Pharmacol.*, 39:158 (1987).

188. E. Chantelan, G. Laoner, M. Gasthaus, M. Boxberger, and M. Berger, *Diabetes Care*, 10:348 (1987).

189. K. G. Hutchinson, *J. Pharm. Pharmacol.*, 37:528 (1985).

190. S. Sato, C. D. Ebert, and S. W. Kim, *J. Pharm. Sci.*, 72:228 (1983).

191. T. D. Rohde, K. H. Kernstine, B. D. Wigness, S. R. Kryjeski, F. D. Dorman, F. J. Goldenberg, and H. Buchwald, *Trans. Am. Soc. Artif. Intern. Organ B.*, 33:316 (1987).

192. J. Bringer, A. Heldt, and G. M. Grodsky, *Diabetes*, 30:83 (1981).

193. U. Grau, *Diabetologia*, 28:458 (1985).

194. H. Thuro and K. Geisen, *Diabetologia*, 27:212 (1984).

195. J. L. Selam, P. Zinnis, M. Mellet, and J. Mirouze, *Diabetes Care*, 10:343 (1987).

196. E. H. Massey and T. A. Sheliga, *Pharm. Res.*, 3:26S (1986).

197. G. A. Truskey, R. Gabler, A. Dileo, and T. Manter, *J. Parenter. Sci. Technol.*, 41:180 (1987).

198. B. Couriel, *Bull. Parenter. Drug Assoc.*, 31:227 (1977).

199. J. Travell, *J. Am. Med. Assoc.*, 158:368 (1955).

200. J. C. Taggart, *Bull. Parenter. Drug Assoc.*, 26:87 (1972).

201. I. Lindmayer, K. Menassa, J. Lambert, A. Moghrabi, L. Legendre, C. Legault, M. Letendre, and J. P. Halle, *Diabetes Care*, 9:294 (1986).

202. J. P. Halle, J. Lambert, I. Lindmayer, K. Menassa, F. Coutu, A. Moghrabi, L. Legendre, C. Legault, and G. Lalumiere, *Diabetes Care*, 9:279 (1986).

203. P. Haycock, *Diabetes*, 35:145A (1986).

204. L. A. Distiller, L. I. Robertson, R. Moore, and F. Bonnici, *S. Afr. Med. J.*, 71:749 (1987).

205. C. Kuhl, B. Edsberg, P. Hildebrandt, and D. Herly, *Diabetologia*, 29:561A (1986).

206. A. E. Kitabchi, J. N. Fisher, R. Matteri, and M. B. Murphy, *Adv. Intern. Med.*, 28:449 (1983).

207. M. V. Sefton, *CRC Crit. Rev. Biomed. Eng.*, 14:201 (1987).

208. M. Franetzki, *Pharm. Res.*, 1:237 (1984).

209. J. Mirouze, *Diabetologia*, 25:209 (1983).

210. C. S. rosenberg, *Mt. Sinai J. Med. N.Y.*, 54:217 (1987).

211. G. R. Chambers, I. A. Sutherland, S. White, P. Mason, and H. S. Jacobs, *Upsala J. Med. Sci.*, 89:91 (1984).

212. N. A. Armar, S. L. Tan, A. Eshel, H. S. Jacobs, J. Adams, and I. A. Sutherland, *Br. J. Hosp. Med.*, 37:429 (1987).

213. P. J. Blackshear, F. D. Dorman, P. L. Blackshear, R. L. Varco, and H. Buchwald, *Surg. Gynecol. Obstet.*, 134:51 (1972).

214. R. L. Reid and E. Sauerbrei, *Am. J. Obstet. Gynecol.*, 148:648 (1984).

215. D. S. Schade and R. P. Eaton, *N. Engl. J. Med.*, 312:1120 (1985).

216. S. E. Chistensen, J. Weeke, H. Orskor, N. Moeller, A. Flyvbjerg, A. G. Harris, E. Lund, and J. Jorgensen, *Clin. Endocrinol.*, 27:297 (1987).

217. M. Cavallini, F. Halberg, G. Gornelissen, F. Enrichens, and G. Margarit, *J. Controlled Release*, 3:3 (1986).

218. N. W. Rodger, J. Dupre, C. L. B. Canny, and W. F. Brown, *Diabetes Care*, 8:447 (1985).

219. H. Bibergeil, I. Huettl, W. Felsing, U. Felsing, I. Seidlein, S. Herfurth, J. Dabels, G. Reichel, C. Lueder, G. Albrecht, W. Brum, and R. Menzel, *Exp. Clin. Endocrinol.* 90:51 (1987).

220. S. B. Leichter, M. E. Schreiner, L. R. Reynolds, and T. Bolick, *Arch. Intern. Med.*, 145:1409 (1985).

221. S. M. Teutsch, W. H. Herman, D. M. Dwyer, and J. M. Lane, *N. Engl. J. Med.*, 310:361 (1984).

222. A. Pietri and P. Raskin, *Diabetes Care*, 4:624 (1981).

223. W. Lougheed and A. M. Albisser, *Int. J. Artif. Organs*, 3:50 (1980).

224. K. Schmidt, C. Rosak, B. Bochm, E. Schifferdecker, P. H. Althoff, and K. Schoeffling, *Klin. Wochenschr.*, 64:804 (1986).

225. F. Theeuwes and S. I. Yum, *Ann. Biomed. Eng.*, *4*:343 (1976).

226. J. W. Fara, *Methods Enzymol.*, *112*:470 (1985).

227. M. V. Sefton, H. M. Lusher, S. R. Firth, and M. U. Waher, *Ann. Biomed. Eng.*, *7*:329 (1979).

228. M. V. Sefton, D. G. Allen, V. Horvath, and W. Zingg, in *Recent Advances in Drug Delivery Systems* (J. M. Anderson and S. W. Kim, eds.), Plenum Press, New York (1984), p. 349.

229. P. K. Walter and M. V. Sefton, *Proc. Int. Symp. Controlled Release Bioact. Mater.*, *14*:231 (1987).

230. M. V. Sefton, *Adv. Chem. Ser.*, *199*:511 (1982).

231. E. Uhlig, W. F. Graydon, and W. Zingg, *J. Biomed. Mater. Res.*, *17*:931 (1983).

232. M. Nalecz, J. Lewadonski, A. Werynski, and I. Zawicki, *Artif. Organs*, *2*:305 (1978).

233. A. D. Bangham, *Prog. Biophys. Mol. Biol.*, *18*:29 (1968).

234. G. Gregoriadis, ed., *Liposome Technology*, Vols. I, II, and III, CRC Press, Boca Raton, Fla. (1984).

235. F. Szoka and D. Papahadjopoulos, in *Liposomes: Physical Structure to Therapeutic Applications* (C. G. Knight, ed.), Elsevier/North-Holland, Amsterdam (1981), p. 51.

236. G. Gregoriadis, *Hindustan Antibiot. Bull.*, *20*:14 (1977).

237. R. L. Juliano, in *Liposomes* (M. Ostro, ed.), Marcel Dekker, New York (1983), p. 53.

238. D. J. A. Crommelin and G. Storm, *Int. Pharm. J.*, *1*:179 (1987).

239. D. A. Eppstein and W. E. Stewart II, *J. Virol.*, *41*:575 (1982).

240. D. A. Eppstein, *J. Interferon Res.*, *2*:117 (1982).

241. A. L. Weiner, S. S. Carpenter-Green, E. C. Soehngen, R. P. Lenk, and M. C. Popescu, *J. Pharm. Sci.*, *74*:922 (1985).

242. R. P. Liburdy and T. S. Tenforde, *Radiat. Res.*, *108*:102 (1986).

243. R. P. Liburdy, *Radiat. Res.*, *103*:266 (1985).

244. M. B. Yatvin, J. N. Weinstein, W. H. Dennis, and R. Blumenthal, *Science*, *202*:1290 (1978).

245. J. N. Weinstein, R. L. Magin, M. B. Yatvin, and D. S. Zaharko, *Science*, *204*:188 (1979).

246. R. L. Magin and M. R. Niesman, *Cancer Drug Delivery*, *1*:109 (1984).

247. C. Weingarten, A. Moufti, J. Delattre, F. Puisieux, and P. Couvreur, *Int. J. Pharm.*, *26*:251 (1985).

248. H. H. Hsieh, M. Schreiber, N. Stowe, A. Novick, S. Deodhar, B. Barna, C. Pippenger, and R. Shamberger, *Transplant. Proc.*, *17*:1397 (1985).

249. G. Gregoriadis, P. D. Leathwood, and B. E. Ryman, *FEBS Lett.*, *14*:95 (1971).

250. G. Gregoriadis and B. E. Ryman, *Biochem. J.*, *129*:123 (1972).

251. G. Gregoriadis and B. E. Ryman, *Eur. J. Biochem.*, *24*:485 (1972).

252. G. Gregoriadis, D. Putman, L. Louis, and E. D. Neerunjun, *Biochem. J.*, *140*:323 (1974).

253. R. L. Juliano and D. Stamp, *Biochem. Biophys. Res. Commun.*, *63*:651 (1975).

254. A. W. Segal, E. J. Willis, J. E. Richmond, G. Slown, C. D. V. Black, and G. Gregoriadis, *Br. J. Exp. Pathol.*, *55*:320 (1974).

255. G. Poste, *Biol. Cell*, *47*:19 (1983).

256. N. C. Phillips, M. L. Moras, L. Chedid, P. Lefrancier, and J. M. Bernard, *Cancer Res.*, *45*:128 (1985).

257. I. Saiki, S. Sone, W. E. Fogler, E. S. Kleinerman, G. Lopez-Berestein, and I. J. Fidler, *Cancer Res.*, *45*:6188 (1985).

258. W. C. Koff, S. D. Showalter, B. Hampar, and I. J. Fidler, *Science*, *228*:495 (1985).

259. M. Fukunga, M. M. Miller, K. Y. Hostetler, and L. J. Deftos, *Endocrinology*, *115*:757 (1985).

260. T. D. Heath and F. J. Martin, *Chem. Phys. Lipids*, *40*:347 (1986).

261. P. A. H. M. Toonen and D. J. A. Crommelin, *Pharm. Weekbl. Sci. Ed.*, *5*:269 (1983).

262. K. K. Mathay, T. D. Heath, C. C. Badger, I. D. Bernstein, and D. Papahadjopoulos, *Cancer Res.*, *46*:4904 (1986).

263. L. D. Leserman, P. Machy, C. Devaux, and J. Barbet, *Biol. Cell*, *47*:111 (1983).

264. K. Hiroshi, S. Junko, Y. Sunju, and K. Yuriko, *Chem. Pharm. Bull.*, *34*:4253 (1986).

265. C. R. Alving, in *Liposomes: From Biophysics to Therapeutics* (M. J. Ostro, ed.), Marcel Dekker, New York (1987), p. 195.

266. G. Scherphof, F. Roerdink, M. Waite, and J. Parks, *Biochim. Biophys. Acta*, *542*:296 (1978).

267. O. O. Petrukhina, N. N. Ivanov, M. M. Feldstein, A. E. Vasil'ev, N. A. Plate, and V. P. Torchilin, *J. Controlled Release*, *3*:137 (1986).

268. J. H. Wiessner and K. J. Kwang, *Biochim. Biophys. Acta*, *689*:490 (1982).

269. J. H. Wiessner, H. Mar, D. G. Baskin, and K. J. Kwang, *J. Pharm. Sci.*, *75*:259 (1986).

270. R. L. Shew and D. W. Deamer, *Biochim. Biophys. Acta*, *816*:1 (1985).

271. L. Stuhue-Sekalec, J. Chudzik, and N. Z. Stanacev, *Biochem. Biophys. Methods*, *13*:23 (1986).

272. L. S. Rao, *J. Parenter. Sci. Technol.*, *37*:72 (1983).

273. M. Singh and L. J. Ravin, *J. Parenter. Sci. Technol.*, *40*:34 (1986).

274. A. T. Florence and D. Whitehill, *Int. J. Pharm.*, *11*:277 (1982).

275. J. Freund and M. V. Bonato, *J. Immunol.*, *48*:325 (1944).

276. J. E. Salk, M. Contakos, A. M. Laurent, M. Sorenson, A. J. Rapalski, I. H. Simmons, and H. Sandbert, *J. Am. Med. Assoc.*, *151*:1169 (1953).

277. J. Freund, *Adv. Tuberc. Res.*, *7*:130 (1956).

278. F. M. Davenport, *J. Allergy*, *32*:177 (1961).

279. B. Zbar, E. Ribi, T. Meyer, T. Azuma, and H. J. Rapp, *J. Natl. Cancer Inst.*, *52*:1571 (1974).

280. E. Yarkoni and H. J. Rapp, *Cancer Res.*, *39*:535 (1979).

281. M. Parant, F. Parant, L. Chedid, A. Yapo, J. F. Patit, and E. Laderer, *Int. J. Immunopharmacol.*, *1*:35 (1979).

282. C. A. McLaughlin, S. M. Schwartzman, B. L. Horner, G. H. Jones, J. G. Moffatt, J. J. Nesto, Jr., and D. Tegg, *Science*, *208*:415 (1980).

283. C. A. McLaughlin, E. E. Ribi, M. B. Goren, and R. Toubiana, *Cancer Immunol. Immunother.*, *4*:109 (1978).

284. E. R. Barnhart, *Physicians' Desk Reference*, 41th ed., Medical Economics Company, Oradell, N.J. (1987), p. 1785.

285. T. Cavanak and H. Sucker, *Prog. Allergy, 38*:65 (1986).

286. D. L. Howrie, R. J. Pitachcinski, B. P. Griffith, R. J. Hardesty, J. T. Rosenthal, G. J. Burchart, and R. Venkataramanan, *Drug Intell. Clin. Pharm., 19*:425 (1985).

287. R. Wassef, Z. Cohen, and B. Langer, *Transplantation, 41*:266 (1986).

288. M. Akimoto and Y. Morimoto, *Biomaterials, 4*:49 (1983).

289. S. S. Davis, L. Illum, J. G. McVie, and E. Tomlinson, eds., *Microspheres and Drug Therapy: Pharmaceutical, Immunological and Medical Aspects*, Elsevier/North-Holland, Amsterdam (1984).

290. E. Tomlinson, *Int. J. Pharm. Technol. Prod. Manuf., 4*:49 (1983).

291. A. J. A. M. Dekker, M. M. Princen, H. deNijs, L. G. J. deLeede, and C. L. E. Brockkamp, *Peptides, 8*:1057 (1987).

292. U. Schroeder, *Methods Enzymol., 112*:116 (1985).

293. P. Artursson, P. Edman, T. Laakso, and I. Sjoholm, *J. Pharm. Sci., 73*:1507 (1984).

294. P. Edman and I. Sjoholm, *J. Pharm. Sci., 71*:576 (1982).

295. A. Rembaum and W. J. Dreyer, *Science, 298*:364 (1980).

296. K. J. Widder and A. E. Senyei, *Pharmacol. Ther., 20*:377 (1983).

297. J. G. Treleavan, F. M. Gibson, J. Ugelstad, A. Rembaum, T. Philip, G. D. Gaine, and J. T. Kemshead, *Lancet, 1*:70 (1984).

298. J. R. DeLoach, *Med. Res. Rev., 6*:487 (1986).

299. G. M. Ihler, *Pharmacol. Ther., 20*:151 (1983).

300. D. McIntosh and A. J. S. Davies, *Proc. Int. Symp. Controlled Release Bioact. Mater., 14*:217 (1987).

301. V. P. Torchilin and A. V. Mazaev, in *Advanced Drug Delivery Systems* (J. M. Anderson and S. W. Kim, eds.), Elsevier/North-Holland, Amsterdam (1986), p. 321.

302. W. R. Gombotz, A. S. Hoffman, G. Schmer, and S. Uenoyama, in *Advanced Drug Delivery Systems* (J. M. Anderson and S. W. Kim, eds.), Elsevier/North-Holland, Amsterdam (1986), p. 375.

303. T. M. S. Chang, *Science, 146*:524 (1964).

304. A. M. Sun and G. M. O'Shea, in *Advanced Drug Delivery Systems* (J. M. Anderson and S. W. Kim, eds.), Elsevier/North-Holland, Amsterdam (1986), p. 137.

305. M. J. Poznansky and L. G. Cleland, in *Drug Delivery Systems* (R. L. Juliano, ed.), Oxford University Press, New York (1980), p. 253.

306. J. R. Uren and R. C. Ragin, *Cancer Res., 39*:1927 (1979).

307. P. S. Ringrose and M. J. Humphrey, in *Targeting of Drugs with Synthetic Systems* (G. Gregoriadis, J. Senior, and G. Poste, eds.), Plenum Press, New York (1986), p. 65.

308. B. L. Ferraiolo and L. Z. Benet, *Pharm. Res., 2*:151 (1985).

309. P. Anderson and J. Vilcek, *Virology, 123*:457 (1982).

310. R. A. Roth, B. A. Maddux, K. Y. Wong, Y. Iwamoto, and I. D. Goldfine, *J. Biol. Chem., 256*:5350 (1981).

311. D. B. Cawley, H. R. Herschman, D. G. Gilliland, and R. J. Collier, *Cell, 22*:563 (1980).

312. T. N. Oeltman and E. C. Heath, *J. Biol. Chem., 254*:1028 (1979).

313. J. Blake, J. Hagman, and J. Ramachandran, *Int. J. Pept. Protein Res., 20*:97 (1982).

314. C.-J. G. Yeh and W. P. Faulk, *Clin. Immunol. Immunopathol., 32*:1 (1984).

315. W. K. Maskimins and N. Shimizu, *Biochem. Biophys. Res. Commun.*, *91*:143 (1979).

316. Y. Watanabe, H. Miyazaki, and T. Osawa, *J. Pharm. Dyn.*, 7:593 (1984).

317. L. Chedid, M. Parant, F. Parant, F. Audibert, F. Lefrancier, J. Choay, and M. Sela, *Proc. Natl. Acad. Sci. USA*, *76*:6557 (1979).

318. H. Sezaki and M. Hashida, *CRC Crit. Rev. Ther. Drug Carrier Syst.*, *1*:1 (1984).

319. E. R. Edelman, J. Kost, H. Bobeck, and R. Langer, *J. Biomed. Mater. Res.*, *19*:67 (1985).

320. J. Kost, R. Noecker, E. Kunica, and R. Langer, *J. Biomed. Mater. Res.*, *19*:935 (1985).

321. J. Kost and R. Langer, *Pharm. Res.*, 7:60 (1986).

322. J. Kost, K. Leong, and R. Langer, *Proc. Int. Symp. Controlled Release Bioact. Mater.*, *14*:186 (1987).

323. T. A. Horbett, B. D. Ratner, J. Kost, and M. Singh, in *Recent Advances in Drug Delivery Systems* (J. M. Anderson and S. W. Kim, eds.), Elsevier/North-Holland, Amsterdam (1985), p. 209.

324. G. Albin, T. A. Horbett, and B. D. Ratner, in *Advances in Drug Delivery Systems* (J. M. Anderson and S. W. Kim, eds.), Elsevier/North-Holland, Amsterdam (1986), p. 153.

325. V. H. L. Lee, *Biopharm. Manuf.*, *1*:24 (1988).

326. M. Browniee and A. Cerami, *Science*, *206*:1190 (1979).

327. S. Y. Jeong, S. W. Kim, M. J. D. Eenink, and J. Feijen, *J. Controlled Release*, *1*:57 (1984).

328. S. Sato, S. Y. Jeong, J. C. McRea, and S. W. Kim, *J. Controlled Release*, *1*:67 (1984).

329. S. Y. Jeong, S. W. Kim, D. L. Holmberg, and J. C. McRea, in *Advances in Drug Delivery Systems* (J. M. Anderson and S. W. Kim, eds.), Elsevier/North-Holland, Amsterdam (1986), p. 143.

330. Y. H. Bae, T. Okano, and S. W. Kim, *Makromol. Chem. Rapid Commun.*, *9*:185 (1988).

331. A. S. Hoffman, A. Afrassiabi, and L. C. Dong, *J. Controlled Release*, *4*:213 (1986).

332. S. W. Kim, T. Okano, Y. H. Bae, R. Hsu, and S. J. Lee, *Proc. Int. Symp. Controlled Release Bioact. Mater.*, *14*:37 (1987).

333. N. Sheppard, Jr., Y. Madrid, and R. Langer, *Proc. Int. Symp. Controlled Release Bioact. Mater.*, *14*:61 (1987).

334. L. C. Dong and A. S. Hoffman, *J. Controlled Release*, *4*:223 (1986).

11

Buccal Routes of Peptide and Protein Drug Delivery

Hans P. Merkle

Swiss Federal Institute of Technology, Zurich, Switzerland

Reinhold Anders

Hoechst AG, Frankfurt, Federal Republic of Germany

Aloys Wermerskirchen, Susanne Raehs, and Gregor Wolany

University of Bonn, Bonn, Federal Republic of Germany

I. INTRODUCTION

Various routes are being considered for peptide and protein drug delivery [1,2]. Of these, the nasal route currently appears to be the route of choice [3,4]. Nonetheless, the nasal route does have distinct limitations. For chronic treatment there is the risk of pathologic changes to the nasal mucosa [4], including the compromising of ciliary activity [5]. Moreover, there is the uncertain effect of vast individual variations in mucus secretion and turnover on the extent and rate of nasal absorption. Finally, proteases and peptidases present in the mucus, or associated with the nasal membrane, may act as an enzymatic barrier to peptide absorption, as found with other mucosal sites [6,7]. It may thus be concluded that, in spite of many promising aspects, the nasal route may not be the optimum answer to peptide absorption.

An alternative is to utilize the oral mucosa—which is comprised of the buccal, sublingual, and gingival mucosae—as a platform for peptide and protein drug delivery. The attractive features of this mucosa include excellent accessibility, high patient acceptance and compliance, and significant robustness of the mucosa. Because of the excellent accessibility of the oral mucosa, appropriate dosage forms can easily be attached and removed without significant associated pain and discomfort. Due to the familiarity of the oral route of drug administration, buccal or sublingual dosage forms are anticipated to be well accepted by patients. Equally as important, since the oral mucosa is routinely exposed to a multitude of

545

different foreign compounds, it may be rather robust and less prone to irreversible irritation or damage by the drug, the dosage form, or additives such as absorption promotors.

Information on the buccal absorption of peptides is still rather scarce, except for oxytocin due to the work of Wespi and Rehsteiner [8], Bergsjö and Jenssen [9], and Sjöstedt [10]. Other peptides that have since been studied include vasopressin analogs, insulin, protirelin, buserelin, and calcitonin [11-18]. By contrast, there is more information on the buccal absorption of conventional drugs, as evident in the reviews by Gibaldi and Kanig [19], Beckett and Hossie [20], Moffat [21], Muzaffar [22], and deBoer et al. [23]. The earliest reports on this subject dated back to those of Karmel in 1873 [24] (Ref. by Tanaka et al. [25]) and of Brunton in 1877 [26] (Ref. by Gimbaldi and Kanig [29]). A survey of the literature since 1966 is given in Table 1.

Table 1 Literature Survey on Buccal and Sublingual Absorption (Peptides Underlined)

Author(s)	Year	Ref.	Drugs or comments[a]
Wespi and Rehsteiner	1966	8	Oxytocin
Beckett and Triggs	1967	27	Basic drugs, model for passive absorption
Beckett et al.	1968	28	Amphetamines
Beckett and Moffat	1968	29	Effect of alkyl substitution in acids
Bickel and Weder	1969	30	Imipramine and metabolites
Beckett and Moffat	1969	31	Phenylacetic acids
Beckett and Moffat	1969	32	Correlation of partition coefficients and buccal absorption, amines, and acids
Bergsjö and Jenssen	1969	9	Oxytocin
Sjöstedt	1969	10	Oxytocin
Beckett and Moffat	1970	33	Carboxylic acids, correlation with partition coefficients
Taraszka	1970	34	Clindamycin, poor absorption
Stojanovic and Milivojevic	1970	35	Radiocesium
Beckett and Moffat	1971	36	Barbiturates
Lien et al.	1971	37	Physicochemical properties and bioavailability

Table 1   (Continued)

Author(s)	Year	Ref.	Drugs or comments[a]
Ho and Higuchi	1971	38	Physical model of buccal absorption
Dearden and Tomlinson	1971	39	p-Substituted acetanilides, buccal absorption, and analgesic activity
Dearden and Tomlinson	1971	40	Buccal absorption model
Beckett and Hossie	1971	20	Review, 50 Refs.
Bogaert and Rosseel	1972	41	Nitroglycerin
Vora et al.	1972	42	Analysis of absorption by physical model approach
Earle	1972	16	Insulin, unsatisfactory oral absorption
Bye et al.	1972	43	Trimethoprim, sulfamethoxazole
Wagner and Sedman	1973	44	Pharmacokinetic model
Mikhailova et al.	1973	45	Phenothiazines, buccal absorption, and partition coefficients
Sutherland et al.	1974	46	Ergotamine
Meyer et al.	1974	47	Procainamide
Beckett et al.	1975	48	Amphetamines
Kaye and Long	1976	49	Accbutolol
Kiddie and Kaye	1976	50	Mexiletin, lignocaine
Ankier and Kaye	1976	51	Disopyramide
Manning and Evered	1976	52	Glucose, 3-O-methyl-D-glucose, fructose, galactose; carrier-mediated mechanism
Lippold and Schneider	1976	53	Benzilic acid esters of homologous dimethyl-(2-hydroxyethyl)-alkyl-ammonium bromides, chain-length effect
McMullan et al.	1977	54	D-glucose, D-xylose, L-xylose, 3-O-methyl-D-glucose, D-fructose, D-arabinose, D-ribose; effect of calcium
Odumosu and Wilson	1977	55	Ascorbic acid

Table 1   (Continued)

Author(s)	Year	Ref.	Drugs or comments[a]
Schürmann and Turner	1977	56	Atenolol, propranolol
Past et al.	1977	57	DL-1-(2',5'-dichlorophenoxy)-3-tert-butylamino-2-propanol-HCl, solution, and sublingual tablets
Kates	1977	58	Propranolol
Schürmann and Turner	1978	59	Propranolol, atenolol; three compartment diffusional model
Temple and Schesmer	1978	60	Fomacaine
Past et al.	1978	61	Barbiturates, diphenyl-hydantoin, salicylic acid, tolbutamide, pyramidon, syncoumar, phenylbutazone: good correlation with dipole moments and buccal absorption
Sprake and Evered	1979	62	D- and L-Cylcoserin, in vitro transport through human buccal tissue
Sadoogh-Abasian and Evered	1979	63	L-ascorbic acid, dihydro-ascorbic acid, D-isoascorbic acid: effects of Na, D-glucose and 3-O-methyl-D-glucose
Henry and Ohashi	1979	64	Propranolol
Kaspi et al.	1979	65	Potential antidepressive drug
Davis and Turner	1979	66	Beta-adrenoceptor antagonist, propranolol
Chan	1979	67	Pethidine and basic metabolites
Muzaffar	1979	22	Review, 34 Refs.
Henry et al.	1980	68	Propranolol
Evered et al.	1980	69	Nicotinic acid, nicotinamide
Chan et al.	1980	70	Indomethacin
Laczi et al.	1980	17	Vasopressin analogues
Tanaka et al.	1980	71	Salicylic acid
Ishida et al.	1981	72	Insulin; mucosal dosage form, absorption in dogs

Table 1 (Continued)

Author(s)	Year	Ref.	Drugs or comments[a]
Evered and Offer	1981	73	D-glucose; inhibition of buccal absorption by aspirin
Evered and Vadgama	1981	74	Amino acids
Muhiddin and Johnston	1981	75	Flecainide
Past et al.	1981	76	Correlation of buccal absorption of 15 drugs with dipole moment in dioxane
Evered et al.	1982	77	Inhibition of D-glucose absorption by sulfite; and D-fructose by aspirin
Milovac et al.	1982	78	Dihydroergocristine, dihydroergocornine, dihydroergocryptine; as methanesulfonates
Erb	1982	79	Nitroglycerin; controlled release buccal preparation
Greenblatt et al.	1982	80	Lorazepam, sublingual tablets
Ishida et al.	1982	81	Lidocaine, mucosal dosage form, absorption by gingiva
Windorfer	1982	82	Nitroglycerin, sublingual versus buccal
Abrams	1983	83	Nitroglycerin, controlled release polymer matrix
Evered and Mallett	1983	84	Thiamine; no effect on valine and tyrosine
Blackett et al.	1983	85	Aspirin
Past et al.	1983	86	Correlation of dipole moment and buccal absorption
Schor et al.	1983	87	Nitroglycerin; controlled release buccal preparation
Woodcock et al.	1983	88	Verapamil; by-passing the first-pass effect
Anders et al.	1983	12	Protirelin (TRH)
Anders et al.	1983	89	Calcitonin, no buccal absorption
Patterson et al.	1983	90	N-dimethylpropranolol, in dogs
Ishida et al.	1983	91	Na salicylate, in hamster cheek pouch

Table 1   (Continued)

Author(s)	Year	Ref.	Drugs or comments[a]
Asthana et al.	1984	92	Verapamil, bioavailability and pharmacokinetics
DeBoer et al.	1984	23	Review of different drugs, including oxytocin and vasopressin
Marty et al.	1984	93	Clonidine, pipamazine, flunitrazepam, sublingual tablets
Schurr et al.	1985	94	Protirelin (TRH), adhesive buccal patches
Pitha et al.	1986	95	Testosterone, progesterone, estradiol; effect of hydrophilic cyclodextrin derivatives on buccal absorption
Hussain et al.	1986	96	Nalbuphine
Merkle et al.	1986	11–15	Protirelin (TRH) in humans and rats; Buserelin, poorly absorbed in rats, adhesive buccal patches
Randhava et al.	1986	97	Medifoxamine
Duchateau et al.	2986	98	Propranolol, comparison of peroral, sublingual and nasal administration, prolonged mean residence time after sublingual absorption
Lee et al.	1987	99	Proteolytic activities against insulin in mucosal homogenates
Veillard et al.	1987	100	Buccal absorption of a model tripeptide in dogs
Ebert et al.	1987	101	Buccal absorption of diclofenac sodium in dogs
Siddiqui et al.	1987	102	Review of nonparenteral administration of peptide and protein drugs
Veillard et al.	1987	103	Buccal absorption of a model tripeptide
Aungst et al.	1988	104	Buccal and sublingual insulin absorption in rats
Banga et al.	1988	105	Review of systemic delivery of peptides and proteins

Table 1   (Continued)

Author(s)	Year	Ref.	Drugs or comments[a]
Le Brun et al.	1988	106	β-Blocking drugs through porcine buccal mucosa
Tavakoli-Saberi et al.	1988	107	Properties of cultured hamster buccal epithelium
Garren et al.	1988	108	In vitro model, hamster cheek pouch
Oh and Ritschel	1988	109	Buccal absorption of insulin in rabbits
Oh and Ritschel	1988	107	Buccal absorption of insulin in rabbits
Kurosaki et al.	1988	111	Hamster cheek pouch, surfactant effects
Aungst et al.	1988	112	Insulin in rats
Ritschel et al.	1988	113	Insulin in dogs
Barsuhn and Ho	1988	114	Studies on factors influencing peptide absorption in dogs
Nakada et al.	1988	115	Calcitonin in rats, effect of additives
Tucker	1988	116	Improved mouth wash method to study oral mucosal absorption
Paulesu et al.	1988	117	Interferon-α_2 in rats, poor absorption
Banerjee and Robinson	1988	118	Electrophysiological characterization of rabbit buccal mucosa
Cassidy and Quadros	1988	119	Buccal aminopeptidase activity
Yamamoto et al.	1988	120	Proteolytic activity against insulin in buccal mucosal homogenates of rabbits
Nakada et al.	1988	121	Degradation of calcitonin in rat's oral mucosa homogenate
Kurosaki et al.	1989	122	Hamster cheek pouch, effect of Azone

[a]All animal studies indicated in text, all other studies in human subjects.

The focus of this chapter will be on the absorption of peptides and proteins across the buccal mucosa. Because of the scarcity of information on this subject mentioned earlier, many of the conclusions drawn in this chapter will be based on reasonable assumptions rather than fully established facts from the literature.

II. ANATOMY AND PHYSIOLOGY OF THE ORAL MUCOSA
RELEVANT TO PEPTIDE AND PROTEIN DRUG ABSORPTION

The oral cavity is lined by a relatively thick, dense, and multilayered mucous membrane of a highly vascularized nature. Drug penetrating into the membrane gains access to the systemic circulation via a network of capillaries and arteries. The arterial flow is supplied by branches of the external carotid artery. The venous backflow goes through capillaries and a venous net and is finally taken up by the jugular veins. The equally well-developed lymphatic drainage runs more or less parallel to the venous vascularization and ends up in the jugular ducts.

The epithelium of the oral cavity is, in principle, similar to that of the skin, with interesting differences in keratinization and the protective and lubricant mucus spread across the surface. The total area is about 100 cm^2 [38]. The buccal part with about one third of the total surface is lined with an epithelium of about 0.5 mm thickness, and the rest by one of 0.25 mm thickness [56]. The multilayered structure of the oral mucosa is formed by cell division that occurs mainly in the basal layer. As reviewed by Jarrett [123], the mucosa of the oral cavity can be divided into three functional zones. First, the mucus-secreting regions (consisting of the soft palate, the floor of the mouth, the under surface of the tongue, and the labial and buccal mucosa) have a normally nonkeratinized epithelium. These regions are supposed to represent the major absorption sites in the oral cavity. Second, the hard palate and the gingiva are the regions of the masticatory mucosa and have a keratinized epidermis. Third, specialized zones are the borders of the lips and the dorsal surface of the tongue with its highly selective keratinization.

An important feature of the oral mucosa is the rapid turnover of the cells in comparison with the skin epidermis, 3–8 days versus 30 days. This is because of the constant replacement of the nonkeratinized or partly keratinized cells that is necessary to stabilize its function and integrity. A reduction of the mucosal mitotic activity would result in a loss of epithelial continuity [123].

Keratinization and average size of the epithelial cells seem to have an inverse relationship. The mean cross-sectional area of the cells of the cheek is about 263 μm^2, as compared to about 133 μm^2 for the cells of the keratinized palate. While the basal cells of the hard palate are not markedly different from those producing the nonkeratinized buccal epithelium, as the cells move toward the surface the differences become increasingly apparent. Palatal cells show a greater concentration of fibrillar and keratohyalin granular structures, while the buccal cells show more glycogen granules and numerous ribosomes [123].

An important difference between the keratinized and the nonkeratinized regions of the oral epithelium is the lipid composition of the cells. The keratinized oral epithelium shows a lipid pattern of mainly neutral lipids, i.e., ceramides, whereas the nonkeratinized epithelium contains few neutral but polar lipids, particularly cholesterol sulfate and glucosylceramides [124,125]. Another feature of the buccal membrane is the presence of numerous elastic fibers in the dermis that provide a typical elastic behavior. These fibers represent another effective barrier against the diffusion of drug molecules to the circulation system.

The nature of the junction between epidermis and dermis in the region of the hard palate is different from that in the buccal mucosa. Whereas the hard palate is a acanthotic-type epidermis, with a large contact area between dermis and epidermis, the buccal mucosa has a much flatter dermo-epidermal junction and, therefore, a much smaller contact area. The collagen fibers of the buccal mucosa are relatively unpolymerized and less dense as compared to those of the hard-palate dermis.

The surface of the mucous membrane is constantly washed by a stream of about 0.5 up to 2 liters of saliva daily produced in the salivary glands. The main glands are the three pairs of the parotid, the submaxillary, and the sublingual glands. The first are located under and in front of each ear, with ducts opening to the inner surface of the cheek. The submaxillary glands lie below the lower jaw releasing saliva through one duct on each side. Finally, the sublingual glands are located below the tongue with its ducts opening to the floor of the mouth under the tongue. In addition to these main glands there is a variety of small glands dispersed on the tongue and the buccal and sublingual mucosa. Minor salivary glands are situated in the buccal, palatal, and retromolar regions of the oral cavity. There are major differences with respect to type of mucins, mucin content, and secretion.

The surface of the oral cavity is the site of a complex microbial flora. Its composition varies widely with the location. Large differences exist between the surface of the teeth, the gingiva, the tongue, and the buccal mucosa. In order to retain health and appearance of the mucosae, each local bacterial composition has to be preserved.

Drug absorption from the oral mucosa most likely occurs in the non-keratinized sections. The first efficient barrier against penetration, however, is the mucus layer covering the oral epithelium. This consists of glycoproteins produced by the nonkeratinized oral mucosae. The drug then diffuses across the epithelia via two possible pathways: by crossing the cell membranes or by using the intercellular space. The latter is presumed to be preferred by ions and very small molecules [123,126]. The main route for most drug molecules is by partitioning into the lipid bilayer of the cells and from there into the cells. Hydrophilic medium and large size molecules like peptides, however, are not likely to cross the lipid bilayers of the cells to a great extent. This might occur through more polar fenestrations in the lipid bilayer. It remains an open question as to how the so-called tight junctions between the cells can be sufficiently opened to allow ready absorption of larger molecules. There is some evidence [117,123,127–129] that even large molecules like horseradish peroxidase and interferon-α_2 may penetrate the oral mucosa to some extent.

A not yet fully known influence on peptide absorption is that of peptidase activity located in the saliva and the mucus layer, the mucosal surface, and the microbial flora. Studies using oral mucosal homogenates revealed substantial peptidase activity [99,119–121]. The significance of studies with cell homogenates, however, is not entirely established, since the enzymatic activity exposed after homogenization may not be identical to the environmental activity which the peptides might encounter during their

passage across the epithelium. Moreover, there is no differentiation between extracellular and intracellular peptidase activity.

III. EXPERIMENTAL TECHNIQUES TO STUDY BUCCAL DRUG ABSORPTION

A. In Vivo Studies

Human Studies

The mouth wash procedure developed by Beckett and Hossie [20], which
yields surprisingly reproducible results, is frequently used in human buccal
drug absorption studies. In this procedure, 25 ml of a drug solution of
appropriate concentration in an appropriate buffer is placed in a subject's
mouth. The solution is circulated 300–400 times around the mouth for
5 min by movement of the cheeks and the tongue. To obtain full time profiles the procedure has to be repeated for each additional time point.
Care must be taken not to swallow any of the solution. This can be
achieved through practice and by inclining the head forward and downward
during the test as well as by performing some manual task to reduce the
temptation to swallow. After the required period the solution is emptied
into a beaker with its volume recorded and pH measured. This is followed
by a 10-sec rinse with 10 ml of water to wash off nonabsorbed drug. The
solutions are combined and analyzed for the drug remaining. In order to
account for volume changes due to dilution by saliva, a nonabsorbable
marker may be added to the initial drug solution. To avoid repeated
mouth wash periods to obtain full time profiles a modification of the original procedure was recently published by Tucker [116] using phenol red
as a nonabsorbable marker. This allows repeated sampling from the drug
solution during a single experiment.

There are at least two drawbacks with the above procedure. First, it
does not provide an accurate measure of the extent or rate of drug absorption because only drug loss is monitored. Second, it may not be suitable
for poorly soluble drugs, for which the absorption kinetics may be governed by dissolution [130]. Of the two drawbacks, the first may be overcome by drug-, blood-, or urine-level monitoring. Pharmacokinetic analysis
of the plasma profiles and/or the urinary excretion data should provide information on the rate and extent of drug absorption. However, no mechanistic insight can be gained.

Aside from the mouth wash procedure, buccal absorption can also be
studied using an absorption chamber [11,12]. In preliminary experiments
in humans, a polytef-disk, 0.5–1 cm in thickness and 3–4 cm in diameter,
was used with a central circular depression to receive a filter paper disk
soaked with a concentrated aqueous peptide solution with or without additives. This design limits the contact area between the device and the
buccal mucosa and, therefore, is able to more realistically simulate drug
absorption from adhesive dosage forms. The rim of the polytef disk is
firmly pressed onto the buccal mucosa, thus preventing excessive leakage
from the system. The effect of leakage and swallowing of the drug may be
minimized by constantly aspirating saliva during the experiment.

Another technique to study oral mucosal absorption is based on adhesive
drug-loaded patches made of an eroding or a noneroding polymer. The
absorption process may be followed by monitoring the amount of drug

remaining in the patch after certain time intervals or by pharmacokinetic
or pharmacodynamic analysis. It is obvious that the amount remaining in
the patch can reflect absorption only if the absorption step is not rate-
limiting. More details about different types of patches will be discussed
in Sec. V.

Animal Studies

Only a small number of absorption studies have been conducted on animals.
As stated by Pitha et al. [95], this may be due to the fact that sublingual/
buccal administration of drugs to animals is difficult and often produces
artifactual results. So far there has been no critical evaluation to determine which animal models could be representative for human absorption.
Rabbits, dogs, hamsters, and rats have been used. Rabbits were used to
study buccal insulin absorption [109,110]. Conscious dogs were used by
Nagai and co-workers [18,131] to investigate adhesive mucosal dosage
forms, as did Patterson et al. [90]. The animals were preconditioned by
receiving placebos for 4 consecutive days. Dogs were also used by other
workers [100,101,103,113,114]. It is interesting to note that, with dogs,
specially designed buccal cells may be used. Veillard et al. [100,103] and
Barsuhn and Ho [114] describe perfusion cells, whereas Ritschel et al.
[113] report on a closed-cell system. Such cells need mechanical attachment to the oral surface, by either cyanoacrylate cement [103] or spring
clips [113]. Experimental dosage forms with a central drug compartment
and a peripheral adhesive ring may be used instead [101]. The cheek
pouch of the male golden hamster was used first by Tanaka et al. [71] and
then by Ishida et al. [91] to study the buccal absorption of Na salicylate.
In this procedure, about 0.5 g of a gel containing the drug was administered through a syringe to the inside of the cheek pouch of anesthetized
animals that had been cleaned with a tampon. Both plasma levels and the
amount of drug remaining in the pouch were investigated. The hamster
cheek pouch was also recommended by Kurosake et al. [111] as a model of
the keratinized oral mucosa.

Merkle and co-workers [11,13–15] conducted animal studies using the
cheek pouch of anesthetized rats. Adhesive erodible type buccal patches
of 5 to 6 mm in diameter were administered to the buccal pouch of the
animals. Absorption was followed by monitoring the drug remaining in the
patch as well as by monitoring plasma and urinary drug levels. Blood was
constantly withdrawn by means of a peristaltic pump. To preserve the
limited total blood volume in the rat, suspensions of the animal's own blood
cells in the preceding blood samples were periodically recycled to the
animals. Additional rat studies were performed by Aungst and co-workers
[104,112]. These studies include experiments investigating possible differences in the local permeability of different sites of the oral rat mucosa,
e.g., buccal versus sublingual. Although the authors state that the volume of drug solution applied did not visibly leak from the site of administration, doubts concerning the significance of this observation may be
raised. This is due to the rather large volumes applied (ca. 50 µl),
which are not likely to be locally contained when taping the rats' mouths
closed. In order to avoid swallowing of excessive drug solution Aungst
and co-workers recommended ligation of the esophagus of the rats.

It has to be pointed out that the state of hydration of the buccal
mucosa of the rat seems to be lower than in humans [104,112]. Correlation

between rats and humans remained to be verified. Preliminary results
from Merkle and co-workers [11,13–15] suggest that such a correlation
may exist.

In Vitro Studies

Galey et al. [132] conducted in vitro permeation studies using a glass diffusion cell with the buccal tissue mounted between the two halves of the
cell which were filled with constantly stirred buffer solutions. The
mucosa was dissected from the cheek tissue of the dog immediately after
sacrifice and was stored in a special skin bank solution. Siegel et al.
[133] used the same procedure with the isolated lingual frenulum of dogs.
The same technique was applied by Veillard et al. [103]. In addition to
excised dog tissue excised hamster cheek pouch mucosa has also been
studied [103,108]. The authors were able to derive a meaningful physical
model to describe the simultaneous transport and metabolism in the tissue.
In vitro penetration of some β-blocking agents was studied using excised
porcine buccal mucosa [106]. So far in most cases no attention has been
paid to the viability of the excised tissues under in vitro conditions.
This aspect has been recently stressed by Banerjee and Robinson [118].
Electrophysiological characterization appears to be a valuable tool to indicate the viability of the tissue. To retain the viability of the tissues
gassing with O_2/CO_2 mixtures and addition of glucose appear to be appropriate [103].

Recently a novel cell culture approach was taken by Tavakoli-Saberi
and Audus [107]. Isolated hamster pouch epithelial cells grown in primary culture form a stratified multilayered tissue that retains the features
of the in vivo hamster pouch epithelium. This technique seems to be a
potentially useful in vitro tool to study simultaneous transport and
metabolism.

IV.   MECHANISM OF BUCCAL PEPTIDE
DRUG ABSORPTION

The mechanism of buccal/sublingual absorption is the subject of several
reviews and original papers [20,29,32,33,37,45,53,61,76,86]. None of the
literature, however, explicitly covers peptides or proteins. Models of
buccal absorption have been described by Ho and Higuchi [38], Wagner
and Sedman [44], and Schürmann and Turner [56]. The first is a
physical model, whereas the other two are based on compartmental analysis.
Unfortunately, the Ho-Higuchi model [38] as such may not be applicable
to buccal peptide absorption since its focus is on the effects of dissociation and chain length on the buccal absorption of a homologous series of
n-alkanoic acids. Nonetheless, the background information for construct-
ing a physical model for peptide absorption may be derived from a review
on the mathematical descriptions of effective membrane permeability coefficients under various absorption conditions [134]. Two of the factors
that must be considered are the conditions and pathways of absorption.

Regarding the first issue, one may assume that buccal peptide absorption occurs via a passive absorption mechanism. It is uncertain
whether the buccal mucosa offers a carrier-mediated transport mechanism,
as existed for dipeptides and tripeptides in the gastrointestinal tract [135].

A companion issue is whether pores are involved in buccal peptide absorption. These can be defined conceptually as regions of the membrane that are more polar than the lipid part and which may include the junctions between cells. While the presence of such pores has been questioned by Schürmann and Turner [59], pore formation is being proposed as one of the mechanisms by which bile salts enhance insulin absorption across the nasal mucosa [136].

In addition to the aqueous pore pathway the lipid pathway must also be considered. This is the dominant pathway for low molecular weight drugs, for which lipophilicity is the major determinant of buccal absorption. Whether this is also valid for peptides is as yet unclear. Studies of Hussain and co-workers [137] on the nasal absorption of various derivatives of L-tyrosine demonstrated that the absorption rate will not necessarily be enhanced by simply increasing lipophilicity. Charge effects appeared to be much more important.

V. DOSAGE FORMS FOR ORAL MUCOSAL PEPTIDE DELIVERY

A multitude of dosage forms are available or being investigated for peptide delivery from the oral mucosa. These include aqueous solutions, conventional buccal and sublingual tablets, adhesive tablets, adhesive gels, adhesive patches, and possibly devices mechanically attached to the teeth or implanted in the tooth enamel [138]. Depending on the pharmacodynamics of the peptides, various buccal dosage forms of different release rates may be designed. In some cases fast release of the peptide may be required; for other peptides, a sustained release may be desirable. To achieve sustained release a number of standard strategies are at hand, such as matrix-diffusion control, membrane control, or polymer-erosion control.

Delivery of a peptide drug to the oral mucosa by conventional means is limited to solutions and buccal or sublingual tablets and capsules. Solutions less than about 1 ml in volume may be filled into capsules with the liquid being released upon chewing. More common are erodible buccal or sublingual tablets or capsules. Their manufacture is based on well-known techniques using appropriate excipients and binders.

The possibility of excluding a major part of the drug from absorption by involuntary swallowing of a drug solution and by continuous dilution due to salivary flow is quite high. Conventional tablets and capsules, on the other hand, do not allow drinking and are, at least, a handicap for speaking. Administration time is therefore limited, so that controlled release is not within the scope of such formulations. However, these limitations can be overcome to some extent by adhesive dosage forms, such as adhesive tablets, gels, and patches.

The heart of these adhesive dosage forms is the adhesive polymer. As seen in Table 2, examples of adhesive polymers include water-soluble and insoluble hydrocolloid polymers from both the ionic and the nonionic type. A very popular polymer is the anionic polyacrylate-type hydrogel [148]. Drug release from soluble polymers is accomplished by gradual erosional dissolution of the polymers, whereas drug release from nonsoluble hydrogels follows Fickian or non-Fickian diffusion kinetics [149].

Close attachment of a dosage form to the buccal, sublingual, or gingival mucosa serves to retain the dosage form in the oral cavity while

Table 2   Mucosal Adhesive Polymers

Author(s)	Year	Ref.	Mucosal adhesive polymer
Machida and Nagai	1978	139	Hydroxypropycellulose
Machida et al.	1980	140	Combination of hydroxy propyl-cellulose and polyacrylic acid
Ishida et al.	1981	72	
Ishida et al.	1982	81	
Ishida et al.	1983	91	Polyacrylic acid
Bremecker et al.	1983	141	Polymethylmethacrylate
Bremecker	1983	142	
Bremecker et al.	1984	143	
Gurny et al.	1984	144	Na carboxymethylcellulose
Anders	1984	11	Methylcellulose, methyl-hydroxyethylcellulose, hydroxyethylcellulose, hydroxypropylcellulose, poly(vinylpyrrolidone), poly(vinyl alcohol), agarose
Anders and Merkle	1987	145	
Nagai	1986	131	Combination of hydroxy-propylcellulose and poly-acrylic acid, or polyethylene glycol
Park and Robinson	1986	146	Isoluble cross-linked poly-acrylic acid polymers (poly-carbophil type polymers)
Peppas and Buri	1986	147	Review on potentially bioad-hesive polymers of all above mentioned classes and others

simultaneously establishing an intimate contact with the absorption site.
Adhesion between the polymer and the underlying mucosa is established by
the entanglement of the polymer chains and the glycoprotein coat of the
mucosa [147]. A comprehensive review on the nature of mucosal interactions with polymers, its mechanisms, experimental methods to evaluate adhesion, and a survey of adhesive polymers has been published by Peppas
and Buri [147]. Fundamental aspects of adhesion to mucus glycoproteins
have been outlined by Robinson and co-workers [146,150,151]. Basic information on adhesion is presented by Manly [152] and by Anderson et al.
[153]. Bioadhesive stages, methods of studying bioadhesion as well as
the existing bioadhesive dosage forms are described by Duchêne et al.
[154].

A. Types of Adhesive Dosage Forms

Adhesive Tablets

Adhesive tablets for buccal or sublingual administration were suggested,
for instance, on the basis of eroding hydrocolloid/filler tablets. An example is given by Davis et al. [155] and Shor et al. [87]. Hydroxypropylcellulose (Synchron) and lactose were mixed with the drug, followed
by compression into tablets. As shown by scintigraphy, the preparation
remained in place for about 3 hr. This was due to the adhesion of the
gradually eroding polymer to the buccal mucosa. Erb et al. [79] evaluated adhesive buccal nitroglycerin tablets based on hydroxypropylcellulose, and observed pharmacodynamic effects for up to 5 hr. In principle
a multitude of other polymers can be used. A small portion of the patent
literature is reviewed by Chien [156], and a thorough review on bioadhesive polymers was given by Peppas and Buri [147]. Unlike conventional
tablets adhesive tablets allow drinking and speaking without major discomfort.

Adhesive Gels

Viscous adhesive gels may be used to deliver drugs to the buccal, sublingual, or gingival mucosa. Examples for local therapy have been given
by Ishida et al. [91,157] and by Bremecker and co-workers [141–143],
using polyacrylic acid and polymethylmethacrylate, respectively, as gel-
forming polymers. Systemic therapy with peptides has not yet been reported but appears feasible. As compared to solutions, gels can significantly prolong residence on the oral mucosa which may improve absorption and/or allow for some degree of sustained release of the active principle.

Adhesive Patches

Adhesive patches are a relatively new technology to pharmacy. The formation of adhesive patches may take a number of different approaches. A
collection of four different set-ups is given in Fig. 1. These range from
simple adhesive disks to laminated systems [13,72,81,155; and patents:
e.g., Offenl. DE 2908847 (1980); Offenl. DE 3237945 A 1 (1983)]. The adhesive polymer may work as the drug carrier itself [Case (a) and (d) in
Fig. 1], or it may act as an adhesive link between a drug loaded layer
and the mucosa as in Case (c). Also, a drug containing disk may be fixed
to the mucosa by using an adhesive shield as in Case (b).

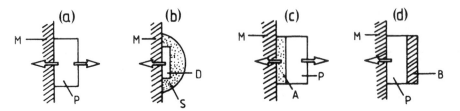

Fig. 1   Schematic view of four different types of adhesive patches for buccal peptide delivery.   Case (a) bi-directional release from adhesive patch by dissolution or diffusion.   Case (b) unidirectional release from patch embedded in adhesive shield.   Case (c) bi-directional release from laminated patch.   Case (d) unidirectional release from laminated patch. M, mucosa; P, polymer with peptide; D, drug depot; S, adhesive shield; A, adhesive layer; B, impermeable backing layer.

An important difference may be seen with respect to the directions available for drug release.   Cases (a) and (c) allow for a bi-directional release of the drug to both the underlying mucosa and the saliva bathing the oral cavity.   This may lead to a substantial loss of the drug when the saliva is swallowed.   Drug loss to the saliva may be decreased by using an adhesive protective shield [Case (b)] or a nonpermeable backing layer [Case (d)].   However, the main absorption site is now confined to the mucosal area in contact with the dosage form.   An additional advantage of these systems carrying a nonpermeable protective shield or layer is that the effect of the additives they contain can be restricted to the very site of application.   These are included to modify mucosal pH or permeability.

An interesting modification of the principal types of adhesive patches (Fig. 1) is given by Veillard et al. [103].   This patch consists of an impermeable backing layer, a matrix for drug release control, and an adhesive layer.

The size of adhesive patches varies, with $10-15 \text{ cm}^2$ being the upper limit.   Depending on the systems' size and shape, several administration sites are possible.   Patches near the maximum size limit can be administered only at the central position of the buccal mucosa; ellipsoid-shaped patches seem to be most suitable for this site.   Small patches may be attached to variable sites on the buccal, sublingual, or gingival mucosa. The sublingual and the gingival sites can only tolerate small patches with a maximum size of $1-2 \text{ cm}^2$.   Drug absorption from various sites of application is anticipated to be different due to intrinsic differences in drug release and dissolution characteristics and in membrane permeability characteristics [79,82,130,148].

The theoretical maximum application time span for adhesive mucosal dosage forms is in the order of several days [131].   In most cases, however, the maximum buccal residence time does not exceed several hours. This is due to the fact that buccal devices may possibly interfere with drinking, eating, and even talking.   Buccal patches for treatments over several hours must be perfectly formulated in order to improve patient compliance.   A smooth surface and good flexibility are prerequisites to prevent mechanical irritation or local discomfort.

B. Evaluation of Adhesive Dosage Forms

Adhesion Studies

In vivo studies. Merkle and co-workers [11,145] studied the effect of
the nature, amount, and processing of polymers on the duration of adhesion
of a patch on the human buccal mucosa. In a typical experiment, an adhesive patch is administered on one side of the buccal mucosa, whereas a
reference nonadhesive patch is administered on the other side. The adhesion time is defined as the time required for the subject to feel that
both patches behave identically. Obviously, this technique requires well-
trained and motivated subjects [11].

Using this technique, ionic polymers such as polyacrylic acid have
been found not to be as effective as nonionic polymers. This conflicts
with the findings that stress the superiority of ionic polymers. Among the
cellulose ethers investigated, hydroxyethyl- and hydroxypropylcellulose
afforded the best adhesion. In contrast, poly(vinylpyrrolidone) and
poly(vinyl alcohol) were less adhesive.

The influence of increasing viscosity grades on adhesion time varies
with the polymer. Hydroxyethylcellulose and poly(vinyl alcohol) show a
maximum function, whereas hydroxypropylcellulose shows a minimum function. Still poly(vinylpyrrolidone) affords an adhesion time which increases exponentailly with viscosity grade. The basis for this behavior is
as yet unclear.

Regardless of the viscosity, increasing the amounts of polymer per
patch results in an increase in the adhesion time. Patches with backing
layers that are permeable to water show shorter adhesion times than
those with impermeable backing layers. This is due to the slower erosion
of the hydrocolloid when one side of the patch is protected against water
uptake. Even the hydrocolloid film forming technique affects the adhesion
time [11].

Human adhesion studies on adhesive force may be also performed directly on the inner side of the lower lip as indicated by Hunt et al. [158].
This involves the measurement of the mechanical force required to remove
a patch or tablet of a defined size from the mucosa.

In vitro studies. In the literature a number of techniques have been
described to measure the adhesion force of bioadhesive polymers to natural
and synthetic membranes. These techniques are thoroughly reviewed by
Peppas and Buri [147]. In a method reported by Smart and Kellaway
[159] and by Smart et al. [160] a glass plate is coated with the adhesive
material to be tested. Then the force required to lift the plate out of a
mucus-containing beaker is measured. The design of the experiment is
analogous to the Wilhelmy-plate method to measure surface tension.

A microbalance system was described by Gurny et al. [144]: The system consisted of a specially designed cell which was mounted to a typical
tensile tester. The set-up was used mainly to measure the adhesiveness of
sublingual controlled-release systems. Another approach was described by
Robinson and co-workers [146,148]. They studied adhesion by measuring
the force required to separate a polymer specimen from freshly excised
rabbit stomach tissue as a model mucous membrane. These studies were
aimed at polymer/mucus interaction as a means to slow GI transit. A similar technique was employed by Robert et al. [161] using frog esophagus
tissue as a model membrane. A more sophisticated Instron-based technique

was applied by Peppas and co-workers [162–164]. This technique allows one to study the dynamic behavior of the adhesive bond, including stress relaxation phenomena, viscoelasticity, and fracture analysis.

A specific method to study the bond strength between adhesive polymers and certain functional groups of the mucus was presented by Park and Robinson [165] using a fluorescence probe technique. The result is restricted to molecular interactions, but no information can be derived for the adhesive force between polymer and mucus on a macroscopic level. The technique of Peppas and Buri [147] utilizes the movement of a bio-adhesive particle across a pipe coated with artificial mucin or natural mucus. The bioadhesive behavior is assumed to relate to the motion of the particle caused by a stream of air or of a second viscoelastic solution flowing over the particle.

Although both of the techniques just described provide valuable information on polymer/mucus interactions, more practical results can only be expected from directly measuring the adhesive forces between a natural or a synthetic membrane and an adhesive dosage form. Preliminary results from this laboratory [130] show that the maximum force may be as high as 100 g/cm^2. The measurements were carried out using a standard electronic balance. Adhesion force curves were recorded on-line to a suitable computer. The polymers covered by this study include initially dry hydrocolloids laminated to an impermeable backing layer.

Release Studies

The results on patches based on water-soluble hydrocolloids and obtained in human subjects will be described. A full experimental description of the work has been reported by Anders and Merkle [145]. The experimental variables were type of polymers used, viscosity grade of the polymers, and amount of polymer per unit square of the patch. So far no local irritation or other signs of damage to the oral mucosa whatsoever have been observed with these patches.

The polymers used in the studies were hydroxyethylcellulose (HEC; Natrosol 250L and Natrosol 250G, from Hercules), and poly(vinyl-pyrrolidone) (PVP; Kollidon 30 and Kollidon 90, from BASF). The release was studied in humans using sub-therapeutic doses of protirelin (TRH) or Na salicylate. The release patterns observed are shown in Fig. 2. All experiments were run on the same subject. As can be seen the release of the marker can be controlled over a wide range. An increase of the viscosity grade sustains the release of the marker from the patch. If the amount of polymer per unit square is increased the release of the marker is slowed down. With low amounts of polymer per unit square the amount of marker remaining in the patch appears to decrease exponentially, whereas higher amounts approach an almost linear release profile, as is the case with PVP 90. Studies currently performed in this laboratory at even higher polymer amounts show that the release profiles are indeed essentially linear. We also found that the rate of polymer erosion is the rate-determining step in drug release in vivo [130].

The efficiency of the two polymers in sustaining drug release is different. The high viscosity HEC is three times more effective than the high-viscosity PVP.

The data in Fig. 2 supply some additional information regarding the intra-subject variability of drug release from this type of buccal patch.

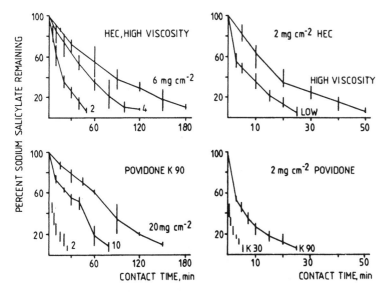

Fig. 2   Effect of polymers, viscosity grades, and amounts of polymer per unit square on the buccal release of Na salicylate from adhesive polymeric patches.   Three independent experiments were run on each of the contact times.   Bars indicate total range of values obtained.

The release profiles are obtained from three independent runs in the same subject, each run being represented by a single data point. The range of variation seen in Fig. 2 represents the within-subject variability of these patches.   Clearly, buccal release from such polymeric laminates appears to be rather reproducible in humans.

Data regarding the inter-subject variation are displayed in Figs. 3 and 4. Fig. 3 shows the fraction of protirelin remaining in the patch after staying in contact with the buccal mucosa for 30 min in 4 to 11 subjects. Three different patch formulations were evaluated. Depending on the amount of polymer employed and the viscosity grade of the polymer, slow, medium, or fast release of the drug could be obtained. This indicates that the release rates of adhesive patches can be individually tailored to meet the specific requirements of a certain peptide or therapy. The variations found between subjects were substantial, but tolerable.

More information on the inter-subject variations of drug release is given in Fig. 4. Here the release of a marker was studied in 5 subjects following complete release profiles. Each data point of the graph represents an independent run. Again, there is a substantial amount of variability in the data. It is more pronounced than the intra-subject variability, but stays within a normally tolerable range. The variations may be explained by individual habits in saliva flow or jaw movements of the subjects, resulting in different dissolution rates of the hydrocolloid polymer. Studies on the optimum site for application are currently under way. From the standpoint of reproducibility of drug release from such dosage forms, it appears that the area facing the 6th or 7th molar of the upper jaw is to be preferred.

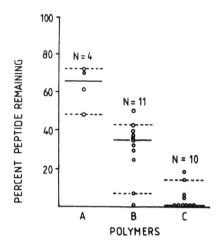

Fig. 3   Percent protirelin remaining in buccal patch after 30 min of
buccal contact in human subjects. Protirelin content of patch: 1–2 mg/
cm^2.   HEC polymer content: (A) 6.2 mg/cm^2 high viscosity; (B) 2.2 mg/
cm^2 high voscosity, (C) 2.1 mg/cm^2 low viscosity.   Solid lines denote the
5% confidence limits.

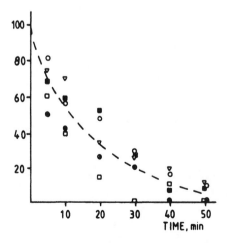

Fig. 4   Release profile of Na salicylate (1 mg/cm^2) from buccal patch
made of 2.1 mg/cm^2 high viscosity HEC in five subjects (symbols indicating subjects).

VI. BUCCAL ABSORPTION OF SELECTED PEPTIDES

A. Human Studies

Preliminary studies in humans [12] using the polytef disk approach (see
Sec. III) demonstrated that protirelin (MW = 362) was readily absorbed
across the buccal mucosa, as shown by significant increases of pituitary
thyrotropin and prolactin. Since it is known to be absorbed to a small
extent from the GI tract, it is necessary to correct for the GI contribution
following involuntary swallowing of the saliva. Buccal absorption was
found to be the major route since the same thyrotropin and prolactin stimulation patterns were obtained when the saliva was continuously aspirated
during the experiment [12].

   The same pharmacodynamic response was obtained when the study was
repeated with buccal adhesive patches [11,13,94]. The result is shown in
Fig. 5. Between 48 and 72% of the peptide were recovered in the patches
and therefore were not available for absorption within the 30 min contact
time. In order to increase the rate of peptide release from the patch the
polymer fraction of the patch was decreased by two-thirds with a concomitant lowering of the protirelin dose from 20 to 10 mg. Both high- and
low-viscosity grade HEC were used as the polymer. The study was run
with healthy female and male subjects. The result of the test is demonstrated in Fig. 6. As expected females showed higher thyrotropin and
prolactin levels than males. The fraction of peptide remaining in the patch
after 30 min mucosal contact was in the range of 0-15% for the low-viscosity
grade polymer and 0-50% for the high-viscosity grade polymer. In spite
of the obviously higher release rate of the low-viscosity polymer patch the
thyrotropin and prolactin stimulation was not statistically different between
the polymers used. But in the case of prolactin found in female subjects
and for the high-viscosity HEC only, there was a clear correlation between
the pharmacodynamic response and the amount of peptide released after
30 min, as indicated by a statistically significant Spearman rank order correlation coefficient (P = 0.05). This is one of several occasions in this
work where a correlation of the peptide release pattern and the pharmacodynamic response upon application was detected.

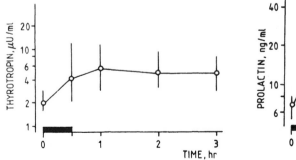 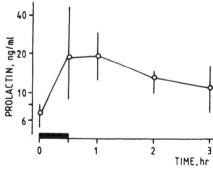

Fig. 5   Plasma thyrotropin and prolactin concentrations in four human
subjects upon buccal application of adhesive protirelin patches. Dose:
20 mg protirelin in high-viscosity HEC-patch; bars indicate time of mucosal
contact.

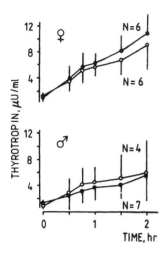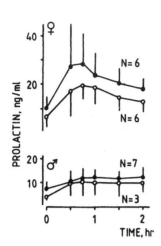

Fig. 6   Plasma thyrotropin and prolactin concentrations in female and male human subjects upon buccal application of adhesive protirelin patches. Dose: 10 mg protirelin in high-viscosity (●) and low-viscosity (○) HEC-patches.

The efficiency of buccal protirelin absorption has been compared to IV, nasal, and peroral administration [94]. With respect to the protirelin dose required to achieve significant thyrotropin and prolactin incremental increases, nasal application was about 10 times more efficient than the buccal route. The effective buccal dose, however, was only about one quarter of the peroral dose. No side effects of the treatment were noticed with the buccal patches, whereas minor effects were caused by the nasal application. Nausea, flushing, and urinary urgency were observed in the majority of the subjects following IV administration. This is another demonstration of the much more moderate pituitary stimulation kinetics of buccal protirelin, as compared to the IV and nasal route. It remains an open question whether this aspect might be a relevant factor for drug formulation of other peptides as well.

B.   Animal Studies

The first attempt to study buccal peptide absorption in animals was made by Nagai and co-workers [81] using beagle dogs. Prior to the experiment the dogs were accustomed to the procedure in order to exclude the effect of stress. This was achieved by placing a placebo tablet on the oral mucosa for 6 hr for 4 days. The results showed that insulin can be absorbed via the buccal mucosa only when Na glycocholate was included as an absorption promotor. Compared to intramuscular injection, only 0.5% of the dose was absorbed.

   Further animal studies were conducted in this laboratory using the rat as a model [11,13-15]. The purpose of this technique is threefold: (1) to determine the buccal permeability to peptides, (2) to optimize possible formulation effects with adhesive buccal peptide patches, and (3) to evaluate potential absorption promotors for oral mucosal transport of peptides.

The stability of the rat as a model to study buccal peptide absorption has recently been assessed by the authors. The peptides so far investigated were protirelin and buserelin. The buccal patches administered to the rats were similar to the two-ply laminate type tested in human studies, except for the size, which was 5–7 mm in diameter. The patches could be easily attached to the mucosa of the buccal pouch of the anesthetized rat. Upon continuous blood sampling, the absorption process was followed by monitoring peptide plasma levels or the pharmacodynamic response.

Protirelin absorption was studied both with and without triiodothyronine pretreatment. Due to negative feedback, triiodothyronine pretreatment led to a strong depression of endogenous protirelin and thyrotropin. The result without such pretreatment is shown in Fig. 7. As can be seen protirelin was readily absorbed via the buccal mucosa, since plasma thyrotropin levels were greatly increased above the physiological level. More experiments were run with triiodothyronine pretreated animals. This set-up was regarded to represent the more reliable and robust test method as compared to the method without pretreatment. Erratic thyrotropin level fluctuations in the test animals under stress and/or anaesthesia may thus be excluded. Again, there was excellent experimental proof that protirelin is bucally absorbed. It is interesting to note that there was a statistically significant rank order correlation (Spearman, P = 0.05) between the individual amounts of drug released to the mucosa of the rats during the experiment and the corresponding pharmacodynamic response in terms of the maximum thyrotropin plasma levels observed (Table 3). Thus, the dosage form used must be further optimized for attainment of faster and more reproducible peptide release rates in order to achieve high pharmacodynamic response at low peptide doses. So far, the vast inter-animal variation in drug release cannot be explained. In humans we were able to show that the release of peptides from such patches is far more reproducible [11,145]. The design of appropriate formulations and experimental assessments by in vitro dissolution studies should help to find a more suitable and reproducible release behavior.

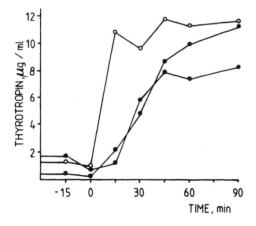

Fig. 7   Thyrotropin stimulation in rats upon buccal application of adhesive buccal patches containing 931 µg protirelin, in high-viscosity HEC patch 5.5 mm in diameter (○) and 6.7 mm in diameter (●).

Table 3   Spearman Rank Order Correlation[a] of Protirelin Release from
Patch and Maximum Pituitary Thyrotropin in Rats

Protirelin, released		Maximum thyrotropin		Patch diameter (mm)
Percent	Rank	µg/ml	Rank	
97.2	1	7.44	1	5.5
74.9	2	5.79	2	5.5
69.7	3	3.50	5	5.5
34.1	4	4.65	4	6.7
32.9	5	5.57	3	6.7
3.2	6	2.03	6	6.7

[a]Spearman rank order correlation coefficient r_s = 0.771 (P = 0.05).

A comparison of the pharmacodynamic effects upon buccal protirelin
and IV protirelin in rats is given in Fig. 8. The graph shows that the
buccal dose required to achieve an equivalent pharmacodynamic response is
about 200 times higher than that for IV injection of the peptide. It must,
however, be kept in mind that this relation does not allow calculation of
the fraction of the peptide absorbed. The efficiency of peptide absorption
with respect to its pharmacodynamic response is not only a matter of the
total dose absorbed but is also influenced by the rate at which the peptide
appears at the target organ.

An IV bolus injection of protirelin is followed by a sharp incremental
rise of the drug's concentration at the target site, i.e., the pituitary
gland, resulting in a pronounced pharmacodynamic response. In case of
the buccal route, however, the mucosal membrane transport and distribution

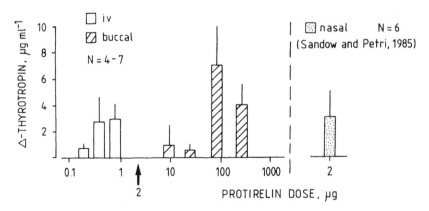

Fig. 8   Dose effects of pharmacodynamics of IV, buccal, and nasal protirelin in rats (nasal data from Sandow and Petri [32]).

impedes the incremental rise of protirelin at the target, and, therefore, buccal doses have to be much higher than IV doses in order to achieve the same pharmacodynamic response. It would have been an erroneous conclusion to regard the data in Fig. 8 as a means to derive first estimations of the total fraction of protirelin buccally absorbed. But it is safe to conclude that the pharmacodynamic efficiency of buccal protirelin was around 200 times less than that of IV protirelin.

Another interesting relation can be observed by comparing the IV and buccal data in Fig. 8 with the results of Sandow and Petri [166] on nasal protirelin absorption. Using the same animal model a 2 μg nasal dose gave a peak thyrotropin concentration of around 4 μg/ml. Based on the pharmacodynamic effects observed, the nasal dose, therefore, had to be about 5 times higher than the IV dose, whereas the buccal dose required was 40 times the nasal one. This result is in close agreement with the results found in human subjects [94] and does indeed strengthen our confidence in the rat model for future studies.

The maximum thyrotropin responses found after buccal protirelin, although requiring higher doses, were about twice as high as those found after IV or nasal protirelin administration. This may be the result of a more sustained type pituitary stimulation in the course of buccal absorption, as compared to the IV and nasal application. At this time no further evidence can be given to support this hypothesis. For a complete understanding more should be known about buccal peptide absorption kinetics and its effects on pituitary pharmacodynamics in man and in rats. This may have interesting therapeutic implications in the future.

In addition to the absorption studies with protirelin, these authors also conducted preliminary rat absorption studies with buserelin, an LHRH agonist [11,13–15]. The same experimental design as with protirelin was used. The outcome of an absorption study with adhesive patches loaded with 0.7 mg of the peptide is given in Fig. 9. Only a minute amount of buserelin was absorbed as indicated by the low plasma buserelin recovered and the resulting lutropin levels from pituitary stimulation. In this composition buccal buserelin patches therefore do not appear to be suitable for

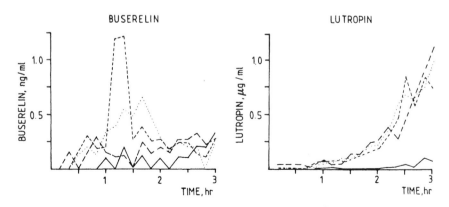

Fig. 9   Plasma buserelin and lutropin levels following buccal buserelin in rats.

Table 4   Rank Order Correlations of Buserelin Amount Released from
Patch and Renally Excreted Buserelin, and Maximum Pituitary Lutropin,
Respectively, in Rats

Buserelin, released		Buserelin,[a] renally excreted		Maximum lutropin[b]	
Percent	Rank	ng	Rank	ng/ml	Rank
49.2	1	12.79	1	1160.5	1
47.4	2	3.09	3	1001.3	2
27.1	3	3.87	2	862.1	3
13.0	4	0.83	4	99.8	4

[a]Spearman correlation coefficient r_s = 0.800 (P = 0.05).
[b]Spearman correlation coefficient r_s = 1.000 (P < 0.05).

efficient buserelin delivery to the systemic circulation. It remains to be
determined whether effective absorption promotors may overcome the
problem.

It is, however, interesting to note that there was a rather constant
lutropin stimulation pattern with three animals, whereas one animal failed
to respond. Detailed inspection of the percentage of buserelin released
during the experiment, and the amount renally excreted and the maximum
lutropin concentration achieved, respectively, for the four rats investigated showed that there is a significant rank order correlation (Spearman,
P = 0.05) between these variables (Table 4). This implies a direct relationship of the amount and/or rate of peptide released from the polymeric
patch and the resulting pharmacokinetics and pharmacodynamics.

As is the case with protirelin, the release of buserelin from the polymeric patches could be improved. The patches should be designed in such
a way that the peptide is released fully and reproducibly after application
of the patch onto the buccal mucosa. It remains to be clarified what the
effect of an optimized patch onto buccal peptide efficiency and efficacy
would be. But further optimization of such patches is currently being
sought in this laboratory.

VII.   EFFECTIVENESS AND MECHANISMS OF ADJUVANTS FOR BUCCAL PEPTIDE ABSORPTION

Of the many factors that can affect buccal peptide absorption, only the
influence of molecular weight has been studied. Protirelin (MW = 362) was
readily absorbed via the buccal mucosa, as shown by significant increases
of pituitary thyrotropin and prolactin, whereas calcitonin (MW = 3500) was
not absorbed [89]. This is possibly due to the large difference in the
molecular weight of the peptides. Nonetheless, the relationship between
peptide molecular weight and membrane permeability is not clear. For
instance, buserelin (MW = 1239) cannot pass the oral mucosa [166], whereas

oxytocin (MW 1007) with about the same molecular weight can [8–10]. The vasopressin analogs (DDAVP and DVDAVP) of similar molecular weight were also absorbed [17]. Even insulin (MW = 5734) was observed to penetrate the oral mucosa to some extent [16]. This information is based on the observation that after administration of a sublingual tablet that also contained a vasodilator the blood glucose concentration fell to lower levels within 10 min of the administration. Collectively, these results suggest that parameters in addition to molecular weight, such as conformation, polarity, dissociation, and enzymatic and chemical stability, also influence the extent of buccal peptide absorption.

In order to improve the permeability of the buccal membrane and the efficacy of buccal peptide administration, highly effective, biocompatible and nontoxic absorption promotors are required. Only a few of these additives have been tested for buccal absorption enhancement. Nagai and co-workers [72] found that Na glycocholate was effective in enhancing the buccal absorption of insulin in beagle dogs to some degree. Earle [16] used a nonspecified vasodilator for absorption enhancement of insulin in human subjects. Interesting data were also reported by Nakada et al. [115] for the enhancement of buccal calcitonin absorption in rats. The study includes bile acids, quillajasaponin, sodium lauryl sulfate, sodium myristate, and sugar esters. Kurosaki et al. [122] reported the effect of Azone on the salicylic acid permeability of the hamster cheek pouch as a keratinized tissue.

In this laboratory two other potential absorption promotors were investigated [11,13–15], citric acid and Na-5-methoxysalicylate. These compounds were previously reported to increase the vaginal absorption of leuprolide [167] and the rectal absorption of insulin [168], respectively. The opening of mucosal pores, stabilized with calcium ions, was reported to be the most probable mechanism. We evaluated both compounds using patches of both high and low viscosity HEC as the polymeric carrier.

In the case of citric acid, an approximate 100% increase of the means of maximum thyrotropin concentrations was found with both high- and low-viscosity HEC polymers. Na 5-methoxysalicylate gave a similar increase with the low-viscosity polymer only, and not with the high-viscosity type. None of these effects, however, was shown to be statistically significant (t-test of means of maximum concentrations). This appears to be due to the small number of animals available in the study and the rather high inter-animal variations of the pharmacodynamic response. In addition, the protirelin dose used for this study, 84 µg, was a dose that already yielded a maximum pharmacodynamic response (see Fig. 8). Further effects by addition of absorption promotors, therefore, are not likely to be detected. Future studies with more animals and a more sensitive experimental design, i.e., testing at lower peptide doses, will clarify these points and help to further evaluate possible promotors.

VIII.   CONCLUSIONS

As stratified epithelia the mucosae of the oral cavity appear to be much less permeable than the nasal mucosa. However, when it comes to accessibility and robustness the buccal and sublingual mucosae have clear advantages over the other mucosal sites. Therefore, the oral epithelia continue to represent an interesting possibility for special drugs. But safe and

effective absorption enhancers have to be developed to overcome the low
permeability of such tissues.

REFERENCES

1. B. W. Barry, *The Transdermal Route for Delivery of Peptides and Proteins*, Proceedings of NATO Advanced Research Workshop: Advanced Drug Delivery Systems for Peptides and Proteins, Copenhagen (S. S. Davis, L. Illum, and E. Tomlinson, eds.), Plenum Press, New York (1986), p. 265.
2. R. R. Burnette and D. Marreiro, *J. Pharm. Sci.*, 75:738 (1986).
3. K. S. E. Su and K. M. Campanale, *Transnasal Systemic Medications* (Y. W. Chien, ed.), Elsevier, Amsterdam (1985), p. 139.
4. K. S. E. Su, *Pharm. Int.*, 7:8 (1986).
5. H. J. M. Van de Donk, A. G. M. Van den Heuvel, J. Zuidema, and F. W. H. M. Merkus, *Rhinology*, 20:127 (1982).
6. R. E. Stratford and V. H. L. Lee, *J. Pharm. Sci.*, 74:731 (1985).
7. V. H. L. Lee, *Enzymatic Barrier to Peptide and Protein Absorption and Use of Penetration Enhancers to Modify Absorption*, Proceedings of NATO Advanced Research Workshop: Advanced Drug Delivery Systems for Peptides and Proteins, Copenhagen (S. S. Davis, L. Illum, and E. Tomlinson, eds.), Plenum Press, New York (1986), p. 87.
8. H. J. Wespi and H. P. Rehsteiner, *Gynaecologia*, 162:414 (1966).
9. P. Bergsjö and H. Jenssen, *J. Obstet. Gynaec. Brit. Cwlth.*, 76:131 (1969).
10. S. Sjöstedt, *Acta Obstet. Gynaec. Scand.*, 48 (Suppl. 7):3 (1969).
11. R. Anders, Ph.D. Thesis, *Selbsthaftende Polymerfilme zur bukkalen Applikation von Peptiden*, Universität Bonn, Bonn (1984).
12. R. Anders, H. P. Merkle, W. Schurr, and R. Ziegler, *J. Pharm. Sci.*, 72:1481 (1983).
13. H. P. Merkle, R. Anders, J. Sandow, and W. Schurr, *Self-Adhesive Patches for Buccal Delivery of Peptides*, Proceedings of the International Symposium on Controlled Release of Bioactive Materials, CH-Geneva (N. A. Pappas and R. J. Haluska, eds.), Controlled Release Society, Lincolnshire, Illinois, 12:85 (1985).
14. H. P. Merkle, R. Anders, and J. Sandow, Abstract: *Buccal Absorption of Peptides in Rats*, Proceedings of 32nd Annual Congress of International Association for Pharmaceutical Technology (APV), NL-Leiden (1986), p. 57.
15. H. P. Merkle, R. Anders, J. Sandow, and W. Schurr, *Drug Delivery of Peptides: The Buccal Route*, Proceedings of NATO Advanced Research Workshop: Advanced Drug Delivery Systems for Peptides and Proteins, Copenhagen (S. S. Davis, L. Illum, and E. Tomlinson, eds.), Plenum Press, New York (1986), p. 159.
16. M. P. Earle, *Isr. J. Med. Sci.*, 8:899 (1972).
17. F. Laczi, G. Mezei, J. Julesz, and F. A. Laszlo, *Int. J. Clin. Pharmacol. Ther. Tox.*, 18:63 (1980).
18. M. Ishida, Y. Machida, N. Nambu, and T. Nagai, *Chem. Pharm. Bull.*, 29:810 (1981).
19. M. Gibaldi and J. L. Kanig, *J. Oral Therap. Pharmacol.*, 1:440 (1965).

20. A. H. Beckett and R. D. Hossie, *Handbook of Experimental Pharmacology*, Vol. 28, 1: Concepts in Biochemical Pharmacology (B. B. Brodie, J. R. Gillette, and H. S. Ackerman, eds.), Springer, Berlin (1971), p. 25.

21. A. C. Moffat, *Topics in Medicinal Chemistry*, Vol. 4: Absorption Phenomena (J. L. Rabinowitz and R. M. Myerson, eds.), Wiley-Interscience, New York (1972), p. 1.

22. N. A. Muzaffar, *Pak. J. Sci.*, *31*:21 (1979).

23. A. G. DeBoer, L. G. J. DeLeede, and D. D. Breimer, *Br. J. Anaesth.*, *56*:69 (1984).

24. J. Karmel, *Dent. Arch. Klin. Med.*, *12*:466 (1873).

25. M. Tanaka, N. Yanagibashi, H. Bukuda, and T. Nagai, *Chem. Pharm. Bull.*, *28*:1056 (1980).

26. R. L. Brunton, *The Gaulstonian Lectures Delivered Before the Royal College of Surgeons*, London, MacMillan and Co. (1977); Ref. by Jarrett [99] as: T. L. Bruton, *Youlstonian Lectures Delivered Before the Royal College of Physicians 1877*, MacMillan and Co., London (1880).

27. A. H. Beckett and E. J. Triggs, *J. Pharm. Pharmacol.*, *19 (Suppl)*: 31S (1967).

28. A. H. Beckett, R. N. Boyes, and E. J. Triggs, *J. Pharm. Pharmacol.*, *20*:92 (1968).

29. A. H. Beckett and A. C. Moffat, *J. Pharm. Pharmacol.*, *20 (Suppl)*: 239S (1968).

30. M. H. Bickel and H. J. Weder, *J. Pharm. Pharmacol.*, *21*:160 (1969).

31. A. H. Beckett and A. C. Moffat, *J. Pharm. Pharmacol.*, *21 (Suppl)*: 139S (1969).

32. A. H. Beckett and A. C. Moffat, *J. Pharm. Pharmacol.*, *21 (Suppl)*: 144 (1969).

33. A. H. Beckett and A. C. Moffat, *J. Pharm. Pharmacol.*, *22*:15 (1970).

34. M. J. Taraszka, *J. Pharm. Sci.*, *59*:873 (1970).

35. D. Stojanovic and K. Milivojevic, *Iugoslav. Physiol. Pharmacol. Acta*, *6*:603 (1970).

36. A. H. Beckett and A. C. Moffat, *J. Pharm. Pharmacol.*, *23*:15 (1971).

37. E. J. Lien, R. T. Koda, and G. L. George, *Drug Intel. Clin. Pharm.*, *5*:38 (1971).

38. N. H. F. Ho and W. I. Higuchi, *J. Pharm. Sci.*, *60*:537 (1971).

39. J. C. Dearden and E. Tomlinson, *J. Pharm. Pharmacol.*, *23 (Suppl)*: 73S (1971).

40. J. C. Dearden and E. Tomlinson, *J. Pharm. Pharmacol.*, *23 (Suppl)*: 68S (1971).

41. M. G. Bogaert and M. T. Rosseel, *J. Pharm. Pharmacol.*, *24*:737 (1972).

42. K. R. M. Vora, W. I. Higuchi, and N. F. H. Ho, *J. Pharm. Sci.*, *61*: 1785 (1972).

43. A. Bye, A. S. E. Fowle, and C. Pullin, *Advan. Antimicrobial. Antineoplastic Chemother.*, *Proc. Int. Congr. Chemother.*, 7th meeting 1971, Vol. 1 (M. Hejzar, ed.), University Park Press, Baltimore, Maryland (1972), p. 27.

44. J. G. Wagner and A. L. Sedman, *J. Pharmacokin. Biopharm.*, *1*:23 (1973).

45. D. Mikhailova, R. Nacheva, and R. Ovcharov, *Pharmazie*, *28*:208 (1973).

46. J. M. Sutherland, W. D. Hooper, M. J. Eadie, and J. H. Tyrer, *J. Neurol. Neurosurg. Psychiatry*, *37*:1116 (1974).

47. W. Meyer, C. M. Kaye, and P. Turner, *Eur. J. Clin. Pharmacol.*, *7*: 287 (1974).

48. A. H. Beckett, O. Grech, and D. Mikhailova, *J. Pharm. Pharmacol.*, *27 (Suppl)*:67P (1975).

49. C. M. Kaye and A. D. Long, *Br. J. Clin. Pharmacol.*, *3*:196 (1976).

50. M. A. Kiddie and C. M. Kaye, *Br. J. Clin. Pharmacol.*, *3*:320 (1976).

51. S. I. Ankier and C. M. Kaye, *Br. J. Clin. Pharmacol.*, *3*:672 (1976).

52. A. S. Manning and D. F. Evered, *Clin. Sci. Mol. Med.*, *51*:127 (1976).

53. B. C. Lippold and G. F. Schneider, *Arzneim.-Forsch./Drug Res.*, *26*: (1976).

54. J. M. McMullan, A. S. Manning, and D. F. Evered, *Biochem. Soc. Trans.*, *5*:129 (1977).

55. A. Odumosu and C. W. Wilson, *Int. J. Vitam. Nutr. Res.*, *47*:135 (1977).

56. W. Schürmann and P. Turner, *Br. J. Clin. Pharmacol.*, *4*:655P (1977).

57. T. Past, Z. Tapsonyi, L. Bodis, and T. Javor, *Zentralbl. Pharm., Pharmakother. Laboratoriumsdiagn.*, *116*:365 (1977).

58. R. E. Kates, *J. Med.* (Westbury, NY), *8*:393 (1977).

59. W. Schürmann and P. Turner, *J. Pharm. Pharmacol.*, *30*:137 (1978).

60. D. J. Temple and K. R. Schesmer, *Arch. Pharm.* (Weinheim), *311*:485 (1978).

61. T. Past, Z. Tapsonyi, L. Nagy, and T. Javor, *Int. J. Clin. Pharmacol. Biopharm.*, *16*:413 (1978).

62. S. A. Sprake and D. F. Evered, *J. Pharm. Pharmacol.*, *31*:113 (1979).

63. F. Sadoogh-Abasian and D. F. Evered, *Brit. J. Nutr.*, *42*:15 (1979).

64. J. A. Henry and K. Ohashi, *Br. J. Clin. Pharmacol.*, *8*:406P (1979).

65. T. Kaspi, A. Blackett, and P. Turner, *Br. J. Clin. Pharmacol.*, *8*: 412P (1979).

66. B. J. Davis and P. Turner, *Br. J. Clin. Pharmacol.*, *8*:405P (1979).

67. K. Chan, *J. Pharm. Pharmacol.*, *31*:672 (1979).

68. J. A. Henry, K. Ohashi, J. Wadsworth, and P. Turner, *Br. J. Clin. Pharmacol.*, *10*:61 (1980).

69. D. F. Evered, F. Sadoog-Abasian, and P. D. Patel, *Life Sci.*, *27*:1649 (1980).

70. K. Chan, J. K. C. Li, N. Baber, E. Ohnhaus, M. Orme, and R. G. Sibeon, *J. Pharm. Pharmacol.*, *32 (Suppl)*:42P (1980).

71. M. Tanaka, N. Yanagibashi, H. Fukuda, and T. Nagai, *Chem. Pharm. Bull.*, *28*:1056 (1980).

72. M. Ishida, Y. Machida, N. Nambu, and T. Nagai, *Chem. Pharm. Bull.*, *29*:810 (1981).

73. D. F. Evered and R. M. Offer, *Biochem. Soc. Trans.*, *9*:133 (1981).

74. D. F. Evered and J. V. Vadgama, *Biochem. Soc. Trans.*, *9*:132 (1981).

75. K. A. Muhiddin and A. Johnston, *Br. J. Clin. Pharmacol.*, *12*:283 (1981).
76. T. Past, Z. Tapsonyi, L. Nagy, T. Horvath, and T. Javor, Drugs Biochem. Metab., Sci. Mater. Pap. Colloq. (I. Klebovich, I. Laszlovszky, and B. Rosdy, eds.), Hung. Chem. Soc., Biochem. Sect., Budapest, p. 155 (1981).
77. D. F. Evered, T. D. Salman, and Y. M. Trought, *Biochem. Soc. Trans.*, *10*:25 (1982).
78. J. Milovac, M. Mohar, and Z. Kopitar, *Farm. Vestn.* (Ljubljana), *33*:21 (1982).
79. R. J. Erb, Controlled release nitroglycerin in buccal and oral form, in *Advanced Pharmacotherapy*, Vol. 1 (W.-D. Bussmann, R.-R. Dries, and W. Wagner, eds.), Karger, Basel (1982), p. 35.
80. D. J. Greenblatt, M. Divoli, J. S. Harmatz, and R. I. Shader, *J. Pharm. Sci.*, *71*:248 (1982).
81. M. Ishida, N. Nambu, and T. Nagai, *Chem. Pharm. Bull.*, *30*:980 (1982).
82. A. Windorfer, Controlled release nitroglycerin in buccal and oral form, in *Advanced Pharmacotherapy*, Vol. 1 (W.-D. Bussmann, R.-R. Dries, and W. Wagner, eds.), Karger, Basel (1982), p. 44.
83. J. Abrams, *Am. Heart J.*, *105*:848 (1983); J. Abrams, Controlled release nitroglycerin in buccal and oral form, in *Advanced Pharmacotherapy*, Vol. 1 (W.-D. Bussmann, R.-R. Dries, and W. Wagner, eds.), Karger, Basel (1982), p. 4.
84. D. F. Evered and C. Mallett, *Life Sci.*, *32*:1355 (1983).
85. A. Blackett, N. Brion, and P. Turner, *Int. J. Clin. Pharmacol. Res.*, *3*:5 (1983).
86. T. Past, Z. Tapsonyi, L. Nagy, T. Horvath, and T. Javor, *Zentralbl. Pharm., Pharmakother. Laboratoriumskiagn.*, *122*:693 (1983).
87. J. M. Schor, S. S. Davis, A. Nigalaye, and S. Bolton, *Drug Dev. Ind. Pharm.*, *9*:1359 (1983).
88. B. G. Woodcock, O. Asthana, M. Wenchel, and N. Rietbrock, *Methods Find. Exp. Clin. Pharmacol.*, *5*:537 (1983).
89. R. Anders, H. P. Merkle, W. Schurr, and R. Ziegler, unpublished results.
90. E. Patterson, P. Stetson, and B. R. Lucchesi, *Pharmacology*, *27*:192 (1983).
91. M. Ishida, N. Nambu, and T. Nagai, *Chem. Pharm. Bull.*, *31*:4561 (1983).
92. O. P. Asthana, B. G. Woodcock, M. Wenchel, K. H. Frömming, L. Schwabe, and N. Rietbrock, *Arzneim.-Forsch./Drug Research*, *34*:493 (1984).
93. P. Marty, F. Rodgriguez, P. Poitou, and R. Rouffiac, *Pharm. Acta Helv.*, *59*:16 (1984).
94. W. Schurr, B. Knoll, R. Ziegler, R. Anders, and H. P. Merkle, *J. Endocrin. Invest.*, *8*:41 (1985).
95. J. Pitha, S. M. Harman, and M. E. Michel, *J. Pharm. Sci.*, *75*:165 (1986).
96. M. A. Hussain, B. J. Aungst, and E. Shefter, *J. Pharm. Sci.*, *75*: 218 (1986).

97. M. A. Randhava, A. N. Blackett, and P. Turner, *J. Pharm. Pharmacol.*, *38*:629 (1986).

98. G. S. M. J. E. Duchateau, J. Zuidema, and F. W. H. M. Merkus, *Pharm. Res.*, *3*:108 (1986).

99. V. H. L. Lee, S. Dodda Kashi, R. M. Patel, E. Hayakawa, and K. Inagaki, *Proceed. Intern. Control. Rel. Bioact. Mater.*, *14*:23 (1987).

100. M. M. Veillard, M. A. Longer, I. A. Tucker, and J. R. Robinson, Buccal controlled delivery of peptides, *Proceed. Intern. Symp. Control. Rel. Bioact. Mater.*, *14*:22 (1987).

101. C. D. Ebert, V. A. John, P. T. Beall, and K. A. Rosenzweig, Transbuccal absorption of diclofenac sodium in a dog model, in *Controlled Telease Technologie* (P. I. Lee, ed.) (1987), p. 310.

102. O. Siddiqui and Y. W. Chien, *CRC Crit. Rev. Ther. Drug Carrier Syst.*, *3*:195 (1987).

103. M. M. Veillard, M. A. Longer, T. W. Martens, and J. R. Robinson, *J. Contr. Rel.*, *4*:123 (1987).

104. B. J. Aungst and N. J. Rogers, *Pharm. Res.*, *5*:305 (1988).

105. A. K. Banga and Y. W. Chien, *Int. J. Pharm.*, *48*:15 (1988).

106. P. P. H. Le Brun, P. L. A. Fox, M. E. de Fries, and H. E. Boddé, *Proceedings of the International Symposium on Controlled Release Bioactive Materials*, CH–Basel, Vol. 15 (1988), p. 336.

107. M. R. Tavakoli-Saberi and K. L. Audus, *Pharm. Res.*, *5 (Suppl)*:99 (1988).

108. K. W. Garren, E. M. Topp, and A. J. Repta, *Pharm. Res.*, *5 (Suppl)*:223 (1988).

109. C. K. Oh and W. A. Ritschel, *Pharm. Res.*, *5 (Suppl)*:100 (1988a).

110. C. K. Oh and W. A. Ritschel, *Pharm. Res.*, *5 (Suppl)*:100 (1988b).

111. Y. Kurosaki, S. Hisaichi, C. Hamada, T. Nakayama, and T. Kimura, *Int. J. Pharm.*, *47*:13 (1988).

112. B. J. Aungst, N. J. Robers, and E. Shefter, *J. Pharmacol. Exp. Ther.*, *244*:23 (1988).

113. W. A. Ritschel, H. Forusz, and M. Kraeling, *Pharm. Res.*, *5 (Suppl)*:108 (1988).

114. C. L. Barsuhn and N. F. H. Ho, *Pharm. Res.*, *5(Suppl)*:103 (1988).

115. Y. Nakada, N. Awata, C. Nakamichi, and I. Sugimoto, *J. Pharmacobiol. Dyn.*, *11*:395 (1988).

116. I. G. Tucker, *J. Pharm. Pharmacol.*, *40*:679 (1988).

117. L. Paulesu, F. Corradeschi, C. Nicoletti, and V. Bocci, *Int. J. Pharm.*, *46*:199 (1988).

118. P. S. Banerjee and J. R. Robinson, *Pharm. Res.*, *5 (Suppl)*:133 (1988).

119. J. Cassidy and E. Quadros, *Pharm. Res.*, *5 (Suppl)*:100 (1988).

120. A. Yamamoto, A. M. Luo, and V. H. L. Lee, *Pharm. Res.*, *5 (Suppl)*:107 (1988).

121. Y. Nakada, N. Awata, C. Nakamichi, and I. Sugimoto, *Yakuzaigaky*, *47*:213 (1987).

122. Y. Kurosaki, S. Hisaichi, L. Hong, T. Nakayama, and T. Kimura, *Int. J. Pharm.*, *49*:47 (1989).

123. A. Jarrett, *The Physiology and Pathophysiology of the Skin*, Vol. 6 (A. Jarrett, ed.), Academic Press, London (1980), p. 1871.

124. C. A. Squier, P. S. Cox, P. W. Wertz, and D. T. Downing, *Arch. Oral. Biol.*, *31*:741 (1986).

125. W. Curatolo, *Pharm. Res.*, *4*:271 (1987).

126. N. H. F. Ho, J. Y. Park, P. I. Ni, and W. I. Higuchi, in *Animals Models for Oral Drug Delivery in Man: In Situ and In Vivo Approaches* (W. Crouthamel and A. C. Sarapu, eds.), APhA, Washington (1983), p. 27.

127. A. A. Rizzo, *J. Periodont.*, *41*:210 (1970).

128. R. Tolo, *Archs. Oral Biol.*, *19*:259 (1974).

129. C. A. Squier, *J. Ultrastruct. Res.*, *43*:160 (1973).

130. A. Wermerskirchen and H. P. Merkle, unpublished data (1986).

131. T. Nagai, *Advances in Drug Delivery Systems* (J. M. Anderson and S. W. Kim, eds.), Elsevier, Amsterdam (1986), p. 121.

132. W. R. Galey, H. K. Lonsdale, and S. Nacht, *J. Invest. Dermatol.*, *67*:713 (1976).

133. I. A. Siegel, K. T. Izutsu, and J. Burkhart, *J. Pharm. Sci.*, *65*:129 (1976).

134. N. F. H. Ho, H. P. Merkle, and W. I. Higuchi, *Drug Dev. Ind. Pharm.*, *9*:1111 (1983).

135. D. M. Matthews, *Physiol. Rev.*, *55*:537 (1975).

136. G. S. Gordon, A. C. Moses, R. D. Silver, J. S. Flier, and M. C. Carey, *Proc. Natl. Acad. Sci.*, *82*:7423 (1985).

137. A. A. Hussain, R. Bawarshi-Nassar, and C. H. Huang, in *Transnasal Systemic Medications* (Y. W. Chien, ed.), Elsevier, Amsterdam (1985), p. 121; C. H. Huang, R. Kimura, R. Bawarshi-Nassar, and A. Hussain, *J. Pharm. Sci.*, *74*:1298 (1985).

138. D. R. Cowsar, T. R. Tice, and D. B. Mirth, in *An Intraoral Fluoride-Releasing Device for Prevention of Dental Caries*, Proceedings of the International Symposium on Controlled Release of Bioactive Materials, CH-Geneva (N. A. Peppas and R. J. Haluska, eds.), Controlled Release Society, Lincolnshire, Illinois, Vol. 12 (1985), p. 310.

139. Y. Machida and T. Nagai, *Chem. Pharm. Bull.*, *26*:1652 (1978).

140. Y. Machida, H. Masuda, N. Fujiyama, M. Iwata, and T. Nagai, *Chem. Pharm. Bull.*, *28*:1125 (1980).

141. K. D. Bremecker, G. Klein, H. Stremple, and A. Rübesamen-Vokuhl, *Arzneim.-Forsch./Drug. Res.*, *33*:591 (1983).

142. K. D. Bremecker, *Pharm. Ind.*, *45*:417 (1983).

143. K. D. Bremecker, H. Stremple, and G. Klein, *J. Pharm. Sci.*, *73*:548 (1984).

144. R. Gurny, J. M. Meyer, and N. A. Peppas, *Biomaterials*, *5*:336 (1984).

145. R. Anders and H. P. Merkle, *Int. J. Pharm.*, *49*:231 (1989).

146. H. Park and J. R. Robinson, in *Advances in Drug Delivery Systems* (J. M. Anderson and S. W. Kim, eds.), Elsevier, Amsterdam (1986), p. 47.

147. N. A. Peppas and P. A. Buri, in *Advances in Drug Delivery Systems* (J. M. Anderson and S. W. Kim, eds.), Elsevier, Amsterdam (1986), p. 257.

148. H. S. Ch'ng, H. Park, P. Kelly, and J. R. Robinson, *J. Pharm. Sci.*, *74*:399 (1985).

149. P. I. Lee, in *Advances in Drug Delivery Systems* (J. M. Anderson and S. W. Kim, eds.), Elsevier, Amsterdam (1986), p. 277.

150. S. S. Leung and J. R. Robinson, *J. Contr. Rel.*, *5*:223 (1988).

151. S. S. Leung and J. R. Robinson, *Pharm. Res.*, *5 (Suppl)*:108 (1988).

152. R. S. Manly (ed.), *Adhesion in Biological Systems*, Academic Press, New York (1970).

153. G. P. Anderson, S. J. Bennet, and K. L. DeVries, *Analysis and Testing of Adhesive Bonds*, Academic Press, New York (1977).

154. D. Duchêne, F. Touchard, and N. A. Peppas, *Drug Dev. Ind. Pharm.*, *14*:283 (1988).

155. S. S. Davis, P. B. Daly, J. W. Kennerley, M. Frier, J. G. Hardy, and C. G. Wilson, in *Controlled Release Nitroglycerin in Buccal and Oral Form, Advanced Pharmacotherapy*, Vol. 1 (W.-D. Bussmann, R.-R. Dries, and W. Wagner, eds.), Karger, Basel (1982), p. 17; J. G. Hardy, J. W. Kennerley, M. J. Taylor, C. G. Wilson, and S. S. Davis, *J. Pharm. Pharmacol.*, *34 (Suppl)*:91S (1982).

156. Y. W. Chien, *Drug. Dev. Ind. Pharm.*, *9*:1291 (1983).

157. M. Ishida, N. Nambu, and T. Nagai, *Chem. Pharm. Bull.*, *31*:1010 (1983).

158. G. Hunt, P. Kearney, and I. W. Kellaway, in *Mucoadhesive Polymers in Drug Delivery Systems* (P. Johnson and J. G. Lloyd-Jones, eds.), E. Horwood, Chichester and VCH Verlagsgesellschaft, Weinheim (1987), p. 180.

159. J. D. Smart and I. W. Kellaway, *J. Pharm. Pharmacol.*, *34*:70P (1982).

160. J. D. Smart, I. W. Kellaway, and H. E. C. Worthington, *J. Pharm. Pharmacol.*, *36*:295 (1984).

161. C. Robert, P. Buri, and N. A. Peppas, *Acta Pharm. Technol.*, *34*:95 (1988).

162. G. Ponchel, F. Touchard, D. Duchêne, and N. A. Peppas, *J. Contr. Rel.*, *4*:129 (1987).

163. N. A. Pappas, G. G. Ponchel, and D. Duchêne, *J. Contr. Rel.*, *5*:143 (1987).

164. G. Ponchel, F. Touchard, D. Wouessidjewe, and D. Duchêne, *Int. J. Pharm.*, *38*:65 (1987).

165. K. Park and J. R. Robinson, *Int. J. Pharm.*, *19*:107 (1984).

166. J. Sandow and W. Petri, in *Transnasal Systemic Medications* (Y. W. Chien, ed.), Elsevier, Amsterdam (1985), p. 183.

167. H. Okada, I. Yamazaki, Y. Ogawa, S. Hirai, T. Yashiki, and H. Mima, *J. Pharm. Sci.*, *71*:1367 (1982).

168. T. Nishihata, J. H. Rytting, T. Higuchi, and L. Caldwell, *J. Pharm. Sci.*, *33*:334 (1981).

12

Rectal Route of Peptide and Protein Drug Delivery

J. Howard Rytting

University of Kansas, Lawrence, Kansas

I. INTRODUCTION

With the development of a number of peptide and protein molecules as potential therapeutic agents, problems of formulation and delivery have emerged
and limited their widespread utility. The rectal route of delivery offers
several advantages over some of the other common dosage forms for peptide
and protein delivery. For example, the pain, inconvenience, and cost in
terms of professional administration, tissue damage, and other difficulties
have limited the parenteral route of administration for many potential drugs.
Poor absorption, degradation in the gastrointestinal (GI) tract, stability,
and other problems place severe limitations on oral administration of peptides
and proteins. Several reviews [1-3] have described a number of potential
advantages of rectal drug delivery in general for many types of drugs.
Many of these considerations also apply to peptides and proteins.
    In this chapter we review a number of studies that have illustrated the
potential use of rectal drug delivery techniques for a number of peptide
and protein drug entities. We discuss the general requirements for effective absorption from the rectum as well as experimental approaches that have
been used to study rectal absorption. We also describe the use of various
types of adjuvants to enhance rectal drug absorption, particularly for peptides and proteins as well as design requirements for rectal peptide delivery
systems.

A. Anatomy and Physiology of the Rectum

The human rectum has a length of 5 in. The rectal epithelium contains a
number of goblet cells. However, unlike the small intestine, it does not
contain villi. The surface area is only about 200 to 400 cm^2, compared
with 2,000,000 cm^2 for the small intestine [4]. Thus just from surface
area considerations, one would expect absorption to be much less from the

rectum than from the upper GI tract. The rectal area has an extensive blood supply from the various rectal arteries. It is drained by three veins [1]. Although there is significant intermingling of these veins due to the presence of anastomoses, it is often reported that the inferior and middle rectal veins drain into the inferior vena cava, thereby bypassing the portal system and presystemic metabolism in the liver. On the other hand, the superior rectal vein drains into the portal vein.

A number of the physiological factors influencing absorption from the rectum have been reviewed by de Blaey and Polderman [3]. Some of the factors cited include the amount of liquid in the rectum (about 3 ml), the pH of the rectum (which is usually reported to be between 7 and 8) [2], the buffer capacity of the rectal fluid (which is small), the surface tension and viscosity of the rectal fluid, and the effects of the luminal pressure exerted on the suppository or other dosage form by the wall of the rectum. This pressure tends to enhance rectal absorption. A number of factors related to the drug and the dosage form are also important. These include: drug solubility, pK_a, surface properties, particle size, partition coefficient, and concentration.

B. Rectal Administration and Presystemic Metabolism

One of the suggested advantages of rectal administration of drugs is the possibility of avoiding, to some extent, first-pass or presystemic metabolism. For example, de Boer and Breimer [5] have presented data which indicate that for lignocaine the systemic bioavailability in humans is about double compared to oral administration. They suggested that about half of the rectally administered drug avoided hepatic first-pass metabolism. Studies in rats suggest that presystemic metabolism can be almost totally avoided by rectal administration [5,6].

For peptides and proteins, there are few data available with respect to avoiding first-pass metabolism. However, it may be expected to be similar to other substances. On the other hand, the fact that in humans rectal administration typically only partially avoids delivery of the drug to the portal system has been cited [7] as an advantage of rectal administration over parenteral administration for insulin. This is because delivery of insulin to the portal blood system is more physiologic than to the peripheral cells from subcutaneous injections.

II. EXPERIMENTAL APPROACHES TO STUDYING RECTAL ABSORPTION

Several approaches have been used to study rectal absorption of drugs. Usually, they involve measurement of plasma levels following rectal administration of the drug or measurement of the amount of drug lost from a solution or other dosage form placed in the rectal compartment or from a perfusing solution allowed to circulate through the rectal compartment.

A. Perfusion Techniques

The use of perfusion techniques has been found in many situations to be more rapid and to present fewer experimental difficulties than sampling and measuring plasma levels of the drug. Indeed, several investigators have used such perfusion techniques [8−10] in their studies of rectal drug absorption.

The approach used in our laboratory involves exposing the rectum of the rat by an abdominal incision, after which a glass cannula was inserted in the distal direction and tied firmly to keep it in position. A second cannula was inserted through the anus about 1 cm inside the rectum and secured by ligation. This exposed about 2 cm of the rectum to the per-fusate. Typically, about 6 ml of a perfusing solution is circulated at a rate of about 2 ml/min at 38°C during the course of the experiment. At appropriate intervals the perfusing liquid is sampled and assayed for drug concentration. The loss of drug from the perfusate is then interpreted as representing the extent of absorption from the rectum. Generally, this assumption is verified by measuring plasma levels as well. In each case where this was done, the plasma levels were found to fit well with the perfusate levels.

One approach to in vivo absorption studies utilized in our laboratories [10] involved measuring plasma drug levels after administration of the drug to the rectum. If the drug was presented as a solution, the rectum was exposed by an abdominal incision, a sample (usually about 0.3 ml) of the drug solution was injected into a 2-cm section of the rectum, and the drug solution was kept in that section by ligating the rectum with thread. Fol-lowing this procedure, blood samples would be collected as a function of time, often from a leg vein using a cannula. Another approach that was used on occasion involved injecting 0.3 ml of the drug solution into the rectum using a cannula; then the anus would be firmly tied to prevent leakage. If the solution was not restricted to a relatively small area of the rectum, the extent of absorption usually was reduced substantially.

B. Use of Enemas

Many studies of rectal absorption have used microenemas as the vehicle. These have the advantage that often the drug is in solution, which makes it readily available for absorption. In the situation where absorption ad-juvants are being used to enhance the absorption of the drug, it is impor-tant to maintain as high a concentration as possible of both the drug and adjuvant at the site of absorption. Thus the use of either ligation or a more viscous vehicle to maintain the solution in a relatively small region of the intestine or rectum has often been employed.

The ionic strength of the solution has often been found to influence the degree of absorption from the rectum. In most situations, the higher the ionic strength, the greater the absorption of the drug. For example, Fig. 1 shows that salicylate is absorbed more readily from a perfusing solu-tion in the rat rectum at an ionic strength of 0.75 than of 0.15 [11]. One suggested reason is that high cation concentrations may reduce the viscosity of the mucin layer, which covers the mucosal epithelium, thereby allowing greater diffusion of the drug and adjuvant to the mucosa. Another sug-gestion involves the uptake of sodium ion to the intercellular space by Na^+, K^+/ATPase, resulting in a higher osmolality and water flux, which may facilitate the transport of drug and adjuvant across the mucosa (solvent drag).

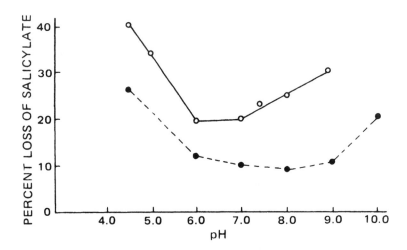

Fig. 1   Percent loss of salicylate from a perfusate as a function of pH
having an initial salicylate concentration of 0.5% and ionic strength of 0.75
(open circle) and 0.15 (filled circle). (From Ref. 11. Reproduced with permission of the copyright owner, the American Pharmaceutical Association.)

C. Suppositories

Presenting the drug to the rectal cavity in a suppository has been found
in a number of situations to have profound effects on the absorption of
the drug compared with the use of a microenema. For example, Nishihata,
et al. [12] found that although both sodium salicylate and 5-methoxysalicy-
late increased the rectal absorption of insulin in dogs using various formulations, microenema formulations containing 4% gelatin showed three times
higher insulin bioavailability than did suppository formulations. On the
other hand, they also found [13] that for gentamicin and cefemtazole much
higher bioavailabilities were observed using suppositiories with Witepsol H-15
as the base than from microenemas containing either normal saline or 4%
gelatin. Other studies [7,14] have indicated that insulin is readily absorbed from many suppositiories. However, there are vast differences in
absorption depending on the formulation chosen for the suppository base.

III. EXAMPLES OF RECTAL DELIVERY OF PEPTIDES
AND PROTEINS

The literature contains a number of examples of rectal absorption of peptides and proteins. In the majority of cases some type of adjuvant is included in the formulation to facilitate absorption. Several examples are
included in the following sections.

A. Insulin

Insulin is probably the most often studied protein with respect to rectal
absorption. Ritschel and Ritschel [7] have presented a comprehensive review of most of the early studies dealing with rectal absorption of insulin.

Studies in the 1920s [15,16] generally showed no lowering of blood sugar
following the administration of insulin rectally. Beginning in the 1930s,
several studies have suggested that the use of adjuvants might enhance
insulin absorption. Many of these studies employ the use of surface-active
agents. For example, Touitou el al. [17] found that a microenema containing PEG 1000 and PEG 400 and 27 units of insulin reduced blood glucose
in the rat about the same amount as intraperitoneal (IP) administration of
4 units of insulin. Several studies in rabbits have been reviewed [7]
which suggest that the use of various surfactants can enhance insulin absorption. The relative bioavailability in most of these cases was rather
low (i.e., a few percent). Surfactant combinations having a HLB value of
12.73 were also found to be optimal for in vitro release and absorption of
insulin from suppositories. The effects of a variety of surfactants, including nonionics, anionics, cationics, and amphoteric materials along with
bile acids and phospholipids, were studied by Ichikawa et al. [18]. The
most effective surfactant in their study was polyoxyethylene(9)lauryl ether.
It was also shown in this and other studies that an increase in the insulin
dose does not always result in a proportionally greater decrease in blood
sugar levels. This may be due to a saturation effect since permeation of
the rectal mucosa appears to be rate limiting and the overall bioavailability
of insulin is low.

Yamasaki et al. [14] found that in humans, diabetics showed a higher
response to insulin than did nondiabetics when insulin was administered as
a suppository containing polyoxyethylene(9)lauryl ether. They found
that an approximately steady glucose level was observed in diabetic subjects who ate regular meals with a 100-unit insulin suppository three times
daily 15 min after meals.

Ritschel and Ritschel [7] have reviewed the use of various absorption
enhancers and compiled a list of the major ones used to promote the absorption of insulin. Their list is extensive and includes a variety of surface-
active agents, phospholipids such as lecithin, saponins, bile acids, organic
alcohols, acids, salts, amines, and fats. They have also suggested that
rectal delivery of insulin may have physiological advantages over parenteral
administration since the rectal anatomy allows some of the drug to be absorbed into the portal circulation and thereby pass through the liver and
be utilized there.

Among the various absorption enhancers that have been examined,
several adjuvants having structural similarities to salicylate have been found
to be particularly effective. For example, Nishihata, et al., [19] reported
that sodium salicylate, sodium 5-methoxysalicylate, sodium 3-methoxysalicy-
late, and sodium homovanilate all enhanced the absorption of insulin and
heparin in rats. A rectal dosage form of insulin including 5-methoxysalicy-
late appears to be particularly promising. Sodium 5-methoxysalicylate was
also found [20] to enhance insulin absorption from the upper gastrointestinal tract in rats. In this study it was found that restricting the movement
of insulin and adjuvant from the point of administration in the intestine by
either ligation or the use of a more viscous vehicle further increased insulin
absorption. As shown in Fig. 2, sodium salicylate and 5-methoxysalicylate
both increased the rectal absorption of insulin in dogs and reduced plasma
glucose levels when coadministered in a variety of formulations, including
suppositories and microenemas containing gelatin [21]. Maximum plasma
levels of insulin occurred within 30 min after rectal administration and correlated closely with the plasma concentrations of the absorption adjuvants

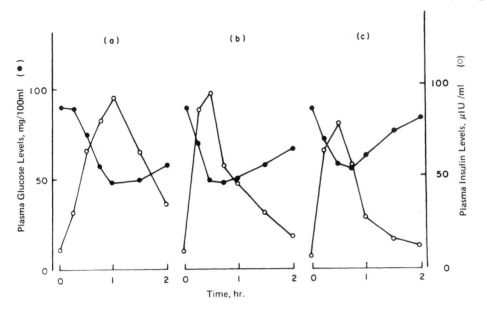

Fig. 2   Concentrations of glucose (mg/100 ml) and insulin (μIU/ml) in the plasma of dogs following (a) an intramuscular injection of 10 IU of insulin, (b) the administration of a 0.5-ml microenema containing 20 IU of insulin with 150 mg of sodium 5-methoxysalicylate in a 0.9% NaCl solution containing 4% gelatin, and (c) the administration of a 0.5-ml microenema containing 20 IU of insulin with 300 mg of sodium salicylate in a 0.9% NaCl solution containing 4% gelatin.   (From Ref. 21.)

employed. The minimum in the plasma glucose concentrations after rectal administration of insulin appeared to lag about 15 min behind the peak insulin levels. When the absorption adjuvant was omitted from the microenema, there was little change in the plasma insulin levels and actually a slight increase in the plasma glucose concentrations. When insulin solutions containing one-half the rectal dose were administered intramuscularly, a somewhat different delivery profile for insulin was observed with maximum plasma levels occurring 1 hr following administration (Fig. 2a).

Liversidge and co-workers [22] observed that modifications in the suppository formulation had profound effects on the release of insulin and on the resulting plasma levels of glucose in dogs. For example, decreasing suppository size while maintaining a constant amount of insulin and adjuvants resulted in increased insulin absorption. Also, increasing the acetic acid content of the suppository increased insulin absorption. The addition of a polypeptide protease inhibitor, aprotinin, to a microenema containing insulin and sodium salicylate resulted in a greater decrease in the plasma glucose level than the use of salicylate alone [23]. Aprotinin in an insulin microenema without sodium salicylate had no effect on the plasma glucose levels. The authors concluded that aprotinin did not enhance rectal insulin absorption, but perhaps inhibited the degradation of insulin in the rectal area before salicylate enhanced absorption of insulin occurred.

Enamine derivatives prepared from the reaction of ethyl acetoacetate
and amino acids have been found to enhance the rectal absorption of insulin in normal rabbits [24]. In this study the enamine of sodium DL-phenyl-
alanate was the most effective adjuvant. Typically, it was found that insulin suppositories containing enamines were effective in lowering plasma
glucose levels for about 2.5 hr. Similar results were obtained in dogs.
However, higher bioavailability of insulin was observed when the insulin
was dissolved and dispersed in a gelatin microenema than when a crystalline suspension of insulin was dispersed in a glyceride suppository base
[25].

B. Lysozyme

Lysozyme is a relatively large polypeptide with a molecular weight of about
14,400 and consisting of a single polypeptide chain of 129 amino acids.
Lysozyme chloride was found [26] to be absorbed from the rectum of rabbits when presented in a triglyceride-base (Witepsol H-15) suppository containing about 5% ethylacetoacetate enamines of phenylalanine and phenyl-
glycine. Glycerine-1,3-diacetoacetate, 1,2-isopropylideneglyceryl-3-aceto-
acetate, and ethylacetoacetylglycolate were also found to enchance the rectal absorption of lysozyme and heparin. Simultaneous administration of
calcium ion inhibited the adjuvant action. Similar inhibition by calcium ion
has also been observed with salicylate type adjuvants [10].

C. [Asu[1,7]]-eel Calcitonin

Calcitonin, which is often used to reduce calcium levels in the blood, is
usually given parenterally since it is digested by the proteolytic enzymes
in the GI tract. [Asu[1,7]]-eel calcitonin is a calcitonin analog and was
found [27] to be absorbed from the rectum of rats when administered in a
0.1% v/v polyacrylic acid gel base at pH 5.5. Under these conditions
plasma calcium level decreased by about 18%, with the maximal effect occurring about 30 min after administration of a 35-fold higher dose than the
intravenous dose. By contrast, rectal administration in vehicles such as
triglyceride suppository bases, polyetheylene glycol 1000 or in saline solution had little, if any, hypocalemic effect at a dose of 5 U/kg.

D. Phenylalanine and Di-, Tri-, and Tetraphenylalanines

Phenylalanine and phenylalanylglycine were found [28] to be poorly absorbed
from a solution perfusing the rat rectum. However, their disappearance
from the perfusate was markedly enchanced in the presence of salicylate or
5-methoxysalicylate at pH 4.5, 7.4, and 8.5. The disappearance of di-,
tri-, and tetraphenylalanine was also found to be facilitated by these adjuvants. Interestingly, their absorption was fairly substantial in the absence of adjuvant. These peptide analogs of phenylalanine were found to
enhance the rectal absorption of cefoxitin and cefmetazole, two highly
water soluble antibiotics. It appears that the enhancement of the absorption of these antibiotics requires the simultaneous absorption of the adjuvant.
Similar observations were reported earlier by Nishihata, et al. [11] for a
number of other drugs.

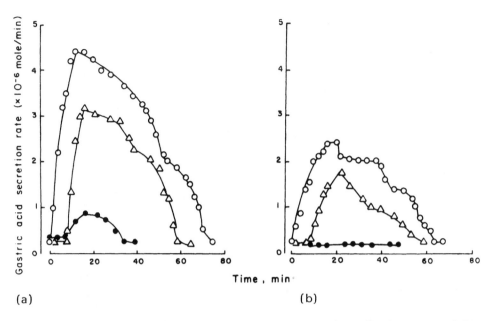

(a)   (b)

Fig. 3   Gastric acid secretion following administration of (a) pentagastrin
and (b) gastrin to the anesthetized rat. The solutions were administered:
IV with 125 µg of polypeptide/kg in rat in 0.1 ml of saline, pH 8.0 (open
circle), and rectally with 125 µg of polypeptide/kg rat in the absence
(filled circle) and presence (triangle) of 25 mg of 5-methoxysalicylate/kg
rat in 0.1 ml of saline, pH 8.0. (From Ref. 29. Reproduced with permission of the copyright owner, the American Pharmaceutical Association.)

E.   Gastrin, Pentagastrin, and Tetragastrin

Rectal absorption of gastrin and pentagastrin was examined by Yoshioka
et al. [29] both in the absence and presence of 5-methoxysalicylate. The
pharmacological effects of these drugs were measured by monitoring the
rate of stomach acid secretion. As can be seen in Fig. 3, the pharmacological response after rectally administered polypeptide was small (pentagastrin)
or nondetectable (gastrin) in the absence of the adjuvant. In the presence
of 5-methoxysalicylate the response was a significant fraction of the response obtained after intravenous (IV) administration. Dose-response
studies indicate that the bioavailability of rectally administered pentagastrin
increased from 6% to 33% when 5-methoxysalicylate was present compared
with the IV administration. In the case of gastrin the bioavailability was
about 18% in the presence of the adjuvant. In an independent study by
Jennewein et al. [30], tetragastrin was found to be absorbed from a number of different sites in rats and dogs. However, rectal administration of
tetragastrin to rats resulted in more distinct increases in acid secretion
than administration to the other sites studied. The bioavailability was found
to be 16% when compared with a single IV injection.

IV. USE OF ADJUVANTS TO ENHANCE THE ABSORPTION OF PEPTIDES AND PROTEINS

As can be seen from the preceding section in which several examples of rectal absorption of peptides and proteins was described, most such substances require the presence of some type of absorption enhancer in order to obtain reasonable absorption of the drug. Ritschel and Ritschel [7] have described the use of several classes of compounds as absorption promoters for insulin. These include surface-active agents, bile acids, saponins, phospholipids, organic slcohols, acids, salts, amines, and fats. In our laboratories we have looked at three classes of absorption enhancers. These include the sodium salts of salicylate-type molecules, EDTA, and surfactants. Each of these compounds seems to act somewhat differently, as described below.

A. Surfactants

A large number of surfactants have been used to enhance the rectal absorption of drugs, particularly insulin. These include cetomacrogol; Tween 40, 60, and 61; Span 40; various polyoxyethylene ethers and esters; glyceryle monostearate; sorbose fatty acid ester; sodium lauryl sulfate; and dioctyl-sulfosuccinate [7]. The surfactants appear to interact with the lipoidal fraction of the membrane, and their effects appear to be irreversible in the short term [31]. Various types of metabolic inhibitors had little, if any effect on the enhancing abilities of a surfactant, polyoxyethylene 23-lauryl ether [32]. Surface-active agents appear to enhance drug absorption through damage to the rectal mucosa.

B. Salicylate Type

Among several adjuvants that are structurally similar to sodium salicylate, four were found to be quite effective in enhancing the rectal absorption of a variety of compounds [33]. These were sodium salicylate, 3-methoxysalicylate, 5-methoxysalicylate, and homovanilate. Binding studies [33] of the potential adjuvants studied and rat rectal tissue showed a good correlation between the amounts of adjuvant and drug absorbed and the binding of the adjuvant with rat rectal tissue (Fig. 4).

The presence of sodium chloride was found [31] to enhance the effectiveness of sodium salicylate, while the presence of several inhibitors such as phlorizin and 4,4'-diisothiocyano-stilbene-2,2'-disulfonic acid (DIDS) reduced the effectiveness of salicylate. A later study [32] showed that the effectiveness of sodium salicylate was reduced by concurrent administration of small amounts (<0.5 mg/ml) of N-ethylamaleimide (NEM) or sodium p-chloromercuriphenylsulfonic acid (p-CMP). However, higher concentrations (>1 mg/ml) resulted in enhanced absorption of cefoxitin. Ouabain and 2,4-dinitrophenol (DNP) also suppressed the effectiveness of salicylate as an absorption enhancer. It has further been suggested [31] that the effects of salicylate occur at the protein fraction of the rectal mucosa through a saturable process.

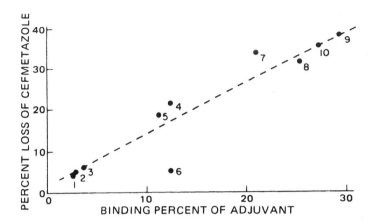

Fig. 4   Percent loss of cefmetazole from perfusate at 60 min versus the percent binding of adjuvant to rat rectal tissue. 1, Benzoate; 2, p-anisate; 3, o-anisate; 4, 2,4-dihydroxybenzoate; 5, 3,5-dihydroxybenzoate; 6, 2,5-dihydroxybenzoate; 7, salicylate; 8, 3-methoxysalicylate; 9, 5-methoxysalicylate; 20, homovanillate. (From Ref. 33. Reproduced with permission of the copyright holder, the American Pharmaceutical Association.)

C. Ethylenediamintetraacetic Acid (EDTA)

EDTA was found [34] to enhance the absorption of salicylate, but to decrease the absorption of m- and p-hydroxybenzoic acids. It has also been found [31] to enhance the rectal absorption of sodium cefoxitin. The effectiveness of EDTA was enhanced by sodium chloride. Several of the inhibitors (phlorizin, DIDS, NEM, and p-CMP) of salicylate-induced enhancement had no effect on the enhancing effects of EDTA. Ouabain and 2,4-dinitrophenol (DNP) suppressed the enhancing effects of EDTA when EDTA was administered at low doses. At higher concentrations of EDTA, DNP and little effect and ouabain only partially suppressed the effects of EDTA.

D. Comparison of the Three Types of Adjuvants

Nishihata et al. [31] found the following differences in the enhancing action of sodium salicylate, disodium EDTA, and polyoxyethylene(23)lauryl ether (POE) relative to cefoxitin absorption. First, the effectiveness of salicylate or EDTA was enhanced by sodium chloride, whereas the activity of POE was not. Second, although the ratios of plasma drug peak values for cefoxitin to the dose of cefoxitin were constant with POE or EDTA, the peak-to-dose ratios with salicylate decreased with increasing drug concentration. Third, phlorizin and DIDS inhibited the effectiveness of salicylate but did not modify the adjuvant action of either POE or EDTA. Fourth, although treatment with salicylate resulted in slightly less protein release than treatment with NaCl, both POE and EDTA increased the release of protein from the rectal mucosa. Based on these observations it was concluded that the effects of salicylate occur at the protein fraction of the rectal mucosa through a saturable process, whereas the adjuvant action of POE and EDTA appears to involve some irreversible disruption of the membrane.

A later study [32] from the same laboratories found that the coadministration of salicylate and POE, or of EDTA and POE, yielded similar plasma
concentrations of cefoxitin as would be expected from summing the plasma
levels resulting from administration of each adjuvant individually at the
same concentrations. However, coadministration of salicylate and EDTA
with cefoxitin yielded plasma cefoxitin concentrations which were much
higher than would be expected from the sum of their individual actions.
Mixtures containing 0.11% EDTA and 3.0% sodium salicylate resulted in protein release to about the same extent as normal saline.

E. Fatty Acid Enhancers

The use of fatty and carboxylic acids as enhancers for rectal absorption
has been reported by Ritschel and Ritschel [7] as well as by Nishimura et
al. [35], Yata et al. [36], and Nishihata et al. [37]. Nishimura et al. [35]
studied the effects of a series of para-substituted benzoic acids and saturated straight-chain fatty acids on the rectal absorption of several β-lactam
antibiotics in rats. They found that the effectiveness of the carboxylic
acid sodium salts was parabolically correlated with their partition coefficient
on a logarithmic basis (log P). The optimal log P values for exerting the
maximal absorption-promoting effect was in the range 4.2 to 4.8. Using
sodium caproate, they found that the extents of rectal absorption of the
β-lactam antibiotics that were coadministered improved from less than 10%
bioavailability to values ranging from about 30 to 90%. These bioavailabilities were found to correlate well with the permeability of the drugs to cellulose membrane.

Yata et al. [36] also have reported that carboxylic acids having metal
ion chelating ability enhance the rectal absorption of poorly absorbed drugs
including the water-soluble antibiotics. They suggested that the carboxylic
acids may serve to make the intercellular space more accessible by temporarily removing calcium ions from the rectal mucosa. Calcium ions are necessary to maintain the integrity of the tight junctions.

Nishihata et al. [37] found that the absorption of sodium cefoxitin from
the small intestine was significantly increased after its administration in a
suppository form which was prepared with a triglyceride base. It was suggested that this may be due to the fatty acids produced by the actions of
lipase on the triglycerides. These fatty acids could then act as surfactants
in their ionized form in the luminal fluid. However, such enhancement was
not found for rectal absorption from triglyceride suppositories, possibly
due to the absence of lipase activity in the rectum. This suggests that
another possible mechanism for the action of fatty acids in enhancing rectal
absorption of water-soluble drugs may be due to their surfactant characteristics in the ionized form.

F. Importance of Lymphatic Transport of Insulin Following
     Rectal Administration in the Presence of Adjuvants

As mentioned above, it appears that a substantial portion of a drug administered rectally may bypass the hepatic portal system and enter the general
circulation directly. This has generally been attributed to the fact that
the lower hemorrhoidal veins do not enter the hepatic portal system, and
therefore a substantial portion of the rectally absorbed drug will enter the

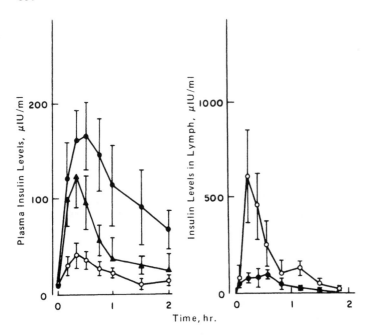

Fig. 5   Concentration of insulin in the plasma and lymph of rats following the intramuscular administration of 0.8 IU/body insulin (filled circle), rectal administration of 2.0 IU/body insulin in the presence of 10 mg/body 5-methoxysalicylate and intact thoracic duct (triangle), and rectal administration of 2.0 IU/body insulin in the presence of 10 mg/body 5-methoxysalicylate with the thoracic duct cannulated for collection of lymph (open circle). The error bars represent standard deviations with n = 6.   (From Ref. 38.)

the general circulation directly.   However, another factor that may account for delivery of drugs directly to the general circulation involves lymphatic uptake of drugs, particularly those that are water soluble.   Nishihata et al. [38] and Caldwell et al. [39] have provided several examples where lymphatic absorption of drugs appears to be very important in rectal absorption.   Insulin was one of the drugs studied by this group.   In their study the thoracic duct of the rat was cannulated and lymph was collected as a function of time.   Insulin was found to be transported primarily via the lymphatic system to the general circulation as shown in Fig. 5 when administered in the presence of 5-methoxysalicylate, at least in this somewhat artificial system.   When administered intramuscularly, plasma insulin levels were significantly higher than were the levels in the lymph.   The concentration of insulin in the lymph after rectal administration to rats with cannulated thoractic ducts was more than 15-fold greater than the corresponding plasma concentrations in the same time frame.   Rectal administration of insulin in the presence of 5-methoxysalicylate in rats with intact thoracic duct produced insulin plasma concentrations about three-fold greater than in the rat in which the lymphatic drainage had been diverted. This suggests that much of the rectally administered insulin reaches the general circulation via the lymphatic system after rectal administration in the presence of adjuvants such as 5-methoxysalicylate.   Furthermore, the peak appearance of insulin in the lymph was somewhat earlier than the rate of appearance in the plasma. Again, rectal administration of insulin without the adjuvant did not produce appreciable concentrations in the plasma or lymph under the conditions of this study.

V. DESIGN REQUIREMENTS FOR RECTAL PEPTIDE DELIVERY SYSTEMS

Many of the considerations that should be taken into account in the formulation of suppositories, retention enemas, or other vehicles for rectal peptide delivery are similar to those found for many other compounds. De Blaey and Polderman [3] have indicated that such factors as the amount, composition, pH, buffer capacity, viscosity, and surface tension of the rectal fluid can be important. Equally important are the solubility, particle size, pK_a, and partition coefficient of the drug as well as the characteristics of the vehicle or delivery device. A recent report [40] has shown that in order to obtain absorption of a series of prodrugs of 5-fluorouracil to better than 50%, the prodrugs should have an octanol/water partition coefficient greater than 0.5 and a solubility in water of pH 7.4 greater than 0.05 M.

In addition to these considerations, the use of absorption enhancers such as those described above have the potential to facilitate greater the absorption of polypeptides and proteins. This is particularly the case for those substances having relatively high aqueous solubilities.

VI. CONCLUSIONS

There are numerous examples of effective delivery of peptide and protein drugs using the rectal route of administration. These include insulin; various gastrins; di-, tri- and tetraphenylalanine; lysozyme; and calcitonin. In most cases it is necessary to use an adjuvant to enhance the absorbtion of peptides and proteins in order to obtain adequate absorption for practical use. This is particularly true for larger molecules and those having relatively high aqueous solubilities. Overall, for a number of situations, the rectal route of administration has advantages over oral and parenteral routes of administration. Rectal drug delivery appears to be a feasible alternative method for administering a number of polypeptide drugs, and for some may be the method of choice.

REFERENCES

1. A. G. de Boer, F. Moolenaar, L. G. J. de Leede, and D. D. Breimer, *Clin. Pharmacokinet.*, 7:285 (1982).
2. A. G. de Boer, L. G. J. de Leede, and D. D. Breimer, *Br. J. Anaesth.*, 56:69 (1984).
3. C. J. de Blaey and J. Polderman, in *Drug Design*, Vol. 9 (E. J. Ariens, ed.), Academic Press, New York (1980), p. 237.
4. T. H. Wilson, *Intestinal Absorption*, W. B. Saunders Company, Philadelphia (1962).

5. A. G. de Boer and D. D. Breimer, in *Drug Absorption* (L. F. Prescott and W. S. Nimmo, eds.), Adis Press, Sydney, (1979), p. 61.
6. A. Kamiya, H. Ogata, and H.-L. Fung, *J. Pharm. Sci.*, *71*:621 (1982).
7. W. A. Ritschel and G. B. Ritschel, in *Rectal Therapy* (B. Glas and C. J. de Blaey, eds.), J. R. Prous Publishers, Barcelona, Spain (1984), p. 67.
8. D. J. A. Crommelin, J. Modderkolk, and C. J. De Blaey, *Int. J. Pharm.*, *3*:299 (1979).
9. T. Nishihata, J. H. Rytting, and T. Higuchi, *J. Pharm. Sci.*, *69*: 744 (1980).
10. T. Nishihata, J. H. Rytting, and T. Higuchi, *J. Pharm. Sci.*, *70*:71 (1981).
11. T. Nishihata, J. H. Rytting, and T. Higuchi, *J. Pharm. Sci.*, *71*: 869 (1982).
12. T. Nishihata, J. H. Rytting, A., Kamada, T. Higuchi, M. Routh, and L. Caldwell, *J. Pharm. Pharmacol.*, *35*:148 (1983).
13. T. Nishihata, J. H. Rytting, T. Higuchi, L. J. Caldwell, and S. J. Selk, *Int. J. Pharm.*, *21*:239 (1984).
14. Y. Yamasaki, M. Shichiri, R. Kawamori, M. Kikuchi, T. Yagi, S. Arai, R. Tohdo, N. Hakui, N. Oji, and H. Abe, *Diabetes Care*, *4*:454 (1981).
15. F. G. Banting and C. H. Best, *J. Lab. Clin. Med.*, *7*:464 (1922).
16. S. Peskind, J. M. Rogoff, and G. N. Stewart, *Am. J. Physiol.*, *68*: 530 (1924).
17. E. Touitou, M. Donbrow, and E. Azaz, *J. Pharm. Pharmacol.*, *30*: 662 (1978).
18. K. Ichikawa, I. Ohata, M. Mitomi, S. Kawamura, H. Maeno, and H. Kawata, *J. Pharm. Pharmacol.*, *32*:314 (1980).
19. T. Nishihata, J. H. Rytting, T. Higuchi, and L. Caldwell, *J. Pharm. Pharmacol.*, *33*:334 (1981).
20. T. Nishihata, J. H. Rytting, A. Kamada, and T. Higuchi, *Diabetes*, *30*:1065 (1981).
21. T. Nishihata, J. H. Rytting, A. Kamada, T. Higuchi, M. Routh, and L. Caldwell, *J. Pharm. Pharmacol.*, *35*:148 (1983).
22. G. G. Liversidge, T. Nishihata, K. K. Engle, and T. Higuchi, *Int. J. Pharm.*, *23*:87 (1985).
23. T. Nishihata, G. Liversidge, and T. Higuchi, *J. Pharm. Pharmacol.*, *35*:616 (1983).
24. S. Kim, A. Kamada, T. Higuchi, and T. Nishihata, *J. Pharm. Pharmacol.*, *35*:100 (1983).
25. S. Kim, T. Nishihata, S. Kawabe, Y. Okamura, A. Kamada, T. Yagi, R. Kawamori, and M. Shichiri, *Int. J. Pharm.*, *21*:179 (1984).
26. M. Miyake, T. Nishihata, N. Wada, E. Takeshima, and A. Kamada, *Chem. Pharm. Bull.*, *32*:2020 (1984).
27. K. Morimoto, H. Akatsuchi, R. Aikawa, M. Morishita, and K. Morisaka, *J. Pharm. Sci.*, *73*:1366 (1984).
28. T. Nishihata, C.-S. Lee, M. Yamamoto, J. H. Rytting, and T. Higuchi, *J. Pharm. Sci.*, *73*:1326 (1984).
29. S. Yoshioka, L. Caldwell, and T. Higuchi, *J. Pharm. Sci.*, *71*:593 (1982).
30. H. M. Jennewein, F. Waldeck, and W. Konz, *Arzneim.-Forsch. (Drug Res.*, *24*:1225 (1974).
31. T. Nishihata, H. Tomida, G. Frederick, J. H. Rytting, and T. Higuchi, *J. Pharm. Pharmacol.*, *37*:159 (1985).

32. T. Nishihata, C.-S. Lee, J. H. Rytting, and T. Higuchi, *J. Pharm. Pharmacol.*, *39*:180 (1987).

33. T. Nishihata, J. H. Rytting, and T. Higuchi, *J. Pharm. Sci.*, *71*:865 (1982).

34. H. Kunze, G. Rehback, and W. Vogt, *Naunyn-Schmiedeberg's Arch. Pharmacol.*, *273*:331 (1972).

35. K. Nishimura, Y. Nozaki, A. Yoshimi, S. Nakamura, M. Kitagawa, N. Kakeya, and K. Kitao, *Chem. Pharm. Bull.*, *33*:282 (1985).

36. N. Yata, Y. Higashi, T. Murakami, R. Yamajo, W. M. Wu, K. Taku, Y. Sasakai, and Y. Hideshima, *J. Pharmacobio-Dyn.*, *6*:s-78 (1983).

37. T. Nishihata, H. Yoshitomi, and T. Higuchi, *J. Pharm. Pharmacol.*, *38*:69 (1986).

38. T. Nishihata, J. H. Rytting, L. Caldwell, S. Yoshioka, and T. Higuchi, in *Optimization of Drug Delivery*, Alfred Benzon Symposium 17 (H. Bundgaard, A. B. Hansen, and H. Kofod, eds.), Munksgaard, Copenhagen (1982), p. 17.

39. L. Caldwell, T. Nishihata, J. H. Rytting, and T. Higuchi, *J. Pharm. Pharmacol.*, *34*:520 (1982).

40. A. Buur and H. Bundgaard, *Int. J. Pharm.*, *36*:41 (1987).

13

Nasal Route of Peptide and Protein Drug Delivery

Kenneth S. E. Su

Eli Lilly and Company, Indianapolis, Indiana

I. BACKGROUND

The popularity of the oral route of drug administration will make it the dominant approach for drug delivery in the foreseeable future, but the nasal route of delivering peptides and proteins is a promising alternative, especially for the young, the very old, the blind, and the debilitated patients. The use of intranasal delivery for peptides dated back to the early 1920s. The administration of pituitary extracts—which contained oxytocin—by the intranasal route to 80 subjects showed that the delivery system was an efficient and acceptable method for accelerating labor already in progress [1]. Subsequently, extensive clinical studies on nasal absorption of oxytocin were explored [2-7]. In 1971, an investigation of the safety of intranasal oxytocin in 1800 patients revealed a total success rate of 88% in terms of induction or stimulation of labor with no major maternal complications [8]. The intranasal administration of oxytocin to dogs resulted in significant increases in plasma insulin and glucagon secretion [9] with no untoward effects.

Vasopressin, an antidiuretic peptide, is another classical example of a peptide drug given by the intranasal route of administration. In the early 1960s, a synthetic vasopressin analog, lypressin, was evaluated for the treatment of diabetes insipidus in humans following intranasal administration [10,11]. Intranasal delivery of this peptide was effective in patients with diabetes insipidus. Various clinical studies on vasopressin analogs were then reported [12-16]. In 1968, the antidiuretic activity of [1-deamino-8-arginine]vasopressin was compared with that of lysine-8-vasopressin in patients following intravenous and intranasal administration. These peptides were safe and effective in the treatment of cranial diabetes insipidus in humans by the intranasal route [17]. These studies have led to the commercialization of synthetic lypressin (Diapid, Sandoz) and desmopressin (DDAVP, Ferring AB) in the intranasal dosage form.

Adrenal corticotropic hormone (ACTH), a 39-peptide (MW = 4500), is used clinically either by intramuscular or intravenous administration. Because of enzymatic degradation in the GI tract, the oral route of administering ACTH was unsuccessful [18]. However, a pharmacological response was observed in subjects following intranasal administration of ACTH in a suitable formulation [19,20]. Similarly, tetracosatrin—a synthetic polypeptide of corticotropin—was well absorbed via nasal mucosa after administration by nasal insufflation in normal subjects [21]. In 1971, the adrenocorticotrophic action of alpha (1-18)-ACTH was compared in normal subjects following intranasal, intramuscular, subcutaneous, and sublingual administration [22]. The ACTH analog was rapidly absorbed intranasally, intramuscularly, and subcutaneously, but not sublingually.

While the evidence above suggests that the nasal route offers a great potential for drug absorption, only recently has it received significant attention. For instance, studies have been reported that the nasal route might be a simple and practical way for administration of hormones such as gonadotropin releasing hormone (GnRH) and luteinizing hormone-releasing hormone (LHRH) and its analogs as contraceptive agents [23–25]. Buserelin, for example, is a synthetic nonapeptide with a biological activity 50 to 70 times that of natural LHRH. After intranasal administration of buserelin in normal subjects, LH secretion peaked at about 2 hr and lasted for 10 hr. The follicle-stimulating hormone and testosterone response following 800 µg of buserelin intranasally were similar to those observed after administration of 5 µg by the intravenous route. The contraceptive efficacy of buserelin in women suggested that the intranasal administration of buserelin at a daily dose of 400 µg or more might control fertility in humans. Although a much higher dose is required than with the parenteral route, synthetic LHRH analogs were effective when administered intranasally. The low bioavailability is obviously of concern.

The analogs of the naturally occurring enkephalins have been evaluated for their analgesic activity over the last 20 years [26–28]. These peptides have to be given parenterally to have measurable activity. Serum concentrations of enkephalin analogs after intranasal administration in rats were not significantly different from those after injection [29]. The data clearly showed that enkephalins can be delivered by the intranasal route, and the absorption was rapid.

Recent studies have shown that the synthetic human pancreas growth hormone-releasing factor (GHRF) or its analogs may be useful therapeutically to stimulate growth hormone (GH) release [30]. Several reports suggest that treatment with GHRF analogs may require repeated administration of multiple doses during the day and night [30–32]. Chronic intravenous or subcutaneous administration would not be practical or ideal. The effect of GHRF-40 administered in normal men, some adult patients, and some children with GH deficiency was investigated [30]. The results showed that GHRF-40 stimulated GH release following intravenous, subcutaneous, and intranasal administration. However, a higher dose was required for the subcutaneous and intranasal routes to achieve responses comparable to that obtained with intravenous administration. The intranasal route of administering GHRF analogs was effective in stimulating GH release and should be more acceptable for chronic therapy than intravenous and subcutaneous injection.

Another example of a peptide that has been studied extensively is insulin. Type I diabetics often require an injection about 45 min to 1 hr before a meal becuase of the slow absorption and/or action of subcutaneously administered insulin. The intranasal delivery of insulin can result in rapid absorption of circulating insulin to control meal-induced hyperglycemia. Peak concentrations of insulin occur within 5 to 10 min following intranasal administration, the onset of action being as rapid as with intravenous injection. The rapid action of insulin mimics the temporal release of insulin by the normal pancreas. For many diabetics, the flexibility and convenience of intranasal dosing are advantageous.

The idea of delivering insulin intranasally began soon after the discovery of insulin [33,34]. In 1922, the intranasal route was compared with intravenous and other routes of insulin administration [33]. Since then, many clinical investigations on nasal absorption of insulin were conducted with pork and beef insulin [35-42]. More recently, studies on the nasal absorption of recombinant human insulin in rats, dogs, and humans were

Table 1   Biopharmaceutical Aspects of Peptides After Nasal Administration

Peptide	Time of peak level (min)	Relative absorption[a] (%)	Approx. mol. wt.	Testing model
Thyrotropin releasing hormone (TRH)	5-15	10-20	360	Rats, humans
Enkephalin analogs	5-10	70-90	600	Rats, humans
Somatostatin analogs	10-30	65-75	800	Rats
Oxytocin	5-10	30-40	1000	Humans
Vasopressin analogs	10-20	4-12	1070	Humans
LHRH agonists and antagonists	10-30	2-10	1200	Monkeys, humans
Tetracosactrin	40-60	—	2900	Humans
Secretin	2-4	4-6	3000	Rats
Glucagon	5-10	70-90	3500	Humans
Adrenal corticotropic hormone (ACTH)	—	10-14	4500	Humans
Growth hormone releasing factor (GHRF)	20-40	2-20	4800	Rats, dogs, humans
Insulin	5-10	10-40	6000	Rats, dogs, humans

Source: Adapted from Ref. 47, reproduced with permission of the copyright owner, Pharm. Int.

[a]Relative to the IV, SC, or IM dose; based on data from different references.

also reported [43,44]. The results have shown that insulin can be delivered across the nasal mucosa either by mixing the peptide with bile salts or other surfactants or by dissolving it in a relatively acid medium. A marked increase in serum insulin concentrations, sufficient to lower the blood glucose concentration, was observed. The results also indicated that insulin can be absorbed nasally with reproducible kinetics and acceptable efficiency. In 1985, a three-month investigation on an intranasal aerosol formulation of insulin in type I diabetes was reported [45]. The feasibility of using intranasal insulin as an adjunct to subcutaneous insulin was demonstrated.

Based on these experimental and clinical experiences, the nasal route seems to offer a practical method for delivering peptides and proteins. The advantages of rapid absorption, fast onset of action, avoidance of hepatic first-pass metabolism, ease of administration, and utility of chronic medication are well demonstrated in the literature. An extensive review of historic evaluation on drugs for intranasal systemic medications was reported [46]. An overall review of the biopharmaceutical aspects of peptides after intranasal administration is summarized in Table 1.

II. ANATOMY AND PHYSIOLOGY OF THE NOSE

The anatomy and physiology of the nose must be understood to evaluate the intranasal route for drug administration. Several comprehensive books on the nose have been published [48,49]. Here only the subjects that are relevant to the intranasal drug delivery systems are reviewed. Figure 1 is a schematic of the lateral wall of the nasal cavity with identification of areas of interest.

A. Nasal Vestibule and Ostium

The nasal vestibule, within 1.5 cm from the nostril, is lined with the skin that carries a zone of short stiff hairs or vibrissae. The vibrissae form a protective screen or filter by trapping large particles from the incoming airstream. In the anterior area of the nose, there is a constriction known as the internal ostium or nasal valve, beyond which the internal ostium, the nasal cavity, is divided into a large respiratory region and a small olfactory region. The internal ostium plays an important role in nasal functions, as it is the narrowest passage of the entire airway. Naturally, the incoming airstream converges at high linear velocity to pass through the small cross section of the internal ostium and then bends at angles of 60 to 130 degrees from the initial direction [50]. The internal ostium has a cross-sectional area of approximately 0.3 to 0.4 cm^2 per nostril, whereas the total surface area of both nasal cavities is about 160 cm^2 with a total volume of about 20 ml [51].

The shape and size of the vestibule and the valvular slit all contribute to vestibular airflow resistance [52]. As a result, turbulence occurs because of the constriction in the nasal airway and the characteristic of the nasal airflow, which, in turn, makes the disposition of drugs most likely in the anterior nose. To achieve a good drug distribution, the nasal actuator or applicator must be placed at the entrance into the narrowest passage near the nasal ostium.

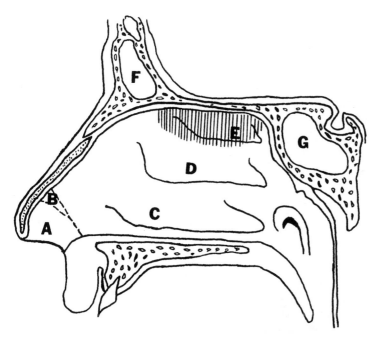

Fig. 1   Lateral wall of the nasal cavity.   (Hatched area) olfactory region;
A, nasal vestibule; B, internal ostium; C, inferior turbinate; D, middle
turbinate; E, superior turbinate; F, frontal sinus; G, sphenoidal sinus.
(Adapted Ref. 49, reproduced with permission of the copyright owner,
Blackwell Scientific Publications, Oxford.)

B. Nasal Epithelium

The anterior one-third of the nasal cavity is covered by a squamous and
transitional epithelium.   The squamous epithelium is essentially lined without microvilli (Fig. 2), while the transitional epithelium is lined with short
microvilli of varying length but uniform within a single cell (Fig. 3).   Beyond this area, the nasal cavity is covered by a mixture of sparsely (Fig.
4) and densely (Fig. 5) ciliated columnar cells.   At the nasopharynx,
which is about 12 to 14 cm from the nostrils, the septum ends and the airway becomes one.   This region gradually shifts to a transition to squamous
epithelium, which continues downward to the larynx [53,54].

C. Nasal Turbinates, Mucus, and Cilia

On the outer sidewall of the nasal cavity are three scroll-shaped bones
called turbinates which are covered and surrounded by soft spongy tissue
with a rich blood supply.   The lateral wall is predominantly covered by
the large inferior turbinate with the orifice of the nasolacrimal duct just
above the nasal floor of cavity.   Deeper in the nasal cavity is the middle
turbinate with the orifices of frontal sinus, anterior ethmoidal sinuses, and
maxillary sinus.   Further back in the nasal cavity is the superior turbinate

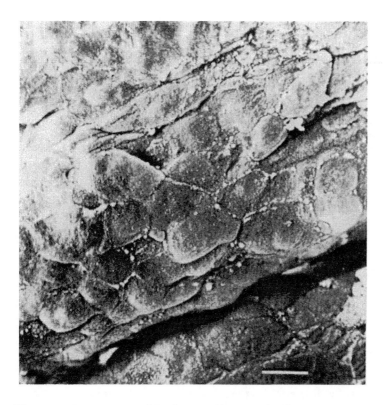

Fig. 2   Squamous epithelium. The bar indicates 10 μm. (From Ref. 48, reproduced with permission of the copyright owner, Elsevier/North-Holland Biomedical Press, Amsterdam.)

with the orifices of posterior ethmoidal sinuses. The superior turbinate is a rudimentary structure of little physiological significance in humans, whereas the middle and inferior turbinates are essential factors in the functional morphology of the nose.

The airway epithelium in the nose consists of four cell types. These are nonciliated columnar cells, goblet cells, basal cells, and ciliated columnar cells (Fig. 6). The goblet cells, together with the nasal glands in the respiratory pathways, produce the mucous secretion. The mucosal surface is covered by a layer of secretions which contain mucous glycoproteins, immunoglobulins (e.g., IgA and IgG), albumin, lysozyme, and other substances. In addition to protecting the nasal mucosa from cold and low humidity and to facilitating olfaction, mucus serves to trap inhaled substances for elimination by mucociliary clearance, coughing, or sneezing. It is believed that cilia beat in the superficial layer of periciliary fluid. The beating of cilia sets up a current such that the upper layer of this fluid is moved in the direction of the forward beat of the cilia, whereas the lower layer remains stationary.

The anatomy of cilia is identical for species from the protozoa to humans based on the studies with transmission electron microscopy [55]. In the human nose, each ciliary cell has about 100 cilia on its surface and each cilium is about 5 μm long and 0.3 μm in diameter [56]. Each beats

with a frequency of about 20 beats per second [57]. The beat rate may
be influenced by the temperature and relative humidity of the ambient air
[58], by the nature of airway secretion [59,60], and by excipients in the
formulation, as discussed in Sec. V.

D. Olfactory Region

The olfactory region, a small patch of tissue containing the smell receptors,
is located at the very top of the nasal cavity near the inner end of the
upper throat. The patch contains several million tiny endings of the olfactory nerve whose bundles pass through the cribriform plate and enter the
farthest forward extension of the brain. The possibility that drugs, absorbed nasally via the olfactory mucosa, enter through the olfactory neurons and the supporting cells directly into the brain as well as into the
systemic circulation in monkeys has been suggested [61].

   The nasal passage is quite different from the rest of the airways in
terms of shape, condition, and surface area. The nasal vascular system is
extremely rich and highly adjustable, making the intranasal administration
of drugs both feasible and desirable.

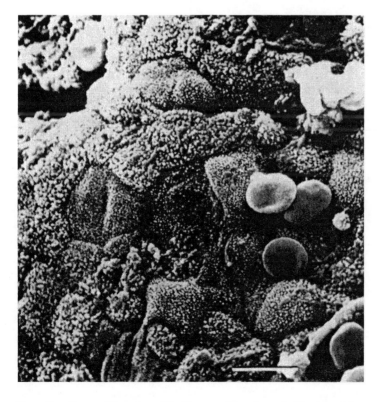

Fig. 3   Transitional epithelium. The bar indicates 10 µm. (From Ref. 48,
reproduced with permission of the copyright owner, Elsevier/North-Holland
Biomedical Press, Amsterdam.)

Fig. 4   Pseudostratified columnar epithelium with few ciliated cells.   The
bar indicates 10 µm.   (From Ref. 48, reproduced with permission of the
copyright owner, Elsevier/North-Holland Biomedical Press, Amsterdam.)

Fig. 5   Pseudostratified columnar epithelium with many ciliated cells.   The
bar indicates 10 μm.   (From Ref. 48, reproduced with permission of the
copyright owner, Elsevier/North-Holland Biomedical Press, Amsterdam.)

E.   Blood Supply of the Nasal Mucosa

The nasal vasculature consists of a rich capillary network in the subepithe-
lium and around the nasal glands and a cavernous plexus deep to the glandular zone [62].   It is characterized by fenestrated endothelium.   The
cavernous plexuses receive blood from the capillary bed and from arteries
by means of arteriovenous anastomoses.   Venous drainage occurs through a
number of venous plexuses, some of which communicate with intracranial
venous sinuses.   Thus the nasal mucosa is obviously well suited for heat
exchange and for potential drug absorption.   In general, drugs absorbed
via the nasal mucosa enter the right side of the heart for distribution to
the systemic arterial circulation prior to traversing the liver.

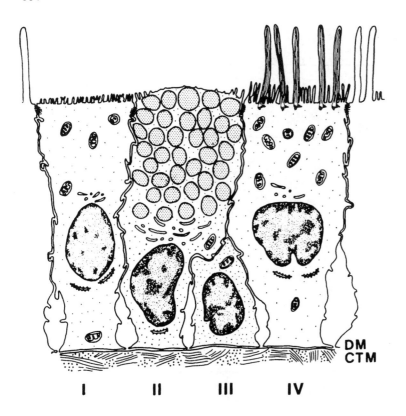

Fig. 6   Transmission electron microscopic diagram of the four cell types
in the nasal epithelium. I, nonciliated columnar cell with microvilli; II,
goblet cell with mucous granules and a well-developed Golgi apparatus; III,
basal cell; IV, ciliated columnar cell with many mitochondria in the apical
part. DM, double membrane; CTM, connective tissue membrane. (From
Ref. 49, reporduced with permission of the copyright owner, Blackwell
Scientific Publications, Oxford.)

III. EVALUATION OF NASAL PEPTIDE ABSORPTION

A. Preformulation Considerations

Peptides and proteins are very much different from the traditional small
organic molecules used as pharmaceuticals. The preformulation data form
the basis for the subsequent consideration of dosage form and/or delivery
system for maximum stability and optimum bioavailability. More broad and
detailed discussion of physical chemistry and physical biochemistry of peptides and proteins is contained in Chap. 4. Several key preformulation
parameters pertinent to nasal peptide and protein drug delivery are discussed briefly below.

Solubility

The solubility of most peptides and proteins is a function of ionic strength,
pH, and the concentration of organic solvents. The solubility of a molecule

often is an extremely important consideration in its biological performance as well as in its incorporation into a specific drug delivery system. For many drugs, the absorption process is limited by solubility [63]. For instance, carboxyhemoglobulin is "salted in" and "salted out" at low and high ionic strength, respectively (Fig. 7), behavior that is typical of peptides and proteins. Moreover, when the solubility of β-lactoglobulin at constant ionic strength is plotted as a function of pH at four different concentrations of NaCl, a U-shaped curve is observed (Fig. 8). In each case, minimal solubility occurs near the pH of 5.2 to 5.4, the isoelectric point. Most organic solvents also affect the solubility of peptides and proteins. Organic solvents usually precipitate proteins. Although very useful and important in protein isolation and purification, this issue is probably less relevant to formulation.

Effects of pH and ionic strength on nasal absorption of peptides and proteins have been demonstrated [40,64]. For instance, insulin is best absorbed at pH 3.0 and least absorbed at pH 6.1 in dogs. Secretin has been observed to be optimally absorbed at an ionic strength of 0.462μ in rats.

Chemical Instability: Hydrolysis, Oxidative and Reductive
Cleavage, and Diketopiperazine Formation

Hydrolysis. Peptides and proteins may be hydrolyzed by heating with acid and base or by the action of proteolytic enzymes. They are hydrolyzed

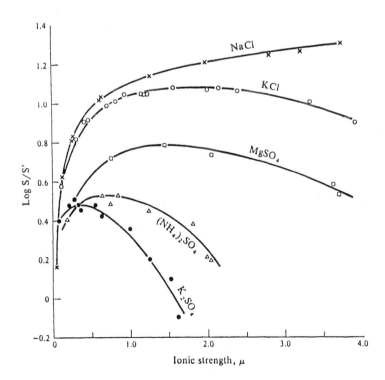

Fig. 7   Solubility of carboxyhemoglobin as a function of ionic strength and ion type. (From Ref. 63, reproduced with permission of the copyright owner, Harper & Row Publishers, New York.)

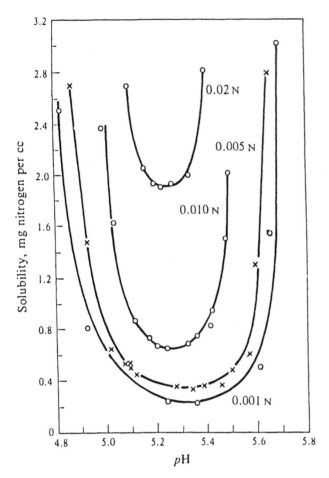

Fig. 8   Solubility of β-lactoglobulin as a function of pH at four different concentrations of sodium chloride. (From Ref. 63, reproduced with permission of the copyright owner, Harper & Row Publishers, New York.)

first to smaller peptides and, ultimately, to the constituent amino acids. As mentioned earlier, the nasal absorption of peptides and proteins is considerably lower than that observed for other nonpeptide drugs administered intranasally. Hydrolysis in the nasal cavity can decrease stability and cause low bioavailability [65].

Oxidative and reductive cleavage.   Polypeptide chains comprising a protein molecule are frequently held together by the disulfide bridge of cystine residues. Oxidation and reduction as well as hydrolysis of chains containing such disulfide linkages may lead to complex mixtures of peptides. For instance, insulin has been reported to break down into its corresponding cysteic acid residues or sulfhydryl form by oxidative and reductive cleavage of the disulfide bridge under certain condtions (Fig. 9) [66].

Diketopiperazine formation.   Alpha-amino acids, upon heating, formed a diketopiperazine, a double amide formation [67]. Although the details of

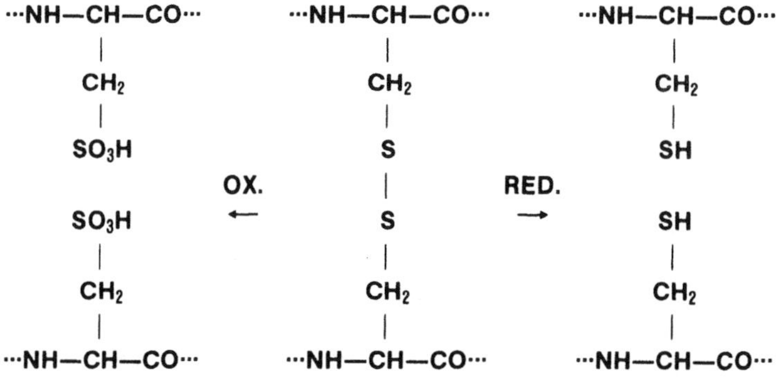

···NH—CH—CO···   ···NH—CH—CO···   ···NH—CH—CO···
     | | |
    CH₂   CH₂   CH₂
     | | |
    SO₃H   S   SH
         OX. | RED.
    SO₃H   ←   S   →   SH
     | | |
    CH₂   CH₂   CH₂
     | | |
···NH—CH—CO···   ···NH—CH—CO···   ···NH—CH—CO···

Fig. 9   Oxidation and reduction of disulfide bridge. (From Ref. 66, reproduced with permission of the copyright owner, *Biochem. J.*)

the mechanism of this reaction are not clear, the reaction can be made to proceed in good yield at room temperature or in the presence of dehydrating agents. This reaction was observed in the stability study on metkephamid, a pentapeptide of enkephalins for nasal absorption [68]. The peptide bond at the second and third amino acid residues of metkephamid molecule was cleaved at room temperature. After cyclization, a diketopiperazine and a tripeptide were found (Fig. 10). It is clear that during the course of

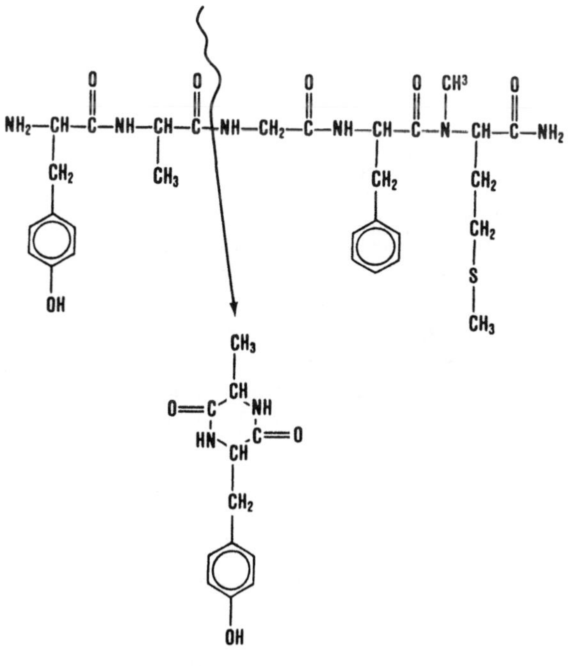

Diketopiperazine product

Fig. 10   Diketopiperazine formation from a pentapeptide.

evaluating nasal absorption of peptides and proteins, conditions that will affect their stability must be considered.

Physical Loss Through Adsorption and Denaturation

Adsorption. The adsorption of peptides and proteins onto glass and plastic surfaces is a common phenomenon and is well documented [69–71]. Adsorption is more significant at low than at high concentration. The adsorption process onto glass surfaces has been ascribed to ionic amine-silanol bonding which can be prevented by silylating the glass with organosilanes such as Prosit 28 [72] or by including additives such as the phosphate and acetate ions [71]. Adsorption to other surfaces, such as filters, syringes, and tubing, varies depending on the nature of the materials used. It is important to determine the loss due to adsorption in the manufacturing process and under storage conditions. The possibility of loss on the delivery of peptide and protein drug substances at low concentration must be considered. The adsorption of insulin onto containers, syringes, and infusion apparatus result in underdosage to patients [73].

Denaturation. Peptides and proteins can be denatured when subjected to extremes of pH, heat, violent shaking and foaming, organic solvent, or high concentration of heavy metal ions at room temperature. The characteristics of denaturation are a decrease in solubility, loss of crystallinity, and loss of specific biological activity. Other characteristics resulting from denaturation include alteration in surface tension, loss of enzyme activity, and perhaps loss or alteration of antigenicity.

B. Experimental Models

Two types of experimental models have been used in intranasal drug absorption studies, in vivo and in situ perfusion.

In Vivo Models: Rat, Dog, and Sheep

Rat model. The rat model used in our laboratory is adapted from the literature with minor modifications [74,75]. The surgical preparation for in vivo nasal absorption is as follows. The rat is anesthetized by an intraperitoneal (IP) injection of sodium pentobarbital. After an incision is made in the neck, the trachea is cannulated with a polyethylene tube. Another tube is inserted through the esophagus toward the posterior part of the nasal cavity (Fig. 11). The passage of the nasopalatine tract is sealed to prevent the drainage of drug solution from the nasal cavity into the mouth. The drug solution is delivered to the nasal cavity through the esophagus cannulation tubing instead of from the nostril as described in the literature. The blood samples are then collected periodically from the femoral vein.

Since all the possible outlets in this rat model are blocked after surgical preparation, the only possible passage for the drug to be absorbed and transported into the systemic circulation is penetration and/or diffusion through the nasal mucosa.

Dog model. The in vivo dog model for nasal absorption studies is as follows. A male dog is anesthetized or maintained in the conscious state depending on the characteristics of the drug and the purpose of the study. Anesthesia is initiated by an intravenous (IV) injection of sodium thiopental and maintained with sodium phenobarbital. A positive pressure pump

Rat Model

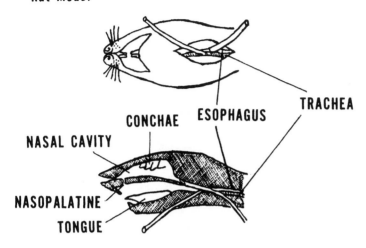

CONCHAE   ESOPHAGUS   TRACHEA

NASAL CAVITY

NASOPALATINE

TONGUE

Fig. 11   Top and side views of the cannulation arrangement in the rat
model.   (From Ref. 74, reproduced with permission of the copyright owner,
J. Pharm. Sci.)

provides ventilation through a cuffed endotracheal tube and a heating pad
keeps the body temperature at 37 to 38°C.   The blood samples are collected
periodically from the jugular vein.

Sheep model.   The in vivo sheep model for nasal absorption is basically
similar to that described for the dog model.   Male in-house-bred sheep free
of nasal infections are used.   Because of their larger nostril and body size
compared to the rat model, both dog and sheep models are suitable and
practical for evaluating nasal delivery parameters with more sophisticated
formulations.   A kinetic study of the absorption profile within a single animal can easily be conducted.

In Situ Nasal Perfusion Technique

Figure 12 shows the experimental setup for the in situ nasal perfusion
studies [76].   The surgical preparation is as described in the in vivo rat
model.   A funnel is provided between the nose and the reservoir to minimize loss of drug solution.   The testing drug solution is placed in the
reservoir and is maintained at 37°C.   The drug solution is circulated
through the nasal cavity of the rat by means of a polystaltic pump.   The
perfusion solution passes out from the nostrils, through the funnel, and
into the reservoir again.   The drug solution in the reservoir is stirred
constantly.   The amount of drug absorbed is then determined by measuring
the drug concentration remaining in the perfusing solution.

C. Assay Methodology

During the evaluation of a peptide or protein molecule for possible nasal
absorption, development of a sensitive analytical method for assaying the

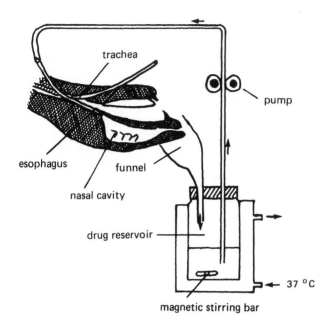

Fig. 12 Experimental setup for in situ nasal perfusion studies in the rat model. (From Ref. 76, reproduced with permission of the copyright owner, *Int. J. Pharm.*)

material and its possible metabolites in the biological fluids is essential. Methods of peptide and protein analysis have been described in Chaps. 5 and 6. The analytical approaches commonly used in nasal peptide and protein absorption studies are described briefly here.

For Radiolabeled Molecules

Radiolabeled molecules are often used in the preliminary assessment of absorption from a new drug delivery system. The disadvantage of this method is that only a gross estimation of absorption due to the sum measurement of parent molecule and metabolites. A detailed quantitative analysis should be performed. The radiolabeled biological fluids can be treated with the following two methods.

Acid digestion method. Samples are weighed and then acid digested in 0.2 ml of 70% $HClO_4$ and 0.4 ml of 30% H_2O_2 at 70°C in scintillation vials for 1 hr. The samples are then cooled down to room temperature and counted after the addition of scintillation fluid.

Combustion method. Samples are weighed and combusted in a sample oxidizer with a fluid mixture to trap $^{14}CO_2$ and then counted on a scintillation spectrometer. Urine samples can be analyzed by liquid scintillation counting.

For a quantitative analysis of peptides and proteins in the biological fluids, a more complex and sensitive method is required. After extraction, isolation, and purification, samples are then analyzed by liquid scintillation counting. A specific detector may be required for enhancing the sensitivity.

An example of qualitative and quantitative studies on nasal absorption of peptides was illustrated in the literature [29].

For Unlabeled Molecules

For quantitative analysis of unlabeled peptides and proteins, a high-performance liquid chromatography method is generally used. After samples are prepared through the processes of extraction, isolation, separation, and derivatization (if necessary), samples are analyzed by HPLC equipped with an ultraviolet, electrochemical, or fluorescent detector. Other methods such as radioimmunological assay (RIA) and colorimetric assay are also used.

IV. MEANS TO FACILITATE NASAL PEPTIDE AND PROTEIN ABSORPTION

Small peptides that have been delivered via the nasal mucosa include oxytocin, vasopressin and its analogs, and enkephalins. Nasal absorption is probably by passive diffusion. This is so at least for metkephamid [29]. As shown in Table 2, the area under the curve of serum metkephamid levels versus time increased with increasing of dose. The extent of absorption (AUC) per unit dose remained relatively constant. Thus the amount of metkephamid absorbed nasally was directly proportional to the dose administered according to the law of diffusion.

Drugs can also be absorbed nasally by special transport mechanisms. For instance, sodium guaiazulene-3-sulfonate (GAS) is absorbed via the nasal membrane by at least two kinds of transport systems [77]. The

Table 2   Effect of Dose on the Nasal Absorption of Metkephamid in Rats[a]

Route of administration	n	AUC[b]	Mean AUC[c]	Bioavailability (%)
Intravenous				
25 mg/kg	3	186.71 ± 34.74	7.47	—
Intranasal				
50 mg/kg	3	273.40 ± 35.57	5.47[d]	73.2
25 mg/kg	4	190.38 ± 65.33	7.61[d]	101.9
12.5 mg/kg	3	60.51 ± 15.57	4.84[d]	64.8

Source: Ref. 29, reproduced with permission of the copyright owner, *J. Pharm. Sci.*

[a]Metkephamid was dissolved in saline, and the volume was maintained at 0.1 ml and was given as drops.

[b]AUC: $\mu g \cdot min/ml$, mean ± SEM, integrated from 0 to ∞.

[c]Mean specific AUC: $(\mu g \cdot min/ml)/(mg/kg)$ = mean AUC/administered dose.

[d]$p > 0.1$, comparison between various doses of nasal administration.

in situ nasal perfusion study in rats showed that the nasal absorption rate changed biphasically depending on the initial GAS concentration in the perfusate. At the low concentration range, the absorption rate of GAS decreased with increasing of concentration. Moreover, the nasal absorption was inhibited by $HgCl_2$ and ouabain. It was suggested that a carrier-mediated transport mechanism was involved. However, at the higher concentration range, the absorption rate from the nasal mucosa increased significantly with increasing concentration, and ouabain had no inhibitory effect. This was attributed to ion-pair formation between GAS and a mucosal constituent on the basis of two lines of evidence: (1) The apparent partition coefficient of GAS increased as the concentration of GAS was increased, and (2) GAS was found to be readily transferred to the organic phase by ion-pair complex formation with cations such as NH_4^+.

Unlike small peptides, larger peptides and proteins do not cross the nasal membrane efficiently. An adjuvant is usually required for facilitating the nasal absorption of larger peptides and proteins. For example, the nasal route of delivering insulin in humans required a promoter to produce the desired pharmacological response. The nasal absorption of nafarelin—a LHRH analog—in monkeys and humans was slow, whereas the bioavailability was greatly improved in the presence of a promoter in the formulation [78].

Various types of adjuvants have been used. These can be classified as surfactants, chelators, bile salts, and others, such as saponins [79—81] and fusidic acid derivatives [82]. Examples of surfactants include polyoxyethylene lauryl ether, polyoxyethylene cetyl ether, polyoxyethylene stearyl ether, polyoxyethylene sorbitan monooleate, and polyoxyethylene monolaurate [79]. Examples of chelators include EDTA and salicylates [83—85]. Examples of bile salts include sodium cholate, sodium glycocholate, sodium deoxycholate, sodium tauracholate, sodium glycodeoxycholate, sodium taurodeoxycholate chenodeoxycholic acid, and ursodeoxycholic acid [79,86]. Examples of fusidic acid derivatives include sodium taurodihydrofusidate and sodium dihydrofusidate [82].

Several mechanisms can be used to facilitate nasal peptide and protein absorption.

A. Modification of pH

Peptides and proteins often exhibit the lowest solubility at their isoelectric point. Adjusting the pH away from the isoelectric point of a peptide can increase the solubility. For example, insulin was demonstrated to cross the nasal membrane in dogs and rats in an acidic medium [40,79]. The least nasal absorption was observed at pH about 6.1, which is close to the isoelectric point of insulin. The final pH of the commerical product DDAVP was adjusted to 4.0 to maintain good solubility [87].

B. Dissociation of Aggregation

Many surfactants are known to promote dissociation and even denaturation of proteins in vitro. Proteins may form higher-order aggregates in solution. For example, at pH 7.0, insulin in solution exists predominatly as hexameric aggregates. Insulin did not cross the nasal membrane without an enhancer in the formulation, whereas good nasal absorption of insulin was observed with sodium deoxycholate as the enhancer [42]. The effect of sodium deoxycholate on the physical properties of insulin was studied

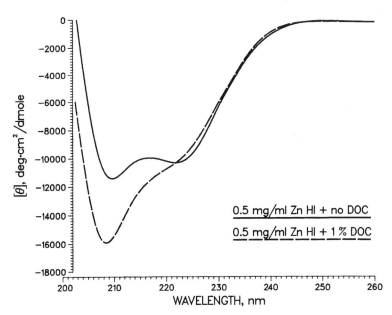

Fig. 13   Effect of 1% sodium deoxycholate on the circular dichroic spectra of human zinc insulin in the far-UV region.   All solutions are formulated at pH 7.3.

using circular dichroism spectral analysis and other techniques [43,44]. Plotted results of mean molar ellipticity versus wavelength are vastly different for zinc insulin in the absence of sodium deoxycholate compared to its presence.   In the far-UV region (Fig. 13), the 209-nm peak is greatly enhanced from $-11,477$ to $-16,042$ deg·cm^2/dmol with a slight blue shift, while the 222 nm peak is reduced to a shoulder.   In the near-UV region (Fig. 14), the maximum for zinc insulin at 276 nm is greatly reduced in magnitude from -265 to -191 deg·cm^2/dmol and blue shifted to 271 nm in the presence of sodium deoxycholate in the formulation.   The circular dichroic properties suggested that sodium deoxycholate may disrupt the formation of insulin hexamers and higher-order aggregates.   Whether sodium deoxycholate prevents the formation of dimers is not known.   The significance of the 5-nm blue shift in the near-UV extremum is presently not understood.   Based on the data, it was assumed that dissociation of insulin hexamers to dimers and monomers in the presence of sodium deoxycholate was partly responsible for enhancing the transport of insulin across the nasal epithelium.

C.   Reverse Micelle Formation

Bile salts affect the micellar properties of biomembranes [88].   Because of these properties, bile salts promote the transmembrane movement of endogenous and exogenous lipids, and other polar substances within the gastrointestinal tract [89–91].   As a result, the potential application of bile salts as an adjuvant for transmucosal delivery of drugs has been widely explored [41,79,92,93].   For instance, the effect of the physicochemical characteristic

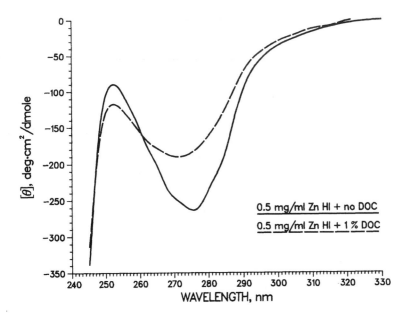

Fig. 14   Effect of 1% sodium deoxycholate on the circular dichroic spectra of human zinc insulin in the near-UV region. All solutions are formulated at pH 7.3.

of bile salts on nasal absorption of insulin in terms of hydrophobicity to adjuvant activity was extensively studied [86]. The more hydrophobic bile salts demonstrated greater adjuvant activity in the order of deoxycholate, chenodeoxycholate, cholate, and ursodeoxycholate. A similar phenomenon was also observed between glycocholate and glycodeoxycholate, or between taurocholate and taurodeoxycholate. The adjuvant potency for nasal insulin absorption correlates positively with increasing hydrophobicity of the bile salts. The data suggested that the hydrophobicity of the steroid nucleus is the major determinant of adjuvant activity. It was also reported that insulin absorption begins at the aqueous critical micellar concentration (CMC) of the bile salts and becomes maximal when micelle formation is well established. Furthermore, the differing adjuvant activities of various bile salt species relates to their differing capacities to penetrate and self-associate as reverse micelles within native membranes as they do in nonpolar solvents or when dispersed in pure phospholipid conditions [89]. In reverse micelles, the hydrophilic surfaces of the molecules face inward and the hydrophobic surfaces face outward from lipid environment (Fig. 15). Therefore, reverse micelles could act as transmembrane channels or mobile carriers for insulin to move down an aqueous concentration gradient through the nasal mucosal cells, into the intercellular space, and into the bloodstream.

D. Membrane Transport and Enzyme Inhibition

Effects of surfactants on drug absorption have been the subject of many studies. For example, the relationship between the absorption-promoting effect of surfactants and their effect on biomembrane was studied in terms of hemolytic activity and protein-releasing effect on the nasal mucosa [79]. The promoting effect of several nonionic surfactants, sodium lauryl sulfate and saponin, paralleled their ability to lyze the rabbit erythrocyte and to release protein from the nasal mucosa.   The data suggested that the effect of these surfactants on the permeability of the nasal mucosa to insulin may be due to the perturbation or creating disorder of the structural integrity of the nasal mucosa.   On the other hand, bile salts showed less effect on biomembrane than those of other surfactants.   A lower hemolytic activity with less protein release from the nasal mucosa was observed.

The effect of bile salts on the activity of proteolytic enzymes in the nasal mucosa was distinctly different from those of the other types of surfactants (Table 3).   The addition of bile salts resulted in a significant inhibition of the enzymatic activity of leucine aminopeptidase and of the enzymatic degradation of insulin.   The results suggested that bile salts affect not only the permeability of the nasal mucosa but also the activity

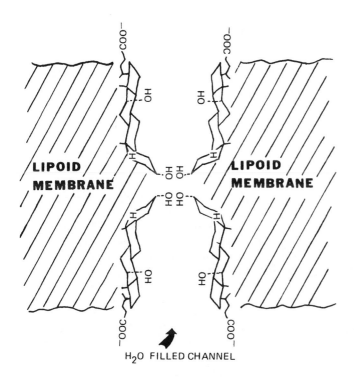

Fig. 15   Schematic molecular model of reverse micelle formation on cell membranes. Two pairs of sodium deoxycholate molecules are shown stacked end to end, spanning a lipid bilayer and forming an aqueous pore for the transport of insulin monomers from the extracellular space, where high concentrations of insulin monomers are solubilized in mixed bile salt and insulin micelles. (From Ref. 86, reproduced with permission of the copyright owner, *Proc. Natl. Acad. Sci. USA.*)

Table 3   Effects of Surfactants on Biological Properties of Nasal Mucosa

Surfactant	Released protein (μg/min/g[a])	Hemolysis (ml/g)	Enzymatic activity[b] inhibition (%)
POE(5) lauryl ether	0.36 ± 0.03	10.0	19.7
POE(9) lauryl ether	0.83 ± 0.13	25.0	22.7
POE(10) monolaurate	0.26 ± 0.03	3.3	16.9
POE(40) monostearate	0.17 ± 0.01	0.1	20.1
POE(20)sorbitan monoolearate	0.17 ± 0.01	0.07	0.8
Sodium lauryl sulfate	1.99 ± 0.07	14.3	98.3
Sodium taurocholate	0.29 ± 0.00	2.5	87.4
Sodium cholate	0.36 ± 0.03	3.3	84.9
Sodium glycocholate	0.21 ± 0.02	3.3	87.2
Saponin	0.77 ± 0.01	20.0	36.9

Source: Ref. 79, reproduced with permission of the copyright owner, *Int. J. Pharm.*

[a]Grams of body weight.

[b]Leucine aminopeptidase activity.

of proteolytic enzyme.   Thus the enhancement of nasal absorption of insulin by bile salts in rats may result from both these mechanisms.

In short, many individual observations on the effect of various adjuvants in promoting nasal absorption of peptides have now been reported.   On the basis of these results, it is already possible to deduce certain phenomena and to draw conclusions with respect to the possible mechanisms of adjuvants in enhancement of absorption.   Because of the great differences between various peptide drug substances and the forms in which they are applied, from the author's experience in this area, it is impossible to find a universal adjuvant for the enhancement of nasal absorption.   Biopharmaceutical studies must be conducted for each peptide and protein to evaluate whether a given adjuvant is suitable for use as an absorption promoter.   Uncontrolled high absorption may increase the toxicity of the peptide and protein drug substances.   On the other hand, a possible interaction with the adjuvant could also influence the activity of the peptides and proteins.

E.   Increased Nasal Blood Flow

Enhancement of nasal peptide absorption by increasing local nasal blood flow was demonstrated [94].   Presumably this is due to an increase in the concentration gradient for passive peptide diffusion.   Agents that possess vasoactive properties and enhance nasal mucosal blood flow include histamine [95], prostaglandin E_1 [96], and the beta-adrenergic agonists [97].

The effects of histamine on nasal blood flow and on nasal absorption of desmopression were evaluated in normal volunteers [94]. The data showed that the combination of histamine and desmopressin resulted in increased nasal blood flow response and the antidiuretic activity of desmopressin, as compared with controls. Furthermore, the increase in duration of the systemic activity of desmopressin is consistent with increased nasal absorption of the peptide. The results suggest that improving nasal blood flow may enhance absorption of peptides and proteins.

V. NASAL PEPTIDE DELIVERY SYSTEMS

A. Types

The efficiency and efficacy of peptides administered intranasally depend on the method of delivering drug to the nasal mucosa. Due to the different experimental procedures, substances, and clinical protocols used in the studies, different conclusions have been drawn with respect to the selection of various delivery systems used for administering drugs to the nasal membrane such as nasal drops, nasal sprays, and nasal inhalers [98—100]. To make a rational judgment on the selection of a delivery system for intranasal administration, the comparative study must be conducted under well-controlled conditions using the same experimental design. More specifically, the drug used for the study must be suitably prepared for all methods of administration to be evaluated. Thus the investigation requires that an agent be formulated either in solution or in volatile form and administered as nasal drops, nasal sprays, and nasal inhalers.

A comparative study on the efficacy of nasal delivery systems in humans was reported by using a vasoconstrictor agent [98]. The results indicate that nasal sprays and nasal inhalers give similar results in terms of intensity and duration of effects produced, while nasal drops are far less effective as a method of medication.

The effects of prolonged administration on nasal mucosa were also conducted in rabbits for 90 days and in a dog for 180 days by comparing these various delivery systems. Repetitive administration with nasal sprays and nasal inhalers produced far less pathologic change than did the use of nasal drops. The severe pathologic changes produced by nasal drops as compared with sprays and inhalers are not entirely clear. Nevertheless, the data suggested that in conditions requiring prolonged and repeated administration, nasal drops should be used with caution. It was also observed that when the nasal spray or the nasal inhaler is used, the drug reaches the nasal mucosa in a more diffuse form and is distributed to greater areas of the mucosa. More recently, the patterns of deposition and the rates of clearance of human serum albumin delivered by nasal sprays and nasal drops were monitored by gamma scintigraphy (Fig. 16) [99—101]. Clearance was faster following administration of the nasal drops. Therefore, nasal sprays or nasal inhalers are suggested as the delivery system of choice.

B. Drug Distribution in the Nose

Because the method of delivery will result in different drug distribution in the nose, it will subsequently influence the sites of deposition, the degree

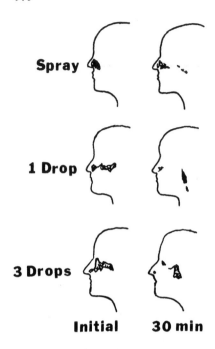

Spray

1 Drop

3 Drops

Initial   30 min

Fig. 16   Sites of deposition and patterns of clearance following administration of nasal sprays and drops. Each pair of images is of the same subject, but the three sets are of different subjects. (From Ref. 99, reproduced with permission of the copyright owner, *J. Pharm. Pharmacol.*)

of absorption, and the efficacy of drug. Using a cast of the human nose (Fig. 17), significant differences in drug distribution from the various delivery devices have been observed [102]. Delivery of a large volume of a solution from the nasal drop device obviously gives a good distribution over the nasal cavity. However, it is probably unacceptable for other reasons. Whereas, after application of a few drops of a solution from a nasal drop device, the drug distribution in the nasal cavity is unsatisfactory (Fig. 18). Thus proper instruction is required to maximize the efficiency of this delivery system.

An atomized metered-dose mechanical pump spray produces a good drug distribution (Fig. 19) and delivers a much more constant dose than nasal drops or other plastic bottle nebulizers. By contrast, the mucosal distribution of drug delivered from a pressurized metered-dose aerosol was not distributed evenly over the nasal cavity (Fig. 20). There was much drug distributed in the septal wall and little in the lateral wall. This drawback can be overcome by delivering drug twice in each nostril, one "puff" in the upper direction and one "puff" in the lower direction.

Given the current state of advanced aerosol technology, there are probably only two types of intranasal delivery devices worthy of consideration: metered-dose pump and metered-dose aerosol. With proper instruction to the patients, both devices produce good distribution in the nasal cavity as well as accurate dose dispension and are commonly used. The selection of either a metered-dose pump or a metered-dose aerosol as the delivery

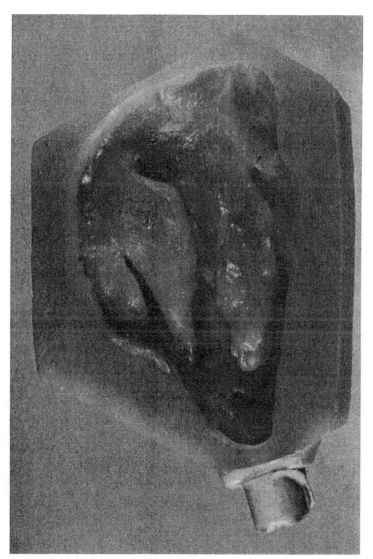

Fig. 17   Cast of the artificial human nose.   (From Ref. 103, reproduced with permission of the copyright owner, Elsevier Science Publishing Co., Amsterdam.)

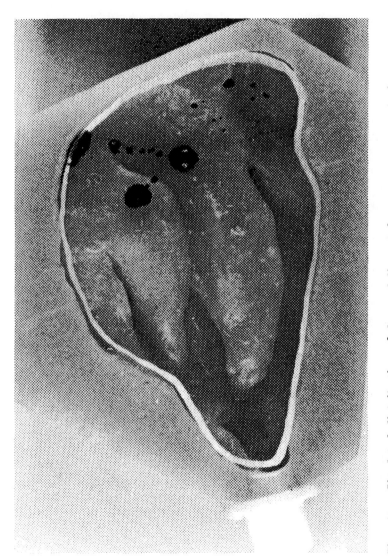

Fig. 18 Simulated distribution of drops delivered from a nasal drop device. (From Ref. 49, reproduced with permission of the copyright owner, Blackwell Scientific Publications, Oxford.)

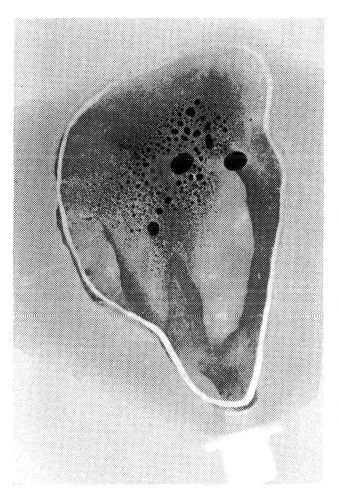

Fig. 19 Simulated distribution of dosage delivered from an automized metered-dose pump. (From Ref. 49, reproduced with permission of the copyright owner, Blackwell Scientific Publications, Oxford.)

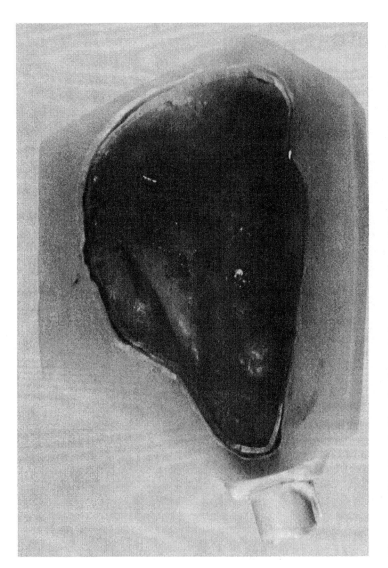

Fig. 20   Simulated distribution of dosage delivered from a pressurized metered-dose aerosol. (From Ref. 103, reproduced with permission of the copyright owner, Elsevier Science Publishing Co., Amsterdam.)

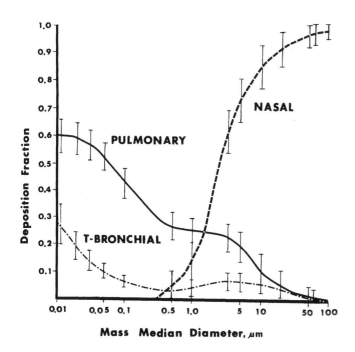

Fig. 21   Relationship between drug deposition fraction and mass medium diameter.

system for intranasal peptides and proteins administration entirely depends on the nature of the drug. An overall review of aerosol formulation has been described elsewhere [103,104].

C. Particle Size Distribution

Particle size is an important parameter determining penetration of the dispensed particles into the respiratory tract. The relationship between drug deposition fraction and mass median diameter is well established (Fig. 21). When mass median diameter is less than 3 μm, it has a good deposition in the pulmonary airways. On the contrary, when the mass median diameter is greater than 10 μm, it has an optimal deposition in the nasal cavity. The reader is referred to the text by Silverman et al. [105] for a detailed treatment in the methods of measuring particle size distribution.

D. Nasal Clearance Process

Nasal mucociliary transport carries materials, including excess secretions or foreign substances, backward to the nasopharynx. This material is transported by a wiping action of cilia to the stomach periodically through swallowing. There is also a small area in the anterior part of the nasal cavity, where ciliary movement carries materials forward to the anterior nares from where they can be removed by nose blowing, wiping, or sneezing. In recent years, nasal mucociliary transport has been studied extensively in this area of ciliary motion and airway secretions [106].

Table 4 Effects of Preservatives on Ciliary Movement

Compound	Freq.[a] t = 20 min	Freq.[b] t = 60 min	$t_{50\%}$[c] (hr)	$t_{0\%}$[d] (hr)	Re[e] (hr)
Lipophilic					
Chlorobutol, 0.5%	0%	0%	0.04	0.08	+
Chlorocresol, 0.05%	0%	0%	0.01	0.02	+
Methyl-p-hydroxybenzoate, 0.15%	0%	0%	0.06	0.33	+
Propyl-p-hydroxybenzoate, 0.02%	0%	0%	0.10	0.33	+
Polar					
Benzalkonium chloride, 0.01%	100%	65%	1.13	>2	−
Benzalkonium chloride, 0.01% + EDTA, 0.05%	75%	58%	1.12	>2	−
Benzalkonium chloride, 0.01% + EDTA, 0.1%	50%	26%	0.33	1.33	−
Others					
EDTA 0.1%	61%	49%	0.86	>2	++

Source: Adapted from Ref. 107, reproduced with permission of the copyright owner, *Rhinology*.

[a]Percentage of frequency of the reference after contact of 20 min.

[b]Percentage of the frequency of the reference after a contact of 60 min.

[c]Exposure time in hours after which the frequency was 50% of the reference frequency.

[d]Exposure time in hours after which motility lacked completely.

[e]Reversibility: ++, complete; +, dependent on exposure time; −, nihil.

Under normal conditions, the major clearance process for the airways is by mucociliary transport. Thus it is important that the integrity of the nasal clearance mechanism must be maintained intact for removing dust, allergens, and bacteria. However, the mechanism can be influenced in many ways, such as by the drug and the additives in its formulation. Van de Donk et al. [107] demonstrated that preservatives can diminish the ciliary movement in the nasal cavity and trachea, which, in turn, can often cause rhinities, sinusitis, and other airway infections. Table 4 shows the effects of some of the preservatives on the onset, speed, and reversibility of ciliary movement. While the long-term effects of even a temporary impediment to the mechanism of nasal clearance are unknown, the design of an intranasal system for administering peptides and proteins should be considered carefully to avoid substantial interference with mucociliary activity.

E. Evaluation of Nasal Irritation

The possibility that good nasal absorption of a compound could be attributed to damage to the mucosa was reported [75]. Irritation associated with lesion

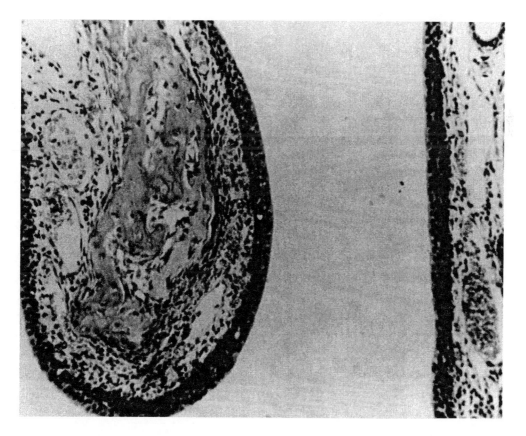

Fig. 22   Nasal mucosa of untreated control rat. (From Ref. 29, reproduced with permission of the copyright owner, *J. Pharm. Sci.*)

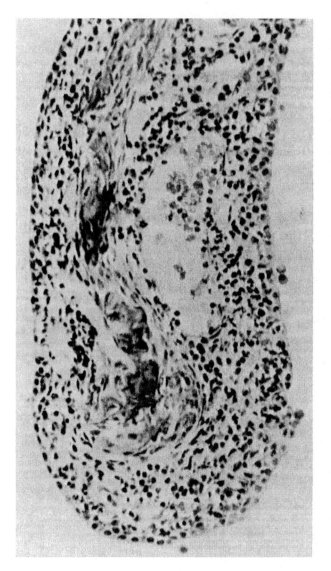

Fig. 23   Nasal mucosa of rat exposed to 25 mg of metkephamid/kg for 5 hr.   (From Ref. 29, reproduced with permission of the copyright owner, *J. Pharm. Sci.*)

of the mucosal membrane or transient sensation of burning and stinging must be evaluated systematically [108, 109]. Typical microscopic specimens of exposed nasal mucosa in rats following nasal administration of peptides are shown in Figs. 22 and 23. Histologically, the surface of the nasal mucosa is smooth in the control group, whereas the debris of sloughed mucosa cells and inflammation can be detected if the testing material is an irritant. Therefore, histopathological studies have to be conducted in animal species to ascertain the integrity of the nasal mucosa before testing in humans [103].

F. Other Factors

Proteins and peptides are subjected to degradation by proteases and peptidases during passage through the mucosal membrane [110] and/or while at the surface of mucosa. For these reasons it is possible that only modest improvements of absorption over the oral delivery have been observed in many reports, even in the presence of absorption enhancers. The issue of enzymatic barriers to peptide absorption appears to have been ignored until recent years [79,111,112]. In general, there are three types of aminopeptidases present in the nasal mucosa. These are plasma membrane-bound peptidases; aminopeptidase N, aminopeptidase A, and a cytosolic enzyme, aminopeptidase B. Their enzymatic activity and effect on hydrolysis of peptides and proteins are discussed in detail in Chap. 7. The implication of enzymatic barriers to nasal absorption and the application of an enzymatic inhibitor and/or competitor into a formulation are not well understood. Other factors need to be considered, such as blood flow, relative humidity, and temperature, which may influence either nasal absorption or nasal clearance.

VI. CONCLUSIONS

The intranasal route of drug administration will undoubtedly be an increasingly important means for delivering drugs to the systemic circulation. Particularly, the intranasal delivery of peptides and proteins is a promising alternative to injection as a route for administration. The highly vascularized surface of nasal mucosa for absorption, bypassing of hepatic first-pass metabolism, rapid onset of action, and avoidance of injections are advantages of intranasal drug delivery. Nonetheless, numerous factors impede the efficiency of intranasal absorption of drugs, such as common cold, allergic rhinitis, and abnormal nasal conditions. However, the application of scientific methodology to the physical, physiological, and clinical issues with respect to these factors could lead to improvement and/or resolution. With the current trend of medical rediscovery of peptides and proteins as future drug candidates, further studies of the intranasal route as the delivery system seem justified.

ACKNOWLEDGMENTS

The author would like to express his appreciation to Dr. R. W. Fuller and Dr. A. Bingham for their valuable discussion and to Ms. A. Simpson for her skillful assistance in the preparation of the manuscript.

REFERENCES

1. J. Hofbauer and J. K. Hoerner, *Am. J. Obstet. Gynecol.*, *14*:137 (1927).
2. D. G. Morton, *Am. J. Obstet. Gynecol.*, *26*:323 (1933).
3. C. Hendricks and R. A. Gabel, *Am. J. Obstet. Gynecol.*, *79*:789 (1960).
4. C. H. Hendricks and S. V. Pose, *J. Am. Med. Assoc.*, *175*:384 (1961).
5. J. E. Clement, V. C. Harwell, and J. R. McCain, *Am. J. Obstet. Gynecol.*, *83*:778 (1962). \
6. E. Talledo, S. F. Adams, and F. P. Zuspan, *J. Am. Med. Assoc.*, *189*:348 (1964).
7. J. Laine, *Acta Obstet. Gynecol.*, *Scand.*, *49*:149 (1970).
8. R. T. Hoover, *Am. J. Obstet Gynecol.*, *110*:788 (1971).
9. N. Altszuler and J. Hampshire, *Proc. Soc. Exp. Biol. Med.*, *168*:123 (1981).
10. V. U. Guhl, *Schweiz. Med. Wochenschr.*, *91*:798 (1961).
11. A. R. Spiegelman, *J. Am. Med. Assoc.*, *184*:657 (1963).
12. D. Barltrop, *Lancet*, *1*:276 (1963).
13. R. Fraser and D. J. Scott, *Lancet*, *1*:1159 (1963).
14. A. M. Dashe, C. R. Kleeman, J. W. Czaczkes, H. Rubinoff, and I. Spears, *J. Am. Med. Assoc.*, *190*:1069 (1964).
15. S. B. Chirman and L. W. Kinsell, *Calif. Med.*, *101*:1 (1964).
16. W. Hung, *Med. Ann. D.C.*, *36*:400 (1967).
17. K. E. Anderson and B. Arner, *Acta. Med. Scand.*, *192*:21 (1972).
18. E. B. Astwood, M. S. Raben, and R. W. Payne, *Recent Prog. Horm. Res.*, *7*:1 (1952).
19. F. Paulson and S. Nordstrom, *Swed. Med. J.*, *49*:2998 (1952).
20. J. B. R. McKendry, H. Schwartz, and M. Hall, *Can. Med. Assoc.*, *70*:244 (1954).
21. J. Keenan and M. A. Chamberlain, *Br. Med. J.*, *4*:407 (1969).
22. J. Keenan, J. B. Thompson, M. A. Chamberland, and G. M. Besser, *Br. Med. J.*, *3*:742 (1971).
23. G. I. Zatuchni, J. D. Shelton, and J. J. Sciarra, *LHRH Peptides as Female and Male Contraceptives*, Harper & Row, New York, (1976).
24. G. Fink, G. Gennser, P. Leidholm, J. Thorrell, and J. Mulder, *J. Endocrinol.*, *63*:351 (1974).
25. S. T. Anik, G. McRae, C. Nerenberg, A. Worden, J. Foreman, J. Y. Hwang, S. Kushinsky, R. E. Jones, and B. Vickery, *J. Pharm. Sci.*, *73*:684 (1984).
26. R. C. A. Frederickson, *Life Sci.*, *21*:23 (1971).
27. M. Motta, *The Endocrine Functions of the Brain*, Raven Press, New York (1980), p. 233.
28. J. D. Leander and C. R. Wood, *Peptides*, *3*:771 (1982).
29. K. S. E. Su, K. M. Campanale, L. M. Mendelsohn, L. G. Kerchner, and C. A. Gries, *J. Pharm. Sci.*, *74*:394 (1985).
30. S. M. Rosenthal, E. A. Schriock, S. L. Kaplan, R. Guillemin, and M. M. Grumback, *J. Clin. Endocrinol. Metab.*, *57*:677 (1983).
31. W. S. Evans, J. L. C. Borges, D. L. Kaiser, M. L. Vance, R. P. Sellers, R. M. Macleod, W. Vale, J. Rivier, and M. O. Thorner, *J. Clin. Endocrinol. Metab.*, *57*:1081 (1983).

32. W. S. Evans, M. L. Vance, D. L. Kaiser, R. P. Sellers, J. L. C. Borges, T. R. Downs, L. A. Frohman, J. Rivier, W. Valve, and M. O. Throner, *J. Clin. Endocrinol. Metab.*, *61*:846 (1985).
33. R. T. Woodyatt, *J. Metab. Res.*, *2*:793 (1922).
34. N. F. Fisher, *Am. J. Physiol.*, *67*:65 (1923).
35. W. Heubner, S. E. deJough, and E. Laquer, *Klin. Wochenschr.*, *3*: 2342 (1924).
36. W. S. Collens and M. A. Goldzieher, *Proc. Soc. Exp. Biol. Med.*, *29*: 756 (1932).
37. R. H. Major, *Am. J. Med. Sci.*, *192*:257 (1936).
38. J. Hankiss and C. S. Hadhazy, *Acta Med. Acad. Sci. Hung.*, *12*:107 (1958).
39. T. Yokosuka, Y. Omori, Y. Hirata, and S. Hirai, *J. Jpn. Diabetes Soc.*, *20*:146 (1977).
40. S. Hirai, T. Ikenaga, and T. Matswzawa, *Diabetes*, *27*:296 (1978).
41. A. E. Pontiroli, M. Alberetto, A. Secchi, G. Dossi, I. Bosi, and G. Possa, *Br. Med. J.*, *284*:303 (1982).
42. A. C. Moses, G. S. Gordon, M. C. Carey, and J. S. Flier, *Diabetes*, *32*:1040 (1983).
43. K. S. E. Su, J. Q. Oeswein, and K. M. Campanale, *APhA Acad. Pharm. Sci.*, *15*:89 (1985).
44. K. S. E. Su, D. C. Howey, K. M. Campanale, and J. Q. Oeswein, *Diabetes*, *35*:64A (1986).
45. R. Salzman, J. E. Manson, G. T. Griffing, R. Kimmerle, N. Ruderman, A. McCall, E. L. Stoltz, C. Mullin, D. Small, J. Armstrong, and J. C. Melby, *N. Engl. J. Med.*, *312*:1078 (1985).
46. Y. W. Chien and S. F. Chang, in *Transnasal Systemic Medications: Fundamental, Developmental Concepts and Biomedical Assessments* (Y. E. Chien, ed.), Elsevier Science Publishing Co., Amsterdam (1985), pp. 1–99.
47. K. S. E. Su, *Pharm. Int.*, *7*:8 (1986).
48. D. F. Proctor and I. B. Anderson, *The Nose: Upper Airway Physiology and the Atmospheric Environment*, Elsevier/North-Holland Biomedical Press, Amsterdam, (1982), pp. 1–464.
49. N. Mygind, *Nasal Allergy*, Blackwell Scientific Publications, Oxford, (1979), pp. 3–333.
50. D. F. Proctor, *Am. Rev. Respir. Dis.*, *115*:97 (1977).
51. D. L. Swift and D. F. Proctor, in *Respiratory Defense Mechanisms* (J. D. Brain et al., eds.), Marcel Dekker, New York (1977), Chapter 3.
52. Haevan Dishoek, *Int. Rhinol.*, *3*:19 (1965).
53. M. Y. Ali, *J. Anat.*, *99*:657 (1965).
54. N. S. Bryant, *J. Anat.*, *50*:172 (1916).
55. P. Satir, *Environ. Health Perspect.*, *35*:77 (1980).
56. G. Ewert, *Acta Oto-Laryngol. Suppl.*, *200*:1 (1965).
57. G. A. Laurenzi, *J. Occup. Med.*, *15*:175 (1973).
58. U. Mercke, C. H. Hakansson, and N. G. Toremalm, *Acta Oto-Laryngol.*, *78*:444 (1974).
59. J. R. Blake, *Respir. Physiol.*, *17*:394 (1973).
60. J. R. Blake and M. A. Sleigh, *Biol. Rev.*, *49*:85 (1974).
61. P. G. Gopinath, G. Gopinath, and T. C. Anand Kumar, *Curr. Ther. Res.*, *23*:550 (1985).

62. E. Zuckerkandl, *Wien. Med. Wochenschr.*, *38*:1121 (1884).

63. H. R. Mahler and E. H. Cordes, *Biological Chemistry*, Harper & Row Publishers, New York, (1971), pp. 99–101.

64. T. Ohwak, H. Ando, S. Watanabe, and Y. Miyake, *J. Pharm. Sci.*, 74:550 (1985).

65. A. Hussain, J. Faraj, and J. E. Truelove, *Biochem. Biophys. Res. Commun.*, *123*:923 (1985).

66. F. Sanger, *Biochem. J.*, *39*:507 (1945).

67. J. March, *Advanced Organic Chemistry: Reactions, Mechanisms, and Structures*, McGraw-Hill Book Company, New York (1968), p. 337.

68. K. S. E. Su, *Proc. Am. Pharm. Assoc., Acad. Pharm. Sci. Basic Symp.*, San Francisco, *16*:17 (1986).

69. T. Mizutani, *J. Pharm. Sci.*, *79*:493 (1981).

70. J. Ogino, K. Noguchi, and K. Terato, *Chem. Pharm. Bull.*, 27:3160 (1979).

71. S. T. Anik and J. Y. Hwang, *Int. J. Pharm.*, *16*:181 (1983).

72. R. A. Messing, *J. Non-Cryst. Solids*, *19*:277 (1975).

73. E. W. Kraegen, L. Lazarus, H. Meler, L. Campbell, and Y. O. Chia, *Br. Med. J.*, *23*:464 (1975).

74. A. A. Hussain, S. Hirai, and R. Bawarshi, *J. Pharm. Sci.*, *68*:1141 (1980).

75. K. S. E. Su, K. M. Campanale, and C. L. Gires, *J. Pharm. Sci.*, 73:1251 (1984).

76. S. Hirai, T. Yashiki, T. Matsuzawa, and H. Mima, *Int. J. Pharm.*, 7:317 (1981).

77. J. Seki, H. Mukai, and M. Sugiyama, *J. Pharmacobio-Dyn.*, *8*:337 (1985).

78. S. T. Anik, *Proc. Land O'Lake*, *86*, Merrimac, Wis., Lecture Note, (1986), p. 85.

79. S. Hirai, T. Yashiki, and H. Mima, *Int. J. Pharm.*, *9*:173 (1981).

80. F. Lasch and S. Brigel, *Arch. Exp. Pathol. Pharmakol.*, *116*:7 (1926).

81. F. O. W. Meyer, *Pharm. Ztg.*, *80*:329 (1935).

82. J. P. Longenecker, A. C. Moses, J. S. Flier, R. D. Silver, M. C. Carey, and E. J. Dubor, *J. Pharm. Sci.*, 76:351 (1987).

83. W. M. Sweeney, S. M. Hardy, A. C. Dornbusch, and J. M. Ruegsegger, *Antibiot. Med. Clin. Ther.*, *4*:642 (1957).

84. E. Windsor and G. E. Gronheim, *Nature (London)*, *190*:263 (1961).

85. D. C. Monkhouse and G. A. Groves, *Australas. J. Pharm.*, *48*:53 (1967).

86. G. S. Gordon, A. C. Moses, R. D. Silver, J. S. Flier, and M. C. Carey, *Proc. Natl. Acad. Sci. USA*, *82*:7419 (1985).

87. E. R. Barnhart, *Physicians' Desk Reference*, 41st ed., Medical Economics Company, Oradell, N.J. (1987), p. 2015.

88. A. Helenius and K. Simons, *Biochim. Biophys. Acta*, *415*:29 (1975).

89. M. C. Carey, in *Liver, Biology and Pathobiology* (Z. M. Arias, H. Popper, D. Schacter, and D. Shafritz, eds.), Raven Press, New York (1982), p. 429.

90. C. Tagesson and R. Sjodahl, *Eur. Surg. Res.*, *16*:274 (1984).

91. J. W. Dobbins and J. H. Binder, *Gastroenterology*, *70*:1096 (1976).

92. E. Ziv, A. Eldor, Y. Kleinman, H. Baron, and M. Kidron, *Biochem. Pharmacol.*, *32*:773 (1983).

93. T. Murakami, Y. Sasaki, R. Yamajo, and N. Yata, *Chem. Pharm. Bull.*, *32*:1948 (1984).

94. L. S. Olanoff and R. E. Gibson, *J. Clin. Invest.*, *80*:890 (1987).

95. M. Bende, A. Elner, and P. Ohlin, *Acta Oto-Laryngol.*, *97*:99 (1984).

96. H. Bisgaard, P. Olsson, and M. Bende, *Prostaglandins*, *27*:599 (1984).

97. P. Malm, *Eur. J. Respir. Dis.*, *64* (*Suppl. 28*):139 (1983).

98. D. B. Butler and A. C. Ivy, *Arch. Otolaryngol.*, *39*:109 (1944).

99. J. G. Hardy, S. W. Lee, and C. G. Wilson, *J. Pharm. Pharmacol.*, *37*:294 (1985).

100. F. Y. Aoki and J. C. W. Crawley, *Br. J. Clin. Pharmacol.*, *3*:869 (1976).

101. A. S. Harris, I. M. Nilsson, Z. G. Wagner, and U. Alkner, *J. Pharm. Sci.*, *75*:1085, (1986).

102. M. Mygind and S. Vesterhange, *Rhinology*, *11*:79 (1978).

103. K. S. E. Su and K. M. Campanale, *Transnasal Systemic Medications: Fundamental, Developmental Concepts and Biomedical Assessments* (Y. E. Chien, ed.), Elsevier Science Publishing Co., Amsterdam (1985), pp. 139–160.

104. S. P. Newman, *Inhal. Ther.*, *76*:194 (1984).

105. L. Silverman, C. E. Billings, and M. W. First, *Particle Size Analysis in Industrial Hygiene*, Academic Press, New York (1971).

106. D. F. Proctor, *Am. Rev. Respir. Dis.*, *115*:97 (1977).

107. J. H. M. van de Donk, I. P. Uller-Plantema, J. Zuidema, and F. W. H. M. Merkus, *Rhinology*, *18*:119 (1980).

108. C. S. Barrow, Y. Alarie, J. C. Warrick, and M. F. Stock, *Arch. Environ. Health*, *32*:68 (1977).

109. J. T. Young, *Fundam. Appl. Toxicol.*, *1*:309 (1981).

110. H. Klostermyer and R. E. Humbel, *Angew. Chem. Int. Ed. Engl.*, *5*:807 (1966).

111. R. E. Stratford and V. H. L. Lee, *J. Pharm. Sci.*, *74*:731 (1985).

112. R. E. Stratford and V. H. L. Lee, *Int. J. Pharm.*, *30*:73 (1986).

14

Vaginal Route of Peptide and Protein Drug Delivery

Hiroaki Okada

Takeda Chemical Industries, Ltd., Osaka, Japan

I. INTRODUCTION

Most peptides and proteins, including endogenous hormones, enzymes, the other biological response modifiers, and their potent modified analogs, have been administered parenterally, owing to their low permeability through mucous membranes and their rapid degradation by local enzymes. However, the congenital and acquired metabolic disorders for which these hormones are needed usually require long-term therapy. Moreover, the rapid disappearance of most peptides from the body requires multiple dosing to maintain the therapeutic effect. Many attempts have therefore been made to develop a convenient and reliable method for nonparenteral self-administration, such as oral, sublingual, buccal, nasal, lung, rectal, and vaginal application.

In this chapter, vaginal application as a rational systemic delivery system for peptides and proteins is considered. The vaginal preparations on the market are restricted exclusively to those that are topically effective: (1) anti-bacterial, -fungal, -protozoal, -chlamydial, and -viral agents; (2) estrogens, which restore the vaginal mucosa; (3) prostaglandins, which induce labor and therapeutic abortion; and (4) spermicidal agents or steroids, which act as contraceptives. Most of these agents act directly on the vaginal membrane, and except for some steroids and prostaglandins, their absorption into the systemic circulation has been considered only from the standpoint of toxicity. Aref et al. [1] and Benziger and Edelson [2] have presented excellent reviews on the vaginal absorption of various kinds of substances. To date, the absorption of peptides and proteins through the vagina has not been studied systematically. Nonetheless, it has been known for some time that certain peptides and proteins, hydrophilic and large molecular compounds, such as insulin [3,4], TSH [5], peanut protein [6], and various antigens [7,8], are definitely absorbed from the vaginal membrane. However, their bioavailability seems to be insufficient and variable and not suitable for systemic therapy.

Substances are generally transported through biological membranes by transcellular and paracellular routes. It is possible that a considerable amount of peptides and proteins is transported through the vagina predominantly by the latter [50]. Vaginal absorption is further characterized by (1) the feasibility of self-insertion, and prolonged retention of the delivery system; (2) partial avoidance of first-pass effects [9–12]; and (3) ovarian dependency with the reproductive cycle.

In this chapter we summarize the anatomical and physiological aspects of the vagina, discuss the current understanding of the absorption of peptides and proteins from the vagina, and point out how to approach rational peptide and protein delivery by the vaginal route.

II. ANATOMY AND PHYSIOLOGY OF THE VAGINA

A. Human

The morphology and physiology of the human vagina have been described in detail [13–21]. It is a complex organ with multiple functions in reproduction. It is the final part of the internal female genitalia, the parturient canal. It serves as a passage for the outflow of cervical fluids and the menstrual flow.

The vaginal tract is divided into two segments: an anterior wall (about 8 cm) and a posterior wall (about 11 cm). The posterior vaginal fornix possesses a peritoneal lining (Fig. 1). Puncturing of the peritoneal cavity is possible by mechanical piercing of the vaginal vault during induced abortion.

The vaginal wall membrane consists of the epithelial layer, the muscular layer, and the tunica adventitia (vaginal fascia) [15]. The vaginal fascia is part of the visceral pelvic fascia and consists of loose connective tissue. The epithelial layer consists of an epithelial lamina and a lamina propria. The epithelium is composed of noncornified, stratified squamous cells, similar to those of the buccal mucosa. It is characterized by distinct changes due to fluctuations in blood ovarian steroid concentrations associated with aging (neonate, juvenile, adult, and senescence), biphasic sexual cycling (follicular and luteal phases), and pregnancy.

The epithelium in the neonate is very thick, owing to high maternal blood levels of estrogen, but flattens within 3 weeks after parturition. Following puberty, it increases in thickness and gains resistance. In the adult stage, the vaginal surface during the follicular phase appears homogeneous, with large superficial polygonal cells with a high degree of proliferation and the presence of cornification (Fig. 2A and B) [16]. The narrow intercellular edges are of dense structure, corresponding to the cellular cluster of a nonkeratinized squamous epithelium, and showing fine webbing and anastomosing intercellular bridges between individual cells. During the luteal phase, desquamation occurs on the superficial epithelial layer extending as far as the intermediate cells. The vaginal surface loses its intact structure once luteinization starts (Fig. 2C). Zipperlike intercellular connections with desmosomes appear on the open-ended microridges

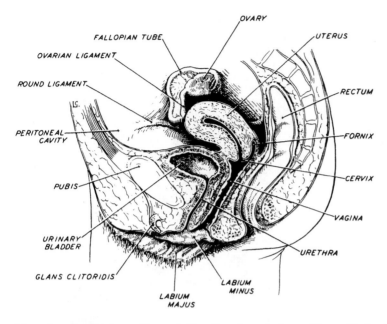

Fig. 1   Sagittal section of the female pelvis. (From Ref. 14.)

of desquamating cells (Fig. 2D). This cyclic desquamation is preceded by
loosening of intercellular grooves (Fig. 2E) as well as a porelike widening
of intercellular crevices following ovulation (Fig. 2F).

The detailed cross-sectional ultrastructure of the human vaginal
epithelium is illustrated in Fig. 3 [17]. It is composed of five different
cell layers: basal (single row), parabasal (two rows), intermediate (about
10 rows), transitional (about 10 rows), and superficial (about 10 rows).
The cyclical variations of the vaginal epithelium generally involve proliferation, differentiation, and desquamation. The intermediate, transitional,
and superficial layers are strongly affected by the cycle and become the
thickest layers at ovulation. During the normal cycle, the number of layers varies from 22 on day 10, to 45 on day 14 (ovulation), to 33 on day
19, and back to 23 on day 24.

During the follicular phase estrogen stimulates mitoses in the basal and
parabasal layers, which are the germinal beds of the epithelium. This proliferation of cells leads to an increase in epithelial thickness as well as in
the number of layers (Fig. 3C). A parallel increase in the number of
intercellular junctions renders the epithelium more cohesive. The number
of desmosomes also increases approximately 10 times from the early to late
follicular phases. The intercellular channels are narrow during early
follicular phase but become widened at ovulation and during the luteal
phase (Fig. 3A and B). Leukocytes migrating from the lamina propria
dilate these channels remarkably. During the luteal phase the proteins of
the lamina propria (i.e., albumin and globulins) transudate through the
dilated intercellular channels.

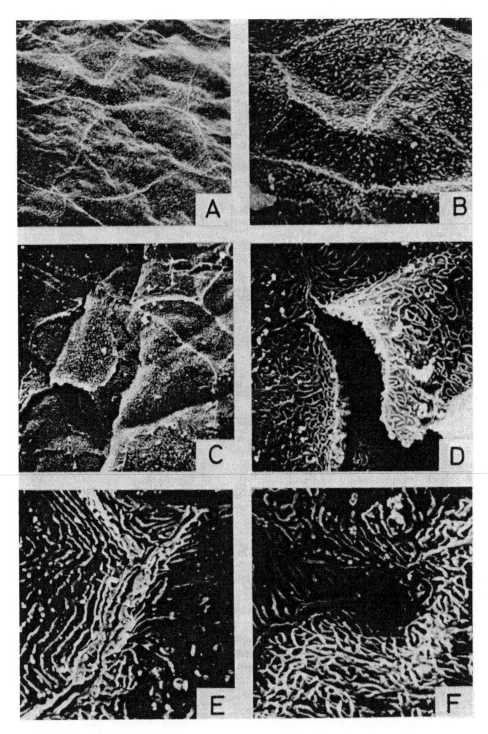

Fig. 2   Epithelial surface of the human vagina during the follicular phase.
(A) × 1260 and (B) × 3150, and, during the luteal phase, (C) × 1260,
(D) × 3150, (E) and (F) × 6300. (From Ref. 16.)

The lamina propria of the vaginal mucosa is a specialized supporting structure for the epithelial cells. It contains a blood supply, a lymphatic drainage system, and a network of nerve fibers (Fig. 3c). It is composed of a dense connective tissue, consisting of collagen fibers, ground substance, and cells. The ground substance represents a pool of proteins for nutrition of the vaginal mucosa. The cells are fibroblasts, macrophages, mast cells, lymphocytes, plasma cells, neutrophils, and eosinophils. Each cell type plays an important role in the physiology of the vagina. Plasma cells are the chief producers of immunoglobulins. Leukocytes penetrate the intercellular channels and separate the vaginal epithelial cells, opening desmosomes and dilating the intracellular space, and finally, mechanically detach the surface epithelium in the desquamation process during the luteal phase [17].

In postmenopausal women the vaginal epithelium becomes extremely thin. The cell boundaries in the surface are less distinct, the microridges are dramatically reduced, and the vagina is often invaded with leukocytes [18]. The incidence of mitosis in the basal and parabasal layers and the number of small blood vessels decreases. By contrast, the concentration of immunoglobulins and the number of plasma cells increase.

During pregnancy the most marked change occurring in the vagina is increased vascularity and venous stasis [19]. The epithelial layer is greatly thickened. On the extended vaginal surface, distended and densely convoluted intercellular microridges and tender cellular borders have been seen [15]. Following delivery, the vagina requires several weeks to reestablish its prepregnancy appearance.

Although the vaginal epithelium is aglandular, it is usually covered with a surface film of moisture, "the vaginal fluids" [20]. In nonmenstrual woman the vaginal fluid is composed mostly of cervical fluid and small amounts of the secretion from Bartholin's glands in the vaginal wall. At ovulation the amount of fluid increases by mixing with the uterine fluid, oviductal fluid, follicular fluid, and even peritoneal fluid. The vaginal fluid contains carbohydrate from the epithelial glycogen, amino acids, aliphatic acids, and proteins.

The pH in the vaginal lumen is controlled primarily by lactic acid produced from the cellular glycogen or carbohydrates by action of the normal microflora, Döderlein's bacilli [19]. Glycogen is deposited in cells of the vaginal epithelium under the influence of estrogen. The vagina of the newborn has a pH of about 5.7. Döderlein's bacilli appear about the fourth day, and the pH falls to about 4.8. It rises to neutrality beginning on the eighth day, and neutral values persist until the onset of puberty. Following the onset of menarche, the pH falls and varies between 4.0 and 5.0, depending on the stage of the cycle. The lowest values are found at midcycle and the highest during menstruation. This change parallels the curves of estrogen excretion for the normal cycle. The lowest values are found near the anterior fornix, intermediate ones in the mid-vagina, and the highest ones near the vestibule. During pregnancy the pH varies between 3.8 and 4.4, owing to the increase in cellular glycogen content. In the postmenopausal state, the pH increases from 4.5–5.5 to 7.0–7.4, owing to a decrease in cellular glycogen content. This acidity plays a clinically important role in preventing the proliferation of pathogenic bacteria.

The arterial blood supply in the vagina is derived from the visceral branches of the internal iliac artery [15]. The main venous drainage from the vaginal plexus occurs via the uterine vein to the internal iliac vein.

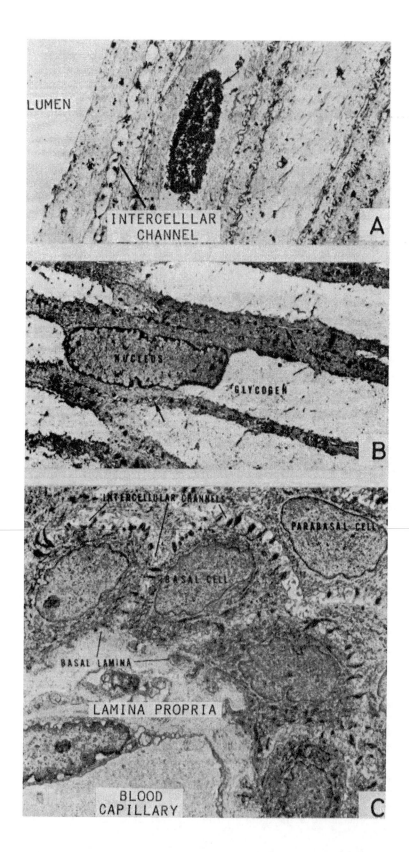

Fine-meshed networks of lymph and blood capillaries are located in the lamina propria and tunica muscularis of the vagina and are connected with those of the uterus, rectum, urinary bladder, and exterior genital organs. Lymph drainage from the vagina takes place to the iliac sacral, gluteal, rectal, and inguinal lymphatic nodes.

B. Experimental Animals

In most animals the vagina is regulated by cyclic alteration of the reproductive system [14]. The cyclic alterations are directly regulated by hormones such as estrogens, progesterone, LH, and FSH derived from the anterior pituitary-gonadal axis. The release of these hormones is also controlled by stimulation from the hypothalamus, which is further conditioned to a large extent by the levels of steroid hormones in the circulation and by external environmental stimuli such as light, temperature, nutritional status, and social relationships.

Vertebrates fall into two categories: those that breed continuously throughout the year, and those that restrict reproductive events to a particular season. Most vertebrates have cyclic alterations of the reproductive system, the estrous cycle, which is fundamentally comparable to the menstrual cycle in primates (about 28 days in length) (Fig. 4) [21]. The latter is characterized by periodic vaginal bleeding that occurs with the breakdown of the endometrium at the end of the cycle. Among nonhuman primates, monkeys and baboons exhibit an ovarian cycle and reproductive system similar to those of the human female. These animals also have a comparable vaginal anatomy and physiology.

The rat, mouse, and guinea pig have short estrous cycles, which are completed in 4 to 5 days; the timing of the cycle is influenced by environmental factors. The changes in the rat vagina are shown in Fig. 5 [14]. During estrus many mitoses occur in the vaginal mucosa, and the superficial layers become squamous and cornified (Fig. 5C). Large numbers of cornified epithelial cells with degenerate nuclei are exfoliated into the lumen as a result of estrogen action. During metestrus many leukocytes appear in the vaginal lumen along with a few cornified cells (Fig. 5D). During diestrus the vaginal mucosa is thin and leukocytes migrate through it (Fig. 5A). During proestrus the vaginal wall is the thickest and consists of fresh cells (Fig. 5B). The vaginal smear is dominated by nucleated epithelial cells. Ovariectomy causes a marked involution of the vagina; the vaginal mucous becomes thin and mitotic divisions are seldom encountered (Fig. 5E). The effect of these changes on vaginal peptide and protein absorption will be amplified subsequently.

Fig. 3   Epithelial membrane of the human vagina. (A) Superficial cell layer during the luteal phase. Asterisk represents the moderate opened intercellular channel with fewer desmosomes; arrow indicates the pyknotic nucleus (× 8300). (B) Transitional cell layers rich in glycogen during the follicular phase. Arrows indicate narrow intercellular channels (× 3800). (C) Basel and parabasal cell layers, and the lamina propria with blood capillary during the follicular phase (× 4500). (From Ref. 17.)

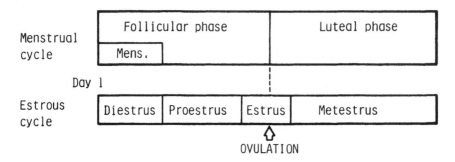

Fig. 4 Timing of estrous cycle and menstrual cycle. Day 1 of a menstrual cycle is the first day of bleeding, menstruation (Mens.), while day 1 of an estrous cycle is the first day of estrus.

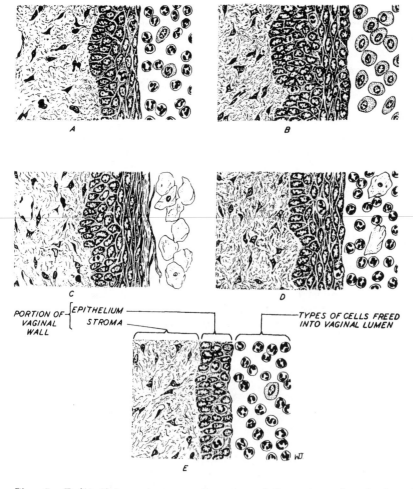

Fig. 5 Epithelial membrane and smears of the rat vagina during different stages of the estrous cycle. (A) Diestrus, (B) proestrus (C) estrus (D) metestrus, (E) ovariectomized rats for 6 months. (From Ref. 14.)

Adult nonpregnant domestic rabbits, often called "induced ovulators," are in a constant state of estrus, and ovulation is induced by coitus or comparative cervical stimulation [14]. The ovaries of wild rabbits, however, are inactive in winter but enlarge in spring. Cats and ferrets exhibit a similar type of estrous cycle.

Ovulation in dogs is spontaneous, and there are generally two estrous periods per year [14]. Proestrus lasts for about 10 days and is followed by a 6- to 10-day estrus. Loss of blood occurs through the vagina during proestrus. Each estrus is followed by a functional luteal phase lasting approximately 60 days.

III. THEORETICAL ASPECTS OF VAGINAL ABSORPTION

Substances are transported through mucous membranes by transcellular and intercellular routes, as illustrated in Fig. 6. Higuchi et al. [22] have proposed a physiological model comprised of an aqueous diffusion layer in series with the membrane that consists of parallel lipoidal and aqueous pore pathways. This model is quite consistent with experimental data for the permeability of aliphatic alcohols and carboxylic acids in the rabbit. The model equations are as follows:

$$P_{app} = \cfrac{1}{\cfrac{1}{P_{aq}} + \cfrac{1}{P_m}} \tag{1}$$

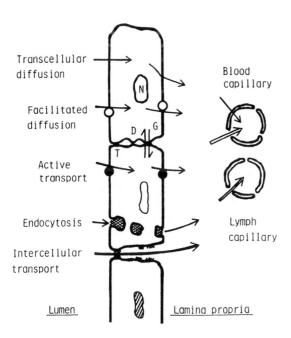

Fig. 6   Diagram of the transport pathway through the vaginal epithelium. T, tight junction; D, desmosome; G, gap junction; N, nucleus.

$$P_{aq} = \frac{D_{aq}}{h} \tag{2}$$

$$P_m = \left(\frac{[H]_s}{K_a + [H]_s} \right) P_l + P_p \tag{3}$$

In these equations, P_{app} is the apparent permeability coefficient and P_{aq} is the permeability coefficient for the aqueous unstirred diffusion layer. The latter is equal to the diffusion coefficient in the diffusion layer (D_{aq}) divided by the thickness of effective diffusion layer (h) on the vaginal epithelium [Eq. (2)]. The permeability coefficient for the membrane (P_m) is equal to the sum of the permeability coefficient through the lipid pathway (P_l) (only for the undissociated acid fraction), and through the pore pathway (P_p). K_a represents the dissociated constant of the acid, and $[H]_s$ the membrane surface hydrogen-ion concentration.

In these experiments, increased chain length resulted in an increased rate of absorption. It is thought that these lipophilic substances are absorbed transcellularly. The absorption of a peptide and protein drug may also be represented by Eqs. (1)–(3), but the apparent permeability coefficient for the membrane predominantly depends on that for the pore pathway. It will be influenced by the physiologic changes of the vaginal epithelium in thickness and number of intercellular pores and aqueous channels. The charge of the membrane surface, pore, and drugs must also be considered in permeating the epithelium. Peptides and proteins are susceptible to self-association and aggregation in the medium due to changes in pH, ionic strength, and concentration and to the presence of additives. It is anticipated that the monomer, oligomer, or aggregated complex may each have its characteristic diffusion and permeation coefficient.

Since the vagina is not the absorption site of nutrients or exogenous substances, its epithelium has no villi or cilia but consists of a keratinized stratified squamous epithelium providing a tight barrier against environmental substances. Nevertheless, many peptide hormones, antigenic proteins, and water-soluble molecules such as penicillin [23] can be absorbed intact. As the vaginal epithelial membrane barrier becomes loose and porous during the luteal and early follicular phases, large and water-soluble substances can be adsorbed in larger quantities by the intercellular route. It should be noted that permeability results in larger fluctuations than occurs with lipophilic compounds transported by the transcellular route. Nevertheless, bioavailability from the vaginal route is expected to be greater than that from the gastrointestinal route, due to high intercellular permeability and reduced first-pass effects. The latter is attributed to the presence of fewer enzymes in the vaginal epithelium and to the avoidance of the vaginal blood drainage into the portal vein and liver. Reduced first-pass effects after vaginal application of estrogens, but not of protein compounds, have been reported [9–12]. We have confirmed less proteolytic activity in 10% homogenate of the rat vaginal membrane than in the small intestine by the agar plate method [38] using casein as a substrate. However, a comparative study of the aminopeptidase activities against enkephalins in the absorptive mucous membranes in the rabbit demonstrated that the supernatants of homogenates of the vaginal, nasal, buccal, rectal, and ileal mucous membranes exhibited similar activity [39]. Nonetheless, the morphological organization of the enzymes and the probability of encounters between the peptides and the enzymes during the penetrating process should be considered in addition to the gross activity. The latter is affected by the presence of proteases on the dimeric IgA-rich glycocalyx mucus barrier [33,40], membraneous or lysosomal proteolysis, diffusion or active transport, receptor- or nonreceptor-mediated endocytosis [30,41,42], and transcellular or intercellular transport.

That carrier-mediated transport systems for di- and tripeptides exist in the intestine is well known [27]. By contrast, there is as yet no evidence for carrier-mediated transport of peptides across the vaginal mucosa, although prostaglandins have been demonstrated to utilize such a mechanism [24-26]. Receptor-mediated endocytosis of peptides and proteins into cells, accompanied by coated pits and vesicles, has recently become recognized as an important transport system in nature [28-30]. The transport of peptides and proteins, such as low-density lipoprotein, insulin, epidermal growth factor, nerve growth factor, transcobalamin II, and transferrin by this route plays an essential role in the growth, nutrition, and differentiation of cells. Furthermore, nonreceptor-mediated endocytosis (bulk fluid pinocytosis) [30-33] and the nonpinocytotic receptor-mediated transport [28,34-37] of bacterial and plant toxins [34,35] and endogenous proteins— "the signal hypothesis"—[36] are well known. In the vagina, seminal and bacterial antigens are assumed to be transported through nonspecifically intercellular channels [50]. There must be active transport of endogenous peptides in an individual epithelial cell of the vagina to regulate proliferation, but no active transport of exogenous peptides, such as receptor-mediated endocytosis or bulk-fluid endocytosis, across the vaginal membrane has been reported.

IV. EXPERIMENTAL APPROACHES TO STUDYING VAGINAL ABSORPTION

A. Method

In determining vaginal absorption, three points must be taken into account: (1) the animal species to be used, (2) whether overall absorption or the transport process is the focus of concern, and (3) the control of experimental or physiological variables.

Vaginal drug absorption has not been investigated to the same degree as gastrointestinal drug absorption. Higuchi et al. [47,48] have described perfusion systems to determine the vaginal permeability to aliphatic alcohols and carboxylic acids. Figure 7 displays one such system for use in the monkey. The rib-cage cell, made of Teflon ending, a stainless steel frame, and perfusion tubing, was inserted into the vagina through the vulval orifice to maintain a constant volume and surface area of the absorption compartment. Since the permeability to peptides and proteins is expected to be low, this method, which monitors the disappearance of a test substance from the perfusate, may not be suitable, as the change in concentration will be minute.

Rapid drainage of a test sample from the vaginal lumen must be prevented if absorption is to be quantitatively assessed. To overcome this, specific vehicles, such as creams, gels, tablets, suppositories, and tampons, are utilized. Figure 8 shows the vaginal application of an antigen solution using three formed plastic sponges in monkey [49]. In experiments with anesthetized rats we have directly closed the vaginal orifice with a surgical

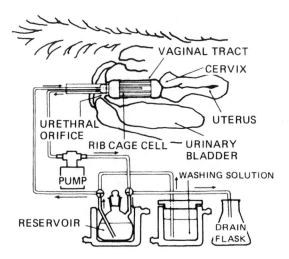

Fig. 7   Vaginal perfusion system.   (From Ref. 48.)

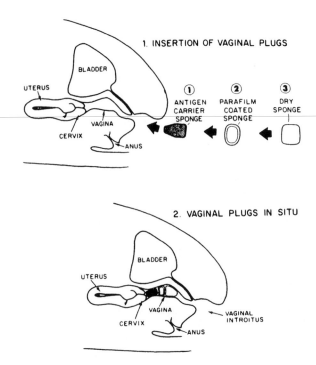

Fig. 8   Diagram of vaginal application by plastic sponges in the monkey.
(From Ref. 49.)

adhesive agent (polyalkylcyanate) after the drug has been applied. Here an auxiliary application device must be utilized to introduce a vaginal preparation accurately and smoothly into the vaginal lumen near the uterine cervix. We have used glass inserters ($4\Phi \times 120$ mm), with or without a tablet-sized dilated neck and having a plastic plunger, to introduce a drug-ladened cotton ball (ca. 12 mg), oleaginous suppository, tablet ($5\Phi \times 8$ mm cylindrical), and jelly into the vaginal lumen. For in situ perfusion of the rat vaginal tract, the base of the V-shaped uterus near the cervix is punctured and cannulated with polyethylene tubing (Intermedic PE50, Clay Adams). A glass tube with a 3 to 4 mm neck is then inserted through the vaginal opening, and the orifice is ligated with a surgical suture. The test solution maintained at 37° is recirculated from the cervical tubing, and the drug levels in blood taken from the cannulated jugular vein or changes in the drug concentration of the perfusate are determined.

Nonhuman primates, such as macaques, rhesus monkeys, and baboons, have comparable anatomy and physiology to the human female, hence providing useful permeability data. However, handling these animals requires considerable skill; they are also expensive. By comparison, it is easier to carry out large numbers of absorption experiments on rats, and more convenient to investigate the cyclic permeation behavior, owing to the rapid ovarian cycle. In selecting an animal model, species differences must be considered, especially with respect to the fundamental permeability of the epithelial membrane, due to ovarian cyclic changes in membrane thickness, the number of intercellular channels, and the nature of the lubricant fluids.

B. Vaginal Absorption of Peptides and Proteins

In 1923, Fisher [3] reported a quick and temporary reduction in blood sugar levels when insulin was administered vaginally to depancreatized dogs in single or consecutive doses. The maximum effect, blood sugar one-sixth that before treatment, was exerted 3 hr after insulin was administered in an aqueous solution. This effect, achieved at a dose of 20 units, was more potent than that achieved by 30 units through an intestinal fistula. The vaginal administration of insulin (40 units and 100 units) in anesthetized cats produces a gradual but definite decline of the blood sugar level that reaches approximately one-fourth of the normal level in 4 to 5 hr [4]. These studies present the first evidence of the absorption of peptides through the vaginal epithelium; they are pioneer efforts in the clinical approach of drug delivery by the vaginal route.

Several investigators have reported on the vaginal absorption of proteins (e.g., antigens and antibodies), as summarized by Aref et al. [1], Benziger and Edelson [2], and Roig de Vagas—Linares [50]. In 1943, Rosenzweig and Walzer used an immunological method to demonstrate that peanut protein can be absorbed through the human vagina and cervix [51]. Vaginal absorption was slow, requiring 40 to 120 min, while absorption from the cervix took only 8 to 25 min. Macromolecular antibodies and bovine milk immunoglobulins are absorbed from the baboon vagina. The vaginal administration of these antibodies inhibited ovarian steroid synthesis by neutralizing the biological activity of endogenous LH [52].

Specific antibodies to a typhoid vaccine containing soluble typhoid bacillus antigen introduced vaginally in human females were demonstrated

by Straus [7].   The vaginal application of a tampon soaked with an antigen
solution for 3 days caused the local secretion of antibodies and exerted
strong local immunological responses.   Antibodies against bacterial antigens,
such as *Candida albicans* and *Escherichia coli*, were also detected [53].
Absorption of bacterial antigens through the vaginal epithelium plays a
fundamentally important role in the local production of antibodies to prevent bacterial infection of the genital organs, the so-called "local host defense mechanism."

The absorption of antigens on the surface of spermatozoa from the
vagina and cervix stimulates the production of spermatozoal antibodies in
the serum and cervicovaginal secretions of women [8,54–58].   The
physiological role of these antibodies is ambiguous but may be involved in
infertility of unknown etiology [57].   It was recently elucidated that antisperm antibodies against the sperm head, not the sperm tail, may prevent
sperm from penetrating the cervical mucus to cause immunological infertility [58].   They also suggested that infertility can be attributed to
abnormal sperm-vaginal interaction and poor sperm survival in the cervicovaginal secretions, possibly resulting from antisperm antibodies.

In these secretory immune systems, the vaginal intercellular channels
play an essential role in the passage of different immunogens from the
vaginal lumen to the lamina propria and finally, the blood or lymphatic
vessels.   The immunoglobulin-producing cells (i.e., lymphocytes and
plasma cells) are located in the lamina propria of the human vagina, and
mature to produce and secrete immunoglobulins (IgA, IgG), depending on
the phase of the ovarian cycle [50].   Thus positive tests for antisperm
antibodies in serum do not always correlate well with immunologic infertility, because the concentration of immunoglobulins in the secretions
decreases to extremely low levels at midcycle when sperm migration and
fertilization take place [59].   This decrease is probably due to the pre-
ovulatory rise of endogenous estrogens, reducing the dimension of the
intercellular channels [17,50] and cervical mucus production [60].

A humoral immune response was induced in cervical secretions and
serum following the repeated vaginal application of two model antigens, a
DNA virus particle (T-4 coliphages) and the lipopolysaccharide of *Sal-
monella typhosa*, in a plastic sponge in rhesus monkeys [49].   However,
the response was weak compared with that achieved by systemic immunization, because cell-mediated immunity was limited after vaginal immunization.
Walker et al. have explicitly demonstrated a local immune system in the
gastrointestinal tract, an extensive area with an abundance of pathological
organisms and toxic agents [29,40,61,62].   Lymphocytes in the Peyer's
patches are stimulated to proliferate and differentiate by the permeated
antigens.   These cells migrate to the lymph nodes for further maturation,
enter the systemic circulation, and ultimately redistribute to the intestinal
lamina propria.   The dimeric IgA produced from these plasma cells is
secreted and interacts with antigens within the glycocalyx compartment to
inhibit antigen permeation on the one hand and to enhance breakdown by
pancreatic enzymes on the other.   Although there is no specially developed
lymphoid tissue in the vagina, it must be noted that the antibodies induced
following chronic dosing may neutralize peptide drugs or prevent their
penetration.

The vaginal application of insulin and LH–RH analogs to obtain systemic effects has recently been investigated.   Humphrey et al. reported
the vaginal absorption of synthetic luteinizing hormone releasing factor

(LRF) in estrogen/progesterone-blocked ovariectomized rats [60], whose vaginal epithelium is thin (Fig. 5). A sustained effect on LH secretion was elicited following the vaginal application of 10 μg or 100 μg of synthetic LRF suspended in Carbowax 1000. The serum LH was significantly elevated by 10 min, reaching a maximum at 30 min, and remaining elevated for at least an additional 90 min.

Schally et al. reported that synthetic LH–RHs, (D–Ala^6, des–Gly^{10}) LH–RH ethylamide and (D–Leu^6, des–Gly^{10}) LH–RH ethylamide, were absorbed from the vaginal lumen in ovariectomized steroid-blocked rats and intact immature female rats [64,65]. Vaginal application of LH–RH and its analogs induced a greater elevation of serum LH and FSH levels than did oral administration of the same doses. In a study using immature rats, the bioavailability of the analogs, estimated by the maximum serum levels of gonadotropins, was assumed to be approximately 1 to 2% (vaginal route) and 0.1% (oral route) of that achieved by the subcutaneous route. A human study was carried out on the vaginal application of (D–Leu^6, des–Gly^{10}) LH–RH ethylamide in 10 women during the early or midfollicular phase of the cycle [66]. A tablet containing 2 mg of the analog was inserted into the posterior fornix of the vagina; a cotton tampon was used to prevent leakage for 9 hr. The analog was absorbed from the vagina and elevated the plasma levels of the gonadotropins and estrogen. The gonadotropin levels peaked 4 to 6 hr after administration; a more prolonged response than that induced by a subcutaneous injection was elicited. The releasing response on LH by the vaginal route was, however, half as great as that induced by subcutaneous injection of 25 μg of the analog, corresponding to about 0.6% bioavailability. These results indicated that the vaginal route was useful for the administration of LH–RH and its analogs, especially when a low but long-lasting release of gonadotropins is required, as is the case for stimulating follicular maturation and estrogen secretion. In this investigation the analog was applied selectively at the early and midfollicular phases, during which the vaginal epithelial membrane is thick and the intercellular channels are narrow, to provide a significant barrier against the penetration of peptides. An increase of vaginal absorption is expected by applying the agents during an earlier follicular phase and the luteal phase, during which porosity of the epithelium is high.

Insulin (27 IU) was vaginally administered in the nonionic surface-active agent, Cetomacrogol 1000 (polyethylene glycol 1000 monocetyl ether) with ethylene glycol 400, to streptozocin-induced diabetic female rats [67]. The blood glucose level was reduced to 66.3, 50.7, and 48.9% of the initial value at 1, 2, and 4 hr after administration. This hypoglycemic effect was less than that achieved by the rectal route (27 IU) in the same base and by the intraperitoneal route (4 IU). By contrast, rapid and pronounced hypoglycemic effects were produced in alloxan diabetic female rats and rabbits by the vaginal administration of insulin (0.5 to 5 IU/kg) suspended in a polyacrylic acid aqueous gel base (0.1%, pH 6.5) [68]. Dose-related changes in the plasma glucose and insulin levels were demonstrated. In these studies, the rate of absorption was influenced directly by the release of hormones from the vaginal preparations: slow but sustained from an oleaginous suppository and rapid from an aqueous gel.

A series of investigations on the vaginal absorption of (D–Leu^6, des–Gly^{10}) LH–RH ethylamide (leuprolide) have been conducted to establish a rational dosage form for therapy of hormone-dependent tumors,

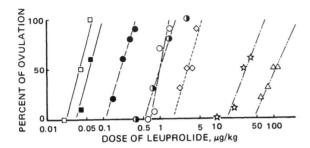

Fig. 9 Ovulation-inducing activity of leuprolide administered by different routes to diestrous rats. Open square, intravenous; filled square,
subcutaneous; open circle, vaginal in saline; half-filled circle, vaginal in
oleaginous base; filled circle, vaginal with 10% citric acid; diamond, rectal;
star, nasal; triangle, oral.

endometriosis, and so on [69–77]. Figure 9 shows the pharmacological
effect, ovulation-inducing activity in diestrous rats, following the administration of leuprolide by the intravenous, subcutaneous, rectal, nasal,
oral, and vaginal routes [71]. The absolute bioavailability of the analog
estimated by the pharmacological effect was 0.05% by the oral route in an
absorption-promoting formulation [78] that contained a mixed micellar solution of monoolein, sodium taurocholate, and sodium glycocholate, 1.2% by
the rectal route, and 0.11% by the nasal route without an absorption promoter and 1.8 to 3% with 1% sodium glycocholate, surfactin, or polyoxyethylene 9-lauryl ether [79,80]. Nasal absorption was probably underestimated due to drainage of the test solution. In contrast, the bioavailability from the vaginal route was relatively large, 3.8% from oleaginous
and aqueous gel bases without the aid of an absorption promoter.
   In rats, the pregnancy terminating effectiveness of leuprolide administered by the vaginal route is almost the same as that by the subcutaneous
route [81–83]. This is possibly due to the persistent blood levels of the
analog by the vaginal route, thereby encouraging the possibility of artificial abortion by this route.
   These studies on the vaginal application of hormones indicate the
feasibility of self-administration by this route. However, the vaginal bioavailability of peptides and proteins may be too low and, at the same time,
too variable to be useful clinically.

C. Absorption Enhancement

As with other routes of peptide and protein drug administration, penetration enhancers, also called absorption promoters, are required. The
screening of promoters for the vaginal absorption of leuprolide has been
carried out by evaluating its ovulation-inducing activity in diestrous rats
using an oleaginous base containing different additives [71]. The vaginal
absorption of this peptide was markedly facilitated by polybasic carboxylic
acids, citric, succinic, tartaric, and glycocholic acids, and slightly increased by hydroxycarboxylic acids and acidic amino acids (Table 1). The
absolute bioavailability increased from about 4% to about 20% by adding

Table 1   Ovulation-Inducing Activity of Leuprolide After
Vaginal Administration with Additives to Diestrus Rats

Additives (10%)	ED_{50} (ng/rat)[a],[b]	Relative potency
None	270 (194−353)	1
Citric acid	56 (38−69)	4.9[c]
Succinic acid	50 (37−63)	5.4[c]
Tartaric acid	82 (69−97)	3.3[c]
Glycocholic acid	47 (32−62)	5.6[c]
Ascorbic acid	113 (80−161)	2.4[c]
Lactic acid (2%)	117 (95−184)	2.3[c]
Aspartic acid	122 (79−167)	2.1[c]
Glutamic acid	177 (133−243)	1.6
Dipotassium EDTA	104 (87−134)	2.6[c]
Taurine	182 (112−373)	1.5
Glycine	755 (570−5904)	0.4[c]
Boric acid	200 (153−261)	1.3
Caproic acid	341 (249−484)	0.8
Oleic acid	358 (244−541)	0.7
Polyoxyethylene 9-lauryl ether	254 (193−348)	1.1
Sodium glycocholate	151 (107−666)	1.9[c]
Sodium oleate	323 (237−461)	0.8
Sodium citrate	245 (160−365)	1.0

[a] The ED_{50} was calculated by Finney's probit analysis [84].

[b] Values in parentheses are fiducial limits (95%).

[c] Significant ($p < 0.05$).

polybasic carboxylic acids. The absorption was poorly enhanced with fatty acids and surfactants such as sodium glycocholate, sodium oleate, and polyoxyethylene 9-lauryl ether, which are known to enhance rectal and nasal absorption of hydrophilic drugs. Dipotassium ethylenediaminetetraacetate (EDTA) significantly promoted absorption, but sodium citrate did not. Citric and succinic acid showed a similar concentration-dependent promoting effect; the activity increased with concentration to attain a maximum activity over 10% in an oleaginous base [71] and an aqueous solution (pH 3.5) [72]. Jellies prepared in highly polar polymer, such as sodium carboxymethylcellulose and polyacrylate, reduced the absorption enhancement effects in preparations containing 5% citric acid (pH 3.5), whereas

those with a less polar polysaccharide afforded sufficient absorption. A
tablet containing 10% citric acid allowed considerable absorption of the
analog and gave a smaller ED_{50} than an oleaginous suppository. Examination of leakage of intravenously administered Evans blue across the vaginal
mucosa revealed that the tablet and oleaginous suppository produced deep
staining of the mucus, possibly due to the higher local concentration of an
organic acid. In contrast, the jelly exhibited only faint staining or none.
Overall, the vaginal jelly appears to be the most suitable practical dosage
form because of its activity with lower concentrations of citric acid, less
local reaction resulting from the homogeneous dissolved promoter and the
easily adjusted pH, and its easy handling due to high water solubility.

The vaginal absorption of synthetic LH−RH and insulin was also markedly
enhanced by adding citric acid in diestrous rats [72]. Then percent citric
acid enhanced the ovulation-inducing activity of LH−RH by 30 times after
vaginal administration in an oleaginous suppository. Decreased plasma glucose
levels of 125% · hr for 6 hr was produced by a 1.1-U/kg dose of insulin by
the intravenous route, and by 30 U/kg and 6.0 U/kg in buffer solution (pH
1.70) and in 10% citric acid solution (pH 1.72), respectively, by the vaginal
route under anesthesia (Fig. 10). Absolute bioavailability following vaginal
administration with citric acid was 18% estimated by the effective dosage.

The mechanism of vaginal absorption enhancement of peptides by
organic acids was attributed to their acidifying and chelating abilities [72].
The vaginal absorption of leuprolide was increased by lowering the pH of
the solution; the ovulation-inducing effects after vaginal administration of
the analog was elevated gradually with a decrease of pH until pH 2.0 was
reached. In contrast, the vaginal absorption of phenol red (as a hydrophilic marker), as assessed by urinary excretion in anesthetized rats, was
decreased with acidification (Fig. 11). However, the vaginal absorption of
phenol red was also increased by adding 5 or 10% citric acid in a pH 3.5
solution, but it was not affected by adding 5% citric acid in a pH 6.6

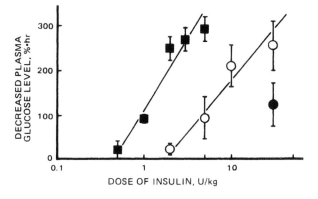

Fig. 10   Decreased plasma levels of glucose after subcutaneous and vaginal
administration of insulin to diestrous rats. The decreased plasma level was
exhibited by integrating the percent decrease against the initial level for
6 hr (mean ± SE, n = 5). Filled square, subcutaneous; open circle, vaginal
in 10% citric acid (pH 1.72); filled circle, vaginal in KCl−HCl buffer (pH
1.70).

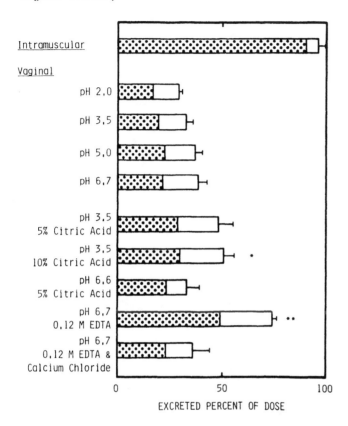

Fig. 11   Effects of pH, citric acid, and EDTA on vaginal absorption of phenol red in diestrous rats. Bars represent the urinary excretion of phenol red after vaginal administration (dotted square, 0 to 3 hr; open square, 3 to 6 hr); (mean ± SE, n = 5). *, p < 0.05, **, p < 0.01.

solution. It seems that the effect of pH on the absorption of leuprolide is due to self-association or conformational changes of the peptide, or to changes in the charge of leuprolide and the epithelial surface. Since the pH of the vaginal lumen is normally 3.5 to 5.0, an acid preparation of pH 3 to 4 should not disturb the normal condition of the vagina. The activity of leuprolide after vaginal administration in aqueous solutions containing 0.238 M of each of several organic acids at pH 3.5 and the chelating ability of their acids are shown in Table 2. These results indicate the enhancing effects of the acids, which correlated somewhat with their chelating ability.

The important roles of Ca^{2+} and its regulatory substances on the stability of biological membranes and the transport of endogenous or exogenous substances are well demonstrated. Recently, several kinds of cell-to-cell adhesion molecules, Ca^{2+}-dependent and Ca^{2+}-independent adhesion systems, have been implicated in the structure of vertebrate tissues. Takeichi et al. have identified the proteins N-cadoherin and E-cadoherin, which are involved in Ca^{2+}-dependent cell-to-cell adhesion systems [85–87]. These

Table 2   Ovulation-Inducing Activity of Leuprolide After Vaginal Administration in Aqueous Solution Containing 0.238 M of Organic Acids to Diestrous Rats

Organic acid	Ovulation[a]	Chelating ability[b]
None	0/10	—
Citric	9/10	1.21
Succinic	10/10	0.80
Tartaric	2/10	0.37
Malonic	7/10	0.25
Malic	3/10	0.13
Acetic	5/10	—
Lactic	1/10	0.06
Ascorbic	0/10	—

[a] Number of rats in which ovulation was induced per number of rats examined. Each solution was adjusted to pH 3.5 with 10 N NaOH or 2 N HCl to be isotonic with sodium chloride. Leuprolide was administered at a dose of 400 ng/kg per 0.2 ml.

[b] Gram ions of calcium ion sequestered by 1 M of organic acids at pH 10 and 30°C. The value of EDTA was 1.75.

proteins are thought to be particularly important in the initial recognition between cells and in maintaining the cell adhesion of various tissues. Thus they lead to the formation of more rigid intercellular structures, such as tight and gap junctions. It is noted that Ca^{2+} can protect such cell surface proteins from protease digestion [85,86,88].

The channel permeability of gap junctions, cell-to-cell channels that provide pathways for the direct flow of hydrophilic molecules up to 1000 to 2000 daltons, is also regulated by the intracellular Ca^{2+} concentration. This is the result of a change in channel charge configuration or in channel pore size [89–92]. Thus junctional pores are closed by high Ca^{2+} concentrations, the junctional pathway being blocked to all molecular species, including low-molecular-weight electrolytes [90]. The pore can be opened gradually by decreasing Ca^{2+} concentration with EDTA [91,92], indicating that the Ca^{2+} effects are reversible.

There appears to be a good correlation between the absorption enhancement of leuprolide and the leakage of intravenously administered Evans blue across the vaginal membrane following treatment with carboxylic acids but not with calcium salt [76]. This observation indicates that (1) the blood-vaginal epithelium barrier is breached by organic acids, and (2) chelation contributes to absorption enhancement probably by widening of the intercellular channel through reversible electrical uncoupling with calcium ions as demonstrated in gap junction or Ca^{2+}-dependent adhesion systems, and/or by cleaving the intercellular junctions through uptake of the calcium ions by Ca^{2+}-containing binding proteins [93].

Substances with a strong chelating ability, such as EDTA [94–96], tetracyclines [97], and enamines [98–100], enhance the intestinal and rectal absorption of hydrophilic compounds. Schuchner et al. have elucidated this phenomenon by ultrastructural observation. In the human vaginal epithelium, the desmosomes open and the size of the intercellular spaces increase after daily treatment for 6 days by a tampon impregnated with a solution of 0.2 M EDTA (pH 3.5). These changes were reversible and were assumed to be due to the depletion of calcium [101]. The chelating ability of organic acids is usually increased by increasing pH, whereas their enhancing effect is decreased. This phenomenon is assumed to be ascribed to the decrease in permeability of organic acids through the membrane channels and/or to the interaction with protein barriers due to the dissociation of the acid molecule. Promoters must penetrate membranes to exert their strong enhancing potency in the rectum [98]. A time-course study on the change of vaginal epithelial membranes with citric acid treatment revealed that faint staining by intravenously administered Evans blue was observed in the early stage after treatment with a 5% solution; while deep staining was observed at 30 min and 1 hr after treatment with a 10% solution, but the stain gradually faded [76]. The change in vaginal epithelium produced by the acids recovered rapidly after the epithelium was washed with physiological saline solution, indicating that the change was transient.

The vaginal absorption of leuprolide with citric acid was assessed in rats by radioimmunoassay of the serum levels of analog and gonadotropins and by anticancer activity in an experimental tumor. Leuprolide disappeared rapidly from the serum in rats after it was injected intravenously. The biological half-life was 8.4 min in the α-phase and 33.2 min in the β-phase (Fig. 12) [77]. After vaginal administration using a cotton ball soaked with a 5% citric acid solution (pH 3.5), high and long-lasting serum levels of leuprolide were observed, with absolute bioavailability (up to 6 hr) being 25.8%. Both a high dose and chronic administration of leuprolide, a potent LH–RH agonist, paradoxically cause down regulation of receptors in the pituitary, inhibition of the gonadotropin releasing response (densensitization), and functional gonadal atrophy ("chemical ovariectomy or castration"). As a result, leuprolide exerts therapeutic effects against hormone-dependent tumors [102], endometriosis [103], and precocious puberty [104]. It is also useful in birth control [105,106].

After continuous administration of leuprolide subcutaneously (above 0.1 µg/kg) or vaginally (1 µg/kg), serum LH levels were elevated once at the beginning of treatment, a "flare-up," and were then drastically inhibited [75]. A stronger inhibitory effect was elicited by the vaginal route; this was ascribed to the longer-lasting serum levels of the analog. Vaginal and subcutaneous constant infusion of the analog by an osmotic minipump (Alzet, Model 1701) maintained high and sustained serum levels of the analog and produced a drastic inhibitory effect on the releasing responses of LH and FSH. These results indicate that continuous stimulation of the target organs with the analog exerts much stronger desensitization than does pulsatile stimulation. Vaginal administration is therefore a preferred self-administration method for medical treatments using the antagonistic

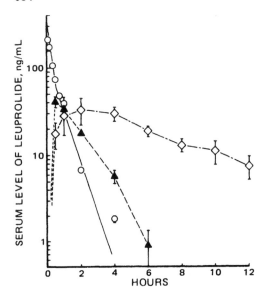

Fig. 12   Serum levels of leuprolide following intravenous (circle), subcutaneous (triangle), and vaginal (diamond) administration to rats.

activity of leuprolide. Recently, we have developed a sustained release injectable microsphere formulation (for 1 month) of the analog using poly(lactic/
glycolic acid) as a biodegradable cell-wall polymer to obtain the more continuous stimulation of the target [107].

Consecutive daily vaginal administration of more than 500 µg/kg
leuprolide in a 5% citric acid jelly (pH 3.5) resulted in a highly significant
regression of hormone-dependent mammary tumors induced in rats by DMBA
(Table 3) [74]. A slight decrease of vaginal absorption after 14 days of
continuous administration was observed. The decrease was caused not by
antibody induction but probably by an increase in surface mucus and by a
two- to threefold thicker vaginal epithelium than that seen in rats treated
subcutaneously during the same period. These changes resulted from persistent stimulation by the glass inserter or cotton ball [75].

Recently, we found that α-cyclodextrin facilitates the vaginal absorption of leuprolide in rats by approximately sixfold (patented), just as it
does in the nasal absorption of leuprolide and insulin in rats and dogs.
This effect may be elicited by the removal of fatty acids, such as palmitic
and oleic acid, which are minor membrane components [108]. The promoters used on the other mucous membranes, such as Azone [109,110],
fatty acid [111], enamines [98–100], and salicylate [112–114], may exert
effects on the vaginal application of peptides and proteins. Mechanisms
of enhancement in nasal insulin absorption by bile salts have recently
been proposed [115]. Two pairs of bile salt are envisioned to stack end
to end to span the lipid bilayers of the nasal mucous membrane, with the
ionized polar groups projecting into the aqueous environment on either
side and forming an aqueous channels for the passage of insulin from the
extracellular space. In the future we may consider substances that promote transcellular diffusion using spiked proteins such as receptors

Table 3 Response of Individual Tumors After Intraperitoneal and Vaginal Administration of Leuprolide for 8 Weeks

Treatment	Dose (μg/kg/day)	Initial tumors	Tumor responses[a]				New tumors
			Growing	Static	Regressing[b]		
Control	–	14	10	0	4 (3)		13
Intraperitoneal	500	15	0	0	15 (13)		2
Vaginal[c]	100	20	4	0	16 (10)		7
	500	19	3	0	16 (10)		2
	1000	20	2	1	17 (10)		1
	2500	17	2	3	12 (10)		0
	5000	18	1	1	16 (9)		2

[a]Response of tumors was defined by comparing the mean diameter as follows: growing, increased by at least 10%; regressing, decreased by at least 10%; and static, changed by less than 10%.

[b]Numbers in parentheses represent number of tumors that disappeared.

[c]The analog was administered in aqueous jelly containing 5% citric acid (pH 3.5) daily except Sundays for 8 weeks. Tumors have been regressed by 1 week and dramatically by 2 to 4 weeks after the onset of treatment.

(endocytosis [29] and proteinaceous tunnels [36]), carrier enzymes, ion
channels [116], and cytoskeletons (band 3, spectrin, actin, etc.).

In mucous application of peptides and proteins, the low bioavailability
and large fluctuation of absorption will remain an important problem.
Thus further attempts at absorption enhancement are required, and the
toxic effects of promoters on the mucous membrane must be studied. Only
then would their clinical utility be decided on practical merits and demerits.

D. Effects of Cyclical Changes of the Reproductive System

Vaginal absorption, especially of hydrophilic compounds, is characterized
by variations associated with the cyclical structural changes of the vaginal
epithelium. We have demonstrated the drastic effects of the estrous cycle
on the vaginal absorption of water-soluble compounds in the rat [73,77].
Mature female rats exhibiting two or more consecutive 4-day estrous cycles
by continuous examination of daily morning vaginal smears (Giemsa's solution staining [117]) were used at each stage. Figure 13 shows the plasma
glucose levels as a percentage of the initial level after vaginal administration of porcine insulin at different stages of the cycle in rats [73]. Insulin
was administered vaginally at a dose of 20 U/rat in an oleaginous suppository containing 10% citric acid. A slight decrease of glucose level was observed at an early period during proestrus, whereas a distinct decrease
was observed during estrus and a more remarkable decrease during
metestrus and diestrus. Vaginal absorption of phenol red, a water-soluble
marker, is markedly affected by the cycle [73]. The excreted percentage
of the dose in 6 hr after vaginal administration of phenol red without any
absorption promoter was 2.4% during proestrus, 5.5% during estrus, 37.5%
during metestrus, and 31.4% during diestrus. The vaginal absorption of
leuprolide is also affected by the estrous cycle to the same degree seen

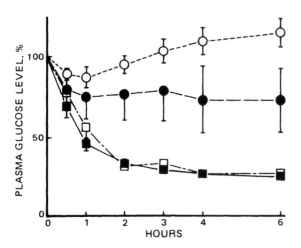

Fig. 13   Effects of estrous cycle on vaginal absorption of insulin in rats.
The plasma glucose level is shown as a percentage of the initial level after
vaginal administration of insulin (20 U) in an oleaginous suppository containing 10% citric acid (mean ± SE, n = 5). Open circle, proestrus; filled
circle, estrus; open square, metestrus; filled square, diestrus.

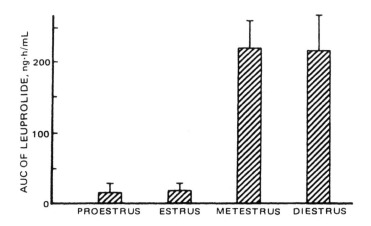

Fig. 14   Effects of the estrous cycle on vaginal absorption of leuprolide
in rats.   AUC of serum leuprolide level over 6 hr after vaginal administration at a dose of 500 µg/kg per 0.2 ml of 5% citric acid solution (pH 3.5)
is shown (mean ± SE, n = 5).

with phenol red (Fig. 14) [77].   The AUC of leuprolide serum levels over
6 hr after vaginal administration (500 µg/kg) with 5% citric acid solution
(pH 3.5) was 15.5 ng · hr/ml during proestrus, 17.4 ng · hr/ml during
estrus, 221.3 ng · hr/ml during metestrus, and 218.8 ng · hr/ml during
diestrus.   The effect of the estrous cycle can be explained by changes in
the porelike pathway of the epithelium.   The apparent porosity during
metestrus and diestrus is presumed to be more than 10 times that during
proestrus and estrus [73].
     To estimate the absorption pathway of hydrophobic and hydrophilic
compounds, vaginal absorption of the ionic and undissociated forms of
salicylic acid was investigated during proestrus and diestrus (Fig. 15)
[73].   The disappearance of the undissociated acid from the low pH buffer
through the vagina was rapid and was almost similar during the two stages.
The disappearance of the ionized form was different: 66% of the dose in
1 hr during diestrus and 29% during proestrus.   This result indicates that
the permeability of the hydrophobic compound is less affected by the
ovarian cycle mainly because the compound was absorbed transcellularly,
whereas the hydrophilic compound was mainly absorbed through porelike
pathways, such as intercellular channels.   The latter is highly dependent
on the stage of the estrous cycle, and, in turn, epithelial porosity.
     The effects of the reproductive cycle on the vaginal absorption of
penicillin, a hydrophilic substance, have been described in humans [23].
High blood levels of penicillin, sufficient to be therapeutic, after a vaginal
suppository was inserted were found near the end of menstrual cycle and
during menopause, but absorption was somewhat diminished during the
ovulation phase and late pregnancy.   These observations can be explained
by the morphological changes of vaginal epithelium during the cycle.   The
vaginal membrane permeability to vidarabine, a hydrophilic antiviral compound, has been reported to be affected by the estrous cycle in the mouse
[118] and guinea pig [119] using a two-chamber diffusion cell apparatus.

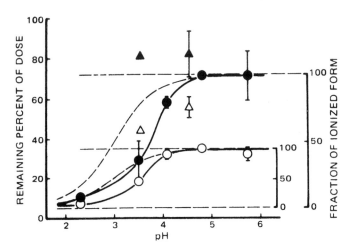

Fig. 15 Percentage of salicylic acid remaining 1 hr after vaginal administration in various pH solutions to diestrous and proestrous rats, and the fraction of the ionized form of the acid. Dose: 2 mg/kg per 0.2 ml in the buffer solutions, except at pH 2.29 (0.8 mg/kg per 0.2 ml). Open circle, diestrus; filled circle, proestrus; dashed lines, fraction of the ionized form. Glycine buffer shows the likely stabilizing effect on the membrane of the diestrus (open triangle) and proestrus (filled triangle).

The permeability coefficients were 5 to 100 times higher during early diestrus or diestrus than during estrus. The thicker superficial epithelial layer, the keratinized layer combined with the mucous layer, appeared to be the major diffusion barrier for the drug during estrus.

Freeze-fracture and thin-sectioning studies of the guinea pig vagina have demonstrated that during estrus keratinized epithelial cells have a tight-junctional network, thereby blocking the paracellular diffusion of such water-soluble tracers as horseradish peroxidase and lanthanum [120]. During metestrus the intercellular space of the epithelium was stained by tracer molecules even though tight-junctional belts could be observed; the junctions seemed to become functionally leaky, although they remained morphologically intact. Poor vaginal permeation of water-soluble markers has been exhibited during the early phase of the menstrual cycle (days 4 to 7) in the rhesus monkey [121]. The upper layers of the vaginal epithelium appear to present a barrier to the penetration of water-soluble molecules through the intercellular channel system. Gap junctions that exist between cells in the lower layers disappear in the upper layers. It may also be possible that the intercellular lamellar sheets, identified as lipids, detected in the rhesus monkey vaginal epithelium might serve as such a barrier [122]. By contrast, the vaginal permeability in the guinea pig is regulated mainly by tight-junctional belts rather than by intercellular lipids, because the latter are extremely rare and cannot therefore form an effective barrier to diffusion [120]; thus this appears to be an evidence for species differences.

The effect of cyclic changes of the reproductive system, while enhancing the vaginal permeability to peptides and proteins, also causes a large

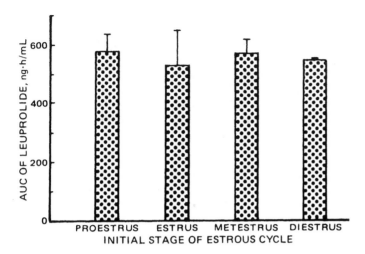

Fig. 16 AUC of leuprolide serum level over 6 hr after vaginal administration in pretreated rats. Rats were pretreated subcutaneously for 10
days with 100 µg/kg of the analog, followed by a single vaginal administration of 500 µg/kg in 5% citric acid solution (pH 3.5) (mean ± SE,
n = 5).

fluctuation in absorption. However, in continuous therapy using leuprolide,
the cycle is halted at diestrus, corresponding to the luteal and early follicular phases in humans. The vaginal mucosa becomes as thin as that
seen in normal diestrus or ovariectomized rats. As a result, the absorption of hydrophilic compounds such as phenol red is increased [73]. Similarly, the absorption of leuprolide from the vagina is enhanced with less
variation by 10-day pretreatment with subcutaneous daily injections of the
analog (Fig. 16) [77]. For this reason, vaginal application of leuprolide
has been proposed as a maintenance dose following continuous parenteral
administration.

In summary, fluctuation of vaginal absorption associated with the
cyclic effects of the reproductive system must be considered in vaginal
application of peptides and proteins in therapy.

V. DESIGN REQUIREMENTS FOR VAGINAL
DELIVERY SYSTEMS

The dosage forms for the vaginal route are usually powders, creams, gels,
solutions, capsules, tablets, suppositories, and aerosol foams, administered
with the aid of an applicator. Their physicochemical characteristics and
drug releasing properties are similar to those of rectal preparations
[123–126].

The desirable attributes of vaginal preparations are as follows [124]:
(1) no adverse reactions (i.e., the absence of tissue irritation), (2) easy
application, (3) even distribution of the drug throughout the vagina (i.e.,
not concentrated in one spot), (4) retention of the drug in the vagina
even when the patient is standing, (5) absence of an offensive odor,

(6) no staining of clothes or skin, and (7) compatibility with other medication or contraception. In continuous multiple dosing, the preparations should contain less base material or vehicle, which could form residues after the drug release was complete.

Bulbocavernosus muscles, sometimes termed the "sphincter vaginae," surround the orifice of the vagina. They are not usually strong enough to retain vaginal preparations as the anal sphincter retains rectal suppositories. It may be preferable to insert a vaginal preparation at bedtime to avoid slipping out or leakage; this is especially the case with preparation involving a relatively large volume of liquid or semisolid.

Adverse reactions, such as local irritation, by vaginal devices must be avoided, particularly when absorption promoters are utilized. The toxic shock syndrome (with several cases resulting in death) is well known [127–130]. This syndrome is caused by the vaginal absorption of a bacterial toxin, derived mainly from *Staphylococcus aureus*, that proliferates after a tampon has been left in the vaginal lumen during menstruation. A blood-soaked tampon at body temperature is a suitable culture medium for the bacteria and may also enhance toxin production. The constituent materials of certain tampons probably stimulate or permit this growth. It is important that the materials used in vaginal preparations should not play the role of a growth medium for the proliferation of pathogenic microorganisms, bacteria, fungi, and protozoan.

Vaginal preparations must not disturb the normal environmental conditions of the vagina (i.e., normal microbial flora, pH, and conditions favoring fertilization). An acidic pH is essential to maintain a healthy vaginal mucosa and to prevent infection caused by pathogenic microorganisms, but is not a favorable milieu for the sperm survival. Menstruation, intercourse, delivery, and the other anatomical or physiological changes in the life cycle of women must also be taken into account when the timing and effectiveness of drug application are being considered.

Continuous long-term vaginal administration of peptides and proteins should be feasible by utilizing a drug delivery device such as a silastic ring [43,44] or collagen sponge [45,46]. Since the vaginal lumen is continuous with the cervical canal into the uterus, some portion of a vaginal preparation may be transferred into the uterus, where it is absorbed. Usually, the cervical mucous barrier prevents the transport of fluids (i.e., the seminal plasma) from the vagina into the uterus; a drug preparation applied to the vagina scarcely invades the uterus.

VI. CONCLUSIONS

The vagina is a complex genital organ with multiple functions. The vaginal epithelium consists of noncornified, stratified squamous cells that are similar to those of the buccal mucosa. Its permeability to peptides and proteins is strongly influenced by serum estrogen concentration throughout a woman's life cycle from birth to menarche and menopause, and during the menstrual cycle in an adult. During the luteal and early follicular phases in humans, and the metestrus and diestrus in the mouse, rat, and guinea pig, the epithelium becomes explicitly porous. This is attributed to loosened intercellular channels and desquamation of the superficial cell layer.

First-pass effects with local and hepatic proteases are almost avoided by using the vaginal route. However, the bioavailability is somewhat insufficient and fluctuates greatly; absorption promoters are required. The polybasic carboxylic acids possessing a chelating ability in an acidic milieu exert a potent absorption-enhancing effect. However, surfactants and the salts of these organic acids are less effective. The cessation of the estrous cycle in rats at diestrus during continuous administration of a potent LH–RH analog, leuprolide, has overcome the large variation in vaginal absorption throughout the estrous cycle. At the same time it increases the absorption of the analog and other hydrophilic compounds.

In summary, the vagina is a possible site for the systemic administration of peptides and proteins. However, variations in absorption resulting from cyclic changes in the reproductive system throughout the life cycle must be taken into account in developing new vaginal preparations.

REFERENCES

1. I. Aref, Z. El-Sheikha, and E. S. E. Hafez, in *The Human Vagina* (E. S. E. Hafez and T. N. Evans, eds.), Elsevier/North-Holland Biomedical Press, Amsterdam (1978), p. 179.
2. D. P. Benziger and J. Edelson, *Drug Metab. Rev.*, *14*:137 (1983).
3. N. F. Fisher, *Am. J. Physiol.*, *67*:65 (1923).
4. G. D. Robinson, *J. Pharmacol. Exp. Ther.*, *32*:81 (1927).
5. C. Aron and R. Aron-Brunetiere, *Ann. Endocrinol.*, *14*:1039 (1953).
6. M. Rosenzweig and M. Walzer, *Am. J. Obstet. Gynecol.*, *42*:286 (1943).
7. E. K. Straus, *Proc. Soc. Exp. Biol. Med.*, *106*:617 (1961).
8. R. R. Franklin and C. D. Dukes, *J. Am. Med. Assoc.*, *190*:682 (1964).
9. C. Longcope and K. I. H. Williams, *J. Clin. Endocrinol. Metab.*, *38*: 602 (1974).
10. I. Schiff, D. Tulchinsky, and K. J. Ryan, *Fertil. Steril.*, *28*:1063 (1977).
11. L. A. Rigg, B. Milanes, B. Villanueva, and S. S. C. Yen, *J. Clin. Endocrinol. Metab.*, *45*:1261 (1977).
12. L. A. Rigg, H. Hermann, and S. S. C. Yen, *N. Engl. J. Med.*, *298*: 195 (1978).
13. F. H. Netter, in *Reproductive System: The Ciba Collection of Medical Illustrations*, Vol. 2 (E. Oppenheimer, ed.), Ciba, Summit, N.J. (1970), p. 89.
14. C. D. Turner and J. T. Bagnara, *General Endocrinology*, 6th ed., W. B. Saunders Company, Philadelphia (1976), pp. 360, 450.
15. W. Platzer, S. Poisel, and E. S. E. Hafez, in *The Human Vagina* (E. S. E. Hafez and T. N. Evans, eds.), Elsevier/North-Holland Biomedical Press, Amsterdam (1978), p. 39.
16. K. A. Walz, H. Metzger, and H. Ludwig, in *The Human Vagina* (E. S. E. Hafez and T. N. Evans, eds.), Elsevier/North-Holland Biomedical Press, Amsterdam (1978), p. 55.

17. M. H. Burgos and C. E. Roig de Vargas-Linares, in *The Human Vagina* (E. S. E. Hafez and T. N. Evans, eds.), Elsevier/North-Holland Biomedical Press, Amsterdam (1978), p. 63.

18. R. W. Steger and E. S. E. Hafez, in *The Human Vagina* (E. S. E. Hafez and T. N. Evans, eds.), Elsevier/North-Holland Biomedical Press, Amsterdam (1978), p. 95.

19. R. W. Kistner, in *The Human Vagina* (E. S. E. Hafez and T. N. Evans, eds.), Elsevier/North-Holland Biomedical Press, Amsterdam (1978), p. 109.

20. G. Wagner and R. J. Levin, in *The Human Vagina* (E. S. E. Hafez and T. N. Evans, eds.), Elsevier/North-Holland Biomedical Press, Amsterdam (1978), p. 121.

21. W. F. Ganong, *Review of Medical Physiology*, 11th ed., Lange Medical Publications, Los Altos, Calif. (1983), p. 336.

22. S. Hwang, E. Owada, L. Suhardja, N. F. H. Ho, G. L. Flynn, and W. I. Higuchi, *J. Pharm. Sci.*, 66:781 (1977).

23. J. Rock, R. H. Barker, and W. B. Bacon, *Science*, 105:13 (1947).

24. L. Z. Bito, *J. Physiol.*, 221:371 (1972).

25. L. Z. Bito and P. J. Spellane, *Prostaglandins*, 8:345 (1974).

26. L. Z. Bito, *Nature*, 256:134 (1975).

27. D. M. Matthews and J. W. Payne, in *Current Topics in Membranes and Transport*, Vol. 14 (F. Bronner and A. Kleinzeller, eds.), Academic Press, New York (1980), p. 331.

28. D. M. Neville, Jr. and T.-M. Chang, in *Current Topics in Membranes and Transport*, Vol. 10 (F. Bronner and A. Kleinzeller, eds.), Academic Press, New York (1978), p. 65.

29. J. L. Goldstein, R. G. W. Anderson, and M. S. Brown, *Nature*, 279:679 (1979).

30. J. G. Lecce, in *Intestinal Toxicology* (C. M. Schiller, ed.), Raven Press, New York (1984), p. 33.

31. D. H. Alpers and K. J. Isselbacher, *J. Biol. Chem.*, 242:5617 (1967).

32. D. E. Bockman and M. D. Cooper, *Am. J. Anat.*, 136:455 (1973).

33. W. A. Walker and K. J. Isselbacher, *N. Engl. J. Med.*, 297:767 (1977).

34. P. Boquet and A. M. Pappenheimer, Jr., *J. Biol. Chem.*, 251:5770 (1976).

35. A. M. Pappenheimer, Jr., *Ann. Rev. Biochem.*, 46:69 (1977).

36. V. R. Lingappa and G. Blobel, *Recent Prog. Horm. Res.*, 36:451 (1980).

37. G. Von Heijne and C. Blomberg, *Eur. J. Biochem.*, 97:175 (1979).

38. A. Masaoka, M. Maeda, T. Mori, S. Ohshima, and T. Ninomiya, *Jpn. J. Clin. Med.*, 33:113 (1975).

39. S. Dodda-Kashi and V. H. L. Lee, *Life Sci.*, 38:2019 (1986).

40. W. A. Walker, M. Wu, K. J. Isselbacher, and K. J. Bloch, *Gastroenterology*, 69:1223 (1975).

41. P. F. Bonventre, C. B. Saelinger, B. Ivins, C. Woscinski, and M. Amorini, *Infect. Immun.*, 11:675 (1975).

42. R. Ducroc, M. Heyman, B. Beaufrere, J. L. Morgat, and J. F. Desjeux, *Am. J. Physiol.*, 245:G54 (1983).

43. D. R. Mishell, Jr., D. E. Moore, S. Roy, P. F. Brenner, and M. A. Page, *Am. J. Obstet. Gynecol.*, 130:55 (1978).

44. S. Meta, U. M. Joshi, G. M. Sankolli, A. Adati, U. M. Donde, and B. N. Saxena, *Contraception*, *23*:241 (1981).

45. M. Chvapil, *Fertil. Steril.*, *27*:1387 (1976).

46. A. Victor, H. A. Nash, T. M. Jackanicz, and E. D. B. Johansson, *Contraception*, *16*:125 (1977).

47. T. Yotsuyanagi, A. Molokhia, S. Hwang, N. F. H. Ho, G. L. Flynn, and W. I. Higuchi, *J. Pharm. Sci.*, *64*:71 (1975).

48. E. Owada, C. R. Behl, S. Hwang, L. Suhardja, G. L. Flynn, N. F. H. Ho, and W. I. Higuchi, *J. Pharm. Sci.*, *66*:216 (1977).

49. S.-L. Yang and G. F. B. Schumacher, *Fertil. Steril.*, *32*:588 (1979).

50. C. D. Roig de Vargas-Linares, in *The Human Vagina* (E. S. E. Hafez and T. N. Evans, eds.), Elsevier/North-Holland Biomedical Press, Amsterdam (1978), p. 193.

51. M. Rosenzweig and M. Walzer, *Am. J. Obstet. Gynecol.*, *42*:286 (1943).

52. L. R. Beck, L. R. Boots, and V. C. Stevens, *Biol. Reprod.*, *13*:10 (1975).

53. J. Govers and J. P. Girard, *Gynecol. Invest.*, *3*:184 (1972).

54. R. R. Franklin and C. D. Dukes, *Am. J. Obstet. Gynecol.*, *89*:6 (1964).

55. D. M. Israelstam, *Fertil. Steril.*, *20*:275 (1969).

56. R. H. Glass and R. A. Vaidya, *Fertil. Steril.*, *21*:657 (1970).

57. R. H. Waldman, J. M. Cruz, and D. S. Rowe, *Clin. Exp. Immunol.*, *12*:49 (1972).

58. C. Wang, H. W. G. Baker, M. G. Jennings, H. G. Burger, and P. Lutjen, *Fertil. Steril.*, *44*:484 (1985).

59. G. F. B. Schumacher, M. H. Kim, A. H. Hosseiman, and C. Dupon, *Am. J. Obstet. Gynecol.*, *129*:629 (1977).

60. J. A. Holt, G. F. B. Schumacher, H. I. Jacobson, and D. P. Schwartz, *Fertil. Steril.*, *32*:170 (1979).

61. W. A. Walker, K. J. Isselbacher, and K. J. Bloch, *Science*, *177*:608 (1972).

62. K. Y. Pang, W. A. Walker, and K. J. Bloch, *Gut*, *22*:1018 (1981).

63. R. R. Humphrey, W. C. Dermody, H. O. Brink, F. G. Bousley, N. H. Schottin, R. Sakowski, J. W. Vaitkus, H. T. Veloso, and J. R. Reel, *Endocrinology*, *92*:1515 (1973).

64. N. Nishi, A. Arimura, D. H. Coy, J. A. Vilchez-Martinez, and A. V. Schally, *Proc. Soc. Exp. Biol. Med.*, *148*:1009 (1975).

65. A. De La Cruz, K. G. De La Cruz, A. Arimura, D. H. Coy, J. A. Vilchez-Martinez, E. J. Coy, and A. V. Schally, *Fertil. Steril.*, *26*:894 (1975).

66. M. Saito, T. Kumasaki, Y. Yaoi, N. Nishi, A. Arimura, D. H. Coy, and A. V. Schally, *Fertil. Steril.*, *28*:240 (1977).

67. E. Touitou, M. Donbrow, and E. Azaz, *J. Pharm. Pharmacol.*, *30*:662 (1978).

68. K. Morimoto, T. Takeeda, Y. Nakamoto, and K. Morisaka, *Int. J. Pharm.*, *12*:107 (1982).

69. I. Yamazaki and H. Okada, *Endocrinol. Jpn.*, *27*:593 (1980).

70. H. Okada, I. Yamazaki, and T. Yashiki, *J. Pharmacobio.-Dyn.*, *4*:S-17 (1981).

71. H. Okada, I. Yamazaki, Y. Ogawa, S. Hirai, T. Yashiki, and H. Mima, *J. Pharm. Sci.*, *71*:1367 (1982).
72. H. Okada, I. Yamazaki, T. Yashiki, and H. Mima, *J. Pharm. Sci.*, *72*:75 (1983).
73. H. Okada, T. Yashiki, and H. Mima, *J. Pharm. Sci.*, *72*:173 (1983).
74. H. Okada, Y. Sakura, H. Kawaji, T. Yashiki, and H. Mima, *Cancer Res.*, *43*:1869 (1983).
75. H. Okada, I. Yamazaki, Y. Sakura, T. Yashiki, T. Shimamoto, and H. Mima, *J. Pharmacobio.-Dyn.*, *6*:512 (1983).
76. H. Okada, *J. Takeda Res. Lab.*, *42*:150 (1983).
77. H. Okada, I. Yamazaki, T. Yashiki, T. Shimamoto, and H. Mima, *J. Pharm. Sci.*, *73*:298 (1984).
78. R. H. Engel and M. J. Fahrenbach, *Proc. Soc. Exp. Biol. Med.*, *129*:772 (1968).
79. S. Hirai, T. Ikenaga, and T. Matsuzawa, *Diabetes*, *27*:296 (1978).
80. S. Hirai, T. Yashiki, T. Matsuzawa, and H. Mima, *Int. J. Pharm.*, 7:317 (1981).
81. I. Yamazaki, *Endocrinol. Jpn.*, *29*:197 (1982).
82. I. Yamazaki, *Endocrinol. Jpn.*, *29*:415 (1982).
83. I. Yamazaki, *J. Reprod. Fertil.*, *72*:129 (1984).
84. D. J. Finney, *Probit Analysis*, Cambridge University Press, Cambridge (1952).
85. C. Yoshida-Noro, N. Suzuki, and M. Takeichi, *Dev. Biol.*, *101*:19 (1984).
86. K. Hatta, T. S. Okada, and M. Takeichi, *Proc. Natl. Acad. Sci. USA*, *82*:2789 (1985).
87. Y. Kanno, Y. Sasaki, Y. Shiba, C. Yoshida-Noro, and M. Takeichi, *Exp. Cell Res.*, *152*:270 (1984).
88. F. Hyafil, C. Babinet, and F. Jacob, *Cell*, *26*:447 (1981).
89. W. R. Loewenstein, Y. Kanno, and S. J. Socolar, *Fed. Proc.*, *37*:2645 (1978).
90. W. R. Loewenstein, in *Current Topics in Membranes and Transport*, Vol. 21 (F. Bronner and A. Kleinzeller, eds.), Academic Press, New York (1984), p. 221.
91. C. Peracchia and A. F. Dulhunty, *J. Cell Biol.*, *70*:419 (1976).
92. C. Peracchia, *Nature*, *271*:669 (1978).
93. B. Alberts, D. Bray, J. Lews, M. Raff, K. Roberts, and J. D. Watson, eds, *Molecular Biology of the Cell*, Garland Publishing, New York (1983), p. 673.
94. E. Windsor and G. E. Cronheim, *Nature*, *190*:263 (1961).
95. L. S. Schanker and J. M. Johnson, *Biochem. Pharmacol.*, *8*:421 (1961).
96. W. D. Erdmann and S. Okonek, *Arch. Toxikol.*, *24*:91 (1969).
97. T. Nadai, K. Nishii, and A. Tatematsu, *Yakugaku Zasshi*, *90*:262 (1970).
98. T. Murakami, N. Yata, H. Tamauchi, and A. Kamada, *Chem. Pharm. Bull.*, *30*:659 (1982).
99. T. Yagi, N. Hakui, Y. Yamasaki, R. Kawamori, M. Shichiri, H. Abe, S. Kim, M. Miyake, K. Kamikawa, T. Nishihata, and A. Kamada, *J. Pharm. Pharmacol.*, *35*:177 (1983).

100. M. Miyake, T. Nishihata, A. Nagano, Y. Kyobashi, and A. Kamada, *Chem. Pharm. Bull.*, *33*:740 (1985).

101. E. B. Schuchner, A. Foix, C. A. Borenstein, and C. Marchese, *J. Reprod. Fertil.*, *36*:231 (1974).

102. E. S. Johnson, J. H. Seely, W. H. White, and E. R. DeSombre, *Science*, *194*:329 (1976).

103. A. Lemay, R. Maheux, N. Faure, C. Jean, and A. T. A. Fazekas, *Fertil. Steril.*, *41*:863 (1984).

104. P. A. Boepple, M. J. Mansfield, M. E. Wierman, C. R. Rudlin, H. H. Bode, J. F. Crigler, Jr., J. D. Crawford, and W. F. Crowley, Jr., *Endocr. Rev.*, *7*:24 (1986).

105. A. Corbin, C. W. Beattie, J. Yardley, and T. J. Foell, *Endocrinol. Res. Commun.*, *3*:359 (1976).

106. K. L. Sheeham, R. F. Casper, and S. S. C. Yen, *Science*, *215*:170 (1982).

107. H. Okada, M. Yamamoto, Y. Ogawa, T. Yashiki, and T. Shimamoto, *105th Ann. Meet. Pharm. Soc. Jpn.*, Kanazawa (abstract) (1985), p. 790.

108. S. Hirai, H. Okada, T. Yashiki, and T. Shimamoto, *105th Ann. Meet. Pharm. Soc. Jpn.*, Kanazawa (abstract) (1985), p. 797.

109. R. B. Stoughton and W. O. McClure, *Drug Dev. Ind. Pharm.*, *9*:725 (1983).

110. P. K. Wotton, B. Mollgaard, J. Hadgraft, and A. Hoelgaard, *Int. J. Pharm.*, *24*:19 (1985).

111. K. Morimoto, E. Kamiya, T. Takeeda, Y. Nakamoto, and K. Morisaka, *Int. J. Pharm.*, *14*:149 (1983).

112. T. Nishihata, J. H. Rytting, A. Kamada, and T. Higuchi, *Diabetes*, *30*:1065 (1981).

113. S. Yoshioka, L. Caldwell, and T. Higuchi, *J. Pharm. Sci.*, *71*:593 (1982).

114. T. Nishihata, J. H. Rytting, A. Kamada, and T. Higuchi, *J. Pharm. Pharmacol.*, *35*:148 (1983).

115. G. S. Gordon, A. C. Moses, R. D. Silver, J. S. Filer, and M. C. Carey, *Proc. Natl. Acad. Sci. USA*, *82*:7419 (1985).

116. A. E. Shamoo and D. A. Goldstein, *Biochem. Biophys. Acta*, *472*:13 (1977).

117. T. Nobunaga and K. Nakamura, *Jpn. J. Anim. Reprod.*, *14*:1 (1968).

118. C. C. Hsu, J. Y. Park, N. F. H. Ho, W. I. Higuchi, and J. L. Fox, *J. Pharm. Sci.*, *72*:674 (1983).

119. M. J. Durrani, K. Kusai, N. F. H. Ho, J. L. Fox, and W. I. Higuchi, *Int. J. Pharm.*, *24*:209 (1985).

120. E. Winterhager and W. Kühnel, *Cell Tissue Res.*, *241*:325 (1985).

121. B. F. King, *J. Ultrastruct. Res.*, *83*:99 (1983).

122. P. M. Elias, *J. Invest. Dermatol.*, *80*:44 (1983).

123. M. A. Lieberman and T. Anschel, in *The Theory and Practice of Industrial Pharmacy* (L. Lachman, H. A. Leiberman, and J. L. Kanig, eds.), Lea & Febiger, Philadelphia (1970), p. 538.

124. A. K. Bhattacharyya and L. J. D. Zaneveld, in *The Human Vagina* (E. S. H. Hafez and T. N. Evans, eds.), Elsevier/North-Holland Biomedical Press, Amsterdam (1978), p. 487.

125. C. J. de Blaey and J. Polderman, in *Drug Design*, Vol. 9 (E. J. Ariens, ed.), Academic Press, New York (1980), p. 237.
126. A. R. Gennaro, ed., *Remington's Pharmaceutical Sciences*, 17th ed., Mack Publishing Company, Easton, Pa. (1986), p. 1580.
127. J. Todd, M. Fishaut, F. Kapral, and T. Welch, *Lancet*, *2*:1116 (1978).
128. J. P. Davis, P. J. Chesney, P. J. Wand, and M. La Venture, *N. Engl. J. Med.*, *303*:1429 (1980).
129. K. N. Shands, G. P. Schmid, B. B. Dan, D. Blum, R. J. Guidotti, N. T. Hargrett, R. L. Anderson, D. L. Hill, C. V. Broome, J. D. Band, and D. W. Fraser, *New Engl. J. Med.*, *303*:1436 (1980).
130. M. W. Kehrberg, R. H. Latham, B. T. Haslam, A. Hightower, M. Tanner, J. A. Jacobson, A. G. Barbour, V. Noble, and C. B. Smith, *Am. J. Epidemiol.*, *114*:873 (1981).

15

Transdermal Route of Peptide and Protein Drug Delivery

Yie W. Chien

Rutgers—The State University of New Jersey, Piscataway, New Jersey

I. HISTORICAL DEVELOPMENT OF TRANSDERMAL DRUG DELIVERY

Continuous intravenous infusion of drug at a programmed rate has been recognized as a superior mode of drug delivery. A closely monitored intravenous infusion can provide both the advantages of direct entry of drug into the systemic circulation and control of circulating drug levels. However, such a mode of drug delivery potentially entails certain risks and therefore necessitates hospitalization of the patients and close medical supervision of the medication. Recently, there is a growing awareness that the benefits of intravenous drug infusion can be closely duplicated, without its potential hazards, by continuous drug delivery through transdermal route [1,2].

The skin of an average adult body covers a surface area of approximately 2 m^2 and receives about one-third of the blood circulating through the body [3]. It is one of the most readily accessible organs on the human body. Microscopically, the skin is a multilayered organ comprised of three main histological layers: the epidermis, the dermis, and the hypodermis (subcutaneous) tissues (Fig. 1). The epidermis is further subdivided into five layers, with the stratum corneum forming the outermost layer.

The stratum corneum consists of multiple layers of horny cells that are compacted, flattened, dehydrated, and keratinized. These cells are physiologically inactive and are continuously shed with constant replacement from the underlying viable epidermal tissue [4]. The stratum corneum has a water content of around 20% as compared to the 70% in a physiologically active tissue, such as in the stratum germinativum (which is the regenerative layer of the epidermis).

An average human skin surface is known to contain an average 40 to 70 hair follicles and 200 to 250 sweat ducts on every square centimeter,

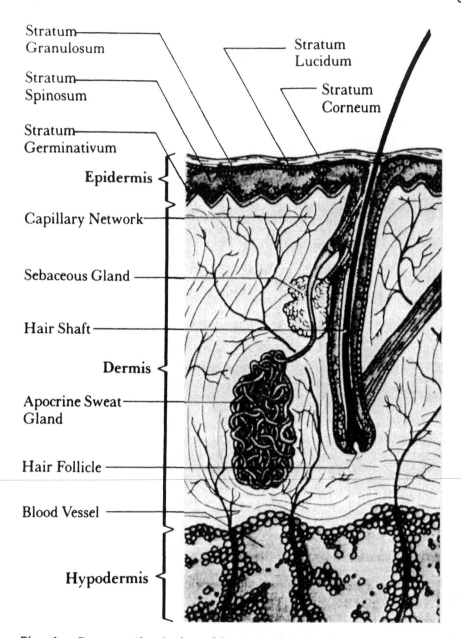

Stratum Granulosum
Stratum Spinosum
Stratum Germinativum
Stratum Lucidum
Stratum Corneum
Epidermis
Capillary Network
Sebaceous Gland
Hair Shaft
Dermis
Apocrine Sweat Gland
Hair Follicle
Blood Vessel
Hypodermis

Fig. 1   Cross-sectional view of human skin, showing various skin tissue layers and appendages.   (From Ref. 4.)

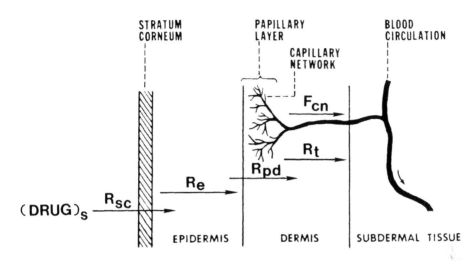

Fig. 2   Simplified multilayer model of the human skin, showing various histologic layers with varying diffusion resistance (R) to the permeation of drug molecules. (From Ref. 5.)

occupying 0.1% of the total human skin surface [5-7]. Even though water-soluble and/or ionic compounds such as peptides and proteins may penetrate the skin via these appendages at a rate faster than through the intact area of the stratum corneum, this transappendageal rate contributes only to a very limited extent to the overall kinetic profile of transdermal permeation. Therefore, the transdermal permeation of most neutral molecules at steady state can be considered as a process of passive diffusion through the intact stratum corneum in the interfollicular region according to the model shown in Fig. 2 [8,9].

For many decades, the skin has been used as the site for topical administration of dermatological drugs to achieve a localized pharmacologic action in the skin tissues, such as the use of hydrocortisone for dermatitis, benzoyl peroxide for acne, and neomycin for superficial infection [10]. In this case, the drug molecules are considered to diffuse to a target tissue in the proximity of drug application to produce its therapeutic effect before they are further distributed to the systemic circulation for elimination (Fig. 3).

In the case of transdermal drug delivery, in which the skin serves as the site for administration of systemically active drugs, the drug applied topically must be distributed following skin permeation, first into the systemic circulation and then transported to the target tissues, which could be relatively remote from the site of drug application (Fig. 3). This new mode of drug delivery is exemplified by the transdermal controlled delivery of nitroglycerin to the myocardium for the treatment of angina pectoris, of scopolamine to the vomiting center for the prevention of motion-induced sickness, and of estradiol to various estradiol-receptor sites for the relief of postmenopausal syndromes [11-13].

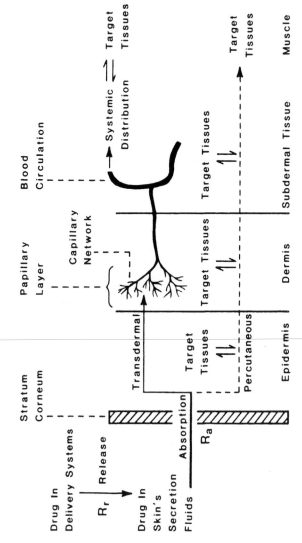

Fig. 3 Schematic illustration of the release of drug from a topically applied delivery system and its subsequent absorption across the skin tissues for localized therapeutic actions (in the tissues that are directly underneath the site of drug administration) or for systemic medications (in the tissues that are remote from the site of topical drug application).

The potential of using the intact skin as the port of drug delivery has been recognized for several decades, as evidenced by the development of medicated plasters and their popularity in medical uses. Plasters are made by blending drugs with a natural adhesive material, spread on a backing support. With proper balance of cohesive strengths, the adhesive material provides good bonding to the skin as the plaster is applied and a clean adhesive break from the skin surface when the plaster is removed [14].

The use of medicated plasters can be traced back to ancient China. They have also been very popular in Japan, and many are available in the over-the-counter pharmaceutical marketplace, commonly called cataplasms [15]. Medicated plasters have also been used in Western medicine for several decades. In the United States, three medicated plasters (belladonna, mustard, and salicylic acid) have been listed in the official compendia for the last 40 years [16,17].

II. TRANSDERMAL DELIVERY OF PHARMACEUTICALS

The potential of using intact skin as the port for continuous transdermal delivery of drugs has recently been recognized beyond the boundary of topical medication. The unexpected development of female syndromes in male operators working in manufacturing areas of estrogen-containing pharmaceutical dosage forms has challenged the traditional belief that the skin is a perfectly protective barrier, and has also triggered intensive research curiosity by biomedical scientists in studying the feasibility of using transdermal drug delivery process for systemic medication. The findings accumulated over the years have practically revolutionized the old theory of impermeable skin barrier and have motivated a number of pharmaceutical scientists to develop rate-controlled drug delivery systems for controlling transdermal administration of drugs to accomplish the objective of systemic medication [2,18–36].

Several transdermal drug delivery (TDD) systems have recently been developed. The potential of transdermal delivery of pharmaceuticals was first demonstrated by development of the Transderm-Scop System, a scopolamine-releasing TDD system, for 72-hr prophylaxis or treatment of motion-induced sickness and nausea [22]. It was followed by the successful marketing of several nitroglycerin-releasing TDD systems, a isosorbide dinitrate-releasing TDD system for once-a-day medication of angina pectoris [23–30], and very recently by regulatory approval of a clonidine-releasing TDD system for weekly therapy of hypertension [25,31,32] and of an estradiol-releasing TDD system for twice-a-week treatment of postmenopausal syndromes [33–36].

These TDD systems can be classified, according to the technological basis of their approach, into four categories [37]: (1) membrane permeation-controlled TDD systems, (2) adhesive dispersion-type TDD systems, (3) matrix diffusion-controlled TDD systems, and (4) microreservoir dissolution-controlled TDD systems. Although these TDD systems have demonstrated usefulness in affording transdermal controlled delivery of pharmaceuticals that are somewhat lipophilic and relatively small in molecular size, they are rather limited in assisting the delivery of peptides and protein drugs.

III. TRANSDERMAL DELIVERY OF PEPTIDES
AND PROTEINS

Since the inception of genetic engineering, therapeutic applications of
peptides/proteins have received increasing attention [38–40]. A number
of peptides and proteins have been approved by the regulatory authorities
for medical use, such as for endocrine disorders, cardiovascular diseases,
cancers, and viral infections, or in preventive medicine, such as in active
and passive immunizations. Numerous studies have been initiated to develop potential techniques to deliver these peptides and proteins by
routes other than parenteral and oral [41–50]. Among these is the transdermal route, which has been shown to provide the possibility of bypassing gastrointestinal degradation and hepatic first-pass elimination as
well as achieving better patient compliance [8,9,32,51–53]. A review of
the literature has revealed an additional advantage: The skin lacks
proteolytic enzymes [54], which can degrade peptide and proteins. All these
factors render the skin an appealing site for the administration of therapeutically important peptides and proteins. Nonetheless, to deliver a peptide
or a protein drug successfully across the intact skin to attain a therapeutic
effect, it is essential to develop a technique to facilitate the transdermal
permeation of peptides and proteins through the stratum corneum and to
overcome the diffusional resistance of various skin tissues (Fig. 2).

A. Mechanisms of Transdermal Iontophoretic
Drug Delivery

Iontophoresis is a process that induces increased migration of ions or
charged molecules in an electrolyte medium following the flow of electric
current [55]. In addition to electrophoresis, it has also been explored as
a potential technique to facilitate the membrane transport of charged molecules based on their ionic characteristics [56–59]. Iontophoresis with
direct current (dc) has recently been proven to be a safe and useful technique to facilitate the penetration of charged anti-inflammatory drugs into
cutaneous tissues for localized effects [60–63].

Skin consists of lipids (15 to 20%), proteins (40%, mostly keratin), and
water (40%) [64–66]. As an electrical potential is applied across the skin,
the electric current may alter the molecular arrangement of skin components, which could yield some changes in the skin permeability. The
"flip-flop gating mechanism" [67] could be an operating model in voltage-
dependent pore formation in the stratum corneum, which is rich in keratin
(an α-helical polypeptide). Under the electrical potential applied, the
flip-flop of polypeptide helices may occur to form a parallel arrangement.
Pores are thus opened up as a result of the repulsion between neighboring
dipoles, and water molecules and ions will flow in the pore channels to
neutralize the dipole moments. This phenomenon should lead to the enhancement of skin permeability for peptide and protein molecules.

The upper layers of the skin are known to have an isoelectric point
at pH 3 to 4, so the pores in the stratum corneum will have a negative
charge when exposed to a solution with pH 4 or higher [55,64,65]. As
the skin is maintained at a negative charge, iontophoresis will affect the
movement of water molecules into the body from the positive electrode
electroosmotically toward the surface of the stratum corneum at the negative electrode. This leads to shrinkage of the skin pores at the positive

electrode and causes swelling of the skin pores at the negative electrode after intensive iontophoresis [55]. This process will facilitate the transdermal permeation of cationic drugs. In addition to ionic conduction and electroosmosis, other phenomena, such as solute–solvent and solute–solute coupling, may account for the observed enhancement in transdermal permeation of drugs under iontophoresis [66].

Although a number of theoretical expressions have been derived to describe the observed rate of iontophoretic delivery of small molecules, an exact relationship has not been defined, primarily because of the difficulties encountered in correlating various studies conducted under different experimental conditions [66]. Some authors [67] attempted to apply Faraday's law to describe the rate of drug deposition at the skin surface. This states that the amount of substance deposited at either electrode is proportional to the quantity of electricity passing through the system. However, because of the complexity of the factors involved in the process of iontophoresis, theoretical predictions based on Faraday's law and their correlations with experimental data are virtually impossible.

On the other hand, Abramson and Gorin [68] derived an equation to correlate the iontophoretic dosage to various components contributed by electrical mobility, electroosmosis, and simple diffusion. Masada et al. [69] also developed equations to define the enhancement of the in vitro skin permeation flux of small molecules by iontophoresis, using a four-compartment diffusion cell electrode system. They found that under a low electrical potential (0 to 0.25 V), the experimental data on the iontophoretic permeation flux of benzoic acid (and its sodium salt) across hairless mouse skin are in agreement with the theoretical calculation. At higher electrical potential, however, the enhancement in the skin permeation flux was greater than the data predicted from the equation.

B. Early Studies on Iontophoretic Delivery of Peptide and Protein Drugs

Feasibility of applying iontophoresis to facilitate the transdermal permeation of peptides was recently investigated in vitro using thyrotropin-releasing hormone (TRH) as a model peptide [70]. The results indicated that the permeation of this tripeptide across the excised hairless mouse skin is facilitated by contributions from an electric term driven by applied current and an electrically induced convective term that may be affected by ionic strength. The steady-state permeation rate of the TRH was found in agreement with the prediction based on the Nernst–Planck flux equation.

Transdermal iontophoretic delivery of insulin was recently attempted by Stephen et al. [71]. They reported that a highly ionized monomeric form of insulin can be delivered through the pig skin, in the presence of iontophoresis, to produce some systemic effects. Kari [49] observed that to control the glucose levels in diabetic rabbits effectively by iontophoresis, the stratum corneum needs to be removed.

C. Recent Advances in Transdermal Iontophoretic Delivery of Peptide and Protein Drugs

The skin is known to produce a large diffusional resistance to the transport of charged molecules driven by an applied electrical field. The

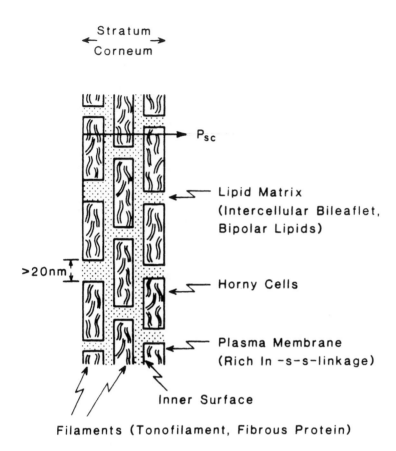

Stratum
Corneum

P_{sc}

Lipid Matrix
(Intercellular Bileaflet,
Bipolar Lipids)

>20nm

Horny Cells

Plasma Membrane
(Rich In -s-s-linkage)

Inner Surface

Filaments (Tonofilament, Fibrous Protein)

Fig. 4   Diagrammatic illustration of the microstructure of stratum corneum showing the dispersion of keratinized horny cells, as multilayered "brick-like" organization, in the fatty matrix.

electrical properties of the skin are also known to be dominated by the least conductive stratum corneum. As discussed in Sec. I, the stratum corneum consists of multilayers of horny cells that are breached by hair follicles and sweat ducts. These skin appendages could potentially act as the pathway for shunt diffusion across the skin. This shunt pathway may be significant, especially for ionic penetrants, which show an extremely poor skin permeation through the transcellular route [72]. Under the influence of electric current, ions or charged molecules are driven across the skin, possibly through the shunt pathway and/or intercellular spacings in the stratum corneum, as the skin is likely to be perturbed during the iontophoresis treatment, which may remove or disrupt the intercellular lipids and activate the formation of artificial "shunts" to facilitate skin permeation (Fig. 4).

The stratum corneum shows two important electrical features. First, it tends to become polarized as an electrical field is continuously applied. Second, its impedance changes with the frequency of the applied electrical field. Therefore, as an electrical field with direct current (dc) is applied

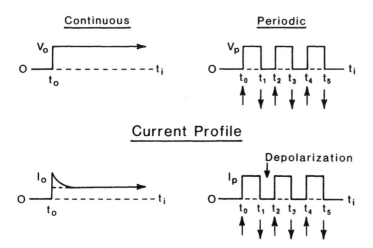

Fig. 5 Current profiles across the skin barrier as a function of the applied electrical field: periodic versus continuous application. Use of periodic waveform minimizes the occurrence of polarization and allows the skin to depolarize during the "off" state, so the intensity of effective current across the skin will not decay exponentially as a function of treatment duration.

in a continuous manner to the stratum corneum to facilitate the permeation of charged molecules, electrochemical polarization may occur in the skin. This polarization often operates against the applied electrical field and greatly decreases the magnitude of effective current across the skin. So the current gradient through the skin decays exponentially (Fig. 5). Consequently, the efficiency of transdermal iontophoretic delivery is reduced as a function of the application time of dc iontophoresis.

To avoid the counterproductive polarization, the current should be applied in a periodic manner, called the pulse current mode. With pulse dc, the electrical field is switched on and off periodically. In the "on" state, charged molecules are delivered iontophoretically into the skin. Within a short period, polarization is expected to occur. By incorporating the "off" state into the system, external stimulation is removed to permit the skin to have a chance to depolarize, a process that is equivalent to the discharging of the current from the skin [73]. Recently, a pulse current with 20% duty (4 μs) followed by an 80% depolarizing period (16 μs) was applied successfully in five human subjects to deliver a beta-blocker systemically without skin irritation caused by polarization [74]. Therefore, every new cycle could start with no residual polarization remaining in the skin from the previous cycle if the proper frequency were selected (Fig. 5).

With these theoretical concepts as the foundation, a transdermal periodic iontotherapeutic system (TPIS) was developed in this laboratory. It is capable of delivering pulse dc with various combinations of waveform,

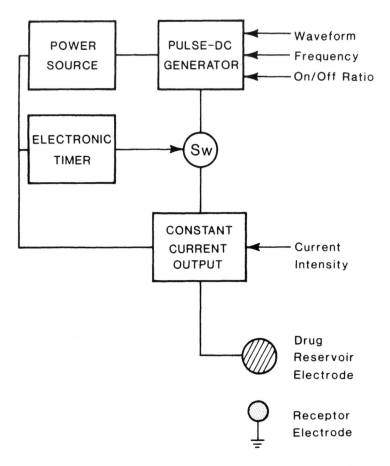

Fig. 6 Diagrammatic illustration of the transdermal periodic iontothera-peutic system (TPIS). The pulse current generated with a preset combination of waveform, frequency, on/off ratio, and intensity is delivered to the drug reservoir electrode.

frequency, on/off ratio, and current intensity for a specified duration of application time (Fig. 6).

In Vitro Evaluation of TPIS

In vitro studies of TPIS-facilitated skin permeation of peptide and protein drugs were conducted using the experimental setup shown in Fig. 7. The results indicated that the skin permeation profile of vasopressin, a peptide hormone with nine amino acid residues (and a molecular weight of 1084 daltons), consists of two phases: activation phase, in which a pulse current with a physiologically acceptable intensity is applied alternatively from the TPIS; and postactivation phase, in which TPIS treatment is terminated (Fig. 8). The results suggested that without TPIS treatment, the rate of skin permeation of vasopressin is relatively low (0.94 ± 0.62 ng/cm^2·hr). During TPIS treatment, the lag time was reduced from 9.12 (± 1.06) hr to

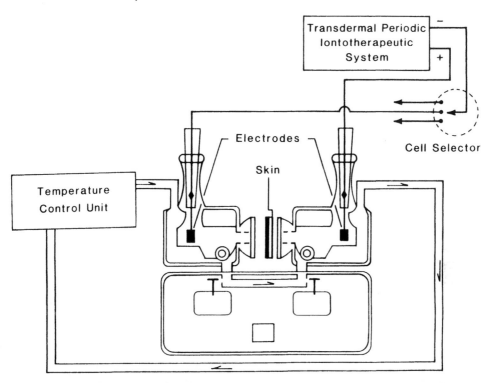

Fig. 7 Experimental setup for the in vitro skin permeation kinetics studies of peptide and protein drugs. The direct current or pulse current generated by TPIS is delivered into the donor solution in the thermostated skin permeation cell and transported across the skin specimen using a pair of platinum electrodes.

less than 0.5 hr and the skin permeation rate was increased nearly 190-fold (178.0 ± 25.0 ng/cm^2·hr). After TPIS treatment, the permeability properties of the skin appeared to recover and the skin permeation rate was reduced to 5.3 (±0.5) ng/cm^2·hr. The enhancement in skin permeation rate profile and the reversibility of skin permeability appeared to be dependent on the intensity and duration of pulse current applied (Table 1).

The in vitro skin permeation profile of insulin, a protein hormone with 51 amino acid residues and a molecular weight of approximately 6000 daltons, was also studied. It has an isoelectric point at pH 5.4 and the degree of enhancement in its skin permeability was found to be pH dependent (Table 2).

The observations outlined in Table 2 can be explained by the fact that a peptide or a protein drug, such as insulin, can be rendered as positively charged molecules or negatively charged molecules by controlling the solution pH below or above the isoelectric point (pH$_{iso}$) of the drug [75]. The solubility and the charge density of peptide (protein) molecules increase when the solution pH is made either higher or lower than its isoelectric point as a result of protonation or dissociation of the amino acid residues in the peptide (protein) molecules. The results suggested that

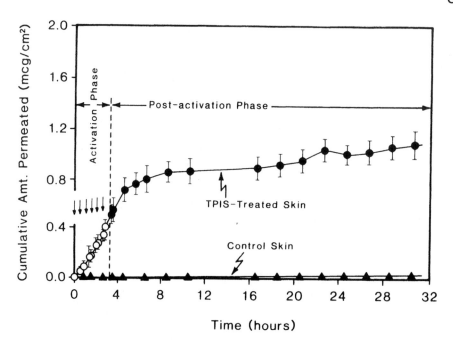

Fig. 8 In vitro skin permeation profiles of vasopressin, from donor solution at pH 5 across the abdominal skin of hairless rats, and the enhancement of skin permeation by TPIS treatment (square waveform; frequency, 2 kHz, on/off ratio, 1:1) at current intensity of 1 mA, which was switched on for 10 min (↑) and then off for 30 min repeatedly for 6 times over 4 hr.

Table 1 Effect of Pulse DC Iontophoresis (TPIS) on Skin Permeation Profiles of Vasopressin

Conditions		Skin permeation profile ($\bar{x} \pm SD$)		
Treatment	Current intensity (mA)	Lag time (hr)	Permeation rate[a] (ng/cm^2·hr)	
Control	0.0	9.12 (±1.06)	0.94 (±0.62)	
TPIS-treated[b]			Activation	Postactivation
	0.5	<0.5	116.2 (±10.7)	0.7 (±0.4)
	1.0	<0.5	178.0 (±25.0)	5.3 (±0.5)

[a]In vitro permeation across freshly excised rat skin mounted in a modified V–C skin permeation cell (Fig. 7).

[b]Pulse dc (frequency, 2 kHz; on/off ratio, 1:1).

Table 2   Enhancement in Skin Permeability of Insulin[a]
by Transdermal Iontophoretic Delivery[b]

Donor solution pH	Enhancement factor[c]
3.7	37.3 ($\pm$2.8)
5.2	12.0 ($\pm$2.3)
7.1	9.5 ($\pm$1.1)

[a]5.2 IU/ml (with 0.3 μCi [I^{125}] insulin) in a donor
solution exposed to the stratum corneum surface of a
freshly excised skin specimen from the abdominal
region of hairless rats.

[b]Application of current intensity at 1 mA (frequency,
0 kHz; on/off ratio, 1:1) for 5 min/hr for 7 hr.

[c]Enhancement factor = $\dfrac{\text{(skin permeation rate)}_{\text{with TPIS}}}{\text{(skin permeation rate)}_{\text{control}}}$

the lower the pH of the insulin solution, the greater the degree of enhancement. This phenomenon may be attributed to the fact that insulin
molecules tend to become aggregated as pH increases, which may explain
the experimental results obtained: that transdermal iontophoretic delivery
has yielded a 37.3-fold enhancement in skin permeation of insulin at pH
3.7, which is 1.7 pH units below its pH$_{iso}$ value (5.4), and only a 9.5-
fold increase at pH 7.1, which is 1.7 pH units above its pH$_{iso}$ value
(Table 2).

In Vivo Evaluation of TPIS

To study the efficiency of transdermal iontophoretic delivery of peptide
and protein drugs, in vivo studies in animal model were conducted. In
the case of insulin, a diabetic animal model was developed in harless rats
by intraperitoneal injection of streptozotocin [76]. After a single injection
of streptozotocin, the beta cells appeared to be extensively damaged with
the Langerhans islet necrotized. The blood glucose level was found to
increase, within 24 hr after the streptozotocin administration, from the
normoglycemic level at around 80 mg/dl to the hyperglycemic level of more
than 200 mg/dl. This hyperglycemic state was found to remain fairly
stable for 3 to 4 days and was unaffected by fasting, anesthesia, and
TPIS treatment with placebo formulation (i.e., no insulin in the reservoir
electrode) (Fig. 9).

After applying the TPIS equipped with insulin-containing reservoir
electrode, which contains insulin formulation at various pH levels, and
the receptor electrode to the abdominal skin surface of diabetic rats (Fig.
10), the blood glucose levels in the treated diabetic rats were observed to
decrease rapidly following the application of dc at 4 mA (current density =
0.67 mA/cm^2) for 80 min [76]. Pronounced hypoglycemia was observed in
the animals treated with insulin formulation, while the blood glucose level
in the untreated diabetic controls remained hyperglycemic (Fig. 11). The

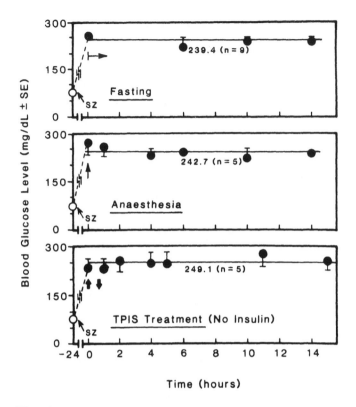

Fig. 9 Hyperglycemia in diabetic hairless rats is stable and is unaffected by fasting and anaesthesia as well as treatment of the transdermal periodic ionto- the peutic system (TPIS) with no insulin in the drug reservoir electrode.

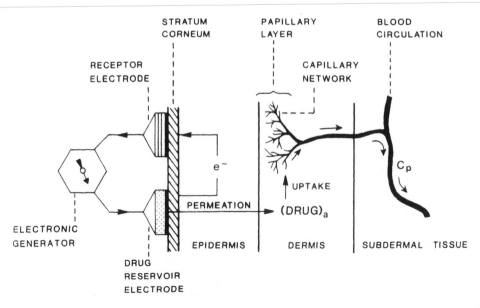

Fig. 10 Diagrammatic illustration of the transdermal iontophoretic delivery of peptide and protein drugs across live animal's skin of systemic administration.

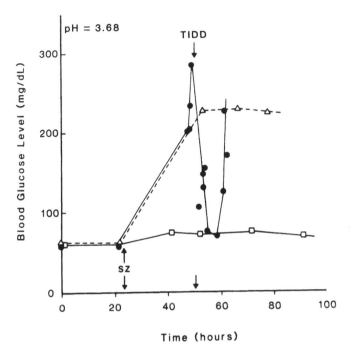

Fig. 11 Reduction of blood glucose level in diabetic hairless rats from the hyperglycemic level down to the normoglycemic level as a result of the transdermal iontophoretic delivery of insulin, at pH 3.68, by application of TPIS with dc mode at current intensity of 4 mA for 80 min.

normoglycemic level in the treated diabetic rats was maintained at the same level as in normal rats for over 2 to 3 hr even after the completion of the treatment (80 min). The reduction of blood glucose levels in diabetic rats, in response to the transdermal iontophoretic delivery of insulin, also appears to be dependent on the pH of insulin solution, with the maximum effect (a 71.2% reduction in blood glucose level) accomplished at pH 3.7 (Fig. 12). This observation could be attributed to the acidic nature of the solution, in which insulin molecules exist predominantly as positively charged molecules and are likely to be less aggregated [75].

As observed in the in vitro studies, the in vivo data on the reduction of blood glucose level also indicated that as solution pH is higher than the isoelectric point of insulin, the efficiency of normalizing the blood glucose level in the diabetic animal is diminished with a 54.5% reduction in blood glucose level at pH 7.1, which is 1.7 pH units above the pH_{iso}, and a 37.5% reduction at pH 8.0, which is 2.6 pH units above the pH_{iso}. Again, this observation could possibly have resulted from the aggregation of insulin molecules at solution pH higher than pH_{iso} [75], which yields insulin aggregates with lower skin permeability.

The experimental results outlined in Fig. 13 illustrate the effect of the delivery mode of current on the efficiency of transdermal iontophoretic delivery of insulin and the control of blood glucose levels. In this study,

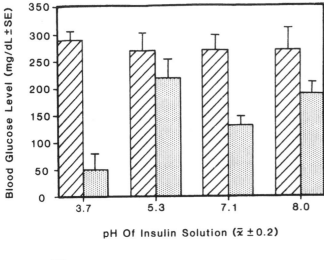

pH Of Insulin Solution ($\bar{x} \pm 0.2$)

▨ Before TIDD (3–4 hours)
▥ After TIDD (4 mA, 80 minutes)

Fig. 12   The pH dependence of the hypoglycemic effect of insulin delivered transdermally by TPIS with dc mode at current intensity of 4 mA
for 80 min.

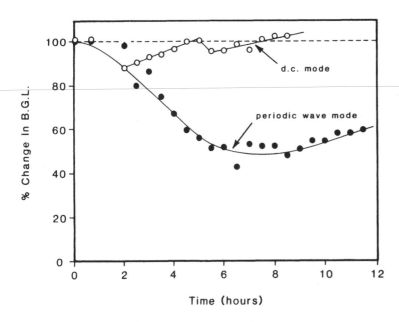

Fig. 13   Effect of the delivery mode of current from the transdermal
periodic iontotherapeutic system (TPIS) on the magnitude and duration of
the percent reduction in blood glucose level (BGL) from hyperglycemic
state (100%) in diabetic hairless rats.

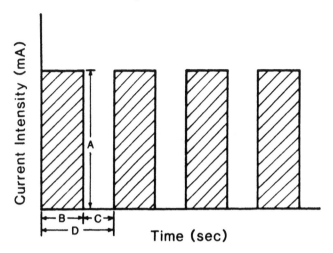

Fig. 14 Diagrammatic illustration of the pulse current delivered by the transdermal periodic iontotherapeutic system (TPIS). A, amplitude of the current intensity (milliamperes); B/C, on/off ratio; D, one complete cycle (seconds); and 1/D, frequency (hertz).

direct current is applied to the diabetic hairless rats from the TPIS's controlling module using either the conventional dc mode or the periodic wave mode using the same current density and for the same duration. The blood glucose levels appear to be much more effectively reduced when pulse dc is applied as compared to the simple dc mode. It is possible to reduce the blood glucose levels effectively with pulse dc for at least 12 hr before the levels go back to the original hyperglycemic state (which is defined as 100%). The percent change in blood glucose level (BGL) in the diabetic animals from the hyperglycemic state is calculated from the following relationship:

$$\% \text{ change in BGL} = \frac{(BGL)_t}{(BGL)_i} \times 100\%$$

where $(BGL)_i$ and $(BGL)_t$ are the blood glucose levels before and after the TPIS treatment. The results in Fig. 13 suggest that the TPIS with pulse dc is much more effective than the conventional dc method in facilitating the transdermal delivery of macromolecular insulin [73].

The pulse dc delivered by TPIS is composed of four features: waveform, intensity, frequency, and on/off ratio (Fig. 14). These system parameters must be characterized in order to optimize TPIS's enhancing effect on the transdermal iontophoretic delivery of peptides and proteins. While the intensity of current controls the amount and rate of charged macromolecules penetrating the skin (Faraday's law), the waveform, frequency, and on/off ratio of current determine the efficiency of overcoming the impedance of the least conductive stratum corneum. The data generated so far have demonstrated that the control of blood glucose levels in

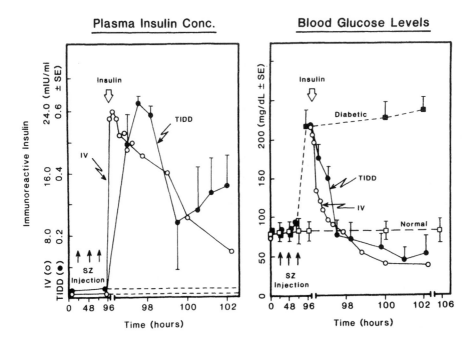

Fig. 15   Comparative plasma profiles of immunoreactive insulin from the transdermal delivery of insulin by TPIS with dc mode (TIDD) and intravenous (IV) administration as well as the correspondent reduction in blood glucose levels in the diabetic rabbits.   TIDD achieved the equivalent efficacy as IV, even though TIDD provided a significantly lower plasma insulin concentration than IV.

the diabetic rats is dependent on the frequency, on/off ratio, and waveform of the input current as well as the duration of TPIS application [73].

New Zealand white rabbits were selected as the animal model to study the relationship between the plasma profiles of insulin following transdermal iontophoretic delivery and its pharmacodynamic responses. By daily intravenous injection of streptozotocin for 2 to 3 consecutive days, diabetes was induced successfully in rabbits. As in the hairless rats, a stable hyperglycemic level was established, with the blood glucose level elevating from the normoglycemic level of approximately 80 mg/dl to a sustained hyperglycemic level of greater than 200 mg/dl.

The feasibility of using this diabetic rabbit model to study the relationship of pharmacodynamic responses, in terms of the percent reduction in blood glucose levels, with the plasma profiles of insulin delivered transdermally was demonstrated experimentally by simultaneously assaying the blood glucose levels by a glucose analyzer and the plasma insulin concentrations by radioimmunoassay.

The results in Fig. 15 demonstrate that using the conventional dc mode, at 4 mA for 80 min, the transdermal iontophoretic delivery of insulin yields a plasma profile of immunoreactive insulin with peak level achieved within 1 to 2 hr. Even though the area under the plasma concentration (AUC) of immunoreactive insulin achieved by the transdermal iontophoretic delivery

Plasma Insulin Conc.

Blood Glucose Levels

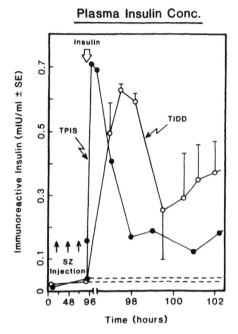
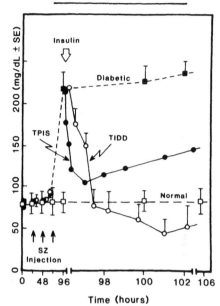

Fig. 16   Comparative plasma profiles of immunoreactive insulin from the
transdermal delivery of insulin by TPIS and the correspondent reduction in
blood glucose levels in the diabetic rabbits.   Transdermal delivery of
insulin using periodic waveform (TPIS) apparently produces a faster transdermal absorption of insulin into the systemic circulation and a better control of blood glucose levels in the diabetic animal than that using dc mode
(TIDD).

was only approximately one-fortieth of that by intravenous administration,
the transdermal iontophoretic delivery was essentially bioequivalent to the
intravenous administration in reducing the hyperglycemia in the diabetic
rabbits.   The observation may be the result of continuous administration
of short-acting insulin ($t_{\frac{1}{2}} \sim 9$ min) by transdermal iontophoretic delivery.
The continuous drug delivery sustains the biological effectiveness of
insulin.

The in vivo studies reported earlier in diabetic hairless rats suggested
that the efficiency of transdermal iontophoretic delivery of insulin can be
improved by using a periodic waveform (Fig. 16).   The results of the
studies in diabetic rabbits further confirmed that by applying pulse dc
with a square waveform at a current intensity of 1 mA, a frequency of
2000 Hz, and an on/off ratio of 1:1 for a duration of 40 min, the peak
plasma level (0.72 mIU/ml) of immunoreactive insulin was reached within
30 min, compared to 1 to 2 hr by conventional dc mode at a condition
which is eight times more intense than that under pulse dc mode.   Both
the plasma insulin concentration and the blood glucose level profiles demonstrated that the pulse mode produces a faster transdermal delivery of
insulin and maintains a shorter duration of action than the conventional
mode.

D. Summary Statement

In summary, insulin and vasopressin can be delivered successfully through the intact skin by iontophoresis. Using diabetic hairless rats and hairy rabbits, the systemic bioavailability of insulin that resulted from transdermal iontophoretic delivery was quantitatively assessed by radioimmunoassay and correlated well with the pharmacodynamic responses as determined by glucose-level measurement. Comparative studies with parenteral administration have demonstrated that insulin delivered transdermally by the application of a specially designed transdermal periodic iontotherapeutic system is able to achieve a therapeutically effective concentration that effectively controls hyperglycemia in diabetic animals. The results further confirmed that transdermal iontophoretic delivery of insulin can be accomplished more efficiently by using pulse rather than conventional dc.

IV. CONCLUSIONS

Development of a transdermal delivery technique for the controlled administration of drugs can potentially yield one or more of the following biomedical benefits: (1) avoid the risks and inconveniences of parenteral therapy; (2) prevent the variation in the absorption and metabolism associated with oral administration; (3) permit continuous drug delivery and the use of a drug with a short biological half-life; (4) increase therapeutic efficacy by bypassing hepatic first-pass elimination; (5) reduce the chance of over- or underdosing as the result of prolonged, preprogrammed delivery of drug at the required therapeutic rate; (6) to provide a simplified therapeutic regimen, leading to better patient compliance; and (7) permit rapid termination of the medication, if needed, simply by removing the transdermal drug delivery system from the skin surface.

The transdermal periodic iontotherapeutic system (TPIS) discussed in Sec. III.C appears to add the capability of modulation (or regulation) to the transdermal controlled delivery of drugs. This capability is particularly important in the systemic administration of peptides and proteins, which are very potent and extremely short-acting and must often be delivered in a rhythmic pattern for simulation of physiological conditions.

ACKNOWLEDGMENTS

The author wishes to express his appreciation to Dr. O. Siddiqui, Dr. Y. Sun, Ms. P. Lelawongs, and Mr. W. M. Shi for sharing their technical data, and to Ms. M. Boslet for her able assistance in preparing this manuscript.

REFERENCES

1. Y. W. Chien, *Novel Drug Delivery Systems—Fundamentals, Developmental Concepts and Biomedical Assessments*, Marcel Dekker, New York (1982), Chaps. 1 and 5.

2. J. E. Shaw, S. K. Chandrasekaran, and P. Campbell, *J. Invest. Dermatol.*, *67*:677 (1976).

3. S. W. Jacob and C. A. Francone, *Structure and Function of Man*, 2nd ed., W. B. Saunders Company, Philadelphia (1970), pp. 55–60.

4. S. C. Laitin, *Handbook of Nonprescription Drugs*, 7th ed., American Pharmaceutical Association, Washington, D.C. (1982), pp. 525–529.

5. J. E. Treherne, *J. Physiol. (London)*, *133*:171 (1956).

6. R. J. Scheuplein, *J. Invest. Dermatol.*, *48*:415 (1967).

7. I. H. Blank, *J. Invest. Dermatol.*, *43*:415 (1964).

8. Y. W. Chien, *Drug Dev. Ind. Pharm.*, *9*:497 (1983).

9. Y. W. Chien, in *Controlled Drug Delivery* (J. R. Robinson and V. H. L. Lee, eds.), Marcel Dekker, New York (1987), Chap. 12.

10. E. K. Kastrip and J. R. Boyd, *Drug: Facts and Comparisons*, 1983 ed., J. B. Lippincott Company, Philadelphia (1983), pp. 1634–1708.

11. J. E. Shaw, W. Bayne, and L. Schmidt, *Clin. Pharmacol. Ther.*, *19*: 115 (1976).

12. P. W. Armstrong, J. A. Armstrong, and G. S. Marks, *Am. J. Cardiol.*, *46*:670 (1980).

13. R. Sitruk-Ware, B. deLignieres, A. Basdevant, and P. Mauvais-Jarvis, *Maturitas*, *2*:207 (1980).

14. A. Osol, *Remington's Pharmaceutical Sciences*, 16th ed., Mack Publishing Company, Easton, Pa. (1980), p. 1534.

15. First Transdermal Therapeutic System Symposium on Evaluation of External Adhesive System, Tokyo, Japan, July 26, 1985.

16. United States Pharmacopeial Convention, *National Formulary*, 8th ed., Mack Publishing Company, Easton, Pa. (1946).

17. United States Pharmacopeial Convention, *United States Pharmacopeia*, 14th ed., Mack Publishing Company, Easton, Pa. (1950).

18. A. S. Michaels, S. K. Chandrasekaran, and J. E. Shaw, *AIChE J.*, *21*:985 (1975).

19. *World Cong. Clin. Pharmacol. Symp. Transdermal Delivery Cardiovas. Drugs*, Washington, D.C. (Aug. 5, 1983); proceedings published in *Am. Heart J.*, *108*(1):195 (1984).

20. *Int. Pharm. R&D Symp. Adv. Transdermal Controlled Drug Admin. Syst. Med.*, Rutgers University, College of Pharmacy (June 20 and 21, 1985).

21. *Neu Ulm Conf. Transdermal Drug Delivery Syst.*, University of Ulm, Neu Ulm, West Germany (Dec. 1–3, 1986).

22. J. E. Shaw and S. K. Chandrasekaran, *Drug Metab. Rev.*, *8*:223 (1978).

23. W. R. Good, *Drug Dev. Ind. Pharm.*, *9*:647 (1983).

24. A. Gerardin, J. Hirtz, P. Fankhauser, and J. Moppert, *APhA/APS 31st Nat. Meet. Abstr.*, *11*:84 (1981).

25. J. E. Shaw, *Am. Heart J.*, *108*:217 (1984).

26. A. D. Keith, *Drug Dev. Ind. Pharm.*, *9*:605 (1983).

27. A. F. Kydonieus and B. Berner, *Transdermal Delivery of Drugs*, Vol. I, CRC Press, Boca Raton, Fl. (1986), Chap. 11.

28. P. K. Noonan, M. A. Gonzalez, D. Ruggirello, J. Tomlinson, E. Babcock-Atkinson, M. Ray, A. Golub, and A. Cohen, *J. Pharm. Sci.*, *75*:688 (1986).

29. A. Karim, *Drug Dev. Ind. Pharm.*, *9*:671 (1983).

30. D. R. Sanvordeker, J. G. Cooney, and R. C. Wester, U.S. patent 4,336,243 (June 22, 1982).

31. M. A. Weber and J. I. M. Drayer, *Am. Heart J.*, *108*:231 (1984).

32. D. Arndts and K. Arndts, *Eur. J. Clin. Pharmacol.*, *26*:79 (1984).

33. W. R. Good, M. S. Powers, P. Campbell, and L. Schenkel, *J. Controlled Release*, *2*:89 (1985).

34. L. F. Prescott and W. S. Nimmo, *Rate Control in Drug Therapy*, Churchill Livingston, Edinburgh (1983), Chap. 31.

35. L. R. Laufer, J. L. De Fazio, J. K. H. Lu, D. R. Meldrum, P. Eggena, M. P. Sambhi, J. M. Hershman, and H. L. Judd, *Am. J. Obstet. Gynecol.*, *146*:533 (1983).

36. L. Schenkel, *Neu Ulm Conf. Transdermal Drug Delivery Syst.*, University of Ulm, Neu Ulm, West Germany, Dec. 1–3, 1986.

37. Y. W. Chien, *Transdermal Controlled Systemic Medication*, Marcel Dekker, New York (1987), Chap. 2.

38. D. H. Hey and D. I. John, *Amino Acids, Peptides and Related Compounds*, Organic Chemistry, Series 1, Vol. 6, University Park Press, Baltimore (1973), Chaps. 1, 2, 3, and 5.

39. J. B. Dence, *Steroids and Peptides: Selected Chemical Aspects for Biology, Biochemistry and Medicine*, John Wiley & Sons, New York (1980), Chap. 4.

40. D. M. Matthews, *Physiol. Rev.*, *55*:537 (1975).

41. O. Siddiqui and Y. W. Chien, *CRC Crit. Rev. Ther. Drug Carrier Syst.*, *3*:195 (1987).

42. M. P. Earle, *Isr. J. Med. Sci.*, *8*:899 (1971).

43. R. P. Walton and C. F. Lacey, *J. Pharm. Exp. Ther.*, *54*:61 (1938).

44. K. Ichikawa, I. Ohata, M. Mitomi, S. Kawamura, H. Maeno, and H. Kawata, *J. Pharm. Pharmacol.*, *32*:314 (1980).

45. Y. Yamaski, M. Schichiri, and R. Kawamari, *Diabetes Care*, *4*:454 (1981).

46. S. W. Lee and J. J. Sciarra, *J. Pharm. Sci.*, *65*:567 (1976).

47. F. M. Wigley, J. H. Londono, and S. H. Wood, *Diabetes*, *20*:552 (1971).

48. G. S. Gordon, A. C. Moses, R. D. Silver, J. S. Flier, and M. C. Carey, *Proc. Natl. Acad. Sci. USA*, *82*:7419 (1985).

49. B. Kari, *Diabetes*, *35*:217 (1986).

50. T. Nishihata, S. Kim, S. Morishita, A. Kamada, N. Yata, and T. Higuchi, *J. Pharm. Sci.*, *72*:280 (1983).

51. Y. W. Chien, *11th Ann. A. R. Granito Mem. Lect.*, Saddlebrook, N.J., Dec. 10, 1985.

52. Y. W. Chien, *Proc. 11th Conf. Pharm. Technol.*, Academy of Pharmaceutical Science and Technology of Japan, Shirakabako, Nagano Prefecture, Japan (1986), pp. 1–56.

53. Y. W. Chien and C. S. Lee, in *Controlled Release Technology—Pharmaceutical Applications* (P. I. Lee and W. R. Good, eds.), ACS Symposium Series #348, American Chemical Society, Washington, D.C., 1987, Chap. 21.

54. P. J. Pannatier, B. Testa, and J. C. Etter, *Drug Metab. Rev.*, *8*: 319 (1978).

55. R. Harris, in *Therapeutic Electricity and Ultraviolet Radiation* (S. Licht, ed.), John Wiley & Sons, New York (1967), Chap. 4.

56. M. Comeau, R. Brummett, and J. Vernon, *Arch. Otolaryngol.*, *98*:114 (1983).

57. D. F. Echols, C. H. Norris, and H. G. Talb, *Arch. Otolaryngol.*, *101*:418 (1975).
58. O. Siddiqui, M. S. Roberts, and A. E. Polack, *J. Pharm. Pharmacol.*, *37*:732 (1985).
59. L. von Sallman, *Am. J. Ophthalmol.*, *25*:1292 (1942).
60. L. E. Bertolucci, *J. Orthop. Sports Phys. Ther.*, *4*:103 (1982).
61. L. P. Gangarosa, N. H. Park, and J. M. Hill, *Proc. Soc. Exp. Biol. Med.*, *154*:439 (1977).
62. J. E. Marchand and N. Hagino, *Exp. Neurol.*, *78*:790 (1982).
63. J. M. Hill, L. P. Gangarosa, and N. H. Park, *Ann. N.Y. Acad. Sci.*, *248*:604 (1977).
64. T. Rosendal, *Acta Physiol. Scand.*, *5*:130 (1942–1943).
65. K. Harpuder, *Arch. Phys. Ther. X-Ray Rad.*, *18*:221 (1937).
66. P. Tyle, *Pharm. Res.*, *3*:318 (1986).
67. G. Jung, E. Katz, H. Schmitt, K. P. Voges, G. Menestrina, and G. Boheim, in *Physical Chemistry of Transmembrane Ion Motion* (G. Spach, ed.), Elsevier Science Publishing Co., New York (1983).
68. H. A. Abramson and M. H. Gorin, *J. Phys. Chem.*, *44*:1094–1102 (1940).
69. T. Masada, U. Rohr, W. I. Higuchi, J. Fox, C. Behl, W. Malick, A. H. Goldberg, and S. Pons, in *39th National Meeting of the Academy of Pharmaceutical Sciences Abstracts*, Vol. 15(2), Minneapolis, Minnesota, 1985, p. 73.
70. R. R. Burnette and D. Marrero, *J. Pharm. Sci.*, *75*:738 (1986).
71. R. L. Stephen, T. J. Petelenz, and S. C. Jacobsen, *Biomed. Biochim. Acta*, *43*:553 (1984).
72. Y. W. Chien, *Novel Drug Delivery Systems*, Marcel Dekker, New York (1982), Chap. 5.
73. Y. W. Chien, O. Siddiqui, Y. Sun, W. M. Shi, and J. C. Liu, in *Ann. N.Y. Acad. Sci.*, *507*:32–51 (1987).
74. K. Okabe, H. Yamaguchi, and Y. Kawai, *J. Controlled Release*, *4*:79 (1986).
75. H. Klostermeyer and R. E. Humbel, *Angew. Chem. Int. Ed. Engl.*, *5*:807 (1966).
76. O. Siddiqui, Y. Sun, J. C. Liu, and Y. W. Chien, *J. Pharm. Sci.*, *76*:341 (1987).

16

Oral Route of Peptide and Protein Drug Delivery

Vincent H. L. Lee

University of Southern California School of Pharmacy, Los Angeles, California

Satish Dodda-Kashi

Ciba-Geigy Pharmaceuticals Division, Summit, New Jersey

George M. Grass

Syntex Research, Inc., Palo Alto, California

Werner Rubas

Institute of Pharmaceutical Sciences, Palo Alto, California

I. INTRODUCTION

Maximizing the bioavailability of orally administered peptide and protein drugs has been an ongoing, yet elusive, goal for many years. The challenge here is to improve the oral bioavailability from less than 1% to at least 30−50%. Throughout the years there have been many reports on the oral activity of peptide and protein drugs. But the doses required are often excessive in comparison to parenteral doses. For instance, doses of 4−8 mg of SMS201-995, a somatostatin analog, given in a glucose sodium acetate −acetic acid buffer (pH 4.0) three times daily were required to lower the mean 24-hr growth-hormone concentration in patients with active acromegaly by over 50% (range, 51−88%) [1]. An oral dose of 10 mg was required of leuprolide, a luteinizing-hormone-releasing hormone analog, to be given in gelatin capsules to elicit a statistically significant increase in mean plasma luteinizing hormone in healthy men [2]. While a mixed micellar solution improved the oral absorption of this peptide, the absolute bioavailability remained low, being less than 0.05% [3]. Much better absorption, amounting to at least 10% of the orally administered dose, was observed for U-71038 (Boc −Pro −Phe −N −MeHis −Leuψ[CHOCH$_2$]Val −Ile −Amp), a renin-inhibitor peptide, when tested in monkeys [4]. The compound was designed to resist proteolytic degradation by carboxypeptidase Y, chymotrypsin, elastase, and the enzymes in a rat liver homogenate.

The purpose of this chapter is fourfold: (1) to review the nature of the barriers to oral peptide and protein absorption, (2) to review the factors influencing the absorption of an orally active peptide, cyclosporine, (3) to review the approaches that could be considered in maximizing oral protein absorption, and (4) to review the factors influencing the lymphatic absorption of peptide and protein drugs.

II. ENZYMATIC AND PENETRATION BARRIERS TO ORAL PEPTIDE AND PROTEIN ABSORPTION

The enzymatic barrier is by far the most important of the multitude of barriers limiting the absorption of natural peptide and protein drugs from the gastrointestinal (GI) tract. Indeed, there is an inverse relationship between the amount of peptide transported across the intestine and its rate of hydrolysis [5]. Absorption is further compromised by the resistance exerted by the intestinal membrane to peptide and protein penetration through simple diffusion, carrier-mediated transport, or endocytosis. For this reason, penetration enhancers are often required to improve oral peptide and protein absorption. This is exemplified by the oral absorption of human calcitonin in rats. Nakada et al. [6] reported that although calcitonin was poorly absorbed, coadministration with surfactants, such as sodium myristate, sodium deoxycholate, sodium taurodeoxycholate, sodium laurel sulfate, quillajasaponin, and sugar esters, resulted in significant hypocalcemia (Table 1).

Perhaps the interplay of enzymatic and penetration barriers in the absorption of macromolecules across the gut is best exemplified in the neonates. During a short period after birth, the GI tract is relatively permeable to macromolecules. The duration of this period depends on the animal species and the macromolecular markers used. On average, it is about 1 day in the pig [7], 7 days in the rabbit [8], and 22–30 days in the rat [9]. Macromolecular closure is assumed to be due to some as yet unidentified factors in the colostrum, since feeding the piglet with glucose instead of colostrum [7] and feeding the rabbit with bottle instead of breast milk

TABLE 1   Effect of Surfactants on Plasma Calcium
Levels After Oral Administration of Human
Calcitonin in Rats at pH 7.4

Surfactant	% Calcium decrease
Control	1.6 ± 1.4
Sodium myristate	7.4 ± 1.2
Sodium deoxycholate	21.6 ± 2.1
Sodium taurodeoxycholate	14.2 ± 2.7
Sodium laurylsulfate	32.2 ± 1.8
Quillayasaponin	15.6 ± 2.9
Sucrose palmitate	8.3 ± 1.2

Source: Ref. 6.

[10] prolong the macromolecular closure somewhat. In the rabbit, macro-molecular closure is not accompanied by changes in intestinal morphology that can be discerned by light and electron microscopy. Interestingly, macromolecular closure does not appear to affect the permeability of the intestine to polyethylene glycols (PEGs) in the 414–942-dalton range [11]. This finding implies that PEG molecules cross the intestine by a mechanism different from that for proteins and that PEGs are not good markers to assess membrane permeability of proteins.

The transient increase in permeability to macromolecules may be attributed to several factors: (1) transient deficiency of intestinal secretory antibodies [12], (2) increased "receptor" sites for antigens, (3) enhanced endocytosis of macromolecules, (4) increased membrane fluidity [13,14], and (5) decreased intestinal proteolysis [15,16]. The interaction of these factors in determining the extent of protein absorption in the newborn is best illustrated in the study by Telemo et al. [9] on the intestinal absorption of IgG, bovine serum albumin, and FITC-dextran 70,000 in young rats of 10, 14, 18, 22, and 30 days of age. In the 10- and 14-day-old rats, absorption of all three macromolecular markers, particularly IgG, was high. Bovine serum albumin, which was susceptible to proteolysis, was absorbed to a greater extent than the proteolytically stable FITC-dextran 70,000. This finding suggests that proteolysis was not a dominant factor limiting macromolecular absorption at this young age. On the other hand, the more extensive absorption of IgG in spite of its molecular size was attributed to the involvement of its specific receptor in its uptake [9]. The proportion of the absorbed IgG appearing intact was dose dependent. Contrary to expectation, at subsaturation dose levels, very little intracellular degradation of IgG occurred and over 80% of the IgG was transported intact. At doses above saturation (>500 µg), a larger proportion of IgG was removed from the intestine as breakdown products [16].

In the rat, absorption of IgG began to decline after 14 days. Indeed, absorption was greatly reduced by 18 days (even though the capacity to absorb proteins by endocytosis was still intact [17,18], and was totally abolished for the protein markers at 22 days and for FITC-dextran 70,000 after 30 days. The decrease in absorption for the protein markers was attributed partly to increases in luminal proteolytic activity as judged by the enhancement of absorption in the presence of such protease inhibitors as soybean trypsin inhibitor and swine colostrum trypsin inhibitor. Udall et al. [16] reported that the basal trypsin-like activity, which together with chymotrypsin comprises about two-thirds of the total proteolytic activity in the duodenal juice [19], was 12.2 ± 1.3 $\Delta OD/hr/ml$ in the newborn rabbits and 22.0 ± 3.0 $\Delta OD/hr/ml$ in the 4-week-old animals. Consequently, aprotinin was more effective in controlling the proteolytic activity in the newborn and 2-week-old rabbits than in the 4-week-old rabbits [16]. Weström et al. [15] speculated that the increase in protein content due to the presence of protease inhibitors may enhance protein uptake by stimulating endocytosis. They based their observation on the enhancement of absorption of FITC-dextran 70,000 in the neonatal pig by bovine serum albumin and IgG.

Without question, the enzymatic barrier is well designed to efficiently digest proteins to a mixture of amino acids and small quantities of peptides consisting of two to six amino acid residues prior to the appearance in portal circulation [20,21]. In the GI tract, hydrolysis of peptides and proteins can occur at several sites —luminally, at the brush border, and intracellularly. Much of the information in this regard was obtained in the 1960s and 1970s from the nutritional point of view [22−24]. In contrast, there is scanty information on the gastrointestinal hydrolysis of biologically active peptides.

Protein digestion is initiated in the gastric juice by a family of aspartic proteinases called pepsins, which are most active at pH 2−3 but which become inactive at a pH above 5. Although they are capable of doing so, pepsins rarely degrade proteins to amino acids. The partial digest that results is then acted upon by pancreatic proteases in the duodenum and beyond. These proteases consist of trypsin, chymotrypsin, elastase, and carboxypeptidase A. The first three are endopeptidases, whereas the last is an exopeptidase.

The three endopeptidases have evolved to complement one another in cleaving almost all the internal peptide linkages likely to be encountered in a wide spectrum of peptides and proteins. α-Chymotrypsin prefers to cleave peptide bonds near hydrophobic amino acids such as leucine, methionine, phenylalanine, tryptophan, and tyrosine. By contrast, trypsin preferentially cleaves peptide bonds near basic amino acids such as arginine and lysine. Elastase complements the other two proteases by cleaving peptide bonds near alanine, glycine, isoleucine, leucine, serine, and valine; in other words, peptide bonds of amino acids bearing smaller, unbranched, nonaromatic side chains [25]. All three proteases have an optimum functional pH of about 8.

Carboxypeptidase A is a well-characterized C-terminal exopeptidase [26]. Substrates for this enzyme must meet two requirements: a free terminal carboxyl group and a C-terminal amino acid bearing a branched aliphatic or an aromatic group are particularly vulnerable to attack by this enzyme. Moreover, the nature of amino acids near the susceptible bond can modify the overall hydrolytic rate.

The combined activity of the pancreatic proteases against dietary proteins is impressive. As much as 200−800 μg of trypsin and chymotrypsin is present in the human duodenum shortly after feeding. This is enough to convert half of the protein in the duodenal contents to trichloroacetic acid-soluble material within 10 min, 60−70% of which is in the form of small peptides with 2−6-amino-acid residues. By the time the content leaves the jejunum, over 50% of it is in the form of amino acids [28,29].

Although the pancreatic proteases are rather active against dietary proteins, their activity toward small peptides is very much restrained. The bulk of luminal fluid activity against small peptides is derived from either the brush border or the cytoplasm of the enterocyte [22]. Luminal activity is less than 3% of cytosolic activity in the jejunum but rises to as high as 40% of brush-border activity in the ileum [22]. Overall, even when luminal degradation occurs, it constitutes less than 5%, and at best 20%, of the total degradation in a given intestinal segment [24,30]. The rest of the degradation occurs upon contact with the brush border or following entry into the cell. Although many of the observations in luminal hydrolysis are based on nonbiologically active peptides, it is anticipated that biologically active peptides would behave similarly.

Proteases in the brush border and the cytosol of the enterocyte are potentially the most important deterrent to the absorption of small biologically active peptidases across the intestinal mucosa. Whether hydrolysis

occurs at the brush border or in the cytosol depends on the resistance of
a given peptide to the proteases at the brush border [31]. These proteases
consist of aminopeptidases A and N, diaminopeptidase IV, endopeptidase
24.11, angiotensin-converting enzyme, Gly−Leu peptidase, and Zn-stable
Asp−Lys peptidase [32−36]. Typically, they are anchored in the apical
membrane with the active site essentially in an extracellular environment
[37,38]. Several of them have been implicated in limiting the absorption
of a nonapeptide renin inhibitor, PHPFHLFVF, across the jejunum [39]. In
addition to the membrane-bound proteases, trypsin, chymotrypsin, and
other pancreatic proteases may be adsorbed from the luminal fluid onto the
brush border of the enterocyte, thereby assisting in proteolysis of oligopeptides and proteins [40,41]. Although the brush-border proteases as a
group are capable of readily hydrolyzing peptides of up to 10 amino acid
residues [42], they tend to prefer tri- and tetrapeptides to dipeptides
[43] and to prefer peptides whose N-terminal amino acid residue possesses
a lipophilic side chain [42]. The cytosolic proteases, by contrast, tend to
prefer dipeptides to larger oligopeptides [44].

Due to the subcellular distribution of the proteases just described,
tripeptides are hydrolyzed principally in the brush border, whereas dipeptides are hydrolyzed principally in the cytosol. Specifically, about 60−90%
of the cellular proteolytic activity against tripeptides can be found in the
brush border [45−48]. In contrast, only 5−20% of the proteolytic activity
against dipeptides can be found in the brush border [45,47,49,50]. Because of this, a significant fraction of the applied dipeptides is absorbed
into the enterocyte, where it is hydrolyzed.

Besides the brush border and the cytosol, lysosomes and other organelles are also potential sites of peptide and protein degradation. So far,
their role in the degradation of orally administered peptides and proteins
has been neglected principally because such organelles are rarely encountered by the peptides that have been studied so far. Nevertheless,
proteolysis in lysosomes and other organelles is expected to assume greater
quantitative importance as a larger fraction of the orally administered protein survives luminal and brush border hydrolysis. This is indeed the
case in the GI absorption of epidermal growth factor (EGF) in suckling
rats [51], in which less than 1% of the administered dose (60 ng of [125]I-
EGF) was found in plasma after up to 3 hr. While little degradation occurred in the lumen of the stomach (9%) or intestine (10%), 84% of the total
radioactivity found in plasma was in degraded form based on size-exclusion
chromatography. Gonnella and co-workers localized specific binding sites
for EGF by electron microscopic autoradiography on apical membranes of
ileal epithelial sheets in vitro [52]. During uptake in vivo, radiolabeled
molecules were concentrated in apical endosomal compartments and were
also associated with lysosomal vacuoles, basolateral cell surfaces, and lamina
propria. Autoradiographic and biochemical data revealed degradation in the
lysosomes. To a certain degree, intact transepithelial transport was also
confirmed, but not quantitated.

III. CYCLOSPORINE: AN ORALLY ACTIVE PEPTIDE

Cyclosporine is an atypical peptide in that it possesses significant oral activity as an immunosuppressant [53,54]. In a mixed transplant patient

population about 26% of an oral dose is absorbed [55]. This rather unusually high bioavailability for a peptide drug can be attributed partly to
its resistance to proteolytic degradation and partly to its physiochemical
properties, which are unusual for a peptide. Both features are probably
related to extensive N-methylation and to rigidity of its skeleton due to hydrogen bonding (Fig. 1).

Cyclosporine is very hydrophobic with a 1-octanol/Ringer's buffer partition coefficient of about 991 [56], primarily due to the presence of seven
hydrophobic N-methylated amino acid residues in a cyclic structure. This
high lipophilicity also results in poor aqueous solubility, about 6 µg/ml at
37°C [57]. Because of this, the commercial oral preparation (Sandimmune)
is formulated in a mixture of peglicol-5-oleate, olive oil, and ethanol in a
ratio of 30:60:10 [58]. Upon dilution in a liquid, peglicol-5-oleate, a nonionic surfactant, disperses the oral solution by its emulsifying action.
Johnston et al. [59] reported that the nature of the diluent had no effect
on cyclosporine absorption in 12 healthy human volunteers who had received either the oral formulation in 150 ml of water or an oral solution
dispersed in 150 ml of white milk, chocolate milk, or orange juice. These
findings were, however, contradicted by recent results obtained in cardiac
transplant patients [60]. There was a 40% increase in the AUC and a 67%
increase in the maximum concentration when cyclosporine was taken in milk
rather than in orange juice. That the vehicle influence on oral cyclosporine
absorption is complex can be seen in two other examples. On the one hand,
Elzinja et al. [61] reported a 67% increase in the AUC when given in de-
methylsulfoxide (DMSO) compared to olive oil alone. On the other hand,
Wagner et al. [62] did not find any difference in the extent of cyclosporine

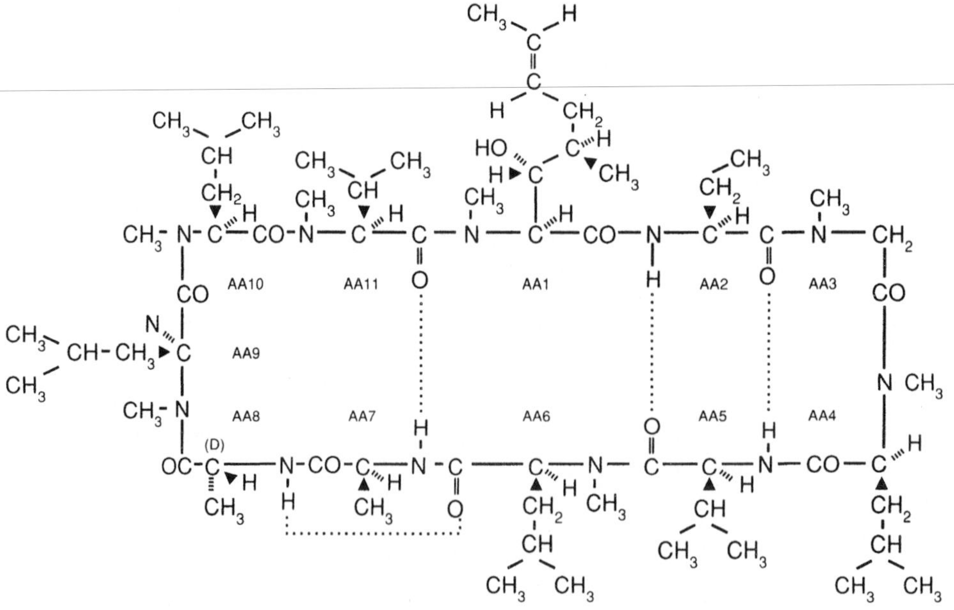

Fig. 1   Primary structure of cyclosporine with four hydrolgen bonds as
deduced from the crystal structure. (From Ref. 68.)

absorption using the commercial preparation and a 2:98 mixture of ethanol and PEG 200.

Largely as a result of poor aqueous solubility, rather slow and incomplete absorption of cyclosporine is observed [57]. Although extensive metabolism of cyclosporine via hydroxylation and N-methylation in the liver undoubtedly contributes in part to the low bioavailability [62–64], recent work tends to focus on improving absorption through formulation optimization. Of the formulations tested, emulsions with a defined droplet size appear to be most promising. Tarr and Yalkowsky [65] showed that reducing the emulsion droplet size from 4 to 2 μm increased the surface area for partitioning of the drug in the surrounding aqueous environment. At the same time, the activity of lipase was enhanced, which led to the production of more fatty acids and monoglycerides for micelle formation. Subsequent solubilization of cyclosporine in such micelles resulted in a greater driving force for penetration across the unstirred aqueous layer of the intestinal mucosa [66].

A major problem in the oral absorption of cyclosporine is the wide interindividual variability. The absolute oral bioavailability ranged from less than 5% to 89% in adult kidney transplant patients [67] and from less than 5% to 19% in pediatric liver transplant patients [69]. In normal individuals and nonliver transplant patients, the mean bioavailability was about 30% [68]. Such a high degree of variability in absorption is not an inherent property of the drug but is due to factors associated with the formulation or with the patient's physiological status [57]. In both liver [69] and kidney [70] transplant patients, the time since surgery affects the extent of oral cyclosporine absorption. For example, the mean bioavailability improved from 24.2% at 6 months to 50.2% at 12 months after kidney transplantation [71]. Although the underlying mechanism for such improvement is not known, it is possible that the disease state of the patient influences the extent of absorption and metabolism through its effect on physiological and biochemical factors controlling the above two processes. Such factors could include gastric emptying, intestinal transit, mesenteric and hepatic blood flow, lymph flow, intestinal secretion and fluid volume, bile secretion and flow, and epithelial cell turnover. Even under optimal physiological conditions, cyclosporine absorption is still incomplete. This is indicated by the negative anatomical reserve length calculated for absorption in rabbits [72], which means that complete absorption of cyclosporine would occur only if there were an additional segment exceeding the normal length of the small intestine.

Conjugated bile acids play an important role in solubilizing cyclosporine, thus affecting its oral bioavailability. It is therefore not surprising that the bioavailability of cyclosporine was very low in liver transplant patients with external bile drainage via a T-tube. A significant increase in the rate and extent of absorption was observed in the same patient following T-tube clamping [73], which drained the bile back into the upper intestine. Since unchanged cyclosporine available for absorption through enterohepatic recycling is insignificant (about 1% or less of the administered dose) [74], the higher blood levels cannot be attributed to reabsorption alone. In rats, the bile flow into the duodenum is normally uninterrupted since the gallbladder, the storage organ for bile, is absent. A certain fraction of the bile pool will always be present in the GI tract. Hence rats may be a poor model for studying the oral absorption of cyclosporine.

The actual absorption mechanism of cyclosporine across the small intestinal mucosa is as yet unclear. In 1983, Ueda et al. [75] invoked carrier-mediated transport to explain the zero-order absorption kinetics and low bioavailability of cyclosporine observed in rats. But such a possibility was rejected in a follow-up study by the same investigators who then observed a significant increase in the bioavailability from 13% to 22% upon increasing the cyclosporine dose from 6 to 23 mg/kg in rats in a vehicle similar to the commercial one [76]. Since there was no decrease in the fraction of the dose absorbed with increasing dose, dose-dependent absorption was attributed to reasons other than carrier-mediated transport. Further evidence supporting the lack of carrier involvement came from the work of Tarr and Yalkowsky [65]. These investigators found that after selective in situ perfusion of the upper part of the rat intestine for 2 hr at increasing concentrations of 0.04-0.18 mg/ml cyclosporine in an emulsion prepared by dilution of commercial preparation, the fraction of dose absorbed did not change.

The phenomenon of dose-dependent absorption reported earlier by Ueda et al. [76] was attributed to the vehicle effect on gastric emptying rate. In that study, the proportion of olive oil was less than 60% (v/v) in the highest dose and more than 80% (v/v) in the lowest dose administered. Since olive oil is known to slow gastric emptying, a characteristic of fatty acids and lipids [77], the absorption rate of cyclosprine is expected to decrease as a result, given that the stomach is not the primary absorption site for cyclosporine. This expectation is borne out of the decrease in the apparent absorption rate constant from 481 to 112 µg/L/hr with an increase in the percent of olive oil in the mixture from about 40% to 82%. Such a negative correlation was observed between the olive oil volume and cyclosporine bioavailability also. Since rate and extent of cyclosporine bioavailability are highly correlated, the volume of olive oil appears to be a factor in dose-dependent absorption.

That gastrointestinal motility influences the bioavailability of cyclosporine is indicated in a clinical study on the effect of metoclopramide on oral cyclosporine bioavailability [79]. Here, metoclopramide was given three times: 10 mg 30 min before, 5 mg with, and 5 mg 30 min after administration of 4–10 mg/kg of cyclosporine. The result was a significant reduction in the peak time, a 56% increase in the mean maximum cyclosporine concentration, and a 29% increase in the mean AUC of cyclosporine. Thus, an increase in gastric emptying rate inproves the absorption and therefore the bioavailability of cyclosporine.

Food, whose retarding effect on gastric emptying is well known, is expected to reduce the oral bioavailability of cyclosporine. Nevertheless, the effect of food on cyclosporine absorption is controversial at best. In a clinical study in renal transplant patients, the ingestion of solid food (standard breakfast) did not affect the rate of absorption, but it did cause increases in maximum plasma concentration from 1120 to 1465 µg/L, minimum plasma concentration from 228 to 267 µg/L, and area under the plasma concentration-time curve from 7881 to 11,430 µg/L/hr [67]. In another clinical study involving cardiac transplant patients, food did not affect the peak time and AUC of orally administered cyclosporine [60]. Aside from its effect on gastric emptying, food may also stimulate bile flow and/or intestinal and/or hepatic blood flow, which in turn may affect the metabolic clearance of cyclosporine in the gut wall and the liver.

In addition to vehicle effects, the residence time at the optimal site of absorption is another factor that contributes to dose-dependent absorption. Sawchuk and Awni [72] demonstrated that upon perfusing the small intestine of the rabbit, more of the cyclosporine, which was formulated in a 35% PEG 400 solution, disappeared from the duodenal region than from the jejunal region of the same animal. Based on the rapid onset of absorption (lag time <0.4 hr) and short absorption phase (2.7 hr) in normal healthy volunteers, Grevel et al. [80] suggested that absorption was mainly confined to the upper part of the intestinal tract. Such a notion of site-dependent absorption of cyclosporine is supported by several other findings. For instance, Takada et al. [81] found little difference in the blood levels of cyclosporine whether the dose was confined to a 15-cm segment of the upper small intestine, a 15-cm segment of the middle small intestine, or a 30-cm segment of the lower small intestine. Thus, a larger fraction of the dose reached the bloodstream from the duodenum than from the other two regions in the small intestine in spite of markedly reduced surface area for absorption (Fig. 2). Moreover, Ritschel et al. [82] found that the time to reach maximal plasma concentration in the beagle dog was longer from a capsule formulation containing cyclosporine and 5 mg sodium taurocholate in a gel (3.6 – 10 hr) than from another formulation containing cyclosporine dissolved in olive oil and peglicol-5-oleate (2.4 hr). The longer peak time for the capsule was attributed to the imposition of an additional disintegration and dissolution step before cyclosporine could be absorbed, thus delaying exposure of the peptide to the upper part of the intestinal tract, where it is best absorbed. The apparent lack of cyclosporine absorption

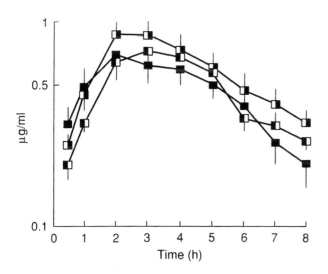

Fig. 2   Effect of absorption site on the systemic blood levels of cyclosporine after oral administration to rats at 10 mg/kg. Each point represents mean ± s.e.m. (n = 4). ▫, 15-cm segment of the upper small intestine; ▫, next 15-cm segment of the upper small intestine; ■, 30-cm segment of the lower small intestine. (From Ref. 81.)

from the dog's ileum, which is very short relative to that in the rat, rabbit, and human, underscores the point that cyclosporine absorption is confined to the upper part of the small intestine.

Further evidence for site-dependent absorption of cyclosporine is provided in a recent experiment by Tarr and Yalkowsky [65], who observed
no significant increase in the fraction of the dose absorbed in the rat intestine by doubling the length of the intestine perfused. The narrow absorption window was assigned to the proximal 8 – 16 cm of the rat intestine.
Whether the existence of such an absorption window is related to differences in the physiological and/or morphological features, such as surface
area, surfactant concentration, pH gradient, as well as lymph and blood
drainage in that region, remains to be investigated.

The proximal small intestine is also where absorption of cyclosporine
into the lymphatic system is maximized [81]. The lymphatic absorption of
cyclosporine is probably unaffected by bile since only marginal decrease in
the lymphatic absorption of cyclosporine was observed in bile-duct-cannu-
lated rats [81]. Lymphatic absorption was favored by administering cyclosporine in mixed micelles of linolic acid (2.5 w/v%) and polyoxyethylated
hydrogenated oil (HCO-60) (8.0 w/v%) in distilled water [83]. The thoracic lymph concentration of cyclosporine was about 20 times higher than
in plasma following intraduodenal administration in mixed micelles in rats
(Fig. 3). Interaction of cyclosporine with the chylomicrons was largely
responsible for its transfer into the lymphatic circulation [83]. A subsequent experiment by the same investigators revealed that about seven
times more cyclosporine was transferred into the major intestinal lymphatics
than into the thoracic lymphatics in the rat [81]. In that study, rats were

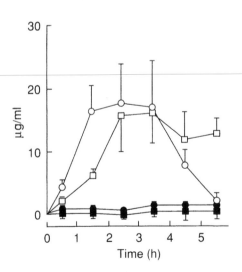

Fig. 3   Cyclosporine concentrations in plasma and lymph of the thoracic
duct after oral administration of cyclosporine at 7 mg/kg in mixed micellar
solution to rats.   Each point is expressed as mean ± s.e.m. (n = 4).   ○,
Lymph concentrations after intragastric administration; □, lymph concentrations after intraduodenal administration; ●, plasma concentrations after intragastric administration; ■, plasma concentrations after intraduodenal administration.   (From Ref. 83.)

administered 1 ml of fresh milk orally prior to surgery to maintain high
lymph flow and also to facilitate lymph duct cannulation.

A site-specific release and selective lymphatic uptake with increased
bioavailability was achieved in rats following oral administration of granular
solid dispersions of cyclosporine and HCO-60 coated with an enteric coating
material, hydroxypropylmethylcellusose phthalate [288]. The enteric coat
was stable at pH 2 and dissolved in 8−12 min at pH 6, thereby causing release of the drug as it entered the proximal region of the intestine. Such
site-specific release in the lumen, combined with mixed micelle formation,
elevated the peak lymph level to about four times more than the commercial
preparation. The percent of dose transferred was about 2−2.5 times better
than the commercial formulation. The above approach, though attractive,
has yet to be tested in humans.

IV. APPROACHES TO MAXIMIZE THE ORAL ABSORPTION OF PEPTIDE AND PROTEIN DRUGS

In order to promote oral absorption of peptides and proteins from the GI
tract, the many components of the enzymatic barrier must be controlled.
This can be achieved to some extent by modifying the peptide or protein
structure, by restricting the release of a peptide or protein drug in a region of the GI tract that favors its absorption, through coadministration of
protease inhibitors, or by using the formulation approach. The last two
approaches have already been discussed in Chapter 7. Thus, only the
first two approaches will be emphasized here.

A. Chemical Modifications of Peptide and Protein Drugs

The strategy of chemically modifying peptide structure to improve enzymatic
stability, membrane penetration, or both has been met with some success.
Examples include enkephalins, thyrotropin-releasing hormone, and vasopressin. So far, there have been few examples of modification of protein drugs
to improve oral activity, although many examples exist for improving formulation stability [84,85], extending circulation half-life [86], and minimizing
immunogenicity [87]. In an attempt to improve the oral activity of insulin,
Hiltbrunner [88] covalently linked insulin to 1,3-dipalmitoylglycerol at the
free amino groups of glycine A1, phenylalanine B1, and lysine B29 to form
mono- and diacyl insulin. While such derivatives retained 60−70% of the
hypoglycemic activity of native insulin, they were orally ineffective.

Enkephalins

Enkephalins are endogenous pentapeptides possessing significant analgesic
properties. The two types that occur naturally, leucine (YGGFL) and
methionine (YGGFM) enkephalins, are derived from the catabolism of
β-endorphin. Being extremely susceptible to rapid hydrolysis by the intestinal peptidases [89], neither peptide is orally active.

Roemer et al. [90] synthesized five pentapeptides structurally related
to YGGFM in order to prolong the parenteral and oral analgesic activity in
the mouse (tail flick and hot plate procedures), the rat (paw pressure
test), and the rhesus monkey (shock titration test). The synthetic

peptides were prepared by (1) altering the carboxyl group to form the car-
binol analog (YGGFM-ol), (2) substituting the glycine residue in position 2
by D-alanine (YAGFM), (3) combining these two modifications (YAGFM-ol),
(4) oxidizing YAGFM-ol to sulfoxide [YAGFM(O)-ol], or (5) N-methylating
YAGFM(O)-ol [YAGF(Me)M(O)-ol or FK 33-824]. With each modification,
the analgesic potency as well as duration increased gradually in the above
order after intracerebroventricular (i.c.v.) administration. The latter two
peptides, YAGFM(O)-ol and FK 33-824, produced significant analgesic activity after oral administration. The oral ED_{50} for analgesic activity at 2
hr postdosing for these two peptides was, respectively, 319 and 102 mg/
kg. On an equimolar basis, the i.c.v. ED_{50} was about 0.0001 and 0.00003
times that of oral ED_{50} for the two peptides, respectively. After i.c.v.
administration, FK 33-824 was about 23, 1000, and 30,000 times more potent
than β-endorphin, morphine, and YGGFM, respectively [90]. Such enhanced potency as a result of improved stability underscores the benefit of
the chemical modification approach once the susceptible bonds have been
identified.

Another analog of methionine enkephalin is metkephamid, which structurally is Tyr −D −Ala −Gly −Phe −N −Me −Met −NH_2. Metkephamid readily
penetrated the nasal mucosa with 54% bioavailability relative to subcutaneous
administration [91]. This peptide was, however, orally inactive. Su et al.
[91] suggested that enzymatic degradation was possibly responsible for the
poor oral activity.

Potentiation and prolongation of biological responses have been achieved
for enkephalins by designing synthetic analogs. L-Tyrosyl-D-methionyl-L-
glycyl-4-nitrophenylalanyl-L-prolinamide S-oxide hydrochloride (BW942C,
nifalatide) is a synthetic enkephalin-like peptide which is resistant to intestinal digestion and metabolism [92]. Nevertheless, this peptide was
poorly absorbed across the isolated rabbit ileum based on potential difference, short-circuit current, and sodium/chloride flux measurements. In
contrast, in a parallel, double-blind, placebo-controlled evaluation of nifalatide in 72 subjects with castor-oil-induced diarrhea, the time to first stool
after castor oil administration was demonstrated to be significantly greater
after the 16- and 48-mg doses of the peptide as compared with placebo dosing [93]. The duration of the pharmacological activity was about 4−6 hr
based on both stool frequency and time to first stool. Whether the effect
was due to local or systemic action of nifalatide was not investigated in the
study.

Chemical modifications followed by conformational analysis will establish
the importance of a particular residue in eliciting a certain biological response. Enkephalin analogs containing β-naphthylalanine at the fourth
position were evaluated to examine the importance of the aromatic side
chains on opiate activity [94]. The findings were in agreement with their
model of an extended structure required for μ selectivity and a folded form
with close aromatic ring placement for δ selectivity. Whether similar modification and analysis can be attempted in terms of altering membrane penetration should be explored since this would lead to orally bioavailable peptides. An example of restricted membrane penetration is L −Tyr −D −Arg −
Gly −Phe($4NO_2$) −Pro −NH_2 (BW443C), which is structurally similar to nifalatide without the S-oxide and with D-arginine instead of D-methionine
[95]. This peptide was designed to penetrate the blood-brain barrier
poorly so as to maximize the peripheral receptor-mediated antinociceptive
activity. Following i.v. infusion in humans, this peptide yielded a plasma

clearance of about 123 ml/min and a half-life of about 2 hr based on double-antibody radioimmunoassay. As expected, there was no evidence of central activity such as sedation, mood change, nausea, vomiting, miosis, or respiratory depression. Lack of penetration into brain was further confirmed by autoradiography studies of the C-14-labeled peptide.

Thyrotropin-Releasing Hormone

Thyrotropin-releasing hormone (TRH), a tripeptide having amino acid composition of p-glutamyl histidyl proline, is one of several hypothalamic factors that exert control over pituitary functions at low concentrations. In clinical practice, TRH is presently administered parenterally because of low levels in the brain following oral administration. In mice, about $0.02-0.2\%$ of TRH reached the brain after i.v., i.m., oral, and rectal administration [96]. The peak brain level of TRH after oral dosing was about five times less than that of parenteral administration. Strangely enough, the highest level, twice that from the parenteral route, was seen after rectal administration.

The oral absorption of TRH in the rat and the dog appears to be site-specific, carrier-mediated, dose-dependent (Fig. 4), and subject to inhibition by oligopeptides and some β-lactam antibiotics [97]. Since TRH is rather resistant to proteolytic degradation in the GI tract [98], its poor oral activity is probably due to poor absorption and rapid clearance in the bloodstream [98]. The absolute bioavailability after oral absorption was less than 2% at all doses studied in the rat and human. In order to overcome poor absorption and to circumvent the rapid metabolism and clearance following absorption, several analogs of TRH have been synthesized and tested for potency.

MK-771, which has a sulfur atom in the pyrrolidine ring at position 2 of the proline residue, was synthesized to obtain a peptide with CNS activity superior to that of TRH but with much slower clearance rate. It was about 200 times more potent than TRH when tested for CNS function but was equipotent in causing the release of TSH [99]. Among others, only

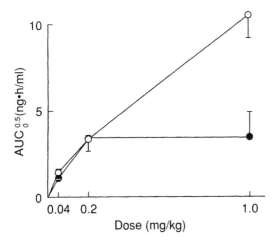

Fig. 4 Relationship between dose and AUC of thyrotropin-releasing hormone (●) and DN-1417 (○) from 0 to 0.5 hr in the in situ rat intestinal loops. Each point represents mean ± s.e.m. ($n = 4$). (From Ref. 102.)

methylhistidyl analog exhibited potencies superior to TRH in causing the release of TSH. MK-771 was more stable than TRH in both whole blood and
plasma. The in vitro half-life of MK-771 in rat, monkey, and human plasma was about 1.1, 8 and 14 min, respectively, which was about 18, 6, and
46 times longer than that of TRH, respectively. In vivo studies demonstrated that the terminal half-life of MK-771 was about 11 min in the rat
[99], compared with 4.6 min for TRH [98]. Improved stability against degradation was confirmed in humans when i.v. MK-771 and TRH yielded terminal half-lives of 46 and 9 min, respectively [99]. In spite of significant
improvement in metabolic stability, the oral bioavailability of MK-771 was
only 3% in monkeys and 2% in humans [99]. This finding suggests that
the poor oral bioavailability of MK-771 was due mainly to inefficient membrane transport rather than to degradation.

A dimethyl analog of TRH, $p-glu-his-(3,3'-dimethyl)-pro-NH2$ (RX
77368), stimulated gastric acid output which was rapid in onset, long-
lasting, and dose-dependent upon parenteral administration [100]. The
analog was, however, less effective than TRH when infused intravenously
but 22 times more potent when given intracisternally, indicating limited
transport of this analog across the blood-brain barrier. The longer-
lasting and more potent action of RX 77368 appears to be related to its
stability to protease action.

Attempts have been made to modify the N-terminus by replacing the
p-glutamic acid residue of TRH with a γ-butyrolactone-γ-carboxyl residue
(DN 1417) [101]. Such modification resulted in a dose proportional absorption in rats (Fig. 4) and dogs up to doses of 500 mg/kg and 100 mg/kg,
respectively. The absolute oral bioavailability was, however, still low,
being about 1% in rats and 10% in dogs. Unlike TRH, DN-1417 was passively and uniformly absorbed from all regions of the small intestine [102].

Vasopressin

Several vasopressin analogs with significant antidiuretic activity have been
synthesized and tested. These include 1-deamino-(8-D-arginine)-vasopres-
sin (desmopressin, DDAVP), (1-deaminopenicillamine, 2-O-methyltyrosine)-
arginine vasopressin [DP-Tyr(OMe)AVP], (1-deaminopenicillamine)-
arginine-vasopressin (dP-AVP), (1-deaminopenicillamine, 4-valine, 8-D-
arginine)-vasopressin (DAVP), (2-O-methyltyrosine)arginine-vasopressin
[Tyr(OMe)AVP], (4-valine, 8-D-arginine)-vasopressin (VAVP), deamino(4-
threonine, 8-D-arginine)-vasopressin (DTDAVP), deamino(4-valine, 8-D-
arginine)-vasopressin (DVDAVP), and [1-(β-mercapto-β,β-cyclopentameth-
ylene propionic acid), 4-valine, 8-D-arginine]-vasopressin(cyclo-DVDAVP)
[103-106].

DDAVP, synthesized by Zaoral et al. [106], differs structurally from
the natural antidiuretic hormone in two positions. DDAVP has β-mercapto-
propionic acid (deaminohemicystine) instead of hemicystine in position 1 and
D-arginine in place of natural L-arginine in position 8. The natural lysine
vasopressin (LVP) is also orally active in the water-loaded rat, albeit at
large doses [104]. Due both to increased membrane penetration and enhanced stability against proteolytic degradation [107], DDAVP is over two
times more orally active than LVP at a 75 times lower dose. Marked differences in potency have, however, been reported, depending on the biological assay used. For instance, Rado et al. [103] reported that DDAVP was
only moderately better than LVP in absolute potency in humans, while

Vavra et al. [104] reported that DDAVP was about 500 times better in rats. This discrepancy can be attributed to the type of biological assay used for quantification of effects. Urinary osmolality was used to quantitate the response in the human study, whereas the half-time to excrete a water load, a poor measure of response intensity, was used in the rat study.

Another analog of vasopressin, DVDAVP, is also orally active. Based on an antidiuretic assay, a 160-pg dose of DVDAVP is equivalent to a 10-μg dose of LVP orally with a potency ratio of 60,000. By comparison, the potency ratio is only 6 by the i.v. route [108]. As is the case with DDAVP, the enhanced activity of DVDAVP is due both to its inherent resistance to proteolytic degradation before and after absorption [108] and to improved membrane penetration [107].

The absorption of DDAVP in anesthetized rabbits, where the dose was administered intraluminally, appears to be site-dependent (Fig. 5) [109]. Absorption was rapid, peaking at 10−20 min. The optimal sites of absorption appeared to be the duodenum and the ileocecal junction. Compared with the gastric, ileal, and colonic regions, the plasma levels of immunoreactive DDAVP were 2−3 times and 4−5 times better from the duodenum and the ileocecal junctions, respectively. The reason for better absorption of DDAVP from these two regions is unclear, but enzymatic activity and porosity of the absorptive mucosa have been implicated [109]. Although DDAVP has been shown to be absorbed passively and largely by

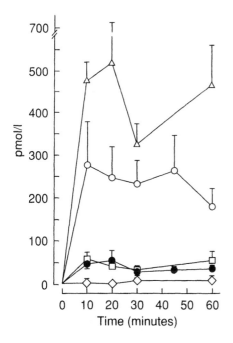

Fig. 5   Plasma DDAVP concentrations after administration of 30 nmol/kg at various sites in the rabbit gastrointestinal tract. Each point represents mean ± s.e.m. △ , Ileocecal function; ○, duodenum (n = 9), □, ileum (n = 4); ●, stomach (n = 3); ◇, colon (n = 4). (From Ref. 109.)

the paracellular route across the rat jejunum [107], it is not known whether
the synthetic modification of a natural peptide affects the pathway of transport, especially when the carrier-mediated transport is not involved.

DDAVP has been used for the treatment of cranial diabetes insipidus
since 1975. To improve patient compliance and decrease unreliable dose
dispensing associated with intranasal administration of DDAVP in solution
or spray form, the peroral route was investigated in humans. Vilhardt and
Bie [110] observed dose-dependent antidiuretic response within 15—30 min
of administration of 20, 40, and 200 µg of DDAVP in hydrated humans.
Administration of 200 µg of DDAVP through a duodenal tube into the lumen
of the small intestine of two subjects caused a prompt antidiuretic response,
with concomitant increase in urine concentration [110]. In an expanded
study involving both adults and children, water-loaded normal subjects,
and patients with cranial diabetes insipidus (CDI), similar antidiuretic effects lasting over 6 hr were seen at all three doses studied, 50, 100, and
200 µg [111]. Based on the plasma levels of immunoreactive DDAVP, absorption appeared to be dose-proportional (Fig. 6). In a long-term study,
nine cranial diabetes insipidus patients were successfully maintained on oral
tablets of DDAVP for over five months with excellent and convenient control of CDI [112]. Based on the plasma AUC of immunoreactive DDAVP in
children with CDI, the doses required to treat CDI patients by the oral
route were approximately 20 times greater than by the nasal route [113].
The relative oral bioavailability was 3.9—5.4%.

In conclusion, the above examples provide grounds for optimism that
the chemical modification approach can be used to meet the challenges in
successfully developing peptides and proteins as therapeutic agents.

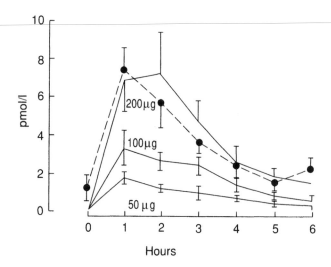

Fig. 6   Plasma DDAVP concentrations in five water-loaded normal subjects
after oral doses of 50, 100, and 200 µg (————) and in seven water-loaded
subjects with cranial diabetes insipidus after 200-µg doses (– – –).
(From Ref   111.)

B. Site-Specific Delivery of Peptide and
    Protein Drugs in the GI Tract

The concept of an optimal site for peptide and protein drug absorption is
appealing. Such a concept is supported by examples of peptide and protein drugs that are best absorbed when confined to a certain region within
the GI tract. For instance, Grass and Morehead [114] have obtained indirect evidence for the absorption of a novel, dipeptide-like angiotensin converting enzyme (ACE) inhibitor from only the upper small intestine in humans, in both the fed and fasted states. The mechanism of absorption is
as yet unknown. A carrier-mediated mechanism has, however, been invoked by Friedman and Amidon [115] to describe the absorption of another
ACE inhibitor, enalapril (an ester prodrug of enalaprilat), in the perfused
rat jejunum. This is indicated by the permeability of enalparil being dependent on a concentration in the range of 0.01−1 mM and its reduction
in the presence of 10 mM Tyr−Gly or 1 mM cephradine. Permeability of
the parent drug was, however, much lower and unchanged over two orders
of magnitude changes in concentration. Therefore, the increased absorption rate of enalapril is due more to its affinity for the peptide carrier than
to increased lipophilicity.

     That some peptide and protein drugs would be nonuniformly absorbed
along the length of the GI tract is to be expected from the gradient of
protease activity that exists. In the rabbit as well as human, the activities
of aminopeptidase N and diaminopeptidase IV, two of the many intestinal
proteases, are higher in the ileum than the jejunum [116−121]. This gradient of activity reflects that of the brush border but not the cytosol,
which shows no regional variation [117]. In contrast, in the rat, aminopeptidase N activity is higher in the jejunum, although diaminopeptidase
IV activity is still higher in the ileum [121−123]. There are, therefore,
species differences in the regional distribution of aminopeptidase activity.
Interestingly, regardless of the animal species studied, these two proteases plus several others collectively render the ileum proteolytically more
active than the jejunum [42,122,123]. This is partly responsible for the
favorable absorption of peptides such as Met−Met, Gly−Gly, and several
others from the jejunum when compared with the otherwise well-absorptive
ileum [124].

     Various techniques are available to delineate the absorption characteris-
tics of drugs from different regions of the GI tract. Staib and co-workers
have used a high-frequency capsule (HF-capsule) for their bioavailability
studies in humans [125]. The device can be triggered to release its contents by means of an external radio signal at a predetermined location.
Such a device allows targeting of a drug to possible absorption windows
and additionally it offers protection from the potentially harmful environment of the intestinal tract.

     The absorption of a theophylline solution containing 80−120-mg doses
delivered to different sites in the gastrointestinal tract has been investigated in male volunteers using the HF-capsule. There was no difference
between the stomach, the ileum, and the colon in the amount of drug absorbed. The absorption half-life of theophylline from the colon was prolonged when compared with that in the upper intestinal tract [125]. Using
the same approach, isosorbide-5-mononitrate was shown to be absorbed
equally well from the stomach, duodenum, jejunum, and ascending colon
[126].

Most recently, Knutson et al. [127] developed a multichannel tube with
two occluding balloons for segmental perfusing studies in jejunum and ileum
in humans (Fig. 7). The tube can be placed at any desired part of the
small intestine under fluoroscopic guidance. In that study, a jejunal segment of healthy volunteers was perfused for 3 hr with ^{14}C-PEG 4000 at a
flow rate of 3 ml/min and an osmolality of 290 mOsm/L. No loss of perfusate was reported, and recovery of PEG 4000 was approximately 97% and
stable during the 3 hr of perfusion. These findings imply that little absorption occurred through aqueous channels in the jejunum. Although drug
absorption was not investigated in the above study, this closed perfusion
technique seems suited for such studies in the small intestine of humans
since it is easy to perform, atraumatic, reproducible, and can continue for
at least 6 hr. Another advantage of the above technique over open or
semiopen systems is the low perfusion rate (3 ml/min), which not only
avoids diluting the perfusate contents, but also avoids disturbing the hydrodynamic mucus layer.

An intubating technique has been used to investigate the absorption of
SMS 201-995, a synthetic somatostatin analog, from the various regions of
the GI tract. This involved positioning a rubber tubing with openings for
discharge of drug solution at various locations: stomach, proximal duodenum, ligament of Tritz, jejunum, and ileum. SMS 201-995 was found to
be equally well absorbed from the various sites, although there was a tendency toward decreased peptide absorption in the ileum. Interindividual
variability and site-to-site variability were quite high. The bioavailability
compared to a s.c. dose was only 0.28%.

Although the stomach is normally the first available site for drug absorption, it is primarily a secretory organ, not an absorptive tissue. For
peptide delivery, however, the stomach is probably the most fatalistic site
for absorption. The enzyme content, low pH, and relatively low tissue

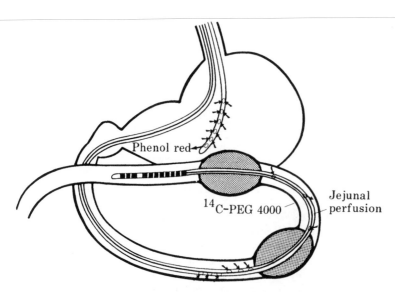

Fig. 7   Apparatus for evaluating drug absorption from various regions of
the GI tract. (From Ref. 127.)

surface area make the stomach a primary site of peptide degradation rather than absorption.

In its simplest form, the intestine is often regarded as a tube-like structure. It is a leaky epithelium, as indicated by its relatively low resistance when compared to the tight epithelia of tissues such as frog skin [128]. Although the area of contact between neighboring cells appears microscopically tight, these junctions are permeable to water, electrolytes, and in a size-dependent manner to charged and uncharged molecules, such as naproxen and mannitol, respectively [129]. An effective pore radius has been estimated at 6 Å in the jejunum and 3 Å in the ileum [130].

The intestinal surface is covered by a layer of mucus in equilibrium with soluble mucus in the luminal juice [131]. In the duodenum, mucus is secreted by Brunner's glands. This is a thick, alkaline mucus [132] whose function is to protect the underlying duodenal tissue from the acidic chyme released from the stomach. In the remainder of the GI tract mucus is secreted by the goblet cells and glands in the crypt cells (crypts of Lieberkühn). Its secretion is stimulated by direct tactile or chemical exposure from the contents of chyme [132]. The unstirred water layer of which the mucus is an integral component can greatly influence the movement of actively transported substrates, such as glucose [132]. Indeed, the excessive mucus stimulated by prostaglandins has been implicated in the reduction in glucose as well as mannitol transport [133]. Of course, the mucus layer can also adversely affect the intestinal absorption of passively absorbed substances, such as ergot peptide alkaloids [134]. This is indicated by the twofold increase in absorption by mucolytic enzymes, such as papain and hyaluronidase.

When considering absorption from the intestine, it is important to address the differences that exist in its various segments. Morphologically, the intestinal segments differ, having progressively fewer and smaller villi in the more distal sections. This creates a smaller surface area, with the colon having the lowest surface area per intestinal length. In the upper jejunum the mucosal area per cm serosal length is approximately 98, while in the lower ileum it is about 20 [135].

In addition to the anatomical differences noted for the various segments of the intestine, there is a differential in pH. The most common method for measurement of intestinal pH is through the use of a radiotelemetry device, the Heidelberg capsule [136]. Youngberg and co-workers conducted such a study in 66 ambulatory subjects. A sample tracing of the results is given in Fig. 8. In a similar study, Lui et al. [137] investigated the differences in pH of human and dog gastrointestinal tract. Their results suggest that, in some cases, test studies carried out in dog may not be predictive in humans because of the potential ionizational differences for some compounds between these two species.

The relative permeability of each of the intestinal segments has not been well explored. In vivo, it is difficult to make such measurements in the absence of confounding factors such as GI transit or pH differences. One method to compare the permeabilities of various intestinal segments in vivo has been to use a steady-state perfusion technique with PEG 400, as described by Chadwick et al. [138,139]. Since PEG is a mixture of low-molecular-weight polymers (M.W. 250–600) and is not metabolized, these perfusion studies allow the examination of a range of molecular-weight fractions. The finite lipid solubility of PEG fractions has, however, prompted the criticism by several investigators on the suitability of such

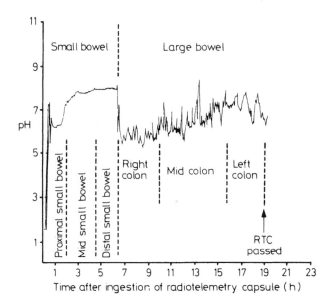

Fig. 8   Gastrointestinal pH profile from a normal subject. The position
of the Heidelberg capsule (RTC) in specific areas is labeled accordingly.
(From Ref. 136.)

fractions as markers [140-144]. But such criticism has now been refuted
by Ma et al. [145], who measured the partition coefficients of PEG fractions, 288-722 in molecular weight, between water and hexane, toluene,
and benzene. They found a progressive decrease in the partition coefficient of PEG with increasing molecular weight, indicating increasing water
solubility of the higher-molecular-weight fractions. For PEG they found
1.5 parts of the compound in hexane per 100,000 parts in equal volume of
water. Clearly, PEG is a strongly hydrophilic molecular probe. The limited intestinal uptake of the higher molecular probes must therefore be due
to increased molecular size.

Ileum as an Absorption Site: Availability
of Specialized Transport Mechanisms

Although the absorbing capacity of the small intestine decreases from the
duodenum to the ileum, the latter is noted to contain highly specialized
mechanisms for absorption of specific macromolecules.
    A specialized transport mechanism exists in the ileum of the rat to ab-
sorb bile acids. The essential molecular requirement of bile acids for active transport is the acidic side chain at the 17-position of the ring system.
C3-derivatives may also undergo active transport. Ho [146] showed that
the absorption of p-toluenesulfonic acid from the ileum was enhanced as a
result of conjugation with cholic acid.
    In addition to the bile acid carrier, a carrier for cyanocobalamin (vi-
tamin B_{12}) also exists in the ileum. Vitamin B_{12} is a dietary coenzyme required for the maintenance of mammalian life. Its molecular weight is 1355,
with a radius that far exceeds the "effective pore size" of intestinal

membranes [147]. Although very small quantities of vitamin B_{12}, on the order of $10-35$ pg/100 mg ileal tissue homogenate, may be absorbed by the jejunum and ileum via a nonspecific mechanism [148], the bulk of vitamin B_{12} is absorbed via a carrier-transport mechanism. Normally, vitamin B_{12} does not exist in its free form but is coupled to dietary proteins, from which it must first be cleaved enzymatically before it can be utilized [149].

Cyanocobalamin binds with high affinity to three distinct transport proteins: intrinsic factor (IF), transcobalamin II, and the nonintrinsic factor cobalamin-binding proteins (R proteins). The three cobalamin-binding transport proteins have different origins. IF, a specific glycoprotein with a molecular weight of $50-70$ kD, is secreted by the parietal cells in the stomach [150]. Transcobalamin II is present in plasma. The R proteins are glycoproteins with a molecular weight of $56-66$ kD, found in plasma, granulocytes, and several glandular secretions, including saliva, gastric juice, and chyme [151,152]. Among the binding proteins the intrinsic factor is the most selective.

Cobalamin transport across the enterocyte involves several steps. The initial step is cobalamin removal from the dietary protein, probably by proteolytic enzymes such as trypsin and chymotrypsin [153]. Cooper and Castle [154] suggested that an acidic environment promotes the transfer of cobalamin from the dietary protein to IF. Allen and co-workers [152], on the other hand, suggested that vitamin B_{12} is bound almost exclusively to R protein rather than to IF in the acid milieu of the stomach and that it remains bound until trypsin and chymotrypsin, in concert with elastase, partially degrade R protein to release the vitamin, which then becomes bound exclusively to IF. The complex formed is very stable over a wide pH range and in the presence of intestinal enzymes [155]. The better absorption of cobalamin via jejunostomy than via the mouth has led Sriram et al. [156] to speculate that administration of unbound cobalamin into the jejunum decreases the possibility of its binding to R proteins due to the greater pH of the ileum thereby maximizing cobalamin binding to IF. Intrinsic factor is not internalized into the enterocyte during cobalamin absorption [150].

Cobalamin binding to its receptor occurs preferentially between pH 6 and 8 [156-158]. Binding is unaffected by changes in temperature from 4 to 37°C and requires no energy, but it requires calcium. Free IF, free R proteins, and free cobalamin will not bind to this receptor [159]. This receptor is present on the surface of the microvillous membranes in the ileum of human and guinea pig [160-163] but in the jejunum of the rat [164]. All receptors examined to date have a molecular weight of approximately 200 kD and consist of two subunits, one of which anchors to the ileal microvillous and the other of which binds intrinsic factor [165,166].

The number of receptors available to the cobalamin-IF complex is remarkably small. Hamsters, for example, have no more than $300-400$ receptors per ileal absorptive cell, or approximately one receptor per microvillous [167]. In human ileum, receptor sites are distributed in a patchy fashion that varies from individual to individual [161,162]. In mice, absorption as well as the number of receptors doubles during pregnancy and this correlates with increases noted during pregnancy in humans [168-170].

The binding of cyanocobalamin-IF to its receptor appears to be species specific. Although hamster IF promotes B_{12} uptake by brush border and microvillous membranes prepared from hamster ileum, human IF does not bind and is ineffective [167]. On the other hand, human IF does enhance the intestinal uptake of B_{12} in the guinea pig [171]. Kolhouse and Allen [172] have examined 14 different cobalamin analogs and only two with minor changes displayed binding similar to intrinsic factor. Transcobalamin II, by contrast, shows moderate selectivity whereas R proteins bind a wide variety of cobamides as well as cobinamid with high affinity. Cobamides are bacterially produced vitamin B_{12} which differ from cobalamin by the absence of the cyano and the 5,6-dimethylbenzimidazol groups. Their ability to compete with the cobalamin-IF complex for ileal receptors is controversial [158,173]. Kanazawa et al. [174] demonstrated binding of cobalamin analogs to hog cobalamin-IF complex receptors. Shaw et al. [175] demonstrated that in humans with inadequate intrinsic factor secretion, the uptake of cobamides was unaffected whereas that of vitamin B_{12} was increased. The conclusion is that vitamin B_{12} analogs are absorbed from the ileum by a mechanism independent of the intrinsic factor. The implication is that it may be possible to couple macromolecules to cobalamin for transport into the enterocyte via the receptor-mediated transport process just described. This hypothesis was confirmed in pilot experiments by Russel-Jones and de Aizpurua [176], who observed stimulation of ovulation in two of five mice to which were administered as small an oral dose as 6 μg of LHRH-vitamin B_{12} conjugate. By contrast, oral administration of an eight to nine times higher dose of the unconjugated hormone resulted only in one animal undergoing ovulation.

Colon as a Site of Drug Delivery

For systemic peptide absorption, the colon is viewed as an area of decreased enzyme concentration, especially peptidases, when compared to the small intestine. Nevertheless, there are several concerns with the colon as a primary drug delivery site. First, the colon has a flat epithelium with a smaller surface/volume ratio available for absorption than other intestinal segments [177]. It is therefore not surprising that the permeability of the colonic epithelium to polar compounds is less than that of the small intestinal epithelium [178]. Although this may be offset by the much longer residence time, colonic transit is somewhat unpredictable. This is indicated by the 17-72-hr range for transit of Heidelberg capsules through the normal human colon [178]. Second, colonic pH can be extremely variable and, depending on the type of food ingested (protein rich or not), can rise to a level as high as pH 8. Third, in the colon, drug is surrounded by almost solid fecal matter, which restricts free diffusion to membranes. Fourth, the concentration of bacteria, largely anaerobic species, is several orders of magnitude higher than in the small bowel. The related rich bacterial enzyme population may potentially lead to faster drug degradation. This may explain the reduced bioavailability of an enteric-coated digoxin formulation which released its contents in the distal small intestine where the drug was degraded by the microflora to inactive dihydrodigoxin [179]. Peppcorn and Goldman [180] also implicated colonic metabolism in the lower efficacy of atropine given via the oral route compared with parenteral application.

On the positive side, microbial enzymes have been exploited to afford colon-targeted drug delivery. This is exemplified by salicyloazosulfapyridine, an antiulcerative compound. In 1980, Parkinson et al. [181] patented a procedure to link 5-aminosalicylic acid via aromatic azo groups to high-molecular-weight polymers, thus rendering the drug inactive and unabsorbable in the GI tract except the colon, where the azo bond is reduced,

liberating 5-aminosalicylic acid (5-ASA). The same concept was the basis
of the design of a 5-ASA prodrug by Willoughby et al. [182]. Their prodrug was comprised of two 5-ASA molecules joined by an azo-bond. In addition to the azoreductase, microbial glycosidases have also been relied
upon to afford colon-specific drug delivery. Friend and Chang [183,184]
demonstrated that in rats up to 60% of an oral dose of steroid glycoside
reached the cecum intact. In the case of the dextran-naproxen ester prodrug examined by Harboe et al. [185], 10–500 kD in molecular weight, a
host of microbial enzymes is involved in its activation. The postulated
mechanism of prodrug activation involved an initial depolymerization step of
the dextran chains by endo- and exodextranases secreted from colonic bacteria yielding small fragments that became substrates for esterases and
other hydrolases, thus releasing the native drug. The bioavailability of
naproxen after oral administration of a 70-kD prodrug in the pig was close
to 100% when compared to an equivalent oral dose of the parent drug [186].

As mentioned in Chapter 7, reduction of the diazovinyl cross-links in a
copolymer of styrene and hydroxyethylmethacrylate by azoreductases is the
basis for the design of a colon-specific delivery system for insulin [187] and
vasopressin [188]. Polysaccharides such as fibers, chitin, and chitosan,
which are degraded by bacteria to short-chain fatty acids, have been suggested as coating materials [169], provided the reduction in pH during digestion of the polysaccharides will not adversely affect absorption of the
drug housed by the system. An alternative system is that based on coating a capsule with Eudragit-S, an acrylic-based resin, to yield coats varying from 95 to 135 μm in thickness [189]. Since the films are insoluble
below pH 6 but are slowly soluble at weakly alkaline pH's, drug release
will be controlled by erosion of the coat.

Peyer's Patches as a Potential Port of Oral Peptide and Protein Absorption: Oral Immunization

Anatomical and physiological aspects. In 1676, Peyer described patches
in the intestinal tract belonging to the gut-associated lymphoid tissue
(GALT) [190], which comprises as much as 25% of the GI mucosa (Fig. 9).
The function of the Peyer's patches is believed to be that of sampling antigens, including poliovirus type I, whether Sabin or Mahoney strain [191].
These two observations have stimulated interest in the Peyer's patches as
a port of entry of lymphokines and other immunomodulating drugs into the
lymphatic circulation. Lymphatic absorption of peptide and protein drugs
will be discussed in the next section.

Peyer's patches are generally rectangular or oval, and they are usually
situated on the mesenteric border of the intestine. In humans they are
more numerous in the distal ileum, whereas in mice they are more uniformly distributed over the small intestine. Their number increases gradually through the end of puberty, and thereafter a gradual and fairly
equal reduction in the number of patches is found (Fig. 10). The number
of Peyer's patches varies widely between species and individuals, from 1
to 10 in rabbits [192], from 26 to 29 in dogs [193], and from 245 to 320 in
horses [194]. The patch size varies from a few mm to about 10 cm in
length before puberty and can be as long as 28 cm in the adult. One
Peyer's patch can contain 5–260 lymphoid follicles, and as many as 980 in
some cases. Patches are only one follicle thick, the follicle being intimately
related to the overlaying epithelium.

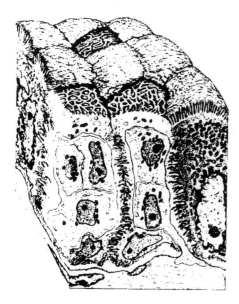

Fig. 9   Surface and longitudinal view of the Peyer's patch showing three
M cells with luminal surface microfolds interdigitating with adjacent columnar
cells.   The mononuclear lymphoid cells directly beneath the M cells can approach to within 0.3 µm of the intestinal lumen.   (From Ref. 213.)

Two distinct types of Peyer's patches are known to exist in sheep
[195].   The ileocecal Peyer's patches are dominated by the lymphoid follicles
that are tightly packed together, whereas the follicles in the jejunal Peyer's
patches are more pear-shaped and in general smaller than those in the ileocecal Peyer's patches.   The rabbit also has two different types of gut-
associated lymphoid tissues.   The appendix is similar to the ileocecal Peyer's
patches, while the jejunal Peyer's patches in the rabbit resemble the jenu-
nal Peyer's patches in the sheep.   The mean size of Peyer's patches in jejunum and ileum is not statistically different.

B and T cells enter Peyer's patches from the recirculating lymphoid
pool via postcapillary venules (Fig. 11) [196].   Follicles with germinal centers host mostly B lymphocytes, which are actively dividing and can bind
high levels of peanut agglutinin [197].   T lymphocytes predominate in the
interfollicular area and are also present in the dome region overlying Peyer's
patches.   In adult mice 11−40% of Peyer's patch lymphocytes are T cells,
whereas 40−70% are B cells [198−200].   Similar numbers were found in
children during ileal endoscopy [201].   Biopsy specimens revealed that the
mean cell yield was 1.7×10^8 lymphocytes/g tissue, 26−48% of which were
B cells and 50% of which were T cells.

Compared to B cells from peripheral lymph nodes, B cells from the
Peyer's patches generate a greater proportion of B cell clones that produce only IgA [202,203].   T cells from Peyer's patches appear to have a
selective role in switching from IgM to IgA commitment and have an amplifying role in the development of mature, antigen-specific IgA B cells [204,
205].   After antigen stimulation, lymphocytes that exit Peyer's patches appear to migrate to the mesenteric lymph nodes, the superior mesenteric

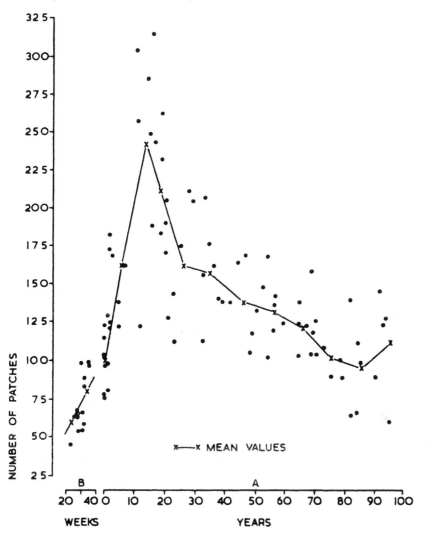

Fig. 10 Age-related changes in number of Peyer's patches in the human small intestine. B, Before term, from 24 to 37 weeks' gestation; A, after term, from 0 to 95 years. (From Ref. 240.)

duct, and the thoracic duct lymph before entering the circulation and returning to the lamina propria and intraepithelial region of the intestine (Fig. 11) [198,206]. In this respect Peyer's patches can be viewed as the initiating site of mucosal immunity.

In their study on the migration behavior of sheep Peyer's patch lymphocytes, Pabst and Reynolds [207] showed that lymphocytes emigrated from the gut and mesenteric lymph nodes via lymphatics and that they did not enter the blood directly in large numbers, as occurred in the pig [208]. The total number of fluorescein isothiocyanate (FITC)-labeled lymphocytes found throughout the lamb after 1 day was about 62 times greater after

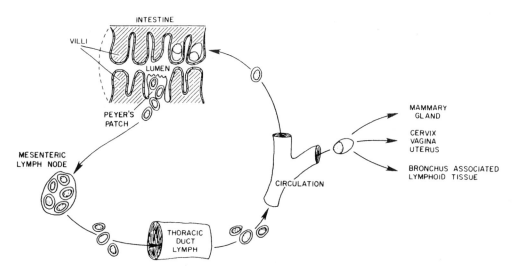

Fig 11   Migration pathway of Peyer's patch cells.   (From Ref. 244.)

ileal Peyer's patch perfusion with FITC than after jejunal Peyer's patch
perfusion, 100×10^9 and 1.6×10^9, respectively, although the amount of
labeled Peyer's patches was only 25 times greater after ileal perfusion than
after jejunal Peyer's patch perfusion, 107 g and 4.2 g, respectively.   Thus,
it was apparent that the number of FITC-labeled cells remote from the perfused site was not directly related to the amount of Peyer's patches perfused.   About 72×10^6 lymphocytes emigrated per g of perfused Peyer's
patch in the jejunum as compared with 17×10^6 cells per g of perfused
Peyer's patches in the ileum.

Mahida et al. [209] investigated the mononuclear phagocyte system
within human Peyer's patches.   The macrophages in the germinal center
were large and stained strongly positive for acid phosphatase.   The macrophages in the dome region were heterogeneous in their size distribution
and also stained for acid phosphatase.   In the interfollicular areas there
were cells with a dendritic morphology, which did not stain for acid phosphatase.   Some macrophages and lymphocytes in the germinal centers and
dome regions expressed interleukin-2 and transferrin receptors.

Microscopic examination of the cells lying over Peyer's patch follicles
has revealed a relatively sparse number of goblet cells, creating a breach
in the mucus blanket, thereby allowing particles to reach the microfold M
cells.   These cells are believed to transport antigenic material from the
lumen into the space between cells where it can be taken up by lymphocytes.

It has been postulated that M cells are present in the lamina epithelia-
lis in every region where it overlies accumulation of lymphoid tissue in the
intestine [210].   These cells have also been identified in human appendix,
tonsils of rabbits, and tonsils and the cecal Peyer's patch of mice [211].
In light microscopy, M cells are characterized by luminal microfolds rather
than microvilli.   The M cell shares tight junctions and desmosomes with
adjacent epithelial cells.   The epithelium, which appears to be disorganized,
consists of a reticulum formed by the attenuated cell processes of M cells.

Such an arrangement is believed to allow lymphoid cells, such as lympho-cytes, lymphoblasts, or macrophages, to approach to within 300 nm of the intestinal lumen [212,213]. Staining with ruthenium red suggests that M cells have a less elaborate glycocalyx than absorptive cells [214]. More-over, the apical plasma membrane of M cells contains less protein and may have a lower protein-to-lipid ratio than the microvillus membrane of ab-sorptive cells [215].

In mice, M cells are distributed evenly throughout the higher regions of the follicle-associated epithelium (FAE). The frequency decreases to-ward the lower regions and no M cells are found in the follicle-associated crypts. M cells account for about 15% and lymphoid cells represent ap-proximately 30% of the total cell population in the highest region of the dome FAE [216]. A similar number was given by Pappo using monoclonal antibodies directed against rabbit FAE [217]. These antibodies labeled a subpopulation of FAE cells that occurred at a frequency of 50% in the epi-thelial cells. Immunostaining was mainly observed on FAE cell basolateral surfaces corresponding to the central hollow or pocket containing clusters of mononuclear leukocytes, a characteristic of M cells. By contrast, the stem cell and proliferative regions facing the lamina propria were devoid of immunologically reactive sites. In an attempt to describe the phagocy-tosis behavior of FAE, Peyer's patch cells were incubated in vitro with FITC-polystyrene microspheres (0.75 μm) at a ratio of 100 particles/cell, which were then immunologically stained. While phagocytotic cells com-prised 51−55% of the Peyer's patch population, only a subpopulation (15−25%) both phagocytosed FITC-microspheres and expressed the antigens that could be recognized by the antibodies.

Cytochemical analysis of M cells in rat, mice, calf, and rabbit Peyer's patches has revealed a strong reduction in alkaline phosphatase activity, supposedly a special characteristic of antigen-sampling epithelial cells. Identical binding for concanavalin A, Ricinus communis agglutinin, wheat germ agglutinin, and peanut agglutinin on M cells and enterocytes indicates that the two cell types share the same common glycoproteins and glycolip-ids [218]. Nevertheless, Neutra and co-workers [219−221] found that al-though ferritinin-conjugated ricin binds in vitro at 4°C almost equally to both enterocyte and M-cell surfaces, only M cells took up ricin and trans-ported it to their basolateral membranes. Moreover, M cells could trans-port cationized ferritin, 12−13 nm in diameter, within vesicles, to the in-tercellular spaces. The molecular weight of ferritin was between that of apoferritin (450 kD) and holoferritin (900 kD), depending on the iron content.

Using light microscopic autoradiography and electron microscopy, Weltzin and co-workers found that all gold or radiolabeled mouse IgA was bound to M cells upon injection into mouse, rat, and rabbit intestinal loops containing Peyer's patches [222]. Competitive experiments with unlabeled IgA demonstrated inhibition, indicating specificity in binding. These re-sults imply that luminal antigens and microorganisms complexed with secre-tory IgA (sIgA) could be targeted to M-cell surfaces and be selectively transported to underlying macrophages and lymphoid cells.

The Peyer's patches are notably deficient in lysosomes [223−225]. Owen et al. [225] examined ileal rat Peyer's patches by electron micro-scopy to identify lysosomes by acid phosphatase activity. The enzyme was found in dense bodies in enterocytes but not in M cells. Comparison between enterocytes and M cells showed that the volume fraction occupied

by dense bodies in M cells was 16 times less than in enterocytes. Both
the absence of acid phosphatase activity and the small amount of dense
bodies in M cells correlated with the absence of lysosomal degradation of
luminal microorganisms during transport into lymphoid follicles by M cells.
Unlike absorptive cells, M cells lacked any evidence for lipid absorption
[220].

McClugage and Low [227] investigated the porosity of the epithelial
basement membrane overlying lymphoid follicles within Peyer's patches of
rats and owl monkeys by scanning electron microscopy. Basement membranes overlying the lymphoid follicles were markedly porous, showing increasing porosity centrifugally from the cap to the periphery of the follicle.
In comparison to the small intestine, porosity of the large intestine is reduced [226]. Such a pattern of porosity could facilitate transcytosis of
luminal macromolecules into the underlying follicles. Nevertheless, the
overlying epithelium remains the rate-limiting barrier [231].

The above features of the Peyer's patches may be responsible for the
threefold increase in transport of HRP across the jejunum Peyer's patches
of 20-day-old piglets [224], although in rabbits, the permeability to PEG
4000 and mannitol was the same in both Peyer's patches and nonpatch tissues [231]. Although the number of compounds studied was small, there
did not appear to be a correlation between the extent of macromolecular uptake by Peyer's patches and the molecular weight of the macromolecules [229].
A comparison between precipitated protein and total recovered radioactivity
revealed that degradation in Peyer's patches was less than in nonpatch tissue [229]. This was also the case for an immunomodulating glycopeptide
extracted from *Klebsiella pneumoniae* in the duodenal Peyer's patches of the
rabbit [230]. Approximately 4% of the glycopeptide remained as a high-molecular-weight fraction after transport across the non-Peyer's patches, while
about 17% was found to be the high-molecular-weight fraction after transport
through Peyer's patches.

Absorption of particulate matter. Particles measuring up to 10 µm
have been shown to accumulate in mouse and rabbit Peyer's patches [232–
235]. Using immunohistochemical studies, Gilley et al. [235] demonstrated
the transport of microspheres less than 5 µm in diameter within macrophages through Peyer's patches into mesenteric lymph nodes. These findings are in agreement with the results of LeFevre and Joel [236], who indicated that the majority of particles that traverse Peyer's patches were
sequestered into macrophages. It has been suggested that particle surface
properties, as well as particle size, govern accumulation in Peyer's patches,
with penetration being favored by particles with hydrophobic surfaces
[237,238]. By counting carboxylated rhodamine B-conjugated polystyrene
latex particles, LeFevre et al. [233] discovered that Peyer's patches of
aged mice accumulated about five times more latex particle (1.8 µm in diameter) than those of young mice. This unexpected result was attributed
to the longer exposure made possible by the slower transit of material
through the small intestine of aged mice.

Carbon particles have been shown to enter intestinal tissue principally
through Peyer's patches [239]. No difference in transport pathway was
discerned with latex particles (2 µm) in germ-free mice, although non-
germ-free mice did accumulate a larger number of particles by virtue of
larger lymphoid follicles [237,238].

Oral immunization. Antigen challenge and exposure to bacterial lipo-
polysaccharides will induce B cells in the Peyer's patches to eventually develop into IgA plasma cells [205]. Effective immunization against proteins
usually demands the coupling of the peptides to suitable carriers and/or
the injection in combination with an adjuvant. Guyon-Gruaz et al. [241]
orally administered synthetic fragments of cholera toxin to mice without
either carrier or adjuvant to induce antibodies recognizing the intact toxin.
Synthetic peptides as small as 20 amino acids were potent enough as immunogens following a priming dose (day 0) and boosters on days 10, 16,
and 22. Wassef et al. [242] found that, when the ileal loops (8 – 10 cm) of
nonimmune rabbits were exposed to both virulent and avirulent shigellae,
about four times more of the virulent strain was recovered in the M cells.
Such avid uptake was reduced to control levels upon heat-treating of the
virulent strain, which also destroyed its ability to elicit a memory mucosal
response.

Several investigators have encapsulated antigens into unilamellar dipal-
mitol phosphatidylcholine liposomes with a negative surface charge in hopes
of enhancing the immune response. Michalek et al. [243] demonstrated M-
cell uptake and a successful immune response in rats from a liposomal antigen preparation. Childers et al. [245] injected gold-labeled solid core
liposomes (AuSCL), made of dioleoylphosphatidylcholine/cholesterol/dio-
leoylphosphatidylglycerol/triolein (4.5:4.5:1:1) into isolated small-intestinal
sections of rats containing Peyer's patches. Electron microscopy revealed
that after 2 hr of incubation, pinocytotic vesicles containing AuSCLs appeared in M cells from the dome region of the Peyer's patches. This finding therefore supports the assumption that liposomes can be taken up by
M cells for induction of mucosal immune resonse.

To facilitate the uptake of liposomes by Peyer's patches, Rubas et al.
[246] coated liposomes composed of soy phosphatidycholine with the reovirus cell attachment protein $\sigma 1$, for which receptors exist on the M cells
[247 – 251]. Incorporation was via the hydrophobic region of the amino-
terminal portion of protein $\sigma 1$ [252]. Competition studies with reovirus on
mouse fibroblasts (L929 cells) grown as a monolayer revealed selective
binding to the reovirus receptor. After in vitro incubation at 4°C for 1
hr, approximately 10-fold greater amounts of coated liposomes were found
associated with the rat Peyer's patches as compared to uncoated liposomes.
These results suggest that selective adherence of a carrier to M cells may
facilitate delivery to mucosal lymphoid tissue, thereby enhancing the immune
response.

V. LYMPHATIC ABSORPTION OF PEPTIDE
AND PROTEIN DRUGS

There is a perception that targeting peptide and protein drugs into the
lymphatic circulation would improve their oral bioavailability since the liver
will be bypassed. This is advantageous for those drugs whose target is in
the lymphatic system. Indeed, some protein drugs are inherently lympho-
tropic. In the case of elastase, a proteolytic enzyme with a molecular
weight of 25,000 about 36% of what was absorbed from doses of 1 mg/rat
or 5 mg/rat, 0.053 – 0.149%, was in the lymphatic circulation [253]. Obviously, this protein was capable of crossing the intestinal wall intact. In

a perfusing experiment using an antielastase rabbit antibody, Tsukii et al. [254] found that elastase was absorbed without decomposition through healthy rat jejunum epithelial cells by the pinocytotic process and transported into lymph as well as blood.

The oral route is not the only means by which peptide and protein drugs can gain access to the lymphatic circulation. Such drugs can also reach the lymphatic circulation from parenteral administration. There appears to be a linear relationship between the extent of lymphatic absorption and the molecular weight of the absorbed species [255]. Molecules with M.W. > 16,000 are predominantly absorbed by the lymphatics that drain the application site. Similar results were obtained by Bocci et al. in rabbits and pigs [256]. Supersaxo and co-workers reported that following the intradermal or subcutaneous administration of 2×10^7 U of recombinant human interferon alpha-2a (rIFN α-2a) to lymphoid cells, very high levels were achieved in lymphatics as compared to intravenous administration in the sheep [255]. Additionally, Bocci and co-workers also demonstrated a vehicle effect [256]. Subcutaneous coadministration of rIFN alpha-2a with a solution containing 12.5% human albumin increased the lymphatic uptake, as the lymph AUC/plasma AUC ratios increased from 0.8 (without albumin) to 1.7 in the presence of albumin. This enhancement effect of albumin was attributed to its property as an interstitial fluid expander and the attendant increase in fluid pressure.

A.   Anatomy and Physiology of the Lymphatics

The lymphatic vessels in the small intestine originate as elongated, blind-ended vessels (lacteals) in the center of each villus (Fig. 12). Finger-like villi host one central lacteal, whereas flattened villi, such as those found in rats, host several vessels. The lacteals lie approximately 50 μm beneath the epithelium and their radius is about 20 μm (Fig. 13). These vessels join a network of capillaries in the glandular layer of the mucosa and are linked to collecting lymphatics. At the mesenteric border the lymphatics leave the intestine in association with the blood vessel [257−259]. In contrast to the small intestine, large-intestinal lymphatic vessels are fewer in number and smaller in diameter and lie deep in the mucosa 300−400 μm beneath the mucosal epithelium (Fig. 14).

Unlike blood vessels, the lacteals are composed of endothelial cells with no fenestrations. Nevertheless, since the junctions are widely separated. macromolecules such as chylomicrons (75−600 nm diameter) can pass readily into the lymphatic lumen. Moreover, the lymphatic capillaries have a fragmented basement membrane offering little barrier to the passage of solutes, fluids, and large particles [260,261]. Consequently, the major pathway for the transport of fluids and particulate components from the interstitium into the lymphatic lumen is the intercellular route. Particles ranging in diameter from 3 to 50 nm have been shown to pass through intermediate junctions of the endothelial cells but not the tight junctions. Substances in this group can be taken up by vesicles approximately 50 nm in diameter for transport across the cell.

The lacteals lie in a gel-like structure (interstitium) through which water and solutes slowly percolate. The interstitium is composed of mainly collagen fibers and hyaluronic acid, a polymer of N-acetylglucosamine and glucuronic acid with a molecular weight ranging from a few thousand to several million daltons. Cross-linking between hyaluronic acid and collagen

Alimentary tract lymph vessels

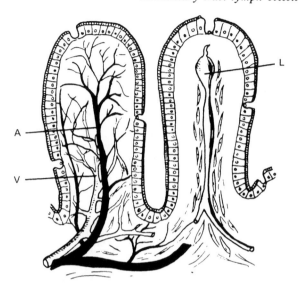

Fig. 12   Vessels of the intestinal villus. A, central artery; V, central
vein; L, lacteal. (From Ref. 265.)

and other proteins creates a fine interstitial mesh, with a pore size of
about 25 nm [262].   Since hyaluronic acid is anionic, the gel has a net
negative charge at physiological pH and ionic strength [263].   When interstitial fluid volume rises during a meal from its normal 25 ml/100 g, the
porosity of the matrix increases from 20 nm to approximately 100 nm. The
net result is reduced frictional resistance to diffusion of macromolecules
and an increased hydraulic conductivity [264].

The interstitial-to-lacteal hydrostatic pressure gradient is the major
driving force for lymphatic filling and therefore determines lymph flow.
When venous outflow pressures (Pv) were increased from 0 to 30 nm Hg
in an in situ cat intestinal loop, the interstitial fluid pressure increased
from −1.8 to 5.3 mm Hg, yielding a plateau at 20 mm Hg (Pv). Lymph
flow increased over the entire range of venous outflow pressures and appeared to approach a plateau at the higher venous outflow pressures (25
and 30 mm Hg). Changes in lymph flow were minimal at the lower venous
outflow pressures (up to 10 mm Hg) but were marked in the range between 10 and 20 mm Hg. The largest lymph flow observed was at Pv = 30
mm Hg, when approximately 7 times the lymph flow of 0 mm Hg venous
outflow pressure was seen [258].

Several investigators have reported an increased intestinal or thoracic
duct lymph flow following a meal or fluid ingestion [265,266]. During net
fluid absorption, the increase can be 5−20 times higher than normal lymph
flow in the nonabsorptive state [264]. The magnitude of increase is quite
variable, due possibly to such factors as tonicity of the fluid ingested,
portal vein pressure, intraenteric pressure, and GI motility. Should these
factors be held constant or eliminated, the rate of fluid absorption would

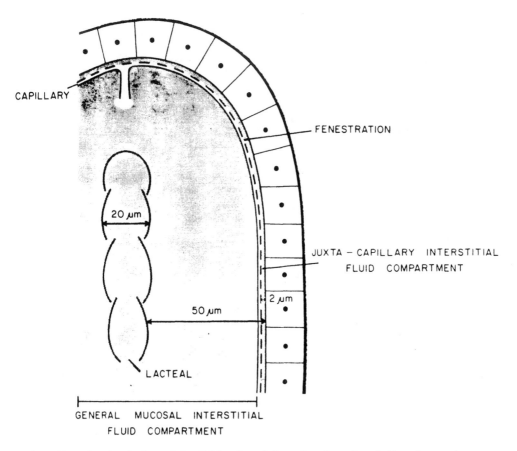

Fig. 13   Anatomical model of blood and lymph microcirculation in an in-testinal villus. (From Ref. 258.)

become a major determinant of intestinal lymph flow. In any event, the overall lymphatic absorption rate in the absorptive state usually does not exceed 15−20% [258].

While glucose and electrolytes do not affect intestinal vascular permea-bility, there is a six- to sevenfold increase in intestinal lymph protein flux during glucose absorption [267,268]. Lymphatic protein flux is also increased during fat absorption [265,269]. Granger and co-workers dem-onstrated that either cream or a mixture of bile and oleic acid produced a four- to sevenfold increase in lymphatic protein flux and a simultaneous increase in vascular permeability. The reflection coefficient (σ), which describes the fraction of the total osmotic pressure generated across a capillary membrane, decreased from a normal value of 0.92 to 0.68 with bile and oleic acid to 0.71 with cream feeding [270]. By way of reference, impermeable proteins yield 100% of their maximum osmotic pressure and yield a σ of 1, whereas freely permeable proteins do not generate an effective os-motic pressure and yield a σ of 0.

The pore size in the intestinal capillary wall can be calculated using the osmotic reflection coefficient values obtained for several endogenous

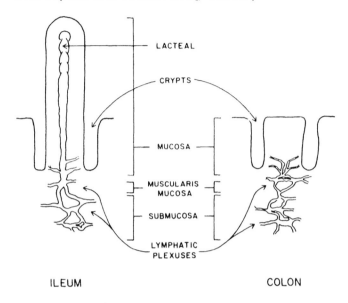

ILEUM   COLON

Fig. 14   Schematic representation of the mucosal-submucosal lymphatic
microcirculation of the canine ileum and colon.   (From Ref. 268.)

proteins of well-defined molecular size (3.7 – 12 nm) [270].   During fat absorption, minimal change was found for small pores (5.1 nm) compared to
the control (4.6 nm), whereas large pores gained in size from 20 to 30 nm.
Because the morphological equivalent of the large pores in intestinal capillaries is considered to reside at the open fenestrae (approximately 20-nm
radius), fat absorption must preferentially affect these structures.   Clearly,
increased vascular permeability during fat absorption may be due to the
dramatic increase in plasma protein flux in perfused capillary surface or to
an increase in capillary permeability [270].   Granger and co-workers [271]
have demonstrated that plasma proteins move from blood to interstitium
across filtering capillaries in the nonabsorptive state (at normal portal
pressure) by both convection and diffusion, while small quantities of fluid
are filtered across the capillary wall into the interstitium.   By contrast, in
the absorptive state (at elevated portal pressure), the movement of fluid is
from the interstitium to blood.   Under these conditions plasma proteins are
moving by diffusion and convection in opposite directions across the capillary wall, and lymphatic protein flux is now the net result of two much
larger and opposing fluxes.   This results in a large entry of plasma proteins into the interstitium in the absorptive state.   Most of the proteins
return to the circulation by convection across the walls of absorbing capillaries, whereas the remainder is returned to lymph.   The existence of a
blood-tissue circulation of plasma proteins might be an important mechanism
for the lymphatic absorption of medium- and long-chain free fatty acids as
well as drugs with a high protein binding capacity that are not incorporated
into chylomicrons but which exist in the circulation in significant amounts
bound to albumin [272].
    Chylomicrons, with diameters ranging from 75 to 600 nm, are barred
from the blood capillaries whose open fenestrae are only 40 – 60 nm in

diameter. Once formed in the enterocyte, they leave the cell by exocytosis and pass through pores in the basement membrane of the epithelial cell.
Chylomicrons traverse approximately 50 μm of interstitium to the lacteal,
where the particles are taken up by cell gaps via an unidentified transport mechanism. Granger et al. [264] studied the effect of net volume absorption rate on interstitial fluid volume, lymph flow, and the excluded
volume fraction for interstitial albumin in an autoperfused cat ileum preparation. Their results indicate that fluid absorption increased the interstitial volume and lymph flow while decreasing the excluded volume fraction
for albumin. This finding suggests that the degree of exclusion of a
given molecule is inversely proportional to matrix hydration (Fig. 15).
This is probably the reason for enhanced transport of chylomicrons when
the matrix was expanded during water absorption and increased lymph
flow.

B. Means to Enhance Lymphatic Absorption

In the late 1960s, Volkenheimer and co-workers introduced a term called
"persorption" to describe a mechanical process by which solid particles are
taken up by the paracellular pathway through the epithelium. They found
that particulate matter such as diatoms, pollens, spores, cellulose particles,
plant cells, starch granules, and others could cross the intestinal mucosa
[273-278]. The best results were obtained with hard particles in the
range 7-70 μm, although persorption would also occur in the size range
5-150 μm. The persorption rate in humans was estimated at 1:50,000
(±50%). This means that 1 of every 50,000 particles will be absorbed, or
approximately 2.5 mg of 100 g cornstarch.

Volkenheimer [277] also found that the persorption rate was dependent
on the amount ingested and was age-related, with younger subjects exhibiting a higher persorption rate. Drugs that increase GI motility, such as
neostigmine and caffeine, were found to increase persorption, whereas the

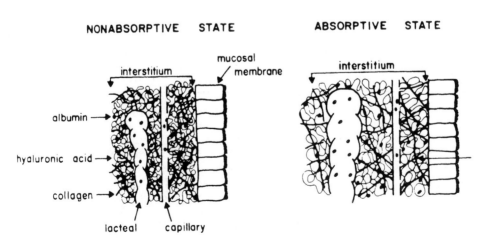

Fig. 15   Intestinal interstitium in nonabsorptive and absorptive states.
(From Ref. 264.)

reverse was true for drugs that reduce motility, such as atropine and bar-
bituric acid [274]. Thus, mechanical factors were thought to be involved
in persorption. Volkenheimer suggested that persorbed particles were
transported mainly by the lymphatic system and that transport was favored
by large particles, although the precise size range was not indicated.

Results similar to those just described were obtained by Sanders and
Ashworth [279]. Specifically, latex particles (200 nm) were taken up into
jejunal absorptive cells of adult rats and shortly thereafter were found in
intact intracellular vesicles. These vesicles were released into the interstitium, where they migrated into both lymphatic lacteals and blood capillaries. Persorption has also been described for *Candida albicans* in human
volunteers and confirmed in dogs and primates [280,281]. Following the
administration of 10^{12} *C. albicans* cells as a suspension to a volunteer, two
Candida colonies were recovered in blood samples collected at 3 and 6 hr,
whereas 31 colonies grew from the sediment of the urine taken at 165 min
and eight colonies from one taken at 195 min. The rest of the blood urine
cultures remained negative up to 14 days. The authors suggested that
C. albicans caused systemic infection in two ways, first by persorption
and then by penetrative growth.

There is some evidence that microparticles may target drugs to the
lymph nodes. This has been investigated by Hagiwara and co-workers
[290] using activated carbon particles, averaging 167 nm in size, to which
were coated pepleomycin, an anticancer agent. Following serosal injection
of such particles into the anterior wall of the greater and lesser curva-
tures in the antrum of the stomach, $6-33$ times higher pepleomycin concentrations were detected in the lymph nodes located along the left gastric artery and the right gastropiploic artery. By comparison, the plasma levels
were very low. Similar results were observed with mitomycin C.

In addition to microparticulate systems, lipid vehicles have also been
considered to promote the lymphatic absorption of orally administered drugs.
Charman and Stella [282] reported that the lipid vehicle in which a drug
was administered influenced the lag time and the rate of lymphatic drug
absorption. The transport of 1,1-bis(p-chlorophenyl)-2,2,2-trichlorethane
(DDT) into lymph was investigated by administration either in peanut oil
(triglyceride carrier) or in oleic acid (fatty acid carrier) [282]. The
quantity of DDT transported lymphatically doubled to about 36% for the
oleic acid carrier compared to the triglyceride carrier. Furthermore, oleic
acid resulted in a 50% reduction in lag time (about 1 hr) compared to that
of peanut oil. The lag times seen in the peanut oil vehicle were attributed
to the extra time required for initial hydrolysis of the triglyceride in peanut oil, in contrast to oleic acid, which was already in the absorbable form.
The differences in lymphatic transport were most likely a function of differences in kinetics of lipid and drug processing, such as digestion or
metabolism within the intestinal lumen, or in the epithelial cells. In a subsequent report, Charman et al. [283] stressed the importance of lipid solubility in influencing lymphatic transport. This was investigated using
hexachlorobenzene (HBC) and DDT as model compounds. These two compounds are similar in octanol/water partition coefficients (log P_{DDT}, 6.19;
log P_{HBC}, 6.53), but are markedly different in solubility in peanut oil
(9.75 g DDT/100 ml lipid, 0.75 g HBC/100 ml lipid). Cumulative lymphatic
transport over 10 hr was 33.5% for DDT and 2.3% for HBC.

An extension of the lipid vehicle approach is to form prodrugs with
α- and β-monoglycerides. So far such an approach has been examined

for low-molecular-weight drugs but has yet to be extended to peptide and protein drugs except for insulin [88]. The drugs that have been studied to date include acetaminophen, naproxen, and nicotinic acid. In each instance, the lymphatic absorption rate was increased. Approximately 28.3% of the orally administered radiolabeled dose was recovered in the lymph with α-[^{3}H]naproxen glycerol octyl ester. Recovery decreased with an increase in the n-alkyl chain length. The opposite was observed with the α-monoglyceride derivatives of [^{3}H]nicotinic acid, where there was an increase in the recovery of radioactivity with the increase in the n-alkyl chain length. Under similar conditions, alkyl ester and triglyceride derivatives were poorly absorbed [284,285].

Unlike lipid vehicles, liposomes were found not to enhance the lymphatic absorption of carboxyfluorescein following intestinal administration [286]. Nevertheless, after intramural injection of liposomal carboxyfluorescein into the intestinal wall, $10-57$ times higher concentration was seen in lymph than in plasma.

By contrast, lipid microspheres, composed of olive and soybean oil, egg lecithin, and glycerol, with an average particle size of 191.3 nm, have been found to promote the absorption of orally administered cyclosporine (CsA) into the thoracic lymph in the rat [287]. Two hours following administration, CsA thoracic lymph concentration afforded by lipid microspheres was approximately 46 times greater than from the conventional preparation (Fig. 16). For the lipid microspheres preparation, maximal lymph levels were achieved within 2 hr (3608.6 ng/ml), while the conventional formulation levels, which were much lower, were still on the rise. As mentioned earlier, Takada et al. [288] administered an enteric solid dispersion of CsA at a dose of 7 mg/kg to rats. The solid dispersion was prepared by stirring CsA and polyoxyethylated, hydrogenated castor oil (HCO-60) in a methanolic solution of enteric coating polymers, such as cellulose acetate phthalate, methacrylic acid and methacrylic acid methylester copolymer (Eudragit L-100), and hydroxypropylmethylcellulose phthalate (HP-55), in a mortar until granules resulted. These granules were dried under vacuum overnight at room temperature, pulverized, screened through a 50-mesh screen, and placed in tubes for insertion into the rat gut via an incision. The highest lymph level (7680 ng/ml) was obtained after administration of a solid dispersion of CsA in hydroxypropyl methylcellulose phthalate (HP-55). The percentage transferred into the lymphatics after 6 hr was approximately 2% of the administered dose.

A bifunctional delivery system for the selective transfer of poorly absorbed drugs into the lymphatics via the enteral route has been described by Yoshikawa and co-workers [289]. Bleomycin (BLM), a glycopeptide, was administered as a BLM−dextran sulfate complex using mixed micelles as an absorption promoter. Approximately half the complex remained intact in the lumen. In the presence of micelles, both complexed and free BLM was able to enter the epithelial barrier, where more BLM dissociated from the complex, reducing the complexed drug to about 15% in the small intestine but to 44% in the large intestine. While free bleomycin entered the blood and lymph in equal proportions, the complexed drug preferentially entered the lymph. Thus, stability of the complex in the tissue is a major factor determining the effectiveness of selective transfer of the delivery system into the lymph.

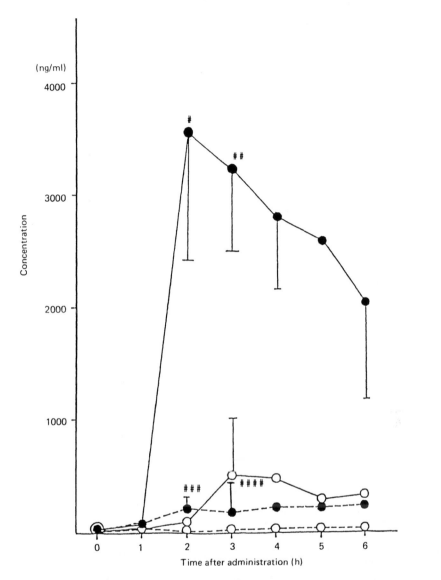

Fig. 16   Time course of cyclosporine concentrations in lymph and blood following oral administration in lipid microspheres (closed symbols, Lipo-CsA) and in conventional preparation (open symbols, CsA) in the rat. —•—, Lipo-CsA in lymph; —○—, CsA in lymph; – –•– –, Lipo-CsA in plasma; – –○– –, CsA in plasma. Levels of significance: lipo-CsA vs. CsA in lymph #, $p < 0.1$; ##, $p < 0.05$; lipo-CsA vs. CsA in plasma ###, $p < 0.02$; ####, $p < 0.05$. (From Ref. 287.)

REFERENCES

1. G. Williams, J. M. Burrin, J. A. Ball, G. F. Joplin, and S. R. Bloom, *Lancet*, October:774–777 (1986).
2. D. Gonzalez-Barcena, M. C. Miller, III, D. H. Coy, A. J. Kastin, D. S. Schalc, and A. V. Schally, *Lancet*, December:1126–1127 (1975).
3. H. Okada, I. Yamazaki, Y. Ogawa, S. Hirai, T. Yashiki, and H. Mima, *J. Pharm. Sci.*, 71:1367–1371 (1982).
4. J. Pless, W. Bauer, U. Briner, W. Doepfner, P. Marchbach, R. Maurer, T. J. Petcher, J.-C. Reubi, and J. Vonderescher, *Scand. J. Gastroenterol.*, 21(suppl. 119):54–64 (1986).
5. L. J. Saidel and I. Edelstein, *Biochim. Biophys. Acta*, 367:75–80 (1974).
6. Y. Nakada, N. Awata, C. Nakamichi, and I. Sugimoto, *J. Pharmacobio-Dyn.*, 11:395–401 (1988).
7. B. R. Weström, J. Svendsen, B. G. Ohlsson, C. Tagesson, and B. W. Karlsson, *Biol. Neonate*, 46:20–26 (1984).
8. J. N. Udall, P. Colony, L. Fritze, K. Pang, J. S. Trier, and W. A. Walker, *Pediatr. Res.*, 15:245–249 (1981).
9. E. Telemo, B. R. Weström, and B. W. Karlsson, *Biol. Neonate*, 52:141–148 (1987).
10. J. N. Udall, K. Pang, L. Fritze, R. Kleinman, and W. A. Walker, *Pediatr. Res.*, 15:241–244 (1981).
11. B. Weström, J. Svendsen, and C. Tagesson, *Gut*, 25:520–525 (1984).
12. B. Taylor, A. P. Norman, M. A. Orgel, C. R. Stoken, M. W. Thomas, and J. F. Soothill, *Lancet*, 2:111–113 (1973).
13. K. Y. Pang, J. L. Bresson, and W. A. Walker, *Pediatr. Res.*, 17:856–861 (1983).
14. K. S. Pang, W. F. Cherry, and E. H. Ulm, *J. Pharmacol. Exp. Ther.*, 233:788–795 (1985).
15. B. R. Weström, B. G. Ohlsson, J. Svendsen, C. Tagesson, and B. W. Karlsson, *Biol. Neonate*, 47:359–366 (1985).
16. J. N. Udall, K. J. Bloch, G. Vachino, P. Feldman, and W. A. Walker, *Biol. Neonate*, 45:289–295 (1984).
17. R. M. Clarke and R. N. Hardy, *J. Anat.*, 108:63–77 (1971).
18. J. G. Lecce, *J. Nutr.*, 103:751–756 (1973).
19. D. M. Goldberg, R. Campbell, and A. D. Roy, *Biochim. Biophys. Acta*, 167:613–615 (1968).
20. M. L. Chen, Q. R. Rogers, and A. E. Harper, *J. Nutr.*, 76:235–241 (1962).
21. S. A. Adibi and D. W. Mercer, *J. Clin. Invest.*, 52:1586–1594 (1973).
22. D. S. A. Silk, J. A. Nicholson, and Y. S. Kim, *Gut*, 17:870–876 (1976).
23. S. A. Adibi and E. L. Morse, *J. Clin. Invest*, 60:1008–1016 (1977).
24. S. A. Adibi, *J. Clin. Invest.*, 50:2266–2275 (1971).
25. M. A. Naughton and F. Sanger, *Biochem. J.*, 78:156–163 (1961).
26. J. R. Whitaker, F. Menger, and M. L. Bender, *Biochemistry*, 5:386–392 (1966).
27. H. Neurath and G. W. Schwert, *Chem. Rev.*, 46:69–153 (1950).
28. S. E. Nixon and G. E. Mawer, *Br. J. Nutr.*, 24:227–240 (1970).
29. S. E. Nixon and G. E. Mawer, *Br. J. Nutr.*, 24:241–258 (1970).

30. R. F. Crampton, M. T. Lis, and D. M. Mathews, *Clin. Sci.*, 44:583–594 (1973).
31. M. G. Gardner, *Clin. Sci.*, 57:217–220 (1979).
32. D. Louvard, M. Semeriva, and S. Maroux, *J. Mol. Biol.*, 106:1023–1035 (1976).
33. E. E. Sterchi and J. F. Woodley, *Clin. Chem. Acta*, 102:49–56 (1980).
34. A. J. Kenny and S. Maroux, *Physiol. Rev.*, 62:91–128 (1982).
35. P. E. Ward and M. A. Sheridan, *Biochem. Pharmacol.*, 32:265–274 (1983).
36. N. Tobey, W. Haizers, R. Yeh, T.-I. Huang, and C. Hoffner, *Gastroenterology*, 88:913–926 (1985).
37. H. Feracci, S. Maroux, J. Bonicel, and P. Desnuelle, *Biochim. Biophys. Acta*, 684:133–136 (1982).
38. M. M. Hussain, *Biochim. Biophys. Acta*, 815:306–312 (1985).
39. K. Takaori, J. Burton, and M. Donowitz, *Biochem. Biophys. Res. Commun.*, 137:682–687 (1986).
40. D. M. Goldberg, R. Campbell, and A. D. Roy, *Biochim. Biophys. Acta*, 167:613–615 (1968).
41. D. M. Goldberg, R. Campbell, and A. D. Roy, *Comp. Biochem. Physiol.*, 38:697–706 (1971).
42. D. M. Matthews and J. W. Payne, *Curr. Top. Memb. Transport*, 14:331–425 (1980).
43. R. J. Kania, N. A. Santiago, and G. M. Gray, *Gastroenterology*, 62:768A (1972).
44. Y. S. Kim, W. Birtwhistle, and Y. W. Kim, *Clin. Invest.*, 51:1419–1430 (1972).
45. T. J. Peters, *Biochem. J.*, 120:195–203 (1970).
46. M. Fujita, D. S. Parsons, and F. Wojnarowska, *J. Physiol.*, 227:377–394 (1972).
47. Y. S. Kim, W. Birtwhistle, and Y. W. Kim, *J. Clin. Invest.*, 51:1419–1430 (1972).
48. M. Maze and G. M. Gray, *Biochemistry*, 19:2351–2358 (1980).
49. W. D. Heizer, R. L. Kerley, and K. J. Isselbacher, *Biochim. Biophys. Acta*, 264:450–461 (1972).
50. J. B. Rhodes, A. Eichholz, and R. K. Crane, *Biochim. Biophys. Acta*, 135:959–965 (1967).
51. W. Thornburg, L. Matrisian, B. Magun, and O. Koldovsky, *Am. J. Physiol.*, 246:G80–G85 (1984).
52. P. A. Gonnella, K. Siminosky, R. A. Murphy, and M. R. Neutra, *J. Clin. Invest.*, 80:22–32 (1987).
53. D. J. G. White, *Drugs*, 24:322–334 (1982).
54. R. M. Ferguson and R. Fidelus-Gort, *Transplant Proc.*, 15:2350–2356 (1983).
55. R. J. Ptachcinski, R. Venkataramanan, and G. J. Burckart, *Clin. Pharmacokinet.*, 11:107–132 (1986).
56. W. T. Cefalu and W. M. Pardridge, *J. Neurochem.*, 45:1954–1956 (1985).
57. J.-P. Reymond, J.-L. Steimer, and W. Niederberger, *J. Pharmacokinet. Biopharm.*, 16:331–353 (1988).
58. J. Grevel, *Transplant Proc.*, 18:9–15 (1986).
59. A. Johnston, J. T. Marsden, K. K. HLA, J. A. Henry, and D. W. Holt, *Br. J. Clin. Pharmacol.*, 21:331–333 (1986).

60. A. Keogh, R. Day, L. Critchley, G. Duggin, and D. Baron, *Transplant Proc.*, *20*:27−30 (1988).

61. L. W. Elzinga, W. M. Bennet, and J. M. Barry, *Transplantation*, *47*:394−395 (1988).

62. O. Wagner, E. Schreier, F. Heitz, and G. Maurer, *Drug. Metabol. Dispos.*, *15*:377−383 (1987).

63. G. Maurer, H. R. Loosli, E. Schreier, and B. Keller, *Drug Metabol. Dispos.*, *12*:120−126 (1984).

64. G. Fabre, P. Bertault-Peres, I. Fabre, P. Maurel, S. Just, and J.-P. Cano, *Drug. Metabol. Dispos.*, *15*:384−390 (1987).

65. B. D. Tarr and S. H. Yalkowsky, *Pharm. Res.*, *6*:40−43 (1989).

66. I. McColl and G. E. Sladen, eds., in: *Intestinal Absorption in Man*, Academic Press, New York (1975), pp. 187−221.

67. R. J. Ptachcinski, R. Venkataramanan, J. T. Rosenthal, G. J. Burckart, R. J. Taylor, and T. R. Hakala, *Transplantation*, *40*:174− 176 (1985).

68. W. Vine and L. Bowers, *Crit. Rev. Clin. Lab. Sci.*, *25*:275−311 (1987).

69. G. Burckart, T. Starzl, L. Williams, A. Sanghvi, C. Gartner, R. Venkataramanan, B. Zitelli, J. Malatack, A. Urbach, W. Diven, R. Ptacheinski B. Shaw, and Irvatsuki, *Transplant Proc.*, *17*:1172−1175 (1985).

70. B. D. Kahan, M. Reid, and J. Newberger, *Transplant Proc.*, *15*:446− 453 (1983).

71. B. D. Kahan, W. G. Kramer, C. Wideman, S. M. Flechner, M. I. Lorber, and C. T. Van Buren, *Transplantation*, *41*:459−463 (1986).

72. R. J. Sawchuk and W. M. Awni, *J. Pharm. Sci.*, *75*:1151−1156 (1986).

73. R. Venkataramanan, G. J. Burckart, and R. J. Ptachcinski, *Semin. Liver Dis.*, *5*:357−368 (1985).

74. R. Venkataramanan, R. J. Ptachcinski, C. J. Burckart, J. Gray, and D. H. Van Thiel, *Drug Intell. Clin. Pharmacol.*, *19*:451−454 (1985b).

75. C. T. Ueda, M. Lemaire, G. Gsell, and K. Nussbaumer, *Biopharm. Drug Dispos.*, *4*:113−124 (1983).

76. C. T. Ueda, M. Lemaire, G. Gsell, P. Misslin, and K. Nussbaumer, *Biopharm. Drug Dispos.*, *5*:141−151 (1984).

77. H. W. Davenport, *Physiology of the Digestive Tract*, 3rd ed., Year Book Medical Publishers, Chicago (1971), pp. 163−170.

78. N. K. Wadhwa, T. J. Schroeder, E. O' Flaherty, A. J. Pesce, S. A. Myre, and M. R. First, *Transplant. Proc.*, *19*:1730−1733 (1987).

79. R. A. Harrington, C. W. Hamilton, and R. N. Brogden, *Drugs*, *25*:452−454 (1983).

80. J. Grevel, E. Nuesch, E. Abisch, and K. Kutz, *Eur. J. Clin. Pharmacol.*, *31*:211−216 (1986).

81. K. Takada, Y. Furuya, H. Yoshikawa, and S. Muranishi, *J. Pharmacobio. Dyn.*, *11*:80−87 (1988).

82. W. A. Ritschel, G. B. Ritschel, A. Sabouni, D. Wolochuk, and T. Schroeder, *Meth. Find. Exp. Clin. Pharmacol.*, *11*:281−287 (1989).

83. K. Takada, H. Yoshimura, N. Shibata, Y. Masuda, H. Yoshikawa, S. Muranishi, T. Yasaumura, and T. Oka, *J. Pharmacobio. Dym.*, *9*:156−160 (1986).

84. S. Barbaric, I. Leustek, B. Pavlovic, V. Cesi, and P. Mildner, *Ann. NY Acad. Sci.*, *542*:173−179 (1985).

85. V. V. Mozhaev, V. A. Siksnis, N. S. Melik-Nubarov, N. Z. Galkantaite, G. J. Denis, E. P. Butkus, B. Y. Zaslavsky, N. M. Mestechkina, and K. Martinek, *Eur. J. Biochem.*, *173*:147−154 (1988).

86. G. R. Larsen, M. Matzger, K. Henson, Y. Blue, and P. Horgan, *Blood*, *73*:1842−1850 (1989).

87. A. Abuchowski, T. van Es, N. C. Palczuk, and F. F. Davis, *J. Biol. Chem.*, *252*:3578−3581 (1977).

88. C. Hiltbrunner, PhD thesis, ETH, Diss. ETH Nr. 9029 (1989).

89. S. Dodda-Kashi and V. H. L. Lee, *Life Sci.*, *38*:2019−2028 (1986).

90. D. Roemer, H. H. Huescher, R. C. Hill, J. Pless, W. Bauer, F. Cardinaux, A. Closse, D. Hauser, and R. Huguenin, *Nature*, *268*:547−549 (1977).

91. K. S. E. Su, K. M. Campanale, L. G. Mendelsohn, G. A. Kerchner, and C. L. Gries, *J. Pharm. Sci.*, *74*:394−398 (1985).

92. H. M. Berschneider, H. Martens, and D. W. Powell, *Gastroenterology*, *94*:127−136 (1988).

93. J. Ryan, J. Leighton, D. Kirksey, and G. McMahon, *Clin. Pharmacol. Ther.*, *39*:40−42 (1986).

94. D. F. Mierke, O. E. Said-Nejad, P. W. Schiller, and M. Goodman, *Biopolymers*, *29*:179−196 (1990).

95. J. Posner, K. Dean, S. Jeal, S. G. Moody, A. W. Peck, G. Rutter, and A. Talekes, *Eur. J. Clin. Pharmacol.*, *34*:67−71 (1988).

96. T. Mitsuma and T. Nogimori, *Experientia*, *39*:620−622 (1983).

97. S. Yokohama, T. Yoshioka, K. Tamashita, and N. Kitamori, *J. Pharmacobio. Dyn.*, *7*:445−451 (1984).

98. S. Yokohama, K. Yamashita, H. Toguchi, J. Takeuchi, and N. Kitamori, *J. Pharmacobio. Dyn.*, *7*:101−111 (1984).

99. M. Hichens, *Drug. Metab. Rev.*, *14*:77−98 (1983).

100. Y. Tache, Y. Goto, M. Lauffenburger, and D. Lesiege, *Regul. Peptides*, *8*:71−78 (1984).

101. S. Yokohama, T. Yoshioka, and N. Kitamori, *J. Pharmacobio. Dyn.*, *7*:527−535 (1984).

102. S. Yokohama, T. Yoshioka, N. Kitamori, T. Shimamoto, and A. Kamada, *J. Pharmacobio. Dyn.*, *8*:278−285 (1985).

103. J. P. Rado, J. Marosi, L. Szende, L. Borbely, J. Tako, and J. Fischer, *Int. J. Clin. Pharmacol.*, *13*:199−209 (1976).

104. I. Vavra, A. Machova, and I. Krejci, *J. Pharmacol. Exp. Ther.*, *188*:241−247 (1974).

105. W. H. Sawyer, M. Acosta, and M. Manning, *Endocrinology*, *95*:140−149 (1974).

106. M. Zaoral, J. Kolc, and F. Storm, *Collect. Czech. Chem. Commun.*, *32*:1250−1257 (1967).

107. H. Vilhardt and S. Lundin, *Acta Physiol. Scand.*, *126*:601−607 (1986).

108. M. Saffran, R. Franco-Saenz, A. Kong, D. Papahadjopoulos, and F. Szoka, *Can. J. Biochem.*, *57*:548−553 (1979).

109. S. Lundin and H. Vilhardt, *Acta Endocrinol.*, *112*:457−460 (1986).

110. H. Vilhardt and P. Bie, *Acta Endocrinol.*, *105*:474−476 (1984).

111. D. M. Williams, D. B. Dunger, C. C. Lyon, R. J. Lewis, F. Taylor, and L. Lightman, *J. Clin. Endocrinol. Metab.*, *63*:129−132 (1986).

112. D. Cunnah, G. Ross, and G. M. Besser, *Clin. Endocrinol.*, *24*:253–257 (1986).

113. A. Fjellestad-Paulsen, N. Tubiana-Rufi, A. Harris, and P. Czernichow, *Acta Endocrinol.*, *115*:307–312 (1987).

114. G. M. Grass and W. M. Morehead, *Pharm. Res.*, *6*:759–765 (1989).

115. D. I. Friedman and G. L. Amidon, *Pharm. Res.*, *6*:1043–1047 (1989).

116. G. J. Leitch, *Arch. Int. Physiol. Biochim.*, *79*:279–286 (1971).

117. T. Lindberg, *Acta Physiol. Scand.*, *66*:437–443 (1966).

118. A. Heringova, O. Koldovsky, V. Jirosova, J. Uher, R. Noack, M. Freidrich, and G. Schenk, *Gastroenterology*, *51*:1023–1027 (1966).

119. S. Auricchio, L. Greco, B. DeVizia, and V. Buonocore, *Gastroenterology*, *75*:1073–1079 (1978).

120. E. E. Sterchi, *Pediatr. Res.*, *15*:884–885 (1981).

121. N. Triadou, J. Bataille, and J. Schmitz, *Gastroenterology*, *85*:1326–1332 (1983).

122. S. Miura, I. S. Song, A. Morita, R. H. Erickson, and Y. S. Kim, *Biochim. Biophys. Acta*, *761*:66–75 (1983).

123. G. F. Vaeth and S. J. Henning, *J. Pediatr. Gastroenterol. Nutr.*, *1*:111–117 (1982).

124. A. M. Asatoor, A. Chadha, M. D. Milne, and D. I. Prosser, *Clin. Sci. Mol. Med.*, *45*:199–212 (1973).

125. A. H. Staib, D. Loew, S. Harder, E. H. Graul, and R. Pfab, *Eur. J. Clin. Pharmacol.*, *30*:691–697 (1986).

126. A. Wildfeuer, H. Laufen, R. Dölling, G. Pfaff, B. Hugemann, H. E. Knoell, and O. Schuster, *Therapiewoche*, *36*:2996–3003 (1986).

127. L. Knutson, B. Odlind, and R. Hällgren, *Am. J. Gastroenterol.*, *84*:1278 (1989).

128. R. A. Frizell and S. A. Schultz, *J. Gen. Physiol.*, *59*:318–346 (1972).

129. G. M. Grass and S. A. Sweetana, *Pharm. Res.*, *5*:372–376 (1988).

130. N. F. Ho, J. Y. Park, P. F. Ni, and W. I. Higuchi, in: *Animal Models for Oral Drug Delivery in Man* (W. Crouthamel and A. C. Sarapu, eds.), Academy of Pharmaceutical Sciences, Washington, DC (1983), pp. 27–107.

131. A. B. R. Thomson and W. M. Weinstein, *Dig. Dis. Sci.*, *24*:442–448 (1979).

132. N. C. Hightower and H. D. Janowitz, in: *Physiological Basis of Medical Practice* (J. R. Brobeck ed.), Williams & Wilkins, Baltimore (1973), pp. 2–81.

133. G. M. Grass, S. A. Sweetana, and C. A. Bozarth, *J. Pharm. Pharmacol.*, *42*:40–44 (1990).

134. J. M. Franz, J. P. Vonderescher, and R. Voges, *Int. J. Pharm.*, *7*:19–28 (1980).

135. W. S. Snyder, M. J. Cook, E. S. Nassat, L. R. Karhausen, G. Parry Howells, and I. H. Tipton, International Commission on Radiological Partition 23, *Report of the Taskgroup on Reference Man*, Pergamon Press, New York, (1974), p. 139.

136. C. A. Youngberg, J. Wlodyga, S. Schmaltz, and J. B. Dressman, *Am. J. Vet. Res.*, *46*:1516–1521 (1985).

137. C. Y. Lui, G. L. Amidon, R. R. Berardi, D. Fleisher, C. Youngberg, and J. B. Dressman, *J. Pharm. Sci.*, *75*:271–274 (1986).

138. V. S. Chadwick, S. F. Phillips, and A. F. Hoffman, *Gastroenterology*, 73:243–246 (1977).

139. V. S. Chadwick, S. F. Phillips, and A. F. Hoffman, *Gastroenterology*, 73:247–251 (1977).

140. B. T. Cooper, *J. Clin. Gastroenterol.*, 6:499–501 (1984).

141. I. S. Menzies, in: *Intestinal Absorption and Secretion* (E. Skadhauge and K. Heintze, eds.), MTP Press, Higham (1984), pp. 527–544.

142. S. O. Ukbam and B. T. Cooper, *Dig. Dis. Sci.*, 29:809–816 (1984).

143. T. J. Peters and I. Bjarnason, *Can. J. Gastroenterol.*, 2:127–132 (1988).

144. E. F. Phillipsen, W. Batsberg, and A. B. Christensen, *Eur. J. Clin. Invest.*, 18:139–145 (1988).

145. T. Y. Ma, D. Hollander, P. Krugliak, and K. Katz, *Gastroeneterology*, 98:39–46 (1990).

146. N. F. H. Ho, *Ann. NY Acad. Sci.*, 507:315–329 (1987).

147. R. Gräsbeck, *Prog. Hematol.*, 6:233–260 (1969).

148. R. Carmel, A. H. Rosenberg, K. S. Lau, R. R. Streiff, and V. Herbert, *Gastroenterology*, 56:548–554 (1969).

149. A. Doscherholem, J. MacMahon, and D. Ripley, *J. Lab. Clin. Med.*, 78:326–339 (1971).

150. J. S. Levine, P. K. Nakane, and R. Allen, *Gastroenterology*, 79:493–502 (1980).

151. K. Simons, *Commentat. Biol. Soc. Sci. Fenn.*, 27:1 (1964).

152. R. H. Allen, B. Seetharam, E. Pdell, and D. H. Alpers, *J. Clin. Invest.*, 61:47–54 (1978).

153. S. G. Schade and R. F. Schilling, *Am. J. Clin. Nutr.*, 20:636–640 (1967).

154. B. A. Cooper and W. B. Castle, *J. Clin. Invest.*, 39:199–214 (1960).

155. I. L. MacKenzie and R. M. Donaldson, Jr., *J. Clin. Invest.*, 51:2465–2471 (1972).

156. K. Sriram, G. A. Gergans, and H. Badger, *J. Am. Coll. Nutr.*, 8:75–81 (1989).

157. R. M. Donaldson, I. L. MacKenzie, and J. S. Trlier, *J. Clin. Invest.*, 46:1215–1228 (1967).

158. V. I. Mathan, B. M. Babior, and R. M. Donaldson, *J. Clin. Invest.*, 54:598–608 (1974).

159. J. S. Levine, P. K. Nakane, and R. Allen, *Gastroenterology*, 82:284–290 (1982).

160. D. C. Hooper, D. H. Alpers, R. L. Burger, C. S. Mehlman, and R. H. Allen, *J. Clin. Invest.*, 52:3074–3083 (1973).

161. C. H. Hagedorn and D. H. Alpers, *Gastroenterology*, 73:1019–1022 (1977).

162. C. H. Hagedorn and D. H. Alpers, *Gastroenterology*, 73:1010–1022 (1977).

163. C. Kapadia and L. K. Esseandoh, *Dig. Dis. Sci.*, 33:1377–1382 (1988).

164. K. Okuda, *Am. J. Physiol.*, 199:84–90 (1960).

165. I. Kouvonen and R. Gräsbeck, *J. Biol. Chem.*, 256:154–158 (1981).

166. B. Seetharam, S. S. Bagur, and D. H. Alpers, *J. Biol. Chem.*, 256:9813–9815 (1981).

167. R. M. Donaldson, D. M. Small, S. Robins, and V. I. Mathan, *Biochim. Biophys. Acta*, 311:477–481 (1973).

168. A. Hellegers, K. Okuda, R. E. L. Nesbitt, D. W. Smith, and B. F. Chow, *Am. J. Clin. Invest.*, 5:327–331 (1957).

169. J. Brown, J. Robertson, and N. Callagher, *Gastroenterology*, 72:881–885 (1977).

170. J. A Robertson and N. D. Gallagher, *Gastroenterology*, 77:511–517 (1979).

171. T. H. Wilson and E. W. Strauss, *Am. J. Physiol.*, 197:926–928 (1959).

172. J. F. Kolhouse and R. H. Allen, *J. Clin. Invest.*, 60:1274–1392 (1977).

173. L. J. Brandt, L. Goldberg, L. H. Bernstein, and G. Greenberg, *Am. J. Clin. Nutr.*, 32:1832–1836 (1979).

174. S. Kanazawa, H. Terada, T. Iseki, S. Iwasa, K. Okuda, H. Kondo, and K. Okuda, *Proc. Soc. Exp. Biol. Med.*, 183:333–338 (1986).

175. S. Shaw, E. Jayatilleke, S. Meyers, N. Colman, B. Herzlich, and V. Herbert, *Am. J. Gastroenterol.*, 84:22–26 (1989).

176. G. J. Russel-Jones and H. J. deAizpurua, *Proc. Intern. Control. Rel. Bioact. Mater.*, 15:142–143 (1988).

177. J. Meier, H. Rettig, and H. Hess, eds., *Biopharmazie*, Georg Thieme Verlag, Stuttgart, New York, (1981), p. 60.

178. N. W. Read and K. Sugden, *CRC Crit. Rev. Ther. Drug Carrier Syst.*, 4:221–263 (1987).

179. J. O. Magnusson, B. Bengalil, C. Goentoft, and U. E. Johnsson, *Br. J. Clin. Pharmacol.*, 14:284–285 (1982).

180. M. A. Peppcorn and P. Goldman, *Rev. Drug Interact.*, 2:75–88 (1976).

181. T. M. Parkinson, J. P. Brown, and R. E. Wingard, U.S. Patent 4,190,716 (1980).

182. C. P. Willoughby, J. K. Aranson, H. Agback, N. O. Bodin, and S. C. Truelove, *Gut*, 23:1081–1087 (1982).

183. D. R. Friend and G. W. Chang, *J. Med. Chem.*, 27:261–266 (1984).

184. D. R. Friend and G. W. Chang, *J. Med. Chem.*, 28:51–57 (1985).

185. E. Harboe, C. Larsen, M. Johansen, and H. P. Olesen, *Pharm. Res.*, 6:919–923 (1989).

186. E. Harboe, C. Larsen, M. Johansen, and H. P. Olesen, *Int. J. Pharm.*, 53:157–165 (1989).

187. M. Saffran, G. S. Kumar, C. Savariar, J. C. Burnham, F. Williams, and D. C. Neckers, *Science*, 123:1081–1084 (1986).

188. M. Saffran, C. Bedra, G. S. Kumar, and D. C. Neckers, *J. Pharm. Sci.*, 77:33–38 (1988).

189. M. J. Dew, P. J. Hughes, M. G. Lee, B. K. Evans, and J. Rhodes, *Br. J. Clin. Pharmacol.*, 14:405–408 (1982).

190. C. L. Ten Cate, *Janus*, 56:22–45 (1969).

191. P. Sicinski, J. Rowinski, J. B. Warcholz, Z. Jarzabek, W. Gut, B. Szczygiel, K. Gielechi, and G. Koch, *Gastroenterology*, 98:56–58 (1990).

192. W. Sackmann, *Acta Anat.*, 97:109–113 (1977).

193. H. HogenEsch, J. M. Housman, and P. J. Felsburg, *Adv. Exp. Med. Biol.*, 216A:249–256 (1987).

194. O. Carlens, *Z. Anat. Entw.*, 86:393–493 (1928).

195. J. Reynolds, R. Pabst, and G. Bordmann, in: *Microenvironments in the Lymphoid System* (G. G. B. Klaus, ed.), Plenum Press, New York (1985), pp. 101–109.

196. M. Schmitz, D. Nunez, and E. Butcher, *Gastroenterology*, 94:576–581 (1988).

197. E. C. Butcher, R. V. Rouse, R. L. Coffman, C N. Nottenburg, R. R. Hardy, and I. L. Wiessman, *J. Immunol.*, *129*:2698–2707 (1982).

198. D. Guy-Grand, C. Griscelli, and P. Vassalli, *Eur. J. Immunol.*, *4*:435–443 (1974).

199. M. McWilliams, M. E. Lamm, and J. A. Phillips-Quagliata, *J. Immunol.*, *113*:1326–1333 (1974).

200. G. E. Roleants, F. Loor, H. von Goehmer, J. Sprent, L. B. Hagg, K. S. Mayor, and A. Ryden, *Eur. J. Immunol.*, *5*:127–131 (1975).

201. T. T. MacDonald, J. Spencer, J. L. Viney, C. B. Williams, and J. A. Walker-Smith, *Gastroenterology*, *93*:1356–1362 (1987).

202. J. J. Cebra, C. A. Crandall, P. J. Gearhart, S. M. Robertson, J. Tseng, and P. M. Watson, in: *Immunology of Breast Milk* (P. Ogra and D. Dayton, eds.), Raven Press, New York (1979), p. 1.

203. P. J. Gearhart and J. J. Cebra, *J. Exp. Med.*, *149*:216–227 (1979).

204. H. Kiyono, J. R. McGhee, L. M. Mosteller, J. M. Eldridge, W. J. Koopman, J. F. Kearney, and S. M. Michalek, *J. Exp. Med.*, *156*:1115–1130 (1982).

205. H. Kawanishi, K. Ozato, and W. Strober, *J. Immunol.*, *134*:3586–3591 (1985).

206. D. Guy-Grand, C. Griscelli, and P. Vassalli, *J. Exp. Med.*, *148*:1661–1677 (1978).

207. R. Pabst and J. D. Reynolds, *J. Immunol.*, *139*:3981–3985 (1987).

208. R. M. Binns, R. Pabst, and S. T. Licence, *Immunology*, *54*:105–111 (1985).

209. Y. R. Mahida, S. Patel, and D. P. Jewell, *Clin. Exp. Immunol.*, *75*:82–86 (1989).

210. H. J. A. Egberts, M. G. M. Brinkhoff, J. M V. M. Mouwen, J. E. van Dijk, and J. F. G. Koninkx, *Vet. Q.*, *7*:333–336 (1985).

211. J. L. Wolf and W. A. Bye, *Annu. Rev. Med.*, *35*:95–112 (1984).

212. R. L. Owen and A. L. Jones, *Gastroenterology*, *66*:189–203 (1974).

213. R. L. Owen and P. Nemanic, *Scanning Electron Microscopy*, *11*:367–378 (1978).

214. L. R. Inman and J. R. Cantey, *J. Clin. Invest.*, *71*:1–8 (1983).

215. J. L. Madara, W. A. Bye, and J. S. Trier, *Gastroenterology*, *87*:1091–1103 (1984).

216. M. W. Smith and M. A. Peacock, *Am. J. Anat.*, *159*:167–175 (1980).

217. J. Pappo, *Cell. Immun.*, *120*:31–41 (1989).

218. R. L. Owen and D. K. Bhalla, *Am. J. Anat.*, *168*:199–212 (1983).

219. H. N. Munro and M. C. Linder, *Physiol. Rev.*, *58*:317–333 (1978).

220. W. A. Bye, C. H. Allan, and J. S. Trier, *Gastroenterology*, *86*:789–801 (1984).

221. M. R. Neutra, T. L. Phillips, E. L. Mayer, and D. J. Fishkind, *Cell Tissue Res.*, *247*:537–546 (1987).

222. R. Weltzin, P. Lucia-Jandris, P. Michetti, B. N. Fields, J. P. Kraehenbuhl, and M. R. Neutra, *J. Cell. Biol.*, *108*:1673–1685 (1989).

223. H. Heyman, R. Ducroc, J.-F. Desjeux, and J. L. Morgat, *Am. J. Physiol.*, *242*:G558–G564 (1982).

224. D. J. Keljo and J. R. Hamilton, *Am. J. Physiol.*, *244*:G637–G644 (1983).

225. R. L. Owen, R. T. Apple, and D. K. Bhalla, *Anat. Rec.*, *216*:521–527 (1986).

226. S. G. McClugage and F. N. Low, *Am. J. Anat.*, *171*:207–216 (1984).

227. S. G. McClugage, F. N. Low, and M. L. Zimny, *Gastroenterology*, *91*:1128–1133 (1986).

228. R. Ducroc, M. Heyman, B. Beaufrere, J. L. Morgat, and J. F. Desjeux, *Am. J. Physiol.*, *245*:G54–G58 (1983).

229. S. J. Beahon and J. F. Woodley, *Biochem. Soc. Trans.*, *12*:1087 (1984).

230. M. Heyman, A. Bonfils, M. Fortier, A. M. Crain-Denoyelle, P. Smets, and J. F. Desjeux, *Int. J. Pharm.*, *37*:33–39 (1987).

231. W. Rubas, N. Jezyk, R. Kos, and G. M. Grass, *Proc. Intern. Symp. Control. Rel. Res. Bioact. Mater.*, *17*:309–310 (1990).

232. M. E. LeFevre, D. C. Hancock, and D. D. Joel, *J. Toxicol. Environ. Health*, *6*:691 (1980).

233. M. E. LeFevre, A. M. Bocci, and D. D. Joel, *Proc. Soc. Exp. Biol. Med.*, *190*:23–27 (1989).

234. J. Pappo and T. H. Ermak, *Clin. Exp. Immunol.*, *76*:144–148 (1989).

235. R. M. Gilley, J. H. Eldrige, J. L. Opitz, L. K. Hanna, J. K. Staas, and T. R. Tice, *Proc. Intern. Symp. Control. Rel. Bioact. Mat.*, *15*:123–124 (1988).

236. M. E. LeFevre and D. D. Joel, in: *Intestinal Toxicology* (C. M. Schiller, ed.), Raven Press, New York (1984), pp. 45–56.

237. M. E. LeFevre, J. B. Warren, and D. D. Joel, *Exp. Cell Biol.*, *53*:121–129 (1985).

238. M. E. LeFevre, D. D. Joel, and G. Schidlovsky, *Proc. Soc. Exp. Biol. Med.*, *179*:522–528 (1985).

239. D. D. Joel, J. L. Laissue, and M. E. LeFevre, *J. Reticuloendothel. Soc.*, *24*:477–487 (1978).

240. J. S. Cornes, *Gut*, *6*:225–233 (1965).

241. A. Guyon-Gruaz, A. Delmas, S. Pedoussaut, H. Halami, G. Milhaud, D. Raulais, and P. Rivaille, *Eur. J. Biochem.*, *159*:525–528 (1986).

242. J. Wassef, D. F. Keren, and J. L. Mailloux, *Infect. Immun.*, *57*:858–863 (1989).

243. S. M. Michalek, N. K. Childers, J. Katz, and R. Curtiss, *Clin. Immun. Lett.*, *8*:158–159 (1987).

244. M. F. Kagnoff, in: *Physiology of the Gastrointestinal Tract*, 2nd ed. (L. R. Johnson, ed.), Raven Press, New York (1987), p. 1699.

245. N. K. Childers, F. Denys, and S. M. Michalek, *Annu. Meet. Am. Soc. Microbiol.*, *88*:124 (1988).

246. W. Rubas, A. C. Banerjea, H. Gallati, P. P. Speiser, and W. K. Joklik, *J. Microencaps.*, *7*:385–395 (1990).

247. J. L. Wolf, D. H. Rubin, R. Finberg, R. S. Kauffman, A. H. Sharpe, J. S. Trier, and B. N. Fields, *Science*, *212*:471–472 (1981).

248. J. L. Wolf, R. S. Kauffman, R. Finberg, R. Dambrauska, B. N. Fields, and J. S. Trier, *Gastroenterology*, *85*:291–300 (1983).

249. R. Bassel-Duby, A. Jayasuriya, D. Chatterjee, N. Sonenberg, J. V. Maizel, and B. N. Fields, *Nature*, *315*:421–423 (1985).

250. D. B. Furlong, M. L. Nibert, and B. N. Fields, *J. Virol.*, *62*:246–256 (1988).

251. P. W. Lee, E. C. Hayes, and W. K. Joklik, *Virology*, *108*:156–163 (1981).

252. L. Nagata, S. A. Masri, C. W. Mah, and P. W. Lee, *Nucleic Acid Res.*, *12*:8699−8710 (1984).

253. K. Katayama and T. Fujita, *Biochem. Biophys. Acta*, *288*:181−189 (1972).

254. T. Tsujii, M. Akita, K. Katayama, S. Yamamoto, and S. Seno, *Histochemistry*, *81*:427−433 (1984).

255. A. Supersaxo, W. R. Hein, and H. Steffen, *Pharm. Res.*, *7*:167−169 (1990).

256. V. Bocci, G. P. Pessina, L. Paulesu, and C. Nicoletti, *J. Biol. Resp. Mod.*, *7*:390−400 (1988).

257. J. A. Borrowan, in: *Physiology of the Gastrointestinal Lymphatic System*, Cambridge University Press, Cambridge (1978).

258. D. N. Granger, *Am. J. Physiol.*, *240*:G343−G349 (1981).

259. D. N. Granger and P. R. Kvietys, in: *Blood Vessels and Lymphatics in Organ Systems* (D. I. Abramson and P. B. Dobrin, eds.), Academic Press, Orlando, FL (1984), pp. 450−455.

260. F. Clementi and G. E. Palade, *J. Cell. Biol.*, *41*:33−58 (1969).

261. L. V. Leak, *Fed. Proc.*, *35*:1863−1871 (1976).

262. D. N. Granger, T. Miller, R. Allen, R. E. Parker, J. C. Parker, and A. E. Taylor, *Gastroenterology*, *77*:103−109 (1979).

263. W. D. Comper and T. C. Laurent, *Physiol. Rev.*, *58*:255−315 (1978).

264. D. N. Granger, N. A. Mortillaro, R. Kvietys, G. Rutili, J. C. Parker, and A. E. Taylor, *Am. J. Physiol.*, *238*:G183−G189 (1980).

265. J. A. Barrowman, *Physiology of the Gastrointestinal Lymphatic System*, Cambridge University Press, Cambridge (1978).

266. J. S. Lee, in: *Physiology of the Intestinal Circulation* (A. P. Shepherd and D. N. Granger, eds.), Raven Press, New York (1984) p. 201.

267. D. N. Granger and A. E. Taylor, *Am. J. Physiol.*, *235*:E429−E436 (1978).

268. P. R. Kvietys, W. H. Wilborn, and D. N. Granger, *Gastroenterology*, *81*:1080−1090 (1981).

269. S. G. Turner and J. A. Barrowman, *J. Exp. Physiol.*, *62*:175−180 (1977).

270. D. N. Granger, M. A. Perry, P. R. Kvietys, and A. E. Taylor, *Am. J. Physiol.*, *242*:G194−G201 (1982).

271. D. N. Granger, M. A. Perry, P. R. Kvietys, and A. E. Taylor, *Am. J. Physiol.*, *241*:G31−G36 (1981).

272. D. N. Granger, M. A. Perry, P. R. Kvietys, and A. E. Taylor, *Am. J. Physiol.*, *241*:G31−G36 (1981).

273. G. Volkenheimer and F. H. Schulz, *Digestion*, *1*:213−218 (1968).

274. G. Volkenheimer and F. H. Schulz, *Pharmacology*, *1*:8−14 (1968).

275. G. Volkenheimer and F. H. Schulz, *Nutr. Diet.*, *11*:13−22 (1969).

276. G. Volkenheimer, F. H. Schulz, H. John, J. Meier Zu Eisen, and K. Niederkorn, *Gynecologia*, *168*:86−92 (1969).

277. G. Volkenheimer, *Ann. NY Acad. Sci.*, *246*:164−171 (1975).

278. G. Volkenheimer, *Adv. Pharmacol. Chemother.*, *14*:163−187 (1977).

279. E. Sanders and E. Ashworth, *Exp. Cell. Res.*, *22*:137−145 (1961).

280. W. Krause, H. Matheis, and K. Wulf, *Lancet*, *1*:598−599 (1969).

281. H. H. Stone, L. D. Kolb, C. A. Currie, C. E. Geheber, and J. Z. Cuzzell, *Ann. Surg.*, *179*:697−711 (1974).

282. W. N. A. Charman and V. J. Stella, *Int. J. Pharm.*, *34*:175–178 (1986).
283. W. N. A. Charman, T. Noguchi, and V. J. Stella, *Int. J. Pharm.*, *33*:155–164 (1986).
284. J. Sugihara, S. Furuuchi, K. Nakano, and S. Harigaya, *J. Pharm. Dyn.*, *11*:369–376 (1988).
285. J. Sugihara, S. Furuuchi, H. Ando, K. Takashima, and S. Harigaya, *Pharm. Dyn.*, *11*:555–562 (1988).
286. M. Hashida, M. Murakami, H. Yoshikawa, K. Takada, and S. Muranishi, *J. Pharmcobio. Dyn.*, *7*:195–203 (1984).
287. A. Yanagawa, T. Iwayama, T. Saotome, Y. Shoji, K. Takano, H. Oka, T. Nakagawa, and Y. Mizushima, *J. Microencapsulation*, *6*:161–164 (1989).
288. K. Takada, M. Oh-hashi, Y. Furuya, H. Yoshikawa, and S. Muranishi, *Chem. Pharm. Bull.*, *37*:471–474 (1989).
289. H. Yoshikawa, S. Muranishi, N. Sugihara, and H. Sezaki, *Chem. Pharm. Bull.*, *31*:1726–1732 (1983).
290. A. Hagiwara, T. Takahashi, T. Ueda, A. Iwamoto, and T. Torii, *Anti-Cancer Drug Design*, *1*:313–321 (1987).

part three
PRACTICAL CONSIDERATIONS

17
Mucoadhesive Dosage Forms for Peptide and Protein Drug Delivery

Sau-Hung S. Leung

Columbia Research Laboratories, Madison, Wisconsin

Tsuneji Nagai and Yoshiharu Machida
Hoshi University, Tokyo, Japan

I. INTRODUCTION

Mucoadhesive drug delivery is an important means of drug delivery. Numerous studies have been conducted to understand the mechanism of mucoadhesion, screen mucoadhesives, and design new dosage forms [1-14].

A mucoadhesive drug delivery system has three desirable features:

1. Localization of the dosage form in specified regions to improve and enhance bioavailability of drugs
2. Promotion of intimate contact of the formulation with the underlying absorbing surface to allow modification of tissue permeability for absorption of macromolecules, e.g., peptides and proteins
3. Prolonged residence time of the dosage form to permit once-a-day dosing

Materials administered to different sites of the body, e.g., ocular, nasal, buccal, and gastrointestinal, are removed from the site of administration by natural clearance mechanisms. In the preocular area, materials are diluted by tears and removed via the drainage system, leaving only one-third of an administered dose remaining. In the nasal cavity, the mucous fluid is moved at a rate of 5 mm/min toward the throat and carries any foreign material with it. In the buccal cavity, the administered dosage form will be washed daily by 0.5-2 liters of saliva secreted by the salivary gland. Furthermore, the housekeeper wave in the IMMC (Interdigestive Migrating Motor Complex) will move any material in the stomach or small intestine distally toward the large intestine.

741

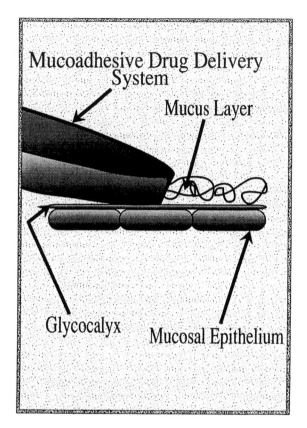

Fig. 1   Interaction of mucoadhesive with mucus and the mucosal membrane.

Thus, in order to localize the dosage form in specified regions for drug delivery, incorporation of mucoadhesive with sufficient adhesive strength is required.

As illustrated in Fig. 1, two different components are involved in mucoadhesive drug delivery: (1) the mucoadhesive drug delivery system itself, which acts as a platform and reservoir for the drug, and (2) the underlying substrate, which may be the mucin layer or the mucosal epithelium, or perhaps some of each. Thus, it is beneficial to review the characteristics of the mucus layer and the mucosal epithelium.

II. ELEMENTS OF MUCOADHESION

A. Characteristics of the Mucus Layer

With the exception of the ears, mucus covers all the internal tracts and orifices of the body, presenting as a continuous unstirred gel layer over the mucosal epithelium. Being the first layer that interacts with foreign materials, the mucus layer is normally the first layer that the mucoadhesive drug delivery system will encounter.

Mucus is a mixture of large glycoproteins, enzymes, water, electrolytes, sloughed epithelial cells, bacteria, and bacterial products. Most of the volume is fluid, saline, with some macromolecules [15]. Mucins are synthesized by goblet cells lining the mucosal epithelium layer and by special exocrine glands, such as salivary glands [16]. During secretion, the mucoprotein is closely associated with proteins via noncovalent bonds, and separation from proteins is difficult even though no permanent bond exists [17,18]. These extra proteins are composed of all other amino acids except serine and threonine [18].

The basic component of all mucus is the mucin glycoprotein with oligosaccharids sidechains [15,19] and terminal sialic acids [20,21] with a pKa of 2.6 [22]. Thus, at physiological pH of 7.4, the mucin network carries a substantial negative charge that affects mucoadhesion significantly. On average, each oligosaccharide unit, which may be linear or branched, consists of eight to ten (ranges from 1–15) sugars, e.g., D-galactose, L-fucose, N-acetyl-D-glucosamine, N-acetyl-D-galactosamine, and sialic acid [23]. The oligosaccharides are linked to the polypeptide backbone via the hydroxyamino acids, serine, and threonine, on the polypeptide chain [23,24].

Mucus is a viscoelastic gel with macromolecules linked together via cross-linking to form an infinite network. Cross-linking in mucus involves disulfide bonds [18] and physical entanglement [15]. The thickness of the mucus layer is the distance between the solution interface and the mucus–mucosa interface. The mean thickness of mucus is 192 μm in the human stomach, 77 μm in the rat stomach, 81 μm in the rat duodenum [25], and 1.4 μm in the human conjunctiva [26].

The continuity and thickness of the mucus layer is important to its protective function against physical and chemical insults. In the gastrointestinal tract, gastric mucin, together with bicarbonate secretion, provides a diffusion barrier for hydrogen ion and pepsin and maintains a near-neutral pH at the mucosal surface. Where there is no discontinuity on the mucin network, the mucoadhesive interacts directly with the mucin layer and localizes the dosage form close to the mucin layer next to the absorbing mucosal epithelium. However, once the continuity of the mucus layer is interrupted or its thickness is reduced, bioadhesion at the epithelial surface may occur. Reduction of mucus thickness and interruption of the continuity of mucin coverage can be accomplished either physically or chemically.

B. Characteristics of the Mucosal Epithelium

It is possible that a good mucoadhesive can adhere to the epithelium directly when the continuity of mucin coverage is interrupted either chemically or physically or it can penetrate across the reduced mucus network for attachment to the epithelial surface. Attachment of adhesives to the mucosal epithelium may serve three different functions [27]:

1. Formation of a continuous layer with the mucus network and thus reduced exposed cell surface area
2. Functioning as a protective covering for the underlying epithelium against physical and chemical insults
3. Functioning as a drug reservoir for sustained drug delivery and facilitating recovery of the damaged or diseased cell layers.

The adhesion of mucoadhesives onto the mucosal epithelium depends significantly on the interactions between the mucoadhesives and specialized structures and/or macromolecules on the epithelial cell surface. Bacteria, cells, and surfaces to which bacteria attach have an overall negative surface charge [28]. Regarding cell-surface charge, there is a growing appreciation that the distrubution rather than the net charge is the important parameter in intercellular adhesion [29,30]. Massa et al. [31] suggested that the increase in adhesion under some circumstances may result from charge redistribution.

Norde and Lyklema [32−34] found that the carboxyl groups of the adsorbed protein tend to accumulate close to the negatively charged polystyrene surface (i.e., the adsorbed proteins tended to accumulate their negative groups toward the negative polystyrene surface). The importance of negatively charged groups in bioadhesion is further supported by Wright et al. [29]. Cells that had a high net rate of spontaneous aggregation showed rearrangement of anionic sites on their surface membrane. Norde and Lyklema [33,35] also found that the amount of protein adsorbed increased with increasing surface charge and decreased with decreasing charge difference between the protein and latex [33].

The presence of macromolecules surrounding the cell has been well established [36]. Examples of these macromolecules are polypeptides [37], ligatin [38], concanavalin A [39], lectins [39], and fibronectin [40]. Fibronectin is a component of the extracellular matrix, which binds certain forms of collagen and glycosaminoglycans and mediates the adhesion and spreading of cells in culture [41]. Lectins are carbohydrate-binding proteins and glycoproteins derived from both plants and animals [42]. The term glycocalyx includes all polysaccharide-containing structures on the external surface of cells and which is maintained and synthesized continuously by the underlying cell [43].

C. Methods to Screen Mucoadhesives and to Evaluate Mucoadhesion

Over the past decade, considerable effort has been devoted to establish suitable screening techniques to evaluate mucoadhesives. Such techniques involve either tensile strength or shear strength measurements.

Tensile strength is the vertical component of the mucoadhesive strength. A Wilhelmy plate method [4] and a modified tensiometer method [9,14] were used to measure the tensile strength in mucoadhesion. The Wilhelmy plate method was used by Smart et al. [4] to measure the tensile strength between mucus and water-soluble polymers. The glass plates were first submerged into a 1% polymer solution and then oven-dried at 60°C to constant weight. The polymer coated plate was then immersed in homogenized mucus solution. The vertical interaction force between polymer and mucus was measured after 7 min of contact. The modified tensiometer method [9,14] was used to measure the vertical force of detachment of a test polymer from mucus layer. In this method, two pieces of rabbit stomach tissues were mounted onto a lower support and an upper stopper. Test polymer was placed on the mucin side of the upper stopper. The stopper was then lowered onto the support. The pressure applied to the interacting interface was varied by varying the weight of the stopper. The tensile strength of mucin−mucin or polymer−mucin interaction was then determined by measuring the force of detachment.

Tensile strength provides only a partial reflection of mucoadhesion, since the majority of mucus-covered surfaces, e.g., the gastrointestinal and buccal mucosae, have some elements of tangential shear motion. In recognition of this fact, a dual tensiometer apparatus was designed by Leung and Robinson [9] to measure the shear strength in mucoadhesion. In this method, two modified tensiometers were used. One of them was used to adjust the position for mucoadhesion and counterbalance the weight of the stopper. The second tensiometer was used to measure the shear strength of mucoadhesion.

In addition to the above screening methods, other methods have also been developed to evaluate the mucoadhesive properties of polymers. For instance, Peppas and Buri [5] have developed a method to study the static and dynamic mucoadhesive properties of polymer particles. Their method utilizes a thin channel made of glass or Plexiglas filled with artificial mucus gel or natural mucus. The artificial mucin solution is usually a pure fraction of glycoprotein from bovine submaxillary glands. The polymeric particles are preswollen in the mucin solution and placed on the mucin surface inside the channel. A laminar flow of air or other viscous solution is directed over the particle. Pictures of the motion of the particles are taken and their velocities determined. The hydrodynamic force needed for detachment of particle is then calculated.

An in-situ method using glass spheres coated with tested polymer placed on rat jejunum or stomach has been used by Ranga Rao and Buri to test the mucoadhesiveness of different polymer [11]. In this method, test polymers are first coated onto glass spheres or drug crystals and then placed on rat jejunum or stomach and kept in a humid environment. The tissues are then washed with phosphate buffer or dilute HCl. The amount of coated particles retained on the tissue is measured and used as an index for mucoadhesion [11].

A falling liquid film system has been developed by Teng and Ho to study the adhesiveness of micron size particles to the mucous surface [10]. The micron size particles used in this method are negatively charged poly(vinyl toluene) and hydroxylated Dynosphere particles with and without a positively charged polybrene coat. An excised intestinal segment is spread on a plastic flute and stand at an incline angle, and a suspension of the test particles is allowed to flow down the flute over the intestinal segment. Fraction of particles adsorbed onto the mucous surface is determined by measuring the particle concentrations entering and leaving the intestinal segment [10].

Recently, a colloidal gold-staining technique has been developed by Park [12] to study mucoadhesion without the use of animal tissue. In this method, mucin gold staining, red colloidal gold particles are stabilized by adsorbed mucin molecules to form mucin gold conjugates. When mucoadhesive hydrogels interact with mucin gold conjugates, a red color is developed on the surface of the hydrogels. The mucoadhesive properties of the hydrogels can then be compared by the intensity of the red color [12]. This technique allows study of interactions between polymer chains and mucin gold conjugate at the molecular level.

Interactions of mucoadhesives with biological substrate surface may be studied using a confocal imaging technique. The laser confocal fluorescence microscopy, an improved light microscopy, may allow biological structures to be visualized with a minimal amount of disturbance [44]. Confocal fluorescence microscopy can image samples within a narrow depth of focus.

Regions outside this depth of focus will appear black. Thus, out-of-focus noise can be rejected and resolution improved. Confocal imaging allows biological structures to be visualized that are completely obscured with conventional imaging techniques. Thus, interactions of polymers with natural mucin network may be studied with the confocal imaging technique.

Alternatively, the fluorescent technique developed by Park and Robinson [3] can be used to study the attachment of soluble polymers to the cell surface. Polymers were added to the surface of pyrene-labeled lipid bilayers of cultured human conjunctival epithelial cells. A change in fluorescence was observed when the polymer was bound to the lipid bilayer and compressed it. A large value of the fluorescence change indicates strong binding potential of the soluble polymer to the lipid bilayer of the cell. Thus bioadhesive strength of different soluble polymers can be compared.

For water-soluble polymers and one water-insoluble polymer, cross-linked polyacrylic acid, acridine orange (3,6-bis-dimethyl amino acridine) was used as the fluorescent dye to determine their charge densities [45]. The assay depends on the interaction of anionic charged sites with the ionized acridine orange. Once bound, the dye shows a fluorescent emission that is shifted to longer wavelengths which can be easily separated from fluorescence due to the unbound dye. In this way, the amount of anionic charge groups on the polymer can be obtained by titration.

For water-insoluble but swellable polymers and copolymers, their charge densities and amount of anionic charge group on the polymer can be obtained by potentiometric titration [46]. In this method, polymers are titrated potentiometrically using sodium hydroxide as a titrant. From the titration curve, the total number of carboxyl groups in the cross-linked polymers can be calculated.

D. Structure–Adhesion Relationships

Charges

In a study of a series of anionic, cationic, and neutral polymers, as exemplified by poly-L-lysine, dextran sulfate, and dextran, respectively, the charge sign and density were found to be important elements for mucoadhesion [6]. Polyanionic polymers were preferred over polycationic and natural polymers when both bioadhesive strength and cellular toxicity were considered.

The mucoadhesive strength of polyacrylic acid cross-linked with divinyl glycol in different pHs was determined by Park [46]. The adhesive force of polyacrylic acid, with an apparent pKa of approximately 4.75, to rabbit stomach tissue was highest at low pHs (1–3) and fell sharply at around pH 4, suggesting that mucoadhesion was favored when the carboxylate groups were in an undissociated form. Thus, hydrogen-bonding through polar groups, e.g., carboxyl, hydroxyl, amide, and sulfate, appears to be involved in mucoadhesion.

Hydration and Expanded Nature of the Polymer Network

When mucoadhesives make contact with an aqueous media, they swell and form a gel. The rate and extent of water uptake by the mucoadhesive may depend on the type and number of hydrophilic groups in the polymer structure and on the pH and ionic strength of the aqueous medium.

For water-insoluble polymers, the polymer network is maintained by covalent bonds formed during polymerization. When in contact with aqueous medium, the mucoadhesives swell and their networks expand. As the degree of hydration increases, the expanded nature of the polymer network increases, and the mucoadhesive strength increases at the same time [9]. The degree of swelling at equilibrium has been found to decrease with increasing degree of cross-linking. Thus, the equilibrium swelling of cross-linked polyacrylic acid decreases almost linearly with increase in the concentration of the cross-linker, 2,5-dimethyl-1,5-hexadiene [46].

Beside degree of cross-linking, the percent of charged groups also influences the degree of hydration of the polymer network significantly. Leung and Robinson [9] found that the degree of hydration and hence expanded nature of the polymer decreased with decreasing percent of charged groups. The average infrastructural mesh size of the polymer network was correlated linearly with the percent of acrylic acid, as shown in Fig. 2 [47]. Clearly the percent of hydrophilic groups controls the degree of hydration and the average infrastructural mesh size of the polymer network. For acrylic polymers, the diffusion coefficient was found to be proportional to the average infrastructural mesh size to the sixth power [48], i.e.,

$$D = \text{constant} \times (\xi)^6 \tag{1}$$

and the tensile strength was proportional to the average infrastructural mesh size of the acrylic polymer's network [48], i.e.,

$$\text{tensile strength} = \text{constant} \times (\xi)^3 \tag{2}$$

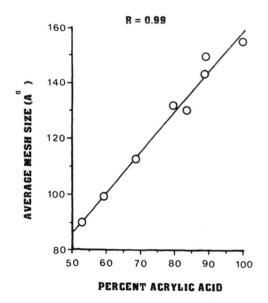

Fig. 2   Effect of acrylic acid composition on average mesh size of the copolymer network. (From Ref. 47.)

Thus, mucoadhesives with desirable tensile strength can be designed by controlling infrastructural mesh size of the polymer network.

As the percent of charged groups increases, the degree of hydration increases and the average infrastructural mesh size of the polymer network increases. Hydration (amount of water uptake) is correlated with the tensile strength as shown in Fig. 3. Therefore, one of the factors which controls the strength of mucoadhesion is expanded nature of the interacting mucus and polymer network. After initial contact is established, the expanded network maximizes physical entanglement and bond formation.

Chain Segment Mobility and Interpenetration

Interpenetration and entanglement of the mucoadhesive polymer to the substrate are partly responsible for their mucoadhesive properties [2,7,9,48]. The ability of the polymer chains to interpenetrate and increase the area for interaction can be approximated by their ability to diffuse.

The exact interpenetrating depth required to achieve adequate mucoadhesion is not known, but the representative mean diffusional path, s, for macromolecules can be estimated using Eq. (3) [49],

$$s = (2tD)^{1/2} \tag{3}$$

For high molecular weight (10^4 to 10^6) polymers, autoadhesion self-diffusion coefficients of 10^{-11} to 10^{-4} cm^2/sec have been observed [49]. Thus, by using Eq. (3), the representative mean diffusional path for bioadhesion of macromolecules for a contacting time of 60 sec can be estimated to be

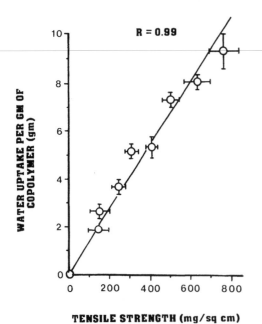

TENSILE STRENGTH (mg/sq cm)

Fig. 3   Correlation of tensile strength with water uptake. (From Ref. 9.)

roughly 11 to 346 Å. The mucin–mucin and polymer–mucin tensile strength is a time-dependent process, i.e., tensile strength increases with an increase in contacting time between polymer and mucin [48].

Mucoadhesion is a temperature dependent process. The temperature dependency of the mucoadhesive tensile strength can be calculated from Eq. (4) [48],

$$\ln \text{(tensile strength)} = \text{constant} - \frac{E}{2R} (1/T) \qquad (4)$$

A plot of ln (tensile strength) versus 1/T was shown to be a straight line [48]. The experimental activation energy of the interpenetration process for cross-linked polyacrylic acid is found to be 30.5 kJ/mole [48].

The importance of interpenetration process suggests the importance of mobility and flexibility of the polymer and mucin chains. More rigid macromolecules with higher cross-linking density have smaller flexibility and smaller chain-segment mobility, thus hindering interpenetration/interdiffusion. Indeed, Park [50] found that increase in cross-linking density of cross-linked polyacrylic acid decreased the mucoadhesive strength of the polymer onto rabbit stomach tissue [46].

An additional factor must be considered when a dry mucoadhesive is placed in contact with the mucous/mucosal epithelial layer, namely, the absorbing properties of the polymer network. As the polymer is hydrating, water is being drawn into the polymer network, and mucin strands are drawn inside the polymer network at the same time. For this reason, administration of dry mucoadhesive to the mucin/mucosal epithelium results in greater mucoadhesive strength as compared with that of fully hydrated mucoadhesive [48].

III. MUCOADHESIVE DOSAGE FORMS FOR VARIOUS ROUTES OF ADMINISTRATION

A. Ocular Route

The lacrimal gland, located at the temporal area above each eye, provides tears that bath the cornea and protect it from getting dry. Tears are removed from the precorneal area via the lacrimal drainage system into the nasopharynx. Mucus forms the inner layer of the tear film, which is in contact with the microvilli of the corneal and conjunctival epithelia.

Normal tear flow facilitates drug release from a precorneal drug delivery system by providing a constant flow of bathing medium. The major problem for conventional ocular delivery systems is, however, excessive drainage of the drug before adequate absorption can occur. When a 50 μl drop is administered, 20–30 μl is lost to the cheek from overflow and immediate drainage and 2 μl lost continuously per blinking [51]. In other words, only one third of the original 50 μl dose remains for absorption after it has been diluted by tear. Thus, a good ocular mucoadhesive delivery system needs to control drug loss via the ocular drainage system. This may be achieved by reducing the drug dissolution rate.

Besides excessive ocular drainage, other problems associated with ocular drug delivery are induced lacrimation, nonproductive absorption, and protein binding. The eye is an extremely sensitive organ, which responds quickly to external stimuli. Application of any foreign matter,

e.g., ointment, inserts, or eye drops, stimulate the eye to increase tear
secretion and increase drug loss. It is therefore essential to carefully adjust the formulation to minimize eye irritation.

In addition to rapid precorneal drug loss, the ocular absorption of topically applied peptides will encounter substantial metabolic and permeation
barriers. Using enkephalins as model peptides, Lee et al. [52] found that
the percent of a topical dose absorbed intraocularly in the albino rabbit
was $0.92 \pm 0.09\%$ for leucine enkephalin (YGGFL), $0.63 \pm 0.04\%$ for
methionine enkephalin (YGGFM) and $0.11 \pm 0.02\%$ for [D-ala^2] metenkephal-
inamide (YAGFM). While such absorption efficiency was comparable to that
seen in small drug molecules, a large fraction of what was absorbed was in
degraded form. As early as 5 min postdosing, only 13% of YGGFL, 1% of
YGGFM, and 74% of YAGFM was intact.

Further complicating the absorption of topically applied peptides into
the eye is their absorption into the bloodstream. Stratford et al. [53]
found that $36.1 \pm 4.4\%$ of a typically applied dose of YAGFM was systemically
absorbed via the blood vessels in the conjunctival and nasal mucosae. These
investigators concluded that the contact time of the instilled dose with the
conjunctival and nasal mucosae, their intrinsic permeability, and extent of
dilution of the instilled dose are key factors determining the vehicle effects
on the extent of systemic absorption of ocularly applied peptides.

To retard rapid drug loss from the precorneal area, various inserts
have been tested. An example is erodible polyvinyl alcohol film for the
ocular delivery of pilocarpine [54,55]. The slow dissolution of the insert
controlled the rate and duration of pilocarpine delivery. The ocular bioavailability of pilocarpine was twice that of an aqueous formulation. Thus,
decreasing drug dissolution affects its ocular bioavailability. Another ocular
product that has achieved prolongation of the therapeutic effect of pilocarpine is Piloplex [56]. It is an aqueous dispersion with limited water solubility. Upon instillation into the eye, the apparent opaque mass adheres to
and stays in the lower fornix for extended periods of time. The slow dissolution of the polymer itself and diffusion of pilocarpine from the polymer probably control the availability of the pilocarpine for ocular absorption.

A mucoadhesive system based on cross-linked polyacrylic acid has been
used to increase the ocular bioavailability of progesterone in the rabbit
[56]. The area under the curve was found to be 4.2 times greater than a
conventional ocular suspension. The increase in ocular bioavailability of
progesterone was probably due to the precorneal retention of the ocular
formulation via mucoadhesion. Similar favorable results were obtained for
pilocarpine by Saettone et al. [57], using hyaluronic acid, polygalacto-
uronic acid, mesoglycan, and carboxymethylchitin as bioadhesive polymers.

Taking the above factors into consideration, it seems that the ocular
bioavailability of peptides and proteins may be improved by reducing the
precorneal drainage loss, decreasing the dissolution of drug and/or formulation, and promoting the precorneal retention of the ocular delivery system via mucoadhesion. Whether this is indeed the case remains to be seen.

B. Nasal Route

The nasal route is an attractive route for systemic drug delivery, especially for drugs that suffer from extensive first-pass metabolism or gastro-

intestinal degradation. The nasal mucosa has a dense vascular network, providing an excellent absorptive surface. Hydrophobic drugs seem to be better absorbed than hydrophilic drugs [58]. The nasal absorption of nafarelin acetate, a highly potent superagonist of luteinizing hormone-releasing hormone, was found to be rapid and very reproducible [59]. Nasal delivery of enkephalins in rats was found to give serum levels comparable to that of intravenous injection [60]. By contrast, the nasal absorption of other peptide drugs such as insulin and secretin is not favorable unless penetration enhancers are included in the formulation or unless the formulation pH and tonicity are manipulated. In the case of insulin, its nasal absorption was enhanced by addition of hydrophobic bile salts [61,62] and nonionic surfactants [63]. It is proposed that bile salts act as absorption adjuvants by solubilization of insulin in mixed bile salt micelles and forming reverse micelles within nasal membranes [62]. Similar favorable results were observed in the nasal administration of β-interferon [64], as shown in Fig. 4. For secretin, its bioavailability was enhanced simply by controlling the medium at an acidic pH or by adding sodium chloride to control the osmolarity [65].

The nasal mucosa is normally covered by a thin layer of mucus. This mucus contains $90-95\%$ water, $1-2\%$ salt, and $2-3\%$ mucin. The mucus layer is moved at a rate of 5 mm/min toward the throat by the ciliated cells present on the surface of the mucosa [66]. Thus, a good mucoadhesive has to hydrate rapidly in the nasal mucous fluid and adhere strongly to the mucus or mucosal epithelium thereby resisting nasal clearance by the cilia.

Different nasal formulations have been developed using mucoadhesive polymers. Nagai et al. [67] developed a powder nasal drug delivery system for insulin without the use of absorption enhancers. This system consisted of Carbopol 934 and hydroxypropyl cellulose mixed in various ratios, as shown in Fig. 5. The powder dosage forms of insulin were administered nasally to five healthy male beagle dogs (3 U/kg). As shown in Fig. 6, resultant hypoglycemic effect was twice that of intravenous administration of 0.5 U/kg insulin. Kuroishi et al. [68] used hydroxypropyl cellulose as the bioadhesive base for a nasal powder spray. Once administered, the powder swelled and adhered to the mucosal membrane for up to 6 hr after application. Illum et al. [69] used bioadhesive microspheres consisted of albumin, starch, and DEAE-dextran for the nasal delivery of sodium cromoglycate. The slow clearance of the microspheres from the nasal cavity was found to prolong contact between the delivery system and the mucin/epithelial layer, thereby increasing drug availability. Morimoto et al. [70] tested the effect of polyacrylic acid gel on the nasal absorption of insulin and calcitonin in rats. Maximum hypoglycemic effect was observed between 30 min and 1 hr following the nasal administration of insulin (1 U/kg) in polyacrylic acid gel (0.1 and 1% w/v). By contrast, carboxymethyl cellulose (1% w/v) solution was found to be ineffective, as shown in Fig. 7. When calcitonin (10 U/kg) was administered nasally in polyacrylic acid gel (0.1% w/v), a prominent hypocalcemic effect was observed in the first 30 min. By contrast, when normal saline was used instead of polyacrylic acid gel, no hypocalcemic effect was observed. Clearly, polyacrylic acid enhanced the nasal absorption of insulin and calcitonin [70].

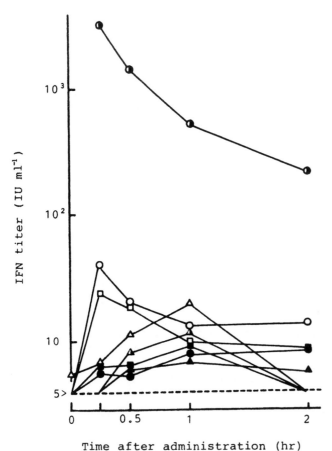

Fig. 4   Effect of surfactants on plasma interferon (IFN) after nasal administration of IFN (Dose: 2×10^6 IU).   Concentration of each surfactant
was 3% w/v. Key: (○) Na glycocholate; (□) Na cholate; (△) saponin;
(■) polyoxyethylene lauryl ether; (▲) polyoxyethylene cetyl ether; (▲)
polyoxyethylene stearyl ether; (●) polyoxyethylene nonylphenyl ether; (◑)
intravenous administration.   (From Ref. 64.)

C. Buccal

The buccal route offers the advantages of lower enzymatic activity [71,72]
and avoidance of first-pass metabolism and gastric enzymatic degradation.
The oral cavity allows precise localization of a mucoadhesive drug delivery
system and removal of the formulation if and when irritation occurs.   Localization of the formulation allows the addition of an absorption enhancer to
modify the underlying absorbing epithelium to enhance drug availability,
e.g., addition of sodium glycocholate increases the buccal absorption of
insulin by 40% [73].
    The total area of the buccal cavity is about 100 cm^2 [74].   The oral
mucosa can be divided into nonkeratinized, keratinized, and highly selective
keratinized regions [75].   The nonkeratinized region consists of the soft

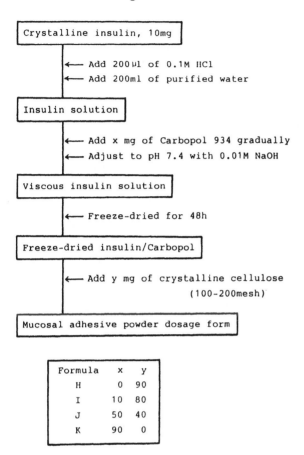

Fig. 5   Method of preparation of mucosal adhesive powder of insulin
(formulae H, I, J, and K.)   (From Ref. 67).

palate, the floor of the mouth, the ventral surface of the tongue, and the
labial and buccal mucosae.   The keratinized region consists of the hard
palate and the gingiva.   The borders of the lips and the doral surface of
the tongue make up the specialized zone of highly selective keratinization.
The complete turnover rate of the oral mucosal cells is 3 to 8 days, which
is much faster than that of the skin epidermis (which is approximately
30 days) [75].

The permeability characteristics of the oral mucosa are similar to those
of the dermis of the skin [76].   Scharaber [77] hypothesized that lipophilic molecules traverse the oral mucosa more easily by moving along or
across the plasma membrane of the epithelial cells, while hydrophilic molecules probably move through the intercellular pathway.

The mucus of the buccal mucosa is derived primarily from salivary
glands.   Thus, buildup of mucus underneath any mucoadhesive drug delivery system is minimized, and interference of interactions between the
adhesive and the underlying mucus/epithelial layer is reduced.   Furthermore, the mucosal surface is constantly washed by 0.5 − 2 liters of saliva

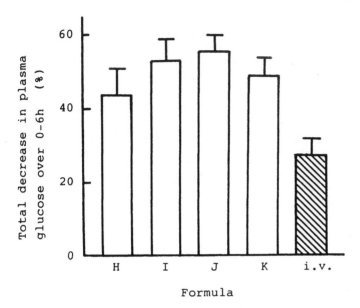

Fig. 6   Total decrease in plasma glucose after the administration of insulin as an adhesive powder, formulations H−K (3 U kg⁻¹); and as an intravenous injection (0.5 U kg⁻¹). Data are expressed as the mean ± S.E. of 4−5 observations. (From Ref. 67.)

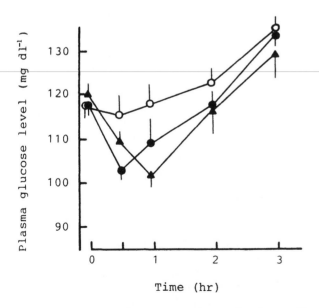

Fig. 7   Effects of concentrations of polyacrylic acid in the gel bases (pH 6.5) on changes in plasma glucose levels after nasal administration of insulin (1 IU kg⁻¹) in rats. Key: (●) 0.1% w/v PAA gel; (▲) 1% w/v PAA gel; (○) CMC solution (1% w/v). Each point represents a mean ± S.E. of 5 animals. (From Ref. 70.)

daily, which provides a sufficient amount of hydrating medium for the mucoadhesive dosage form. However, voluntary swallowing of the saliva may remove dissolved drug from the buccal cavity.

An example of a peptide that is absorbed via the buccal mucosa is protirelin [78]. Crystalline protirelin was spread onto a filter pater, which was placed at the center of a ∿3.5 cm diameter polytef disk. The disk was put in contact with the buccal mucosa of ten clinically healthy volunteers (five males and five females) and was removed after 30 min. Protirelin dissolved immediately and was absorbed via the buccal mucosa. Significant stimulation of both thyrotropin and prolactin was observed.

Buccal or sublingual tablets using hydroxypropyl cellulose [79,80] or hydroxypropyl cellulose/Carbopol [4,13] as adhesive layers have been developed. The dosage form is a double-layered tablet [4,13]. The upper layer consists of lactose. The lower layer is the adhesive layer (hydroxypropyl cellulose/Carbopol) with triamicinolone entrapped in it. Once administered the surface of the tablet swelled and eroded slowly. The adhesive tablet remained in place for at least 30 minutes. The entrapped drug was released as polymer eroded or by diffusion through the hydrating matrix.

A two-layer tablet with an adhesive peripheral layer and an oleaginous core for the buccal administration of insulin has been described by Ishida et al. [4], as illustrated in Fig. 8. The core consisted of insulin, sodium glycocholate, and cacao butter. Upon administration of this dosage form to the buccal mucosa of beagle dogs, an increase in plasma insulin levels and a decline in blood glucose levels were observed. This is shown in Fig. 9. Unfortunately, the bioavailability of insulin was only 0.5% when compared to an intramuscular dose.

Other buccal adhesive dosage forms that have been explored are adhesive gels [81] or ointments [82]. Freeze-dried hydroxypropyl cellulose and Carbopol were used as mucoadhesives to deliver a local anesthetic, lidocaine [81]. Adhesive ointment consisting of neutralized polymethacrylic acid methyl ester was formulated to deliver tretinoin and reduce irritation to the mucous membranes [82]. The adhesiveness of the formulation was found to increase with decrease in the degree of neutralization. Bremecker et al. [82] also found that partially neutralized gel gave good mucoadhesion without causing mucosal irritation.

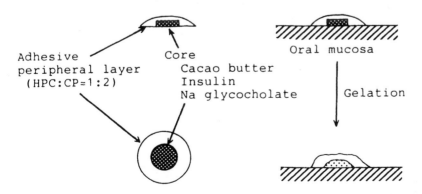

Fig. 8   Buccal mucosal dosage form of insulin. (From Ref. 4.)

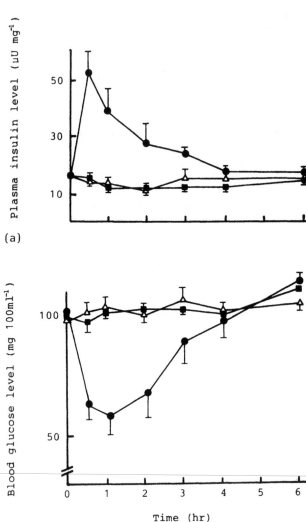

(a)

(b)

Time (hr)

Fig. 9   Change of (a) plasma insulin levels and (b) blood glucose levels
after mucosal administration of an adhesive dosage form of insulin to beagle
dogs. Key: (△) control; (■) insulin 10 mg; (●) insulin 10 mg + Na
glycocholate. Each symbol represents the mean ± S.E. of 9 determinations.
(From Ref. 4.)

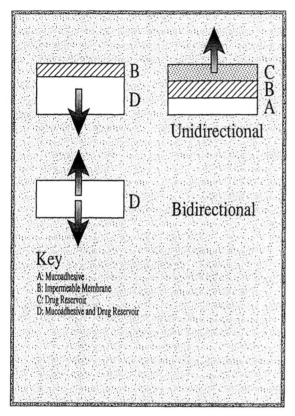

Fig. 10   Mono- or multilaminated mucoadhesive patches.

In all the previous systems direction of drug release was not controlled. Drug was released in all directions and large amounts of drug
would be swallowed. Thus, better control of the direction and rate of drug
release is desirable. To this end, bioadhesive buccal films [83] and
patches [84] have been designed to release drug unidirectionally to the
mucosa or buccal cavity or bidirectionally to both the buccal mucosa and
buccal cavity for local and/or systemic effects [85]. This is illustrated in
Fig. 10. The direction of drug release can be controlled by introducing a
impermeable membrane, e.g., polyvinylacetate, as a diffusion barrier.

D. Gastrointestinal Route

An important factor to consider in oral drug delivery is gastrointestinal
(GI) motility. GI motility in man and animal follows a cyclic pattern, which
can be divided into the digestive (fed) mode and the interdigestive (fasted)
mode. In the fed mode, the stomach is in a state of continuous motility
with substantial retropulsive forces. In the fasted mode, the motility pattern is divided into four distinct phases [86]. Phase I is a quiescent
period, phase II has random spikes of electrical activity or intermittent
contractions, phase III (housekeeper wave) is a period of regular spike

bursts of regular contractions at maximal frequency that migrate distally, and phase IV is the transition period between phase III and
phase I. The GI motility cycle in the fasted mode is commonly known
as the interdigestive migrating motor complex (IMMC) with an average
duration of 90–120 min. The IMMC is interrupted when food is administered. A normal meal converts GI motility to a fed mode and
maintains it up to 8 hr depending on the caloric content of the food
[87]. It seems that a minimum amount of gastric content is required to
change GI motility from IMMC to postprandial. Gupta and Robinson
[88] found that during phase I, 150 ml of water was required to change
GI motility from a fasted mode to a fedlike mode in the dog.

This GI motility cycle originates in the foregut and propagates
distally to the terminal ileum. The housekeeper wave (phase III) serves
to clear all indigestible materials from the stomach and small intestine.
Any mucoadhesive drug delivery system located in the stomach and small
intestine may be dislodged or loosened by the distally migrating contractile forces. Thus, a good oral mucoadhesive drug delivery system
needs to overcome the cleaning action of the housekeeper wave.

Another factor that potentially may affect the performance of mucoadhesives is substantial mucin turnover in the stomach. Once administered, the external surface of mucoadhesive drug delivery system will
interact with "soluble" mucin and all of the active adhesive sites will be
covered by mucin. As a result, the apparent mucoadhesive strength
may be reduced from polymer–mucin to mucin–mucin interaction, and
the dosage form may be removed from the stomach and small intestine
in the phase III housekeeper wave. This effect may be more significant
when adhesive dosage form is administered to the dog, which is known
to have a higher soluble mucin content in the stomach than the rat [91].

The adult small intestine, approximately 12 feet in length, is composed of the duodenum, jejunum, and ileum. Absorption takes place
predominantly in the small intestine because of its large surface area
and high permeability. Small peptides with 2–3 [89] and up to 6 [90]
amino acid residues can be taken up by the microvillus membrane of
the intestine. This microvillus membrane contains several peptidases
which are capable of hydrolyzing tetrapeptides, pentapeptides, and
hexapeptides. Some macromolecules, e.g., insulin, horseradish peroxidase [91], and bovine serum albumin [92] have been shown to
traverse the small intestine in trace amount. The mechanism of absorption of macromolecules may involve pinocytosis in specialized cells
found in the Peyer's patches.

Peyer's patches are organized mucosal lymphoid tissues of the small
intestine. Their size and number vary from species to species [92],
and are generally larger and more numerous in the distal small intestine.
Peyer's patches are capable of internalizing particulate matter [93],
bacteria [94], and marker proteins [95]. Microfold or "M" cells, covering Peyer's patches, have been demonstrated to participate in antigen
uptake [97], thus providing a possible explanation for the uptake of
high molecular weight soluble and colloidal proteins. Localization of
mucoadhesive drug delivery system at or around the Peyer's patches
has the potential of favoring the absorption of peptides and proteins.

E. Colonic and Rectal Routes

The colon is about 56 inches in length and divided into ascending, transverse, descending, and sigmoid sections. The rectum is about 5 inches in length and is continuous with the sigmoid colon. The mucous layer of the large intestine is thicker than that of the small intestine. The mucosal layer of the large intestine is smooth and devoid of villi, and is covered with a large number of goblet cells [96]. The primary function of the rectum is water absorption. Rectal drug absorption is primarily via simple diffusion through the lipid membrane [97] and is consistent with the pH partition theory [98].

Mucoadhesive drug delivery system targeted for the colon area must be protected from its environment as it traverses the upper G.I. tract, as the viscosity of fecal matter is usually high and can interfere with absorption of drugs. One possible approach is to control the rate of erosion of a protective coat, so that the mucoadhesives will not be exposed and hydrated until the formulation reaches the colon. Once the mucoadhesive dosage form reaches the colon, the bioavailability of the drugs may be enhanced by using penetration enhancers [99] or prodrugs [100,101]. Colonic absorption of cefmetazole and inulin was shown to increase by addition of 1 and 0.25% of enhancers, respectively. The enhancers used were sodium caprate, sodium laurate, and mixed micelles composed of sodium oleate and sodium taurocholate [99]. Colon-targeted naproxen delivery has been attempted using dextran ester prodrugs [100,101]. Naproxen was found to be liberated 15−17 times faster in pig cecum and colon homogenates than in aqueous pH 7.4 buffer or in homogenates of the small intestine [101]. During incubation of dextran prodrug in colon homogenates, the average molecular weight of the dextran prodrug decreased. The mechanism of a drug release may involve an initial degradation of the dextran by dextranases secreted from bacteria in the pig colon.

Another possible approach for colonic delivery of peptides is to coat the drug particles with an impervious layer, which will be degraded by the enzymes in the colon and release the entrapped drug. For example, vasopressin and insulin were coated with copolymers of styrene and hydroxyethylmethacrylate (either 1:6 or 100% hydroxyethylmethacrylate) and cross-linked with 1 or 2% of divinylazobenzene or substituted divinylazobenzenes [102]. The coated peptides were then given orally to rats. The peptides were protected from enzymatic digestion in the stomach and small intestine. However, when the azopolymer-coated drug reached the colon, the microflora in the colon reduced the azo bonds and broke the cross-links, thereby releasing the entrapped peptides for colonic absorption or local action. The use of azopolymer-coated insulin was shown to have a hypoglycemic effect in diabetic rats [102].

The mucous membrane in the rectum is thicker and more vascular than the mucous membrane in the colon. The proximal one-third of the rectum is drained via the superior hemorrhoidal vein to the inferior mesenteric vein and finally to the hepatic portal system. By contrast, the middle and distal one-third of the rectum are drained by the middle and inferior hemorrhoidal veins, respectively, and enter the general circulation via the iliac veins. Thus, dosage formulation located in the lower two-thirds of the rectum can be absorbed and enter the systemic circulation thereby avoiding the hepatic first-pass effect. In this light, the incorporation of mucoadhesives in rectal formulations can serve three different but related functions. First, mucoadhesive attaches to the mucin and/or epithelial layer and localizes the rectal suppository in the lower two-thirds of the rectum and delay its upward migration toward the upper one-third of the rectum, thereby avoiding first-pass metabolism. Second, localization of the rectal formulation allows more intimate contact of the drug with the absorbing epithelium and increases the bioavailability of the medicament. Third, close contact of the formulation with the mucosal layer promote action of adjuvants that are present in the formulation and possibly reduce the amount of adjuvant needed. Thus, irritation of these adjuvants to the mucosal layer may be reduced.

Penetration enhancers, e.g., sodium salicylate [103,104], 5-methoxy-salicylate [103], and enamine [105], were shown to enhance the rectal bioavailability of insulin in depancreatized dog. The effect of enhancers, e.g., salicylate, was prolonged (>1.5 hr) by addition of lecithin to the suppository base [104]. The sustaining effect may be due to slow release of sodium salicylate. Slow release of insulin from the suppositories, however, may decrease the availability of insulin [106].

There have been some studies in peptidase activity in the rectum of rabbits [107,108]. Aprotinin is a polypeptide of 58 amino acids obtained from the lumg or parotid glands. Aprotinin has a broad spectrum of protease inhibitory activity. When aprotinin was added to a microenema containing insulin and sodium salicylate, a higher plasma insulin level (187.4 ± 47.6 i.u./ml) was observed in 20 min after administration as compared to insulin and sodium salicylate alone (83.3 ± 20.8 i.u./ml) [109]. Thus, addition of a protease inhibitor may enhance rectal peptide absorption.

Beside penetration enhancers and protease inhibitors, the use of polyacrylic acid gel, a mucoadhesive, containing long-chain fatty acids was shown to promote the rectal absorption of insulin in rats [110]. Maximum hypoglycemic effects of insulin (1 IU/kg) in polyacrylic acid gel base containing each long-chain fatty acids (oleic acid, linolic acid, and linolenic acid) at 1% v/v occurred at 1 hr after administration [110]. As shown in Fig. 11, peak plasma insulin level was reached during the first 30 min after administration and was dose-dependent for the polyacrylic acid gel base containing 1% v/v of long-chain fatty acids [110].

F. Cervical and Vaginal Routes

The uterus is divided structurally and functionally into the body (proximal region) and the cervix (distal region). The mucous membrane in the upper two-thirds of the cervix, is responsible for mucus secretion. The epithelium there is cylindrical and ciliated whereas the lower one-thirds consists of stratified squamous epithelium. The vaginal portion of the cervix (exocervix) projects into the vaginal cavity. The mucous membrane of the vagina is continuous with that of the uterus, and the epithelium is the stratified squamous epithelium.

During a woman's lifetime, there are cyclic changes and variations in endogenous estrogen level, which affects the environment of the cervix and vagina. During childbearing age, the squamous epithelium of the vagina is thick and stratified, and an acidic environment is maintained via metabolism of glycogen [111]. During the menstrual cycle, the level of estrogen and progesterone changes. When estrogen dominates, the cervical mucus

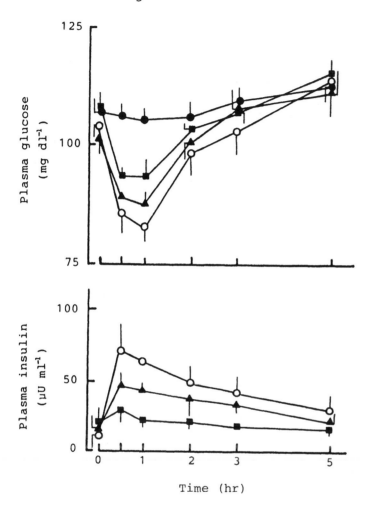

Fig. 11   Dose dependency of plasma insulin levels and plasma glucose levels following rectal administration of insulin in gel base containing 1% v/v linolic acid in rats. Key: (●) none; (■) 0.5 IU kg^{-1}; (▲) 1 IU kg^{-1}; (○) 3 IU kg^{-1}. Each point is a mean ± S.E. of 5 animals. (From Ref. 110.)

has low viscoelasticity, whereas when progesterone dominates, the cervical mucus has high viscoelasticity.

The cyclic changes in the mucus coat and the underlying cervical and vaginal tissues have made the design of mucoadhesive dosage forms difficult. Most of the cervical or vaginal formulations developed were for local delivery of drugs. For example, Machida et al. [112] developed a disklike formulation for treatment of vaginal carcinoma colli, where hydroxypropyl cellulose and Carbopol were used as mucoadhesive base. With this formulation, the amount of bleomycin released increased with increase in amount of mucoadhesive. Furthermore, Machida et al. [113] developed a sticklike formulation to deliver bleomycin hydrochloride,

carboquone, or 5-fluorouracil cervically for treatment of carcinoma colli. In
this formulation a mixture of hydroxypropyl cellulose and Carbopol was
used as mucoadhesive. Once administered to the cervical canal, the formulation swelled and the shape of the dosage form was fairly well maintained.

In postmenopausal women, the endogenous estrogen level decreases.
The cells of reproductive organs and surrounding tissues shrink in size or
atrophy slowly. The shrinkage of the cervix results in a decline in
cervical mucus production, and the vagina may appear dry. The vaginal
mucosa may be only 3-4 cells in thickness and with a decrease in vascularity. Furthermore, as the vaginal epithelium atrophies, the level of glycogen decreases, with resulting increases in vaginal pH to alkaline or slightly acidic. This increase in pH promotes bacterial growth and infection.

For postmenopausal women, there is less variation in the cervical and
vaginal tissues. However, the amount of mucus and moisture is greatly
reduced, affecting the hydration and mucoadhesiveness of the dosage form.
Postmenopausal vaginal dryness was successfully treated by using a bioadhesive moisturizer [114]. The formulation is a cream containing hydrated polycarbophil. Polycarbophil, polyacrylic acid cross-linked with
divinyl glycol, acts as a bioadhesive that attaches to the vaginal mucosa.
In addition to moisturization, this bioadhesive maintains the desiccated
vaginal tissue at an acidic pH. The increase in moisturization of the vaginal mucosa may probably enhance the adhesiveness of mucoadhesive drug
delivery system.

IV. CONCLUSION

Mucoadhesive drug delivery systems can localize the formulation to specified site of delivery and improve bioavailability of peptide or protein.
The localization of the formulation to specified regions also allows modification of tissue permeability by addition of penetration enhancer, or decreases enzymatic degradation of peptide and protein by addition of protease inhibitors. The potential of mucoadhesive dosage forms in peptide
and protein drug delivery is only beginning to be realized.

REFERENCES

1. R. Gurny, J. M. Meyer, and N. A. Peppas, *Biomaterial*, 5:336-340
    (1984).
2. J. D. Smart, I. W. Kellaway, and H. E. C. Worthington, *J. Pharm.
    Pharmacol.*, 36:295-299 (1984).
3. K. Park and J. R. Robinson, *Int. J. Pharm.*, 19:107-127 (1984).
4. M. Ishida, Y. Machida, N. Nambu, and T. Nagai, *Chem. Pharm. Bull.*,
    29:810-816 (1981).
5. N. A. Peppas and P. A. Buri, *J. Controlled Release*, 2:257-275
    (1985).
6. K. Park, H. S. Ch'ng, and J. R. Robinson, Alternative approaches to
    oral controlled drug delivery: Bioadhesives and in-situ systems, in
    Recent Advances in Drug Delivery Systems (J. M. Anderson and S. W.
    Kim, eds.), Plenum Press (1984), pp. 163-183.

7. G. Ponchel, F. Touchard, D. Duchene, and N. Peppas, *J. Controlled Release, 5*:129−141 (1987).
8. N. Peppas, G. Ponchel, and D. Duchene, *J. Controlled Release, 5*: 143−149 (1987).
9. S. H. S. Leung and J. R. Robinson, *J. Controlled Release, 5*:223−231 (1988).
10. C. L. C. Teng and N. F. H. Ho, *J. Controlled Release, 6*:133−149 (1987).
11. K. V. Ranga Rao and P. Buri, *Int. J. Pharm., 52*:265−270 (1989).
12. K. Park, *Int. J. Pharm., 53*:209−217 (1989).
13. T. Nagai, *Medicinal Res. Revs., 6*:227−242 (1986).
14. H. S. Ch'ng, H. Park, P. Kelly, and J. R. Robinson, *J. Pharm. Sci., 74*:399−405 (1985).
15. A. Silberberg and F. A. Meyer, Structure and function of mucus, in Mucus in health and disease — II, *Advances in Experimental Medicine and Biology, 144* (E. N. Chantler, J. B. Elder, and M. Elstein, eds.), Plenum Press, New York (1982), pp. 53−174.
16. H. Schachter and D. Williams, *AEMB, 144*:3−28 (1982).
17. F. A. Meyer, *Biorheology, 13*:49−58 (1976).
18. B. J. Starkey, D. Snary, and A. Allen, *Biochem. J., 141*:633−639 (1974).
19. A. Allen and A. Garner, *Gut, 21*:249−262 (1980).
20. A. Gottschalk, *The Chemistry and Biology of Sialic Acid and Related Substances*, Cambridge University Press, London (1960), pp. 1−115.
21. R. W. Jeanloz, in *Glycoprotein, Their Composition, Structure and Function* (A. Gottschalk, ed.), Elsevier, Amsterdam (1972), pp. 403−449.
22. P. M. Johnson and K. D. Rainsford, *Biochim. Biophys. Acta 286*: 72−78 (1972).
23. J. F. Forstner, *Digestion, 17*:234−263 (1978).
24. R. Kornfeld and S. Kornfeld, *Annu. Rev. Biochem, 45*:217−237 (1976).
25. A. Allen, D. A. Hutton, J. P. Pearson, and L. A. Sellers, Mucus glycoprotein structure, gel formation and gastrointestinal mucus function, in *Mucus and Mucosa, Ciba Foundation Symposium 109*, J. Nugent and M. O'Connor, eds.), Pitman, London (1984), pp. 137−156.
26. B. A. Nichols, M. L. Chiappino, and C. R. Dawson, *Invest. Ophthalmol. Vis. Sci., 26*:464−473 (1985).
27. S. H. S. Leung and J. R. Robinson, *ACS Symposium Series* (1990), in press.
28. G. W. Jones, The attachment of bacteria to the surface of animal cells, in *Microbial Interaction* (J. L. Ressig, ed.), Chapman and Hall, London (1977), pp. 130−176.
29. T. C. Wright, B. Smith, B. R. Ware, and M. J. Karnousky, *J. Cell Sci., 45*:99−117 (1980).
30. J. J. Deman and E. A. Bruyneel, *Biochem. Biophys. Res. Commun., 62*:895−900 (1975).
31. S. Massa and H. B. Bosmann, *Pharm. Ther., 21*:101−124 (1983).
32. W. Norde and J. Lyklema, *J. Colloid Interface Sci., 66*:266−276 (1978).
33. W. Norde and J. Lyklema, *J. Colloid Interface Sci., 66*:285−294 (1978).

34. W. Norde and J. Lyklema, *J. Colloid Interface Sci.*, *71*:350−366
    (1979).
35. W. Norde and J. Lyklema, *J. Colloid Interface Sci.*, *66*:257−265
    (1978).
36. G. I. Bell, M. Dembo, and P. Bongrand, *Biophys. J.*, *45*:1051−1064
    (1984).
37. R. Bertolotti, U. Rutishauser, and G. M. Edlemann, *Proc. Natl.
    Acad. Sci. USA*, *77*:4831−4835 (1980).
38. R. B. Marchase, P. Harges, and E. R. Jakoi, *Dev. Biol.*, *86*:
    250−255 (1981).
39. G. M. Edelman, *Prog. Clin. Biol. Res.*, *17*:467−480 (1977).
40. M. Pierschbacher, E. G. Hayman, and E. Ruoslahti, *Proc. Natl.
    Acad. Sci. USA*, *80*:1224−1227 (1983).
41. W. Dessau, F. Jilek, B. C. Adelman, and H. Hormann, *Biochem.
    Biophys. Acta*, *533*:227−237 (1978).
42. J. P. McCoy, Jr., *Biotechniques*, *4*:252−262 (1986).
43. I. Ito, *Fed. Proc. Fed. Am. Soc. Exp. Biol.*, *28*:12−25 (1969).
44. Y. Rujanasakul, Mechanistic studies of ocular peptide absorption and
    its enhancement by various penetration enhancers. Ph.D. Thesis,
    University of Wisconsin−Madison (1989).
45. S. H. S. Leung, The determination of charge density for water-
    soluble and water-insoluble anionic bioadhesives. M.S. Thesis,
    University of Wisconsin−Madison (1985).
46. H. Park, On the mechanism of bioadhesion. Ph.D. Thesis, University
    of Wisconsin−Madison (1986).
47. S. H. S. Leung, Polymer structural features contributing to muco-
    adhesion. Ph.D. Thesis, University of Wisconsin−Madison (1987).
48. S. H. S. Leung and J. R. Robinson, *J. Controlled Release* (1990),
    in press.
49. R. P. Campion, *J. Adhesion*, *7*:1−23 (1974).
50. N. A. Peppas and C. T. Reinhart, *J. Membrane Sci.*, *15*:275−287
    (1983).
51. D. M. Maurice and S. Mishima, Ocular pharmacokinetics, in *Handbook
    of Experimental Pharmacology*, Vol. 69 (M. L. Sears, ed.), Springer-
    Verlag, Basel (1984), pp. 19−38.
52. V. H. L. Lee, L. W. Carson, S. D. Kashi, and R. E. Stratford, Jr.,
    J. Ocular Pharmacol., *2*:345−351 (1986).
53. R. E. Stratford, Jr., L. W. Carson, S. Dodda-Sashi, and V. H. L.
    Lee, *J. Pharm. Sci.*, *77*:838−842 (1988).
54. M. F. Saettone, B. Giannaccini, and P. Chetoni, *J. Pharm. Pharmacol.*,
    36:229−234 (1984).
55. G. M. Grass, J. Cobby, and M. C. Makoid, *J. Pharm. Sci.*, *73*:
    618−621 (1984).
56. J. R. Robinson and V. H. K. Li, Ocular disposition and bioavailability
    of pilocarpine from Piloplex® and other sustained release drug delivery
    system, in *Recent Advances in Glaucoma* (U. Ticho and R. David,
    eds.), Excerpta Medica, Amsterdam (1984), pp. 231−236.
57. M. F. Saettone, D. Monti, M. T. Torracca, P. Chetoni, and
    B. Giannaccini, *Drug Develop. Ind. Pharm.*, *15*:2475−2489 (1989).
58. G. S. M. J. E. Duchateau, J. Zuidema, W. M. Albers, and F. W. H. M.
    Merkus, *Int. J. Pharm.*, *34*:131−136 (1986).

59. S. T. Amil, G. McRae, C. Nerenberg, A. Worden, J. Foreman,
     J. Hwang, S. Kushinsky, R. E. Jones, and B. Vickery, *J. Pharm.
     Sci.*, *73*:684–685 (1984).
60. K. S. E. Su, K. M. Campanale, L. G. Mendelsohn, G. A. Kerchner,
     and C. L. Gries, *J. Pharm. Sci.*, *74*:394–398 (1985).
61. A. C. Moses, G. S. Gordon, M. C. Carey, and J. S. Flier, *Diabetes*,
     32:1040–1047 (1983).
62. G. S. Gordon, A. C. Moses, R. D. Silver, J. S. Flier, and M. C.
     Carey, *Proc. Natl. Acad. Sci. USA*, *82*:7419–7423 (1985).
63. R. Salzman, J. E. Manson, G. T. George, T. Griffing, R. Kimmerle,
     N. Ruderman, A. McCall, E. I. Staltz, C. Mullin, D. Small,
     J. Armstrong, and J. Melby, *New Engl. J. Med.*, *312*:1078–1081
     (1985).
64. T. Ohwaki, H. Ando, S. Watanabe, and Y. Miyake, *J. Pharm. Sci.*,
     74:550–552 (1985).
65. T. Ohwaki, H. Ando, S. Watanabe, and Y. Miyake, *J. Pharm. Sci.*,
     74:550–552 (1985).
66. N. Mygind, *Nasary Allergy*, Blackwell Scientific, Oxford (1978),
     pp. 39–56.
67. T. Nagai, Y. Nishimoto, N. Nambu, Y. Suzuki, and K. Sekine, *J.
     Controlled Rel.*, *1*:15–22 (1984).
68. T. Kuroishi, H. Aska, and M. Okamoto, *Jpn. Pharmacol. Ther.*, *27*:
     4055–4059 (1984).
69. L. Illum, H. Jorgensen, H. Bisgaard, O. Krogsgaard, and N. Rossig,
     Int. J. Pharm., *39*:189–199 (1987).
70. K. Morimoto, K. Morisaka, and A. Kamada, *J. Pharm. Pharmacol.*,
     37:134–136 (1985).
71. K. W. Garren, E. M. Topp, and A. J. Repta, *Pharm. Res.*, *6*:
     966–970 (1989).
72. V. H. L. Lee, K. Dodda, R. M. Patel, E. Hayakawa, and
     K. Inagaki, *Proc. Int. Symp. Controlled Release Bioact. Mater.*, *14*:
     23–24 (1987).
73. D. M. Matthew and S. A. Alibi, *Gastroenterology*, *71*:151–161 (1976).
74. N. H. F. Ho and W. I. Higuchi, *J. Pharm. Sci.*, *60*:537–541 (1971).
75. A. Jarrett, *The Physiology and Pathophysiology of the Skin*, Vol. 6,
     Academic Press, London (1980), pp. 1871–2155.
76. W. R. Gale, H. K. Lonsdale, and S. Nacht, *J. Invest. Dermatology*,
     67:713–717 (1976).
77. L. S. Schanber, *Adv. Drug Res.*, *1*:71–106 (1964).
78. R. Anders, H. P. Merkle, W. Schurr, and R. Ziegler, *J. Pharm.
     Sci.*, *72*:1481–1483 (1983).
79. J. M. Schor, S. S. Davis, A. Nigaloyem, and S. Bolton, *Drug Dev.
     Ind. Pharm.*, *9*:1359–1377 (1983).
80. R. J. Erb, Bioavailability of controlled release buccal and oral nitro-
     glycerin by digital plethysmography, in *Controled Release Nitro-
     glycerine in Buccal and Oral Form*, *Advances in Pharmacotherapy*,
     Vol. I (W. D. Bussmann, R. R. Drics, and W. Wagner, eds.),
     S. Karger, Basel (1982), pp. 35–43.
81. M. Ishida, N. Nambu, and T. Nagai, *Chem. Pharm. Bull.*, *30*:980–984
     (1982).
82. K. D. Bremecker, H. Strempel, and G. Klein, *J. Pharm. Sci.*, *73*:
     548–552 (1984).

83. T. Yotsuyanagi, K. Yamamura, and Y. Akao, *Lancet*, *14*:613–614 (1985).

84. I. M. Brook, G. T. Tucker, E. C. Tuckley, and R. N. Boyes, *J. Controlled Rel.*, *10*:183–188 (1989).

85. H. P. Merkle, R. Anders, and A. Wermerskirchen, Mucoadhesive buccal patches for peptide delivery, in *Bioadhesive Drug Delivery Systems* (V. Lenaerts and R. Gurny, eds.), CRC Press, Inc., Boca Raton, Florida (1990), pp. 93–104.

86. J. H. Szurszewski, *Am. J. Physiol.*, *217*:1757–1763 (1969).

87. P. Kerlin and S. Phillips, *Gastroenterology*, *82*:694–700 (1982).

88. P. K. Gupta and J. R. Robinson, *Int. J. Pharm.*, *43*:45–52 (1988).

89. S. A. Adibi and E. Phillips, *Clin. Res.*, *16*:446 (1968).

90. M. Semeriva, L. Varesi, and D. Gratecos, *Eur. J. Biochem.*, *122*: 619–626 (1982).

91. R. L. Smith, C. L. Slough, K. L. Buckingham, R. W. Boyce, J. E. McOsker, and A. F. Franks, *Proceed. Intern. Symp. Control. Rel. Bioact. Mater.*, *16*:215–216 (1989).

92. R. A. Good and J. Finstad, The phylogenetic development of the immune responses and the germinal center system, in *Germinal Centers in Immune Response* (H. Cottier, N. Odartchanko, R. Schindler, and C. C. Congdon, eds.), Springer-Verlag, Berlin (1967), pp. 4–27.

93. D. E. Bockman and M. D. Cooper, *Am. J. Anat.*, *136*:455–461 (1973).

94. Y. Rhimizu and W. Andrew, *J. Morphol.*, *123*:231–237 (1967).

95. R. L. Owen, *Gastroenterology*, *72*:440–451 (1977).

96. H. Gray, *Anatomy of the Human Body* (C. D. Clemente, ed.), Lea and Febiger, Philadelphia, 30th Ed. (1985), pp. 1402–1506.

97. M. J. Binder, *Biochim. Biophys. Acta*, *219*:503–506 (1970).

98. S. Muranishi, *Methods Findings Exp. Clin. Pharmacol.*, *6*:763–772 (1984).

99. K. Tomita, M. Shiga, M. Hayashi, and S. Awazu, *Pharm. Res.*, 5: 341–346 (1988).

100. E. Harboe, C. Larsen, M. Johansen, and H. P. Olesen, *Pharm. Res.*, *6*:919–923 (1989).

101. C. Larsen, E. Harboe, M. Johansen, and H. P. Olesen, *Pharm. Res.*, *6*:995–999 (1989).

102. M. Saffran, S, G. Kumar, C. Savariar, J. C. Burnham, F. Williams, and D. C. Neckers, *Science*, *223*:1081–1084 (1986).

103. T. Nishihata, J. H. Rutting, A. Kamada, T. Higuchi, M. Routh, and L. Caldwell, *J. Pharm. Pharmacol.*, *35*:148–151 (1983).

104. T. Nishihata, M. Sudoh, H. Inagaki, A. Kamada, T. Yagi, R. Kawamori, and M. Shichiri, *Int. J. Pharm.*, *38*:83–90 (1987).

105. T. Nishihata, Y. Okamura, A. Kamada, T. Higuchi, T. Yagi, R. Kawamoni, and M. Shichiri, *J. Pharm. Pharmacol.*, *37*:22–26 (1985).

106. S. Kim, T. Nishihata, S. Kawabe, Y. Okamura, A. Kamada, T. Yagi, R. Kawamori, and M. Shichiri, *Int. J. Pharm.*, *21*:179–186 (1984).

107. R. E. Stratford and V. H. L. Lee, *Int. J. Pharm.*, *30*:73–82 (1986).

108. S. D. Kashi and V. H. L. Lee, *Life Sci.*, *38*:2019–2028 (1986).

109. T. Nishihata, G. Liversidge, T. Higuchi, *J. Pharm. Pharmacol.*, *35*: 616–617 (1983).

110. K. Morimoto, E. Kamiya, T. Takeeda, Y. Nakamoto, and K. Moriska, *Int. J. Pharm.*, *14*:149−157 (1983).
111. A. Bergman and P. F. Brenner, Alterations in the urogenital systems, in *Menopause-Physiology and Pharmacology* (D. R. Mishell, Jr., ed.), Year Book Medical Publishers, Inc., Chicago and London (1987), pp. 67−75.
112. Y. Machida, H. Masuda, N. Fujiyama, S. Ito, M. Iwater, and T. Nagai, *Chem. Pharm. Bull.*, *27*:93−100 (1979).
113. Y. Machida, H. Masuda, N. Fujiyama, M. Iwater, and T. Nagai, *Chem. Pharm. Bull.*, *28*:1125−1130 (1980).
114. Replens by Columbia Laboratories, Miami, Florida.

18

Preformulation and Formulation Considerations of Peptide Drugs: Case History of an LHRH Analog

Shabbir T. Anik and David M. Johnson

Syntex Research, Inc., Palo Alto, California

I. INTRODUCTION

Peptides are small and relatively simple molecules compared to proteins. In addition to size, peptides also differ from some proteins in that only the primary structure is responsible for biological activity. This further constraint is not generally used to define peptides, but it is useful from a drug development point of view when one considers the requirements for drug characterization, stability, and formulation development. Specifically, it eliminates a complex topic—dependence of biological activity on conformation; that topic is left to proteins exclusively. Peptides do, however, clearly possess a diverse and interesting array of physical/chemical and biological properties that provide special challenges for formulation development.

In this chapter we examine a case history of the preformulation and formulation development of nafarelin acetate, a highly potent analog of the naturally occurring hormone luteinizing hormone releasing hormone (LHRH) [1]. Aqueous formulations for parenteral and nasal administration of this peptide were developed for use in clinical trials. Although nafarelin acetate has its own set of unique properties, it is hoped that the approach used and the results found for this compound may provide some general guidance in the pharmaceutical development of other peptides.

II. PREFORMULATION CONSIDERATIONS

The initial objective in the formulation development of nafarelin was to define its physical and chemical properties. Although there is an enormous literature on the chemistry of peptides, little of it is devoted to the *quantitative* characterization of the physical properties of these compounds.

Additionally, quantitative data on chemical reactivity in aqueous solution
have generally been limited to the pH extremes, with little information
available in the pH region (pH ca. 3 to 9) of pharmaceutical interest
[2-5]. Thus, armed with very limited reference data, we undertook preformulation studies to explore the properties of nafarelin acetate.

A. Purity

Nafarelin acetate drug substance contains three predominant types of impurities: (1) peptides, (2) acetic acid and its salts, and (3) water. The
peptide impurities result from the inevitable carryover of small amounts of
intermediates and side products into the final product. Acetic acid is
present because narafelin is isolated by precipitation or lyophilization of an
acetate buffer solution. In addition to incorporation of 1 mol of acetate
ion in nafarelin as the counterion of the charged arginine residue, up to 1
mol of acetate is also present as acetic acid. Thus the total acetate content varies from ca. 4.3 to 8.3% in the peptide. Due to the hygroscopic
nature of nafarelin acetate, water is also present. Water is not usually
considered an impurity, but was included because the water content varies
from about 2.5 to 9.4% in batches of the peptide.

The quantity and nature of peptide impurities can vary in early development when the synthetic methods are changing and can cause problems
during formulation development when analytical methodology is not sufficiently refined. For example, in developing injectable solutions for
nafarelin acetate, solutions of relatively low concentrations (50 to 200 µg/
ml) showed large adsorptive losses to glass when measured by ultraviolet
(UV) spectrophotometry. However, on further investigation it was determined that UV-absorbing impurities in the initial batch of drug raw material were adsorbing on the glass, resulting in a false indication of adsorption. Hence the availability of an analytical method that is specific for the
drug substance is essential before formulation development is initiated.

In early development, batch-to-batch variations in the peptide impurities present in nafarelin acetate were carefully monitored by high-performance liquid chromatography (HPLC). Identification and quantitative analysis of nonpeptide impurities in the drug substance, however, required additional development time. During this time, each analytical standard solution was compared to previous standard solutions to assure consistent results in stability studies. Ultimately, a sensitive ion-chromatography assay was developed to monitor acetate concentration, thus allowing the concentration of nafarelin acetate to be corrected for residual acetic acid in the drug substance. The analytical controls used to deal with problems of batch-to-batch water variation in nafarelin acetate is discussed subsequently in the hygroscopicity section.

B. Solid-State Properties

Crystallinity

Many peptides do not exhibit a strong tendency to crystallize and are isolated as amorphous powders. Isolation techniques for nafarelin have utilized lyophilization and precipitation methods that favor formation of amorphous material. This lack of crystallinity is characterized by low intensity peaks or by broad bands in the powder x-ray pattern, as shown in Fig. 1. Differential scanning calorimetry (DSC) and thermogravimetric analysis (TGA) plots are also shown in Fig. 1. The DSC wave is essentially featureless except for a minor endotherm at 178° C that is probably due to salt decomposition and release of acetic acid. The TGA curve is consistent with loss of about 7% water up to 140°C, followed by loss of acetic acid and decomposition.

Hygroscopicity

Nafarelin acetate is moderately hygroscopic, absorbing up to about 20% water at 93% relative humidity. In Fig. 2 the percent moisture absorbed after equilibration at a given humidity for 1 week is plotted for three batches of drug substance. Water sorption increases nonstoichiometrically and shows a dramatic increase between 79 and 93% relative humidity (RH). The figure shows that the three batches give similar water sorption profiles, except at very high humidity (93% RH). This variation may be a result of differences in the morphology or particle size distribution between the three batches.

The rate of sorption-desorption of water in nafarelin acetate is shown in Fig. 3 and indicates that large changes in moisture content can occur rapidly. This is especially true for desorption (93% RH → 44% RH) where in only 10 min the sample (10 mg) lost about 5% moisture. Although this initial water loss was rapid, the plateau reached after 2 hr indicates that the sample still contained about 3% more water than the original sample. The reversibility of the sorption and desorption of moisture for this amorphous solid on a longer time scale is shown in Fig. 4. In this experiment, all samples were initially dried under P_2O_5 to 0% water content. Three samples were stored at 20%, 47%, and 93% RH for approximately 3 days. They were then placed at an intermediate humidity (79% RH), where they equilibrated to similar, but not identical water content, indicating that sorption-desorption was not completely reversible.

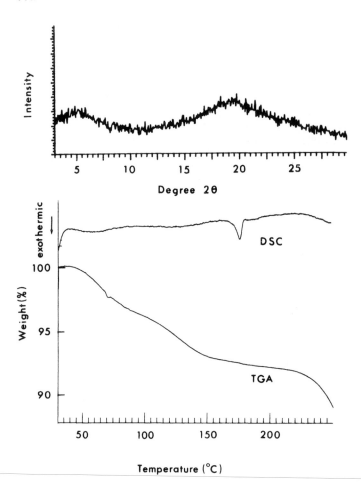

Fig. 1   Powder x-ray diffraction (top), DSC (middle), and TGA (bottom)
plots of nafarelin acetate.

The rapid rate and lack of reversible water sorption–desorption are
annoying problems when preparing analytical standards from solid drug
and also in weighing operations during manufacturing. Thus for analytical
standards, Karl Fischer analysis was used each time a solid sample was
weighed to correct for the true water content. This may be impossible to
achieve early in development due to the small quantities of drug available.
To achieve the target drug concentration in manufacturing, measuring
water content on each batch of drug substance was not feasible. In this
case, in-process controls using an HPLC method to monitor nafarelin
acetate concentration were developed to ensure that the bulk solution from
each batch would meet specifications.

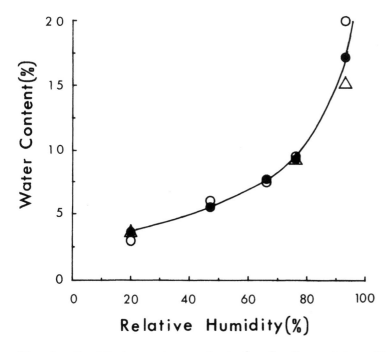

Fig. 2   Equilibrium water content of nafarelin acetate.   Three batches are shown denoted by open circles, triangles, and filled circles.

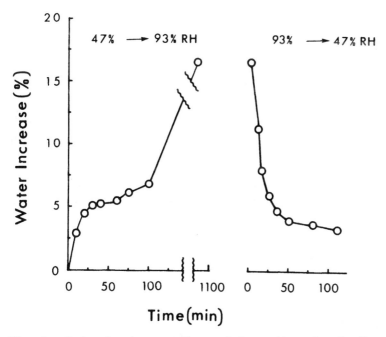

Fig. 3   Rate of water sorption and desorption of nafarelin acetate between 47 and 93% RH.

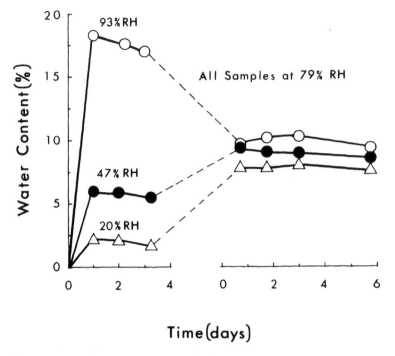

Fig. 4   Plot showing water content as a function of time for three nafarelin acetate samples. Initially, all samples were dried to 0% water content and then placed at the relative humidity (RH) shown. After about 3 days all samples were placed at 79% RH.

C. Solution-Phase Properties

Ionization Constants

Nafarelin contains three ionizable amino acid residues: histidine, tyrosine, and arginine. The pK_a of the tyrosine residue was 9.92 by UV spectrophotometry. The pK_a of the histidine residue was determined to be 5.9 by potentiometric titration. Arginine was not determined but is expected to be similar to that of free arginine (ca. 12.5). Thus nafarelin exists predominantly as a monocation above pH 6 and as a dication below pH 6. The state of ionization of nafarelin in solution can affect its ability to ion-pair, aggregate, and bind to surfaces, which in turn can affect the solution properties described below.

Adsorption

Loss of peptides and proteins from solution by adsorption to various surfaces is a common phenomenon and should be addressed early in the preformulation activities. Losses are particularly significant at low concentrations. In the case where adsorption is due to ionic interaction of the peptide with the silanol groups on the glass surface, it can be prevented by silylating the glass (flasks, syringes, etc.) with organosilanes such as Prosil 28 [6]. This procedure is acceptable only for experimental purposes.

Table 1   Percent Adsorbed at Equilibrium of D-Nal (2)^6 LHRH onto
Borosilicate Solid Glass Beads (ca. 100 cm^2) in the Presence of Different
Additives[a]

Additive	Nafarelin concentration (μg/ml)			
	20	10	5	1
None (pH 7)	29	40	43	59
Sodium dihydrogen phosphate				
0.02 M	13	20	35	62
0.05 M	8	15	23	35
0.1 M	4	7	10	18
0.16 M sodium acetate	4	7	12	36
0.2% sodium carboxymethylcellulose	11	–	16	57
0.1 M glycine	18	–	43	60
0.5% bovine serum albumin	11	–	17	32
0.1% gelatin	10	–	14	19

[a]All solutions were adjusted to pH 5 except where indicated. All values
are ±2%.

Although it is very difficult to generalize what agents may help prevent
adsorption of peptides, carrier proteins [7], ionic species [8], and
surfactants [9] appear to work in some cases. The effects of various additives on the adsorption of nafarelin acetate on glass are given in Table 1.
For injectable formulations of nafarelin acetate at concentrations of 5 to
100 μg/ml, 0.1 M phosphate buffer was effective in preventing glass adsorption. Ion pairing or complexation of the phosphate ion with nafarelin
may be a possible explanation, but the mechanism has not been confirmed
[6].

Adsorption to other materials, such as syringes, tubing, and filters,
was also significant, as shown in Table 2. Membrane fiber filters showed
large adsorption losses of nafarelin due to the tortuous channels and large
surface areas of the filter. Although presaturation of these filters is a
possibility, this was not a desirable solution and was not explored. Filters
that function by size exclusion pores such as Nuclepore were preferred and
showed minimal losses of nafarelin.

Solubility

One of the major problems in solubility measurements is determining whether
the system has reached equilibrium. Peptides, due to their conformational
mobility, may not reach their most stable equilibrium state immediately.
This is particularly true if an aggregated form is the most stable state. In
the case of nafarelin acetate, it was often possible to dissolve enough drug

Table 2   Adsorption of D-Nal (2)^6 LHRH onto Various Surfaces[a]

Surface	Initial concentration of D-Nal(2)^6 LHRH (μg/ml)	Adsorption (%)
High-density polyethylene bottle (49 cm^2)[b]	20	2.4
1-ml Stylex plastic syringe 7020D	5	5.5
	20	2.4
1-ml B–D Glaspak syringe	5	12.0
	20	5.7
	50	2.0
	100	0.6
100-μl gastight Hamilton syringe	50	2.5
Rubber testing, amber latex (0.45 cm diameter, 40 cm length)	20	3.7
Tygon tubing (0.6 cm diameter, 40 cm length)	20	0
Millex-GS filter (25 mm diameter, 0.22 μm)	20	93.2
Nucleopore filter (13 mm diameter, 0.2 μm)	20	4.9

[a]The solution used was an isotonic 0.1 M phosphate buffer at pH 4.4, except where indicated. Adsorption was measured by scintillation counting of nafarelin solutions contained in the device. For filters, a fixed volume of solution was passed through and counted. For complete experimental details, see Ref. 6.

[b]Aqueous solution without additives.

to give a 10-mg/ml aqueous solution. These concentrated solutions, however, were weakly birefringent when viewed through crossed polars, indicating the existence of a liquid–crystal phase [10]. On standing, some solutions eventually developed turbidity, a precipitate, or a gel. Solutions were also cycled between 5 and 40°C for several months to gain added assurance of the long-term physical stability of the system.

The solubility of nafarelin acetate is very sensitive to ionic strength, pH, and buffer, as shown in Table 3. The solubility drops rapidly with increasing ionic strength (sodium chloride), and the solution gels at higher concentrations of sodium chloride, a rather surprising occurrence for a decapeptide at these low concentrations. The solubility also shows marked dependence on the nature of the buffer. It decreases dramatically on going from acetic to succinic to citric acid. Equilibrium solubility values could be determined for nafarelin in citrate and phosphate buffers, whereas the solution gelled in succinate buffer.

Table 3   Effect of Ionic Strength, pH, and Buffers on Nafarelin Acetate
Solubility

Vehicles	pH	Ionic strength (M)	Solubility (mg/ml)
NaCl			
0.14%	5.8	0.024	>5
0.5%	5.8	0.086	Opaque gel at 5 mg/ml[a]
0.9%	5.8	0.154	Opaque gel at 10 mg/ml[a]
Na acetate, 0.01 M	4.0	0.0015	>10
	6.0	0.0095	>10
Na succinate, 0.01 M	4.0	0.0038	>10
	6.0	0.024	Opaque gel at 5 mg/ml[a]
Na citrate, 0.001 M	4.0	0.0012	1.71[b]
	6.0	0.0054	0.58[b]
Sodium phosphate buffer, 0.02 M	6.0	0.022	1.19[b]

[a]Concentration where gelling was first observed.

[b]Values and measurements similar to conventional solubility data. No
gelling observed.

Although pH-solubility data are limited, the solubility was found to decrease with increasing pH in succinate and citrate buffers. In the former
case the solution gels at 5 mg/ml at pH 6. It is possible that the di- and
tri-acid salts of nafarelin form ordered structures in solution, resulting in
aggregation of the peptide leading to low solubility and/or gelation.

The sensitivity of solubility to these factors requires that excipients
for formulation development be chosen judiciously. However, the ability to
vary solubility by choice of excipient can be useful in those cases where
it is necessary to target to a desired solubility to maximize thermodynamic
activity.

Chemical Stability

The degradation rate of narafelin acetate was studied in the pH range 1 to
11 in aqueous solution from 40 to 80°C [11]. For solutions of intermediate
pH and low temperature, sterilization was required to prevent bacterial
degradation of nafarelin. In some buffer media, microbiological degradation
was much more rapid than chemical degradation. In sterile solutions, the
disappearance of nafarelin was followed using a reverse-phase HPLC method (Fig. 5) and obeyed pseudo-first-order kinetics. The log k_{obs} versus

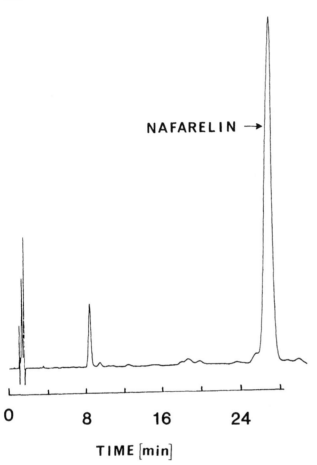

NAFARELIN →

0   8   16   24

TIME [min]

Fig. 5   Reverse-phase HPLC chromatogram of a pH 5.3 solution of
nafarelin acetate (ca. 94% remaining) after 14 months at 40°C. The small
peaks appearing after 2 min represent unidentified degradation products.

pH profile for the degradation of nafarelin at 80°C is defined by the open
circles in Fig. 6. The data points are connected by line segments to illustrate the U-shaped curve of the profile.

The acid portion of the profile between pH 1 and 3 exhibits a slope of
−1, indicating that the reaction is specific-acid catalyzed. The remainder
of the pH profile includes a base-catalyzed region above pH 7 and a
region of maximum stability from pH 4 to 6.

Further kinetic investigations of this region at lower temperatures
showed that the pH of maximum stability shifts somewhat with temperature
and is ca. 5.3 at 40°C. An Arrhenius plot of the data at pH 5.4 is reasonably linear (Fig. 7) and gives a predicted mean t_{90} of 3.8 years at
25°C. In the acidic region of the profile (<pH 3), specific acid hydrolysis
of the terminal glycinamide group to give nafarelin free acid predominates.
With the exception of a small amount of free acid, the degradation products

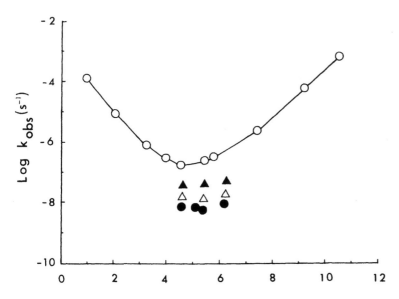

Fig. 6   Log k_{obs} versus pH profile for the degradation of nafarelin in aqueous solution at 80°C (open circle), 60°C (filled triangle), 50°C (open triangle), and 40°C (filled circle).

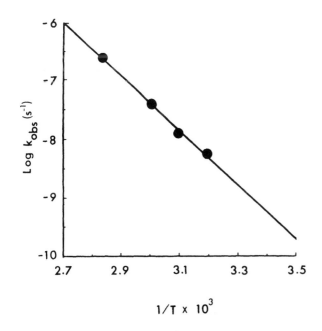

Fig. 7   Arrhenius plot of log k_{obs} versus 1/T for the degradation of nafarelin in aqueous solution at pH 5.4

at intermediate pH values are numerous (Fig. 5) but as yet have not been identified.

III. FORMULATION CONSIDERATIONS

Injectable aqueous buffer solutions of nafarelin acetate were used to define the early toxicology and clinical profile of nafarelin. However, for daily administration over longer periods, an alternative route had to be established. The nasal route was considered to be the most promising for nafarelin acetate. Reports in the literature had established nasal absorption of native LHRH and some of its analogs. Furthermore, since the solution properties of nafarelin were well characterized, it was relatively straightforward to develop an aqueous nasal solution for metered dose delivery.

To study the nasal absorption of nafarelin acetate and optimize the formulation, the rhesus monkey was used as the animal model [12]. Two points became evident from the monkey experiments: nasal absorption from aqueous buffer solutions of nafarelin acetate was quite low, with a

Table 4 Effect of Penetration Enhancers on the Nasal Absorption of Nafarelin Acetate in Monkeys[a]

Enhancer	Nafarelin concentration (mg/ml)	Dose (μg)	Peak height relative to buffer solution[b]	AUC 8 hr relative to buffer solution[b]
None[c]	0.625	133	1	1
1% Sodium glycocholate	0.625	120	48	22.4
2% Sodium glycocholate	0.625	120	48	26.0
1% Polyoxyethylene-10-cetyl ether	0.625	137	18.7	14.0
None[c]	1.25	272	1	1
1% Polyoxyethylene-10-cetyl ether	1.25	229	5.5	4.2
1% Polysorbate 20	1.25	232	1.0	0.9
1% Azone +1% polysorbate 60	1.25	256	3.9	3.8
0.05% EDTA Na$_2$	1.25	212	2.5	2.0

[a] n = 6; standard errors on peak heights and AUC range from 15 to 30%.

[b] Ratio of data for solutions with enhancer relative to aqueous buffer of the same concentration and dose.

[c] Aqueous buffered solution.

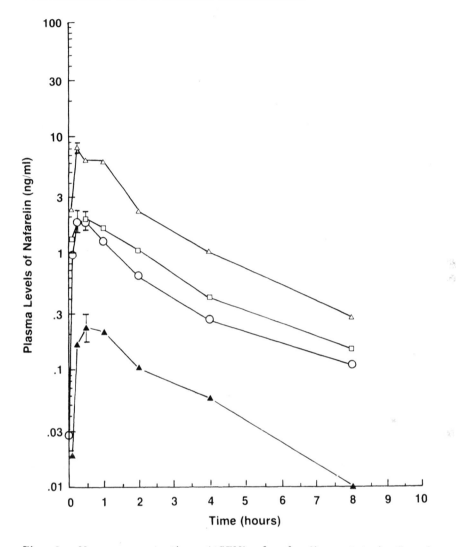

Fig. 8   Mean concentrations (±SEM) of nafarelin acetate in the plasma of monkeys after a single subcutaneous or intranasal dose. Open triangle, 431 µg (2.5 mg/ml), nasal; circle, 272 µg (1.25 mg/ml), nasal; filled triangle, 133 µg (0.625 mg/ml), nasal; square, 5 µg, subcutaneous.

bioavailability of 2 to 3%; there was a marked concentration effect on the extent of absorption as shown in Fig. 8. Improvement in bioavailability by increasing the concentration of nafarelin was limited due to restriction on the solubility and practical considerations for the volume delivered. The 2 to 3% bioavailability was insufficient to obtain the desired therapeutic levels for prostatic carcinoma and had to be improved.

     The monkey served as a useful animal for studying the relative effects of various additives on the extent of absorption. The results are listed in Table 4. Of the various absorption enhancers studied in the monkey,

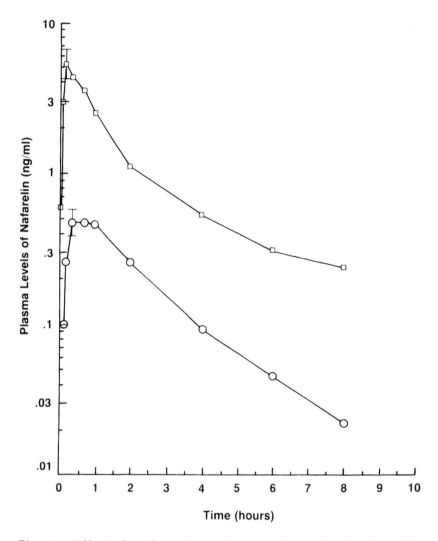

Fig. 9   Effect of sodium glycocholate on plasma levels of nasally administered nafarelin acetate in humans.   Circle, 347 µg, buffered solution;
square, 347 µg, buffered solution + 2% sodium glycocholate.

sodium glycocholate was found to be the most effective in enhancing absorption.   Bioavailability studies in humans with a 1.75-mg/ml nafarelin
acetate solution containing 2% sodium glycocholate showed a sevenfold increase in absorption and a bioavailability of approximately 20%.   Blood-
level profiles were similar to those observed in the monkey and are shown
in Fig. 9.   With this increase in bioavailability and blood levels, it is now
feasible to treat patients with prostatic carcinoma with an intranasal dosing
regimen.   Although a 1:1 correlation between the monkey and human absorption data was lacking, the rank correlation among the enhancers was
extremely useful in choosing formulations for human testing.

IV. CONCLUSIONS

The individual physical and chemical properties of nafarelin acetate described in this chapter are not unique to peptides. However, some of these properties when grouped together form a set of characteristics that are common to many peptides. Often the solid peptides exist in an amorphous state. Amorphous compounds are usually more soluble, in some cases more hygroscopic [13–15], and can be less pure and less stable [16] than their crystalline forms. Peptides have a tendency to form ordered structures in concentrated solutions and at interfaces that can result in aggregation, surface activity, and gel formation. They also possess a binding affinity for hydrophobic and/or hydrophilic surfaces, leading to adsorption, which can be especially significant at low drug concentration. Finally, peptides are very susceptible to bacterial degradation, and all aqueous drug solutions under study must be sterile.

Peptides contain both ionic and hydrophobic regions and can have strong inter- and/or intramolecular hydrogen bonding forces as well. Due to their multifunctional nature, the physical properties of peptides in solution, such as aggregation, surface properties, and solubility, are dependent on ionic strength, ionic species, pH, cosolvents, and surfactant content of the formulation. Thus although there are general approaches to understanding these complex drugs, there may be few general solutions to formulation problems that can be applied to peptides. These challenges may require that the scope of preformulation and formulation activities be broader for peptides than is usual for small organic drug molecules.

The delivery of peptides provides exciting opportunities for the pharmaceutical scientist. Because the injectable route with daily or more frequent administration is not desirable, alternative routes such as nasal, rectal, buccal, and vaginal should be explored carefully. At this time the nasal route appears to be the most feasible. However, low bioavailabilities on the order of 5 to 30% will probably be the norm. The judicious use of enhancers should help in this regard. Delivery of peptides orally by manipulations of formulations or appropriate peptide prodrugs and analogs will always be the preferred mode of administration. At this point, however, only modest progress has been made in solving the difficult problems associated with poor oral absorption of peptides. Thus the design of controlled-release injectable systems will probably play a pivotal role in the successful use of peptides as therapeutic agents in the years to come.

REFERENCES

1. B. H. Vickery, J. J. Nestor, Jr., and E. S. E. Hafez, eds., *LHRH and Its Analogs. Contraceptive and Therapeutic Applications*, MTP Press, Lancaster, England (1984).
2. A. B. Robinson and C. J. Rudd, in *Curr. Top. Cell. Regul.*, 8:247 (1974).
3. S. M. Stenberg and J. L. Bada, *J. Org. Chem.*, 48:2295 (1983).
4. H. Hartmann and R. Brussau, *Z. Naturforsch.*, 22b:380 (1967).
5. G. G. Smith and B. S. DeSol, *Science*, 207:765 (1980).
6. S. T. Anik and J.-Y. Hwang, *Int. J. Pharm.*, 16:181 (1983).
7. C. Petty and N. L. Cunningham, *Anesthesiology*, 40:400 (1974).

8. T. Mizutani and A. Mizutani, *J. Chromatogr.*, *111*:214 (1975).

9. T. Mizutani and A. Mizutani, *J. Non-Cryst. Solids*, *27*:437 (1978).

10. A. Rogerson and L. M. Sanders, *Proc. Int. Symp. Controlled Release Bioact. Mater.*, *14*:97 (1987).

11. D. M. Johnson, R. A. Pritchard, W. F. Taylor, D. Conley, G. Zuniga, and K. G. McGreevy, *Int. J. Pharm.*, *31*:125 (1986).

12. S. T. Anik, G. McRae, C. Nerenberg, A. Worden, J. Foreman, J. Y. Hwang, S. Kushinsky, R. E. Jones, and B. Vickery, *J. Pharm. Sci.*, *73*:684 (1984).

13. M. Otsuka and N. Kaneniwa, *Chem. Pharm. Bull.*, *32*:1071 (1984).

14. M. J. Pikal, A. L. Lukes, J. E. Lang, and K. Gaines, *J. Pharm. Sci.*, *67*:767 (1978).

15. R. Huttenrauch, *Pharmazie*, *32*:240 (1977).

16. T. W. Chan and A. R. Becker, *Pharm. Res.*, *5*:523 (1988).

19

Controlled Delivery Systems for Peptides

Lynda M. Sanders

Syntex Research, Palo Alto, California

I. INTRODUCTION

The present rapid expansion of the design and development of peptides and proteins as pharmaceutical compounds is precedented only by the analogous, though less dramatic growth of the field of steroids some several decades ago. Endogenous peptides and proteins are, however, proving to be more numerous and ubiquitous than steroid hormones, and consequently, the pharmaceutical and medical fields they are opening promise to have depths as yet unplumbed. There are two clear paths of approach being taken by different groups. Each involves identification, isolation, and sequencing of a native compound. The worker may then develop synthetic proprietary analogs of this compound, with desired characteristics such as optimal hydrophobicity, stability to proteases and consequent longer half-life, or chemical or physical stability. Alternatively, it may be preferable to synthesize the actual native compound, frequently by biotechnological techniques as described in Chapter 2. One limitation here is that biotechnology does not lend itself to preparation of peptides or proteins containing synthetic (including D-) amino acids, so the preparation of synthetic analogs of a peptide found endogenously will usually be by chemical means. The capability of modification of a compound for desired effect and properties results in compounds with varying physicochemical properties and biological potencies and consequently presenting different challenges in their formulation for controlled release.

Compared with conventional drug compounds, peptides have unique requirements and restrictions for delivery. Controlled delivery technologies of various types have been known for many years, but the field has experienced recent massive advancement in scope and sophistication, with the thrust provided by the needs of peptides.

Peptides have in general low oral activity. Although the goal is elusive, this means of delivery is sufficiently attractive to support continuing

efforts to achieve it [1], but with limited success to date. A simple injectable solution, usually aqueous, is usually a necessity for in vivo characterization of the compound, and may well be useful for early clinical
studies, but has limitations in utility for long-term therapy. The limitations will be a function of the patient acceptability/benefit ratio, acceptability including such diverse factors as physical discomfort, noncompliance
and consequent nonefficacy, adverse effects from highly variable plasma
levels, and difficulty in maintaining a personal inventory of supplies of
sufficient stability. Even for the extreme acceptability/benefit ratio of
treatment of an acute life-threatening condition, if similar benefits will
accrue from either a conventional injectable or controlled-release system,
the latter will be preferred and will potentially provide superior efficacy,
pharmacological profiles being equal. Nasal delivery is an increasingly
useful means of delivery of these compounds, but this again presents
problems with patient compliance and high fluctuations in plasma levels.
In the majority of cases of low bioavailability by this route, the typical
high cost of peptide drugs makes this route undesirable. Even though
there have been major advances in the design of systems to optimize nasal
delivery, as discussed in detail in Chapter 13, the inherent drawbacks of
this and the other routes outlined may often be surmounted by use of a
well-designed controlled-release system.

   Controlled release is, however, far from being a panacea for the problems inherent in delivery of peptides. Of significant practical consideration is the need to at least partially characterize a candidate peptide drug
compound in terms of safety and efficacy, to justify the significant developmental effort required for a controlled-release system, as well as providing
an indication of potency, critical to the feasibility and design of such a
system. A simple injectable solution, being the most rapidly developable
dosage form, will provide this information, but may well introduce a significant confounding factor. The pharmacological profiles of a peptide
compound given continuously versus intermittently may well be dissimilar
both qualitatively and quantitatively. Experiences in the author's laboratory have highlighted this phenomenon with analogs of luteinizing hormone releasing hormone (LHRH) in that nafarelin, an agonistic analog of
LHRH, is more potent and elicits more profound pharmacological effects in
both male and female primates when delivered continuously versus intermittently [2]. It is anticipated that this may be an issue with any
endogenous peptide or analog thereof in which the native compound is
secreted in a pulsatile manner. The phenomenon mandates early assessment of the properties of such a compound delivered continuously (e.g.,
by implantable osmotic pump) for a well-informed decision on the profile,
or even indeed of the overall practical viability, of a particular peptide.
It also renders the traditional parameter of bioavailability—area under the
curve of plasma level versus a reference—unamenable to traditional interpretation. Clearly, bioavailability as thus defined does not differentiate
between regimens of delivery and will not itself be expected to correlate
directly with efficacy. Design of controlled-release systems for peptides
therefore poses challenges for the pharmacokineticist, pharmacologist, and
clinician, as well as the pharmaceutical scientist.

   In the design of controlled-release systems, peptides offer some major
feature permitting flexibility in system design. In general, they are highly
potent, at least compared with the majority of conventional drug compounds.

The theoretical maximum duration of release of a given system will be directly proportional to potency of compound. For a parenteral controlled-release system, the mass of material administered will again be related to potency of the compound. Various mechanisms of release are possible with different "loading levels" of active contents. A highly potent peptide will therefore have latitude in maximum duration of release possible, size or mass of the dosage form, and release mechanism for optimum release kinetics. Each of these parameters becomes more restricted with decreased potency.

The small molecular size of peptides, usually defined as fewer than 20 amino acid residues, usually renders them water soluble, although they may have significant hydrophobic character. They usually do not have secondary or tertiary structure, and therefore conformational integrity is not an issue in processing the compounds to prepare a controlled-release system. However, their molecular size, except for tripeptides or smaller, renders them poor candidates for membrane permeability, and consequently the major route of controlled delivery envisaged for peptides to date is parenteral. In this chapter we focus on this route, with some speculation on opportunities to develop other routes.

Even though peptides as a class have broad characteristics common to the group, each compound has its own distinct properties, and it is not possible except in the broadest sense to design a generic controlled-release system that can be applied to all or even most peptides. The synthetic flexibility in design of the compounds makes possible an enormous range of physicochemical properties. For example, a decapeptide has been (indirectly) observed to display secondary conformational structure, in that it forms an ordered structure of liquid crystals, analogous to β-sheet formation, in a suitable aqueous environment. The compound [D$-$Nal(2)^2, D$-$pCl$-$Phe^2, D$-$Trp^3, D$-$hArg(Et$_2$)^6, D$-$Ala^{10} LHRH was designed for high hydrophobicity, with high receptor binding and long circulatory half-life [3]. Its hydrophobicity also provides a high degree of nonaffinity for an aqueous environment (although the polar components confer good aqueous solubility). In aqueous solution, particularly in the presence of electrolytes, birefringence is observed, and this ordering in solution exhibits itself in the extreme case as gelation. This is perceived to be an interesting and unusual case of solution structure formation, but from a practical viewpoint makes the compound problematic to handle in solution, frequently a required processing step in, for example, microencapsulation of the compound.

The example above exemplifies the difficulty of the use of model compounds. Another factor to be considered in the selection of a model compound is its physicochemical similarity to the active compound, in terms of solubility and partitioning properties, and particle size and morphology where applicable. If the limitations of such models are borne in mind, however, they may provide some utility, particularly in early broad screening of system types prior to availability of sufficient quantities of the actual compound. A model should in this case be selected on the basis of similarity of molecular size and physicochemical properties, as well as ready and inexpensive availability, ease of detection by conventional methodologies (e.g., good ultraviolet absorption), and ease of handling. The task is easier than that with proteins, since in that case the criterion of comparable conformational stability may also apply, which adds another level of complexity.

The challenge of design and development of controlled-release systems for peptides opens new questions and demands skills beyond those traditionally used in conventional dosage form design. Expertise is required in a wide range of fields, including polymer chemistry as well as physical chemistry, pharmaceutics, and biological sciences. The continued growth in this technically sophisticated field, stimulated by the successes already achieved, will take pharmaceutical scientists into new areas and require them to bring about a convergence of creativity from a breadth of these previously insular areas.

In this chapter we review the considerations, limitations, and potential in the design of controlled-release delivery systems for peptides and give the reader an overview of the present scope of this field.

II. CONTROLLED-RELEASE DOSAGE FORM TYPES

A. Microencapsulated Systems

The concept of a biodegradable microencapsulated system that will afford an injectable controlled-release product for a peptide is very attractive. It allows the patient compliance and convenience that can be a limiting factor in practical utility of a system, and puts control of therapy closer to the physician. It may improve efficiency of use of a possibly expensive material and may be no more intrusive in administration than a simple daily injectable system. The major disadvantage is commitment to therapy for the duration, which requires a high degree of confidence that the dosage will not require adjustment for that period of time. Another disadvantage is the relatively complex fabrication technology involved, which may be slowly optimized for a particular compound and may require specialized equipment and expertise. A problem common to the development of all controlled-release systems is that of the time frame of development being directly related to the duration of release required. For example, if a one-year system is under study, every screening of a candidate formulation will take up to one year to determine its performance. The variability in release kinetics also increases with target duration of release. These factors all contribute to the strategic decision on required duration of the system—longest not always being preferred.

Microencapsulation as a technology has been available since the development of "carbonless carbon paper" by NCR [4]. More recently it has been applied to controlled delivery of pharmaceuticals and the technology modified in many different ways to accommodate the requirements of different compounds [5]. Prerequisites that remain, however, are sufficient potency of the compound, an acceptable therapeutic ratio, and ruggedness of the compound to the microencapsulation process, which may involve conditions of high shear, organic solvents, or nonambient temperatures.

As an example of such a system, nafarelin (D-[Na](2)6 LHRH), has been prepared as a controlled-release injectable system by microencapsulation in poly(d,l-lactic-co-glycolic) acid (PLGA) [6,7]. The compound has high potency, having a plasma half-life five to seven times that of the native compound. A dose of 0.5 µg/day is sufficient to maintain suppression of estrus in the female rat [8], and 5 µg/day has been shown to inhibit ovulation in the female rhesus monkey [9]. Nasal administration of a daily dose as low as 300 µg has achieved castrate levels of testosterone in prostatic cancer patients within 4 weeks of initiation of therapy [10].

The compound is a decapeptide with no conformational structure that would render it delicate, and it has a high therapeutic ratio.

The carrier selected, PLGA, is a biodegradable polyester that has a long history of use in controlled-release pharmaceuticals as well as other biomedical applications. Although many biodegradable carriers are available, PLGA is frequently the one of choice, for reasons elaborated in Sec. III. An optimum encapsulation process for a compound of this type is phase separation of a water-in-oil (w/o) emulsion. The process and some critical variables are described in more detail elsewhere [6]. Briefly, the compound is prepared as an aqueous solution and the polymer dissolved in a suitable organic solvent such as methylene chloride. These two immiscible solutions are coemulsified by high-speed stirring, with a suitable surfactant if necessary, the phase ratios being appropriate for formation of a water-in-oil emulsion. A nonsolvent for the polymer is then added, causing precipitation of the polymer around the aqueous droplets to form embryonic microspheres. This suspension is added to a larger volume of a nonsolvent to harden and complete the formation of the microspheres. The process conditions may be modified to optimize the overall yield and efficiency of encapsulation, as well as providing a good, spherical, fairly monodisperse product in the required size range and of the required loading level. By these means, microspheres of nafarelin in PLGA of up to 10% loading have been prepared, of mean sizes in the range 10 to 50 μm. The product is collected, dried in vacuo, and prepared immediately before use as an injectable suspension by agitation with a suitable suspending vehicle.

Selection of Size Range

The generally preferred gauge needle for intramuscular (IM) injection is 18 G, which has an internal diameter of 838 μm. One would therefore expect a great deal of latitude in the size range of microspheres to form an injectable suspension. In practice, however, there are limitations. Above about 100 μm, depending on other properties of the suspension, interparticulate interactions are seen to be a factor in occasional clogging in the vicinity of the syringe tip and needle. Although use of a more dilute suspension and/or modification of the suspending vehicle may ameliorate this, it is inadvisable to target at sizes above this.

At the other extreme, very small microspheres present their own problems. Even though they might be anticipated to provide a very high quality suspension, their high surface area may be difficult to wet with the suspending vehicle, and may require concentrations of surfactant in the vehicle that are unacceptable for other reasons. Also, even if a good suspension is readily formed, non-Newtonian rheological properties, particularly dilatancy, may ensue [11]. This may be increasingly pronounced with reduction in particle size. Dilatancy exhibits itself as an increase in viscosity with increase in applied shear. For an injectable suspension, there may be good flow properties in the syringe barrel, but as the shear rate rapidly increases at the needle junction the viscosity may increase momentarily and block all flow. On relaxation of pressure or slight withdrawal of the plunger the effects are lost, until another attempt to eject the contents is made. Again, modification of the vehicle may ameliorate the effects. Other factors against the use of extremely small microspheres include poor dry flow properties, required for ease of filling vials, and frequently, poor handling properties, with high electrostatic charges being

generated too readily. For several reasons, therefore, the use of extremely small microspheres is usually ill advised, and selection of a target
size range should compromise between the pros and cons outlined. There
may, however, be overriding considerations in cases where particle size
(i.e., surface area) is a factor in determining the rate of release of the
compound from the microspheres. This is usually not the case with peptides and is discussed further in Sec. IV.

Selection of Suspending Vehicle

Development of an optimum suspending vehicle is perhaps the only aspect
of the development of controlled-release technology for peptides that requires classical pharmaceutical formulation studies. Criteria for a vehicle
include toxicological properties, wetting properties, isotonicity, specific
gravity, foaming properties, and properties of suspension formed.

Toxicological properties include local irritancy and metabolic profiles.
Some surfactants that may otherwise be very acceptable toxicologically may
have membrane activity that renders them unacceptably locally irritating.
Of the other potential excipients, cellulose derivatives are not considered
acceptable in a parenteral vehicle, since they are neither metabolized nor
sufficiently small to be excreted renally.

The vehicle must have sufficiently low surface tension to wet the
microsphere mass rapidly and completely, but not foam, even with the displaced air from the mass. If necessary, an antifoaming compound may be
considered in conjunction with a wetting agent.

Isotonicity, or isoosmoticity with serum, is a requisite for a suspending
vehicle. This may be achieved by the addition of an appropriate amount
of a suitable osmotically active ingredient. Isotonicity is usually designed
into a solution on a theoretical basis, knowing the osmotic pressures of the
components, and is verified by demonstrating compatibility with whole
blood. Erythrocytes will neither swell nor shrink when exposed to an isotonic medium. However, the presence of surfactants may confound this
study if they lyse erythrocyte membranes; such a phenomenon should not
be interpreted as nonisotonicity but should be explored further. The
choice of an ionic or nonionic tonicifier may also profoundly influence the
nature of the final suspension, particularly when there is a net charge on
the surface of the microspheres.

A suspending vehicle should ideally have specific gravity close to that
of the microspheres, to minimize sedimentation rate. This requires determination of the specific gravity of the microspheres—which may be accomplished by floatation of the microspheres in glycerol of varying concentrations and densities. Appropriate additives (such as glycerol) may
then be used to adjust the specific gravity of the vehicle.

The properties of the suspension formed are the most difficult to
quantify in terms of acceptability. The required maximum working concentration of microspheres in the product should first be defined. A simulated use test on candidate vehicles should then be performed, with
suspensions being prepared and tested for syringeability. This may be
performed using a syringe of the type to be used clinically, and a needle
one size smaller than that desired clinically as a more discriminating challenge. The suspension should be injected into a material providing some
back-pressure, as will be encountered in the clinical situation.

Examples of Microencapsulated Peptides

There are several examples of luteinizing hormone releasing hormone analogs having been successfully microencapsulated for parenteral controlled
release. As mentioned previously, nafarelin has been encapsulated in
PLGA, and its release kinetics and pharmacological properties assessed
clinically [12]. A similar system of tryptorelin encapsulated in PLGA has
also been developed and has been shown to have utility in prostatic carcinoma [13]. Leuprolide, another LHRH analog, has been encapsulated in
PLGA by a modified phase separation technique [14] and has been shown
to be pharmacologically active in vivo [15].

   To touch on a related area, that of microencapsulation of proteins, the
challenges here are much more profound. The frequent conformational
delicacy of proteins renders them susceptible to denaturation in the encapsulation process. In particular, high shear stirring and organic solvents are known to be problematic. Despite this, there are examples of
the successful microencapsulation of larger molecules, including insulin in
PLGA [16,17]. The technology will no doubt evolve to surmount the problems presented in microencapsulation of proteins.

B. Implantable Systems

Implants, or single-unit systems, offer a number of advantages over multi-
particulate systems. Even if degradable, they offer a degree of retrieva-
bility, at least in the early stages of degradation, which permits contingency
action in the event of a short-term adverse response. They are significantly more readily prepared, often being prepared by a simple melt extrusion
of a coarse blend of the compound and the carrier. This requires a degree
of thermal stability of the compound, usually not a problem with peptides.
Such a process will not raise questions of solvent residues, or syringeability,
or any other of the myriad of issues arising in the more sophisticated microsphere system. The major disadvantage of the single-unit implant is that
of means of administration: it will require subcutaneous delivery with a
device such as a trochar. It may also require closed medical supervision
and may involve a suture. However, there is a great deal more flexibility
in the design of a controlled-release implant than of a microencapsulated
system, both in choice of materials and methods of manufacture and in
loading levels achievable. This is highlighted by the multitude of carriers
and systems described in the literature.

Biodegradable Systems

Such systems have the advantage of not requiring removal while still providing the possibility of removal, at least in the earlier phase of release.
An increasingly commonly used carrier for these systems is again PLGA,
and the peptide nafarelin has been shown to be readily formulated in this
polymer for long-term controlled release [18]. In this system the influence of composition and molecular weight of the polymer has been elucidated
systematically in terms of mechanism of release of the compound, properties
discussed further in Sec. IV. The method of manufacture of these devices
was a simple comelt extrusion of a blend of the compound and milled
polymer, a method that proved useful across the range of formulation
parameters spanned. The exception is with a high-molecular-weight polymer

requiring heating to become malleable beyond that consistent with compound stability. It is possible that the use of equipment applying pressure would have eliminated this problem; certainly, force driving in an extruder rather than passive feeding would reduce the temperature requirement.

Another LHRH analog developed as a single-unit controlled-release parenteral system is Zoladex [D−Ser, (t-Bu)^6, AzaGly^{10}−LHRH] formulated in PLGA [19]. This subdermal depot has qualitative release properties similar to those of the nafarelin system, although of shorter duration, consistent with different physicochemical parameters of the polymer. This system has shown clinical utility in the treatment of prostatic cancer [20].

Leuprolide [Des−Gly^{10}-(D−Leu^6) LHRH ethylamide] has also been prepared as a single-unit device in PLGA by a hot pressing technique [21], and its performance in a rat model assessed. Buserelin, another analog of LHRH, has been formulated for the same purpose in poly(hydroxybutyric acid) [22], a much less commonly used carrier. The fabrication procedure involves a coarse combination of the compound and the polymer in the presence of methanol, followed by compression into "tablets" containing 0.7% compound. Long-term release in rats of over 200 days was provided by this system. An analogous system providing one month's release has also been clinically evaluated [23].

Poly(ortho esters) are biodegradable polymers with some unique features, described further in Sec. III. They have been shown to have good utility in the long-term zero-order controlled release of steroids (levonorgestrel) [24,25], and more recently have shown more utility for controlled release of [D−(Nal^2)^6, AzaGly^{10}] LHRH in vitro [26]. These systems again are readily prepared by melt blending and extrusion of a mix of the components.

All of the examples cited above are illustrative of the relative ease of preparation of controlled release implants with a peptide drug compound. The commonality of LHRH analogs in these systems points to the rapid development of these compounds for therapeutic purposes and the strong requirement for controlled release technology to realize their full utility. Other compounds have, however, been studied in such a system, examples covering peptides of a range of molecular weights, including tetragastrin, epidermal growth factor, and bovine prolactin [27].

Again, a primary criterion for the feasibility of preparation of a controlled release implant of a given compound is stability to the fabrication process. This frequently involves subjecting the peptide to elevated temperatures. If this is problematic, the temperature requirement may be reduced either by the concomitant application of pressure, or by the use of plasticizers to reduce the glass transition temperature or the malleability point of a polymer. The latter alternative bears with it, however, the complication of the potential influence of the additive on the release profile of the compound, as well as raising toxicological questions.

Nondegradable Systems

Use of a nondegradable carrier material opens many options in system design. The advantage of removability in the event of need is balanced by the additional surgical manipulation required. Therefore, the decision for a degradable versus a nondegradable system must be an individual one based on an assessment of the needs and limits of acceptability of dosage form for a given clinical scenario. A precedent has been set by the development and use of Norplant, a long-term (5-year) nondegradable implant for controlled release of levonorgestrel for contraception [28]. The apparent acceptability of this system in this particular patient population indicates that nondegradable subdermal implants have broader acceptability than previously believed.

Many materials have been used in preparing nondegradable implants of peptides and similar compounds. Silicone rubber (Silastic) has been shown to have utility for at least short-term controlled release of LHRH in vivo [29] and for a variety of other macromolecules in vitro, including bovine serum albumin, pepsin, insulin, and chymotrypsin [30]. Silicone rubber has features of ease of device fabrication and well-recognized local tolerance and toxicological acceptability, and could well be an as yet underutilized means of delivery of peptides. There are some limitations on system design due to the inherent physicochemical properties of peptides, discussed further in Sec. IV.

A great deal of work has been done exploring the utility of ethylene vinyl acetate (EVA) for controlled release of polypeptides and other macromolecules, and the reader is referred to the extensive literature on the subject [31–34], including the work with insulin [35] culminating in a patented system [36]. EVA is a biocompatible polymer with some history of clinical use for parenteral controlled release. The broad information base available on it also renders it particularly useful as a matrix material. Preparation of devices, however, is a little more complex [31] than with silicone rubber, and EVA may be prepared with a wide range of hydrophobicity whereas silicone rubber does not offer this flexibility. Silicone rubber and EVA may be contrasted in many respects to hydrogels, hydrophilic polymers that will absorb water to a controllable extent and become selectively permeable to hydrophilic substances, including peptides. In this respect, the behavior of peptides in a hydrogel carrier is analogous to the behavior of steroids in a silicone rubber carrier, each class of compounds having useful permeability through the continuum of the carrier.

Hydrogels have been recognized as being useful for the design of controlled release implants of peptides for several years, since Davis recognized their utility for the delivery of such compounds as bovine serum albumin, luteinizing hormone, and insulin [37], using polyacrylamide/polyvinylpyrrolidone devices. A more common nondegradable hydrogel is poly(hydroxyethyl methacrylate) (methyl methacrylate) (HEMA/MMA), which may be cross-linked. The ratio of HEMA, the more hydrophilic component, to MMA, the more hydrophobic component, and the degree of cross-linking, determines the overall hydrophilicity, and permeability of the matrix. The physical nature of the water taken up by such systems has been studied [38]. It is believed that there are three classes of such water: bound, interfaced, and bulk. This model was expanded upon by Kim and co-workers [39], who concluded that the diffusion coefficients for hydrophilic solutes through hydrogel membranes depends on the molecular size of the solute and pore size of the membrane, diffusion occuring primarily in the bulk water. Similar studies with poly(vinyl alcohol) (PVA) showed this also to be a hydrophilic polymer with controllable permeability to water-soluble solutes, theophylline being in this case the model compound [40].

Polyacrylamide has been shown to be useful in vivo for controlled delivery of insulin [41], as has also the more widely used HEMA [42]. Similar acrylate hydrogels have been used to prepare a composite implant of leuprolide, an LHRH analog [43].

Hydrogels may also be designed to be degradable, offering another avenue of approach to system design. Heller and co-workers have designed degradable hydrogels of fumaric acid and poly(ethylene glycol) cross-linked with vinylpyrrolidone and related polymers [44], and have studied the release of bovine serum albumin as a model peptide from these carriers. Appropriate adjustment of critical parameters resulted in a system providing near-zero-order release in vitro. A more sophisticated system has been developed by Churchill and Hutchinson [45], in which the hydrogel has a biodegradable hydrophobic component and a hydrophilic component that may or may not be degradable. This has been shown to have potential utility for the controlled release of such compounds as bovine growth hormone, immunoglobulin A, epidermal growth factor, and the LHRH analog Zoladex. The field of hydrogels will undoubtedly offer opportunities for the future as the demands for delivery methodologies for these compounds continue to increase.

Another potential matrix, relatively unsophisticated, is cholesterol. Contrary to expectations, this is nondegradable. Prills blended with an appropriate amount of peptide, such as the LHRH analogs (D–Ala^6, Pro^9–NHEt) LHRH and (D–Trp^6, Pro^9–NHEt) LHRH, will compress to form implantable pellets [46,47]. These will then provide classical diffusion-controlled release from the porous matrix. This has seen limited application, probably due to practical difficulty in controlling the prilling and compaction processes for sufficiently tight control of the porosity of the matrix.

C. Precipitated and Absorbed Gels

The extended availability of insulin from an insoluble zinc complex has long been known and utilized. The same approach may be taken with other peptide compounds, and the LHRH analog (D–Trp^6, Pro^9–NHEt) LHRH has been thus formulated as a zinc tannate salt suspended in an aluminum monostearate/sesame oil gel [47]. This low-solubility salt system provided efficacy in the female rat for a median of 24 days. If close control of kinetics of release is not an issue, such a system may be developed much more rapidly than a more complex system and may very well have utility.

D. Viscosity-Controlled Systems

The temperature-dependent rheological properties of poloxamers (block copolymers of propylene and ethylene oxides), or pluronics, have been well characterized [48,49]. These compounds have the unusual property that their aqueous solutions increase in viscosity with increased temperature. Poloxamers and the related poloxamines may be designed such that an aqueous solution is liquid and injectable at room temperature, but gels postinjection at body temperature. This has recently been applied to the design of an injectable controlled-release system [50] useful for a wide variety of pharmaceuticals, including insulin. This is again a simple sustained-release (as opposed to controlled-release) system that may be very suitable for a particular delivery system. The only word of caution in the utilization of these polymers for delivery of peptides is their reported adjuvant activity in antibody formation [51], a phenomenon that would be intolerable for any system except perhaps a vaccine.

It is undoubtedly clear from the discussion and examples above that there are many possible approaches to the design of controlled-release systems for the delivery of peptides. The field of immobilization of enzymes provides some precedent for approaches to system design [52], and provides background in such areas as the use of liposomes as a carrier for enzymes and proteins, the use of erthrocytes as carriers, and covalent linkage of enzymatic substances to carriers. The field is developing at such a rate that the general literature is unavoidably well behind actual events in the field; the patent literature is in fact a much better source of information.

III. SELECTION OF CARRIER TYPE

Many balancing factors will contribute to the decision on selection of a particular carrier type. These will include considerations of the merits of a degradable versus nondegradable system, and the requirements on release kinetics. For a compound of relatively low therapeutic ratio, for example, close to zero-order release will probably be a requirement, and removability will certainly be an advantage, at least in early studies. The finite possibility of catastrophic failure of a reservoir or membrane-bounded device will probably be unacceptable for such a compound. The potency of the drug compound also influences the choice of carrier: a compound of low potency, required as a long-term controlled-release injectable, will require microencapsulation or other incorporation into a carrier at a relatively high loading level. A risk/benefit assessment of the clinical indication may also define bounds of choices of carrier available.

Other considerations include physicochemical compatibility of the compound with the carrier, including stability, ease of device preparation, ease and reproducibility of synthesis and availability of the carrier material, toxicological profile and data base available on the carrier, and extent of precedent and background information available. The latter factor may critically influence the developmental time of a proposed system. There are many carrier materials partially characterized that probably have advantages over others with proven performance. However, the task of actually demonstrating their acceptability may be prohibitive, at least until there is a compelling requirement for the particular features of that carrier. Some features under consideration among the major groups of carriers follow. A more extensive review of the preparation and properties of carrier materials for controlled release is provided by Baker [60].

A. Degradable Carriers

Biodegradable polymers may be designed around one of many types of labile bond. The relative rates of hydrolysis of these bonds under neutral condition is

$$\underset{\substack{\text{Poly-}\\\text{carbonates}}}{-O-\overset{\overset{\displaystyle O}{\|}}{C}-O-} \;>\; \underset{\text{Polyesters}}{-\overset{\overset{\displaystyle O}{\|}}{C}-O-} \;>\; \underset{\substack{\text{Poly-}\\\text{urethanes}}}{-NH-\overset{\overset{\displaystyle O}{\|}}{C}-O} \;>\; \underset{\substack{\text{Polyortho}\\\text{esters}}}{-O-\overset{\overset{\displaystyle |}{\underset{\displaystyle |}{CH}}}{}-O} \;>\; \underset{\text{Polyamides}}{-\overset{\overset{\displaystyle O}{\|}}{C}-NH-}$$

However, polymer morphology and the presence of substituents may also influence hydrolysis rates significantly.

Polyesters

Linear polyesters are by far the most widely characterized and utilized group of biodegradable polymers, the most significant among them being copolymers of lactic and glycolic acids (PLGA). A major factor in this is their long history of use and acceptability as erodible sutures [53], bone plates, and other temporary prostheses. A wide variety of pharmaceuticals have been so formulated [18].

The unique demands of peptides have provided a decisive thrust for the advancement of technology utilizing carriers such as PLGA. This polymer allows both injectable (microsphere) systems and single-unit implants to be prepared, is soluble in a wide range of solvents, and is malleable and ductile at moderate temperatures. The monomer ratio and the choice of l- or d,l-lactide provide a wide choice of hydrophilicity and degree of crystallinity, two factors paramount in determining hydrolytic rate. The degradation products, l-lactic acid and glycolic acid, are found endogenously, and any d-lactic acid will readily be excreted renally. In common with most degradable polymers, it has the feature of homogeneous or bulk erosion, in that when exposed to an aqueous environment, the polymer will hydrate and degrade more or less uniformly throughout.

Peptides, at least decapeptide LHRH analogs, appear to have minimal diffusivity through the intact polymer and are released primarily on degradation of the polymer matrix. This does not make the design of an optimally releasing system obvious since the polymer does not inherently provide zero-order release. It does, however, give gross independence from surface area effects, so that precise reproducibility of size of a microencapsulated system is a noncritical vairable. In practice, an early phase of limited diffusional release of compound from the surface of the system is seen, followed by a later phase of erosional release [8,18,27]. Appropriate adjustment of monomer ratio and molecular weight may provide sufficient overlap of these phases for continuous efficacy, and these factors have been elucidated in some detail [2]. The synthesis of PLGA to tightly reproducible specifications is a specialized technology that has been studied in depth [54]. The combination of history of acceptability of use, degradation to simple products with no toxicological questions, increasingly broad information base on which to draw, and flexibility in handling make this polymer uniquely useful. In the space of a few years, it has emerged as the polymer of general choice for a candidate carrier.

Other polyesters include polycaprolactone, reported to have surface-eroding properties, which has shown utility for controlled release of steroids [55] but has not to date been reported in application to peptides or proteins.

Poly (ortho esters) may be designed to be primarily surface eroding, as has been demonstrated in the long-term controlled release of steroids [56]. This provides control of release kinetics by control of surface area of the system, and zero-order release is undoubtedly achievable with appropriate system design. Short-term linear release in vitro and prolonged efficacy in vivo has been demonstrated with two such systems of an LHRH analog [26].

Polyhydroxybutyric acid has recently emerged as another candidate carrier material. It may be synthesized by fermentation from *Alcaligenes eutrophus*, novel methodology resulting in material of high purity. The polymer has been shown both in rats [22] and clinically [23] to be useful for controlled release of buserelin. This polymer is limited to not having the flexibility of hydrophilicity conferred by a copolymer and, being a novel material, by requiring demonstration of acceptability.

Poly(amino Acids)

Poly(glutamic acid) copolymers were developed by Sidman et al. [57] and have the potential for utility for controlled-release devices. However, they have not found wide application, because of the obvious question of potential immunogenicity of a poly(amino acid) carrier. Even through no such phenomenon has been reported, the consideration has been sufficient to limit their use to research studies rather than developmental use.

Polyanhydrides

Maleic anhydride/methyl vinyl ester copolymers have been well studied for controlled-release applications [58], but have not yet been reported in the context of delivery of peptides. These polymers are again believed to be surface eroding, with the advantages and disadvantages which that presents. More recently, another group of surface-eroding polyanhydrides has been developed [59], opening another avenue of opportunity in controlled-release system development.

Bioerodible Hydrogels

Bioerodible hydrogels have been mentioned in the context of fumaric acid/ poly(ethylene glycol) copolymers which have been used for controlled release of the model protein bovine serum albumin [44]. In the present state of the art hydrogels are the only class of polymer that may be designed such that the peptide will be permeable through the continuum of the carrier. A bioerodible hydrogel will therefore have the combination of features of a diffusion-controlled release, possibly augmented by later erosional release, without the requirement of removal of the system on depletion.

B. Nondegradable Carriers

Nondegradable carriers are restricted by the need for device removal and by the fact that only diffusional (rather than erosional) release is available for system design. Advantages include removability in case of need and relative simplicity of device design. The chemistry of the carrier is significantly less critical to the release profile of the compound than in the case of degradable polymers providing erosion-controlled release.

Silicone Rubber

This has proven utility for controlled release of macromolecules, as mentioned previously [29,30] and is well recognized as being acceptable for parenteral use. Peptides appear to have minimal diffusivity through these polymers, and the only release mechanism available is matrix leaching, or channeling. If this is not a limitation, and if a noninjectable implantable product is acceptable, a high loading level of soluble solids may well provide a useful diffusion-controlled (nonzero-order) delivery system (see Sec. IV).

Ethylene Vinyl Acetate

Ethylene vinyl acetate has seen a little less utilization than silicone rubber, but even so, has potential for similar design of single-unit systems. Its fabrication is a little more complex. However, it does offer a degree of flexibility in hydrophilicity by variation of the monomer ratio, which may be used to tailor the design of the system.

Hydrogels

Hydrogels offer properties opposite those of more hydrophobic polymers, and are appropriate when the peptide is required to diffuse through the intact polymer. Hydrogels of different types have been well characterized for their utility as drug delivery systems [61], and the various approaches to system design have been explored in depth.

   To summarize, the choice of carrier material is dictated by a number of considerations as follows: (1) requirement for an encapsulated injectable product versus an implant, (2) requirements on linearity of release kinetics, (3) requirements for removability of system, and (4) potency of compound versus loading level feasible. It should be borne in mind that in vivo, a release profile may be modified by formation of a capsule around this system. For this reason, in vitro data should be viewed with caution until validated by parallel in vivo studies.

IV. DESIGN OF CONTROLLED DELIVERY SYSTEMS

A. Monolithic Systems

Monolithic systems are characterized by having a more or less homogeneous and uniform dispersion of the compound within the carrier matrix, defined on either a macro or a molecular scale. The system may then range from a solid solution (molecular dispersion) to a matrix of discrete particles throughout the carrier. Release of the compound will be by diffusion and will be controlled by the physicochemical properties of the components, the loading level of compound within the system, and system geometry. The simplicity of design of such a system, together with the absence of disadvantages of reservoir devices (where minor defects may result in complete system failure) and low fabrication costs, often outweigh the main disadvantage of a monolithic system, that of diminishing release rate with time.

Low-Loading Systems

The two main types of rate control from a monolithic system may be achieved by adjusting the loading level of the system. At low loading levels, the majority of the particles will be isolated, each being completely surrounded by carrier material. The only mechanism for release of such particles is by dissolution and diffusion through the continuum of the carrier, which must be selected for good permeability properties. Hydrogels are a good candidate, as discussed earlier. The primary factors then controlling release are the permeability of the peptide through the polymer and geometry (surface area, volume, shape, and membrane thickness) of the device.

High-Loading System

As the loading level increases, particles occasionally become sited such that
they are in immediate contact with a similar particle. This phenomenon increases with increase in loading level, to a point where most particles have
such contact. The average number of contacts made by a particle as a function of loading level of a binary system have been calculated (for a model
system of soft spheres) [62], and this has been applied to the design of
controlled-release systems [63]. A particle at the surface will dissolve and
diffuse away, leaving a cavity and an exposed surface on any connected
particle, which will then also be free to dissolve and diffuse out. A system
of channels will thus be created as compound is lost. The connectivity of
the channels, depending on the initial loading level of compound and tortuosity, depends on the size and shape of the particles originally occupying the
space. The loading level and compound morphology therefore determine diffusional resistance of the matrix of channels or pores, and consequently,
control the release of compound by this mechanism.

Release from these monolithic dispersions may also be described by
percolation theory, which again analyzes interparticulate contact and consequent channel formation as a function of loading level [64]. For both models,
the critical loading level at which a continuous network is formed (i.e., each
particle has at least two contact points) is about 15%. Below this level, network connectivity will be reduced and diffusional resistance will be increased.
Duration of release will be increased, at the expense of reduction in rate of
release and reduction in total amount available for release. Those particles
having zero contact points will be "entombed." Converse features occur at
higher loading levels.

Clearly, the size and shape of particles will also be critical factors. It
has been shown that for a high-loading matrix of albumin in silicone rubber
the particle size has profound effects, as does the dose's loading level [30].
Also, the points of particulate contact may provide constrictions in the channels, which then impede diffusional flow of peptide [31]. Close physicochemical characterization of a system will therefore be required for reproducible fabrication and performance.

Carriers that have low permeability to the peptide are useful in the
design of high-loading matrix systems. These would include EVA [31],
silicone rubber [30], and degradable polymers such as the polyesters. The
properties of the peptide are, as described, the main controlling parameters.
The properties of the carrier are relatively noncritical unless they independently contribute degradability or diffusion through the polymer
continuum.

B. Erosion-Controlled Systems

In an erosion-controlled system, the nature of the carrier is paramount and
the characteristics of the peptide are relatively insignificant. A wide variety
of degradable polymers are available, but as described in Sec. III, polyesters, particularly PLGA, are particularly widely used for peptides (as
well as for other pharmaceuticals). The majority of these materials provide
bulk degradation, with onset of erosion when the molecular weight profile

has diminished to a critical level. As the diffusional resistance of the degrading matrix decreases in time, the rate of compound release increases,
which tends to increase linearity of release and provides abrupt termination.
The flexibility of adjustment of release kinetics, which may make possible
sufficiently close to linear kinetics, together with complete availability of
compound, relatively abrupt termination of release, and no requirement for
device removal, will frequently make an erosion-controlled system the system of choice.

C. Reservoir Systems

A reservoir or membrane-limited system is characterized by having a com-
pound-rich central core and a rate-limiting external membrane. The rate of
release is controlled by the geometry (area) of the device, the thickness
and permeability of the membrane to the compound, and the concentration
gradient across the membrane. If the internal reservoir is maintained at
saturation with excess solid, zero-order or linear release is theoretically
possible. With the choice of appropriately permeable membrane material, a
device providing optimum release kinetics with maximum efficiency of use of
compound may be designed.

Two reservoir system designs are possible: (1) utilizing a polymer such
as a hydrogel for the external membrane, or (2) using an impermeable polymer
made with a microporous structure. The latter concept has been put into
practice in the case of vasopression or other peptides contained within a
microporous polypropylene membrane [65]. The nature of the pores (diameter, length, tortuosity, and number) is a determining factor in the release
rate provided by the reservoir.

There are major disadvantages to a reservoir system, however, since a
crack or pinhole will result in complete system failure, and the complexity
and difficulty of manufacture will increase manufacturing costs. These disadvantages contribute to the present dearth of such systems for peptides.

D. Pumps

Mechanical pumps, invaluable as a tool to study the properties of a compound
when delivered continuously, provided the earliest means of such delivery.
Implantable pumps have been developed [66] that operate on the principle of
osmotic imbibition of water generating an internal pressure that expels the
contents of a reservoir containing the compound, at a rate determined by the
design of the pump. This technology has made possible the ready characterization of the profiles of drug substances as a function of the delivery
regimen. This is particularly important in the case of hormones and their
synthetic analogs, in which rate and nature of profile of release are critical
to the response, in particular in the case of peptide hormones secreted in a
pulsatile manner.

It is also possible to delivery compounds clinically at a predetermined
rate by use of a pump worn externally. This technology has been applied to
insulin and has also been used for the pulsatile delivery of LHRH [67].

V. TOXICOLOGICAL AND BIOCOMPATIBILITY ISSUES

Degradable polymers as well as nondegradables that may contain leachable
components present questions as to the toxicological acceptability of those

materials that become available systemically. A related issue is that of biocompatibility, or acceptability, in terms of possible local reactions to the implanted or injected system.

The toxicological issue may be addressed by conventional animal studies over an appropriate period. A problem that has arisen in the development of a number of potential carrier systems is that of prohibitive requirements in demonstrating toxicological acceptability of a material, even though the outcome is judged likely to be positive. This is a practical as well as a very real constraint in the selection of a carrier material that separates systems of academic interest for use in model systems from those that have real practical potential. This issue has led to an expansion of the use of polymers that have a proven history of use, such as PLGA, at the expense of more novel systems.

Local tissue response is a ubiquitous phenomenon that may or may not have significance. In its simplest form, a capsule may form around the parenterally-administered system. Although this may have no toxicological consequences, typically being resolved on removal of the device, the capsule may modify compound release from the system [68]. This phenomenon has not yet been reported for peptides, but it would be anticipated that with the typically low permeability through biological membranes, such a capsule may provide a rate-limiting barrier for a peptide. In such a case, rate control would not be provided by the device but by a variable biological phenomenon, a less than ideal situation. In vivo release rates would also not then parallel in vitro release rates.

Models have been established for biocompatibility testing; in particular an in vitro method [69] has been shown to parallel an in vivo model [70] for a wide variety of polymeric materials. The rabbit cornea [71] has been shown to be a useful in vivo model in evaluating the biocompatibility of EVA, HEMA, and other less innocuous polymers.

Of the degradable polymers, a series of copoly(amino acids) have been shown to have biocompatibility correlating with hydrophilicity of polymer [72]. PLGA has been characterized in terms of tissue reaction and has been reported to give a totally nonadverse reaction [73]. On the other hand, HEMA, which has been generally recognized to be similarly biocompatible [71,74], has more recently been suggested not to be suitable as an implant material [75]. The compatibility with blood, or thrombogenic properties, of a polymer also contributes to biocompatibility profile, and this has been studied for a series of polymers used in controlled-release devices [76].

The factors that determine biocompatibility of a polymer have not been fully elucidated, although it is suggested that they may include hydrophilicity, surface charge, interfacial tension, and mechanical features. As work continues it may become possible to predict biocompatibility from a knowledge of the physicochemical properties of the system, but in the interim, each system will require independent evaluation.

VI. FUTURE TRENDS

The field of controlled delivery of peptides will undoubtedly continue its rapid expansion for as long as the fields of genetic engineering and peptide hormone isolation and sequencing continue theirs. The trend has been toward a narrowing of focus as certain system types have proven more feasible than others, and also toward more developmental and applications-oriented work rather than solely theoretical design studies. A number of areas are clearly

ripe for rapid growth. These include pulsatile delivery systems and biofeedback systems, as well as systems that release on demand under external
control.

A. Biofeedback Systems

Biofeedback or self-regulated systems that regulate drug output according
to need are in a sense the ideal system type. The concept represents a
tremendously more complex developmental technology, in that a sensor is
required and a means of control of release rate dependent on the sensor
output. Systems that release insulin in response to a change in glucose
level have already been designed and are under study [77]. Applications of
this and similar technologies for compounds where availability is required
contingent only on another biological event will provide one major avenue of
research for the future [78].

B. Pulsatile Systems

It is well recognized that intermittent or pulsatile delivery may be a necessity
for particular peptide hormones for particular applications. This is especially true of the releasing factors, where the endogenous compound is secreted
in a pulsatile manner and the response is dependent on this. Pulsatile delivery may be achieved by use of a programmable pump worn externally [67].
Although there is presently much discussion and interest around the concept
of a self-contained parenteral system providing pulsatile release, such a system has not yet been realized. This will undoubtedly be an avenue of activity for the future.

C. On-Demand Systems

An example of an on-demand system is that of EVA/bovine serum albumin
containing magnetic beads [33,79]. When the device is implanted, an externally applied magnetic field greatly increases the rate of release of the
compound. There is opportunity for expansion of this concept to design
devices for this type of on-demand release of peptides as well as other compounds, and this will also provide an area of potential future growth.

D. Liposomes

Another potential means of controlled delivery of peptides is by entrapment
in liposomes. This has been shown to afford protection to enzymatic degradation for both insulin [80] and proteins [81]. Although this technology has
not yet seen a general application for delivery of peptides, it is an area of
possible future growth.

E. Nonparenteral Routes

The foregoing discussion has focused on parenteral controlled-release systems
since the parenteral route is currently the one of general choice for peptides.
However, the future will no doubt see other routes of delivery being explored
for controlled release. For example, the nasal route is becoming well established

for the delivery of peptides (Chap. 13), and it is conceivable that a system could be developed for nasal delivery that provides prolonged availability through diffusion-controlled release to the nasal mucosa. Another route of delivery under study is the rectal route (Chap. 12); with the limitations of absorption by this route, a controlled-release suppository could certainly be designed to give the required release profile.

The transdermal route has been examined for delivery of peptides, but so far with limited success (Chap. 15). It is generally recognized that peptides beyond about 1000 daltons will not permeate the stratum corneum by conventional means. Promising data seen in animals models [82] have not been reproduced in primates when care was taken to maintain the integrity of the stratum corneum (i.e., avoiding shaving or depilation). Another report on the transdermal availability of LHRH in dimethyl sulfoxide was also developed from a model in which the animal was shaved [83]. Insulin and vasopressin have been reported to be iontophoretically absorbed across the skin [84], but the practical utility of the system, particularly in light of the current required, remains to be determined. These data are, however, sufficiently thought provoking to stimulate further work, and the transdermal route may well become a portal of entry for some peptides.

Of other opportunities, the vaginal route has been shown to be useful for the LHRH analog leuprolide [85], and this may have applicability for controlled-release suppositories of other compounds (Chap. 14). Curiously, the buccal membrane appears to be effectively impermeable to peptides, and at the current state of knowledge a controlled-release buccal tablet has little promise (Chap. 11).

In conclusion, the state of the art is presently very dynamic and will undoubtedly continue its rapid multidisciplinary growth for years to come. The directions of this growth will depend on the ingenuity of pharmaceutical scientists designing and developing the systems. The level of technology required is far beyond that required in conventional dosage form design, and the possibilities for system development will increase as the basic material technology advances.

REFERENCES

1. M. Saffran, G. S. Kumar, C. Savariar, J. C. Burnham, F. Williams, and D. C. Neckers, *Science, 223:*1081 (1986).
2. L. M. Sanders, K. M. Vitale, G. I. McRae, and P. B. Mishky, in *Proceedings of NATO Workshop on Advanced Drug Delivery Systems for Peptides and Proteins*, Plenum Press, New York (1986), p. 125.
3. J. J. Nestor, Jr., R. Tahilramani, T. L. Ho, G. I. McRae, and B. H. Vickery, in *Peptides, Structure and Function, Proceedings of the 8th American Peptide Symposium* (V. J. Hruby and D. H. Rich, eds.), Pierce Chemical Company, Rockford, Ill. (1983), p. 861.
4. B. K. Green and L. Schleicher, U.S. Patent 2,800,457 (July 1957).
5. C. Thies, *CRC Crit. Rev. Biomedical Eng., 8:*335 (1982).
6. J. S. Kent, L. M. Sanders, T. R. Tice, and D. H. Lewis, in *Long Acting Contraceptive Delivery Systems* (G. I. Zatuchni, A. Goldsmith, J. D. Shelton, and J. J. Sciarra, eds.), Harper & Row Publishers, New York (1984), p. 169.

7. J. S. Kent, L. M. Sanders, D. H. Lewis, and T. R. Tice, European
    Patent 52,510 (Cl A61K9/50) (1986).
8. L. M. Sanders, J. S. Kent, G. I. McRae, B. H. Vickery, T. R. Tice,
    and D. H. Lewis, *J. Pharm. Sci.*, *73*:1294 (1984).
9. B. H. Vickery and G. I. McRae, in *LHRH and Its Analogs: Contra-
    ceptive and Therapeutic Applications* (B. H. Vickery, J. J. Nestor,
    Jr., and E. S. E. Hafez, eds.), MTP Press, Lancaster, England
    (1984), p. 91.
10. P. G. Hoffman, M. R. Henzl, M. D. Chaplin, and C. A. Nerenberg,
    J. Androl., *8*:ST7 (1987).
11. A. H. Martin, J. Swarbrick, and A. Cammarata, eds., *Physical
    Pharmacy*, 2nd ed., Lea & Febiger, Philadelphia (1969), Chap. 18,
    p. 520.
12. B. H. Vickery, G. I. McRae, L. M. Sanders, P. Hoffman, and S. N.
    Pavlou, in *Proceedings of International Symposium on Hormonal
    Manipulation of Cancer: Peptides, Growth Factors and New Anti-
    steroidal Agents*, Raven Press, New York (1986).
13. H. Parmar, S. L. Lightman, L. Allen, R. H. Phillips, L. Edwards,
    and A. V. Schally, *Lancet*, p. 1201 (Nov. 30, 1985).
14. H. Okada, Y. Ogawa, and T. Yashiki, European Patent Application
    0145240, (Cl A61K9/52) Application 84307506 to Takeda Chemical
    Industries, Ltd. (2nd November 1984).
15. T. Shimamoto, *3rd Int. Cong. Androl.*, Boston (Apr. 1985).
16. T. M. S. Chang, *J. Bioeng.*, *1*:25 (1976).
17. M. F. A. Goosen and A. M. F. Sun, European Patent Application
    134,318 (Cl A61K9/22) to Connaught Labs, Ltc. (Mar. 20, 1985).
18. L. M. Sanders, B. A. Kell, G. I. McRae, and G. W. Whitehead,
    J. Pharm. Sci., *75*:356 (1986).
19. F. G. Hutchinson and B. J. A. Furr, *Biochem. Soc. Trans.*, *13*:520
    (1985).
20. A. D. Turkes, W. B. Peeling and K. Griffiths, *J. Steroid Biochem.*,
    27:543 (1987).
21. M. Asano, M. Yoshida, I. Kaetsu, K. Imai, T. Mashimo, H. Yuasa,
    H. Yamanaka, K. Suzuki, and I. Yamazaki, *Makromol. Chem. Rapid
    Commun.*, *6*:509 (1985).
22. W. Koenig, H. R. Seidel, and J. K. Sandow, European Patent Applica-
    tion 133,988 (Cl A61K37/02) to Hoechst A.G. (Mar. 13, 1985).
23. J. H. Waxman, J. Sandow, P. J. Magill, and R. T. D. Oliver, *Symp.
    Treatment Adv. Prostatic Cancer Role LHRH—Superagonists*, Baden
    (June 1985).
24. J. Heller, B. K. Fritzinger, S. Y. Ng, and D. W. H. Penhale,
    J. Controlled Release, *1*:225 (1985).
25. J. Heller, B. K. Fritzinger, S. Y. Ng, and D. W. H. Penhale,
    J. Controlled Release, *5*:173 (1985).
26. J. Heller, L. M. Sanders, P. Mishky, and S. Y. Ng, *Proc. Int.
    Symp. Cont. Rel. of Bioact. Mater.*, *13* (1986) p. 69.
27. F. G. Hutchinson, European Patent Application 58,481 (Cl A61K37/02)
    to Imperial Chemical Industries PLC (1982).
28. S. Roy, D. R. Mishell, D. N. Robertson, R. M. Krauss, and M. J.
    Duda, *Am. J. Obstet. Gynecol.*, *148*:1006 (1984).
29. W. Lotz and B. Syllwasscy, *J. Pharm. Pharmacol.*, *31*:649 (1979).
30. D. S. T. Hsieh, C. C. Chiang, and D. S. Desai, *Pharm. Technol.*,
    9:39 (1985).

31. R. A. Siegel and R. Langer, *Pharm. Res. N.Y.*, *1*:2 (1984).
32. R. Langer, D. S. T. Hsieh, W. Rhine, and J. Folkman, *J. Membr. Sci.*, 7:333 (1980).
33. R. Langer, D. S. T. Hsieh, A. Peil, R. Bawa, and W. Rhine, *AIChE Symp. Ser. 206*, 77:10 (1981).
34. R. Langer, *Chemtech*, *12*:98 (1982).
35. H. M. Creque, R. Langer, and J. Folkman, *Diabetes*, *29*:37 (1980).
36. M. J. Folkman and R. Langer, U.S. Patent 4,164,560, Application 756,892 (Aug. 14, 1979).
37. B. K. Davis, *Proc. Natl. Acad. Sci. USA*, *71*:3120 (1974).
38. H. B. Lee, M. S. John, and J. D. Andrade, *J. Colloid Interface Sci.*, *51*:225 (1975).
39. S. W. Kim, J. R. Cardinal, S. Wisniewski, and G. M. Zentner, in *Water in Polymers* (S. P. Rowland, ed.), ACS Symposium Series, Vol. 127, No. 20 (1980), p. 347.
40. R. W. Korsmeyer and N. A. Peppas, *J. Membr. Sci.*, *9*:211 (1981).
41. B. K. Davis, *Experientia*, *28*:348 (1972).
42. K. Ishihara, K. Mineko, and I. Shinohara, *Polym. J.*, *16*:467 (1984).
43. M. Yoshida, M. Asano, I. Kaetsu, K. Nakai, H. Yamanaka, K. Shida, and A. Shiraishi, German Patent DE 3,421,065 (Cl A61K37/43) to Japan Atomic Energy Research Institute, Takeda Chemical Industries, Ltd. (Dec. 13, 1984).
44. J. Heller, R. F. Helwing, R. W. Baker, and M. E. Tuttle, *Biomaterials*, *4*:262 (1983).
45. J. R. Churchill and F. G. Hutchinson, European Patent Application 92,918 (Cl A61K9/22) to Imperial Chemical Industries PLC (Mar. 1983).
46. J. S. Kent, B. H. Vickery, and G. I. McRae, *Proc. 7th Symp. Controlled Release Bioact. Mater.*, Fort Lauderdale, Fla. (1980) p. 67.
47. B. H. Vickery, G. I. McRae, L. M. Sanders, J. S. Kent, and J. J. Nestor, Jr., in *Long Acting Contraceptive Delivery Systems* (G. I. Zatuchni, A. Goldsmith, J. D. Shelton, and J. J. Sciarra, eds.), Harper & Row Publishers, New York (1984), p. 180.
48. I. R. Schmolka and L. R. Bacon, *J. Am. Oil Chem. Soc.*, *44*:559 (1967).
49. M. Vadnere, G. Amidon, S. Lindenbaum, and J. L. Haslam, *Int. J. Pharm.*, *22*:207 (1984).
50. J. L. Haslam, T. Higuchi, and A. R. Mlodozeniec, U.S. Patent 4,474,752 (Cl A61K31/415) to Merck & Company (Oct. 2, 1984).
51. R. L. Hunter and B. Bennett, *J. Immunol.*, *133*:3167 (1984).
52. T. M. S. Chang, ed., *Biomedical Applications of Immobilized Enzymes and Proteins*, Vol. 1, Plenum Press, New York (1977).
53. A. K. Schneider, U.S. Patent 3,636,956 (1972).
54. D. K. Gilding and A. M. Reed, *Polymer*, *20*:1459 (1979).
55. C. G. Pitt, R. W. Hendren, A. Schindler, and S. C. Woodward, *J. Controlled Release*, *1*:3 (1984).
56. J. Heller, *J. Controlled Release*, *2*:167 (1985).
57. K. R. Sidman, A. D. Schwope, W. D. Steber, S. E. Rudolph, and S. B. Poulin, *Polym. Prepr.*, *20*:27 (1979).
58. J. Heller, *Biomaterials*, *1*:51 (1980).
59. K. W. Leong, B. C. Brott, and R. J. Langer, *Biomed. Mater. Res.*, *19*:941 (1985).

60. R. W. Baker, *Controlled Release of Biologically Active Agents*, John Wiley & Sons, New York (1987).

61. W. E. Roorda, H. E. Bodde, A. G. DeBoer, and J. E. Junginger, *Pharm. Weekbl. Sci. Ed.*, *8*:165 (1986).

62. G. A. Moore, *Proceedings Fourth International Congress for Sterology*, NBS Spec. Publ. National Bureau of Standards, 431, Gaithersburg, Md. (1976), p. 41.

63. R. G. Wheeler and P. G. Friel, in *Controlled Release of Pesticides and Pharmaceuticals* (D. H. Lewis, ed.), Plenum Press, New York (1981), p. 111.

64. T. D. Gierke and W. Y. Hsu, in *Perfluorinated Ionomer Membranes* (A. Eisenbery and H. L. Yaeger, eds.), ACS Sympossium Series (1982), p. 283.

65. J. Kruisbrink and G. J. Boer, *J. Pharm. Sci.*, *73*:1713 (1984).

66. F. Theeuwes, U.S. Patent 3,760,984 (Sept. 1973).

67. G. R. Chambers, I. A. Sutherland, S. White, P. Mason, and H. S. Jacobs, *Upsala J. Med. Sci.*, *89*:91 (1984).

68. J. M. Anderson, H. Niven, J. Palagalli, L. S. Olanoff, and R. D. Jones, *J. Biomed. Mater. Res.*, *15*:889 (1981).

69. R. M. Rice, A. F. Hegyeli, S. J. Gourlay, C. W. R. Wade, J. B. Dillon, H. R. Jaffe, and R. K. Kulkarni, *J. Biomed. Mater. Res.*, *12*:43 (1978).

70. S. J. Gourlay, R. M. Rice, A. F. Hegyeli, C. W. R. Wade, J. B. Dillon, H. R. Jaffe, and R. K. Kulkarni, *J. Biomed. Mater. Res.*, *12*:219 (1978).

71. R. Langer, H. Brem, and D. Tapper, *J. Biomed. Mater. Res.*, *15*:267 (1981).

72. K. W. Marck, C. R. H. Wilvuur, W. L. Sederal, A. Bantjes, and J. Feijen, *J. Biomed. Mater. Res.*, *11*:405 (1977).

73. G. E. Visscher, R. L. Robinson, H. V. Maulding, J. W. Fong, J. E. Pearson, and G. J. Argentieri, *J. Biomed. Mater. Res.*, *20*:667 (1986).

74. R. A. Abrahams and S. H. Ronel, *Polym. Prep. 16*:535 (1975).

75. Y. Imai and E. Masuhara, *J. Biomed. Mater. Res.*, *16*:609 (1982).

76. J. M. Anderson and D. F. Gibbons, *Biomater. Med. Devices Artif. Organs*, *2*(3) (1974).

77. J. Heller, D. W. H. Penhale, and B. K. Fritzinger, *Proc. Int. Symp. Controlled Release Bioact. Mater.*, *13*:37 (1986).

78. J. Heller, *Med. Device Diagn. Ind.*, *6*:61 (1984).

79. D. S. T. Hsieh, R. Langer, and J. Folkman, *Proc. Natl. Acad. Sci. USA*, *78*:1863 (1981).

80. C. Weingarten, A. Moufti, J. Delattre, F. Puisieux, and P. Couvuer, *Int. J. Pharm.*, *26*:251 (1985).

81. G. Adrian and L. Huang, *Biochemistry*, *18*:5610 (1979).

82. Y. Uda and M. Yamada, European Patent Application 127,426 (Cl A61K47/00) (May 22, 1984).

83. J. R. Ree, R. R. Humphrey, J. W. Vaitkus, and W. C. Dermody, *Endocr. Res. Commun.*, *2*:327 (1975).

84. Y. Sun, O. Siddiqui, J. C. Liu, Y. W. Chien, W. Shi, and J. Li, *Proc. Int. Symp. Controlled Release Bioact. Mater.*, *13*:175 (1986).

85. H. Okada, I. Yamazaki, Y. Ogawa, S. Hiai, T. Yashiki, and H. Mima, *J. Pharm. Sci.*, *71*:136 (1982).

20

Characterization of the Toxicity Profile of Peptide and Protein Therapeutics

B. J. Marafino, Jr., and J. R. Kopplin

Cetus Corporation, Emeryville, California

> All substances are poisons; there is none which is
> not a poison. The right dose differentiates a
> poison and a remedy.
>
> Paracelsus (1493–1541)

I. INTRODUCTION

The advent of recombinant DNA and hybridoma technologies has caused an explosion in the number of potential peptide and protein therapeutics in the armamentarium of both the human and veterinary clinician. As well, these new fields of biomedical research have provided novel substances for use in the food, beverage, and agricultural industries. In the less than 20 years since the twin discoveries of the construction of biologically active plasmids [1] and of the fusion process necessary for murine hybridoma development [2], a burgeoning "cloning" technology has developed that is revolutionizing the pharmaceutical and other industries.

Today's developmental scientists in the biotechnology field, as distinguished from some of their counterparts in the traditional pharmaceutical field, often start by choosing a disease or metabolic process, and then use the techniques of recombinant research to identify the peptide or protein which can treat or modify that process. Only then is the substance manufactured.

Where previously available, many of these substances, such as the interferons (IFNs), were only crude biologic preparations, sometimes of less than 1% purity and contaminated with unidentified and potentially interactive agents [3]. Gone are the days when these agents were of questionable purity and potency and were assayable only in tests of biologic activity of little more sophistication than that of the tests used to evaluate the potency of the galenicals a century ago. With the development of appropriate immuno- and receptor-binding assays, these substances can now be accurately quantified in picogram amounts in physiologic substrates. Further, these agents can often be produced in dekagram quantities in a relatively short period of time; the proteins are often expressed at levels which are a significant fraction of the total protein synthesized by the cell [4]. The pharmacologic agents produced by recombinant DNA technology can rightfully be called drug entities, frequently with 99+% purity [5].

All this notwithstanding, many of these polypeptide entities are powerful, potent biologic agents. Seldom are there sufficient a priori physiologic and pharmacologic bases of knowledge to allow their appropriate use in a clinical setting. Some of these substances are clearly of therapeutic value to correct a deficiency state (e.g., human growth hormone, human insulin, erythropoietin, procoagulant Factor VIII, and albumin). Others, such as many of the more newly discovered cytokines, are only inferred to be of therapeutic value, as, for example, antineoplastic agents, that inference originating from in vitro demonstrations of their biologic activity. Further, knowledge of the appropriate dose level, route of administration, frequency of dosing, and indeed the dosing schedule usually cannot be obtained until in vivo biologists are provided with sufficient quantities of the drug. Often these agents are administered systemically, as though they were endocrine hormones, when in fact they may eventually be discovered to be paracrine or even autocrine substances. Often too, there exists a gray region where it is difficult to decide whether or not a polypeptide or protein is actually a xenobiotic in the species to which it is being administered.

It is therefore important to begin a discussion of practical problems that may be encountered in the characterization of the toxicity profile of protein and peptide therapeutics, as a portion of the development process of these substances as therapeutic agents. The ideas put forth and the arguments made in this chapter are a blend of our experiences in developing more than a dozen proteins for clinical use and the experiences reported in the literature.

II. TOXICITY OF NATURAL SUBSTANCES

A. Fallacy of the Safety of Natural Substances

It is inappropriate to assume that products normally found in nature are inherently safer than xenobiotics, or that agents normally found in the human body in physiologic quantities are therefore safe in pharmacologic quantities. There are a host of naturally occurring substances that have been demonstrated to exert toxic responses in various animal and clinical settings (Tables 1 and 2). Thus a real need exists to perform toxicologic testing of these agents in preclinical studies. Some of the preliminary information concerning the safety of these drug candidates will fall naturally from the early work investigating their in vivo efficacy and biodistribution in various animal models. Nonetheless, formal studies of the toxicity of these compounds must be performed in animals for a variety of very different reasons.

B. Reasons for Studying Protein and Peptide Toxicity

Frequently, there are no data to suggest the possible mechanism(s) of pharmacologic and/or toxicologic actions of these often novel agents in intact

Table 1   Naturally Occurring Botanical and
Microbial Toxins

Cardiac glycosides	Veratrum alkaloids
Botulinum toxin	Ricin
Aflatoxins	4-Ipomeanol
Dicoumarin	Pyrrolizidines
Ergot alkaloids	Goitrogenic glycosides

mammalian systems. Nor are there any data on the maximally tolerated dose
by a given route of administration, and on the dose-limiting toxicity and
target organs involved, with dosing in the clinical setting.   These data can
frequently be obtained from preclinical studies.   Additionally, data are needed (1) to confirm the intrinsic and interactive safety of the proposed formulation to be used in the clinic, and (2) to ensure that the manufacturing
and formulation processes will provide material which is biologically active,
stable, and free from contaminating substances that might decrease efficacy
and/or increase toxicity.

Also inherent in the performance of preclinical studies is the necessity
to satisfy the regulatory requirements of the U.S. Food and Drug Administration (FDA) with regard to the testing of proposed pharmaceutical agents.
To date the FDA has considered the extent and magnitude of testing of protein therapeutics to be decided on a case-by-case basis.   In our experience,
we have often found it sufficient to perform studies, under Good Laboratory
Practice conditions (according to Title 21, Code of Federal Regulations, Section 58, [6]), in two species (rodent and nonrodent) and as acute and subacute dosings by the route proposed to be used in the clinic.   The duration
of subacute dosing is often related to the anticipated duration of clinical
therapy.   Special studies are then dictated by the findings in these first-
tier examinations.   Further, it is most prudent to consider performing a
parenteral irritation study if the route of administration is to be other than
intravenous.   But again, the FDA considers the testing to be performed on
a case-by-case basis and has no rigid regulations in this regard.   However,
it has from time to time issued various "Points to Consider" documents, to
which the reader may wish to refer (see Ref. 7).   Other reasons, concerning bioethics and product liability considerations, should be self-evident and
will not be discussed here.

Table 2   Naturally Occurring Animal Toxins

Estrogen	Vitamins A and D
Ciguatoxin	Various snake and toad venoms
Tetrodotoxin	Cantharidin
Corticosteroids	Adrenocorticotrophic hormone
Insulin	Arthropod venoms

III. OVERVIEW OF THE TOXICITY STUDY DESIGN

A. Rationale for Testing

The overriding concerns behind all preclinical testing are threefold:

1. To provide an estimate of the starting dose for clinical trials
2. To set an escalation increment and schedule in dosing
3. Where possible, to alert the clinician to toxic clinical and laboratory signs that should be anticipated in phase 1 human safety studies

The prime purposes of any preclinical testing are to maximize the probability of detecting potential human clinical toxic responses, and to alert the the clinician to these signs. In clinical studies, the Latin maxim *primum non nocere* applies; that is, "first, do no harm."

B. Species Selection

Of unquestionably the most importance in the design, conduct, and interpretation of any toxicity study is the choice of species, whether the compound to be tested is proteinaceous or xenobiotic in character. With the testing of human proteins and peptides, the choice takes on an even higher level of significance. This is because frequently there are no precedent-setting compounds that might be useful in predicting the appropriate test species. However, there are steps to be followed from the more well-traveled road of preclinical pharmacologic and toxicologic assessment of xenobiotics. Similar to the responses of these traditional compounds, a sigmoidal log dose response curve is commonly observed with increasing doses of protein therapeutics in responsive species. The inference, then, is that the population of that particular test species displays a log-normal (i.e., log-Gaussian) distribution of receptor numbers (or affinities) for the test protein, as it does for various low-molecular-weight, lipid-soluble xenobiotic compounds.

Often the toxicities seen in the clinic at maximally tolerated doses are *exuberant pharmacologic responses*: effects that are predictable as one moves to the right on a particular dose response curve. It then becomes axiomatic that one toxicologic criterion involved in the choice of species is that the species be capable of mounting a pharmacologic response to the test article. To conduct a toxicity study in a pharmacologically unresponsive species is as inappropriate as it is to conduct a toxicity study at sub-pharmacologic doses.

For some substances, this species choice does not present a concern. Insulin (INS) and interleukin-2 (IL-2) are proteins that have been shown to demonstrate the same response in laboratory animals as in humans. INS is capable of reducing blood glucose in all species tested [8], and IL-2 is capable of evincing responses with treatment of a number of different species [9]. Tissue-type plasminogen activator (t-PA) has been shown to lyse experimentally induced fibrin clots very well in rabbits [10], while the human growth hormone (GH) bioassay employs somatic growth of hypophysectomized rats to gauge therapeutic potency [11]. The IFNs, on the other hand, are much more species selective, and generally show neither antiviral nor antineoplastic efficacy in nonprimate species [12]. However, this is not always the case [13,14], and this species selectivity has been abolished by encapsulation of an IFN in liposomes [15]. Thus, prior to toxicologic testing in earnest, with the possible exception of the agents based on or consisting of monoclonal antibodies (MAbs), it is prudent to perform cross-species efficacy comparisons to evaluate relative species responsiveness to these proteins in various common laboratory animal species (i.e., rat, mouse, hamster, rabbit, guinea pig, dog) or in other species [16]. It would not be necessary, however, to demonstrate antitumor efficacy in a nonhuman primate by administering an IFN, since at present no good experimental tumor model exists in a nonhuman primate. Testing in one of various antiviral models would be appropriate in this case [17–19].

C. Proof of Efficacy Insufficient

A note of caution should probably be introduced at this point. Although it is generally a *necessary* condition in the species selection process that efficacy be demonstrated, usually that is not a *sufficient* condition. The appropriateness of the animal model should be investigated in light of comparative species data available. Differences in pharmacokinetics and/or pharmacodynamics, as well as in metabolism and tissue distribution, may tend to under- (or rarely over-) estimate quantitative aspects of the toxicity assessment. For example, murine MAbs can be produced against human tumors and are intended for clinical diagnostic and/or antineoplastic therapy. These substances can easily be shown to bind to human xenografts in athymic nude mice [20]. However, the pharmacokinetics of these agents has been shown to be dramatically different when examined in nonhomologous systems [21,22]. As the toxicity of MAb-based therapeutics can reasonably be expected to correlate with certain pharmacokinetic parameters related to exposure (i.e., half-life and volume of distribution), the differences in pharmacokinetics in different species may lead to differences in toxicity in those species [22]. The converse, however, need not be the case. In practice, then, it is probably wisest to look to the animal models only for qualitative information concerning toxic signs, target organs, and hematologic and physiologic responses to the test material.

D. Advice Against Testing of Homologous Agents

Another cautionary note may be appropriate at this juncture. It is certainly possible, by recombinant or other means, to generate the homologous protein and test it in an animal system (e.g., mouse IFN in mice). This is the so-called "simulation modeling" technique. In our opinion, it is imprudent to use this technique to attempt to predict the *human* clinical toxicity of the *human* protein under study. Although it is certainly of scientific value to study proteins in homologous efficacy test systems [23], and is most definitely a recombinant tour de force to produce a protein for this purpose [24], for all the reasons listed above it is potentially disastrous to employ these agents as the exclusive mode of testing in toxicologic assessments. Further, the protein may be expressed in a different expression system, or it may have different physicochemical properties which will necessitate a different isolation and/or purification scheme and a different formulation in order to produce a stable preparation. Insofar as these factors influence the toxicity of the human protein, we feel that the more prudent course of action is to concentrate on the toxicologic evaluation of the protein to be used in the clinic. The species-species and

protein-protein extrapolations are too great a leap of scientific faith to make
for the purposes of predicting the effects of the human protein in humans.

IV. PATHOLOGY

A. Toxicologic Pathology and the New Protein
    Therapeutics

The toxicologic pathologist is a key player in the characterization of the
toxicity of novel polypeptides and proteins, and is of premier importance
in the safety assessment of the cytokines, where immune cells are the major
effector organs. Thus it is important to assess the role of the toxicologic
pathologist in determining the impact, if any, of the new protein therapeutics on the study design and conduct of toxicity evaluations.

The specific functions of a toxicologic pathologist differ somewhat from
those of a diagnostic pathologist. Central to the more obvious differences
in the level of data management and experimental manipulation, there are
four basic functions that define the role of toxicologic pathologists. They
must:

1. Intelligently create experimental designs to answer pertinent ques-
    tions of pathogenesis, using interpretable techniques
2. Accurately recognize the morphologic states corresponding to
    "normal" and "abnormal" functions
3. Correctly interpret the relationship between the test material and
    those morphologic changes, by integrating antemortem observations,
    clinical laboratory findings, postmortem macroscopic changes, organ
    weight data, and all other available experimental parameters
4. Clearly and concisely communicate the interpretation of that relation-
    ship in writing, characterizing the morphologic and biochemical
    changes as to their toxicologic consequence, their secondary or re-
    sponsive nature, their biologic significance, or their spontaneous
    nature

These activities of the toxicologic pathologist are unchanged by the advent of recombinant protein therapeutics. The necessity for recognition
of morphologic change remains unaltered. Cells are injured and die in
much the same manner as they have in the past. There are, however,
new terms that perhaps should be applied to old changes, and more emphasis that should be placed on cytology than on histology. New partnerships
must be forged among the toxicologist, the pharmacologist, the physiologist,
the cell biologist, and the pathologist. In reality, many of these new approaches reflect our lack of detailed knowledge of the immune system and
of other humoral interactions, and our lack of knowledge of the full range
of effects of these potent compounds, rather than something unique to
these new therapeutics. In their native form, most are naturally occurring
substances that are integral components of a complex network that maintains
the heart and mind, perpetuates the species, and customizes the soul. Despite this lofty origin, these proteins and peptides are often administered
in such a way as to be "foreign" to the body and entertain the same relationship between "cure" and "poison" as any xenobiotic.

B. True Morphologic Nomenclature

The new therapeutics belonging to the group classified as biological response modifiers (BRMs) generate perhaps the greatest need for modification of histopathologic nomenclature.   The BRMs pharmacologically manipulate the immune system.   A previously undesirable test-article-related focus
of inflammation, when induced by a BRM, becomes evidence of a pharmacologic response (and a desirable one) in an anticancer drug.   Extraordinary
proliferation of lymphocytes becomes the desirable "lymphoid hyperplasia"
instead of the less desirable "lymphoma."   The proliferation and relocation
of cells of the immune system should be viewed as an expected pharmacologic response.   Far more attention should be paid to the "effect" on the
parenchyma and the function of the organ in which the immune cells have
relocated.   Attention should be directed toward identification of the morphologic and biochemical changes induced by the relocating cells, as those
changes may be the only ones of toxicologic consequence.   Polymorphonuclear leukocytes may be numerous in the sinusoids of the liver in the absence of any necrosis, altered permeability, vascular obstruction, and/or
alterations in liver function.   Under the influence of a BRM, this response
may well be characterized as extramedullary hematopoiesis and not "hepatitis."   Cataloging this morphologic change as "hepatitis" confuses the nature of the biologic response and misses the scientific point.

C. Cytopathology Versus Histopathology

Linked to the necessity to alter nomenclature to a more morphologically de-
scriptive and less process-associated terminology is the need to emphasize
cytology over histology.   Previously the domain of the clinical pathologist,
the cytologic examination of solid and liquid tissues is central to the identification of cells of the immune system that may relocate in unusual sites
and may be present in developmental stages which are synchronous and
not readily identifiable.   The well-used terminology of "mononuclear cell
infiltration" is no longer sufficient, if indeed it ever was.

D. Special Histocytochemistry and Cell Sorting

The standard histology laboratory, which mass-produces hematoxylin-and-
eosin stained sections, with an occasional periodic acid—Schiff's, Prussian
Blue, Trichrome, or Congo Red stain, is insufficient for the toxicologic
assessment of BRMs or the related therapeutics, such as antibodies, antibody-directed effector proteins, hormones, and enzymes.   Special histocyto-
and immunocytochemistry are the tools of the toxicologic pathologist working
with proteins and peptides.   Phenotypic identification of cell subclasses,
immature forms, and stem cells is commonplace and indispensable.   The
combination of cell-sorting techniques and in situ identification of target
cells or molecules is commonplace and necessary to an understanding of the
effects of therapeutic administration of proteins.

E. Mechanisms of Cell Injury and Cell Death

The toxicologic pathologist dealing with BRMs is faced with the challenge
of examining tissue injury from the aspect of each and every cell type

present in the lesion. To understand the nature of the injury and, with
that knowledge, be enabled to modify the "inflammatory" response necessitates distinguishing bystander or inactive cells from those that are releasing
mediators of cell alteration and cell injury. It is imperative to understand
the known spectrum of biologic response for each cell type involved in a
lesion, as well as to recognize those cells that proliferate without inducing
a local morphologic change and appear only coincidentally with the marked
physiologic changes in the host animal.

F. Cause of Death

Determining the cause of death in an animal and providing prognostic information on a biopsy specimen are basic activities for diagnostic pathologists.
Their training as clinicians and their experience as pathologists, correlating
their clinical observations with morphologic and biochemical changes, provide them with unique insight into the degree of severity of a clinical observation, as well as a unique understanding of the clinical manifestations
of a morphologic or biochemical change. Toxicologic pathologists are trained
in exactly the same way. However, for various reasons, many have strayed
from the function of determining the cause of death for an animal on study
in all cases except those involving tumors. In studying BRMs and other
therapeutics, it is essential to evaluate those animals that die, from the
standpoint of *identifying the underlying morphologic change responsible for
the death of the animal*. It is equally important to declare that the cause
of death is *unknown*, for by so declaring, the pathologist is clearly com-
municating that there are no morphologic changes present capable of inducing a fatal response in the host. This type of condition is a clear signal for further investigation by the physiologist, the immunologist, or the
toxicologist, in an attempt to identify systemic or perhaps local humoral
factors that may be responsible for irreversible alterations in physiologic
homeostasis.

G. Conserving a Precious Resource, the Nonhuman Primate

There is an old saying that of all the clinicians involved in the treatment
of a patient, the pathologist "knows everything, but knows it too late."
In the case of nonclinical drug testing that must be amended to read that
the toxicologic pathologist "knows everything, but has to sacrifice the animal to obtain the information." This is an accepted strategy for drug testing in rodent and other nonprimate species, but is one that needs careful
scrutiny and reconsideration in the case of nonhuman primates. Their
close phylogenetic relationship and strong protein-structural homology to
humans renders them of supreme importance in the safety evaluation of
potential human protein and MAb-based therapeutics. These animals must
be preserved rather than being needlessly sacrificed to satisfy established
morphologic tradition.

Nonhuman primates can and, in our opinion, should be reused as experimental subjects, with the appropriate prestudy screening for health
status and antibody presence. Pathologists must ask themselves if there
are sufficient morphologic data available or obtainable in nonprimate species,
if biopsy data are sufficient, whether radiologic or electrophysiologic examinations can provide sufficient data, or if sampling cerebrospinal fluid
will provide enough data to replace euthanasia and standard histopathologic

evaluation, in the absence of clinical signs with protein administration in
the nonhuman primate. We have also found specific organ function testing
(e.g., clearance tests of liver or kidney, electrophysiologic monitoring of
heart or brain, and arterial gas measurements as a test of lung function)
to be extremely valuable, and this testing has, under certain circumstances,
supplanted morphologic examinations of whole organs from euthanized nonhuman primates. Other nontraditional testing of the intact animal, such
as blood pressure or body temperature measurements or antigen challenge,
is an excellent source of information about the functional status of the animals, without sacrificing the animal. This preservation strategy complicates the task of preclinical scientists but is well within their training and
capabilities, and is most certainly a resource-effective mode of safety assessment.

Proteins intended for single dosing or very short term administration,
or therepeutics intended only for use in life-threatening disease states,
are prime candidates for consideration of the preservation strategy. In
short, each study using a nonhuman primate should be reviewed from the
standpoint of the intended clinical application of the protein and the existing data base in other species before automatically scheduling sacrifice of
this precious resource. The reader is referred to an excellent discussion
by Hobson and Fuller [25] of the use of primates in the preclinical safety
assessment of proteins.

H. Antibody Specificity and Cross-Reactivity Screening

Exploring MAb specificity and cross-reactivity by using cryostat preparations of normal human tissue has become an accepted technique in toxicity
characterization, and the pathologist should be involved in the process.
Therapeutic antibodies or antibody-directed molecules pose a serious problem in host selection for the preclinical process of safety assessment. Obviously, a specific human antigenic site is the therapeutic target, and that
target may exist only in the human; this exclusivity would eliminate the
use of laboratory animals as expositors of toxicity. Such toxicities may
be expressed as either an exuberant pharmacologic response at the intended
tissue or as damage due to cross-reactivity with other unintentional target
tissues. The support for a claim of efficacy with these products is based
upon, among other things, the reactivity of the antibody with the intended
antigen in the target tissue. One thus cannot ignore cross-reactivity with
unintended tissues as "artifactual" or of minor importance. We must note
here that to date, we have been unable to establish a good correlation
between in vitro cross-reactivity with unintended tissues and clinically detectable toxicities in those tissues. Further, we are not aware of any such
correlation in the scientific literature. There have been published demonstrations, however, that antibodies and antibody-directed molecules do
find their targets and can produce the desired therapeutic or diagnostic
effect. Examples may be found in antitumor antibodies which have demonstrated clinical effects [26,27] and in tumor-imaging antibodies [28,29].

When normal tissues are examined for cross-reactivity with MAbs directed at specific abnormal target cells, these tissues should be evaluated
and documented in the same manner and with the same degree of data management as are used for any other tissue section examined histopathological-
ly. Just as one would have a protocol-specified tissue list for each species
under study, one should have a protocol-specified cell-type list for each

tissue examined. This is especially important when using external laboratories with banks of normal human tissues. It is not sufficient to know
that the pancreas is negative for cross-reactivity if one cannot be assured
that there were several islets, a pancreatic duct, and perhaps a sizable
blood vessel in the plane of section with the acini, and to know whether
or not these structures were also negative for cross-reactivity. Care
should also be taken to distinguish between staining of surface antigens
and staining of those within the cytoplasm of the cell. The interior antigenic site may never be exposed to a parenterally administered antibody.

Thus the role of the toxicologic pathologist is much altered by the advent of the new protein therapeutics. In reality, the role has returned to
the more investigative mode of the experimental pathologist and is not
simply focused on the cataloging of lesions.

V. CLASSES OF TOXICITIES OF PROTEINS AND PEPTIDES

A. Exuberant Pharmacologic Responses

The most frequently encountered toxic manifestations of protein or peptide
toxicity are those simply related to dose. With movement to the right on
a particular dose response curve, a greater fraction of the protein receptors are occupied, and a greater physiologic or biochemical response to
the drug is seen. The reduction in blood glucose with INS administration,
and the vasoconstriction subsequent to dosing with angiotensin II or vasopressin, each may have very desirable therapeutic benefits. In neither
instance, however, will it be the case that *"more is better."* Profound hypoglycemia and the resultant convulsions with an overdose of INS, and
the arteriospasm and hypertension possible after angiotensin II or vasopressin treatment in large-dose infusions, certainly qualify as toxicities of
this type [30,31] (i.e., as exuberant pharmacologic responses).

Endocrine Peptide and Protein Hormones

Other endocrine protein hormones for which toxic responses can occur
from either large acute or cumulative doses, and whose toxicities are actually exuberant pharmacologic responses, include adrenocorticotrophic hormone
(ACTH, responsible for Cushing's disease), luteinizing hormone and its releasing hormone (receptor down-regulation can lead to infertility), oxytocin
(fetal distress often occurs), thyroid stimulating hormone and its releasing
hormone (causing hyperthyroidism), and parathyroid hormone (bone de-
mineralization and nephrocalcinosis result). Further examples are provided
in a review by Robbins and Cotran [32]. Interestingly enough, calcitonin
does not appear to be associated with toxicities of this type, owing to a
reflex increase in the circulating parathyroid hormone levels and the rapid
tachyphylaxis that occurs with continued calcitonin administration [33].

Protein hormones associated with somatic growth—namely GH and its
releasing factor (GRF), somatostatin (GH release inhibiting factor), and
the somatomedins, or insulin-like growth factors (IGFs) induced by GH in
the liver—all have at least potential toxicities associated with their use,
and all are involved extensively in both positive and negative feedback
loops [34]. GH itself may cause gigantism or acromegaly in the pre- and
postpubertal animal, respectively, as might GRF [32]. The potential toxicity of the IGFs should be obvious to the reader. Somatostatin, which inhibits not only GH secretion but also ACTH, glucagon, cholecystokinin

(CCK), and INS, among other hormones, would theoretically have potential toxicities associated with reduced blood and tissue concentrations of all these hormones were it not for its extremely short half-life in the blood [35].

Blood and Cardiovascular Proteins

To a greater or lesser extent, the doses of various fibrinolytic proteins used to treat certain thrombotic disorders are also accompanied by toxicities associated with exuberant pharmacologic responses. Streptokinase (SK), urokinase (UK), and t-PA are each capable of lysing fibrin clots at various sites within the vascular system [36–38]. Each will lyse clots wherever found, at venipuncture or arteripuncture sites as easily as at the pathologic clot site in leg, lung, heart, or brain. These agents share the common gross mechanism of action of converting the inactive plasminogen to plasmin, doing so in a more (t-PA) or less (SK and UK) specific manner [39]. Plasmin is a quite nonspecific plasma protease, and once formed is capable of degrading many different plasma proteins, fibrinogen probably being most important from a toxicologic standpoint. Relative overdoses of each of these thrombolytic agents are therefore followed by decreases in plasma fibrinogen, ranging from noticeable to severe [40]. Plasminogen activation after t-PA administration occurs for the most part only in the vicinity of a fibrin clot, and thus plasmin is produced almost exclusively at that site. This local production of plasmin is responsible for the much lower systemic plasminogen activation observed with that agent [41], and thus the greater therapeutic index of t-PA than that of SK or UK [42]. The latter agents, without fibrin-binding specificity, cause much greater fibrinogen depletion, due to systemic plasmin generation.

Three other blood proteins have been produced recombinantly, procoagulant Factor VIII, erythropoietin, and human serum albumin. Factor VIII would naturally produce increased and possibly nonphysiologic clotting when dosed to excess. As anticipated, large doses of erythropoietin lead to polycythemia, although the toxicologic consequences of this are not clear. The iron deficiency that results from large-dose erythropoietin administration is simply a result of the increased demand for erythrocyte production [43]. Excess intravascular dosing with albumin could conceivably raise the plasma oncotic pressure and increase the plasma volume. We speculate that this would lead to an increased load on the heart, and over the long term might lead to cardiac hypertrophy. However, no data exist to support this contention.

Cytokines

When one enters into a discussion of the toxicities of the cytokines, it must be realized that little is known about their mechanisms of action in vivo. A case in point is that of the IFNs, some of which have been studied in humans for over 10 years, but for which little is known about their in vivo antiviral and antiproliferative mechanisms. Although blood concentrations of IFN-α seem to correlate roughly with efficacy in human and/or animal systems [44], this is not the case for IFN-β [45] or for IFN-γ [46], suggesting indirect mechanisms for the action of the latter two agents [12].

These BRMs, in general, are the most poorly understood of all the protein therapeutics in terms of their in vivo efficacy and toxicity mechanisms. We have attempted to synthesize various data available to suggest the

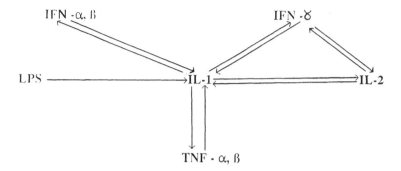

Fig. 1   Generation of, and interactions among, the major cytokines.   This
diagram is based on the IRIS (interleukin regulation of the immune system)
scheme of Palladino et. al. [47]; we have modified and expanded it to reflect unpublished observations and more recently published work [48,49].

complexity of the interactions among some of the more common and better
studied proteins of this class (Fig. 1).   This schema shows that IL-1 and
IL-2, the IFNs, and tumor necrosis factors α and β (TNF-α,β) may all be
generated in vivo.   The introduction of endotoxin (lipopolysaccharide, or
LPS) into a biologic system triggers this cascade of BRMs, all of which
are capable of interacting with one another directly or indirectly.

Monoclonal Antibodies and Their Derivatives

The technology of MAb development has progressed greatly in the last five
years.   Unmodified MAbs and those covalently conjugated to toxins, radio-
nuclides, or conventional chemotherapeutic agents have been, and are
being used as diagnostic and therapeutic anticancer agents.   These substances are also of at least potential usefulness against a broad range of
disease states (Table 3).   This subject has recently been reviewed by
Larrick and Bourla [50].

     Obviously, exuberant pharmacologic responses to MAbs and their con-
jugates used as therapeutic agents are a direct consequence of their binding to antigen which is either free in the circulation or on a target cell
in solid tissue [51].   However, other cells in the body may also present
the antigen, and thus be subject to the same therapeutic action as the target cell and/or tissue.   The FDA routinely requires in vitro binding studies
for this class of candidate therapeutics prior to initiating clinical trials [52].
This knowledge will then allow at least a *prediction* of possible target organs of toxicity in humans.   Further, these data can be used to survey
tissues from various species, in attempts to determine which test species
might be the best in vivo predictors of human exposure and toxicity.   Then
the appropriate toxicity studies can be initiated (see Sec. III.B).

B.   Generic Toxicities

The next most frequently encountered toxic manifestations of protein and
peptide toxicity are those that we term *generic toxicities*.   We classify as
generic any adverse reactions which are exhibited by most, if not all,

members of a class of proteins. Among these are the fever, rigor, chills, and diaphoresis seen when MAbs are infused that react with circulating cells [53], the inhibitory effects of the IFNs on the function of various cytochrome P-450 enzymes (or "mixed-function oxidases") in the liver of certain species [54–56], and the fever, arthralgias, myalgias, fatigue, and malaise that develop with clinical administration of any of the IFNs [57,58]. But perhaps the single most distinguishing generic toxicity of the protein and peptide therapeutics is a result of their propensity for immunogenicity.

Immunogenicity

Proteins have been known for many years to evoke a host response, when administered by various routes, that may be either desirable or undesirable. Their provocative nature has been used therapeutically to protect much of the world from devastating disease through the process of vaccination. This characteristic has also lead to seasonal discomforts, incapacitating asthmas, dermal exfoliations, and ignominious deaths from the stings of errant bees, when sensitized individuals are exposed.

As of this writing, the new protein therapeutics have been spared the role of perpetrators of widespread immunologic disease. These new agents have not been reported to produce dose-limiting immunologic injury, for reasons discussed below. Certainly, it is not customary to report such findings in the scientific literature, but in larger measure this lack of immunogenic effect may be because the proteins are of sufficient purity and are being dosed for a relatively short period. Consequently, serious immunologic injury is avoided. However, some have been reported to generate an antibody response (e.g., the IFNs, GH, and INS) [57,59,60]; fewer yet are known to be inactivated or neutralized by that antibody response [57]. But the field is growing rapidly and the data are accumulating, and thus it is really too early to assess the impact of immunogenicity on this new class of drugs. Nonetheless, a discussion of the substance of this potential toxicity is still warranted.

Table 3   Present and Potential Uses of MAbs
and Their Conjugates

Antineoplastic therapy (liquid and solid tumors)

Discrete lesion imaging (e.g., tumors, abscesses)

Clearance of circulating antigens (viral, bacterial, proteal)

Prophylaxis of graft versus host disease (e.g., OKT3 therapy)

Treatment of xenobiotic and other intoxications (e.g., antidigoxin)

Allergy/autoimmune therapy

Immune system reconstitution

Diagnosis and reduction of infarct size (e.g., antimyosin, antineutrophil)

The immunogenicity of any molecule, whether xenobiotic or protein, is dependent on a complex relationship between the state of the host immune system, the characteristics of the antigen, the amount of the substance administered, the route of administration, and the theoretical and technical nature of the detection methods used to quantify the immune response. Those characteristics that place proteins at risk of immunogenic behavior include their molecular weight, degree of structural complexity (e.g., secondary, tertiary, and quaternary structure) and hydrophobicity, aromatic amine content, extent of aggregation, sequence homology to the native protein in the test species, terminal glycosylation, and various qualities of the absorption, distribution, metabolism, and excretion peculiar to the test species and the specific protein moiety [61]. In Darwinian terms, the host has evolved a manner of response that aids in its survival against invasion by foreign molecular networks variously known as viruses, fungi, bacteria, and the like. That response is expressed through mechanisms that lead to precipitation, agglutination, immobilization, neutralization, complement fixation, and cell lysis.

The toxicologist working with the new protein therapeutics will generally be supplied with a pharmaceutically elegant preparation of a molecule already selected for its paucity of characteristics favoring immunogenicity. The primary concern of the toxicologist then becomes not a documentation of immunologic disease resulting from protein administration, but rather the antibody-mediated removal of the molecule and the possible toxic sequelae, or the neutralization of its pharmacologic effects. In either case, the toxicologist is looking for immune responses that would invalidate or obscure any conclusions reached in toxicity experiments.

It is important to realize that the presence of circulating antibodies to a protein does not de facto dictate the elimination or neutralization of a protein or its effects. We have found in unpublished work that antibodies may be present in the bloodstream, yet fail to "neutralize" the antigen in classic in vitro assays of antigen function. On the contrary, antibodies may also be present in the circulation that demonstrate in vitro "neutralization," yet the host will continue to demonstrate an unaltered pharmacologic response. A careful analysis of the significance of the presence of antibody is therefore crucial to the interpretation of study validity.

If the conditions are correct, and the host and protein properties favor a major immune response, immune effector mechanisms can lead to disease.

Mechanisms of Immunologic Injury

Mechanisms of host immunologic injury are mediated chiefly through the action of lymphocytes, either directly in a T-lymphocyte response to an antigen, or indirectly through B-lymphocyte production of antibody. However, although the reactions are complex and are not solely mediated by lymphocytes, these cells do form the major scaffolding upon which both injurious and protective immunologic responses develop. Repeated exposure to an unrecognized or undesirable organism or molecule may set up a series of responses that culminate in injury through four general pathogenetic mechanisms.

Type I reaction in immunologic injury. This form of immunologic injury is the result of active mediators released from mast cells through an antibody-based mechanism. These active mediators include histamine, numerous arachidonic acid metabolites, serotonin, bradykinin, and heparin.

They all are potent compounds that variously alter vascular permeability, affect smooth muscle contractility, aggregate platelets, stimulate mucus secretion, initiate nociception, alter cyclic nucleotide synthesis, act as chemoattractants, and potentiate further active mediator production [62].

The sequence of events involves immunoglobulin E (IgE) antibody binding through its Fc portion to receptors located on circulating basophils or tissue mast cells. The exposure of the fixed antibody to antigen leads to degranulation of the basophil or mast cell, releasing the active mediators. The toxic result depends on the type of provoking antigen and the site of the reaction. The spectrum of responses includes allergic rhinitis, allergic dermatitis, urticaria, angioneurotic edema, asthma, gastrointestinal hypermotility, acute respiratory distress, esophageal varices, and acute hypotension. IgM and IgA can also precipitate the release of mast cell/basophil-derived active mediators [62].

Type II reaction in immunologic injury. This mechanism of immunologic injury is based on the interaction between antibody and antigenic sites at the cell surface. Usually, these cells are within the vascular bed, although extravascular cells may be targeted and exhibit either cell stimulation and mediator release, or inhibition of cell function. The intravascular antigen may be a naturally occurring constituent of the cell surface or an exogenous immunogenic substance which has adsorbed to the cell surface. In either event, complement is usually involved in the intravascular process, and its activation leads to a series of events culminating in cytotoxicity and cell lysis [63]. Complement may be directly cytotoxic, or it may initiate an inflammatory response, attracting neutrophils and their destructive enzymes. It is also possible that the antibody may attract killer T lymphocytes, leading to antibody-directed cell cytotoxicity (ADCC) of the antigen-presenting cell [64].

Cytotoxicity mechanisms generally lead to anemias, hemorrhagic conditions, neutropenias, and related secondary disorders. Although the antigen may be located in the basement membrane instead of the cell membrane, the mechanism of damage remains the same, involving antibody, complement, and inflammation. Kidney, lung, and skin basement membrane are not uncommon sites for such a response [65].

Antibodies may either stimulate or inhibit the natural activities of a cell through receptor binding. Although not of themselves cytotoxic, they can lead to disease. Long-acting thyroid stimulator (LATS) is an antibody to the thyroid cell that brings about increased production of thyroid hormone and all the manifestations associated with that elevation [66]. In a similar manner, antibodies may inhibit a cell function by blocking a receptor or inactivating/neutralizing an important cell product. Examples of this type of immune disease may be found among patients with autoimmune hemolytic anemia, idiopathic thrombocytopenia, and Goodpasture's disease, among others [67].

Type III immunologic injury. This mechanism of immunologic injury is mediated through preformed antigen-antibody complexes, usually in antibody excess, which produce damage upon their deposition on blood vessel walls or in the interstitium of vital organs. The deposition within blood vessel walls is also mediated through the action of complement. Once in place, further complement activation leads to a destructive inflammatory response. Examples include vasculitis, arthritis, glomerulonephritis, hepatitis, and serum sickness [68,69].

Type III immunologic injury may be the most likely mechanism for immunologic injury resultant from protein therapeutic administration, as a result of untoward and excessive immunogenicity.

Type IV immunologic injury. This form of immunologic injury occurs in the absence of antibody and complement. It is mediated by sensitized T lymphocytes and is termed "delayed hypersensitivity." Damage is mediated through direct T-cell cytotoxicity and/or cytokine release, which mobilizes and attracts monocytes, other lymphocytes, and neutrophils, leading to further tissue injury. Examples of type IV reactions include the classical tuberculin skin reaction, allograft rejection, skin reactions to diphtheria toxoid and poison ivy, graft versus host response, cell-mediated lympholysis, and mixed-lymphocyte response [70]. For a discussion of means to reduce the immunogenicity of peptides and proteins, see Chap. 21.

C. Idiopathic Toxicities

Toxic reactions of this type are usually unrelated to the mechanism of action of the test article (at least at the time of categorization), and are not generally caused by all members of a protein therapeutic class. These are the unexpected toxicities. Documentation of and research into the causes of these toxicities may provide insight into the mechanism of action of the therapeutic moiety in vivo. Examples of this type of toxicity include the hypotension seen acutely with TNF-α administration [22,71], the so-called "vascular leak syndrome" associated with long-term IL-2 infusion [72,73], the retinopathy reportedly linked to insulin-like growth factor [74], the finding that IFN-α is an abortifacient in the rhesus monkey [75], the shortening of bleeding time in subjects administered 1-deamino-8-D-arginine vasopressin [76], and the inhibitory effects of TNF-α on the function of various cytochrome P-450 enzymes [77]. This last result may be due to the IFNs induced indirectly by TNF-α (see Sec. V.A).

VI.   TOXICITIES OF GENETICALLY ENGINEERED PRODUCTS

A. Expression System Contaminants

Proteins and peptides have been produced successfully by recombinant techniques in three different phylogenetic groupings of cells: the prokaryotic gram-positive and gram-negative bacteria, and the eukaryotic yeast and mammalian cells. The latter group includes both human and nonhuman cell types. We have only limited knowledge of recombinant work done in gram-positive and yeast cell expression systems, so we will confine our discussion to contaminants of proteins and peptides from the two most common expression systems, gram-negative bacteria and mammalian cells.

Gram-Negative Bacteria

Without doubt the most important substance responsible for differences in the toxicity of recombinant proteins and peptides from their nonrecombinant forms is LPS, a cell wall component of gram-negative bacteria. Falling into this class of organisms is *Escherichia coli*, the most frequently employed cell type in the production of recombinant proteins. As the majority of proteins are expressed directly into the cytoplasm of the *E. coli* cell, rather than being secreted into the medium, this cloning technique demands that the cells be lysed to liberate the recombinant protein. The cell lysate is then subjected to an elaborate purification process to obtain a pharmaceutical-grade compound.

Minor amounts of LPS that contaminate the therapeutic material are sufficient to cause an acute-phase response and fever in man, at doses of about 1 ng/kg [78,79]. Assuming that the recombinant drug is administered at 1 mg/kg, this implies that the LPS contamination must be less than 1 part per million. While this level of purity is now routinely met, pyrogen contamination in some form had been a problem associated with at least one recombinant protein administered to humans in an early phase 1 clinical trial [80]. Further, since only the rabbit is as sensitive to the pyrogenic effects of LPS as is the human, the presence of LPS must be routinely tested for in rabbits, on a lot-by-lot basis [81]. Although LPS contamination is not usually a problem with final product, it is of more than theoretical concern if crude bacterial-cell supernatant solutions are tested in animals during early preclinical experimentation.

Another confounding problem associated with the administration of LPS to various species stems from the triggering of a cytokine cascade, as mentioned previously (Fig. 1). These cytokines may lead to synergistic and antagonistic interactions, as well as producing frank toxicities in their own right. It may be as a result of this cascade phenomenon that LPS or bacterial protein contaminants exert the commonly observed "adjuvant effect," wherein the immunogenicity of proteins is increased by coadministration of lysed gram-negative bacteria.

Also falling loosely into the category of bacterial system contaminants are the N-formyl-derivatized proteins produced when recombinant proteins are expressed in bacterial systems (and not secreted); this occurs as a natural consequence of the metabolic process whereby bacteria make all proteins [82]. These derivatized proteins are by themselves quite immunogenic.

Mammalian Systems

Certain proteins must be produced in mammalian cells for the following reasons: (1) foreign gene expression in bacteria may either inhibit the growth of or actually kill the bacterial cell; (2) to ensure proper folding of the protein of interest; and (3) due to the need to have various posttranslational modifications which the bacterial cell is incapable of performing (see Levinson et. al. in Ref. 83). Here, too, there are potential problems associated with expression system contaminants. Because these cell lines are generally continuous ones and are to a large extent immortal, there is a concern that the cells may be contaminated with retrovirus or oncogenes. Again, the FDA has set rigid guidelines on the amount of contaminating DNA in products from transformed cells. That quantity is currently set at 10 pg per patient dose [52], and the final product is tested on a lot-by-lot basis. The entire matter of nucleic acid and protein contamination from mammalian cells has been addressed at length [83]; the reader is directed to that work for details on this subject, as well as a discussion of the possible presence and concerns over adventitious agents.

B. Amplification /Induction Chemicals

In many cases, recombinant protein expression systems require the presence
of various molecular weight chemical or protein entities at one stage or
another in the process of producing the protein of interest. These substances are necessary for the selective growth of the gene-containing cells,
or to prime these cells to produce the recombinant protein. These substances include various antibiotics, such as tetracycline, ampicillin, or
penicillin, as well as butyrate, transferrin, indoleacetic acid, methotrexate,
and β-propiolactone. These substances are all routinely tested for in the
final product. They may cause untoward reactions in sensitized individuals,
but may also be of specific concern if crude cell supernatants are used in
early preclinical testing.

C. Process and Formulation Effects

Often, the protein of interest is present in solution with a myriad of other
proteins of similar molecular weight, electrophoretic mobility, and/or carbohydrate content, and must be separated from them by numerous separation techniques. In the separation process, the protein will sometimes, and
variously, be subject to denaturation, oxidation, deglycosylation, proteolysis, deamidation and decarboxylation, dimerization, and aggregation. Also
possible is the so-called "disulfide shuffling" reaction, wherein intramolecular disulfide bonds are cleaved and re-formed at new sites, sometimes on
different molecules. When the protein does not have any internal methionine, it is sometimes expressed in *E. coli* as the fusion protein, which is
then converted to the protein of interest by treatment with cyanogen bromide. This process can also result in degradative reactions of the amino
acid side chains. In all cases, closely related but subtly different contaminating peptides and proteins can appear as contaminants (a condition
referred to as "microheterogeneity" [4]), and sophisticated analytic techniques are necessary to identify these contaminants [5]. Abrogation of
therapeutic effects, and toxicities of various forms (especially immunogenicity), may result from these contaminating moieties.

The use of solubilizing agents, antioxidants, chaotropic agents, protease inhibitors, chelating agents, and detergents in the purification
process and in final product formulations can potentially lead to toxicity
problems. These may occur directly through the intrinsic toxicity of these
substances, or by means of some physiologic interaction of the aforementioned agents with the protein in vivo [84], or through the ability of
these compounds to increase the immunogenicity of the protein therapeutic
[85].

VII. CLOSING COMMENTS

The expanded use of proteins and peptides as pharmacologic agents has
lead to the expanded study of their toxicity. The novel classes of therapeutic agents generated in the past 15 years demand increased awareness,
by the toxicologist and toxicologic pathologist, of many scientific disciplines
heretofore outside their ken. But the novelty of many of these agents and
their actions does not imply that the long-followed, traditional precepts of
toxicology and pathology are no longer valid. As discussed and/or alluded
to, intelligent experimental design and proper species selection remain

critical to the proper assessment of potential risk in protein and peptide administration in the clinical setting. Traditional toxicology testing, when strengthened by the inclusion of nontraditional endpoints and enlightened interpretation of all the various signs observed, will expand the scope and sharpen the vision of the preclinical scientists charged with safety assessment, and allow them to make better predictions of the risk/benefit ratio of a potential protein or peptide therapeutic.

Nevertheless, the words of Paracelsus remain as true today as when they were pronounced five centuries ago. As paraphrased by another, contemporary toxicologist, the late former chancellor of the University of California, Davis:

> There are no harmless substances; there are only harmless ways of using substances.
>
> Emil Mrak (1902–1987)

ACKNOWLEDGMENTS

The authors wish to thank Drs. David Mark, Scott Burchiel, Mike Bjorn, and Mike Palladino and Messrs. M. Refaat Shalaby and Art Blum for providing certain information used in preparing this manuscript, Drs. Joseph Nachtman and Gene Fuller for critically reviewing the manuscript, and Eric Ibsen for preparing the figure. Special thanks are accorded to Ms. Colleen Hanley for her work in helping to generate this document.

REFERENCES

1. S. N. Cohen, A. C. Y. Chang, H. W. Boyer, and R. B. Helling, *Proc. Natl. Acad. Sci. USA, 70*:3240 (1973).
2. G. Kohler and C. Milstein, *Nature, 256*:495 (1975).
3. I. Gresser, *Tex. Rep. Biol. Med., 35*:394 (1977).
4. D. F. Mark, M. V. Doyle, and K. Koths, in *Recombinant Lymphokines and Their Receptors* (S. Gillis, ed.), Marcel Dekker, New York (1987), p. 1.
5. W. S. Hancock, *Chromatog. Forum, 1*:57 (1986).
6. *Fed. Reg., 43*:60013 (1978).
7. W. M. Galbraith, in *Preclinical Safety of Biotechology Products Intended for Human Use* (C. E. Graham, ed.), Alan R. Liss, New York (1987), p. 3.
8. R. E. Humbel, H. R. Bosshard, and H. Zahn, in *Endocrinology*, Vol. 1, *Handbook of Physiology* (D. F. Steiner and N. Freinkel, eds.), American Physiological Society, Washington, D.C. (1972), p. 139.
9. Y. Harada, in *Preclinical Safety of Biotechnology Products Intended for Human Use* (C. E. Graham, ed.), Alan R. Liss, New York (1987), p. 127.
10. D. Collen, J. M. Stassen, and M. Verstraete, *J. Clin. Invest., 71*: 368 (1983).
11. C. H. Li, *Hormonal Proteins and Peptides*, Vol. 4 (C. H. Li, ed.), Academic Press, London (1977), p. 1.

12. N. Stebbing and P. K. Weck, in *Recombinant DNA Products: Insulin, Interferon and Growth Hormone* (A. P. Bollon, ed.), CRC Press, Boca Raton, Fla. (1984), p. 75.

13. P. K. Weck, E. Rinderknecht, D. A. Estell, and N. Stebbing, *Infect. Immunol.*, *35*:660 (1982).

14. G. Grabner, G. Smolin, M. Okumoto, and N. Stebbing, *Curr. Eye Res.*, *2*:785 (1983).

15. I. J. Fidler, W. E. Fogler, E. S. Kleinerman, and I. Saiki, *J. Immunol.*, *135*:4289 (1985).

16. H. R. Lijnen, B. J. Marafino, Jr., and D. Collen, *Thromb. Haemostasis*, *52*:308 (1984).

17. H. Rabin, R. H. Adamson, R. H. Neubauer, J. L. Cicamanec, and W. C. Wallen, *Cancer Res.*, *36*:715 (1976).

18. H. Schellekens, A. Dereus, R. Bolhuis, M. Fountoulakis, C. Schein, J. Ecsodi, S. Nagata, and C. Weissmann, *Nature*, *292*:775 (1981).

19. N. Stebbing, P. K. Weck, J. T. Fenno, D. A. Estell, and E. Rinderknecht, *Arch. Virol.*, *76*:365 (1983).

20. S. E. Halpern, P. L. Hagan, P. R. Garver, J. A. Koziol, A. W. N. Chen, J. M. Frincke, R. M. Bartholomew, G. S. David, and T. H. Adams, *Cancer Res.*, *43*:5347 (1983).

21. S. W. Burchiel, R. R. Pollock, D. G. Covell, G. Fuller, and M. D. Scharff, *Int. J. Immunopharmacol.*, *9*:913 (1987).

22. B. J. Marafino, Jr., J. D. Young, I. L. Greenfield, and J. R. Kopplin, in *Therapeutic Peptides and Proteins: Assessing the New Technologies*, Banbury Report 29 (D. R. Marshak and D. T. Liu, eds.), Cold Spring Harbor Laboratory, Cold Spring Harbor, N.Y., (1988), p. 175.

23. M. R. Shalaby, E. B. Hamilton, A. H. Benninger, and B. J. Marafino, Jr., *J. Interferon Res.*, *5*:339 (1985).

24. P. W. Gray and D. V. Goeddel, *Proc. Natl. Acad. Sci. USA*, *80*: 5842 (1983).

25. W. C. Hobson and G. B. Fuller, in *Preclinical Safety of Biotechnology Products Intended for Human Use* (C. E. Graham, ed.), Alan R. Liss, New York (1987), p. 55.

26. H. Sears, D. Herlyn, Z. Steplewski, and H. Koprowski, *J. Biol. Response Modif.*, *3*:138 (1984).

27. A. N. Houghton, D. Mintzer, C. Cordon-Cardo, S. Welt, B. Fliegel, S. Vadhan, E. Carswell, M. R. Melamed, H. F. Oettgen, and L. J. Old, *Proc. Natl. Acad. Sci. USA*, *82*:1242 (1985).

28. F. H. DeLand, E. E. Kim, and D. M. Goldenberg, *Cancer Res.*, *40*: 2997 (1980).

29. C. H. Thompson, M. Lichtenstein, S. A. Stacker, M. J. Leyden, N. Salehi, J. T. Andrews, and I. F. C. McKenzie, *Lancet*, *2*: 1245 (1984).

30. K. E. Sussman, J. R. Crout, and A. Marble, *Diabetes*, *12*:38 (1963).

31. I. L. Slotnick and J. D. Tiegland, *J. Am. Med. Assoc.*, *146*:1126 (1951).

32. S. L. Robbins and R. S. Cotran, in *Pathologic Basis of Disease*, 2nd ed. (S. L. Robbins and R. S. Cotran, eds.), W.B. Saunders Company, Philadelphia (1979), p. 1336.

33. D. H. Copp, *Recent Prog. Horm. Res.*, *20*:59 (1964).

34. P. M. Plotsky and W. Vale, *Science*, *230*:461 (1985).

35. A. V. Schally and C. A. Meyers, *Mater. Med. Pol.*, *12*:28 (1982).
36. R. N. Brogden, T. M. Speight, and G. S. Avery, *Drugs*, *5*:357 (1973).
37. R. Paoletti and S. Sherry, eds., *Thrombolysis and Urokinase*, Academic Press, London (1977).
38. C. Korninger, O. Matsuo, R. Suy, J. M. Stassen, and D. Collen, *J. Clin. Invest.*, *69*:573 (1982).
39. D. Collen, J. M. Stassen, B. J. Marafino, Jr., S. Builder, F. DeCock, J. Ogez, D. Tajiri, D. Pennica, W. F. Bennett, J. Salwa, and C. F. Hoyng, *J. Pharmacol. Exp. Ther.*, *231*:146 (1984).
40. B. Wiman and D. Collen, *Nature*, *272*:549 (1978).
41. M. Hoylaerts, D. C. Rijken, H. R. Lijnen, and D. Collen, *J. Biol. Chem.*, *257*:2919 (1982).
42. M. Verstraete, R. Bernard, M. Bory, R. W. Brower, D. Collen, D. P. DeBono, R. Erbel, W. Huhmann, R. J. Lennane, J. Lubsen, D. Mathey, J. Meyer, H. R. Michels, W. Rutsch, M. Schmidt, R. Uebis, and R. von Essan, *Lancet*, *1*:845 (1985).
43. B. J. Payne, in *Preclinical Safety of Biotechnology Products Intended for Human Use* (C. E. Graham, ed.), Alan R. Liss, New York (1987) p. 107.
44. H. Schellekens, P. M. C. A. van Eerd, A. de Reus, P. K. Weck, and N. Stebbing, *Antiviral Res.*, *2*:313 (1982).
45. P. K. Weck, R. N. Harkins, and N. Stebbing, *J. Gen. Virol.*, *64*: 415 (1983).
46. J. U. Gutterman, M. G. Rosenblum, A. A. Rios, A. F. Herbert, and J. Quesada, *Cancer Res.*, *44*:4164 (1984).
47. M. A. Palladino, L. P. Svedersky, H. M. Shepard, K. M. Pearlstein, J. Vilcek, and M. P. Scheid, in *Interferon: Research, Clinical Application, and Regulatory Consideration* (K. C. Zoon, P. D. Noguchi, and T. Y. Liu, eds.), Elsevier Science Publishing Co., New York (1984), p. 137.
48. J. Vilcek, D. Henriksen-DeStefano, D. Siegel, and J. Le, in *The Interferon System* (F. Dianzani and G. B. Rossi, eds.), Raven Press New York (1985), p. 43.
49. D. K. Blanchard, T. W. Klein, J. J. Djeu, H. Friedman, and W. E. Stewart II, in *The Interferon System* (F. Dianzani and G. B. Rossi, eds.), Raven Press, New York (1985), p. 49.
50. J. W. Larrick and J. M. Bourla, *J. Biol. Response Modif.* *5*:379 (1986).
51. M. J. Bjorn, D. Ring, and A. Frankel, *Cancer Res.*, *45*:1214 (1985).
52. *Points to Consider in the Manufacture and Testing of Monoclonal Antibody Products for Human Use*, Center for Drugs and Biologics, Food and Drug Administration, Bethesda, Md. (1987).
53. R. O. Dillman, J. C. Beauregard, S. E. Halpern, and M. Clutter, *J. Biol. Response Modif.*, *5*:73 (1985).
54. J. A. Moore, B. J. Marafino, Jr., and N. Stebbing, *Res. Commun. Chem. Pathol. Pharmacol.*, *39*:113 (1983).
55. G. Taylor, B. J. Marafino, Jr., J. A. Moore, V. Gurley, and T. F. Blaschke, *Drug Metab. Dispos.*, *13*:459 (1985).
56. M. R. Franklin and B. S. Finkle, *J. Interferon Res.*, *5*:265 (1985).

57. J. R. Quesada, M. Talpaz, A. Rios, R. Kurzrock, and J. U. Gutterman, *J. Clin. Oncol.*, *4*:234 (1986).

58. R. Kurzrock, M. G. Rosenblum, S. A. Sherwin, A. Rios, M. Talpaz, J. R. Quesada, and J. U. Gutterman, *Cancer Res.*, *45*:2866 (1985).

59. Food and Drug Administration, *Summary Basis of Approval, Protropin*, NDA 19-107, FDA, Rockville, Md. (1985).

60. Food and Drug Administration, *Summary Basis of Approval, Humulin R*, NDA 18-780, FDA, Rockville, Md. (1982).

61. J. W. Goodwin, in *Basic and Clinical Immunology*, 2nd ed. (H. H. Fundenberg, D. P. Stiles, J. P. Caldwell, and J. V. Wells, eds.), Lange Medical Publications, Los Altos, Calif. (1978), p. 39.

62. O. L. Frick, in *Basic and Clinical Immunology*, 2nd ed. (H. H. Fundenberg, D. P. Stiles, J. P. Caldwell, and J. V. Wells, eds.), Lange Medical Publications, Los Altos, Calif. (1978), p. 246.

63. D. T. Fearon and W. W. Wong, *Annu. Rev. Immunol.*, *1*:243 (1983).

64. K. F. Trofather, Jr., and C. A. Daniels, *J. Immunol.*, *122*:1363 (1979).

65. R. K. Root, in *Pathophysiology: The Biological Principles of Disease*, 2nd ed. (L. H. Smith and S. O. Thier, eds.), W.B. Saunders Company, Philadelphia (1985), p. 139.

66. A. C. Allison, *N. Engl. J. Med.*, *295*:821 (1976).

67. F. H. Bach, in *Autoimmunity: Genetic, Immunologic, Virologic, and Clinical Aspects* (N. Talal, ed.), Academic Press, New York (1977), p. 3.

68. M. C. Gelfand, M. M. Frank, and I. Green, *J. Exp. Med.*, *142*:1029 (1975).

69. S. E. G. Fligiel, P. A. Ward, K. J. Johnson, and G. O. Till, *Am. J. Pathol.*, *115*:375 (1984).

70. V. Kumar, in *Pathologic Basis of Disease*, 2nd ed. (S. L. Robbins and R. S. Cotran, eds.), W.B. Saunders Company, Philadelphia (1979), p. 262.

71. K. J. Tracey, B. Beutler, S. F. Lowry, J. Merryweather, S. Wolpe, I. W. Milsark, R. J. Hariri, T. J. Fahey III, A. Zentella, J. D. Albert, G. T. Shires, and A. Cerami, *Science*, *234*:470 (1986).

72. S. A. Rosenberg, in *Important Advances in Oncology 1986* (V. T. DeVita, S. Hellman, and S. A. Rosenberg, eds.), J.B. Lippincott Company, Philadelphia (1986), p. 55.

73. M. Rosenstein, S. E. Ettinghausen, and S. A. Rosenberg, *J. Immunol.*, *137*:1735 (1986).

74. T. J. Merimee, J. Zapf, and E. R. Froesch, *N. Engl. J. Med.*, *309*: 527 (1983).

75. P. W. Trown, R. J. Wills, and J. J. Kamm, *Cancer*, *57*:1648 (1986).

76. N. L. Kobrinsky, E. D. Israels, J. M. Gerrard, M. S. Cheang, C. M. Watson, A. J. Bishop, and M. L. Schroeder, *Lancet*, *1*:1145 (1984).

77. P. Ghezzi, B. Saccardo, and M. Bianchi, *Biochem. Biophys. Res. Commun.*, *136*:316 (1986).

78. S. E. Greisman and R. B. Hornick, *J. Infect. Dis.*, *128*:S265 (1973).

79. C. A. Dinarello and S. M. Wolff, *N. Engl. J. Med.*, *298*:607 (1978).

80. R. L. Hintz, in *Recombinant DNA Products: Insulin, Interferon and Growth Hormone* (A. P. Bollon, ed.), CRC Press, Boca Raton, Fla. (1984), p. 146.

81. United States Pharmacopeial Convention, *United States Pharmacopeia*, 21st ed., Mack Publishing Company, Easton, Pa. (1985), p. 1181.

82. B. F. C. Clark and K. A. Marcker, *J. Mol. Biol.*, *17*:394 (1966).

83. H. E. Hopps and J. C. Petricciani, eds., *Abnormal Cells, New Products and Risk*, Tissue Culture Association, Rockville, Md. (1985).

84. H. R. Strausser, R. A. Bucsi, and J. A. Shillcock, *Proc. Soc. Exp. Biol. Med.*, *128*:814 (1968).

85. A.-M. Lompre, P. Bouveret, J. Leger, and K. Schwartz, *J. Immunol. Methods*, *28*:143 (1979).

21

Reduction of Immunogenicity and Extension of Circulating Half-Life of Peptides and Proteins

Frank F. Davis, Glenn M. Kazo, Mary L. Nucci, and Abraham Abuchowski

ENZON, Inc., South Plainfield, New Jersey

I. INTRODUCTION

Historically, most therapeutic peptides and proteins were obtained from nonhuman sources. Applications of the many techniques developed through biotechnology have resulted in the production of an increasing diversity of enzymes, peptide hormones, kinins, and factors, many of which may become effective therapeutic agents. Unfortunately, repeated parenteral administration of a heterologous (nonhost) protein may result in an immunogenic response that accelerates the clearance of the protein from the blood. Immunogenicity (the ability to induce the formation of antibodies), when combined with antigenicity (the ability to react with specific antibodies), can potentially result in the development of hypersensitivity reactions. In these instances, severe allergic reactions are common and anaphylaxis-induced deaths are sometimes observed.

Some investigators, in attempts to ameliorate this problem, have established screening programs to identify protein sources which would yield particular proteins that were more compatible with the host's immune system. These investigations have had little success in resolving the problem.

The advent of recombinant DNA technology has allowed the large-scale production of human-derived peptides and proteins. The immunogenicity problem has decreased with the use of these proteins, but by no means has the problem been eliminated. In fact, it has yet to be proven that a human-derived protein is truly nonimmunogenic.

It is now recognized that immunogenicity is not the primary obstacle to the effective clinical use of proteins. Poor pharmacokinetics and pharmacodynamics are frequently the major impediments to efficacy. This is due primarily to rapid protein clearance from the blood by the liver, kidneys, or other organs, degradation by plasma proteases, and in some cases, as in the plasminogen activators, through inactivation by specific inhibitors.

In nonsensitized individuals, the plasma half-life of a human-derived protein is rarely better and usually worse than that of its nonhuman counterpart. The human origin of an injected protein is no guarantee that it will remain in the blood long enough to produce satisfactory pharmacokinetics and function as an effective therapeutic agent. Also, if the human-derived protein is produced by a cell other than the human cell in which it is normally expressed, it may lack the proper posttranslational modifications, and probably will be immunogenic.

Considerable research has been undertaken on therapeutic peptides and proteins to develop methods to control immunogenicity, minimize inactivation by plasma proteases and inhibitors, and improve pharmacokinetics. Success in these endeavors should result in the realization of the full therapeutic potential of these compounds. In this chapter we review techniques that have been developed to address problems inhibiting the use of proteins and peptides in clinical situations.

II. TECHNIQUES FOR MINIMIZING IMMUNOGENICITY AND ENHANCING CIRCULATING HALF-LIFE

A. Cloned Peptides and Proteins

It has been well documented by a number of investigators that both heterologous and homologous proteins suffer from the twin problems of short circulating half-life and immunogenicity when administered to humans. Human growth hormone (hGH) has been the most widely studied of all the proteins presently in use in the clinic. Thirty to sixty percent of children treated with acetone-preserved glandular hGH were reported to develop antibodies to the hormone and as a result, 5 to 10% of the patients develop growth attenuation secondary to antibody formation [1]. Recombinant human growth hormone (r-hGH) has also proven to be immunogenic, as antibody formation was observed in up to 50% of the patients. The immunogenicity was believed to be caused by the aggregates or impurities contaminating the preparations [2]. Recent reports about a "natural human growth hormone" (i.e., sequence homology with naturally occurring human growth hormone) indicated that antibody formation was reduced but not eliminated. The effect, if any, on growth attenuation due to antibody formation has yet to be determined for this preparation. In any case, no clear relationship has been established between the titer or affinity of the antibodies produced and the growth rate in response to therapy.

Luteinizing hormone releasing hormone (LHRH) is also commonly used in the clinic. Brown et al. found that drug failure following chronic administration of LHRH was correlated to neutralization of the LHRH by antibodies present in the patient's sera [3]. LHRH is rapidly metabolized and it has been suggested that these metabolites may also play a role in antibody formation [3].

A number of investigators have also demonstrated antibody formation to different insulins derived from a variety of sources [4,5]. Heading and Larkins have separately shown that semisynthetic human insulin is no less antigenic than purified procine insulin [6,7]. Significant titers of insulin binding antibodies developed in patients treated with either human or porcine insulin [7]. Fireman et al. identified antibodies in patients treated with human recombinant DNA insulin [8]. Velcovsky et al. also observed an insulin-specific antibody response in type I diabetic subjects treated with recombinant human insulin [9]. From a practical perspective, clinical experience indicates that even small amounts of circulating antibody may have negative effects on acute glycemic control [4], recovery from hypoglycemia [5], beta cell function, and metabolic control [6].

Dietrich et al. found resistance in patients to synthetic human calcitonin and suggested that antibody formation was responsible [10]. Subsequent studies using porcine calcitonin, synthetic calcitonin, and salmon calcitonin indicated that all sources, including the homologous (human) form, were capable of eliciting an antibody response [10,12,13].

A monoclonal antibody (Orthoclone OKT3, a registered trademark of the Ortho Pharmaceuticals Corporation) is currently being marketed to prevent kidney rejection following transplantation. Antibodies have been detected against the murine immunoglobulin following treatment [14,15].

Considering these results, it is highly probable that proteins produced by recombinant DNA technology will elicit an immune response in patients. One explanation for this immunogenicity may be that most r-DNA proteins are not truly identical to their human counterparts. Posttranslational changes are not well characterized and the addition of a terminal methionine by cloned bacterium is common, as is lack of proper glycosylation. In eukaryotic cells, improper glycosylation may also occur.

Although these may be part of the explanation, it should again be emphasized that even human-derived proteins may be immunogenic in humans and it is unlikely that recombinant techniques alone will yield proteins that are nonimmunogenic. Most important, the problem of short circulating half-life associated with recombinant proteins has not been adequately addressed.

B. Conjugation of Peptides and Proteins to Low-Molecular-Weight Compounds

A number of small molecules have been covalently attached to enzymes for the purpose of improving their therapeutic performance. Asparaginase has been subjected to a series of modifications in attempts to alter its clearance rate from the blood. Marsh et al. [16] coupled lactose and N-acetylneuraminyl lactose to asparaginase. The N-acetylneuraminyl lactose-modified asparaginase had a half-life about twice that of native asparaginase, while the lactosylated enzyme was cleared very rapidly, presumably by galactose-specific receptors. Other modifications of asparaginase, such as succinylation, acetylation, and attachment of N,N-dimethyl-1,3-propanediamine and glucuronic acid, either decreased the half-life or produced only a slight enhancement [17]. Blazek and Benbough, using a rat model, showed that compared to succinylation, guanidation, acetimidation, and reductive alkylation were more likely to increase the half-lives of asparaginases from *Escherichia coli* and *Erwinia caratovora*, and the glutaminase-asparaginase from *Achromobacter* [18].

Wagner et al. [19] prepared partially deaminated asparaginase (*E. coli*) by treating the enzyme with nitrous acid. This preparation exhibited a half-life of 24 hr in cancer patients compared to 11 hr for the native enzyme [20]. Rutter and Wade modified asparaginase from *E. carotovora* and produced a series of enzymes with isoelectric points ranging from 3.3 to 9.7. In rabbits, a comparison of half-lives revealed a parabolic relationship between half-life and isoelectric point (pI), with the longest half-lives in the pI range 5.6 to 6.0 [21]. A phase I study in cancer patients using a succinylated glutaminase-asparaginase was reported [22]. Daily doses were required to achieve and maintain continuous glutamine depletion. Significant side effects were observed.

Although some of the modified asparaginases showed extended circulating lives on initial injection, detailed immunological studies following repetitive injections were not reported. There is a real likelihood that enzymes modified with low-molecular-weight compounds will prove to be immunogenic; hence they may not become practical therapeutic agents.

An especially useful modification is the acylation of the active serine in the catalytic sites of plasmin and the streptokinase-plasminogen complex [23]. Acylation inactivates these enzymes and largely prevents their attack on plasma proteins. However, the attached acyl groups, such as the anisoyl group, exhibit hydrolytic rates such that significant reactivation of the plasminogen activators occurs after binding to fibrin, resulting in the subsequent lysis of the fibrin.

C. Conjugation of Peptides and Proteins to Various Polymers and the Effects of Conjugation

An approach to minimize immunogenicity and extend circulating half-life is the conjugation of proteins to various hydrophilic polymers. Using both natural and synthetic polymers, investigators developed a variety of covalent coupling chemistries to produce polymer–protein conjugates. It was theorized that conjugation of such polymers would mask determinant sites on the protein surface which are responsible for recognition by the target organ. Such a conjugate would thus demonstrate decreased clearance and an extended circulating life following injection.

Because of the wide array of polymers and coupling chemistries available, covalent attachment offers numerous advantages over other forms of immobilization schemes, such as entrapment, ionic binding, or noncovalent cross-linking. In terms of immunogenicity, covalent coupling ensures that a stable linkage is formed between the polymer and the protein. With other schemes, the possibility exists that a noncovalent linkage will be broken in vivo. This would, of course, release free protein and free polymer. This could further result in an immune response against both the free and conjugated protein as well as the polymer itself [24,25].

To be therapeutically effective, polymer–protein conjugates should be water-soluble, exhibit reduced immunogenicity, have extended circulating half-life in vivo, and still retain a substantial portion of the original biological activity. In some situations, as in certain storage diseases, targeting of the conjugate to specific locations in the body is desirable. To date, the majority of work with protein–polymer conjugates has focused on enzyme modifications, and a number of excellent reviews are available on this topic [24,25].

In general, protein–polymer conjugates exhibit properties quite different from those of the native protein. This approach has been used with varying degrees of success to alter certain physical characteristics of proteins, including stability, pH optimum, and solubility. Although some general principles have emerged as a result of these studies, the effect of any modification procedure is specific to both the polymer and the protein involved.

Dextran

Dextran has been used as a blood expander and was reported to be of low toxicity [26]. It has been covalently coupled to a variety of proteins, including insulin [29,30], pullulanase [31], lysozyme, β-glucosidase [27], nicotinamide adenine dinucleotide [32,33], hemoglobin [34–36], α- and β-amylase [37], trypsin, chymotrypsin [38], carboxypeptidase G, arginase [39], asparaginase [26,40], superoxide dismutase [41], and catalase [42]. Generally, the characteristics of thermal stability, resistance to proteolytic digestion, half-life, and immunogenicity are improved following modification of proteins with dextran, but some loss in biological activity is often noted [27,28].

Dextran–protein conjugates are prepared by coupling the protein via both the ε-amino groups of lysine and the terminal amino groups to the activated monosaccharide units of the carbohydrate [43]. The most common preparatory methods involved the activation of dextran by either alkylation with cyanogen bromide to yield an N-bromoacetylaminoethylamino-dextran [36], or by oxidation with either sodium periodate or sodium bisulfite to yield a dialdehyde–dextran [44]. The protein was then added and coupling was allowed to proceed overnight at 4°C. Uncoupled protein was removed by chromatography [36]. After modification, the adduct consisted of a heterogeneous mixture of high-molecular-weight molecules resulting from the interaction of lysine residues from different protein molecules with the same molecule of dextran [45]. Immunological analysis indicated that for dextran-modified antigen E, products were heterogeneous relative to the site of attachment and number of dextran molecules [46].

The method of preparation and the characteristics of the polymer and the protein affect the properties of the final product [31]. By changing reaction temperature, pH, and coupling time, one can alter the activity and extent of conjugation of the product [47]. The activating agent used to prepare the dextran can also alter the nature of the dextran–protein conjugate. For example, synthesis of dextran–hemoglobin by the alkylation method resulted in an adduct with a different oxygen binding curve than when prepared by the oxidation method [36]. When dextran–hemoglobin was prepared at a high pH, the resulting conjugate was more stable and of lower molecular weight than when prepared at low pH [35].

The molecular weight of the dextran used can also affect the characteristics of the adduct. Generally, as the molecular weight of the dextran is increased, the stability of the adduct is also increased. In addition, the circulating half-life of some proteins were also increased with higher-molecular-weight dextran (Table 1). However, a dextran of too high molecular weight may not increase the circulating half-life to the same extent as a lower-molecular-weight dextran [28], but either preparation had a longer circulating half-life than that of the native protein.

The stability of naturally occurring glycoproteins is generally superior to that of noncarbohydrate proteins [29,49–51]. When dextran was conjugated to noncarbohydrate proteins, the resulting adduct had stability properties superior to those of the unmodified protein. Stability was

Table 1 Circulating Half-Lives of Various Proteins Modified with Dextran

Protein	MW dextran	Host	Route	Circulating half-life		Reference
				Native	Modified	
α-Amylase	—	Rat	IV	16%	75%	47
				(% remaining after 2 hr)		
Catalase	—	Acatalasemic mice	IV	17 min	140 min	47
Carboxypeptidase G	40,000	DBA/2 mice	IP	3.5 hr	17.0 hr	39
	40,000	Tumor-bearing DBA/2	IP	7.0 hr	18.0 hr	39
Arginase	40,000	DBA/2	IP	1.4 hr	12.0 hr	39
	40,000	Tumor-bearing DBA/2	IP	2.5 hr	17.0 hr	39
Asparaginase	40,000	NZW rabbits (nonimmune)	IV	8.0 hr	46 hr	48
	70,000	NZW rabbits (nonimmune)	IV	8.0 hr	46 hr	48
	250,000	NZW rabbits (nonimmune)	IV	8.0 hr	56 hr	48
	40,000	Man	—	6-10 days	—	26
	40,000	Rabbits	IV	6-10 hr	30-40 hr	48

generally increased with increased molecular weight of dextran [28]. The
heat stability of lysozyme, chymotrypsin, β-glucosidase [27], α- and
β-amylase, and trypsin [37] was increased following modification. The
half-life for α-amylase increased from 2.5 min to 63 min, while the half-
life for β-amylase increased from 5 min to 175 min [37].

Modification also decreased the susceptibility of lysozyme, chymotrypsin,
casein, and serum albumin to proteolytic inactivation [27]. In addition,
the autolytic digestion of trypsin was inhibited following modification [37,
45]. Enzymes modified with dextran may also demonstrate less susceptibility
to inactivation by enzyme inhibitors, as was the case with dextran–trypsin.
When the dextran was removed from this adduct by enzymatic digestion,
the enhanced stabilization was reversed [45].

The stabilization of proteins after modification with dextran is believed
to be due to changes in the degree of hydration of the protein and conformational stabilization provided by multipoint attachment of the protein to
the same dextran molecule [28]. It is believed that multiple covalent links
form between the polymer and the protein, resulting in an adduct with
rigid conformation. This makes it resistant to inactivation and offers
steric protection from inhibitors and proteases. When the number of covalent links is relatively low, there is a smaller increase in stability.
Conjugates obtained early in the reaction sequence were of lower stability
than were those allowed a longer time to react [31].

Some activity loss is noted following modification (Table 2), the degree
of which can depend on the molecular weight and amount of dextran

Table 2   Effect of Dextran Modification on Protein Activity

Protein	MW dextran	Percent activity remaining	Reference
Arginase	40,000	45	39
Asparaginase	11,000	34	40
	40,000	50	48
	250,000	50	48
	250,000	60	149
	2,000,000	35	40
Carboxypeptidase	40,000	52	39
Catalase	17,000	49–67	42
	40,000	77–82	42
Chymotrypsin	10,000	55	27
β-Glucosidase	10,000	30–32	27
Lysozyme	10,000	24 (17% carbohydrate)	27
	10,000	16 (40% carbohydrate)	27

attached to the protein. Residual activity following modification was also
affected by the coupling method used and whether attachment occurred at
or near an active site [27]. By-products of the coupling reaction utilizing
cyanogen bromide may also be responsible for this loss in activity [42].
Marshall et al. [47] showed that the major loss of activity occurred during
the attachment of the protein to the activated dextran, and that the
coupling reaction methodology was not solely responsible for protein
inactivation.

In 1958, Kabat and Bezer demonstrated that dextrans of molecular
weight of less than 51,300 daltsons had a decreased ability to induce antibody formation in humans compared to dextrans of molecular weight 91,700
or greater [52]. Dextran alone was also nonimmunogenic in rabbits, mice,
and guinea pigs [53]. However, when coupled to protein, dextran was immunogenic in rabbits, sheep, and guinea pigs [53,54].

In a more recent study, repetitive intradermal injections of dextran-
asparaginase conjugates into rabbits did not result in the production of
antibodies specific for this conjugate [48]. In tests using asparaginase-
sensitized guinea pigs, injections of the conjugate caused local tissue damage. Injections of dextran (MW 70,000) alone also caused local tissue
damage in similarly sensitized animals but of less severity. Evaluations of
the conjugate or dextran alone in virgin animals were not performed.
Tissue damage was fully reversed within 72 hr in the dextran- or conjugate-
injected animals. In contrast, sensitized guinea pigs challenged with
asparaginase showed increased tissue damage at 24 hr which did not resolve by the 72-hr examination. In human studies, no hypersensitivity
reactions were noted when a single injection of dextran–asparaginase was
given to three lymphoblastic leukemia patients [26].

The attachment of dextran to other proteins resulted in a decrease in
immunogenicity rather than the total loss seen with asparaginase.
Terrilytine [55], α-amylase [56], and catalase conjugates [47] were all
capable of eliciting antibody formation, although at a lower level than
that of the nonconjugated protein.

Antigen E, when modified with dextran, lost some but not all of its
ability to produce and react with specific antibody. In skin tests on
human subjects, the coupled allergen was tenfold less reactive than unmodified antigen E, and in histamine release assays the coupled allergen
was sixfold less reactive than the native antigen. In rabbits, antibodies
to both the dextran and the antigen E were produced following immunization with the conjugate [46].

Albumin

Albumin has also been conjugated to a variety of therapeutically useful
proteins. As a natural constituent of plasma, serum albumin exhibits a
prolonged plasma half-life when administered to humans. It was thought
that the conjugation of this naturally occurring protein to a foreign protein would allow the adduct to be accepted as a normal component of the
plasma. In 1974, Paillot et al. [57] developed procedures for preparing
soluble cross-linked albumin-enzyme conjugates. Albumin has been attached to uricase [58], superoxide dismutase [59], α-1,4-glucosidase [60,
61], asparaginase [62,63], insulin [64], heparin [65], and wasp venom
[66]. In all cases, the attachment of albumin resulted in an adduct with
improved therapeutic characteristics.

According to Poznansky [24], there is no fixed methodology for preparing albumin–protein conjugates. Generally, the protein of interest is added to a reaction mixture consisting of albumin in buffer and a cross-linking agent. Glutaraldehyde, carbodiimide, or sodium periodate have all been used as cross-linking agents. It should be recognized that the agent used can affect the characteristics of the final product, such as activity, stability, immunogenicity, antigenicity, and circulating half-life. The protein–albumin complex was purified from the reaction mixture by chromatography or ultrafiltration [24].

Following modification, albumin was covalently attached to the primary amino groups of the protein [67]. The molecular weight of the conjugate was dependent on the cross-linking agent and the length of time that the reaction was allowed to proceed. Activity recovery rates for enzymes modified with albumin range from 4% for hog liver uricase prepared in the absence of uric acid to 84% for human α-glucosidase. The recovery rate for enzymes may be increased by running the coupling reaction in the cold or in the presence of substrate to protect the active site [24].

There was some loss of activity following the attachment of albumin (Table 3). Activity levels as a percent of original activity varied from 35% for albumin-α-1,4-glucosidase [60] to 70% for albumin–superoxide dismutase [59]. The addition of substrate to the reaction medium may increase the activity of the final product. In the case of asparaginase, omitting L-asparagine from the reaction mixture resulted in a loss of at least 90% of the original activity, while 50% was retained when the amino acid was present [63]. Although considerable activity was lost, the increase in stability, resistance to proteolytic inactivation, and increase in blood circulating half-life more than offset the loss of activity following modification [61].

Cross-linked albumin conjugates were much more resistant to proteolytic degradation and heat inactivation than was nonconjugated protein. In the presence of serum, albumin–uricase retained over 90% of activity over a period of 8 days, while nonconjugated uricase lost 40% of activity in the first 3 days [68]. Albumin–asparaginase adducts had an in vitro half-life of 20 hr in the presence of serum, and 10 hr when incubated with a proteolytic enzyme. This was compared to half-lives of 1.5 and 0.5 hr, respectively, for the native asparaginase under similar conditions [57].

The albumin adducts did retain the ability to act in vivo on their specific substrates. Albumin–uricase lowered blood uric acid levels in Dalmatian dogs by 38% with 1 hr of injection and up to 50% for 48 hr following administration. Repeated injections did not result in the development of hypersensitivity reactions and continued to lower blood uric acid levels effectively [68]. Albumin–superoxide dismutase (SOD) retained the ability to act as an anti-inflammatory agent in carrageenan-induced paw swelling in rats. Albumin–SOD inhibited up to 65% of the inflammation compared to 10 to 14% for native SOD, although given at a twofold-higher dose. Only one of three rabbits immunized with albumin–SOD developed antibodies to the antigen, while all the rabbits injected with native SOD mounted an immune response to the protein [59].

Cross-linked albumin–asparaginase adducts were more effective than native asparaginase in inhibiting the in vivo cell growth of the 6C3HED lymphosarcoma in mice. One hundred percent survival was increased from 16 days for the native enzyme to approximately 23 days following injection of albumin–asparaginase [63].

Table 3 Effect of Albumin Modification on Circulating Half-Life, Proteolytic and Temperature Inactivation, and Activity of Proteins

Protein	Percent original activity	Half-life at 37°C		Proteolysis half-life		Circulating half-life		Reference
		Native	Modified	Native	Modified	Native	Modified	
Superoxide dismutase	60–70	–	–	–	–	6 min	2.5–15 hr	59
Asparaginase	60	3–4 hr	40 hr	1.5 hr	20 hr	–	–	63
				(serum)				
Uricase	2–30	<5 min	>120 min	–	–	–	–	57
α-1,4-Glucosidase	35–60	1 hr	>15 hr	<1 min	15 min	–	–	60
α-1,4-Glucosidase	50	–	–	–	–	15 min	45 min	61
Uricase	–	3 day	>8 day	6.5 day	>8 day	4 hr	21 hr	68

The antigenicity of proteins decreased after modification with albumin. Antisera from rabbits immunized with rabbit albumin–uricase did not react with other uricase, rabbit albumin–uricase, rabbit albumin, dog albumin, or dog albumin–uricase. However, the source of albumin can alter the resulting immunogenicity. When albumin from dogs was attached to uricase and injected into rabbits, the rabbits were able to mount an immune response to the antigen. The antisera reacted with both dog albumin and dog albumin–uricase, but not with uricase. Although the complex retained immunogenicity, the antibodies produced were directed toward the albumin and not the uricase [58]. Such a response was probably due to the use of a heterologous albumin in the rabbit host.

DL-Amino Acids

Amino acid polymers have long been used to study the immune response. In 1962, Sela reported that multichain poly(DL-alanine) preparations were nonimmunogenic [69]. It was later observed that in contrast to the unmodified protein, the attachment of poly(DL-alanine) to proteins would yield a conjugate with both reduced antigenicity and immunogenicity. Asparaginase, trypsin, and ribonuclease were modified using this process and the conjugates exhibited both extended circulating half-life and reduced immunogenicity [70–72].

The method of conjugate preparation involved the addition of poly-D-L-alanine-N-carboxyanhydride to the enzyme solution. The complex was then purified by either gel filtration or hydroxylapatite chromatography. The modification process resulted in a 40 to 50% loss in specific activity [73]. However, there was no change in the K_m value of the enzyme. Slight changes in the optimal pH were seen after modification; however, at physiological pH both the native and modified enzymes were close to their maximal activity.

Poly(DL-alanine)-protein conjugates have increased stability and increased resistance to proteolytic digestion. Modified asparaginase (*E. carotovora*) was approximately three times more heat stable than was the native protein. Modified asparaginases from either of two sources were totally resistant to tryptic digestion and slightly more than fivefold more resistant to digestion by chymotrypsin [73].

After modification, asparaginase exhibited an extended circulating half-life. Native *E. coli* asparaginase had a half-life of 3 hr in mice, whereas the poly(DL-alanyl) derivative had a half-life of 20 hr. Modified *E. corotovora* asparaginase exhibited a half-life of 36 hr in mice as opposed to a half-life of 5 hr for the native enzyme. The clearance was biphasic; the first phase had a half-life of 4 hr and the second phase, 13 hr. When administered to rhesus monkeys, modified asparaginase exhibited a half-life of 254 hr, over 100 times longer than that of native enzyme (1.6 hr). In animals immunized to the native enzyme, the *E. coli* conjugate exhibited an extended circulating half-life, but a similar effect was not seen in hyperimmune animals [73].

Polyvinylpyrrolidone

Polyvinylpyrrolidone (PVP) is a water-soluble polymer that is available in a molecular weight range from a few thousand to several hundred thousand. It was used extensively in World War II was a plasma expander. A half-million people received infusions of PVP. Subsequent observations that

PVP was stored in the reticuloendothelial system led to the gradual aban-
donment of its use as a plasma expander in favor of other polymers, such
as dextran and gelatin derivatives [75].

PVP has been attached to trypsin, chymotrypsin [76], and D-N-acetyl-
hexosaminidase A [77]. PVP—trypsin exhibited in vitro properties of enhanced esterase activity, decreased proteolytic activity, resistance to
inhibition by trypsin inhibitor, and resistance to autolysis. PVP-chymo-
trypsin exhibited similar properties, except that its esterase activity was
reduced relative to chymotrypsin.

PVP/β-D-N-acetylhexosaminidase A exhibited a resistance to proteases
and a significantly slower clearance rate from rabbit blood than that of the
unmodified enzyme [77]. The conjugate was capable of reacting with antibody developed against the native enzyme. Extended immunogenicity
studies, or repetitive injections that might elicit an immune response, were
not reported.

There appear to be two major drawbacks to the use of PVP as a polymer
for peptide conjugation and therapeutic use. First, PVP lacks a specific
site for activation and coupling. Sites must be created by drastic hydrolytic reactions to achieve the opening of a few percent of the lactam rings,
which then are activated by conversion to the N-hydroxysuccinimide ester.
Multiple and random activation sites on PVP may produce multipoint attach-
ments, and the attached polymer will contain free carboxyl groups following
conjugation. The difficulty in reproducibility in obtaining a defined conjugate by this approach may make it difficult to obtain regulatory approval.
Also, PVPs of higher molecular weights are definitely immunogenic [78].
Thus a PVP-peptide conjugate probably will also be immunogenic, which
may present difficulties in developing effective enzyme drugs.

Polyethylene Glycol

The most extensively studied method of protein modification involves the
use of the polymer monomethoxypolyethylene glycol (PEG). This methodology arose from the desire to find a compound which when attached to
proteins would enhance their circulating half-life and render them nonimmunogenic. PEG is a linear, uncharged, hydrophilic, nonimmunogenic
molecule that has been used to modify a large number of compounds, including trypsin [79], superoxide dismutase [80], catalase [81], adenosine
deaminase [82], bovine serum albumin [83], asparaginase [84], uricase
[85], NAD [86], antigen E [87], streptokinase [88], urokinase [89],
D-α-tocopherol [90], lipase [91], hemoglobin [92], interleukin-2 [93],
and arginase [94]. As a general rule, PEG-modified proteins exhibit increased stability, decreased to nonexistent immunogenicity, increased circulatory half-lives, and low toxicity. PEG-proteins were shown to be soluble
in organic solvents [95], yet retain activity.

In 1977, Abuchowski et al. described a method for the covalent attachment of PEG to proteins. Several other methods have since been described
for the preparation of PEG-proteins, but all involve the preparation of an
activated PEG having a reactive functional group that can be coupled to the
lysine groups on the protein. The most common reagents used for the
activation of PEG are trichloro-s-triazine (cyanuric chloride) [81], carbonyldiimidazole [96], or succinic anhydride [97—99]. These yield, respectively,
2-o-methoxypolyethylene glycol-4,6-dichloro-s-triazine [100], carbonyl-
imidazyl monomethoxypolyethylene glycol, and methoxypolyethylene

glycolyl-succinimidyl succinate. For further discussion on these and other means of activating PEG, see the review by Harris [101].

Each method is reported to have its disadvantages. The attachment of cyanuryl-PEG resulted in substantial inactivation of certain enzymes [96, 99]. The third chloride of the reagent may react further and cross-link the protein, although this is highly unlikely due to the short intramolecular distance between the adjacent chlorides. The cyanuryl moiety may modify a reactive —SH group in the active site of a protein and it causes interference in spectroscopy due to the s-triazine ring. There is the question of whether the carbonyldiimidazole activation of PEG results in the attachment of PEG to tyrosine residues as well as lysine residues (unpublished observation).

The ester bond of succinyl PEG was reported to be hydrolyzed spontaneously at physiological pH and in vivo by nonspecific hydrolases in the blood [102]. Apparently, the ester linkage is somehow stabilized by the presence of the PEG, as PEG-proteins have long shelf lives when stored under physiological conditions and circulate for extended periods in the blood. The succinic anhydride method does offer certain advantages over the other methods, in that it is a rapid reaction at physiological pH and mild conditions, which minimizes any loss in activity. To date, this procedure has been used successfully to prepare over a dozen PEG-proteins, including enzymes, peptides, and hormones.

Boccu et al. [98] reported the preparation of PEG derivatives by attachment through active esters. Two methods for producing carboxyl terminated PEG were described, the first involving the conversion of the terminal hydroxyl group to an amine group which was then reacted with succinic anhydride to yield PEG-succinate. The other proceeds by a two-step oxidation of the terminal hydroxyl to a carboxylic acid. The extent of modification was a nonlinear function of the amount of reagent added. Both reagents produced adducts with identical properties.

The reactions noted above involve a single strand of PEG attached to an available ε-amino and α-amino group in the protein. Several investigators have reported a PEG modification in which two PEG strands are attached to the protein at one site. It was thought that the extra PEG molecule added to each site would reduce the percent modification required to alter immunogenicity. Cyanuric chloride was used to activate PEG to yield 2,4-bis(O-methoxypolyethylene glycol)-6-chloro-s-triazine (PEG2). The immunoreactivity was indeed eliminated at a lower degree of modification, but the resulting conjugate suffered a substantial loss in activity [103].

The molecular weight of the PEG and the specific properties of the protein may affect the characteristics of the final product [104]. With the percent modification held constant and as the molecular weight of the PEG used to modify asparaginase increased from 750 daltons to 5000 daltons, the activity dropped from 12% to 7% of original; however, the binding ability to antiserum specific for asparaginase fell from 83% to 0% [84].

In the case of allergens, the molecular weight of the PEG used had an distinctive effect on the tolerability of the adduct and the extent of reduction of allergenicity [105]. The coupling agent used may affect the resulting immunogenicity of the adduct. For example, at similar degrees of modification, the antibody-combining ability of PEG-antigen E was higher when modified by the dithiocarbonate method than by the cyanuric chloride method [106]. Reaction pH and the ratio of the reactants used can also affect the activity of the modified protein. When phenylalanine ammonia-lyase was modified with cyanuric chloride-activated PEG, increased pH also increased the number of amino groups modified, but the percent of original activity was decreased. As the ratio of PEG was increased, an increase in the percentage of amino groups modified and a decrease in activity was also noted for β-glucuronidase [107], phenylalanine ammonia-lyase [108], and uricase from *Candida utilis* [109]. It is necessary to determine the optimum molecular weight of PEG, degree of modification, and reaction sequence for each protein [101].

Attachment of PEG to proteins increased their resistance to proteolytic inactivation. Ashihara et al. (1978) showed that PEG–asparaginase retained 80% of its original activity after incubation for 30 min with trypsin. Native asparaginase was quickly digested, in contrast, losing more than 80% of its activity in the same period [84]. In the presence or absence of the competitive inhibitor, cinnamic acid, phenylalanine ammonia-lyase was digested by trypsin at equal rates. The corresponding PEG conjugate, in the absence of the inhibitor, was digested more slowly than the native enzyme, and in the presence of cinnamic acid showed increased resistance to digestion by trypsin [108]. Catalase [79], trypsin [79], β-glucuronidase [107], acyl-plasmin-streptokinase [100], and PEG2-asparaginase [84] also showed resistance to proteolytic digestion following modification. PEG-superoxide dismutase did not demonstrate increased resistance to proteolytic digestion after modification, probably because native superoxide dismutase is already very stable to such degradation [110]. PEG also protected trypsin against autolysis. Trypsin with 24% of its amine groups modified with PEG showed a complete loss of autolysis under conditions that would inactivate native trypsin [79].

Resistance to thermal inactivation does not appear to be enhanced in most proteins to the same extent as resistance to proteolytic inactivation is after modification. Superoxide dismutase [110] and lipase [111] demonstrated increased thermal stability after attachment of PEG, while the thermal stability of catalase modified with PEG was not increased. Both native and PEG-catalase (MW PEG = 1900 daltons) denatured at 55°C, but the maximum activity of PEG-catalase (MW PEG = 5000) occurred at a temperature several degrees lower than the 40°C demonstrated for catalase. The authors suggested that this may be due to some destabilization of the catalase structure following PEG attachment [81].

Long-term stability studies indicate that PEG-proteins are stable for extended periods in solution. Adenosine deaminase modified with PEG is stable for more than 9 months at 4°C [112]. PEG derivatives of catalase, superoxide dismutase, and asparaginase are also stable for longer than 18 months in solution at 4°C (unpublished).

Polyethylene glycol attachment also affects other properties of proteins, including solubility, K_m value, and activity. Following modification, proteins often become soluble over an extended pH range. Bovine serum albumin was found to be soluble from pH 1 to 12 [113]. Proteins that are normally insoluble under physiological conditions can be solubilized by PEG attachment [114]. PEG2-modified horseradish peroxidase [115], lipoprotein lipase [116,117], catalase [95], and chymotrypsin [118] have all been shown to be soluble and active in organic solvents such as benzene, toluene, acetone, ethanol, and dimethylformamide [119]. Generally, native proteins are not soluble in organic solvents, but the amphipathic nature of the PEG moiety allows the modification of protein in an aqueous solution and the retention of activity in a hydrophobic environment [116]. The necessity for a trace amount of water in the solvent [120] implies that water molecules are necessary for protein activity. The retention of activity in organic solvents may be due to the proposed ability of the attached PEG strands to concentrate the water around the protein and thus create a microenvironment in which it can function.

The K_m value is generally higher following modification (Table 4), and activity decreases. As modification is increased, most proteins, including acyl-plasmin-streptokinase [100], batroxobin [121], α-galactosidase [122], and elastase [123], demonstrate decreased activity. The fibrinolytic activity of acyl-plasmin-streptokinase decreased from 71% at a 26 modification to 10% at a 68% modification [100], and α-galactosidase decreased from 30% at a 17% modification to less than 1% at a 52% modification. The loss of activity is believed to be due to steric hindrance or denaturing effects of the strands of PEG that reduce the ability of the substrate to interact with the protein [104]. However, PEG modification of batroxobin did allow it to degrade proteolytically its large macromolecular substrate, fibrinogen [121].

Investigators using PEG2 to modify proteins found that a lower percentage of modification was required to decrease the immunogenicity of the protein than PEG and that subsequently, more activity was retained (Table 5). The relative activity of the conjugate remained low; however, the decreased activity due to PEG modification was more than offset by the concomitant increase in circulating half-life after attachment of PEG. All proteins modified with PEG demonstrated an increased half-life regardless of the route of injection (Table 6). Increasing the amount of attached PEG also resulted in a concomitant increase in the circulating half-life. In rats, as the percentage of amino groups in superoxide dismutase modified with PEG was increased from 13 to 90, the circulating half-life increased from 4 hr to 20 hr. This was in marked contrast to the 5-min circulating half-life for unmodified superoxide dismutase [110]. Uricase from either hog liver or *Candida utilis* also circulated longer when modified to a greater extent [113].

In addition, PEG compounds appear to cross biological membranes freely, without loss of activity. After intraperitoneal injection, PEG-adenosine deaminase was detected in the bloodstream and showed a maximal level (48%) at 6 hr which was close to the theoretical maximal amount. Native adenosine deaminase at 1 hr was only 6 to 9% of the theoretical calculated level [82].

PEG-proteins injected into animals immunized to the corresponding native protein circulated as if injected into virgin animals. PEG2-asparaginase [124], PEG-asparaginase [99], PEG-arginase [94], PEG-phenylalanine ammonia-lyase [108], PEG-bovine serum albumin [83], and PEG-catalase [81] all circulated similarly both in immunized and in virgin animals, indicating that PEG modification had rendered them nonantigenic.

The circulating half-life of proteins in humans was also increased after PEG attachment. The half-life for PEG-uricase increased from less than 3 hr for the native enzyme to over 8 hr for the conjugate [82], from 20 hr to 357 hr for PEG-asparaginase [126], from 15 min to 37 hr for PEG-superoxide dismutase (unpublished), and from minutes to 5 days for PEG-adenosine deaminase [112].

As a general rule, PEG proteins demonstrate a decrease in immunogenicity. Examples are catalase [81], ovalbumin, ragweed pollen extract [127], superoxide dismutase [80, β-glucosidase [122], β-glucuronidase modified greater than 33% [107], bovine serum albumin [83], and asparaginase

Table 4 Effect of PEG Modification on Protein K_m Value

Protein	Percent modification	PEG MW	Percent activity	$K_m{}^a$	Reference
Arginase	53	5000	65	N = 6.0×10^{-3}	94
	—	—	—	P = 1.2×10^{-2}	—
Asparaginase	29	350 (PEG2)	9.2	—	144
	34	2000 (PEG2)	—	—	—
	35	5000	30	N = 7.0×10^{-3}	144
Catalase	40	5000	95	—	81
	43	1900	93	—	81
α-Galactosidase	17	—	—	N = 3.4×10^{-4}	122
	—	—	—	P(17%) = 4.0×10^{-4}	—
	35	—	—	P(35%) = 5.9×10^{-4}	122
β-Glucosidase	40	—	—	N = 1.2×10^{-3}	122
	—	—	—	P = 3.0×10^{-4}	—

60	5000	20	$N = P$	107
L-Asparaginase 56	5000 (PEG2)	–	5.0×10	159
70	5000	52	–	159
Glutaminase-asparaginase 71	5000	12	–	151
Phenylalanine ammonia lyase 27	5000	42	$N = 3.9 \times 10^{-4}$	108
–	–	–	$P = 1.04 \times 10^{-3}$	–
52	5000	15	$P = 1.06 \times 10^{-3}$	108
Superoxide dismutase 95	5000	51	–	80
Uricase 36	5000 (PEG2)	45	–	109
57	5000	23	$N = 5.0 \times 10^{-5}$	113
–	–	–	$P = 5.6 \times 10^{-5}$	–
58	5000	28	$N = 2.0 \times 10^{-5}$	113
–	–	–	$P = 6.0 \times 10^{-5}$	–

[a]N, native enzyme; P PEG-modified enzyme.

Table 5   Effect of PEG Modification on Activity Using Cyanuric Chloride as the Coupling Agent

Protein	PEG	Percent modification	Percent original activity	Reference
Adenosine deaminase	PEG1	60	28	82
Asparaginase	PEG1	71	12	150
	PEG2	35	30	144
	PEG2	56	8	124
Batroxobin	PEG2	29	93 (esterolytic)	121
Uricase	PEG2	37	45	124
	PEG1	58	28	113
	PEG1	57	23	113

[129]. Modification also resulted in a decrease in antigenicity. Uricase
[128], elastase [123], batroxobin [121], asparaginase [129], superoxide
dismutase, and catalase [130] all showed a decrease in antigenicity. Modification of antigen E or ovalbumin resulted in both a decrease in allergenicity and an increase in the ability to inhibit an IgE response to the
PEG antigen. In addition, the PEG antigen was capable of abrogating an
ongoing IgE response to the corresponding native antigen [127,131].

The decreased immunogenicity of proteins following the attachment of
PEG was thought to be due to either steric hindrance of antigenic determinants on the surface of the protein or to an active effect on the immune
system that induces the formation of T suppressor (Ts) cells specific for
the PEG antigen [132]. Radioimmunoassay studies on PEG β-glucuronidase
conjugates indicated that with increased modification, accessibility to determinants on the surface of the protein became limited. When 48, 60,
and 73% of available amine groups on the protein were modified with PEG,
an 18-fold, 15-fold, and 300-fold excess, respectively, of each antigen
was required to displace the native antigen bond to antiserum against PEG
β-glucuronidase (60%). The ability of PEG-ovalbumin conjugates to induce
tolerability against ovalbumin was eliminated following treatment with
cyclophosphamide [132], which specifically eliminates Ts cells [133]. One
method of producing specific Ts cells for an antigen is to use an antigen
that avoids macrophage processing, such as might be the case for PEG
antigens [132].

When a nonimmunogenic compound (hapten) is attached to an immunogenic compound (carrier), the hapten generally becomes capable of inducing an immune response specific for its own antigenic determinants and
the new determinants formed by attachment. When administered alone,
PEG was shown to be nonimmunogenic in rabbits and mice [134]. After attachment to a protein and testing by repeated injections into rabbits in the
presence of complete Freund's adjuvant, PEG may become slightly hapto-
genic. In clinical trials with PEG-modified allergens, PEG reactive antibodies did appear in the sera. Antibodies were demonstrated in 50% of the
patients, with a decrease in titer occurring after the second year of therapy.
No secondary response was noted after receipt of booster immunizations, and
most of the antibodies consisted of the IgM isotype. The authors considered the response moderate and of no clinical significance [135]. In clinical trials on PEG-modified asparaginase, PEG-superoxide dismutase, and
PEG-adenosine deaminase, antibodies reactive to the PEG-proteins did develop, but were of a very low level and of no clinical significance (unpublished).

The degree of immunogenicity acquired by PEG after its attachment to
a protein varies with each protein to which it is attached and the percent
of conjugation (Table 7). When six PEG molecules (MW 11,000) were attached to ovalbumin, precipitating antibodies specific for both PEG and
ovalbumin were detected. In comparison, the attachment of 20 PEG molecules resulted in an adduct with greatly decreased immunogenicity [134].
Phenylalanine ammonia-lyase with 23% of its amine groups modified with PEG
reacted strongly with antisera specific for either PEG-phenylalanine ammonia-
lyase or the native antigen. When 44% of the amine groups of the same
protein were modified, the conjugate did not react as strongly against antiserum specific for PEG-phenyalanine ammonia lyase. Antiserum against a
48% PEG-phenylalanine ammonia lyase adduct had the same complement-fixing

Table 6 Circulating Half-Lives of Various Proteins Modified with Polyethylene Glycol

Enzyme	Source	Route	Host	Half-life		Ref.
				Native	Modified	
Adenosine deaminase	Calf intestine	IV	Balb/c mice	30 min	28 hr	82
		IP	—	40 min	32 hr	—
Asparaginase	Achromobacter	IV	BDF1 mice	2 hr	24 hr	150
	Achromobacter	IV	Human	<30 min	72 hr	151
	Escherichia coli	IP	Rats	2.9 hr	56 hr	152
	Escherichia coli	IV	Mice	3.35 hr	32.9 hr	153
	Escherichia coli	IP	—	<6 hr	>3 days	99
	Vibrio succinogenes	IP	—	—	>3 days	159
Arginase	Beef liver	IV	B6D2 F1/J mice	1 hr	12 hr	94
	Beef liver	IV	B6D2 F1/J mice	24 hr: 1%	24 hr: 45%	154
		—	—	—	72 hr: 16%	—
Catalase	Bovine liver	IV	Long-Evans mice	10 min	>4 hr	156
β-Glucuronidase	Bovine liver	IV	Swiss-Webster mice	5 min	4–8 hr	107
Phenylalanine ammonia-lyase	Rhodotorula glutinis	IV	Swiss-Webster mice	1st: 6 hr	1st: 20 hr	155
				5th: 1 hr	7th: 4 hr	—

Enzyme	Source	Route	Species			
Superoxide dismutase	Bovine erythrocytes	IV	Swiss–Webster mice	—	16 hr	80
	Bovine erythrocytes	IV	Lewis male rats	5 min	PEG37000: 1.5 hr	110
					PEG46000: 3 hr	—
					PEG43000: 11 hr	—
					PEG66000: 25 hr	—
					PEG121000: 25 hr	—
	Bovine erythrocytes	IV	Lewis male rats	5 min	4 hr	157
					18 hr	—
					20 hr	—
	Bovine erythrocytes	—	Lewis male rats	—	4 hr	98
				20 hr	20 hr	—
	Bovine liver	IV	Long-Evans rats	56 min	24 hr	156
Uricase	Candida utilis	IV	Human	<3 hr	8 hr	125
Urokinase		IV	Beagle dog	37 min	167 min	89
		IV	Rabbit	4–5 min	30–40 min	158
					80–100 min	—
					110–150 min	—

Table 7   Immunogenicity of Various Proteins as Related to Percent
Modification with Polyethylene Glycol

Protein	Percent modification	PEG type	Percent binding	Reference
Asparaginase	18	PEG1 (5000)	90	84
	23		94	
	38		83	
	48		39	
	73		0	
Batroxobin	29	PEG2 (5000)	1:10	121
	45		1:19	
	46		1:90	
Elastase	25	PEG1 (5000)	100	123
	50		91	
	75		9	
	100		0	
Uricase	22	PEG2 (5000)	86	109
	25		49	
	37		0	
	47		0	
	51		0	

ability as that of anti-phenylalanine ammonia lyase antiserum when tested
against the native antigen [122].

The modification of antigen E also resulted in a reduction in both immunogenicity and antigenicity, but the protein was still able to induce
histamine release from the leukocytes of ragweed patients and completely
inhibit agglutination of antigen E-coated cells. As multiple surface antigens
must interact with immunoglobulin E (IgE) in order to induce the cellular
release of the chemical mediators of the immune response, the antigenic
determinants must have been available for interaction [87]. The antigenic
valency of the PEG adduct must be greater than 1 for histamine release to
occur [136]. The inability of PEG modification to greatly reduce or eliminate the immunogenicity of some proteins, such as antigen E and phenylalanine ammonia lyase, may be due to insufficient modification of the antigenic determinants on the surface of the protein, to dissociation of subunits caused by the bulk of the PEG molecules [108], or to the inherent

Table 8 Effect of PEG Modification on Protein K_m Value

Protein	Percent modification	PEG MW	Percent original activity	K_m	Reference
Glutaminase-asparaginase	71	5000	12	—	151
L-Asparaginase	70	5000	52	—	159
	56	5000 (PEG2)	8	$N = P \ 5.0 \times 10$	124
Arginase	53	5000	65	$N = 6.0 \times 10^{-3}$	94
				$P = 1.2 \times 10^{-2}$	94
Phenylalanine ammonia lyase	27	5000	42	$N = 3.9 \times 10^{-4}$	108
				$P = 1.04 \times 10^{-3}$	
	52	5000	15	$P = 1.06 \times 10^{-3}$	108
Uricase	57	5000	23	$N = 5.0 \times 10^{-5}$	113
				$P = 5.6 \times 10^{-5}$	
	58	5000	28	$N = 2.0 \times 10^{-5}$	113
				$P = 6.0 \times 10^{-5}$	
	36	5000 (PEG2)	45	—	109
Catalase	43	1900	93	—	81
	40	5000	95	—	81
Superoxide dismutase	95	5000	51	—	80

characteristics of the protein, which may allow antigenic determinants to
remain exposed after modification.

III. CLINICAL USES OF MODIFIED PEPTIDES
   AND PROTEINS

Of the various conjugates available, PEG conjugates have been the most
extensively studied clinically. Conjugates of PVP, DL-amino acids, and
albumin have not yet been tested clinically, and dextran-asparaginase
(E. coli) for the treatment of acute lymphoblastic leukemia (ALL) has been
tested in only three patients.

A. Dextran

Dextran-asparaginase was administered to three adult ALL patients to assess its therapeutic potential. Each patient had relapsed and was given a
single injection (300 U/kg) of the conjugate. None showed any hypersensitivity to the drug; one went into partial remission and survived 14
months, the others developed metabolic acidosis and died within 4 weeks of
commencement of the trial. Plasma half-lives ranged from 6 to 11 days and
plasma asparagine levels remained undetectable for 12 weeks in the one patient. Native asparaginase depleted plasma asparagine for only 7 days.
Severe metabolic effects, which may have been associated with the enzyme's
inherent glutaminase activity, were noted following the injection of the
adduct. This led the author to state that further use of this conjugate for
the therapy of leukemia was not recommended without further study [137].

B. Polyethylene Glycol

Six different PEG-enzyme conjugates have reached clinical trials: PEG-
adenosine deaminase (PEG-ADA) for severe combined immunodeficiency
disease (SCID); PEG-antigen E for ragweed hay fever; PEG-asparaginase
E. coli) of both the PEG and PEG2 variety for acute lymphoblastic, lymphocytic, and undifferentiated leukemias (ALL, AUL) and for lymphomas; PEG-
honeybee venom (PEG-HBV) for reperfusion injury associated with organ
transplantation; and PEG-uricase for hyperuricemia associated with chemotherapy or gout. One of these six conjugates was approved by the Federal
Food and Drug Administration in 1990.

PEG-Adenosine Deaminase (PEG-ADA)

PEG-ADA (PEG MW 5000) is being used for enzyme replacement therapy in
ADA-deficient severe combined immunodeficiency disease (SCID), a disease
in which the immune system does not develop due to the buildup of
adenosine and 2'-deoxyadenosine [138]. After administration to SCID patients at approximately 15 U/kg, PEG-ADA had a half-life of 48 to 72 hr
[112]. In mice, PEG-ADA had a half-life of 24 hr compared to minutes for
the unmodified form [82]. The conjugate reversed the biochemical effects
of the disease by normalizing the erythrocyte concentrations of adenosine,
S-adenosylhomocysteine hydrolase, and deoxyadenosine, thereby causing an
increase in the mitogen responsiveness of blood mononuclear cells. In one
patient, E-rosetting cells increased from 40% to 60% of blood mononuclear
cells after only 6 months of treatment. No patients in the trial experienced

any serious infections since the start of therapy, nor have they developed any allergic or immune complex-related side effects [112].

PEG-Antigen E

Immunotherapy with native allergens was used to suppress IgE responses to allergens that occur naturally, such as ragweed pollen. PEG-antigen E was used in an attempt to induce the formation of Ts cells, thereby abrogating ongoing IgE responses to those naturally occurring allergens.

Norman et al. [106] reported that although the coupling of PEG (PEG MW 2000 or 5000) to the antigen did not carry any additional risk of toxicity, no specific suppression of IgE antibody formation was noted. However, PEG-antigen E was able to induce a significant IgG response, which was suggested as being more important than IgE for efficacy of immunotherapy [139]. It was noted that the method of conjugation affected the immunogenic properties of the conjugate. At a similar degree of modification, PEG-antigen E (MW 5000) prepared by the dithiocarbonate method [140] had a higher antibody-combining activity than when prepared by the cyanuric chloride method.

In patients with seasonal rhinoconjunctivitis, the use of PEG-antigen E (0.2 to 2.0 mg per protein) was associated with a reduction in symptoms of ragweed pollen-induced rhinoconjunctivitis. Side effects were mild and no toxicologic changes were observed. An increase in ragweed-specific IgG and mild increase in IgE were noted. The author concluded that PEG-antigen E in sufficiently high doses reduced the severity of ragweed pollen rhinoconjunctivitis [141].

Two regimens were compared in a more recent study of the efficacy of PEG-antigen E in seasonal rhinoconjunctivitis. The first regimen consisted of weekly injections of the antigen and the second of daily injections (the rush group) until there was a wheal reaction >4 cm in diameter or systemic symptoms developed. Subsequently, all injections were given weekly. The rush group had a higher number of adverse responses, which due to cessation of therapy, resulted in the weekly group having a higher total cumulative dose. A greater increase in ragweed-specific IgG and a greater reduction in skin sensitivity to ragweed pollen extract was also noted in the weekly group.

Five patients showed mildly positive skin tests to PEG after therapy with the conjugate, and systemic reactions to the conjugate developed in only 8% of the rush group and 4% of the weekly group. As before, the adduct was able to induce a significant IgG response and a mild IgE response [142].

PEG-Asparaginase

PEG-asparaginase (PEG MW 5000) is currently in phase II clinical trials for the assessment of efficacy against acute lymphocytic and lymphoblastic leukemias and lymphomas. In phase I trials, patients received biweekly intravenous injections ranging from 500 to 8000 U/m^2. PEG-asparaginase was given as the sole agent. The mean plasma half-life of the adduct (357 ± 243 hr) was not dose related and varied from patient to patient [126]. The unmodified form of the enzyme had a half-life of 20 hr [143].

Apparent anaphylactic reactions were noted in 3 of 37 patients, though only one had identifiable antibodies to the enzyme. This patient had received native asparaginase before and had experienced a mild allergic reaction approximately 6 months prior to receiving PEG-asparaginase. In another patient who experienced an anaphylactic reaction, PEG-asparaginase remained in the plasma for at least 9 days. Clinical responses included four complete remissions and one stable disease, three of whom received low doses of the conjugate [126]. Fifty percent of all patients had some response to the drug.

Clinical trials with this form of PEG-asparaginase are continuing. One phase II program is evaluating the clinical activity of PEG-asparaginase in lymphoma. Patients receive biweekly injections of 54 U/kg of the adduct as a single agent. Another phase II protocol is studying the effect of incorporating PEG-asparaginase in the standard chemotherapeutic regimen for therapy of ALL and AUL in place of asparaginase. PEG-asparaginase will shortly enter phase III to evaluate efficacy in newly diagnosed children.

Asparaginase modified with PEG2 was tested in three patients with leukemia who had developed allergic reactions to the unmodified form of the enzyme. Asparaginase was then replaced by PEG2-asparaginase in a regimen consisting of three injections every 2 weeks. Injections of 200 U of the adduct were administered intravenously over a 60-min period [144].

The half-life of PEG2-asparaginase was on the order of days compared to a few hours for asparaginase given at a much higher dose of 10,000 U. Two patients tolerated more than 10 courses of the regimen without side effects; data on the third patient were not given. No immunological responses to the conjugate were noted in any of the patients [144].

PEG-Honeybee Venom (PEG-HBV)

PEG-HBV was compared with HBV in a study on venom immunotherapy of honeybee stings. Twenty-four patients were enrolled in this double-blind study, and injected according to a cluster regimen with a total weekly maintenance dose of 100 µg of PEG-HBV. Therapy with PEG-HBV was tolerated well, with mild systemic reactions being noted in only one patient. Systemic reactions caused two dropouts in the HBV group. Patients receiving HBV showed increased levels of HBV-specific IgG and IgE, while only HBV-specific IgG levels increased in those receiving PEG-HBV. After sting challenge, one patient receiving HBV had a systemic reaction, which consisted of urticaria-angiodema. Three patients receiving PEG-HBV had systemic reactions: respiratory reactions in one and shock in the other two [145].

PEG-Superoxide Dismutase (PEG-SOD)

PEG-SOD is currently in phase I trials to examine its efficacy in reperfusion injury associated with organ transplantation. Safety and tolerance are currently being investigated in healthy volunteers. Preliminary data indicate that the adduct circulates at its peak level for up to a week.

PEG-Uricase

PEG-uricase was used in humans to test its efficacy against hyperuricemia associated with malignancy. Five patients were injected intravenously with 120 U/m^2 over a period of 30 min. Serum uric acid levels became undetectable within 60 min of injection and remained undetectable for at least

32 hr. The serum half-life of PEG-uricase was calculated to be 8 hr. Although the half-life of the native enzyme was not calculated in this study, in other studies it was reported as less than 3 hr. The same dose of PEG-uricase given to all patients 30 days later had the same activity in vivo, with no toxicity being noted. No precipitating antibodies were detected during the 60-day study period [82]. Currently, PEG-uricase is in phase I trials and will be examined as a therapeutic agent for hyperuricemia associated with chemotherapy or gout. The study will evaluate the response in both adults and children.

IV. PRACTICAL CONSIDERATIONS

A. Limitations of the Conjugate Approach

Regulatory

Interestingly, requirements imposed by the Food and Drug Administration may limit the type of conjugate used, the coupling methodology, and finally the approval of an NDA to market the drug. The polymer itself must be prepared according to Good Manufacturing Practices, and meet appropriate standards for homogeneity, freedom from pyrogens, and so on. Activation and coupling of the polymer to the bioactive peptide must be by a process that is reproducible and in which the chemistry is clearly described. For example, a polymer such as dextran activated with cyanogen bromide or periodate will have multiple activation sites of largely random location on the polymer. The peptide may be attached through one or more of these sites, but it cannot be said with certainty which of the activated groups reacted with the peptide or, in fact, if more than one did. Such uncertainty may greatly increase the difficulty of obtaining regulatory approval.

In addition to the need for clearly described and reproducible chemistry, the preparations must show minimal immunogenicity. Again, random, possibly multipoint attachment may lead to structures that elicit an immune response. Although the polymer and the peptide individually may not be immunogenic, the conjugate may show immunogenicity. Procedures that modify the structure of the polymer, such as periodate oxidation or cyanogen bromide treatment in the case of dextran, may also create new immunogenic sites.

Cost

There is substantial expense in the preparation of a soluble polymer–enzyme conjugate that must be justified by the improved therapeutic performance of the conjugate. Costs include the preparation or acquisition of medical-grade polymer, activation and cleanup of the polymer, reaction of the activated polymer with the medical-grade peptide, and purification and evaluation of the conjugate. The modification procedure must yield essentially reproducible products from each run. If the conjugate is not homogeneous, chromatographic purification may be required. The added expense of chromatography is usually compounded by the inability to salvage conjugate fractions that do not fall within specifications. This can be a substantial loss if the biopolymer is expensive and the yield of acceptable conjugate is low.

B. Peptides and Proteins That Lend Themselves
to the Conjugate Approach

Attachment of hydrophilic polymers to peptides enhances blood circulation
half-life, alters the immunological response, protects against protease inactivation (or, in the case of protease conjugates, against inactivation by plasma
protease inhibitors), and may enhance the stability of the peptide. In addition, relatively insoluble peptides show enhanced solubility.

Peptides of various sizes and complexities are amenable to polymer modification. For example, small peptides such as insulin [146] and inter-
leukin-2 [93] are active after PEG modification and show extended in vivo
activity. Small enzymes such as superoxide dismutase show greatly enhanced plasma life after attachment of polymers such as albumin, dextran,
ficoll, and PEG [41,59,80]. Oligomeric enzymes such as asparaginase,
uricase, and catalase also show greater plasma persistence, and altered immunological responses, following polymer attachment. PEG-modified hemoglobin is an excellent oxygen-carrying blood substitute [148,149]. The
upper limits of peptide size and complexity that can be successfully modified have not yet been reached.

V. CONCLUSIONS

The attachment of various soluble polymers to enzymes and other peptides
with biological activity increases their blood circulating half-life while,
usually, allowing retention of significant activity. Extended circulating
half-life for most peptides and proteins is desirable or even essential for
efficacious therapeutic performance. Extension of circulating half-life by
polymer attachment occurs even with glycosylated peptides.

Although reduction in antigenic response commonly occurs after polymer
attachment, the long-term immunological effects of chronic administration of
the various polymer–peptide conjugates have not been extensively explored.
The possible exception is PEG-conjugated peptides. Such studies must be
concluded successfully before a polymer can be considered useful. Continuing developments in coupling technology may provide other activated
polymers which yield conjugates with peptides and proteins that meet the
requirements for long-term nonimmunogenicity. The coming years may see
the emergence of a soluble polymer technology that yields a broad spectrum
of polymer–peptides and proteins, with therapeutic efficacies far beyond
those to be achieved with the native peptides and proteins.

Of the five polymers discussed for use in modifying proteins for clinical applications—dextran, albumin, poly(DL-amino acids), PVP, and PEG—
only PEG has been studied in extensive clinical trials. This does not eliminate the potential of the other polymers for modifying proteins for therapeutic use, as in order to choose a polymer successfully, one must take
into consideration the application for which the protein is intended and the
effect that the chemical manipulation will have on the final product. The
reason for modifying the proteins with polymers is to obtain a defined
chemical entity with decreased immunogenicity, increased circulating life,
and nonexistent or low-level toxicity which at the same time retains sufficient biological activity to warrant its use in humans and approval by the
FDA. As each protein will respond to modification in a manner as unique

as is the protein, one must take into account the potential effect that the polymer will have on the protein in order to choose the modification scheme that logically, will result in a compound with the characteristics desired.

ACKNOWLEDGMENTS

We would like to thank Sylvia Greenspan for preparation of the tables and Friedericke Fuertges for her help in rewriting several sections of the chapter, in compiling references, and for her general support.

REFERENCES

1. W. V. Moore and P. Leppert, *J. Clin. Endocrinol.*, *51*:691 (1980).
2. L. A. Retegui, P. L. Masson, and A. C. Paladini, *J. Clin. Endocrinol. Metab.*, *60*:184 (1985).
3. G. M. Brown, G. R. Van Loon, B. C. W. Hummel, L. J. Grota, A. Arimura, and A. V. Schally, *J. Clin. Endocrinol. Metab.*, *44*:784 (1977).
4. G. Bolli, P. D. Feo, P. Compagnucci, M. G. Cartechini, G. Angeletti, F. Santeusanio, P. Brunetta, and J. E. Gerich, *Diabetes*, *32*:134 (1983).
5. G. B. Bolli, G. D. Dimitriades, P. Brunetti, P. E. Cryer, P. De Feo, and J. E. Garich, *Diabetologia*, *27 (Suppl.)*:74 (1984).
6. L. G. Heding, M. O. Marshall, B. Persson, G. Dahlquist, B. Thalme, F. Lindgren, H. K. Akerblom, A. Rilva, M. Knip, J. Ludvigsson, L. Stenhammar, L. Stromberg, O. Sovik, H. Baivre, K. Wefring, J. Vidnes, J. J. Kjaergard, P. Bro, and P. H. Kaad, *Diabetologia*, *27*:96 (1984).
7. R. G. Larkins, J. Zajac, R. Saunders, A. Read, and J. L. Hopper, *Aust. N.Z. J. Med.*, *16*:206 (1986).
8. P. Fireman, S. E. Fineberg, and J. A. Galloway, *Diabetes Care*, *5 (Suppl. 2)*:119 (1982).
9. H. G. Velcovsky and K. F. Federlin, *Diabetes Care*, *5 (Suppl. 2)*:126 (1982).
10. F. M. Dietrich, J. A. Fischer, and O. L. M. Bijvoet, *Acta Endocrinol.*, *92*:468 (1979).
11. M. O. Thorner, J. Reschke, J. Chitwood, A. D. Rogol, R. Furlanetto, J. Rivier, W. Vale, and R. M. Blizzard, *New Engl. J. Med.*, *312*:994 (1985).
12. J. G. Haddad and J. G. Caldwell, *J. Clin. Invest.*, *51*:3133 (1972).
13. F. R. Singer, J. P. Aldred, R. M. Neer, S. M. Krane, J. T. Potts, and K. J. Bloch, *J. Clin. Invest.*, *51*:2331 (1972).
14. F. L. Demnico and A. B. Cosimi, *Surg. Gynecol. Obstet.*, *166*:89 (1988).
15. J. F. Bach and L. Chatenoud, *Transplant. Proc.*, *19*:17 (1987).
16. J. W. Marsh, J. Denis, and J. C. Wriston, Jr., *J. Biol. Chem.*, *252*:7678 (1977).
17. E. C. Nickle, R. D. Solomon, T. E. Torchia, and J. C. Wriston, Jr., *Biochim. Biophys. Acta*, *704*:345 (1982).

18. R. Blazek and J. E. Benbough, *Biochim. Biophys. Acta*, 677:220
     (1981).
19. O. Wagner, E. Irion, A. Arens, and K. Bauer, *Biochem. Biophys.
     Res. Commun.*, 37:383 (1969).
20. J. Putter and G. Gehrman, *Klin. Wochenschr.*, 47:1324 (1969).
21. D. A. Rutter and H. E. Wade, *Br. J. Exp. Pathol.*, 52:610 (1971).
22. R. P. Warrell, Jr., T. C. Chou, C. Gordon, C. Tan, J. Roberts,
     S. S. Sternberg, F. S. Philips, and C. W. Young, *Cancer Res.*, 40:
     4546 (1980).
23. R. A. G. Smith, R. J. Dupe, P. D. English, and J. Green, *Nature,
     290*:505 (1981); R. A. G. Smith, U.S. Patent 4,285,932 (Aug. 25,
     1981).
24. M. J. Poznansky, *Pharmacol. Ther.*, 21:53 (1983).
25. J. D. Holcenberg, in *Enzymes as Drugs* (J. C. Holcenberg, ed.),
     John Wiley & Sons, New York (1981).
26. T. Wileman, M. Bennett, and J. Lilleymann, *J. Pharm. Pharmacol.*,
     35:762 (1983).
27. G. Vegarud and T. B. Christensen, *Biotechnol. Bioeng.*, 17:1391
     (1975).
28. J. J. Marshall, *Trends Biochem. Sci.*, 3:79 (1978).
29. F. Suzuki, Y. Daikuhara, O. Masayaski, and Y. Takeda,
     Endocrinology, 90:1220 (1972).
30. Y. Sakamoto, Y. Akanuma, K. Kosaka, and B. Jeanrenaud, *Biochim.
     Biophys. Acta*, 498:102 (1977).
31. J. P. Lenders and R. R. Crichton, *Biotechnol. Bioeng.*, 26:1343
     (1984).
32. Y. Sakaguchi and T. Murachi, *J. Appl. Biochem.*, 2:117 (1980).
33. P. O. Larsson and K. Mosbach, *FEBS Lett.*, 46:119 (1974).
34. F. Bonneaux, P. Labrude, and E. Dellacherie, *Experientia*, 37:884
     (1981).
35. E. Dellacherie, F. Bonneaux, P. Labrude, and C. Vigneron, *Biochim.
     Biophys. Acta*, 749:106 (1983).
36. S. Tam, J. Blumenstein, and J. Wong, *Proc. Natl. Acad. Sci. USA*,
     73:2128 (1976).
37. J. J. Marshall and M. L. Rabinowitz, *Arch. Biochem. Biophys.*, 167:
     777(1975).
38. R. Axen and J. Porath, *Nature, 210*:367 (1966).
39. R. F. Sherwood, J. K. Baird, T. Atkinson, C. N. Wiblin, D. A.
     Rutter, and D. C. Ellwood, *Biochem. J.*, 164:461 (1977).
40. J. E. Benbough, C. N. Wiblin, T. N. A. Rafter, and J. Lee,
     Biochem. Pharmacol., 28:833 (1979).
41. J. M. McCord and K. Wong, in *Oxygen Free Radicals and Tissue
     Damage*, Ciba Foundation Symposium 65 (J. M. McCord, ed.),
     Excerpta Medica, New York (1979), p. 343.
42. J. J. Marshall and J. D. Humphreys, *Biotechnol. Bioeng.*, 19:1739
     (1977).
43. R. Axen and S. Ernback, *Eur. J. Biochem.*, 18:351 (1971).
44. A. Jeanes and C. A. Wilham, *J. Am. Chem. Soc.*, 72:2655 (1950).
45. J. J. Marshall and M. L. Rabinowitz, *J. Biol. Chem.*, 251:1081 (1976).
46. T. P. King, L. Kochoumian, K. Ishizka, L. M. Lichtenstein, and
     P. S. Norman, *Arch. Biochem.*, 169:464 (1975).

47. J. J. Marshall, J. D. Humphreys, and S. L. Abramson, *FEBS Lett.*, *83*:249 (1977).

48. T. E. Wileman, R. L. Foster, and P. N. C. Elliott, *J. Pharm. Pharmacol.*, *38*:264 (1986).

49. J. H. Pazur and N. N. Aronson, *Adv. Carbohydr. Chem. Biochem.*, *27*:301 (1972).

50. J. W. Coffey and C. DeDuve, *J. Biol. Chem.*, *243*:3255 (1968).

51. J. H. Pazur, H. R. Knull, and D. L. Simpson, *Biochem. Biophys. Res. Commun.*, *40*:110 (1970).

52. E. A. Kabat and A. E. Bezer, *Arch. Biochem.*, *78*:306 (1958).

53. W. Richter and L. Kagedal, *Int. Arch. Allergy Appl. Immunol.*, *42*:885 (1972).

54. K. Himmelspach and J. Wrede, *FEBS Lett.*, *118*:118 (1971).

55. A. P. Kashkin, G. M. Lindenbaum, L. V. Makhovenko, I. Z. Konshina, B. V. Maskvichev, and I. M. Tereshin, *Khim. Farm. Zh.*, *12*:27 (1978).

56. J. D. Humphreys, S. L. Abramson, and J. J. Marshall, *Fed. Proc. Fed. Am. Soc. Exp. Biol. Med.*, *36*:865 (1977).

57. B. Paillot, M. H. Remy, D. Thomas, and G. Brown, *Pathol. Biol.*, *22* (1974).

58. M. H. Remy and M. J. Poznansky, *Lancet*, 68 (1978).

59. K. Wong, L. G. Cleland, and M. J. Poznansky, *Agents Actions*, *10*:231 (1980).

60. M. J. Poznansky and D. Bhardwaj, *Can. J. Physiol. Pharmacol.*, *58*:322 (1980).

61. M. J. Poznansky and D. Bhardwaj, *Biochem. J.*, *196*:89 (1981).

62. T. Yagura, Y. Kamisaki, H. Wada, and Y. Yamamura, *Int. Arch. Allergy Appl. Immunol.*, *64*:11 (1981).

63. M. J. Poznansky, M. Shandling, M. A. Salkie, J. Elliott, and E. Lau, *Cancer Res.*, *42*:1020 (1982).

64. M. J. Poznansky, R. Singh, and B. Singh, *Science*, *223*:1304 (1984).

65. W. E. Hennink, C. D. Ebert, S. W. Kim, W. Breemhaar, A. Bantjes, and J. Feijen, *Biomaterials*, *5*:264 (1984).

66. A. Gewurz, L. C. Grammer, M. A. Shaughnessy, and R. Patterson, *J. Allergy Clin. Immunol.*, *77*:521 (1986).

67. T. M. Allen, L. Murray, D. Bhardwaj, and M. J. Poznansky, *J. Pharmacol. Exp. Ther.*, *234*:250 (1985).

68. M. J. Poznansky, *Life Sci.*, *24*:153 (1979).

69. M. Sela, S. Fuchs, and R. Arnon, *Biochem. J.*, *85*:223 (1962).

70. R. Arnon and H. Neurath, *Immunochemistry*, *7*:241 (1970).

71. R. K. Brown, A. Trzpis, M. Sela, and C. B. Anfinsen, *J. Biol. Chem.*, *238*:3876 (1963).

72. M. Sela, *Adv. Immunol.*, *5*:29 (1966).

73. J. R. Uren and R. C. Ragin, *Cancer Res.*, *39*:1927 (1979).

74. J. R. Uren, B. J. Hargis, and R. Beardsley, *Cancer Res.*, *42*:4068 (1982).

75. W. Wessel, M. Schoog, and E. Winkler, *Arzneim. Forsch.*, *21*:1468 (1971).

76. B. U. Von Specht and W. Brendel, *Biochim. Biophys. Acta*, *484*:109 (1977).

77. B. Geiger, B. U. Von Specht, and R. Arnon, *Eur. J. Biochem.*, *73*:141 (1977).

78. B. Andersson and H. Blomgren, *Cell Immunol.*, *2*:411 (1971).

79. A. Abuchowski and F. F. Davis, *Biochim. Biophys. Acta*, *587*:41 (1979).

80. P. S. Pyatak, A. Abuchowski, and F. F. Davis, *Res. Commun. Chem. Pathol. Pharmacol.*, *29*:113 (1980).

81. A. Abuchowski, J. R. McCoy, N. C. Palczuk, T. van Es, and F. F. Davis, *J. Biol. Chem.*, *11*:3582 (1977).

82. S. Davis, A. Abuchowski, Y. P. Park, and F. F. Davis, *Clin. Exp. Immunol.*, *46*:649 (1981).

83. A. Abuchowski, T. van Es, N. C. Palczuk, and F. F. Davis, *J. Biol. Chem.*, *11*:3578 (1977).

84. Y. Ashihara, T. Kono, S. Yamazaki, and Y. Inada, *Biochem. Biophys. Res. Commun.*, *93*:385 (1978).

85. A. Abuchowski, D. Karp, and F. F. Davis, *J. Pharmacol. Exp. Ther.*, *219*:352 (1981).

86. N. Katayama, K. Hayakawa, I. Urabe, and H. Okada, *Enzyme Microb. Technol.*, *6*:538 (1984).

87. T. P. King, L. Kochoumian, and L. M. Lichtenstein, *Arch. Biochem. Biophys.*, *178*:442 (1977).

88. J. Newmark, A. Abuchowski, and G. Murano, *J. Appl. Biochem.*, *4*:185 (1982).

89. N. Sakuragawa, K. Shimizu, K. Kondo, S. Kondo, and M. Niva, *Thromb. Res.*, *41*:627 (1986).

90. R. J. Sokol, J. E. Heubi, N. A. Butler-Simon, and H. J. McClung, *Pediatr. Res.*, *20*:249A (1986).

91. T. Yoshimoto, K. Takahashi, H. Nishimura, A. Ajima, Y. Tamaura, and Y. Inada, *Biotechnol. Lett.*, *6*:337 (1984).

92. K. Ajisaka and Y. Iwashita, *Biochem. Biophys. Res. Commun.*, *97*:1076 (1980).

93. N. V. Katre, M. J. Knauf, and W. J. Laird, *Proc. Natl. Acad. Sci. USA*, *84*:1487 (1987).

94. K. V. Savoca, A. Abuchowski, T. van Es, F. F. Davis, and N. C. Palzcuk, *Biochim. Biophys. Acta*, *578*:47 (1979).

95. K. Takahashi, A. Ajima, T. Yoshimoto, and Y. Inada, *Biochem. Biophys. Res. Commun.*, *125*:761 (1984).

96. C. O. Beauchamp, S. L. Gonias, D. P. Menapace, and S. V. Pizzo, *Anal. Biochem.*, *131*:25 (1983).

97. F. M. Veronese, E. Boccu, O. Schiavon, G. P. Velo, A. Conforti, L. Franco, and R. Milanino, *Clin. Res.*, *32*:530A (1984).

98. E. Boccu, R. Largajolli, and F. M. Veronese, *Z. Naturforsch.*, *38*:94 (1984).

99. A. Abuchowski, G. M. Kzao, C. R. Verhoest, T. van Es, D. Kafkewitz, M. L. Nucci, A. T. Viau, and F. F. Davis, *Cancer Biochem. Biophys.*, *7*:175 (1984).

100. N. Tomiya, K. Watanabe, J. Awaya, M. Kurono, and S. Fujii, *FEBS Lett.*, *193*:44 (1985).

101. J. M. Harris, *Rev. Macromol. Chem. Phys.*, *C25*:325 (1985).

102. J. B. Stenlake, in *Foundations of Molecular Pharmacology*, Vol. II (J. B. Stenlake, ed.), Athlone Press, London (1979), p. 213.

103. A. Matsushima, H. Nishimura, Y. Ashchara, Y. Tokata, and Y. Inada, *Chem. Lett.*, *7*:773 (1980).

104. K. A. Sharp, M. Yalpani, S. J. Howard, and D. E. Brookes, *Anal. Biochem.*, *154*:110 (1986).
105. S. I. Wie, C. W. Wie, W. Y. Lee, L. G. Filion, A. H. Sehon, and E. Akerblom, *Int. Arch. Allergy Appl. Immunol.*, *64*:84 (1981).
106. P. S. Norman, T. P. King, J. F. Alexander, A. Kagey-Sabotka, and L. M. Lichtenstein, *J. Allergy Clin. Immunol.*, *73*:782 (1984).
107. P. J. Lisi, T. van Es, A. Abuchowski, N. C. Palczuk, and F. F. Davis, *J. Appl. Biochem.*, *4*:19 (1982).
108. K. J. Wieder, N. C. Palczuk, T. van Es, and F. F. Davis, *J. Biol. Chem.*, *254*:12579 (1979).
109. H. Nishimura, A. Matsushima, and Y. Inada, *Enzyme*, *26*:49 (1981).
110. E. Boccu, R. Velo, and F. M. Veronese, *Pharmacol. Res. Commun.*, *14*:113 (1982).
111. K. Takahashi, Y. Kodera, T. Yoshimoto, A. Ajima, A. Matsushima, and Y. Inada, *Biochem. Biophys. Res. Commun.*, *131*:532 (1985).
112. M. S. Hershfield, R. H. Buckley, M. L. Greenberg, A. L. Melton, R. Schiff, C. Hatem, J. Kurtzberg, M. L. Markert, R. H. Kobayashi, A. L. Kobayashi, and A. Abuchowski, *New Engl. J. Med.*, *316*:589 (1987).
113. R. H. L. Chen, A. Abuchowski, T. van Es, N. C. Palczuk, and F. F. Davis, *Biochim. Biophys. Acta*, *660*:293 (1981).
114. A. Abuchowski and F. F. Davis, in *Enzymes as Drugs* (J. C. Holcenberg, ed.), John Wiley & Sons, New York (1981), p. 367.
115. K. Takahashi, H. Nishimura, T. Yoshimoto, Y. Saito, and Y. Inada, *Biochem. Biophys. Res. Commun.*, *121*:261 (1984).
116. K. Takahashi, L. T. Yoshimoto, A. Ajima, Y. Tamaura, and Y. Inada, *Enzyme*, *32*:235 (1984).
117. A. Ajima, K. Takahashi, A. Matsushima, Y. Saito, and Y. Inada, *Biotechnol. Lett.*, *8*:547 (1986).
118. A. Matsushima, N. Okada, and Y. Inada, *FEBS Lett.*, *178*:275 (1984).
119. K. Takahashi, A. Ajima, T. Yoshimoto, N. Okada, A. Matsushima, Y. Tamaura, and Y. Inada, *J. Org. Chem.*, *50*:3414 (1985).
120. K. Takahashi, H. Nishimura, T. Yoshimoto, M. Okada, A. Ajima, A. Matsushima, Y. Tamaura, Y. Saito, and Y. Inada, *Biotechnol. Lett.*, *6*:765 (1984).
121. H. Nishimura, K. Takahashi, K. Sakurai, K. Fujinuma, Y. Imamura, M. Ooba, and Y. Inada, *Life Sci.*, *33*:1467 (1983).
122. K. J. Wieder and F. F. Davis, *J. Appl. Biochem.*, *5*:337 (1983).
123. A. Koide and S. Kobayashi, *Biochem. Biophys. Res. Commun.*, *111*:659 (1983).
124. Y. Kamisaki, H. Wada, T. Yagura, A. Matsushima, and Y. Inada, *J. Pharmacol. Exp. Ther.*, *216*:410 (1981).
125. S. Davis, Y. K. Park, F. F. Davis, and A. Abuchowski, *Lancet*, *2*:281 (1981).
126. D. H. Ho, N. S. Yen, R. Holmes, M. Keating, A. Abuchowski, R. A. Newman, and I. H. Krakoff, *Drug Metab. Dispos.*, *14*:349 (1986).
127. W. Y. Lee and A. H. Sehon, *Int. Arch. Allergy Appl. Immunol.*, *56*:159 (1978).
128. J. I. Tsuju and K. Hirose, *Int. J. Immunopharmacol.*, *7*:725 (1985).

129. K. Kawamura, T. Igarashi, T. Fujii, Y. Kamisaki, H. Wada, and
     S. Kishimoto, *Int. Arch. Allergy Appl. Immunol.*, *76*:324 (1985).
130. M. L. Nucci, J. Olejarczyk, and A. Abuchowski, *J. Free Radicals
     Biol. Med.*, *2*:321 (1986).
131. W. Y. Lee and A. H. Sehon, *Int. Arch. Allergy. Appl. Immunol.*,
     56:193 (1978).
132. W. Y. Lee, A. H. Sehom, and E. Akerblom, *Int. Arch. Allergy
     Appl. Immunol.*, *64*:100 (1981).
133. N. Chiorazzi, D. A. Fox, and D. H. Katz, *J. Immunol.*, *117*:1629
     (1976).
134. A. W. Richter and E. Akerblom, *Int. Arch. Allergy Appl. Immunol.*,
     70:124 (1983).
135. A. W. Richter and E. Akerblom, *Int. Arch. Allergy Appl. Immunol.*,
     74:36 (1984).
136. K. Ishizaka, in *The Antigens* (M. Sela, ed.), Academic Press, New
     York (1973), p. 479.
137. T. Wileman, M. Bennett, and J. Lilleyman, *J. Pharm. Pharmacol.*,
     35:762 (1983).
138. N. M. Kredich and M. S. Hershfield, in *The Metabolic Basis of
     Inherited Disease* (J. B. Stanbury, J. B. Wyngarden, D. S.
     Fredrickson, J. L. Goldstein, and M. S. Brown, eds.), McGraw-Hill
     Book Company, New York (1983), p. 1157.
139. P. S. Norman, K. Ishizaka, L. M. Lichtenstein, and N. F.
     Adkinson, *J. Allergy Clin. Immunol.*, *66*:336 (1980).
140. T. P. King and C. Weiner, *Int. J. Pept. Protein Res.*, *15*:147
     (1980).
141. E. F. Juniper, R. S. Roberts, M. Tech, L. K. Kennedy,
     J. O'Conner, M. Syty-Golda, J. Dolovich, and F. E. Hargreave,
     J. Allergy Clin. Immunol., *75*:578 (1985).
142. E. F. Juniper, J. O'Conner, R. S. Roberts, M. Tech, S. Evans,
     F. E. Hargreave, and J. Dolovich, *J. Allergy Clin. Immunol.*, *78*:
     851 (1986).
143. R. L. Capizzi, J. R. Bertino, and R. E. Handchumacher, *Annu.
     Rev. Med.*, *21*:433 (1970).
144. T. Yoshimoto, N. Nishimura, Y. Saito, K. Sakurai, Y. Kamisake,
     H. Wada, M. Sako, G. Tsujino, and Y. Inada, *Jpn. J. Cancer Res.
     (Gann)*, *77*:1264 (1986).
145. U. Muller, A. Lanner, P. Schmid, M. Bischof, S. Dreborg, and
     R. Hoigne, *Int. Arch. Allergy Appl. Immunol.*, *77*:201 (1985).
146. F. F. Davis, N. C. Palczuk, and T. Van Es, U.S. Patent 4,179,337
     (Dec. 18, 1979).
147. M. Leonard and E. Dellacherie, *Biochim. Biophys. Acta*, *791*:219
     (1984).
148. K. Iwasaki, Y. Iwashita, K. Ikeda, and T. Uematsu, *Artif. Organs*,
     10:470 (1986).
149. R. L. Foster and T. Wileman, *J. Pharm. Pharmacol.*, *31*:37 (1979).

22

Regulatory Perspective of Peptide and Protein Drug Development

John L. Gueriguian, Yuan-Yuan H. Chiu, Alex Jordan, Carlos A. Schaffenburg,* and Solomon Sobel

Division of Metabolism and Endocrine Drug Products, Food and Drug Administration, Rockville, Maryland

I. INTRODUCTION

For almost a decade, the Division of Metabolism and Endocrine Products, Center for Drug Evaluation and Research, U.S. Food and Drug Administration (FDA), has participated, however vicariously, in the introduction in the marketplace of a number of drugs of peptidic and proteinaceous origin, particularly those obtained through recombinant DNA technology [1]. The division has also been asked to reflect on its experience and to offer sometimes speculative comments on questions not directly under its purview [2]. During that same period, old drugs have been improved, new drugs have been introduced in general practice, and novel delivery routes have been actively investigated.

In the 1970s, peptides and proteins in clinical use were of natural origin. Insulins, for example, were extracted from bovine and porcine pancreases, gonadotropic hormones were concentrated from human urine, and human growth hormone was obtained from cadaveric human pituitaries. These and other protein-containing products—used exclusively through parenteral administration—were, by and large, rather impure and sometimes dramatically so. For example, rare insulin batches contained as much as 30,000 ppm of proinsulin, while the norm was 300 to 3000 ppm of proinsulin. Human growth hormone preparations showed several bands when studied with the available protein separation systems. Such impurities are known to present toxic potential, unfortunately realized from time to time.

The scientific advances of recent years have dramatically changed this landscape. New methodologies—particularly solid-phase synthesis, recombinant DNA cloning, and high-performance liquid chromatography (HPLC; analytical and preparative)—have permitted the improvement of old drugs and the introduction of new ones. Finally, efforts have been made to eliminate the burden of daily injections.

Current affiliation: Consultant, NICHD, National Institutes of Health, Bethesda, Maryland.

II. INTERACTIVE NATURE OF DRUG DEVELOPMENTAL EFFORTS

A. Fundamental Issues

The complex, costly, and time-consuming process of developing new drugs and formulations is dependent on an organized interplay between multi-disciplinary teams from academia, industry, and government. An integrated effort is needed for a prompt, cost-effective development of safe and useful therapeutic agents. Under these modern constraints, it is imperative that the regulatory arm of government join the mainstream of developmental efforts very early in the process, and with four major purposes in mind: (1) to understand the significant theoretical and practical parameters of any given situation, (2) to inform industry as clearly and quickly as possible on the nature of its statutory and regulatory requirements, (3) to ensure that the developmental process proceeds smoothly in a manner consistent with these requirements, and (4) to expend whatever effort is required to increase the probability of success of the drug development process.

B. Networking

The kind of interaction that we have just described entails a large network of connections between the regulatory arm of government and other institutions. In the particular case of interest, it was thought that the following contacts were particularly needed: (1) the research arms of government and industry (to seek from them early information about promising new discoveries); (2) the pioneering basic and clinical scientists (to learn from them the essentials of a situation and, in turn, advise them during their investigative efforts); (3) the developmental arms of government and industry (to make sure that all their efforts are consistent with all essential regulatory requirements); and (4) the international scientific community (to attempt a standardization of at least some common efforts and to share scientific and medical information).

C. The Demands of Novelty

Early on, it was felt by all parties concerned that there was no reason to single out the products obtained through modern methodologies—just because they happened to be obtained through techniques that seemed exotic—and subject them to unreasonable and discriminatory hurdles prior to their eventual approval. On the other hand, these technologies did entail special problems that had to be addressed. The basic approach was, therefore, to determine, on a case-by-case basis, the conditions that were felt to be necessary and sufficient to prove the safety and efficacy of a given moiety obtained by a given method [3,4].

D. Obtaining a Consensus

Given the overall novelty of the situation, it was important to attempt to obtain, among all experts involved, a scientific consensus on the appropriateness and validity of the criteria that were developed. Such attempts

were numerous and eventually culminated during two public scientific
forums where, finally, some of the more important elements of the needed
consensus emerged and were recorded [5,6].

E. Reassessment of Quality Assurance

By definition, quality assurance is an effort expended to ensure the
ultimate quality and uniformity of a manufactured product. In established
pharmaceutical endeavors, quality assurance has generally meant adherence
to the so-called Good Manufacturing Practices, although it is well understood that these two concepts afford considerable overlap without actually
being synonymous. A new situation, on the other hand, requires a reas-
sessment of quality assurance above and beyond traditional practices.
Also, since risk cannot be exhaustively assessed in uncharted territory, it
becomes necessary to define an ordered and sequential array of precautions which, when taken together, will synergistically reduce the overall
risk to almost infinitesimal proportions. It is these new approaches that
we will now present and discuss in a categorical fashion, while illustrating
each concept with examples.

III. CHEMISTRY AND MANUFACTURING CONTROLS

A. Generalities

Chemistry is the first discipline to deliberate on the appropriateness of
protein preparations for clinical investigations and, eventually, therapeutic
use. Its role is absolutely crucial. If properly understood and imple-
mented, it can considerably simplify and reduce the burden of scrutiny by
the other disciplines.

B. Sources of Therapeutic Proteins

Qualitatively and quantitatively, the testing of a peptidic drug is very
largely dependent on its origin.

Peptides Extracted from Natural Sources

Peptides extracted from natural sources may include products of human as
well as animal origin (e.g., gonadotropic hormones from human urine and
porcine insulin from pork pancreas). The purification of these peptidic
drugs, usually by extraction, salt fractionation, and sequential column
chromatography, should be designed to remove all potential contaminants.
It is always desirable to include viral deactivation steps in the process,
especially when human tissues are the source of the drug. If, for whatever reason, deactivation is impractical or incomplete, the purification
process must be rigorously validated for its ability to remove significant
pathogens.

Semisynthetic Peptides

Semisynthetic peptides may be obtained by chemically treating natural
source peptides. For example, **porcine** insulin differs from its human

counterpart by one amino acid. It is extracted from pork pancreases, then converted into human insulin through the action of specific enzymes. In such a case, additional purification steps are needed to remove reaction by-products, unreacted proteins, and added reagents (e.g., enzymes and organic solvents).

Biosynthetic Peptides

Biosynthetic peptides are usually manufactured through recombinant DNA technology. To ensure that a correct and safe product is manufactured, the vector-host system must be critically characterized, the sterility of the cell banks proven, and the fermentation process properly designed and monitored. When mammalian cell lines are used in production, it is essential to document, as extensively as possible, the presence or absence of viral pathogens in the tissue culture system. Elaborate purification processes may be needed to remove contaminants that are specific to this particular methodology.

Full Chemical Synthesis

Both solution- and solid-phase synthesis can and are being used to obtain such products. The automation of solid-phase synthesis has made this manufacturing method relatively easy and reliable. Purifying a peptide containing about 40 amino acids to 99% purity can usually be achieved by using reverse-phase, ion-exchange, affinity, and size-exclusion preparative HPLC. It is important to maintain high coupling yield during synthesis, and to avoid racemization and chemical modification of the product during synthesis and purification. These issues have been discussed in detail in Chap. 2.

C. Structure of the Final Product

It is always important to determine the chemical structure of a protein. Regulatory efforts are meaningless if the structure has not been determined satisfactorily. The following minimal assurances are always necessary: correct amino acid sequence, formation of the proper disulfide bridges and of the appropriate folding of the chain, and absence of any telling structural divergence from the native structure. A synthetic product will have to be tested more intensively, the more so the more structurally different it is from its natural counterpart. In such cases one must prove that one has achieved the desired chemical modifications.

Natural Sequence Peptides

We are referring to molecules with amino acid sequences identical to human endogenous proteins, not to those obtained directly from prokaryotic cells through recombinant DNA technology. The latter peptides should be considered as analogs of natural products since they usually contain an additional methionine residue at their N-terminus.

Natural sequence peptides can be obtained from any of the four sources mentioned previously. In certain instances, elaborate chemical modifications may be required to convert the direct gene products obtained through biotechnology into natural peptides. For example, human insulin contains two chains, A and B. The A and B chains can be prepared separately by

fermentation of cloned *Escherichia coli* cells and then chemically combined
to yield human insulin. Alternatively, proinsulin can also be made by
E. coli and the treatment of proinsulin by appropriate enzymes can produce human insulin. In any event, if the manufactured products are
ultimately shown to be structurally identical to the natural ones, toxicological and clinical studies using such products would be expected to provide satisfactory results, provided that extraneous contamination is kept at
very low levels. Under the circumstances, it can be well understood that
minimal safety studies are needed prior to the initiation of clinical trials.

Agonistic Analog

Amino acid sequences of agonists are usually highly, but sometimes remote-
ly homologous to that of the corresponding natural (model) moieties. Or
they may fall somewhere between these two extremes. In general, it is
desirable to obtain analogs with the following traits: high biological potency, minimal impurities, and very low or even nonexistent intrinsic immunogenicity. The agonistic analogs can also be obtained from any of the
four sources mentioned above.

Natural-source agonistic analogs. As stated earlier, porcine insulin differs from human insulin by a single amino acid. Once extracted and purified, it contains low (ppm) levels of extraneous protein and its immunogenicity is nominal. Some consider it as good a therapeutic agent as
human insulin itself. Salmon calcitonin is more potent than its human
counterpart and is thus an excellent addition to the armamentarium, even
though it is immunogenic.

Biosynthetic agonistic analog. Methionyl-human growth hormone (hGH)
is equipotent with hGH. But it is now clear that it possesses high intrinsic immunogenicity, probably directly or indirectly attributable to its
additional amino acid residue. Other things being equal, it is a less satisfactory product than a safe preparation of hGH.

Fully synthetic agonistic analog. There are many good reasons to synthesize analogs whose structures are at a distance from the native model
structure. Replacing certain L-amino acids of luteinizing hormone releasing
hormone (LHRH) by D-amino acids or related structures, for example, can
drastically increase relative potency, through protection from circulating
proteolytic enzymes, through a better fit with receptors, or by a combination of both effects.

Antagonistic Analogs

The structure of an antagonist is usually quite different from that of its
natural agonist. This difference may reside in a considerably modified
amino acid sequence, due to significant chemical modifications of the
residues affected. Compared to the natural agonist, antagonists show lower
binding constants to the common receptors, and consequently, lower relative potency on a molar basis. It therefore stands to reason that such
products be tested much more rigorously than more potent moieties.
Theoretically at least, antagonists can be obtained from any of the four
sources mentioned above. In practice, antagonists with considerable chemical modifications of its amino acid residues can only be obtained through
full chemical synthesis. Conceivably, biotechnology may allow the economical

preparation of large antagonistic proteins, as long as they did not contain "unnatural" amino acids. It may also be possible to produce small peptides using such technology. This may be achieved by cloning a lengthy message consisting of repeats of the hormonal message, interspersed with suitable "dividers." The expressed protein would then be treated appropriately to separate and liberate the numerous small peptides from the expressed "dividers." Although this method would increase yields dramatically, it presents obvious, although perhaps surmountable theoretical difficulties.

D. Purity

Regardless of the method used to obtain therapeutically useful proteins, residual contamination of the final product should be kept as low as possible to avoid toxicity to the recipient. The level of concern is different depending on the nature, amount, and origin of these moieties.

Extraneous Proteins

Such contamination is present in all products, regardless of their method of manufacture. Contaminants may comprise natural analogs (e.g., molecular variants of hGH), chemically modified native proteins (e.g., deaminated insulins), natural precursors (e.g., proinsulin), other physiologically or pharmacologically natural active moieties (e.g., pancreatic polypeptide in glandular insulin), inactive fragments of the active moiety, failure peptides obtained during solid-phase synthesis, unmodified proteins during semisynthetic processes (e.g., porcine insulin in semisynthetic human insulin), and extraneous host proteins from cloned microorganisms or cell lines. When human tissues or fluids are used in production, it is wise to test the purified product for the absence of marker proteins of known human viruses (e.g., those of hepatitis). Regardless, the power of modern purification methods allows impressive reductions of any and all contaminating proteins, usually to ppm levels. Also, powerful analytical techniques (e.g., radioimmunoassays) allow precise identification and measurement of these residual levels. Unless there is a specific reason for concern (and there rarely is), one is not justified to harbor any significant amount of concern about the potential toxicity of these very low amounts of impurities.

Viral and Cellular DNA

When eukaryotic cells, especially mammalian tumor cells, are used to produce therapeutically useful proteins, residual DNA should be kept exceedingly low, to reduce to a minimum all conceivable potential for toxicity. Oncogenicity is of concern here, since, conceivably, exogenous DNA might induce the expression of abnormal genetic information in the treated subject. However, known facts and conservative assumptions assure us that a residual cellular DNA of about 1 pg per dose can claim a safety factor of at least 10 million; that is, there is in the amount of DNA one 10 millionth or less of the amount needed to induce a single transformation event in the recipient [7]. Few products claim such safety. Technically, the detection of DNA in a drug product at picogram levels can be achieved readily by using DNA probe hybridization.

Infectious Contaminants

For years, live yellow fever vaccines were produced in eggs containing
avian leukosis virus, yet a large retrospective study failed to show an increased risk of cancer in the vaccinated populations [8]. On the other
hand, a few cases of Creutzfeldt-Jakob disease were recently detected in
a young population treated for long periods with certain preparations of
hGH obtained from human pituitaries. These extreme instances frame well
the nature and the dimensions of the problem. Under the circumstances,
and presumably for some years to come, many experts will choose to expend every conceivable effort to eliminate, through state-of-the-art methodologies, the presence of all known infectious contaminants (e.g., retroviruses). While the true nature and properties of a given causative agent
(e.g., that of Creutzfeldt-Jakob disease) that might contaminate a drug
(e.g., hGH from human pituitaries) are as yet unknown or uncertain, it
would seem prudent to refuse to accept even the smallest theoretical risk
entailed by the use of such medications, but then only if relatively safer
replacement products are available. Obviously, the therapeutic use of a
drug will be acceptable as long as the known and perceived risks are well
justified by the documented and postulated benefits.

E. Stability

Most formulated proteins are stable for long periods. Under certain conditions, however (e.g., when insulin solutions are used in external or internal pumps), they may be denatured readily. It is therefore reasonable
to require that the stability of specially formulated products (e.g., insulin
containing ad hoc additives) be tested in vitro, in the potentially destabilizing equipment (pumps, catheters, etc.), and under conditions mimicking
actual use as closely as possible. Once such products have been proven
to be stable under these circumstances, they may then be tested in properly conducted clinical trials consisting of pharmacokinetic, safety, and
efficacy studies.

IV. ANIMAL STUDIES

The extent of pharmacological and toxicological studies required to ascertain
the relative safety of peptidic drugs prior to their clinical use depends on
a number of considerations. In general, it may reasonably be felt that a
well-defined natural human product requires very little preclinical scrutiny,
regardless of its origin, provided that it has been purified to a high
degree. Similarly, a peptide analog which is at least as potent as its
natural model will require very little preclinical testing. However, as the
analog resembles the natural peptide less and less, it tends to become a
true xenobiotic.

A. Preclinical Studies

The potential for human toxicity of novel molecular entities is traditionally
appreciated during animal studies. Such studies often require a significant
commitment in time and resources. Classical subchronic toxicity tests require the study of at least two species for a duration of up to one year,

with standard hematology, clinical chemistry, and histology endpoints.
Such studies would not be expected to provide much usable information in
the case of natural human peptides and proteins, for empirical as well as
scientific reasons. Indeed, we do not know of any case where such a
molecule was found to be overly toxic, other than through its known biological activity; any toxicity that may be noted could be due to a reaction
in the animal to the human (i.e., "foreign") protein. Thus we either do
not expect evidence of toxicity or if some toxicity does exist, it would be
difficult, if not impossible, to interpret it correctly. By extension, non-
exotic peptides, even if identical in structure to human products, should
not have to undergo the same rigorous tests as frank xenobiotics. It
would thus seem reasonable to tailor individual requirements to the merits
of each case.

Natural Sequence Human Proteins

As stated, once well-defined standards for structure, purity, and pyro-
genicity have been met, the usual preclinical toxicological requirements
for xenobiotic drugs are more or less unnecessary in this case. It is thus
possible to minimize the reliance on animals for the testing of natural and
endogenous human peptides, even when the peptides have been obtained
synthetically.

Acute toxicity test. Minimal tests are deemed necessary, primarily to
detect the totally unforeseen. Any given preparation is better tested in
several species. It is fairly useful to administer the preparation on two
consecutive days, at fairly high multiples of the expected human dose.
The animals should be observed for 7 days after the last injection.

Acute physiology test. Certain human peptides may possess undesirable side effects when injected systemically (e.g., hypotension induced
by growth hormone releasing factor or corticotropin releasing hormone).
This may be due to the introduction of large quantities of an otherwise
natural peptide in an "unnatural" bodily compartment, where "latent" receptors could be activated. It is therefore prudent to use these peptides
and test important physiological functions to determine if untoward effects
are obtained at the human doses contamplated. There are, for example, a
number of useful tests of drug-induced effects on cardiovascular function.
A fairly straightforward and useful one is a physiology test in catheterized
dogs. Specifically, after implantation of an arterial catheter, each dog
should receive successive daily intravenously doses of vehicle as well as 1,
10, and 25 times the expected human therapeutic dose over 4 days. Pertinent
cardiovascular parameters are measured prior to dosing and at hourly
intervals for 5 hr after each administration.

Frank to Relative Xenobiotics

There is no doubt that frankly xenobiotic peptides and proteins should be
tested as rigorously as all the traditional drugs. The only difficulty resides in recognizing a xenobiotic peptide and defining its relative degree
of foreignness. In the area of peptide drugs, circumstances force us to
discard a strictly binary (yes or no) decision-making process and replace
it with a more complex system of graded and shaded responses. In our
experience, the following practical criteria have been found to be most
helpful to define the degree of foreignness of a compound: (1) the moiety

is structurally very different from the original native model (e.g., through extensive amino acid replacements, particularly with D and other "unnatural" amino acids); (2) it has a much lower binding affinity for in vitro "receptor" preparations; (3) it is much less potent than the native protein; (4) it can penetrate or even concentrate in bodily compartments relatively inaccessible to the native moiety; and (5) its metabolism is qualitatively and quantitatively different from that of the native product.

B. Additional Tests

Any number of additional animal studies may be needed in some special cases, either before or after the initiation of clinical trials. Preclinically, special considerations may dictate the performance of studies to resolve, ahead of time, certain perceived difficulties. Later, as clinical studies are unfolding, careful scrutiny of the data generated may be needed to resolve or at least alleviate such concerns. Immunological tests in animals are an illustrative case in point. For a number of reasons (e.g., alterations in peptide folding during synthesis or purification of natural peptides), proteins administered as drugs may induce immune-related effects in the recipient. At first blush, it would seem that such problems are only poorly addressed by studies in animals, particularly when using a human protein with structural differences with the homolog protein of the recipient animal. However, satisfactory animal model systems can be developed to address specific needs, at least in some instances. For example, protocols can be devised whereby primates treated with a protein preparation of acceptable purity (e.g., one that has not been shown in the past to be immunogenic in humans) would not form antibodies, whereas less pure preparations of the same protein would still be immunogenic. In such carefully crafted model systems, undue immunogenicity, or perhaps even other adverse effects, can probably be recognized and addressed.

V. CLINICAL STUDIES

A. Medical Involvement During the Preclinical Period

It may be simplistically assumed that the medical staff of the FDA (and of industry, for that matter) ought only to be interested in clinical trials. Nothing could be further from the truth. In fact, every drug developmental phase should be assessed by medically competent personnel for any and all clinical significance or implications.

Extramural Interactions

Interactions with industry. Industry–government contacts should be initiated very early, preferably before any studies, preclinical or clinical, are begun. Such interactions are the foundation of all subsequent traditional "quality assurance" efforts. A prior agreement between the parties as to which trials to conduct is consistent with the new drug approval process in the United States. Thus useless testing of drugs is not permitted inasmuch as no risk, however small, may be tolerated if the proposed studies have little or no chance of proving the efficacy and safety of the therapeutic moiety. Medical input is, of course, essential in this type of determination.

Other extramural interactions. The ability to chart a proper course of development for a drug entity is largely dependent upon precise self-educational efforts on the part of the medical personnel involved. Those efforts comprise, among other things, a number of interactions with basic and clinical scientists, themselves directly or indirectly involved in the various aspects of the developmental process.

Intramural Interactions

Protocols for peptidic drug studies are reviewed at the FDA by a team of nonmedical professionals. Examples will follow as to typical communications between them and the medical officers.

With chemists. As stated earlier, the role of the chemist is fundamental in this area. It is the chemist who sets (1) the acceptable criteria of identity, purity, and sterility; (2) the permissible levels of contamination by established and unknown foreign substances; (3) the proper reference standards for purity and potency; (4) the criteria of method and process validations; and (5) the overall batch or lot rejection standards. It is imperative that medical officers understand the rationale of these chemical decisions. In turn, they may assist the chemist in distinguishing between essential and reasonable requirements and excessive or unnecessary ones. For example, a certain amount of immunogenicity seems to be acceptable from the various insulin formulations, given our ample and reassuring medical experience in this area. On the other hand, immunogenicity is less acceptable in growth hormone preparations since neutralizing antibodies (which inhibit growth) have been observed in the past, however rarely.

With pharmacologists and toxicologists. A true collaboration between these professionals and the medical officers will often result in the recognition of the studies that are needed to attempt a prediction of the general toxicity profile of the drug being studied. It is also important collectively to understand and accept the limits of such enquiry. Special questions may also arise, requiring collaborative answers. For example, during solid-phase synthesis, failure peptides and other contaminants are readily formed, and this contamination may vary from batch to batch, both qualitatively and quantitatively. Under the circumstances, how could one conduct meaningful preclinical studies? With the proper consultations, the answer to this somewhat vexing problem was as follows: (1) several drug batches should be admixed once and for all, to be used for all projected preclinical studies; (2) the results of such studies, if acceptable, would then permit clinical trials, provided that such trials used subsequent batches with a degree of purity at least equal to that of the initial admixture; (3) the probability of occurrence of toxic effects in humans would be much diminished by further purification of the clinical batches; and (4) "final" (i.e., to be marketed) formulations should meet certain purity specifications, and phase III clinical trials should be conducted only with such preparations.

With pharmacokineticists. Many therapeutically useful peptides and proteins are extremely potent biologicals. Consequently, it is difficult to measure the circulating levels of many of them. Thus classical pharmacokinetic studies may be impossible to perform, and medical input is necessary to identify specific and sensitive endpoints that would allow adequate measurement of bioavailability or bioequivalency. LHRH agonists, for example, are used in the palliative treatment of prostatic cancer, for their castrating effect. Various analogs do not seem to be equipotent in this regard. Therefore, it becomes imperative to prove that any tested drug and regimen are able to bring down, and maintain, blood testosterone at castrate levels.

With statisticians and epidemiologists. It is most important that statisticians and clinicians agree on all essential aspects of the protocols of clinical trials prior to their initiation. Here, prospective decisions are of the essence, and retrospective analyses are rarely adequate or acceptable, particularly if they purport to address substantive issues. During preliminary discussions, the clinicians must present to the statisticians the important parameters of the clinical situation, to allow the latter to identify appropriate data analysis methods. In turn, the medical officer has to learn the essentials of the methods proposed, in order to understand their true scope, assumptions, and limitations. For example, in the case of the treatment of a prostatic cancer, a given trial intended to compare the effect of an LHRH analog to that of DES. After numerous discussions, it was agreed that the purpose of the trial was to prove that the two treatments were not significantly different from each other, not to show that they yielded strictly identical results. Similarly, when planning phase IV efforts, full consultations ought to be conducted with staff epidemiologists since their input may be absolutely necessary for proper performance of the needed studies.

B. Novel Conceptual Approaches to Drug Development

Finding the Initial Indication

When one proposes to test a totally novel drug, it is most important to choose the correct lead indication, one that may establish or destroy its credibility. In our case, this process was simplified since we were dealing with highly potent endogenous hormonal substances and their analogs. The ideal lead indication, in general terms, is one that fulfills all of the following criteria: (1) an indication allowing the use of the least chemically alien moiety available (e.g., LHRH itself); (2) one where very low doses of the moiety chosen are expected to be effective to achieve the intended purpose (e.g., physiological amounts of GRF administered for diagnostic purposes); (3) one based on a sound rationale (e.g., endometriosis results from the estrogenic stimulation of ectopic target cells. Since LHRH analogs reversibly shut off the gonads that produce estrogens, they are expected to correct endometriosis); and (4) the subjects that are going to receive the new drug initially belong to a category for which the benefit/risk ratio is unquestionably comfortable (i.e., the expected benefits are well worth the conceivable risks).

Defining a Developmental Algorithm

During the development of LHRH analogs [9], the FDA gradually adopted the following general algorithm of drug development:

1. List all conceivable indications (this obviously requires considerable informational gathering followed by numerous critical discussions).
2. Eliminate those that are not based on correct rationale. Specifically, it is our opinion that LHRH analogs act as reversible medical

castration agents. Therefore, their contraceptive efficacy is highly questionable inasmuch as the categorical concept of "castration"—a state usually followed by loss of libido and potency—
does not overlap with the concept of "contraception." We are not
saying that it is impossible to develop contraceptive agents that
contain LHRH analogs, but we are stating that such an outcome is
highly improbable.

3. Choose from the remaining indications the so-called "lead indica-
    tion" (see above).
4. Initiate clinical trials with that indication, thus allowing for a high
    probability of initial success with the least expenditure in time and
    resources.
5. After this initial use (during which the drug's preliminary safety
    profile has emerged, e.g., the reversibility of LHRH-induced ef-
    fects), adopt a second indication where relatively high doses are
    safe to use, to allow for a careful collection of human safety data.
6. After that, develop one conceivable indication after another, ac-
    cording to a descending list of their estimated probability of
    success.

Additional Useful Concepts

Other techniques may be able to improve the process of drug development,
allowing for an expeditious, yet safe evaluation of the therapeutic potential of entirely novel agents.

Shuttling. This is defined as a continuum of back-and-forth movements from animal to human studies. Initially, minimal animal studies will
allow the initiation of limited and preliminary human trials. If the data obtained during the latter are promising, clinical trials may continue. During such trials, safety data are carefully collected and continuously evaluated to determine whether feasible additional animal studies are needed to
refine certain elements of the safety picture. Shuttling, as a concept, is
partly justified by the pragmatic observation that (1) most toxicities wit-
nessed in animals are seldom predictive of untoward reactions later seen in
humans, and (2) side effects eventually detected in the human had rarely
been forecasted by prior observations in animals. In any event, an essential element in shuttling is to avoid halting human studies while waiting
the completion of needed or traditionally required animal studies.

Lead groups. It is impossible to forecast the one-in-a-million long-term
side effect of an as yet untested drug. Nonetheless, we require a very
high level of assurance of safety in certain patient groups, such as preadolescent children whose reproductive functions we feel compelled to
manipulate chemically in order to alleviate an intolerable condition. In
such cases, additional approaches are needed to lower the risk to socially
acceptable levels. The concept of the "lead group" is one such modality,
and it is defined as the establishment of a small group of patients initially
treated with the novel entity. All aspects of the long-term safety issues
are carefully assessed in that group. Larger groups of patients are then
introduced into clinical trials, at prescribed times, when the lead group
has safely attained a defined chronological milestone. The clinical trials
may consist of a lead group, followed by a single large definitive trial
group. Alternatively, one may find it necessary to use several groups,
one following another, each larger than the one it immediately succeeds. In

any case, if an untoward reaction is observed in the lead group(s), all
necessary decisions can be made, in full cognizance of the facts, to protect the "lagging" groups.

Dovetailing of animal and clinical studies. Animal and clinical studies
should be performed in such a manner as to allow orderly and efficient
testing of new products or formulations. Preclinical studies should consist of the smallest number of meaningful tests needed to ensure a reasonable level of safety during the initial (phase I) clinical studies. Phase
I studies will then help determine the therapeutic potential of the tested
peptide or protein. If the results are negative, the test product is discarded, quickly and inexpensively. If, on the other hand, the results
are positive, further animal and clinical studies may be justified. At all
times, it seems beneficial to adhere to the following principle: No animal
studies are to be performed unless they are needed and meaningful. This
most simple and basic concept ought not to be discarded even if an im-
passe is reached in the drug developmental process. This can occur, for
example, when a certain type of serious human toxicity is indirectly suspected during the clinical trials, but it cannot be studied adequately either
because of the absence of a suitable animal model system, or because we
are technically unable to define clearly the nature and the course of such
a toxicity in humans.

Dovetailing of clinical trials. If the proper resources are available,
one can also choose to dovetail the clinical trials for the various credible
indications of a given drug, provided that all such indications are based
on a common rationale. Thus the various indications may be tested in a
sequentially interconnected fashion. The lead indications should be studied
ahead of all the others, and by a safe margin. Otherwise, one is taking
an unnecessary risk: if the drug is found to be ineffective for a lead
indication, all other indications based on the same rationale may be doomed
as well. Again, if one possesses the necessary resources, one may choose
to test groups of indications simultaneously, each based on a respective
rationale. Under the proper conditions, such risks may well be worth
taking and may result in distinct competitive advantages.

C. Various Phases of Human Investigations

Clinical investigations are classically performed in three phases. In phase I
investigate general tolerance and clinical pharmacology is studied. In
phases II and III the safety and effectiveness of the tested drug for a defined indication is studied. A new drug is approved for public use when
all these trials (including well-controlled ones) have conclusively proven
its general usefulness.

D. Special Problems

Safety Issues

A safety concern may arise, based either on deductive reasoning or on
pragmatic observation, or on both. For example, LHRH analogs appear to
be useful in the treatment of endometriosis by reversibly suppressing
ovarian function (i.e., they induce an artificial menopausal state in young
women). This gain may be at the expense of possible bone loss. The
duration of ovarian suppression will therefore have to be balanced against

the intensity of the observed bone loss, such that the maximum benefit is
derived at a minimum and acceptable cost. Safety concerns, once raised,
ought to be addressed clearly and forcefully, not diffusely or excessively.
In the example above, we want to find out whether a given regimen of
LHRH analog achieves its intended result without unacceptable bone loss;
we do not need to determine what amount of LHRH analog is needed to obtain osteoporosis and bone fractures. In the realm of peptidic drugs, a
number of other safety issues are notable. For example, even endogenous
peptides (e.g., LHRH) and proteins (e.g., insulin) can be immunogenic.
As the tremendous volume of experience with insulins attest, such immunogenicity is devoid of clinical significance in the great majority of cases.
However, a number of potential toxicities are not to be excluded. These
include reactions, sometimes anaphylactoid in nature (e.g., with certain
LHRH antagonists), circulation of neutralizing antibodies (as seen with
various preparations of hGH and methionyl-hGH), and perhaps even the
possible initiation in the long run of immune-mediated diseases.

Efficacy Considerations

Route of administration. Most protein drugs are currently administered
through either intramuscular or, preferably, subcutaneous daily injections.
Needless to say, this is cumbersome, and a number of alternative routes
are in the process of being studied. Satisfactory monthly depot formulations of proteins can be achieved. As discussed in Chap. 13, the intranasal route, without the proper absorption enhancers, appears to be
erratic and wasteful, particularly for larger molecules, since usually only a
small and variable fraction of the administered dose is absorbed. Other
more convenient routes (e.g., oral and rectal) have been studied but without much success to date. Of course, the definitive solution to this rather
vexing situation requires novel and bold forays, including the synthesis of
a small molecule effectively reproducing the receptor-binding and/or the
the receptor-activating functions of the therapeutic protein or interest.
This has been achieved in the past, albeit in an easier situation, when
captopril was developed to imitate the angiotensin converting-enzyme inhibitor activity of a group of peptides (e.g., the nonapeptide teprotide),
known as bradykinin potentiating factors.

Efficacy of proteins. Many proteins (e.g., insulin) are used in a so-
called replacement modality. In that case, it is reasonable to accept that
the normalization of an important abnormality (e.g., hyperglycemia) proves
the efficacy of the exogenously administered insulin. However, such a
normalization does not necessarily imply a cure of the disease process,
nor does it mimic the normal route of initial endogenous insulin delivery to
the liver through the portal circulation. Through collaborative efforts between industry and government, insulin pumps are now available for careful study of the therapeutic results of glycemic control that is as tight as
is reasonably feasible. All other proposals to cure diabetes have to correct the biochemical (hyperglycemia) as well as the clinical (e.g., neuro-
pathy, retinopathy, and/or nephropathy) complications of the disease. In
general terms, objective clinical improvement of a pathological condition is
usually essential to prove the efficacy of a given drug.

Combination therapies. It is often useful to use peptidic drugs in
combination with other peptidic drugs or other more traditional moieties.

For example, one may need to treat growth hormone-deficient preadolescent children with LHRH analogs to delay puberty, thus preventing the closure of epiphyses, while administering hGH and giving the latter protein a better chance to maximize final stature. In another example, LHRH analogs might be tested in conjunction with anti-estrogens and aromatase inhibitors for the palliative treatment of breast cancer. To have such combinations approved, the FDA usually adheres to the following policy: The combination therapy must be proven more effective or safer than either agent used alone; in other words, both additive and synergistic interactions are acceptable in the case of proposed combination therapies. Therefore, two general strategies may be used in such a case: Either each agent is proven to be safe and effective during independent studies, or each element of the combination is proven to offer an additive or synergistic advantage. Drugs may be sold as a combination, but only when the latter argument has been made conclusively. Otherwise, the various drugs must be sold separately, but the medical practitioner may still prescribe them in combination.

VI. CONCLUSIONS

We have endeavored to present a number of theoretical and practical approaches thought to optimize proper development of peptides and proteins with therapeutic potential. It is our perception that the complex and difficult difficult process of drug development can be improved on by the case-by-case use of these and other concepts and methods. A number of these (custom-tailoring of preclinical studies, dovetailing of toxicological and clinical trials) will result in considerable economies. Other suggested techniques (adequate purification of proteins, careful definition of toxicity parameters in the human, use of lead group in particularly delicate situations) will improve the safety aspects of newly introduced drugs. Even so, there are quite a few unsolved problems (development of more practical dosage forms) that need our attention. The only way to improve our collective chances of addressing these residual issues is to continue to work collectively.

REFERENCES

1. S. O. Waife and L. Lasagna, *Regul. Toxicol. Pharmacol.*, 5:212 (1985).
2. J. L. Gueriguian and Y. Y. H. Chiu, in *Opioid Peptides: Molecular Pharmacology, Biosyntheses, and Analysis*, NIDA Research Monograph 70 (R. S. Rapaka and R. L. Hawks, eds.), U.S. Department of Health and Human Services, Washington, D.C. (1986).
3. Y. Y. H. Chiu, D. J. Kertexa, and C. S. Kumkumian, in *Insulins, Growth Hormone, and Recombinant DNA Technology* (J. L. Gueriguian, H. I. Miller, C. Schaffenburg, A. T. Gregoire, and S. Sobel, eds.), Raven Press, New York (1981), pp. 207–213.
4. H. I. Miller, J. L. Gueriguian, G. Troendle, and S. Sobel, in *Insulins, Growth Hormone, and Recombinant DNA Technology* (J. L. Gueriguian, H. I. Miller, C. Schaffenburg, A. T. Gregoire, and S. Sobel, eds.), Raven Press, New York (1981), pp. 215–222.

5. J. L. Gueriguian, H. I. Miller, C. Schaffenburg, A. T. Gregoire, and S. Sobel, eds., *Insulins, Growth Hormone, and Recombinant DNA Technology*, Raven Press, New York (1981).
6. J. L. Gueriguian, E. D. Bransome, and A. S. Outschoorn, eds., *Hormone Drugs*, U.S. Pharmacopeial Convention, Rockville, Md. (1982).
7. Y.-Y. H. Chiu, J. L. Gueriguian, and S. Sobel, in *Biotechnologically Derived Medical Agents: The Scientific Basis of Their Regulation* (J. L. Gueriguian, V. Fattorusso, and D. Poggiolini, eds.), Raven Press, New York (1988).
8. T. D. Waters, G. W. Beebe, and R. W. Miller, *Science, 177*:76 (1972).
9. J. L. Gueriguian, C. A. Schaffenburg, Y. Y. H. Chiu, and V. Berlinger, in *LHRH and Its Analogs* (F. Labrie, A. Belanger, and A. Dupont, eds.), Excerpta Medica, Amsterdam (1984).

Index